The transition from the Eocene to the Oligocene epochs from approximately 47 to 27 million years ago was one of the most dramatic episodes of climatic and biotic change since the demise of the dinosaurs. The mild tropical climates of the Paleocene and early Eocene were replaced by modern climatic conditions and extremes, including glacial ice in Antarctica. The best terrestrial record of the Eocene-Oligocene transition is found in North America, including the spectacular cliffs and spires of Big Badlands National Park in South Dakota.

The first part of the book summarizes the latest information in dating and correlation of the strata of late middle Eocene through early Oligocene age in North America, including the latest insights from argon/argon dating and magnetic stratigraphy. The second part reviews almost all the important terrestrial reptiles and mammals found near the Eocene-Oligocene boundary in the White River Chronofauna, from the turtles, snakes, and lizards, to the common rodents, carnivores, artiodactyls, and perissodactyls.

This is the first comprehensive treatment of these rocks and fossils in over sixty years. The *Terrestrial Eocene-Oligocene Transition in North America* will be an invaluable resource to vertebrate paleontologists, geologists, mammologists, and evolutionary biologists.

T0185614

Morris F. Skinner in the Big Badlands of South Dakota during the 1950s. (Photo courtesy Marie Skinner).

Morris F. Skinner
1906-1989

Our present understanding of the White River Group and its fossils would never have been possible without the enormous contributions of Morris Skinner. Although he was originally hired by Childs Frick in 1927 to collect late Cenozoic mammals of western North America, he spent much time in the White River Group as well. Beginning in 1938 and continuing through the 1940s and 1950s, Skinner made large, stratigraphically zoned collections from the Big Badlands of South Dakota. In the 1950s and 1960s, he made some of the most important collections from White River outcrops in Wyoming, Nebraska, and North Dakota as well. Unlike many collectors, however, Morris was a dedicated geologist and stratigrapher. He measured hundreds of stratigraphic sections, and made sure that every specimen he and his field parties collected had the best possible stratigraphic information. Thus, the Frick White River collections made by Skinner and parties have the best biostratigraphic data available (frequently zoned to the nearest foot from marker ashes), allowing the first detailed range-zone biostratigraphy after 150 years of study in the White River Group. Most of the systematic paleontological studies in this volume are based largely on the Frick Collections, and their insights would not be possible without the excellent stratigraphic data provided by Skinner and crew.

Skinner was more than one of the best fossil collectors of this century, however. He was also an excellent field geologist, and his many insights into White River stratigraphy were largely unappreciated because they remained in his field notebooks, unpublished. Only his brief summary of the North Dakota sequence (Skinner, 1951) was published in his lifetime, but his contributions to the stratigraphy of White River deposits at Flagstaff Rim (Emry, 1992) and the Lusk and Douglas areas of eastern Wyoming (see Evanoff et al., 1992; partially summarized in this volume, Chapter 13, and Prothero and Whittlesey, in press), have been the basis of many later publications by others. For example, Skinner recognized that the divisions of the Chadron Formation used by his contemporaries were inadequate, and described a latest Chadronian unit he called the "Trunk Butte Member." This concept was revived 30 years later as the Big Cottonwood Creek Member of the Chadron Formation (Terry et al., 1995). As early as 1953, Skinner collected brontothere bones from Brule Formation equivalents (see Prothero and Whittlesey, in press), and recognized that the classic definitions of the Chadronian and Orellan needed revision. This, too, is finally occurring 40 years after Skinner's insight. Based on his knowledge of biostratigraphy, Skinner realized that the Chadronian and earliest Orellan were poorly represented in the Big Badlands, but much thicker and more completely exposed in eastern Wyoming. This has major implications for interpretations of the Chadronian or Orellan based on the less complete Big Badlands sequence (e.g., Clark et al., 1967; Retallack, 1983). Skinner's stratigraphic concepts about the White River Group only reached print as an illustration in Mellett (1977, fig. 71, pp. 128-129). His profound understanding of White River stratigraphy and paleontology is finally being appreciated. At the White River Group symposium held at the North Central–South Central section meeting of the Geological Society of America in Lincoln, Nebraska, on April 27, 1995, the prevailing refrain from the speakers and all their "new" research was, "Morris Skinner knew this years ago!"

Much of the best science in this book represents Morris Skinner's collections or ideas finally seeing publication. It is appropriate that it be dedicated to his memory.

The Terrestrial Eocene-Oligocene Transition in North America

The Terrestrial Eocene-Oligocene Transition in North America

Edited by

DONALD R. PROTHERO
Occidental College

ROBERT J. EMRY
Smithsonian Institution

CAMBRIDGE
UNIVERSITY PRESS

CAMBRIDGE UNIVERSITY PRESS
Cambridge, New York, Melbourne, Madrid, Cape Town, Singapore, São Paulo

Cambridge University Press
The Edinburgh Building, Cambridge CB2 2RU, UK

Published in the United States of America by Cambridge University Press, New York

www.cambridge.org
Information on this title: www.cambridge.org/9780521433877

First published 1996
This digitally printed first paperback version 2005

A catalogue record for this publication is available from the British Library

Library of Congress Cataloguing in Publication data
The terrestrial Eocene-Oligocene transition in North America / edited
by Donald R. Prothero, Robert J. Emry.
p.   cm.
Includes bibliographical references and index.
ISBN 0-521-43387-8 (hc)
1. Eocene-Oligocene boundary – North America.   2. Geology,
Stratigraphic.   3. Geology – North America.   4. Paleontology,
Stratigraphic.   5. Paleontology – North America.   I. Prothero,
Donald R.   II. Emry, Robert J.
QE692.8.T47   1996
551.7'84 – dc20   95-40903
                                       CIP

ISBN-13 978-0-521-43387-7 hardback
ISBN-10 0-521-43387-8 hardback

ISBN-13 978-0-521-02109-8 paperback
ISBN-10 0-521-02109-X paperback

The original hardback volume was prepared by Donald Prothero as camera-ready copy on an Apple
Macintosh Quadra 630 using Microsoft Word 6.0, and printed on a 800 dpi laser printer.
All layouts and paste-ups were done by Prothero.

Front cover caption: The strata of the Big Badlands of South Dakota preserve one of the best terrestrial records of terrestrial Eocene-Oligocene transition in North America.   This rendition is from Henry Fairfield Osborn's 1929 titanothere monograph, showing the upper Eocene Chadron Formation overlain by the lower Oligocene Brule Formation.   Superimposed on this figure are side views of the skulls of three of the most common oreodonts from the Eocene-Oligocene transition (from bottom to top): the middle Chadronian *Merycoidodon presidioensis,* the late Chadronian – early Orellan *Merycoidodon culbertsoni,* and the late Orellan *Merycoidodon bullatus* (from Stevens and Stevens, this volume, Chapter 25).

CONTENTS

PART II: Common Vertebrates of the White River Chronofauna

CONTRIBUTORS

Jon A. Baskin
Department of Geological Sciences
Texas A & I University
Kingsville, TX   78363

Harold N. Bryant
Mammalogy Program
Provincial Museum of Alberta
12845 102nd Avenue
Edmonton, Alberta T6G 2E9 Canada

Emily A. CoBabe
Department of Geological Sciences
University of Massachusetts
Amherst, MA   01003

T. H. Huxley Dozier
Department of Geology
Occidental College
Los Angeles, CA   90041-3392

Robert J. Emry
Department of Paleobiology
NHB-E207 MRC 121
Smithsonian Institution
Washington, DC   20560

John J. Flynn
Department of Geology
Field Museum of Natural History
Roosevelt Road at Lakeshore Drive
Chicago, IL   60605-2496

Daniel Garcia
8252 126th Avenue NE, D301
Kirkland, WA   98033

C. Bruce Hanson
5505 Sierra Avenue
Richmond, CA   94805

Timothy H. Heaton
Department of Earth Sciences
University of South Dakota
Vermilion, SD   57069

J. Alan Holman
The Museum
Michigan State University
East Lansing, MI   48823

Jeffrey L. Howard
Department of Geology
Wayne State University
Detroit, MI   48202

Robert M. Hunt, Jr.
Vertebrate Paleontology
W436 Nebraska Hall
University of Nebraska
Lincoln, NE   68588-0541

J. Howard Hutchison
Museum of Paleontology
University of California
Berkeley, CA   94720

Steven A. King
Department of Geology
Occidental College
Los Angeles, CA   90041-3392

William W. Korth
928 Whalen
Penfield, NY   14526

Walter G. Lohr II
Department of Geology
Occidental College
Los Angeles, CA   90041-3392

Spencer G. Lucas
New Mexico Museum of Natural History
1801 Mountain Road NW
Albuquerque, NM   87104

David J. Lundquist
Department of Geology
Occidental College
Los Angeles, CA   90041-3392

Steven M. McCarroll
Department of Geology
Field Museum of Natural History
Roosevelt Road at Lakeshore Drive
Chicago, IL   60605-2496

Donald R. Prothero
Department of Geology
Occidental College
Los Angeles, CA   90041-3392

Jennifer Snyder
Department of Earth Sciences
Denver Museum of Natural History
2001 Colorado Boulevard
Denver, CO 80205

James B. Stevens
Department of Geology
Lamar University
Box 10031, LU Station
Beaumont, TX 77710

Margaret S. Stevens
Department of Geology
Lamar University
Box 10031, LU Station
Beaumont, TX 77710

John E. Storer
Royal Saskatchewan Museum
2340 Albert Street
Regina, Saskatchewan S4P 3V7 Canada

Richard K. Stucky
Department of Earth Sciences
Denver Museum of Natural History
2001 Colorado Boulevard
Denver, CO 80205

Robert M. Sullivan
State Museum of Pennsylvania
Third and North Streets
P.O. Box 1026
Harrisburg, PA 17108-1026

James B. Swinehart
Nebraska Conservation and Survey Division
113 Nebraska Hall
University of Nebraska
Lincoln, NE 68588-0541

Carl C. Swisher III
Berkeley Geochronology Center
2455 Ridge Rd.
Berkeley, CA 94709

Alan R. Tabrum
Section of Vertebrate Fossils
Carnegie Museum of Natural History
4400 Forbes Avenue
Pittsburgh, PA 15213-4080

Richard H. Tedford
Department of Vertebrate Paleontology
American Museum of Natural History
Central Park West at 79th St.
New York, NY 10024

Timothy E. Tierney
Department of Geology
Occidental College
Los Angeles, CA 90041-3392

William D. Turnbull
Department of Geology
Field Museum of Natural History
Roosevelt Road at Lakeshore Drive
Chicago, IL 60605-2496

Edward H. Vance, Jr.
Department of Geology
Occidental College
Los Angeles, CA 90041-3392

Steve Walsh
Department of Paleontology
Natural History Museum
P.O. Box 1390
San Diego, CA 92112

Wang Xiaoming
Department of Vertebrate Paleontology
American Museum of Natural History
Central Park West at 79th Street
New York, NY 10024

Xu Xiaofeng
Department of Geology
Southern Methodist University
Dallas, TX 75275

PREFACE

The transition from the Eocene to the Oligocene epochs (from about 47 to 27 million years ago) was one of the most dramatic episodes of climatic and biotic change since the demise of the dinosaurs. The mild tropical climates that characterized the Paleocene and early Eocene were replaced by the beginning of modern climatic extremes, including glacial ice in Antarctica and modern deep-water oceanic circulation (summarized in Prothero, 1994). These changes were seen in plants and animals worldwide, both in the oceans and on land. Land floras changed from dense forests (found even at polar latitudes) to a mixture of woodland and scrubland. The land faunas responded with extinction of many forest-dwelling and leaf-eating animals, and replacement by snails, reptiles, and mammals tolerant of drier conditions and the more open vegetation.

The best terrestrial record of the Eocene-Oligocene transition is found in North America, in deposits which include the spectacular cliffs and spires of Big Badlands National Park in South Dakota, world-famous for its scenery and abundant fossils. Although the fossils and deposits have been studied since 1846, much critical new information has accumulated in the last twenty years. Enormous collections of fossil mammals from these beds were made by the Frick Laboratory of the American Museum of Natural History in New York, but only a small fraction of the studies on these fossils has been published. On the 150th anniversary of the discovery and description of the first fossil mammal from the Badlands, we hope to bring the subject up to date. The last comprehensive monographs on the White River mammals were published in 1936-1941 by William Berryman Scott, Glenn Lowell Jepsen, and Albert E. Wood, and many of these mammals have not been reviewed since then. This volume is, in part, a long-overdue update of the classic White River monographs.

There has also been a vast improvement in our chronological understanding of these beds. With the new tools of magnetic stratigraphy and ^{40}Ar/^{39}Ar dating, we are gaining our first high-resolution correlation of the terrestrial North American section with the global climatic record. New correlations have already radically changed our understanding of the time scale, and even changed the position of the North American Eocene-Oligocene boundary itself. Uintan and Duchesnean fossils, long thought to be late Eocene, are now considered middle Eocene; Chadronian fossils, long thought to be early Oligocene, are now considered late Eocene; the Orellan and Whitneyan land mammal "ages," once thought to be middle and late Oligocene, are now considered early Oligocene. As important as these new data are, very little was published in sufficient detail. In addition to correlations of the classic White River and Uinta Basin deposits, this volume also summarizes the geology of relevant deposits in Saskatchewan, Montana, Oregon, Texas, New Mexico, and California, with detailed faunal lists, magnetics, and geochronology not previously published.

Contrary to widespread misconceptions, there was no singular, catastrophic "Terminal Eocene Event." Instead, the transition was marked by a series of extinctions, beginning at the end of the middle Eocene (about 37 Ma). Consequently, our stratigraphic coverage spans about 20 million years (47-27 Ma), beginning with the late middle Eocene (Uintan and Duchesnean, from 47-37 Ma) through the late Eocene (Chadronian, 37-33.5 Ma), and the early Oligocene (Orellan-Whitneyan-earliest Arikareean, from 33.5-27 Ma). The chapters in the first part of the book review the chronostratigraphy of nearly all the important areas where terrestrial mammal fossils of these ages are found.

The second part of the book summarizes the systematic paleontology of nearly all the common land vertebrates of the White River Chronofauna. These include the most common mammalian taxa, as well as the reptiles (turtles, lizards, snakes, and amphisbaenians). Unfortunately, it was not possible to include every taxon. Some have been recently revised elsewhere (hyaenodonts by Mellett, 1977; horses by Prothero and Shubin, 1989; rabbits by Korth and Hageman, 1988). The taxonomy of other groups (e.g., brontotheres, enteledonts) was not ready for publication. Nevertheless, the updated systematics of the mammals in this volume covers all the taxa which are critical to the biostratigraphy of the Chadronian through Whitneyan interval.

In the final chapter, we summarize the chronostratigraphic and biostratigraphic information presented in this book, and suggest outlines of a biostratigraphic zonation for the entire interval. Much work remains to be done, of course, but this summary presents a framework for further refinement of the chronostratigraphy, biostratigraphy, and systematic paleontology of this important interval in Earth history.

ACKNOWLEDGMENTS

This book would never have been possible without the cooperation and support of many people. First, we thank the many authors, who worked so hard to produce polished manuscripts on disk and finished art, and the reviewers acknowledged in each chapter, who gave freely of their time to ensure the scientific accuracy of each contribution. We thank Clifford R. Prothero for all his help with the production of this volume. We thank the editoral staff at Cambridge University Press for their support of this project. They include Catherine Flack, developmental editor; and Elizabeth Avery, copy editor. Much of the support for Prothero's research over the last 20 years published herein was provided by NSF grants EAR87-08221, 91-17819, 94-05942, grants from the Donors of the Petroleum Research Fund of the American Chemical Society, a Guggenheim Fellowship, and a Columbia University Department of Geological Sciences research fellowship. Emry's research was supported by grants from the Charles Walcott Fund, Smithsonian Research Foundation, the Research Opportunities Fund, and other sources within the Smithsonian Institution.

LITERATURE CITED

Clark, J., J. R. Beerbower, and K. K. Kietzke. 1967. Oligocene sedimentation, stratigraphy, paleoecology and paleoclimatology of the Big Badlands of South Dakota. Fieldiana: Geology Memoir 5:1-158.

Emry, R. J. 1992. Mammalian range zones in the Chadronian White River Formation at Flagstaff Rim, Wyoming; pp. 106-115 *in* D. R. Prothero and W. A. Berggren (eds.), Eocene-Oligocene Climatic and Biotic Evolution, Princeton University Press, Princeton, N. J.

Evanoff, E., D. R. Prothero, and R. H. Lander. 1992. Eocene-Oligocene climatic change in North America: the White River Formation near Douglas, east-central Wyoming; pp. 116-130 *in* D. R. Prothero and W. A. Berggren (eds.), Eocene-Oligocene Climatic and Biotic Evolution, Princeton University Press, Princeton, N. J.

Korth, W. W., and J. Hageman. 1988. Lagomorphs (Mammalia) from the Oligocene (Orellan and Whitneyan) Brule Formation, Nebraska. Transactions of the Nebraska Academy of Sciences 16:141-152.

Mellett, J. S. 1977. Paleobiology of North American *Hyaenodon* (Mammalia, Creodonta). Contributions to Vertebrate Evolution 1:1-134.

Prothero, D. R. 1994. The Eocene-Oligocene Transition: Paradise Lost. Columbia University Press, New York.

Prothero, D. R., and N. Shubin. 1989. The evolution of Oligocene horses; pp. 142-175 *in* D. R. Prothero and R. M. Schoch (eds.), The Evolution of Perissodactyls. Oxford University Press, New York.

Prothero, D. R., and K. E. Whittlesey. In press. Magnetostratigraphy and biostratigraphy of the Orellan and Whitneyan land mammal "ages" in the White River Group. Geological Society of America Special Paper (in press).

Retallack, G. 1983. Late Eocene and Oligocene fossil paleosols from Badlands National Park, South Dakota. Geological Society of America Special Paper 193.

Scott, W. B., G. L. Jepsen, and A. E. Wood. 1936-1941. The mammalian fauna of the White River Oligocene, Parts I-V. Transactions of the American Philosophical Society 28:1-980.

Skinner, M. F. 1951. The Oligocene of western North Dakota; pp. 51-58 *in* J. D. Bump (ed.), Society of Vertebrate Paleontology Guidebook, 5th Annual Field Conference, Western South Dakota, August-September 1951.

Terry, D. O., H. LaGarry, and W. B. Wells. 1995. The White River Group revisited: vertebrate trackways, ecosystems, and stratigraphic revision, reinterpretation, and redescription. Nebraska Conservation and Survey Division Guidebook 10:43-57.

PART I:

THE CHRONOSTRATIGRAPHY OF THE UINTAN THROUGH ARIKAREEAN

1. Magnetic Stratigraphy and Biostratigraphy of the Middle Eocene Uinta Formation, Uinta Basin, Utah

DONALD R. PROTHERO

ABSTRACT

The Uinta Formation in northeastern Utah was the original basis of the Uintan land mammal "age." Magnetostratigraphic studies were conducted in four sections in the northeastern, north-central, and northwestern Uinta Basin. The uppermost Evacuation Creek Member of the Green River Shale, and all of unfossiliferous Uinta Formation unit "A," was of normal polarity. This normal interval probably correlates with Chron C21n (46.3-47.8 Ma), as originally suggested by Prothero and Swisher (1992). Most of Uinta "B" was reversed (= Chron C20r, 43.8-46.3 Ma). A short normal zone spanning upper Uinta "B" and lower Uinta "C" probably correlates with Chron C20n (42.5-43.8 Ma). The upper part of the Uinta "C" and the lowermost portion of the Duchesne River Formation were also reversed (= C19r, 41.4-42.5 Ma), with normal (= C19n, 41.1-41.4 Ma) and reversed (= C18r, 40.0-41.1 Ma) magnetozones in the higher part of the Brennan Basin Member.

Although the original biostratigraphic data for most Uinta Basin collections are very poor, distinctions between the faunas of Uinta "B1," "B2," and "C" are possible. Uinta "B1" (the "*Metarhinus* zone" of Osborn, 1929) spans the interval 45-46 Ma, and is characterized by overlapping ranges of the brontotheres *Sthenodectes* and *Metarhinus*, the rhinocerotoids *Hyrachyus eximius*, *Forstercooperia grandis* and *Triplopus obliquidens*, and the agriochoerid oreodont *Protoreodon parvus*. Uinta "B2" (the "*Eobasileus-Dolichorhinus* zone" of Osborn, 1929), including White River Pocket, spans the interval 43-45 Ma, and is characterized by the overlapping ranges of the brontotheres *Sphenocoelus*, *Metarhinus*, *Eotitanotherium*, the chalicothere *Eomoropus*, the horse *Epihippus gracilis*, the creodont *Oxyaenodon*, and the artiodactyls *Diplobunops*, *Oromeryx*, and *Leptotragulus*. The taeniodonts (*Stylinodon*), uintatheres (*Uintatherium* and *Eobasileus*), achaenodont artiodactyls, and protoptychid and sciuravid rodents last appear in Uinta B2.

Uinta "C" (the "*Diplacodon-Protitanotherium* zone" of Osborn, 1929), including Myton Pocket, Kennedy Hole, and Leota Quarry, spans the interval from 42.5-43 Ma, and is characterized by the first occurrences of the brontotheres *Protitanotherium*, *Metatelmatherium*, the lagomorph *Mytonolagus*, the primate *Mytonius*, the rodent *Janimus*, and the last appearance of numerous taxa, including the rodents *Thisbemys*, *Ischyrotomus*, and *Reithroparamys*, the creodont *Oxyaenodon*, the carnivorans *Procynodictis* and *Uintacyon*, the artiodactyls *Poebrodon*, *Oromeryx*, *Auxontodon*, *Bunomeryx*, and *Mytonomeryx*, and the rhinocerotoids *Amynodon* and *Triplopus*. The upper part of Uinta "C" (spanning the interval 42.0-42.5 Ma) is unfossiliferous. Sparse but distinctive fossils characterize the latest Uintan Brennan Basin Member and the Duchesnean Lapoint Member of the Duchesne River Formation.

INTRODUCTION AND GEOLOGIC SETTING

The Uinta Basin in northeastern Utah (Dane, 1954; Untermann and Untermann, 1964; Cashion, 1967) is an asymmetric synclinal structure about 7000 square miles in area (Fig. 1). Its axis trends roughly east-west, and the north limb is inclined more steeply than the shallow-dipping south limb. The Uinta Basin is about 135 miles wide along its east-west axis and 100 miles across in the north-south direction. It is bounded by the Uinta Mountains to the north, the Douglas Creek arch to the east (which separates it from the similar Piceance Basin of Colorado), the Wasatch Range on the west, and the Roan Cliffs to the south.

Thick sequences of Paleozoic and Mesozoic rocks are found along the edges of the basin and plunge into the subsurface. These were deformed during the latest Cretaceous-middle Eocene Laramide Orogeny, which created the basin in its present configuration. During Laramide orogenesis, over 15,000 feet of Eocene sediments accumulated in this rapidly subsiding structure. The bulk of the sedimentary package consists of the fluvial lower Eocene Wasatch Formation (up to 4100 feet thick) and the lacustrine lower middle Eocene Green River Formation (up to 7000 feet thick). The latter is part of an extensive system of middle Eocene lakes that once covered much of northeastern Utah, southwestern Wyoming, and western Colorado. Because of its importance as source rock for oil and oil shale, the Green River Formation has been studied in great detail (see Bradley, 1929, 1931, 1964; Dane, 1954; Cashion, 1967; Ryder et al., 1976; Surdam and Stanley, 1979, 1980; Johnson, 1985; and Roehler, 1992b, for some of the key features of the Green River lake system).

Toward the end of the early middle Eocene

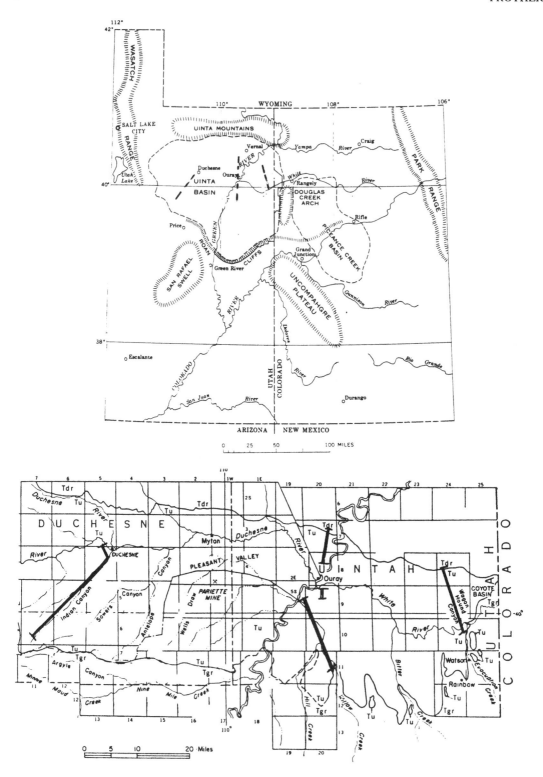

Figure 1. Index map of the Uinta Basin, showing the location of the stratigraphic sections and key localities. Tdr = Duchesne River Formation; Tgr = Green River Formation; Tu = Uinta Formation. (Top map after Cashion, 1967; bottom map modified after Dane, 1954).

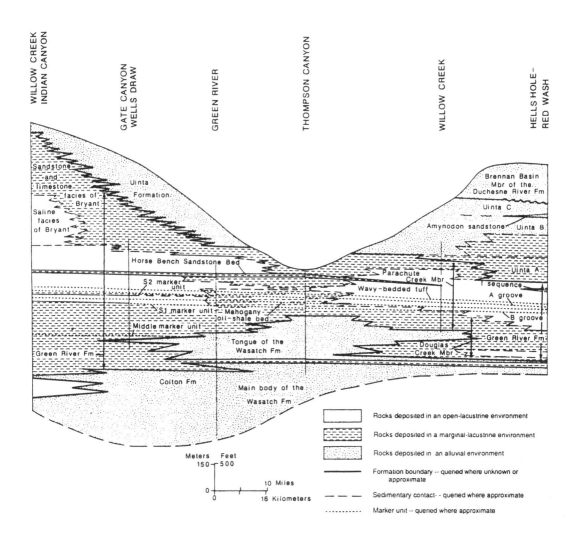

Figure 2. Stratigraphic cross-section of the Uinta Basin, showing the interfingering relationships between the Green River, Uinta, and Duchesne River formations (after Franczyk et al., 1989, fig. 14).

(Bridgerian, about 47-49 Ma), the Green River lake system began to recede. Lacustrine shales were replaced by fluvial-deltaic mudstones and sandstones which entombed many terrestrial fossil vertebrates. The Bridger Basin (Bridger Formation) and Washakie Basin (Kinney Rim Member of the Washakie Formation–Roehler, 1973, 1992a) began to dry up first, with Bridgerian-aged fluvial sediments capping and interfingering with the Green River shales. Much of this fluvial sediment was supplied by a large influx of volcaniclastic debris from the Absaroka volcanic field of northwestern Wyoming and the Challis volcanics of Idaho, forming the "volcanic lithic sandstone petrofacies" of Surdam and Stanley (1979, 1980; see also Johnson, 1985). The Piceance Basin began to dry up slightly later as volcanic

debris spilled south from the Washakie-Sand Wash Basins (Surdam and Stanley, 1979, 1980; Dickinson et al., 1988; see Stucky et al., this volume, Chapter 3).

In the Uinta Basin, the Green River lake system was gradually replaced by the fluvial Uinta Formation prograding westward from the east end of the basin (Fig. 2). Thus, the lower fluvial sandstones of the eastern Uinta Formation are laterally equivalent to lacustrine evaporites, and sandstones and limestones in the western Uinta Basin, with complex interfingering between the two units (Dane, 1954, 1955; Ray et al., 1956; Cashion, 1967; Ryder et al., 1976). Most lower Uinta Formation sandstones have west-trending channels and paleocurrents (Stagner, 1941; Cashion, 1967), and apparently had arkosic source areas to the southeast in the Laramide uplifts, especially the Uncompahgre uplift in west-central Colorado, and the Park Ranges of central Colorado (Stagner, 1941; Bruhn et al., 1983; summarized by Dickinson et al., 1986). By the Duchesnean, these sources were swamped by quartzite debris and recycled Paleozoic sedimentary clasts from the Uinta Mountains to the north as these ranges experienced renewed late Laramide uplift (Andersen and Picard, 1972, 1974; Picard and Andersen, 1975; Dickinson et al., 1986).

Compared to the Green River Formation, the Uinta Formation has been much less studied. The unit was first named by Comstock (1875), and fossils from the upper part of the formation were first reported by Marsh (1870) and Scott and Osborn (1887). The history of collecting in the Uinta Basin is reviewed by Black and Dawson (1966, pp. 326-328). Peterson (in Osborn, 1895) first used the terms "Uinta A" and "Uinta B" for the lower fossiliferous sequence, and "Uinta C" for the upper fossiliferous beds. Most of the early collections follow this terminology. However, Osborn (1929) confused matters by reshuffling the names "Uinta A" and "B." The unfossiliferous sandstones at the base of the sequence (the lower half of Peterson's "Uinta A") became the totality of Osborn's "Uinta A," and the upper half of Peterson's "Uinta A" was renamed "Uinta B1." Peterson's "Uinta B" was renamed "Uinta B2." This unfortunate recycling of similar terminology led to much confusion, and every museum specimen label has to be read carefully to determine when and where the stratigraphic data were determined. Paleontologists who understood these changes (e.g., Krishtalka et al., 1987, p. 83) were fooled by older museum labels, and erroneously attributed collections from Uinta "B1" to Osborn's Uinta "A" (e.g., Krishtalka et al., 1987, p. 89; see Prothero and Swisher, 1992, p. 54, for discussion). Wood (1934) renamed Uinta "A" and "B" the Wagonhound Member and Uinta "C" the Myton Member. Wood et al. (1941) based their Uintan land mammal "age" on the faunas from the Uinta Formation, and considered it late Eocene in age.

Osborn (1929) also introduced another change in terminology which can cause confusion. Peterson (in Osborn, 1895) originally drew the boundary between his Uinta "B" and "C" at the color change from greenish-gray mudstones in the Devil's Playground area of Kennedy's Hole and the reddish beds overlying them. But Osborn and Matthew (1909, fig. 8) and Osborn (1929, fig. 63 and p. 92) redefined the Uinta B-C boundary as the *Amynodon* sandstone, placing the overlying greenish-gray mudstones of Peterson's Uinta B in their Uinta C. This confusion means that some specimens which are called "Uinta B, Kennedy's Hole" (such as the type of *Protoptychus hatcheri*) are actually from Uinta C as currently understood. For this reason, it is important to go back to the original locality data on every specimen,

and not trust assignments such as Uinta A, B, or C.

Overlying and interfingering with the Uinta Formation is the Duchesne River Formation. It was originally named Duchesne Formation by Scott (1932), and then renamed Duchesne River Formation by Kay (1934) when the original name was found to be preoccupied. Cropping out along the northern flank of the Uinta Basin, it locally consists of more than 3,000 feet of fluvial sandstones and conglomerates with lesser floodplain mudstones (Andersen and Picard, 1972, 1974). Although the distinction is subtle, the reddish color of the Duchesne River Formation is typically used to distinguish it from the greenish-gray-tan Uinta Formation. The Wood Committee (1941) used the Duchesne River Formation as the basis for their Duchesnean land mammal "age," and considered it late Eocene.

The age of the Duchesne River Formation, and of the Duchesnean land mammal "age," has been the subject of considerable dispute ever since the Wood Committee report. The fossils of the lower two members, Andersen and Picard's (1972) Brennan Basin and Dry Gulch Creek Members (source of the Randlett and Halfway faunas of Peterson, 1934) are now considered latest Uintan in age (Clark et al., 1967, p. 59; Tedford, 1970, pp. 690-692; Emry, 1981; Krishtalka et al., 1987, p. 84). Only the fauna of the third member, or Lapoint Member of Andersen and Picard (1972), is presently used as the basis for the Duchesnean. The fourth, or uppermost member, the Starr Flat Member of Andersen and Picard (1972) is unfossiliferous (Krishtalka et al., 1987). Because of the limited nature of the Lapoint fauna, some authors have suggested making the Duchesnean a subage of the Chadronian (Wilson, 1978, 1984; Emry, 1981), but more recently the value of retaining the Duchesnean as a distinct land mammal "age" has been reaffirmed (Krishtalka et al., 1987; Kelly, 1990; Lucas, 1992).

In addition to this confusion between rocks, faunas, and time terms, the Lyellian epoch assignment of the Duchesnean has also been controversial. Scott (1945) and Clark et al. (1967) regarded it as early Oligocene, whereas Simpson (1946), Black and Dawson (1966), and Krishtalka et al. (1987) considered it late Eocene. Clearly, a better chronostratigraphic framework for the Uinta and Duchesne River formations is critical to understanding the middle-late Eocene transition.

MAGNETIC ANALYSIS

In the summers of 1986, 1987, and 1988, we sampled the Uinta Formation, parts of the underlying Evacuation Creek Member of the Green River Formation, and the overlying Brennan Basin Member of the Duchesne River Formation. The final laboratory work on these samples was completed and presented in 1990 (Prothero, 1990; Prothero and Swisher, 1990). A summary of this research was published (Prothero and Swisher, 1992).

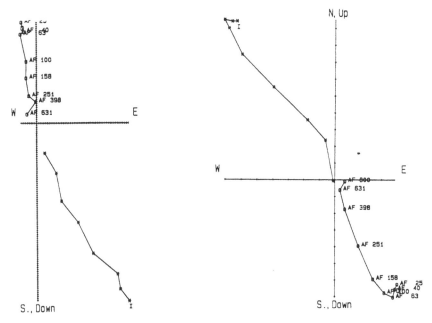

Figure 3. Vector demagnetization ("Zijderveld") plots of AF demagnetization of representative samples from the Uinta Formation. AF intensity in Gauss shown at each step. Horizontal component indicated by circles, vertical component by asterisks. I = NRM direction of vertical component. Note that both samples declined rapidly in intensity, indicating that a low-coercivity mineral such as magnetite is a significant component of the remanence.

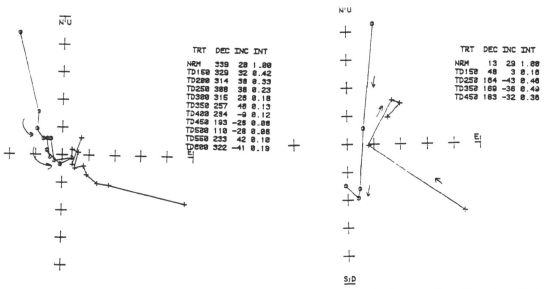

Figure 4. Vector demagnetization plots of selected thermal demagnetization results from the Uinta Formation. Circles indicate horizontal component, + symbol is the vertical component. Sample on left had a normal overprint that was removed by 400°C. Sample on right had a normal overprint removed by 250°C.

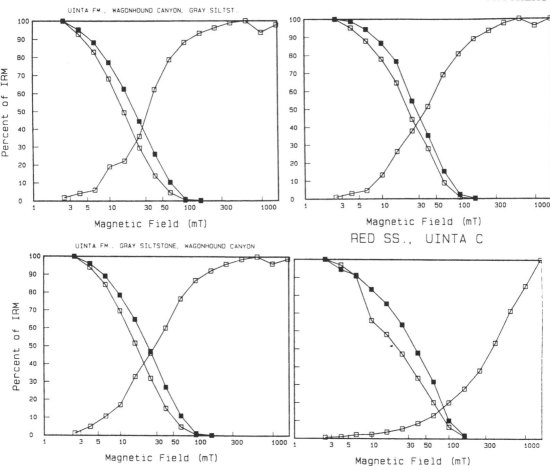

Figure 5. IRM acquisition and Lowrie-Fuller tests of selected samples from the Uinta Formation (see text and Pluhar et al., 1991, for further details). Solid boxes are ARM intensities at each AF demagnetization step, open boxes are IRM intensities. Note that during IRM acquisition (ascending curve on right), most samples reached saturation at about 300 mT, indicating the presence of magnetite; only the red sandstone from Uinta C (lower right) failed to saturate, showing that hematite is a primary component of the remanence in that sample. In the modified Lowrie-Fuller tests (descending curves on left), the ARM is more resistant to AF demagnetization than the IRM, showing that the remanence is carried by single-domain or pseudo-single domain grains.

Over 350 sites (each containing a minimum of three samples) were collected using simple hand tools, resulting in over 1200 individual samples. Four sections, representing the northeastern, north-central, and northwestern parts of the basin, were taken to see if the magnetic pattern could be correlated across the basin, and to tie in as many fossil localities as possible (Fig. 1). The routes of each section are described in the Appendix. Given the time-transgressive nature of the formational boundaries, and the lateral facies changes, such parallel sampling was critical (Fig. 2).

After measurement of NRM (natural remanent magnetization), a suite of pilot samples was demagnetized using alternating field (AF) and thermal demagnetization. Under AF demagnetization (Fig. 3), nearly all samples decreased rapidly in intensity with increasing applied fields, suggesting that the carrier of the rema-

nence is a low-coercivity mineral such as magnetite. Thermal demagnetization (Fig. 4) typically showed a normal overprint which was removed by 250-350°C; this overprint was probably due to an iron hydroxide such as goethite. A stable reversed component was typically obtained between 350-500°C, and that component was used for further analysis in all samples. Above the Curie point of magnetite (580°C), less than 10% of the magnetization remained, suggesting that very little of the remanence is carried by hematite.

This interpretation was corroborated by IRM (isothermal remanent magnetization) acquisition studies (Fig. 5). Most samples (Fig. 5A-C) from the Uinta Formation reached saturation IRM values at 100-300 mT (millitesla); the remanence in these rocks is carried mostly by magnetite. However, some red sandstones from Uinta "C" and the Duchesne River Formation

(Fig. 5D) showed no evidence of IRM saturation, even at fields of 1300 mT; these samples clearly contained hematite. A modified Lowrie-Fuller ARM (anhysteretic remanent magnetization) test (e.g., Johnson et al., 1975) was also conducted along with the IRM analysis (see Pluhar et al., 1991, for details). This test compares the resistance of AF demagnetization of both an IRM acquired in a 100 mT peak field, and an ARM gained in a 100 mT oscillating field. In almost all samples, the ARM (black squares) demagnetizes at higher peak fields than does the IRM (open squares), indicating that the remanence is carried by single-domain or pseudo-single domain grains.

The stable sample directions were then clustered by site, and statistically analyzed by the methods of Fisher (1953; see Butler, 1992). Class I sites of Opdyke et al. (1977) showed a clustering that differed significantly from random at the 95% confidence level. In Class II sites, one sample was lost or crumbled, but the remaining samples gave a clear polarity indication. In Class III sites of Opdyke et al. (1977), two samples showed a clear polarity preference, but the third sample was divergent because of insufficient removal of overprinting. A few samples were considered indeterminate if their magnetic signature was unstable, or their direction uninterpretable.

The beds of the Uinta Basin ranged in dip from 45° to horizontal, so it was possible to conduct a modified fold test for stability. Before the dip correction, the cleaned mean inclination (I) for normal sites was 340.7°, and declination (D) was 56.3°; the precision parameter (k) was 2.4 and the ellipse of confidence (α_{95}) was 40.8°. After dip correction, the directions were much less scattered [D = 2.9, I = 51.7, k = 14.6, α_{95} = 5.6] and much closer to the Eocene pole position for the region, showing that the remanence was acquired before tilting. The cleaned but uncorrected mean for reversed sites [D = 155.2, I = −35.6, k = 1.5, α_{95} = 44.1] is much more scattered than the corrected mean of reversed sites [D = 176.9, I = −57.8, k = 8.9, α_{95} = 13.1], also suggesting that the magnetization was acquired prior to tilting.

These statistics also provide a reversal test. The mean direction of all cleaned and corrected normal sites shown above was antipodal to the mean direction of all cleaned and corrected reversed samples, indicating that the magnetization was probably acquired during deposition of the beds.

MAGNETIC CORRELATIONS

The magnetic pattern for all the Uinta Basin sections is shown in Fig. 6. In the northeastern Uinta Basin (Wagonhound Canyon, Bonanza area, Coyote Basin, plus Kennedy Wash of the topographic maps, or Kennedy Hole of Peterson and Kay), the uppermost Evacuation Creek Member of the Green River Shale, and all of Uinta "A" were of normal polarity. Most of Uinta "B"

was reversed, except for the top 100 feet, which was of normal polarity; this normal polarity persisted through the lower half of Uinta "C." The rest of Uinta "C" was of reversed polarity, as was nearly all of the lower part of the Duchesne River Formation in Kennedy Hole (except for the last two sites at the top, which were of normal polarity).

In the central Uinta Basin, the Willow Creek-Ouray-Brennan Basin section showed a similar pattern. The Green River section below the base of the Uinta Formation was of normal polarity, but most of what was called "Uinta A" and "Uinta B" in Willow Creek by Kay (1934; Peterson and Kay, 1931) was reversed. This is not really surprising, since there are rapid facies changes across the basin, and the migrating fluvial Uinta "A" channel complex would be expected to be time-transgressive. This reversed magnetozone ended low in Uinta "C," just above White River Pocket, and most of the rest of Uinta "C" in the Leota Bottom area was of normal polarity. After a reversed magnetozone in upper Uinta "C," the base of the type Brennan Basin Member of the Duchesne River Formation was of normal polarity, and the top three sites in that section were reversed. As discussed in the Appendix, the rest of the Duchesne River Formation type section of Andersen and Picard (1972) was not sampled due to difficulties with chemical overprinting by iron oxides and hydroxides.

The other section in the central Uinta Basin, Myton Pocket, was entirely of normal polarity. This is not surprising, since the fauna of Myton Pocket is usually correlated with Uinta "C" faunas, such as those of Leota and Skull Pass quarries (Hamblin, 1987).

In the western Uinta Basin section at Indian Canyon, comparable results were obtained. If the Horse Bench Sandstone correlates with the rocks just below Uinta "A" of the eastern part of the basin (Dane, 1954; Franczyk et al., 1989), then the normal magnetozone in the upper Evacuation Creek Member probably correlates with the Uinta "A" normal zone in Wagonhound Canyon. Assuming no discontinuities, the second normal magnetozone in the middle of the saline facies probably correlates with the lower Uinta "C" magnetozone. Finally, the zone of normal polarity in the sandstone and limestone facies at the top probably correlates with the normal magnetozone near the base of the Duchesne River Formation to the east.

Geochronological calibration of this magnetic pattern has been problematic. Mauger (1977) reported a number of K-Ar dates from the Indian Canyon section, including a date of 43.1 ± 1.3 Ma from an ash about 70 m above the Horse Bench Sandstone (in the middle of the lowest normal magnetozone), and 42.8 ± 1.0 Ma for a tuff 19 m below the contact between the saline facies and the limestone and sandstone facies (in the middle of the higher reversed magnetozone). On the current magnetic polarity time scale (Cande and Kent, 1993; Berggren et

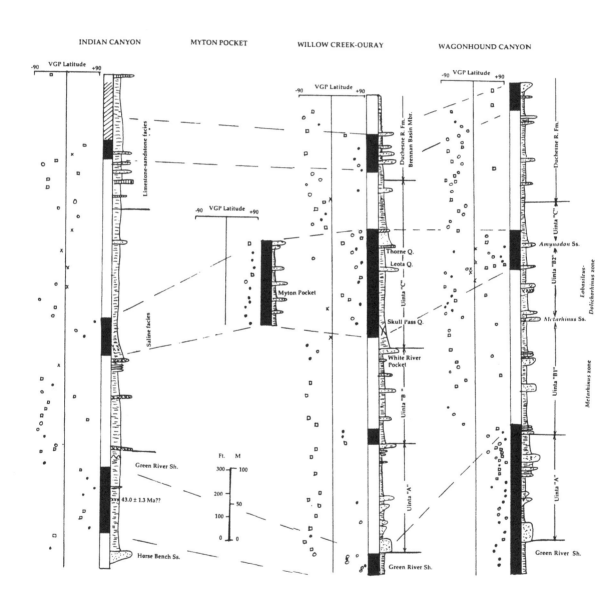

Figure 6. Magnetic stratigraphy of four sections in the Uinta Basin, Utah (see Fig. 1). Positive virtual geomagnetic pole (VGP) latitudes indicate normal polarity; negative VGP latitudes indicate reversed polarity. Solid circles are Class I sites of Opdyke et al. (1977); open squares, Class II sites; open circles, Class III sites; X = indeterminate sites. Stratigraphy of Wagonhound Canyon after Osborn (1929); Willow Creek-Ouray area after Kay (1934); Indian Canyon after Dane (1954) and Dyni and Cashion (unpublished). See Appendix for location of sections.

al., 1995), the lower date would place the lowest Uinta normal magnetozone in Chron C20n (42.5-43.7 Ma), but the upper date is also within the age span of Chron C20n—yet it occurs in reversed rocks that should correlate with Chron C18r if the lowest normal magnetozone is truly C20n. Either one date or the other (or both) must be wrong. Because a number of authors (e.g., Krishtalka et al., 1987; Prothero and Swisher, 1992) have questioned the quality of Mauger's (1977) dates, I will not rely on them too heavily here.

Ideally, ^{40}Ar/^{39}Ar dating would provide much better chronostratigraphic tie points for the magnetostratigraphy, but the date of 47.4 ± 0.24 Ma in upper Uinta "B2" in Coyote Basin reported by Prothero and Swisher (1990) appears to be in error, since the date probably came from detrital contaminants. Swisher (personal communication) has tried dating other volcanic materials from the same Indian Canyon ashes dated by Mauger (1977), but so far without success. Hence, other correlation tie points must be sought.

Prothero and Swisher (1992, pp. 54-55) attempted to calibrate the Uinta Basin section based on similiarities of the faunas in other magnetically calibrated areas (Fig. 7). Two correlations were possible, and discussed in that paper. In one interpretation, the Uinta "A" normal magnetozone was correlated with Chron C21n, and the Uinta Basin sequence continues through C20n (lower Uinta "C") and C19n (lower Duchesne River Formation). We thought this interpretation was the most reasonable, because early Uintan faunas do not occur until Uinta "B1" in the Uinta Basin (which was reversed, and thought to be C20r), and early Uintan faunas are associated with upper Chron C20r in the Washakie Basin (Flynn, 1986), and also in Trans-Pecos Texas (Walton, 1992).

The other alternative was that the Uinta "A" normal magnetozone corresponds to Chron C20n, the lower Uinta "C" normal magnetozone correlates with C19n, and the lower Duchesne River Formation records C18n. Prothero and Swisher (1992, p. 55) rejected that interpretation because it conflicts with the K-Ar date of 42.7 ± 1.6 Ma just above the late Uintan Serendipity l.f. (Walton, 1992). It also conflicts with the ^{40}Ar/^{39}Ar date of 39.74 ± 0.07 Ma on the Lapoint tuff, which marks the contact between the Dry Gulch Creek and Lapoint members (Prothero and Swisher, 1992, p. 49, tables 2.1-2.2). This date would fall in Chron C18n of the Berggren et al. (1995) time scale. If the upper Dry Gulch Creek and lower Lapoint members of the Duchesne River Formation are correlative with Chron C18n, then our reversed-normal-reversed magnetozones in the Brennan Basin Member cannot also be C18n, but must correlate with C19r-C18r.

Prothero and Swisher (1992, p. 55) also suggested that the alternative correlation would conflict with the supposed occurrence of late Uintan faunas in the upper

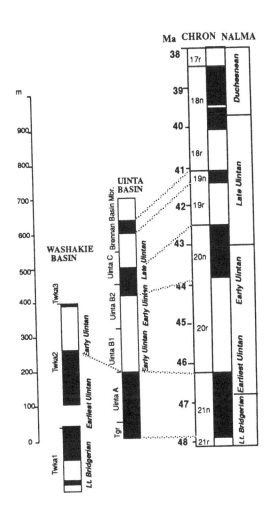

Figure 7. Magnetostratigraphic correlation of Uinta Basin sections with those of the Washakie Basin (after Flynn, 1986, and McCarroll et al., this volume, Chapter 2) and the magnetic polarity time scale (after Berggren et al., 1995). Abbreviations: Tgr = Green River Formation; Twkk = Kinney Rim Member; Twka1-3 = Adobe Town Member, levels 1-3.

part of the Washakie Basin section. Flynn (1986, p. 347 and fig. 9) briefly mentions upper Washakie faunal level "D," which he correlated (Flynn, 1986, fig. 9) with late Chrons C20n and early C19n. Although Flynn does not specifically describe this fauna as late Uintan, we followed earlier workers (Roehler, 1973; Turnbull, 1978), who speculated that the upper levels of the Washakie Formation might be late Uintan based on lithologic changes. As McCarroll et al. (1993; this volume, Chapter 2) point out, however, recent detailed study shows that there are no late Uintan (Uinta "C") fossils in the Washakie Basin; the sequence spans only the late Bridgerian, earliest Uintan (Shoshonian), and early Uintan (= Uinta "B"). However, this is a minor point. The preponderance of the evidence still favors the original interpretation of Prothero and Swisher that the Uinta Basin sequence spans Chrons C21n-C18r (see Walsh, this volume, Chapter 4, and the summary chapter to this volume).

BIOSTRATIGRAPHY

Judging from these calibrations, the Uintan is one of the longest of the North American land mammal "ages." It spans late Chron C21n (46.5 Ma) to early C18n (40 Ma), with a duration of about 6.5 million years. Unfortunately, it has never been finely divided into biostratigraphic zones, but instead the lithostratigraphically based "zones" of Uinta "B1," "B2," and "C" have been used. Osborn (1929, pp. 92-95) gave biostratigraphic names to these lithostratigraphic units. "B1" was called the *Metarhinus* zone, "B2" the *Eobasileus-Dolichorhinus* zone, and "C" the *Diplacodon* zone. Except for the detailed position of a few brontothere fossils (Osborn, 1929, figs. 65, 66) and some other specimens, however, no detailed range information was provided to facilitate biostratigraphic subdivision of these units.

Osborn's biostratigraphic zonal terminology was not adopted by later workers, including Kay and Peterson of the Carnegie Museum, who used lithostratigraphic terms, or Wood, who named the Wagonhound and Myton members. In spite of the fact that Osborn's zonal names were established long before, Gunnell (1989) named two "zones" (Ui1, the *Epihippus* assemblage "zone," and Ui2, the camelid-canid appearance "zone") for the early and late Uintan. However, these are not based on recent biostratigraphic work, but simply rename the distinction between the faunas of Uinta "B" and "C." In addition, both names are now inappropriate, as discussed below.

It would be expected that the 6.5 million years of the Uintan should be divisible by biostratigraphy into more than three zones. However, in most cases the original collections do not have adequate stratigraphic data, since the collectors (mostly Peterson, Kay, Clark, Riggs, and other early twentieth-century parties) rarely recorded the position of their fossils beyond Uinta "B" or "C." I have made an extensive survey of the collections and field notes (primarily in the Carnegie Museum and American Museum) and it is clear that more detailed information cannot be obtained from most of these old collections. Unfortunately, the best specimens that had weathered out over the centuries from the Uinta Formation were removed by these early collectors, and numerous parties since then have not obtained significant new collections that would help refine the biostratigraphy. Only Myton Pocket (Hamblin, 1987), and occasionally some of the other classic localities, such as White River

Pocket (e.g., Dawson, 1966), are still producing. Tab Rasmussen (personal communication) has recently collected in the Coyote Basin/Bonanza area.

Given these limitations, I re-examined the museum collections and their original labels to tie down as many specimens as possible to the new magnetic sections. Fortunately, many of the specimens come from named quarries and localities whose position in the section is known. I kept the biostratigraphy of the eastern (Wagonhound Canyon-Bonanza-Coyote Basin-Kennedy Hole) and central (Willow Creek-Ouray-Leota Bottom-Brennan Basin, plus Myton Pocket) areas separate, since the "Uinta A" rocks of Willow Creek are temporally equivalent to Uinta "B" in the Wagonhound-Bonanza area (Fig. 6).

The results are shown in Table 1. As is clear from the table, most specimens come from a single local level in only one area, so the ranges are not very long or overlapping. In addition, these are only partial range zones, or teilzones, for a single basin. Many of these ranges would surely be extended if other regions were taken into account (e.g., Black and Dawson, 1966; Stucky, 1992). This is particularly true of the small mammals. Most of the insectivores, primates, rabbits, and rodents are known from only two localities: White River Pocket and Myton Pocket. Very little collecting for small mammals has occurred in higher or lower beds in the Uinta Basin, so many of these smaller mammals would have longer ranges if there were not such a size bias in the collections.

Nevertheless, some overlapping ranges of biostratigraphic utility can be detected from the existing collections. These ranges can also be combined with stratigraphic data from the adjacent Washakie-Sand Wash Basins of Colorado and Wyoming (McCarroll et al., this volume, Chapter 2; Stucky et al., this volume, Chapter 3), to produce a composite biostratigraphy for the Uintan in the greater Green River Basin (Fig. 8). Of course, the teilzones for this basin would be extended if the biostratigraphic data from other regions (especially Texas, San Diego, the Sespe Formation, the Badwater area, and others) were added to them (see Krishtalka et al., 1987).

Uinta A

As discussed above, approximately 500 feet of fluvial sandstones at the base of Wagonhound Canyon were part of Peterson's Uinta A (in Osborn, 1895), but Osborn (1929) split off the upper fossiliferous part of Peterson's Uinta A and renamed it Uinta B1. Hence, most of the pre-1929 museum labels that say "Uinta A" should be read "Uinta B1," and very few specimens can be traced positively to Uinta A *sensu* Osborn (1929). I have double-checked the specimen labels and the museum records, and the specimens of *Amynodon, Triplopus, Metarhinus, Dolichorhinus,* and

Table 1. Stratigraphic distribution of mammals from the Uinta Basin. Most ranges after Osborn (1929), Peterson (1931, 1934) Peterson and Kay (1931), Kay (1934), Black and Dawson (1966), Hamblin (1987), Emry (1981), and Stucky (1992). Abbreviations: WRP, White River Pocket, R/H/L, Randlett, Halfway, Lapoint faunas.

TAXON	WAGONHOUND MBR.			MYTON MBR.			DUCHESNE R.
Eastern area	"B1"	"B2"		"C" (Kennedy Hole)			
Ma	46.0			44.0	43.0	40.0	
Central area	Willow Creek	WRP		Myton Pocket	Leota Q.	R/H/L	
INSECTIVORES							
Talpavus duplus				MP			
Nyctitherium sp.				MP			
Micropternodus sp.				MP			
Simidectes magnus				MP			
Simidectes medius			WRP				
Apatemys uintensis		B2					
Protictops alticuspidens							L
PRIMATES							
Ourayia uintensis		B2, WRP		KH, MP			
Mytonius hopsoni				MP			
Macrotarsius jepseni			WRP				
RODENTS							
"Paramyids"							
Thisbemys uintensis		B2					
Thisbemys medius				MP, KH			
Leptotomus mytonensis				MP			
Leptotomus leptodus				KH			
Leptotomus sciuroides				KH			
Leptotomus kayi							R
Janimus rhinophilus				MP			
Microparamys dubius			WRP				
Ischyrotomus petersoni		B2		KH			
Ischyrotomus compressidens			WRP	MP			
Ischyrotomus eugenei				MP			
Reithroparamys gidleyi				KH (base)			
Prosciurines							
Mytonomys robustus			WRP	MP			R
Sciuravids							
Sciuravus altidens			WRP				
Sciuravus popi			WRP				
Protoptychids							
Protoptychus hatcheri				KH			
Cylindrodonts							
Pareumys troxelli			WRP				
Pareumys grangeri			WRP				
Pareumys milleri			WRP				
Pareumys guensbergi							L
Eomyids							
Protadjidaumo typus						R, L	
LAGOMORPHS							
Mytonolagus petersoni				MP			
Mytonolagus robustus							R
CREODONTS							
Oxyaenodon dysclerus			WRP	KH			
Oxyaenodon wortmani		B2					
Limnocyon douglassi		B2		MP			
Apataelurus kayi		B2					
Hyaenodon vetus							L

TAXON	WAGONHOUND MBR.			MYTON MBR.		DUCHESNE R.
Eastern area	"B1"	"B2"		"C" (Kennedy Hole)		
Ma	46.0			44.0 43.0	40.0	
Central area	Willow Creek	WRP		Myton Pocket	Leota Q.	R/H/L

TAXON	B1/WC	B2	WRP	MP, KH	LQ	R/H/L
CARNIVORES						
Miacis gracilis		B2		MP		
Miacis uintensis		B2				
Uintacyon robustus				MP		
Procynodictis vulpiceps				MP, KH (base)		
Mimocyon longipes				MP		
Miocyon scotti		B2		MP		
Eosictis avinoffi						H
ARTIODACTYLS (Gazin, 1955)						
"Dichobunids"						
Auxontodon pattersoni				MP		
Bunomeryx montanus				KH		
Bunomeryx elegans				MP, KH		
Hylomeryx annectens			WRP			
Hylomeryx quadricuspis				MP, KH		
Mesomeryx grangeri		B2				
Pentacemylus leotensis					LQ	
Pentacemylus progressus				MP		R
Mytonomeryx scotti				MP		
Simimeryx minutus						L
Achaenodonts						
Achaenodon uintensis	WC	B2				
Entelodonts						
Brachyhyops wyomingensis						L
Agriochoeres						
Protoreodon petersoni				MP, KH		
Protoreodon pumilus				MP, KH		R
Protoreodon parvus	B1	B2				
Protoreodon paradoxicus			WRP			
Protoreodon medius		B2,	WRP			
Protoreodon primus					LQ	R
Protoreodon minor				KH		
Diplobunops matthewi				MP	LQ	
Diplobunops vanhouteni			WRP			
Diplobunops crassus						R
Agriochoerus maximus						L
Oromerycids						
Oromeryx plicatus	B1?	B2		MP		
Protylopus petersoni		B2		KH		
Protylopus annectens				MP		
Camelids						
Poebrodon kayi				MP		
Protoceratids						
Leptotragulus proavus		B2		MP, KH		
Leptotragulus clarki				MP		
Leptotragulus medius				MP		
Leptoreodon marshi		B2		MP, KH		
Poabromylus kayi						L
MESONYCHIDS						
Harpagolestes uintensis	B1					
Harpagolestes leotensis				MP	LQ	
Mesonyx obtusidens	B1	B2				
Hessolestes ultimus						L

TAXON	WAGONHOUND MBR.		MYTON MBR.		DUCHESNE R.
Eastern area	"B1"	"B2"	"C" (Kennedy Hole)		
Ma	46.0		44.0	43.0	40.0
Central area	Willow Creek WRP		Myton Pocket	Leota Q.	R/H/L

PERISSODACTYLS

Equids

Epihippus gracilis		B2, WRP			
Epihippus parvus			MP, KH		
Epihippus (Duchesnehippus) intermedius					H
"Tapiroids"					
Isectolophus annectens		B2, WRP	MP		
Dilophodon leotanus				LQ	R
Chalicotheres					
Eomoropus amarorum	B1	B2			
Rhinocerotoids					
Hyrachyus eximius	B1				
Amynodon reedi		B2			
Amynodon advenus		B2, WRP	MP, KH		
Megalamynodon regalis					R
Triplopus implicatus	B1,WC	B2	KH (base)		
Triplopus obliquidens	B1	B2, WRP	MP, KH	LQ	
Triplopus rhinocerinus			?MP		
Forstercooperia grandis	B1				
Epitriplopus uintensis		B2	MP		
Epitriplopus medius					R, L
Hyracodon primus					L
Brontotheres					
Sthenodectes priscus	WC (just above Green River Fm.)				
Sthenodectes incisivum	B1				
Dolichorhinus	B1, WC	B2			
Eotitanotherium osborni		WC (just below WRP)			
Diplacodon			MP, KH	LQ	
Metatelmatherium ultimum		KH			
Rhadinorhinus	B1, WC	B2			
Metarhinus	B1				
Protitanotherium			KH	LQ	R
Duchesneodus uintensis					L

UINTATHERES

Eobasileus cornutus	B1, WC	B2			
Uintatherium anceps	B1	B2			

TAENIODONTS

Stylinodon mirus		B2			

Forstercooperia reported by Krishtalka et al. (1987, p. 89) from Uinta "A" are actually from Uinta "B1" in Osborn's (1929) terminology (see Prothero and Swisher, 1992, p. 54).

Peterson (in Osborn, 1895), Riggs (1912), and Osborn (1929) all found the lower 500 feet of the Uinta Formation (all of Uinta "A" in modern parlance) in Wagonhound Canyon to be unfossiliferous. Kay (1934, plate XLVI) indicates that _Sthenodectes priscus_ comes from a horizon about 30 feet above the base of the Green River Formation in Willow Creek, which he correlated with Uinta A. However, this sequence is reversed in polarity and apparently correlates with Uinta B1 in the Bonanza area, not with the Uinta A sandstones. Hence, there are no well documented mammal fossils from Uinta A as it is now understood.

Uinta B1

The lower part of Uinta B is not particularly fossiliferous, either, so most of the stratigraphic data must be inferred from Riggs (1912, fig. 1) and Osborn (1929, fig. 65) in the Wagonhound Canyon-Bonanza area, and from Kay (1934, plate XLVI) for the Willow Creek-Ouray area. In addition, there are important localities like "Well no. 2" (Peterson and Kay, 1931, plate IX) just north of Bonanza that can be placed high in Uinta B1, allowing the determination of the exact level of a number of specimens.

Most of the specimens positively known from Uinta B1 are brontotheres, but among them _Metarhinus fluviatalis_ last appears in Uinta B1, and _Sthenodectes_ (both _priscus_ and _incisivum_, if they are both valid) appears to be restricted to this level. Mader (1989) indicates that

"*Sthenodectes*" *australis* from the Pruett Formation of Texas (Wilson, 1978) is not referable to that genus. A few taxa (*Protoreodon parvus, Triplopus obliquidens*) first appear at the B1 level, and a few more (*Pareumys grangeri, Harpagolestes uintensis, Hyrachyus eximius, Forstercooperia grandis*) last occur at this level. Osborn (1929) called this interval the "*Metarhinus* zone," and that name could still be used, since *Metarhinus* is one of the commonest mammals in the interval. However, *Metarhinus* is no longer restricted to Uinta B1, but first occurs in the middle Adobe Town Member in the Washakie Basin (McCarroll et al., this volume, Chapter 2), which correlates with Chron C21n. Mader (1989) has also synonymized *Rhadinorhinus* with *Metarhinus*, extending the range of *Metarhinus* into Uinta B2. Hence, the Uinta B1 interval could be redefined by the overlapping ranges of *Protoreodon parvus* and *Triplopus obliquidens* (first occurrence), and *Hyrachyus eximius, Forstercooperia grandis, Pareumys grangeri*, and *Harpagolestes uintensis* (last occurrence). *Sthenodectes* is the only taxon restricted to this interval.

Uinta B2

By contrast with Uinta B1, the upper part of the Wagonhound Member is very fossiliferous due to the extensive collections of small mammals from White River Pocket. Osborn (1929) called this the "*Eobasileus-Dolichorhinus* zone," and both of those taxa are still characteristic, and last occur in the interval (although both have earlier occurrences). A long list of taxa (mostly small mammals from White River Pocket) is restricted to the upper Wagonhound Member (Fig. 8), including *Simidectes medius, Apatemys uintensis, Ourayia uintensis, Thisbemys uintensis, Microparamys dubius, Pareumys troxelli, Pareumys milleri, Oxyaenodon wortmani, Apataelurus kayi, "Miacis" uintensis, Hylomeryx annectens, Mesomeryx grangeri, Protoreodon paradoxicus, Protoreodon medius, Diplobunops vanhouteni*, and *Amynodon reedi*. Numerous other taxa (*Ischyrotomus petersoni, Oxyaenodon dysclerus, Limnocyon douglassi, Miocyon scotti, Oromeryx plicatus, Leptotragulus proavus, Leptoreodon marshi, Isectolophus annectens*) first occur in Uinta B2 and continue into Uinta C. There is a major faunal break between the Wagonhound and Myton members, with taxa which persisted from the Bridgerian and earliest Uintan (*Sciuravus, Eobasileus, Uintatherium, Stylinodon, Eomoropus, Protoptychus hatcheri, Achaenodon, Sphenocoelus, Epihippus gracilis, Pareumys grangeri*, and *Metarhinus*) disappearing at the top of Uinta B.

Uinta C

Again, there is a long list of small mammals from the lower part of the Myton Member because of the extraordinarily diverse collections from Myton Pocket. However, most of the other quarries and levels in the Myton Member are much less rich, so the local ranges of Uinta C mammals are often restricted to this single locality. Nearly all the fossil localities appear to be located in the lower part of the member (except for Leota Quarry), and the upper part of the Myton Member above Leota Quarry is virtually unfossiliferous.

Taxa restricted to the lower Myton Member (mostly small mammals from Myton Pocket) include *Talpavus duplus, Simidectes magnus, Mytonius hopsoni, Thisbemys medius, Leptotomus mytonensis, Leptotomus sciuroides, Janimus rhinophilus, Ischyrotomus eugenei, Reithroparamys gidleyi, Uintacyon robustus, Procynodictis vulpiceps, Mimocyon longipes, Auxontodon pattersoni, Bunomeryx montanus, Bunomeryx elegans, Hylomeryx quadricuspis, Pentacemylus progressus, Mytonomeryx scotti, Protoreodon petersoni, Protoreodon pumilis, Protoreodon minor, Leptotragulus clarki, Leptotragulus medius, Epihippus parvus*, and *Metatelmatherium ultimum*. In addition, *Diplacodon, Protitanotherium, Harpagolestes leotensis*, and *Diplobunops matthewi* are restricted to the interval spanning Myton Pocket to Leota Quarry. *Pentacemylus leotensis, Protoreodon primus, Dilophodon leotanus*, and *Pentacemylus progressus* first occur at the Leota Quarry level; the first two taxa are also restricted to that locality.

In addition to this uniquely Uinta C fauna, a number of taxa from the early Uintan (*Amynodon advenus, Poebrodon, Leptotomus leptodus, Protylopus petersoni, Triplopus implicatus, Triplopus obliquidens, Ischyrotomus petersoni, Ischyrotomus compressidens, Protoreodon parvus, Oxyaenodon dysclerus, Limnocyon douglassi, Miocyon scotti, Oromeryx plicatus, Leptotragulus proavus, Leptoreodon marshi, Isectolophus annectens*) last occur at the Myton Pocket level. Only one taxon (*Mytonolagus*) that first appears in Uinta C continues into the Duchesnean. Thus, there is a big drop in diversity and change in the fauna in the middle of the late Uintan (between Leota Quarry and the Randlett horizon, around 42 Ma), as numerous authors (Black and Dawson, 1966; Krishtalka et al., 1987; Stucky, 1990, 1992) have noted.

Osborn (1929) called Uinta C the "*Diplacodon-Protitanotherium* zone," and that name is still appropriate, since both of these brontotheres are restricted to Uinta C. However, the names coined by Gunnell (1989) are no longer useful for the Uintan. His "*Epihippus* assemblage zone" (Ui1) for Uinta A-B is not very descriptive, since *Epihippus parvus* occurs in Uinta C, and *Epihippus (Duchesnehippus) intermedius* is found in the Halfway fauna. With the discovery of the camelid *Poebrodon* in the early Uintan middle Adobe Town Member of the Washakie Basin (McCarroll et al., this volume, Chapter 2), the "camelid-canid appearance zone" (Ui2) is no longer appropriate for Uinta C. Nor does Wang (1994) regard any Uintan "miacids" as canids. The earliest true canid is from the Duchesnean.

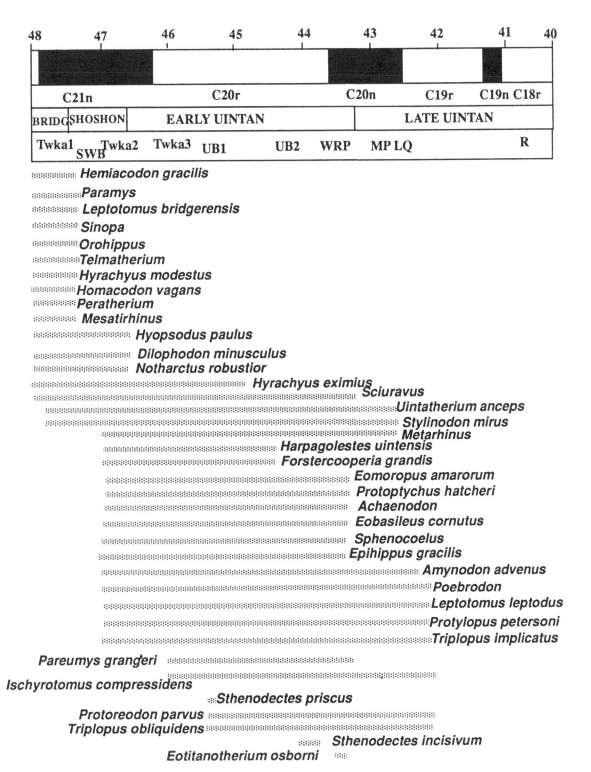

Figure 8. Local stratigraphic ranges (teilzones) of Uintan mammals in the Washakie, Sand Wash, and Uinta Basins, based on data from Table 1 and McCarroll et al. (this volume, Chapter 2) and Stucky et al. (this volume, Chapter 3). Time scale after Berggren et al. (1995); other abbreviations as in Fig. 7.

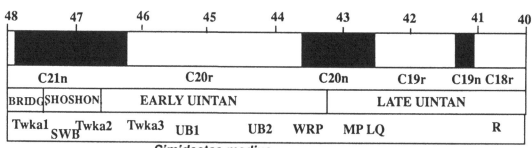

Simidectes medius
Apatemys uintensis
Ourayia uintensis
Thisbemys uintensis
Microparamys dubius
Pareumys troxelli
Pareumys milleri
Oxyaenodon wortmani
Apataelurus kayi
Miacis uintensis
Hylomeryx annectens
Mesomeryx grangeri
Protoreodon paradoxicus
Protoreodon medius
Diplobunops vanhouteni
Amynodon reedi
Ischyrotomus petersoni
Oxyaenodon dysclerus
Limnocyon douglassi
Miocyon scotti
Oromeryx plicatus
Leptotragulus proavus
Leptoreodon marshi
Isectolophus annectens
Mytonomys robustus
Talpavus duplus
Simidectes magnus
Mytonius hopsoni
Thisbemys medius
Leptotomus mytonensis
Leptotomus sciuroides
Janimus rhinophilus
Ischyrotomus eugenei
Reithroparamys gidleyi

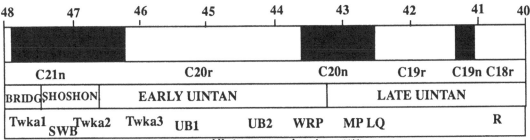

48	47	46	45	44	43	42	41	40

C21n		C20r			C20n		C19r	C19n	C18r

BRIDG	SHOSHON	EARLY UINTAN				LATE UINTAN			

Twka1	SWB	Twka2	Twka3	UB1	UB2	WRP	MP LQ	R

Uintacyon robustus

Procynodictis vulpiceps

Mimocyon longipes

Auxontodon pattersoni

Bunomeryx montanus

Bunomeryx elegans

Hylomeryx quadricuspis

Pentacemylus progressus

Mytonomeryx scotti

Protoreodon petersoni

Protoreodon pumilis

Protoreodon minor

Leptotragulus clarki

Leptotragulus medius

Epihippus parvus

Metatelmatherium ultimum

Diplacodon

Protitanotherium

Harpagolestes leotensis

Diplobunops matthewi

Mytonolagus

Pentacemylus leotensis

Protoreodon primus

Dilophodon leotanus

Pentacemylus progressus

Leptotomus kayi

Protadjidaumo typus

Diplobunops crassus

Megalamynodon regalis

Epitriplopus medius

Duchesne River Formation and Duchesnean

As discussed above (and is apparent from Table 1 and Fig. 8), the faunas of the Duchesne River Formation are so scarce, and their biostratigraphic levels are so poorly constrained, that it is pointless to suggest a range-zone biostratigraphy in the Uinta Basin. This is much better done in the Sespe Formation of California (see Kelly, 1990, and Prothero et al., this volume, Chapter 8) and the Trans-Pecos Texas region (see Wilson, 1986; Lucas, 1992; and the summary chapter to this volume).

CONCLUSIONS

Although confusing stratigraphic problems, and the lack of precise biostratigraphic data associated with many specimens, have long hampered our understanding of the Uintan, magnetic stratigraphy and detailed stratigraphic re-examination of the collections makes refinements possible. It seems that the Uintan can be subdivided into at least five discrete intervals:

Earliest Uintan ("Shoshonian") spans late Chron C21n-early Chron C20r (about 46.5-46 Ma), and is best represented by faunas in the Tepee Trail Formation of northwest Wyoming, the middle Adobe Town Member of the Washakie Formation in the Washakie and Sand Wash basins, the Friars Formation in the San Diego area (Walsh et al., this volume, Chapter 6; Flynn, 1986, pp. 379-380), and the Whistler's Squat l.f. in Trans-Pecos Texas.

Uinta B1 ("*Metarhinus* zone" of Osborn, 1929) spans early Chron C20r (45-46 Ma). It is sparsely fossiliferous, but has some distinctive taxa. The fossils of the upper Adobe Town Member of the Washakie Formation may also correlate with this interval.

Uinta B2 ("*Eobasileus-Dolichorhinus* zone" of Osborn, 1929) spans late Chron C20r and early Chron C20n (43.5-45.0 Ma), and includes many distinctive taxa (especially the collections of smaller mammals from White River Pocket). Most "Uinta B" correlatives (Black and Dawson, 1966; Krishtalka et al., 1987) appear to fall in this time interval.

Uinta C ("*Diplacodon* zone" of Osborn, 1929) spans most of Chron C20n (42.5-43.5 Ma), including the extensive faunas of Myton Pocket, Kennedy Hole, and Skull Pass, Leota, and Thorne quarries. Most "Uinta C" faunas of other authors (Black and Dawson, 1966; Krishtalka et al., 1987) probably correlate with this interval, or with the unfossiliferous upper Myton Member (which spans early Chron C19r, 42.0-42.5 Ma).

Randlett horizon (in the Brennan Basin Member of the Duchesne River Formation of Andersen and Picard, 1972) includes latest Uintan faunas, and correlates with early Chron C19n-C18r (40.0-41.0 Ma). Correlative faunas might include the Candelaria l.f. of Trans-Pecos Texas.

ACKNOWLEDGMENTS

I thank the Donors of the Petroleum Research Fund of the American Chemical Society, and the National Science Foundation (grant EAR87-08221) for supporting this research. Joe Kirschvink graciously allowed me to use the Caltech paleomagnetics laboratory to run the samples. Preliminary samples were run by Annie Walton in the paleomagnetics laboratory of the University of Texas at Austin. Field sampling would never have been possible without the hard work, enthusiasm, and good spirits of several field crews. In 1986, they included Dana Gilchrist, Kecia Harris, and Annie Walton. In 1987, they included Jill Bush, Susan Briggs, John Foster, and Steve King. In 1988, they included Jeff Amato, Jennifer Chean, and Jim Finegan. I thank Alden Hamblin and J. R. Dyni for all their help with Uinta Basin stratigraphy, and Carl Swisher for valiantly trying to get a decent date out of the Uinta Basin rocks. I thank Craig Black, Alden Hamblin, Tom Kelly, Jay Lillegraven, Bryn Mader, Steve McCarroll, Malcolm McKenna, John Storer, Steve Walsh, and Annie Walton for reviewing the manuscript.

LITERATURE CITED

Andersen, D. W., and M. D. Picard. 1972. Stratigraphy of the Duchesne River Formation (Eocene-Oligocene?), northern Uinta Basin, northeastern Utah. Utah Geological and Mineralogical Survey Bulletin 97:1-23.

Andersen, D. W., and M. D. Picard. 1974. Evolution of synorogenic clastic deposits in the intermontane Uinta Basin of Utah. SEPM Special Publication 22:167-189.

Berggren, W. A., D. V. Kent, M.-P. Aubry, C. C. Swisher III, and K. G. Miller. 1995. A revised Paleogene geochronology and chronostratigraphy. SEPM Special Publication 54:129-212.

Black, C. C., and M. R. Dawson. 1966. A review of the late Eocene mammalian faunas from North America. American Journal of Science 264:321-349.

Bradley, W. H. 1929. Algae reefs and oolites of the Green River Formation. U.S. Geological Survey Professional Paper 154-G, 21 pp.

Bradley, W. H. 1931. Origin and microfossils of the oil shale of the Green River Formation of Colorado and Utah. U.S. Geological Survey Professional Paper 168, 58 pp.

Bradley, W. H. 1964. Geology of Green River Formation and associated Eocene rocks in southwestern Wyoming and adjacent parts of Colorado and Utah. U.S. Geological Survey Professional Paper 496-A, 86 pp.

Bruhn, R. L., M. D. Picard, and S. L. Beck. 1983. Mesozoic and early Tertiary structure and sedimentology of the central Wasatch Mountains, Uinta Mountains, and Uinta Basin. Utah Geological and Mineralogical Survey Special Studies 59:63-105.

Butler, R. F. 1992. Paleomagnetism. Blackwell, Boston.

Cande, S. C., and D. V. Kent. 1993. A new geomagnetic polarity time scale for the late Cretaceous and Cenozoic. Journal of Geophysical Research 97 (B10):13917-13951.

Cashion, W. B. 1967. Geology and fuel resources of the Green River Formation, southeastern Uinta Basin, Utah and Colorado. U.S. Geological Survey Professional

Paper 548:1-48.

Clark, J., J. R. Beerbower, and K. K. Kietzke. 1967. Oligocene sedimentation, stratigraphy, paleoecology, and paleoclimatology in the Big Badlands of South Dakota. Fieldiana Geology Memoirs 5.

Comstock, T. B. 1875. Report upon the reconnaissance of northwestern Wyoming including Yellowstone National Park, for 1873, by Wm. A. Jones, House Rep. Exec. Doc. 285, 43d Congress, 1st session, Jan. 1875.

Dane, C. H. 1954. Stratigraphic and facies relationships of the upper part of the Green River Formation and lower part of the Uinta Formation in Duchesne, Uintah, and Wasatch counties. Bulletin of the American Association of Petroleum Geologists 38:405-425.

Dane, C. H. 1955. Stratigraphic and facies relationships of upper part of Green River Formation and lower part of Uinta Formation in Duchesne, Uintah, and Wasatch Counties, Utah. U.S. Geological Survey Oil and Gas Chart OC-52.

Dawson, M. R. 1966. Additional late Eocene rodents (Mammalia) from the Uinta Basin, Utah. Annals of the Carnegie Museum 38 (4):97-114.

Dickinson, W. R., M. A. Klute, M. J. Hayes, S. U. Janecke, E. R. Lundin, M. A. McKittrick, and M. D. Olivares. 1988. Paleogeographic and plate tectonic setting of the Laramide sedimentary basins in the central Rocky Mountain region. Geological Society of America Bulletin 100:1023-1039.

Dickinson, W. R., T. F. Lawton, and K. F. Inman. 1986. Sandstone detrital modes, central Utah foreland region: stratigraphic record of Cretaceous-Paleogene tectonic evolution. Journal of Sedimentary Petrology 56:276-293.

Emry, R. J. 1981. Additions to the mammalian fauna of the type Duchesnean, with comments on the status of the Duchesnean "Age." Journal of Paleontology 55:563-570.

Fisher, R. A. 1953. Dispersion on a sphere. Proceedings of the Royal Astronomical Society A217:295-305.

Flynn, J. J. 1986. Correlation and geochronology of middle Eocene strata from the western United States. Palaeogeography, Palaeoclimatology, Palaeoecology 55:335-406.

Franczyk, K. J., et al. 1989. Evolution of resource-rich foreland and intermontane basins in eastern Utah and western Colorado. 28th International Geological Congress Field Trip Guidebook T324, American Geophysical Union, Washington, D.C.

Gazin, C. L. 1955. A review of the upper Eocene artiodactyls of North America. Smithsonian Miscellaneous Collections, 131 (7):1-96.

Gunnell, G. F. 1989. Evolutionary history of the Microsyopoidea (Mammalia, ?Primates) and the relationship between Plesiadapiformes and Primates. University of Michigan Papers in Paleontology 27:1-157.

Hamblin, A. H. 1987. Paleogeography and paleoecology of Myton Pocket, Uinta Basin, Utah (Uinta Formation—Upper Eocene). Brigham Young University Geology Studies 34(1):33-60.

Johnson, H. P., W. Lowrie, and D. V. Kent. 1975. Stability of anhysteretic remanent magnetization in fine and coarse magnetite and maghemite particles. Geophysical Journal of the Royal Astronomical Society 41:1-10.

Johnson, R. C. 1985. Early Cenozoic history of the Uinta and Piceance Creek Basins, Utah and Colorado, with special reference to the development of Eocene Lake Uinta; pp. 247-276 in R. M. Flores and S. S. Kaplan (eds.), Cenozoic Paleogeography of the West-Central United States. SEPM Rocky Mountain Paleogeography Symposium 3.

Kay, J. L. 1934. The Tertiary formations of the Uinta Basin. Annals of the Carnegie Museum, 23:357-371.

Kay, J. L. 1957. The Eocene vertebrates of the Uinta Basin, Utah. Intermountain Association of Petroleum Geologists Guidebook, 8th Annual Field Conference, pp. 110-114.

Kay, J. L., and V. Garwood. 1953. Guidebook to the 6th Annual Field Conference of the Society of Vertebrate Paleontology in northeastern Utah. 34 pp.

Kelly, T. S. 1990. Biostratigraphy of Uintan and Duchesnean land mammal assemblages from the middle member of the Sespe Formation, Simi Valley, California. Contributions to Science of the Natural History Museum of Los Angeles County 419:1-42.

Krishtalka, L., R. K. Stucky, R. M. West, M. C. McKenna, C. C. Black, T. M. Bown, M. R. Dawson, D. J. Golz, J. J. Flynn, J. A. Lillegraven, and W. D. Turnbull. 1987. Eocene (Wasatchian through Duchesnean) biochronology of North America; pp. 77-117 in M. O. Woodburne (ed.), Cenozoic Mammals of North America, Geochronology and Biostratigraphy. University of California Press, Berkeley.

Lucas, S. G. 1992. Redefinition of the Duchesnean land mammal "age," late Eocene of western North America; pp. 88-105 in D. R. Prothero and W. A. Berggren (eds.), Eocene-Oligocene Climatic and Biotic Evolution. Princeton University Press, Princeton, N. J.

Mader, B. J. 1989. The Brontotheriidae: a systematic revision and preliminary phylogeny of North American genera; pp. 458-484 in D. R. Prothero and R. M. Schoch (eds.), The Evolution of Perissodactyls. Oxford University Press, New York.

Marsh, O. C. 1870. Professor Marsh's Rocky Mountain expedition: discovery of the Mauvaises Terres formation in Colorado. American Journal of Science, 2nd series, 59:292.

Mauger, R. L. 1977. K-Ar ages of biotites from tuffs in Eocene rocks of the Green River, Washakie, and Uinta basins, Utah, Wyoming, and Colorado. Contributions to Geology of the University of Wyoming 15:17-41.

McCarroll, S. M., J. J. Flynn, and W. D. Turnbull. 1993. Biostratigraphy and magnetic polarity correlations of the Washakie Formation, Washakie Basin, Wyoming. Journal of Vertebrate Paleontology 13 (3):49A.

Opdyke, N. D., E. H. Lindsay, N. M. Johnson, and T. Downs. 1977. The paleomagnetism and magnetic polarity stratigraphy of the mammal-bearing section of Anza-Borrego State Park, California. Journal of Quaternary Research 7:316-329.

Osborn, H. F. 1895. Fossil mammals of the Uinta Basin: Expedition of 1894. Bulletin of the American Museum of Natural History 7:71-105.

Osborn, H. F. 1929. The titanotheres of ancient Wyoming, Dakota, and Nebraska. United States Geological Survey Monograph 55:1-953 (2 vols.).

Osborn, H. F., and W. D. Matthew. 1909. Cenozoic mammal horizons of western North Americ. United States Geological Survey Bulletin 361:1-138.

Peterson, O. A. 1919. Report upon the material discovered in the upper Eocene of the Uinta Basin by Earl Douglass in the years 1908-1909, and by O. A. Peterson in 1912. Annals of the Carnegie Museum 12:40-169.

Peterson, O. A. 1924. Osteology of *Dolichorhinus longiceps* Douglass, with a review of the species of

Dolichorhinus in the order of their publication. Memoirs of the Carnegie Museum 9(4):405-472.

Peterson, O. A. 1931. New species from the Oligocene of the Uinta. Annals of the Carnegie Museum 21:58-78.

Peterson, O. A. 1934. List of species and description of new material from the Duchesne River Oligocene, Uinta Basin, Utah. Annals of the Carnegie Museum 23:373-389.

Peterson, O. A., and J. L. Kay. 1931. The Upper Uinta Formation of northeastern Utah. Annals of the Carnegie Museum 20:293-306.

Picard, M. D., and D. W. Andersen. 1975. Paleocurrent analysis and orientation of sandstone bodies in the Duchesne River Formation (Eocene-?Oligocene), northern Uinta Basin, northeastern Utah. Utah Geology 2:1-15.

Pluhar, C. J., J. L. Kirschvink, and R. W. Adams. 1991. Magnetostratigraphy and clockwise rotation of the Plio-Pleistocene Mojave River Formation, central Mojave Desert, California. San Bernardino County Museum Association Quarterly 38 (2):31-42.

Prothero, D. R. 1990. Magnetostratigraphy of the middle Eocene Uinta Formation, Uinta Basin, Utah. Journal of Vertebrate Paleontology 10(3):38A.

Prothero, D. R., and C. C. Swisher III. 1990. Magnetostratigraphy and ^{40}Ar/^{39}Ar dating of the middle Eocene Uinta Formation, Utah. Geological Society of America Abstracts with Programs 22 (7):A364.

Prothero, D. R., and C. C. Swisher, III. 1992. Magnetostratigraphy and geochronology of the terrestrial Eocene-Oligocene transition in North America; pp. 74-87 *in* D. R. Prothero and W. A. Berggren (eds.), Eocene-Oligocene Climatic and Biotic Evolution. Princeton University Press, Princeton, N. J.

Ray, R. G., B. H. Kent, and C. H. Dane. 1956. Stratigraphy and photogeology of the southwestern part of the Uinta Basin, Duchesne and Uintah Counties, Utah. U.S. Geological Survey Oil and Gas Investigations Map OM-171.

Riggs, E. S. 1912. New or little-known titanotheres from the lower Uintah formations. Field Museum of Natural History, Geology Series 4(2):17-41.

Roehler, H. W. 1973. Stratigraphy of the Washakie Formation in the Washakie Basin, Wyoming. U.S. Geological Survey Bulletin 1369:1-40.

Roehler, H. W. 1992a. Description and correlation of Eocene rocks in stratigraphic reference sections for the Green River and Washakie Basins, southwest Wyoming. U.S. Geological Survey Professional Paper 1506-D.

Roehler, H. W. 1992b. Correlation, composition, areal distribution, and thickness of Eocene stratigraphic units, greater Green River Basin, Wyoming, Utah, and Colorado. U.S. Geological Survey Professional Paper 1506-E.

Rowley, P. D., W. R. Hansen, O. Tweto, and P. E. Carrara. 1985. Geologic map of the Vernal 1° x 2° quadrangle, Colorado, Utah, and Wyoming. U.S. Geological Survey Map I-1526.

Ryder, R. T., T. D. Fouch, and J. H. Elison. 1976. Early Tertiary sedimentation in the western Uinta Basin, Utah. Geological Society of America Bulletin 87:496-512.

Scott, W. B. 1932. An introduction to geology, 3rd ed. Macmillan, New York, 441 pp.

Scott, W. B. 1945. The Mammalia of the Duchesne River Oligocene. Transactions of the American Philosophical Society 34:209-253.

Scott, W. B., and H. F. Osborn. 1887. Preliminary report on the vertebrate fossils of the Uinta formation, collected by the Princeton expedition of 1886. Proceedings of the American Philosophical Society, 24:255-264.

Scott, W. B., and H. F. Osborn. 1895. The Mammalia of the Uinta Formation. Transactions of the American Philosophical Society 15:461-572.

Simpson, G. G. 1946. The Duchesnean fauna and the Eocene-Oligocene boundary. American Journal of Science 244:52-57.

Stagner, W. L. 1941. The paleogeography of the eastern part of the Uinta Basin during Uinta B (Eocene) time. Annals of the Carnegie Museum 28:273-308.

Stucky, R. K. 1990. Evolution of land mammal diversity in North America during the Cenozoic. Current Mammalogy, 2:375-432.

Stucky, R. K. 1992. Mammalian faunas in North America of Bridgerian to early Arikareean "ages" (Eocene and Oligocene); pp. 463-493 *in* D. R. Prothero and W. A. Berggren (eds.), Eocene-Oligocene Climatic and Biotic Evolution. Princeton University Press, Princeton, N. J.

Surdam, R. C., and K. O. Stanley. 1979. Lacustrine sedimentation during the culminating phase of Eocene Lake Goshiute, Wyoming (Green River Formation). Geological Society of America Bulletin 90:93-110.

Surdam, R. C., and K. O. Stanley. 1980. Effects of changes in drainage-basin boundaries on sedimentation in Eocene Lake Goshiute and Uinta of Wyoming, Utah, and Colorado. Geology 8:135-139.

Tedford, R. H. 1970. Principles and practices of mammalian geochronology in North America. Proceedings of the North American Paleontological Convention 2F:666-703.

Turnbull, W. D. 1978. The mammalian faunas of the Washakie Formation, Eocene age, of southwestern Wyoming. Part 1: Introduction: The geology, history, and setting. Fieldiana Geology 33:569-601.

Untermann, G. E., and B. R. Untermann. 1964. Geology of Uintah County. Utah Geological and Mineralogical Survey Bulletin 72:1-112.

Walton, A. H. 1992. Magnetostratigraphy and the ages of Bridgerian and Uintan faunas in the lower and middle members of the Devil's Graveyard Formation, Trans-Pecos Texas; pp. 74-87 *in* D. R. Prothero and W. A. Berggren (eds.), Eocene-Oligocene Climatic and Biotic Evolution. Princeton University Press, Princeton, N. J.

Wang, X. 1994. Phylogenetic systematics of the Hesperocyoninae (Carnivora: Canidae). Bulletin of the American Museum of Natural History 221:1-207.

Wilson, J. A. 1978. Stratigraphic occurrence and correlation of early Tertiary vertebrate faunas, Trans-Pecos Texas, Part 1: Vieja area. Texas Memorial Museum Bulletin 25:1-42.

Wilson, J. A. 1984. Vertebrate fossil faunas 49 to 36 million years ago and additions to the species of *Leptoreodon* found in Texas. Journal of Vertebrate Paleontology 4:199-207.

Wilson, J. A. 1986. Stratigraphic occurrence and correlation of early Tertiary vertebrate faunas, Trans-Pecos Texas: Agua Fria-Green Valley areas. Journal of Vertebrate Paleontology 6:350-373.

Wood, H. E. II. 1934. Revision of the Hyrachyidae. Bulletin of the American Museum of Natural History 67:182-295.

Wood, H. E., R. W. Chaney, J. Clark, E. H. Colbert, G. L. Jepsen, J. B. Reeside, Jr., and C. Stock. 1941. Nomenclature and correlation of the North American continental Tertiary. Bulletin of the Geological Society of America 52:1-48.

APPENDIX

The routes of the various stratigraphic sections taken in the Uinta Basin are described below. The general geology of these areas has been mapped by Cashion (1967) and Rowley et al. (1985). Further details will be deposited in the Archives of the Department of Vertebrate Paleontology of the American Museum of Natural History.

Northeastern basin.—The first section was taken in the eastern Uinta Basin, and ran through Riggs's (1912, fig. 1), Peterson's (1924, fig. 1), and Osborn's (1929, fig. 63) sections at Wagonhound Canyon, near Bonanza, Utah, and then up through Coyote Basin to Kennedy Hole. At first glance these sections appear to be clearly diagrammed, but in actuality the thicknesses and the route of the section given by these authors are only approximate. In the field it was difficult to trace the route of the original sections, or determine the exact position of the key beds that separate Uinta "A" from "B1," or "B1" from "B2." However, after much double-checking and backtracking from known landmarks, a section was obtained which closely approximated the original sections in thicknesses and lithologic details.

The Wagonhound Canyon section began at the level of the White River at the mouth of the canyon, taking the bottom three sites in the uppermost part of the Evacuation Creek Member of the Green River Formation. The section then proceeded up the dirt road through the Uinta "A" sandstones in Wagonhound Canyon (NE and SE of NE NW Sec. 2, T10S R24E, Southam Canyon 7.5' quadrangle, Uintah County, Utah). At the top of the roadcuts, the route of section then proceeded westward up the ridges (SE SE SW Sec. 35, T9S R24E) to the highest point in the area (NE SW SE Sec. 34). Over 900 feet of section, including all of Uinta "A" and "B1" was measured. Projecting the highest beds of this section down the dip slope to the north, the next part of the section was taken in a northeasterly direction in the flats around Bonanza and Coyote Basin. The upper part of "B1" was measured in SW SW SW NW Sec. 23, T9S, R24E, Bonanza 7.5' quadrangle, and then continued through "B2" in ridges located in SW SW NW Sec. 13, SW SW SW Sec. 12, SE NW NE Sec. 11, culminating in the prominent ledge known as the *"Amynodon* Sandstone" (see Riggs, 1912, fig. 2; Osborn, 1929, fig. 64) in NE SW SW Sec. 34, T8S R24E and Center NW SW Sec. 24, just east of road from benchmark 5350. Moving down the dip slope of the *"Amynodon* Sandstone," the section was resumed in Uinta "C" mudstone badlands (the "Devil's Playground") just to the north (NE SW SW to SE SW NE Sec. 21, T8S R24E, Bonanza 7.5' quadrangle). This section culminated just below the color change (at the 5388 benchmark) from gray bentonitic "popcorn" claystones of Uinta "C" to the red and yellow clays of the Duchesne River Formation.

Over 400 feet of the Brennan Basin Member of the Duchesne River Formation was taken in several sections in the Kennedy Hole area (SW NW SE Sec. 17, T8S R24E Bonanza 7.5' quadrangle and finishing in SW SW NE Sec. 17, T8S R24E, Dinosaur NW 7.5' quadrangle), then continuing from NW SE SE Sec. 6, T8S R24E, Dinosaur NW 7.5' quadrangle, to NE SE SE of the same section. This was the highest continuous exposure of Duchesne River Formation in the area, and the rest of the formation could not be sampled because the exposures disappear completely under the sod-covered bench for miles to the north.

North-central basin.—Many important Carnegie Museum localities, including White River Pocket, and Leota, Thorne, and Skull Pass quarries, are located along a north-south transect through the central Uinta Basin. Kay

(in Peterson and Kay, 1931, Plate IX, and Kay and Garwood, 1953, p. 19) gave the location of these quarries and a route of section on a very crudely drawn, large-scale map of the area. When that route was traced, however, much of it proved to have no exposures whatsoever, and other parts were completely inaccessible. Like the cartoonish stratigraphic sections of Riggs and Osborn in the eastern basin, the generalized stratigraphic sections of Kay (in Kay, 1934, Plate XLVI, and Peterson and Kay, 1931, Plate X) were very inaccurate and all the details had to be carefully remeasured and redocumented in the field. Consequently, I picked a route that gave better exposures and closely paralleled the route indicated by Kay (1953). Rather than follow the road north to Ouray, which had no continuous outcrop along the top of the soil-covered bench, I followed the Willow Creek drainage, starting 150 feet below the contact with the Green River Formation. A series of sections was patched together along the east bank of Willow Creek, starting with SE NE SW Sec. 12, T11S R20E, Big Pack Mountain 7.5' quadrangle. The second leg was taken in SE SE SW Sec. 10 along the road up "Turkey Trail Hill," then traced west along the bed which caps the bench in the area to another outcrop along Willow Creek where the next 100 feet of Uinta Formation were exposed (NE SW SE Sec. 30, T9S R20E, Ouray 7.5' quadrangle). The Willow Creek section concluded in SW NE NE Sec. 19 in the same quadrangle. The highest level was then projected northeast to the area of White River Pocket (SE SW SE Sec. 5, T9S R20E, Ouray 7.5' quadrangle, just west of road), where the main White River Pocket section was taken.

The sandstone capping the Uinta "B" rocks at White River Pocket was then projected to the north, and the section was resumed in SE SE Sec. 19, T4S R3E, Ouray 7.5' quadrangle, where the lower part of Uinta "C" could be measured. The next segment was taken in SW SW NW NW Sec. 8, T4S R3E, Pelican Lake 7.5' quadrangle. A series of sections was then taken along the east-facing bluffs west of Leota Bottom (site of Leota Quarry), in segments located in SW SE SE Sec. 2, T8S R20E, Pelican Lake 7.5' quadrangle, then in NW SW NW Sec. 1, T8S R20E, Brennan Basin 7.5' quadrangle, followed by NE NW SW Sec. 36, T7S R20E in the same quadrangle, ending in the base of the Brennan Basin Member of the Duchesne River Formation ("unit 1" of the type section of the Brennan Basin Member of Andersen and Picard, 1972, p. 26). An additional 300 feet of that member (spanning Andersen and Picard's units 2-5) were measured and sampled in NW NE NE Sec. 18 and NW SW SE Sec. 7, T7S R21E, Brennan Basin 7.5' quadrangle. At this point, the section was discontinued, because the sandstones were too hard for hand sampling, the shales were too poorly exposed, and every unit had a deep red hematitic stain. In our 1986 reconnaissance sampling, almost all of our deep red Duchesne River samples proved to have an intractable chemical overprinting due to hematite. Hence, further sampling in the Duchesne River Formation was abandoned.

A third section in the central Uinta Basin ran through the Myton Pocket area, 15 miles east of Ouray, and 7 miles west of the town of Myton. Fortunately, this locality has been carefully measured and studied in recent times by Hamblin (1987). Approximately 360 feet of section of Uinta "C" ("Myton Member") were taken in SW SE NE Sec. 12 (lower half) and concluded in NW SW SW Sec. 6 (upper half), T4S R1E of the Uintah Special Meridian, Windy Ridge 7.5' quadrangle, Uintah County, Utah.

Western basin.—In the western portion of the Uinta

Basin, the fluvial sandstones and mudstones are replaced by the relict lacustrine facies of Eocene Lake Uinta. Instead of Uinta "A" sandstones, the Evacuation Creek Member of the Green River Formation is overlain by the "saline facies" and "sandstone and limestone facies" of the Uinta Formation (Dane, 1954, 1955; Ray et al., 1956). The most complete section through the region is in Indian Canyon, southwest of Duchesne, in Duchesne County, Utah. A brief sketch of the section was published by Dane (1954, fig. 2, column 5), and the stratigraphic relationships were summarized in Franczyk et al. (1989, fig. 14). A detailed, unpublished stratigraphic section through Indian Canyon made by J. R. Dyni and W. B. Cashion was graciously provided for our research by Dr. Dyni, and our stratigraphic section followed theirs closely. This section and its nomenclature are also cited by Mauger (1977), who attempted to date several ash layers in the Indian Canyon.

The section began on the distinctive "Horse Bench Sandstone Bed" (bed 221 of Dyni and Cashion) of Dane (1954, 1955), which can be traced eastward through the Evacuation Creek Member across most of the basin, and is apparently equivalent to the base of Uinta "A" in the east (Francyzk et al., 1989, fig. 14). The first segment (section 6 of Dyni and Cashion) was located in west side of the left fork of Indian Canyon, east half of the NW Sec. 22, T6S R7W, Jones Hollow 7.5' quadrangle, Duchesne County, Utah.

This segment ran from the Horse Bench Sandstone to the base of the saline facies of the Uinta Formation (unit 297 of Dyni and Cashion), covering the upper 460 feet of the Evacuation Creek Member. The second segment (Dyni and Cashion, section 7, units 298-350) along the bluffs on the west side of Indian Canyon, was located in NW SW Sec. 12 in the same map; it covered 360 feet of the lower saline facies. The third segment (section 8 of Dyni and Cashion), also on the west side of the left fork of Indian Canyon in SE SE Sec. 1, T6S R7W, Lance Canyon 7.5' quadrangle, covered units 351-390 of Dyni and Cashion. The fourth segment (Dyni and Cashion section 9, units 390-428) was located in the NW SW Sec. 22, T5S R6W, Buck Knoll 7.5' quadrangle; units 429-456 were collected in SW NE Sec. 22. The fifth segment (Dyni and Cashion, section 10, units 457-484) covers the uppermost 330 feet of the saline facies, and was located in NW NW NW Sec. 28, T4S R5W, Duchesne SW 7.5' quadrangle. The sixth segment covered the basal 200 feet of the limestone and sandstone facies of the Uinta Formation (units 485-503 of Dyni and Cashion), and was located in a west-trending dry wash 1 mile east of Indian Canyon in NE SW Sec. 14, T4S R5W, Duchesne 7.5' quadrangle. The remaining portion of the sandstone and limestone facies was not sampled because of difficulty of access and poor continuity of exposures.

2. Biostratigraphy and Magnetostratigraphy of the Bridgerian-Uintan Washakie Formation, Washakie Basin, Wyoming

STEVEN M. McCARROLL, JOHN J. FLYNN, AND WILLIAM D. TURNBULL

ABSTRACT

We summarize and add to the biostratigraphy and magnetic polarity stratigraphy of the Washakie Formation, Washakie Basin, Wyoming. Previously the Washakie Formation (divided into the lower Kinney Rim Member and the upper Adobe Town Member) was thought to contain rocks of early Bridgerian through late Uintan age. Continuing collection efforts in the Washakie Basin by the Field Museum of Natural History (FMNH) allow us to revise the biochronologically determined age of the Washakie Formation to late Bridgerian through early Uintan age.

A late Bridgerian age for the poorly fossiliferous Kinney Rim Member of the Washakie Formation is indicated by the presence of *Hyrachyus eximius* (a taxon with a late Bridgerian first occurrence elsewhere). In addition, the following taxa are also known from the Kinney Rim Member: *Peratherium* cf. *P. knighti*, cf. *Apatemys bellus*, *Hyopsodus* sp., *Orohippus* sp., *Mesatirhinus* sp., *Helaletes nanus*, and *Hyrachyus modestus* (all known from the Bridgerian elsewhere, but none restricted to the late Bridgerian, except possibly *Mesatirhinus*). In addition, taxa restricted to the early Bridgerian of the Bridger Basin (e.g., *Smilodectes*) have not been recovered from the Washakie Formation, except for a possible new species of tillodont from the Kinney Rim Member.

An early Uintan age for the upper unit of the Adobe Town Member, the uppermost unit in the Washakie Formation, is indicated by the occurrence of *Pareumys grangeri* (restricted to the early Uintan elsewhere). In addition, *Paramys compressidens* and *Epihippus gracilis* (both known from the Uintan elsewhere, but neither restricted to the early Uintan) are also known from the upper unit. Taxa with late Uintan first occurrences or late Uintan index taxa have not been recovered from the Washakie Formation.

The presence of earliest Uintan (Shoshonian "Subage") faunas in the Washakie Formation, indicated by the co-occurrence of smaller-bodied taxa characteristic of the late Bridgerian (e.g., *Notharctus robustior*, *Hyopsodus*, and *Dilophodon minusculus*) and larger-bodied taxa characteristic of the early Uintan (e.g., *Dolichorhinus*, *Eobasileus cornutus*, *Amynodon advenus*, and *Achaenodon*), is documented for the first time in the middle unit of the Adobe Town Member.

INTRODUCTION

This paper contains an overview of the biostratigraphic and magnetostratigraphic information currently available from the Washakie Formation in the Washakie Basin, southwest Wyoming (see Fig. 1). We provide a species list for the four stratigraphic units that make up the formation (Table 1), discuss current conclusions regarding the biostratigraphy and magnetostratigraphy of the formation, and discuss its correlation with other formations that temporally overlap the Washakie Formation, particularly the Uinta Formation. This paper is an expansion of an earlier abstract (McCarroll et al., 1993).

To avoid confusion over terminology we will start by giving a few definitions. We use Bridger A-B, Bridger C-D, Uinta A-B, Uinta C, Washakie A, and Washakie B as informal lithostratigraphic terms. These terms correspond to rock units of the Bridger, Uinta, and Washakie Formations. The North American Land Mammal "Ages" of Wood et al. (1941) (e.g., Uintan, Bridgerian, etc.) are used as biochronologic terms. "Ages" and "Subage" are in quotes because the NALMAs are not based on corresponding time-stratigraphic Stages (see discussions in Savage, 1962, and Tedford, 1970). Subdivisions of the NALMAs are indicated either by a temporal modifier (e.g., earliest, early, middle, late) or by a name given specifically to that "Subage" (e.g., Shoshonian for earliest Uintan or Gardnerbuttean for earliest Bridgerian). We avoid "Subage" names informally derived from formational subdivisions (e.g., there is no formally defined "Black's Forkian Subage" of the Bridgerian). The definitions of the NALMAs used in this paper can be found in Wood et al. (1941) and Woodburne (1987). The definition for the Shoshonian "Subage" of the Uintan can be found in Flynn (1986).

At present the Washakie Formation is divided into two members, the lower Kinney Rim Member and the upper Adobe Town Member. The Adobe Town Member is divided into three informal lithostratigraphic units:

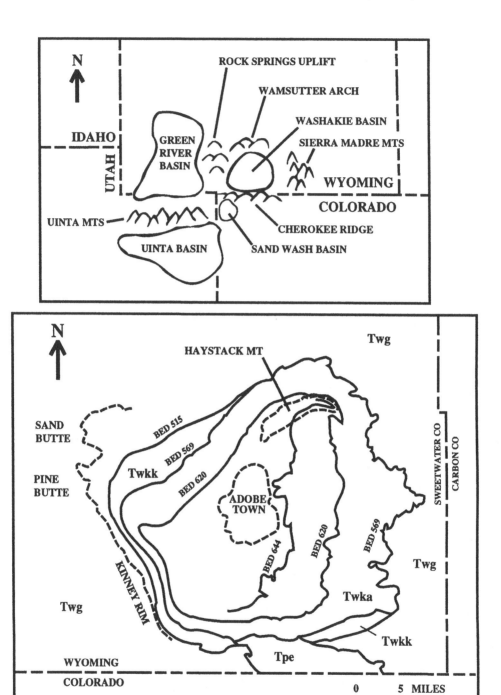

Figure 1. Above, general location map showing geographic relationship of the Washakie, Green River (containing the Bridger Basin) and Uinta basins (From Roehler, 1973). Below, detailed map of study area showing outcrop extent of the Washakie Formation members and informal units. Major physiographic features are also shown. Abbreviations as follows: Twg = Wasatch and Green River formations; Twkk = Washakie Formation, Kinney Rim Member; Twka = Washakie Formation, Adobe Town Member; Tpe = post-Eocene rocks. (From Roehler, 1973)

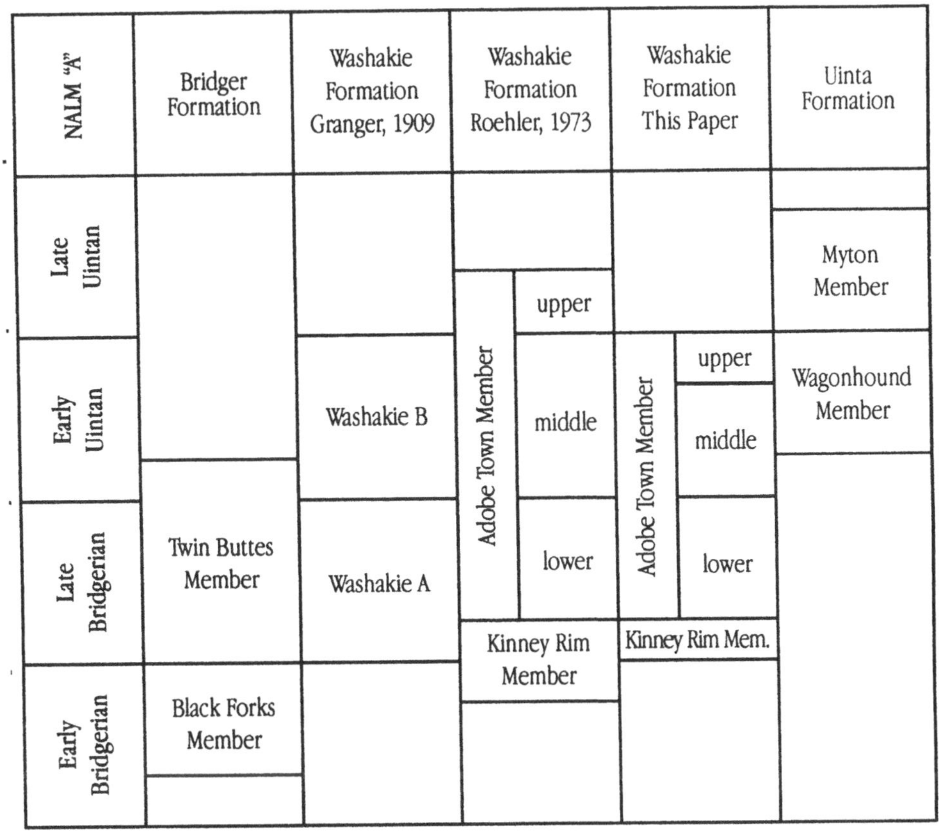

NALM "A"	Bridger Formation	Washakie Formation Granger, 1909	Washakie Formation Roehler, 1973	Washakie Formation This Paper	Uinta Formation
Late Uintan			Adobe Town Member — upper	Adobe Town Member — upper	Myton Member
Early Uintan		Washakie B	Adobe Town Member — middle	Adobe Town Member — middle	Wagonhound Member
Late Bridgerian	Twin Buttes Member	Washakie A	Adobe Town Member — lower	Adobe Town Member — lower	
			Kinney Rim Member	Kinney Rim Mem.	
Early Bridgerian	Black Forks Member				

Figure 2. Generalized correlation of the Washakie, Bridger and Uinta Formations to the North American Land Mammal "Ages."

lower, middle, and upper (Roehler, 1992; Turnbull, 1978; see Figs. 2 and 3).

Historically the formation has been considered Bridgerian-Uintan in age. Granger (1909) divided the formation into two lithostratigraphic units (Granger, 1909, p. 20, did not explicitly call them lithostratigraphic units but did define them in lithologic terms): a lower Washakie A, corresponding faunally to the Bridger C-D, and an upper Washakie B, corresponding faunally to the Uinta A-B. Roehler (1973) formally divided the Washakie Formation into the Kinney Rim and Adobe Town members and speculated on the age of the formation. Roehler (1973, p. 23) tentatively placed the early/late Bridgerian boundary at the top of bed 540 within the Kinney Rim Member. Roehler cited the stratigraphic position of bed 540, its widespread distribution, and the suggested age of the fossil mammals from the Kinney Rim Member as his reasons for doing so. (Roehler, 1973, did not use the terms early or late Bridgerian but instead used Bridger A-B and Bridger C-D as chronologic terms; as stated previously, Bridger A-B, Uinta C, and Washakie B, etc., are used here as lithostratigraphic terms.) The fossil mammals listed by Roehler (1973, table 2) from the Kinney Rim Member (*Hyrachyus* cf. *H. eximius* and *Sciuravus* cf. *S.*

nitidus) do not necessarily indicate an early Bridgerian age, as they occur in early and late Bridgerian deposits elsewhere (Krishtalka et al., 1987).

Roehler (1973, p. 24) speculatively assigned a late Uintan age to the upper unit of the Adobe Town Member, "based on lithologic differences in the rocks signifying a slight change in sedimentary regime." We wish to emphasize this was a speculation and should be treated as such; its subsequent use (e.g., Prothero, 1990; Prothero and Swisher, 1992) as an accepted, valid age for the upper unit has been reconsidered and is no longer being used (Prothero, this volume, Chapter 1). We discount a late Uintan age assignment for the upper unit based solely on lithologic differences and a change in sedimentary regime. An age assignment for the unit must be based on direct positive evidence of temporal distribution, such as biochronology, magnetic stratigraphy, and radioisotopic dating.

Fossil mammal specimens from FMNH collections allow a revision of the age of the Washakie Formation. The majority of specimens from the Washakie Formation at FMNH were collected with this goal in mind and thus have associated geographic and stratigraphic data (Turnbull, 1972). Unfortunately many older collections of fossil specimens from the Washakie

Formation were given only general geographic localities, so their exact stratigraphic position is difficult to determine or unknown (see Radinsky, 1963, pp. 54-56 for an example). In this chapter we revise and restrict the age of the Washakie Formation to be only late Bridgerian through early Uintan. This is essentially the age assigned by Granger (1909), although the Washakie Formation as defined by him represented a stratigraphically more restricted subset of the strata included in the formal formational definition (used here) of Roehler (1973). Our age assignments represent a shorter duration than those of Roehler (1973), but again many of his age assignments were speculative (this is especially true for the Kinney Rim Member and the upper unit of the Adobe Town Member).

BIOSTRATIGRAPHY

In this section we discuss the stratigraphic occurrences within the Washakie Formation of selected taxa listed in Table 1, and their biochronologic significance. We also discuss possible correlations of the Washakie Formation to other middle-late Eocene formations of North America, including the Washakie Formation of the Sand Wash Basin, Colorado. Stratigraphic occurrences listed in Table 1 have been determined from FMNH specimens as well as from specimens from other collections and the published literature, provided they have associated stratigraphic information. Taxa listed in Table 1, but not discussed, are either poor biochronologic indicators (i.e., long temporal distribution) or from taxonomic groups with poor generic and/or species-level resolution. Temporal ranges are taken from Stucky (1992), Krishtalka et al. (1987), Gazin (1976), and other references cited for a particular taxon.

Hyrachyus eximius—Radinsky (1967) discounted all previous reports of *Hyrachyus* from the Washakie Formation giving either misidentification or poor locality data as the reason. His statement is discounted based on our identification of both *H. eximius* and *H. modestus* from the Washakie Formation (McCarroll et al., in press).

In the Bridger Formation, *H. eximius* is restricted to the upper Twin Buttes Member, except for one specimen (not examined by us) reported from just below the Sage Creek White Layer (Gazin, 1976). *Hyrachyus* cf. *eximius* is also reported from the upper Bridger Formation at Tabernacle Butte (McGrew, 1959). *Hyrachyus* sp. cf. *H. eximius* is reported from an unnamed sequence of beds stratigraphically above the Wapiti Formation in Park County, Wyoming (Eaton, 1982) as well as from the Blue Point marker (not assigned to a formation by Eaton, 1982) stratigraphically above the unnamed sequence of beds. These occurrences are consistent with a late Bridgerian first occurrence for *H. eximius*, as the faunas from the unnamed sequence of beds appear to be transitional between faunas from the Bridger B and Bridger C units

of the Bridger Formation. However, as long as the age of the unnamed sequence is thought to be transitional between those of the Bridger B and Bridger C, this occurrence of *H. eximius* may represent a late early Bridgerian occurrence of the taxon. Eaton (1982) did not, however, report *H. eximius* from the underlying Wapiti Formation, whose fauna compares to that from the Bridger B unit of the Bridger Formation. *Hyrachyus eximius* was not reported from the Tepee Trail Formation in its type are in Fremont County, Wyoming by MacFadden (1980), but *Hyrachyus* cf. *H. eximius* was subsequently reported from the type area (Flynn, 1983, 1986) and from the Tepee Trail Formation in Hot Springs County, Wyoming, by Eaton (1985). Both occurrences in the Tepee Trail Formation were thought to be late Bridgerian or earliest Uintan in age and are consistent with a late Bridgerian first occurrence for *H. eximius*.

Two reported occurrences of *H. eximius* may require a change in age of the first occurrence of this taxon. The first is the report of *H. eximius* from the ?Tepee Trail Formation of Togwotee Pass by McKenna (1980). The second is the report of *Hyrachyus* sp. cf. *H. eximius* by Stucky (1984) from the Lost Cabin Member of the Wind River Formation. Both reports are preliminary (we have not examined the specimens) and the specimens have not been described in detail in the subsequent literature. Both occurrences would extend the first occurrence of *H. eximius* into either the early Bridgerian (McKenna, 1980) or the earliest Bridgerian (Gardnerbuttean) (Stucky, 1984). Other faunas of earliest/early Bridgerian age have not yielded *H. eximius* (Morris, 1954; Robinson, 1966; West, 1973; West and Dawson, 1973; Bown, 1982; Emry, 1990). At present, the known temporal range of *H. eximius* is late Bridgerian through earliest or possibly early Uintan. The occurrence of *H. eximius* in the Kinney Rim Member (at bed 527) places the early/late Bridgerian boundary below bed 527 and it probably lies within the underlying Laney Shale Member of the Green River Formation.

Pareumys grangeri—This taxon is presently restricted to the Uinta B of the Uinta Formation (Burke, 1935) and the Friars and Mission Valley formations of southern California, thought to be earliest/early Uintan only in age (Lillegraven, 1977; Flynn, 1986). Its occurrence in the upper unit of the Adobe Town Member is thought to preclude a late Uintan age assignment for that unit.

Protoptychus hatcheri—This rare, highly specialized taxon was named by Scott (1895) based upon a skull from Kennedy's Hole, Wagonhound Member of the Uinta Formation (upper Uinta B), Uinta Basin. Now known from about 70 specimens from the middle unit of the Adobe Town Member (Turnbull, 1991) the two localities are assumed to be time equivalent on this basis as well as other faunal elements.

Notharctus robustior—Flynn (1986) recognized *N.*

robustior as a small-bodied Bridgerian holdover taxon within the earliest Uintan elsewhere. The occurrence of *N. robustior* in the middle unit of the Adobe Town Member, specifically its co-occurrence with larger-bodied early Uintan taxa (e.g., *Dolichorhinus*, *Amynodon*), indicates the presence of an earliest Uintan age fauna (see discussion in Flynn, 1986) from some lower portion of the middle unit. *Notharctus robustior* is common in the lower unit of the Adobe Town Member, especially the middle red bed sequence of Roehler (1973) in the southern part of the Washakie Basin. Plentiful *N. robustior* indicates a late Bridgerian age elsewhere (Krishtalka et al., 1987).

Hyopsodus sp.—Specimens of *Hyopsodus* from the Kinney Rim Member and the middle unit of the Adobe Town Member are similar to *H. paulus*. Unfortunately the material is fragmentary, currently preventing unambiguous assignment to that species. The presence of *H. paulus* from the middle unit would indicate an earliest Uintan age, rather than Bridgerian, for the middle unit as it occurs with larger-bodied earliest/early Uintan taxa (*Dolichorhinus*, *Amynodon*).

Uintatheres--Turnbull (1993; in prep) has called into question the existence of the genus *Tethiopsis*. He has evidence suggesting a rapid transition from a relatively large *Bathyopsis* to relatively small *Uintatherium anceps* between 25 and 50 meters above the Sage Creek White Layer in the Bridger Formation. In the Washakie Formation, large *U. anceps* is replaced by small *Eobasileus cornutus* between the top of the lower and the bottom of the middle units of the Adobe Town Members. This transition indicates that *U. anceps* may be restricted to the late Bridgerian and that *E. cornutus* may be restricted to the earliest and early Uintan.

Dilophodon minusculus—*Dilophodon* was listed by Flynn (1986) as a characteristic earliest Uintan taxon. The species *D. minusculus* is Bridgerian and earliest Uintan elsewhere and its co-occurrence with taxa typical of the early Uintan within the middle unit of the Adobe Town Member indicates an earliest Uintan age fauna from some lower portion of that unit, near Bed 630. Schiebout (1977) reports *Dilophodon* sp. indet. from the Friars Formation of southern California, thought to be earliest/early Uintan in age.

Epihippus gracilis—The occurrence of *E. gracilis* from the upper unit of the Adobe Town Member reconfirms a Uintan age for that unit but does not refine the age further, as *E. gracilis* is known elsewhere from the early through late Uintan.

Amynodon advenus—Flynn (1986, p. 380) designated the first appearance of *Amynodon* as the defining taxon (after Woodburne, 1977) for the base of the earliest Uintan (Shoshonian "Subage"). Thus, if single taxon boundaries are preferred to identify limits of biochrons, the lowest stratigraphic occurrence of *Amynodon* in a given local section should be used to mark the base of the earliest Uintan in that section. The lowest stratigraphic occurrence of *Amynodon* within the

Washakie Formation (from FMNH specimens) is from approximately bed 630, within the middle unit of the Adobe Town Member. This is stratigraphically above bed 620 (= Granger's bed 11) which Granger (1909) and Roehler (1973) used to mark the Bridgerian/Uintan boundary within the Washakie Formation. In light of the generally poor fossil record from this portion of the Washakie Formation (beds 620-630) and for the sake of consistency with previous usage we tentatively place the Bridgerian/Uintan boundary within the Washakie Formation at bed 620. This may be revised if future collection efforts do not yield *Amynodon* specimens or other Uintan characterizing taxa below bed 630.

Tillodonts—The occurrence of a possible new species of tillodont from the Kinney Rim Member is of considerable interest. Tillodonts are known elsewhere from the early Bridgerian (Gazin, 1976) and have not to our knowledge been found in younger deposits (late Bridgerian). This occurrence requires either an early Bridgerian age assignment for some portion of the Kinney Rim Member or a range extension of tillodonts into the late Bridgerian. The specimen in question is smaller than *Tillodon fodiens* from the early Bridgerian and may represent a new late Bridgerian species. To assume an early Bridgerian age for some lower portion of the Kinney Rim Member would require an earlier first occurrence of *Hyrachyus eximius* than is currently known elsewhere. For the time being we prefer to extend the last occurrence of the tillodonts into the late Bridgerian. Justifications for doing so are to favor the temporal distribution of the much more common and more widely distributed taxon, *H. eximius*. This occurrence may extend the known temporal range of the tillodonts upward into the late Bridgerian.

Poebrodon—This genus is previously known only from the Uinta C of the Uinta Formation (Gazin, 1955) and the Laguna Riviera local fauna from the Santiago Formation of southern California (thought to be a Uinta C equivalent and thus late Uintan in age; Golz, 1976, p. 14). Its occurrence in the middle unit of the Adobe Town Member extends the temporal range of the genus into the earliest to early Uintan, as the diverse fauna from the middle unit of the Adobe Town Member is clearly earliest to early Uintan in age, not late Uintan.

To summarize, at present we can only tentatively exclude an early Bridgerian age assignment for the Kinney Rim Member. The presence of *H. eximius* indicates a late Bridgerian age. Preliminary reports of *H. eximius* from older deposits may, however, cast some doubt on a late Bridgerian first occurrence for this taxon. At present we consider the Kinney Rim Member to be late Bridgerian in age.

The evidence precluding a late Uintan age for the upper unit of the Adobe Town Member is more conclusive. The occurrence of *P. grangeri* indicates an early Uintan age; the other six taxa from the unit indicate an undifferentiated Uintan age. The fauna from the upper unit of the Adobe Town Member will be

Table 1. Stratigraphic distribution of mammalian taxa from the Washakie Formation. Abbreviations for stratigraphic units: Twkk = Washakie Formation, Kinney Rim Member; Twka1 = Washakie Formation, Adobe Town Member, lower unit; Twka2 = Washakie Formation, Adobe Town Member, middle unit; Twka3 = Washakie Formation, Adobe Town Member, upper unit. Question marks indicate either uncertain taxonomic identification or uncertain stratigraphic locality data.

	TWKK	TWKA1	TWKA2	TWKA3
Marsupials				
Peratherium cf. *P. knighti*	X	X		
Insectivores				
Pantolestes cf. *P. longicaudus*		X		
Pantolestes cf. *P. elegans*		X		
Pantolestes cf. *P. natans*		X		
Microsyops annectens		X		
cf. *Apatemys bellus*	X	X		
Primates				
Notharctus robustior		X	X	
Omomys carteri		X		
Ourayia uintensis			X	
Hemiacodon gracilis		X		
Tillodonts				
new species?	X			
Taeniodonts				
Stylinodon mirus		X		
Stylinodon inexplicatus		X		
Rodents				
Paramys compressidens				X
Paramys sp.		X		
Leptotomus leptodus			X	
Leptotomus bridgerensis		X		
Thisbemys sp.		X		
Protoptychus hatcheri			X	
Mysops parvus		X		
Sciuravus nitidus	X	X		
Sciuravus sp.		X		
Pareumys grangeri				X
Creodonts-Carnivorans				
Viverravus sp.		X		
Miacis sp.				X
Miacis medius		X		
Miacis washakius		X		
Miacis cf. *M. sylvestris*		X		
Patriofelis ferox		X		
Limnocyon potens			X	
Thinocyon cledensis		X		
Sinopa sp.		X		
Harpagolestes sp.		X	X	
Hyopsodontids				
Hyopsodus sp.	X		X	
Hyopsodus paulus		X		
Uintatheres				
Uintatherium anceps		X		
Eobasileus cornutus			X	

TABLE 1 (continued)	TWKK	TWKA1	TWKA2	TWKA3
Perissodactyls				
Orohippus sp.	X	X		
Epihippus gracilis				X
Palaeosyops sp.		X		
Telmatherium sp.		X		
Mesatirhinus sp.	X	X		
Metarhinus sp.			X	
Dolichorhinus sp.			X	
Isectolophus latidens		X		
Eomoropus amarorum			X	
Helaletes nanus	X			
Dilophodon minusculus			X	
Hyrachyus modestus	X	X		
Hyrachyus eximius	X	X	X?	
Triplopus cubitalis			X	
Triplopus implicatus			X	
Forstercooperia grandis			X	
cf. *Forstercooperia minuta*			X	
Amynodon advenus			X	X?
Artiodactyls				
Homacodon vagans		X		
Achaenodon insolens			X	
Achaenodon robustus			X	
Poebrodon sp.			X	
Protylopus petersoni			X	
cf. *Mesomeryx grangeri*				X
?Pentacemylus			X	
Artiodactyl—new taxon	X			

described in a subsequent publication.

These interpretations of the age of the Washakie Formation allow us to re-examine previously proposed correlations between the Washakie and other formations. Recent work by Prothero and Swisher (1992) in the Uinta Formation has called into question the age of that formation. Confusion over the age of the lower Uinta Formation (= Uinta A) has resulted from the changing stratigraphic nomenclature early this century and the persistent lack of fossils from the lower portion of the formation. Using the lithostratigraphic divisions of Osborn (1929) we assume Uinta A is unfossiliferous (see Prothero and Swisher, 1992), that Uinta B (divided into two units, B1 and B2 by Osborn, 1929) has yielded early Uintan faunas, and that Uinta C has yielded late Uintan faunas. Osborn (1929, p. 91) correlated Uinta A with Washakie A but gave no reason for doing so. As long as the Uinta A lacks vertebrate fossils, accurate temporal correlation of this portion of the Uintan section will be done only through magnetic polarity stratigraphy, radioisotopic dating, and possibly pollen/ invertebrate biostratigraphy. No mammalian faunal evidence exists which allows Uinta A to be correlated with Washakie A.

Prothero and Swisher (1992, p. 54) suggested that "the lowermost part of Uinta B1 is Bridgerian." They based this on the presence of *Metarhinus*, *Rhadino-rhinus*, and *Hyrachyus eximius* from within the lower half of Uinta "B1." They did not recognize diagnostic Uintan taxa (e.g., *Triplopus*, *Amynodon*, *Eobasileus*, and *Dolichorhinus*) until halfway through Uinta "B1." At present, however, *Metarhinus* and *Rhadinorhinus* are known only from the Uintan, and their presence within known Bridgerian deposits has not been demonstrated (Mader, 1989, and references therein). *Hyrachyus eximius*, previously thought to be restricted to the late Bridgerian, has now been recovered from localities thought to be earliest and possibly early Uintan in age (Flynn, 1986; Eaton, 1985).

Faunal correlation of the Washakie Formation with part of the Devil's Graveyard Formation of southwest Texas is possible, even though some typical small Bridgerian taxa appear to have survived there into the late Uintan (Walton, 1992). The first appearance of *Amynodon* in the Devil's Graveyard Formation is within the Whistler Squat Local Fauna (Wilson, 1986) which therefore would correlate with the faunas from the middle unit of the Adobe Town Member. The Devil's Graveyard Formation also contains the Serendipity Local Fauna, a late Uintan equivalent (Wilson, 1986; Walton, 1992). Definitive late Uintan faunas are not currently known from the Washakie Formation, thus no faunal correlation between the Serendipity Local Fauna and the Washakie Formation is possible.

West and Dawson (1975) reported a Bridgerian and Uintan age fauna from the Washakie Formation of the Sand Wash Basin, northwest Colorado. They reported 27 taxa, mostly perissodactyls, with Bridgerian and

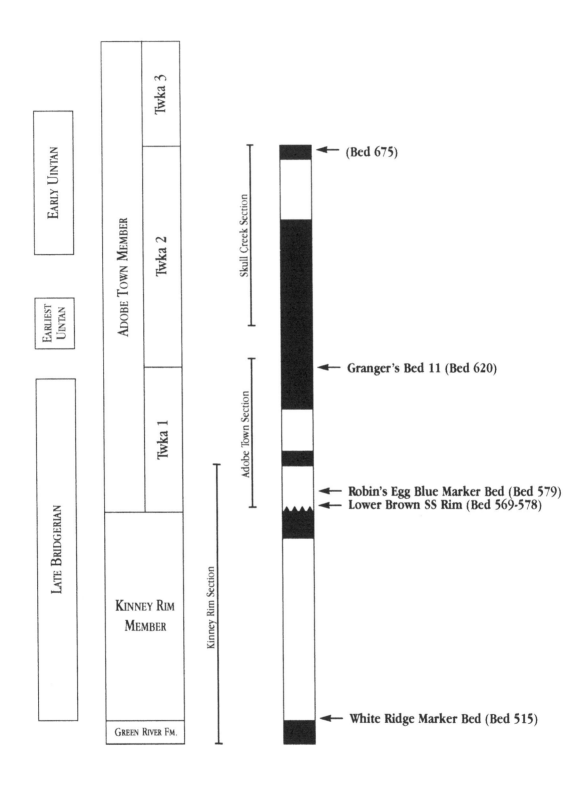

Figure 3. To the left are stratigraphic subdivisions and known faunas of the Washakie Formation. To the right are the paleomagnetic sections and the composite magnetostratigraphic section for the Washakie Formation, both from Flynn (1986), and distinctive basin-wide marker beds (bed numbers in parenthesis are those of Roehler, 1973).

Uintan age assignments elsewhere, and found a generalized geographic distribution of larger-bodied Uintan taxa in the northwest portion of the basin and smaller-bodied Bridgerian taxa in the southern and eastern portions of the basin. To what extent this geographic distribution represents a stratigraphic difference (i.e., Bridgerian rocks to the south and east and Uintan rocks to the northwest) is still not clearly understood. West and Dawson (1975) speculated that some of these differences may be environmental rather than stratigraphic. In contrast to West and Dawson's (1975) observations, Stucky and Snyder (1992) and Stucky et al. (this volume, Chapter 3) indicate no taxonomic differentiation within the Sand Wash Basin fauna. If true, this would indicate to us that the fauna from the Sand Wash Basin represents only a subset of that found in the Washakie Basin, the fauna from the Washakie Basin having a distinct stratigraphic and taxonomic differentiation reflecting a broader temporal range. Based on their interpretation of the age of the Sand Wash Basin fauna (earliest Uintan), the fossiliferous rocks of that basin should correlate with some part of the lower and middle units of the Adobe Town Member of the Washakie Formation.

Stucky and Snyder (1992) assigned the small *Hyrachyus* species of West and Dawson (1975) to *Triplopus* sp. and used its presence (as well as *Protoreodon and Protylopus*) to assign a Shoshonian (earliest Uintan) age to the Sand Wash Basin sediments. If indeed these specimens represent *Triplopus* it is interesting to note they come from West and Dawson's (1975) localities 1 and 3, both of which are from the southern portion of the basin. The presence of *Triplopus* (a characteristic Uintan taxon, but not listed by Flynn, 1986, as a Shoshonian characterizing taxon) in what was previously thought by West and Dawson (1975) to be in a Bridgerian portion of the section, supports the assertion of Stucky and Snyder (1992) and Stucky et al. (this volume, Chapter 3) that no faunal stratification exists within the Sand Wash Basin section.

IMPLICATIONS FOR MAGNETIC POLARITY STRATIGRAPHY AND TEMPORAL CORRELATION

A change in age assignment for portions of the Washakie Formation based on the mammalian faunas has implications for previous interpretations of the magnetic polarity stratigraphy of the formation. Flynn's (1986) integration of mammalian biostratigraphy, magnetic polarity stratigraphy, and radioisotopic chronology of the Washakie Formation might find confirmation or require adjustment with the addition of more extensive collections, more detailed faunal descriptions, and a finer scale mammalian biostratigraphy. In this section we review present interpretation of the Washakie Basin magnetic polarity stratigraphy in relation to the biostratigraphic conclusions of this chapter, and briefly discuss possible interbasinal correlations with other middle Eocene paleomagnetic polarity sections.

The previously proposed magnetic polarity stratigraphy of the Washakie Formation (Flynn, 1986) consists of a basal normal polarity interval, a long reversed polarity interval, a normal polarity interval, and a final reversed polarity interval (Figs. 3 and 4, section A). Flynn's (1986) data indicated the possibility of another normal interval at the very top of his section, but he showed this as a short duration normal interval within his last reversed interval. Flynn's (1986) preferred correlation with the Geomagnetic Polarity Time Scale (magnetic polarity time scale) indicated that Chron C21n through Chron C19r are present within his sampled sections. The intervals at the bottom, Chron C21n, and top, Chron C19r, are of unknown duration and may extend down and up section, respectively, into rocks he did not sample. Flynn's (1986) correlation of the magnetic polarity sections and the Bridgerian-Uintan mammalian faunas from the Washakie Basin, East Fork Basin, and San Diego sections, to the magnetic polarity time scale placed the Bridgerian/Uintan boundary within Chron C20r.

Flynn (1986) discussed two alternatives to his preferred correlation which we review here. Flynn's (1986) magnetostratigraphic sections showed two short duration normal polarity events (B2+ and B4+) within the interval he correlated to Chron C20r. The first Washakie Formation alternative correlation (Fig. 4, section B) would be to correlate the short duration normal polarity interval B2+ with Chron C21n and correlate polarity intervals B3- through B5- with Chron C20r. Polarity interval B1- is then correlated to Chron C21r.

This alternative correlation (Fig. 4, section B) is very unlikely as it would place the early/late Bridgerian boundary within the upper 200 feet of the Kinney Rim Member. As discussed in more detail below, Jerskey (1981) placed the early/late Bridgerian boundary within Chron C21n. As we have previously shown, the Washakie Formation does not contain an early Bridgerian fauna and the early/late Bridgerian boundary may lie within the underlying Green River Formation, or at the very highest, within the lowest 100 feet of the Kinney Rim Member.

In the second Washakie Basin alternative correlation (Fig. 4, section C), Chron C20r is assigned to polarity intervals B3- through B5- and Chron C21n is assigned to polarity intervals A+ through B2+. This is a viable alternative, as the older boundary of Chron C21n within the Washakie Basin section is unknown. In this alternative the early/late Bridgerian boundary could be placed within a Chron C21n correlative and still lie within the underlying Green River Formation. The data at present do not allow a definitive choice of preference between alternatives A and C shown in Figure 4.

At present the only magnetic polarity stratigraphy of

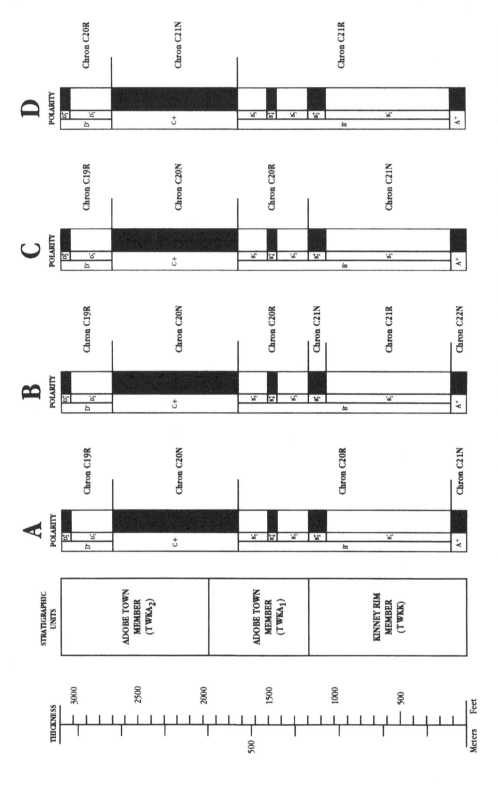

Figure 4. Four alternative correlations of the Washakie Formation magnetic polarity stratigraphy (from Flynn, 1986) with the magnetic polarity time scale. Stratigraphic divisions of the Washakie Formation shown to the left. Alternative correlations A and B allow the early/late Bridgerian boundary to be placed in the underlying Green River Formation as well as in Chron C21n. Alternative correlation C would require the presence of the early/late Bridgerian boundary within the Washakie Formation. Alternative correlation D is supported by biostratigraphic evidence presented by Walsh et al. (this volume, Chapter 6).

the Bridger Formation is the unpublished master's thesis of Jerskey (1981). We briefly discuss the implications of this study here, as these are the only results currently available, but we note that the data must be interpreted with great caution, as the study has not been formally published and there are serious analytical weaknesses (e.g., lack of thermal demagnetization treatment, insufficient statistical analysis, etc.). Jerskey (1981) found one long normal interval at the base of his section which he correlated with Chron C21n, and a shorter reversed interval at the top of the section which he correlated with Chron C20r. This correlation placed the early/late Bridgerian boundary, marked in the Bridger Basin by the Sage Creek White Layer, within Chron C21n. A Shoshonian fauna was recently reported from the Bridger E of the Bridger Formation (Evanoff et al., 1994). This finding may support Jerskey's (1981) correlation of the upper Bridger reversed interval to Chron C20r.

Prothero (1990) and Prothero and Swisher (1992) have made preliminary correlations (Fig. 5, section A) of magnetic polarity sequences from the Uinta Basin with Chrons C21n-C19n. Their correlations indicated that the Chron C20r correlative contains portions of the Uinta A and Uinta B of the Uinta Formation (Prothero and Swisher, 1992). In addition the Uinta C spans strata correlated to Chrons C20n through some part of Chron C19r. Their preferred correlation (Fig. 5, section A) places the early/late Uintan boundary within either Chron C21n, or the earliest portion of Chron C20r.

Prothero and Swisher's (1992, p. 55) interpretation is based in part on the supposed occurrence of late Uintan faunas from Chron C20n within the Washakie Basin, which they attributed to Flynn (1986). No statements in Flynn (1986) indicate the presence of late Uintan faunas from the Washakie Formation. As stated previously, a late Uintan age has been attributed tentatively to the upper unit of the Adobe Town Member (Turnbull, 1978; Roehler, 1973), but only on the basis of lithologic changes, not on observed faunal distributions.

Prothero and Swisher (1992, p. 55) also noted the presence of late Uintan taxa from an interval in the Devil's Graveyard Formation of Trans-Pecos Texas assigned to Chron C20n (Walton, 1992) and used this occurrence to support their correlation of the Uinta Formation to the magnetic polarity time scale. We view Walton's (1992) correlation as tentative, because of the many short reversals found in her composite magnetic polarity stratigraphy (suggesting possible incomplete isolation of primary remanences, making it difficult to precisely correlate the magnetostratigraphic pattern) and the lack of convincing independent temporal data to correlate the Texas sections to the magnetic polarity time scale.

In the biostratigraphic section of this paper, we showed that Prothero and Swisher's (1992) recognition of a late Bridgerian fauna in the Uinta Formation is not supported based on faunal evidence. Prothero and Swisher's (1992) tentative inclusion of the Bridgerian/Uintan boundary within the Uinta Formation relied solely on the correlation of their magnetic polarity stratigraphy to Chron C21n through Chron C19n. That correlation placed Chron C20r within the lower Uinta Formation. Flynn (1986) proposed that the Bridgerian/Uintan boundary lies within Chron C20r. If Chron C20r is present in the Uinta Formation it may be reasonable to assume, as Prothero and Swisher (1992) have done, that the Bridgerian/Uintan boundary is also present. Prothero and Swisher (1992) used the supposed, correlated presence of a time interval equivalent to the Bridgerian/Uintan boundary, not faunas, to infer the presence of Shoshonian (earliest Uintan) age rocks within the Uinta A of the Uinta Formation. At present, the Uinta Formation is not known to preserve a Shoshonian fauna (earliest Uintan). In addition, the Uinta "A" cannot be characterized based on a fauna (Prothero and Swisher, 1992; Krishtalka et al., 1987).

Within the Washakie Formation, early Uintan faunas are known from the upper portion of Chron C20n and some portion of Chron C19r (the younger boundary of Chron C19r is not preserved in the Washakie Basin section) (Flynn, 1986). Prothero and Swisher's (1992, p. 52, Wagonhound Canyon section) correlation placed the known early Uintan faunas of the Uinta Formation within Chron C20r and the earliest portion of Chron C20n. These correlations would make early Uintan faunas heterochronous between these two geographically proximate basins. There is no evidence supporting significant heterochrony of faunal assemblages characterizing any NALMA (Flynn et al., 1984).

Prothero and Swisher (1992) detailed an alternative to their preferred correlation in which correlation of the sequence is shifted up by one complete polarity chron. In that correlation (Fig. 5, section B) Chron C20n through Chron C18n correlatives are present within the Uinta Formation. This correlation places the known early Uintan faunas of the Uinta Formation within Chron C19r and possibly the very earliest portion of Chron C19n, yielding temporal continuity between the mammalian biostratigraphy and magnetic polarity stratigraphies of the Uinta and Washakie formations. Early Uintan faunal heterochrony is not a problem in this alternative correlation. Prothero and Swisher's (1992) hesitation in accepting that correlation rested in part on the occurrence of late Uintan faunas in the upper portion of the Washakie Formation. As we have shown, no definitive late Uintan faunas are known from the Washakie Formation.

At present, a Shoshonian (earliest Uintan) fauna from the Washakie Formation has been found in rocks correlated to Chron C20n. This is younger than, and currently not known to be temporally overlapping with, Flynn's (1986) placement of Shoshonian faunas and the Bridgerian/Uintan boundary within Chron C20r in the East Fork Basin of Wyoming and the San Diego area of California. The lowest part of the middle unit of the

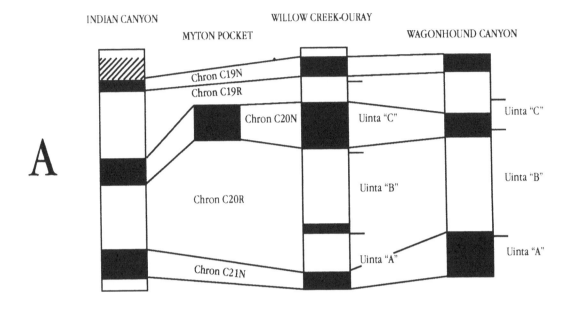

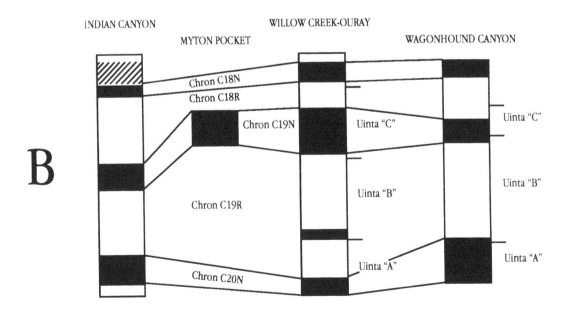

Figure 5. Two alternative correlations of the Uinta Formation magnetic polarity stratigraphy (from Prothero and Swisher, 1992) with the magnetic polarity time scale.

Washakie Formation currently is not known to produce fossils. The lack of a Shoshonian fauna from the upper part of the lower unit of the Adobe Town Member may, however, be a taphonomic artifact, as larger taxa (specifically perissodactyls) are rare, and less common in the lower unit than in the middle unit of the Adobe Town Member. At present we have no data allowing us to confidently place the older boundary of the Shoshonian within the lower unit of the Adobe Town Member, and our available positive evidence indicates the oldest Washakie Formation Shoshonian faunas lie within a normal polarity interval. Alternatively, taphonomic biases may still permit the future discovery of a Shoshonian fauna in the lower unit as suggested by Flynn (1986, p. 379).

Paleomagnetic correlation of the Washakie Formation with the Eocene deposits of the San Diego County Eocene seems straightforward as outlined by Flynn (1986). New data published by Walsh (this volume, Chapter 4) and Walsh et al. (this volume, Chapter 6), however, complicate the situation. Walsh reports *Amynodon* (the defining taxon for the Bridgerian/Uintan boundary) and *Leptoreodon* (a characteristic earliest Uintan taxon) in a horizon that has been interpreted from museum records to lie within the top of a normal interval near the base of the Friars Formation (although lack of faunas lower in this normal could permit the boundary to be even older). This normal interval was correlated to Chron C21n by Flynn (1986) who placed the Bridgerian/Uintan boundary within the overlying Chron C20r. The new biostratigraphic data indicate that the Bridgerian/Uintan boundary may be slightly older, lying within some portion of Chron C21n rather than C20r. It should be noted that the lowest stratigraphic occurrence of *Amynodon* in the Washakie Formation is also within a normal magnetic polarity interval (interpreted by Flynn, 1986, as Chron C20n). The lowest stratigraphic occurrence of *Amynodon* in the East Fork Basin area of northwest Wyoming is within a reversed magnetic polarity interval. The next stratigraphically lowest fauna discussed by Flynn (fauna from horizon B, Flynn, 1986, fig. 9, p. 387) is late Bridgerian and occurs in the lower portion of a normal magnetic polarity interval. Thus an earliest Uintan fauna could potentially be discovered from the upper portion of this same normal magnetic polarity interval. (Flynn, 1986, fig. 9, p. 387, shows another faunal horizon near the top of this normal magnetic polarity interval, but this fauna is not age diagnostic.) Thus, earliest Uintan (Shoshonian) faunas occur in a normal magnetic interval in two of the three areas (and possibly could occur in the third). The San Diego area correlations, based on marine biostratigraphic evidence, strongly support identification of this normal as Chron C21n. If, however, faunal horizon C of Flynn (1986) is late Bridgerian in age, then the lower boundary of the earliest Uintan in the East Fork Basin must lie within a reversed magnetic polarity interval.

Alternatively, the earliest Uintan faunas may be heterochronous between the San Diego area and those from the Washakie Formation and the East Fork Basin area of Wyoming, first appearing in Chron C21n in San Diego and in Chron C20r in Wyoming, or there might be problems with correctly determining polarity of strata in one or more areas. This would, however, be the first instance of demonstrated heterochrony of a NALMA fauna (Flynn et al., 1984). We could also postulate that the Shoshonian is not a real temporal interval and that the observed co-occurrence of smaller-bodied late Bridgerian and larger-bodied early Uintan taxa has another explanation.

Another possible alternative would be the presence of a previously undetected reversed interval between Flynn's (1986) Adobe Town and Skull Creek paleomagnetic sections. Flynn (1986, fig. 7) shows a magnetic sampling gap of about 300 feet (90 m) between these two sections. If a reversed interval is present within this gap we would interpret it as Chron C20r, with Flynn's (1986) C+ interval being Chron C20n above the reversed interval and Chron C21n below the reversed interval. This would allow earliest Uintan faunas to be present in normal and reversed intervals in the Washakie Formation, as they are known to occur in the Tepee Trail Formation and the San Diego region. It should be noted that Stucky et al. (this volume, Chapter 3) show a short reversed interval within their long normal interval which they have correlated to Flynn's C+ interval.

This alternative may appear *ad hoc*, but in light of the new biostratigraphic data from San Diego and the Washakie Basin, we must at least question Flynn's Washakie Basin section. In Flynn's (1986) section the long normal interval C+ contains apparently late Bridgerian (but depauperate in large-bodied taxa) faunas at its base, earliest Uintan faunas above that, and an early Uintan fauna above that. Thus, the earliest Uintan in the Washakie Basin is constrained to fall entirely (or in part if there is a hiatus or a reversal in the unsampled strata) within a normal interval. This is in conflict with the Tepee Trail and San Diego sections (Walsh, this volume, Chapter 4, and Walsh et al., this volume, Chapter 6), and possibly with those from the Sand Wash Basin (Stucky et al., this volume, Chapter 3).

The Sand Wash Basin paleomagnetic section and fauna (Stucky et al., this volume, Chapter 3) are also problematic. Stucky et al. (this volume, Chapter 3) show earliest Uintan faunas within a reversed magnetic interval as well as from the overlying normal interval. They correlate this pattern with magnetic intervals B+3 through C+ of Flynn (1986). At present, earliest Uintan faunas are unknown from reversed intervals in the Washakie Formation. One explanation of this apparent discrepancy is to correlate the Sand Wash Basin reversed interval (and earliest Uintan faunas) to the previously postulated, but as yet unsampled, reversed interval within Flynn's (1986) C+ interval.

In summary, there are a number of plausible correlations of the Washakie Formation to the magnetic polarity time scale (Fig. 4). Unfortunately, no single correlation is currently consistent with all presently available data sets. The correlation by Prothero (this volume, Chapter 1) of the Uinta Formation to the magnetic polarity time scale is consistent with the preferred correlation of Prothero and Swisher (1992) and would require changing at least the upper part of the Washakie Basin magnetic column by shifting it down one polarity chron (Fig. 4, section D). That correlation would predict that Shoshonian faunas should be found in the top of Chron C21n and the base of Chron C20r in the Washakie Formation. This portion of the Washakie section preserves early Uintan faunas. This alternative may be tested by paleomagnetic analysis and $^{40}Ar/^{39}Ar$ dating of the Bridgerian/Uintan boundary in the Bridger Basin, currently under study by the authors. In addition, further paleomagnetic and biostratigraphic research in the Washakie Formation may help clarify the current situation. In addition, a more rigorous delineation of the relationship between the Bridgerian/Uintan boundary and the magnetic polarity stratigraphy in the San Diego and Washakie Basin sections is needed, as well as further detailed collecting of fossil mammals.

ACKNOWLEDGMENTS

We thank the editors for the invitation to contribute this chapter. For typing and proofreading we thank Elaine Zeiger. We thank Andrew H. Leman for the computer-generated figures. We thank Jason A. Lillegraven and Richard K. Stucky for reviewing the manuscript and improving its content and clarity. We thank Donald R. Prothero, Stephen L. Walsh, and John Alroy for comments on portions of the manuscript.

LITERATURE CITED

Bown, T. M. 1982. Geology, paleontology, and correlation of Eocene volcaniclastic rocks, southeast Absaroka Range, Hot Springs County, Wyoming. United States Geological Survey Professional Paper 1201-A:1-75.

Burke, J. J. 1935. Fossil rodents form the Uinta Eocene series. Annals of Carnegie Museum 25:5-12

Eaton, J. G. 1982. Paleontology and correlation of Eocene volcanic rocks in the Carter Mountain area, Park County, southeastern Absaroka Range, Wyoming. Contributions to Geology, University of Wyoming 21(2):153-194.

Eaton, J. G. 1985. Paleontology and correlation of the Eocene Tepee Trail and Wiggins formations in the north fork of Owl Creek area, southeastern Absaroka Range, Hot Springs County, Wyoming. Journal of Vertebrate Paleontology 5(4):345-370.

Emry, R. J. 1990. Mammals of the Bridgerian (middle Eocene) Elderberry Canyon Local Fauna of eastern Nevada. Geological Society of America Special Paper 243:187-210.

Evanoff, E., P. Robinson, P. C. Murphey, D. G. Kron, D. Engard, and P. Monaco. 1994. An early Uintan fauna from Bridger E. Journal of Vertebrate Paleontology, 14, supplement to #3:24A.

Flynn, J. J. 1983. Correlation and geochronology of middle Eocene strata from the western United States. Thesis, Columbia University, 496 pp. (unpublished)

Flynn, J. J. 1986. Correlation and geochronology of middle Eocene strata from the western United States. Palaeogeography, Palaeoclimatology, Palaeoecology 55:335-406.

Flynn, J. J., B. J. MacFadden, and M. C. McKenna. 1984. Land-Mammal ages, faunal heterochrony, and temporal resolution in Cenozoic terrestrial sequences. Journal of Geology 92:687-705.

Gazin, C. L. 1955. A review of the upper Eocene Artiodactyla of North America. Smithsonian Miscellaneous Collections 128(8):1-96.

Gazin, C. L. 1976. Mammalian faunal zones of the Bridger middle Eocene. Smithsonian Contributions to Paleobiology 26.

Golz, D. J. 1976. Eocene Artiodactyla of southern California. Bulletin, Natural History Museum of Los Angeles County 26:1-85.

Granger, W. 1909. Faunal horizons of the Washakie Formation of southern Wyoming. Bulletin, American Museum of Natural History 26(3):13-24.

Jerskey, R. G. 1981. A paleomagnetic study of the Bridger Formation, southern Green River Basin, Wyoming. M.S. thesis, University of Wisconsin at Milwaukee, 60 pp.

Krishtalka, L., R. K. Stucky, R. M. West, M. C. McKenna, C. C. Black, T. M. Bown, M. R. Dawson, D. J. Golz, J. J. Flynn, J. A. Lillegraven, and W. D. Turnbull. 1987. Eocene (Wasatchian through Duchesnean) biochronology of North America; pp. 77-117 in M. O. Woodburne (ed.), Cenozoic Mammals of North America, Geochronology and Biostratigraphy. University of California Press, Berkeley.

Lillegraven, J. A. 1977. Small rodents (Mammalia) from Eocene deposits of San Diego County, California. Bulletin, American Museum of Natural History 158(4):223-261.

MacFadden, B. J. 1980. Eocene perissodactyls from the type section of the Tepee Trail Formation of northwestern Wyoming. Contributions to Geology, University of Wyoming 18(2):135-143.

Mader, B. J. 1989. The Brontotheriidae: a systematic revision and preliminary phylogeny of North American genera; pp. 458-484 in D. R. Prothero and R. M. Schoch (eds.), The Evolution of Perissodactyls. Oxford University Press, New York.

Matthew, W. D. 1909. The Carnivora and Insectivora of the Bridger Basin, middle Eocene. Memoir, American Museum of Natural History 9:291-567.

McCarroll, S. M. in prep. An early Uintan fauna from the upper unit of the Adobe Town Member, Washakie Formation, Washakie Basin, Wyoming. (to be submitted to Fieldiana, Geology)

McCarroll, S. M., J. J. Flynn, and W. D. Turnbull. 1993. Biostratigraphic and magnetic polarity correlations of the Washakie Formation, Washakie Basin, Wyoming. Journal of Vertebrate Paleontology 13, supplement to #3:49A.

McCarroll, S. M., J. J. Flynn, and W. D. Turnbull.. In press. The mammalian faunas of the Washakie Formation, Eocene age, of southern Wyoming. Part III. The Perissodactyls. Fieldiana: Geology.

McGrew, P. O. 1959. The geology and paleontology of the Elk Mountain and Tabernacle Butte area, Wyoming. American Museum of Natural History Bulletin 117(3):121-176.

McKenna, M. C. 1980. Late Cretaceous and early Tertiary vertebrate paleontological reconnaissance, Togwotee

Pass area, northwestern Wyoming; pp. 321-343 in L. L. Jacobs (ed.), Essays in Honor of Edwin Harris Colbert. Museum of Northern Arizona Press, Flagstaff.

Morris, W. J. 1954. An Eocene fauna from the Cathedral Bluffs Tongue of the Washakie Basin, Wyoming. Journal of Paleontology 28(2):195-203.

Osborn, H. F. 1929. The titanotheres of ancient Wyoming, Dakota and Nebraska. United States Geological Survey Monograph 55 (2 vols.):1-953.

Prothero, D. R. 1990. Magnetostratigraphy of the middle Eocene Uinta Formation, Uinta Basin, Utah. Journal of Vertebrate Paleontology, 10, supplement to #3:38A.

Prothero, D. R., and C. C. Swisher, III. 1992. Magneto-stratigraphy and geochronology of the terrestrial Eocene-Oligocene transition in North America; pp. 74-87 in D. R. Prothero and W. A. Berggren (eds.), Eocene-Oligocene Climatic and Biotic Evolution. Princeton University Press, Princeton, N. J.

Radinsky, L. 1963. Origin and early evolution of North American Tapiroidea. Bulletin of the Peabody Museum of Natural History, Yale University 17:1-106.

Radinsky, L. 1967. Hyrachyus, Chasmotherium, and the early evolution of helaletid tapiroids. American Museum of Natural History Novitates no. 2313:1-23.

Robinson, P. C. 1966. Fossil Mammalia of the Huerfano Formation, Eocene, of Colorado. Bulletin of the Peabody Museum of Natural History, Yale University 21:1-95.

Roehler, H. W. 1973. Stratigraphy of the Washakie Formation in the Washakie Basin, Wyoming. United States Geological Survey Bulletin 1369:1-40.

Roehler, H. W. 1992. Description and correlation of Eocene rocks in stratigraphic reference sections for the Green River and Washakie basins, southwest Wyoming. United States Geological Survey Professional Paper 1506-D:1-83.

Savage, D. E. 1962. Cenozoic geochronology of the fossil mammals of the Western Hemisphere. Revista del Museo Argentina de Ciencias Naturales "Bernardino Rivadavia." Zoologia, 8:53-67.

Schiebout, J. A. 1977. Eocene Perissodactyla from the La Jolla and Poway groups, San Diego County, California. Transactions of the San Diego Society of Natural History, 18(13):217-228.

Scott, W. B. 1895. Protoptychus hatcheri, a new rodent from the Uinta Eocene. Proceedings of the Academy of Natural Sciences of Philadelphia 1895:269-286.

Stucky, R. K. 1984. The Wasatchian-Bridgerian Land Mammal Age boundary (early to middle Eocene) in western North America. Annals of Carnegie Museum 53:347-382.

Stucky, R. K. 1992. Mammalian faunas in North America of Bridgerian to early Arikareean "ages" (Eocene and Oligocene); pp. 463-493 in D. R. Prothero and W. A. Berggren (eds.), Eocene-Oligocene Climatic and Biotic Evolution. Princeton University Press, Princeton, N. J.

Stucky, R. K., and J. R. Snyder. 1992. Mammalian fauna of the Sand Wash Basin, Colorado (Washakie Formation, Middle Eocene, Earliest Uintan). Journal of Vertebrate

Paleontology 12, supplement to #3:54A.

Tedford, R. H. 1970. Principles and practices of mammalian geochronology in North America. Proceedings of the North American Paleontological Convention, Part F:666-703.

Turnbull, W. D. 1972. The Washakie Formation of Bridgerian-Uintan ages, and the related faunas; pp. 20-31 in R. M. West (ed.), Guidebook for Field Conference on Tertiary Biostratigraphy of Southern and Western Wyoming.

Turnbull, W. D. 1978. The mammalian faunas of the Washakie Formation, Eocene age, of southern Wyoming. Part I. Introduction: the geology, history, and setting. Fieldiana: Geology 33:569-601.

Turnbull, W. D. 1991. Protoptychus hatcheri Scott, 1895. The mammalian faunas of the Washakie Formation, Eocene age of southern Wyoming. Part II. The Adobetown [sic] Member, middle division (= Washakie B), Twka2 (in part). Fieldiana: Geology, New Series, no. 21.

Turnbull, W. D. 1993. Additions to knowledge of the Uintatheres of the Washakie Formation and aspects of their biology. Journal of Vertebrate Paleontology 13, supplement to #3:60A.

Walton, A. H. 1992. Magnetostratigraphy of the lower and middle members of the Devil's Graveyard Formation (Middle Eocene), Trans-Pecos, Texas; pp. 74-87 in Prothero, D. R., and W. A. Berggren (eds.), Eocene-Oligocene Climatic and Biotic Evolution. Princeton University Press, Princeton, N. J.

West, R. M. 1973. Geology and mammalian paleontology of the New Fork-Big Sandy area, Sublette County, Wyoming. Fieldiana, Geology 29:1-193.

West, R. M., and M. R. Dawson. 1973. Fossil mammals from the upper part of the Cathedral Bluffs Tongue of the Wasatch Formation (early Bridgerian), northern Green River Basin, Wyoming. University of Wyoming, Contributions to Geology 12(1):33-41.

West, R. M., and M. R. Dawson. 1975. Eocene fossil Mammalia from the Sand Wash Basin, northwestern Moffat County, Colorado. Annals of Carnegie Museum 45:231-253.

Wilson, J. A. 1986. Stratigraphic occurrence and correlation of early Tertiary vertebrate faunas, Trans-Pecos Texas: Agua Fria-Green Valley areas. Journal of Vertebrate Paleontology 6(4):350-373.

Wood, H. E., R. W. Chanety, J. Clark, E. H. Colbert, G. L. Jepsen, J. B. Reeside, Jr., and C. Stock. 1941. Nomenclature and correlation of the North American continental Tertiary. Bulletin, Geological Society of America 52:1-48.

Woodburne, M. O. 1977. Definition and characterization in mammalian chronostratigraphy. Journal of Paleontology 51(2):220-234.

Woodburne, M. O. (ed.) 1987. Cenozoic Mammals of North America, Geochronology and Biostratigraphy. University of California Press, Berkeley.

3. Magnetic Stratigraphy, Sedimentology, and Mammalian Faunas of the Early Uintan Washakie Formation, Sand Wash Basin, Northwestern Colorado

RICHARD K. STUCKY, DONALD R. PROTHERO, WALTER G. LOHR, AND JENNIFER R. SNYDER

ABSTRACT

The Sand Wash Basin in northwestern Colorado is a southern sub-basin of the Washakie Basin of Wyoming. It contains several hundred meters of the middle Eocene Washakie Formation, overlying a thick sequence of the lacustrine Green River Formation. Late Bridgerian and earliest Uintan mammalian faunas have been recovered from several localities within the basin. Key biochronological indicators of the earliest Uintan age of the fauna include the earliest agriochoerids (*Protoreodon*), oromerycids (*Oromeryx*), as well as hyracodontids (*Triplopus*) and eomyids (*Namatomys*).

Abundant petrified wood, nonmarine stromatolites, gastropods, bivalves, fish, turtles, and crocodilians show that the Washakie Formation in the Sand Wash Basin was deposited in a marginal lacustrine-fluvial setting. Paleocurrents indicate sediment transport from the north, and the composition of the sandstones is mostly devitrified volcaniclastics derived from the Absaroka volcanic field of northwest Wyoming, or possibly the Challis volcanic field of Idaho. No sediments appear to be derived from the nearby Uinta uplift to the west.

The lower part of the sequence is all of reversed magnetic polarity, and the upper part is of normal polarity. Based on correlations with the revised magnetic stratigraphy of the Washakie Basin, the Sand Wash Basin sequence was deposited during Chrons C21r and C21n (47-48 Ma).

INTRODUCTION

During the latest Cretaceous through early Eocene, the Laramide Orogeny caused Rocky Mountain intermontane basins to subside and collect thick piles of terrestrial sediments (Dickinson et al., 1988). Many of these basins also yield our most prolific assemblages of Eocene fossil mammals (Krishtalka et al., 1987). During the early and early middle Eocene (Wasatchian and Bridgerian North American land mammal "ages"), thick lacustrine deposits of the Green River Formation accumulated in many of these basins (especially the greater Green River Basin of Wyoming, the Uinta Basin of Utah, and the Piceance Basin of Colorado). As the Green River lake system contracted, it was replaced by fluvial deposits which continued to fill many of the basins (e.g., the Bridger Formation in the Bridger Basin of Wyoming, the Uinta Formation in the Uinta Basin of Utah, and the Washakie Formation in the Washakie Basin of Wyoming).

Most of these basins have been collected intensively for fossil vertebrates, and some have been studied in considerable detail. Until recently, however, the Sand Wash Basin in northwestern Colorado received much less attention. The first fossils were collected by Earl Douglass and J. LeRoy Kay of the Carnegie Museum in 1922, 1923, and 1924, and by Denver Museum parties in 1924 and 1925; three short papers were published on these collections (Abel and Cook, 1925; Cook, 1926a, 1926b). Little further collecting or publication on the fossils of the Sand Wash Basin occurred for about fifty years until parties from the Carnegie Museum and University of Colorado Museum began to revisit the old localities and discover new ones in the 1960s (summarized by West and Dawson, 1975). From 1972-1976, the Sand Wash Basin was again collected by Denver Museum parties. Recent work by the Carnegie Museum in 1988, and by Denver Museum parties from 1989-1994, produced major new collections.

The collections from the Sand Wash Basin are particularly important because they supplement and extend the collections from other basins. In particular, the "type" sections in the Uinta Basin of Utah yield few earliest Uintan fossils from Uinta Formation "A." This span of time seems to be better represented in the Washakie Basin of Wyoming, where Flynn (1986) recognized the "Shoshonian" land mammal "subage" for the earliest Uintan (this volume, Chapter 2). Other areas of this age include those from southern California and the Togwotee Pass and Lysite Mountain areas in northern Wyoming. As discussed below, the Sand Wash collections are apparently the same age as the earliest Uintan ("Shoshonian") faunas from the Washakie Basin to the north, so they augment our understanding of this poorly sampled interval of the mammalian record.

GEOLOGIC SETTING

The Sand Wash Basin is a Laramide structure of about 600 square miles in Moffat County, Colorado (Fig. 1). Although it is presently separated from the larger Washakie Basin to the north in Wyoming by Cherokee Ridge (a younger structural arch), it is considered a sub-basin of the Washakie Basin (Roehler, 1973, 1992a; Dickinson et al., 1988), since the two were originally a single basin. The stratigraphic sequences of the two areas are so similar that the marker bed nomenclature coined by Roehler (1973) for the Washakie Basin in Wyoming can be used south of the border in Colorado. The only published geologic studies of the Sand Wash Basin were U.S.G.S. maps of Lone Mountain (McKay, 1974) and Maybell (McKay and Bergin, 1974) 7.5' quadrangles.

Most of the limited research in the Sand Wash Basin focused on the Green River Formation because of its economic importance as a source of oil. McKay (1974) and McKay and Bergin (1974) mapped about 1200 feet of the Laney and Tipton members of the Green River Formation, which interfinger with the underlying lower Eocene Wasatch Formation. These same maps show about 400 feet of "Bridger" Formation overlying the Green River Formation, especially in Lone Mountain Quadrangle. Since the Sand Wash mammalian faunas were relatively poorly known at the time, and the rocks do bear some resemblance to the type Bridger Formation, this assignment was not unreasonable. Since that work, Roehler (1973, 1992a) has distinguished the Uintan rocks of the Washakie Basin from the type Bridger Formation and formally named them the Washakie Formation, following Granger (1909). However, he did not formally designate the same rocks in the Sand Wash Basin as "Washakie Formation" (Roehler, 1992b, p. E26). Nevertheless, we believe that the "Bridger" Formation in the Sand Wash Basin should also be referred to the Washakie Formation because of its lithological similarity to the Washakie Formation in the adjacent sub-basin in Wyoming.

Although McKay (1974) and McKay and Bergin (1974) recognize a lower and upper part of their "Bridger" Formation, only the upper part yields mammalian fossils. The "lower Bridger" Formation, as mapped by these geologists, is actually much closer in lithology to the laminated Green River shales, and interfingers with the Green River Formation in the area. In Wyoming, Roehler (1992a, fig. 6) identifies the lateral equivalent of these Colorado beds (beds 519-620 in the Washakie Basin) as Washakie Formation, so there may be some justification for distinguishing the "lower Bridger" Formation of the Sand Wash Basin from the classic Green River shales. However, for our study, only the "upper Bridger" (= upper Washakie) yielded mammalian fossils, and so our paleomagnetic studies focused on this part of the formation.

Some time after the upper Washakie Formation was deposited in the Sand Wash Basin, it underwent structural deformation, because the beds along the southern rim of the basin are faulted and tilted as much as 66° to the south (McKay, 1974). Middle Eocene beds are overlain by the middle-upper Miocene Brown's Park Formation with an angular unconformity. Thus, most of the deformation must have taken place after the late middle Eocene. It was probably a result of the final phase of Laramide deformation, which caused late Uintan and Duchesnean uplift to the west in the Uinta Basin (Andersen and Picard, 1972; Dickinson et al., 1986, 1988). The Brown's Park Formation, in turn, was also slightly deformed and faulted by Plio-Pleistocene tectonic activity related to the renewed uplift of the Rocky Mountains.

SEDIMENTOLOGY

Like the Washakie Formation to the north (Roehler, 1992a), the upper Washakie Formation in the Sand Wash Basin is full of sedimentary evidence of a marginal lacustrine-fluvial depositional environment. It consists of about 450 feet (150 m) of grayish-brown and purple tuffaceous mudstones, with greenish and lavender cross-bedded sandstones. The mudstones are poorly bedded, and do not show the fine-scale lamination characteristic of the Green River Formation in this area. Consequently, these probably represent floodplain and lake margin mudstones, rather than the classic finely laminated lake shales found lower in the section.

The shallow depth of the water in this basin is indicated by a number of beds of silicified lacustrine stromatolites found in the mudstones; some of the domed stromatolites are 10-50 cm across. Such stromatolites are also common in the Washakie (Roehler, 1992a) and Uinta basins (Bradley, 1929; Ryder et al., 1976). Further evidence of the habitat is provided by the abundant fossils of freshwater bivalves and gastropods (*Goniobasis, Biomphalaria*), fish such as garpike (*Lepisosteus*), bowfin (*Amia*), and ictalurid catfish (*Astephus, Rhineastes*), aquatic turtles (*Baena, "Trionyx," Plastomenus*), crocodiles (*Allognathosuchus, Pristichampsus*), varanid (*Saniwa*) and anguid lizards, and snakes.

The abundance of sandstones in the section also attests to the higher energies represented by the upper Washakie Formation (compared to the Green River shales). Both tabular and trough cross-stratification were observed in the sandstones, and some units show striking aggrading ripples ("climbing ripple drift") indicative of large sediment volumes settling rapidly out of the stream flow. Most sandstone bodies are narrow and lenticular, although in some places they form large channel complexes with an architecture of anastomosing incised channels.

Taken together, all of this evidence strongly indicates that the upper Washakie Formation in the Sand Wash Basin was deposited on the margins of a receding Green

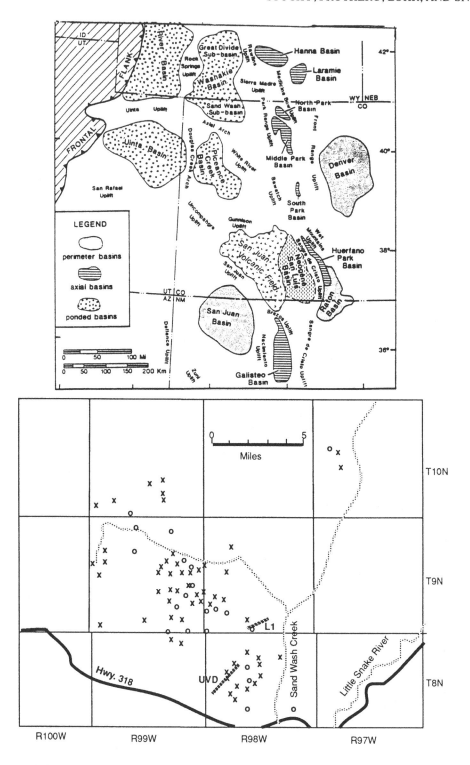

Figure 1. (*Top*) Index map showing location of Sand Wash Basin with respect to other Laramide structures in the Rocky Mountains (after Dickinson et al., 1988). (*Bottom*) Detail of fossiliferous portion of Sand Wash Basin, modified from West and Dawson (1975). "o" indicates localities shown by West and Dawson (1975, fig. 2); "x" indicates recently discovered DMNH localities. Exact location of these fossil sites is availabe to qualified investigators. Bold patterned lines show routes of measured magnetostratigraphic sections: L1 = "locality 1"; UVD = Upper Vaughn Draw.

River lake system (Lake Gosiute). In fact, the combination of channel sandstone complexes, gray-green mudstones, freshwater mollusks, and stromatolites is typical of the "marginal lacustrine-deltaic" facies of Ryder et al. (1976, p. 505), described from the margins of the Green River Formation in the Uinta Basin.

Paleocurrents were measured on 45 different sets of cross-beds and all showed a strong southerly trend (mean direction = 179.7° ± 50.9°); this is concordant with the directions reported by Surdam and Stanley (1980, fig. 1) for the same area. Several thin sections of the sandstones were examined and point counted (minimum 300 points). All contained between 40-80% devitrified glassy volcanic rock fragments, by far the dominant component in the sandstone. Most of the quartz (15-30% of most slides) is also angular and apparently volcanic in origin. In this respect, they are very similar to the volcaniclastic sandstones described by Roehler (1992a) from the Adobe Town Member (= upper member) of the Washakie Formation in the Washakie Basin. Many of our thin sections closely resembled the tuffaceous channel sandstone described by Roehler (1992a, fig. 19) from Bed 677 of the Adobe Town Member. These rocks are part of the upper middle Eocene "volcanic lithic sandstone petrofacies" of Surdam and Stanley (1979, 1980). Clearly, the bulk of the sediments supplied to the Sand Wash Basin came from volcanic fields to the north. The likeliest source is the massive Absaroka eruptions occurring in northwestern Wyoming at that time (Smedes and Prostka, 1973; Surdam and Stanley, 1979, 1980; Johnson, 1985).

According to Surdam and Stanley (1979, 1980) and Johnson (1985), the "volcanic lithic sandstone petrofacies" apparently spread southward in such volume that it swamped local sources. From paleocurrents and petrologic evidence, Surdam and Stanley (1980, fig. 1) and Dickinson et al. (1988, fig. 12) inferred that the river systems in the early Uintan drained south out of the southeastern Lake Gosiute (= Sand Wash Basin) into the Piceance Basin, then westerly into the Uinta Basin. Indeed, paleocurrents and river channels in the Uinta "A" member of the Uinta Formation trend westerly (Stagner, 1941; Dane, 1954; Cashion, 1967). From its sources in north-central Wyoming to the greater Green River Basin (forming the Washakie Formation), the "volcanic lithic sandstone petrofacies" can be traced over the divide into the Piceance Creek Basin, and ultimately eastward into the Uinta Basin of Utah (depositing the Uinta Formation "A-B" sandstones). As it did so, it contributed to the infilling and the final desiccation of the Green River lake system.

It is surprising, however, that there is no indication of sediments from the nearby Uinta Mountain uplift to the southwest of the Sand Wash Basin. No Proterozoic quartzite fragments or reworked Paleozoic sediments ("Paleogene quartzolithic petrofacies" of Dickinson et al., 1986) so characteristic of these mountains (Andersen and Picard, 1972, 1974; Dickinson et al., 1986; Bruhn et al., 1983; Picard and High, 1972) were observed. Because the Uinta Mountains apparently did not begin shedding sediment into the Uinta Basin to the south until the Duchesnean (Andersen and Picard, 1972, 1974), they must not have been emergent enough in the early Uintan to compete with the flood of volcanic debris from the north.

MAMMALIAN FAUNA

Since the most recent review of the mammalian fauna from the Washakie Formation in the Sand Wash Basin by West and Dawson (1975), field parties from the Denver Museum and Carnegie Museum have discovered an additional 18 mammalian species (Table 1). West and Dawson considered the fauna to be intermediate between the Bridgerian and Uintan land mammal "ages." They suggested that the fauna from the northwestern part of the basin, which consisted of predominately large mammals, was different in age and composition from that recovered in the southeastern part, which consisted of smaller mammals. Essentially, the larger mammals were typical of the Uintan and the smaller ones of the Bridgerian. Three possible explanations for these differences were offered: (1) inadequate collecting of small mammals in the northwestern part of the basin; (2) stratigraphic distance separating the strata in the two regions; and (3) recovery of taxa from different facies and paleoenvironments. They favored a composite explanation of (2) and (3), suggesting that "perhaps there is a time difference accompanied by ecologic variation" (West and Dawson, 1975, p. 252).

Recent collecting expeditions have recovered both small and large mammals from the northwestern and southeastern parts of the basin in both channel sandstones and overbank mudstones modified into paleosols. There is no question that the larger animals are well represented in channel sandstones and conglomerates exposed more commonly in the northwestern part of the basin. The smallest mammal known from the channel sandstones is *Notharctus robustior* (West and Dawson Locality 2) but this species is also known from the southeastern part as well, and occurs in both the Bridgerian and Uintan sequence of the Bridger Basin (Gazin, 1976; Evanoff et al., 1994; P. Robinson, personal communication to RKS) and the early Uintan sequence at Lysite Mountain, Wyoming (RKS, unpublished data, CMNH collections). Finer-grained overbank mudstones have yielded specimens of large mammals such as uintatheres and titanotheres, and small mammals in both parts of the basin also, thus suggesting that bias due to depositional regime is not a factor.

Almost all of the specimens reported by West and Dawson (1975) and the newer collections come from above the Robin's Egg Blue Tuff (REBT) of Roehler (1973), which crops out in the northern part of the basin (Sec. 30, T10N, R98W—see Fig. 1). Only one

Table 1. Vertebrate fauna of the Washakie Formation, Sand Wash Basin, northwestern Colorado. Taxa added to the fauna since the most recent synopsis are marked with an asterisk (*).

OSTEICHTHYES
SEMIONOTIFORMES
  Lepisosteidae
    *Lepisosteus sp.
AMIIFORMES
  Amiidae
    *Amia sp.
CYPRINIFORMES
  Catostomidae
    *Catostomidae sp.
SILURIFORMES
  Ariidae
    *Rhineastes peltatus
  Ictaluridae
    *"Rhineastes" smithi
    *Astephus sp.
REPTILIA
CHELONIA
  Baenidae
    *Baena arenosa
  Dermatemydidae
    *Baptemys wyomingensis
  Testudinidae
    *Echmatemys septaria
    *Echmatemys sp.
SQUAMATA
  Anguidae
    *Anguidae sp.
  Varanidae
    *Saniwa sp.
OPHIDIA
  *Ophidia sp.
CROCODILIA
  Crododylidae
    *Allognathosuchus polydon
    *Crocodylus sp.
    *Pristichampsus sp.
MAMMALIA
MARSUPIALIA
  Didelphidae
    Peratherium knighti
    (Krishtalka and Stucky, 1983)
    *Peratherium innominatum
LEPTICTIDA
  Leptictidae
    *Leptictidae sp.
PANTOLESTA
  Pantolestidae
    Pantolestes natans
INSECTIVORA
  Geolabididae
    *Centetodon sp.
  Apternodontidae
    *Apternodontidae sp.
DINOCERATA
  Uintatheriidae
    Eobasileus cornutus
DERMOPTERA
  Microsyopidae
    *Microsyops annectens
    *Microsyops sp.
PRIMATES
  Adapidae
    Notharctus robustior
  Omomyidae
    Hemiacodon sp. (=Washakius
    sp. of West and Dawson, 1975)

    *Omomyidae sp.
CARNIVORA
  Miacidae
    Uintacyon vorax
    Viverravus minutus
    Viverravus sp.
  Amphicyonidae
    *?Amphicyonidae sp.
RODENTIA
  Ischyromyidae
    Leptotomus bridgerensis
    *Leptotomus sp.
    *Thisbemys sp.
    Ischyromyidae sp.
  Sciuravidae
    *Sciuravus sp.
    Tillomys sp. cf. T. senex
  Eomyidae
    *Namatomys sp.
ARCHAIC UNGULATES
  Hyopsodontidae
    Hyopsodus markmani (=H. despicians
    of West and Dawson, 1975)
ARTIODACTYLA
  Homacodontinae
    Hylomeryx sp. (=Homacodon sp.
    cf. H. vagans of West and Dawson, 1975)
  Helohyidae
    Helohyidae, n. gen., n. sp. (=Parahyus sp.
    of West and Dawson, 1975)
  Agriochoeridae
    *Protoreodon sp.
    *cf. Agriochoeridae, n. gen., n. sp.
  Oromerycidae
    *Oromeryx sp.
MESONYCHIA
  Meonychidae
    *Mesonyx sp.
PERISSODACTYLA
  Equidae
    *Eppihippus parvus
    Orohippus sylvaticus
  Brontotheriidae[1]
    Metarhinus sp.
    Tanyorhinus blairi
    Tanyorhinus bridgeri
    Tanyorhinus harundivorax
    Tanyorhinus sp.
    Telmatherium acola
    Telmatherium advocata
    Manteoceras foris
    Manteoceras pratensis
  Isectolophidae
    Isectolophus annectens
  Helaletidae
    *Dilophodon sp.
  Hyracodontidae
    Hyrachyus eximius or
    Forstercooperia grandis
    Triplopus implicatus (= Hyrachyus
    small species of West and Dawson, 1975)

---

[1]The plethora of taxa listed here are greater than the actual diversity (See West and Dawson, 1975; Mader, 1989). The group is currently under revision by Bryn Mader.

locality reported by West and Dawson, their Locality 20, lies below the REBT and preserves specimens of *Peratherium* and *Hemiacodon* (= *Washakius* of West and Dawson, 1975). Several additional specimens in Denver Museum collections from this area near Lang Springs are identified as *Hyopsodus*, a tapir, *Microsyops*, and a uintathere. The fauna from the Sand Wash Basin from below the REBT is too incomplete to determine if that part of the sequence is Bridgerian or Uintan in age, but it should be noted that no taxa characteristic of the Uintan are known.

The fauna from horizons above the REBT (Fig. 1) is typical of the earliest Uintan, or "Shoshonian" from California and Wyoming (Flynn, 1986; Stucky and Snyder, 1992). Especially significant are new records of *Epihippus parvus* (West and Dawson Loc. 16), *Protoreodon* (West and Dawson Loc. 3), *Oromeryx* (West and Dawson Loc. 9 and 12), and *Namatomys* (West and Dawson Loc. 12). New identifications include *Hylomeryx* and *Triplopus*, which were identified by West and Dawson (1975, pp. 247, 251) as *Homacodon* sp. cf. *H. vagans*, and *Hyrachyus* small species, respectively. Additional specimens in Denver Museum and University of Colorado Museum collections are referred to these taxa. We agree with West and Dawson (1975) that the titanotheres and *Eobasileus cornutus* (Wheeler, 1961) are also suggestive of a Uintan age.

We have some reservations with regard to the biostratigraphic significance of the hyracodontid taxa. Many characters which define Uintan hyracodontids are also indicated as transitional between late Bridgerian *H. modestus* and Uintan *Triplopus* (Radinsky, 1967a, 1967b). Specimens assigned here to *Triplopus implicatus* are based on a relatively higher crown height of the lower molars, and a reduced metacone and lengthened postmetacrista on M2. A larger hyracodontid representing either *Hyrachyus eximius* (typical of the late Bridgerian and early Uintan) or *Forstercooperia grandis* (typical of the Uintan) is also present, but the isolated teeth assigned here are insufficiently diagnostic. A revision of Eocene hyracodontids is needed to clarify the utility of genera in this family for biostratigraphic interpretations. This is also true of the brontotheres, which are currently under study (see Mader, 1989).

Several other new records may eventually prove to be significant for age determination upon further study. An associated maxilla and mandible are tentatively identified as an amphicyonid, which would be the earliest appearance of this family. A mandible with m1-2 appears to represent a new agriochoerid. The *Parahyus* sp. (DMNH 1764) of West and Dawson (1975, p. 25) represents a new genus of helohyid artiodactyl (Snyder, 1993). Other taxa reported from the Sand Wash fauna for the first time here are typical of both the Uintan and Bridgerian.

The new identifications and records from the localities that lie above the REBT in the Sand Wash Basin include taxa which typify the earliest Uintan or

"Shoshonian" (Flynn, 1986; Stucky 1992; Evanoff et al., 1994; see also Gazin, 1955, and Radinsky, 1967a). The fauna from both the northwestern and southeastern Sand Wash Basin is comparable in taxonomic content to that from Bridger E (Evanoff et al., 1994), Bone Bed A of the Tepee Trail Formation, Togwotee Pass area, Wyoming (McKenna 1980; Flynn, 1986), Washakie B, and the Poway assemblage of southern California (Flynn, 1986), and potentially Whistler Squat, Texas (Wilson, 1986). *Epihippus*, *Triplopus*, *Namatomys*, *Protoreodon*, and *Oromeryx* characterize the Uintan land mammal "age" and their first appearance marks the beginning of the Uintan and the Shoshonian. Taxa such as *Auxontodon*, *Diplobunops*, *Bunomeryx*, and *Leptotragulus* are absent from the Sand Wash fauna but do occur in Uinta B2. This suggests that, within the Rocky Mountain region, the Shoshonian is distinct.

MAGNETIC STRATIGRAPHY

Two main stratigraphic sections were collected for magnetic analysis (Fig. 1). The lower sequence (200 feet of section) was collected near Denver Museum "Locality 1" (DMNH 296), and represented the localities in this area and at nearby "Turtle Hill" (DMNH 276). This section started at the Tbl/Tbu contact (McKay, 1974) (NE NW SW Sec. 9, T8N R98W, Lone Mountain 7.5' quadrangle, Moffat County, Colorado) and ran up section and due west to the crest of the hill with benchmark 6297 (NE SW NW of the same section). From the crest of this hill, the dip of the bedding was projected to the southwest, and the section was resumed in the bottom of Upper Vaughn Draw. This upper section (about 220 feet thick) started in SE SW SW Sec. 8, and ran southwesterly along the exposures at the east edge of the grassy bench, culminating in NE NW SE Sec. 18. Because of obvious faulting, the section was terminated at this point, since we could not determine if the remaining sequence repeated what we had already sampled, or might be separated from our previous section by a gap of unknown length.

In addition, two paleomagnetic sites were collected from isolated exposures of the distinctive Robin's Egg Blue Tuff (REBT) found in both the Sand Wash and Washakie basins (Bed 579 of Roehler, 1973). These sites were located about nine miles due north of the main section (in NW NW NE Sec. 30, T10N R98W, Sheepherder Spring 7.5' quadrangle, Moffat County, Colorado). According to Roehler (1973, pp. D62-63), approximately 470 feet of section separate the Robin's Egg Blue Tuff (Bed 579) with the "Rim above Adobe Town Rim" (Bed 600, which is approximately equivalent to the base of our "Locality 1" section) in the Washakie Basin. Unfortunately, in the Sand Wash Basin the exposures between these two areas were so poor and covered that it was impossible to connect them with continuous sampling. Based on the projected dip of the REBT and overlying tuffaceous mudstones and

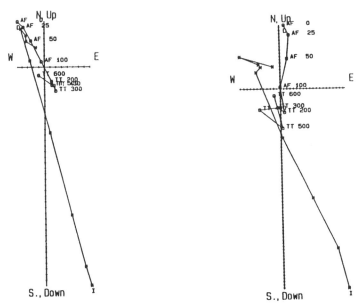

Figure 2. Vector demagnetization plots of selected AF amd thermal demagnetization results from the Washakie Formation. Circles indicate horizontal component, + symbol is the vertical component. Sample on left had a normal overprint that was removed by 300°C. Sample on right had a normal overprint removed by 200°C.

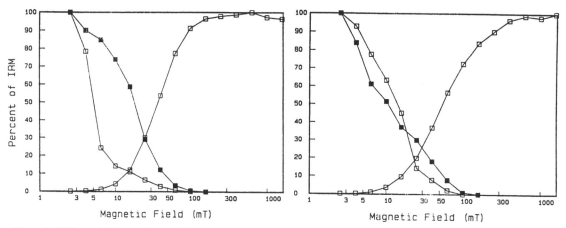

Figure 3. IRM acquisition and Lowrie-Fuller tests of selected samples from the Washakie Formation (see text and Pluhar et al., 1991, for further details). Solid boxes are ARM intensities at each AF demagnetization step, open boxes are IRM intensities. Note that during IRM acquisition (ascending curve on right), both samples reached saturation at about 300 mT, indicating the presence of magnetite. In the modified Lowrie-Fuller tests (descending curves on left), the ARM is more resistant to AF demagnetization than the IRM, showing that the remanence is carried by single-domain or pseudo-single domain grains.

sandstones, there are probably no more than 200 feet of section that remained unsampled between the REBT and the lowest fossiliferous Uintan beds where our section began.

Twenty-five magnetic sites (three samples per site) were collected with simple hand tools, stratigraphically spaced about 7 m (22 feet) apart. Samples were trimmed on a band saw with a tungsten-carbide blade and run at the paleomagnetic laboratory of the California Institute of Technology. After measurement of NRM (natural remanent magnetization), samples were demagnetized in alternating fields (AF) at 25, 50, and 100 Gauss to remove any possible multidomain components. Each sample was then subjected to thermal demagnetization from 200-600°C in steps of 100°C.

The behavior of representative samples is shown in Figure 2. Under AF demagnetization, the magnetic intensity of most samples decreased significantly, showing that a low-coercivity mineral such as magnetite is a major carrier of the remanence. Thermal demagnetization was used to remove overprints caused by high-coercivity, low-blocking-temperature iron hydroxide mineral, such as goethite, which dehydrates by 200°C. Indeed, overprinted reversed sites first revealed reversed directions at about 200°C, and a stable reversed component was typically isolated at 400-500°C. This was the component used in further analysis. Very little of the remanence was left above 580°C (the Curie point of magnetite), indicating that most of the magnetic signal is carried by magnetite and not by hematite.

Analyses of IRM (isothermal remanent magnetization) acquisition further confirmed that magnetite was the primary carrier of the remanence. Most rocks (Fig. 3) reached saturation IRM values at 100-300 mT (millitesla), a characteristic of magnetite. A modified Lowrie-Fuller ARM (anhysteretic remanent magnetization) test (e.g., Johnson et al., 1975) was also conducted during the IRM analysis (see Pluhar et al., 1991, for details). This test compares the resistance of AF demagnetization of both an IRM acquired in a 100 mT peak field, and an ARM gained in a 100 mT oscillating field. In most samples, the ARM (black squares in Fig. 3) demagnetizes at higher peak fields than does the IRM (open squares), indicating that the remanence is carried by single-domain or pseudo-single-domain grains. However, in a few samples, the IRM was more resistant to AF demagnetization than the ARM, showing that multidomain grains were also present. Thin sections of representative samples were examined under reflected light. In several slides, euhedral magnetite grains with oxidized rims of hematite could be seen.

Once a characteristic magnetic component had been obtained for each sample, the vectors were averaged using the methods of Fisher (1953; see Butler, 1992). Class I sites of Opdyke et al. (1977) showed a clustering that differed significantly from random at the 95% confidence level. In Class II sites, one sample was lost

or crumbled, but the remaining samples gave a clear polarity indication. In Class III sites of Opdyke et al. (1977), two samples showed a clear polarity preference, but the third sample was divergent because of insufficient removal of overprinting. A few samples were considered indeterminate if their magnetic signature was unstable or their direction uninterpretable.

To determine if the stable remanence was primary or secondary, a reversal test for stability was conducted. The Fisher mean of the normal sites (I = 349.0, D = 46.4, k = 18.2, α_{95} = 11.6) was antipodal to the mean for reversed sites (I = 181.0, D = -50.2, k = 7.8, α_{95}= 19.7), showing that the magnetization is primary depositional remanence.

The magnetic stratigraphy of the Sand Wash sections is shown in Figure 4. Both sites at the "Robin's Egg Blue Tuff" locality were reversed in polarity. Except for two sites, all of the rocks in the "Locality 1" section were also reversed. Most of the "Upper Vaughn Draw" section was of normal polarity (with a two-site reversed zone in the middle).

The interpretation of this polarity pattern is shown in Figure 5. Because the Sand Wash mammals can be correlated to the earliest Uintan faunas in the nearby Washakie Basin, that section (Flynn, 1986; McCarroll et al., this volume, Chapter 2) was used as a standard. As shown by McCarroll et al. (this volume, Chapter 2, fig. 3), the REBT occurs in an interval of reversed polarity originally labeled "B3–" by Flynn (1986). Earliest Uintan ("Shoshonian") faunas occur at the base of Twka2, and near the base of Flynn's (1986) "C+" normal magnetozone. Typical early Uintan faunas occur higher in Twka2.

Thus, it appears that our two sites of reversed polarity around the REBT match the pattern reported for the Washakie Basin. However, the presence of "Shoshonian" fossils in rocks of reversed polarity in the Sand Wash Basin seems to contradict the pattern seen in the Washakie Basin. This may be resolved in one of two ways:

1) Twka1 (the lower Adobe Town Member of the Washakie Formation) is indicated as being "Late Bridgerian" by McCarroll et al. (this volume, Chapter 2, fig. 3). However, these authors point out that the upper part of Twka1 is sparsely fossiliferous, especially with respect to the larger mammal taxa that are critical to recognizing the Uintan. Thus, our "Locality 1" section, with its reversed polarity could actually correlate with the upper part of Twka1 (magnetozone "B5-" of Flynn, 1986), and the Vaughn Draw section would correlate with the base of normal magnetozone C+.

2) McCarroll et al. (this volume, Chapter 2) point out that there is a 90 m (300 foot) gap between the Adobe Town and Skull Creek sections of Flynn (1986), so that the polarity of the lower part of Flynn's (1986) magnetozone C+ is not actually known. This interval also yields a "Shoshonian" fauna. It is possible that the

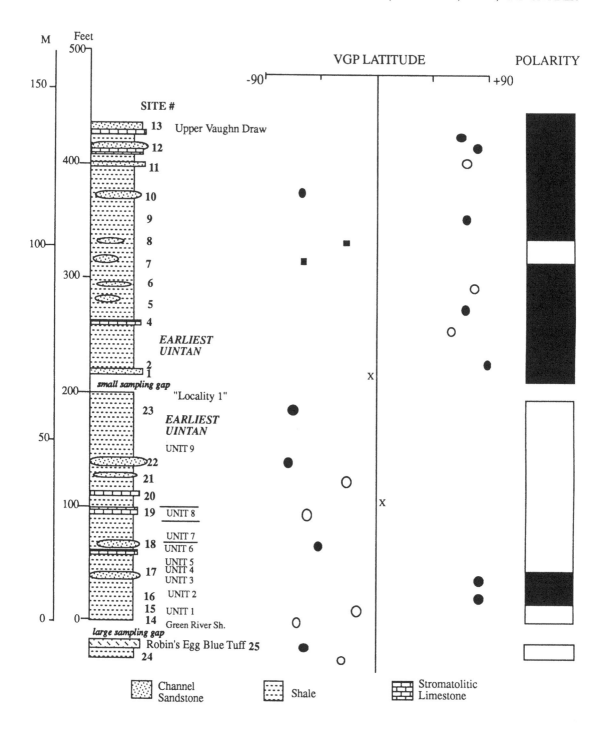

Figure 4. Magnetostratigraphy of the Locality 1 and Upper Vaughn Draw sections in the Sand Wash Basin (see Fig. 1). Solid circles = Class I sites (statistically significant) of Opdyke et al. (1977); solid squares = Class II sites (one sample missing); open circles = Class III sites (one sample divergent); x = indeterminate sites.

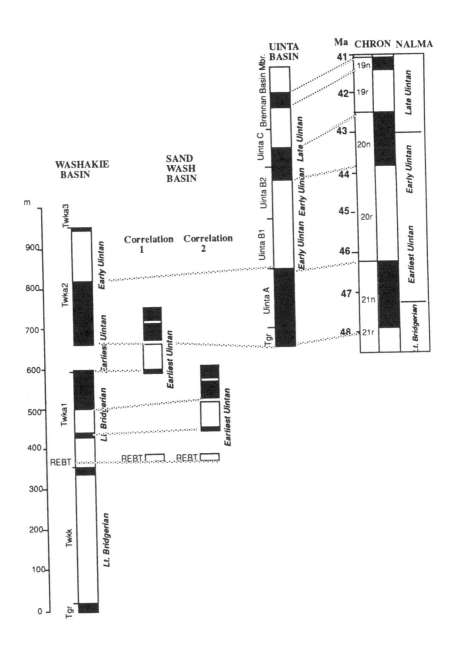

Figure 5. Possible correlations of the Sand Wash Basin magnetostratigraphy to other Uintan strata and to the magnetic polarity time scale of Berggren et al. (1995). Washakie Basin after McCarroll et al. (this volume, Chapter 2); Uinta Basin after Prothero (this volume, Chapter 1). REBT = Robin's Egg Blue Tuff; Twkk = Kinney Rim Member of the Washakie Formation; Twka = Adobe Town Member of the Washakie Formation, units 1, 2, and 3.

unsampled gap in magnetozone C+ is actually of reversed polarity and thus could correlate with out "Locality 1" reversed section. The Vaughn Draw section would again correlate with the higher part of magnetozone C+.

Whichever interpretation is adopted, the correlation of our sections with the magnetic polarity time scale hinges on the interpretation of the Washakie Basin section. As discussed by Walsh (this volume, Chapter 4), Walsh et al. (this volume, Chapter 6), and Prothero and Emry (summary chapter of this volume), the evidence is now quite strong that the normal magnetozone C+ in the Washakie Basin correlates with Chron C21n (46.3-47.9 Ma). This places the REBT in late Chron C21r, and the rest of the section in either C21r or C21n. Thus, the Sand Wash Basin section appears to span the interval from about 48.2 Ma to about 47 Ma.

ACKNOWLEDGMENTS

The paleomagnetic aspects of this project were supported by NSF grant EAR91-17819 to Prothero. Curation and storage of the DMNH fossil material was supported by the Bureau of Land Managment and by NSF grant DEB92-1817 to Stucky. We thank Erin Wilson and Joby Campbell for their help with sampling, and Dr. Joseph Kirschvink for graciously allowing access to his paleomagnetic laboratory. We would especially like to thank the students in the DMNH Certification Program in Paleontology for all their help with collecting, and Harley Armstrong and Hal Keesling of the Bureau of Land Management for their support of the field work. The manuscript was reviewed by Emmett Evanoff, John Flynn, Jay Lillegraven, Steve McCarroll, Malcolm McKenna, and Steve Walsh.

LITERATURE CITED

Abel, O., and H. J. Cook. 1925. A preliminary study of early mammals in a new fauna from Colorado. Proceedings of the Colorado Museum of Natural History 5 (4):33-36.

Andersen, D. W., and M. D. Picard. 1972. Stratigraphy of the Duchesne River Formation (Eocene-Oligocene?), northern Uinta Basin, northeastern Utah. Utah Geological and Mineralogical Survey Bulletin 97:1-23.

Andersen, D. W., and M. D. Picard. 1974. Evolution of synorogenic clastic deposits in the intermontane Uinta Basin of Utah. SEPM Special Publication 22:167-189.

Berggren, W. A., D. V. Kent, M. -P. Aubry, C .C. Swisher III, and K. G. Miller. 1995. A revised Paleogene geochronology and chronostratigraphy. SEPM Special Publication 54:129-212.

Bradley, W. H. 1929. Algal reefs and oolites of the Green River Formation. U.S. Geological Survey Professional Paper 154-G, 21 pp.

Bruhn, R. L., M. D. Picard, and S. L. Beck. 1983. Mesozoic and early Tertiary structure and sedimentology of the central Wasatch Mountains, Uinta Mountains, and Uinta Basin. Utah Geological and Mineralogical Survey Special Studies 59:63-105.

Butler, R. F. 1992. Paleomagnetism. Blackwell, Boston.

Cashion, W. B. 1967. Geology and fuel resources of the Green River Formation, southeastern Uinta Basin, Utah and Colorado. U. S. Geological Survey Professional Paper 548:1-48.

Cook, H. J. 1926a. A new genus of uintatheres from Colorado. Proceedings of the Colorado Museum of Natural Histor, 6 (2):7-11.

Cook, H. J. 1926b. New Eocene titanotheres from Moffat County, Colorado. Proceedings of the Colorado Museum of Natural History 6 (3):12-18.

Dane, C. H. 1954. Stratigraphic and facies relationships of the upper part of the Green River Formation and lower part of the Uinta Formation in Duchesne, Uintah, and Wasatch counties. Bulletin of the American Association of Petroleum Geologists 38:405-425.

Dickinson, W. R., T. F. Lawton, and K. F. Inman. 1986. Sandstone detrital modes, central Utah foreland region: stratigraphic record of Cretaceous-Paleogene tectonic evolution. Journal of Sedimentary Petrology 56:276-293.

Dickinson, W. R., M. A. Klute, M. J. Hayes, S. U. Janecke, E. R. Lundin, M. A. McKittrick, and M. D. Olivares. 1988. Paleogeographic and plate tectonic setting of the Laramide sedimentary basins in the central Rocky Mountain region. Geological Society of America Bulletin 100:1023-1039.

Evanoff, E., P. Robinson, P. C. Murphey, D. G. Kron, D. Engard, and P. Monaco. 1994. An early Uintan fauna from Bridger E. Journal of Vertebrate Paleontology 14 (Supplement to no. 3):24A.

Fisher, R. A. 1953. Dispersion on a sphere. Proceedings of the Royal Society A217:295-305.

Flynn, J. J. 1986. Correlation and geochronology of middle Eocene strata from the western United States. Palaeogeography, Palaeoclimatology, Palaeoecology 55:335-406.

Gazin, C. L. 1955. A review of the upper Eocene Artiodactyla of North America. Smithsonian Miscellaneous Collections 128(8):1-96.

Gazin, C. L. 1976. Mammalian faunal zones of the Bridger middle Eocene. Smithsonian Contributions to Paleobiology 26.

Granger, W. 1909. Faunal horizons of the Washakie Formation of southern Wyoming. Bulletin of the American Museum of Natural History 26(3):13-24.

Hartl, P., L. Tauxe, and C. Constable. 1993. Early Oligocene geomagnetic field behavior from Deep Sea Drilling Project Site 522. Journal of Geophysical Research 98 (B11):19649-19665.

Johnson, H. P., W. Lowrie, and D. V. Kent. 1975. Stability of anhysteretic remanent magnetization in fine and coarse magnetite and maghemite particles. Geophysical Journal of the Royal Astronomical Society 41:1-10.

Johnson, R. C. 1985. Early Cenozoic history of the Uinta and Piceance Creek basins, Utah and Colorado, with special reference to the development of Eocene Lake Uinta; in R. M. Flores and S. S. Kaplan (eds.), Cenozoic Paleogeography of the West-Central United States. SEPM Rocky Mountain Paleogeography Symposium 3:247-276.

Krishtalka, L. and R. K. Stucky. 1983. Paleocene and Eocene marsupials of North America. Annals of the Carnegie Museum 52:229-263.

Krishtalka, L., R. K. Stucky, R. M. West, M. C. McKenna, C. C. Black, T. M. Bown, M. R. Dawson, D. J. Golz, J. J. Flynn, J. A. Lillegraven, and W. D. Turnbull. 1987. Eocene (Wasatchian through Duchesnean) biochronology of North America; pp. 77-117 in M. O. Woodburne (ed.), Cenozoic Mammals of North America, Geochronology and Biostratigraphy. University of California

Press, Berkeley.

Mader, B. J. 1989. The Brontotheriidae:a systematic revision and preliminary phylogeny of North American genera; pp. 458-484 *in* D. R. Prothero and R. M. Schoch (eds.), The Evolution of Perissodactyls. Oxford University Press, New York.

McKay, E. J. 1974. Geologic map of the Lone Mountain Quadrangle, Moffat County, Colorado. U.S. Geological Survey Map GQ-1144.

McKay, E. J., and M. J. Bergin. 1974. Geologic map of the Maybell Quadrangle, Moffat County, Colorado. U.S. Geological Survey Map GQ-1145.

McKenna, M. C. 1980. Late Cretaceous and early Tertiary vertebrate paleontological reconnaissance, Togwotee Pass area, northwestern Wyoming; pp. 321-343, *in* L. L. Jacobs (ed.), Aspects of Vertebrate History: Essays in Honor of Edwin Harris Colbert. Museum of Northern Arizona Press, Flagstaff, Arizona.

Opdyke, N. D., E. H. Lindsay, N. M. Johnson, and T. Downs. 1977. The paleomagnetism and magnetic polarity stratigraphy of the mammal-bearing section of Anza-Borrego State Park, California. Journal of Quaternary Research 7:316-329.

Picard, M. D., and L. R. High, Jr. 1972. Paleoenvironmental reconstructions in an area of rapid facies change, Parachute Creek Member of the Green River Formation (Eocene), Uinta Basin, Utah. Geological Society of America Bulletin 83:2689-2708.

Pluhar, C. J., J. L. Kirschvink, and R. W. Adams. 1991. Magnetostratigraphy and clockwise rotation of the Plio-Pleistocene Mojave River Formation, central Mojave Desert, California: San Bernardino County Museum Association Quarterly 38 (2):31-42.

Prothero, D. R., and W. G. Lohr. 1994. Magnetostratigraphy and sedimentology of the earliest Uintan Sand Wash Basin, Colorado. Journal of Vertebrate Paleontology 14 (supplement to no. 3):42A.

Radinsky, L. R. 1967a. A review of the rhinocerotoid family Hyracodontidae (Perissodactyla). Bulletin of the American Museum of Natural History 136:1-45.

Radinsky, L. R. 1967b. *Hyrachyus, Chasmotherium*, and the early evolution of helaletid tapiroids. American Museum Novitates 2313:1-23.

Roehler, H. W. 1973. Stratigraphy of the Washakie Formation in the Washakie Basin, Wyoming. U.S. Geological Survey Bulletin 1369:1-40.

Roehler, H. W. 1992a. Description and correlation of Eocene rocks in stratigraphic references sections for the Green River and Washakie Basins, southwest Wyoming. U.S. Geological Survey Professional Paper 1506-D.

Roehler, H. W. 1992b. Correlation, composition, areal distribution, and thickness of Eocene stratigraphic units, greater Green River Basin, Wyoming, Utah, and Colorado. U.S. Geological Survey Professional Paper 1506-E.

Ryder, R. T., T. D. Fouch, and J. H. Elison. 1976. Early Tertiary sedimentation in the western Uinta Basin, Utah. Geological Society of America Bulletin 87:496-512.

Smedes, H. W., and H. J. Prostka. 1973. Stratigraphic framework of the Absaroka volcanic supergroup in the Yellowstone National Park region. U.S. Geological Survey Professional Paper 729-C:C1-C33.

Snyder, J. R. 1993. A new genus of Helohyidae (Mammalia, Artiodactyla) from the Sand Wash Basin, Colorado (Washakie Formation, Eocene, Earliest Uintan). Journal of Vertebrate Paleontology 14 (Supplement to no. 3):58A.

Stagner, W. L. 1941. The paleogeography of the eastern part of the Uinta Basin during Uinta B (Eocene) time. Annals of the Carnegie Museum, 28:273-308.

Stucky, R. K. 1992. Mammalian faunas in North America of Bridgerian to early Arikareean "ages" (Eocene and Oligocene); pp. 464-493 *in* D. R. Prothero and W. A. Berggren (eds.), Eocene-Oligocene Climatic and Biotic Evolution. Princeton University Press, Princeton, N. J.

Stucky, R. K., L. Krishtalka, and M. R. Dawson. 1989. Paleontology, geology, and remote sensing of Paleogene rocks in the northeastern Wind River Basin, Wyoming, USA; pp. 34-44 *in* J. J. Flynn and M. C. McKenna (eds.), Mesozoic/Cenozoic Vertebrate Paleontology: Classic Localities, Contemporary Approaches. 28th International Geological Congress Field Trip Guidebook T322.

Stucky, R. K., and J. R. Snyder. 1992. Mammalian fauna of the Sand Wash Basin, Colorado (Washakie Formation, Middle Eocene, Earliest Uintan). Journal of Vertebrate Paleontology 12 (supplement to no. 3):54A.

Surdam, R. C., and K. O. Stanley. 1979. Lacustrine sedimentation during the culminating phase of Eocene Lake Gosiute, Wyoming (Green River Formation). Geological Society of America Bulletin 90:93-110.

Surdam, R. C., and K. O. Stanley. 1980. Effects of changes in drainage-basin boundaries on sedimentation in Eocene Lakes Gosiute and Uinta of Wyoming, Utah, and Colorado. Geology 8:135-139.

West, R. M., and M. R. Dawson. 1975. Eocene fossil Mammalia from the Sand Wash Basin, northwestern Moffat County, Colorado. Annals of the Carnegie Museum 45 (11):230-253.

Wheeler, W. H. 1961. Revision of the Uintatheres. Peabody Museum of Natural History, Yale University, Bulletin 14:1-93.

Wilson, J. A. 1986. Stratigraphic occurrence and correlation of early Tertiary vertebrate faunas, Trans-Pecos Texas: Agua Fria-Green Valley areas. Journal of Vertebrate Paleontology 6:350-373.

4. Theoretical Biochronology, the Bridgerian/Uintan Boundary, and the "Shoshonian Subage" of the Uintan

ABSTRACT

Operational definitions of biochron boundaries refer to a stratigraphic record; either the lowest known occurrence of the defining taxon in a given section (Lowest Stratigraphic Datum), or the presumed Earliest Known Record of the defining taxon in a geographic area (EKR). Operational definitions in this context are subject to a variety of geological, paleoecological, taphonomic, and sampling biases, and will often result in biochron "boundaries" that are inappropriate for consistent magnetobiochronological correlations. Theoretical definitions of biochron boundaries refer to historical events, such as the immigration or evolution of the defining taxon (First Historical Appearance; FHA), and are recommended here. In order to demonstrate that such a boundary occurs within a particular magnetozone, (1) the defining or characterizing taxa for the beginning of this biochron *and* (2) an assemblage confidently referable to the next older biochron must both be known from this same magnetozone.

Flynn (1986) proposed that the Bridgerian/Uintan North American Land Mammal "Age" boundary occurred during Chron C20r. However, evidence from the La Jolla Group in San Diego and the Aycross and Tepee Trail formations in Wyoming now suggests that this boundary occurred either during the latest part of Chron C21r or during C21n, and given the time scale of Berggren et al. (1995), its age is estimated at between 46.3 and 48.3 Ma.

Flynn's (1986) suggestion that the beginning of the Uintan is older than the geochron of Uinta B is probably correct, but the status of the "Shoshonian Subage" of the Uintan is problematical. While the term "Shoshonian" can be used in the sense of a "faunal facies" to designate those early Uintan assemblages that contain certain "Bridgerian holdover taxa," it should not connote that all such assemblages are necessarily older than the geochron of Uinta B. Intensive screenwashing of Uinta B1 and B2 should be done in order to establish the stratigraphic ranges of various micromammals, and to determine the presence or absence of "Bridgerian holdover taxa." This task will be critical to the faunal definition and characterization of putative "earliest Uintan" and "late early Uintan" biochrons, and until it is accomplished, use of the term "Shoshonian" in a temporal sense should be suspended.

INTRODUCTION

Flynn (1983, 1986) integrated mammalian biochronology and paleomagnetic and radiometric data in a bold correlation project that was designed to provide a chron correlation and numerical age for the boundary between the Bridgerian and Uintan North American Land Mammal "Ages" (NALMA). Flynn's study may be summarized in five major steps as follows: (1) Marine microfossil assemblages (e.g., Bukry and Kennedy, 1969) from the lower middle Eocene Ardath Shale in San Diego were used to correlate the normal polarity interval in this formation with Chron C21n (Berggren et al., 1985); (2) Upsection extrapolation of the San Diego Eocene magnetostratigraphic pattern resulted in an assignment of the reversed polarity interval in the Friars Formation to Chron C20r; (3) Early Uintan mammal assemblages from the Friars Formation and the Tepee Trail and Washakie formations of Wyoming were proposed to be significantly older than the early Uintan assemblages from Uinta B of Utah, and were assigned to Flynn's (1986) newly recognized "Shoshonian Subage"; (4) Since these "Shoshonian" assemblages were assumed to be essentially synchronous, the reversed polarity intervals in which each assemblage and the Bridgerian/Uintan (B/U) boundary allegedly occurred were assigned to Chron C20r; (5) Several radiometric dates from the Aycross and Wiggins formations of Wyoming were used to derive estimates of 49 Ma for the B/U boundary, and 49.5 Ma for the end of Chron C21n.

Flynn's (1986) conclusion that the B/U boundary occurred during Chron C20r was reiterated by Krishtalka et al. (1987), and has been generally accepted by other workers (e.g., Prothero and Swisher, 1992). However, re-examination of Flynn's data and other magnetostratigraphic studies, in conjunction with the new Geomagnetic Polarity Time Scale, or GPTS (Berggren et al., 1995) indicates that some of his conclusions should be modified. The purposes of this chapter are: (1) to discuss two methods by which terrestrial mammalian biochrons are defined, and the suitability of these methods to continent-wide magnetobiochronological studies; (2) to examine the current evidence for the chron correlation of the B/U boundary; and (3) to discuss several issues regarding the "Shoshonian Subage" of the Uintan.

THEORETICAL BIOCHRONOLOGY
LSD's, FHA's, and FAD's

Before examining the chron correlation of the B/U boundary, an extended discussion of some basic principles is necessary. I largely accept the distinctions made by Murphy (1977) between definition, characterization, and identification (used here in the context of mammalian biochrons). I also endorse the recommendation of Woodburne (1977, 1987) that the beginnings of mammalian biochrons should be defined by using a single taxon, but there are several ways to interpret this recommendation. For example, Flynn (1986, p. 380) defined the beginning of the Uintan NALMA by "the first occurrence of the amynodontoid rhinoceros *Amynodon*." Without further discussion, however, this definition is ambiguous, because the word "occurrence" (or the interchangeable word "appearance") can be interpreted in either a theoretical or operational context, as follows:

Operational Definition: "The beginning of the Uintan NALMA is defined by the stratigraphically lowest known occurrence of any fossil of *Amynodon*."

Theoretical Definition: "The beginning of the Uintan NALMA is defined by the first historical appearance of *Amynodon* in North America, whether this event is the result of immigration or *in situ* evolution."

Amynodon was suggested to be an Asian immigrant by Wall (1982a), and this conclusion is assumed to be correct for the following discussions. In the present context, theoretical definitions are designed to reflect our understanding of what (unobserved) events and processes actually took place in the past. These definitions provide the explanatory basis for interpreting the existence and distribution of operational data, such as fossils in a given section. Continued collection of facts serves to either corroborate or refute the theoretical ideas used to explain them. The fundamental differences between operational and theoretical definitions in this context have not always been fully appreciated. Even workers who are familiar with these concepts have nevertheless sometimes phrased their discussions in such a way as to invite confusion. For example, Evander (1986, p. 380) states:

The Hemingfordian-Barstovian boundary is redefined *at the immigration* [a theoretical definition referring to a historical event; italics mine] of gomphotheriid proboscideans into North America. Specifically, *a boundary stratotype* [an operational definition referring to an actual fossil occurrence; italics mine] is set in the abundantly fossiliferous and well-studied Mud Hills section of the Barstow Formation, at the base of New Year Quarry.

Especially in terrestrial sections, however, the point in time represented by even the oldest known fossil record of a particular taxon does not necessarily correspond closely to the time of immigration, evolution, or local historical appearance of that taxon (e.g., Pickford and Morales, 1994, p. 306), and these and several related concepts must be kept separate. Since the commonly used terms "First Appearance Datum" and "Lowest Stratigraphic Datum" do not cover all of these concepts, additional ones are necessary. The term "Earliest Known Record" was proposed by Pickford and Morales (1994) and will be adopted here. Some new terms will also be coined below for several concepts that are relevant to this discussion.

Operational Terms:

LSD = Lowest Stratigraphic Datum (Opdyke et al., 1977; Lindsay et al., 1987)—In a specified local stratigraphic section, the lowest level from which fossil remains confidently identified as pertaining to a particular taxon have been recognized.

EKR = Earliest Known Record (Pickford and Morales, 1994)—Of all known LSD's of a particular taxon in a specified geographic area, the LSD that is the oldest.

LSD's have no necessary temporal significance; they are simply local stratigraphic records whose positions are initially described in terms of meters above or below a given geological horizon or magnetozone boundary. Lindsay et al. (1987, p. 278) discuss the concept of the LSD as follows:

Our indication of . . . *lowest stratigraphic datum* rather than *first appearance datum* indicates a local stratigraphic event rather than a widespread temporal event. The . . . LSD events are restricted to a local section and geographic region; they are explicitly biostratigraphic events.

Marine micropaleontologists often use the term "Lowest Occurrence" (LO) for the same concept as the LSD (e.g., Berggren et al., 1995). I prefer the latter term for two reasons. First, "LO" has also been used to abbreviate "Last Occurrence" (Berggren and Prothero, 1992), which is a temporal (not stratigraphic) term. Second, "LSD" is self-defining, because "stratigraphic" and "datum" clearly designate an observed record of a fossil in a specific stratigraphic section. The term "First Appearance Datum" (FAD) was not rigorously defined when first used ("changes in the fossil record with extraordinary geographical limits"; Berggren and Van Couvering, 1974, p. ix), and so has been interpreted in a variety of ways. Van Couvering and Berggren (1977, p. 284) used this term in a theoretical context, referring to "the most widespread and distinctive *events in biological history* [italics mine]." Lindsay et al. (1987,

p. 278) also used this term in a theoretical context, when they stated that "An FAD . . . event is recognized as a significant *biological and biochronological event* prior to evaluating its chronological limits [italics mine]." Nevertheless, the term "FAD" has often been used as a synonym of the LSD (e.g., Wing et al., 1991, fig. 2). Yet another meaning was proposed by Lindsay and Tedford (1990, p. 609) who state:

> The LSD represents the lowest known record of [a] taxon in a stratigraphic sequence; the FAD represents an **interpretation** of the earliest record of that taxon among its numerous fossil records in any biogeographic region.

Lindsay and Tedford's (1990) concept of the FAD does not refer to a historical event, but rather represents the LSD that is believed to be the oldest known, and I adopt Pickford and Morales's (1994) term "Earliest Known Record" (EKR) for essentially the same concept. Although the EKR still refers to an operational datum (a known fossil occurrence in a specific stratigraphic section), it has a theoretical element in that this particular occurrence is actually the oldest known record of that taxon (whether or not any evidence exists to demonstrate this). I will give a revised definition of the "FAD" after discussing the following new terms.

Theoretical Terms:

FHA_O = Oldest First Historical Appearance—The point in time at which individuals of a particular taxon first appeared in a specified geographic area, either as a result of immigration or *in situ* evolution.

FHA_l = Local First Historical Appearance—Within a specified subarea of a larger geographic area, the point in time at which individuals of a particular taxon first appeared in that subarea.

FHA_{cd} = Time of Completed Dispersal—The point in time at which disperal of a particular taxon throughout a specified geographic area was essentially complete.

FHA = First Historical Appearance—Used instead of FHA_O, FHA_l, and FHA_{cd} when the distinction between them is unnecessary.

FHA's have no necessary stratigraphic significance; they are historical events whose occurrence in time is described (at least conceptually) in terms of millions of years before present (Ma), or as occurring during particular biochrons or geomagnetic chrons. I use the term "First *Historical* Appearance" here in order to give this phrase a built-in theoretical connotation (use of the term FAD in this context has been confusing, because the operational term "datum" implies that a stratigraphic record and not a historical event is being used for the definition). The use of events such as evolutionary origins and immigrations to provide the basis for the

definitions (if not the definitions themselves) of the beginnings of terrestrial mammal biochrons has been discussed by several workers (Repenning, 1967; Tedford, 1970; Woodburne, 1977, 1987). As implied above, the beginning of a particular mammal biochron could be defined to be "instantaneous" by using the FHA_O, or it could be defined to have a certain amount of intrinsic diachrony by using the interval between the FHA_O and the FHA_{cd}. The temporal difference between the FHA_O and FHA_{cd} of a taxon will be negligible when used to define biochrons applicable only to local depositional basins and small provinces, but may become significant when the biochron is meant to apply throughout large continents. Although an accurate estimate for the duration of most Tertiary mammal dispersal events may never be possible, studies of the Pliocene, Pleistocene, and Recent record have led some workers to conclude that these events often take place in less than a few tens of thousands of years (Kurten, 1957; Savage and Russell, 1983; Flynn et al., 1984), which is insignificant relative to the duration of most chrons. Nevertheless, some large-scale mammal dispersal events apparently take much longer than this (e.g., Lindsay et al., 1987; Woodburne and Swisher, 1995), and this subject will be addressed below in the discussion of the FAD.

The concepts represented by the terms LSD, EKR, FHA_O, FHA_l, and FHA_{cd} are clarified upon inspection of Figure 1, which illustrates several problems typically encountered in mammal-based biochronologic correlation of widely separated areas. Five hypothetical stratigraphic sections are depicted (1-5), spanning three different land mammal ages (the "Aian," "Bian," and "Cian"), and very accurate numerical age control for each LSD is assumed for the purposes of discussion. Fossil taxa diagnostic of each land mammal age are represented by corresponding letters in each section. Individuals of Taxon B (whose FHA defines the beginning of the Bian land mammal age) first appeared on the western edge of the continent at 47.5 Ma, and dispersed slowly eastward over a period of 0.5 m.y., which represents the intrinsic diachrony of the Aian/Bian (A/B) boundary (compare with "dispersal lag" of Woodburne and Swisher, 1995). In Section 1, the LSD of Taxon B occurs above a significant disconformity (e.g., a sequence boundary in a coastal basin). Thus, even though individuals of Taxon B were present in this area for a million years after the FHA_O, no strata exist in this basin to record that fact. Section 2 spans the A/B boundary, but has not been explored for vertebrate fossils. In Section 3, the LSD of Taxon B occurs at 47 Ma. In Section 4, only strata of late Bian age occur in this basin, and the LSD of Taxon B here occurs long after the A/B boundary. Finally, in Section 5, a continuous section across the A/B boundary is present, but the strata in this interval are unsuitable for the preservation of vertebrate fossils, and the LSD of

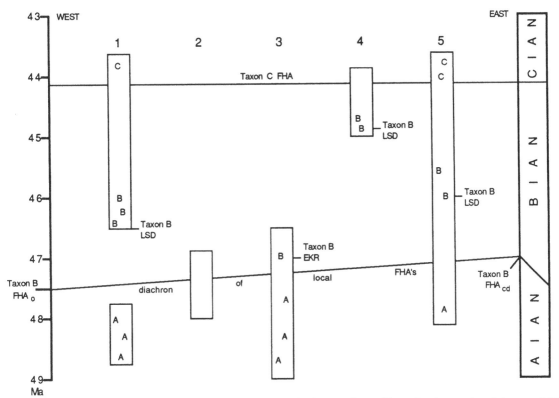

Figure 1. Five hypothetical sections spanning three hypothetical land mammal ages, illustrating the meaning of the terms LSD, EKR, FHA_o, FHA_l, and FHA_{cd}. See text for discussion.

Taxon B does not occur until more favorable strata began to be deposited at about 46 Ma.

In principle, since the LSD in Section 3 represents the EKR of Taxon B, some workers would designate it as the "boundary stratotype," and then use this particular record to define the beginning of the Bian land mammal age (this example is simplistic, since advocates of boundary stratotypes would presumably require more continuously fossiliferous sections before choosing the "best" one). However, there is no theoretical significance in the EKR as such. Like any other fossil occurrence, it owes its existence in and collection from that particular section to various geologic, paleoecologic, taphonomic, and sampling biases, and it is understood that future fossil discoveries may render its status as the EKR obsolete (e.g., Lindsay, 1990, p. 6). EKR's are often difficult to recognize as such anyway (Lindsay and Tedford, 1990, p. 609), so their identification prior to the definition of a chronostratigraphic stage will often not be possible. Herein lies an important conceptual difference between operationally-defined chronostratigraphic stages (and their corresponding geochronologic ages) and theoretically-defined biochrons. For example, in Fig. 1, given a chronostratigraphic definition of the Bian Stage based on the EKR in Section 3, if a 47.3 Ma record of Taxon B were subsequently discovered in Section 2, this new EKR would not be of Bian *age*, precisely because it is older than the designated boundary in the stratotype. This conclusion is required by the definitions of the categories "chronostratigraphic stage" and "geochronologic age" as given by the North American Stratigraphic Code (North American Commission on Stratigraphic Nomenclature, 1983, Articles 66, 67, 80, and 81). In contrast, if we used a theoretical definition for the beginning of the Bian Age (i.e., the immigration of Taxon B), any pre-Cian assemblage containing Taxon B would be of Bian age.

Finally, I agree with Pickford and Morales (1994) that the term "FAD" has come to be loaded with unfortunate connotations. However, a better definition for this ambiguous term may clarify matters. I endorse the general concept of the FAD and its relationship to the LSD as discussed by Woodburne (in press). The FAD as envisioned by many workers seems to encompass a combination of ideas, which I define in terms of a hypothetical Taxon X as follows:

First Appearance Datum of Taxon X—The immigrational or evolutionary appearance of Taxon X, which, together with its associated dispersal throughout a specified geographic region, has a duration short enough such that in conformable, reasonably fossiliferous, well-sampled sections, the LSD's of Taxon X will be synchronous relative to the duration of other specified temporal units that are essentially coeval.

The FAD refers to a combination of historical (origin and dispersal) events, and is thus a theoretical term, and a subcategory of the First Historical Appearance. However, when attempting to determine whether a given FHA qualifies as a FAD, only conformable, relatively fossiliferous, well-sampled sections can be considered, because only in them can the absence of Taxon X be regarded as significant, and only in them can the LSD's of Taxon X be assumed to correspond closely in time to the FHA₁'s of Taxon X (Fig. 1). If precise radiometric dating of the LSD's in such sections throughout a large geographic area is then accomplished, we can begin to evaluate the synchrony of a proposed FAD (Woodburne, in press). If the requirements of the theoretical definition are met, then a given FHA can be considered as a "datum" relative to our ability to detect significant diachrony.

Finally, the biochronologic utility of each origin and dispersal event must be considered in relation to the duration of certain other temporal units (e.g., other biochrons and geomagnetic chrons) with which it is temporally associated. For example, the objectively diachronous set of LSD's resulting from an evolutionary and dispersal event that takes two million years may nevertheless be relatively synchronous when compared to the duration of the chrons in the late Cretaceous (see Cande and Kent, 1992), and so this event may be considered as a FAD in this context. However, another immigration and dispersal event that takes only a few hundreds of thousands of years may be too long to allow its corresponding LSD's to be used for consistent identification of certain subchrons that occur in the Pliocene and Miocene portions of the GPTS (Berggren et al., 1995), and therefore such a FHA would not qualify as a FAD in this context.

To summarize the conceptual relationships between the general categories of the LSD, EKR, FHA, and FAD, if dozens of richly fossiliferous sections spanning a given land mammal age boundary were available for study, each with its own LSD of the defining taxon for the beginning of that age, the oldest of these records (the EKR) would correspond very closely in time to the initial immigration or evolution of that taxon (the FHA₀). In reality, however, many land mammal age boundaries are known to occur in only a handful of sparsely fossiliferous sections. Further, since many mammal dispersal events take considerable amounts of time, the duration of a particular FHA and its associated dispersal event can either be synchronous or diachronous relative to other important temporal units, and the practice of consistent temporal correlation requires that our defining FHA's must in fact be FAD's (whether or not we know them to be in a particular case). Each one of these concepts must be labeled by its own unambiguous term if mammalian biochronologists are to communicate effectively with one another.

Definition vs. Identification

Lindsay and Tedford (1990, p. 609) state: "Probably the most significant problem in application of any mammal chronology is the definition and recognition of the boundary between two units." However, these authors did not recommend a specific way to define biochron boundaries. Lindsay (1995) designated the cricetid rodent *Copemys* as "the single taxon *identifier* for the Barstovian/Hemingfordian boundary event" (italics mine). However, identification (recognition) is not the same as definition (Murphy, 1977), and Lindsay's (1995) discussion still leaves the Hemingfordian/Barstovian temporal boundary undefined. That is, he has provided operational criteria with which we can agree to call a given assemblage Barstovian, but he has not defined what the Hemingfordian/Barstovian "event" *actually refers to*. The distinction between these two concepts is simple: The Hemingfordian/Barstovian boundary may be theoretically *defined* by the historical event of the immigration of *Copemys* into North America. A given geological horizon can therefore be operationally *identified* (recognized) as being of Barstovian or younger age if it contains *Copemys*.

An example from archaeology provides a useful analogy here. Do we define the beginning of the "Bronze Age" by compiling all radiocarbon dates of sites with bronze artifacts, finding the oldest of these, and then declaring that the Bronze Age began at X years before present (requiring us to change our definition every time an older date is obtained on a bronze artifact?) Obviously not. Instead, we use the actual historical event that makes it possible to recognize a Bronze Age in the first place. The precise numerical age of this event may never be known, but it is the occurrence of the historical event itself that is of importance, and which is therefore appropriately used for the definition. Thus, the beginning of the Bronze Age can be theoretically *defined* as the time of the invention of the alloy of copper and tin, and the resulting spread of bronze technology throughout Eurasia. A given archaeological site can therefore be operationally *identified* as being of Bronze Age (or younger) if it contains bronze artifacts.

In endorsing the use of FHA's to define the beginnings of mammalian biochrons, I see no need to establish these biochrons on the basis of formally-defined chronostratigraphic stages (see discussions in Tedford, 1970; Murphy, 1977; Woodburne, 1977, 1987; and Evander, 1986). Although well-corroborated biostratigraphic data provide the essential raw material for the selection of useful "biochron boundary-defining taxa," I do not believe that a particular stratigraphic section should play any role in the temporal definition of these intervals. The current North American Stratigraphic Code (NASC) lacks a discussion of biochronologic units and their theoretical definition based upon

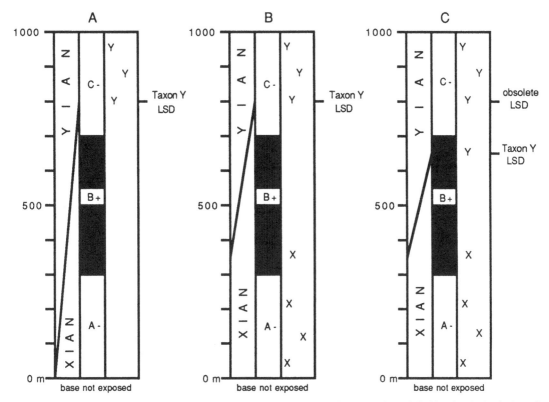

Figure 2. Three hypothetical sections showing how the choice of a theoretical or operational definition for the beginning of a mammalian biochron affects our determination of the magnetostratigraphic position of that boundary. See text for discussion.

historical events (Lindsay, 1990; Lindsay and Tedford, 1990). In my view, the failure of the NASC to offer guidelines for the theoretical definition, practical characterization, use, revision, and abandonment of biochronologic units, encourages confusion of the concepts discussed above. In addition, if the NASC emphasized that theoretical definitions of biochron boundaries should only be proposed (and, perhaps, that any new biochron names would only become valid) after the establishment and mutual corroboration of a given minimum number of local biostratigraphies, there is little doubt that workers would place more emphasis on faunal succession and superposition before casually defining new biochrons.

Implications for Magnetobiostratigraphy

Implications of the concepts of the FHA and LSD for magnetobiostratigraphy are clarified in Figure 2, which portrays three versions of a hypothetical 1000 m thick magnetostratigraphic section. In Figure 2A, Taxon Y (whose fossils are represented by "Y's") defines the beginning of the Yian land mammal age. The LSD of Taxon Y occurs at 800 m in this section, and fossils have not been found below this level. From this evidence, where in the section and in which magnetozone does the boundary between the Xian (the next-older land mammal age) and the Yian occur? If we are using an operational (LSD-based) definition for the beginning

of the Yian, the answer is simple: the boundary occurs at the 800 m level, in magnetozone C-. However, if we are using a theoretical definition, based on the immigration or evolution of Taxon Y, we can only say that the boundary occurs no higher than the 800 m level, that it may even occur below the base of the section, and therefore it cannot be assigned to a particular magnetozone at this time. In Figure 2B, diagnostic Xian taxa ("X's"; i.e., taxa that in all other known sections are found stratigraphically below, but never above or in association with diagnostic Yian taxa) are known from as high as the 350 m level. Given an operational definition, the Xian/Yian boundary remains at 800 m, while the theoretically-defined Xian/Yian boundary can now be constrained to occur somewhere between 350 m and 800 m, and therefore either within magnetozone B+ or C-. Finally, in Figure 2C, a new specimen of Taxon Y has been discovered at the 650 m level, and the theoretically-defined Xian/Yian boundary occurs within magnetozone B+.

It may be observed that the criteria outlined above for unambiguously assigning a mammal biochron boundary to a given magnetozone (and ultimately a given geomagnetic chron) will seldom be met in practice. That is precisely the point of this discussion, for if the statement "the beginning of the Yian land mammal age occurred in Chron Z" is made in the literature, such a correlation soon assumes the role of a primary piece of data. It is then used by other workers to physically correlate other sections, to assign ages to strata barren of fossils, and to otherwise support certain hypotheses over others. But if the initial magnetobiochronologic correlations of biochron boundaries in widely separated, unevenly fossiliferous, and often disconformity-ridden terrestrial sections are to be meaningful, I contend that they must meet the requirements discussed above. If they do not, the resulting chron correlations and all other implied temporal conclusions based upon them will be suspect. This is especially true when attempting to correlate biochron boundaries with those portions of the GPTS where each chron is composed of several short subchrons of alternating polarity.

Assumptions Involved in the Correlation of Middle Eocene Mammalian Biochrons to the GPTS

Finally, before discussing the magnetostratigraphy of specific Bridgerian/Uintan sections, it is necessary to review some of the assumptions involved in the practice of correlating the middle Eocene mammal biochrons of North America with the GPTS. There are currently two methods by which this can be done. The first is applicable only to coastal sections where the mammal-bearing magnetozones also encompass marine strata assignable to certain microfossil biochrons. In such cases, chron assignments can be made on the basis of information from deep-sea cores and other relatively complete sections wherein the relationship between the microfossil biochrons and geomagnetic chrons has already been established (e.g., Berggren et al., 1995). This was the method used by Flynn (1986) to assign the middle Eocene magnetozones of San Diego to specific chrons. Unfortunately, non-mammalian biochronologies capable of correlating middle Eocene magnetozones in the coastal and interior portions of North America do not exist. Sporomorph biostratigraphies have been developed for the Eocene of southern California and the Gulf Coast (Frederiksen, 1989; 1991; 1995), and for the Paleocene and earliest Eocene of the western interior (Nichols and Ott, 1978; Nichols, 1994), but similar studies have not been completed for the middle Eocene of the latter region (see Newman, 1974, for a preliminary effort). Thus, the floral events observed by Frederiksen (1989) in southern California currently cannot be recognized in the Bridger, Uinta, and Washakie basins. Furthermore, provincial differences and local physiographic effects in the western interior may hamper efforts to identify synchronous sporomorph events common to both areas (N. O. Frederiksen, written communication, 1995).

The second method of identifying terrestrial magnetozones independently of mammalian biochronology involves direct radiometric dating of the magnetozones, and their assignment to particular chrons on the basis of current time scales (e.g., Berggren et al., 1995). However, numerical calibration of the Eocene chrons is

a difficult task in itself. An important aspect of Flynn's (1986) study concerned his attempt to provide an estimate for the age of Chron C21n(y), which could then be used as a tie-point for the calibration of the GPTS. Several K-Ar dates have been obtained on the Aycross and Wiggins formations in the Absaroka Range of Wyoming, and Flynn (1986, pp. 392-393; fig. 10) used some of these dates to assign an age of about 49.5 Ma to the top of a magnetozone that he believed to represent Chron C21n. As discussed below, however, it now appears that the magnetozone in question represents (in part) C22n (Sundell et al., 1984; Harland et al., 1990, p. 157). As a result, the estimates of Berggren et al. (1985) and Flynn (1986) for the age of C21n(y) are probably too old (Berggren et al., 1995).

More recent estimates for the age of Chron C21n have been based on radiometrically-dated deep-sea horizons yielding paleomagnetic and biochronologic data (Bryan and Duncan, 1983). Berggren et al. (1995) give estimates for the age of C21n(o) and C21n(y) of 47.9 Ma and 46.3 Ma, respectively, using the older of the two K-Ar dates of Bryan and Duncan (1983) as one of their tie-points. However, as discussed by Berggren et al. (1995), $^{40}Ar/^{39}Ar$ dates of about 45.6 Ma obtained by Swisher and Montanari (in preparation) for the horizon of Bryan and Duncan (1983) imply a somewhat younger age for C21n(y). Additional imprecision in these figures is introduced by uncertainty regarding the precise position of the dated horizon within the magnetozone that represents Chron C21n in Hole 516F (D. V. Kent, personal communication). These factors, combined with other assumptions (Cande and Kent, 1992), contribute to inherent potential errors of about ±1 m.y. in the current numerical calibration of the middle Eocene chrons (C18, C19, C20, and C21; S. C. Cande and D. V. Kent, personal communication; see also Harland et al., 1990, p. 154). In view of this degree of uncertainty, a given radiometrically-dated magnetozone may not be confidently identifiable if the calibrated age difference between chrons of the same polarity is less than the sum of 1 m.y. and the analytical error in the date. In the context of the B/U boundary, this problem is mitigated by the fact that the early middle Eocene chrons (C21r-C20n) are of relatively long duration. However, in other parts of the GPTS (e.g., C19r-C19n-C18r), even relatively precise radiometric dates may not allow the assignment of a given magnetozone to a particular chron.

In practice, since the most common situation in the western interior is one in which the Eocene magnetozones are not associated with precise radio-metric dates, the only way to identify them is by using mammalian biochrons that have already been correlated with the GPTS on the basis of one of the independent methods discussed above. However, the validity of this technique may depend upon the assumption that the biochrons in question encompass only one chron of a given polarity. For example, since the early Uintan interval in California has been independently correlated (in part) with C20r, certain reversed magnetozones in Wyoming and Utah that contain early Uintan assemblages are also assumed to pertain to C20r (e.g., Flynn, 1986; Prothero, this volume, Chapter 1; this chapter). In this case, the assumption may be reasonable owing to the relatively long combined duration of C21n and C20r (see Berggren et al., 1995). However, for other mammal biochrons in magnetically-complex parts of the GPTS, these assumptions may not be justifiable. Although it would be convenient if certain mammal biochrons could be used as "chron-specific" indicators, if these correlations are not to become circular, they must be continually tested by independent methods over wide geographic areas (as has been the case for marine microfossil biochrons—Berggren et al., 1995).

Having discussed the necessary biochronologic and magnetostratigraphic framework, concrete examples of these ideas will be addressed in the following examination of the magnetostratigraphic occurrence of the Bridgerian/Uintan boundary.

MAGNETOBIOSTRATIGRAPHY OF THE BRIDGERIAN/UINTAN BOUNDARY
La Jolla Group, San Diego County, California

Flynn (1986) discussed the magnetobiostratigraphy of the Eocene formations in southwestern San Diego County, California (Fig. 3). Of relevance to the location of the B/U boundary in this section is the middle Eocene La Jolla Group, which includes the Delmar Formation, Torrey Sandstone, Ardath Shale, Scripps Formation, and Friars Formation (Kennedy and Moore, 1971; Walsh et al., this volume, Chapter 6).

Given a theoretical definition for the beginning of the Uintan (i.e., the immigration of *Amynodon* into North America), Flynn's (1986, p. 386) statement that the "Shoshonian (Earliest Uintan) faunas and the Bridgerian/Uintan boundary consistently occur within a reversed polarity interval in the three Wyoming and California sequences" is unsupported by biostratigraphic evidence in San Diego. There is no faunal basis for the recognition of the B/U boundary within the reversed polarity interval in the upper part of the Friars Formation (magnetozone C- of Flynn, 1986, fig. 8). In order to conclude this, it is necessary to demonstrate that an undoubted Bridgerian assemblage occurs in the Friars within the lower part of this magnetozone, or perhaps within the uppermost part of the normal polarity interval in the base of the Friars and the underlying Scripps Formation (an "undoubted Bridgerian assemblage" would be one containing taxa that in all other known sections are found stratigraphically below, but never above or in association with characteristic Uintan taxa; or, lacking these diagnostic Bridgerian taxa, an assemblage numerically well-represented and taxonomically diverse enough

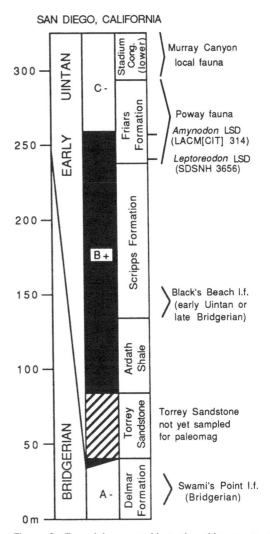

Figure 3. Essential magnetobiostratigraphic aspects of Bridgerian through early Uintan strata in southwestern San Diego County. Compiled from Golz and Lillegraven (1977), Flynn (1986), Walsh et al. (this volume, Chapter 6), and LACM(CIT) records.

such that the *absence* of *Amynodon* and other characteristic Uintan taxa provides compelling *negative* evidence for the Bridgerian age of that assemblage; see Flynn, 1986, p. 380 for taxa characteristic of the beginning of the Uintan). Flynn (1986, fig. 8) did not indicate the position of *Amynodon*-bearing localities or formational contacts on his San Diego sections, so it is difficult to determine where his magnetozone boundaries occur relative to these crucial data points. Nevertheless, when combined with a knowledge of the stratigraphic occurrence of two important fossil localities, Flynn's paleomagnetic sections from the Friars Formation do help to constrain the magnetostratigraphic position of the B/U boundary in San Diego (Fig. 3).

The 35 m-thick type section of the Friars Formation (Kennedy and Moore, 1971) contains the vertebrate locality UCMP V-68156, which has yielded *Amynodon reedi* and *Metarhinus*? *pater* (Golz and Lillegraven, 1977, table 1). According to the UCMP locality description, V-68156 occurs roughly 20 m above the base of the Friars Formation. Inspection of Flynn's (1986, fig. 8) magnetostratigraphic chart for the San Diego area shows that approximately the lower half of the type section of the Friars is of normal polarity. The exact location of the normal-to-reversed transition in the Friars type section is unclear, because there is roughly a 20 m stratigraphic gap between Flynn's highest normal site and his only reversed site. However, these facts indicate that *Amynodon* occurs either in or less than a few meters above the top of a normal polarity interval that Flynn correlated with C21n. Flynn's (1986) Genessee Avenue section is another important outcrop exposing about 22 m of the ?Scripps and Friars formations. As discussed by Walsh et al. (this volume, Chapter 6), the LSD of *Amynodon* in San Diego occurs here within the uppermost part of a normal magnetozone near the base of the Friars Formation. This magnetozone may be confidently correlated with Chron C21n based on discussions in Flynn (1986) and Walsh et al. (this volume, Chapter 6). Finally, in the State Route 52 East section of Walsh et al. (this volume, Chapter 6), the LSD of the characteristic Uintan artiodactyl *Leptoreodon major* occurs in the lower tongue of the Friars Formation at least 17 m below the top of a normal interval that is reasonably correlated with Chron C21n.

The evidence summarized above indicates that the normal polarity interval in the lower part of the Friars Formation is at least in part of early Uintan age. But if this interval correlates with Chron C21n, as suggested by Flynn (1986, p. 384), and supported by Walsh et al. (this volume, Chapter 6), then the B/U boundary cannot occur in C20r, but must occur in C21n or some older chron. Unfortunately, evidence for the stratigraphic position of the B/U boundary in San Diego is meager. The current lack of a record of *Amynodon* and other characteristic Uintan taxa from the Scripps Formation cannot be regarded as significant. This formation largely represents shelfal deposition in open marine waters (Kennedy and Moore, 1971; Givens and Kennedy, 1979), and therefore would not be expected to produce terrestrial mammals. The presence of rare, abraded vertebrate fossils ("Horizon A" of Flynn, 1986, fig. 9; = Black's Beach local fauna of Walsh, this volume, Chapter 5) in the basal conglomerate of the type Scripps Formation is an exceptional occurrence, as these strata probably represent shallow-marine deposition on a low-stand delta (May et al., 1991). Only one mammal taxon tentatively identified to genus level is known from the Black's Beach local fauna (cf. *Uintatherium* sp.; Hutchison, 1971), and this assemblage can

only be assigned a late Bridgerian or early Uintan age (Walsh, this volume, Chapter 5). The lower constraint for the stratigraphic position of the B/U boundary in San Diego is provided by the Swami's Point local fauna, which occurs in the upper part of the Delmar Formation (Walsh, this volume, Chapter 5). This small assemblage has produced only two mammal taxa identified to generic level, but is nevertheless confidently assigned to the Bridgerian based on the presence of the Bridgerian index taxon *Trogosus*. Therefore, the B/U boundary in San Diego may occur in either the uppermost part of the Delmar Formation (Chron C21r; Flynn, 1986; Walsh et al., this volume, Chapter 6), the Torrey Sandstone (of unknown magnetic polarity), or the Ardath Shale, Scripps Formation, or basal Friars Formation (all assigned to C21n; Flynn, 1986; Walsh et al., this volume, Chapter 6).

Aycross and Tepee Trail Formations, Absaroka Range, Wyoming

Flynn (1983, 1986) discussed the magnetobiostratigraphy of the Aycross and Tepee Trail formations in the East Fork Basin area, northwestern Wyoming. Flynn's 640-m-thick composite section consisted of two subsections; the "Aycross Formation section" (Aycross Formation) and the stratigraphically higher "Tepee Creek section" (Tepee Trail Formation). For the Aycross Formation section, Flynn (1986, fig. 6) recorded a short basal reversed magnetozone (A-), an overlying normal magnetozone (B+), an overlying reversed magnetozone (C-), another normal magnetozone (D+), and an uppermost reversed magnetozone (E). The Tepee Creek section is composed of a basal normal interval and a long overlying reversed interval. Flynn (1986, pp. 376-77) correlated the basal normal interval in the Tepee Creek section with the top of magnetozone D+ in the Aycross Formation section, and then correlated the Aycross Formation magnetozone E– with the base of the reversed interval in the Tepee Creek section. For the moment, I will assume that these magnetozone correlations are correct, and Flynn's resulting composite section for the East Fork Basin is shown in Figure 4.

Flynn (1983, p. 94) reported that the LSD of *Amynodon* in the East Fork Basin occurs in the Tepee Trail Formation at the Bare A Locality, about 198 m-229 m above the base of the Tepee Creek section. Given the data shown by Flynn (1986, fig. 6), this would translate to a level about 168-199 m above the top of magnetozone D+. The early Uintan Bone Bed A locality ("Horizon D" of Flynn, 1986, fig. 9), was reported to occur about 61 m above this level (Flynn, 1983, p. 94) and contains *Amynodon* and other characteristic early Uintan taxa (McKenna, 1980, 1990). In addition, Flynn (1986, fig. 9) showed a faunal horizon "?C" occurring at the base of magnetozone E–. According to Flynn (1983; personal communication),

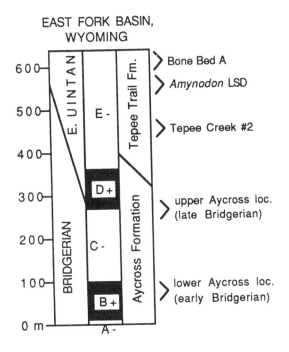

Figure 4. Flynn's (1983, 1986) interpretation of the magneto-biostratigraphy of the Aycross and Tepee Trail formations in the East Fork Basin area, southwestern part of the Absaroka Range, Wyoming.

Horizon "?C" represents a composite assemblage of two taxa from two localities ("Flynn's Folly" and "Tepee Creek #2") consisting of *Hyopsodus* sp. cf. *H. paulus* and *Hyrachyus* sp. cf. *H. eximius*. Flynn (1983, p. 95) implied that the Tepee Creek #2 locality occurs about 107 m above the top of magnetozone D+ (Fig. 4). The assemblage from Horizon "?C" is of uncertain age, because *Hyopsodus paulus* is known from the early Bridgerian to the early Uintan (Flynn, 1983, p. 83), and *Hyrachyus eximius* is known from the late Bridgerian to the early Uintan (Flynn, 1983, p. 98; Sundell et al., 1984, p. 46).

Based on the available evidence, the B/U boundary in the East Fork Basin must occur below the LSD of *Amynodon* (Flynn, 1986, p. 382), but cannot be assigned to either magnetozone D+ or E- (Fig. 4). Nevertheless, Flynn (1986, p. 382 and fig. 9) asserted that the B/U boundary occurs within magnetozone E-. As discussed above, however, in order to demonstrate that the theoretically-defined (and therefore biochrono-logically meaningful) beginning of the Uintan occurs in magnetozone E-, an undoubted Bridgerian assemblage must be known from the lower part of this magnetozone. In the absence of this biostratigraphic control, it cannot be assumed that the LSD of *Amynodon* in the Tepee Creek section corresponds closely in time to the FHA₁ of this taxon. Indeed, as discussed below, Sundell et al. (1984) and Eaton (1985) present evidence to show that elsewhere in the

Absaroka Range, the LSD's of *Amynodon* and two other characteristic Uintan taxa occur much lower in the Tepee Trail Formation than they do in the East Fork Basin.

In addition, Flynn's (1983, 1986) correlation of the magnetozones in the Aycross Formation section and Tepee Creek section was questioned by Sundell et al. (1984, p. 5):

Flynn's polarity reversal from normal to reverse in the upper Aycross section should not be correlated with the normal to reverse polarity change in the Tepee Trail Formation. A substantial amount of stratigraphic section (250-300 m) was not sampled by Flynn because of poor exposures and local un-conformities, including all of the lower Tepee Trail (of Sundell, 1982), the Tepee Trail/Aycross contact and the Blue Point marker. Our sampling of a more complete section in the Bighorn Basin indicates two polarity changes occur within the sequence missed by Flynn's sampling. We agree that the normal episode at the base of the Tepee Trail Formation correlates with anomaly 21 (Shive et al., 1980) but the major normal interval within the upper Aycross Formation is believed to be anomaly 22, not 21 as Flynn (1983) suggests.

Due to the problems involved in magnetobiostrati-graphic correlation in the East Fork Basin area recognized by Sundell et al. (1984), I will use the data of these authors and Eaton (1985) to constrain the position of the B/U boundary in the Absaroka Range, which is summarized for the Aycross and Tepee Trail formations in Figure 5. This composite, but superpositionally-controlled magnetostratigraphic section may contain the most continuous mammalian fossil record across the B/U boundary in North America. Readers should consult Eaton (1982), Bown (1982), and Sundell et al. (1984) for stratigraphic interpretations and information on the fossil localities shown here. Decker (1990) also proposes a stratigraphic model for these units similar to that of Sundell et al. (1984).

Sundell et al. (1984) report that the strata in the low-ermost 54 m of the Aycross Formation are of reversed polarity. This interval represents the upper part of a reversed magnetozone spanning the underlying Tatman? Formation and the upper part of the Willwood Formation (not shown in Fig. 5). The overlying 47 m of the Aycross Formation are of uncertain paleomag-netic significance, as they consist of a short interval of uncertain polarity, a short interval of normal polarity, and a short interval of reversed polarity. A long normal interval occurs between 101 m and 277 m above the base of the Aycross. The uppermost 56 m of the Aycross is of reversed polarity, and this reversed inter-val continues into the overlying, 30 m-thick Blue Point marker. This horizon is overlain by the Tepee Trail

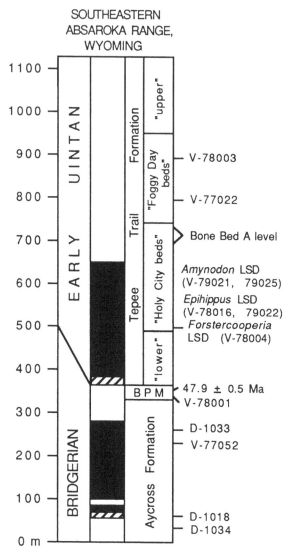

Figure 5. Essential magnetobiostratigraphic aspects of the Aycross and Tepee Trail formations in the southeastern part of the Absaroka Range, Wyoming. Compiled from Eaton (1982), Sundell et al. (1984), and Eaton (1985). BPM = Blue Point Marker.

Formation, the lowermost 19 m of which are of uncertain polarity. A long normal interval then occurs between 19 and 282 m above the base of the Tepee Trail, extending into the upper part of the "Holy City beds." The remaining 508 m of the Tepee Trail are of reversed polarity, and include the uppermost part of the "Holy City beds," the "Foggy Day beds," and the "upper Tepee Trail." The Tepee Trail Formation is overlain by about 1450 m of early Uintan and possibly late Uintan rocks of the Wiggins Formation (not shown in Fig. 5). The lower part of the Wiggins is of reversed polarity, while the "upper Wiggins" is of normal polarity (Sundell et al., 1984).

Diverse mammal assemblages from the Aycross Formation are of undoubted Bridgerian age, and have

been described by Bown (1982) and Eaton (1982). The uppermost Bridgerian locality in this formation ("Fantasia" = V-78001) is situated in the Blue Point marker, which is a distinctive (and probably essentially synchronous) volcaniclastic horizon recognizable throughout much of the Absaroka Range (Eaton, 1982; Sundell et al., 1984, p. 47; Decker, 1990, p. 18). The unit designated by Sundell et al. (1984) as the "lower Tepee Trail" has not produced identifiable mammal fossils, but the LSD of the characteristic Uintan perissodactyl *Forstercooperia* occurs in the base of the "Holy City beds" in the lower part of the long normal interval previously noted (Sundell et al., 1984, p. 32). The LSD's of *Amynodon* and *Epihippus* also occur in the "Holy City beds" in the upper part of this normal magnetozone (Eaton, 1985; Eaton, personal communication, 1995). Strata in the type section of the Tepee Trail Formation that yield the diverse, early Uintan "Bone Bed A assemblage" (McKenna, 1980, 1990) are correlated by Sundell et al. (1984, p. 27) with the upper part of the "Holy City beds." The Bone Bed A horizon occurs in the lower part of the long reversed magnetozone comprising the remainder of the Tepee Trail Formation. Early Uintan assemblages are recorded in this reversed interval from several localities in the "Foggy Day beds" and "upper Tepee Trail" (Eaton, 1985).

Correlation of the Aycross and Tepee Trail magnetozones with the GPTS is aided by relatively precise K-Ar dates of 47.9 ± 0.5 Ma and 48.5 ± 0.6 Ma for the Blue Point marker on Carter Mountain (J. D. Obradovich, oral communication *in* Bown, 1982, p. 24; see discussion in Eaton, 1982, pp. 158–159). These dates correspond well to the age of Chron C21r (47.9–49.0 Ma) as estimated by Berggren et al. (1995), so a correlation of the overlying normal interval in the lower part of the Tepee Trail Formation with C21n seems likely. Sundell et al. (1984, fig. 2) show an unconformity between the Tepee Trail Formation and the Blue Point marker, but this feature is unlikely to be of chron-encompassing magnitude for the following reasons. First, if the normal interval in the lower part of the Tepee Trail Formation correlated with Chron C20n, then the reversed interval in the upper part of the Tepee Trail and the lower part of the Wiggins Formation would correlate with C19r, and these horizons would be predicted to yield a late Uintan fauna (see discussions in Walsh et al., this volume, Chapter 6; Prothero, this volume, Chapter 1; and Prothero et al., this volume, Chapter 8). Instead, they produce a diagnostic early Uintan assemblage, and late Uintan taxa are only doubtfully known from the uppermost part of the Wiggins Formation (Eaton, 1985). Second, the K-Ar date of 46.0 ± 1.7 Ma from reversed strata in the lower part of the Wiggins Formation reported by Sundell et al. (1984, p. 17) is consistent with the age of C20r (43.8–46.3 Ma) and much older than the age of C19r (41.5–42.5 Ma) as estimated by Berggren et al. (1995).

Therefore, the data suggest that the normal magneto-zone in the lower part of the Tepee Trail Formation correlates with Chron C21n, and that the upper reversed interval in this formation correlates with Chron C20r (Sundell et al., 1984).

Bridgerian and Uintan assemblages do not occur in superposition within the same magnetozone in the reference section of Sundell et al. (1984), so current evidence from the Absaroka Range can only constrain the B/U boundary to either the later part of Chron C21r, or some part of C21n. While corroborating Uintan fossils should be sought from the lower, normally-magnetized part of the Tepee Trail Formation in the East Fork Basin area (probably C21n), the data pro-vided by Sundell et al. (1984) are consistent with the evidence previously discussed from San Diego. Both sections indicate that the B/U boundary occurred either during late C21r or C21n, and not during C20r as pro-posed by Flynn (1986). This consistency is significant because the identity of the chrons in which the B/U boundary probably occurs was determined by marine biochronology in San Diego, and by radiometric dating in Wyoming.

Uinta Formation, Uinta Basin, Utah

Prothero (this volume, Chapter 1) discusses the mag-netostratigraphy of the Uinta Formation in the Uinta Basin, Utah, which has traditionally produced the "type" faunas for the Uintan NALMA. In the Wagon-hound Canyon-Kennedy Hole section (the "classic" eastern Uinta Basin section portrayed by Riggs, 1912, and Osborn, 1929), Prothero reports that the uppermost part of the Green River Formation and all of Uinta A are of normal polarity. The extreme base of Uinta B1 is also normal, while the rest of Uinta B1 is reversed. Finally, the lower half of Uinta B2 is reversed, while the upper half of Uinta B2 and the lower part of Uinta C is normal (see Fig. 6). Correlation of this section with Prothero's Willow Creek-Ouray section (east-central part of the basin) is problematical, owing to the westward pinchout of true Uinta A, and the time-equivalence of the upper part of the Green River Formation in the Willow Creek area to parts of Uinta A and B1 in the Wagonhound Canyon area (Cashion, 1967; Cashion and Donnell, 1974; Cashion, written communication, 1995). For these reasons, I will largely confine my discussion of the B/U boundary in the Uinta Basin to the better understood and more fossiliferous Wagonhound Canyon-Kennedy Hole section. However, it is important to note that Prothero's determination of reversed polarity for Peterson and Kay's (1931) Uinta "A" along Willow Creek is consistent with Cashion's interpretation that these strata are time-equivalent to part of Uinta B1 (and possibly the lower part of Uinta B2) in the Bonanza and Coyote Basin area.

Independent correlation of the magnetozones in the Wagonhound Canyon-Kennedy Hole section to the

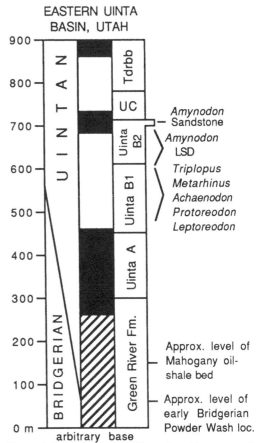

Figure 6. Essential magnetobiostratigraphic aspects of the Uinta and Green River formations in the eastern part of the Uinta Basin, Utah. Compiled from Prothero (this volume, Chapter 1), Cashion (in Dawson, 1968), and Cashion (1986). UC = Uinta C; Tdrbb = Brennan Basin Member of Duchesne River Formation.

GPTS is not yet possible, because reliable radiometric dates are unavailable from the Uinta Formation (Prothero and Swisher, 1992). Identification must therefore be made on the basis of biochronological correlation of radiometrically-dated mammal assembla-ges in other areas. According to Prothero (this volume, Chapter 1), the Uinta B/C lithostratigraphic boundary (which may approximate but should not define the early Uintan/late Uintan temporal boundary; see Walsh, this volume, Chapter 5) occurs within the middle of a normal polarity interval. Based on the estimate of 42.6 Ma given by Berggren et al. (1995) for the age of C20n(y), identification of this "earliest late Uintan" normal interval as Chron C20n is consistent with the following radiometric dates on late Uintan assemblages: 1) An ^{40}Ar/^{39}Ar date of 42.83 ± 0.24 Ma from the Mission Valley Formation in San Diego (Obradovich and Walsh, in prep.; Walsh et al., this volume, Chapter 6); 2) Recalculated K-Ar dates of 43.9 ± 0.9 Ma from stratigraphically below, and 42.7 ± 1.6 Ma from stratigraphically above the Serendipity local fauna in Texas (McDowell, 1979; Wilson, 1986); 3) A series of recalculated K-Ar dates averaging 42.2 Ma, and a K-Ar date of 42.0 ± 0.8 Ma from the Cook Mountain Formation, which approximates the age of the Casa Blanca local fauna of Texas (Westgate, 1988; Berggren et al., 1992, table 1.6). Given a correlation of the early Uintan/late Uintan boundary with Chron C20n, the reversed interval comprising most of Uinta B would correlate with C20r, and the normal interval in Uinta A and the upper part of the Green River Formation would correlate with C21n (Prothero, this volume, Chapter 1). Identification of the Uinta B reversed interval as C20r is consistent with an assignment of the reversed interval in the upper part of the early Uintan Friars Formation to C20r (Flynn, 1986; Walsh et al., this volume, Chapter 6).

The occurrence of the B/U boundary within the Uinta Formation has long been unclear, largely owing to the absence of identifiable mammals from both Uinta A and the lacustrine shales of the (late Bridgerian?) upper part of the Green River Formation (Parachute Creek Member of Cashion and Donnell, 1974). Prothero and Swisher (1992) pointed out that the records of *Amynodon, Triplopus, Metarhinus, Dolichorhinus*, and *Forstercooperia* reported from Uinta A by Krishtalka et al. (1987) actually came from Uinta B1 of Osborn (1929). Riggs (1912, pp. 19, 23) reported that the type specimens of *Metarhinus fluviatilus* and *Sphenocoelus uintensis* were collected from the lowermost levels of the Uinta Formation from somewhere along the White River, but Osborn (1929, fig. 65) assigned these specimens to Uinta B1. Peterson (1934, p. 356) and Peterson and Kay (1931, p. 301 and Plates X, XI) indicate that the type specimen of *Sthenodectes priscus* was collected in the Willow Creek area, from the lowermost 20 feet of "Uinta A" sandstones. However, this specimen was probably obtained from reversed strata which are presumably younger than the normally-magnetized rocks of Uinta A in Wagonhound Canyon (Prothero, this volume, Chapter 1; W. B. Cashion, personal communication).

The precise position of the LSD of *Amynodon* in the Uinta Formation is also unclear. This genus is common in Uinta B2 ("*Amynodon* Beds" of Riggs, 1912), but its occurrence below this level is uncertain. Osborn (1895, p. 75) listed one specimen of *Amynodon* (AMNH 1878) from "Horizon A" (presumably Uinta B1 of Osborn, 1929). However, this genus was not recorded by Riggs (1912) from the "Upper *Metarhinus* Zone," nor by Osborn (1929, fig. 65) from Uinta B1. Wood (1941, p. 87) recognized *Amynodon advenus* from "Uinta A" (presumably Uinta B1; see Wilson and Schiebout, 1981, p. 10). Wall (1982b, p. 436) recognized *Amynodon reedi* from Uinta B, but it is unclear whether the relevant specimens are from Uinta B1 or B2 (Wood, 1941, apparently recognized *A. reedi* only from Uinta B2). Finally, Prothero (this volume, Chapter 1) does not record *Amynodon* from Uinta B1. In any case, the occurrence of *Triplopus, Metarhinus, Achaenodon, Protoreodon* (Prothero, this volume, Chapter 1), and now *Leptoreodon* (Walsh, this volume, Chapter 5) demonstrate an early Uintan age for most or all of Uinta B1. Prothero and Swisher (1992, pp. 54-55) suggested that the "Bridgerian-Uintan transition occurs midway through Uinta B," but this proposal was influenced by their acceptance of Flynn's (1986) assignment of this transition to C20r, and their assumption that *Hyrachyus eximius* (reportedly from the base of Uinta B1) was a Bridgerian index fossil. The early Bridgerian Powder Wash mammal locality at Raven Ridge (e.g., Krishtalka and Stucky, 1984) has been determined to occur within the Douglas Creek Member of the Green River Formation, about 270 feet (82 m) stratigraphically below the Mahogany oil-shale bed (W. B. Cashion, *in* Dawson, 1968). Accordingly, if the B/U boundary occurred during latest C21r or during C21n, as argued above, then it must occur in the Uinta Basin either within the normally-polarized strata of Uinta A, or the upper part of the Green River Formation (Fig. 6).

Washakie Formation, Washakie Basin, Wyoming

Flynn (1986, fig. 7) reported the presence of nine magnetozones in his composite, 950 m-thick section of the Kinney Rim and Adobe Town members of the Washakie Formation in the Washakie Basin of southwestern Wyoming. The basal 30 m of the section pertained to the Laney Shale Member of the Green River Formation (Flynn, 1986, p. 354), and was reported to be of normal polarity (magnetozone A+). The Kinney Rim Member was reported by Flynn to be entirely of reversed polarity (magnetozone B1-), with the exception of a relatively thin normal interval (B2+) at the top of this member. According to Roehler (1973) and Flynn (1986, p. 377 and fig. 9), the contact between the Adobe Town Member and the underlying Kinney Rim Member is an unconformity, so the original extent of magnetozone B2+ is unknown. According to Flynn (1986), the lower unit of the Adobe Town Member is composed of a basal reversed magnetozone (B3-), an overlying short normal magnetozone (B4+), an overlying reversed magnetozone (B5-), and the lower part of a long normal magnetozone (C+). The latter is divided here into two parts (C1+ and C2+) by a magnetic sampling gap of about 90 m that McCarroll et al. (this volume, Chapter 2) show within the base of the middle unit of the Adobe Town Member. The upper part of the middle unit of the Adobe Town Member contains a reversed magnetozone (D1-) and a short normal magnetozone (D2+). The upper unit of the Adobe Town Member was not sampled by Flynn (1986). For the purposes of this discussion, I have approximated the position of Roehler's (1973) mammal localities and important stratigraphic horizons relative to the magnetozones of Flynn (1986, fig. 7) by observing Turnbull's (1978, fig. 2) placement of the boundaries between the Kinney Rim Member and the lower and middle units of the Adobe Town Member relative to Roehler's bed numbers. The resulting simplified magnetobiostratigraphic section for the Washakie Formation is shown in Figure 7, and is consistent with that presented by McCarroll et al. (this volume, Chapter 2).

Identification of the magnetozones in the Washakie Formation is difficult. First, it is uncertain if Flynn's short magnetozones B2+ and B4+ are anomalous, or if they are erosionally-truncated representatives of *bona fide* normal chrons. Second, the radiometric evidence required to independently correlate these magnetozones with the GPTS is suspect owing to sample contamination by older biotite grains (Mauger, 1977, p. 33; Flynn, 1986, p. 343; Prothero and Swisher, 1992). Nevertheless, since Mauger's work contains the only published radiometric data for the Washakie Formation, it is briefly discussed here. A K-Ar date of 49.6 ± 1.0 Ma was reported by Mauger (1977; recalculated value from Krishtalka et al., 1987, table 4.1) on the Robin's Egg Blue Tuff, of probable Bridgerian age (Bed 579 of Roehler, 1973; see Fig. 6). Given the reversed polarity of this horizon (Flynn, 1986), the date may suggest deposition during the early part of Chron C21r (see Berggren et al., 1995), since Mauger (1977, p. 33) believed that this date was too old. The younger dates of 45.7 ± 0.9 Ma and 46.7 ± 0.9 Ma reported by Mauger (1977; recalculated values from Krishtalka et al., 1987, table 4.1) were obtained from Bed 664 of Roehler (1973), and thus from magnetozone D1- of Flynn (1986). Mauger's (1977, p. 33) adjustment of these dates to compensate for contamination by older biotites would result in a value for the recalculated ages about the same as his original estimates of 44-45 Ma. Given the time scale of Berggren et al. (1995), these figures suggest a correlation of magnetozone D1- with Chron C20r. Taken at face value, therefore, Mauger's dates would suggest that Flynn's magnetozone C+ correlates at least in part with Chron C21n.

Unfortunately, the position of the B/U boundary within the Washakie Formation is also uncertain. Flynn (1986, p. 385) concluded that this boundary occurred within magnetozone B5- (probably because the LSD of *Amynodon* occurred in a reversed polarity interval in the East Fork Basin). However, McCarroll et al. (this volume, Chapter 2) indicate that no definite Uintan faunas are known from the lower unit of the Adobe Town Member. The lower constraint for the position of the B/U boundary is provided by the Bridgerian age of at least the lower part of the lower unit of the Adobe Town Member, as determined by the presence of the Bridgerian index taxa *Patriofelis*, *Thinocyon*, *Palaeosyops*, and *Homacodon* (McCarroll et al., this volume, Chapter 2). These taxa all occur below the "sandstone rim below the Adobe Town Rim" (S. McCarroll, personal communication). The upper constraint for the position of this boundary is provided by

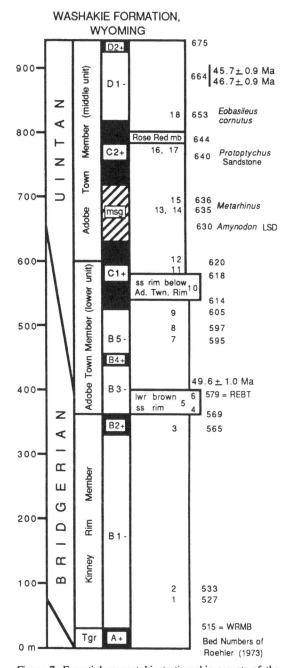

Figure 7. Essential magnetobiostratigraphic aspects of the Washakie Formation in the Washakie Basin, Wyoming. Compiled from Roehler (1973), Turnbull (1978, 1991), Flynn (1986), and McCarroll et al. (this volume). Numbers 1-18 correspond to vertebrate localities of Roehler (1973). Tgr = Green River Formation; REBT = Robin's Egg Blue Tuff; WRMB = White Ridge Marker Bed; msg = magnetic sampling gap.

the LSD of *Amynodon*, which occurs at about the level of Roehler's Bed 630 (McCarroll et al., this volume, Chapter 2), and so apparently within the lower part of the 90 m magnetic sampling gap between magnetozones C1+ and C2+ (Fig. 7). The reported provenance of the type specimen of the early Uintan uintathere *Eobasileus cornutus* from Granger's (1909) Bed 21 (Locality 18 of Roehler, 1973), and the reported presence of the characteristic early Uintan brontothere *Metarhinus* from Roehler's (1973) localities 13 and 15, support an early Uintan age for magnetozone C2+. Nevertheless, if the 90 m sampling gap proves to be of reversed polarity, it is possible that magnetozone C2+ could correlate with Chron C20n, which could imply a late Uintan age for the upper part of the middle unit of the Adobe Town Member. The report of McCarroll et al. (this volume, Chapter 2) that the otherwise late Uintan camelid *Poebrodon* occurs in the middle unit may hint at this possibility.

On the basis of the faunal evidence summarized above, the B/U boundary can only be constrained to occur within magnetozone B3-, B4+, B5-, C1+, or the lower part of the 90 m sampling gap (Fig. 7), so the Washakie Formation in the Washakie Basin currently provides no independent evidence for the chron correlation of the B/U boundary. Further biostratigraphic, paleomagnetic, and radiometric work will be required before Flynn's (1986) magnetozones can be uniquely correlated with particular chrons.

Washakie Formation, Sand Wash Basin, Colorado

Stucky et al. (this volume, Chapter 3) discuss the Washakie Formation in the Sand Wash Basin of northwestern Colorado. The magnetobiostratigraphic evidence currently available from the Sand Wash Basin is limited, owing to the relatively short and discontinuous section available. In particular, it is difficult to determine the exact stratigraphic distance between the Bridgerian? Robin's Egg Blue Tuff and the main composite section. Therefore, it is difficult to be sure whether the main reversed interval of reportedly earliest Uintan age correlates with Flynn's (1986) Washakie magnetozones B3- and/or B5-, or with the 90 m sampling gap of uncertain polarity in magnetozone C+. If the former alternative is correct, Flynn's conclusion that the B/U boundary occurred in magnetozone B5- could be corroborated. The significance of the paleomagnetic results of Stucky et al. (this volume, Chapter 3) must be clarified by sampling the 90 m gap in Flynn's (1986) Washakie Basin section, and by the documentation of the "earliest Uintan" taxa occurring in their primary reversed interval. Until then, the evidence from the Sand Wash Basin sheds no independent light on the identity of the chron in which the B/U boundary occurs.

Devil's Graveyard Formation, Trans-Pecos Texas

Wilson (1986) reported that the LSD of *Amynodon* in the Devil's Graveyard Formation of Trans-Pecos Texas occurs at the early Uintan Whistler Squat quarry and the Boneanza locality, both of which were assigned to the Whistler Squat local fauna. Below the level of these localities is a unit known as the "basal Tertiary conglomerate," which has yielded *Leptoreodon* and a micromammal assemblage very similar to that from the level of the Whistler Squat quarry (Wilson, 1986, p. 354, table 1).

Walton (1992) presented paleomagnetic data for the lower and middle members of the Devil's Graveyard Formation, and reported that the Whistler Squat quarry occurred within strata of reversed polarity, while the underlying basal Tertiary conglomerate was questionably of normal polarity. The occurrence of *Leptoreodon* in the basal Tertiary conglomerate indicates a Uintan age for this unit, so the B/U boundary is probably not represented within the Devil's Graveyard Formation. This boundary may occur within the Canoe Formation (Runkel, 1988), but no paleomagnetic data are currently available from these strata.

Wilson (1986) and Walton (1992, fig. 3.8) note that two recalculated K-Ar dates of 46.9 ± 1.1 Ma and 49.7 ± 1.3 Ma have been obtained from a tuff directly below the Whistler Squat quarry, and the younger date is regarded as the more reliable (McDowell, 1979; Henry et al., 1986). Given the younger date and the time scale of Berggren et al. (1995), the Whistler Squat reversed interval could correlate with either late C21r or early C20r. Taken alone, the magnetostratigraphic pattern for the lower member of the Devil's Graveyard Formation does not allow a confident choice between these alternatives, and re-dating of the Whistler Squat Quarry Tuff by the ^{40}Ar/^{39}Ar method is called for. The presence of three different species of *Leptoreodon* in addition to *Amynodon* is weak evidence suggesting that assignment of the Whistler Squat local fauna to C20r is preferable to C21r. If so, this local fauna may be approximately coeval with the Poway fauna of San Diego, the assemblage from Uinta B1 in Utah, and the Bone Bed A assemblage from the Tepee Trail Formation of Wyoming.

Amynodon: Other LSD's, the EKR, and the FAD

The LSD's of *Amynodon* discussed above were selected because they all occur in strata of undisputed early Uintan age, and therefore bear on the question of the B/U boundary. However, the LSD of this genus occurs in many areas in decidely younger rocks, reflecting some of the biases discussed with reference to Figure 1. For example, the LSD's of *Amynodon* in the Santiago Formation of northwestern San Diego County, California, the Sespe Formation of Ventura County, California, the Laredo area of Texas, southwestern Montana, and northwestern New Mexico respectively occur in the late Uintan Laguna Riviera, Brea Canyon, Casa Blanca, and Douglass Draw local faunas, and the Duchesnean part of the Galisteo Formation (Golz and Lillegraven, 1977; Westgate, 1990; Tabrum et al., this volume, Chapter 14; Prothero

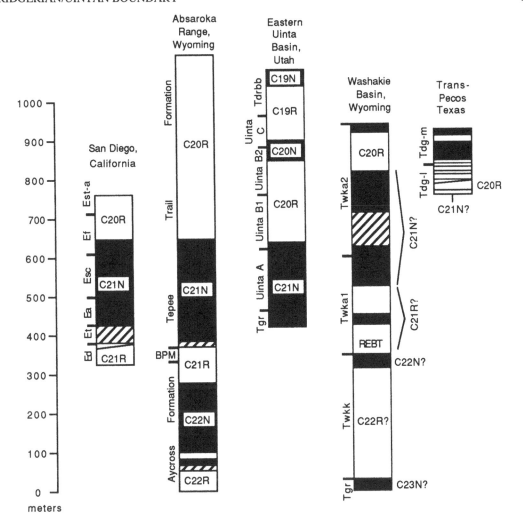

Figure 8. Proposed correlations of Bridgerian and early Uintan magnetozones from San Diego, the Absaroka Range, the Uinta Basin, the Washakie Basin, and Trans-Pecos Texas, based on a variety of evidence discussed in this chapter. Abbreviations: Ed = Delmar Formation, Et = Torrey Sandstone, Ea = Ardath Shale, Esc = Scripps Formation, Ef = Friars Formation, Est-a = lower member of Stadium Conglomerate, Tgr = Green River Formation, Tdrbb = Brennan Basin Member of Duchesne River Formation, Twkk = Kinney Rim Member, Twka1 and Twka2 = lower and middle units of Adobe Town Member, Tdg-l and Tdg-m = lower and middle members of Devil's Graveyard Formation.

and Lucas, this volume, Chapter 10). None of these relatively young LSD's can be ascribed to significant intrinsic diachrony of the B/U boundary, because early Uintan records of *Amynodon* are known from the same general areas.

The EKR of *Amynodon* in the early Uintan sections discussed above is unclear. This genus is not known from the Sand Wash Basin of Colorado (Stucky et al., this volume, Chapter 3). The LSD's of *Amynodon* in the Uinta Formation and the Devil's Graveyard Formation occur within reversed polarity intervals correlated with C20r. The LSD's in San Diego and the Absaroka Range occur in the uppermost part of normal magnetozones correlated with C21n, and would therefore be about the same age. The LSD of *Amynodon* in the Washakie Formation of Wyoming occurs in the middle part of a composite normal magnetozone, all of which may

correlate with C21n. However, the identity of this magnetozone is too uncertain to permit even a tentative designation of the Washakie Basin record as the EKR.

Finally, owing to the small number of available sections containing the B/U boundary and the lack of precise radiometric control on any of the *Amynodon* LSD's within them, it is uncertain to what extent the FHA of this genus can be considered as a FAD. The available evidence suggests that the FHA of *Amynodon* in both San Diego and the Absaroka Range occurred no earlier than late C21r, and no later than latest C21n. It therefore appears that any intrinsic diachrony of the B/U boundary between California and Wyoming did not transgress more than one chron of a given polarity. In this context, the immigration and dispersal of *Amynodon* is assumed to represent a FAD until a great deal of future evidence indicates otherwise.

Summary of the Chron Correlation and Numerical Age of the Bridgerian-Uintan Boundary

Suggested correlations of the magnetostratigraphic sections discussed above are shown in Figure 8. Undoubted Bridgerian and undoubted Uintan mammal assemblages are not known to occur in superposition within the same magnetozone anywhere in North America. This fact, together with the presence of unsampled strata and/or unconformities of unknown duration between Bridgerian and Uintan horizons in both San Diego and the Absaroka Range, entails that an unequivocal chron correlation for the B/U boundary cannot be made at this time. Nevertheless, the boundary probably occurred either during the later part of C21r, or during some part of C21n. This hypothesis will soon be tested by magnetostratigraphic studies of the Bridger Formation (McCarroll et al., this volume, Chapter 2; E. Evanoff, personal communication), where the stratigraphic position of the B/U boundary may be more precisely located than in any other section in North America (Evanoff et al., 1994). Finally, magnetostratigraphic evidence from San Diego (Walsh et al., this volume, Chapter 6) and the Uinta Basin (Prothero, this volume, Chapter 1), together with radiometric dates on three different late Uintan localities in San Diego and Texas, further suggests that the early Uintan interval ends either during late C20r or during C20n, with a duration of somewhere between 3 and 5 m.y. (Fig. 9).

For reasons previously discussed, the derivation of an age for the B/U boundary from present estimates of the numerical age of Chron C21n is hazardous, but a range of between 46.3 and 48.3 Ma is consistent with the time scale of Berggren et al. (1995) and the date of 47.9 ± 0.5 Ma from the Bridgerian Blue Point marker (Eaton, 1982). Direct $^{40}Ar/^{39}Ar$ dating of Bridgerian and early Uintan horizons in the western interior will provide the best constraints for the age of this boundary (Prothero and Swisher, 1992). Finally, given the correlation of the reversed magnetozone in the upper part of the Aycross Formation with Chron C21r, the middle normal and lower reversed magnetozones in the Aycross are logically correlated with C22n and C22r (Sundell et al., 1984). Since the latter chrons are of early Eocene age (Berggren et al., 1995), and since the Aycross Formation mammal assemblages discussed by Bown (1982) and Eaton (1982) are of undoubted Bridgerian age, it would appear that the older part of this NALMA should be assigned to the late early Eocene (Fig. 9). Given the estimate of 49 Ma by Berggren et al. (1995) for the early Eocene/middle Eocene boundary, this conclusion is consistent with estimates for the age of the end of the Wasatchian of ~51 Ma and 49.7-50.7 Ma, proposed by Wing et al. (1991) and Clyde et al. (1995), respectively (see also Woodburne and Swisher, 1995, fig. 2).

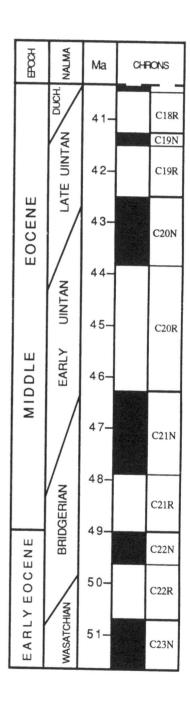

Figure 9. Proposed correlations of Bridgerian and Uintan NALMA's with the Geomagnetic Polarity Time Scale of Berggren et al. (1995).

THE "SHOSHONIAN SUBAGE" OF THE UINTAN
Previous Characterization

Prior to Flynn's (1983, 1986) study, Lillegraven (1979) referred the assemblage from the Friars Formation in San Diego to the "earliest Uintan," and suggested a temporal correlation with the assemblage from the type section of the Tepee Trail Formation in Wyoming. McKenna (1980) proposed that the early Uintan mammal assemblage from Bone Bed A of the Tepee Trail Formation was more primitive than that known from Uinta B, and speculated that it might correlate with the (unknown) fauna of Uinta A.

Flynn (1986) reviewed the early Uintan mammal assemblages from San Diego and the Tepee Trail and Washakie formations of Wyoming, and noted that they were characterized by the co-occurrence of certain small-bodied taxa (micromammals; largely marsupials, insectivores, primates, and rodents) typical of Bridgerian rocks with large-bodied Uintan index taxa. Flynn (1986, pp. 379-380) asked:

> What is the resolution of this apparent anomaly in temporal distribution of taxa in these areas? Given the apparent intermediate nature of several of these faunas, is it possible to define rigorously and recognize a discrete chronostratigraphic or chronologic interval that lies temporally between the previously (Wood et al., 1941; West et al., in press) defined Bridgerian and Uintan mammal ages?

One possible explanation for this apparent anomaly had already been suggested by Lillegraven (1980). After documenting species-level similarities between the small mammal assemblages from the early Uintan of San Diego and the Bridgerian of the western interior, Lillegraven (1980, p. 201) stated:

> The Bridgerian comparisons should not be perplexing, however, because the general vertebrate fauna of the early Uintan (equivalent of Uinta A and B) of the Rocky Mountain region is poorly known. Possibly many of these species, or closely related forms, eventually will be discovered in rocks of this age in the western interior.

As discussed by Prothero (this volume, Chapter 1), most micromammal taxa from the "type" early Uintan of the Uinta Formation were collected from the upper part of Uinta B in the Willow Creek-Ouray area (i.e., White River Pocket), and none are recorded from the lower part of Uinta B in this area. In the eastern part of the Uinta Basin, Prothero records only five micromammal taxa from Uinta B2, and none from Uinta B1 (to illustrate what these numbers imply as to the magnitude of our ignorance of the "type" early Uintan fauna, of the approximately 61 mammal species known

from the intensively screenwashed early Uintan Poway fauna of San Diego County, 43 of them are micromammals). In view of the rarity of recorded micromammal taxa from the lower part of Uinta B, the most economical hypothesis concerning the age of the San Diego and Wyoming early Uintan assemblages might be to consider them as approximate temporal equivalents of Uinta B1 (and possibly in part B2), and to ascribe the apparent lack of small-bodied "Bridgerian holdover taxa" in these horizons to preservational and/or collecting bias. However, Flynn (1986) does not consider this possibility, and instead resolves the apparent anomaly by recognizing the Shoshonian Land Mammal Subage. Flynn (1986, p. 380) stated:

> This unit may be *defined* by the first occurrence of the amynodontoid rhinoceros *Amynodon*. It is characterized, and can be recognized, by the occurrences of *Amynodon*, *Ischyrotomus*, *Metarhinus*, *Protylopus*, *Apatemys*, *Herpetotherium* (= *Peratherium*), *Leptoreodon*, *Protoreodon*, *Dilophodon*, *Epihippus*, *Sciuravus*, *Microsyops*, *Uintasorex*, *Nyctitherium*, *Notharctus robustior*, *Achaenodon*, *Trogolemur myodes*, and *Hyopsodus paulus* and *Hyopsodus* n. sp. . . . First appearances include *Amynodon*, selenodont artiodactyls, (*Leptoreodon*, *Protoreodon*, *Protylopus*), *Epihippus*, and *Achaenodon*. Last occurrences include *Uintasorex*, *Northarctus* [sic] *robustior*, *Trogolemur myodes*, *Microsyops annectens*, and *Hemiacodon*. No genera are presently known to be restricted to this temporal interval.

First, it should be noted that *Uintasorex* is known from the late Uintan of Wyoming (Robinson, 1968), Texas (Runkel, 1988), and California (Walsh, this volume, Chapter 5; Kelly and Whistler, 1994), while *Microsyops* is now recorded from the late Uintan Casa Blanca and Serendipity local faunas of Texas by Westgate (1990). These two taxa can therefore no longer be regarded as having their last occurrences in the "earliest Uintan." More importantly, Flynn's definition of this interval is incomplete, because it only addresses the beginning and not the end of the "Shoshonian." This would preferably have been accomplished by faunal criteria, using the first appearances of appropriate new taxa (e.g., Woodburne, 1987; Stucky et al., 1989). Instead, Flynn (1986, pp. 380-381) seems to have intended the end of the "Shoshonian" to correspond to the beginning of the geochron of Uinta B (although this was never stated explicitly). The "Shoshonian" as originally envisioned is therefore not a biochronologic unit, but seems to be a "hybrid" temporal unit whose beginning is defined and whose end is characterized by a biological and geological event, respectively.

If the preceding analysis is correct, several different

issues concerning the "Shoshonian" must be kept separate: (1) Is the beginning of the Uintan (as defined by the immigration of *Amynodon*) older than the geochron of Uinta B? (2) Can some portion of the early Uintan be characterized by the presence of certain "Bridgerian holdover taxa," and if so, are such assemblages necessarily older than the geochron of Uinta B? (3) Can the early Uintan interval be otherwise subdivided by using certain taxa whose first occurrences are consistently recognizable throughout North America? Hereafter, I will use the term "Shoshonian" informally, in the sense of a "faunal facies," to designate only those early Uintan assemblages that are characterized by certain "Bridgerian holdover taxa," but without any implication that all such assemblages are necessarily older than the geochron of Uintan B.

1. Is the beginning of the Uintan older than the geochron of Uinta B?

Although it was not unreasonable for Flynn (1986) to speculate that a significant faunal change within the early Uintan might correspond approximately to the Uinta A-Uinta B boundary, this hypothesis is virtually untestable owing to the barren nature of Uinta A. Nevertheless, as discussed above, Flynn's (1986) suggestion that the earliest part of the Uintan is older than the geochron of Uinta B is substantiated by the fact that if the B/U boundary occurred during latest Chron C21r or C21n (as argued above), and if Uinta B correlates with C20r (Prothero, this volume, Chapter 1), then this boundary must occur within or below the normally-polarized strata of Uinta A (assuming no significant disconformity between Uinta A and B). Ironically, however, if Flynn's original assignment of the B/U boundary to C20r had been correct, then this boundary would occur within Uinta B, in which case the "Shoshonian" interval would be entirely of "Uinta B age."

2. Can some portion of the early Uintan interval throughout North America be characterized by the presence of certain "Bridgerian holdover taxa," and if so, are all such assemblages necessarily older than the geochron of Uinta B?

A significant difficulty with Flynn's faunal characterization of the "Shoshonian" lies in the question of what taxa are and are not considered to be significant "Bridgerian holdover taxa" (BHT's). In his original characterization, Flynn (1986, p. 380) stated that the "Shoshonian" could be recognized in part by the occurrences of *Peratherium*, *Apatemys*, *Nyctitherium*, *Uintasorex*, and *Dilophodon*, which were said to be characteristic of the Bridgerian. Since all of these taxa extend into the late Uintan or younger, however (Stucky, 1992), they have little relevance to the characterization of the boundaries of a putative "earliest Uintan" interval. In this context, the only relevant BHT's would

seem to be those that apparently still do become extinct during or at the end of this time, namely, *Notharctus*, *Trogolemur myodes*, and *Hemiacodon* (other taxa recently recognized in the Poway fauna of San Diego might be added to this list, such as *Antiacodon* and cf. *Lophiohyus* sp.). Although the presence of one or more of these taxa in association with early Uintan mammals may arguably be used to characterize an "earliest Uintan" interval, in the absence of our knowledge of the micromammal fauna of the lower part of Uinta B, it cannot be assumed that these BHT's do not occur in these strata.

Since Flynn's (1986) study, several workers have noted the presence of BHT's in certain early Uintan assemblages, and have assigned them to the "Shoshonian" or "earliest Uintan." Eaton (1985) suggested that the assemblage from the "Holy City beds" (Owl Creek area, Wyoming; lower part of Tepee Trail Formation) might correlate temporally with the unfossiliferous strata of Uinta A, based on the presence of the Uintan index taxa *Epihippus* and *Forstercooperia* in association with several taxa typical of the Bridgerian. Krishtalka et al. (1987) suggested a possible "Shoshonian" age for the Whistler Squat local fauna of Texas (Wilson, 1986). Stucky et al. (this volume, Chapter 3) refer the Sand Wash Basin fauna of Colorado to the "Shoshonian" on the basis of the primitiveness of some of the Uintan taxa, and formal description of this material is in progress. Finally, Evanoff et al. (1994) assigned an assemblage from Bridger E to the "Shoshonian" on the basis of the co-occurrence of the rodent "*Namatomys*" and a "Uintan artiodactyl" (assigned to *Oromeryx* sp. by R. K. Stucky, personal communication) with the BHT's *Hemiacodon* and *Omomys*. Note, however, that the status of *Omomys* as a BHT capable of resolving an "earliest Uintan" age is questionable owing to its occurrence in the late Uintan Serendipity and Casa Blanca local faunas of Texas (Wilson, 1986; Westgate, 1990), and the Duchesnean Lac Pelletier Lower Fauna of Saskatchewan (Storer, 1990).

Present evidence suggests that some "Shoshonian" assemblages may be older than the geochron of Uinta B, whereas others are not. If the above assignment of the B/U boundary to late C21r or C21n is correct, then, given the magnetostratigraphic results of Prothero (this volume, Chapter 1), the early Uintan assemblages from magnetozone C+ in the Washakie Formation in Wyoming (McCarroll et al., this volume, Chapter 2) and from Colorado (Stucky et al., this volume, Chapter 3) may be entirely older than Uinta B. Similarly, based on magnetostratigraphic evidence (Sundell et al., 1984; Eaton, 1985) the early Uintan assemblage from the lower part of "Holy City beds" of the Tepee Trail Formation of Wyoming is older than that from Uinta B. Finally, the possibly Uintan assemblage from Bridger E (Evanoff et al., 1994) may be tentatively assumed to be older than Uinta B based on its stratigraphic position only a short distance above "type" Bridgerian strata. However, the "Shoshonian" assemblages from the Friars Formation, the upper part of the "Holy City beds" and "Foggy Day beds" of the Tepee Trail Formation, and apparently the Devil's Graveyard Formation occur largely or entirely within reversed magnetozones that are best correlated with C20r, and are therefore approximately coeval with the Uinta B1 assemblages correlated with C20r in Utah (Prothero, this volume, Chapter 1). If so, "Bridgerian holdover taxa" should eventually be recovered from Uinta B1.

3. Can the early Uintan interval be otherwise subdivided by using taxa whose first occurrences are consistently recognizable throughout North America?

Stucky et al. (1989) proposed a characterization for the beginning of the "late early Uintan" by using the first occurrences of *Hessolestes*, *Auxontodon*, *Bunomeryx*, *Diplobunops*, and *Leptotragulus*. First, the reported presence of *Hessolestes* in the "Shoshonian" Whistler Squat local fauna (Wilson, 1986; Gustafson, 1986) casts doubt on the validity of this genus as a characterizing taxon for the beginning of the "late early Uintan," while the rarity of this genus (Gustafson, 1986, p. 14) renders it unsuitable as a practical biochronological indicator. *Auxontodon* is also relatively uncommon. It is known in the Uinta Basin only from strata of apparent late Uintan age (Prothero, this volume, Chapter 1), although Eaton (1985) identified several specimens of "cf. *Auxontodon* sp." from strata of possible early Uintan age in Wyoming. Stucky's (1992, table 24.1) record of *Auxontodon* from the "middle Uintan" ("late early Uintan" of present usage) was based on specimens from the Wind River Basin in Wyoming (R. Stucky, personal communication), but these have not been observed to occur in superposition above an "earliest Uintan" assemblage. The artiodactyl *Leptotragulus* is known from the "late early Uintan" of Uinta B2 (Prothero, this volume, Chapter 1), but fragmentary material of this genus and *Leptoreodon* are difficult to distinguish (e.g., Golz, 1976), so the utility of *Leptotragulus* as a biochronological indicator is open to question. As a result of these considerations, perhaps only *Diplobunops* and *Bunomeryx* can be regarded as potentially useful indicators for the beginning of the "late early Uintan."

Assuming that these doubts can be resolved, the fact that none of the genera listed by Stucky et al. (1989) are known from Uinta B1 (see taxon range chart in Prothero, this volume, Chapter 1) suggests that their characterization of the beginning of the "late early Uintan" might correspond to the Uinta B1-B2 boundary better than the Uinta A-B boundary. Finally, a test of the continent-wide applicability of this characterization will be provided by continued collecting from the Murray Canyon local fauna of San Diego County, which occurs in superposition above the "earliest Uintan" Poway fauna (Walsh, this volume, Chapter 5). Unfortunately, none of the genera used by Stucky et al. (1989) to recognize the beginning of the "late early Uintan" are currently known from the Murray Canyon local fauna, nor from the diverse late Uintan mammal assemblages of southern California.

CONCLUSIONS

Operational definitions of biochron boundaries are based upon a stratigraphic record of a fossil, and refer to either the lowest known occurrence of the defining taxon in a given section (Lowest Stratigraphic Datum = LSD), or upon the presumed Earliest Known Record (EKR) of the defining taxon in a given geographic area. Operational definitions in this context are subject to a variety of geological, paleoecological, taphonomic, and sampling biases, and will often result in highly diachronous "boundaries" of biochrons that are inappropriate for consistent biochronological and magnetochronological correlations. Theoretical definitions of the beginnings of mammalian biochrons refer to historical events such as the actual immigration or evolution of the defining taxon (First Historical Appearance = FHA). Given a theoretical definition for the lower boundary of a mammalian biochron, in order to demonstrate that this boundary occurs within a particular magnetozone, (1) the defining or characterizing taxa for the beginning of this biochron *and* (2) an assemblage confidently referable to the next older biochron must both be known from this same magnetozone. Consistent temporal correlation requires that the particular FHA we designate to define a given boundary must in reality be a First Appearance Datum (FAD). A FAD is a FHA that, together with its associated dispersal event, is synchronous relative to the duration of other specified temporal units that are approximately coeval with it (such as other biochrons and geomagnetic chrons).

Flynn (1986) concluded that the Bridgerian-Uintan boundary occurred within Chron C20r, and estimated its age at about 49 Ma. However, evidence from the La Jolla Group in San Diego, the Aycross-Tepee Trail-Wiggins sequence in Wyoming, and the Uinta Formation in Utah suggests that the B/U boundary occurred either during the latest part of Chron C21r, or during C21n, which would imply a younger numerical age of about 46.3-48.3 Ma given the time scale of Berggren et al. (1995). If this conclusion is correct, then, given the paleomagnetic work of Prothero (this volume, Chapter 1), together with several radiometric dates from late Uintan strata, the B/U boundary is older than the geochron of Uinta B, and the early Uintan interval would span most or all of C20r and end within late C20r or within C20n, with a duration of about 3-5 m.y.

Flynn's (1986) "Shoshonian Subage" of the Uintan NALMA is problematical. Although the term

"Shoshonian" can be used in the sense of a "faunal facies," to informally designate early Uintan assemblages that contain certain "Bridgerian holdover taxa" (e.g., *Notharctus*, *Trogolemur*, and *Hemiacodon*), it should not connote that all such assemblages are necessarily older than those from Uinta B. This conclusion stems from the fact that "Shoshonian" assemblages from the Friars Formation (San Diego), the "middle" part of the Tepee Trail Formation (Wyoming), and the Devil's Graveyard Formation (Texas) probably occurred largely or entirely during the earlier part of C20r, as does the assemblage from Uinta B1 (Prothero, this volume, Chapter 1). It is therefore predicted that "Bridgerian holdover taxa" will eventually be recovered from Uinta B1.

Formal subdivision of the early part of the Uintan NALMA should be based on the first historical appearances of certain taxa that are recognizable within early Uintan assemblages throughout North America (e.g., Stucky et al., 1989). The characterization of the beginning of the "late early Uintan" offered by these authors suggests that a possible faunal break between Uinta B1 and B2 might provide a better basis for the subdivision of the early Uintan interval than that implied by Flynn (1986) to occur between Uinta A and B. This possibility must be tested by intensive screenwashing of Uinta B1 and B2. The stratigraphic ranges of the micromammal taxa thus obtained, and the presence or absence of any "Bridgerian holdover taxa," will be crucial to the proper faunal definition and characterization of putative "earliest Uintan" and "late early Uintan" intervals. Until this is accomplished, use of the term "Shoshonian" in a temporal sense should be suspended.

ACKNOWLEDGMENTS

I thank the editors for their patience and the invitation to contribute to this volume. I thank W. A. Berggren, S. C. Cande, W. B. Cashion, M. R. Dawson, K. Doi, J. G. Eaton, R. J. Emry, E. Evanoff, J. J. Flynn, N. O. Frederiksen, D. V. Kent, P. Murphey, D. G. Nichols, J. D. Obradovich, D. R. Prothero, P. Robinson, P. N. Shive, R. K. Stucky, W. D. Turnbull, and W. P. Wall for discussions and information. I thank T. A. Deméré, E. H. Lindsay, J. A. Lillegraven, S. M. McCarroll, M. C. McKenna, D. T. Rasmussen, A. H. Walton, and M. O. Woodburne for helpful reviews of all or parts of earlier drafts of this chapter. Many of these persons directed my attention to references I would otherwise have missed. I also thank M. O. Woodburne for making available a manuscript which discusses topics similar to those addressed here. However, none of these individuals necessarily agree with any of my interpretations.

LITERATURE CITED

Berggren, W. A., D. V. Kent, and J. J. Flynn. 1985. Paleogene geochronology and chronostratigraphy; pp. 141-95 *in* N. J. Snelling (ed.), The Chronology of the Geologic Record. Geological Society of London Memoir 10.

Berggren, W. A., D. V. Kent, J. D. Obradovich, and C. C. Swisher III. 1992. Toward a revised Paleogene geochronology; pp. 29-45 *in* D. R. Prothero and W. A. Berggren (eds.), Eocene-Oligocene Climatic and Biotic Evolution. Princeton University Press, Princeton, N.J.

Berggren, W. A., D. V. Kent, C. C. Swisher III, and M.-P. Aubry. 1995. A revised Cenozoic geochronology and chronostratigraphy. SEPM Special Publication 54:129-212.

Berggren, W. A., and D. R. Prothero. 1992. Eocene-Oligocene climatic and biotic evolution: An overview; pp. 1-28 *in* D. R. Prothero and W. A. Berggren (eds.), Eocene-Oligocene Climatic and Biotic Evolution. Princeton University Press, Princeton, N.J.

Berggren, W. A., and J. A. Van Couvering. 1974. The late Neogene. Palaeogeography, Palaeoclimatology, Palaeoecology 16:1-216.

Bown, T. M. 1982. Geology, paleontology, and correlation of Eocene volcaniclastic rocks, southeast Absaroka Range, Hot Springs County, Wyoming. United States Geological Survey Professional Paper 1201-A:1-75.

Bryan, W. B., and R. A. Duncan. 1983. Age and provenance of clastic horizons from Hole 516F. Initial Reports of the Deep Sea Drilling Project 15:475-77.

Bukry, D., and M. P. Kennedy. 1969. Cretaceous and Eocene coccoliths at San Diego, California. California Division of Mines and Geology Special Report 100:33-43.

Cande, S. C., and D. V. Kent. 1992. A new geomagnetic polarity time scale for the late Cretaceous and Cenozoic. Journal of Geophysical Research 97:13917-13951.

Cashion, W. B. 1967. Geology and fuel resources of the Green River Formation, southeastern Uinta Basin, Utah and Colorado. United States Geological Survey Professional Paper 548:1-48.

Cashion, W. B. 1986. Geologic map of the Bonanza quadrangle, Uintah County, Utah. United States Geological Survey Miscellaneous Field Studies Map 1865.

Cashion, W. B., and J. R. Donnell. 1974. Revision of nomenclature of the upper part of the Green River Formation, Piceance Creek Basin, Colorado, and eastern Uinta Basin, Utah. United States Geological Survey Bulletin 1394-G:G1-G9.

Clyde, W. C., J. Stamatakos, J.-P. Zonneveld, G. F. Gunnell, and W. S. Bartels. 1995. Timing of the Wasatchian-Bridgerian (early-middle Eocene) faunal transition in the Green River Basin, Wyoming. Journal of Vertebrate Paleontology 15, supplement to no. 3: 24A (abstract).

Dawson, M. R. 1968. Middle Eocene rodents (Mammalia) from northeastern Utah. Annals of Carnegie Museum 39:327-370.

Decker, P. L. 1990. Style and mechanics of liquifaction-related deformation, lower Absaroka Volcanic Supergroup (Eocene), Wyoming. Geological Society of America Special Paper 240:1-72.

Eaton, J. G. 1985. Paleontology and correlation of the Eocene Tepee Trail and Wiggins formations in the North Fork of Owl Creek area, southeastern Absaroka Range, Hot Springs County, Wyoming. Journal of Vertebrate Paleontology 5:345-370.

Evander, R. L. 1986. Formal redefinition of the Hemingfordian-Barstovian land mammal age boundary. Journal of Vertebrate Paleontology 6:374-381.

Evanoff, E., P. Robinson, P. C. Murphey, D. G. Kron, D. Engard, and P. Monaco. 1994. An early Uintan fauna from Bridger E. Journal of Vertebrate Paleontology 14, supplement to no. 3:24A.

Flynn, J. J. 1983. Correlation and geochronology of middle Eocene strata from the western United States. Ph.D. dissertation, Columbia University, New York, 496 pp.

Flynn, J. J. 1986. Correlation and geochronology of middle Eocene strata from the western United States. Palaeogeography, Palaeoclimatology, Palaeoecology 55:335-406.

Flynn, J. J., B. J. MacFadden, and M. C. McKenna. 1984. Land-mammal ages, faunal heterochrony, and temporal resolution in Cenozoic terrestrial sequences. Journal of Geology 92:687-705.

Frederiksen, N. O. 1989. Eocene sporomorph biostratigraphy of southern California. Palaeontographica, Abteilung B 211:135-179.

Frederiksen, N. O. 1991. Age determinations for Eocene formations of the San Diego, California, area, based on pollen data; pp. 195-199 in P. L. Abbott and J. A. May (eds.), Eocene Geologic History San Diego Region, Pacific Section, SEPM Volume 68.

Frederiksen, N. O. 1995. Latest Cretaceous and Tertiary spore/pollen biostratigraphy; in J. Jansonius and D.C. McGregor (eds.), Palynology: principles and applications. American Association of Stratigraphic Palynologists Foundation, Volume 2 (in press).

Golz, D. J. 1976. Eocene Artiodactyla of southern California. Natural History Museum of Los Angeles County Science Bulletin 26:1-85.

Golz, D. J., and J. A. Lillegraven. 1977. Summary of known occurrences of terrestrial vertebrates from Eocene strata of southern California. University of Wyoming Contributions to Geology 15:43-65.

Granger, W. 1909. Faunal horizons of the Washakie Formation of southern Wyoming. Bulletin of the American Museum of Natural History 26:13-24.

Gustafson, E. P. 1986. Carnivorous mammals of the late Eocene and early Oligocene of Trans-Pecos Texas. Texas Memorial Museum Bulletin 33:1-66.

Harland, W. B., R. L. Armstrong, A. V. Cox, L. E. Craig, A. G. Smith, and D. G. Smith. 1990. A Geologic Time Scale 1989. Cambridge University Press, Cambridge.

Henry, C. D., F. W. McDowell, J. G. Price, and R. C. Smyth. 1986. Compilation of K-Ar ages of Tertiary igneous rocks, Trans-Pecos, Texas. University of Texas Bureau of Economic Geology Circular 86-2:1-34.

Hutchison, J. H. 1971. Cf. *Uintatherium* (Dinocerata, Mammalia) from the Uintan (middle to late Eocene) of southern California. Paleobios 12:1-8.

Kelly, T. S., and D. P. Whistler. 1994. Additional Uintan and Duchesnean (middle and late Eocene) mammals from the Sespe Formation, Simi Valley, California. Contributions to Science of the Natural History Museum of Los Angeles County 439:1-29.

Kennedy, M. P., and G. W. Moore. 1971. Stratigraphic relations of Upper Cretaceous and Eocene formations, San Diego coastal region, California. American Association of Petroleum Geologists Bulletin 55:709-722.

Krishtalka, L., and R. K. Stucky. 1984. Middle Eocene marsupials (Mammalia) from northeastern Utah and the mammalian fauna from Powder Wash. Annals of Carnegie Museum 53:31-45.

Krishtalka, L., R. K. Stucky, R. M. West, M. C. McKenna, C. C. Black, T. M. Bown, M. R. Dawson, D. J. Golz, J. J. Flynn, J. A. Lillegraven, and W. D. Turnbull. 1987. Eocene (Wasatchian through Duchesnean) biochronology of North America; pp. 77-117 in M. O. Woodburne (ed.), Cenozoic Mammals of North America, Geochronology and Biostratigraphy. University of California Press, Berkeley.

Kurtén, B. 1957. Mammal migrations, Cenozoic stratigraphy, and the age of Peking Man and the australopithecines. Journal of Paleontology 31:215-227.

Lillegraven, J. A. 1979. A biogeographical problem involving comparisons of later Eocene terrestrial vertebrate faunas of western North America; pp. 333-347 in J. Gray and A. J. Boucot (eds.), Historical Biogeography, Plate Tectonics, and the Changing Environment. Oregon State University Press, Corvallis.

Lillegraven, J. A. 1980. Primates from later Eocene rocks of southern California. Journal of Mammalogy 61:181-204.

Lindsay, E. H. 1990. The setting; pp. 1-14 in E. H. Lindsay, V. Fahlbusch, and P. Mein (eds.), European Neogene Mammal Chronology. NATO Advanced Science Institute Series 180. Plenum Press, New York.

Lindsay, E. H. 1995. *Copemys* and the Hemingfordian/Barstovian boundary. Journal of Vertebrate Paleontology 15:357-365.

Lindsay, E. H., N. D. Opdyke, N. M. Johnson, and R. F. Butler. 1987. Mammalian chronology and the Magnetic Polarity Time Scale; pp. 269-284 in M. O. Woodburne (ed.), Cenozoic Mammals of North America: Geochronology and Biostratigraphy. University of California Press, Berkeley.

Lindsay, E. H., and R. H. Tedford. 1990. Development and application of land mammal ages in North America and Europe, a comparison; pp. 601-624 in E. H. Lindsay, V. Fahlbusch, and P. Mein (eds.), European Neogene Mammal Chronology. NATO Advanced Science Institute Series 180. Plenum Press, New York.

Mauger, R. L. 1977. K-Ar ages of biotites from tuffs in Eocene rocks of the Green River, Washakie, and Uinta basins, Utah, Wyoming, and Colorado. University of Wyoming Contributions to Geology 15:17-41.

May, J. A., J. M. Lohmar, J. E. Warme, and S. Morgan. 1991. Field trip guide: early to middle Eocene La Jolla Group of Black's Beach, La Jolla, California; pp. 27-36 in P. L. Abbott and J. A May (eds.), Eocene Geologic History San Diego Region. Pacific Section, SEPM Guidebook 68.

McDowell, F. W. 1979. Potassium-argon dating in the Trans-Pecos volcanic field. University of Texas Bureau of Economic Geology Guidebook 19:10-18.

McKenna, M. C. 1980. Late Cretaceous and early Tertiary vertebrate paleontological reconnaissance, Togwotee Pass area, northwestern Wyoming; pp. 321-343 in L. L. Jacobs (ed.), Aspects of Vertebrate History, Essays in Honor of Edwin Harris Colbert. Museum of Northern Arizona Press, Flagstaff.

McKenna, M. C. 1990. Plagiomenids (Mammalia: ?Dermoptera) from the Oligocene of Oregon, Montana, and South Dakota, and middle Eocene of northwestern Wyoming. Geological Society of America Special Paper 243:211-234.

Murphy, M. A. 1977. On time-stratigraphic units. Journal of Paleontology 51:213-219.

Newman, K. R. 1974. Palynomorph zones in early Tertiary formations of the Piceance Creek and Uinta basins, Colorado and Utah; pp. 47-55 in D.K. Murray (ed.), Energy Resources of the Piceance Creek Basin, Colorado. Rocky Mountain Association of Geologists 25th field conference guidebook.

Nichols, D. J. 1994. Palynostratigraphic correlation of Paleocene rocks in the Wind River, Bighorn, and Powder River Basins, Wyoming; pp. 17-29 in R. M. Flores, K. T. Mehring, R. W. Jones, and T. L. Beck (eds.), Organics and the Rockies Field Guide. Wyoming State Geological Survey Public Information Circular no. 33.

Nichols, D. J., and Ott, H. L. 1978. Biostratigraphy and evolution of the *Momipites-Caryapollenites* lineage in the early Tertiary in the Wind River Basin, Wyoming. Palynology 2:93-112.

North American Commission on Stratigraphic Nomenclature. 1983. North American Stratigraphic Code. American Association of Petroleum Geologists Bulletin 67:841-75.

Opdyke, N. D., E. H. Lindsay, N. M. Johnson, and T. Downs. 1977. The paleomagnetism and magnetic polarity stratigraphy of the mammal-bearing section of Anza-Borrego State Park, California. Journal of Quaternary Research 7:316-329.

Osborn, H. F. 1895. Fossil mammals of the Uinta Basin: Expedition of 1894. Bulletin of the American Museum of Natural History 7:71-105.

Osborn, H. F. 1929. The titanotheres of ancient Wyoming, Dakota, and Nebraska. United States Geological Survey Monograph 55:1-953.

Peterson, O. A. 1934. New titanotheres from the Uinta Eocene in Utah. Annals of Carnegie Museum 22:351- 361.

Peterson, O. A., and J. L Kay. 1931. The upper Uinta Formation of northeastern Utah. Annals of Carnegie Museum 20:293-306.

Pickford, M., and J. Morales. 1994. Biostratigraphy and palaeogeography of East Africa and the Iberian peninsula. Palaeogeography, Palaeoclimatology, Palaeoecology 112:297-322.

Prothero, D. R., and C. C. Swisher. 1992. Magnetostratigraphy and geochronology of the terrestrial Eocene-Oligocene transition in North America; pp. 46-73 in D. R. Prothero and W. A. Berggren (eds.), Eocene-Oligocene Climatic and Biotic Evolution. Princeton University Press, Princeton, N.J.

Repenning, C. A. 1967. Palearctic- Nearctic mammalian dispersal in the late Cenozoic; pp. 288-311 in D. M. Hopkins (ed.), The Bering Land Bridge. Stanford University Press, Stanford.

Riggs, E. S. 1912. New or little known titanotheres from the lower Uintah formations. Field Museum of Natural History, Geology Series 4:17-41.

Robinson, P. 1968. The paleontology and geology of the Badwater Creek area Central Wyoming. Part 4. Late Eocene Primates from Badwater, Wyoming, with a discussion of material from Utah. Annals of Carnegie Museum 39:307-326.

Roehler, H. W. 1973. Stratigraphy of the Washakie Formation in the Washakie Basin, Wyoming. United States Geological Survey Bulletin 1369:1-40.

Runkel, A. C. 1988. Stratigraphy, sedimentology, and vertebrate paleontology of Eocene rocks, Big Bend region, Texas. Ph.D. dissertation, University of Texas at Austin, 310 pp.

Savage, D. E., and D. A. Russell. 1983. Mammalian paleofaunas of the world. Addison-Wesley, Reading, Ma.

Shive, P. N., K. A. Sundell, and J. Rutledge. 1980. Magnetic polarity stratigraphy of Eocene volcaniclastic rocks from the Absaraka Mountains of Wyoming. EOS 61:945 (abstract).

Stock, C. 1939. Eocene amynodonts from southern California. Proceedings of the National Academy of Sciences 25:270-275.

Storer, J. E. 1990. Primates of the Lac Pelletier Lower Fauna (Eocene: Duchesnean), Saskatchewan. Canadian Journal of Earth Science 27:520-524.

Stucky, R. K. 1992. Mammalian faunas in North America of Bridgerian to early Arikareean "ages" (Eocene to Oligocene); pp. 464-493 in D. R. Prothero and W. A. Berggren (eds.), Eocene-Oligocene Climatic and Biotic Evolution. Princeton University Press, Princeton, N.J.

Stucky, R. K., L. Krishtalka, and M. R. Dawson. 1989. Paleontology, geology and remote sensing of Paleogene rocks in the northeastern Wind River Basin, Wyoming, USA; pp. T322:34-44 in J. J. Flynn (ed.), Mesozoic/Cenozoic Vertebrate Paleontology: Classic Localities, Contemporary Approaches. 28th International Geological Congress, field trip guide.

Sundell, K. A., P. N. Shive, and J. G. Eaton. 1984. Measured sections, magnetic polarity and biostratigraphy of the

Eocene Wiggins, Tepee Trail and Aycross formations within the southeastern Absaroka Range, Wyoming. Earth Science Bulletin of the Wyoming Geological Association 17:1-48.

Tedford, R. H. 1970. Principles and practices of mammalian geochronology in North America. Proceedings of the North American Paleontological Convention, Part F:666-703.

Turnbull, W. D. 1978. The mammalian faunas of the Washakie Formation, Eocene age, of southern Wyoming. Part I. Introduction: the geology, history, and setting. Fieldiana (Geology) 33:569-601.

Van Couvering, J. A., and W. A. Berggren. 1977. Biostratigraphical basis of the Neogene time scale; pp. 283-306 in E. G. Kauffman and J. E. Hazel (eds.), Concepts and Methods of Biostratigraphy. Dowden, Hutchison, and Ross, Inc., Stroudsburg, Penn.

Wall, W. P. 1982a. Evolution and biogeography of the Amynodontidae (Perissodactyla, Rhinocerotoidea). Proceedings of the Third North American Paleontological Convention 2:563-567.

Wall, W. P. 1982b. The genus Amynodon and its relationship to other members of the Amynodontidae (Perissodactyla, Rhinocerotoidea). Journal of Paleontology 56:434-443.

Walton, A. H. 1992. Magnetostratigraphy of the lower and middle members of the Devil's Graveyard Formation (Middle Eocene), Trans-Pecos Texas; pp. 74-87 in D. R. Prothero and W. A. Berggren (eds.), Eocene-Oligocene Climatic and Biotic Evolution. Princeton University Press, N. J.

Westgate, J. W. 1988. Biostratigraphic implications of the first Eocene land-mammal fauna from the North American coastal plain. Geology 16:995-998.

Westgate, J. W. 1990. Uintan land mammals (excluding rodents) from an estuarine facies of the Laredo Formation (middle Eocene, Claiborne Group), Webb County, Texas. Journal of Paleontology 64:454-468.

Wilson, J. A. 1986. Stratigraphic occurrence and correlation of early Tertiary vertebrate faunas, Trans-Pecos Texas: Agua Fria-Green Valley areas. Journal of Vertebrate Paleontology 6:350-373.

Wilson, J. A., and J. A. Schiebout. 1981. Early Tertiary vertebrate faunas, Trans-Pecos Texas: Amynodontidae. Pearce-Sellards Series no. 33:1-62.

Wing, S. L., T. M. Bown, and J. D. Obradovich. 1991. Early Eocene biotic and climatic change in interior western North America. Geology 19: 1189-1192.

Wood, H. E. 1941. Trends in rhinoceros evolution. Transactions of the New York Academy of Sciences Series II, Vol. 3:83-96.

Woodburne, M. O. 1977. Definition and characterization in mammalian chronostratigraphy. Journal of Paleontology 51:220-34.

Woodburne, M. O. 1987. Principles, classification, and recommendations; pp. 9-17 in M. O. Woodburne (ed.), Cenozoic Mammals of North America, Geochronology and Biostratigraphy. University of California Press, Berkeley.

Woodburne, M. O. In press. Precision and resolution in mammalian chronostratigraphy: Principles, practices, examples. Journal of Vertebrate Paleontology.

Woodburne, M. O., and C. C. Swisher III. 1995. Land-mammal high resolution geochronology, intercontinental overland dispersals, sea level, climate, and vicariance. SEPM Special Publication 54:329-358.

5. Middle Eocene Mammal Faunas of San Diego County, California

STEPHEN L. WALSH

ABSTRACT

This chapter discusses a series of stratigraphically superposed middle Eocene mammal assemblages from coastal San Diego County. The Swami's Point local fauna (Delmar Formation) is the first fossil vertebrate assemblage of demonstrably Bridgerian age from southern California. The assemblage from the basal Scripps Formation (Black's Beach local fauna) is either of late Bridgerian or early Uintan age. Uintan biochronology is problematical, due to the lack of a rigorous biostratigraphic framework and other unsolved problems in the Uinta Basin. However, the early Uintan/late Uintan boundary can be characterized in southern California by the extinction of the common early Uintan genera *Crypholestes*, an unnamed dormaaliid insectivore, *Omomys*, *Washakius*, *Metarhinus*?, and *Merycobunodon*, and by the evolution and/or immigration of the common late Uintan genera *Sespedectes*, *Proterixoides*, *Dyseolemur*, *Rapamys*, *Griphomys*, and *Simimys*.

The early Uintan Poway fauna is redefined as the entire set of vertebrate taxa occurring in the Friars Formation. The Mesa Drive local fauna and the assemblage from SDSNH Locality 3486 demonstrate an early Uintan age for the upper part of Member B of the Santiago Formation in northwestern San Diego County. The Murray Canyon local fauna (late early Uintan) is named for a distinctive assemblage from the lower member of the Stadium Conglomerate. The late Uintan Stonecrest local fauna (undifferentiated Stadium Conglomerate) represents the lowest stratigraphic occurrence in San Diego County of several taxa characteristic of the Sespe Formation of Ventura County. The late Uintan Cloud 9 fauna is defined as the entire set of vertebrate taxa present in all outcrops of the Mission Valley Formation. Jeff's Discovery and Rancho del Oro are local faunas of late Uintan age from Member C of the Santiago Formation in northwestern San Diego County. The Laguna Riviera, Camp San Onofre, and Mission del Oro local faunas (Member C, Santiago Formation) are of latest Uintan and/or Duchesnean age.

The global late middle Eocene cooling event recognized by several workers appears to have occurred after the distinctive late Uintan assemblages of California were established. Faunal interchange between California and the western interior was relatively free during the Wasatchian, Bridgerian, and earliest Uintan. Some interchange must also have occurred during the latest Uintan and early Duchesnean. No special taxonomic similarities can be demonstrated for the Uintan assemblages of California and Texas, and the hypothesis that these areas pertained to a distinct Middle American faunal province cannot be supported at this time.

INTRODUCTION

Eocene vertebrates were first described from southern California by Chester Stock and Robert Wilson in the 1930s and 1940s (see Golz and Lillegraven, 1977, for a complete bibliography). Most of these early collections were obtained from the middle member of the Sespe Formation in Ventura County. Several local faunas have been recognized from the Sespe Formation, and are assigned to the later part of the Uintan and the Duchesnean North American land mammal "ages" (NALMA; Wood et al., 1941; Krishtalka et al., 1987). Recent studies of the Sespe faunas have been undertaken by Lindsay (1968), Golz (1976), Golz and Lillegraven (1977), Novacek (1985), Mason (1988, 1990), Kelly (1990, 1992), Kelly et al. (1991), and Kelly and Whistler (1994).

Stock and Wilson also described the first Eocene mammals from San Diego County. Their specimens were obtained from several localities in what is now known as the Friars Formation (Kennedy and Moore, 1971). This assemblage was referred to as the "Poway vertebrate fauna" by Stock (1948), because the fossil-bearing sandstones were generally mapped by Hanna (1926) as part the Poway Conglomerate of Ellis (1919). Wood et al. (1941) and Stock (1948) assigned the Poway fauna to the early part of the Uintan NALMA.

In the late 1960s and early 70s, renewed exploration in San Diego County by personnel of the University of California Museum of Paleontology (Berkeley), San Diego State University, the Natural History Museum of Los Angeles County, and the University of California, Riverside, resulted in the discovery of many significant vertebrate localities, from four major districts. The Scripps Formation (Kennedy and Moore, 1971) yielded a few specimens of late Bridgerian or early Uintan age from the seacliffs of La Jolla (Hutchison, 1971; Golz and Lillegraven, 1977; Walsh, 1991b). Fluvial deposits in the Rancho Peñasquitos area that were mapped as the Mission Valley Formation by Kennedy and Peterson (1975) yielded an early Uintan mammal assemblage, virtually identical to that obtained from the type area of the Friars Formation (Golz and Lillegraven, 1977; Walsh, 1991b). In contrast, terrestrial outcrops of the undoubted Mission Valley Formation in the eastern part of the greater San Diego area yielded a late Uintan assemblage dominated by the micromammals *Simimys* and *Sespedectes* (Lillegraven and Wilson, 1975; Golz, 1976; Walsh, 1987, 1991b). Finally, Member "C" of the Santiago Formation (Wilson, 1972) produced the latest Uintan or Duchesnean Laguna Riviera, Chestnut Avenue, and Camp San Onofre local faunas of northwestern San Diego County (Golz, 1976; Golz and Lillegraven, 1977; Kelly, 1990; Walsh, 1991b). Additional studies addressing particular mammalian taxa from the Uintan of San Diego County include those of Lillegraven (1976, 1977, 1980), Lillegraven et al. (1981), Novacek (1976, 1985), Chiment (1977), Schiebout (1977), and Colbert (1993). Lillegraven (1979) discussed the paleobiogeography of the Eocene faunas of southern California, while Novacek and Lillegraven (1979) addressed aspects of the paleoecology of the Uintan assemblage of San Diego County.

Since the early 1980s, major collections of Eocene vertebrates from San Diego County have been made by personnel of Paleoservices, Inc. (a private paleontological resource firm), and personnel of the Department of Paleontology, San Diego Natural History Museum. Walsh (1991a, 1991b) discussed some of these new collections, most of which were obtained during grading for construction projects. This paper concentrates on the Bridgerian, Uintan, and Duchesnean mammal assemblages of San Diego County. The Morena Boulevard local fauna (Wasatchian; unnamed formation) was discussed by Walsh (1991b) and will not be further addressed here.

Walsh et al. (this volume, Chapter 6) provide a preliminary stratigraphic revision of the Friars Formation and Poway Group. This new framework is essential for understanding the mammalian succession of the southwestern part of San Diego County. The geographic distribution of the local faunas discussed herein is shown in Figure 1, while Figure 2 summarizes the stratigraphic position of all Eocene mammal assemblages from southern California. The goals of this paper are to: 1) Provide faunal lists and biostratigraphic context for several important middle Eocene mammal assemblages from San Diego County; 2) Compare the late Uintan assemblages from San Diego County with those from Ventura County; and 3) compare the Uintan assemblages from southern California with those from the western interior of North America.

ABBREVIATIONS

AMNH, American Museum of Natural History, New York; CM, Carnegie Museum of Natural History, Pittsburgh; LACM, Natural History Museum of Los Angeles County; LACM (CIT), California Institute of Technology collections, now owned by LACM; SDSNH, San Diego Natural History Museum; UCMP, University of California Museum of Paleontology,

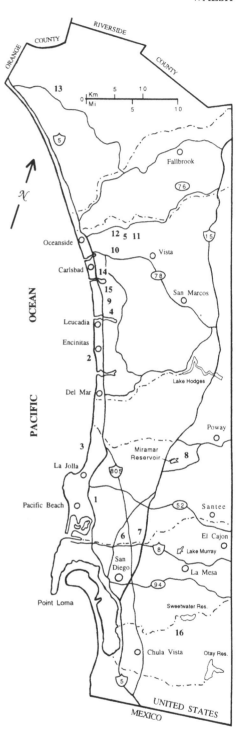

Figure 1. Western part of the coastal plain of San Diego County showing locations of the following Eocene local faunas. 1. Morena Boulevard. 2. Swami's Point. 3. Black's Beach. 4. SDSNH Loc. 3486. 5. Mesa Drive. 6. Murray Canyon. 7. Stonecrest. 8. Eastview. 9. SDSNH Loc. 3378. 10. Jeff's Discovery. 11. Rancho del Oro. 12. Mission del Oro and SDSNH Loc. 3235. 13. Camp San Onofre. 14. Laguna Riviera. 15. SDSNH Loc. 3495. 16. Bonita.

Berkeley; UCR, University of California, Riverside; USNM, United States National Museum, Washington, DC; YPM-PU, Princeton University collections, now owned by Yale Peabody Museum.

THE FIRST BRIDGERIAN ASSEMBLAGE FROM CALIFORNIA
Swami's Point Local Fauna

The Swami's Point local fauna is named here for a small assemblage of fossil vertebrates collected in 1991 from a single locality in the upper part of the Delmar Formation in Encinitas (SDSNH Loc. 3542). The specimens were discovered in beach cobbles of well-cemented green sandstone that may have been derived from the footings for a stairway at Seacliff County Park. These were excavated into the modern marine abrasion platform about 200 m south of the small promontory known locally as Swami's Point. The fossiliferous horizon is not currently exposed, and its exact stratigraphic level is unknown. Taxa recognized to date include the tillodont *Trogosus* sp. (T. A. Deméré, personal communication), the crocodilian *Pristichampsus* sp. (M. W. Colbert, personal communication), and the rhinocerotoid *Hyrachyus* sp. (M. W. Colbert, personal communication). The *Hyrachyus* specimen is intermediate in size between the early and late Bridgerian *H. modestus* and the late Bridgerian and early Uintan *H. eximius* (as diagnosed by Radinsky, 1967). *Trogosus* and other tillodonts have been recognized from strata no younger than early Bridgerian (e.g., Woodburne, 1987, p. 308), but McCarroll et al. (this volume, Chapter 2) report a possible late Bridgerian tillodont from the Washakie Formation of Wyoming. These facts do not allow an assignment of the Swami's Point local fauna to either the early or late Bridgerian, but an undifferentiated Bridgerian age seems secure. The Delmar Formation is the first rock unit in California to produce fossil mammals pertaining to this NALMA. Two other localities in the Delmar Formation have yielded an unidentifiable fragment of a rodent cheek tooth (SDSNH Loc. 3671), and an upper molar of an unidentified sciuravid? or myomorph? rodent (SDSNH Loc. 3793). Crocodile and fish remains are also present in the Delmar at SDSNH Locs. 3547 and 3793.

The Delmar Formation presents the potential opportunity for direct correlation of the Bridgerian NALMA with standard marine biochronologies. Unfortunately, this unit has not yielded biochronologically significant marine microfossils, and the molluscan assemblage from this formation is dominated by brackish-water taxa with limited biostratigraphic value (Givens and Kennedy, 1979). A post-Paleocene age for at least the upper part of the Delmar Formation is indicated by the presence of the tiger shark *Galeocerdo* from SDSNH Loc. 3791 (M. A. Roeder, personal communication; Cappetta, 1987, p. 123). The Delmar can be no younger than the earliest middle Eocene

coccolith Subzone CP12b, based on its stratigraphic position below the oldest-reported strata of the Ardath Shale (Bukry and Kennedy, 1969; Okada and Bukry, 1980; May et al., 1991). The most recent evidence for the age of the Delmar was reported by Frederiksen (1989, 1991a), who assigned a pollen assemblage from this formation to the earliest middle Eocene (Subzone CP12b equivalent), based on his correlation of pollen events and coccolith zones in the Transverse Ranges of California.

THE BRIDGERIAN/UINTAN BOUNDARY IN SAN DIEGO COUNTY

The stratigraphic position of the Bridgerian/Uintan boundary has not been precisely located in San Diego County. Based on the Bridgerian age of the Swami's Point local fauna and the early Uintan age of the lower part of the Friars Formation (see below), this boundary may occur in either the uppermost part of the Delmar Formation, the Torrey Sandstone, the Mount Soledad Formation, the Ardath Shale, the Scripps Formation, or the lowermost part of the Friars Formation. As discussed by Walsh (this volume, Chapter 4) and Walsh et al. (this volume, Chapter 6), Flynn's (1986a) conclusion that the Bridgerian/Uintan boundary occurred during Chron C20r appears to be incorrect, because the lower part of the Friars Formation is of normal polarity (Chron C21n; see Flynn, 1986a, and Walsh et al., this volume, Chapter 6), but has yielded *Amynodon* and *Leptoreodon*, which have been used to define and characterize the beginning of the Uintan (Flynn, 1986a; Krishtalka et al., 1987).

Black's Beach Local Fauna

Of relevance to the determination of the lithostratigraphic position of the Bridgerian/Uintan boundary in San Diego is the Black's Beach local fauna, named by Walsh (1991b) for the small vertebrate assemblage from the basal part of the type outcrops of the Scripps Formation (Kennedy and Moore, 1971), as exposed in the seacliffs north of La Jolla. These fossils were obtained from UCMP Locs. V-6874, V-6884, and V-6978; LACM (CIT) Loc. 456, LACM Locs. 1401, 1778, 3803, and 6673; an unnumbered USNM locality, and SDSNH Locs. 3381, 3567, and 3830. Golz and Lillegraven (1977) assigned several of these localities to the Ardath Shale, but inspection of LACM and UCMP records by T. A. Deméré and the author indicates that all of these fossils were obtained from the conglomeratic sandstones of the basal part of the Scripps Formation, 0-20 m above the top of the Ardath Shale. A detailed measured section of the Scripps Formation that includes some of the strata yielding this local fauna was provided by Lohmar et al. (1979, p. 32). Vertebrate taxa from the Black's Beach local fauna include the following (from Golz and Lillegraven, 1977, and recent SDSNH collections):

Ma: 38 39 40 41 42 43 44 45 46 47 48 49 50

SW SAN DIEGO COUNTY

Ep-uc ? ? Eastview l.f. Ep-lc Eastview l.f. Emv Cloud 9 fauna Epm Est-b Stonecrest l.f.? Stonecrest l.f.? Murray Canyon l.f. Est-a Poway fauna Esc-ut Ef-ut Ef-cg Ef-lt Ef Esc Black's Beach l.f. Ea ? ? Ems Morena Blvd. l.f. unnamed formation

NW SAN DIEGO COUNTY

SANTIAGO FM. (Member C): SDSNH Loc. 3235, SDSNH Loc. 3495, Mission del Oro l.f., Chestnut Ave l.f., Laguna Riviera l.f., Camp San Onofre l.f., SDSNH Loc. 3487, SDSNH Loc. 3378, Rancho del Oro l.f., Jeff's Discovery l.f.

SANTIAGO FM. (Member B): Mesa Drive l.f., SDSNH Loc. 3486, Ea, Et, Swami's Point l.f. ? ? Ed ? ?

VENTURA COUNTY

SESPE FORMATION — Upper Member: Simi Valley Landfill l.f., Pearson Ranch l.f., Strathern l.f.; Middle Member: Brea Canyon l.f.; L. Mbr.: Tapo Canyon l.f., unnamed l.f.

N. Amer. L. Mamm. "Ages": CHAD., DUCHESNEAN, UINTAN (late / early), BRIDGERIAN, WASATCHIAN

W. Coast Mollusc "Stages": "TEJON", "TRANSITION", "DOMENGINE", "CAPAY"

Benthic Foram Stages: NARIZIAN, ULATISIAN, "PENUTIAN"

Planktic Foram Zones: P 15, P 14, P 13, P 12, P 11, P 10, P 9, P 8

Calc. Nanno. Zones: CP 14 (b / a), CP 13 (c / b / a), CP 12 (b / a), CP 11, CP 10

Ma: 38 39 40 41 42 43 44 45 46 47 48 49 50

AGE: BARTONIAN, LUTETIAN, YPRESIAN

EPOCH: EOCENE — MIDDLE (LATE / MIDDLE / EARLY), EARLY

Carcharias macrota
Trionychidae
Crocodilia
Boidae
cf. *Uintatherium* sp.
Mesonychidae? or Arctocyonidae?
    (J. H. Hutchison, personal communication)
Brontotheriidae
Rhinocerotoidea
Artiodactyla?

The uintathere specimens described by Hutchison (1971) from the Black's Beach local fauna were fragmentary, and although they were provisionally assigned to cf. *Uintatherium*, he pointed out that they could pertain to either *Uintatherium* (late Bridgerian-early Uintan), *Tetheopsis* (late Bridgerian-early Uintan), or *Eobasileus* (early Uintan). In any case, the presence of a relatively large uintathere suggests a late Bridgerian or early Uintan age for the Black's Beach local fauna (Wheeler, 1961).

A brontothere specimen from the Black's Beach local fauna (USNM 214887, maxilla fragment with M2-3) was briefly discussed by Schiebout (1977), who noted that it resembled the type specimen of *Metarhinus? pater* (the common, small early Uintan brontothere from San Diego) in size and in the absence of a posterolingual cingulum on M3. If this specimen does pertain to *Metarhinus*, an early Uintan age would be indicated for the Black's Beach local fauna (Mader, 1989). Unfortunately, it is difficult to assign such fragmentary dental material to particular genera, because diagnoses of the several late Bridgerian and early Uintan small brontotheres of the western interior are based largely on skull characters (Mader, 1989).

The assertion of Prothero (Prothero et al., 1993) that

(Facing page) Figure 2. Correlation chart showing geographic and stratigraphic position of southern California mammal assemblages discussed in the text. Abbreviations: Ea = Ardath Shale, Ed = Delmar Formation, Ef = Friars Formation, Ems = Mount Soledad Formation, Emv = Mission Valley Formation, Ep-lc, Ep-m, and Ep-uc = lower conglomerate member, Miramar Sandstone Member, and upper conglomerate member of Pomerado Conglomerate, respectively. Esc = Scripps Formation, Est-a = lower member of Stadium Conglomerate, Est-b = upper member of Stadium Conglomerate, Et = Torrey Sandstone. Vertical ordering of the Camp San Onofre, Laguna Riviera, and Mission del Oro local faunas is arbitrary. The same is true for SDSNH Loc. 3378 and the Rancho del Oro and Jeff's Discovery local faunas, and SDSNH Loc. 3486 and the Mesa Drive local fauna. Compiled and in some cases slightly modified from Almgren et al. (1988), Berggren et al. (1995), Bukry and Kennedy (1969), Givens and Kennedy (1979), Kelly (1990), Kelly et al. (1991), Obradovich and Walsh (in prep.), Prothero and Swisher (1992), Squires (1988), and Steineck et al. (1972). Numerical ages of most local faunas and biochron and lithostratigraphic boundaries shown here should not be taken literally to within at least ± 1 m.y.

the Scripps Formation "produces late Bridgerian mammals" may eventually prove to be correct. However, this statement was not based on the collection of Bridgerian index fossils from the Scripps, but rather on the paleomagnetic correlations of Flynn (1986a), whose implied placement of the Bridgerian/Uintan boundary within the Friars Formation "forced" the Scripps Formation to be of Bridgerian age. Collection of more age-diagnostic material from the Black's Beach local fauna will be relevant to the correlation of the Bridgerian/Uintan boundary with standard marine biochronologies and the geomagnetic polarity time scale (Walsh, this volume, Chapter 4), as well as the testing of competing hypotheses on the stratigraphic relationships of certain middle Eocene units in southwestern San Diego County (Walsh et al., this volume, Chapter 6).

BIOSTRATIGRAPHIC PROBLEMS IN THE UINTA BASIN

In order to provide a meaningful discussion of the Uintan faunas of California, it is necessary to first examine certain issues concerning the mammalian faunas from the Uinta Formation of Utah, which have traditionally represented the "type" faunas of the Uintan NALMA (Wood et al., 1941; Krishtalka et al., 1987). Precise age correlation of Uintan faunas in North America is hampered by several problems, including our ignorance of the precise stratigraphic location of both the Bridgerian/Uintan (Walsh, this volume, Chapter 4), and Uintan/Duchesnean boundaries in the Uinta Basin. Other problems relevant to continent-wide correlations within the Uintan are discussed as follows.

1. Confusion caused by the changing definition of the Uinta B-C boundary in the eastern part of the basin.

Just as Osborn's (1929) change in the definition of the Uinta A-Uinta B boundary has caused confusion in the stratigraphic provenance of certain specimens (Prothero, this volume, Chapter 1), his change in the definition of the Uinta B/Uinta C boundary seems to have achieved the same result. Peterson's (in Osborn, 1895) original definition of this boundary corresponded to the gradational, but nevertheless striking transition between the mostly gray and green clays that he assigned to "Horizon B," and the overlying brownish and reddish sandstones and clays of his "Horizon C." In the vicinity of Utah State Highway 45, this transition generally occurs in the middle to the north side of Kennedy Wash (includes "Kennedy's Hole"; e.g., Hatcher, 1895; Kay, 1934, plate XLV; Stagner, 1941, pp. 282-283). However, Osborn and Matthew (1909, fig. 8) and Osborn (1929, fig. 63 and p. 92) redefined the base of Uinta C to correspond to the top of the *Amynodon* Sandstone, so that the contact between Uinta B and C no longer occurred within Kennedy

Wash, but now occurred on the north rim of Coyote Basin. The result of this change was that the uppermost gray and green clays of Peterson's "Horizon B" ("The Devil's Playground") were redefined into the base of Uinta C (see geologic mapping of Rowley et al., 1985, and Cashion, 1986). Osborn's redefinition may partly account for the difference in the thicknesses of Uinta B (420 feet) and Uinta B2 (285 feet) as given by Douglass (1914) and Osborn (1929, p. 92), respectively. See also Douglass's (1914, p. 420) implication that the position of the Uinta B-C boundary was indeed a matter of controversy among these workers. Presumably, the Devil's Playground beds would pertain to the lower part of Uinta C1 of Osborn (1929), while Uinta C2 of Osborn now corresponds at least in part to the Brennan Basin Member of the Duchesne River Formation (Kay, 1934, p. 361; Andersen and Picard, 1972). Osborn's (1929) redefinition of the Uinta B-C boundary is problematical, because the *Amynodon* Sandstone is not a widely mappable lithologic unit, even in the eastern part of the Uinta Basin (Cashion, 1986). As discussed by Stagner (1941), this sandstone channel deposit pinches out rapidly to the west, and is "replaced" by several thinner, non-continuous sandstones which often cannot be readily correlated with the *Amynodon* Sandstone proper. A consequence of this situation for local biostratigraphy (at least until detailed geologic mapping of the Uinta B-C transition in the eastern part of the basin is accomplished) is that certain fossil localities in the western part of Kennedy Wash and Coyote Wash cannot be confidently assigned to either Uinta B or Uinta C of Osborn (1929).

Finally, Untermann et al. (1964) and Andersen and Picard (1972, fig. 9) retained the lower part of the redbeds comprising Peterson's "Horizon C" in the Red Wash and Kennedy Wash area within the Uinta Formation. In Kennedy Wash, however, Rowley et al. (1985) and Cashion (1986) generally map these strata as the Duchesne River Formation (Brennan Basin Member). Prothero (this volume, Chapter 1) accepts the latter interpretation, and the result is that Uinta C is restricted here to only the approximately 60 m of gray and green strata comprising the Devil's Playground beds. These differing placements of the Uinta C-Duchesne River Formation contact constitute another semantic problem that must be resolved, if only for the sake of a consistent biostratigraphic terminology.

2. Adherence to an unsubstantiated chronostratigraphic definition of the early Uintan/late Uintan boundary.

The shifting position of the Uinta B-C contact in the eastern part of the basin should have no bearing on the definition of the early Uintan-late Uintan faunal transition. By default, however, most workers have essentially defined the latter to correspond to the former. Thus, in the eastern part of the basin, fossils from below the top of the *Amynodon* Sandstone are "early Uintan," and fossils from above the *Amynodon* Sandstone are "late Uintan." The result of such a chronostratigraphic definition is that our concept of the early Uintan/late Uintan temporal boundary is held hostage by the difficulty of recognition of the Uinta B-C lithostratigraphic boundary. In addition, both the Uinta B-C and the Uinta C-Duchesne River Formation contacts are significantly time-transgressive across the Uinta Basin (Andersen and Picard, 1972; Cashion, 1974, 1986, 1994; Cashion, personal communication), so the phrases "Uinta B time" and "Uinta C time" are misleading at best. For this reason alone, the common temporal assignments of "Uinta B" and "Uinta C" to faunas far removed from the Uinta Basin should be viewed with caution.

Osborn and Matthew (1909) and Osborn (1929) may have redefined the Uinta B-C contact to correspond to a significant change in the large-bodied mammal faunas that they recognized in the upper part of the Uinta Formation, i.e., the "*Eobasileus-Dolichorhinus*" zone (Uinta B2) vs. the "*Diplacodon -Epihippus* zone" (Uinta C). This redefinition is not objectionable in itself (there being no established type sections either then or now), but together with the subsequent chronostratigraphic definition of the early Uintan-late Uintan boundary noted above, it may lead to confusion. Thus, some significant specimens collected by early workers from the green and gray clays on the south side of Kennedy's Hole, and which were originally recorded as coming from "Horizon B" of Peterson (and were "therefore" early Uintan), must now be assigned to Uinta C (and are "therefore" late Uintan). For example, the type specimen of the rodent *Protoptychus hatcheri* (YPM-PU 11235) and the much-discussed paired lower jaws of the primate *Ourayia uintensis* (YPM-PU 11236) may fit this category. Both specimens were collected by J. B. Hatcher in 1895, and both are recorded as coming from the "gray clays of upper B, Kennedy's Hole" (Turnbull, 1991; Krishtalka, 1978; M. A. Turner, personal communication), which pertain to Uinta C of Osborn (1929). In view of this situation, the precise stratigraphic provenance of other specimens recorded from "upper Uinta B," "Kennedy's Hole," and "the Devil's Playground" should be re-evaluated.

3. Absence of a rigorous biostratigraphy for the Uinta Basin.

A major difficulty involved in Uintan biochronology is that a detailed biostratigraphy for the Uinta Formation has not been established. Prothero (this volume, Chapter 1) has compiled the stratigraphic distributions of currently known taxa, but some of these ranges may be questionable in view of the problems discussed above. In particular, the Uinta Formation has never been subjected to large scale screenwashing, especially in the eastern part of the basin (M. R.

Dawson, personal communication). Not a single micromammal taxon is recorded in this area by Prothero (this volume, Chapter 1) from Uinta B1, only five are recorded from Uinta B2, only six from Uinta C, and none from the Brennan Basin Member of the Duchesne River Formation (micromammals in this context include all Uinta marsupials, insectivores, and "proteutheres" excluding *Simidectes*, and all primates, "plesiadapiforms," and rodents). Therefore, it has not been convincingly demonstrated that the greatest overall faunal change within the Uinta Formation corresponds to the top of the *Amynodon* Sandstone. This ignorance also hampers precise temporal correlations with the geographically more isolated districts of White River Pocket and Myton Pocket in the east-central part of the basin. Collection of a diverse micromammal assemblage from Uinta B1 and B2 will also be necessary before formal "earliest Uintan" and "late early Uintan" biochrons are defined for North America (Walsh, this volume, Chapter 4). Once these tasks are accomplished, and the stratigraphic provenance of existing specimens is re-evaluated, a revised faunal definition for the early Uintan/late Uintan boundary may also be appropriate. The establishment of a detailed biostratigraphy for the Uinta Formation will be of major importance for the correlation of Uintan faunas throughout North America. Until this scheme is presented and corroborated in other areas, use of the terms "early Uintan" and "late Uintan" is unavoidable, but strict, continent-wide isochrony of the intervals currently designated by these terms should not be casually assumed in either biochronological or magnetochronological studies. Although the quotation marks will be largely dispensed with for the rest of this paper, designations such as "*nominally* early Uintan" and "*nominally* late Uintan" may often be advisable in this context (Woodburne and Swisher, 1995).

UINTAN BIOCHRONOLOGY IN CALIFORNIA
Subdivision of the Early Uintan
The Uintan has been recognized as one of the longest of the NALMAs (about 7 m.y. according to Krishtalka et al., 1987), and subdivsion of this interval into early and late parts, although less than adequately defined, has been generally accepted. Recently, some workers have proposed a further subdivision of the early Uintan interval into "earliest" ("Shoshonian") and "late early" parts (Flynn, 1986a), or "Shoshonian" and "middle" parts (Stucky et al., 1989; Stucky, 1992). Various aspects of this subject are discussed by McCarroll et al. (this volume, Chapter 2), Stucky et al. (this volume, Chapter 3), and Walsh (this volume, Chapter 4), and their opinions will not be repeated here. However, for immediate clarification, the terms "Shoshonian" and "middle Uintan" are not used in this paper. Instead, the terms "earliest Uintan" and "late early Uintan" are used as local, relative, and faunally-undefined subdivisions of a longer "early Uintan" interval. It should not be

assumed that these terms as currently used in California and the western interior represent strictly isochronous biochrons.

The Early Uintan/Late Uintan Boundary
As in the Uinta Basin, recognition of the early Uintan/late Uintan boundary in California is problematical. According to Krishtalka et al. (1987, p. 89), the end of early Uintan time is characterized by the disappearance of *Protoptychus*, *Metarhinus*, *Telmatherium*, taeniodonts, and uintatheres. Of these taxa, only *Metarhinus*? and a uintathere are known from southern California, and both apparently do become extinct at what has been traditionally regarded as the end of "early Uintan time" in this area.

Krishtalka et al. (1987, pp. 89-90) characterized the beginning of the late Uintan by the first appearance of camelids (*Poebrodon*), eomyid rodents, *Domnina*, *Thylacaelurus*, *Colodon*, *Prodaphoenus*, *Simidectes*, *Procyonodictis*, and *Epitriplopus*. However, three of these taxa are apparently known from the early Uintan. *Poebrodon* has been recognized from putatively early Uintan strata of Wyoming by McCarroll et al. (this volume, Chapter 2). A new genus of rodent ("*Namatomys*") assigned to the Eomyidae by Chiment (1977) and Chiment and Korth (in press) is abundant in the early Uintan of San Diego County, and two different eomyids were reported from the early Uintan Tepee Trail Formation by McKenna (1980). Finally, *Simidectes* is recorded from the nominally late early Uintan White River Pocket locality in the Uinta Formation (Prothero, this volume, Chapter 1). Of the remaining taxa cited by Krishtalka et al. (1987) with a "late Uintan" first appearance, only *Procyonodictis* is known from California.

Stucky et al. (1989) characterized the Uintan 2/Uintan 3 boundary (equivalent to the early Uintan/late Uintan boundary of Krishtalka et al., 1987) by the first occurrences of *Domnina*, *Thylacaelurus*, *Chumashius*, *Mytonius*, *Tapocyon*, *Colodon*, *Simimeryx*, *Mytonolagus*, and *Pseudocylindrodon*. Of these, *Tapocyon*, *Chumashius*, *Mytonius*, *Mytonolagus*, and *Simimeryx* are known from California. However, *Tapocyon* is now recorded from the early Uintan Friars Formation of San Diego. *Mytonolagus* is known only from the latest Uintan or Duchesnean Laguna Riviera local fauna, while *Chumashius* is known only from the latest Uintan or Duchesnean Camp San Onofre local fauna and the Duchesnean Pearson Ranch local fauna. Finally, *Mytonius* ("*Yaquius*" of Mason, 1990; Rasmussen and Walsh, in prep.) and *Simimeryx* are known in the late Uintan only from the Tapo Canyon local fauna (Mason, 1988).

Despite these difficulties, students of California Eocene mammals have recognized "early Uintan," "early late Uintan," "late Uintan," and "latest Uintan" assemblages (Golz and Lillegraven, 1977; Kelly, 1990;

Table 1. Temporal distribution of Uintan and Duchesnean mammal genera of California. Compiled from Golz and Lillegraven (1977), Kelly (1990; 1992), Kelly et al. (1991), and this paper. Genera marked with an asterisk may in fact be restricted to the Duchesnean.

Known only from the early Uintan.	Known from both the early Uintan, *and* the late Uintan and/or Duchesnean.	Known from the late Uintan and/or Duchesnean, but not the early Uintan.	Known only from strata of definite Duchesnean age.
Crypholestes	*Peratherium*	*Sespedectes*	*Paradjidaumo*
Dormaaliidae, n. gen.	*Peradectes*	*Proterixoides*	*Heliscomys*
Palaeoryctine gen.	*Apatemys*	*Simidectes*	*Simiacritomys*
Micropternodontine gen.	*Aethomylos*	*Chumashius**	*Hyaenodon*
cf. *Pantolestes* sp.	*Scenopagus*	*Dyseolemur*	*Duchesneodus*
Palaeictops sp.	*Nyctitherium* (cf.)	*Macrotarsius*	*Eotylopus*
Leptictidae, n. gen.	*Centetodon*	*Mytonius*	
Adapidae, n. gen.	*Batodonoides*	*Craseops*	
Omomys	*Oligoryctes*	*Phenacolemur*	
Hemiacodon	*Ourayia*	*Tapomys*	
Washakius	*Uintasorex*	*Mytonomys*	
Stockia	*Microparamys*	*Rapamys*	
Omomyidae, unident. gen.	*Leptotomus*	Manitshini, unident. gen.	
Microsyops	*Pseudotomus*	*Presbymys**	
Palaeanodonta? unident. gen.	*Sciuravus*	*Griphomys*	
Uriscus	*Pareumys*	cf. *Paradjidaumo*	
Pauromys	*Eohaplomys*	*Simimys*	
cf. *Harpagolestes*	"*Namatomys*"	*Mytonolagus**	
Uintatherium	*Miacis*	cf. *Apataelurus*	
Hyrachyus	*Tapocyon*	*Limnocyon*	
Metarhinus?	Tapiroidea, n. gen.	*Miocyon*	
cf. *Lophiohyus*	*Amynodon*	*Procyonodictis*	
Achaenodon	*Antiacodon*(?)	*Triplopus?*	
cf. *Parahyus*	*Protoreodon*	*Amynodontopsis**	
Merycobunodon	(cf.) *Protylopus*	*Epihippus**	
	Leptoreodon	*Ibarus*	
		Tapochoerus	
		Simimeryx	
		*Poebrodon**	
25 genera	26 genera	29 genera	6 genera

Walsh, 1991b). Due to the absence or rarity of most of the genera used to characterize the early Uintan/late Uintan boundary in the western interior, Walsh (1991b) concluded that a practical characterization of this boundary in southern California must make use of several endemic taxa. As a result, however, there can be even less confidence that these terms designate the same intervals of time in both areas, and precise age assignments for California assemblages such as "early late Uintan" should be understood only in a local, relative sense. With this caveat in mind, the early Uintan/late Uintan boundary as recognized in southern California can be characterized by the extinction and/or emigration of the common early Uintan genera *Crypholestes*, an unnamed dormaaliid insectivore, *Omomys*, *Washakius*, *Pauromys*, *Metarhinus?* and *Merycobunodon*, and by the *in situ* evolution and/or immigration of the common late Uintan genera *Sespedectes*, *Proterixoides*, *Dyseolemur*, *Rapamys*, *Griphomys*, and *Simimys*. Of the latter group, only *Rapamys* and *Sespedectes* are known from the western

interior; the possible Badwater occurrence of *Griphomys* mentioned by Sutton and Black (1975) has not been corroborated (M. R. Dawson, personal communication). Several other genera from southern California are known only from either the early or late Uintan, and may be used as supplementary age indicators in the absence of other evidence (Table 1). In addition, some relatively common species of certain genera are also restricted to one side of the boundary or the other. In the early Uintan, these include *Centetodon aztecus*, *Microparamys* sp. cf. *M. minutus*, *Pareumys* sp. near *P. grangeri*, *Protoreodon* new sp. 2, cf. *Protylopus* sp., *Leptoreodon major*, and *L.* sp. cf. *L. marshi*. Species restricted to the late Uintan (and/or Duchesnean) include *Centetodon* sp. probably new, *Microparamys tricus*, *Pareumys* sp. cf. *P. milleri*, *Protoreodon* new sp. 1, *Protoreodon annectens*, *Protylopus stocki*, *P.* sp. cf. *P. petersoni*, *Protylopus* sp. (207/1592 morph), *Protylopus robustus*, *Leptoreodon edwardsi*, *L. pusillus*, *L. leptolophus*, and *L. stocki*. Finally, the rodent *Sciuravus powayensis* is abundant in the early Uintan, but only three isolated teeth of this genus are known from the late Uintan of southern California (Jeff's Discovery local fauna).

That the early Uintan and late Uintan assemblages of southern California are indeed temporally distinct and not facies-controlled is demonstrated by their consistent superpositional relationship in at least two different paleoenvironments. Thus, in northwestern San Diego County, the early Uintan micromammal assemblage (Mesa Drive local fauna) occurs within estuarine strata, and is overlain by a late Uintan mammal assemblage occurring entirely within fluvial strata (Rancho del Oro local fauna). In contrast, in the greater San Diego area, the early Uintan assemblage (Poway fauna) occurs largely in terrestrial floodplain deposits, and is locally overlain by shallow marine strata of the Mission Valley Formation containing a typical late Uintan assemblage.

Based on the above discussion, Table 1 distinguishes between four categories of Uintan and Duchesnean mammal genera of southern California; 1) those known only from the early Uintan; 2) those known from the early Uintan, *and* the late Uintan and/or Duchesnean; 3) those known from the late Uintan and/or Duchesnean, but not from the early Uintan; and 4) those known only from strata of undoubted Duchesnean age. The "late Uintan and/or Duchesnean" category is required because several local faunas of southern California cannot be confidently assigned to either age (see below). Thus, six taxa in category (3) are marked with an asterisk to indicate that their occurrence in the late Uintan of southern California is uncertain. If these taxa are momentarily excluded from consideration, the early Uintan assemblage from southern California (51 genera) would be slightly more diverse than the late Uintan assemblage (49 genera). Note that a few taxa of bats, ischyromyid rodents, and a creodont are excluded from Table 1 because it is uncertain whether their early Uintan representives are congeneric with their late Uintan representatives.

Specific Uintan and Duchesnean assemblages of San Diego County are discussed below in order of decreasing age. Table 2 lists all known more or less identifiable mammalian taxa from named Uintan and/or Duchesnean assemblages of San Diego County (excluding Chestnut Avenue local fauna; see below). Complete references for the authorship of most southern California species are given in Golz and Lillegraven (1977), and many of the specific localities that have yielded these taxa can be obtained from Golz and Lillegraven (1977) and Appendix 1.

EARLY UINTAN ASSEMBLAGES OF SAN DIEGO COUNTY

Poway Fauna

The name "Poway fauna" was previously used by Stock (1948) and Savage and Downs (1954) to refer to the relatively small collection of mammals then known

from what is now the type area of the Friars Formation. Based on the stratigraphic revisions of Walsh et al. (this volume, Chapter 6), the lithostratigraphic basis for the definition of the Poway fauna discussed by Walsh (1991b) is modified here to refer to the entire set of vertebrate taxa occurring in the type outcrops of the Friars Formation, the lower tongue of the Friars, the conglomerate tongue of the Friars, and the upper tongue of the Friars. No significant stratigraphic or geographic variations in faunal content have been detected within any of these units. Due to Kennedy and Moore's (1971) assignment of the Friars Formation to the La Jolla Group, it should be emphasized that none of the stratigraphic units that yield the Poway fauna are presently assigned to the Poway Group. Nevertheless, the former term is well-established for the early Uintan assemblage of southwestern San Diego County, and I prefer to retain it rather than coin a new name for the same concept.

The Poway fauna is currently represented by more than 7000 identified mammal specimens in UCMP, LACM, and SDSNH collections, and is composed of about 53 genera, containing about 61 species (Table 2). As noted by Novacek and Lillegraven (1979), the micromammal assemblage of the Poway fauna is dominated numerically by the marsupial *Peratherium* sp. cf. *P. knighti*, the insectivore *Crypholestes vaughni*, and the rodents *Microparamys* sp. cf. *M. minutus*, *Sciuravus powayensis*, and "*Namatomys*" new sp. 1 (the latter taxon was erroneously identified by Walsh, 1991b as "cf. *Pauromys* sp."). These five species make up about 65% of the individuals represented in the Poway fauna as estimated by Walsh (1987), using the weighted abundance of elements method of Holtzman (1979). The Poway fauna is correlated with the Mesa Drive local fauna and SDSNH Loc. 3486 (Member B, Santiago Formation) of northwestern San Diego County. The stratigraphic occurrence of several localities yielding the Poway fauna is provided by Walsh et al. (this volume, Chapter 6).

Mesa Drive Local Fauna

Walsh (1991b) named the Mesa Drive local fauna for an early Uintan assemblage from SDSNH Locs. 3440, 3443, 3447, 3448, 3450, 3465, and 3571 (Loc. 3447 was erroneously assigned to the late Uintan Rancho del Oro local fauna in that paper). These were all collected from 1988 to 1991, during grading for the Rancho del Oro and Mission del Oro housing developments in Oceanside, north of Oceanside Boulevard and south of State Route 76. All of these sites occurred in the upper part of Member B of the Santiago Formation (Wilson, 1972), within a 7 m-thick estuarine sequence composed of mollusc-and-bone-bearing yellowish very fine-grained sandstones, interbedded with barren reddish siltstones and mudstones (Figure 3). The base of the vertebrate-bearing interval begins 1 m below a widespread, 0.2m–

Table 2. Mammalian faunal lists for selected Uintan and/or Duchesnean assemblages in San Diego County. Sources for taxonomic identifications are given in brackets, abbreviated using the identifier's initials, and if published, are provided in "References Cited".

	Early Uintan			Late Uintan						Latest Uintan and/or Duch.			
	Mesa Drive l.f.	Poway fauna	Murray Cyn. l.f.	Stonecrest l.f.	Cloud 9 fauna	Eastview l.f.	Jeff's Discovery l.f.	Rancho del Oro l.f.	SDSNH Loc. 3378	Lag. Riviera l.f.	Cmp. San Ono. l.f.	Miss. del Oro l.f.	Bonita l.f.
Marsupialia													
Didelphidae													
Peratherium cf. *P. knighti* McGrew, 1959 [JAL 1976]	X	X	–	–	–	–	–	–	–	–	X	–	–
Peratherium cf. *P. innominatum* Simpson, 1928 [SLW]	–	X	X	–	X	–	X	–	–	–	–	–	–
Peradectidae													
Peradectes californicus (Stock, 1936) [JAL 1976; LK&RS 1983]	–	–	–	X	X	X	X	X	–	–	X	X	X
Peradectes sp(p). unidentified [SLW]	X	X	X	–	–	–	–	–	–	–	–	–	–
Apatotheria													
Apatemys cf. *A. bellus* Marsh, 1872 [MJN 1976]	–	X	–	–	–	–	–	–	–	–	–	–	–
Apatemys downsi Gazin, 1958 [SLW]	–	X	–	–	X	–	X	–	–	–	–	–	–
Apatemys uintensis (Matthew, 1921) [West, 1973]	–	–	–	–	–	–	–	–	–	X	–	–	–
Aethomylos simplicidens Novacek, 1976	–	X	–	X	X	–	X	–	–	–	–	–	–
Aethomylos new small sp. [SLW]	–	X	–	–	–	–	–	–	–	–	–	–	–
Pantolesta													
cf. *Pantolestes* sp. [SLW]	–	X	–	–	–	–	–	–	–	–	–	–	–
Simidectes merriami Stock, 1933 [MCC 1971]	–	–	–	–	–	–	–	–	–	X	–	–	–
Simidectes cf. *S. medius* (Peterson, 1919) [DJG & JAL 1977]	–	–	–	–	X	–	–	–	X	–	–	–	–
Leptictida													
Palaeictops sp. [SLW]	–	X	–	–	–	–	–	–	–	–	–	–	–
Leptictidae gen. and sp. prob. new [SLW]	–	X	–	–	–	–	–	–	–	–	–	–	–
Insectivora													
Dormaaliidae													
Sespedectes singularis Stock, 1935	–	–	–	X	X	X	X	X	–	X	–	–	–
Sespedectes stocki Novacek, 1985	–	–	–	–	–	–	–	–	–	–	X	–	–
Sespedectes sp(p).	–	–	–	–	–	–	–	–	–	–	–	X	X
Proterixoides davisi Stock, 1935	–	–	–	X	X	X	X	X	–	X	X	X	–
cf. *Proterixoides* sp.	–	–	–	–	–	–	–	–	X	–	–	–	–
Crypholestes vaughni (Novacek, 1976) [MJN 1985]	X	X	X	–	–	–	–	–	–	–	–	–	–
Crypholestes new sp. (large) [SLW]	–	–	X	–	–	–	–	–	–	–	–	–	–
Scenopagus cf. *S. priscus* (Marsh, 1872) [SLW 1991b]	–	X	–	–	–	–	–	–	–	–	–	–	–
Scenopagus sp. [SLW]	–	–	–	–	–	–	X	–	–	–	–	–	–
Dormaaliidae, new gen. and sp. [MJN et al. 1985]	X	X	–	–	–	–	–	–	–	–	–	–	–
Palaeoryctidae													
Genus and species indet., probably new [MJN 1976]	–	X	–	–	–	–	–	–	–	–	–	–	–
Geolabididae													
Centetodon aztecus Lillegraven et al., 1981	–	X	–	–	–	–	–	–	–	–	–	–	–
C. cf. *C. bembicophagus* JAL et al., 1981 [SLW 1991b]	–	X	–	–	–	–	–	–	–	–	–	–	–
Centetodon, sp. prob. new [SLW]	–	–	X	X	X	X	X	–	–	–	–	–	–
Batodonoides powayensis Novacek, 1976	X	X	X	X	X	–	X	X	–	–	–	–	–
Batodonoides? sp. [SLW 1987]	–	–	–	–	X	–	–	–	–	–	–	–	–
Nyctitheriidae													
Nyctitherium sp. [MJN 1976; DJG & JAL 1977]	–	X	–	–	–	–	–	–	–	–	–	–	–
cf. *Nyctitherium* sp. [DJG & JAL 1977]	–	–	–	–	–	–	–	–	–	X	–	–	–
Micropternodontidae													
micropternodontine-like gen. and sp. [MJN 1976]	–	X	–	–	–	–	–	–	–	–	–	–	–
Apternodontidae													
Oligoryctes sp. 1 [MJN 1976; SLW]	–	X	–	–	X	–	X	–	–	–	–	–	–
Oligoryctes sp. 2 [SLW]	–	–	–	–	–	–	X	–	–	–	–	–	–
Chiroptera, unident. genera and spp. [JAL & SLW]	–	X	X	–	X	–	X	–	–	–	X	–	–
Primates													
Adapidae													
Adapidae, new gen. and sp. [G. Gunnell, in press]	–	X	–	–	–	–	–	–	–	–	–	–	–
Omomyidae													
Omomys cf. *O. carteri* Leidy, 1869 [JAL 1980; JGH 1990]	–	X	–	–	–	–	–	–	–	–	–	–	–
Hemiacodon sp. near *H. gracilis* Marsh, 1872 [JAL 1980]	–	X	–	–	–	–	–	–	–	–	–	–	–
Chumashius balchi Stock, 1933 [JAL 1980]	–	–	–	–	–	–	–	–	–	–	X	–	–
Ourayia new sp. [DTR & SLW, in prep.]	–	–	–	–	–	–	X	–	–	–	–	–	–
Ourayia sp(p). unidentified [SLW]	–	X	X	X	X	X	–	X	X	–	–	–	–
Washakius woodringi (Stock, 1938) [JAL 1980]	X	X	–	–	–	–	–	–	–	–	–	–	–
Dyseolemur pacificus Stock, 1934 [JAL 1980]	–	–	–	X	X	X	X	X	–	X	X	–	–
Stockia powayensis Gazin, 1958 [JAL 1980; JGH 1990]	–	X	–	–	–	–	–	–	–	–	–	–	–

Table 2 (continued).

	Early Uintan			Late Uintan						Latest Uintan and/or Duch.			
	Mesa Drive l.f.	Poway fauna	Murray Cyn. l.f.	Stonecrest l.f.	Cloud 9 fauna	Eastview l.f.	Jeff's Discovery l.f.	Rancho del Oro l.f.	SDSNH Loc. 3378	Lag. Riviera l.f.	Cmp. San Ono. l.f.	Miss. del Oro l.f.	Bonita l.f.
Omomyidae, unident. gen. and sp. [SLW]	–	X	–	–	–	–	–	–	–	–	–	–	–
Dermoptera?													
Microsyopidae													
Microsyops kratos Stock, 1938 [JAL 1980]	–	X	–	–	–	–	–	–	–	–	–	–	–
Microsyops cf. *M. annectens* (Marsh, 1872) [JAL 1980]	–	X	–	–	–	–	–	–	–	–	–	–	–
Uintasorex montezumicus Lillegraven, 1976	X	X	–	–	–	–	–	–	–	–	–	–	–
Uintasorex sp(p). [SLW]	–	–	X	X	X	–	X	X	–	–	–	–	–
Paromomyidae													
Phenacolemur cf. *P. shifrae* Robinson, 1968 [MM 90; SLW]	–	–	–	–	X	–	X	–	–	–	–	–	–
Palaeanodonta													
Palaeanodonta?, unident. gen. and sp. [SLW 1991b]	–	X	–	–	–	–	–	–	–	–	–	–	–
Rodentia													
Ischyromyidae													
Microparamys cf. *M. minutus* (Wilson, 1937) [JAL 1977]	X	X	X	–	–	–	–	–	–	–	–	–	–
Microparamys woodi Kelly and Whistler, 1994	–	–	–	X	X	X	X	X	–	–	–	–	–
Microparamys cf. *M. woodi* Kelly and Whistler, 1994	–	X	–	–	–	–	–	–	–	–	–	–	–
Microparamys tricus (Wilson, 1940) [JAL 1977; SLW 1991b]	–	–	–	–	–	–	–	–	–	X	X	–	–
Microparamys sp(p). [SLW 1991a]	–	–	–	–	–	–	–	–	–	–	–	X	X
Pseudotomus californicus (Wilson, 1940) [WWK 1985]	–	X	–	–	–	–	–	–	–	–	–	–	–
Pseudotomus littoralis (Wilson, 1949) [WWK 1985]	–	X	–	–	–	–	–	–	–	–	–	–	–
Pseudotomus sp. [DJG & JAL 1977; WWK 1985]	–	–	–	–	–	–	–	–	–	X	–	–	–
Manitshini?, unident. gen. and sp. [SLW]	–	–	–	–	–	–	X	–	–	–	–	–	–
Leptotomus? *caryophilus* Wilson, 1940	–	X	–	–	–	–	–	–	–	–	–	–	–
Leptotomus sp(p). [DJG & JAL 1977; SLW]	–	–	–	–	X	–	–	–	–	X	–	–	–
cf. "*Leptotomus* near *L. burkei*" [Wilson, 1940]	–	–	–	–	–	–	–	–	–	–	–	X	–
Uriscus californicus Wood, 1962	–	X	–	–	–	–	–	–	–	–	–	–	–
Rapamys fricki Wilson, 1940 [SLW 1991b]	–	–	–	–	–	–	X	–	–	–	–	–	–
cf. *Rapamys* sp. [SLW 1991b]	–	–	–	X	X	–	–	–	–	–	–	–	–
Eohaplomys cf. *E. tradux* Stock, 1935	–	–	–	–	X	–	X	X	–	X	–	–	–
Eohaplomys cf. *S. serus* Stock, 1935 [SLW]	–	–	–	–	–	–	–	–	–	–	–	X	–
Eohaplomys new sp. [SLW]	–	–	X	X	X	–	–	–	–	–	–	–	–
Ischyromyidae, unident. gen. and sp. 1	–	X	–	–	–	–	–	–	–	–	–	–	–
Ischyromyidae, unident. gen. and sp. 2	–	X	–	–	–	–	–	–	–	–	–	–	–
Ischyromyidae, unident. gen. and sp. 3	–	–	–	X	–	–	–	–	–	–	–	–	–
Sciuravidae													
Sciuravus powayensis Wilson, 1940 [JAL 1977]	X	X	X	–	–	–	X	–	–	–	–	–	–
Sciuravidae, unidentified [DJG & JAL 1977]	–	–	–	–	–	–	–	–	–	X	–	–	–
Cylindrodontidae													
Pareumys sp. near *P. grangeri* Burke, 1935 [JAL 1977]	–	X	–	–	–	–	–	–	–	–	–	–	–
Pareumys sp(p). [JAL 1977; SLW]	–	–	X	–	X	X	X	X	X	X	X	X	X
Presbymys sp. [SLW 1991a]	–	–	–	–	–	–	–	–	–	–	–	–	X
Eomyidae													
"*Namatomys*" new sp. 1 [JJC 1977; JJC & WWK in press]	X	X	X	–	–	–	–	–	–	–	–	–	–
"*Namatomys*" new sp. 2 [JJC 1977; JJC & WWK in press]	–	–	–	–	–	–	–	–	–	–	X	–	–
"*Namatomys*" sp(p).	–	–	–	–	–	–	X	X	–	–	–	X	X
Griphomys alecer Wilson, 1940 [JAL 1977]	–	–	–	–	–	–	X	X	–	–	X	–	–
Griphomys toltecus Lillegraven, 1977	–	–	–	–	–	–	–	–	–	–	X	–	–
Griphomys sp.	–	–	–	X	X	–	–	–	–	X	–	X	–
cf. *Paradjidaumo* sp. [SLW]	–	–	–	–	X	–	X	–	–	–	–	–	–
Rodentia incertae sedis													
Pauromys cf. *P. perditus* Troxell, 1923 [SLW]	–	–	X	–	–	–	–	–	–	–	–	–	–
Simimys simplex (Wilson, 1935) [JAL & RWW 1975]	–	–	–	X	X	X	X	X	–	X	X	X	X
Lagomorpha													
Mytonolagus sp. [DJG & JAL 1977]	–	–	–	–	–	–	–	–	–	X	–	–	–
Mesonychia													
cf. *Harpagolestes* sp. [DJG & JAL 1977]	–	X	–	–	–	–	–	–	–	–	–	–	–
Mesonychidae, unident. gen. and sp. [SLW]	–	–	–	–	–	–	X	▭—▭▭—▭		–	–	–	–
Creodonta													
Hyaenodontidae													
Limnocyon sp. [SLW]	–	–	–	–	–	X	–	–	–	–	–	–	–
cf. *Apataelurus* sp. [T.A. Demeré pers. comm.]	–	–	–	–	–	X	–	–	–	–	–	–	–
Creodonta, unident. gen. and sp.	–	X	–	–	–	–	–	–	–	–	–	–	–

Table 2 (continued).

	Early Uintan			Late Uintan						Latest Uintan and/or Duch.			
	Mesa Drive l.f.	Poway fauna	Murray Cyn. l.f.	Stonecrest l.f.	Cloud 9 fauna	Eastview l.f.	Jeff's Discovery l.f.	Rancho del Oro l.f.	SDSNH Loc. 3378	Lag. Riviera l.f.	Cmp. San Ono. l.f.	Miss. del Oro l.f.	Bonita l.f.
Carnivora													
Procyonodictis progressus (Stock, 1935) [JJF & HG 1982]	–	–	–	–	X	–	X	–	–	–	–	–	–
Tapocyon occidentalis Stock, 1934 [TAD pers. comm.]	–	X	–	–	–	–	X	X	–	–	–	–	–
Miocyon cf. *U. scotti* (JLW & WDM 1899) [HNB 1992]	–	–	–	–	X	–	X	X	–	–	–	–	–
Miacis? cf. *M. hookwayi* Stock, 1934 [SLW]	–	–	–	–	–	X	X	–	–	–	–	–	–
Miacis sp(p). unidentified [DJG & JAL 1977; SLW]	–	X	–	–	–	–	–	–	–	X	–	–	–
Dinocerata													
Uintatheriidae													
cf. *Uintatherium* sp. [JHH 1971; TAD pers. comm.]	–	X	–	–	–	–	–	–	–	–	–	–	–
Perissodactyla													
Brontotheriidae													
Metarhinus? pater Stock, 1937 [JAS 1977]	X	X	–	–	–	–	–	–	–	–	–	–	–
cf. *Duchesneodus* sp.	–	–	–	–	–	–	–	–	–	–	–	X	–
Brontotheriidae, unident. gen. and sp. [SLW]	–	–	–	–	–	–	–	X	–	–	–	–	–
Tapiroidea													
new genus, sp. A [M.W. Colbert 1993]	–	–	–	–	X	X	X	X	X	–	–	–	–
new genus, sp. B [M.W. Colbert 1993]	–	X	–	–	–	–	–	–	–	–	–	–	–
Rhinocerotoidea													
Hyrachyus sp. [M.W. Colbert, pers. comm.]	–	X	–	–	–	–	–	–	–	–	–	–	–
Triplopus ? woodi Stock, 1936 [SLW]	–	–	–	–	–	–	X	X	–	–	–	–	–
Amynodon reedi Stock, 1939	–	X	–	–	–	–	–	–	–	–	–	–	–
A. cf. *A. advenus* (Marsh, 1875) [G&L 1977; W&S 1981]	–	–	–	X	–	–	X	–	–	X	X	–	–
Amynodontopsis bodei Stock, 1933 [W. Wall, pers. comm.]	–	–	–	–	–	–	–	–	–	X	–	–	–
rhinocerotoid? unident. gen. and sp.	–	–	–	–	–	–	–	–	–	–	–	X	–
Equidae													
Epihippus sp. [M.W. Colbert, pers. comm.]	–	–	–	–	–	–	–	–	–	–	–	X	–
Artiodactyla													
"Dichobunidae"													
Antiacodon sp. 1 [SLW]	–	X	–	–	–	–	–	–	–	–	–	–	–
Antiacodon? sp. 2 [SLW]	–	–	–	–	X	–	X	–	–	–	–	–	–
Achaenodon sp. [SLW]	–	X	–	–	–	–	–	–	–	–	–	–	–
cf. *Parahyus* sp. [SLW]	–	X	–	–	–	–	–	–	–	–	–	–	–
cf. *Lophiohyus* sp. [SLW]	–	X	–	–	–	–	–	–	–	–	–	–	–
Ibarus cf. *I. ignotus* Storer, 1984 [SLW]	–	–	–	–	X	–	X	–	–	–	–	–	–
Hypertragulidae													
Simimeryx cf. *S. hudsoni* Stock, 1934 [DJG & JAL 1977]	–	–	–	–	–	–	–	–	–	–	X	–	–
Agriochoeridae													
Protoreodon annectens (Marsh, 1875) [TSK et al. 1991]	–	–	–	–	–	–	–	–	–	X	–	–	–
Protoreodon new sp. 1 [B. Lander, pers. comm.]	–	–	–	–	X	X	X	X	X	–	–	–	–
Protoreodon new sp. 2 [T.A. Deméré pers. comm.]	–	X	–	–	–	–	–	–	–	–	–	–	–
Protoreodon sp(p). unidentified [DJG & JAL 1977; SLW]	–	–	–	–	–	–	–	–	–	–	X	X	–
Oromerycidae													
Merycobunodon littoralis Golz, 1976	–	X	–	–	–	–	–	–	–	–	–	–	–
Protylopus cf. *P. petersoni* Wortman, 1898	–	–	–	–	–	–	–	–	–	X	–	–	–
Protylopus stocki Golz, 1976	–	–	–	–	–	–	X	X	X	X	–	–	–
Protylopus sp. (207/1592 morph) [SLW]	–	–	–	–	X	X	X	–	–	–	–	–	–
Protylopus robustus Golz 1976 [Kelly, 1990]	–	–	–	–	–	–	–	–	–	X	–	–	–
Protylopus cf. *P. robustus* Golz, 1976 [SLW]	–	–	–	–	–	–	–	–	–	–	–	X	–
Protylopus? sp. [DJG 1976]	–	–	–	–	X	–	–	–	–	–	–	–	–
cf. *Protylopus* sp. [SLW]	–	X	X	–	–	–	–	–	–	–	–	–	–
Camelidae													
Poebrodon californicus Golz, 1976	–	–	–	–	–	–	–	–	–	X	–	–	–
Protoceratidae													
Leptoreodon major Golz, 1976	–	X	X	–	–	–	–	–	–	–	–	–	–
Leptoreodon cf. *L. marshi* Wortman, 1898 [DJG 1976]	–	X	–	–	–	–	–	–	–	–	–	–	–
Leptoreodon leptolophus Golz, 1976	–	–	–	–	–	–	–	–	–	X	–	–	–
Leptoreodon pusillus Golz, 1976	–	–	–	–	–	–	–	–	–	X	–	–	–
Leptoreodon sp(p). unidentified [SLW]	–	–	–	–	X	X	X	X	X	–	–	–	–
Total number of genera/species:	11	53	15	16	32	14	39	21	8	22	15	15	7
	11	61	16	16	34	14	41	21	8	24	16	15	7

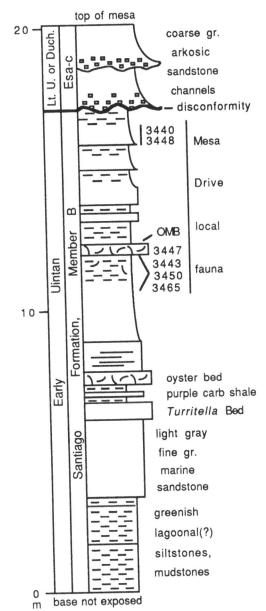

Figure 3. Measured section showing stratigraphic occurrence of the early Uintan Mesa Drive local fauna. OMB = Oyster Marker Bed. Esa-c = Member C of Santiago Formation.

0.5m thick "Oyster Marker Bed" (OMB in Fig. 3) that occurs throughout the Rancho del Oro area, and which is traceable from east of College Boulevard at least 3 km to the west to the Mission del Oro housing development (B. O. Riney, field notes on file at SDSNH). The OMB in turn occurs about 4 m stratigraphically above another laterally persistent, resistant sandstone bed ("Turritella Bed"; includes SDSNH Loc. 3446) containing the molluscs Turritella uvasana, Miltha packi, and Glyptoactis sp. (T. A. Deméré, personal communication). The top of the interval yielding the Mesa Drive local fauna is the disconformity separating

Member B from the overlying Member C of the Santiago Formation. The section shown in Figure 3 was measured about 500 m east-southeast of the intersection of College Boulevard and Mesa Drive, and has since been graded away. In this section, the latest? Uintan and/or Duchesnean coarse-grained arkosic sandstones of the upper part of Member C lie directly upon the early Uintan strata of Member B, so that the lower, finer-grained part of Member C (which produced the late Uintan Rancho del Oro local fauna; see below) is absent.

A revised list of mammal taxa from the Mesa Drive local fauna is given in Table 2. Like the Poway fauna, the Mesa Drive local fauna is dominated numerically by the rodents Sciuravus powayensis and "Namatomys" new sp. 1. On the basis of these taxa, Crypholestes vaughni, the unnamed dormaaliid insectivore, Washakius, Microparamys sp. cf. M. minutus, and Metarhinus? pater, this local fauna is clearly of early Uintan age.

SDSNH Loc. 3486

A sparse but biochronologically significant early Uintan assemblage was collected in 1989 from Member B of the Santiago Formation, during grading for the Aviara development in Carlsbad, north of Bataquitos Lagoon. Stratigraphic context for this assemblage is given in Figure 4. Fossils occurred throughout a 3-4 m-thick bed of light gray, fine- to medium-grained, fluvial? to shallow-marine sandstone (SDSNH Loc. 3486). This bed also contained Poway-type rhyolite pebbles and rare shark teeth, and may correlate stratigraphically with some part of the Friars Formation. Pervasive faulting complicates the stratigraphic relationships at Aviara, but the vertebrate-bearing horizon occurred in apparent conformable contact above cross-bedded arkosic sandstones of the Torrey Sandstone, and was in turn disconformably overlain by arkosic gritstones of Member C of the Santiago Formation. The vertebrate assemblage from SDSNH Loc. 3486 is composed of the following taxa:

Carcharias macrota
Abdounia sp.
Trionychidae
Crocodilia
Uintatherium sp.
Metarhinus? pater

The assemblage from SDSNH Loc. 3486 contains the first record of Uintatherium from the Santiago Formation (T. A. Deméré, personal communication), is dominated by the small brontothere Metarhinus? pater, and corroborates the evidence from the Mesa Drive local fauna that the upper part of Member B is of early Uintan age.

Member C of the Santiago Formation at Aviara was

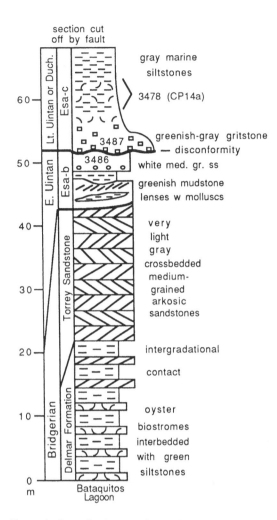

Figure 4. Generalized composite section showing stratigraphic occurrence of SDSNH Locs. 3486, 3487, and 3478.
Esa-b and Esa-c = Members B and C of Santiago Formation,
respectively. Measured by B. O. Riney, T. A. Demére, and
D. R. Swanson.

characterized by a basal "green gravel unit" composed of
metavolcanic and granitic pebbles and cobbles in a
matrix of arkosic gritstone. This unit (SDSNH Loc.
3487) produced a lower jaw fragment of *Amynodon* sp.
cf. *A. advenus*, and an upper molar fragment of cf.
Duchesneodus sp., suggesting a late Uintan or
Duchesnean age. The basal gravel of Member C graded
upward into a shelfal marine siltstone containing well-
preserved molluscs and a diverse assemblage of
coccoliths (SDSNH Loc. 3478). Bukry (1991) assigned
this assemblage to low-latitude Subzone CP14a, which
is correlated with the middle to late part of the middle
Eocene by Berggren et al. (1995). It is possible that the
marine siltstone in Member C at Aviara reflects the
same eustatic sea-level highstand as the Mission Valley
Formation, which is tentatively assigned to CP14a by
Walsh et al. (this volume, Chapter 6).

Murray Canyon Local Fauna

Five SDSNH vertebrate localities are currently known
from the lower member of the Stadium Conglomerate
(Appendix 1; see also Walsh et al., this volume,
Chapter 6). These have produced an unusual mammal
assemblage which is here named the Murray Canyon
local fauna (Table 2). It contains important elements of
both the early and late Uintan faunas of California, but
is characterized by the abundance of the small
myomorph? rodent *Pauromys* sp. cf. *P. perditus*, and by
common specimens of a large new species of the
insectivore *Crypholestes*. This local fauna contains four
species that are common in the early Uintan Poway
fauna (*Crypholestes vaughni*, *Microparamys* sp. cf. *M.
minutus*, "*Namatomys*" new sp. 1, and *Leptoreodon
major*). The Murray Canyon local fauna also contains
the lowest stratigraphic record in San Diego of the
rodent *Eohaplomys*, a genus common in the late Uintan
of southern California. Finally, the Murray Canyon
local fauna is characterized by the rarity of the rodent
Sciuravus (which is abundant in the Poway fauna), and
by the apparent absence of *Sespedectes*, *Proterixoides*,
Simimys, and other taxa characteristic of the late Uintan
of southern California.

Faunal equivalents of the Murray Canyon local fauna
have not been recognized from either the Santiago
Formation of northwestern San Diego County or the
Sespe Formation of Ventura County. Assuming that
the Poway fauna is roughly time-correlative with
assemblages from the lower part of Uinta B in Utah (see
discussion in Walsh, this volume, Chapter 4), then,
based solely on superposition, the Murray Canyon local
fauna may be roughly time-correlative with the
assemblages from the upper part of Uinta B. However,
no species are currently known to occur in both (see
faunal list in Prothero, this volume, Chapter 1). Further
collecting from the Murray Canyon local fauna may
shed light on the origin of certain endemic elements of
the late Uintan assemblages of southern California.

LATE UINTAN MAMMAL ASSEMBLAGES OF
SAN DIEGO COUNTY
Stonecrest Local Fauna

Several fossil mammal localities in the Stadium
Conglomerate were collected in late 1989 and early
1990, during grading for the Stonecrest development on
the west rim of Murphy Canyon, south of Aero Drive
(Stonecrest local fauna of Walsh, 1991b; SDSNH Locs.
3526, 3527, 3528, 3530, 3536, and 3538; see revised
faunal list in Table 2 and MS-9 *in* Walsh et al., this
volume, Chapter 6). The fossils were obtained largely
from a red and green mudstone bed occurring locally at
an elevation of 100 m, and from mudstone rip-up clasts
almost certainly derived from this bed. Unfortunately,
these mudstones have not been observed in the type
section of the Stadium Conglomerate, and it is
uncertain whether this local fauna should be assigned to

the lower or upper member of this formation (MS-10 in Walsh et al., this volume, Chapter 6). If only the 5-10 m of clast-supported conglomerate in the uppermost part of the Stadium Conglomerate at Stonecrest is correlative with the type upper member, then the Stonecrest local fauna would be assignable entirely to the lower member, in which case the latter would span the early Uintan-late Uintan boundary. Alternatively, a 7-m-thick interval of the Stadium Conglomerate was not observed at Stonecrest, and it is possible that the contact between the lower and upper members occurs below the Stonecrest local fauna. In either case, the Stonecrest local fauna is of late Uintan age, and represents the lowest known stratigraphic occurrence in San Diego of several micromammal taxa typical of the late Uintan part of the Sespe Formation in Ventura County (e.g., *Sespedectes*, *Proterixoides*, *Microparamys woodi*, and *Simimys*).

Cloud 9 Fauna

Fossil mammals from the Mission Valley Formation (as restricted stratigraphically and geographically by Walsh et al., this volume, Chapter 6) have been previously collected from several localities in the eastern part of the greater San Diego area, in La Mesa, San Carlos, Fletcher Hills, and East San Diego (Lillegraven and Wilson, 1975; Golz and Lillegraven, 1977; Walsh, 1987). Walsh (1987) used the term "Cloud 9 Fauna" to designate the late Uintan assemblage collected from these "southern outcrops of the Mission Valley Formation" (also mentioned by Kelly, 1990). Walsh (1991b) proposed a more inclusive geographic term for the "Cloud 9 Fauna," and reduced it in rank to a local fauna. However, in 1992 and 1994, diagnostic late Uintan mammals were finally collected from the Mission Valley Formation in the Miramar Reservoir area, from the "Scripps Ranch North" housing development (Appendix 1). These were obtained from several localities in fluvio-deltaic and shallow marine lithofacies, and include *Sespedectes singularis*, *Proterixoides davisi*, *Simimys simplex*, *Eohaplomys* sp. cf. *E. tradux*, *Procyonodictis progressus*, *Protoreodon* new sp. 1, and *Protylopus* sp. (207/1592 morph).

Since the mammal assemblage now known from the Mission Valley Formation in the Poway area is indistinguishable from that in the La Mesa area, the term used by Walsh (1991b) to refer to the latter no longer has any geographic significance, and is therefore abandoned. In addition, the rank of local fauna is no longer appropriate for the combined assemblage from both the southern and northern outcrops of the Mission Valley Formation. In view of these factors, the term "Cloud 9 fauna" of Walsh (1987) is resurrected here and used to designate the entire set of vertebrate taxa contained in all outcrops of the Mission Valley Formation (Table 2). The Cloud 9 fauna correlates

faunally (and presumably temporally) with SDSNH Loc. 3378 and the Jeff's Discovery and Rancho del Oro local faunas of northwestern San Diego County (Member C, Santiago Formation).

An ^{40}Ar/^{39}Ar date of 42.83 ± 0.24 Ma has been obtained by J. D. Obradovich on bentonite from the Mission Valley Formation (Obradovich and Walsh, in prep.; the previous figure of 42.18 Ma cited in Berry, 1991, and Walsh, 1991b, is incorrect). This bentonite occurred 2 m stratigraphically above the now-inaccessible SDSNH Loc. 3428 in La Mesa, which yielded isolated teeth of the characteristic late Uintan taxa *Sespedectes singularis* and *Simimys simplex*. Presumably the same bentonite bed is present in the Mission Valley Formation in a now-landscaped roadcut on the north side of Murray Drive in La Mesa, immediately east of State Route 125 and immediately south of Briercrest Park. Mudstones from about 2 m stratigraphically below the bentonite have yielded numerous teeth of *Sespedectes* and *Simimys* (SDSNH Loc. 3539). These are the only known sites in California in which Eocene mammals are directly associated with a radiometric date. The relevance of this date to the chron correlation of the Mission Valley Formation is discussed by Walsh et al. (this volume, Chapter 6).

Tentative assignment of the late Uintan Mission Valley Formation to coccolith Subzone CP14a (Walsh et al., this volume, Chapter 6) is compatible with Westgate's (1988) correlation of the late Uintan Casa Blanca local fauna (Laredo Formation, Texas) with nannoplankton Zone NP16 (Berggren et al., 1995). Westgate also noted that stratigraphic units in Texas equivalent to the Laredo Formation contain bentonites that yield average K-Ar ages of 42.2 Ma. Thus, two nominally late Uintan assemblages from California and Texas appear to be synchronous within the resolution of one nannoplankton subzone, and nearly synchronous on the basis of radiometric evidence.

Eastview Local Fauna

The first identifiable mammals from the Pomerado Conglomerate were collected in 1989 from the Eastview housing development, in the northeastern part of the community of Scripps Ranch (SDSNH Loc. 3493). This locality was previously assigned to the upper part of the Stadium Conglomerate by Walsh (1991b). However, extensive grading operations at the nearby "Scripps Ranch North" development in 1991-1992 and 1994-1995 have clarified the stratigraphy of this area. It is now evident that SDSNH Loc. 3493 occurred in the upper part of the lower conglomerate member of the Pomerado Conglomerate, and that the white sandstone unit immediately overlying this locality, which Walsh (1991b) had assumed to be the Mission Valley Formation, is in fact a previously unmapped (Kennedy and Peterson, 1975, plate 2B) eastern extension of the

Miramar Sandstone Member of the Pomerado Conglomerate.

The small assemblage from SDSNH Loc. 3493 was designated the Eastview local fauna by Walsh (1991b). In 1992, a taxonomically more diverse vertebrate locality (SDSNH Loc. 3755) in the lower member of the Pomerado Conglomerate was discovered at the Scripps Ranch North housing development, about 1.5 km west of Loc. 3493. The fossils from both of these localities were obtained from greenish and rust-stained (Loc. 3493) and reddish-brown (Loc. 3755) rip-up clasts of sandy mudstone, in sandstones lenses in the upper part of the lower conglomerate member of the Pomerado Conglomerate (MS-6 in Walsh et al., this volume, Chapter 6). In view of their similar geologic and stratigraphic occurrence, and in the absence of any evidence of faunal heterogeneity, Loc. 3755 is included in the Eastview local fauna. The fossil-bearing rip-up clasts are lithologically similar to locally extensive beds of greenish and reddish siltstones and mudstones that crop out in the lower part of the Mission Valley Formation at Scripps Ranch North (e.g., SDSNH Loc. 3870), and it is possible that the former were derived from the latter. A combined mammalian faunal list is given in Table 2. These taxa indicate a late Uintan age, but due to the possibility of reworking, the Eastview local fauna does not necessarily reflect the age of the lower member of the Pomerado Conglomerate. Nevertheless, the occurrence in the overlying Miramar Sandstone Member of *Protoreodon* new sp. 1, *Leptoreodon* sp., and *Miacis* sp. cf. *M. hookwayi* at SDSNH Loc. 3757 (included here in the Eastview local fauna) demonstrates that the lowermost two members of the Pomerado Conglomerate are of late Uintan age, and probably older than the Laguna Riviera local fauna, which is characterized by *Protoreodon annectens* (Golz, 1976; Kelly et al., 1991).

Jeff's Discovery Local Fauna

The Jeff's Discovery local fauna was named by Walsh (1991b) for the fossil assemblage from SDSNH Locs. 3276 and 3560-3564, from Member C of the Santiago Formation. This important late Uintan locality in Oceanside was discovered by Mr. Jeff Dahlgren and brought to the attention of SDSNH personnel in 1986. The locality was further exposed by highway construction in 1991. Thanks to the financial support of the California Department of Transportation (CALTRANS), thousands of vertebrate specimens were salvaged from this site. About 15,000 kg of fossiliferous matrix has been screenwashed and floated using heavy liquids, and the resulting concentrates have been picked and curated. The beds yielding the Jeff's Discovery local fauna fill a broad paleochannel cut into at least 16 m of light gray, mostly fine- to very fine-grained, marine(?) sandstones (Figure 5). The base of this channel probably represents the disconformable

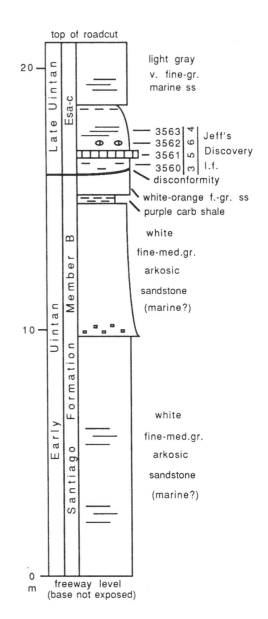

Figure 5. Measured section showing stratigraphic occurrence of the late Uintan Jeff's Discovery local fauna. Measured by S. L. Walsh and B. O. Riney.

contact between Members B and C of the Santiago Formation (Wilson, 1972).

Jeff's Discovery is the most taxonomically-diverse mammal site of late Uintan age known in southern California (Table 2). It contains two creodonts not previously recorded from southern California (cf. *Apataelurus* and *Limnocyon*), and shares with the Cloud 9 fauna the only southern California late Uintan records of the artiodactyls *Antiacodon*? and *Ibarus*. It has produced a variety of interesting small taxa, such as a large new species of the primate *Ourayia*, two species of

Oligoryctes, two species of bats, and the first record of *Sciuravus powayensis* from the late Uintan of California.

Rancho del Oro Local Fauna

The Rancho del Oro local fauna was named by Walsh (1991b) for the mammal assemblage collected in 1988 from SDSNH Locs. 3431-3437, 3441-3442, 3444-3445, 3449, 3451, 3463-3464, and 3466-3467 (Locs. 3438, 3439, and 3447 were erroneously included in this list by Walsh, 1991b). All of these late Uintan sites were collected from Member C of the Santiago Formation in Oceanside, during grading for the Rancho del Oro housing development (Villages II and III) in 1988. These now-destroyed localities occurred in various superposed and complexly interfingered terrestrial lithofacies of Member C, from 0 to 30 m above the disconformable Member B-Member C contact (B. O. Riney, field notes on file at SDSNH). A revised mammalian faunal list from the Rancho del Oro local fauna is given in Table 2. This local fauna shares with the Cloud 9 fauna the first record of the carnivore *Miocyon* from southern California, and is the only undoubted late Uintan local fauna in southern California that has produced a brontothere. This specimen (SDSNH 3433/40815; maxilla fragment with damaged teeth) pertains to a species about the same size as *Duchesneodus uintensis*, but differs from *D. uintensis* in the relative proportions of the upper premolars. Specimens of the primate *Dyseolemur pacificus* from Rancho del Oro are described by Rasmussen et al. (in press).

SDSNH Loc. 3378

A taxonomically limited mammal assemblage was collected in Carlsbad in 1987 from the base of Member C of the Santiago Formation, during grading for the extension of College Boulevard between El Camino Real and Palomar Airport Road. The fossils were obtained from several large reddish and purplish mudstone clasts (SDSNH Loc. 3378) that occurred in a greenish-gray, sharktooth-bearing arkosic gritstone unit of deltaic origin. This gritstone also contains large granitic boulders, and lies disconformably upon white, fine-grained marine sandstones of Member B of the Santiago Formation. The stratigraphic section shown in Figure 6 occurs in fault contact with at least 50 m of greenish marine siltstones of uncertain age. Although the rip-up clasts that comprise SDSNH Loc. 3378 are apparently of late Uintan age, the enclosing gritstone may be late Uintan and/or Duchesnean in age. The latter unit is very similar in lithology to the basal gritstone of Member C at Aviara (SDSNH Loc. 3487), which is located only 3.5 km south of SDSNH Loc. 3378. Stratigraphic equivalence of these coarse-grained horizons seems likely, but cannot be demonstrated due to faulting and lack of continuous outcrops.

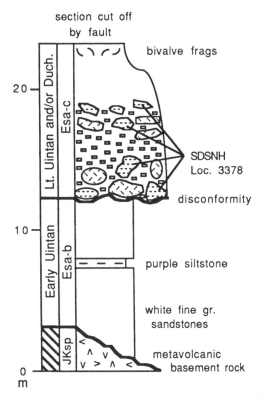

Figure 6. Generalized section showing stratigraphic occurrence of SDSNH Loc. 3378 (late Uintan). Measured by B. O. Riney and T. A. Deméré. JKsp = Santiago Peak Volcanics.

A faunal list from SDSNH Loc. 3378 is given in Table 2. This assemblage is assigned to the late Uintan (rather than early Uintan) on the basis of the presence of *Protoreodon* new sp. 1, *Protylopus stocki*, and a relatively small species of *Leptoreodon*, intermediate in size between *L. pusillus* and *L. leptolophus* (Golz, 1976). This locality shares a record of *Simidectes* sp. cf. *S. medius* with the Cloud 9 fauna, contains a species of *Ourayia* much smaller than the species of this genus known from Jeff's Discovery, and has yielded the most complete skull of a new tapiroid genus from the southern California Eocene (Colbert, 1993).

LATEST UINTAN AND/OR DUCHESNEAN MAMMAL ASSEMBLAGES OF SAN DIEGO COUNTY

Kelly (1990), Kelly et al. (1991), and Lucas (1992) discussed the faunal characterization of the Uintan/Duchesnean boundary in California and throughout North America. Of the several taxa used by these authors to characterize the beginning of the Duchesnean, only *Duchesneodus*, *Amynodontopsis*, and the rodent *Presbymys* are known from San Diego County. Lucas (1992) regarded the brontothere *Duchesneodus* as an index taxon for the Duchesnean, and this genus is definitely known from two localities in San Diego

County (SDSNH Locs. 3235 and 3495). Fragmentary remains of a brontothere about the same size as *D. uintensis* are known from two other localities (SDSNH Loc. 3487, discussed above, and the Mission del Oro local fauna, discussed below), so the strata in which they occur may prove to be of Duchesnean age.

Laguna Riviera Local Fauna

The Laguna Riviera local fauna was named by Golz (1976) for the mammals collected from several associated localities in the Laguna Riviera housing district in Carlsbad. These were all obtained from Member C of the Santiago Formation, but apparently from a higher stratigraphic level within this member than the Jeff's Discovery or Rancho del Oro local faunas. Due to lack of continuous outcrops, however, direct stratigraphic superposition of the Laguna Riviera local fauna above these other local faunas cannot be demonstrated. The faunal list in Table 2 was compiled from Golz and Lillegraven (1977), with modifications discussed below.

Golz (1976) suggested that the Laguna Riviera local fauna was correlative with the late Uintan Brea Canyon local fauna of the Sespe Formation. However, the geographical and taxonomic similarity between the Laguna Riviera local fauna and the assemblage from SDSNH Loc. 3495 (see below) suggests the possibility of a Duchesnean age for the former. Corroboration of this hypothesis must be based on the presence of the brontothere *Duchesneodus* or other Duchesnean indicators (Kelly, 1990; Lucas, 1992). Unidentified brontothere material is known from Laguna Riviera (Golz and Lillegraven, 1977, table 1), but is currently unavailable for study. Most of the UCMP specimens from Laguna Riviera identified as "*Amynodon* sp." by Golz and Lillegraven (1977, Table 1) also have not been examined. However, a partial skull of *Amynodontopsis bodei* from UCMP V-68102 (UCMP 124385; W. Wall, personal communication) is consistent with a Duchesnean age for Laguna Riviera, as this species is otherwise known in southern California only from the Duchesnean of Ventura County (Kelly, 1990) and another Duchesnean locality in San Diego County (SDSNH Loc. 3235). A Duchesnean age for Laguna Riviera is further suggested by the presence of *Simidectes merriami*, which is otherwise known in southern California only from the Pearson Ranch local fauna (Coombs, 1971; Kelly, 1990). *Microparamys tricus* was not recorded from Laguna Riviera by Golz and Lillegraven (1977), but four isolated teeth (UCR 14096 and 22201-22203) from RV-6830 clearly pertain to this species, which is otherwise known only from Pearson Ranch (Kelly, 1990) and the latest Uintan or Duchesnean Camp San Onofre local fauna of the Santiago Formation (Lillegraven, 1977; however, Kelly and Whistler, 1994, record *Microparamys* sp. cf. *M. tricus* from the presumably late Uintan Brea Canyon

local fauna). Following Kelly et al. (1991), the previous record of *Protoreodon pumilus* (Golz, 1976) from Laguna Riviera is modified here to *Protoreodon annectens*. It is interesting to note that Kelly et al. (1991) interpreted the Laguna Riviera specimens of *P. annectens* to be more derived than those from the Brea Canyon local fauna of the Sespe. Finally, the Laguna Riviera local fauna is noteworthy in that it contains the only confident southern California Eocene records of *Mytonolagus* and *Poebrodon* (Golz, 1976).

LACM 3881 ("Carlsbad Palisades") occurs in the Laguna Riviera area (unpublished LACM locality description), but was not specifically included in the Laguna Riviera local fauna by Golz (1976, p. 5). It has produced only one identifiable mammal specimen, a lower molar assigned by Golz and Lillegraven (1977) to *Amynodon* sp. cf. *A. intermedius* (*A. intermedius* was regarded as a junior synonym of *A. advenus* by Wilson and Schiebout, 1981). This specimen (LACM 71218) appears to be conspecific with the large, deep-jawed amynodont from SDSNH Loc. 3487 at Aviara, and may indicate that both *Amynodon* and *Amynodontopsis* coexisted in southern California during the latest Uintan/early Duchesnean. The Chestnut Avenue local fauna (Golz, 1976, table 1, p. 5) occurs a short distance to the north of the Laguna Riviera local fauna, and perhaps at a stratigraphically somewhat higher level within Member C (Kelly, 1990). It has produced only two identifiable taxa (cf. *Amynodon* and *Simimeryx* cf. *S. hudsoni*), and is either of latest Uintan or early Duchesnean age (Golz, 1976, fig. 4; Kelly, 1990, pp. 32-33).

A small collection of vertebrates from Member C of the Santiago Formation was obtained from a series of arkosic sandstones and mudstones similar to those at Laguna Riviera, but about 1.5 km south of the latter (SDSNH Loc. 3495). Although an approximate stratigraphic equivalence for these two localities seems probable, the presence of Agua Hedionda Lagoon between them does not allow direct corroboration of this hypothesis. The mammal assemblage from SDSNH Loc. 3495 includes the following mammal taxa.

Tapiroidea, new genus, species A
Duchesneodus uintensis
Protoreodon sp. cf. *P. annectens*
Protylopus stocki
Leptoreodon leptolophus
Leptoreodon sp. cf. *L. stocki*

The presence of *Duchesneodus uintensis* suggests, by definition (Lucas, 1992), a Duchesnean age for SDSNH Loc. 3495. Lucas and Schoch (1989) regard *D. californicus* Stock, 1935 as a junior synonym of *D. uintensis* (Peterson, 1931), and I follow this interpretation here. SDSNH Loc. 3495 has at least two, and perhaps three artiodactyl species in common with the Laguna Riviera local fauna, suggesting a close temporal equivalence, and therefore the possibility of a Duchesnean age for the latter. If so, SDSNH Loc. 3495 could be readily included in the definition of the Laguna Riviera local fauna.

Camp San Onofre Local Fauna

The Camp San Onofre local fauna (Golz and Lillegraven, 1977; Walsh, 1987; Kelly, 1990) was collected by Lillegraven and his colleagues in the early 1970's from UCMP Loc. V-72088. The fossil-bearing strata crop out on the Camp Pendleton Marine Corps base in northwestern San Diego County, and presumably pertain to Member C of the Santiago Formation. The faunal list in Table 2 was compiled from Golz and Lillegraven (1977), with the following modifications. The previous record of *Sespedectes singularis* from V-72088 (Golz and Lillegraven, 1977) is changed to *S. stocki* given the taxonomic revision of Novacek (1985). Chiment (1977) assigned several isolated teeth from V-72088 to "*Namatomys*" new sp. 2. Finally, the previous record of *Amynodon reedi* from this locality (Golz and Lillegraven, 1977, Table 1) is changed here to *Amynodon* sp. cf. *A. advenus* based on examination of the relevant specimens (e.g., UCMP 98651, 113188, and 113190). The precise age of the Camp San Onofre local fauna is unclear, but is probably no older than latest Uintan based on the presence of *Chumashius*, *Microparamys tricus*, and *Simimeryx* sp. cf. *S. hudsoni* (e.g., Walsh, 1987; Kelly, 1990). Specific identification of the *Protoreodon* material from Camp San Onofre may clarify the age of this local fauna.

Mission del Oro Local Fauna

An interesting vertebrate assemblage was collected in 1991 from Member C of the Santiago Formation, during grading for the Mission del Oro housing project, southeast of the intersection of Rancho del Oro Drive and State Route 76, in the eastern part of Oceanside. The assemblage from SDSNH Locs. 3570, 3572, and 3574 is here named the Mission del Oro local fauna. SDSNH Loc. 3573 is tentatively excluded, because it has produced only *Simimys* sp., and it occurs stratigraphically below the lowermost of the distinctive coarse-grained arkosic sandstone channels and associated brownish gritty mudstones that produce the Mission del Oro local fauna. SDSNH Loc. 3571 occurs within Member B of the Santiago Formation, immediately below the "Oyster Marker Bed," and has produced *Sciuravus powayensis* and "*Namatomys*" new sp. 1 (this locality was assigned above to the early Uintan Mesa Drive local fauna). The stratigraphic context of the Mission del Oro local fauna is provided in Figure 7, and a composite faunal list is given in Table 2. The presence of *Protylopus* sp. cf. *P. robustus* suggests a correlation with the Laguna Riviera local fauna, and

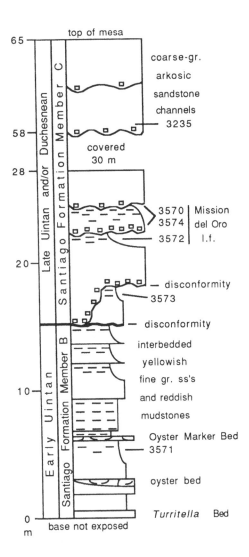

Figure 7. Measured section showing stratigraphic occurrence of the latest Uintan or Duchesnean Mission del Oro local fauna, and SDSNH Loc. 3235 (Duchesnean). Measured by S. L. Walsh and B. O. Riney.

together with cf. *Duchesneodus* sp., a latest Uintan or Duchesnean age for the Mission del Oro local fauna. A single specimen assigned to *Epihippus* sp. (M. W. Colbert, personal communication) represents the first confident record of Equidae from the southern California Eocene.

About 30-35 m stratigraphically above the level of the Mission del Oro local fauna, specimens of *Duchesneodus uintensis* and *Amynodontopsis bodei* have been obtained from SDSNH Loc. 3235 (Figure 7), which is approximately equivalent to LACM Loc. 4022 ("Oceanside East"; no mammals identifiable to generic level). Future collecting from this Duchesnean horizon could allow it to be included within the Mission del Oro local fauna, or established as a new local fauna.

Bonita Local Fauna

Walsh (1991a) named the Bonita local fauna for the assemblage recovered from SDSNH Localities 3471, 3550, and 3608, in the lower part of the "Sweetwater" Formation in southwestern San Diego County. The name "Sweetwater" is preempted (Keroher et al., 1966, p. 3791), so a new lithostratigraphic name will have to be proposed for this unit. Walsh (1991a) assigned a latest Uintan and/or Duchesnean age to the Bonita local fauna on the basis of the presence of *Sespedectes*, *Simimys*, *Presbymys*, and a small unidentified ?cylindrodontid rodent. Since then, two new localities in the lower part of the "Sweetwater" Formation (SDSNH Locs. 3821 and 3823) have produced isolated teeth of a relatively large species of *Pareumys* and an unidentified species of "*Namatomys*." A revised list of mammal taxa from the Bonita local fauna is given in Table 2. The presence of the cylindrodontid rodent *Presbymys* from the Bonita local fauna represents the first record of this genus from San Diego County, although the Bonita species may be significantly smaller than *P. lophatus* from the Duchesnean Pearson Ranch local fauna of the Sespe Formation. There is insufficient evidence to decide whether the Bonita local fauna should be assigned to the latest Uintan or the Duchesnean.

COMPARISON AND CORRELATION OF UINTAN/DUCHESNEAN FAUNAS OF CALIFORNIA

Correction of Previous Statements

Before discussing faunal comparisons and age correlations of the mammal taxa of southern California, it is necessary to correct some errors in the age ranges of certain California taxa that have been caused by Kennedy and Peterson's (1975) erroneous (but understandable) assignment of certain strata in San Diego to the Mission Valley Formation (see Walsh et al., this volume, Chapter 6). First, the early Uintan vertebrate localities from the "Mission Valley Formation" in the Rancho de los Peñasquitos district that were discussed by Golz and Lillegraven (1977) are actually located in the upper tongue of the Friars Formation. In contrast, the type Mission Valley Formation and equivalent outcrops of this unit in the La Mesa-Fletcher Hills area are of late Uintan age. Golz and Lillegraven (1977) and Novacek and Lillegraven (1979) were fully aware of the profound faunal differences between the Rancho de los Peñasquitos and La Mesa districts. At that time, however, the superpositional evidence required to demonstrate that these districts pertained to two different stratigraphic units was largely unavailable (Walsh et al., this volume, Chapter 6).

In the hope of permanently ending this confusion, I will cite and then correct some examples of the unintentional equivocations evident in recent discussions of the Uintan faunas of San Diego. These include statements such as:

Golz and Lillegraven (1977) record *Simidectes* sp. cf. *S. medius* at LACM 65190 in the Mission Valley Formation, which they correlate with the Wagonhound or early Uintan (Wilson, 1986, p. 361).

Younger records of *Hemiacodon* have been reported by Lillegraven (1980) from the early Uintan Mission Valley Formation of San Diego County (Beard et al., 1992, p. 53).

Although Golz and Lillegraven's (1977) record of *Simidectes* sp. cf. *S. medius* from the Mission Valley Formation is still valid, this lithostratigraphic unit is of *late* Uintan age, and the San Diego records of *Hemiacodon* cited by Beard et al. (1992) are from the early Uintan upper tongue of the Friars Formation.

Stucky's (1992, table 24.1) "earliest Uintan" records of *Simidectes*, *Eohaplomys*, and *Simimys* are explained by the occurrence of these genera in the (now restricted) Mission Valley Formation, combined with the early Uintan age of the Rancho de los Peñasquitos outcrops that were formerly assigned to this formation (as noted above, however, a new species of *Eohaplomys* is now known from the late early Uintan Murray Canyon local fauna). Finally, it should be noted that the genera *Tapochoerus* and *Craseops* are listed by Stucky (1992, table 24.1) as being known only from the Duchesnean, when in fact these taxa are known only from the late Uintan Tapo Canyon and Brea Canyon local faunas of the Sespe Formation (Kelly, 1990).

Faunal Comparison of Late Uintan/Early Duchesnean Assemblages from San Diego and Ventura Counties

Lillegraven (1979) concluded that no significant barriers to the dispersal of land vertebrates existed in southern California during the middle Eocene, and considered the Uintan/Duchesnean mammal faunas of San Diego and Ventura counties to belong to a single biogeographical entity. This is a reasonable assumption, since both areas are located west of the San Andreas fault, and are currently separated by a distance of at most 300 km. Due to the absence of known early Uintan assemblages from Ventura County, only the late Uintan and Duchesnean assemblages of the two areas can be compared. As discussed below, the precise position of the Uintan/Duchesnean boundary in the Sespe Formation is unclear, and several local faunas of San Diego County can only be assigned a "latest Uintan or Duchesnean" age. Therefore, a meaningful comparison of "strictly late Uintan" and "strictly Duchesnean" assemblages from San Diego and Ventura counties is difficult. Accordingly, Table 3 lists all known taxa from the late Uintan and/or early Duchesnean of both regions. All Sespe Formation taxa from the Tapo Canyon and Brea Canyon areas from stratigraphic levels up to and

Table 3. Geographic distribution of late Uintan and early Duchesnean species-level mammal taxa between San Diego and Ventura counties, California. Compiled from Golz and Lillegraven (1977), Lillegraven (1980), Kelly (1990, 1992), Kelly et al. (1991), Kelly and Whistler (1994), Mason (1988, 1990), and this paper.

Species-level taxa known from the late Uintan and/or early Duchesnean of San Diego County, but not from the late Uintan and/or early Duchesnean of Ventura County.	Species-level taxa known from the late Uintan and/or early Duchesnean of both San Diego and Ventura counties.	Species-level taxa known from the late Uintan and/or early Duchesnean of Ventura County, but not from the late Uintan and/or early Duchesnean of San Diego County.
Peratherium sp. cf. *P. knighti*	*Peratherium* sp. cf. *P. innominatum*	*Macrotarsius roederi*
Apatemys uintensis	*Peradectes californicus*	*Mytonius travisi*
Sespedectes stocki	*Apatemys downsi*	*Craseops sylvestris*
Scenopagus sp.	*Aethomylos simplicidens*	*Paramys compressidens*
cf. *Nyctitherium* sp.	*Sespedectes singularis*	*Tapomys tapensis*
Oligoryctes sp. 1	*Proterixoides davisi*	*Leptotomus* sp. cf. *L. leptodus*
Oligoryctes sp. 2	*Centetodon* sp., prob. new	*Mytonomys burkei*
Simidectes sp. cf. *S. medius*	*Batodonoides powayensis*	*M.* sp. cf. *M. mytonensis*
Chiroptera, unident. gen. and sp. 1	*Simidectes merriami*	*Eohaplomys matutinus*
Chiroptera, unident. gen. and sp. 2	*Dyseolemur pacificus*	*Presbymys lophatus*
Ourayia new sp. 1	*Chumashius balchi*	*"Namatomys" fantasmus*
Ourayia sp. 2	*Uintasorex* sp. cf. *U. montezumicus*	*Hyaenodon vetus*
Leptotomus sp. unident.	*Phenacolemur* sp. cf. *P. shifrae*	*Hyaenodon venturae*
Manitshini? unident. gen. and sp.	*Microparamys woodi*	*Tapochoerus egressus*
Sciuravus powayensis	*Microparamys tricus*	*Protoreodon pacificus*
Eohaplomys new sp.	*Rapamys fricki*	*Protylopus pearsonensis*
Presbymys sp., poss. new	*Eohaplomys tradux*	*Eotylopus* sp.
Griphomys toltecus	*Eohaplomys serus*	*Leptoreodon edwardsi*
"Namatomys" new sp.	*Pareumys* sp. cf. *P. milleri*	*Simimeryx* new sp.
cf. *Paradjidaumo* sp.	*Griphomys alecer*	
Mytonolagus sp.	*Simimys simplex*	
Limnocyon sp.	*Tapocyon occidentalis*	
cf. *Apataelurus* sp.	*Procyonodictis progressus*	
Miocyon sp. cf. *M. scotti*	*Miacis? hookwayi*	
mesonychid, unident. sp.	Tapiroidea, new gen., sp. A	
Epihippus sp.	*Triplopus? woodi*	
Ibarus sp. cf. *I. ignotus*	*Amynodon* sp. cf. *A. advenus*	
Antiacodon? sp. 2	*Amynodontopsis bodei*	
Protoreodon new sp. 1	*Duchesneodus uintensis*	
Protylopus stocki	*Protoreodon annectens*	
Protylopus? sp.	*Protylopus* sp. cf. *P. petersoni*	
Leptoreodon leptolophus	*P. sp.* (207/1592 morph)	
Poebrodon californicus	*P. robustus*	
	Leptoreodon pusillus	
	Leptoreodon stocki	
	Simimeryx hudsoni	

18 genera, 33 species	35 genera, 36 species	9 genera, 19 species

San Diego County Totals: 53 genera, 69 species.   Ventura County Totals: 44 genera, 55 species.

Simpson Coefficients: Genus level = 35/44 X 100% = 80%. Species level = 36/55 X 100% = 65%.

Total southern California late Uintan + early Duchesnean diversity: 62 genera, 88 species.

including the early Duchesnean locality LACM (CIT) 150 (Pearson Ranch) are listed in Table 3. Taxa from the late Duchesnean Simi Valley Landfill local fauna (Kelly et al., 1991) are not included here (although strata of this age may be present in San Diego County in the upper part of the "Sweetwater" Formation; Walsh, 1991a).

The left hand column of Table 3 shows that 18 genera and 33 species are known from the late Uintan and/or early Duchesnean of San Diego County, but not

Ventura County. However, most of these taxa are known only from one or a few specimens, and their apparent absence from the Sespe Formation can probably be attributed to sampling error. Exceptions to this rule may include *Protoreodon* new sp. 1 and *Leptoreodon leptolophus*, which are common in certain horizons in San Diego County. The central column in Table 3 shows that 35 genera and 36 species are known from the late Uintan and/or early Duchesnean of both areas. The right hand column shows that 9 genera and

19 species are known from the late Uintan and/or early Duchesnean of Ventura County, but not San Diego County. Again, most of these taxa are known from only a few specimens each, although exceptions include *Craseops sylvestris*, *Tapochoerus egressus*, and *Leptoreodon edwardsi* (late Uintan Tapo Canyon local fauna), and *Hyaenodon* and *Protoreodon pacificus* (early Duchesnean Pearson Ranch local fauna). Nevertheless, it seems likely that some of these taxa will eventually be detected in San Diego County, especially since faunas of unequivocal Duchesnean age are poorly known from this area.

A semiquantitative estimate of the overall taxonomic similarity between the late Uintan-early Duchesnean assemblages of San Diego and Ventura counties is desirable. Simpson (1960) advocated an index of faunal resemblance for two assemblages based on the number of taxa (of a given rank) in common, expressed as a percentage of the total number of taxa in the least diverse assemblage (for a discussion of the "Simpson Coefficient," see Flynn, 1986b, and references therein). Given the distributions in Table 3, a Simpson Coefficient of 80% (35/44 X 100%) is obtained for the generic level, and 65% (36/55 X 100%) is obtained for the species level. These values indicate a high taxonomic similarity of the composite late Uintan/early Duchesnean assemblages from San Diego and Ventura counties.

Age Correlation of Uintan Assemblages of San Diego and Ventura Counties

Age correlations of the Uintan local faunas of San Diego and Ventura counties have been discussed by Golz and Lillegraven (1977), Walsh (1987), Kelly (1990), and Walsh (1991b). It may eventually be useful to recognize provincial mammalian biochrons for the Uintan of California, defined on the basis of the local temporal ranges of particularly common and distinctive taxa. This would afford a terminology for precise correlation in California, without invoking the implications of continent-wide isochrony inherent in the use of such awkward terms as "late early" and "early late" Uintan. The superposed mammal assemblages from the lower and middle members of the Sespe Formation of Ventura County (Kelly, 1990; Kelly et al., 1991) would be ideal for defining the later Uintan biochrons of California, but early Uintan and "late early Uintan" assemblages have not been recognized from the Sespe. Mammal assemblages from the disconformity-ridden middle Eocene sequence of San Diego County would therefore be used to define the older Uintan biochrons.

Before these biochrons are formally recognized, however, several problems must be solved. First, a faunal equivalent of the Murray Canyon local fauna should be recognized in either northwestern San Diego County or Ventura County before a biochron based on

this assemblage is proposed. Second, the age of the typical late Uintan assemblages of San Diego County (e.g., Cloud 9 fauna and Jeff's Discovery and Rancho del Oro local faunas) relative to the Tapo Canyon and Brea Canyon local faunas must be established. The reported occurrence of *Protoreodon annectens* in the Sespe assemblages, versus the occurrence of the morphologically more primitive *Protoreodon* new sp. 1 in the San Diego local faunas, may indicate a somewhat older age for the latter (E. B. Lander, personal communication; cf. Kelly, 1990). Indeed, Walsh et al. (this volume, Chapter 6) correlate the late Uintan magnetozones in the type section of the Mission Valley Formation in San Diego with Chrons C20n and C19r, while Prothero et al. (this volume, Chapter 8) assign the late Uintan assemblages from the Sespe Formation to C18r. These correlations imply that the typical late Uintan assemblages of San Diego County are one chron older (and at least 1.3 m.y. older, according to the time scale of Berggren et al., 1995) than the Ventura County local faunas. I am skeptical of this conclusion, since many species co-occurring in these assemblages suggest a close temporal equivalence (Table 3). However, there is no *a priori* reason to suppose that the late Uintan assemblages of southern California could not remain largely unchanged for more than 1.3 m.y. The poorly known "unnamed local fauna" from the top of the lower member of the Sespe Formation (Kelly et al., 1991) may be of critical importance in determining the relative age of the Ventura County and San Diego County late Uintan assemblages.

Third, it is not clear whether the faunal differences between the Tapo Canyon and Brea Canyon local faunas of the Sespe Formation are significant enough to recognize two different biochrons based upon each. It seems probable that some of the taxa known only from the Tapo Canyon local fauna (e.g., *Apatemys*, *Craseops*, *Phenacolemur*, and *Leptoreodon edwardsi*) are absent from the Brea Canyon local fauna due to collecting bias and/or paleoecological factors (e.g., Golz, 1976). Similarly, some of the taxa known only from the Brea Canyon local fauna (*Macrotarsius*, *Mytonomys*, *Rapamys*, *Protylopus* spp., *Leptoreodon pusillus*) may be absent from the Tapo Canyon local fauna for the same reasons. If so, it is possible that a more significant faunal break occurs within the Brea Canyon local fauna (as designated by Kelly et al., 1991) than between the lower part of this local fauna and the Tapo Canyon local fauna.

Finally, the position of the Uintan/Duchesnean boundary within the Sespe Formation should be better established before any "latest Uintan" biochron based on the Sespe assemblages is recognized. Previous workers have correlated the Laguna Riviera local fauna of San Diego County with the late Uintan Brea Canyon local fauna (e.g., Golz, 1976). As discussed above, however, Laguna Riviera may be of Duchesnean age. If so, the

Uintan/Duchesnean boundary could occur within the Sespe Formation as low as the upper part of the stratigraphic interval currently included within the Brea Canyon local fauna. Further collecting from the Sespe Formation may resolve these questions.

PALEOCLIMATE OF THE SOUTHERN CALIFORNIA EOCENE

Can a significant change in the Eocene climate of southern California be correlated with the early Uintan/ late Uintan faunal change discussed above? The sedimentological and pedological evidence for Eocene paleoclimates of San Diego was summarized by Peterson and Abbott (1979). Based on their analysis of a pre-middle Eocene, kaolinite-dominated oxisol developed on the Cretaceous and older rocks of southern California and northern Baja California, these authors concluded that the Paleocene-early Eocene climate was warm (mean annual temperature 20-25°C) and humid (at least 125-190 cm of average annual rainfall). In contrast, based in part on the common caliche horizons in the Friars and Mission Valley formations and the immaturity of the clay mineral suite within these units, Peterson and Abbott (1979) concluded that the Uintan climate of southern California was warm (18-20°C average annual temperature) and semiarid (less than 63 cm of average annual rainfall). Unfortunately, terrestrial floodplain deposits are unknown in the Wasatchian and Bridgerian strata of southern California, so it cannot be determined if pedogenic caliche horizons were capable of forming during this time. Since Peterson and Abbott (1979) assumed the Friars and Mission Valley formations to be of late Eocene age (the Uintan was generally regarded as "late Eocene" at that time), they correlated the climate change recognized in San Diego with the late middle Eocene global decrease in ocean temperatures reported by Savin et al. (1975).

Peterson and Abbott's (1979) conclusions were discussed in the context of the Uintan vertebrate faunas of southern California by Schiebout (1979), Lillegraven (1979), and Novacek and Lillegraven (1979). The latter authors observed that these faunas must have required a substantial amount of water, and speculated that (p. 77):

the overall composition of the Friars-Mission Valley assemblage suggests a complex environment similar to that of present-day East Africa, where riparian habitats grade into gallery forests and/or savanna within a few kilometers. Such regions today are characterized by: (1) very high species densities for small and large mammals (Bouliére, 1963); (2) average annual rainfall and temperature ranges of 50-100 cm and 20°-22°C, respectively; and (3) marked seasonality with intense periods of aridity. It is also noteworthy that droughts on recent savannas occasionally extend through several years, and some of the ancient

caliche horizons in the San Diego deposits may represent transgressions of aridity beyond normal seasonality.

Since 1979, paleobiological evidence for the Eocene paleoclimates of southern California has increased, and the timing of its changes has improved. Elsik and Boyer (1977) studied palynomorphs from the Bridgerian Delmar Formation and the Bridgerian or earliest Uintan Torrey Sandstone, and reported that their assemblage contained several elements indicating a subtropical to tropical climate. This conclusion was corroborated by a recent study of plant megafossils from the Torrey Sandstone. Myers (1991, p. 201) states:

Living taxa similar to Torrey flora species inhabitant warm temperate to paratropical forests of southeastern China, Mexico, and the southeastern United States. The taxonomic composition and foliar physiognomy of Torrey flora fossils indicate a warm (20°C mean annual temperature), equable (less than 8°C mean annual temperature range), frostless climate, with annual rainfall of between 120 and 150 cm, concentrated during the months of April through September.

The same climate appears to have persisted into Ardath Shale time (Bridgerian or earliest Uintan), based on the sporomorph assemblage described by Lowe (1974; quoted in Lillegraven, 1979). Transported land snails from the marine Scripps Formation (late Bridgerian or earliest Uintan) are also consistent with a tropical, equable climate dominated by summer rainfall (Roth and Pearce, 1988; Roth, 1988; fossils described in the latter study were assigned to the Friars Formation, but the outcrop in question is assigned to the upper part of the Scripps Formation by Walsh et al., this volume, Chapter 6).

Paleobotanical evidence is now available for the Uintan climate of San Diego. Estimates of annual rainfall for this time period are somewhat uncertain, because plant megafossils from the Friars Formation (early Uintan) and Mission Valley Formation (late Uintan) have not been studied. No pollen evidence is available from the early Uintan terrestrial strata of San Diego County (e.g., Friars Formation), but a late Uintan pollen assemblage from marine strata near the base of the Mission Valley Formation (locality of Kern, 1978) has been studied by Frederiksen (1989, 1991a,b). Frederiksen (1991b) assigned the Mission Valley Formation to the middle Lutetian, based in part on his correlation of this unit with coccolith Subzone CP13c. However, if the Mission Valley Formation is actually assignable to Subzone CP14a (Walsh et al., this volume, Chapter 6), then, given the time scale of Berggren et al. (1995), Frederiksen's (1989) original late Lutetian/early Bartonian age for the Mission Valley would be correct. Frederiksen (1991b) noted the paleoenvironmental interpretation of Novacek and Lillegraven (1979), and stated (p. 568):

> The angiosperm pollen flora from the Mission Valley Formation is quite rich in diversity, but elements of savanna origin (using the definition of MacGinitie) cannot be separated from elements that may have lived along the streams. Therefore it is difficult to determine on the basis of the pollen flora whether narrow gallery forests existed, perhaps having low diversity, or whether the savanna was 'traversed by streams whose floodplains supported a relatively luxuriant mesic forest of many species' (MacGinitie, 1969, p. 49). There is some indication from angiosperm pollen data that the mean annual temperature (or minimum winter temperature) in the middle Lutetian [Mission Valley time] was at least as high as at the beginning of the Lutetian; the generally paratropical to tropical monosulcate (mainly palm) and Bombacaceae-Sterculiaceae-Tiliaceae pollen groups actually increased slightly in diversity from Delmar-Ardath time (7 and 7 pollen species, respectively) to Mission Valley time (10 and 9 pollen species, respectively).

Based on the pollen evidence, Frederiksen (1991b) suggested that the late Uintan climate (Mission Valley time) was "no dryer than moist subhumid," and noted that relatively little climatic change seems to have occurred from Delmar/Ardath time to Mission Valley time. This conclusion is consistent with the persistence into the latest Uintan of taxa that are associated today with relatively warm, wet climates, such as crocodilians, certain primates, and tapiroids (Colbert, 1993). As seen from Tables 2 and 3, known primate and "plesiadapiform" diversity in southern California did not decrease significantly from the early Uintan (9 genera and 10 species) to the late Uintan (8 genera and 9 species). Unfortunately, samples of Duchesnean micromammals from southern California are relatively small, and do not give an adequate picture of primate diversity during this time.

Frederiksen (1991b) proposed that a significant climate change occurred during the first two-thirds⊐of the Bartonian (roughly Duchesnean; probably soon after deposition of the Mission Valley Formation). He states (p. 569):

> According to angiosperm pollen data, [the "Middle Eocene Diversity Decline"] was marked by far the greatest climatic change during the middle to late Eocene interval in southern California and the Gulf Coast. . . The distinctness of the MEDD is important because it appears to offer a precise timing for the peak of middle Eocene climatic

deterioration. This deterioration was previously known in southern California on the basis of vertebrate fossils and sedimentary features mainly from older rocks.

The MEDD of Frederiksen (1991b) may correspond with the late middle Eocene (approximately planktonic foram Zone P13) oceanic temperature drop inferred by Savin et al. (1975, fig. 6) on the basis of oxygen isotope studies. More recent oxygen/carbon isotope and marine microfaunal studies have provided indications of a slight, gradual cooling from early middle Eocene to late middle Eocene time, punctuated by a relatively abrupt cooling event near the middle Eocene/late Eocene boundary. The latter event is variably correlated with foram Zones P13/14/15 and coccolith Zones NP16/17/18 (e.g., Gaskell, 1991; Miller, 1992; Aubrey, 1992; Keller et al., 1992). McGowran (1990) compiled these and other data and recognized a global late middle Eocene cooling event ("Level E"), which he tied to changes in plate motions in the Indian and Pacific oceans associated with Anomaly 19. Note that the ~44 Ma age of Anomaly 19 implied by McGowran (1990, fig. 6) should be decreased to about 41.4 Ma, according to the time scale of Berggren et al. (1995).

As noted above, Bukry (1991) assigned a coccolith assemblage from Member C of the Santiago Formation to Subzone CP14a (~ Zone NP16), and suggested a correlation with the Cook Mountain Formation of the Gulf Coast (deposited during the late middle Eocene cooling event according to Gaskell, 1991, fig. 8). However, it has not been established that the Aviara strata and/or the Mission Valley Formation (possibly assignable to Subzone CP14a) were also deposited during this event. Subzone CP14a is thought to have lasted about 3 m.y. (Berggren et al., 1995), and deposition of the San Diego County units could have taken place during the early part of the Subzone, before the onset of significant cooling. This scenario is consistent with 1) a correlation of the Mission Valley Formation with Chron C20n/C19r based on an $^{40}Ar/^{39}Ar$ date of 42.83 ± 0.24 Ma (Obradovich and Walsh, in prep.; Walsh et al., this volume, Chapter 6); 2) a somewhat younger age of about 42.0-42.2 Ma for the Cook Mountain Formation (Westgate, 1988; Berggren et al., 1992); 3) the continued presence from Ardath Shale time into Mission Valley time of several molluscan taxa typical of tropical waters (Givens, 1974; Givens and Kennedy, 1979); 4) the occurrence in the late Uintan Jeff's Discovery local fauna of several land snail species consistent with tropical to subtropical conditions (Roth, 1991); and 5) the continued presence of paratropical to tropical pollen groups in southern California during Mission Valley time (Frederiksen, 1991b).

From the available evidence, it appears that the global late middle Eocene cooling event discussed above occurred largely or entirely after the deposition of the late Uintan Mission Valley Formation. Therefore, although this cooling and its possibly associated drying may have contributed to the endemism of the late Uintan mammalian fauna of California (see below), these events apparently had no direct role in the origin of this fauna. Nevertheless, the semi-arid conditions postulated by Peterson and Abbott (1979) for the Uintan of San Diego were cited by Sloan and Barron (1992) as being consistent with certain aspects of their Paleogene climate model for the southwestern margin of North America. The importance of plant megafossil studies in providing independent rainfall estimates for the Uintan of southern California is therefore underscored.

EARLY UINTAN MAMMAL ASSEMBLAGES OF NORTH AMERICA
Faunal Comparison of Seven Early Uintan Assemblages

Lillegraven (1979) compared the Uintan mammal faunas of southern California with those of the western interior. Since then, several more taxa have been recognized in the early Uintan of both regions, and it is appropriate to re-examine the faunal differences between them. The following comparisons are made at the generic level, since species-level identifications are more subjective and frequently unavailable. Table 4 gives mammalian faunal lists for seven early Uintan assemblages from western North America. It includes only those taxa for which at least a tentative generic identification has been made, which are known to represent new genera, or which must otherwise represent genera not known from the other assemblages. For comparisons of this kind to detect significant faunal variations (i.e., variations caused by differences in geologic age, paleoecology, or paleogeography), each assemblage should represent similar amounts of geologic time, should be collected with the same techniques, and should be studied with the same amount of comprehensiveness. That not all of these require- ments have been met is demonstrated by the wide variation between the number of genera in the least and most diverse assemblages (19 vs. 47). Part of this variation is due to the scarcity of micromammal taxa from some assemblages due to limited screenwashing (e.g., Uinta B1 and Twka2). Certain assemblages are also relatively diverse because they probably represent a significant amount of geologic time and/or several paleohabitats (e.g., Poway fauna; Uinta B2 assem- blage), or have been taxonomically oversplit (Uinta B2 assemblage). Despite these difficulties, Table 4 summarizes our knowledge of most of the well-known early Uintan assemblages of North America, and indicates where further study is required.

Table 5 organizes the data in Table 4 into two faunal comparisons; the Simpson Coefficient (Simpson, 1960; Flynn, 1986b), and the number of genera known from two given assemblages, but not from any of the other considered assemblages. Comparisons such as these are also problematical in that certain poorly represented and/or insufficiently studied taxa are assumed (for the sake of obtaining a number) to be "congeneric" with specimens from another assemblage that may have been identified just as tentatively. Thus, most of the Simpson Coefficients and exclusive common occurrences listed in Table 5 cannot be taken literally, but are presented in an effort to detect broad trends in faunal similarity. Unfortunately, it is difficult to attribute much significance to the considerable variation in the Simpson Coefficients. For example, the relatively high coefficient of 58 for the Uinta B1 and Twka2 assemblages probably reflects the low diversity of each sample, and cannot yet be attributed to significant age or paleoecological affinities for the faunas they represent. Other coefficients range from 29 to 55, and show a moderate amount of similarity between most of these early Uintan assemblages (as expected).

Some observations are possible regarding the taxa known only from one assemblage (central diagonal in Table 5). For example, the Poway fauna contains 14 genera (many of them new) not known from any of the other assemblages, and this figure reflects a significant amount of early Uintan endemism for southern California. A few taxa known only from the Whistler Squat local fauna may also indicate a degree of isolation for Trans-Pecos Texas. The presence of 9 genera known only from the Sand Wash Basin fauna (Stucky et al., this volume, Chapter 3) may reflect the best sample yet of a putative "earliest Uintan" interval. Finally, the 11 genera known only from the Tepee Trail Formation of Wyoming include a mixture of "archaic" (neoplagiau- lacid genus, cf. *Tetrapassalus*, and ?*Didymictis*) and "derived" forms (cf. *Ocajila* and *Protadjidaumo*?). McKenna (1980, 1990) noted that the Bone Bed A assemblage (which constitutes most of the taxa listed from the Tepee Trail Formation in Table 4) may represent an unusual paleocommunity, situated at a higher paleoaltitude than other early Uintan assemblages.

The Simpson Coefficients given above are based on a "phenetic" concept of overall faunal similarity. In contrast, the number of genera shared only by two given assemblages may potentially be a better indicator of temporal, paleoecological, or paleogeographic affinity, but comparisons of this type must be evaluated with care. For example, of the three genera known only from the Poway fauna and the Uinta B2 assemblage (*Pseudotomus*, *Pareumys*, and *Miacis*), none of them indicate a significant affinity, because these taxa are known from older and/or younger rocks in other areas. Thus, their absence in the other assemblages is probably attributable to sampling error. The same conclusion holds for the three taxa shared exclusively by

Table 4. Mammal genera from selected early Uintan assemblages of North America. Pow = Poway fauna of San Diego (Golz and Lillegraven, 1977; this paper). WS = Whistler Squat local fauna of Texas (Wilson, 1986; Runkel, 1988). UB1 and UB2 = "members" B1 (includes Willow Creek localities) and B2 (includes White River Pocket localities) of the Uinta Formation of Utah (Prothero, this volume, Chapter 1). SWB = Sand Wash Basin fauna of Colorado (Stucky et al., this volume, Chapter 3). Twka2 = assemblage from middle unit of Adobe Town Member of Washakie Formation, Wyoming (McCarroll et al., this volume, Chapter 2). Ttt = combined assemblage from Tepee Trail Formation of Wyoming, from Bone Bed A (McKenna, 1980, 1990) and the area of the North Fork of Owl Creek (Eaton, 1985). "Cf." and "?" indicate uncertainty in taxonomic identification; "( )" indicates occurrence of taxon in this horizon is uncertain.

	Pow	WS	UB1	UB2	SWB	Twka2	Ttt
MULTITUBERCULATA							
neoplagiaulacid gen.	–	–	–	–	–	–	X
MARSUPIALIA							
Peratherium	X	X	cf.[1]	–	X	–	X
Peradectes	X	–	–	–	–	–	X
INSECTIVORA							
Crypholestes	X	–	–	–	–	–	–
Scenopagus	X	X	cf.[1]	–	–	–	–
dormaaliid, n. gen.	X	–	–	–	–	–	–
Entomolestes	–	?	–	–	–	–	–
Ocajila	–	–	–	–	–	–	**cf.**
Centetodon	X	X	–	–	X	–	–
Batodonoides	X	–	–	–	–	–	–
Nyctitherium	X	X	–	–	–	–	X
palaeoryctine gen.	X	–	–	–	–	–	–
micropternodontine gen.	X	–	–	–	–	–	–
Oligoryctes	X	–	–	–	–	–	X
"PROTEUTHERIA"							
Apatemys	X	–	–	X	–	–	X
Aethomylos	X	–	–	–	–	–	–
Pantolestes	cf.	cf.	–	–	X	–	–
Simidectes	–	–	–	X	–	–	–
Palaeictops	X	–	–	–	–	–	X
leptictid, n. gen.	X	–	–	–	–	–	–
PALAEANODONTA							
Tetrapassalus	–	–	–	–	–	–	cf.
PRIMATES							
Notharctus	–	X	–	–	X	X	–
adapid, n. gen.	X	–	–	–	–	–	–
Omomys	X	X	–	–	–	–	–
Hemiacodon	X	–	–	–	X	–	–
Ourayia	X	X	–	X	–	–	–
Macrotarsius	–	X	–	X	–	–	–
Washakius	X	–	–	–	–	–	–
Stockia	X	–	–	–	–	–	–
"PLESIADAPIFORMES"							
Microsyops	X	X	–	–	X	–	–
Uintasorex	X	X	–	–	–	–	X
Tarka	–	–	–	–	–	–	X
Trogolemur	–	–	–	–	–	–	X
Phenacolemur	–	–	–	–	–	–	cf.

TABLE 4 (continued)

	Pow	WS	UB1	UB2	SWB	Twka2	Ttt
RODENTIA							
Paramys	–	–	–	X[2]	–	–	?
Thisbemys	–	X	–	X	X	–	–
Leptotomus	X	X	–	–	X	X	–
Pseudotomus	X	–	–	X	–	–	–
Reithroparamys	–	–	–	–	–	–	?
Microparamys	X	X	–	X	–	–	X
Lophiparamys	–	X	–	–	–	–	–
Uriscus	X	–	–	–	–	–	–
Mytonomys	–	–	–	X	–	–	–
Sciuravus	X	–	X[1]	X	X	–	X
Tillomys	–	–	–	–	X	–	–
Prolapsus	–	X	–	–	–	–	–
Pareumys	X	–	–	X	–	–	–
Mysops	–	X	–	–	–	–	–
Protoptychus	–	–	–	X	–	X	–
cylindrodontid, n. gen.	–	X	–	--	–	–	–
Pauromys	–	?[3]	cf.[1]	–	–	–	–
Namatomys	–	–	–	–	X	–	–
"*Namatomys*"	X	?[3]	–	–	–	–	–
Protadjidaumo	–	–	–	–	–	–	?
MESONYCHIDAE							
Harpagolestes	cf.	–	X	–	–	X	–
Mesonyx	–	–	X	X	X	–	–
Hessolestes	–	X	–	–	–	–	–
CREODONTA							
Oxyaenodon	–	–	–	X	–	–	–
Limnocyon	–	–	–	X	–	X	–
Apataelurus	–	–	–	X	–	–	–
CARNIVORA							
Protictis	–	–	–	–	–	–	X
Didymictis	–	–	–	–	–	–	?
Viverravus	–	–	–	–	X	–	X
Proviverra	–	?	–	–	–	–	–
Miocyon	–	X	–	X	–	–	–
Miacis	X	–	–	X	–	–	–
Tapocyon	X	–	–	–	–	–	–
Uintacyon	–	–	–	–	X	–	–
?amphicyonid, n. gen.	–	–	–	–	X	–	–
DINOCERATA							
Uintatherium	cf.	cf.	X	X	–	–	–
Eobasileus	–	–	X	X	X	X	–
TAENIODONTA							
Stylinodon	–	X	–	X	–	–	–
HYOPSODONTIDAE							
Hyopsodus	–	X	X	X	X	X	X

TABLE 4 (continued)

	Pow	WS	UB1	UB2	SWB	Twka2	Ttt
PERISSODACTYLA							
Hyrachyus	X	?	X	–	?	?	X
Triplopus	–	?	X	X	X	X	–
Epitriplopus	–	–	–	X	–	–	cf.
Forstercooperia	–	–	X	–	?	X	X
Isectolophus	–	–	–	X	X	–	–
Amynodon	X	X	( )	X	–	X	X
Helaletes	–	–	–	–	–	–	?
Dilophodon	–	–	–	–	X	X	X
Tapiroidea, n. gen.	X	–◻	–	–	–	–	–
Orohippus	–	–	–	–	X	–	–
Epihippus	–	–	–	X	X	–	X
Telmatherium	–	–	–	–	X	–	–
Metarhinus	?	–	X	–	X	X	?
Mesatirhinus	–	–	–	–	–	–	?
Dolichorhinus	–	–	X	X	–	X	–
Rhadinorhinus	–	–	X	X	–	–	–
Tanyorhinus	–	–	–	–	X	–	–
Manteoceras	–	–	–	–	X	–	–
Sthenodectes	–	–	X	–	–	–	–
"*Sthenodectes*" [4]	–	X	–	–	–	–	–
Eomoropus	–	–	X	X	–	X	–
chalicothere, n. gen.	–	–	–	–	–	–	X
ARTIODACTYLA							
Helohyus	–	X	–	–	–	–	–
Lophiohyus	cf.	X	–	–	–	–	–
helohyid, n. gen.	–	–	–	–	X	–	–
Parahyus	cf.	X	–	–	–	( )[5}	–
Achaenodon	X	–	X	X	–	X	X
Antiacodon	X	–	–	–	–	–	–
Bunomeryx	–	–	–	X	–	–	–
Hylomeryx	–	–	–	X	X	–	X
Mesomeryx	–	–	–	X	–	–	–
Pentacemylus	–	–	–	–	–	?	–
Protoreodon	X	–	X	X	X	–	–
Diplobunops	–	–	–	X	–	–	–
agriochoerid, n. gen.	–	–	–	–	X	–	–
Merycobunodon	X	–	–	–	–	–	–
Protylopus	X	–	–	X	–	X	–
Oromeryx	–	–	( )	X	X	–	–
Malaquiferus	–	?	–	–	–	–	–
Poebrodon	–	–	–	–	–	X	–
Leptoreodon	X	X	X[1]	X	–	–	–
Leptotragulus	–	–	–	X	–	–	–
Total genera: 115	47	35	20	40	32	19	33

[1] These taxa collected by the author from Uinta B1 in association with a Washington University
  field crew led by D.T. Rasmussen, summer of 1995.

[2] This record based on Korth's (1988) generic reassignment of "*Ischyrotomus*" *compressidens*.

[3] The record of *Pauromys* from Whistler Squat (Walton, 1993a) may represent "*Namatomys*."

[4] Mader (1989) removed "*S*." *australis* (Wilson, 1977) from the genus *Sthenodectes*.

[5] The type specimen of *Parahyus* probably came from the Washakie Formation of Wyoming, but whether it came from the Kinney Rim
Member or Adobe Town Member is uncertain (Gazin, 1955, pp. 42-43).

the Poway fauna and the Tepee Trail assemblage
(*Peradectes*, *Oligoryctes*, and *Palaeictops*), and the three
taxa shared exclusively by the Poway fauna and the
Whistler Squat local fauna (*Omomys*, *Lophiohyus*, and
Parahyus).

Age Correlation of Early Uintan Assemblages

Walsh (this volume, Chapter 4) discusses the age
correlation of the Poway fauna of San Diego with early
Uintan mammal assemblages in Wyoming, Utah, and
Texas, in relation to the problem of the Bridgerian/

Simpson   Coefficients

		POWAY FAUNA	WHISTLER SQUAT L.F.	UINTA B1	UINTA B2	SAND WASH BASIN FAUNA	TWKA2	TEPEE TRAIL FORMATION
	POWAY FAUNA	Aethomylos Crypholestes dormaaliid, n.g. Batodonoides palaeoryctine micropternodont leptictid, n.g. adapid, n.g.*	18/35 **51**	11/20 **55**	13/40 **33**	16/32 **50**	7/19 **37**	13/33 **39**
	WHISTLER SQUAT L.F.	Omomys Lophiohyus Parahyus	Entomolestes? Lophiparamys Prolapsus Mysops Hessolestes Proviverra? "Sthenodectes" Helohyus Malaquiferus? (cylindrodont, n.g.)	8/20 **40**	11/35 **31**	9/32 **28**	6/19 **32**	8/33 **24**
	UINTA B1	0	0	cf. Pauromys Sthenodectes	11/20 **55**	10/20 **50**	11/19 **58**	8/20 **40**
Genera Known Only From These Two Assemblages	UINTA B2	Pseudotomus Pareumys Miacis	Macrotarsius Stylinodon Miocyon	Rhadinorhinus	Simidectes Mytonomys Oxyaenodon Apataelurus Leptotragulus Bunomeryx Mesomeryx Diplobunops	10/32 **31**	10/19 **53**	9/33 **27**
	SAND WASH BASIN FAUNA	Hemiacodon	0	0	Isectolophus	Tillomys Namatomys ?amphicyonid, n.g. Orohippus Telmatherium Tanyorhinus Manteoceras helohyid, n.g. agriochoerid, n.g.	10/19 **53**	9/32 **28**
	TWKA2	0	0	0	Protoptychus Limnocyon	0	Pentacemylus? Poebrodon	7/19 **37**
	TEPEE TRAIL FORMATION	Peradectes Oligoryctes Palaeictops	0	0	Paramys Epitriplopus	Viverravus	0	neoplagiaulacid cf. Ocajila cf. Tetrapassalus Tarka Trogolemur Phenacolemur Reithroparamys? Protadjidaumo? Protictis Didymictis? chalicothere, n.g. Helaletes

* plus Washakius, Stockia, Uriscus, Tapocyon, tapiroid, n.gen., and Antiacodon.

Table 5. Faunal similarities of selected early Uintan assemblages from North America, derived from data in Table 4. Lower left part of table shows genera known only from two given assemblages. Central diagonal shows genera known only from each assemblage. Upper right part of table shows Simpson coefficients; upper part of each square shows the number of genera in common, divided by the number of genera in the less diverse assemblage. The resulting fraction is expressed as a percentage in the lower part of each square.

Uintan boundary and the "Shoshonian Subage" of Flynn (1986a). Until more detailed species-level comparisons of these assemblages are undertaken, age correlations should be regarded as tentative. Briefly, however, given the magnetobiostratigraphic results of Flynn (1986a) and Walsh et al. (this volume, Chapter 6), the Poway fauna may correlate approximately with: 1) the early Uintan assemblages from the upper part of the "Holy City beds" and lower part of the "Foggy Day beds" of the Tepee Trail Formation in Wyoming (Eaton, 1985); 2) the poorly known early Uintan assemblages from the

lower part of Uinta B in Utah (Prothero, this volume, Chapter 1); and 3) the early Uintan Whistler Squat local fauna from the Devil's Graveyard Formation of Texas (Wilson, 1986; Runkel, 1988; Walton, 1992). If the early Uintan assemblages from the lower part of the middle unit of the Adobe Town Member of the Washakie Formation, Wyoming (McCarroll et al., this volume, Chapter 2), and the Sand Wash Basin of Colorado (Stucky et al., this volume, Chapter 3) accumulated during Chron C21n, then both assemblages would be largely or entirely older than the Poway fauna.

Table 6. Mammal genera from seven late Uintan assemblages of western North America. SD = combined assemblage from Cloud 9 fauna and Jeff's Discovery and Rancho del Oro local faunas, San Diego County (this paper). VENT = combined assemblage from Tapo Canyon and Brea Canyon local faunas, Ventura County (Golz and Lillegraven, 1977; Kelly, 1990; Kelly et al., 1991; Kelly and Whistler, 1994). SER = Serendipity local fauna, Texas (Wilson, 1986; Runkel, 1988; Walton, 1993a). CB = Casa Blanca local fauna, Texas (Westgate, 1988, 1990, 1994a,b; Wilson and Westgate, 1991; Walton, 1993a, b). UC = combined assemblage from "Member C" of Uinta Formation, Utah (Prothero, this volume). BAD = combined assemblage from localities 5, 6, 7, and associated late Uintan sites at Badwater, Wyoming (Black and Dawson, 1966; see numerous reports in the Annals of Carnegie Museum cited by Stucky (1992, p. 485). SCC = Swift Current Creek local fauna, Saskatchewan (Storer, this volume, Chapter 12). "Cf." and "?" indicate uncertainty in taxonomic identification; "( )" indicates occurrence of taxon in this horizon is uncertain.

	SD	VENT	SER	CB	UC	BAD	SCC
MULTITUBERCULATA							
Ectypodus	-	-	-	-	-	X	X
MARSUPIALIA							
Peratherium	X	X	X	X	-	X	X
Peradectes	X	X	-	-	-	X	X
INSECTIVORA							
Sespedectes	X	X	-	-	-	-	X
Proterixoides	X	X	-	-	-	-	-
Macrocranion	-	-	-	-	-	X	-
Scenopagus	X	-	-	-	-	-	-
Talpavus	-	-	-	-	X	X	-
Ankylodon	-	-	-	-	-	X	-
Centetodon	X	X	-	X	-	X	X
Batodonoides	X	X	-	-	-	-	-
Nyctitherium	-	-	-	-	X	X	X
Oligoryctes	X	-	-	-	-	X	X
Micropternodus	-	-	-	-	X	-	-
Domnina	-	-	-	-	-	X	X
Wallia	-	-	-	-	-	-	X
"PROTEUTHERIA"							
Apatemys	X	X	-	-	-	X	X
Aethomylos	X	X	-	-	-	-	-
Palaeictops	-	-	-	-	-	-	X
Leptictis	-	-	-	-	-	-	X
Didelphodus	-	-	-	-	-	-	X
Simidectes	X	-	X	X	X	-	-
PRIMATES							
Omomys	-	-	X	X	-	-	X
Chumashius	-	-	-	-	-	X	-
Dyseolemur	X	X	-	-	-	-	-
Macrotarsius	-	X	-	X	-	X	-
Ourayia	X	-	-	-	X	-	-
Mytonius	-	X[1]	-	-	X	-	-
"PLESIADAPIFORMES"							
Microsyops	-	-	-	X	-	-	-
Craseops	-	X	-	-	-	-	-
Uintasorex	X	X	X	-	-	X	-
Trogolemur	-	-	-	-	-	X	-
Phenacolemur	X	X	-	-	-	X	-
Thylacaelurus	-	-	-	-	-	X	X
RODENTIA							
Paramys	-	X[2]	X	-	X[2]	-	-
Thisbemys	-	-	X	-	X	-	-
Leptotomus	X	X	X	-	X	X	X
Pseudotomus	-	-	-	-	X	X	X
Tapomys	-	X	-	-	-	-	-
Mytonomys	-	X	-	X	X	-	-
Rapamys	X	X	-	-	-	X	-
Eohaplomys	X	X	-	-	-	-	-
Janimus	-	-	-	-	X	-	X
Spurimus	-	-	-	-	-	X	-
Reithroparamys	-	-	-	-	X	-	-
Microparamys	X	X	X	X	-	X	X
Anonymus	-	-	-	-	-	-	X
Microeutypomys	-	-	-	X	-	-	-
Sciuravus	X	-	-	-	X[3]	-	-
Prolapsus	-	-	X	-	-	-	-
Protoptychus	-	-	-	-	( )	-	-
Mysops	-	-	X	-	-	-	-
Pareumys	X	X	-	-	X	X	-
Pseudocylindrodon	-	-	-	-	-	?	X
cylindrodontid, n. gen.	-	-	X	X	-	-	-
Namatomys"	X	X	?[4]	?[4]	-	-	X

TABLE 6 (continued)	SD	VENT	SER	CB	UC	BAD	SCC
Protadjidaumo	–	–	–	–	–	–	X
cf. *Paradjidaumo*	X	–	–	–	–	–	–
Griphomys	X	X	–	–	–	–	–
Pauromys	–	–	?[4]	?[4]	–	–	–
Laredomys	–	–	–	X	–	–	–
Simimys	X	X	–	–	–	–	–
Nonomys	–	–	–	cf.	–	–	–
LAGOMORPHA							
Mytonolagus	–	–	–	–	X	X	–
Procaprolagus	–	–	–	–	–	–	X
MESONYCHIDAE							
Harpagolestes	–	–	–	–	X	–	X
Hessolestes	–	–	X	–	–	–	–
CREODONTA							
Oxyaenodon	–	–	–	–	X	–	–
Limnocyon	X	–	–	–	X	X	–
Apataelurus	cf.	–	–	–	–	–	–
CARNIVORA							
Viverravus	–	–	–	–	–	–	X
Miocyon	X	–	?	cf.	X	–	X
Miacis	X	X	–	–	X	–	–
Tapocyon	X	X	–	–	–	X	–
Uintacyon	–	–	–	–	X	–	–
Procyonodictis	X	X	–	X	X	X	–
SIRENIA							
Sirenian, gen.	–	–	–	X	–	–	–
HYOPSODONTIDAE							
Hyopsodus	–	–	X	X	X	X	X
PERISSODACTYLA							
Triplopus	X	–	?	–	X	–	–
Epitriplopus	–	–	–	–	X	X	–
Isectolophus	–	–	–	–	X	–	–
Amynodon	X	X	–	X	X	X	–
"Metamynodon"	–	–	X	–	–	–	–
Colodon	–	–	–	–	–	X	X
Dilophodon	–	–	–	–	X	X	X
Tapiroidea, n. gen.	X	X	–	–	–	–	–
Epihippus	–	–	X	X	X	X	X
Diplacodon	–	–	–	–	X	–	–
Metatelmatherium	–	–	–	–	X	–	–
Protitanotherium	–	–	–	–	X	–	–
"Sthenodectes"[5]	–	–	X	–	–	–	–
ARTIODACTYLA							
Antiacodon	?	–	–	–	–	–	–
Auxontodon	–	–	–	–	X	–	X
Tapochoerus	–	X	–	–	–	–	–
Ibarus	X	–	–	–	–	–	X
Laredochoerus	–	–	–	X	–	–	–
Bunomeryx	–	–	–	–	X	–	–
Hylomeryx	–	–	–	–	X	–	–
Pentacemylus	–	–	–	–	X	–	–
Mytonomeryx	–	–	–	–	X	–	X
Apriculus	–	–	–	–	–	X	–
Toromeryx	–	–	–	X	–	–	–
Texodon	–	–	X	–	–	–	–
Hendryomeryx	–	–	X	–	–	–	–
Protoreodon	X	X	X	X	X	X	X
Diplobunops	–	–	–	–	X	X	–
Protylopus	X	X	?	–	X	–	–
Oromeryx	–	–	–	–	X	–	–
Malaquiferus	–	–	?	–	–	X	–
Poebrodon	–	–	–	–	X	–	–
Poabromylus	–	–	–	–	–	X	–
Simimeryx	–	X	–	–	–	–	–
Leptoreodon	X	X	X	X	X	X	X
Leptotragulus	–	–	–	–	X	X	–
Total genera: 113	39	35	25	23	45	41	35

[1] *Yaquius travisi* (Mason, 1990) is here regarded as a species of *Mytonius* Robinson, 1968 (Rasmussen and Walsh, in prep.).
[2] These records based on Korth's (1988) generic reassignment of "*Ischyrotomus*" *compressidens* Peterson.
[3] M.R. Dawson and D.T. Rasmussen, personal communication.
[4] At least some of Walton's (1993a) specimens of *Pauromys* from the Serendipity and Casa Blanca local faunas may be referable to "*Namatomys*" (A.H. Walton, personal communication).
[5] Mader (1989) removed "*S.*" *australis* (Wilson, 1977) from the genus *Sthenodectes*.

Simpson   Coefficients

	SAN DIEGO COUNTY	VENTURA COUNTY	SERENDIP-ITY LOCAL FAUNA	CASA BLANCA LOCAL FAUNA	UINTA C ASSEMB.	BAD-WATER ASSEMB.	SWIFT CURRENT CREEK L.F.
SAN DIEGO COUNTY	Scenopagus cf. Paradjidaumo cf. Apataelurus Antiacodon?	27/35 77	11/25 44	10/23 43	14/39 36	17/39 44	13/36 36
VENTURA COUNTY	Proterixoides Batodonoides Aethomylos Dyseolemur Eohaplomys Griphomys Simimys tapiroid, n. gen.	Craseops Tapomys Tapochoerus Simimeryx	9/25 36	9/23 39	11/35 31	16/35 46	10/35 29
SERENDIP-ITY LOCAL FAUNA	0	0	Prolapsus Mysops Hessolestes "Sthenodectes" "Metamynodon" Texodon Hendryomeryx	11/23 48	11/25 44	9/25 36	10/25 40
CASA BLANCA LOCAL FAUNA	0	0	Pauromys? cylindrodont, n.g.	Microsyops Micro-eutypomys Laredomys Nonomys sirenian Laredochoerus Toromeryx	8/23 35	10/23 43	10/23 43
UINTA C ASSEMB.	Ourayia Sciuravus	Mytonius	Thisbemys	0	Micropternodus Oxyaenodon Isectolophus Diplacodon Pentacemylus Bunomeryx Hylomeryx Oromeryx Protitanotherium Metatelmatherium	17/41 41	13/36 36
BAD-WATER ASSEMB.	0	0	Malaquiferus	0	Talpavus Epitriplopus Diplobunops	Macrocranion Ankylodon Trogolemur Spurimus Poabromylus Apriculus	19/36 53
SWIFT CURRENT CREEK L.F.	Ibarus	0	0	0	Janimus Harpagolestes Mytonomeryx Auxontodon	Ectypodus Domnina Pseudocylin-drodon Colodon Thylacaelurus	Palaeictops Leptictis Didelphodus Wallia Anonymus Protadjidaumo Procaprolagus Viverravus

(Left axis label: Genera Known Only From These Two Assemblages)

Table 7. Faunal similarities of selected late Uintan assemblages from North America, derived from data in Table 6. Lower left part of table shows genera known only from two given assemblages. Central diagonal shows genera known only from each assemblage. Upper right part of table shows Simpson coefficients; upper part of each square shows the number of genera in common, divided by the number of genera in the less diverse assemblage. The resulting fraction is expressed as a percentage in the lower part of each square.

LATE UINTAN MAMMAL ASSEMBLAGES OF NORTH AMERICA

Faunal Comparison of Seven Late Uintan Assemblages

Table 6 gives mammalian faunal lists at the generic level for seven late Uintan assemblages from western North America (see discussion of Table 4 for criteria of construction, and see Krishtalka et al., 1987, for more information on some of these assemblages). As with the early Uintan comparisons, there is a wide range of variation between the number of genera in the least and most diverse assemblages (23 vs. 45). The relatively high diversity of the Uinta C assemblage (despite the lack of large-scale screenwashing) is attributable to significant time-averaging and paleoecological variation represented in the entire fauna of the Myton Member of the Uinta Formation (and a certain amount of taxonomic oversplitting). Significant time-averaging is also likely for the late Uintan assemblages of San Diego and Ventura counties, California, and several associated localities from Badwater, Wyoming. The relatively low diversity of the Serendipity and Casa Blanca local faunas is attributable to the fact that these assemblages were obtained from a restricted temporal interval from closely associated localities (Serendipity) or just one quarry (Casa Blanca). In addition, the insectivores from the Serendipity local fauna have not been studied (Runkel, 1988).

Simpson Coefficients for these assemblages are given in Table 7, and range from 31 to 53, with one notable exception. The very high figure of 77 for the late Uintan assemblages from San Diego County and Ventura County reflects their occurrence in a distinct paleobiogeographic region during late Uintan time (Lillegraven, 1979). The large number of genera known only from San Diego and Ventura counties further corroborates their faunal affinity. Several taxa from the late Uintan Serendipity and Casa Blanca local faunas appear to be endemic to Texas, affirming a degree of isolation previously suggested by the early Uintan fauna. A relatively high Simpson Coefficient of 53 is obtained for the Badwater and Swift Current Creek assemblages, and 5 genera are known only from them. Five genera are also known only from Swift Current Creek and Uinta C, despite a relatively low Simpson Coefficient of 36. Other exclusive shared occurrences in Table 7 are intriguing, but are perhaps best explained by sampling bias in the other assemblages.

Age Correlation of Late Uintan Assemblages

Golz (1976), Lillegraven (1979), Wilson (1984, 1986), and Kelly (1990) discussed age correlation of the late Uintan and Duchesnean assemblages of southern California with those from the western interior. As noted above, a major difficulty in this endeavor is caused by the fact that the California assemblages are dominated numerically by only a few endemic micromammal taxa. Wilson (1984) recognized several species of *Leptoreodon* in Texas that Golz (1976) described from California. *L. edwardsi* and *L. leptolophus* are apparently restricted to the late Uintan of both California and Texas, and may prove to be biochronologically useful taxa. However, the intracontinental age significance of other species is questionable, since *L. major* is known only from the early Uintan in California, but was recorded by Wilson (1984) from the late Uintan Serendipity local fauna in Texas. Conversely, *L. pusillus* is known only from the late Uintan in California, but was recorded by Wilson (1984) from the early Uintan Whistler Squat local fauna in Texas. Kelly (1990) reviewed previous correlations of the Tapo Canyon, Brea Canyon, and Laguna Riviera local faunas with late Uintan assemblages from the western interior, and suggested that on the basis of the faunal evidence available, the California assemblages correlated best with the fauna of Uinta C and the late Uintan assemblages from Badwater, Wyoming (Krishtalka et al., 1987, p. 98). Kelly (1990) also suggested that the Cloud 9 fauna was of early late Uintan age, and correlative with the Serendipity local fauna of Texas (Wilson, 1986).

In principle, the paleomagnetic evidence now available from southern California and the western interior could refine temporal correlations of late Uintan assemblages, but this approach is hampered by a significant problem. In contrast to the early Uintan interval, the age differences between chrons of a given polarity within the late Uintan (presumably C20n and C19r, and perhaps C19n and C18r as well) range from only 0.3-1.0 m.y. (Berggren et al., 1995). This means that when the errors in most late Uintan K-Ar dates are combined with the potential errors in the numerical calibration of the Eocene chrons (± 1 m.y.; e.g., Harland et al., 1990, p. 154), the assignment of a particular late Uintan assemblage to a given chron can rarely be considered conclusive. One possible exception to this rule involves the Mission Valley Formation of San Diego, where a precise $^{40}Ar/^{39}Ar$ date is directly associated with late Uintan mammals (Obradovich and Walsh, in prep.; Walsh et al., this volume, Chapter 6). Finally, the occurrence of the Uintan/Duchesnean boundary within a particular chron is not well-established, making the duration of the late Uintan interval relative to the GPTS uncertain.

Chron correlations for late Uintan faunas have been proposed for only three places in the western interior, namely, northeastern Utah (Prothero, this volume, Chapter 1), Trans-Pecos Texas (Walton, 1992; Prothero, this volume, Chapter 9), and western Montana (Tabrum et al., this volume, Chapter 14). Rather than engage in a detailed discussion of these sections, I will only note that none of the correlations are based on precise radiometric dates on the magnetozones in question (nor on other evidence independent of the assumption that the "late Uintan" is an essentially isochronous biochron), and therefore must be viewed with caution. The fact that the very similar late Uintan assemblages from California are nevertheless assigned to different chrons (C20n/C19r in San Diego vs. C18r in Ventura County; Walsh et al., this volume, Chapter 6; Prothero et al., this volume, Chapter 8) casts doubt on our ability to identify late Uintan magnetozones in the western interior with much confidence. I therefore await corroboration of the chron correlations proposed by the above workers based on detailed biostratigraphic work and precise radiometric dates, especially in the Uinta Basin.

PALEOBIOGEOGRAPHIC ASPECTS OF EOCENE FAUNAS OF SOUTHERN CALIFORNIA

Wasatchian-Bridgerian

Discussion of the paleobiogeography of the Eocene mammal faunas of southern California begins with the Wasatchian Lomas las Tetas de Cabra local fauna, which is located about 550 km south of San Diego in Baja California, Mexico. Study of this assemblage reveals that many of the genera also occur in the Wasatchian of the western interior (Novacek et al., 1991). The undescribed Morena Boulevard local fauna (Wasatchian, unnamed formation; Walsh, 1991b) contains typical Wasatchian taxa such as *Knightomys, Hyopsodus*, cf. *Tubulodon*, and *Meniscotherium*, and does not contradict the evidence from the Lomas las Tetas de Cabra local fauna that relatively open migration routes between Baja California-San Diego and the western interior must have been available during the early Eocene.

The Bridgerian Swami's Point local fauna of northwestern San Diego County has produced only two identified mammal genera (*Trogosus* and *Hyrachyus*), but both are typical of the Bridgerian of the western interior. The early Bridgerian Elderberry Canyon local fauna of northeastern Nevada (Emry and Korth, 1989; Emry, 1990) also contains *Hyrachyus*, as does the late Bridgerian Clarno fauna (Hanson, this volume, Chapter 11). Certain rodents from Elderberry Canyon show similarities to taxa in the early Uintan of San Diego (e.g., *Reithroparamys* cf. *R. huerfanensis* and *Leptotomus*? *caryophilus*; *Microparamys sambucus* and *M.* cf. *M. woodi*; certain teeth of *Pauromys exallos* and "*Namatomys*" new sp. 1). Perhaps description of the remaining insectivores from the Elderberry Canyon local fauna (Emry, 1990, p. 190) will reveal additional similarities with the San Diego fauna. The evidence from the Swami's Point and Elderberry Canyon local faunas is consistent with Lillegraven's (1979) suggestion that relatively free faunal interchange between California and the western interior must also have occurred during Bridgerian time (Emry, 1990). Equids have yet to be collected from the Elderberry Canyon l.f., but the presence of *Orohippus* in the Clarno suggests that this genus will eventually be recognized in the Bridgerian of San Diego. On the other hand, equids have yet to be collected from Elderberry Canyon, so their presence in San Diego during the Bridgerian cannot be predicted.

Early Uintan

Uintan faunas have not been recognized in Nevada (R. J. Emry, personal communication). Only a few fossils from the Baca Formation suggest the possibility of a late? Uintan record in New Mexico (Lucas, 1992; Lucas and Williamson, 1993), and the small, endemic early Tertiary assemblage from Guanajuato (central Mexico) is of uncertain age (Ferrusquia-Villafranca, 1984; Westgate, 1990). Therefore, paleobiogeographic comparisons of the San Diego Uintan assemblages can be made only with those from the Rocky Mountains and Trans-Pecos Texas. Lillegraven (1979) compared mammal taxa from the early Uintan Poway fauna of San Diego with those from Bridgerian and early Uintan rocks of the Rocky Mountains. He demonstrated that the two regions showed a high degree of species-level similarity, especially among the small-bodied taxa. Table 8 examines the distribution of early Uintan records of genera from California, Texas, and the Rocky Mountains. The California faunal list consists of the

combined assemblage from the "earliest Uintan" Poway fauna and the "late early Uintan" Murray Canyon local fauna of San Diego. The faunal list for southern Texas consists of the combined assemblage from the early Uintan Whistler Squat local fauna (Wilson, 1986; Runkel, 1988) and the early Uintan "Devil's Graveyard Assemblage" and "Canoe Assemblage A" of Runkel (1988). The faunal list for the early Uintan of the Rocky Mountains reflects the combined assemblage from Uinta B1 and B2, the Sand Wash Basin, the middle unit of the Adobe Town Member of the Washakie Formation, the Tepee Trail Formation, and a few additional taxa that were listed by Stucky (1992). Table 8 excludes most taxa that have not been at least tentatively identified to genus level, but includes a few unnamed taxa that have been determined by various workers to pertain to new genera.

A certain amount of faunal interchange between California, Texas, and the Rocky Mountain area at the beginning of the Uintan is demonstrated by the presence in both areas of similar species of characteristic large-bodied early Uintan taxa such as *Amynodon, Achaenodon, Protoreodon*, and *Leptoreodon*. On the other hand, the apparent absence of *Hyopsodus* and equids in San Diego during the early Uintan remains a mystery, since these taxa are common in the early Uintan of the western interior (Golz, 1976; Lillegraven, 1979). Lillegraven (1980) suggested that southern California may have been an early Uintan refuge, because characteristic Wasatchian and Bridgerian taxa such as *Pelycodus, Notharctus*, and *Hemiacodon* survived there into the Uintan after apparent extinction in the western interior. However, Lillegraven's (1980) specimen of "*Pelycodus* sp. near *P. ralstoni*" is now regarded as a macrotarsiine primate (Gunnell, in press), while previous San Diego records of *Notharctus* (Lillegraven, 1980; Walsh, 1991b) are assigned to a new adapid genus (Gunnell, in press). *Notharctus* and *Hemiacodon* have also been reported from early Uintan deposits in the Rocky Mountain and Trans-Pecos Texas areas (e.g., West, 1982; Eaton, 1985; Flynn, 1986a; Wilson, 1986; Stucky et al., this volume). Nevertheless, additional "Bridgerian holdover taxa" recognized in the early Uintan of California since 1979 include *Scenopagus*, cf. *Pantolestes, Palaeictops, Pauromys*, and cf. *Lophiohyus*. At least three other San Diego genera are currently known only from Bridgerian or older rocks of the western interior. *Antiacodon* is newly recognized here from the Poway fauna, *Aethomylos* and *Crypholestes* were reported from the early Bridgerian of Utah (Krishtalka and Stucky, 1984), and *Batodonoides* may be present in the Wasatchian of Wyoming (Bloch, 1995). Comments on the biochronological utility of "Bridgerian holdover taxa" are provided by Walsh (this volume, Chapter 4). The early Uintan faunas of San Diego and Trans-Pecos Texas existed at comparable paleolatitudes, and might

Table 8. Geographic distribution of mammal genera between California, Texas, and the Rocky Mountains, showing genera that are known in the early Uintan:

only from California.	only from Texas.	only from the Rocky Mtns.	from Calif. and Texas, but not the Rocky Mtns.	from Calif. and the Rocky Mtns., but not Texas.	from Texas and the Rocky Mtns., but not Calif.	from all three areas.
Aethomylos	*Entomolestes?*	neoplagiaulacid	*Omomys*	*Peradectes*	*Notharctus*	*Peratherium*
Crypholestes	*Lophiparamys*	cf. *Ocajila*	*"Namatomys"*	*Apatemys*	*Macrotarsius*	*Scenopagus*
dormaaliid, n. gen.	*Prolapsus*	*Simidectes*	*Lophiohyus*	*Oligoryctes*	*Thisbemys*	*Nyctitherium*
Scenopagus	*Mysops*	*Trogolemur*	*Parahyus*	*Palaeictops*	*Miocyon*	*Centetodon*
Batodonoides	cylindrodont, n. gen.	*Tarka*		*Hemiacodon*	*Stylinodon*	*Pantolestes*
palaeoryctine, gen.	*Hessolestes*	*Phenacolemur*		*Washakius**	*Hyopsodus*	*Uintasorex*
micropternodont	*Proviverra?*	*Paramys*		*Ourayia*	*Isectolophus*	*Microsyops*
leptictid, n. gen.	*"Sthenodectes"*	*Reithroparamys?*		*Pseudotomus*	*Triplopus*	*Microparamys*
adapid, n. gen.	*Helohyus*	*Mytonomys*		*Sciuravus*		*Leptotomus*
Stockia	*Texodon*	*Tillomys*		*Pareumys*		*Pauromys?*
Uriscus	*Malaquiferus?*	*Protoptychus*		*Harpagolestes*		*Uintatherium*
Eohaplomys		*Namatomys*		*Miacis*		*Hyrachyus*
Tapocyon		*Protadjidaumo?*		*Metarhinus?*		*Amynodon*
tapiroid, n. gen.		*Mesonyx*		*Achaenodon*		*Leptoreodon*
Antiacodon		*Oxyaenodon*		*Protoreodon*		
Merycobunodon		*Limnocyon*		*Protylopus*		
		Apataelurus				
		Protictis				
		Didymictis?				
		Viverravus				
		Uintacyon				
		?amphicyonid n.gen.				
		Eobasileus				
		Epitriplopus				
		Forstercooperia				
		Dilophodon				
		Helaletes?				
		Orohippus				
		Epihippus				
		Mesatirhinus?				
		Telmatherium				
		Dolichorhinus				
		Manteoceras				
		Tanyorhinus				
		Rhadinorhinus				
		Sthenodectes				
		Eomoropus				
		chalicothere, n. gen.				
		Auxontodon				
		helohyid, n. gen.				
		Hylomeryx				
		Mesomeryx				
		Bunomeryx				
		Pentacemylus?				
		Diplobunops				
		agriochoerid, n. gen.				
		Oromeryx				
		Poebrodon				
		Leptotragulus				
16	11	48	4	16	8	14

Total genera for California = 50. Total genera for Texas = 37. Total genera for Rocky Mtns. = 86.
Simpson Coefficient for California and Texas = 18/37 X 100% = 49%.
Simpson Coefficient for California and the Rocky Mtns. = 30/50 X 100% = 60%.
Simpson Coefficient for Texas and the Rocky Mtns. = 22/37 X 100% = 59%.
* Rocky Mtn. record of *Washakius* from Honey (1990, p. 206).

therefore be expected to have some significant similarities (e.g., Flynn, 1986b), but this has not been demonstrated. Although three species of *Leptoreodon* are apparently known only from these areas (Wilson, 1984), it appears that the early Uintan assemblages of California and Texas are approximately as distinct from one another as each of them are from the fauna of the Rocky Mountains.

One interesting aspect of the still incompletely known "late early Uintan" Murray Canyon local fauna of San Diego concerns the appearance and abundance of the early myomorph? rodent *Pauromys*, represented by a species quite similar to *P. perditus* from Bridger B or C in Wyoming (Troxell, 1923; Walton, 1993a). Undoubted specimens of *Pauromys* have not been detected in the well-sampled "earliest Uintan" Poway fauna, suggesting that migration of this rodent from the western interior to California may have been delayed by

2-3 m.y. However, *Pauromys* seems to have disappeared in San Diego almost as soon as it arrived. Perhaps it was a victim of competition from (or evolved into!) one or more new taxa that characterize the beginning of the late Uintan in southern California (e.g., *Simimys*).

Late Uintan to Arikareean

Table 9 examines the distribution of late Uintan records of genera from California, Texas, and the Rocky Mountains. The California faunal list consists of the combined assemblage from San Diego and Ventura counties as previously defined for Table 6, but adds a few taxa from the latest Uintan and/or Duchesnean Laguna Riviera, Camp San Onofre, and Mission del Oro local faunas. The faunal list for southern Texas consists of the combined assemblage from the Serendipity and Casa Blanca local faunas, and that for the Rocky Mountains reflects the combined assemblage from Uinta C, the Badwater assemblage, and the Swift Current Creek local fauna (see references in Table 6), plus a few additional late Uintan taxa that were listed by Stucky (1992).

Wood (1974) postulated that the late Uintan to Chadronian assemblages of southern California and the Vieja area of Texas represented the northern range of a distinct Middle American rodent fauna. This hypothesis is intuitively appealing, but as shown in Table 9, there are no genera known from the late Uintan of California and Texas that are not also known from the Rocky Mountains, and the Simpson Coefficient for California and the Rocky Mountains (69%) is notably higher than the coefficient for California and Texas (50%). The only taxon cited by Wood (1974) as potentially occurring in both southern California and Texas (and nowhere else) was the rodent *Simimys*. However, Wood's record of "cf. *Simimys*" from the Chadronian Porvenir local fauna was based on only a single lower molar, and in view of the morphological variability of this genus and other small rodents (Lillegraven and Wilson, 1975), must be regarded with skepticism. Novacek (1976) speculated that some of the Uintan insectivores apparently endemic to southern California could have been part of a Central American radiation, but there is still no Eocene record from this region (Westgate, 1990). The collection of identifiable insectivore material from Guanajuato in central Mexico (Ferrusquia-Villafranca, 1984, p. 189) may be important for testing the existence of a distinct Middle American faunal province during later Eocene time.

Lillegraven (1979) demonstrated that species-level similarities between California and the western interior decreased during the late Uintan (although as noted above, overall genus-level similarities remain high). After compiling a variety of tectonic and paleoclimatic evidence, Lillegraven (1979) suggested that increasing aridity in the continental interior of North America during the late Uintan may have resulted in vegetational and edaphic barriers to continuous dispersal, thus contributing to the endemism of the late Uintan and Duchesnean assemblages of southern California. Storer (1989) divided North America into eight early Tertiary faunal provinces, and compared the numbers of rodent genera and species shared by adjacent provinces from Uintan through Hemphillian time. He concluded that rodent faunal provinces were most distinct during the Duchesnean. The hypotheses of Lillegraven (1979) and Storer (1989) are consistent with Frederiksen's (1991b) recognition of a Bartonian (approximately Duchesnean NALMA) climatic deterioration in southern California and the Gulf Coast. Nevertheless, as discussed below, some communication between these areas must have occurred during the early Duchesnean.

Walsh (1991b) speculated that the early Uintan/late Uintan disconformity in San Diego County reflected the global sea-level drop shown by Haq et al. (1987) as occurring at about 43-44 Ma. Walsh (1991b) also suggested that some of the endemic taxa that characterize the beginning of the late Uintan in southern California may have been immigrants whose dispersal reflected this sea-level lowstand, citing Vianey-Liaud's (1985) proposal that *Simimys* immigrated to North America from Asia. While this remains possible, the recent discovery of *Elymys* from the Bridgerian of Nevada (Emry and Korth, 1989) and *Pauromys* from the early Uintan of southern California (this chapter) suggests that *Simimys* could have evolved in North America. Interestingly, Woodburne and Swisher (1995) recognize the late Uintan as a time of major immigrational input to the mammalian faunas of North America from Asia, and correlate this event with the global sea-level lowstand TA 3.4 (see Haq et al., 1987). Toward the end of the Uintan or at the beginning of the Duchesnean (Laguna Riviera and Mission del Oro local faunas), the first records in southern California of lagomorphs (*Mytonolagus*), equids (*Epihippus*), and camelids (*Poebrodon*) are represented by only one specimen each. Nevertheless, the absence of these taxa in the well-sampled, older late Uintan rocks of southern California, together with the first records of *Duchesneodus*, *Hyaenodon*, and *Eotylopus* in the Pearson Ranch local fauna (Kelly, 1990), suggest that the beginning of the Duchesnean represented a time of partial interchange with the western interior (at least for medium- to large-bodied forms).

Mammal assemblages assigned to the Chadronian, Orellan, and Whitneyan NALMAs (latest Eocene and early Oligocene) are unknown from southern California. The absence of a record for this interval seems to reflect a regional unconformity that may correspond to a large drop in global sea-level at about 30 Ma (Haq et al., 1987; Howard, 1989; Mason and Swisher, 1989; Walsh and Deméré, 1991; Woodburne and Swisher, 1995; Prothero et al., this volume, Chapter 8). Above the unconformity, diverse early Arikareean assemblages are

Table 9. Geographic distribution of mammal genera between California, Texas, and the Rocky Mountains, showing genera that are known in the late Uintan:

only from California.	only from Texas.	only from the Rocky Mtns.	from Calif. and Texas, but not the Rocky Mtns.	from Calif. and the Rocky Mtns., but not Texas.	from Texas and the Rocky Mtns., but not Calif.	from all three areas.
Aethomylos	Omomys	Ectypodus	(none)	Peradectes	Thisbemys	Peratherium
Proterixoides	Microsyops	Didelphodus		Apatemys	Hyopsodus	Centetodon
Scenopagus	Microeutypomys	Palaeictops		Sespedectes	Malaquiferus	Simidectes
Batodonoides	Prolapsus	Leptictis		Nyctitherium		Macrotarsius
Dyseolemur	Mysops	Talpavus		Oligoryctes		Uintasorex
Craseops	cylindrodont, n. gen.	Macrocranion		Chumashius		Paramys
Tapomys	Pauromys?	Ankylodon		Ourayia		Microparamys
Eohaplomys	Laredomys	Micropternodus		Mytonius		Mytonomys
cf. Paradjidaumo	cf. Nonomys	Wallia		Phenacolemur		Pseudotomus
Simimys	Hessolestes	Domnina		Rapamys		Leptotomus
Griphomys	sirenian, gen.	Trogolemur		Sciuravus		"Namatomys"
cf. Apataelurus	Laredochoerus	Ignacius		Pareumys		Procyonodictis
Amynodontopsis	Texodon	Thylacaelurus		Mytonolagus		Miocyon
tapiroid, n. gen.	Toromeryx	Manitsha		Limnocyon		(?)Triplopus
Antiacodon?	Hendryomeryx	Reithroparamys		Tapocyon		Amynodon
Tapochoerus	"Sthenodectes"	Spurimus		Miacis		Epihippus
Simimeryx		Janimus		Ibarus		Protoreodon
		Pseudocylindrodon		Poebrodon		Protylopus
		Protadjidaumo				Leptoreodon
		Procaprolagus				
		Mesonyx				
		Harpagolestes				
		Hemipsalodon				
		Oxyaenodon				
		Uintacyon				
		Viverravus				
		Prodaphoenus				
		Hyrachyus				
		Forstercooperia				
		Epitriplopus				
		Isectolophus				
		Metamynodon				
		Colodon				
		Dilophodon				
		Manteoceras				
		Metarhinus				
		Telmatherium				
		Metatelmatherium				
		Diplacodon				
		Protitanotherium				
		Grangeria				
		Apriculus				
		Auxontodon				
		Bunomeryx				
		Hylomeryx				
		Mytonomeryx				
		Pentacemylus				
		Mesomeryx				
		Diplobunops				
		Oromeryx				
		Poabromylus				
		Leptotragulus				
17	16	52	0	18	3	19

Total genera for California = 54. Total genera for Texas = 38. Total genera for Rocky Mtns. = 92.
Simpson Coefficient for California and Texas = 19/38 X 100% = 50%.
Simpson Coefficient for California and the Rocky Mtns. = 37/54 X 100% = 69%.
Simpson Coefficient for Texas and the Rocky Mtns. = 22/38 X 100% = 58%.

known from the upper member of the Sespe Formation in Ventura County (Tedford et al., 1987, and references cited therein), and the Otay Formation in San Diego County (Deméré, 1988). These assemblages show many generic-level (and possibly some species-level) similarities with the early Arikareean assemblages of Oregon, Nebraska, and South Dakota (Deméré, 1988), and there is no doubt that relatively unencumbered interchange must have taken place between the west coast and the western interior during the late Oligocene.

CONCLUSIONS

The Swami's Point local fauna (Delmar Formation) is the first fossil vertebrate assemblage of demonstrably Bridgerian age from southern California, and contains the first record of the tillodont *Trogosus* from California. The vertebrate assemblage from the basal Scripps Formation (Black's Beach local fauna) may be critical in constraining the position of the Bridgerian/ Uintan boundary in San Diego. Unfortunately, the few fossils available indicate only a late Bridgerian or early Uintan age.

Many biostratigraphic problems must be solved in the Uinta Basin before meaningful subdivisions of the Uintan NALMA are defined throughout North America. Recognition of the early Uintan/late Uintan boundary in southern California is problematical for several reasons, but can be characterized by the extinction of the common early Uintan genera *Crypholestes*, an unnamed dormaaliid insectivore (Novacek et al., 1985), *Omomys*, *Washakius, Sciuravus, Metarhinus*? and *Merycobunodon*, and by the evolution and/or immigration of the common late Uintan genera *Sespedectes, Proterixoides*, *Dyseolemur, Rapamys, Griphomys*, and *Simimys*.

The early Uintan Poway fauna is redefined as the entire set of vertebrate taxa occurring in the Friars Formation, as the latter is redefined by Walsh et al. (this volume, Chapter 6). It includes taxa from strata that were previously mapped as the Mission Valley Formation, but are now regarded as the upper tongue of the Friars Formation. The Mesa Drive local fauna and the assemblage from SDSNH Loc. 3486 indicate an early Uintan age for the upper part of Member B of the Santiago Formation in northwestern San Diego County. The Mesa Drive local fauna in Oceanside is composed largely of micromammals, while the mammal assemblage from Loc. 3486 in Carlsbad contains only the large-bodied taxa *Uintatherium* sp. and *Metarhinus*? *pater*. Both are correlated with the Poway fauna. The Murray Canyon local fauna (late early Uintan) is named for a distinctive assemblage from the lower member of the Stadium Conglomerate. This local fauna is dominated by the myomorph rodent *Pauromys* sp. cf. *P. perditus*, contains a new large species of *Crypholestes*, and the lowest stratigraphic record of the rodent *Eohaplomys* in San Diego.

The late Uintan Stonecrest local fauna was collected from either the lower or upper member of the Stadium Conglomerate, and represents the lowest stratigraphic occurrence in San Diego of several genera typical of the late Uintan of the Sespe Formation of Ventura County. The late Uintan Cloud 9 fauna is defined as the entire set of vertebrate taxa present in all outcrops of the Mission Valley Formation. Jeff's Discovery and Rancho del Oro are local faunas of late Uintan age from Member C of the Santiago Formation in northwestern San Diego County. The Laguna Riviera, Camp San Onofre, and Mission del Oro local faunas (Member C,

Santiago Formation) and the Bonita local fauna ("Sweetwater" Formation) are of latest Uintan and/or Duchesnean age. SDSNH Locs. 3235 and 3495 (Member C, Santiago Formation) have produced the brontothere *Duchesneodus*, and so are the first assemblages of demonstrably Duchesnean age to be documented from San Diego County.

The early Uintan Poway fauna of San Diego County correlates approximately in time with the early Uintan assemblages from the middle part of the Tepee Trail Formation in Wyoming, the sparse assemblages from the lower part of Uinta B in Utah, and the Whistler Squat local fauna of Texas. Based solely on stratigraphic position, the late early Uintan Murray Canyon local fauna may be roughly coeval with the assemblage from the upper part of Uinta B in Utah. The precise age of the late Uintan assemblages of California relative to those from the western interior is difficult to determine. This is due in part to the numerical dominance of endemic taxa in California, the lack of precise radiometric dates from the western interior, and to resulting uncertainties in the paleomagnetic correlation of strata in the two regions.

Although some climatic deterioration may have begun during the Uintan in southern California, most of it apparently occurred after the distinctive late Uintan mammal assemblages were already established. Faunal interchange between California and the western interior appears to have been relatively free during the Wasatchian and Bridgerian, and perhaps the early Uintan, corroborating the predictions of Lillegraven (1979). In addition, the tectonic and climatic factors invoked by Lillegraven to explain the increased species- level endemism in the late Uintan/Duchesnean faunas of California are consistent with Bartonian (~Duchesnean) climatic deterioration recognized by Frederiksen (1991b). Nevertheless, some interchange of larger- bodied taxa did occur between California and the western interior during the latest Uintan and early Duchesnean. No special taxonomic similarities can be demonstrated for either the early Uintan or late Uintan assemblages of California and Texas, and Wood's (1974) hypothesis that these areas pertained to a distinct Middle American faunal province cannot be supported on the basis of current evidence.

ACKNOWLEDGMENTS

I thank the editors for their invitation to contribute to this volume. I thank J. D. Archibald for logistical support. Many of the specimens studied for this paper were collected on construction projects by B. O. Riney, R. A. Cerutti, M. A. Roeder, D. R. Swanson, M. W. Colbert, C. P. Majors, M. W. Cerutti, R. Q. Gutzler, and D. J. McGuire during paleontological salvage operations conducted by the San Diego Natural History Museum and PaleoServices, Inc. Collection of these fossils was made possible by the support of American

Pacific Builders, the Buie Corporation, Brehm Homes, Carmel Mountain Ranch, Centre Development, the City of San Diego, the City of Carlsbad, Gatlin Development, Hillman Properties, Kaufman and Broad, the Koll Company, McMillin Communities, Mission Terrace Associates, Pardee Construction, Shea Homes, Stonecrest Development, and the Waxie Corporation. Collection of fossils from Jeff's Discovery local fauna and State Routes 52, 54, and 56 was made possible by Joyce Corum and Chris White (California Department of Transportation). C. P. Majors helped to collect the Murray Canyon local fauna, and Mr. Louis Cerutti discovered the Swami's Point local fauna. T. A. Deméré, M. W. Colbert, and J. H. Hutchison provided identifications for several larger-bodied taxa, and M. A. Roeder identified the shark teeth. The geological observations of B. O. Riney were critical in compiling some of the measured sections. I thank W. B. Cashion, M. R. Dawson, N. O. Frederiksen, T. S. Kelly, E. B. Lander, D. R. Prothero, P. D. Rowley, D. T. Rasmussen, R. D. Squires, R. K. Stucky, W. Wall, A. H. Walton, and J. W. Westgate for helpful discussions of Eocene rocks and faunas, and I thank D. T. Rasmussen for introducing me to the Uinta Basin. I thank D. P. Whistler and S. McLeod (LACM), J. H. Hutchison (UCMP), M. O. Woodburne (UCR), M. R. Dawson (CM), R. J. Emry (USNM), and M. J. Novacek and J. P. Alexander (AMNH) for allowing me to borrow specimens in their care. M. A. Turner (YPM) provided locality information on certain YPM-PU specimens. M. W. Colbert, T. A. Deméré, T. S. Kelly, M. C. Maas, S. M. McCarroll, M. C. McKenna, and J. W. Westgate reviewed the manuscript and made helpful suggestions for its improvement. None of these individuals necessarily agree with any of the views expressed herein.

LITERATURE CITED

Almgren, A. A., M. V. Filewicz, and H. L. Heitman. 1988. Lower Tertiary foraminiferal and calcareous nannofossil zonation of California: An overview; pp. 83-105 in M. V. Filewicz and R. L. Squires (eds.), Paleogene Stratigraphy, West Coast of North America. Pacific Section SEPM Volume 58.

Andersen, D. W., and M. D. Picard. 1972. Stratigraphy of the Duchesne River Formation (Eocene-Oligocene?), northern Uinta Basin, northeastern Utah. Utah Geological and Mineralogical Survey Bulletin 97:1-23.

Aubry, M.-P. 1992. Late Paleogene calcareous nannoplankton evolution: A tale of climatic deterioration; pp. 272-309 in D. R. Prothero and W. A. Berggren (eds.), Eocene-Oligocene Climatic and Biotic Evolution. Princeton University Press, Princeton, N.J.

Beard, K. C., L. Krishtalka, and R. K. Stucky. 1992. Revision of the Wind River faunas, early Eocene of central Wyoming. Part 12. New species of omomyid primates (Mammalia: Primates: Omomyidae) and omomyid taxonomic composition across the early-middle Eocene boundary. Annals of Carnegie Museum 61:39-62.

Berggren, W. A., D. V. Kent, J. D. Obradovich, and C. C.

Swisher. 1992. Toward a revised Paleogene geochronology; pp. 29-45 in D. R. Prothero and W. A. Berggren (eds.), Eocene-Oligocene Climatic and Biotic Evolution. Princeton University Press, Princeton, N.J.

Berggren, W. A., D. V. Kent, C. C. Swisher III, and M.-P. Aubry. 1995. A revised Cenozoic geochronology and chronostratigraphy. SEPM Special Publication 54:129-212.

Berry, R. W. 1991. Deposition of Eocene and Oligocene bentonites and their relationship to Tertiary tectonics, San Diego County; pp. 107-113 in P. L. Abbott and J. A. May (eds.), Eocene Geologic History San Diego Region. Pacific Section SEPM Volume 68.

Black, C. C. 1971. Paleontology and geology of the Badwater Creek area, central Wyoming. Part 7. Rodents of the Family Ischyromyidae. Annals of Carnegie Museum 43:179-217.

Black, C. C., and M. R. Dawson. 1966. Paleontology and geology of the Badwater Creek area, central Wyoming. Part 1. History of field work and geological setting. Annals of Carnegie Museum 38:297-307.

Bloch, J. I. 1995. A diminutive insectivore (Mammalia: Lipotyphla) from a Wasatchian early Eocene limestone of the Clarks Fork Basin, Wyoming. Journal of Vertebrate Paleontology 15, supplement to no. 3:19A.

Bryant, H. N. 1992. The Carnivora of the Lac Pelletier Lower Fauna (Eocene: Duchesnean), Cypress Hills Formation, Saskatchewan. Journal of Paleontology 66:847-855.

Bukry, D. 1991. Transoceanic correlation of middle Eocene coccolith Subzone CP14a at Bataquitos Lagoon, San Diego County; pp. 189-193 in P. L. Abbott and J. A. May (eds.), Eocene Geologic History San Diego Region. Pacific Section SEPM Volume 68.

Bukry, D., and M. P. Kennedy. 1969. Cretaceous and Eocene coccoliths at San Diego, California. Califonia Division of Mines and Geology Special Report 100:33-43.

Cappetta, H. 1987. Chondrichthyes II: Mesozoic and Cenozoic Elasmobranchii; pp. 1-193 in H.-P. Schultze (ed.), Handbook of Paleoichthyology, Volume 3B. Gustav Fischer Verlag, Stuttgart.

Cashion, W. B. 1974. Geologic map of the Southam Canyon quadrangle, Uintah County, Utah. United States Geological Survey Miscellaneous Field Studies Map MF-579.

Cashion, W. B. 1986. Geologic map of the Bonanza quadrangle, Uintah County, Utah. United States Geological Survey Miscellaneous Field Studies Map MF-1865.

Cashion, W. B. 1994. Geologic map of the Nutter's Hole quadrangle, Uintah and Carbon counties, Utah. United States Geological Survey Miscellaneous Field Studies Map MF-2250.

Chiment, J. J. 1977. A new genus of eomyid rodents from the later Eocene (Uintan) of southern California. Unpub. Master's thesis, Department of Biology, San Diego State University, 83 p.

Chiment, J. J., and W. W. Korth, in press. A new genus of eomyid rodent (Mammalia) from the Eocene (Uintan-Duchesnean) of southern California. Journal of Vertebrate Paleontology.

Colbert, M. W. 1993. New species of tapiroids from the Eocene of San Diego County, California, and their implications to tapiroid phylogeny and evolution. Unpub. M.S. thesis, Department of Biology, San Diego State University, 271pp.

Coombs, M. C. 1971. Status of Simidectes (Insectivora, Pantolestoidea) of the late Eocene of North America.

American Museum Novitates 2455:1-41.

Deméré, T. A. 1988. Early Arikareean (late Oligocene) vertebrate fossils and biostratigraphic correlations of the Otay Formation at Eastlake, San Diego County, California; pp. 35-43 in M. V. Filewicz and R. L. Squires (eds.), Paleogene Stratigraphy, West Coast of North America. Pacific Section, SEPM Volume 58.

Douglass, E. 1914. Geology of the Uinta Formation. Bulletin of the Geological Society of America 25:417-420.

Eaton, J. G. 1985. Paleontology and correlation of the Eocene Tepee Trail and Wiggins formations in the North Fork of Owl Creek area, southeastern Absaroka Range, Hot Springs County, Wyoming. Journal of Vertebrate Paleontology 5:345-370.

Ellis, A. J. 1919. Geology, western part of San Diego County, California. United States Geological Society Water- Supply Paper 446:50-76.

Elsik, W. C., and J. E. Boyer. 1977. Palynomorphs from the middle Eocene Delmar Formation and Torrey Sandstone, coastal southern California. Palynology 1:173 (abstract).

Emry, R. J. 1990 Mammals of the Bridgerian (middle Eocene) Elderberry Canyon Local Fauna of eastern Nevada. Geological Society of America Special Paper 243:187-210.

Emry, R.J., and W. W. Korth. 1989. Rodents of the Bridgerian (middle Eocene) Elderberry Canyon Local Fauna of Eastern Nevada. Smithsonian Contributions to Paleobiology 67:1-14.

Ferrusquia-Villafranca, I. 1984. A review of the early and middle Tertiary mammal faunas of Mexico. Journal of Vertebrate Paleontology 4:187-198.

Flynn, J. J. 1986a. Correlation and geochronology of middle Eocene strata from the western United States. Palaeogeography, Palaeoclimatology, Palaeoecology 55:335-406.

Flynn, J. J. 1986b. Faunal provinces and the Simpson Coefficient. Contributions to Geology, University of Wyoming, Special Paper 3:317-338.

Flynn, J. J., and H. Galiano. 1982. Phylogeny of early Tertiary Carnivora, with a description of a new species of Protictis from the middle Eocene of northwestern Wyoming. American Museum Novitates 2725:1-64.

Frederiksen, N. O. 1989. Eocene sporomorph biostratigraphy of southern California. Palaeontographica, Abteilung B 211:135-179.

Frederiksen, N. O. 1991a. Age determinations for Eocene formations of the San Diego, California, area, based on pollen data; pp. 195-199 in P. L. Abbott and J. A. May (eds.), Eocene Geologic History San Diego Region. Pacific Section SEPM Volume 68.

Frederiksen, N. O. 1991b. Pulses of middle Eocene to earliest Oligocene climatic deterioration in southern California and the Gulf Coast. Palaios 6:564-571.

Gaskell, B. A. 1991. Extinction patterns in Paleogene benthic foraminiferal faunas: Relationship to climate and sea-level. Palaios 6:2-16.

Gazin, C. L. 1955. A review of the upper Eocene Artiodactyla of North America. Smithsonian Miscellaneous Collections 128(8):1-96.

Givens, C. R. 1974. Eocene molluscan biostratigraphy of the Pine Mountain area, Ventura County, California. University of California Publications in Geological Sciences 109:1-107.

Givens, C. R., and M. P. Kennedy. 1979. Eocene molluscan stages and their correlation, San Diego area, California; pp. 81-95 in P. L. Abbott (ed.), Eocene Depositional Systems, San Diego, California. Pacific

Section, SEPM Field Trip Guidebook.

Golz, D. J. 1976. Eocene Artiodactyla of southern California. Los Angeles County Natural History Museum Bulletin 26:1-85.

Golz, D. J., and J. A. Lillegraven, 1977. Summary of known occurrences of terrestrial vertebrates from Eocene strata of southern California. Contributions to Geology, University of Wyoming 15:43-65.

Gunnell, G. F. In press. New notharctine (Primates: Adapiformes) skull from the early Uintan (middle Eocene) of southern California. American Journal of Physical Anthropology.

Gustafson, E. P. 1986. Carnivorous mammals of the late Eocene and early Oligocene of Trans-Pecos Texas. Texas Memorial Museum Bulletin 33:1-66.

Hanna, M. A. 1926. Geology of the La Jolla quadrangle, California. University of California Publications in Geological Sciences Bulletin 16:187-246.

Haq, B. U., J. Hardenbol, and P. R. Vail. 1987. Chronology of fluctuating sea levels since the Triassic. Science 235:1156-1167.

Hatcher, J. B. 1895. On a new species of Diplacodon, with a discussion of the relations of that genus to Telmatotherium. American Naturalist 29:1084-1090.

Holtzman, R. C. 1979. Maximum likelihood estimation of fossil assemblage composition. Paleobiology 5:77-89.

Honey, J. G. 1990. New washakiin primates (Omomyidae) from the Eocene of Wyoming and Colorado, and comments on the evolution of the Washakiini. Journal of Vertebrate Paleontology 10:206-21.

Howard, J. L. 1989. Conglomerate clast populations of the upper Paleogene Sespe formation, southern California; pp. 269-280 in I. P. Colburn, P. L. Abbott, and J. A. Minch (eds.), Conglomerates in Basin Analysis: A Symposium Dedicated to A. O. Woodford. Pacific Section SEPM Volume 62.

Hutchison, J. H. 1971. Cf. Uintatherium (Dinocerata, Mammalia) from the Uintan (middle to late Eocene) of southern California. PaleoBios 12:1-8.

Kay, J. L. 1934. The Tertiary formations of the Uinta Basin, Utah. Annals of Carnegie Museum 23:357-371.

Keller, G., N. MacLeod, and E. Barrera. 1992. Eocene-Oligocene faunal turnover in planktonic foraminifera, and Antarctic glaciation; pp. 218-244 in D. R. Prothero and W. A. Berggren (eds.), Eocene-Oligocene Climatic and Biotic Evolution. Princeton University Press, Princeton, N.J.

Kelly, T. S. 1990. Biostratigraphy of the Uintan and Duchesnean land mammal assemblages from the middle member of the Sespe Formation, Simi Valley, California. Los Angeles County Natural History Museum Contributions in Science 419:1-42.

Kelly, T. S. 1992. New Uintan and Duchesnean (middle and late Eocene) rodents from the Sespe Formation, Simi Valley, California. Bulletin of the Southern California Academy of Science. 91:97-120.

Kelly, T. S., E. B. Lander, D. P. Whistler, M. A. Roeder, and R. E. Reynolds, 1991. Preliminary report on a paleontologic investigation of the lower and middle members, Sespe Formation, Simi Valley Landfill, Ventura County, California. Paleobios 13:1-13.

Kelly, T. S., and D. P. Whistler. 1994. Additional Uintan and Duchesnean (middle and late Eocene) mammals from the Sespe Formation, Simi Valley, California. Los Angeles County Natural History Museum Contributions in Science 439:1-29.

Kennedy, M. P., and G. W. Moore. 1971. Stratigraphic relations of Upper Cretaceous and Eocene formations, San Diego coastal region, California. American

Association of Petroleum Geologists Bulletin 55:709-722.

Kennedy, M. P., and G. L. Peterson. 1975. Geology of the Eastern San Diego Metropolitan area, California. California Division of Mines and Geology Bulletin 200-B:43-56.

Kern, J. P. 1978. Paleoenvironment of new trace fossils from the Eocene Mission Valley Formation, California. Journal of Paleontology 52:186-194.

Keroher, G. C., et al. 1966. Lexicon of geologic names of the United States for 1936-1960. United States Geological Survey Bulletin 1200, Part 3:2887-4341.

Korth, W. W. 1985. The rodents *Pseudotomus* and *Quadratomus* and the content of the Tribe Manitshini (Paramyinae, Ischyromyidae). Journal of Vertebrate Paleontology 5:139-152.

Korth, W. W. 1988. *Paramys compressidens* Peterson and the systematic relationships of the species of *Paramys* (Paramyinae, Ischyromyidae). Journal of Paleontology 62:468-471.

Krishtalka, L. 1976. Early Tertiary Adapisoricidae and Erinaceidae (Mammalia, Insectivora) of North America. Bulletin of Carnegie Museum of Natural History 1:1-40.

Krishtalka, L. 1978. Paleontology and geology of the Badwater Creek area, central Wyoming. Part 15. Review of the late Eocene primates from Wyoming and Utah, and the Plesitarsiiformes. Annals of Carnegie Museum 47:335-359.

Krishtalka, L., and R. K. Stucky. 1983. Paleocene and Eocene marsupials of North America. Annals of Carnegie Museum 52:229-263.

Krishtalka, L., and R. K. Stucky. 1984. Middle Eocene marsupials (Mammalia) from northeastern Utah and the mammalian fauna from Powder Wash. Annals of Carnegie Museum 53:31-45.

Krishtalka, L., R. K. Stucky, R. M. West, M. C. McKenna, C. C. Black, T. M. Bown, M. R. Dawson, D. J. Golz, J. J. Flynn, J. A. Lillegraven, and W. D. Turnbull. 1987. Eocene (Wasatchian through Duchesnean) biochronology of North America; pp. 77-117 in M. O. Woodburne (ed.), Cenozoic Mammals of North America, Geochronology and Biostratigraphy. University of California Press, Berkeley.

Lillegraven, J. A. 1976. Didelphids (Marsupialia) and *Uintasorex* (?Primates) from later Eocene sediments of San Diego County, California. San Diego Society of Natural History Transactions 18:85-112.

Lillegraven, J. A. 1977. Small rodents (Mammalia) from Eocene deposits of San Diego County, California. Bulletin of the American Museum of Natural History 158:221-262.

Lillegraven, J. A. 1979. A biogeographical problem involving comparisons of later Eocene terrestrial vertebrate faunas of western North America; pp. 333-347 in J. Gray and A. J. Boucot (eds.), Historical biogeography, plate tectonics, and the changing environment. Oregon State University Press, Corvallis.

Lillegraven, J. A. 1980. Primates from later Eocene rocks of southern California. Journal of Mammalogy 61:181-204.

Lillegraven, J. A., M. C. McKenna, and L. Krishtalka. 1981. Evolutionary relationships of middle Eocene and younger species of *Centetodon* (Mammalia, Insectivora, Geolabididae), with a description of the dentition of *Ankylodon* (Adapisoricidae). University of Wyoming Publications 45:1-115.

Lillegraven, J. A., and R. W. Wilson. 1975. Analysis of *Simimys simplex*, an Eocene rodent (?Zapodidae). Journal of Paleontology 49:856-874.

Lindsay, E. H. 1968. Rodents from the Hartman Ranch local fauna, California. PaleoBios 6:1-20.

Lohmar, J. M., J. A. May, J. E. Boyer, and J. E. Warme. 1979. Shelf edge deposits of the San Diego Embayment; pp. 15-33 in P. L. Abbott (ed.), Eocene Depositional Systems, San Diego, California. Pacific Section, SEPM Field Trip Guidebook.

Lucas, S. G. 1992. Redefinition of the Duchesnean Land Mammal "Age," late Eocene of western North America; pp. 88-105 in D. R. Prothero and W. A. Berggren (eds.), Eocene-Oligocene Climatic and Biotic Evolution. Princeton University Press, Princeton, N.J.

Lucas, S. G., and R. M. Schoch. 1989. Taxonomy of *Duchesneodus* (Brontotheriidae) from the late Eocene of North America; pp. 490-503 in D. R. Prothero and R. M. Schoch (eds.), The Evolution of Perissodactyls. Oxford University Press, New York.

Lucas, S. G., and T. E. Williamson. 1993. Eocene vertebrates and late Laramide stratigraphy of New Mexico; pp. 145-158 in S. G. Lucas and J. Zidek (eds.), Vertebrate Paleontology in New Mexico. New Mexico Museum of Natural History and Science Bulletin 2.

Mader, B. J. 1989. The Brontotheriidae: A systematic revision and preliminary phylogeny of North American genera; pp. 458-484 in D. R. Prothero and R. M. Schoch (eds.), The Evolution of Perissodactyls. Oxford University Press, New York.

Mason, M. A. 1988. Mammalian paleontology and stratigraphy of the early to middle Tertiary Sespe and Titus Canyon formations, southern California. Ph.D. dissertation, University of California, Berkely, 257 pp.

Mason, M. A. 1990. New fossil primates from the Uintan (Eocene) of southern California. PaleoBios 13:1-7.

Mason, M. A., and C. C. Swisher. 1989. New evidence for the age of the South Mountain Local Fauna, Ventura County, California. Los Angeles County Natural History Museum Contributions in Science no. 410:1-9.

May, J. A., J. M. Lohmar, J. E. Warme, and S. Morgan. 1991. Field trip guide: early to middle Eocene La Jolla Group of Black's Beach, La Jolla, California; pp. 27-36 in P. L. Abbott and J. A. May (eds.), Eocene Geologic History San Diego Region. Pacific Section, SEPM Volume 68.

McGowran, B. 1990. Fifty million years ago. American Scientist 78:31-39.

McKenna, M. C. 1980. Late Cretaceous and early Tertiary vertebrate paleontological reconnaissance, Togwotee Pass area, northwestern Wyoming; pp. 321-343 in L. L. Jacobs (ed.), Aspects of Vertebrate History. Museum of Northern Arizona Press, Flagstaff.

McKenna, M. C. 1990. Plagiomenids (Mammalia: ?Dermoptera) from the Oligocene of Oregon, Montana, and South Dakota, and middle Eocene of northwestern Wyoming. Geological Society of America Special Paper 243:211-234.

Miller, K. G. 1992. Middle Eocene to Oligocene stable isotopes, climate, and deep-water history: The Terminal Eocene Event?; pp. 160-177 in D. R. Prothero and W. A. Berggren (eds.), Eocene-Oligocene Climatic and Biotic Evolution. Princeton University Press, Princeton, N.J.

Novacek, M. J. 1976. Insectivora and Proteutheria of the later Eocene (Uintan) of San Diego County, California. Los Angeles County Natural History Museum Contributions in Science 283:1-52.

Novacek, M. J. 1985. The Sespedectinae, a new subfamily of hedgehog-like insectivores. American Museum Novitates 2822:1-24.

Novacek, M. J., T. M. Bown, and D. M. Schankler. 1985. On the classification of the early Tertiary

Erinaceomorpha (Insectivora: Mammalia). American Museum Novitates 2813:1-22.

Novacek, M. J., I. Ferrusquia-Villafranca, J. J. Flynn, A. R. Wyss, and M. Norell. 1991. Wasatchian (early Eocene) mammals and other vertebrates from Baja California, Mexico: The Lomas Las Tetas de Cabra fauna. Bulletin of the American Museum of Natural History 208:1-88.

Novacek, M. J., and J. A. Lillegraven. 1979. Terrestrial vertebrates from the later Eocene of San Diego County, California: a conspectus; pp. 69-79 in P. L. Abbott (ed.), Eocene Depositional Systems, San Diego, California. Pacific Section SEPM Field Trip Guidebook.

Okada, H., and D. Bukry. 1980. Supplementary modification and introduction of code numbers to the low-latitude coccolith biostratigrapic zonation (Bukry, 1973; 1975). Marine Micropaleontology 5:321-324.

Osborn, H. F., and W. D. Matthew, 1909. Cenozoic mammal horizons of western North America. United States Geological Survey Bulletin 361:1-138.

Osborn, H. F. 1895. Fossil mammals of the Uinta Basin: Expedition of 1894. Bulletin of the American Museum of Natural History 7:71-105.

Osborn, H. F. 1929. The titanotheres of ancient Wyoming, Dakota, and Nebraska. United States Geological Survey Monograph 55:1-953.

Peterson, G. L., and P. L. Abbott. 1979. Mid-Eocene climatic change, southwestern California and northwestern Baja California. Palaeogeography, Palaeoclimatology, Palaeoecology 26:73-87.

Prothero, D. R., and C. C. Swisher, 1992. Magnetostratigraphy and geochronology of the terrestrial Eocene-Oligocene transition in North America; pp. 46-73 in D. R. Prothero and W. A. Berggren (eds.), Eocene-Oligocene Climatic and Biotic Evolution. Princeton University Press, Princeton, N.J.

Prothero, D. R., D. J. Lundquist, T. A. Deméré, and S. L. Walsh. 1993. Magnetostratigraphy of middle Eocene deposits from the San Diego area, California. Geological Society of America Abstracts with Programs 25:A-473.

Radinsky, L. B. 1967. Hyrachyus, Chasmotherium, and the early evolution of helaletid tapiroids. American Museum Novitates 2313:1-23.

Rasmussen, D. T., M. Shekelle, S. L. Walsh, and B. O. Riney. In press. The dentition of Dyseolemur, and comments on the use of the anterior teeth in primate systematics. Journal of Human Evolution.

Roth, B. 1988. Camaenid land snails (Gastropoda: Pulmonata) from the Eocene of southern California and their bearing on the history of the American Camaenidae. San Diego Society of Natural History Transactions 21:203-220.

Roth, B. 1991. Tropical "physiognomy" of a land snail faunule from the Eocene of southern California. Malacologia 33:281-288.

Roth, B., and T. A. Pearce. 1988. "Micrarionta" dallasi, a helicinid (prosobranch), not a helminthoglyptid (pulmonate) land snail: paleoclimatic implications. Southwestern Naturalist 33:117-119.

Rowley, P. D., W. R. Hansen, O. Tweto, and P. E. Carrara. 1985. Geologic map of the Vernal 1° X 2° quadrangle, Colorado, Utah, and Wyoming. United States Geological Survey Map I-1526.

Runkel, A. C. 1988. Stratigraphy, sedimentology, and vertebrate paleontology of Eocene rocks, Big Bend region, Texas. Ph.D. dissertation, University of Texas, Austin, 310 pp.

Savage, D. E., and T. Downs. 1954. Cenozoic land life of southern California; pp. 43-58 in R. H. Jahns (ed.), Geology of Southern California. California Division of Mines and Geology Bulletin 170.

Savin, S. M., R. G. Douglass, and F. G. Stehli. 1975. Tertiary marine paleotemperatures. Geological Society of America Bulletin 86:1499-1510.

Schiebout, J. A. 1977. Eocene Perissodactyla from the La Jolla and Poway Groups, San Diego County, California. San Diego Society of Natural History Transactions 18:217-227.

Schiebout, J. A. 1979. An overview of the terrestrial early Tertiary of southern North America—fossil sites and paleopedology. Tulane University Studies in Geology and Paleontology 15:75-94.

Simpson, G. G. 1960. Notes on the measurement of faunal resemblance. American Journal of Science 258-A:300-311.

Sloan, L. C., and E. J. Barron. 1992. Paleogene climatic evolution: A climate model investigation of the influence of continental elevation and sea-surface temperature upon continental climate; pp. 202-217 in D. R. Prothero and W. A. Berggren (eds.), Eocene-Oligocene Climatic and Biotic Evolution. Princeton University Press, Princeton, N.J.

Squires, R. L. 1988. Geologic age refinements of West Coast Eocene marine molluscs; pp. 107-112 in M. V. Filewicz and R. L. Squires (eds.), Paleogene Stratigraphy, West Coast of North America. Pacific Section SEPM Volume 58.

Stagner, W. L. 1941. The paleogeography of the eastern part of the Uinta Basin during Uinta B (Eocene) time. Annals of Carnegie Museum 28:273-308.

Steineck, P. L., J. M. Gibson, and R. W. Morin. 1972. Foraminifera from the middle Eocene Rose Canyon and Poway formations, San Diego, California. Journal of Foraminiferal Research 2:137-144.

Stock, C. 1948. Pushing back the history of land mammals in western North America. Geological Society of America Bulletin 59:327-332.

Storer, J. E. 1984. Mammals of the Swift Current Creek Local Fauna (Eocene, Uintan, Saskatchewan). Saskatchewan Museum of Natural History Contributions 7:1-158.

Storer, J. E. 1989. Rodent faunal provinces, Paleocene-Miocene of North America; pp. 17-29 in C. C. Black and M. R. Dawson (eds.), Papers on Fossil Rodents in Honor of Albert Elmer Wood. Natural History Museum of Los Angeles County Science Series 33.

Stucky, R. K. 1992. Mammalian faunas in North America of Bridgerian to early Arikareean "ages" (Eocene to Oligocene); pp. 464-493 in D. R. Prothero and W. A. Berggren (eds.), Eocene-Oligocene Climatic and Biotic Evolution. Princeton University Press, Princeton, N.J.

Sutton, J. F., and C. C. Black. 1975. Paleontology of the earliest Oligocene deposits in Jackson Hole, Wyoming. Part 1. Rodents exclusive of the Family Eomyidae. Annals of Carnegie Museum 45:299-315.

Tedford, R. H., M. F. Skinner, R. W. Fields, J. M. Rensberger, D. P. Whistler, T. Galusha, B. E. Taylor, J. R. McDonald, and S. D. Webb. 1987. Faunal succession and biochronology of the Arikareean through Hemphillian interval (late Oligocene through earliest Pliocene Epochs) in North America; pp. 153-210 in M. O. Woodburne (ed.), Cenozoic Mammals of North America, Geochronology and Biostratigraphy. University of California Press, Berkeley.

Troxell, E. L. 1923. Pauromys perditus, a small rodent. American Journal of Science, series 5, 5:155-156.

Turnbull, W. D. 1991. Protoptychus hatcheri Scott, 1895. The mammalian faunas of the Washakie Formation, Eocene age, of southern Wyoming. Part II. The

Adobetown Member, middle division (= Washakie B), Twka2 (in part). Fieldiana: Geology new series no. 21:1-33.

Untermann, G. E., B. R. Untermann, and D. M. Kinney. 1964. Geologic map of Uintah County, Utah (south half). Utah Geological and Mineralogical Survey Bulletin 72 (supplement).

Vianey-Liaud, M. 1985. Possible evolutionary relationships among Eocene and lower Oligocene rodents of Asia, Europe, and North America; pp. 277-309 *in* W. P. Luckett and J. L. Hartenberger (eds.), Evolutionary Relationships Among Rodents: A Multidisciplinary Analysis. NATO Advanced Science Series, ser. A, Life Sciences, Volume 92. Plenum Press, New York.

Walsh, S. L. 1987. Mammalian paleontology of the southern outcrops of the Mission Valley Formation, San Diego County, California. Unpub. undergraduate thesis, Department of Geology, San Diego State University, 171 pp.

Walsh, S. L. 1991a. Late Eocene mammals from the Sweetwater Formation, San Diego County, California; pp. 149-159 *in* P. L. Abbott and J. A. May (eds.), Eocene Geologic History San Diego Region. Pacific Section SEPM Volume 68.

Walsh, S. L. 1991b. Eocene mammal faunas of San Diego County; pp. 161-178 *in* P. L. Abbott and J. A. May (eds.), Eocene Geologic History San Diego Region. Pacific Section SEPM Volume 68.

Walsh, S. L., and T. A. Deméré. 1991. Age and stratigraphy of the Sweetwater and Otay formations, San Diego County, California; pp. 131-148 *in* P. L. Abbott and J. A. May (eds.), Eocene Geologic History San Diego Region. Pacific Section SEPM Volume 68.

Walton, A. H. 1992. Magnetostratigraphy and the ages of Bridgerian and Uintan faunas in the lower and middle members of the Devil's Graveyard Formation, Trans-Pecos Texas; pp. 74-87 *in* D. R. Prothero and W. A. Berggren (eds.), Eocene-Oligocene Climatic and Biotic Evolution. Princeton University Press, Princeton, N.J.

Walton, A. H. 1993a. *Pauromys* and other small Sciuravidae (Mammalia: Rodentia) from the middle Eocene of Texas. Journal of Vertebrate Paleontology 13:243-261.

Walton, A. H. 1993b. A new genus of eutypomyid (Mammalia: Rodentia) from the middle Eocene of the Texas Gulf Coast. Journal of Vertebrate Paleontology 13:262-266.

West, R. M. 1973. Review of the North American Eocene and Oligocene Apatemyidae (Mammalia: Insectivora). Museum of Texas Tech. University Special Publication 3:1-42.

West, R. M. 1982. Fossil mammals from the lower Buck Hill Group, Eocene of southwest Texas. Marsupicarnivora, Primates, Taeniodonta, Condylarthra, bunodont Artiodactyla, and Dinocerata. Pearce-Sellards Series 35:1-20.

Westgate, J. W. 1988. Biostratigraphic implications of the first Eocene land-mammal fauna from the North American coastal plain. Geology 16:995-998.

Westgate, J. W. 1990. Uintan land mammals (excluding rodents) from an estuarine facies of the Laredo Formation (middle Eocene, Claiborne Group), Webb County, Texas. Journal of Paleontology 64:454-468.

Westgate, J. W. 1994a. A new leptochoerid from middle Eocene (Uintan) deposits of the Texas Coastal Plain. Journal of Vertebrate Paleontology 14:296-299.

Westgate, J. W. 1994b. Stratigraphic and paleoecologic analysis of land mammal-bearing Eocene strata on Gulf Coastal Plain. Petroleum Research Fund, 38th Annual Report on Research, p. 161-162 (abstract).

Wheeler, W. H. 1961. Revision of the uintatheres. Yale Peabody Museum of Natural History Bulletin 14:1-93.

Wilson, J. A. 1977. Early Tertiary vertebrate faunas, Big Bend area, Trans-Pecos Texas: Brontotheriidae. Pearce-Sellards Series 25:1-17.

Wilson, J. A. 1984. Vertebrate faunas 49 to 36 million years ago and additions to the species of *Leptoreodon* (Mammalia: Artiodactyla) found in Texas. Journal of Vertebrate Paleontology 4:199-207.

Wilson, J. A. 1986. Stratigraphic occurrence and correlation of early Tertiary vertebrate faunas, Trans-Pecos Texas: Agua Fria- Green Valley areas. Journal of Vertebrate Paleontology 6:350-373.

Wilson, J. A., and J. A. Schiebout. 1981. Early Tertiary vertebrate faunas, Trans-Pecos Texas: Amynodontidae. Pearce-Sellards Series 33:1-62.

Wilson, J. A., and J. W. Westgate. 1991. A lophodont rodent from the middle Eocene of the Gulf Coastal Plain, Texas. Journal of Vertebrate Paleontology 11:257-260.

Wilson, K. L. 1972. Eocene and related geology of a portion of the San Luis Rey and Encinitas quadrangles, San Diego County, California. M.S. thesis (unpub.), University of California, Riverside, 135 pp.

Wood, A. E. 1974. Early Tertiary vertebrate faunas, Vieja Group, Trans-Pecos Texas: Rodentia. Texas Memorial Museum Bulletin 21:1-112.

Wood, H. E. II, R. W. Chaney, J. Clark, E. H. Colbert, G. L. Jepsen, J. B. Reeside Jr., and C. Stock. 1941. Nomenclature and correlation of the North American continental Tertiary. Geological Society of America Bulletin 52:1-48.

Woodburne, M. O. (ed.). 1987. Cenozoic Mammals of North America, Geochronology and Biostratigraphy. University of California Press, Berkeley.

Woodburne, M. O., and C. C. Swisher. 1995. Land mammal high-resolution geochronology, intercontinental overland dispersals, sea level, climate, and vicariance. SEPM Special Publication 54:329-358.

Appendix 1: SDSNH Bridgerian, Uintan, and Duchesnean mammal localities, grouped by subage and stratigraphic unit.

BRIDGERIAN
Delmar Formation
3542 Swami's Point
3671 State Route 56 West site 1-A
3793 I-5 and State Route 56 West site 4

LATE BRIDGERIAN OR EARLIEST UINTAN
Scripps Formation
3278 Computer Media
3381 Dike Rock
3567 Waterfall Locality
3830 Black's Canyon Road

EARLIEST UINTAN
Friars Formation (type outcrops and undiff.)
2964 Murphy Canyon
2999 Murphy Canyon
3145 Murphy Canyon
3414 SDSU Parking Lot 1
3430 SDSU Parking Lot 2
3488 Unocal Cut Slope (= UCMP V-6888)
3637 Friars Road
3639 Hotel Circle South
3649 San Diego Mission 1
3669 San Diego Mission 2
3770 Mission Gorge and Margerum
3781 Stonecrest Square site 1
3784 Stonecrest Square site 4
3785 Stonecrest Square site 5
3786 Stonecrest Square site 6
3787 Stonecrest Square site 7
3788 Stonecrest Square site 8
3789 Stonecrest Square site 9
3790 Stonecrest Square site 10
3831 Mission Terrace sites 1+5
3832 Mission Terrace site 4
3833 Mission Terrace site 7
3849 Waring Road Friars Hole
3851 Stonecrest Square site 11
3898 Del Mar Heights Road Extension site 3

Friars Formation, lower tongue
3494 Westview site 4
3496 Westview site 6
3502 South Creek site 2
3503 South Creek site 3
3505 South Creek site 5
3654 State Route 52 East site 4
3655 State Route 52 East site 5
3656 State Route 52 East site 6
3657 State Route 52 East site 7
3658 State Route 52 East site 8
3659 State Route 52 East site 9
3660 State Route 52 East site 10
3661 State Route 52 East site 11
3662 State Route 52 East site 12
3664 State Route 52 East site 14
3665 State Route 52 East site 15
3666 State Route 52 East site 16
3667 State Route 52 East site 17
3727 Black Mountain Road
3828 4950 Murphy Canyon Road
3892 Silver Country Estates site 2
3893 Silver Country Estates site 3
3894 Silver Country Estates site 4

Friars Formation, conglomerate tongue
3504 South Creek site 4
3507 South Creek site 7
3509 South Creek site 9
3510 South Creek site 10
3615 Scripps Ranch North site 15
3616 Scripps Ranch North site 16
3617 Scripps Ranch North site 17
3618 Scripps Ranch North site 18
3619 Scripps Ranch North site 19
3620 Scripps Ranch North site 20
3621 Scripps Ranch North site 20-B
3824 Scripps Ranch North site 20-R
3730 Clairemont Mesa Boulevard
3732 Scripps Ranch North site 32
3736 Scripps Ranch North site 36
3737 Scripps Ranch North site 37
3738 Scripps Ranch North site 38
3739 Scripps Ranch North site 39
3753 Scripps Ranch North site 53

Friars Formation, upper tongue
2789 Rancho de los Peñasquitos (= V-73138)
2974 Gabacho Drive
3254 Azuaga
3373 Carmel Mountain Ranch Unit 15 site 2
3380 Carmel Mountain Ranch Unit 16 site 3
3391 Carmel Mountain Ranch Unit 16 site 1
3410 Azuaga II site 1
3411 Azuaga II site 3
3412 Azuaga II site 4
3413 Azuaga II site 5
3482 Carmel Mountain Ranch Unit 20-B site 2
3483 Carmel Mountain Ranch Unit 20-B site 3
3484 Carmel Mountain Ranch Unit 20-B site 4
3498 Eastview site 8
3591 Carmel Mountain Ranch Unit 20-A site 1
3592 Carmel Mountain Ranch Unit 20-A site 2
3611 State Route 52 West site 1
3612 State Route 52 West site 2
3623 Scripps Ranch North site 23
3681 State Route 56 East site 1
3682 State Route 56 East site 2
3685 State Route 52 West site 5
3693 Hill 781 Saddle Cut 1
3735 Scripps Ranch North site 35
3771 Carmel Mountain Ranch Unit 19 site 1
3772 Carmel Mountain Ranch Unit 19 site 2
3773 Carmel Mountain Ranch Unit 19 site 3
3777 CMB and MCR Intersection
3825 Waxie Business Park site 1
3862 Scripps Ranch North site 62
3883 Fiesta Island Replacement Project

Santiago Formation, Member B
3440 Rancho del Oro Village III site 10
3443 Rancho del Oro Village III site 13
3447 Rancho del Oro Village III site 17
3448 Rancho del Oro Village III site 10-S
3450 Rancho del Oro Village III site 13-N
3465 Rancho del Oro Village II site 5
3486 Aviara (lower bronto bed)
3571 Mission del Oro site 1

LATE EARLY UINTAN
Stadium Conglomerate, lower member
3691 Murray Canyon 1
3692 Murray Canyon 2
3701 Waring Road
3731 Fenton Quarry
3852 Stonecrest Square site 12

LATE UINTAN
Stadium Conglomerate, lower or upper member
3526 Stonecrest site 7
3527 Stonecrest site 8
3528 Stonecrest site 9
3530 Stonecrest site 13
3536 Stonecrest site 27
3538 Stonecrest site 29

Mission Valley Formation
3273 Cloud 9
3383 Aqueduct Tower
3426 Collwood South
3428 Jackson Offramp
3429 *Cylindracanthus* Spot
3537 Stonecrest site 28
3539 Briercrest Park
3627 Scripps Ranch North site 27
3629 Scripps Ranch North site 29-N
3638 Witherspoon Way
3694 Hill 781 Saddle Cut 2
3715 40th Street
3741 Scripps Ranch North site 41-D
3742 Scripps Ranch North site 42-C
3760 Scripps Ranch North site 60
3822 State Route 54 East site 2
3866 Scripps Ranch North site 66
3867 Scripps Ranch North site 67
3870 Scripps Ranch North site 70
3874 Scripps Ranch North site 74

Pomerado Conglomerate, lower member
3493 Eastview site 3
3755 Scripps Ranch North site 55

Pomerado Conglomerate, Miramar Sandstone Member
3756 Scripps Ranch North site 56
3757 Scripps Ranch North site 57

Santiago Formation, Member C
3276 Jeff's Discovery

3378 College Boulevard site 14
3431 Rancho del Oro Village III site 1
3432 Rancho del Oro Village III site 2
3433 Rancho del Oro Village III site 3
3434 Rancho del Oro Village III site 4
3436 Rancho del Oro Village III site 6
3437 Rancho del Oro Village III site 7
3441 Rancho del Oro Village III site 11
3442 Rancho del Oro Village III site 12
3444 Rancho del Oro Village III site 14
3445 Rancho del Oro Village III site 15
3449 Rancho del Oro Village III site 11-W
3451 Rancho del Oro Village III site 6-W
3463 Rancho del Oro Village II site 3
3464 Rancho del Oro Village II site 4
3466 Rancho del Oro Village II site 6
3467 Rancho del Oro Village II site 7
3560-3564 Jeff's Discovery (Caltrans Quarry)
3573 Mission del Oro site 3

UINTAN UNDIFFERENTIATED
Santiago Formation
3512 Airport Business Center II site 2

LATEST UINTAN AND/OR DUCHESNEAN
Santiago Formation, Member C
3279 Tamarack Avenue
3487 Aviara (green gravel unit)
3570 Mission del Oro site 5
3572 Mission del Oro site 2
3574 Mission del Oro site 4

"Sweetwater" Formation
3470 Bonita Long Canyon site 5
3471 Bonita Long Canyon site 12
3474 Bonita Long Canyon site 17
3550 Sweetwater Regional Equestrian Park
3608 Bonita Valley Christian Center
3821 State Route 54 East site 1
3823 State Route 54 East site 3

Pomerado Conglomerate, upper member
3759 Scripps Ranch North site 59

DUCHESNEAN
Santiago Formation, Member C
3235 Watertower Hill
3495 Kelly's Ranch

6. Stratigraphy and Paleomagnetism of the Middle Eocene Friars Formation and Poway Group, Southwestern San Diego County, California

STEPHEN L. WALSH, DONALD R. PROTHERO, AND DAVID J. LUNDQUIST

ABSTRACT

The stratigraphy of the Friars Formation and Poway Group in southwestern San Diego County is informally revised. The Friars Formation is locally divisible into three units: a lower sandstone and mudstone tongue, a middle conglomerate tongue, and an upper sandstone and mudstone tongue. As far as can be determined, the Friars Formation is entirely of early Uintan age. The Stadium Conglomerate is stratigraphically and geographically restricted, but is nevertheless locally divisible into a lower member of late early Uintan and possibly late Uintan age, and an upper member of late Uintan age. The Mission Valley Formation of late Uintan age is stratigraphically and geographically restricted only to those strata that are correlative with the type section.

We corroborate previous reports that the lower and upper parts of the type outcrops of the Friars Formation are of normal and reversed polarity, respectively, and concur in the assignment of these polarity intervals to Chrons C21n and C20r. However, the polarity of the lower, conglomerate, and upper tongues of the Friars as recognized here is variable, and may reflect significant time-transgression of these units. The lower member of the Stadium Conglomerate is tentatively assigned to C20r. The upper member of the Stadium may straddle the C20r/C20n boundary. The lower normal and upper reversed magnetozones of the Mission Valley Formation in the type section of this formation are tentatively correlated with C20n and C19r, based on the ^{40}Ar/^{39}Ar date of 42.83 ± 0.24 Ma.

INTRODUCTION

Along the coastal plain of San Diego County, middle Eocene mammal-bearing fluvial strata assigned to the Uintan North American Land Mammal "Age" (NALMA; e.g., Krishtalka et al., 1987) interfinger with fossiliferous marine deposits (Kennedy and Moore, 1971; Golz and Lillegraven, 1977). San Diego County is the only place in North America that currently offers the potential of direct correlation of the Bridgerian and early Uintan NALMA's with various marine biochronologies, and is one of only two places in North America allowing such a correlation for the late Uintan (Westgate, 1988).

Due to the general lack of natural outcrops in San Diego County, the excellent but short-lived artificial exposures that are produced by construction activity are significant sources of stratigraphic information. Of particular importance are sections that display several lithostratigraphic units in superposition, and contain age-diagnostic fossils as well. Such sections are ideal for testing stratigraphic models, and are mandatory for the confident correlation of local magnetozones. Locations of the measured sections studied for this paper are shown in Figure 1, which illustrates the major cultural and geographic features of southwestern San Diego County.

A review of the middle Eocene mammal faunas of San Diego County is provided by Walsh (this volume, Chapter 5). Fossil mammals have proven to be useful in the correlation of local stratigraphic units that have rapid lateral facies changes, but few continuous outcrops. The main purpose of this paper is to discuss the stratigraphy and paleomagnetism of the Friars Formation and Poway Group, as these units have yielded most of the Uintan mammal assemblages of southwestern San Diego County. We have collected additional paleomagnetic samples from several important mammal-bearing Uintan sections, and will comment on recent paleomagnetic studies of the local Eocene (Flynn, 1986; Bottjer et al., 1991).

EOCENE STRATIGRAPHY OF SOUTHWESTERN SAN DIEGO COUNTY

Although the Eocene rocks of southwestern San Diego County are generally undeformed, the stratigraphic terminology for these deposits has had a complex history (Hanna, 1926; Milow and Ennis, 1961; Kennedy and Moore, 1971). Factors contributing to this complexity include: (1) the rapid lateral facies changes and obscure disconformities inherent in sedimentary deposits formed along oscillating coastlines; (2)

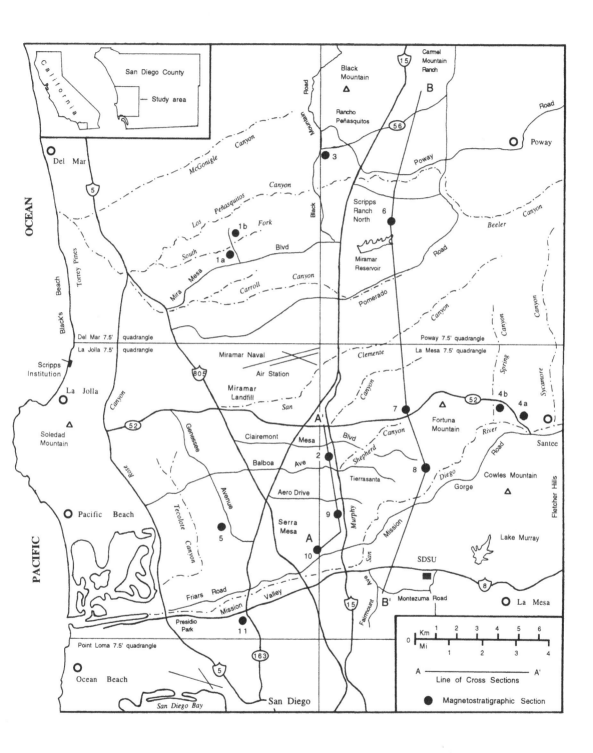

Figure 1. Location map of southwestern San Diego County showing relevant geographic and cultural features. Magnetostratigraphic sections 1-11 are described in Appendix 1.

lithological similarity of the numerous, relatively thin sandstone and conglomerate units that occur at different stratigraphic levels; (3) removal of much of the Eocene section by Plio-Pleistocene erosion (Peterson and Kennedy, 1974, fig. 2); (4) relatively poor outcrops due to the cover of the Eocene strata by Pleistocene terrace deposits, vegetation, and urbanization; and (5) locally significant faulting. Due to these factors, correlation of many isolated outcrops with named lithostratigraphic units must often be regarded as tentative.

Kennedy and Moore (1971), Peterson and Kennedy (1974), Kennedy (1975), and Kennedy and Peterson (1975) discussed the geologic setting of the coastal plain of San Diego County, and substantially modified the stratigraphic terminology of Hanna (1926) and Milow and Ennis (1961). Kennedy and Moore (1971) and Kennedy and Peterson (1975) organized the Eocene strata of southwestern San Diego County into the lower, dominantly marine La Jolla Group (consisting of the Mount Soledad Formation, Delmar Formation, Torrey Sandstone, Ardath Shale, Scripps Formation, and Friars Formation), and the overlying, dominantly nonmarine Poway Group (consisting of the Stadium Conglomerate, Mission Valley Formation, and Pomerado Conglomerate). Their model envisioned essentially continuous deposition during middle Eocene time, beginning with the Mount Soledad Formation and ending with the Pomerado Conglomerate. Three gradual marine transgressions (represented at their maxima by the Ardath Shale, Mission Valley Formation, and Miramar Sandstone Member of the Pomerado Conglomerate) and three gradual marine regressions (represented at their maxima by the Stadium Conglomerate, the lower and upper members of the Pomerado Conglomerate) were invoked to explain the lateral facies changes occurring perpendicular to the Eocene shoreline. No significant disconformities were recognized by Kennedy and Moore (1971) between any of the middle Eocene units, and all contacts between them were described as conformable.

Recently, certain aspects of the model of Kennedy and Moore (1971) have been modified or questioned. Kies (1982) and Walsh (1991) discussed the fact that some of the outcrops mapped by Kennedy (1975) as the Upper Cretaceous Cabrillo Formation were actually of Tertiary age. These distinctive strata were informally referred to by Milow and Ennis (1961) as the "greenstone lithic sandstone and conglomerate" unit. Kies (1982) and Kies and Abbott (1983) considered these strata to be a facies of the Mount Soledad Formation. Walsh (1991) demonstrated a Wasatchian (early Eocene) age for these deposits, and on the basis of their lack of Poway rhyolite clasts (Kies and Abbott, 1983) and stratigraphic position disconformably below the Mount Soledad Formation, he preferred to regard them as an unnamed formation. At the eastern base of Soledad Mountain, the unnamed formation disconfor-

mably overlies the Upper Cretaceous Cabrillo Formation of Kennedy and Moore (1971).

Several unsolved problems regarding the stratigraphy of the lower part of the La Jolla Group are potentially relevant to the stratigraphy, age, and paleomagnetic correlation of the Friars Formation and Poway Group. The most important of these concerns the stratigraphic position of the Mount Soledad Formation relative to the Delmar and Torrey formations. The latter two units are known to occur only on the north side of the Rose Canyon fault, while the Mount Soledad Formation is known to occur only on the south side of the fault (Kennedy, 1975, plates 1A and 2A). Kennedy's (1975, fig. 6, block diagrams 5 and 6) illustrations of the Delmar and Torrey formations overlying the Mount Soledad Formation in the subsurface east of Rose Canyon were hypothetical, and not based upon well data or other objective evidence (M. P. Kennedy, personal communication to SLW, 1995). Much has been written about these units and their possible stratigraphic relationships (Milow and Ennis, 1961; Kennedy, 1975; Lohmar et al., 1979; Flynn, 1983; May et al., 1991; Frederiksen, 1991), but the available evidence is inconclusive. If the type Mount Soledad Formation does occur above the Delmar and Torrey formations, the possibility also exists that this unit could be stratigraphically identical to the type Scripps Formation of Kennedy and Moore (1971), as maintained by Milow and Ennis (1961) and discussed by Lohmar et al. (1979). If so, the "Ardath Shale" as mapped by Kennedy (1975) could actually be composed of two different stratigraphic units (as maintained by Milow and Ennis, 1961). We raise these questions to impress upon others the fact that little can be assumed regarding the stratigraphy of the La Jolla Group, and considerable work remains to be done.

In addition to the complexities in the lower part of the Eocene section, the need for significant changes in Kennedy and Moore's (1971) model of the stratigraphy of the Friars Formation and overlying Poway Group has recently become apparent. The revision of these units as discussed below is informal, because new geologic mapping has not been accomplished, and related stratigraphic questions remain unresolved. In the meantime, use of the informal terminology proposed here is justifiable, because erroneous stratigraphic and age relationships of the units under study will result in erroneous biostratigraphic and paleomagnetic correlations of those units. Since the San Diego section occupies an important place in the Eocene global correlation network, it is essential that the stratigraphic model used for the local units reflect historical reality as closely as possible. Accordingly, the evidence indicating the need for a revision of Kennedy and Moore's (1971) model and Kennedy and Peterson's (1975) geologic mapping of the Friars Formation and Poway Group is discussed here in some detail.

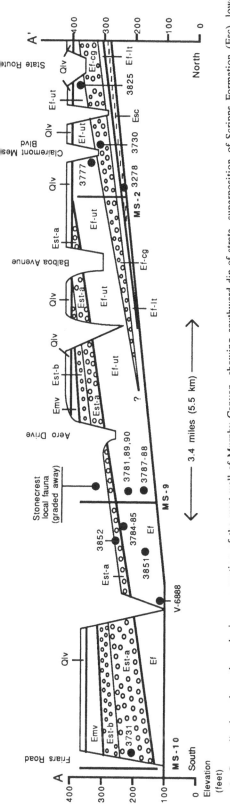

Figure 2. Generalized north-south geologic cross-section of the west wall of Murphy Canyon, showing southward dip of strata, superposition of Scripps Formation (Esc), lower tongue (Ef-lt), conglomerate tongue (Ef-cg), and upper tongue (Ef-ut) of Friars Formation, lower (Est-a) and upper (Est-b) members of Stadium Conglomerate, and Mission Valley Formation (Emv), all capped by the Pleistocene Lindavista Formation (Qlv). Some lithologic contacts from Kennedy and Peterson (1975, plate 3B). Numbered black dots represent significant fossil vertebrate localities. Measured sections 2, 9, and 10 indicated by bold vertical lines. Base of cross-section represents the surface of the west side of the bottom of Murphy Canyon. Vertical exaggeration 10X.

Recent Work on the Friars Formation and Poway Group

Mapping subdivisions of the Friars Formation and Poway Group is difficult, because the various conglomerate and sandstone horizons that occur within these units are generally very similar to one another in lithology. The typical "Poway-type sandstone" is a very light gray to white, friable, moderately sorted, medium-grained, feldspathic volcanic litharenite (Abbott et al., 1979), containing abundant reddish and purplish rock fragments derived from the Poway rhyolite clasts. These sandstones comprise most of the finer-grained strata of the Friars Formation and Poway Group, and hand samples from the lower tongue of the Friars, the upper tongue of the Friars, the Mission Valley Formation, and the Miramar Sandstone Member of the Pomerado Conglomerate are generally indistinguishable. As a result, the stratigraphic identification of "which sandstone is which" often depends on the observation of the superpositional relationships between the sandstone units and the conglomerate units with which they are interbedded. Unfortunately, the determination of "which conglomerate is which" is also difficult when dealing with isolated outcrops, and understandable errors in geologic mapping that have resulted from these problems will be discussed below. Fortunately, in accordance with the North American Stratigraphic Code (North American Commission on Stratigraphic Nomenclature, 1983, Article 22d), fossil mammal evidence can often aid in the assignment of a given outcrop in San Diego to a particular unit (Walsh, this volume, Chapter 5).

Walsh (1991) proposed that certain strata cropping out in the Poway quadrangle that were mapped as the Mission Valley Formation by Kennedy and Peterson (1975) were not correlative with the type Mission Valley Formation of late Uintan age. These fluvial sandstones and mudstones yielded the early Uintan mammal assemblage from the Rancho Peñasquitos district discussed by Golz and Lillegraven (1977), and graded westward into shallow marine deposits in the Del Mar quadrangle. Walsh (1991) assigned these strata to a new lithostratigraphic unit, which was observed to gradationally overlie early Uintan mammal-bearing conglomerates that were mapped as Stadium Conglomerate by Kennedy and Peterson (1975). This alleged new unit was subsequently observed to be disconformably overlain by late Uintan marine sandstones and gritstones of the Mission Valley Formation (e.g., MS-6, MS-7). Walsh's (1991) recognition of the Rancho Peñasquitos strata as a new unit rested on a major assumption, namely, that the early Uintan conglomerates that occurred immediately below them represented the same stratigraphic unit as the lower member of the type Stadium Conglomerate as exposed on the north side of Mission Valley. Given this assumption, Walsh (1991) believed that the allegedly new unit occupied a stratigraphic position between the lower and upper members of the Stadium Conglomerate. If this assumption were correct, one would predict that a typical early Uintan mammal assemblage (Poway fauna of Walsh, this volume, Chapter 5) would be found in the lower member of the type Stadium Conglomerate. However, when fossil mammals were discovered in the latter unit in 1993, they unexpectedly pertained to a younger and unusual micromammal assemblage (Murray Canyon local fauna of Walsh, this volume, Chapter 5) dominated by the early myomorph? rodent *Pauromys* sp. cf. *P. perditus*. The significant faunal differences between the "lower member of the Stadium Conglomerate" in the Mission Valley and Poway areas raised the possibility that these outcrops might in fact pertain to two stratigraphically different conglomerate units.

Field work in the Murphy Canyon and Tierrasanta area has now corroborated this hypothesis. Briefly, an early Uintan conglomerate about 15 m thick crops out in the northern end of Murphy Canyon and in Shepherd Canyon, and was correctly mapped by Kennedy and Peterson (1975, plate 3B) as occurring within the Friars Formation. As exposed on the west side of Murphy Canyon (Fig. 2 and MS-2), this conglomerate erosionally overlies about 7 m of fluvial sandstones, which in turn overlie at least 8 m of concretionary, mollusc-bearing marine siltstones and sandstones of provincial "Transition" age (SDSNH Loc. 3278; Squires and Deméré, 1991; these authors followed Kennedy and Peterson (1975) in assigning these open marine strata to the Friars Formation, but assignment to the Scripps Formation is more consistent with Kennedy and Moore's (1971) and Kennedy's (1975) lithological and environmental concepts of these units). Throughout the Tierrasanta area, the early Uintan conglomerate unit is gradationally overlain by an early Uintan mammal-bearing, fluvial, light gray, medium-grained sandstone and green and reddish mudstone unit up to 30 m thick (SDSNH Locs. 3611, 3612, 3685, 3777, and 3825; University of California Museum of Paleontology (UCMP) Locs. V-68117, V-71056, V-71175, V-71176, and V-71223). On the west side of Murphy Canyon, this sandstone unit is overlain by light gray cobble conglomerates of the lower member of the Stadium Conglomerate (Fig. 2). Lithologically, faunally, and in its stratigraphic position gradationally overlying a fluvial early Uintan conglomerate, the 30 m-thick sandstone unit is identical to the allegedly new stratigraphic unit of Walsh (1991) that crops out in the Poway quadrangle. However, recognizing the small (<2°) but significant southwestward dip of these strata, this sandstone unit can be continuously mapped from Balboa Avenue southward into typical outcrops of the Friars Formation as exposed at the southern end of Murphy Canyon (as correctly shown by Kennedy and Peterson, 1975, plate 3B).

Given these observations, the allegedly new

stratigraphic unit of Walsh (1991) is now recognized to correlate partly or entirely with the Friars Formation, and the early Uintan conglomerate cropping out in the Poway and Murphy Canyon-Shepherd Canyon areas is recognized as a widespread tongue, thickest in the Poway quadrangle, that divides the eastern outcrops of the Friars into lower and upper sandstone-mudstone tongues. This middle conglomerate tongue of the Friars Formation presumably pinches out in the subsurface below the communities of Lindavista and Serra Mesa, because it does not occur in the type section of the Friars in Mission Valley. Consequently, a stratigraphic distinction between the lower and upper tongues is not possible in the type area of the Friars (Fig. 2). It should be emphasized that although fossil mammal evidence was instrumental in first raising doubts about the stratigraphic model of Kennedy and Moore (1971), the revisions proposed here are justifiable entirely on lithostratigraphic and superpositional grounds. That is, the removal of certain strata (conglomerate tongue of the Friars Formation) from the Stadium Conglomerate is justified precisely because they do not represent the same lithostratigraphic unit as the type section of the Stadium Conglomerate. Similarly, removal of certain strata (upper tongue of the Friars Formation) from the Mission Valley Formation is justified precisely because they do not represent the same lithostratigraphic unit as the type section of the Mission Valley Formation.

To summarize these new interpretations in relation to the work of Kennedy and Peterson (1975), their mapping of the Shepherd Canyon-Murphy Canyon area was largely correct when they showed a conglomerate unit of substantial thickness within the Friars Formation, which was in turn overlain by the Stadium Conglomerate, and then the Mission Valley Formation. However, only three km to the northeast of the confluence of Murphy and Shepherd canyons (Fig. 3 and MS-7), Kennedy and Peterson inconsistently mapped the upper tongue of the Friars Formation as the Mission Valley Formation, because they were unaware of the northern pinchout of the Stadium Conglomerate, and did not recognize the resulting marine sandstone-on-fluvial sandstone disconformity between the Mission Valley Formation and the underlying upper tongue of the Friars Formation. This understandable error was compounded in their mapping of the northern part of the La Mesa quadrangle, the Poway quadrangle, and the eastern part of the Del Mar quadrangle, when Kennedy and Peterson mapped the conglomerate tongue of the Friars as the Stadium Conglomerate.

Revised Stratigraphic Model for the Eocene of Southwestern San Diego County

Figure 4 shows a revised hypothesis for the Eocene stratigraphy of southwestern San Diego County. It is still based largely on the transgressive-regressive model of Kennedy and Moore, but differs in the following

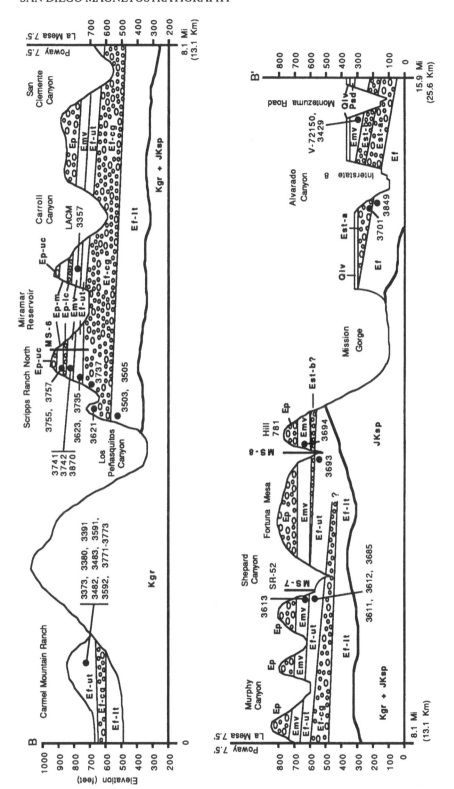

Figure 3. Generalized north-south geologic cross-section from Carmel Mountain Ranch to east San Diego, showing proposed stratigraphy of Friars Formation and Poway Group. Abbreviations: JKsp = Jurassic and Cretaceous metavolcanic basement rocks of the Santiago Peak Volcanics; Kgr = Cretaceous granitic basement rocks of the Peninsular Ranges Batholith; Ep-lc, Ep-m, and Ep-uc = lower, Miramar Sandstone, and upper members of Pomerado Conglomerate, respectively; Psd = Pliocene San Diego Formation. Other abbreviations as in Figure 2. Some lithologic contacts taken from Kennedy and Peterson (1975, plates 2B and 3B). Subsurface topography of Friars Formation-basement rock contact is hypothetical. Numbered black dots represent significant fossil vertebrate localities. Measured sections 6, 7, and 8 indicated by bold vertical lines. Vertical exaggeration 10X.

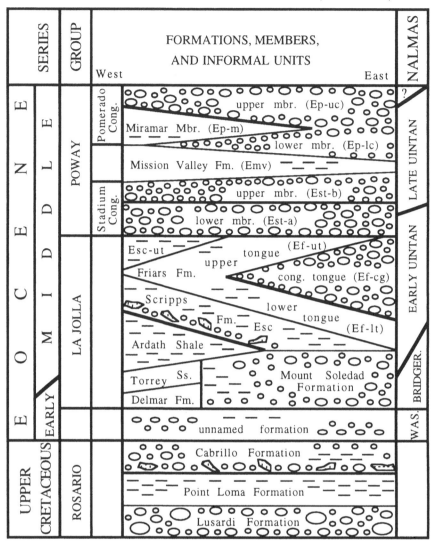

Figure 4. Proposed stratigraphic relationships of Upper Cretaceous and Eocene units in southwestern San Diego County. Substantially modified from Kennedy and Moore (1971).

respects: (1) An unnamed formation of Wasatchian age is recognized between the Cabrillo Formation and the Mount Soledad Formation (Walsh, 1991); (2) The stratigraphic position of the Mount Soledad Formation relative to the Delmar and Torrey formations is left in doubt, because Kennedy and Moore's (1971) placement of the type Mount Soledad entirely below the latter two units has not been convincingly demonstrated; (3) The Friars Formation is stratigraphically expanded, and subdivided into a lower sandstone-mudstone tongue, a middle conglomerate tongue, and an upper sandstone-mudstone tongue. The upper tongue of the Friars is hypothesized to be the dominantly fluvial, lateral equivalent of the upper tongue of the Scripps Formation; (4) The Stadium Conglomerate is subdivided into two locally distinguishable members (Milow and Ennis, 1961) and restricted in its stratigraphic and geographic extent only to those strata that are correlative with the type section; (5) The Mission Valley Formation is restricted in its stratigraphic and geographic extent only to those strata that are correlative with the type section; (6) Finally, as shown by bold lines in Figure 4, more disconformable contacts are recognized between certain units of the La Jolla and Poway Groups than were envisioned by Kennedy and Moore (1971). These include the contact between the lower tongue and the conglomerate tongue of the Friars Formation; the contact between the Friars Formation and the Stadium Conglomerate; the contact between the lower and upper members of the Stadium; the contact between the Friars Formation and Mission Valley Formation (where the Stadium is absent); and the contact between the Miramar Sandstone Member of the Pomerado Conglomerate and the upper conglomerate member of the Pomerado.

Due to the difficulties of portraying complex three-dimensional facies relationships in a single two-dimensional diagram, it is necessary to clarify certain stratigraphic relationships implied by Figure 4. First, a contact between the unnamed formation and the Delmar Formation has not been observed. If this contact exists, it presumably occurs in the subsurface north of Mount Soledad. Second, a contact between the Scripps Formation and the Mount Soledad Formation has not been observed. If it exists, it would probably occur in the subsurface below Clairemont Mesa. Finally, a contact between the Stadium Conglomerate and the upper tongue of the Scripps Formation has not been observed. If it ever existed, it has probably been eroded away. Information on the lithology, thickness, areal distribution, and depositional environments of the units newly recognized, restricted, or resurrected here is summarized as follows.

Friars Formation

The Friars Formation (Ef) was named by Kennedy and Moore (1971) for the vertebrate-bearing, largely fluvial white sandstones and reddish and greenish siltstones and mudstones that occurred above the marine sandstones and siltstones of the Scripps Formation, and below the fluvial conglomerates of the Stadium Conglomerate. The type section of the Friars Formation is located on the north wall of Mission Valley, and typical outcrops of the Friars occur in Tecolote Canyon, Alvarado Canyon, and the southern end of Murphy Canyon. These sandstone-dominated strata will hereafter be referred to as the "type outcrops" of the Friars. As discussed above, our conception of the Friars Formation differs from that of Kennedy and Peterson's (1975) in that the conglomerate lens which they mapped within the Friars in the Shepherd Canyon area is interpreted here to be the southern edge of an extensive conglomerate body that thickens to the north, where it was erroneously mapped by Kennedy and Peterson (1975) as the Stadium Conglomerate.

Milow and Ennis (1961) recognized an "unnamed formation" similar in stratigraphic position and scope to the Friars Formation, and subdivided it into three units ("blocky sandstone," "massive sandstone," and "green mudstone" members), showing complex stratigraphic relationships. As noted above, we also divide the Friars Formation into three units, but our model differs from that of Milow and Ennis (1961) as follows. First, a distinct "mudstone member" cannot be recognized in the Friars, because significant thicknesses of mudstone occur in this formation only in close proximity to buried paleohills of the granitic and metavolcanic basement complex. These mudstones occur locally in both the lower and upper tongue of the Friars as recognized here, and are best regarded as lithofacies rather than lithostratigraphic units. Second, the middle unit of the Friars Formation in our model is a widespread conglomerate tongue. The resulting interpretation of the Friars Formation as a conglomerate tongue enclosed by underlying and overlying sandstone-mudstone tongues is fundamentally different from Milow and Ennis's (1961) concept of this unit.

Friars Formation, lower tongue

Lithology and Depositional Environment—The lower tongue of the Friars Formation (Ef-lt) is dominated by white, medium-grained, fluvial sandstones, and interbedded, caliche-bearing, greenish and reddish-brown, terrestrial sandy siltstones and mudstones. Lenses of Poway-type pebble and cobble conglomerate are locally common in the eastern outcrops of the Ef-lt (e.g., MS-4). The mudstones and siltstones are thickest and most common in the vicinity of buried paleohills of the crystalline basement complex. Adjacent to granitic basement rocks, these fine-grained strata are commonly interbedded with thin tongues of coarse-grained arkosic sandstones. Brackish water molluscs including *Ostrea* occur locally in the western outcrops of the Ef-lt (e.g., SDSNH Loc. 3727; MS-3).

Distribution and Thickness—The Ef-lt crops out from Rancho Bernardo in the north (Kennedy and Peterson, 1975), at least to Santee in the south, and from Poway Valley and Lakeside in the east to Murphy Canyon and Mira Mesa in the west. During excavations for State Route 52 northwest of Santee, a maximum composite thickness of about 55 m was measured for the Ef-lt (MS-4). Apparently due to substantial erosion at the base of the conglomerate tongue of the Friars, the Ef-lt is only 7 m thick in the north end of Murphy Canyon (MS-2), and is absent in the gravel quarry of H. G. Fenton Materials Company in Carroll Canyon (SW1/4 Sec. 2 T15S R3W).

Contacts-The Ef-lt rests nonconformably upon metavolcanic and granitic rocks of Mesozoic age in most areas. In Los Peñasquitos Canyon and in the northern part of Murphy Canyon, this unit rests upon shelfal marine siltstones and sandstones assigned here to and/or mapped by Kennedy (1975) as the Scripps Formation. The Ef-lt is often overlain with as much as 5 m of local erosional relief by the conglomerate tongue of the Friars. This relationship is documented in MS-3, and was also seen in the Westview housing development in Mira Mesa (SDSNH Loc. 3496), and the South Creek housing development southwest of Poway (SDSNH Loc. 3505).

Fossils and Age—The upper part of the Ef-lt yields a typical early Uintan mammal assemblage, which has been collected from several localities in the Rancho Peñasquitos, Poway, and Santee areas. This assemblage is faunally indistinguishable from that collected from the type outcrops, the conglomerate tongue, and the upper tongue of the Friars (Walsh, this volume, Chapter 5). Given Flynn's (1986) faunal characterization of the beginning of the Uintan NALMA, an early Uintan age for at least the upper part of the Ef-lt is corroborated by the presence of *Leptoreodon major* at SDSNH Locs. 3655-3657 in Santee, and by the presence of tooth fragments of selenodont artiodactyls from SDSNH Locs. 3505 and 3496 (Poway and Mira Mesa areas). The Ef-lt also contains occasional fossil leaves (SDSNH Locs. 3653 and 3663) and pulmonate gastropods (e.g., SDSNH Loc. 3656).

Friars Formation, Conglomerate Tongue

Lithology and Depositional Environment—The conglomerate tongue of the Friars Formation (Ef-cg) is dominated by purplish metarhyolite clasts of the Poway suite. In the eastern part of its outcrop area (e.g., Scripps Ranch and upper Murphy Canyon), the Ef-cg is a light gray to dark rusty brown, moderately sorted, matrix-to-clast-supported, pebble to boulder conglomerate of fluvial origin. These outcrops also contain common lenses of light gray and yellowish-brown medium-grained sandstone, and channeled beds and rip-up clasts of greenish and reddish mammal-bearing siltstone, with common multicolored caliche horizons and nodules. In the western part of its outcrop area (e.g., western Carroll Canyon and south fork of Los Peñasquitos Canyon), the Ef-cg is a rust-stained, clast-supported, pebble-to-boulder conglomerate containing light gray and rust-stained, medium-grained sandstone lenses and occasional lenses and rip-up clasts of greenish-gray and rust-stained, mollusc-bearing siltstones. These western outcrops of the Ef-cg are of probable deltaic origin (e.g., May, 1985; referred to the Stadium Conglomerate).

The Ef-cg differs lithologically from the lower member of the Stadium Conglomerate in that the former unit does not have the pale greenish, muddy sandstone matrix that typifies the latter unit in the Mission Valley area. The Ef-cg is more similar to the upper member of the Stadium Conglomerate in color, lithology, and the range of depositional environments it represents, but these two units differ in maximum observed thickness, and are not known to occur in the same geographic area.

Distribution and Thickness—The maximum observed thickness of the Ef-cg is about 60 m at the South Creek, Eastview, and Scripps Ranch North housing developments southwest of Poway. The Ef-cg thins rapidly to the northwest, away from the Poway Fan (Howell and Link, 1979). For example, this unit is 21 m thick in the community of Carmel Mountain Ranch (Figs. 1 and 3; stratigraphic section by B. O. Riney on file at SDSNH), is locally only 11 m thick in the community of Rancho Peñasquitos (MS-3), and apparently pinches out within the Friars Formation in the northern part of the Del Mar quadrangle and the southern part of the Rancho Santa Fe quadrangle (mapped as Stadium Conglomerate by Kennedy, 1975, and Eisenberg, 1985). To the west, the Ef-cg can be

traced at least as far as the western ends of Los Peñasquitos Canyon and Carroll Canyon, where Kennedy (1975) shows this unit (mapped as Stadium Conglomerate) to pinch out between the Scripps Formation and what he maps as the upper tongue of the Scripps. 3-5 km east of this mapped pinchout, the Ef-cg is still at least 50 m thick in the gravel quarries of H. G. Fenton Materials Company and CalMat Company in Carroll Canyon. The Ef-cg also thins to the southwest, where it is about 20 m thick in the City of San Diego Miramar Landfill (northern rim of San Clemente Canyon), and is 15 m thick in the Shepherd Canyon area (Figs. 2-3 and MS-2). South of Santee, the extent of the Ef-cg and its relationship with the Stadium Conglomerate has not been determined. The Ef-cg apparently pinches out within the Friars Formation below Serra Mesa and Tierrasanta, because it has not been identified in Mission Valley and San Carlos (Figs. 2 and 3).

Contacts—In the eastern part of the study area, the Ef-cg generally rests erosionally upon the Ef-lt (MS-2, MS-3, MS-4). However, in the south fork of Los Peñasquitos Canyon, western Carroll Canyon, and western San Clemente Canyon, erosion at the base of the Ef-cg has locally removed the lower tongue of the Friars Formation, probably because this area was located in the main east-west corridor through which the bulk of the conglomerate was transported toward the Eocene coastline during early Uintan time. Here, the Ef-cg overlies shelfal marine sandstones and siltstones mapped as Scripps Formation by Kennedy (1975). The Ef-cg is generally gradationally overlain by the upper tongue of the Friars Formation. This contact can be seen along Camino Santa Fe (MS-1), at Scripps Ranch North (MS-6), in the Shepherd Canyon area (MS-2), and in the Carroll Canyon quarry of H. G. Fenton Materials Company.

Fossils and Age—The Ef-cg has yielded a typical early Uintan mammal assemblage from several SDSNH localities at the Scripps Ranch North housing development, and SDSNH Loc. 3730 in the Shepherd Canyon area (Walsh, this volume, Chapter 5). Pulmonate gastropods and fossil leaves have been recovered from fluvial outcrops of the Ef-cg at Scripps Ranch North (SDSNH Locs. 3621, 3624), and sparse marine molluscs have been collected from deltaic outcrops of the Ef-cg on the south rim of San Clemente Canyon (SDSNH Locs. 3881-3882).

Overall Stratigraphic Position—Figure 4 illustrates the stratigraphic position of the Ef-cg as observed in the eastern part of the study area, where this unit occurs as a relatively thick lens occurring entirely (or nearly entirely) within a stratigraphically-expanded Friars Formation, all of which is assumed to occur stratigraphically above the Scripps Formation. However, the lateral relationship of the Ef-cg to the marine strata exposed along the present-day coastline is uncertain. It

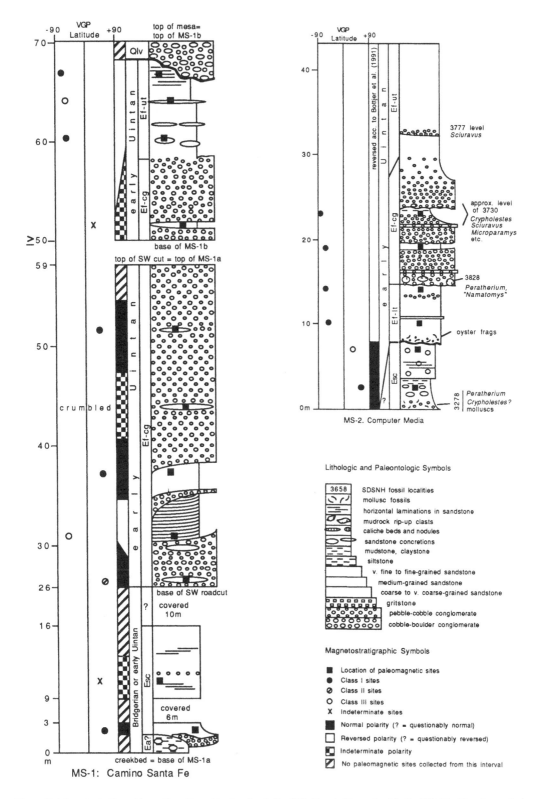

Figure 5. Magnetostratigraphic sections from Camino Santa Fe (MS-1) and Computer Media (MS-2). Key to symbols used in this and the following figures is also shown.

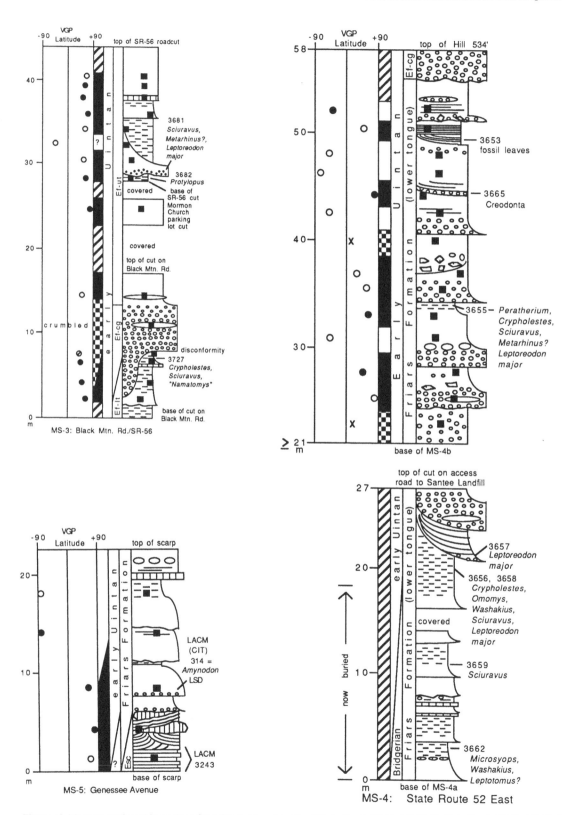

Figure 6. Magnetostratigraphic sections from Black Mountain Road/State Route 56 East (MS-3), State Route 52 East (MS-4), and Genessee Avenue (MS-5). Symbols as in Figure 5.

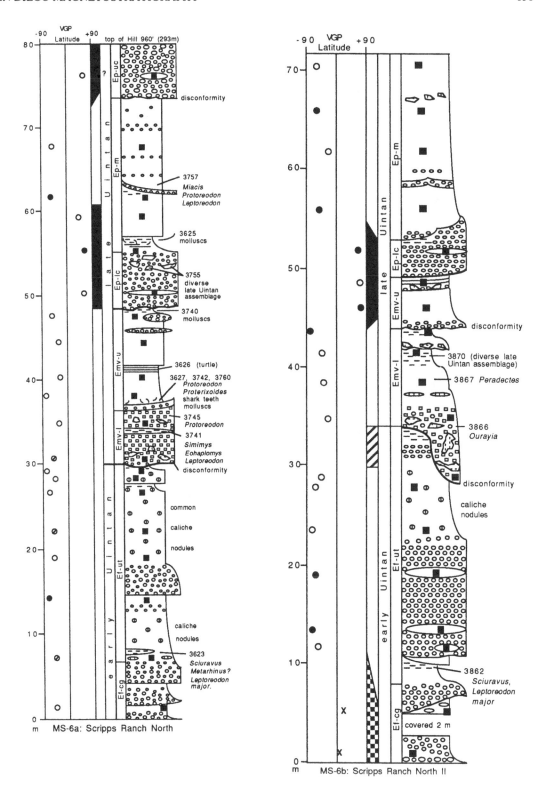

Figure 7. Magnetostratigraphic sections from Scripps Ranch North (MS-6a, b). Symbols as in Figure 5.

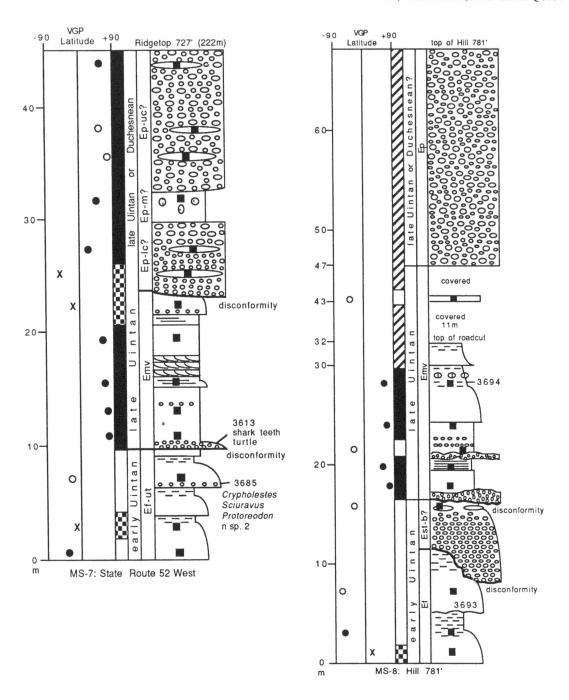

Figure 8. Magnetostratigraphic sections from State Route 52 West (MS-7) and Hill 781' (MS-8). Symbols as in Figure 5.

is tempting to speculate that the Ef-cg represents the coeval, eastern, largely fluvial facies of the marine conglomerates in the lower part of the Scripps Formation as exposed in the seacliffs north of Scripps Institution of Oceanography (Kennedy and Moore, 1971). If so, the Black's Beach local fauna (Walsh, this

volume, Chapter 5) may prove to be of early Uintan age. Unfortunately, due to structural complexities in the southwestern part of the Del Mar quadrangle (Kennedy, 1975), it is difficult or impossible to directly correlate the Ef-cg as exposed in Carroll Canyon with the seacliff outcrops of the Scripps Formation.

Friars Formation, upper tongue

Lithology and Depositional Environment—The upper tongue of the Friars Formation (Ef-ut) is very similar to the Ef-lt in overall lithology, lithological variations, and in the east-to-west geographic distribution of the depositional environments it represents. Fluvial sandstones of the Ef-ut as exposed in the Poway quadrangle grade westward into brackish water strata in the eastern part of the Del Mar quadrangle. These deposits in turn appear to grade westward into shelfal marine strata that were mapped by Kennedy (1975, plate 1A) as the upper tongue of the Scripps Formation.

The Ef-ut was previously mapped as the Mission Valley Formation in the Poway and Delmar quadrangles (Kennedy, 1975; Kennedy and Peterson, 1975). The Ef-ut differs lithologically from the latter in that at a given longitude, the Ef-ut contains a greater proportion of medium-grained fluvial sandstones and floodplain siltstones and mudstones, while the Emv contains a greater proportion of fine- to very fine-grained marine sandstones. The Ef-ut also apparently lacks bentonite beds, and contains a greater proportion of reddish terrestrial mudstones and siltstones than the Mission Valley Formation, whose terrestrial mudrocks are usually green or brownish in color.

Thickness and Distribution—The maximum observed thickness of the Ef-ut is about 60 m in the Rancho Peñasquitos and Carmel Mountain Ranch areas. In the northern part of Murphy Canyon and Tierrasanta, the Ef-ut is about 30 m thick (MS-2, MS-7). The Ef-ut crops out to the northwest as far as the northwest part of the Del Mar quadrangle and southern part of the Rancho Santa Fe quadrangle (Kennedy, 1975; Eisenberg, 1985; mapped as Mission Valley Formation).

Contacts—The Ef-ut locally rests nonconformably upon metavolcanic and granitic rocks of the crystalline basement, but usually it gradationally overlies the Ef-cg (MS-1, MS-2, MS-3, MS-6). The Ef-ut is variably overlain by the lower member of the Stadium Conglomerate (Fig. 2), the upper member of the Stadium Conglomerate (MS-8), or the Mission Valley Formation (Fig. 3; MS-6, MS-7). The upper contact of the Ef-ut is erosional wherever it has been observed.

Fossils and Age—Early Uintan fossil mammals are common in the Ef-ut, and have been collected from many localities in the greater San Diego area, from Murphy Canyon and Tierrasanta north to Scripps Ranch North, Rancho Peñasquitos, and Carmel Mountain Ranch (Golz and Lillegraven, 1977; Walsh, this volume, Chapter 5). Fossils leaves (SDSNH Loc. 3590) and pulmonate gastropods (SDSNH Loc. 3648) have been collected from the Ef-ut in the Carmel Mountain Ranch and Rancho Peñasquitos areas, while brackish-water and freshwater molluscs have been collected from the Ef-ut in McGonigle Canyon (SDSNH Locs. 3641-3642 and localities 4231 and 4233 of Hanna, 1927, p. 266).

Stadium Conglomerate

Milow and Ennis (1961, pp. 27, 29, 30) recognized two "submembers" ("Ep-cg[a]" and "Ep-cg[b]") of their "conglomerate member of the Poway Formation." In discussions of their newly-named Stadium Conglomerate, Kennedy and Moore (1971) and Kennedy (1975) did not recognize this distinction. Although Milow and Ennis's "submembers" are probably not consistently separable throughout the entire outcrop area of the Stadium Conglomerate, two lithologically distinct conglomerate units do exist within this formation in the Mission Valley-Alvarado Canyon area, and their informal recognition calls attention to details of geologic history that would otherwise be ignored. Milow and Ennis's "submembers" are therefore resurrected here as the lower (Est-a) and upper (Est-b) members of the Stadium Conglomerate.

The type section of the Stadium Conglomerate of Kennedy and Moore (1971) is located in the now idle gravel quarry of H. G. Fenton Materials Company on the north side of Mission Valley, north of Friars Road and east of Interstate 805. Two large abandoned quarry faces currently expose the lower and upper members of the Stadium Conglomerate, the overlying Mission Valley Formation, and the Pleistocene Lindavista Formation (Fig. 2). Since 1971, the type section of the Stadium Conglomerate has been modified by quarrying activity, so it is redesignated here as the large south-facing quarry face about 800 m west of the intersection of Friars Road and Mission Village Drive (MS-10).

The Stadium Conglomerate is now recognized to occur over a much smaller geographic area than as mapped by Kennedy (1975) and Kennedy and Peterson (1975). As discussed above, this conclusion stems from the realization that most or all of the outcrops of the "Stadium" mapped by these authors in the Poway and Del Mar quadrangles pertain to a stratigraphically lower unit (regarded here as the conglomerate tongue of the Friars Formation). Neither member of the Stadium Conglomerate has been recognized north of State Route 52, where the upper tongue of the Friars Formation is disconformably overlain by the late Uintan Mission Valley Formation (MS-6, MS-7). Two scenarios that would explain the limited distribution and northward-thinning of both members of the Stadium Conglomerate are as follows: (1) Toward the end of the early Uintan, the Ballenas River (Steer and Abbott, 1984) may have flowed around the southern end of present-day Cowles Mountain. If so, a maximum thickness of the Stadium Conglomerate would have been deposited along the Alvarado Canyon-Mission Valley axis, with rapid thinning to the north resulting from a "clast shadow" effect caused by the granitic barrier of Cowles and Fortuna mountains. (2) From Tierrasanta northward, both members of the Stadium were completely removed by erosion associated with the marine transgression represented by the Mission Valley Formation. Further

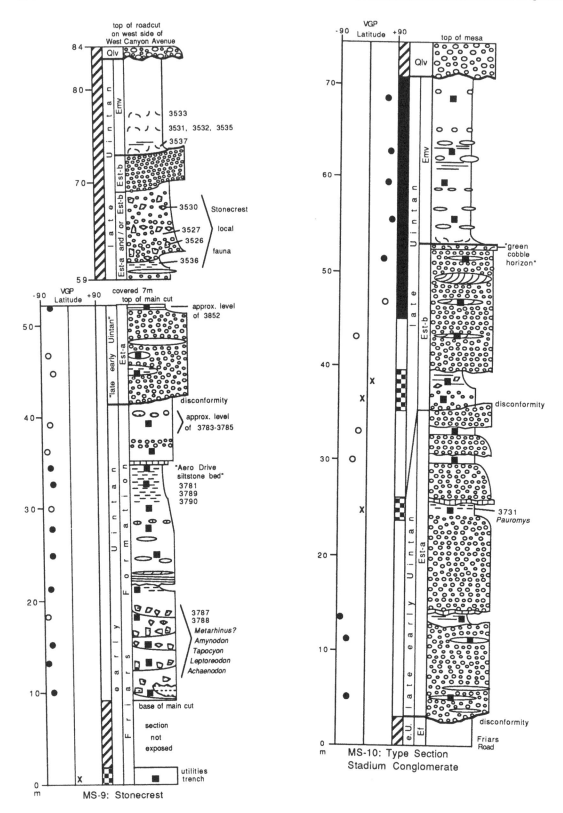

Figure 9. Magnetostratigraphic sections from Stonecrest (MS-9) and the type section of the Stadium Conglomerate (MS-10). Symbols as in Figure 5.

mapping, clast population, and paleocurrent work on the Ef-cg and both members of the Stadium Conglomerate must be done in order to test these scenarios.

Stadium Conglomerate, lower member

Type Section—At the type section of the Stadium Conglomerate, the lower member (Est-a) is composed of 34 m of light yellowish-brown and light gray and pale greenish-gray, moderately sorted, matrix-to-clast-supported, pebble to boulder conglomerate, with several lenses of light gray and pale greenish gray, medium-to-very fine-grained, moderately sorted, friable sandstone (MS-10). The matrix of the conglomerate is a very light gray to pale greenish gray, poorly sorted, muddy, fine-to medium-grained sandstone. This lithology typifies the Est-a in the Mission Valley, Murphy Canyon, and Alvarado Canyon areas. About 22 m above the base of the Est-a, a lens of medium gray, very fine-grained sandstone and caliche nodule-bearing siltstone contains sparse micromammals of late early Uintan age, including *Pauromys* sp. cf. *P. perditus* (SDSNH Loc. 3731). This bed may be stratigraphically equivalent to a laterally persistent bed of micro-mammal-bearing bluish gray and greenish gray siltstone up to 2 m thick ("*Pauromys* bed"; SDSNH Locs. 3691-3692) that occurs within the upper part of the Est-a in Murray Canyon.

Lithology and Depositional Environment—The general lithology of the Est-a is as described above for the type section. Clast assemblages have not been studied in detail, but are dominated by Poway rhyolites (Kies and Abbott, 1983). Milow and Ennis (1961) implied that the Est-a contained a greater variety of less resistant clasts than the Est-b, but detailed clast counts have not been done to corroborate these differences. In general, the light gray, "bleached" appearance of the Poway clasts in the lower member contrasts sharply with the darker, purplish and reddish colors of the unweathered Poway clasts in the upper member of the Stadium. The light gray and pale greenish gray, generally massive sandstone lenses within the lower member also contrast with the iron oxide-stained and generally cross-bedded sandstone lenses within the upper member.

Depositional environments of the Stadium Conglomerate have been studied by Howell and Link (1979), who did not recognize a division of this conglomerate into distinct lithostratigraphic units. However, Howell and Link concluded that in the Mission Valley area, the lower part of the Stadium Conglomerate was largely or entirely of fluvial origin. A non-marine setting for most of the Est-a is supported by the occurrence of caliche layers and local concentrations of fossil micromammals in this member. However, Milow and Ennis (1961) reported a sparse hyposaline microfossil assemblage from near the base of this member.

Thickness and Distribution—North of the 34 m-thick type section, the Est-a decreases rapidly in thickness before being either truncated by the Pleistocene Lindavista Formation (Fig. 2), or pinching out between the upper tongue of the Friars Formation and the Mission Valley Formation (Fig. 3). The Est-a has been recognized as far east as the campus of San Diego State University, and as far south as the intersection of Montezuma Road and Fairmount Avenue in East San Diego (Fig. 1), where it dips below the mesa surfaces south of Mission Valley.

Contacts—The Est-a erosionally overlies the type outcrops of the Friars Formation on the north side of Mission Valley and in the southern end of Murphy Canyon, and erosionally overlies the upper tongue of the Friars in the northern end of Murphy Canyon. A maximum of 5 m of relief was observed on the Est-a/Ef contact at the Stonecrest development on the west wall of Murphy Canyon (MS-9). The Est-a is locally erosionally overlain by the Est-b, with a maximum relief of about 2 m observed in the gravel pit of H. G. Fenton Materials Company, immediately east of Interstate 805. East of Murphy Canyon, the nature of the contact between the lower and upper members of the Stadium is unclear. Milow and Ennis (1961) believed this contact to be entirely disconformable, with the lower member pinching out below the upper member in the San Diego State University area. Outcrops of the Stadium Conglomerate between Alvarado Canyon and El Cajon Valley must be studied in order to resolve this question.

Fossils and Age—As discussed by Walsh (this volume, Chapter 5), the distinctive fossil mammal assemblage definitely known to occur in the Est-a (Murray Canyon local fauna) is assigned a late early Uintan age on the basis of the presence of *Microparamys* sp. cf. *M. minutus*, "*Namatomys*" new sp. 1, and *Leptoreodon major*, and on the apparent absence of *Sespedectes*, *Proterixoides*, *Simimys*, and other late Uintan indicators. However, the late Uintan Stonecrest local fauna (Walsh, this volume, Chapter 5) may have been collected from the upper part of the Est-a, in which case this unit would span the early Uintan-late Uintan boundary as recognized in southern California.

Stadium Conglomerate, Upper Member

Type section—At the type section of the Stadium Conglomerate, the Est-b is 17 m thick, and disconformably overlies the Est-a with about 1 m of erosional relief (MS-10). Here, the upper member is composed largely of rusty brown, clast-supported, Poway-type pebble and cobble conglomerate with a rusty brown, friable, well-sorted, medium-grained sandstone matrix. Two meters above the base of this member, a 2 m-thick, light gray, laminated, very fine-grained sandstone lens contains plant debris and red siltstone rip-up clasts up to 0.3 m in diameter. Several light rusty brown, massive to cross-laminated, well-sorted, friable-to-well-cemented, medium-grained sandstone lenses from 0.2 to 1 m thick occur in the type section. The top of the Est-b is marked by a distinctive, 0.6 m-thick bed of pale yellow, concretionary, matrix-supported, oyster-bearing, pebble-to-cobble conglomerate, whose clasts are all stained a dark green color. A similar (diagenetic?) horizon is also present at the top of the Est-b at the type section of the Mission Valley Formation (MS-11) and along Montezuma Road.

Lithology and Depositional Environment—Throughout the Mission Valley area, the general lithology of the Est-b is as described above for the type section. At and west of the type section, this unit is largely of deltaic origin (Howell and Link, 1979). East of Murphy Canyon, the Est-b is apparently largely of fluvial origin.

Thickness and Distribution—The Est-b displays a maximum observed thickness of 17 m on the north side of Mission Valley. North and east of the type section, the Est-b thins rapidly before either being eroded out by the Pleistocene Lindavista Formation (Fig. 2), or pinching out between the Ef-ut and the Mission Valley Formation (Fig. 3). The southernmost known outcrops of the Est-b occur in the National City quadrangle, in Chollas Valley and along Imperial Avenue (Kennedy and Tan, 1977).

Contacts—In the Mission Valley area, the Est-b erosionally overlies the Est-a. East of San Diego State University, the relationship between these two units is unclear. The Est-b is abruptly overlain by the Mission Valley Formation in the Mission Valley and La Mesa areas (Kennedy and Moore, 1971).

Fossils and Age—As discussed by Walsh (this volume, Chapter 5), the late Uintan Stonecrest local fauna may have been collected from the lower part of the Est-b. If this local fauna was actually collected from the upper part of the Est-a, however, the Est-b would still be of late Uintan age on the basis of superposition. Fossil marine molluscs were documented to occur within what is here regarded as the Est-b by Howell and Link (1979) and Givens and Kennedy (1979). The latter authors assigned a "Tejon" age to the upper member in part because of its gradational contact with the overlying Mission Valley Formation of undoubted "Tejon" age. Some of the foraminifer localities of Dusenbury (1932), Cushman and Dusenbury (1934), Gibson (1971), and Steineck et al. (1972) (variously assigned to the Poway Conglomerate or the upper part of the Stadium Conglomerate) may have been in the basal Mission Valley Formation, which contains common microfossils throughout the Mission Valley area (E. D. Milow, personal communication). However, according to M. P. Kennedy (personal communication), the calcareous nannofossil assemblage discussed by Kennedy and Moore (1971, p. 719) was definitely collected from the Stadium Conglomerate (presumably upper member).

Mission Valley Formation

The Mission Valley Formation (Emv) was named by Kennedy and Moore (1971) for a sandstone unit that had previously been recognized to occur within the old "Poway Conglomerate" by Hanna (1926) and Milow and Ennis (1961). As discussed above, the fluvial early Uintan strata in the Poway quadrangle that were mapped by Kennedy (1975) and Kennedy and Peterson (1975) as the Mission Valley Formation are here assigned to the upper tongue of the Friars Formation. Certain strata in the La Mesa quadrangle mapped by Kennedy and Peterson (1975) as the "lower tongue of the Mission Valley Formation" are probably also referable to the upper tongue of the Friars Formation. Accordingly, the stratigraphic and areal extent of the Mission Valley Formation is substantially restricted here.

Lithology and Depositional Environment—In its western outcrops, the Emv is of shelfal marine origin, and is composed almost entirely of very light gray, fine-to-very fine-grained sandstone with common concretionary horizons containing fossil molluscs (Kennedy and Moore, 1971; Givens and Kennedy, 1979). In the eastern part of its area of outcrop, the Emv is predominantly of fluvial origin, and is composed largely of very light gray, medium-grained, friable "Poway-type" sandstones with common caliche-bearing, light brown and greenish mudstone and siltstone beds, often containing terrestrial vertebrates (Lillegraven and Wilson, 1975). A distinctive lithofacies in the lower part of the Emv occurs at Scripps Ranch North, and consists of light gray, poorly sorted, arkosic coarse-grained sandstones and gritstones, interbedded with vertebrate-bearing, greenish bentonitic mudstones. This lithofacies reaches a maximum thickness of 16 m.

A 25 cm-thick, pink bentonite bed in the Emv was recently exposed at two different places during freeway construction near the intersection of Interstate 8 and State Route 125 in La Mesa. *Simimys* and *Sespedectes* occurred at both sites (SDSNH Loc. 3428 and 3539), about 2 m stratigraphically below the bentonite. Samples of the bentonite from SDSNH Loc. 3428 have yielded a single-crystal $^{40}Ar/^{39}Ar$ age of 42.83 ± 0.24 Ma (Obradovich and Walsh, in prep.; the age of 42.18 Ma reported for this bentonite in Berry, 1991, and Walsh, 1991, is incorrect). It is possible that a laterally persistent, 1-2 m-thick reddish mudstone bed occurring in the upper part of the Mission Valley Formation on the south wall of Mission Valley (e.g., MS-11 and roadcuts along Interstate 15) represents the reworked deposits of this bentonite. However, this bentonite was of normal polarity at SDSNH Loc. 3428 (Prothero, 1991), whereas the red mudstone bed in MS-11 is reported below to be of reversed polarity.

Distribution and Thickness—The Emv crops out discontinuously from Otay Valley in the south to

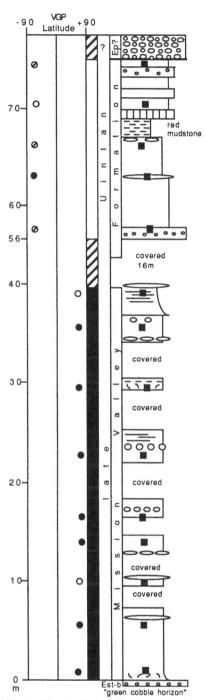

MS-11: Mission Valley Fm. Type Section

Figure 10. Magnetic stratigraphy of the type section of the Mission Valley Formation (MS-11). Symbols as in Figure 5.

Mission Valley, La Mesa, and Scripps Ranch in the north (Kennedy, 1975; Kennedy and Peterson, 1975; Kuper, 1977). The 75 m-thick type section of the Emv represents the maximum preserved thickness of this unit, due to its position in the core of the southern continuation of the gently south-plunging syncline mapped by Kennedy (1975) on the north side of Mission Valley. North of Mission Valley and east and west of the syncline axis, the preserved thickness of the Emv decreases rapidly, due to erosion at the base of the Pleistocene Lindavista Formation. At the type section of the Stadium Conglomerate (MS-10), the Emv is only 20 m thick, and is eroded out completely between Aero Drive and Balboa Avenue, on the west rim of Murphy Canyon (Fig. 2). East of the Lindavista Terrace, variable thicknesses of the Mission Valley Formation are preserved below the Pomerado Conglomerate. For example, at Hill 781 (MS-8), State Route 52 West (MS-7), and Scripps Ranch North (MS-6), the Emv is 30 m, 15 m, and 22 m thick, respectively.

Contacts—In the Mission Valley area, the Emv rests abruptly upon the Est-b. East of San Diego State University, the Mission Valley rests both abruptly and gradationally upon fluvial conglomerates of the undivided Stadium Conglomerate. In the northern part of Tierrasanta, as exposed along roadcuts on State Route 52, 1.5 km and 2.5 km east of Santo Road, white, fine- to very fine-grained marine sandstones of the Emv rest disconformably upon white, medium-grained fluvial sandstones and greenish and reddish mudstones of the Ef-ut (MS-7). The disconformity is locally marked in this area by large granitic boulders in the base of the Emv (roadcut on State Route 52, 2.5 km east of Santo Road), or by a basal conglomerate of well-polished Poway pebbles set in a matrix of white, very fine-grained sandstone (MS-7). At Scripps Ranch North, a distinctive "arkosic gritstone facies" of the Emv disconformably overlies the Ef-ut with as much as 6 m of relief (MS-6a, 6b). The Emv is overlain gradationally and erosionally by the Pomerado Conglomerate in the Scripps Ranch and La Mesa area. From Spring Valley and Encanto south to Otay Valley (south of the area covered by Fig. 1), the Emv is overlain by reddish mudstones of the latest Uintan and/or Duchesnean "Sweetwater" Formation (Walsh and Deméré, 1991).

Fossils and Age—Mammals from many localities in the Emv in the greater San Diego and Miramar Reservoir areas are assignable to the late Uintan, based on a variety of distinctive taxa such as *Simimys*, *Sespedectes*, *Proterixoides*, and *Protoreodon* new sp. 1 (Walsh, this volume, Chapter 5). Pulmonate gastropods are also common in the Emv (SDSNH Locs. 3273, 3870). Marine molluscan assemblages from the Emv have been assigned to the Pacific Coast "Tejon" stage (Givens and Kennedy 1979). Squires (1988) presented a chart wherein the "Tejon" stage was correlated with low-latitude coccolith subzones CP13B through CP15B (Okada and Bukry, 1980). As noted above, benthonic foraminifera described by Dusenbury (1932), Cushman and Dusenbury (1934), and Gibson (1971) from the "Poway Conglomerate" may have been collected from the base of the Mission Valley Formation. These shelfal assemblages were assigned to the California Narizian "Stage" by Milow and Ennis (1961) and Gibson (1971). Almgren et al. (1988) have correlated an emended concept of the Narizian with the middle Eocene low-latitude coccolith subzones CP12b through CP14b. Steineck and Gibson (1971) and Steineck et al. (1972) assigned planktonic foraminifera from the "Poway Formation" (possibly from the base of the Emv) to the late middle Eocene, and suggested an equivalence with the *Orbulinoides beckmanni* Zone of Bolli, 1957 (now the *Globigerapsis beckmanni* Zone = P13; Berggren et al., 1995). Precision of age assignments and other aspects of the work of these authors was questioned by McWilliams (1972) and Phillips (1972). Nevertheless, a correlation of no older than Zone P13 for the Mission Valley Formation planktonic foram assemblage is supported by the fact that the first appearance of *Globorotaloides suteri* and the last appearance of *Truncorotaloides collactea* (both recorded by Steineck et al., 1972) are shown by Toumarkine and Luterbacher (1985, fig. 5-6) to occur at the base of Zone P13 and the top of Zone P14, respectively. Kennedy (1975, p. 28) reported on a sample of the Emv collected from the Miramar Reservoir filtration plant (his locality 15 = LACM 3357), which contained the coccoliths "*Reticulofenestra umbilica* (Levin) and *Discoaster distinctus* Martini, which together suggest either a late middle Eocene or early [late] Eocene age" (Flynn, 1983, p. 185). According to Bukry (1973) and Okada and Bukry (1980), the base of low-latitude coccolith Subzone CP14a is defined by the first appearance of *R. umbilica*. According to Perch-Nielson (1985), *D. distinctus* "has its [last occurrence] in the upper part of Zone NP16," which is approximately equivalent to CP14a. Although the biochronological utility of *R. umbilica* is problematical (see discussion below), these data suggest that the Emv may be assignable to Subzone CP14a, which is correlated by Berggren et al. (1995) with the middle part of the middle Eocene.

Pomerado Conglomerate

The Pomerado Conglomerate (Ep) was named by Peterson and Kennedy (1974) for the cobble conglomerate unit that overlies the Mission Valley Formation in the Poway and La Mesa quadrangles. These authors also named the Miramar Sandstone Member of the Pomerado Conglomerate, which occurs between unnamed lower and upper conglomerate members of the Pomerado in the Miramar Reservoir area.

Pomerado Conglomerate, lower member

Lithology and Depositional Environment—The lower member of the Pomerado Conglomerate (Ep-lc) is composed mainly of iron oxide-stained, Poway clast-bearing pebble to cobble conglomerate. Lenses of light rusty brown, medium-grained friable sandstone are common in this unit, and occasionally contain greenish and reddish rip-up clasts of mammal-bearing siltstone. The Ep-lc appears to be mostly of fluvial origin, but local facies of probable shallow-marine origin were noted at Scripps Ranch North.

Distribution and Thickness—The Ep-lc is known to crop out only in the Miramar Reservoir area, where it is 7-10 m thick (Peterson and Kennedy, 1974; MS-6). Only a few kilometers south of the type section, the Ep-lc has not been recognized, and may pinch out rapidly within the sandstone lithosome composed of the Mission Valley Formation and Miramar Sandstone Member of the Pomerado Conglomerate (Peterson and Kennedy, 1974). Exactly where this pinchout occurs, however, is unknown.

Contacts—The contact between the Ep-lc and the Emv was erosional at the Eastview development on the west rim of Beeler Canyon, and was variably erosional and gradational at the Scripps Ranch North development north of Miramar Reservoir. The contact between the Ep-lc and the overlying Miramar Sandstone Member of the Pomerado Conglomerate was variably erosional and gradational at the Eastview and Scripps Ranch North developments.

Fossils and Age—Walsh (this volume, Chapter 5) discusses the late Uintan Eastview local fauna, which was collected from the upper part of the Ep-lc at the Eastview and Scripps Ranch North housing developments. Most of these specimens were collected from siltstone rip-up clasts, and the sandstone lenses that contained them (e.g., SDSNH Loc. 3755; MS-6a). The siltstone clasts are lithologically similar to the extensive beds of siltstone that occur in the lower part of the Mission Valley Formation at Scripps Ranch North, so it is possible that the fossils from the Eastview local fauna were derived from the Mission Valley Formation. Nevertheless, presence of *Protoreodon* new sp. 1 from the Miramar Sandstone Member (see below) independently suggests that the Ep-lc is of late Uintan age.

Pomerado Conglomerate, Miramar Sandstone Member

Lithology and Depositional Environment—The Miramar Sandstone Member (Ep-m) consists almost entirely of very light gray and pale greenish gray, fine- to medium-grained, friable sandstone, with rare interbeds of reddish siltstone, common lenses of Poway-type conglomerate, and rip-up clasts of green sandy mudstone, often oxidized to a bright rust color. The Ep-m is of both fluvial and shallow-marine origin.

Distribution and Thickness—The Ep-m is definitely known to crop out only in the Miramar Reservoir area, where it is 20 m thick (Peterson and Kennedy, 1974). It extends eastward at least as far as the Eastview housing development on the west rim of Beeler Canyon. Assuming that the lower member of the Pomerado pinches out rapidly to the south and southwest of Miramar Reservoir, it is possible that the Ep-m is stratigraphically correlative with the upper part of the Mission Valley Formation in the type area of the latter. Unfortunately, paleomagnetic results from these units are difficult to interpret, and can neither support nor refute this hypothesis.

Contacts—In general, the Ep-m appears to gradationally overlie the Ep-lc, although an erosional contact was observed at MS-6b. The Ep-m is erosionally overlain by the upper member of the Pomerado Conglomerate at Scripps Ranch North (MS-6a).

Fossils and Age—A small collection of internal molds of marine molluscs has been obtained from the Ep-m from Scripps Ranch North. The fossils occurred in a 0.5 m-thick bed of reddish laminated siltstone (SDSNH Locality 3625; MS-6a), and include *Crassatella* sp. indet., Lucinidae gen. indet., Tellinidae gen. indet., and one indeterminate gastropod (T. A. Deméré, personal communication). Preservation of these fossils is inadequate for species-level identifications, so the biostratigraphic utility of this collection is limited.

Two fossil mammal localities were also discovered within the Ep-m at Scripps Ranch North (SDSNH Locs. 3756 and 3757; MS-6a). Only SDSNH Loc. 3757 has yielded identifiable specimens, which include *Miacis* cf. *M. hookwayi*, *Protoreodon* new sp. 1, and a small unidentified species of *Leptoreodon*. These specimens suggest that the Ep-m is of late Uintan age, but still older than the latest Uintan or Duchesnean Laguna Riviera local fauna of northwestern San Diego County, which is characterized by *Protoreodon annectens* (Kelly et al., 1991; Walsh, this volume, Chapter 5).

Pomerado Conglomerate, Upper Member

Lithology—The upper member of the Pomerado Conglomerate (Ep-uc) is a poorly-sorted, light gray to pale greenish gray, Poway-type, pebble to boulder conglomerate. The Ep-uc contains common lenses of light gray medium-grained sandstone, and occasional thin lenses and rip-up clasts of green sandy mudstone. The mudstone clasts are often oxidized to a bright rust color.

Distribution and Thickness—The Ep-uc reaches a maximum preserved thickness of 32 m north of Miramar Reservoir (Peterson and Kennedy, 1974), and extends southward to Tierrasanta, Santee, and apparently La Mesa, where it thins to less than 10 m (Kennedy and Peterson, 1975). The light gray cobble to boulder conglomerate unit overlying the type section of the Mission Valley Formation may also be referable to the Ep-uc (Kennedy and Moore, 1971).

Contacts—In the Miramar Reservoir area, the Ep-uc erosionally overlies the Ep-m with less than 1 m of observed relief. In the Fletcher Hills and La Mesa areas, the Pomerado Conglomerate (upper member?) gradationally(?) overlies late Uintan, mammal-bearing, white fluvial sandstones mapped by Kennedy and Peterson (1975) as Mission Valley Formation.

Fossils and Age—Only one fossil is known from the Ep-uc. It was collected on the Scripps Ranch North development, from a green sandy mudstone bed (SDSNH Loc. 3759) within gray cobble-to-boulder conglomerate, at an elevation of 293 m. The fossil is a heavily damaged dentary of a small artiodactyl, and is not identifiable. The Ep-uc is of late Uintan or younger age, based on its stratigraphic position above the late Uintan Ep-m.

PALEOMAGNETIC CORRELATION OF EOCENE STRATA IN SOUTHWESTERN SAN DIEGO COUNTY

Previous Work

Flynn (1986) integrated paleomagnetic data with mammalian biochronology in an attempt to correlate the early Uintan rocks and faunas of San Diego with those of Wyoming. The key correlation point in Flynn's analysis was the Ardath Shale, which had been assigned an early middle Eocene age (low-latitude coccolith Subzone CP12b; Bukry and Kennedy, 1969; Okada and Bukry, 1980). Given Flynn's finding that the type section of the Ardath Shale was almost entirely of normal polarity, and given the placement by Berggren et al. (1985) of the early Eocene/middle Eocene boundary at the Chron C22n/C21r boundary, Flynn correlated the Ardath Shale normal polarity interval with Chron C21n. Bottjer et al. (1991) sampled the Ardath Shale along the seacliffs between Scripps Institution of Oceanography and Torrey Pines, reported that all sites were of normal polarity, and followed Flynn (1986) in assigning this interval to C21n.

The type section of the Delmar Formation was reported by Flynn (1986) to be of reversed polarity, with the exception of a single normal site at the top of the section. Flynn correlated this normal site with Chron C21n, and the lower reversed sites with C21r. Bottjer et al. (1991) sampled the Delmar Formation at Torrey Pines State Beach, and found this section to be entirely of reversed polarity. The significance of Flynn's (1986) normal site at the top of the Delmar is therefore uncertain, especially since the overlying Torrey Sandstone has not been sampled.

Flynn (1986) reported that the type section of the Scripps Formation was entirely of normal polarity. Flynn also reported that the lower part of the Friars Formation type section (uppermost Scripps and basal Friars Formation) was of normal polarity, while the upper part of the Friars Formation here was of reversed polarity. Flynn's Genesee Avenue section in the Friars produced similar results. We resampled this section because it contains the significant mammal locality LACM (CIT) 314, the results of which are discussed below (Appendix 1). Flynn (1986) correlated the normal polarity interval in the Scripps and basal Friars formations with Chron C21n, and the reversed polarity interval in the upper part of the Friars Formation with Chron C20r. Bottjer et al. (1991) sampled the Friars Formation at an outcrop about 3 km northwest of Flynn's Genessee Avenue section, and reported that all eight of their sites from the base of the Friars were of normal polarity, which they also correlated with C21n (however, some of the lower sites reported by Bottjer et al. (1991) may have been taken from the upper part of the Scripps Formation). Bottjer et al. (1991) also sampled the Friars Formation at Boyd Road and at Clairemont Mesa Boulevard, found both sections to be entirely of reversed polarity, and correlated these intervals with C20r. The Clairemont Mesa Boulevard section of Bottjer et al. (1991) is located in the upper tongue of the Friars as recognized here, immediately above the conglomerate tongue of the Friars (Fig. 2, MS-2).

Flynn (1986, fig. 8 and p. 378) briefly described two paleomagnetic sections from Murphy Canyon, collected from the Friars Formation and ?Stadium Conglomerate. Flynn shows the stratigraphically lowest section to be entirely of normal polarity, and the stratigraphically highest section to be entirely of reversed polarity. Based on Flynn's (1983, p. 191) statement that his Murphy Canyon section was collected from cuts along Interstate 15 between Friars Road and Clairemont Mesa Boulevard, it is probable that some of his "?Stadium Conglomerate" sites were actually taken from the conglomerate tongue of the Friars Formation. On the basis of the Murphy Canyon sections, Flynn (1986, p. 384) raised the possibility of an upper normal polarity interval in the Friars Formation that might correlate with Chron C20n. However, as he pointed out, correlation of these non-superposed sections in the Friars Formation is difficult, and there is presently no convincing evidence for the recognition of an upper normal polarity interval in the Friars Formation in this area. Flynn (1986, p. 353) indicated that three paleomagnetic sections (two in the Miramar Reservoir area and one near Fairmount Avenue) were collected from the Mission Valley Formation, but only two of them are shown in his summary of magnetostratigraphic data (Flynn, 1986, fig. 8), one of which is entirely of normal polarity, and the other entirely of reversed polarity. Unfortunately, it is not possible to determine which of Flynn's three sections are illustrated.

Of the four paleomagnetic sections collected in the Mission Valley Formation by Bottjer et al. (1991), the Presidio Park section is actually located in the Scripps Formation. The Eocene strata forming the south wall of

Mission Valley between Presidio Park and State Route 163 were assigned entirely to the Stadium Conglomerate and Mission Valley Formation by Kennedy (1975, plate 2A), but the anticline and syncline mapped by Kennedy on the north side of Mission Valley continue through to the south, such that the Scripps (SDSNH Loc. 3834) and Friars formations (SDSNH Loc. 3639) crop out on the south wall of Mission Valley in and east of Presidio Park (T. A. Deméré, personal communication). Bottjer et al. (1991) reported that the Presidio Park section is entirely of normal polarity, which is consistent with Flynn's (1986) report that the type section of the Scripps is entirely of normal polarity. The Miramar Reservoir section of Bottjer et al. (1991) clearly pertains to the Mission Valley Formation, but this section yielded only two sites of unequivocal polarity, both of which were normally magnetized (Powers, 1988, p. 71). The Pomerado Road section of Bottjer et al. (1991) apparently pertains to the Mission Valley Formation, but it produced only three sites of unequivocal polarity, all of which were normally magnetized (Powers, 1988, p. 69). The Montezuma Road section of Bottjer et al. (1991) clearly pertains to the Mission Valley Formation. They report that two sites occurring near the base of their section are of reversed polarity, whereas five sites from the middle and upper part of their section are of normal polarity. However, the original data presented by Powers (1988, p. 67) shows two reversed and two normal magnetozones from Montezuma Road, so the significance of this section is uncertain. Bottjer et al. (1991) correlated the putative reversed and putatively overlying normal magnetozones in the Mission Valley Formation with Chrons C18r and C18n, based largely on the tentative correlation (Steineck et al., 1972) of the upper part of the Stadium Conglomerate (possibly basal Mission Valley Formation) with planktonic foramini-feral Zone P13, which was correlated with Chron C18 by Berggren et al. (1985).

NEW PALEOMAGNETIC STUDIES
Methods

In an effort to clarify some of the ambiguities evident in previous paleomagnetic work on the Friars Formation and Poway Group, we collected samples from several new sections containing both early and late Uintan mammal assemblages (MS 1-11; Appendix 1). All sections were measured by Walsh with the exception of MS-11, which was measured by T. A. Deméré and Walsh. Block samples were collected using simple hand tools, with the top surface planed horizontal. Many samples were hardened with dilute sodium silicate to prevent them from crumbling. At least three samples were collected per site so that site statistics could be calculated. Sites were typically spaced 5-10 m apart stratigraphically, depending upon exposure and availability of suitable lithologies. Samples from MS-3,

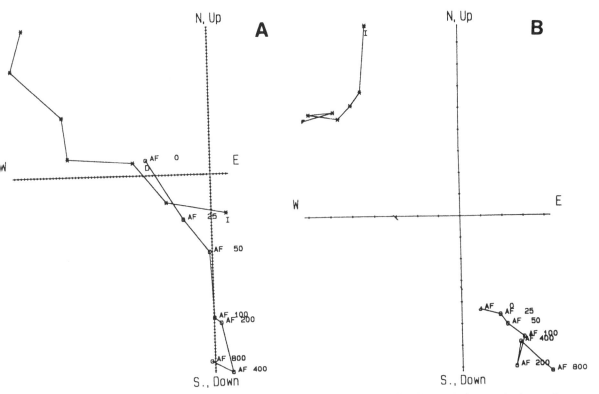

Figure 11. Representative orthogonal demagnetization ("Zijderveld") plots of alternating field (AF) demagnetization results. Solid circles indicate horizontal component, "x" indicates vertical component. AF intensity in Gauss is shown on each step. Each division = 10^{-7} emu. (A) This sample shows a rapid decline in intensity in AF demag, indicating that magnetite is a major carrier of remanence. It also changes from north and down at NRM ("AF0") to south and up (reversed polarity) after the first demagnetization step. (B) This sample does not respond to AF demagnetization, indicating that hematite is a primary carrier of the remanence.

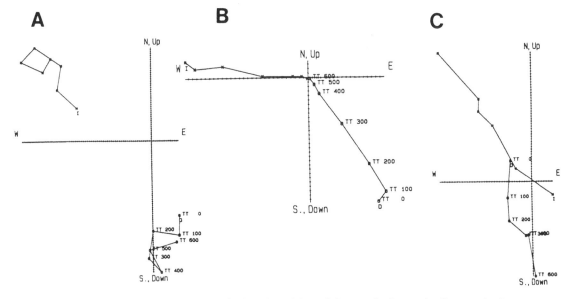

Figure 12. Representative orthogonal demagnetization plots of thermal demagnetization results. Demagnetization temperature ("TT") shown at each step. All other conventions as in Figure 11. (A) This sample shows clear reversed polarity from NRM ("TT0") , but did not lose intensity even at 600°C, indicating that hematite is the primary carrier of remanence. (B) This sample also shows reversed polarity, but completely demagnetizes by 600°C, indicating that all the remanence is carried by magnetite. (C) This sample contained a remanence in hematite, and a normal overprint was removed by 200°C, revealing a reversed polarity.

MS-4, MS-6a, MS-7, MS-10, and MS-11 were collected by Prothero, Lundquist, and their colleagues at Occidental, while samples from the remaining sections were collected by Walsh. Samples were trimmed into 2.5-cm cubes on a band saw with a tungsten carbide blade, and analyzed at the California Institute of Technology paleomagnetics lab by Prothero and Lundquist.

A suite of samples of representative lithologies was subjected to alternating field (AF) and thermal demagnetization. Under AF demagnetization (Fig. 11A), most samples showed a rapid drop in intensity, suggesting that the main carrier of the remanence is a low-coercivity mineral such as magnetite. However, some samples (Fig. 11B) were relatively resistant to AF demagnetization, so their remanence appears to be carried by a high-coercivity mineral such as hematite or goethite.

Under thermal demagnetization (Fig. 12), most samples showed no remanence above the Curie point of magnetite at 580°C; however, a few samples still had significant intensity at 600°C, showing that some of their remanence is carried by hematite. Overprints were typically removed by 100-200°C. Most samples produced a stable vectorial component between 400-500°C, and that component was used in the statistical analysis.

Analyses of IRM (isothermal remanent magnetization) acquisition further confirmed that both magnetite and hematite were present in many samples. Some rocks (Fig. 13B) reached saturation IRM values at 100-300 mT (millitesla); the remanence in these rocks is carried mostly by magnetite. Other specimens (Fig. 13A) showed no evidence of IRM saturation, even at fields of 1300 mT; these samples clearly contained much hematite. A modified Lowrie-Fuller ARM (anhysteretic remanent magnetization) test (e.g., Johnson et al., 1975) was also conducted during the IRM analysis (see Pluhar et al., 1991, for details). This test compares the resistance of AF demagnetization of both an IRM acquired in a 100 mT peak field, and an ARM gained in a 100 mT oscillating field. In almost all samples, the ARM (black squares) demagnetizes at higher peak fields than does the IRM (open squares), indicating that the remanence is carried by single-domain or pseudo-single domain grains. Finally, thin sections of representative samples were examined under reflected light. In several slides, euhedral magnetite grains with oxidized rims of hematite could be seen.

The stable sample directions were then clustered by site, and statistically analyzed by the methods of Fisher (1953; see Butler, 1992). Class I sites of Opdyke et al. (1977) showed a clustering that differed significantly from random at the 95% confidence level. In Class II sites, one sample was lost or crumbled, but the remaining samples gave a clear polarity indication. In Class III sites of Opdyke et al. (1977), two samples

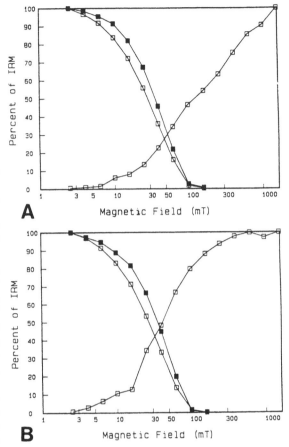

Figure 13. Typical results of IRM acquisition analysis and the modified Lowrie-Fuller test (see Pluhar et al., 1991, for details of methods). IRM values shown in open boxes, ARM in solid boxes. The IRM acquisition test is shown by the ascending curve on the right; the modified Lowrie-Fuller test by the two descending curves on the left. (A) The IRM fails to saturate in this sample, showing that hematite is the predominant magnetic mineral. The ARM is more resistant to AF demagnetization than the IRM, suggesting that the grains are single-domain or pseudo-single-domain. (B) IRM saturation occurs at about 300 millitesla (mT), showing that magnetite predominates in this sample. This sample also shows a single-domain Lowrie-Fuller test.

showed a clear polarity preference, but the third sample was divergent because of insufficient removal of overprinting. A few samples were considered indeterminate if their magnetic signature was unstable, or their direction uninterpretable.

Means for all Class I normal and reversed sites were calculated using the methods of Fisher (1953). The mean for all normal sites was D = 349.2°, I = 47.5°, k = 22.0, and α_{95} = 8.7; the mean for all reversed sites was D = 181.1, I = -26.2, k = 4.4, α_{95} = 19.9. These two means are antipodal within their margins of error, so this positive reversal test suggests that the magnetization is primary depositional remanence and not a later overprint.

Based on these inclinations, the Eocene paleolatitude for the area was about 28.6° (present latitude = 33°), suggesting about 4-5° of northward translation since the Eocene. This is within the range of values reported by Flynn (1986), Prothero (1991), and Powers (1988) for the area. However, given the large error bars, this result is not very definitive for paleolatitudinal studies.

Correlation of Magnetostratigraphic Sections

Simplified versions of the magnetostratigraphic sections studied for this paper are illustrated in Figure 14, and arranged from left to right in general order of decreasing age. The relative age of the sections is inferred from the lithostratigraphic correlations discussed above, and the mammalian biostratigraphy discussed by Walsh (this volume, Chapter 5).

Early Uintan

Magnetostratigraphic interpretation of the Friars Formation in the type area appears to be straight-forward. Thus, the type section and the three Genessee Avenue sections are consistent in that the lowermost part of the Friars is of normal polarity, while the upper part is reversed (Flynn, 1986; Bottjer et al., 1991; this chapter). This reversed interval is reasonably correlated with the reversed interval in the Friars Formation at MS-9, and with the reversed interval in the upper tongue of the Friars at MS-2, MS-6, MS-7, and MS-8.

Magnetostratigraphic interpretation of the Friars Formation becomes more difficult north and east of the type area, where the formation is divided into lower and upper tongues by the conglomerate tongue. The thickest sampled section in the Ef-lt (MS-4b) shows mostly normal polarity in the lower part, and mostly reversed polarity in the upper part of the section. This normal-to-reversed transition may correlate with the similar transition observed a short distance above the Friars/Scripps contact in the type area (e.g., type section, MS-2, and MS-5). If so, the relative thickness of the normal magnetozone at MS-4 suggests that the lower part of the lower tongue of the Friars Formation is indeed coeval with the upper part of the Scripps Formation, in accordance with the time-transgressive model of Kennedy and Moore (1971). The short sections of the Ef-lt at MS-2 and MS-3 were entirely of reversed and entirely of normal polarity, respectively. However, since erosional relief at the base of the Ef-cg may be considerable over several kilometers, the significance of these results is uncertain.

More discrepancies are evident in the results obtained from the Ef-cg and Ef-ut. The Ef-cg was found to be of reversed polarity at MS-2, but was largely of normal polarity at MS-1. The Ef-ut was found to be entirely of reversed polarity at MS-1, MS-2, MS-6, MS-7, and MS-8, but was found to be almost entirely normal at MS-3. Conceivably, the "Ef-cg" and "Ef-ut" in certain sections may be different stratigraphic units than the

"Ef-cg" and "Ef-ut" in other sections. A more parsimonious explanation would invoke lithostratigraphic time-transgression (e.g., Kennedy, 1975), such that deposition of these units as preserved in one area began earlier than deposition of the same units as preserved in another area. If so, the normal magnetozones in MS-1 and MS-3 may correlate with the normal magnetozone in the lower part of the type area of the Friars Formation. Finally, the normal magnetozone at MS-3 could represent a distinct upper normal polarity interval in the Friars. This seems unlikely, since no definite reversed polarity interval was detected in this section, but it is also possible that a reversed interval was originally present in the Ef-lt here, only to be eroded away prior to the deposition of the Ef-cg.

Late Uintan

A potential tiepoint between the early and late Uintan magnetozones is represented by the reversed interval that occurs in the late early Uintan Est-a (MS-9 and MS-10). If the contact between the Est-a and Est-b is assumed not be a chron-encompassing disconformity, then the two reversed sites in the lower part of the Est-b (MS-8 and MS-10) would represent the same chron as the reversed interval in the Est-a. The late Uintan normal interval in MS-10 (upper part of Est-b and Emv) likely correlates with the normal intervals in the Mission Valley Formation at the type section and at MS-7 and MS-8. Unfortunately, other Mission Valley Formation sections are difficult to interpret. Most of the type section (MS-11) is of normal polarity, with the uppermost five sites being reversed. In contrast, the lower part of the Montezuma Road section of Bottjer et al. (1991) was reported to be of reversed polarity. We found the Mission Valley to be of variable, but largely reversed polarity in two adjacent sections at MS-6 (see Appendix 1). However, only 1.5 km to the south and 3 km to the southeast of MS-6a, all five of Powers's (1988, p. 71) determinable Mission Valley Formation sites were of normal polarity. These discrepancies cannot be resolved at this time. Assuming that the polarity has been correctly determined, it is also unclear whether the reversely-magnetized lower unit of the Mission Valley Formation at MS-6 correlates with the reversed magnetozone in the upper part of the Mission Valley type section, or whether the lower unit at MS-6 is entirely older than the type section. In view of these uncertainties, no satisfactory overall magnetostratigraphic hypothesis can be proposed for the Mission Valley Formation at this time.

Stratigraphic uncertainties and limited samples also make interpretation of the Pomerado Conglomerate magnetozones difficult. First, it is uncertain whether the entire thickness of the Pomerado Conglomerate at MS-7 correlates with the Ep-lc, Ep-m, and Ep-uc at MS-6a, or just with the Ep-uc at MS-6a. Second, as noted previously, it cannot be assumed that the lower

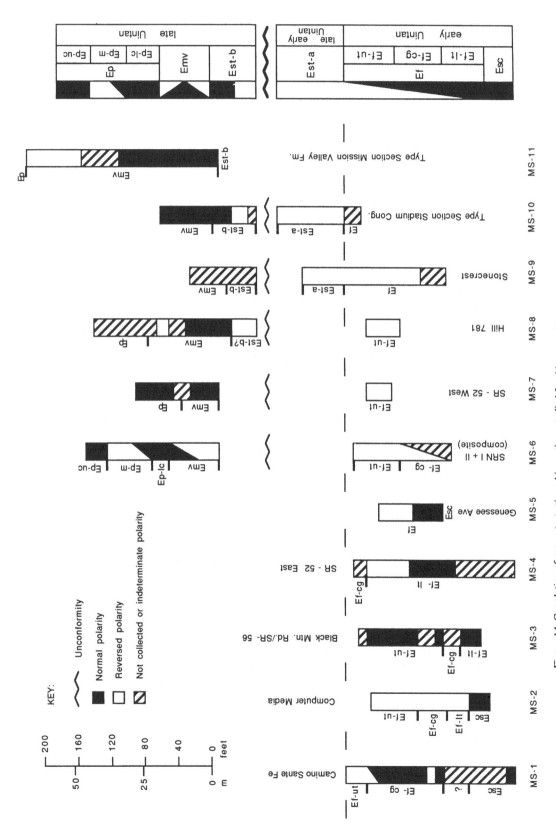

Figure 14. Correlation of magnetostratigraphic sections studied for this paper. Vertical scale applies to the individual sections, not the composite section. Abbreviations as in Figs. 2-3. See text for discussion.

conglomerate and Miramar Sandstone members of the Pomerado are entirely younger than the type outcrops of the Mission Valley Formation. In the absence of a convincing demonstration of the correct stratigraphic and/or temporal relationships between these units, correlation of the magnetozones within them is inadvisable.

Chron Correlations of San Diego Eocene Polarity Intervals

Given the difficulties in magnetozone correlation discussed above, the attempt to identify many of them with specific chrons is questionable. Nevertheless, tentative chron correlations for various lithostratigraphic units of southwestern San Diego County are shown in Figure 15 (after Berggren et al., 1995), and the evidence for these assignments is reviewed below. We emphasize that numerical calibration of the boundaries of the Eocene chrons rests upon a large number of assumptions involving many different sources of error, and that the values given for these boundaries cannot be taken literally to within about ± 1 m.y. (Harland et al., 1990, p. 154).

Delmar Formation

As noted, Flynn (1986) and Bottjer et al. (1991) assigned the Delmar Formation reversed polarity interval to the earliest middle Eocene Chron C21r. This assignment appears to be corroborated by Frederiksen's (1991) correlation of a pollen assemblage from the Delmar with the earliest middle Eocene coccolith Subzone CP12b (Okada and Bukry, 1980; Berggren et al., 1995).

Ardath Shale

The normal polarity interval in the Ardath Shale both at the type section and at the seacliff outcrops north of La Jolla was assigned by Flynn (1986) and Bottjer et al. (1991) to Chron C21n. This assignment was based on the early middle Eocene age of the Ardath as determined by planktonic foraminifera (Zone P10 and/or P11; Steineck et al., 1972) and coccoliths (Subzone CP12b; Bukry and Kennedy, 1969; Okada and Bukry, 1980), in conjunction with the correlations of Berggren et al. (1985). The report by May et al. (1991) that the upper part of the Ardath Shale contains coccoliths assignable to Subzones CP13a and CP13b further supports identification of the Ardath normal polarity interval as C21n (Berggren et al., 1995). Although a literal reading of the time scale of Berggren et al. (1995) would indicate that Subzone CP13b is entirely younger than C21n, these authors note that magnetochronological control on certain middle Eocene microfossil biochrons is poorly constrained. The beginning of CP13b as shown by Berggren et al. (1995) will have to be moved downward into Chron C21n if the presence of Subzone CP13b in the Ardath Shale can

be corroborated (assuming that the normal polarity obtained by Bottjer et al. (1991) for the Ardath Shale at Black's Beach is correct).

Scripps Formation

The normal polarity interval comprising the entire thickness of the type section of the Scripps Formation was assigned by Flynn (1986) to C21n, based on the reasonable assumption that the contact between the Scripps and the Ardath Shale was not a chron-encompassing disconformity. If May et al. (1991) are correct that the Ardath Shale is as young as coccolith Subzone CP13b, and if the normal polarity interval in the type Scripps does correlate with Chron C21n, and if the correlations between the coccolith zones of Okada and Bukry (1980) and chron boundaries illustrated by Berggren et al. (1995) are taken literally, then the entire thickness of the type Scripps must have been deposited extremely rapidly during the latest part of Chron C21n and during the earliest part of Subzone CP13b. Marine mollusc assemblages from the lower and upper parts of the Scripps are assigned by Givens and Kennedy (1979) to the "Domengine" and "Transition" stages, respectively. Since Squires (1988) placed the "Transition" stage entirely within Subzone CP13a, correlation of the Scripps normal polarity interval with Chron C21n would again seem to be indicated (Berggren et al., 1995). However, a partial correlation of the "Domengine" stage with Subzone CP13b would be required if the upper part of the Ardath Shale is in fact assignable to this Subzone (May et al., 1991).

Alternatively, the Scripps-Ardath contact was regarded as a sequence boundary by May et al. (1991), and it is conceivable that the disconformity is of such magnitude that the normal interval in the type Scripps is a full chron younger than the normal interval in the Ardath, in which case the former would correlate with C20n. If so, the Scripps should contain coccoliths assignable to Subzones CP13c and/or CP14a, and planktonic forams assignable to Zones P11 and/or P12, according to the correlations of Berggren et al. (1995). Obviously, a stratigraphically-controlled micropaleontological study of the Scripps Formation should be undertaken in order to clarify the age range of this unit.

Despite these difficulties, we prefer to correlate the Scripps normal interval with Chron C21n for the following reasons. First, if this interval did pertain to C20n, the Scripps-Ardath disconformity would have to represent at least the entire duration of C20r, and the beginning of the Friars Formation reversed interval would then correlate with the beginning of C19r (42.6 Ma according to Berggren et al., 1995). However, given the $^{40}Ar/^{39}Ar$ date of 42.83 ± 0.24 Ma from the Mission Valley Formation (Obradovich and Walsh, in prep.), this would in turn require that most of the Friars Formation, both members of the Stadium Conglomerate, and the lower part of the Mission Valley

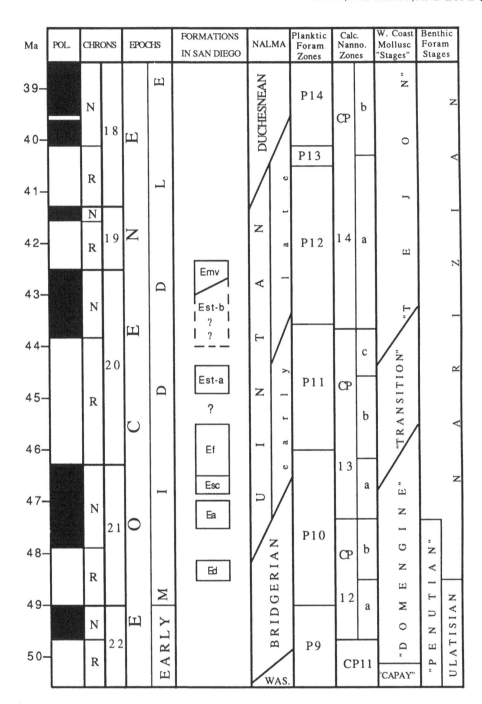

Figure 15. Tentative correlations of certain San Diego lithostratigraphic units with the Geomagnetic Polarity Time Scale (Berggren et al., 1995) and various biochronologies. Based largely on Flynn (1986), Bottjer et al. (1991), Almgren et al. (1988), Squires (1988), Frederiksen (1991), Obradovich and Walsh (in prep.), Walsh (this volume, Chapter 5), and this paper. Ed = Delmar Formation, Ea = Ardath Shale. Other abbreviations as in Figures 2-3. Note that the numerical ages of the chron boundaries cannot be taken literally to within about ± 1 m.y. (Harland et al., 1990, p. 154).

Formation were all deposited in an interval of "negative" 0.2 m.y. Further, an age of about 42.6 Ma for the early Uintan Friars Formation would conflict with the fact that several mammal assemblages in North America of about this age have all been assigned to the late Uintan (Walsh, this volume, Chapter 4). These implications suggest that the Scripps Formation normal polarity interval probably correlates with C21n rather than C20n.

Friars Formation

Because the Friars Formation either conformably overlies the Scripps Formation (Kennedy and Moore, 1971), or locally overlies the Scripps with minor erosional relief, the normal polarity interval in the base of the Friars Formation in the type area can be confidently assumed to pertain to the same chron as the normal interval in the Scripps Formation. As discussed above, the latter is best correlated with C21n, so the Friars reversed interval probably correlates with C20r, as concluded by Flynn (1986) and Bottjer et al. (1991).

Stadium Conglomerate, lower member

The contact between the lower member of the Stadium and the Friars Formation is gradational according to Kennedy and Moore (1971), but was erosional wherever we have seen it, and the amount of time represented by this contact is unknown. Nevertheless, the reversed interval in the lower member of the Stadium is tentatively assigned to Chron C20r on the basis of its late early Uintan mammal assemblage. If the lower member reversed interval does in fact pertain to C19r, then the Friars-Stadium contact would be a disconformity representing at least the entire duration of C20n (1.3 m.y. according to Berggren et al., 1995).

Stadium Conglomerate, upper member

Paleomagnetic evidence from the Est-b is limited. Two sites from the base of the type section of this unit (MS-10) were of indeterminate polarity, and one site each from MS-8 and MS-10 were reversed. If the above assignment of the Est-a to C20r is correct, and if the Est-b/Est-a contact is not a chron-encompassing disconformity, then the lower part of the Est-b may also pertain to C20r, and the normally-magnetized upper part of the Est-b (MS-10) may pertain to C20n. If so, based on the late Uintan age of the Stonecrest local fauna (Walsh, this volume, Chapter 5), the early Uintan/late Uintan boundary as recognized in San Diego would occur during late C20r. However, these scenarios must be regarded as tenuous in view of the small number of paleomagnetic sites from the Est-b, and the discussion of the Mission Valley Formation below.

Mission Valley Formation

As noted above, a planktonic foraminiferal assemblage from either the upper member of the Stadium Conglomerate or the Mission Valley Formation has been correlated with Zone P13 (Steineck et al., 1972). Berggren et al. (1985) correlated Zone P13 with Chron C18n, which led Bottjer et al. (1991) to assign the Mission Valley Formation to the latter. However, Berggren et al. (1995) now correlate Zone P13 with the upper part of C18r. Although little confidence can be placed in the meager planktonic foraminiferal evidence available, if the Mission Valley Formation is indeed approximately of "Zone P13 age," then the lower normal and upper reversed magnetozones in the type section of the Mission Valley Formation could correlate with C19n and C18r, or perhaps C18n.2n and C18n.1r.

As discussed above, a sparse coccolith assemblage from the Mission Valley Formation at the Miramar Reservoir filtration plant may be assignable to Subzone CP14a, based in part on the reported occurrence of *Reticulofenestra umbilica*, which is the defining taxon for the base of this subzone (Bukry, 1973; Okada and Bukry, 1980). Unfortunately, there are problems with the use of this species as a biochronologic marker. Backman and Hermelin (1986) concluded that the "appearance" (defined as a size increase) of *R. umbilica* took place over a period of about 2.3 m.y. According to their redefinition of this taxon (using a lower size limit of 14 microns), the first appearance of *R. umbilica* took place during Chron C19r. As discussed by Berggren et al. (1995, table 15), Wei and Wise (1989) report examples of this species greater than 14 microns from the upper part of a normal magnetozone interpreted as Chron C20n, and Berggren et al. (1995) correlate Subzone CP14a with all of C20n, all of C19, and almost all of C18r. Unfortunately, the original samples collected from Miramar Reservoir that were analyzed by D. Bukry cannot be located (Bukry, personal communication to SLW), so it is impossible to determine if the specimens that Bukry identified in the 1960s were larger than 14 microns (several matrix samples from the Mission Valley Formation recently submitted to Bukry for analysis proved to be barren of coccoliths). To further complicate matters, the coccolith-bearing strata at Miramar Reservoir are of uncertain polarity, and cannot be confidently correlated with the magnetozones in the type section of the Mission Valley Formation. Nevertheless, assuming that Kennedy's (1975) coccolith sample does pertain to CP14a, and that the Miramar Reservoir strata are time-equivalent to the lower part of the type Mission Valley Formation, and that the correlations of Berggren et al. (1995) are correct, then the type Mission Valley magnetozones could correlate with either C20n and C19r, or C19n and C19r, but not C18n.

Frederiksen (1991) correlated a pollen assemblage from an outcrop near the base of the Mission Valley Formation in the type area with coccolith Subzone CP13c, suggesting a correlation of the Mission Valley magnetozones with C20r-C20n (Berggren et al., 1995). However, Frederiksen's age assignment should be regarded as tentative, because it is based largely on weak negative evidence, namely, the apparent absence of three taxa (*Psilodiporites*, *Corsinipollenites thiergartii*, and *Juglans* type) from only one locality.

Prothero and Swisher (1992) report an $^{40}Ar/^{39}Ar$ date of 39.74 ± 0.07 Ma on the LaPoint Ash which, following the time scale of Berggren et al. (1995), establishes at least a partial correlation between the Duchesnean NALMA and the older part of Chron C18n. Correlation of the late Uintan Mission Valley Formation with C18n therefore seems unlikely, especially since an $^{40}Ar/^{39}Ar$ date of 42.83 ± 0.24 Ma has been obtained on bentonite from the Mission Valley Formation at SDSNH Loc. 3428 (Obradovich and Walsh, in prep.). Strata of the Mission Valley Formation at this locality were reported by Prothero (1991) to be of normal polarity, which, given the time scale of Berggren et al. (1995), is consistent with an assignment of the type Mission Valley magnetozones to C20n and C19r.

In summary, chron correlation of the Mission Valley Formation is made difficult by the contradictory correlations (C20n-C18n) indicated by a variety of radiometric and biostratigraphic evidence in conjunction with current time scales. We prefer to place more weight on the radiometric evidence, and tentatively correlate the type Mission Valley Formation lower normal and upper reversed polarity intervals with C20n and C19r. However, it should be emphasized that the temporal difference between the C20n-C19r boundary and the C19n-C18r boundary is only about 1.3 m.y. according to Berggren et al. (1995), which is about the same magnitude as the potential errors in the numerical calibration of the Eocene chrons (~ ± 1 m.y.; e.g., Harland et al., 1990, p. 154). Other workers must be aware of these uncertainties before using our tentative chron correlation of the Mission Valley Formation as a primary piece of data upon which to base other conclusions. Further micropaleontological work on the Mission Valley Formation should be done in order to corroborate the suggested correlation.

Pomerado Conglomerate

Chron correlation of the Pomerado Conglomerate is made difficult by the uncertain stratigraphic and temporal relationship of the lower and Miramar Sandstone members of the Pomerado to the type Mission Valley Formation, and by the lack of a well-corroborated magnetostratigraphic pattern for these units. Accordingly, no chron correlations are proposed for the Pomerado Conglomerate here.

DISCUSSION

When the paleomagnetic part of this project was begun, it was hoped that with the revised stratigraphic model discussed above, and better age control provided by mammalian biostratigraphy (Walsh, this volume, Chapter 5), some of the ambiguities in the results obtained by Flynn (1986) and Bottjer et al. (1991) from the Friars, Stadium, and Mission Valley formations could be resolved. This goal has been only partially achieved. Consistent results were obtained from the lower member of the Stadium Conglomerate, and the

Mission Valley Formation in the type area. However, the variable polarities indicated for the Friars and Mission Valley formations in the composite magneto-stratigraphic section in Figure 14 reflect the large proportion of apparently contradictory paleomagnetic data involved. Additional magnetic sections might be collected in an effort to resolve these discrepancies, and continued field work is being done in order to test the stratigraphic interpretations proposed here. However, it is our opinion that the facies relationships, inherent time-transgressivness, and internal disconformities of the Friars Formation and Poway Group are so complex that a completely consistent magnetostratigraphic pattern for every unit will probably never be obtained. The nature of the outcrops in San Diego also hampers these efforts, since it is difficult to determine the superpositional relationships of the widely separated and relatively short sections available. For perspective, the *thickest* single section studied in San Diego (80 m; MS-6a) is thinner than the 90 m *sampling gap* in Flynn's (1986) Washakie Formation section. From this standpoint, it is highly questionable whether a given magnetozone in a given section of the Friars Formation or Poway Group corresponds to a worldwide geomagnetic event, and if it does, whether it represents a formally recognized chron, or an unrecognized "cryptochron" (Cande and Kent, 1992). Indeed, recent work has shown that many of the short polarity events long ignored by magnetostratigraphers may be real (Hartl et al., 1993).

Despite these difficulties, consistent paleomagnetic results have been obtained in San Diego from the composite, but superposed sections of the fine-grained marine strata of the Delmar, Ardath, and Scripps formations, as exposed in the seacliffs north of La Jolla, and less than a few kilometers inland from the coast (Flynn, 1986; Bottjer et al., 1991). Consistent results have also been obtained by these authors and ourselves from the Friars Formation in the type area. As noted above, the normal polarity interval in the lower part of the Friars Formation in the type area is confidently assignable to Chron C21n (Flynn, 1986; Bottjer et al., 1991; this paper). Given this correlation, (1) the occurrence of *Amynodon* and *Leptoreodon* in the Friars Formation in the upper part of this polarity interval at Genessee Avenue (MS-5); and (2) the presence of *Leptoreodon* in the lower tongue of the Friars Formation, stratigraphically well below the top of a normal interval correlated with C21n at State Route 52 East (MS-4), indicate that the Bridgerian/Uintan NALMA boundary is older than the beginning of Chron C20r. As discussed by Walsh (this volume, Chapter 4), this conclu-sion is relevant to the stratigraphic position of the Bridgerian/Uintan boundary in the Aycross/ Tepee Trail sequence of Wyoming, the Washakie Formation of Wyoming and Colorado, the Uinta Formation of Utah, and the Devil's Graveyard Formation of Texas.

CONCLUSIONS

The stratigraphy of the Friars Formation and Poway Group in southwestern San Diego County is informally revised. The Friars Formation is locally divisible into three units. The lower tongue of the Friars consists largely of sandstone and mudstone, is of fluvial origin in the east, and is partly of lagoonal origin in the west. The middle unit of the Friars is an extensive conglomerate tongue (formerly mapped as part of the Stadium Conglomerate) that erosionally overlies the lower tongue. This conglomerate is of fluvial origin in the east, and of deltaic origin in the west. The upper tongue of the Friars (previously mapped in part as the Mission Valley Formation) gradationally overlies the conglomerate tongue, and consists largely of sandstone and mudstone. The upper tongue is again of fluvial origin in the east, and partly of lagoonal origin in the west. All three units of the Friars produce early Uintan mammal taxa that are collectively referred to as the Poway fauna (Walsh, this volume, Chapter 5).

The type Stadium Conglomerate is locally divisible into two members. The lower member is generally a light yellowish gray and pale greenish gray, matrix to clast-supported conglomerate, is largely of fluvial origin, and produces a distinctive late early Uintan micromammal assemblage dominated by the rodent *Pauromys* sp. cf. *P. perditus*. The upper part of the lower member may be of late Uintan age. The upper member of the Stadium Conglomerate is an iron oxide-stained, clast-supported conglomerate, of marine origin in the western part of its outcrop area, and apparently of fluvial origin in the eastern part of its outcrop area. The upper member of the Stadium is late Uintan in age. The Mission Valley Formation is restricted only to those rocks that are stratigraphically correlative with the type section. The Mission Valley Formation is of late Uintan age, and disconformably overlies the early Uintan upper tongue of the Friars Formation north of the northern pinchout of the Stadium Conglomerate.

The reports of Flynn (1986) and Bottjer et al. (1991) of normal and reversed polarity intervals in the lower and upper part of the type outcrops of the Friars are corroborated here, and we agree with their assignment of these intervals to Chrons C21n and C20r. The polarity of the lower, conglomerate, and upper tongues of the Friars as recognized here is variable, and may reflect significant time-transgression of these units. The lower member of the Stadium Conglomerate is tentatively assigned to C20r. The upper member of the Stadium may straddle the C20r-C20n boundary. Radiometric and biochronologic evidence from the Mission Valley Formation indicate partly contradictory chron correlations. However, more weight is given to an $^{40}Ar/^{39}Ar$ date of 42.83 ± 0.24 Ma from a bentonite in this formation (Obradovich and Walsh, in prep.), which suggests a correlation of the lower normal and upper reversed magnetozones in the type section of this formation with Chrons C20n and C19r.

The existence of *Amynodon* and *Leptoreodon* during Chron C21n in San Diego has important implications for the occurrence of the Bridgerian/Uintan North American land mammal "age" boundary in other sections in the western interior.

ACKNOWLEDGMENTS

We thank Tom Deméré, Brad Riney, Paul Majors, Dean Milow, and Michael Kennedy for discussions of the stratigraphy and paleontology of the region. David Bukry examined several barren samples for nannofossils, and David Whistler provided access to LACM(CIT) locality records. Jon Erskine, Erin Wilson, and Karen Whittlesey helped with the magnetic sampling. Joyce Corum and Chris White (CALTRANS) provided access to the SR-52, SR-56, and SR-163 sections. Gatlin Development and McMillin Communities provided access to the Stonecrest and Scripps Ranch North sections, respectively. Kevin Everly and Lynn Swanson provided access to the Mission Valley and Carroll Canyon quarries of H. G. Fenton Materials Company, while Severo Chavez and Mike Mills provided access to the Mission Valley and Carroll Canyon quarries of CalMat Company. Joe Corones provided access to outcrops in the Miramar Landfill of the City of San Diego, and Coralie Cobb provided access to outcrops on Miramar Naval Air Station. Prothero acknowledges the support of NSF grant EAR91-17819 during this research. We thank Dr. Joseph Kirschvink for allowing us access to the Caltech paleomagnetics lab. We appreciate reviews of the manuscript by Patrick Abbott, Thomas Deméré, Steve McCarroll, Thomas Kelly, Michael Kennedy, Gary Peterson, and Michael Woodburne. None of these individuals necessarily agree with any of the views expressed herein.

REFERENCES CITED

Abbott, P. L., B. M. Smith, N. A. Briedis, and T. E. Moore. 1979. Petrology of some Eocene sandstones, San Diego California; pp. 111-114 in P. L. Abbott (ed.), Eocene Depositional Systems, San Diego, California. Pacific Section, SEPM.

Almgren, A. A., M. V. Filewicz, and H. L. Heitman. 1988. Lower Tertiary foraminiferal and calcareous nannofossil zonation of California: An overview; pp. 83-105 in M. V. Filewicz and R. L. Squires (eds.), Paleogene Stratigraphy, West Coast of North America. Pacific Section, SEPM Volume 58.

Backman, J., and J. O. R. Hermelin. 1986. Morphometry of the Eocene nannofossil *Reticulofenestra umbilicus* lineage and its biochronological consequences. Palaeogeography, Palaeoclimatology, Palaeoecology 57:103-116.

Berggren, W. A., D. V. Kent, J. J. Flynn, and J. A. Van Couvering. 1985. Cenozoic geochronology. Geological Society of America Bulletin 96:1407-1418.

Berggren, W. A., D. V. Kent, C. C. Swisher III, and M.-P. Aubry. 1995. A revised Cenozoic geochronology and chronostratigraphy. SEPM Special Publication 54:129-212.

Berry, R. W. 1991. Deposition of Eocene and Oligocene

bentonites and their relationship to Tertiary tectonics, San Diego County; pp. 107-113 *in* P. L. Abbott and J. A. May (eds.), Eocene Geologic History San Diego Region. Pacific Section, SEPM Volume 68.

Bolli, H. M. 1957. The genera *Globigerina* and *Globorotalia* in the Paleocene-lower Eocene Lizard Springs Formation of Trinidad, B.W.I. United States National Museum Bulletin 215:51-81.

Bottjer, D. J., S. P. Lund, J. E. Powers, and M. C. Steele. 1991. Magnetostratigraphy of Paleogene strata in San Diego and the Simi Valley, southern California; pp. 115-124 *in* P. L. Abbott and J. A. May (eds.), Eocene Geologic History San Diego Region. Pacific Section, SEPM Volume 68.

Bukry, D. 1973. Low-latitude coccolith biostratigraphic zonation. Initial Reports of the Deep Sea Drilling Project 15:817-832.

Bukry, D., and M. P. Kennedy. 1969. Cretaceous and Eocene coccoliths at San Diego, California. California Division of Mines and Geology Special Report 100:33-43.

Butler, R. F. 1992. Paleomagnetism. Blackwell, London.

Cande, S. C., and D. V. Kent. 1992. A new geomagnetic polarity time scale for the late Cretaceous and Cenozoic. Journal of Geophysical Research 97:13917-13951.

Cushman, J. A., and A. N. Dusenbury, Jr. 1934. Eocene foraminifera of the Poway Conglomerate of California: Contributions to the Cushman Laboratory for Foraminiferal Research 10:51-65.

Dusenbury, A. N. Jr. 1932. A faunule from the Poway Conglomerate, upper middle Eocene of San Diego County, California. Micropaleontology Bulletin 3:84-95.

Eisenberg, L. I. 1985. Geologic Map of the Rancho Santa Fe quadrangle; Plate 3 *in* P. L. Abbott (ed.), On the Manner of Deposition of the Eocene Strata in Northern San Diego County. San Diego Association of Geologists Field Trip Guidebook.

Fisher, R. A. 1953. Dispersion on a sphere. Proceedings of the Royal Society A217:295-305.

Flynn, J. J. 1983. Correlation and geochronology of middle Eocene strata from the western United States. Ph.D. Thesis, Columbia University, 496 pp.

Flynn, J. J. 1986. Correlation and geochronology of middle Eocene strata from the western United States. Palaeogeography, Palaeoclimatology, Palaeoecology 55:335-406.

Frederiksen, N. O. 1991. Age determinations for Eocene formations of the San Diego, California, area, based on pollen data; pp. 195-199 *in* P. L. Abbott and J. A. May (eds.), Eocene Geologic History San Diego Region. Pacific Section, SEPM Volume 68.

Gibson, J. M. 1971. Benthonic foraminifera of the Ardath Shale and Stadium Conglomerate (Eocene), San Diego Basin, California. Bulletin of the Southern California Academy of Sciences 70:125-130.

Givens, C. R., and M. P. Kennedy. 1979. Eocene molluscan stages and their correlation, San Diego area, California; pp. 81-95 *in* P. L. Abbott (ed.), Eocene Depositional Systems, San Diego, California. Pacific Section, SEPM.

Golz, D. J., and J. A. Lillegraven. 1977. Summary of known occurrences of terrestrial vertebrates from Eocene strata of southern California. University of Wyoming Contributions to Geology 15:43-65.

Hanna, M. A. 1926. Geology of the La Jolla quadrangle, California. University of California Department of Geological Sciences Bulletin 16:187-246.

Hanna, M. A. 1927. An Eocene invertebrate fauna from the La Jolla quadrangle, California. University of California Department of Geological Sciences Bulletin 16:247-398.

Harland, W. B., R. L. Armstrong, A. V. Cox, L. E. Craig, A. G. Smith, and D. G. Smith. 1990. A Geologic Time Scale 1989. Cambridge University Press, Cambridge.

Hartl, P., L. Tauxe, and C. Constable, 1993, Early Oligocene geomagnetic field behavior from Deep Sea Drilling Project Site 522. Journal of Geophysical Research 98(B11): 19649-19665.

Howell, D. G., and M. H. Link. 1979. Eocene conglomerate sedimentology and basin analysis, San Diego and the southern California borderland. Journal of Sedimentary Petrology 49:517-540.

Johnson, H. P., W. Lowrie, and D. V. Kent. 1975. Stability of anhysteretic remanent magnetization in fine and coarse magnetite and maghemite particles. Geophysical Journal of the Royal Astronomical Society 41:1-10.

Kennedy, M. P. 1975. Geology of the Western San Diego Metropolitan area, California. Califonia Division of Mines and Geology Bulletin 200-A:1-39

Kennedy, M. P., and G. W. Moore. 1971. Stratigraphic relations of Upper Cretaceous and Eocene formations, San Diego coastal region, California. American Association of Petroleum Geologists Bulletin 55:709-722.

Kennedy, M. P., and G. L. Peterson. 1975. Geology of the Eastern San Diego Metropolitan area, California. California Division of Mines and Geology Bulletin 200-B:43-56.

Kennedy, M. P., and S. S. Tan. 1977. Geology of the National City, Imperial Beach, and Otay Mesa quadrangles, southern San Diego metropolitan area, California. California Division of Mines and Geology Map Sheet 29.

Kies, R. P. 1982. Paleogeography of the Mt. Soledad Formation west of the Rose Canyon Fault; pp. 1-11 *in* P. L. Abbott (ed.), Geologic Studies in San Diego. San Diego Association of Geologists Field Trip Guidebook.

Kies, R. P., and P. L. Abbott. 1983. Rhyolite clast populations and tectonics in the California Continental Borderland. Journal of Sedimentary Petrology 53:461-475.

Krishtalka, L., R. K. Stucky, R. M. West, M. C. McKenna, C. C. Black, T. M. Bown, M. R. Dawson, D. J. Golz, J. J. Flynn, J. A. Lillegraven, and W. D. Turnbull. 1987. Eocene (Wasatchian through Duchesnean) biochronology of North America; pp. 77-117 *in* M. O. Woodburne (ed.), Cenozoic Mammals of North America, Geochronology and Biostratigraphy. University of California Press, Berkeley.

Kuper, H. T. 1977. Geological maps of the Jamul Mountains, Otay Mesa, Imperial Beach, and National City quadrangles; *in* G. T. Farrand (ed.), Geology of Southwestern San Diego County, California and Northwestern Baja California. San Diego Association of Geologists Field Trip Guidebook.

Lillegraven, J. A., and R. W. Wilson. 1975. Analysis of *Simimys simplex*, an Eocene rodent (?Zapodidae). Journal of Paleontology 49:856-874.

Lohmar, J. M., J. A. May, J. E. Boyer, and J. E. Warme. 1979. Shelf edge deposits of the San Diego embayment; pp. 15-33 *in* P. L. Abbott (ed.), Eocene Depositional Systems, San Diego, California. Pacific Section, SEPM.

May, J. A., 1985. Submarine canyon system of the Eocene San Diego Embayment; pp. 1-17 *in* P. L. Abbott (ed.), On the Manner of Deposition of the Eocene Strata in Northern San Diego County. San Diego Asssociation of Geologists Field Trip Guidebook.

May, J. A., J. M. Lohmar, J. E. Warme, and S. Morgan. 1991. Field trip guide: Early to Middle Eocene La Jolla Group of Black's Beach, La Jolla, California; pp. 27-36 *in* P. L. Abbott and J. A. May (eds.), Eocene Geologic History San Diego Region. Pacific Section, SEPM Volume 68.

McWilliams, R. G. 1972. Age and correlation of the Eocene Ulatisian and Narizian Stages, California: Discussion. Geological Society of America Bulletin 83:533-534.

Milow, E. D., and D. B. Ennis. 1961. Guide to geologic field trip no. 2, southwestern San Diego County; pp. 23-43 *in* B. E. Thomas (ed.), Geological Society of America

Cordilleran Section 57th Annual Meeting, Field Trip Guidebook.

North American Commission on Stratigraphic Nomenclature. 1983. North American Stratigraphic Code. American Association of Petroleum Geologists Bulletin 67:841-375.

Okada, H., and D. Bukry. 1980. Supplementary modification and introduction of code numbers to the low-latitude coccolith biostratigrapic zonation (Bukry, 1973; 1975). Marine Micropaleontology 5:321-324.

Opdyke, N. D., E. H. Lindsay, N. M. Johnson, and T. Downs. 1977. The paleomagnetism and magnetic polarity stratigraphy of the mammal-bearing section of Anza-Borrego State Park, California. Quaternary Research 7:316-329.

Perch-Nielson, K. 1985. Cenozoic calcareous nannofossils; pp. 427-554 in H. M. Bolli, J. B. Saunders, and K. Perch-Nielson (eds.), Plankton Stratigraphy. Cambridge University Press, Cambridge.

Peterson, G. L., and M. P. Kennedy. 1974. Lithostratigraphic variations in the Poway Group near San Diego, California. San Diego Society of Natural History Transactions 17:251-258.

Philips, F. J. 1972. Age and correlation of the Eocene Ulatisian and Narizian Stages, California: Discussion. Geological Society of America Bulletin 83:2217-2224.

Pluhar, C. J., J. L. Kirschvink, and R. W. Adams. 1991. Magnetostratigraphy and clockwise rotation of the Plio-Pleistocene Mojave River Formation, central Mojave Desert, California. San Bernardino County Museum Association Quarterly 38(2):31-42.

Powers, J. E. 1988. Paleomagnetic analysis of Eocene rocks from the Peninsular Ranges Terrane, San Diego, California. M.S. thesis, University of Southern California, Department of Geological Sciences, 118 pp.

Prothero, D. R. 1991. Magnetic stratigraphy of Eocene-Oligocene mammal localities in southern San Diego County; pp. 125-130 in P. L. Abbott and J. A. May (eds.), Eocene Geologic History San Diego Region. Pacific Section SEPM Volume 68.

Prothero, D. R., and C. C. Swisher. 1992. Magnetostratigraphy and geochronology of the terrestrial Eocene-Oligocene transition in North America; pp. 46-73 in D. R. Prothero and W. A. Berggren (eds.), Eocene-Oligocene Climatic and Biotic Evolution. Princeton University Press, N.J.

Squires, R. L. 1988. Geologic age refinements of West Coast Eocene marine molluscs; pp. 107-112 in M. V. Filewicz and R. L. Squires (eds.), Paleogene Stratigraphy, West Coast of North America. Pacific Section SEPM Volume 58.

Squires, R. L., and T. A. Deméré. 1991. A middle Eocene marine molluscan assemblage from the usually nonmarine Friars Formation, San Diego County, California; pp. 181-187 in P. L. Abbott and J. A. May (eds.), Eocene Geologic History San Diego Region. Pacific Section SEPM Volume 68.

Steer, B. L., and P. L. Abbott. 1984. Paleohydrology of the Eocene Ballenas Gravels, San Diego County, California. Sedimentary Geology 38:181-216.

Steineck, P. L., and J. M. Gibson. 1971. Age and correlation of the Eocene Ulatisian and Narizian stages, California. Geological Society of America Bulletin 82:477-480.

Steineck, P. L., J. M. Gibson, and R. W. Morin. 1972. Foraminifera from the middle Eocene Rose Canyon and Poway formations, San Diego, California. Journal of Foraminiferal Research 2:137-144.

Stock, C. 1939. Eocene amynodonts from southern California. National Academy of Sciences Proceedings 25:270-275.

Toumarkine, M., and H. Luterbacher. 1985. Paleocene and Eocene planktonic foraminifera; pp. 87-154 in H. M. Bolli, J. B. Saunders, and K. Perch-Nielson (eds.), Plankton Stratigraphy. Cambridge University Press, Cambridge.

Walsh, S. L. 1991. Eocene mammal faunas of San Diego County; pp. 161-178 in P. L. Abbott and J. A. May (eds.), Eocene Geologic History San Diego Region. Pacific Section SEPM Volume 68.

Walsh, S. L., and T. A. Deméré. 1991. Age and stratigraphy of the Sweetwater and Otay formations, San Diego County, California; pp. 131-148 in P. L. Abbott and J. A. May (eds.), Eocene Geologic History San Diego Region. Pacific Section, SEPM Volume 68.

Wei, W., and S. W. Wise. 1989. Paleogene calcareous nannofossil magnetobiochronology: Results from South Atlantic DSDP Site 516. Marine Micropaleontology 14:119-152.

Westgate, J. W. 1988. Biostratigraphic implications of the first Eocene land-mammal fauna from the North American coastal plain. Geology 16: 995-998.

APPENDIX 1. DESCRIPTION OF PALEOMAGNETIC SECTIONS

MS-1. Camino Santa Fe

A 70 m-thick composite section consisting of two subsections was measured in the community of Mira Mesa. The base of the first subsection (MS-1a) is located in a cutbank outcrop on the south side of the south fork of Los Peñasquitos Canyon, about 300 m west of the south abutment of the Camino Santa Fe bridge. The lowest paleomagnetic site was collected from rust-stained, medium-grained, conglomeratic sandstones of the Esc, from 1 m above the creekbed. These sandstones erosionally overlie at least 2 m of medium gray, concretionary siltstones that may pertain to the Ardath Shale. Above a covered interval of 6 m, a second site was collected from light brown fine-grained sandstones of the Esc as exposed in a landslide scarp 50 m east of the first site. Above a covered interval of 10 m, five sites were collected from light yellowish-gray, fine-to-medium-grained sandstone lenses within the Ef-cg, as exposed in a large roadcut on the west side of Camino Santa Fe, immediately south of the bridge.

The second subsection (MS-1b) is located on the second, third, and fourth roadcuts extending from about 100-250 m south of Sorrento Valley Boulevard, on the east side of Camino Santa Fe, north of the bridge. One site was collected from a light yellowish-gray, medium-grained sandstone lens in the upper part of the Ef-cg, and three sites were collected from light gray to light rusty brown, locally concretionary, fine-to-very-fine grained sandstones of the gradationally-overlying Ef-ut. Since the Ef-ut does not crop out in MS-1a, the indicated level of the Ef-ut/Ef-cg contact in meters above the base of the composite section is a minimum value.

For MS-1a, the lowest Scripps Formation site was of normal polarity, while the overlying Scripps site showed indeterminate polarity. Three sites from the Ef-cg were of normal polarity, one was reversed, and the samples from one site disintegrated before they could be analyzed. The lowest site from MS-1b was of indeterminate polarity, and all three sites from the Ef-ut were reversed.

MS-2. Computer Media

A 42 m-thick continuous section was measured from the bottom to the top of the east-facing cut slope behind 4760 Murphy Canyon Road, on the west side of Murphy Canyon, about 0.6 km south of Clairemont Mesa Boulevard. The lower 8 m of the section consists of shallow-marine sandstones and siltstones assigned here to the Scripps Formation. SDSNH Loc. 3278 occurs at the base of the section and has yielded a marine mollusc assemblage of "Transition" age, and isolated teeth of *Peratherium* sp. and *Crypholestes*? sp. (Squires and Deméré, 1991). These marine strata are erosionally overlain by fluvial white sandstones and yellowish-gray conglomerates of the Ef-lt and Ef-cg, respectively. The upper part of this section consists of white, medium-grained sandstones of the Ef-ut, and is stratigraphically equivalent to the roadcut outcrop of this unit on Clairemont Mesa Boulevard determined to be of reversed polarity by Bottjer et al. (1991). About 0.5 km southwest of the top of the Computer Media section, in an artificial cut slope 100 m north of View Ridge Avenue, about 10 m of the lower member of the Stadium Conglomerate (not sampled) overlies the Ef-ut, and is in turn disconformably overlain by the Pleistocene Lindavista Formation.

The lower two sites from the Scripps Formation were found to be of normal polarity, and all samples from the Ef-lt and Ef-cg were of reversed polarity.

MS-3. Black Mountain Road and State Route 56

A 44 m-thick composite section consisting of two subsections was measured in the community of Rancho Peñasquitos. The first subsection (paleomagnetic sites 1-6) was located in a roadcut on the east side of Black Mountain Road, about 100-200 m south of the intersection with State Route 56, and spanned the uppermost part of the Ef-lt, the Ef-cg, and the base of the Ef-ut. Site 7 was collected from the Ef-ut as exposed in a short artificial cut slope immediately north of the parking lot for the Mormon Church complex at 12835 Black Mountain Road.

The second subsection (sites 8-15; all from the Ef-ut) was located in the now-filled median of State Route 56, and the north-facing roadcut on the south side of this highway, about 0.5 km east of Black Mountain Road. Correlation of the two subsections is based on elevation, and the essentially horizontal attitude of the beds in this area.

The four sites from the Ef-lt consisted of fine-grained sandstones and greenish siltstones, and all were of normal polarity. The single site from a rust-stained sandstone lens in the Ef-cg disintegrated before it could be analyzed. The ten sites collected from the Ef-ut consisted of fine-to-medium-grained sandstones and multicolored siltstones. Nine of these were of normal polarity, and one was questionably of reversed polarity.

MS-4. State Route 52 East

A 58 m-thick composite section consisting of two subsections was measured west of the community of Santee. The base of the stratigraphically lowest subsection (MS-4a) was located in the bottom of a now-filled shear key about 15-20 m below the present surface of State Route 52. The upper part of this subsection is located on the east side of the mouth of Little Sycamore Canyon, north of SR-52, on a roadcut on the north side of the access road to the Santee Landfill. No paleomagnetic samples were collected from this subsection. The stratigraphically highest subsection (MS-4b) is located on a large roadcut on the south side of State Route 52, in the saddle on the ridge between Spring Canyon and Little Sycamore Canyon, and ends at the summit of Hill 534'. The base of MS-4b is roughly correlated with the upper part of MS-4a on the basis of elevation, the assumption of essentially horizontal strata, and the first appearance of conglomeratic sandstones in the latter section.

A total of 14 sites were collected from MS-4b, from strata assigned entirely to the Ef-lt, and consisting of conglomeratic sandstones, medium-grained sandstones, and mudstones. These strata are erosionally overlain by several tens of meters of yellowish-brown cobble conglomerate mapped as Stadium Conglomerate by Kennedy and Peterson (1975). No fossils or paleomagnetic samples were collected from this unit. These conglomerates are assigned here to the Ef-cg, because they represent the stratigraphically lowest conglomerate unit of significant thickness in the area, and they occur at an appropriate elevation to pertain to this tongue.

Since two single-site normal and three single-site reversed intervals occur in MS-4b, the magnetostratigraphic pattern obtained cannot be taken literally. However, the lower part of this subsection is largely of normal polarity, while the upper part is mostly reversed.

MS-5. Genessee Avenue

Flynn's (1986) Genessee Avenue section is located north of Mission Valley, about 3.5 km north of the type section of the Friars Formation. This outcrop exposes about 22 m of strata, most or all of which pertain to the undifferentiated Friars Formation (as mapped by Kennedy, 1975, plate 2A). As shown by Flynn (1986, fig. 8), however, the basal few meters of ledgy and cross-bedded sandstones here could also be assigned to the Scripps Formation, as these strata appear to represent a transitional Scripps-Friars facies. LACM locality 3243 occurs at the base of the section (LACM[CIT] Photograph 894) and has yielded the type specimen of the rodent *Pseudotomus littoralis* (Wilson, 1949; Korth, 1985), the mesonychid cf. *Harpagolestes* sp., and a lower molar of a small brontothere tentatively assigned here to *Metarhinus? pater* (Golz and Lillegraven, 1977, table 1). This small assemblage cannot be assigned with confidence to either the Bridgerian or the Uintan at this time, although *P. littoralis* is known to occur together with definite Uintan assemblages in the conglomerate tongue (SDSNH Loc. 3621) and upper tongue (SDSNH Loc. 3883) of the Friars Formation. The Genessee Avenue outcrop also contains the vertebrate locality LACM(CIT) 314 (= UCMP V-6882), which is located in the Friars Formation about 8-9 m above the base of the section. This locality has produced the Lowest Stratigraphic Datum of *Amynodon reedi* in San Diego, as well as the selenodont artiodactyl *Leptoreodon major* and a micromammal assemblage typical of the San Diego early Uintan (see Stock, 1939; Golz and Lillegraven, 1977, table 1).

Flynn (1986, fig. 8) showed that one site collected from the base of the Genesee Avenue outcrop is of normal polarity, while the remaining four sites from the upper part of the outcrop are of reversed polarity. The location of the normal-to-reversed transition in this section relative to LACM (CIT) 314 is unclear from Flynn's paper, because there is a 13 m sampling gap between the only normal site and the lowest reversed site. Therefore, we re-collected this section with the following results. The lower three sites were found to be of normal polarity, and the highest of these was taken from the bed in which LACM(CIT) 314 appears to be located, based on LACM (CIT) Photograph 893. The overlying two sites were of reversed polarity, so the general pattern obtained corroborates that reported by Flynn. The importance of this section is that the normal magnetozone in the lower part of the Friars Formation is of early Uintan age. Therefore, if this magnetozone correlates with Chron C21n, as suggested by Flynn (1986) and supported here, the Bridgerian/Uintan boundary cannot occur within Chron C20r, as proposed by Flynn (1986).

MS-6a. Scripps Ranch North I

A composite stratigraphic section about 80 m thick was produced from two subsections in the Scripps Ranch North housing development, north of Miramar Reservoir and south of Cypress Canyon. The stratigraphically lowest subsection was located on the west side of the small hill in the center of Sec. 28, T14S R2W, and begins in the upper part of the Ef-cg. The lower part of this subsection is now heavily graded and suburbanized, but the upper part is still exposed on an abandoned jeep road on the SW side of the hill. The top of the lower subsection is the disconformity between the Ef-ut and the lower "arkosic gritstone unit" of the Emv ("Emv-l" in MS-6a). This same disconformity occurs about 3 m above the base of the upper subsection, which is located on the large roadcut (now partially buttressed and landscaped) on the south side of Spring Canyon Road (NW SE Sec. 28). The upper subsection encompassed the uppermost part of the Ef-ut, the Emv-l, the Emv-u, and the Ep-lc, Ep-m, and Ep-uc.

Most sites consisted of light gray to light rusty brown, fine-to-medium-grained sandstones. All three sites from the Ef-cg were of reversed polarity, as were all five sites from the Ef-ut. All sites from both the lower arkosic gritstone unit and the upper marine sandstone unit of the Emv were of reversed polarity. The single site from the Ep-lc was of normal polarity. The two lowest sites from the Ep-m were of normal polarity, while the two highest sites were of reversed polarity. Finally, the single site from the Ep-uc was of normal polarity.

MS-6b. Scripps Ranch North II

A continuous stratigraphic section about 72 m thick was measured along the entire length of the prominent NW-trending ridge that straddles the boundary between Sections 27 and 28, T14S R2W, on the south side of Cypress Canyon. This section is located about 1 km east of MS-6a, and covers essentially the same stratigraphic interval. MS-6b began in the upper part of the Ef-cg at the NW base of the ridge, and ended in the top of the Ep-m, as exposed at the SE end of the ridge, on the now-graded hill immediately north of Spring Canyon Road, in the center W1/2 SW1/4 Sec. 27. The outcrops on this ridge are now suburbanized. Lateral facies changes in this area are demonstrated by the rapid thinning of a prominent conglomerate body within the Ef-ut from 12 m thick in MS-6b to only 5 m thick in MS-6a. In addition, the "lower gritstone unit" of the Emv is thicker than the upper unit of the Emv in MS-6b, whereas these relative thicknesses are reversed in MS-6a.

Two sites were collected from medium-grained sandstone lenses within the Ef-cg, three sites were collected from medium-grained sandstone lenses within a conglomerate body within the Ef-ut, two sites were collected from medium-grained sandstones of the upper part of the Ef-ut, five sites were collected from medium-grained sandstones and siltstones of the "lower arkosic gritstone unit" of the Emv, two sites were collected from fine-grained sandstones of the upper unit of the Emv, one site was collected from a medium-grained sandstone lens within the Ep-lc, and four sites were collected from medium-grained sandstones of the Ep-m.

Comparison of the magnetic results from MS-6a and MS-6b is interesting. First, the Ef-ut was entirely reversed in both sections. In MS-6a, both units of the Mission Valley Formation were of reversed polarity, but in MS-6b, the upper unit of the Mission Valley was normal. The single site from the Ep-lc in both sections was normal. However, the Ep-m was entirely of reversed polarity in MS-6b, whereas the lower two sites from the Ep-m in MS-6a were normal. The discrepancies between two proximate sections traversing the same lithostratigraphic units demonstrate that the magnetic pattern obtained from a single section in the Friars Fomation and Poway Group may not be meaningful. Therefore, proposed correlations of short magnetozones in different sections must be viewed with skepticism.

MS-7. State Route 52 West

A continuous 45 m-thick section was measured on a large roadcut on the north side of State Route 52, 200 m east of the Second San Diego Aqueduct, and 1.5 km east of Santo Road. This section spanned white, fluvial, medium-grained sandstones of the upper part of the Ef-ut, the disconformably overlying, white, fine-to-very fine-grained marine sandstones of the Emv, and light gray and rust-stained cobble-to-boulder conglomerates of the Ep (undifferentiated). The 3 m-thick white sandstone lens within the Ep may represent the Ep-m, or the entire thickness of the Ep here may correlate only with the Ep-uc as exposed at Scripps Ranch North. Approximately 25 m of the Ef-ut is present below the base of this section. 600 m WSW of MS-7, the Ef-ut was observed to gradationally overlie the Ef-cg in a small south-draining canyon immediately north of State Route 52 and immediately east of Hill 518.

Two of the three sites from the Ef-ut were of reversed polarity, while one was of indeterminate polarity. The lower four sites from the Emv were of normal polarity, while the uppermost site from this formation was indeterminate. The basal site from the Ep was also indeterminate, while the remaining five sites from this unit were of normal polarity.

MS-8. Hill 781

A substantially covered, 69 m-thick section was measured east of Tierrasanta, starting at the base of a roadcut on the south side of a dirt access road for the Second San Diego Aqueduct, located in the saddle immediately north of Hill 781. The section begins in the upper part of the Ef-ut and spans 8 m of yellowish-gray conglomerates tentatively assigned to the Est-b, 30 m of the Emv, and 22 m of the Ep. The upper 37 m of this section is poorly exposed along the jeep road up to the summit of Hill 781.

The lowest site from the Ef-ut was indeterminate, while the upper two sites from this unit were reversed. The single site from a sandstone lens in the top of the Est-b? was reversed. Four sites from the lower part of the Emv were normal, one was reversed, and the uppermost site from the Emv was reversed. No sites were collected from the Ep.

MS-9. Stonecrest

An 84 m-thick composite section was measured in the undifferentiated Friars Formation, the Est-a, Est-b, and Emv, during grading for two different phases of the Stonecrest development on the west wall and rim of Murphy Canyon, south of Aero Drive. Paleomagnetic sites were collected only from the lower 52 m of this section (Ef and Est-a). The upper 32 m of the measured section have been largely graded away, but the Est-b and Emv are currently exposed in artificial cuts behind commercial buildings at the NW corner of the intersection of Aero Drive and Ruffin Road, and in a landscaped roadcut on the west side of West Canyon Avenue. If the Ef-cg is present below the surface of Murphy Canyon at Stonecrest, then the sampled Friars strata would pertain entirely to the upper tongue. However, it is equally likely that the Ef-cg pinches out north of Aero Drive (Fig. 2), in which case the Friars here would be equivalent to all or part of both lower and upper tongues.

The stratigraphically lowest paleomagnetic site was obtained from the wall of a large utilities trench immediately west of Murphy Canyon Road. This site was located about 300 m east of and at least 9 m stratigraphically below the base of the main section, and was of indeterminate polarity. The main section is located on the large east-facing cut on the west wall of Murphy Canyon, about 500 m south of Aero Drive. All sites from the Ef were of reversed polarity, as were all sites from the overlying Est-a.

MS-10. Type Section of Stadium Conglomerate

A composite 73 m-thick section consisting of two subsections was measured in the idle gravel quarry of H. G. Fenton Materials Co., on the north side of Mission Valley, north of Friars Road and east of Interstate 805. This section is essentially the same as Kennedy and Moore's (1971) type section of the Stadium Conglomerate. The lower subsection was located in a roadcut on the north side of Friars Road, about 200 m east of Northside Drive, directly opposite Friars Road from the northwesternmost entrance to the parking lot for San Diego Jack Murphy Stadium. Two sites were collected from light yellowish-brown, fine-to-medium-grained sandstone lenses within the Est-a, from 5 m and 11 m above the Est-a/Ef contact. The base of the upper subsection is located about 400 m northwest of the first section, on the surface of the graded lot immediately below the large south-facing quarry face in the NE part of the Fenton property. The two subsections are correlated on the basis of elevation. Samples from the upper subsection consisted of light gray and pale greenish-gray medium-grained sandstones from lenses in the Est-a, light rusty brown medium-grained sandstones from lenses in the Est-b, and very light gray, very-fine-grained sandstones of the Emv.

All six sites from the Est-a were of reversed polarity. Five sites were collected from the Est-b, of which the lower two were indeterminate, the next-highest was reversed, and the uppermost two were normal. All four sites from the Emv were of normal polarity.

MS-11. Type Section of Mission Valley Formation

A continuous, but substantially covered section spanning the maximum preserved thickness of the Emv (approx. 75 m) was collected from roadcuts on the west side of State Route 163, between Interstate 8 and the Sixth Avenue exit. This section is located immediately east and southeast of the type section of the Emv (Kennedy and Moore, 1971), which is currently obscured by vegetation.

Most sites were collected from light gray, fine-to-very fine-grained, often concretionary sandstones, while some of the uppermost sites consisted of light gray, medium-grained, poorly sorted sandstones. Of the 14 sites collected, the lower 9 were of normal polarity, and the uppermost 5 were reversed.

7. Magnetostratigraphy of the Upper Middle Eocene Coldwater Sandstone, Central Ventura County, California

DONALD R. PROTHERO AND EDWARD H. VANCE, JR.

ABSTRACT

A magnetostratigraphic study was undertaken to improve the age control of the middle Eocene Coldwater Sandstone in southern California. The results show that, although magnetic overprinting is common in Coldwater rocks, most samples yield a stable characteristic remanence that passes reversal and fold tests and shows a clockwise rotation of about 100° ± 17° (consistent with other pre-Miocene units in the western Transverse Ranges). Magnetic stratigraphy and refined chronostratigraphy of the late Uintan and Duchesnean mammals found within redbeds of the Coldwater show that the upper Cozy Dell Shale-Coldwater Sandstone-Sespe Formation succession in central Ventura County spans Chrons C19r-C18n (approximately 39.5-42.5 Ma). These data do not corroborate the sequence stratigraphic interpretations of Campion et al. (1994) or Clark (1994). We suggest that local tectonic forces may have been more important than eustatic sea level fluctuations in causing sequence boundaries in the Coldwater and Sespe formations.

INTRODUCTION

The Transverse Ranges of southern California (Fig. 1) contain an important succession of strata that records a complex history of transgressions and regressions during the Eocene. In the upper Sespe Creek area of central Ventura County, over 4,800 m of Eocene strata were deposited, and some of these formations are even thicker to the east or west. The Eocene succession starts with the lower Eocene deep-water Juncal Shale, overlain by the middle Eocene shallow- to deep-marine Matilija Sandstone, and then by the deep-water Cozy Dell Shale (Kerr and Schenk, 1928; Page et al., 1951; Vedder, 1972; Dibblee, 1966, 1982; Jestes, 1963; Ingle, 1980; Campion et al., 1994). After Cozy Dell deposition, a shallowing-upward trend is marked by the nearshore marine Coldwater Sandstone (Clark, 1994; Jiao and Fritsche, 1994) that is transitional with the overlying nonmarine Sespe Formation (McCracken, 1969, 1972; Howard, 1987, 1989; Prothero et al., this volume, Chapter 8).

Despite almost a century of study, however, the dating and correlation of these beds have never been very precise. For example, parts of the Matilija Sandstone, all of the Cozy Dell Shale and Coldwater Sandstone, and the lower part of the Sespe Formation are middle Eocene in age. Following current time scales (e.g., Berggren et al., 1995), the middle Eocene spans 12 million years (49-37 Ma), so a "middle Eocene" age assignment is not very helpful for high-resolution chronostratigraphy. Similarly, the local biostratigraphic units employed in coastal California are very long in duration, so they do not provide much resolution or precision. For example, the "Tejon" molluskan stage is about 10 million years in duration (about 34-44 Ma), and the "Narizian" benthic foraminiferan stage is about 8 million years in duration (39-48 Ma) (Fig. 2). In a few places, planktonic foraminifera allow more precise correlation, but they are rare, even in the deep-water units.

Several important fossil mammal localities are known from rocks assigned to the Coldwater Sandstone by some authors (or the Sespe Formation by others). With the newly refined dating of the Sespe Formation in Simi Valley to the east (Prothero et al., this volume, Chapter 8), these fossils turn out to be more age-diagnostic than any other kind. The mammalian biostratigraphic framework established in Simi Valley, tied to the global magnetic time scale, allows us to extend our magnetostratigraphic correlation down into the underlying Coldwater Sandstone in Ventura County and greatly improve the age constraints of the rock unit.

Why is such fine-scale dating important? Of course, there is the obvious value of improving the chronostratigraphy of the California Eocene. In recent years, however, the Eocene strata in the Transverse Ranges have been studied in the context of sequence stratigraphy (Thompson and Slatt, 1990; Campion et al., 1994; Clark, 1994). Many of the Eocene sequence boundaries have very imprecise age constraints, so some sequence stratigraphers (e.g., Campion et al., 1994) have chosen to assume that they matched the Exxon global onlap-offlap chart (Haq et al., 1987, 1988). These assumptions and correlations can be tested with biostratigraphically controlled magnetic stratigraphy to see if sequence boundaries are really eustatically controlled, or whether they might be due to local tectonism.

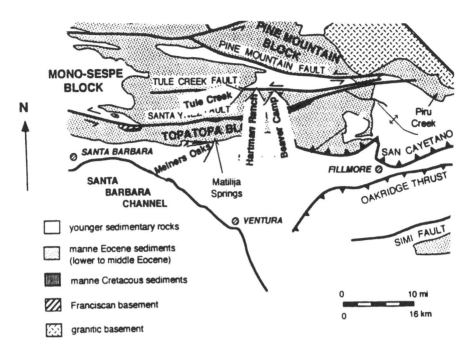

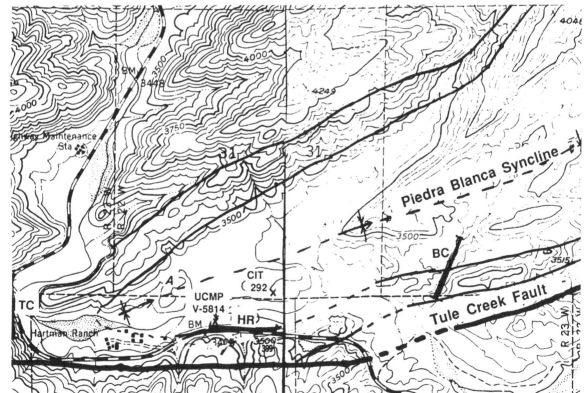

Figure 1. (*Top*) Index map showing location of four sections sampled in this study (after Clark, 1994). (*Bottom*) Details of localities in the Hartman Ranch area. Shaded area shows the outcrop of the marine Coldwater Sandstone plunging northeasterly along the nose of the Piedra Blanca syncline. The area inside the syncline includes the Coldwater redbeds and conglomerates mapped as Sespe Formation by Dibblee (1985, 1987). Location of stratigraphic sections and main fossil localities are shown as follows: A, *Amynodontopsis* tooth fragments; BC, Beaver Camp; HR, Hartman Ranch; TC, Tule Creek. Geology modified from Dibblee (1985, 1987).

THE COLDWATER SANDSTONE

The Coldwater Sandstone was first described by Watts (1897, 1900) and Taliaferro (1924) and named by Kew (1924). Taliaferro (1924) designated the type section as Coldwater Creek in the Topatopa Mountains north of Fillmore (Vedder, 1972; Clark, 1994). The name "Coldwater" has been widely used in the region (e.g., Page et al., 1951; Dibblee, 1966), so that even though the name has been used at least twice elsewhere in North America, Vedder (1972) argued that it should be retained because it had become so well established.

The Coldwater Sandstone (Fig. 1) crops out almost continuously from its type area north of Fillmore to west of Santa Barbara, a distance of more than 80 km (Merrill, 1954; Dibblee, 1966; Kleinpell and Weaver, 1963). As the formation is traced west of Santa Barbara, it interfingers with the mudstones and silt-stones of the upper part of the deeper-water Sacate Formation (Weaver, 1965, fig. 1; Dibblee, 1966; O'Brien, 1972). The Sacate Formation extends all the way to Point Conception, another 40 km to the west. At its type area, the Coldwater is about 150 m thick, but it thickens to 750 m in our study area in upper Sespe Creek. It reaches about 1000 m in thickness north of Santa Barbara, and then thins rapidly to the west as it interfingers with the Sacate Formation. Outcrops of the Coldwater Sandstone also occur in the Pine Mountain area and in the Santa Maria Basin.

At the type locality, the Coldwater gradationally overlies the Cozy Dell Shale. Elsewhere, the contact is gradational or represented by a small disconformity (Vedder, 1972). In some places, the Coldwater/Cozy Dell contact has been confused because a similar-looking shallow marine sandstone body also occurs in the upper part of the Cozy Dell. Some authors (e.g., Dibblee, 1985, 1987a; Clark, 1994) have extended the Coldwater downward to include this sandstone, and called the rocks in between the "Coldwater Shale." Others (e.g., Jestes, 1963; Fritsche, 1994; Squires, 1994) refer to this sandstone as the "Circle B sandstone member" of the Cozy Dell Formation. In this chapter, we will follow the latter usage, since it has been clearly delineated in the sections we measured (Fritsche, 1994; Jiao and Fritsche, 1994).

The upper contact of the Coldwater with the Sespe Formation is even more confusing. Traditionally, most terrestrial redbeds in the region have been mapped as Sespe Formation. However, in several sections in the study area, shallow marine Coldwater sandstones interfinger with these redbeds. Thus, some authors draw the Coldwater/Sespe contact at the base of the first redbed (e.g., Dibblee, 1985; Howard, 1989; Clark, 1994), whereas others place it at the top of the last marine sandstone (e.g., Dibblee, 1987a), or at the base of the first conglomerate in the Sespe (e.g., Campion et al., 1994). The contact is so confusing that in the Meiners Oaks section in the Matilija Quadrangle, Dibblee

(1987a) mapped the redbeds as "red siltstone" layers within the Coldwater, but at Hartman Ranch in the Wheeler Springs Quadrangle (the next quadrangle to the north), Dibblee (1985) mapped the same redbeds as Sespe Formation. As we shall show below, these controversial redbeds are equivalent in age to the Sespe red-beds in the Simi Valley area. As long as one is aware of the complications of the Coldwater/Sespe contact, it really makes no difference whether one includes the red-beds in the Coldwater or the Sespe. Such a complex situation should be expected of transitional, interfingering contacts.

To add to the confusion, the internal stratigraphy of the Sespe Formation is also very complex. Howard (1989) suggested that it is composed of two discrete depositional cycles, a middle Eocene pulse, and a late Oligocene pulse (Prothero et al., this volume, Chapter 8). In some areas, conglomerates in the upper part of the Sespe Formation (which are upper Oligocene) are deeply incised into the lower units, so they cut out much of the section. Prothero et al. (this volume, Chapter 8) showed that an erosional event removed about 8 million years of upper Eocene and lower Oligocene strata in Simi Valley. In the Sandstone Camp (= Pine Mountain Inn) area along Highway 33 near upper Sespe Creek, conglomerates of the upper part of the Sespe cut out the lower part of the Sespe completely, and are incised into the Coldwater (or ?Sespe) redbeds (Howard, 1989).

The depositional interpretation of the Coldwater Sandstone in the study area was most recently summarized by Jiao and Fritsche (1994). As part of his thesis, in early 1994 the junior author of this chapter (EHV) undertook his own independent analysis of the depositional environments of the Coldwater, unaware of the earlier, unpublished work of Jiao (1989). Most of his conclusions corroborated those of Jiao and Fritsche (1994), which were published after his work was completed. We concur with Jiao and Fritsche (1994) that the Coldwater Sandstone represents wave-dominated and storm-dominated nearshore facies, with minor tidal influences. After correction for 90° of Miocene clockwise tectonic rotation, provenance studies and paleocurrents indicate that sediments were eroded from highlands in the Pine Mountain area and transported southeasterly. The Coldwater also progrades to the southeast (after correction for tectonic rotation).

Dating of the Coldwater Sandstone has long been very imprecise (Fig. 2). The only marine fossils found in it are mollusks assigned to the "Tejon Stage" (Squires, 1994), which has about a 10-million-year duration. Link and Welton (1982) reported planktonic microfossils from the upper Cozy Dell Shale that placed it in planktonic foraminiferal Zones P11 and P12. Just south of our study area, Berman (1979, p. 34) reported planktonic foraminiferans which last occur in Zone P12, constraining the age of the upper Cozy Dell to

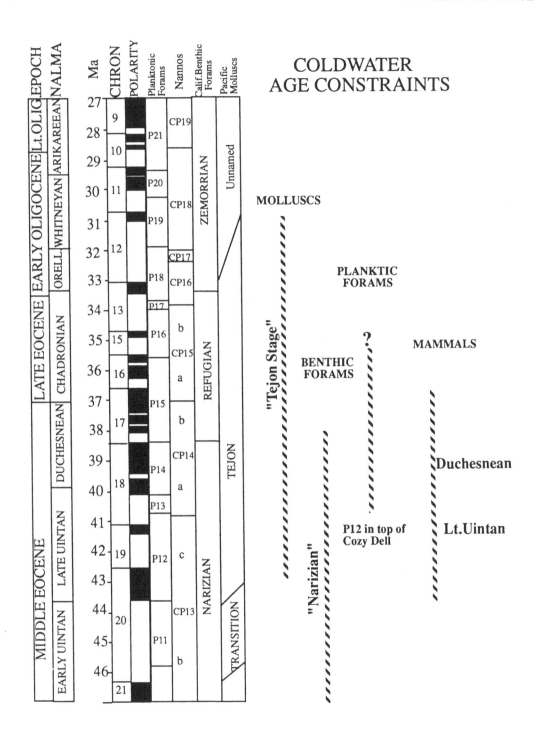

Figure 2. Biostratigraphic age constraints on the Coldwater Sandstone before this study. Time scale after Berggren et al. (1995). California benthic foraminiferal stages after Almgren et al. (1988) and Clark (1994); molluskan stages after Squires (1994).

between 40.5-44.0 Ma, or magnetic Chrons C18r-C20n (Berggren et al., 1995). Benthic foraminiferans from the upper Cozy Dell are assigned to the A-1/A-2 transition (Kelley, 1943; Almgren et al., 1988), which is less age-diagnostic than the planktonic foraminiferans. Fossil mammals from the redbeds between the marine sandstones at Hartman Ranch (Lindsay, 1968; Kelly, 1990; Lander, 1994) are assigned to the late Uintan, but the exact chronostratigraphic position of the late Uintan did not become clear until recent work on the type Uintan in the Uinta Basin of Utah (Prothero, this volume, Chapter 1), and the Uintan faunas of the Sespe Formation in Simi Valley (Prothero et al., this volume, Chapter 8). The redbeds above the marine sandstones have produced Duchesnean mammals in several places. Together with the magnetic stratigraphy, these new biostratigraphic constraints greatly improve the precision of correlation.

MAGNETIC ANALYSIS

Only a few stratigraphic sections of the Coldwater Sandstone produce age-diagnostic mammals that make magnetic sampling worthwhile. Four sections were sampled in this study (Fig. 1). Sections were measured using a Brunton compass and Jacob's staff. Each section was sampled with a coring drill or hand tools, producing 40 sites spaced at about 10-30 meter intervals, depending upon exposure. Three samples were taken per site so that site statistics could be calculated (Fisher, 1953). The hand samples were trimmed into 2.5-cm cubes on a band saw with a tungsten-carbide blade, and then analyzed at the paleomagnetics laboratory at the California Institute of Technology.

Previous work on rocks of the Transverse Ranges (e.g., Hornafius, 1985; Prothero et al., this volume, Chapter 8) has shown that they often contain complex overprinted components that make the primary component hard to recognize and extract. For this reason, each sample was measured at NRM (natural remanent magnetization), and then demagnetized at alternating fields (AF) of 25, 50, and 100 Gauss to remove multidomain components, followed by thermal demagnetization at 200, 300, 400, 500, and 600°C, for a total of nine demagnetization steps. Typical results are shown in Figures 3 and 4. Most normally magnetized samples (i.e., those with downward inclinations) were pointed almost due east, and the component of magnetization between 300 and 500°C showed a stable normal direction with about 100° clockwise rotation (Fig. 3). Likewise, reversed samples (i.e., those with upward inclinations) showed stable directions that were pointed about 280° in declination, which is consistent with about 100° clockwise rotation (Fig. 4). A few samples had intractable normal overprints due to the present-day earth's magnetic field that could not be removed at any demagnetization step. A number of samples yielded highly scattered, random results that could not be inter-

preted at all. These were omitted from the analysis.

The vector demagnetization ("Zijderveld") plots also show that a complex magnetic mineralogy is responsible for the remanence. In some samples (e.g., Fig. 4), the magnetization declined rapidly in intensity under AF demagnetization, showing that magnetite is the primary carrier of the remanence; the intensity of these samples was very weak above the Curie point of magnetite (i.e., the 600°C step). In others, the AF demagnetization had no effect on the intensity, showing that a high-coercivity phase, such as goethite or hematite, was the primary carrier; these samples also retained significant intensity above 600°C, which is also consistent with hematite.

To test these conclusions, several IRM (isothermal remanent magnetization) acquisition tests were conducted on different lithologies (Fig. 5). Some samples (Fig. 5B) showed saturation at 500 mT (millitesla), consistent with magnetite, but others (Figs. 5A, C) did not saturate even at 1300 mT, which indicates that they contained significant hematite. Along with the IRM acquisition analyses, the same samples were subjected to a modified Lowrie-Fuller test (Johnson et al., 1975; Pluhar et al., 1991), which compares the resistance to AF demagnetization of both the IRM produced in a 100 mT peak field, and the ARM (anhysteretic remanent magnetization) gained in a 100-mT oscillating field. As can be seen in Figure 5, the ARM (solid squares) demagnetizes at higher fields than the IRM (open squares) in most samples, so much of the remanence appears to be carried by single-domain or pseudo-single domain grains.

If a stable component was apparent, its vector was averaged for those of the other two samples at that site using the methods of Fisher (1953; see Butler, 1992). These yielded site statistics, which were classified following the system of Opdyke et al. (1977). Class I sites are significantly separated from random at the 95% confidence level, and are shown by the solid circles in Figure 6. Class II sites (circles with diagonal pattern) had one sample missing, so site statistics could not be calculated, but a clear polarity was apparent from the remaining samples. Class III sites (open circles) had one sample divergent, so they did not cluster at the 95% confidence level, but their polarity was clear from the two remaining sites.

The means for the normal and reversed sites were averaged using the methods of Fisher (1953). The mean for all Class I normal sites (N = 17) was D = 99.0°, I = 57.0°, k = 5.2, α_{95} = 17.3; the mean for all Class I reversed sites (N = 10) was D = 293.8, I = -57.3, k = 13.1, α_{95} = 13.9. These directions are almost exactly antipodal, so the magnetic vectors pass the reversal test and are probably due to primary magnetization, and not overprints. The mean declinations clearly show that this block has been rotated clockwise by about 100° since the Eocene, which is consistent (within the α_{95}

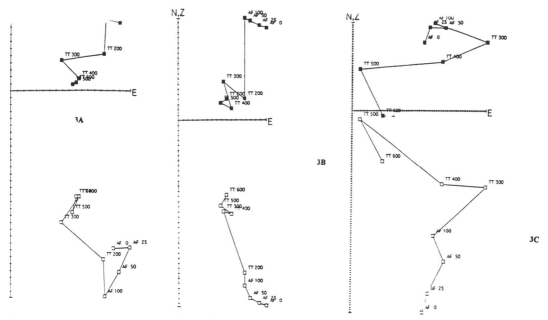

Figure 3. Vector demagnetization ("Zijderveld") plots of representative normal (east and down) samples. Solid squares indicate declination; open squares indicate inclination. AF and thermal ("TT") demagnetization steps are labeled. Each increment is 10^{-6} emu. Note that in most samples, an overprint was removed at 200-300°C, leaving a stable component pointed east and down that decayed to the origin. In samples 3A and 3B, the lack of response to AF demagnetization and the high intensity remaining at 600°C shows that significant hematite was present; in 3C, the sample responded to AF demagnetization and was nearly demagnetized by 500°C, showing that magnetite is the main carrier of the remanence.

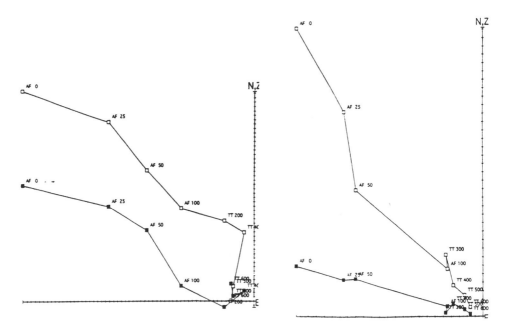

Figure 4. Vector demagnetization plots of typical reversed (west and up) samples. All conventions as in Fig. 3. Note that nearly all samples responded to AF demagnetization and were nearly demagnetized by 500°C, suggesting that magnetite is the main carrier of the remanence.

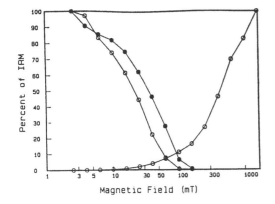

COLDWATER TAN SILTSTONE

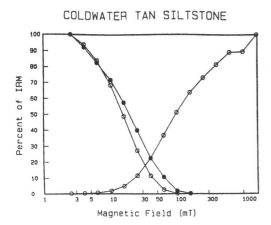

COLDWATER YELLOW SANDSTONE

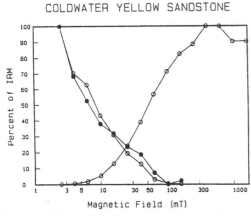

Figure 5. IRM (isothermal remanent magnetization) and Lowrie-Fuller tests for typical lithologies. IRM = open circles, ARM = solid circles. Note that in most samples (except the yellow sandstone), the IRM fails to saturate, indicating the presence of hematite. In most samples, the ARM is more resistant to AF demagnetization than the IRM, indicating that the remanence is carried by single-domain or pseudo-single domain grains (see Pluhar et al., 1991, for further details of the experimental methods).

error estimates) with the approximately 90° clockwise rotation reported for this tectonic block by Luyendyk et al. (1980, 1985).

The mean inclinations would produce an Eocene paleolatitude of 38° ± 17.3° (present latitude = 34.0°), which is actually north of its present position, and further north than would be expected from the shallow paleolatitudes of most Eocene rocks in the region (Lund et al., 1991). However, the error estimates are large enough to accommodate this possibility.

A fold test was also conducted to determine the stability of the magnetic directions. The scatter of the uncorrected normal mean ($\alpha 95$ = 32.5) and reversed mean ($\alpha 95$ = 25.6) was significantly greater than the means of the sites after dip correction. This shows that the magnetization must have been acquired before folding took place.

MAGNETIC STRATIGRAPHY
Meiners Oaks

One of the thickest sections of the Coldwater Sandstone in the region is exposed in west-facing roadcuts along Highway 33 just north of the town of Meiners Oaks (center NE Sec. 33, T5N, R23W, Matilija 7.5' quadrangle). These exposures lie stratigraphically above the type Cozy Dell Shale at Cozy Dell Creek, and span almost 500 m of section. The basal 100 m of section above the Cozy Dell contains lithologies typical of the Coldwater Sandstone, with wave-ripple cross-beds, herringbone cross-beds, glauconitic layers, flaser beds, storm-battered shell lags, and other shoreface-intertidal features. The upper part of the section consists of 400 m of redbeds and fluvial sandstones. However, Dibblee (1987a) mapped the redbeds as "red siltstone" tongues within the Coldwater Sandstone, not as Sespe Formation, as he did in other exposures. Undisputed Sespe Formation occurs to the south of the roadcut (above the top of the section) in the backyards of north Meiners Oaks, but no good sections were available for sampling. This section was also discussed by Campion et al. (1994, p. 20, and figs. 31, 35, and 36).

The basal 100 m of marine Coldwater is entirely of reversed polarity (Fig. 6). The next 100 m of redbeds are of normal polarity. From about 200 m in the section until the covered interval begins at 330 m, the section is again of reversed polarity. An edentulous rodent jaw (University of California, Berkeley, Museum of Paleontology, or UCMP locality V-81116) occurs at about the 330-meter level, but it is not diagnostic of age. The only other exposures (between 380 and 460 m) are of normal polarity. Just above where the exposures disappear in slumped material and vegetation, a *Duchesneodus* (brontothere) jaw was collected (UCMP V-82372). This specimen would indicate that the upper part of the section transitional between Coldwater and Sespe is Duchesnean in age.

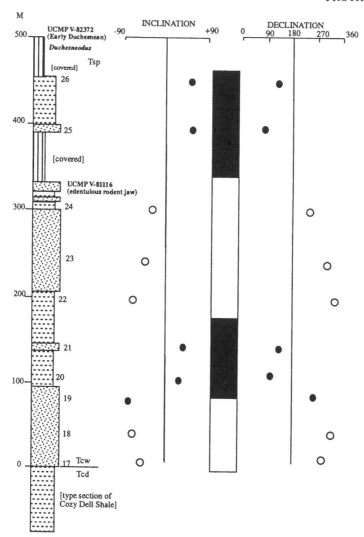

Figure 6. Magnetostratigraphic plots of the Meiners Oaks section. Details of sections described in text. Tcw = Coldwater Sandstone; Tcd = Cozy Dell Shale; Tsp = Sespe Formation. UCMP V = University of California (Berkeley) Museum of Paleontology vertebrate fossil locality. Site numbers shown in small numbers to right of lithostratigraphy. R = redbeds. Solid circles = Class I sites of Opdyke et al. (1977), which are significantly separated from a random distribution at the 95% confidence level; circles with diagonal slash = Class II sites (one sample missing, so no statistics could be performed); open circles = Class III sites (two samples show a polarity preference, but one sample was divergent). Note that most normal sites have a positive inclination and declinations around 90°; reversed sites have a negative inclination and declinations around 270°.

Hartman Ranch

The most important exposures in the area occur in north-facing roadcuts along Highway 33 (N 1/2 NE Sec. 2, T5N, R23W, Wheeler Springs 7.5' quadrangle). Campion et al. (1994, p. 18, fig. 31) gave a brief description of this section, and it was described as the "lower Sespe" by Howard (1989, fig. 4). It begins with about 80 m of Coldwater Sandstone characterized by wave-ripple cross-beds, *Ophiomorpha* burrows, abraded shell beds and other wave- and storm-dominated shoreline features. This unit is entirely of reversed polarity (Fig. 7). Between 80 and 140 m, there are poorly exposed redbeds (called Coldwater by Campion et al., 1994, but mapped as Sespe Formation by Dibblee,

1985, and considered to be Sespe Formation by Howard, 1989). Near the top of this unit in a green sandstone on a north-facing roadcut just east of milepost 28.77, is the Hartman Ranch locality (UCMP V-5814. shown by Squires, 1994, fig. 3), that produces a late Uintan fauna correlative with the Tapo Canyon and Brea Canyon local faunas in the middle member of the Sespe Formation in Simi Valley (Lindsay, 1968; Kelly, 1990; Lander, 1994). The remaining 40 m of section are intertidal marine sandstone (with flaser beds, larger rip-up clasts, and interference ripples) typical of the Coldwater (yet mapped as Sespe by Dibblee, 1985). The entire upper 100 m of this section is of normal polarity.

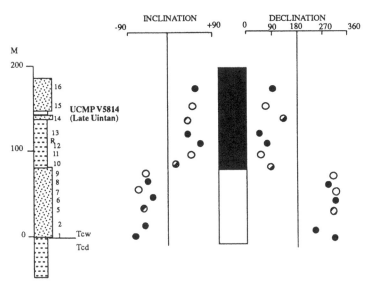

Figure 7. Magnetostratigraphy of the Hartman Ranch section. Conventions as in Fig. 6.

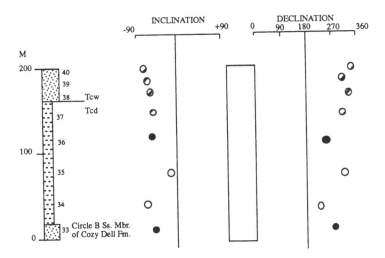

Figure 8. Magnetostratigraphy of the Tule Creek section. Conventions as in Fig. 6.

Tule Creek

To supplement the Hartman Ranch section, another section was taken on the same side of Highway 33 in the east-facing roadcuts where the highway crosses Tule Creek (NE NE NE Sec. 3, T5N, R23W, Wheeler Springs 7.5' quadrangle). Jiao and Fritsche (1994, fig. 3, "Highway 33 roadcut") previously measured and described the upper part of this section. We began the section with exposures of the Circle B sandstone member of the Cozy Dell north of the creek, and sampled some 120 m of Cozy Dell Shale, before concluding at the base of typical Coldwater Sandstone on the south side of the creek. The entire section (Fig. 8) was of reversed polarity, from the lowest exposures of the Circle B sandstone to the typically reversed lower Coldwater Sandstone.

Beaver Camp

The Eocene beds in the Hartman Ranch area are folded into the Piedra Blanca syncline, which plunges west (Dibblee, 1985, 1987b). Just north of Beaver Campground (SE SW SW Sec. 32, T6N, R22W, Lion Canyon 7.5' quadrangle), a small southeast-flowing tributary of Sespe Creek cuts through the cuestas on the south limb of the syncline, exposing a section of upper Coldwater Sandstone, plus several tens of meters of redbeds, capped by typical lower Sespe Formation conglomerate (Fig. 9). This section is near the one labeled "West of Circle B Ranch" as described and measured by Jiao and Fritsche (1994, fig. 3).

The lower 120 m consist of typical marine Coldwater Sandstone with low-angle shoreface cross-beds marked by heavy mineral lags, convolute bedding, tabular and trough cross-bedding, wave-ripple cross-beds, flaser

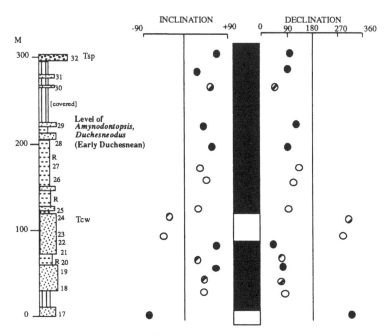

Figure 9. Magnetostratigraphy of the Beaver Camp section. Conventions as in Fig. 6.

bedding, and *Ophiomorpha* burrows. The remaining 180 m of section consists of poorly exposed redbeds, capped by Sespe conglomerate with sandstone clasts characteristic of the middle Eocene phase of Sespe deposition (Howard, 1989, fig. 4). Except for the stratigraphically lowest Coldwater site (which is an unknown height above the base of the Coldwater), and two sites between 100 and 120 m in the section, the entire section was of normal polarity.

Although no fossil mammal localities are known from the Beaver Camp section, a number of important localities occur in nearby areas of the syncline. Kelly (1990, p. 34) reported that *Amynodontopsis* (rhino) teeth were recovered from the correlative redbeds on the north limb of the syncline just northeast of the main (former) Hartman Ranch buildings. This collection was deposited in the Los Angeles County Museum of Natural History (LACM), but never accessioned or catalogued. Apparently, these specimens have been misplaced, because recent attempts by Kelly to locate them have failed (Kelly, personal communication). Kelly (1990, p. 34) also located the exact spot of locality LACM (CIT) 292, from which a *Duchesneodus* palate was collected by Chester Stock and field parties (Stock, 1938). Although it is also on the north limb of the syncline, it was collected from the basal Sespe sandstones and conglomerates correlative with our highest Beaver Camp magnetic sites. Together, these two localities were called the "Sespe Creek local fauna" by Kelly (1990) and Lander (1994), and are correlative with the early Duchesnean Pearson Ranch local fauna in Simi Valley (Kelly, 1990; Lander, 1994). These correlations, in turn, suggest that the upper part of the Beaver Camp section is also early Duchesnean.

Summary of correlations

Correlation of all the Coldwater sections discussed above is shown in Figure 10. The uppermost Cozy Dell Shale (starting with the Circle B sandstone member) and the lower 100 m of typical marine Coldwater Sandstone are of reversed polarity. Above this is a zone of normal polarity in the lowest redbed interval (whether it is called Coldwater or Sespe); this zone also yields the late Uintan Hartman Ranch local fauna. In the Meiners Oaks and Beaver Camp sections, the upper redbeds and fluvial-intertidal Coldwater sandstones are of reversed polarity. Finally, the remaining redbed sequence, up to and including the Sespe sandstones and conglomerates, is entirely of normal polarity; this zone produces early Duchesnean mammals in two sections.

By comparison to the magnetically calibrated Sespe sections in eastern Ventura County (Prothero et al., this volume, Chapter 8), it is possible to correlate this Coldwater sequence to the magnetic polarity time scale. The late Uintan Hartman Ranch local fauna is most similar to the Tapo Canyon and Brea Canyon local faunas in Simi Valley, which occur early in Chron C18r (about 41 Ma); this suggests that the normal zone at the top of the Hartman Ranch section and elsewhere is probably Chron C19n. The presence of P12 planktonic foraminiferans in the upper Cozy Dell means that the reversed zone in the upper Cozy Dell and lower Coldwater must be Chron C19r, and cannot be an older reversed Chron (Fig. 2). The early Duchesnean specimens at Meiners Oaks and Beaver Camp occur in a long zone of normal polarity. This correlates with the early Duchesnean Pearson Ranch local fauna in Simi Valley, which occurs in the early part of Chron C18n (about

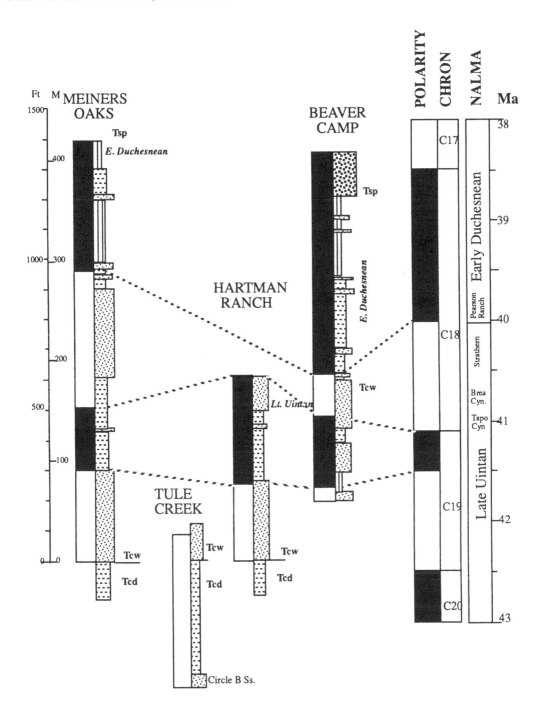

Figure 10. Correlation of Coldwater sections to the magnetic polarity time scale (after Berggren et al., 1995). NALMA = North American land mammal "ages." Chronostratigraphic position of key Sespe local faunas from Simi Valley are also shown (after Prothero, Howard, and Dozier, this volume, Chapter 8).

39.5 Ma). Thus, the entire upper Cozy Dell-Coldwater-Sespe sequence in central Ventura County appears to span C19r (base of the Circle B sandstone), which begins about 42.5 Ma, to somewhere in C18n (lower Sespe), or about 39.0 Ma.

These correlations and calibrations also clear up an apparent contradiction discussed by Golz and Lillegraven (1977). They found it perplexing that "late Eocene" brontotheres such as those from the Sespe Formation at locality LACM (CIT) 292 occurred immediately above middle Eocene mollusks in the Coldwater Sandstone. This anomaly is now resolved by the fact that the Uintan and Duchesnean are both middle Eocene, not late Eocene, as was thought in 1977 (Prothero and Swisher, 1992).

IMPLICATIONS FOR SEQUENCE
STRATIGRAPHY

In recent years, sequence stratigraphy (Vail et al., 1977; Posamentier et al., 1988; Van Wagoner et al., 1988, 1990) has become one of the dominant tools of sedimentary geology. Sequence stratigraphic concepts have been applied to a wide variety of sedimentary settings, although its assumptions are not without critics (e.g., Miall, 1991, 1992; Christie-Blick, 1991; Hallam, 1992). The thick, well exposed transgressive-regressive Eocene rocks of the western Transverse Ranges are ideally suited for sequence stratigraphic analysis, and recently two detailed interpretations have appeared (Campion et al., 1994; Clark, 1994; Thompson and Slatt, 1990, is an abstract).

One of the central tenets of sequence stratigraphy is that unconformable surfaces (sequence boundaries) represent time-synchronous surfaces. Many sequence stratigraphers believe these surfaces can be correlated to the Exxon-Vail onlap-offlap curve (Haq et al., 1987, 1988). The predictions of sequence stratigraphy have been successfully tested in passive margin settings by the biostratigraphy of key sections with good exposures and abundant microfossils (e.g., Poag and Schlee, 1984; Olsson, 1988). However, other critics (Aubry, 1991; Miall, 1992) have pointed out that the Exxon-Vail curve has so many events, and the biostratigraphic control is so poor in many cases, that *any* sequence of rocks could be fit to the curve. Moreover, applications of sequence stratigraphy in active tectonic settings provide a much more dynamic context, with the attendant potential for miscorrelation and misidentification of sequence boundaries. Clearly, fine-scale biostratigraphy, magnetostratigraphy, or chronostratigraphy are critical in testing sequence stratigraphic correlations.

In the western Transverse Range Eocene succession, however, time control has been notoriously poor and the dating based on benthic foraminiferans has long been controversial (McDougall, 1980; Almgren et al., 1988). This poor time control has forced sequence stratigraphers to make a leap of faith and *assume* that

their sequence boundaries matched the global eustatic curve of Haq et al. (1987, 1988). Now that much finer chronostratigraphic resolution is possible for the Coldwater and Sespe formations, it is possible to test these sequence stratigraphic predictions (Fig. 11).

The Campion, Lohmar, and Sullivan (1994) interpretation

Campion et al. (1994) based their sequence stratigraphic interpretations on the Haq et al. (1987, 1988) curve, which was based on the old Berggren et al. (1985) time scale. Unfortunately, the numerical age assignments in the Haq et al. (1987, 1988) curve must be considerably revised since the Eocene time scale has changed so radically in the last few years (Berggren et al., 1995). Although this makes all the numerical dates in Campion et al. (1994) off by several million years, it is still possible to test their predictions by matching their sequence boundary assignments to the original Haq et al. (1987, 1988) curves, and finding out what magnetic chron these events fall in. The correlation between the planktonic microfossil zonation, the Exxon-Vail curve, and the magnetic polarity time scale has not changed over the last decade, even though numerical age assignments have.

Campion et al. (1994) placed a sequence boundary at the contact between the lower marine sandstone and the first redbeds in the Coldwater (which they label the "39.5 Ma unconformity" or the "Ta3.6/Ta4.1" boundary of Haq et al., 1988). This was based on the assumption that the Coldwater can be traced westward in the Santa Ynez Range as a correlative of the Sacate and lower Gaviota formations, which are late Eocene (planktonic foraminiferal Zones P15 and P16, according to Tipton, 1980). Haq et al. (1987, 1988) place the Ta3.6/Ta4.1 boundary in Chron C17n. However, the presence of P12 planktonic foraminifera in the underlying Cozy Dell Shale, and late Uintan mammals just above this sandstone, plus the magnetic pattern, shows that this boundary occurs in the transition between C19r and C19n (about 41.5 Ma), which corresponds to no sequence boundary on the Haq et al. (1987, 1988) chart.

Campion et al. (1994) recognized a second sequence boundary (their "38 Ma unconformity," or the "Ta4.1/Ta4.2" boundary of Haq et al., 1987, 1988) at the top of the redbed sequence at Hartman Ranch. According to Haq et al. (1987, 1988), this sequence boundary should fall within Chron C15r. However, the occurrence of late Uintan mammals below this level, and early Duchesnean mammals above it, plus the magnetic pattern, show that the top of the redbeds occurs within early Chron C18n (about 40 Ma), which is in the middle of the Ta3.5 cycle and thus is not associated with any sequence boundary on the Haq et al. (1987, 1988) curve.

Campion et al. (1994) placed a third sequence boundary at the base of the Sespe conglomerates, which they

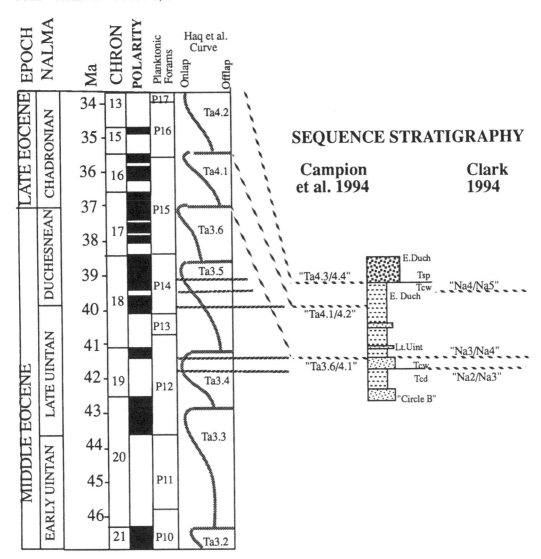

Figure 11. Comparison of chronostratigraphy of Cozy Dell-Coldwater-Sespe section in central Ventura County with sequence stratigraphic correlations of Campion et al. (1994) and Clark (1994), showing the lack of correspondence to the eustatic curve of Haq et al. (1987, 1988). Time scale after Berggren et al. (1995).

interpret as the "36 Ma unconformity" or the "Ta4.3/Ta4.4" boundary of Haq et al. (1987, 1988). In the original global cycle chart, this boundary falls late in Chron C12n, close to the Eocene/Oligocene and Chadronian/Orellan boundaries. However, the presence of Duchesnean mammals in the Sespe conglomerate, such as the *Duchesneodus* in LACM (CIT) 292, and the early Duchesnean *Amynodontopsis* just below it (which first occur in Simi Valley in Chron C18n, about 39.5 Ma) clearly falsifies this correlation. The unconformity below the Sespe conglomerate may correspond to the Ta3.5/Ta3.6 boundary, which does occur late in Chron C18n, although Chron C18n is so long in duration that it is hard to say with confidence that they match.

In summary, the age assignments of Campion et al. (1994) are off by at least four or five cycles on the Haq

et al. (1987, 1988) chart, and only one of the three sequence boundaries they recognize (the unconformity below the Sespe) might match the predictions of the Exxon chart. Clearly, this does not confirm the interpretations of eustatic control assumed by Campion et al. (1994).

The Clark (1994) interpretation

Six months later, Clark (1994) published another sequence stratigraphic interpretation of the same rocks. Unlike Campion et al. (1994), he had access to the new Berggren et al. (1995) time scale, and to some of the information in this chapter, and thus could adjust his interpretations accordingly. Clark (1994, fig. 7) placed a sequence boundary (his "Na2/Na3" boundary) at the Cozy Dell/Coldwater contact, which he correlated with the base of Chron C18n (Clark, 1994, fig. 12). As he pointed out (p. 25), this age assignment does not match any sequence boundary on the Haq et al. (1987, 1988) curve. We concur that the Cozy Dell/Coldwater contact does not match any of the Exxon regressions. However, our data show that this boundary is in the middle of Chron C19r (not C18n), and corresponds to no sequence boundary on the Exxon chart, because it would be in the middle of cycle Ta3.4.

Clark (1994) placed a second sequence boundary ("Na3/Na4" boundary) at the contact between the marine sandstones and the redbeds in the Meiners Oaks section (as did Campion et al., 1994, who labeled it the "39.5 Ma" unconformity). As discussed above, this unconformity does not match any on the Haq et al. (1987, 1988) curve, unless one assumes that this curve is not precise to within a polarity zone and a million years.

Clark's (1994) third sequence boundary ("Na4/Na5" boundary) is more difficult to interpret, because his "Matilija Springs" section (which corresponds in part to our Meiners Oaks section) shows a thick sequence of coarse sandstone (Clark, 1994, fig. 7, unit O) where we found only redbeds with a few sandstone beds (Fig. 6). However, he (p. 25) suggested that this boundary correlates with the "39.5 Ma" unconformity of Campion et al. (1994), which is not consistent with our interpretation of the section. If our reading of Clark's (1994, fig. 7) section is correct, the Na4/Na5 boundary falls somewhere in Chron C18n, so it is impossible to tell if this corresponds to any of the unconformities of the Haq et al. (1987, 1988) chart. Clearly, however, it is not the "39.5" unconformity, which falls within Chron C17n.

In summary, none of Clark's (1994) sequence boundaries appear to match the Exxon cycle chart. To his credit, Clark (1994) does not automatically assume eustatic control of every unconformity, but attributes some to local tectonic forces.

CONCLUSIONS

The Coldwater Sandstone in central Ventura County has been rotated clockwise by about 100°, consistent with other results in the same tectonic block. It is magnetically and biostratigraphically correlated to Chrons C19r-C18n (39.5-42.5 Ma), or late Uintan to early Duchesnean. Detailed examination of the sequence stratigraphic correlations of Campion et al. (1994) and Clark (1994) show that few if any sequence boundaries clearly correlate to the global cycle chart of Haq et al. (1987, 1988), suggesting that local tectonism was more important than eustasy in causing unconformities in the Cozy Dell-Coldwater-Sespe sections in this region.

ACKNOWLEDGMENTS

This research was supported by NSF grant EAR94-05942 to Prothero. We thank Dr. Joseph Kirschvink for access to the Caltech paleomagnetics laboratory.

The Meiners Oaks section was measured in 1988 by Dr. Mark Mason, who showed Prothero the position of the fossil localities in that section and at Hartman Ranch. Greg Brown and Steve King helped with the sampling at Meiners Oaks. The remaining sections were measured in 1994 by Vance; Chris Jaquette and Jeff Norville helped with the field work and sampling. We thank Tom Kelly for all his help with the biostratigraphy. We thank Kirt Campion, Mike Clark, Jeff Howard, Jim Ingle, Tom Kelly, Bruce Lander, Tor Nilsen, and Richard Squires for reviewing the manuscript.

LITERATURE CITED

Almgren, A. A., M. V. Filewicz, and H. L. Heitman. 1988. Lower Tertiary foraminiferal and calcareous nannofossil zonation: overview and recommendation; pp. 83-106 *in* M. B. Filewicz and R. L. Squires (eds.), Paleogene Stratigraphy, West Coast of North America. Pacific Section SEPM Guidebook 58.

Aubry, M.-P. 1991. Sequence stratigraphy: eustasy or tectonic imprint? Journal of Geophysical Research 96: 6641-6670.

Berggren, W. A., D. V. Kent, and J. J. Flynn. 1985. Paleogene geochronology and chronostratigraphy. Geological Society of London Memoir 10:141-195.

Berggren, W. A., D. V. Kent, M.-P. Aubry, C. C. Swisher, III, and K. G. Miller. 1995. A revised Paleogene geochronology and chronostratigraphy. SEPM Special Publication 54:129-212.

Berman, B. H. 1979. Biostratigraphy of the Cozy Dell Formation in Ventura and Santa Barbara Counties. M.S. Thesis, California State University, Long Beach, 116 pp.

Butler, R. F. 1992. Paleomagnetism. Blackwell, Boston.

Campion, K. M., J. M. Lohmar, and M. D. Sullivan. 1994. Paleogene sequence sratigraphy, western Transverse Ranges, California: Pacific Section, American Association of Petroleum Geologists 1994 Annual Convention, Fieldtrip Guidebook, 29 pp.

Christie-Blick, N. 1991. Onlap, offlap, and the origin of unconformity-bounded depositional sequences. Marine Geology 97:35-56.

Clark, M. S. 1994, Sedimentation response to earthquake-related events, middle Eocene Ventura Basin, California; pp. 9-38 *in* A. E. Fritsche (ed.), Sedimentology and paleontology of Eocene rocks in the Sespe Creek area, Ventura County, California. Pacific Section SEPM Guidebook 74.

Dibblee, T. W., Jr. 1966. Geology of the central Santa Ynez Mountains, Santa Barbara County, California. California Division of Mines and Geology Bulletin 186, 91 pp.

Dibblee, T. W., Jr., 1982. Geology of the Santa Ynez-Topatopa Mountains, southern California; pp. 41-56 *in* D. C. Fife and J. A. Minch (eds.), Geology and mineral wealth of the California Transverse Ranges. South Coast Geological Society Guidebook.

Dibblee, T. W., Jr. 1985. Geologic map of Wheeler Springs quadrangle, Ventura County, California. Thomas W. Dibblee, Jr., Foundation, Santa Barbara, CA.

Dibblee, T. W., Jr. 1987a. Geologic map of Matilija quadrangle, Ventura County, California. Thomas W. Dibblee, Jr., Foundation, Santa Barbara, CA.

Dibblee, T. W., Jr., 1987b. Geologic map of Lion Canyon quadrangle, Ventura County, California. Thomas W.

Dibblee, Jr., Foundation, Santa Barbara, CA.

Fisher, R.A. 1953. Dispersion on a sphere. Proceedings of the Royal Society A217:295-305.

Fritsche, A. E. 1994. Field guide to Eocene rocks in the upper Sespe Creek area, Ventura County, California; pp. 89-106 in A. E. Fritsche (ed.), Sedimentology and paleontology of Eocene rocks in the Sespe Creek area, Ventura County, California. Pacific Section SEPM Guidebook 74.

Golz, D. J., and J. A. Lillegraven. 1977. Summary of known occurrences of terrestrial vertebrates from Eocene Strata of southern California. University of Wyoming Contributions in Geology 15:43-65.

Hallam, A. 1992. Phanerozoic Sea Level Changes. Columbia University Press, New York, 266 pp.

Haq, B. U., J. Hardenbol, and P. R. Vail. 1987. Chronology of fluctuating sea levels since the Triassic. Science 235:1156-1167.

Haq, B. U., J. Hardenbol, and P. R. Vail. 1988. Mesozoic and Cenozoic chronostratigraphy and cycles of sea level change; pp. 71-108 in C. K. Wilgus, B. S. Hastings, H. Posamentier, J. Van Wagoner, C. A. Ross and C. G. S. Kendall (eds.), Sea level changes: An integrated approach: SEPM Special Publication 42.

Hornafius, J. S. 1985. Neogene tectonic rotation of the Santa Ynez Range, Western Transverse Ranges, California, suggested by paleomagnetic investigation of the Monterey Formation. Journal of Geophysical Research 90 (B14):12503-12522.

Howard, J. L. 1987. Paleoenvironments, provenance and tectonic implications of the Sespe Formation, southern California. Ph. D. dissertation, University of California, Santa Barbara, CA, 306 pp.

Howard, J. L. 1989. Conglomerate clast populations of the Upper Paleogene Sespe Formation, southern California; pp. 269-280 in I. P. Colburn, P. L. Abbott, and J. A. Minch (eds.), Conglomerates in basin analysis: A symposium dedicated to A. O. Woodford. Pacific Section SEPM Guidebook 62.

Ingle, J.C., Jr. 1980. Cenozoic paleobathymetry and depositional history of selected sequences within the southern California continental borderland. Cushman Foundation Special Publication 19:163-195.

Jestes, E. C. 1963. A stratigraphic study of some Eocene sandstones, northeastern Ventura Basin, California. Ph.D. dissertation, University of California, Santa Barbara, CA, 252 pp.

Jiao, Z. S. 1989. Depositional environments and paleogeography of the Coldwater Formation, upper Sespe Creek, Ventura County, California. M.S. Thesis, California State University, Northridge, 102 pp.

Jiao, Z. S., and A. E. Fritsche. 1994. Depositional environments and paleogeography of the Coldwater Formation, upper Sespe Creek, Ventura County, California; pp. 57-78 in A. E. Fritsche (ed.), Sedimentology and paleontology of Eocene rocks in the Sespe Creek area, Ventura County, California. Pacific Section SEPM Guidebook 74.

Johnson, H. P., W. Lowrie, and D. V. Kent. 1975. Stability of anhysteretic remanent magnetization in fine and coarse magnetite and maghemite particles. Geophysical Journal of the Royal Astronomical Society 41:1-10.

Kelley, F. R. 1943. Eocene stratigraphy of western Santa Ynez Mountains, Santa Barbara County, California. American Association of Petroleum Geologists Bulletin 27:1-19.

Kelly, T. S. 1990. Biostratigraphy of Uintan and Duchesnean land mammal assemblages from the middle member of the Sespe Formation, Simi Valley, California. Contributions to Science of the Los Angeles County Museum 419:1-43.

Kerr, P. F., and H. G. Schenk. 1928. Significance of the Matilija overturn. Geological Society of America Bulletin 39:1087-1102.

Kew, W. S. W. 1924. Geology and oil resources of a part of Los Angeles and Ventura Counties, California. U.S. Geological Survey Bulletin 753, 202 pp.

Kleinpell, R. M., and D. W. Weaver. 1963. Oligocene biostratigraphy of the Santa Barbara embayment, California. University of California Publications in Geological Sciences 43:1-250.

Lander, E. B. 1994. Recalibration and causes of marine regressive-transgressive cycle recorded by middle Eocene to lower Miocene nonmarine Sespe Formation, southern California continental plate margin; pp. 79-88 in A. E. Fritsche (ed.), Sedimentology and paleontology of Eocene rocks in the Sespe Creek area, Ventura County, California. Pacific Section SEPM Guidebook 74.

Lindsay, E. H. 1968. Rodents from the Hartman Ranch local fauna, California. Paleobios 6:1-22.

Link, M. H., and J. E. Welton. 1982. Sedimentology and reservoir potential of Matilija Sandstone: an Eocene sand-rich deep-sea fan and shallow-marine complex, California. American Association of Petroleum Geologists Bulletin 66:1514-1534.

Lund, S. P., D. J. Bottjer, K. J. Whidden, J. E. Powers, and M. C. Steele. 1991. Paleomagnetic evidence for Paleogene terrane displacements and accretion in southern California; pp. 99-106 in P. L. Abbott and J. A. May (eds.), Eocene Geologic History, San Diego Region. Pacific Section SEPM Guidebook 68.

Luyendyk, B. P., M. J. Kamerling, and R. R. Terres. 1980. Geometric model for Neogene crustal rotations in southern California. Geological Society of America Bulletin 91:211-217.

Luyendyk, B. P., M. J. Kamerling, R. R. Terres, and J. S. Hornafius. 1985. Simple shear of southern California during Neogene time suggested by paleomagnetic declinations. Journal of Geophysical Research 90:12454-12466.

McCracken, W. A. 1969. Sespe Formation on upper Sespe Creek; pp. 41-48 in W. R. Dickinson (ed.), Geologic setting of upper Miocene gypsum and phosphorite deposits, upper Sespe Creek and Pine Mountain, Ventura County, California. Pacific Section SEPM, Upper Sespe Creek Field Trip.

McCracken, W. A. 1972. Paleocurrents and petrology of Sespe sandstones and conglomerates, Ventura Basin, California. Ph. D. dissertation, Stanford, University, Stanford, CA, 192 pp.

McDougall, K. 1980. Paleoecological evaluation of late Eocene biostratigraphic zonations of the Pacific Coast of North America. Paleontological Society Monograph 2, 46 pp.

Merrill, W. R. 1954. Geology of Sespe Creek-Pine Mountain area, Ventura County, California. California Division of Mines Bulletin 170, map sheet 3.

Miall, A. D. 1991. Stratigraphic sequences and their chronostratigraphic correlation. Journal of Sedimentary Petrology 61:497-505.

Miall, A. D. 1992. Exxon global cycle chart: an event for every occasion? Geology 20:787-790.

O'Brien, J. M. 1972. Narizian-Refugian (Eocene-Oligocene) marine and marginal marine sedimentation, western Santa Ynez Mountains, Santa Barbara County, California; pp. 37-43 in D. W. Weaver (ed.), Guidebook, Central Santa Ynez Mountains, Santa Barbara County,

California. Pacific Section SEPM and American Association of Petroleum Geologists Guidebook.

Olsson, R. K. 1988. Foraminiferal modeling of sea-level change in the Late Cretaceous of New Jersey. SEPM Special Publication 42:289-298.

Opdyke, N. D., E. H. Lindsay, N. M. Johnson, and T. Downs. 1977. The paleomagnetism and magnetic polarity stratigraphy of the mammal-bearing section of Anza-Borrego State Park, California. Journal of Quaternary Research 7:316-329.

Page, B. M., J. G. Marks, and G. W. Walker. 1951. Stratigraphy and structure of mountains northeast of Santa Barbara, California. American Association of Petroleum Geologists Bulletin 35:1727-1780.

Pluhar, C. J., J. L. Kirschvink, and R. W. Adams. 1991. Magnetostratigraphy and clockwise rotation of the Plio-Pleistocene Mojave River Formation, central Mojave Desert, California. San Bernardino County Museum Association Quarterly 38(2):31-42.

Poag, C. W., and J. S. Schlee. 1984. Depositional sequences and stratigraphic gaps on the submerged United States Atlantic Margin. American Association of Petroleum Geologists Memoir 36:165-182.

Posamentier, H. W., M. T. Jervey, and P. R. Vail. 1988. Eustatic controls on clastic deposition I—conceptual framework. SEPM Special Publication 42:109-124.

Prothero, D. R., and C. C. Swisher III. 1992. Magnetostratigraphy and geochronology of the terrestrial Eocene-Oligocene transition in North America; pp. 46-73 in D. R. Prothero and W. A. Berggren (eds.), Eocene-Oligocene Climatic and Biotic Evolution. Princeton University Press, Princeton, N.J.

Squires, R. L. 1994. Macropaleontology of Eocene marine rocks, upper Sespe Creek area, Ventura County, California; pp. 39-56 in A. E. Fritsche (ed.), Sedimentology and paleontology of Eocene rocks in the Sespe Creek area, Ventura County, California. Pacific Section SEPM Guidebook 74.

Stock, C. 1938. A titanothere from the type Sespe of California. Proceedings of the National Academy of Sciences 24:507-512.

Taliaferro, N. L. 1924. Notes on the geology of Ventura County, California. American Association of Petroleum Geologists Bulletin 8(6):789-810.

Thompson, P. R., and R. M. Slatt. 1990. Depositional

sequence stratigraphy of middle to upper Eocene units, Santa Ynez Mountains (abstract). American Association of Petroleum Geologists Bulletin 74:778.

Tipton, A. 1980. Foraminiferal zonation of the Refugian stage, latest Eocene of California. Cushman Foundation Special Publication 19:258-277.

Vail, P. R., R. M. Mitchum, Jr., and S. Thompson III. 1977. Seismic stratigraphy and global changes of sea level, Part 4:Global cycles of relative changes of sea level; pp. 83-97 in C. E. Payton (ed.), Seismic stratigraphy—Applications to hydrocarbon exploration: American Association of Petroleum Geologists Memoir 26.

Van Wagoner, J. C., H. W. Posamentier, R. M. Mitchum, Jr., P. R. Vail, J. F. Sarg, T. S. Loutit, and J. Hardenbol. 1988. An overview of the fundamental definitions of sequences stratigraphy and key definitions. SEPM Special Publication 42:39-48.

Van Wagoner, J. C., R. M. Mitchum, K. M. Campion, and V. D. Rahmanian. 1990. Siliciclastic sequence stratigraphy in well logs, cores, and outcrops: concepts for high resolution correlation of time and facies. American Association of Petroleum Geologists Methods in Exploration 7, 55 pp.

Vedder, J. G. 1972. Revision of stratigraphic names for some Eocene formations in Santa Barbara and Ventura counties, California. U.S. Geological Survey Bulletin 1354-D:1-12.

Vedder, J. G., T. W. Dibblee, Jr., and R. D. Brown, Jr. 1973. Geologic map of the upper Mono Creek-Pine Mountain area, California. U. S. Geological Survey Miscellaneous Geologic Investigations Map I-752, scale 1:48,000.

Watts, W. L. 1897. Oil- and gas-yielding formations of Los Angeles, Ventura and Santa Barbara counties, California. Bulletin of the California State Mining Bureau 11:22-28.

Watts, W. L. 1900. Oil- and gas-yielding formations of California. California State Mining Bureau Bulletin 19:1-236.

Weaver, D. W. 1965. Summary of Tertiary stratigraphy, western Santa Ynez Mountains; pp. 16-30 in D. W. Weaver and T. W. Dibblee, Jr. (eds.), Western Santa Ynez Mountains, Santa Barbara County, California. Coast Geological Society and Pacific Section SEPM.

8. Stratigraphy and Paleomagnetism of the Upper Middle Eocene to Lower Miocene (Uintan to Arikareean) Sespe Formation, Ventura County, California

DONALD R. PROTHERO   JEFFREY L. HOWARD, AND T. H. HUXLEY DOZIER

ABSTRACT

The paleomagnetism and heavy mineral assemblages of sandstones from the middle and upper members of the Sespe Formation were studied to clarify stratigraphic relations. Magnetostratigraphic data, and an upsection provenance change reflected in heavy mineral suites, indicate a major intraformational unconformity within the upper member of the Sespe Formation. The hiatus represented by this unconformity apparently spans much or all of the late Eocene and early Oligocene (approximately 29.5-36 Ma). In Simi Valley, the late Uintan and Duchesnean (Chrons C19n-C16r, 41.2-36.5 Ma) and the early Arikareean (Chrons C10r-C9r, 28.0-29.5 Ma) are the only intervals represented by strata. These relations indicate that the Sespe Formation was deposited much more rapidly and discontinuously than previously thought. The most prominent unconformity is attributed to a middle Oligocene eustatic sea level drop. The paleomagnetic results support previous interpretations that the western Transverse Ranges province was tectonically rotated clockwise by ~90° during the late Cenozoic.

INTRODUCTION

The Sespe Formation is a sequence of terrestrial clastic rocks typically sandwiched between middle Eocene and lower Miocene marine rocks. It consists of braided fluvial, deltaic, and alluvial fan deposits that crop out in several fault-bounded blocks in southwestern California (Fig. 1). The Sespe Formation is also found in the subsurface offshore and crops out locally on Santa Rosa Island south of the Santa Barbara Channel. Thus, its geographic distribution defines a large non-marine basin elongated in an east-west direction. The Sespe Formation thins gradually away from the central part of the basin where it has its maximum thickness of about 1700 m.

Simi Valley is located in the central part of the Sespe basin. One of the thickest and presumably most complete stratigraphic sections of the Sespe Formation is found in a north-dipping homocline located north of the Simi fault (Fig. 1, locality 1). This Sespe section is also the most richly fossiliferous and has produced important Eocene and Oligocene terrestrial vertebrate faunas (Fig. 2). The only Sespe Formation locality that has been isotopically dated is at South Mountain (Fig. 1, locality 3). These data make it possible to establish a chronostratigraphic framework, so the Simi Valley section is perhaps the best place to examine the depositional history of the Sespe Formation. In addition, this section is the only place in North America where Uintan and Duchesnean (middle Eocene) beds are overlain by strata containing Oligocene land mammals (Savage and Russell, 1983; Krishtalka et al., 1987; Tedford et al., 1987). Because either Eocene or Oligocene strata are missing in the classic sequences in the Rocky Mountains and High Plains, the Simi Valley section presents a unique opportunity to study the late Paleogene fossil record of continental vertebrates.

Previous sedimentological and paleontological studies have documented the depositional and faunal characteristics of the Simi Valley section. However, the internal chronostratigraphic relations of the Sespe Formation and the details of correlation with other strata remain to be determined. These are significant problems because there is some geological evidence that a major, and possibly regional, unconformity is present within the Sespe Formation (Howard, 1987, 1989, 1995). The purpose of this study was to investigate the magnetostratigraphy of the Simi Valley section of the Sespe Formation. Heavy mineral studies were used to further define the depositional character of Sespe sandstones, and paleomagnetic data were collected at several neighboring locations. This chapter focuses on lithostratigraphic, biostratigraphic, and magnetostratigraphic relations, but some of the tectonic implications of our data are also discussed briefly.

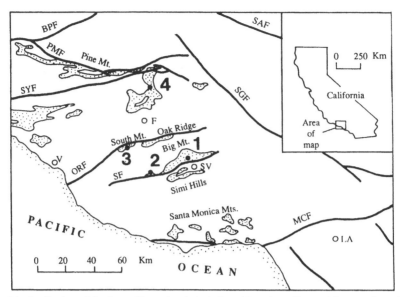

Figure 1. Geographic distribution of the Sespe Formation (stippled) and locations discussed in text. Location names (cities indicated by open circles): LA, Los Angeles; SV, Simi Valley; F, Fillmore; V, Ventura. Fault names: MCF, Malibu Coast fault; ORF, Oak Ridge fault; SYF, Santa Ynez fault; PMF, Pine Mountain fault; BPF, Big Pine fault; SF, Simi fault; SGF, San Gabriel fault; SAF, San Andreas fault. Localities discussed in text: 1, Simi Valley section (Brea Canyon-Alamos Canyon area); 2, Las Posas Hills section (Kew Quarry); 3, South Mountain section; 4, Type location of Sespe Formation.

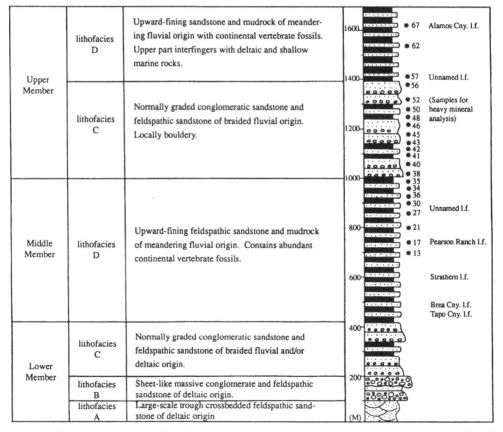

Figure 2. Composite stratigraphic section (from Brea and Alamos canyons) of the Sespe Formation north of Simi Valley, showing lithostratigraphic nomenclature, and location of terrestrial vertebrate faunas and sandstone samples.

PREVIOUS WORK

The Sespe Formation was named and described by Watts (1897) from a type locality on Sespe Creek in the Topatopa Mountains (Fig. 1, locality 4). In the Simi Valley area it was first mapped by Kew (1919, 1924). Reed (1929) briefly mentioned the Sespe sequence in Simi Valley in an early regional sedimentological study. The first paleontological studies were those of Stock (1932a, 1932b, 1948), who amassed an extensive collection of vertebrate fossils. Savage and Downs (1954) also discussed these vertebrate faunas in some detail. Flemal (1966, 1969), McCracken (1972), Yeats et al. (1974), Black (1982), and Hathon and Houseknecht (1993) documented some of the sedimentological aspects of the Simi Valley section as part of their regional investigations. Golz (1976), Golz and Lillegraven (1977), and Mason (1983, 1988, 1989, 1990) have reported on the Eocene vertebrate faunas in the study area, and Lander (1983) summarized the Oligocene faunas. The most detailed stratigraphic studies are those of Taylor (1983, 1984). Howard (1987, 1988, 1989) also discussed stratigraphic and sedimentological relations in a regional context. Mason and Swisher (1989) reported an isotopic age date from the Sespe Formation in South Mountain, west of Simi Valley. New specimens from renewed excavations in the Simi Valley Landfill near Brea Canyon in northwestern Simi Valley have led to much recent paleontologic work (Kelly, 1990, 1992; Kelly et al., 1991; Kelly and Whistler, 1994).

METHODS

A composite stratigraphic section of Sespe Formation was established at locality 1 (Fig. 1) using the geologic mapping of Weber et al. (1973) and Dibblee (1992), and the stratigraphic framework of Taylor (1983, 1984) and Kelly (1990). Site numbers in this study correspond to the stratigraphic numbering scheme of Kelly (1990, fig. 4). The lower and middle parts of the formation were sampled in Brea Canyon, and the upper part in Alamos Canyon. Other nearby locations of Sespe Formation were sampled under the guidance of Mark Mason.

Sampling for paleomagnetic analysis was carried out with simple hand tools, planing the top surface of the sample horizontal for consistency. A minimum of three samples was collected at each site. Samples were spaced at least 3 to 5 m apart stratigraphically in thinner sections (e.g., South Mountain) to as much as 15 to 20 m apart in thicker sections (e.g., Brea Canyon). Poorly consolidated samples were hardened with sodium silicate solution. After removal from the field, the samples were trimmed into 2.5-cm cubes for laboratory analysis using a band saw with a tungsten carbide blade. All measurements were made at the paleomagnetic laboratory at the California Institute of Technology (see Pluhar et al., 1991, for methods).

After measurement of natural remanent magnetization (NRM), several different tests were conducted. One suite of samples was subjected to stepwise alternating field (AF) demagnetization, and another suite was subjected to a detailed stepwise thermal demagnetization. Powdered samples (about 0.1 g) of representative lithologies were placed in epindorph tubes, and subjected to rock magnetic analysis. These samples were then AF demagnetized twice, once after having acquired an isothermal remanent magnetization (IRM) produced in a 100 mT (millitesla) peak field, and once after having acquired an anhysteretic remanent magnetization (ARM) in a 100 mT oscillating field. Such data are useful in conducting a modified Lowrie-Fuller ARM test (see Johnson et al., 1975). These same powdered samples were also subjected to IRM acquisition experiments up to 1300 mT.

The AF demagnetization demonstrated a spectrum of behaviors (Fig. 3). Most samples showed a rapid decline in intensity, indicating that some of the remanence resides in a low-coercivity mineral such as magnetite. Other samples, however, were relatively resistant to AF demagnetization, indicating that a high-coercivity phase, such as hematite or goethite, was dominant. During thermal demagnetization (Fig. 4), most samples had little or no remanence left above the Curie point of magnetite (580°C); only a few samples showed significant remanence at 600°C indicting hematite was present.

Results of the IRM acquisition (ascending curve on right) and Lowrie-Fuller tests (descending curves on left) are shown in Figure 5. Many samples reached IRM saturation at about 300 mT, indicating that their remanence is carried by magnetite. However, a few of the more reddish samples did not reach IRM saturation, showing that hematite is a significant carrier of the remanence. In most samples, the ARM (solid boxes) was more resistant to AF demagnetization than the IRM (open boxes). This suggests that the remanence is carried by fine-grained (<10 micron) single-domain, or pseudo-single-domain grains. A few samples (Fig. 5), however, did not show this behavior, suggesting that multidomain grains are also present.

Heavy-mineral separations also allowed direct examination of the opaque minerals. As first reported by Flemal (1966, 1969) and confirmed by our analysis, hematite pseudomorphs after magnetite are the dominant opaque component. Most of the opaques have the distinctive crystal habit of magnetite, yet were not extractable with a hand magnet. However, the replacement by hematite must not be complete, because the IRM saturation and AF and thermal demagnetization results clearly indicate the presence of significant magnetite in most samples.

The AF demagnetization results were also replotted as an NRM:IRM(s) [saturation IRM] plot, following

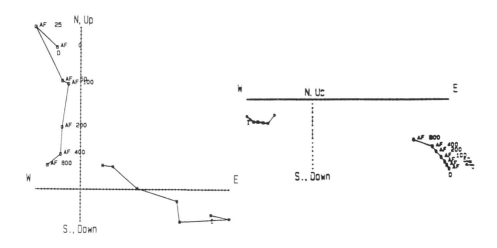

Figure 3. Orthogonal demagnetization ("Zijderveld") plots of AF demagnetization behavior of representative samples of the Sespe Formation. Each division is 10^{-6} emu. Open circles indicate horizontal component; asterisks show vertical component. AF demagnetization field in Gauss shown. Some samples declined in intensity during AF demagnetization, indicating the presence of significant magnetite. Others had a high coercivity, probably due to hematite.

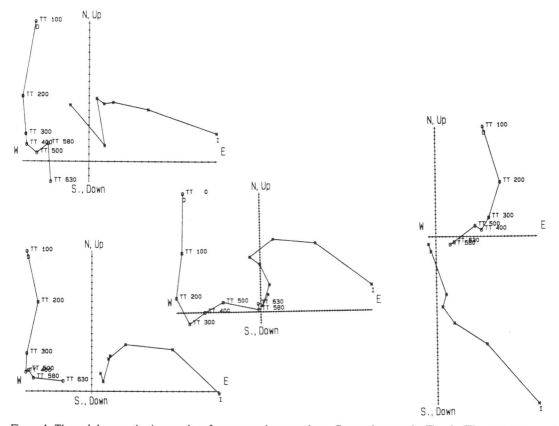

Figure 4. Thermal demagnetization results of representative samples. Conventions as in Fig. 2. TT = temperature of demagnetization step in degrees Centigrade; TT0 = NRM. Note that an overprinted vector was typically removed by 300°C, and then a rotated component was apparent at temperatures between 300 and 500°C. Three samples are reversed and rotated to the west; the fourth is normal and rotated to the east.

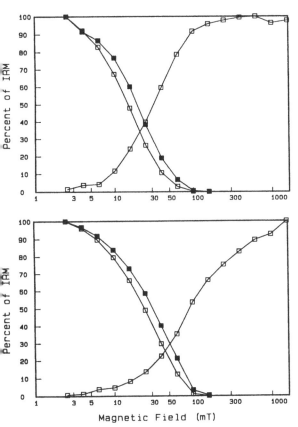

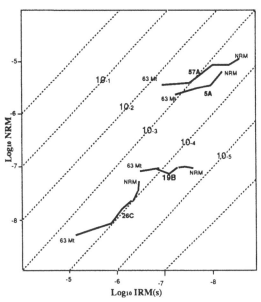

Figure 6. NRM:IRM(s) results of AF demagnetization from NRM to 63 Mt of four representative samples. In all four, the ratio stays below the 10^{-2} threshold, indicating that most of the remanence is primary, rather than a secondary chemical overprint.

Figure 5. IRM saturation test (curve ascending to right) and Lowrie-Fuller test (curves descending to left) of representative samples. AF of IRM shown by open boxes; AF of ARM shown by solid boxes. Some samples reached IRM saturation by 100-300 mT, indicating that magnetite is the main carrier of remanence; others did not saturate even at 1300 mT, which suggests that hematite is dominant. In most samples, the ARM was more resistant to AF demagnetization than the IRM, suggesting that the carrier of the remanence is fine-grained (<10 micron) single-domain or pseudo-single domain magnetite; in others, the IRM was more resistant, suggesting multi-domain remanence. For details, see Pluhar et al. (1991).

Fuller et al. (1988) and Cisowski et al. (1990). These plots are useful in distinguishing primary remanence from a secondary chemical magnetization, because primary magnetizations typically have an NRM:IRM(s) ratio on the order of 10^{-2} or less; secondary chemical remanence will have ratios greater than this value. Figure 6 shows the typical AF demagnetization trajectories of four representative samples on an NRM: IRM(s) plot. All four trajectories are clearly below the 10^{-2} threshold, indicating that most of the remanence is primary.

Because these results clearly indicate the presence of high-coercivity minerals such as goethite and hematite,

all remaining samples were then subjected to stepwise thermal demagnetization. As can be seen in Figure 4, many samples had an overprinted component that was removed by temperatures of 200-300°C; it was probably due to a chemical remanence from an iron hydroxide, such as goethite. Temperatures between 300 and 500°C typically produced a stable magnetic component which was rotated about 80-90° clockwise (Fig. 4). This component was used for all further analyses.

The presence of this rotated component makes it easy to distinguish younger overprints from the characteristic remanence, and reject samples which failed to show this component. In addition to this test, a reversal test for stability was also conducted. Using the statistical methods of Fisher (1953; see Butler, 1992), the mean declination (D) for 40 normal sites was 89.2°; the mean inclination (I) was 30.3°, with a precision parameter (k) of 6.0 and an ellipse of confidence (α_{95}) of 11.1°. For 106 reversed sites, D = 282.3, I = -27.9, k = 4.2, and α_{95}= 7.7. These results are antipodal within the error margin of their α_{95}, indicating that the remanence is primary.

Petrologic methods

Sandstone samples weighing about 500 g were used for the heavy mineral analyses. All but two samples were so weakly indurated that they could be disaggregated under gentle pressure in a mullite mortar and pestle; the harder samples were disaggregated by soaking in 0.02 M Na_2CO_3. The disaggregated samples were wet sieved to obtain the -35+100 mesh (0.15-0.50 mm) fraction and dried at 40°C. Heavy mineral fractions were recovered by gravitational settling in bromoform, and iron oxides removed using the dithionite-citrate-bicarbonate method of Mehra and Jackson (1960). Bulk heavy mineral samples were microsplit by cone-and-quartering and a subsample mounted with Permount for point counting. Another subsample grain mount, prepared using Petropoxy 154, was polished for cathodoluminescence work. The heavy mineral fractions of selected samples were fractionated further by magnetic separation using a Franz magnetic separator according to the method of Hess (1956). Polished grain mounts were also prepared of these subfractions. The polished grain mounts were studied by cathodoluminescence microscopy using a Nuclide ELM-3R luminoscope under a standard atmosphere at 0.5 mA and 25 kV.

SIMI VALLEY AREA
Lithostratigraphy

In the fault block bounded on the south by the Simi fault and on the north by the Oak Ridge fault (Fig. 1), the Sespe Formation contains three informal members, composed of four lithofacies (Fig. 2). Lithofacies A and B in the lower member are interpreted to be delta distributary channel and river mouth bar deposits, respectively. These facies grade upsection into braided fluvial deposits of lithofacies C that compose the upper part of the lower member, and meandering fluvial deposits of lithofacies D that compose the middle member, to form an upward-fining succession about 1000 m in thickness. The upper member is composed of braided fluvial deposits (lithofacies C) in the lower part that grade upward into meandering fluvial deposits (lithofacies D) to form another upward-fining megasequence about 700 m thick. The uppermost part of the Sespe Formation grades upward into, or interfingers with, the Vaqueros Formation (Blundell, 1981; Taylor, 1984) forming a transitional succession deposited in a lower delta plain setting (Taylor, 1984).

In the Simi Hills south of the Simi fault (Fig. 1), the middle part of the Sespe Formation is covered by Quaternary alluvium and the upper part is unconformably overlain by Miocene volcanic rocks. The lower part of the Sespe Formation in this area is lithologically similar to the lower member north of the fault, but the upper part (apparently corresponding to lithofacies C of the upper member) includes boulder conglomerate not found north of the fault. Upsection changes in texture and in the types of clasts found in the conglomerates locally south of the Simi fault, and elsewhere in the Los Angeles basin area, led Howard (1987, 1989) to suspect that an unconformity exists between the middle and upper members.

Biostratigraphy

Figure 2 shows the stratigraphic distributions of seven distinct local vertebrate faunas that were recognized previously in the Sespe Formation of the Simi Valley area. Five faunas, all of late middle Eocene age, are found in the middle member. These include the Tapo Canyon and Brea Canyon local faunas of late Uintan age, the Strathern local fauna of latest Uintan or earliest Duchesnean age, the Pearson Ranch local fauna of early Duchesnean age, and the Simi Valley Landfill local fauna of late Duchesnean age (Kelly, 1990; Kelly et al., 1991).

Lander (1983) recognized three local faunas of Oligocene age in the upper member near Simi Valley: the Alamos Canyon l.f., the Kew Quarry l.f., and an unnamed l.f. However, only two are shown in Figure 2 because there is disagreement over the age and stratigraphic position of the Kew Quarry (Las Posas Hills) local fauna, which is separated from Simi Valley by considerable distance and several faults. The unnamed local fauna consists of a supposed leptauchenid oreodont fossil reported from a locality near the western end of Big Mountain (Fig. 1) approximately 300 m below the top of the Sespe Formation (Stock, 1932b; Wilson, 1949). Although the specimen has been lost, Lander (1983) suggested that it was probably the early Arikareean (late Oligocene) oreodont *Sespia nitida* because this is the only leptauchenine known from the Sespe Formation. This oreodont is also common in the upper part of the Sespe Formation both at South Mountain and Oak Ridge (Fig. 1). The Alamos Canyon local fauna is similar to other early Arikareean (late Oligocene) faunas found elsewhere in the Sespe Formation (e.g., South Mountain), and found in rocks correlative with Chrons C9-10 in the High Plains (see Tedford et al., this volume, Chapter 15). Wilson (1949) reported that the Alamos Canyon fauna was found in association with marine vertebrate remains in the Sespe-Vaqueros transitional sequence north of Alamos Canyon. Lander (1983) also noted that marine mollusks and continental vertebrate fossils are found at nearly the same stratigraphic position near the west end of Big Mountain.

Magnetostratigraphy

The paleomagnetic results for the upper part of the lower member and for the middle member of the Sespe Formation in the Brea Canyon section are shown in Figure 7. About 100 m of the lower member were sampled along the entrance road to the Simi Valley landfill; the lowest 60 m (sites M4 through M6) are of normal polarity. Between site M3 and the Strathern fault gap at about 350 m above the base of the composite section (site 12), the uppermost part of the lower member and the lower part of the middle member are entirely of reversed polarity. These rocks contain the Tapo Canyon, Brea Canyon and Strathern local faunas.

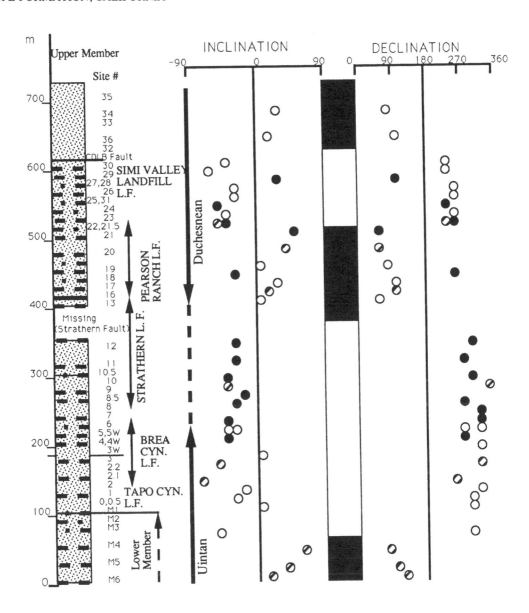

Figure 7. Magnetostratigraphy of the lower and middle members of the Sespe Formation in Brea Canyon north of Simi Valley. Site numbers correspond to the stratigraphic numbering scheme of Kelly (1990). Uppermost 100 m of the lower member and the entire middle member (rest of the section) are shown. Solid circles = Class I sites (significantly clustered) of Opdyke et al. (1977); diagonal patterned circles = Class II sites (one sample missing); open circles = Class III sites (one sample divergent). Note that the declinations are clustered around 90° and 270°, indicating a 90° clockwise rotation.

Above the Strathern fault gap, the interval from sites 13 to 21 (400-515 m above base of section) is of normal polarity; this part of the middle member contains the Pearson Ranch local fauna. From site 21.5 to the Canada de la Brea fault (just above site 30, 515-615 m above base of section), the rocks have reversed polarity. Due to faulting, there is also some repeated section in this area (Kelly, 1990); the sampling takes this into account. The uppermost 100 m of the middle member

(sites 32-35) appear to be of normal polarity, although only two of these sites produced suitable paleomagnetic results.

Most of the 800 m of the upper member studied in Alamos Canyon are of reversed polarity (Fig. 8). Intervals of normal polarity characterize the lowest 20 m of section (sites 37 and 38), and from 250 to 380 m (sites 48 to 54) and 510 to 540 m (sites 60 to 62) above the base of the section. The Alamos Canyon local fauna,

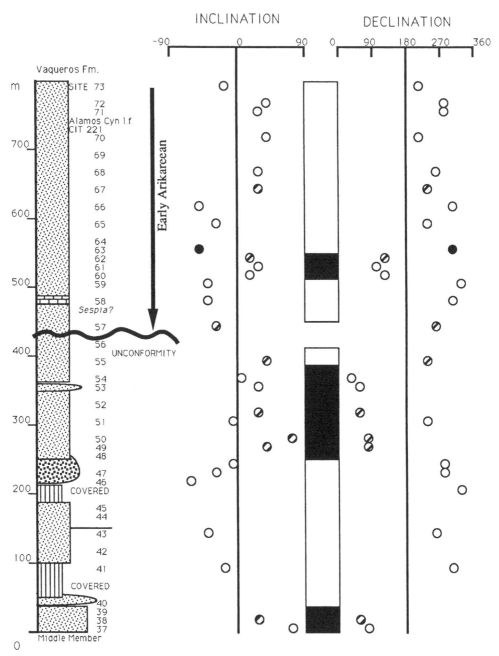

Figure 8. Magnetostratigraphy of the upper member of the Sespe Formation in Alamos Canyon, north of Simi Valley Landfill. Symbols as in Fig. 7.

which dates the top of the section, is found about 720 m above the base of this section, and the unnamed local fauna of Lander (1983)—presumably containing *Sespia nitida*—is approximately 300 m below the top of the section.

The magnetic polarity patterns observed in the Sespe Formation (Figs. 7 and 8) are correlated with the magnetic polarity time scale of Berggren et al. (1995) based on the biostratigraphic control afforded by fossil verte-brates (Fig. 9). This section contains fewer reversals than are present in the global time scale, so one or more unconformities must be present within the Sespe Formation. In the Uinta basin of Utah, the Duchesnean begins in Chron C18n, or about 40 Ma; late Uintan faunas are found in Chrons C18r and C19n (Prothero and Swisher, 1992; Prothero, this volume, Chapter 1). On this basis, the reversely magnetized interval of Sespe Formation in Brea Canyon containing the late

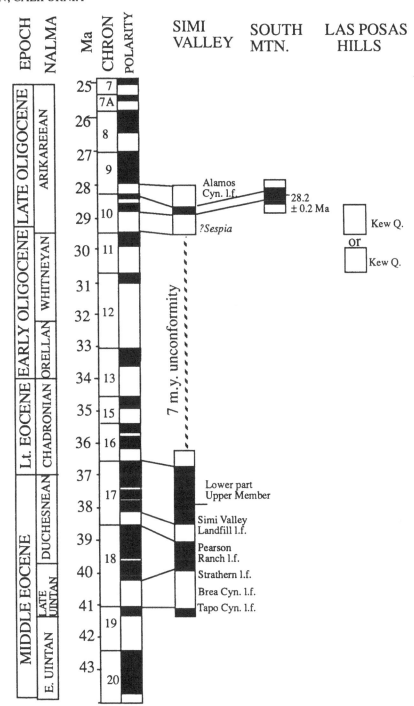

Figure 9. Magnetic correlations of the key Sespe sections discussed in text. Time scale after Berggren et al. (1995).

Uintan Tapo Canyon and Brea Canyon local faunas, and the Uintan-Duchnesnean Strathern local fauna, probably correlates with Chron C18r (40.2 to 41.0 Ma); the normally magnetized interval containing the early Duchesnean Pearson Ranch fauna probably correlates with Chron C18n (38.5 to 40.2 Ma). The interval containing the late Duchesnean Simi Valley Landfill local

fauna higher in the section probably correlates with Chron C17r (38.1 to 38.4 Ma), and the unfossiliferous upper zone of normal polarity possibly correlates with part of Chron C17n.

In the Alamos Canyon Sespe section, the unnamed local fauna characterized by *Sespia nitida*, and the Alamos Canyon local fauna, are both most likely of early Arikareean age. Early Arikareean faunas in the Gering Formation of the High Plains region are well dated and found in rocks correlated with Chron C9r (Tedford et al., this volume, Chapter 15). Thus, if the Sespe strata of reversed polarity containing the Alamos Canyon l.f. correlate with Chron C9r (Fig. 8), then the polarity zones below may represent Chron C10n and C10r (28.0-29.5 Ma). If this correlation is correct, then a major unconformity, representing a hiatus of about seven million years, must exist within the Sespe Formation that separates upper middle Eocene rocks below from upper Oligocene strata above. Furthermore, if the latest Oligocene to earliest Miocene age (i.e., about 23-24 Ma) identification (based on mollusks and benthic foraminifera) of the Vaqueros Formation in this area is correct, another unconformity and a hiatus representing about 4 million years may exist in the uppermost part of the Sespe Formation.

However, this scenario is at odds with evidence indicating that the upper Sespe Formation is conformable or interfingers with the Vaqueros Formation. An alternative explanation is that the entire section is much younger (i.e., correlative with Chrons C7 or C7a). This can be ruled out, because the Alamos Canyon l.f. contains taxa (*Archaeolagus, Leidymys, Hypertragulus*) which are restricted to Chron C9 (Tedford et al., 1987; Tedford et al., this volume, Chapter 15). A third possibility is that the Vaqueros Formation in Alamos Canyon is much older than suggested by Blake (1983), i.e., early late Oligocene (about 27 Ma) rather than latest Oligocene (about 24 Ma). Unfortunately, there are no planktonic microfossils from the Vaqueros in this area, so it is not possible to precisely correlate this section with the global time scale. The Zemorrian benthic foraminifera from the section range through most of the Oligocene, so they offer less chronostratigraphic resolution than the fossil mammals. In addition, Blake (1983) points out that many of the benthic foraminifera are highly depth- and facies-controlled. Their apparent "early Miocene" affinities might be due to the fact that shallow marine rocks of the Vaqueros Formation (which yield shallow-water benthic foraminifera) are common in southern California only during the early Miocene, whereas almost all known late Oligocene strata in the region are deep marine.

Heavy Mineral Assemblages of Sandstones

Although the above data clearly indicate that at least one major unconformity is present within the Simi Valley section of the Sespe Formation, the exact position of this feature cannot be located magnetostratigraphically because of limited paleontologic control in the upper member. Hence, we examined the heavy mineral assemblages of sandstones with the objective of identifying any stratigraphic discontinuities that might be present. Heavy minerals are sensitive indicators of

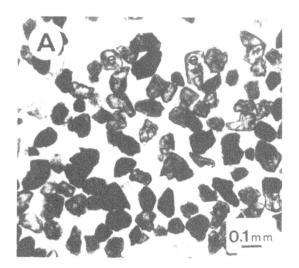

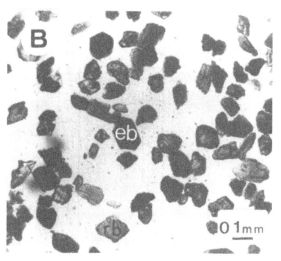

Figure 10. Photomicrographs comparing heavy mineral assemblages in sandstones of the Sespe Formation near Simi Valley. A, lower part (lithofacies C) of upper member; B, upper part (lithofacies D) of upper member (see Fig. 2). Mineral types: e, epidote; rb, pleochroic red biotite; eb, euhedral biotite with pseudohexagonal symmetry.

source rock type and intrastratal solution, so unconformities are liable to be indicated by upsection changes in provenance and/or dissolution of unstable mineral species. As noted previously, conglomerate is scarce in the upper member north of the Simi fault, but lithologic relations south of the fault suggest that a discontinuity might lie between the middle and upper members. We studied eighteen samples spanning this stratigraphic interval (Fig. 2). Except for samples 21 and 27, which are dark greenish-gray wacke, all of the samples studied are medium grained feldspathic arenites with minor mud. The arenite samples 13 through 35 are light or dark gray and contrast sharply in color with

Table 1. Selected results of heavy mineral analysis of Sespe Formation sandstones near Simi Valley, California. Percent abundance of non-opaque, non-mica fraction indicated; n = 125. O = opaque; NO = non-opaque minerals; n = 200. See Figure 2 for location of samples.

MINERAL TYPE	SAMPLES Middle				Member		Upper			Member	
	13	21	27	30	36	34	38	40	42	45	46
Zircon (Z)	4.0	3.2	1.6	1.6	4.8	2.4	8.8	14.4	10.4	5.6	7.2
Apatite (A)	12.0	11.2	2.4	12.0	12.8	1.6	15.2	6.4	5.6	4.0	6.4
Garnet (G)	2.4	5.6	1.6	0.8	0.8	0	4.0	3.2	1.6	0.8	4.8
Tourmaline	1.6	1.6	1.6	0	0.8	0	0	1.6	0	0.8	2.4
Epidote (E)	48.0	48.8	64.8	41.6	54.4	67.2	52.0	61.6	61.6	77.6	68.8
Clinozoisite (C)	30.4	26.4	34.8	27.2	16.8	24.8	13.6	11.2	19.2	7.2	7.2
Zoisite	1.6	1.6	1.6	8.0	4.0	2.4	4.0	1.6	0.8	0.8	0
Rutile	0	0	0	0	0	0	0	0	0	0.8	0.8
Sphene	0	1.6	1.6	8.8	4.8	1.6	2.4	0	0.8	2.4	2.4
Piemontite	0	0	0	0	0.8	0	0	0	0	0	0
E + C	78.4	75.2	89.6	68.8	71.2	92.0	65.6	72.8	80.8	84.8	76.0
G + A + Z	18.4	20.0	6.5	14.4	18.4	4.0	28.0	24.0	17.6	10.2	18.4
			MINERAL		RATIOS						
O/NO	0.45	1.88	0.60	0.28	0.34	0.23	0.46	0.63	0.54	0.26	1.04
G+A+Z/E+C	0.23	0.27	0.06	0.21	0.26	0.04	0.43	0.33	0.22	0.12	0.24
Z/A	0.33	0.29	0.67	0.13	0.38	1.50	0.58	2.25	1.86	1.40	1.12
Z/E+C	0.05	0.02	0.02	0.02	0.07	0.03	0.13	0.20	0.13	0.07	0.09

Table 2. Heavy minerals in Sespe Formation sandstones after magnetic separation. Samples in parentheses.

	MAGNETIC	FRACTION Sideslope Angle = 20°		Sideslope angle = 5°	
A/B	C	D	E	F	
Magnetic at 0.4 Amps	Magnetic at 0.8 Amps	Magnetic at 1.2 Amps	Magnetic at 1.2 Amps	Non-magnetic at 1.2 Amps	
	Upper part of upper member (sample 57)				
Garnet	Chloritoid	Sphene	Sphene	Zircon	
Chlorite	Red biotite		Apatite	Apatite	
Red biotite	Chlorite				
Chloritoid	Epidote				
	Rutile				
	Lower part of upper member (samples 41-43)				
Garnet	Epidote	Zoisite	Sphene	Apatite	
Epidote	Zoisite	Sphene	Apatite	Zircon	
Biotite	Schorlite				
Chlorite					
	Middle Member (samples 13-36)				
Garnet	Epidote	Zoisite	Sphene	Apatite	
Epidote	Schorlite	Sphene	Zoisite	Zircon	
Chlorite	Zoisite				

the light yellowish brown samples 38 through 52.

Modal analysis indicates that the heavy mineral suites of the middle member and the lower part of the upper member are virtually identical (Table 1) despite the change in inferred depositional settings. The mineral varieties found in various fractions obtained by magnetic separation are also the same (Table 2), and they have the same cathodoluminescence features. None of the rocks, not even sample 35 located just below the top of the middle member, were found to contain minerals showing significant etching or other signs of dissolution. Hence, these data do not support the inference based on conglomerate clast assemblages that an unconformity is located between the middle and upper members. After these results were obtained, further studies were made and magnetic separations revealed a

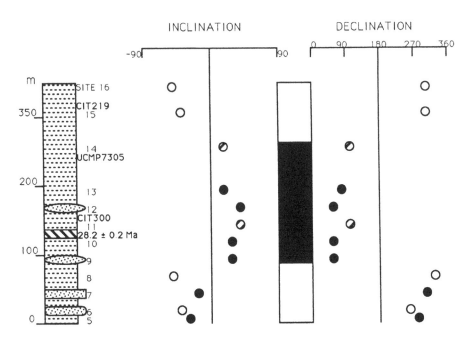

Figure 11. Magnetostratigraphy of the Sespe Formation in South Mountain. Stratigraphy after Mason and Swisher (1989, fig. 2). Symbols as in Fig. 7.

significantly different mineral suite in sample 57 (Table 2) at the base of the upper part of the upper member (Fig. 2), and in samples 62 and 67. The most distinctive feature is the presence of an abundance of euhedral green and brown biotite and a strongly pleochroic red biotite (Fig. 10). Chloritoid and euhedral sphene also are common and epidote minerals, especially zoisite, are much less abundant. These upsection mineralogical changes suggest that the unconformity indicated by paleomagnetics is located at the base of lithofacies D, made up of the upper part of the upper member. Sample 56, located just below the inferred unconformity, contains a mineral suite transitional to that above. This is similar to relations noted elsewhere in the Sespe Formation where upsection changes in provenance first become evident just below intraformational unconformities (Howard, 1987, 1995). Sample 62 contains severely etched epidote; however, because other unstable minerals (e.g., apatite) in the same sample are not etched, the significance of this feature is problematic.

OTHER SECTIONS
South Mountain

About 600 m of the upper member of the Sespe Formation are well exposed in an anticlinal section at South Mountain, 25 km northwest of Simi Valley (Fig. 1, locality 3). The section contains an extensive continental vertebrate fauna first described by Stock (1930) and later referred to as the South Mountain local

fauna (Lander, 1983; Mason and Swisher, 1989). The main section was sampled northward along roadcuts in Willard Canyon, where most of the localities (such as CIT 300) and the dated tuff were clearly tied to the section. Only about 330 m of section were sampled during our study; the remainder was too poorly exposed or disturbed by faulting. Thicker sections elsewhere on South Mountain were available, but because of structural complications and uncertain correlations between structural blocks, they cannot be positively tied to the fossils or dated tuff.

The lowest 75 m of section are reversely magnetized, the next 175 m are of normal polarity and the upper 80 m are reversely magnetized (Fig. 11). Mason and Swisher (1989) obtained a ^{40}Ar/^{39}Ar isotopic age of 28.2 ± 0.2 Ma on a tuff bed located about 140 m above the base of the section in rocks of normal polarity. The date correlates this normally magnetized interval with Chron C10n (28.2-28.7 Ma) (Fig. 9). Thus the total section sampled apparently spans Chrons C9r to C10r (28.0 to 29.3 Ma). According to Lander (1983) and Mason and Swisher (1988), early Arikareean faunas (including *Sespia nitida*) are found throughout the section sampled. Hence, these magnetostratigraphic data are in good agreement with the ^{40}Ar/^{39}Ar dates for the type Arikareean of the High Plains, which produces a similar fauna in Chron C10r (Tedford et al., 1987; Tedford et al., this volume, Chapter 15). This further reinforces the correlation of the Alamos Canyon section in Figure 9.

Las Posas Hills section (Kew Quarry)

The Las Posas Hills section, which produces the Kew Quarry l.f., is located in an isolated outcrop belt of Sespe Formation located 18 km west of the Simi Valley section (Fig. 1, locality 2). The vertebrate faunas were first described by Stock (1932b, 1933a, 1933b, 1935) and include a puzzling mixture of mammals typical of the Whitneyan and earliest Arikareean (Lander, 1983; Tedford et al., 1987, p. 157). The Kew Quarry section is faulted and folded extensively, but four successive sites covering about 500 m of section were analyzed. All were found to be of reversed polarity except for the stratigraphically lowest site (Fig. 9). Because of the equivocal age assignment of the fauna, the magnetic correlation of this section is problematic. Based on the mixture of primitive and advanced faunal elements, it seems reasonable that the section correlates with the earliest Arikareean Chron C10r (28.7 to 29.3 Ma) (Tedford et al., this volume, Chapter 15). However, if the archaic mammals are weighted more heavily (Lander, 1983), then it may correlate with Chron C11r (30.0 to 30.5 Ma).

TECTONIC ROTATIONS AND TRANSLATIONS

All of the Simi Valley samples studied showed anomalous paleomagnetic declinations and inclinations. The clockwise rotation of 89.2 ± 11.1° for the normal sites, and 102.3 ± 7.7° for the reversed sites are consistent with the 93° clockwise rotation reported by Luyendyk (1991) for the Miocene rocks of the western Transverse Range block, and with the 80.1 ± 5.4° rotation for the Sespe Formation in the Santa Ynez Range reported by Liddicoat (1990). Lund et al. (1991, table 1) report some additional results from the Sespe Formation which show a 71.6 ± 7.0° clockwise tectonic rotation. From these data, it is apparent that most of the tectonic rotations in the Transverse Range have taken place during the Neogene, because the Sespe shows no more rotation than the older Miocene rocks reported by Luyendyk (1991).

To determine the northward translation of the Simi Valley block, the sites were clustered by age, with reversed sites inverted by 180° so they could be averaged with normal sites. Middle Eocene Sespe samples gave a mean D = 98.6 and I = 30.1 [k = 4.7, α95 = 7.5, n = 97], and late Oligocene samples gave a mean D = 91.4 and I = 32.1 [k = 4.2, α95 = 11.4, n = 49]. These results are about 17-18° further south in paleolatitude than their present position [34.5° north latitude], or about 19-20° further south than the North American cratonic pole of Eocene or Oligocene age calculated from Irving and Irving (1982). Our latitudinal deviations are much greater than the 10° further south reported by Liddicoat (1990) for the Sespe Formation in the Santa Ynez Range, and the 10° further south reported by Lund et al. (1991, Table 1). Although such

data have been used by some (e.g., Howell et al., 1987; Beck, 1991) to infer large-scale tectonic translations of suspect terranes, others (e.g., Luyendyk et al., 1985; Gastil, 1991; Butler et al., 1991) have questioned the reliability of such data. However, given the large error estimates, these results are within the range of translations reported by Lund et al. (1991) for other exotic blocks of coastal California.

DISCUSSION

Stratigraphic relations in the Sespe Formation at Simi Valley are apparently similar to those in the Santa Ynez Range (west of the area shown in Fig. 1) where lateral facies relations defining an upper Paleogene nonmarine to marine transition are well exposed. Near Santa Barbara, for example, braided fluvial deposits of late Oligocene (early Arikareean) age containing *Sespia nitida* rest with pronounced unconformity on deltaic deposits inferred to be of late Eocene (Refugian or Chadronian) age based on lateral facies relations (Howard, 1995). Because the suspected unconformity at Simi Valley was thought to lie between the middle and upper members, upper Eocene rocks were thought to be missing (Figs. 9, 12) and the lower part of the upper member was thought to correlate physically with those rocks lying above the unconformity in the Santa Ynez Range (Howard, 1989). However, the results of this study indicate the possibility that the lower part of the upper member is upper Eocene and thus correlative with rocks below the unconformity in the Santa Ynez Range (Fig. 12). This reinterpretation implies that either the braided fluvial rocks above the unconformity in the Santa Ynez Range are correlative with meandering fluvial rocks composing the upper part of upper member at Simi Valley, or contemporaneous braided fluvial sediments were never deposited at Simi Valley. The results of this study also suggest that part of the Sespe section in the Santa Monica Mountains (Fig. 1) may be older than previously thought based on physical correlations with the Simi Valley section as tentatively reinterpreted here (Fig. 12).

The difficulties of correlating Sespe rocks have led to some confusion. For example, Lander (1980, 1983) suggested radical changes in the North American time scale on the basis of a single oreodont from the Sespe. Lander (1983) argued that the early Arikareean oreodont *Sespia* occurs in a basal Sespe conglomerate (the Calvary Conglomerate Member of Lian, 1952) in the East Fork of Maria Ygnacio Creek (UCMP locality V-5813) in the Santa Ynez Range near Santa Barbara. This conglomerate was supposedly traceable into the marine Gaviota Formation, which much further to the west has Refugian benthic foraminifera (Kleinpell and Weaver, 1963), and late Eocene planktonic foraminifera (Tipton, 1980) and nannofossils (Warren and Newell, 1980). From this *Sespia* specimen, Lander argued that

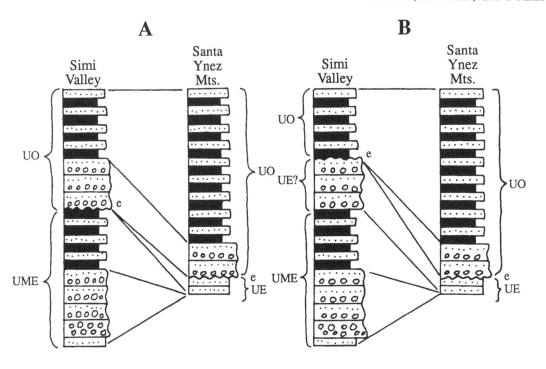

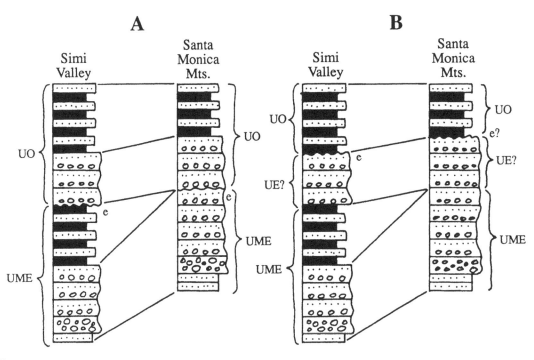

Figure 12. Schematic stratigraphic sections (not drawn to scale) of the Sespe Formation showing tentative chronostratigraphic correlations. A, previous correlations; B, revised correlations. Chronostratigraphic units: UO, upper Oligocene; UE, upper Eocene; UME, upper middle Eocene; e, erosional unconformity.

the early Arikareean was late Eocene, and implied that all the chronostratigraphic control and dozens of radio-isotopic dates on Eocene and Oligocene rocks from North America (e.g., Evernden et al., 1964; Obradovich et al., 1973) were in error. The chronological framework of Evernden et al. (1964) and Obradovich et al. (1973) has since been substantiated and updated by Krishtalka et al. (1987), Emry et al. (1987), and Tedford et al. (1987), further contradicting Lander's conclusions.

What is the reason for this apparent discrepancy? Did *Sespia* have an anomalously early first occurrence on the West Coast? Given the complicated faulting, poor exposures, and problems with tracing time-transgressive units across the Santa Ynez Range, many alternative explanations come to mind. As Howard (1989, fig. 5) and Mason and Swisher (1989, Addendum, p. 8) have noted, a simpler explanation is available. Lander (1983) was unaware of a major unconformity near the base of the Sespe in the disputed area. The base of the section is indeed middle or late Eocene, but the *Sespia* specimen probably comes from just above the unconformity. Above this level, all the rocks are late Arikareean and grade westward into the late Oligocene marine Alegria Formation. Dibblee (1987) shows that the conglomerate in question ("Tspcg" on the geologic map) is not at the base of the Sespe, but disconformably overlies a basal Sespe sandstone ("Tspss" on map). According to Dibblee's map, both units pinch out to the west, where they are affected by a number of faults. The Gaviota Formation first appears at the base of the undifferentiated Sespe Formation in the extreme western edge of the Goleta Quadrangle. According to the most recent mapping, there is no direct interfingering between the *Sespia*-bearing rocks and the marine Eocene, and thus no basis for Lander's correlations.

No correlation was found between paleoburial depth and either the abundances of heavy minerals (Table 1) or the degree of etching. Hence, intrastratal solution does not appear not be a significant factor affecting the types of heavy minerals present. Stratigraphic changes documented in this study probably accurately reflect changes in provenance over time. This interpretation is supported by the fact that sphene and epidote, both of which are abundant in Sespe sandstones, are unstable during burial diagenesis (Morton, 1985). In addition, Taylor (1984) found an abundance of highly unstable minerals such as pyroxene and amphibole even farther downsection in the lower member. It is also consistent with clay mineralogical evidence indicating that the Simi Valley section has not undergone deep burial (Hathon and Houseknecht, 1993). Euhedral biotite with a pseudohexagonal habit is typically of pyroclastic origin (Mange and Maurer, 1992), and biotite high in Ti is commonly pleochroic red and of volcanic origin (Deer et al., 1992). The upsection appearance of a volcanic provenance is interesting because clasts of andesitic tuff are common in conglomeratic parts of the upper member of the Sespe Formation, especially at Simi Valley (Howard, 1987, 1989). Andesitic volcanic rocks with K-Ar isotopic ages generally in the 26 to 29 Ma age range are voluminous in southwestern Arizona (Olmsted, 1968; Lipman et al., 1972; Eberly and Stanley, 1978), that is, in the previously postulated source terrane of the Sespe fluvial system (Howard and Lowry, 1995).

SUMMARY

The paleomagnetism and heavy mineral assemblages of the middle and upper parts of the Sespe Formation near Simi Valley were studied to clarify stratigraphic relations. The paleomagnetics of several nearby localities (Las Posas Hills, South Mountain) of Sespe Formation were also investigated. Many more magnetic anomalies are present in the global time scale than in the Sespe Formation at Simi Valley, so much or all of the upper Eocene to lower Oligocene section is apparently missing. The presence and stratigraphic position of the unconformity in the middle of the upper member is also indicated by an upsection provenance change revealed by heavy mineral assemblages. Magnetostratigraphic data suggest that part of the upper Oligocene section is also probably missing, although the exact location of this unconformity is unknown.

These relations indicate that the Sespe Formation was deposited much more rapidly and discontinuously than previously thought. Because the upper Oligocene sequence contains *Sespia nitida*, deposition following the major hiatus probably resumed about 30 Ma. If so, the intraformational unconformity may represent the drastic eustatic sea level drop that occurred during middle Oligocene time (Haq et al., 1988). The paleomagnetic results from the Sespe Formation also support previous interpretations that the western Transverse Ranges province was tectonically rotated clockwise by ~90° during late Cenozoic time.

ACKNOWLEDGMENTS

Prothero was supported by NSF grants EAR87-08221 and EAR91-17819 during this research. We thank Simi Valley Landfill and Recycling Center, a division of Waste Management of California, Inc., for access to their land, Unocal for their permission to work at South Mountain and Alamos Canyon, and Tom Kelly, Bruce Lander, Mark Mason, and Mark Roeder for their advice and help in the field. Steve King and Greg Brown helped with the 1988 sampling, and Tim Tierney with the 1991 sampling. Joe Kirschvink graciously allowed us access to the California Institute of Technology paleomagnetics laboratory. We thank Mike Clark, Tom Kelly, Joe Liddicoat, Tor Nilsen, Richard Squires, Richard Tedford, and Michael Woodburne for their helpful comments on this manuscript.

LITERATURE CITED

Beck, M. E., Jr. 1991. Case for northward transport of Baja and coastal southern California: Paleomagnetic data, analysis, and alternatives. Geology 19:506-509.

Berggren, W. A., D. V. Kent, M.-P. Aubry, C. C. Swisher III, and K. G. Miller. 1995. A revised Paleogene geochronology and chronostratigraphy. SEPM Special Publication 54:129-212.

Black, B. A. 1982. Excerpts from a report on the Sespe Formation of the Los Angeles and Ventura basins; pp. 1-24 in G. Nulty (ed.), Geologic guide of the central Santa Clara Valley, Sespe and Oak Ridge trend oil fields, Ventura County, California. Coast Geological Society and Pacific section AAPG Guidebook 53.

Blake, G. H. 1983. Benthic foraminiferal paleoecology and biostratigraphy of the Vaqueros Formation, Big Mountain area, Ventura County, California; pp. 173-182 in R. L. Squires and M. V. Filewicz (eds.), Cenozoic geology of the Simi Valley area, southern California. Pacific Section SEPM Guidebook 35.

Blundell, M. C. 1981. Depositional environments of the Vaqueros Formation in the Big Mountain area, Ventura County, California. M. S. thesis, California State University, Northridge, 102 pp.

Butler, R. F. 1992. Paleomagnetism. Blackwell, Boston.

Butler, R. F., W. R. Dickinson, and G. E. Gehrels. 1991. Paleomagnetism of coastal California and Baja California: alternatives to large-scale northward transport. Tectonics 10:561-576.

Cisowski, S. M., J. R. Dunn, M. Fuller, and P. J. Wasilewski. 1990. NRM:IRM(s) demagnetization plots of intrusive rocks and the origin of their NRM. Tectonophysics 184(1990):35-54.

Deer, W. A., R. A. Howie, and J. Zussman. 1992. An introduction to the rock-forming minerals. Longman Scientific and Technical Publishers, England, 696 pp.

Dibblee, T. W., Jr. 1987. Geologic map of Goleta quadrangle, Santa Barbara County, California. Dibblee Foundation, Santa Barbara. Scale 1:24,000.

Dibblee, T. W., Jr. 1992. Geologic map of the Simi quadrangle, Ventura County, California. Dibblee Foundation, Santa Barbara. Scale 1:24,000.

Eberly, L. D., and T. B. Stanley, Jr. 1978. Cenozoic stratigraphy and geologic history of southwestern Arizona. Geological Society of America Bulletin 89:921-940.

Emry, R. J., P. R. Bjork, and L. S. Russell. 1987. The Chadronian, Orellan, and Whitneyan land mammal ages; pp. 118-152 in M. O. Woodburne (ed.), Cenozoic Mammals of North America, Geochronology and Biostratigraphy. University of California Press, Berkeley.

Evernden, J. F., D. E. Savage, G. H. Curtis, and G. T. James. 1964. Potassium-argon dates and the Cenozoic mammalian chronology of North America. American Journal of Science 262:145-198.

Fisher, R. A. 1953. Dispersion on a sphere. Proceedings of the Royal Society A217:295-305.

Flemal, R. C. 1966. Sedimentology of the Sespe Formation, southwestern California. Ph. D. dissertation, Princeton University, Princeton, N. J., 230 pp.

Flemal, R. C. 1969. Iron-bearing heavy minerals in the Sespe Formation (California). Journal of Sedimentary Petrology 39:1616-1618.

Fuller, M., S. Cisowski, M. Hart, R. Haston, and E. Schmidtke. 1988. NRM:IRM(s) demagnetization plots: an aid to the interpretation of remanent magnetization. Geophysical Research Letters 15(5):518-521.

Gastil, G. 1991. Is there a Oaxaca-California megashear?

Conflict between paleomagnetic data and other elements of geology. Geology 19:502-505.

Golz, D. J. 1976. Eocene Artiodactyla of southern California. Natural History Museum of Los Angeles County Science Bulletin 26:1-85

Golz, D. J., and J. A. Lillegraven. 1977. Summary of known occurrences of terrestrial vertebrates from Eocene strata of southern California. University of Wyoming Contributions in Geology 15:43-65.

Haq, B. U., J. Hardenbol, and P. R. Vail. 1988. Mesozoic and Cenozoic chronostratigraphy and cycles of sea level change, in C. K. Wilgus, B. S. Hastings, H. Posamentier, J. Van Wagoner, C. A. Ross, and C. G. S. Kendall (eds.), Sea level changes: An integrated approach. SEPM Special Publication 42:71-108.

Hathon, L. A., and D. W. Houseknecht. 1993. Origin and diagenesis of clay minerals in the Oligocene Sespe Formation, Ventura basin. SEPM Special Publication 47:185-195.

Hess, H. H. 1956. Notes on the operation of Franz isodynamic magnetic separator. Unpublished pamphlet, Princeton University.

Howard, J. L. 1987 Paleoenvironments, provenance and tectonic implications of the Sespe Formation, southern California. Ph. D. dissertation, University of California, Santa Barbara, 306 pp.

Howard, J. L. 1988. Sedimentation of the Sespe Formation in southern California; pp. 53-69 in A. G. Sylvester and G. C. Brown (eds.), Santa Barbara and Ventura basins—Tectonics, structure, sedimentation and oilfields along an east-west transect: Ventura, California, Coast Geological Society, Field Guide 64.

Howard, J. L. 1989. Conglomerate clast populations of the Upper Paleogene Sespe Formation, southern California; pp. 269-280 in I. P. Colburn, P. L. Abbott and J. A. Minch (eds.), Conglomerates in basin analysis: A symposium dedicated to A. O. Woodford. Pacific Section SEPM Guidebook 62.

Howard, J. L. 1995. Conglomerates of the upper middle Eocene to lower Miocene Sespe Formation along the Santa Ynez fault: Implications for the geologic history of the eastern Santa Maria basin area, California. USGS Bulletin 1995H:1-37.

Howard, J. L., and W. D. Lowry. 1995. Medial Cenozoic paleogeography of the Los Angeles area, southwestern California, and adjacent parts of the United States; pp. 22-41 in A. E. Fritsche (ed.), Cenozoic Paleogeography of the Western United States. Pacific Section SEPM Volume 75.

Howell, D. G., D. E. Champion, and J. G. Vedder. 1987. Terrane accretion, crustal kinematics, and basin evolution, southern California; pp. 81-123 in R. V. Ingersoll and W. G. Ernst (eds.), Cenozoic basin development of coastal California, Rubey Volume 6. Prentice-Hall, Englewood Cliffs, N. J.

Irving, E., and G. A. Irving. 1982. Apparent polar wander paths, Carboniferous through Cenozoic, and the assembly of Gondwana. Geophysical Surveys, 5:141-188.

Johnson, H. P., W. P. Lowrie, and D. V. Kent. 1975. Stability of anhysteretic remanent magnetization in fine and coarse magnetite and maghemite particles: Geophysical Journal of the Royal Astronomical Society 41:1-10.

Kelly, T. S. 1990. Biostratigraphy of Uintan and Duchesnean land mammal assemblages from the middle member of the Sespe Formation, Simi Valley, California.

Contributions to Science of the Los Angeles County Museum 419:1-43.

Kelly, T. S. 1992. New Uintan and Duchesnean (middle and late Eocene) rodents from the Sespe Formation, Simi Valley, California. Bulletin of the Southern California Academy of Sciences 91(3):97-120.

Kelly, T. S., and D. P. Whistler. 1994. Additional Uintan and Duchesnean (middle and late Eocene) mammals from the Sespe Formation, Simi Valley, California. Contributions to Science of the Los Angeles County Museum 439:1-29.

Kelly, T. S., E .B. Lander, D. P. Whistler, M. A. Roeder, and R. E. Reynolds. 1991. Preliminary report on a paleontologic investigation of the lower and middle members, Sespe Formation, Simi Valley Landfill, Ventura County, California. PaleoBios 13 (50):1-13.

Kew, W. S. W. 1919. Geology of part of the Santa Ynez River district, Santa Barbara County, California. University of California Publications in Geological Sciences 12:1-121.

Kew, W. S. W., 1924, Geology and oil resources of a part of Los Angeles and Ventura Counties, California. U.S. Geological Survey Bulletin 753, 202 p.

Kleinpell, R. M., and D. W. Weaver. 1963. Oligocene biostratigraphy of the Santa Barbara embayment, California. University of California Publications in Geological Sciences 43:1-250.

Krishtalka, L., R. K. Stucky, R. M. West, M. C. McKenna, C. C. Black, T. M. Bown, M. R. Dawson, D. J. Golz, J. J. Flynn, J. A. Lillegraven, and W. D. Turnbull. 1987. Eocene (Wasatchian through Duchesnean) biochronology of North America; pp. 77-117 in M. O. Woodburne (ed.), Cenozoic Mammals of North America, Geochronology and Biostratigraphy. University of California Press, Berkeley.

Lander, E. B. 1980. Recorrelation and recalibration of the Duchesnean and Arikareean North American land-mammal Ages with the European time scale. Geological Society of America Abstracts with Programs, 12 (7):468.

Lander, E. B. 1983. Continental vertebrate faunas from the upper member of the Sespe Formation, Simi Valley, California, and the Terminal Eocene Event; pp. 142-153 in R. L. Squires and M. V. Filewicz (eds.), Cenozoic geology of the Simi Valley area, southern California. Pacific section SEPM Guidebook 35.

Lian, H. M. 1952. The geology and paleontology of the Carpinteria district, Santa Barbara County, California. Ph.D. dissertation, University of California, Los Angeles, 178 pp.

Liddicoat, J. C. 1990. Tectonic rotation of the Santa Ynez range, California, recorded in the Sespe Formation. Geophysical Journal International 102:739-745.

Lipman, P. W., H. J. Prostka, and R. L. Christiansen. 1972, Cenozoic volcanism and plate-tectonic evolution of the western United States: I. Early and middle Cenozoic. Royal Society of London Philosophical Transactions Series A 271:217-248.

Lund, S. P., D. J. Bottjer, K. J. Whidden, J. E. Powers, and M. C. Steele. 1991. Paleomagnetic evidence for Paleogene terrane displacements and accretion in southern California; pp. 99-106 in P. L. Abbott and J. A. May (eds.), Eocene Geologic History, San Diego Region. Pacific Section SEPM Guidebook 68.

Luyendyk, B. P. 1991. A model for Neogene crustal rotations, transtensions, and transpressions in Southern California. Geological Society of America Bulletin 103:1528-1536.

Luyendyk, B. P., M. J. Kamerling, and R. R. Terres. 1980. Geometric model for Neogene crustal rotations in southern California. Geological Society of America Bulletin 91:211-217.

Luyendyk, B. P., M. J. Kamerling, R. R. Terres, and J. S. Hornafius. 1985. Simple shear of southern California during Neogene time suggested by paleomagnetic declinations. Journal of Geophysical Research 90:12454-12466.

Mange, M. A., and H. F. W. Maurer. 1992. Heavy minerals in colour. Chapman and Hall, London, 147 p.

Mason, M. A. 1983. Fossil land mamals from the Sespe Formation, Simi Valley area, Ventura County, California; p. 141 in R. L. Squires and M. V. Filewicz (eds.), Cenozoic geology of the Simi Valley area, southern California: Pacific section SEPM Guidebook 35.

Mason, M. A. 1988. Mammalian paleontology and stratigraphy of the early to middle Tertiary Sespe and Titus Canyon Formations, southern California. Ph.D. dissertation, University of California, Berkeley, 257 pp.

Mason, M. A. 1989. Mammalian biostratigraphy of the Sespe Formation north of Simi Valley, Ventura County, California and revision of the Duchesnean North American land mammal age (abs.): Geological Society of America Abstracts with Programs 21:112.

Mason, M. A. 1990. New fossil primates from the Uintan (Eocene) of southern California. PaleoBios 13(49):1-7.

Mason, M. A., and C. C. Swisher III. 1989. New evidence for the age of the South Mountain local fauna, Ventura County, California. Contributions in Science of the Los Angeles County Museum 410:1-9.

McCracken, W. A. 1972. Paleocurrents and petrology of Sespe sandstones and conglomerates, Ventura Basin, California. Ph. D. dissertation, Stanford University, Stanford, CA, 192 pp.

Mehra, O. P., and M. L. Jackson. 1960. Iron oxide removal from soils and clays by a dithionite-citrate system buffered with sodium bicarbonate. Clays and Clay Minerals 7:317-327.

Morton, A. C. 1985. Heavy minerals in provenance studies; pp. 249-277 in G. G. Zuffa (ed.), Provenance of arenites. Reidel, Dordrecht.

Obradovich, J. D., G. A. Izett, and C. W. Naeser. 1973. Radiometric ages of volcanic ash and pumice beds in the Gering Sandstone (earliest Miocene) of the Arikaree Group, southwestern Nebraska. Geological Society of America Abstracts with Programs 5:499-600.

Olmsted, F. H. 1968. Tertiary rocks near Yuma, Arizona (abs): Geological Society of America Special Paper 101:153-154.

Opdyke, N. D., E. H. Lindsay, N. M. Johnson, and T. Downs. 1977. The paleomagnetism and magnetic polarity stratigraphy of the mammal-bearing section of Anza-Borrego State Park, California. Journal of Quaternary Research 7:316-329.

Pluhar, C. J., J. L. Kirschvink, and R. W. Adams. 1991. Magnetostratigraphy and clockwise rotation of the Plio-Pleistocene Mojave River Formation, central Mojave Desert, California. San Bernardino County Museum Association Quarterly 38(2):31-42.

Prothero, D. R., and C. C. Swisher III. 1992. Magnetostratigraphy and geochronology of the terrestrial Eocene-Oligocene transition in North America; pp. 46-73 in D. R. Prothero and W. A. Berggren (eds.), Eocene-Oligocene Climatic and Biotic Evolution. Princeton University Press, Princeton, N. J.

Reed, R. D. 1929. Sespe Formation, California. American

Association of Petroleum Geologists Bulletin 13:489-507.

Savage, D. E., and T. Downs. 1954. Cenozoic land life in southern California. California Division of Mines and Geology Bulletin 170(3):43-58.

Savage, D. E., and D. E. Russell. 1983. Mammalian Paleofaunas of the World. Addison-Wesley, Reading, MA.

Stock, C. 1930. Oreodonts from the Sespe deposits of South Mountain, Ventura County, California. Carnegie Institute of Washington Publication 404:27-42.

Stock, C. 1932a. Eocene land mammals on the Pacific coast. Proceedings of the National Academy of Sciences 18(7):518-523,

Stock, C. 1932b. An upper Oligocene mammalian fauna from southern California. Proceedings of the National Academy of Sciences 18(8):550-554.

Stock, C. 1933a. Perissodactyls from the Sespe of the Las Posas Hills, California. Carnegie Institute of Washington Publication 440:16-27.

Stock, C. 1933b. Carnivora from the Sespe of the Las Posas Hills, California. Carnegie Institute of Washington Publication 440:30-41.

Stock, C. 1935. Artiodactyls from the Sespe of the Las Posas Hills, California. Carnegie Institute of Washington Publication 453:121-125.

Stock, C. 1948. Pushing back the history of land mammals in western North America. Geological Society of America Bulletin 59:327-332.

Taylor, G. E. 1983. Braided-river and flood-related deposits of the Sespe Formation, northern Simi Valley, California; pp. 128-140 in R. L. Squires and M. V. Filewicz (eds.), Cenozoic geology of the Simi Valley area, southern California. Pacific section SEPM Guidebook 35.

Taylor, G. E. 1984. Depositional environments of the Eocene through Oligocene Sespe Formation in the northern Simi Valley area, Ventura County, southern California. M.S. thesis, California State University Northridge, 74 pp.

Tedford, R. H., T. Galusha, M. F. Skinner, B. E. Taylor, R. W. Fields, J. R. Macdonald, J. M. Rensberger, S. D. Webb, and D. P. Whistler. 1987. Faunal succession and biochronology of the Arikareean through Hemphillian interval (late Oligocene through earliest Pliocene Epochs) in North America; pp. 152-210 in M. O. Woodburne (ed.), Cenozoic Mammals of North America, Geochronology and Biostratigraphy. University of California Press, Berkeley.

Tipton, A. 1980. Foraminiferal zonation of the Refugian stage, latest Eocene of California. Cushman Foundation Special Publication 19:258-277.

Warren, A. D., and S. H. Newell. 1980. Plankton biostratigraphy of the Refugian and adjoining stages of the Pacific Coast Tertiary. Cushman Foundation Special Publication 19:233-251.

Watts, W. L. 1897. Oil- and gas-yielding formations of Los Angeles, Ventura and Santa Barbara Counties, California. Bulletin of the California State Mining Bureau 11:22-28.

Weber, F. H., Jr., G. B. Cleveland, J. E. Kahle, E. F. Kiessling, R. V. Miller, M. F. Mills, D. M. Morton, and B. A. Cilweck. 1973. Geology and mineral resources study of southern Ventura County, California. California Division of Mines and Geology Preliminary Report 14. Scale 1:48,000.

Wilson, R. W. 1949. Rodents and lagomorphs of the upper Sespe. Carnegie Institute of Washington Publication 584:51-65.

Yeats, R. S., M. R. Cole, W. R. Merchat, and R. M. Parsley. 1974. Poway fan and submarine cone and rifting of the inner southern California continental borderland. Geological Society of America Bulletin 85:293-302.

9. Magnetostratigraphy of the Eocene-Oligocene Transition in Trans-Pecos Texas

DONALD R. PROTHERO

ABSTRACT

Previous studies of the Uintan, Duchesnean, and Chadronian rocks and their mammalian faunas of Trans-Pecos Texas are supplemented by more recent magnetostratigraphic research to provide a more up-to-date chronostratigraphic framework. In the Vieja area, reinterpretations of the work of Testarmata (1978; Testarmata and Gose, 1979), plus new magnetics, show that the Vieja Group (Colmena Chambers-Capote Mountain Tuffs) spans the interval from Chron C19r (42 Ma) to C13n (33 Ma). The early Duchesnean Candelaria l.f. occurs in Chron C19n (about 41.5 Ma), the late Duchesnean Porvenir l.f. and early Chadronian Little Egypt l.f. in Chron C17n (37-38 Ma), and the mid-Chadronian Airstrip l.f. in C16n (35.5 Ma). In the Agua Fria area, the Devil's Graveyard Formation spans the interval from C21n (46.5 Ma) to C13n (33 Ma), with a gap between C16n (35.5 Ma) and C18n (39 Ma) (wherein lie the Duchesnean faunas of the Skyline and Cotter channels). Revisions of Walton's (1992) magnetic interpretations place the late Bridgerian Junction l.f. in C21n (46.5 Ma), the early Uintan Whistler Squat l.f. in early C20r (45-46 Ma), the late Uintan Serendipity and Purple Bench l.f. in C19r (42 Ma) and C18r (41 Ma), respectively. The mid-Chadronian Coffee Cup l.f. occurs in Chron C15r (about 35 Ma).

INTRODUCTION

The Trans-Pecos region of Texas is the only place in North America where the middle-late Eocene is recorded by superposed mammalian faunas datable by both radiometrics and magnetic stratigraphy. There is no place in North America that contains interfingering marine/nonmarine sequences of this age, so this area is the key to unlocking the details of the Uintan-Duchesnean-Chadronian transition in land mammals.

The Trans-Pecos rocks (Fig. 1) contain a number of well-studied faunas of middle Eocene (Uintan and Duchesnean) and late Eocene (Chadronian) age (Wilson et al., 1968; Wilson 1978, 1980, 1984, 1986) with extensive radioisotopic dating (McDowell, 1979; Henry et al., 1986; Henry and McDowell, 1986; Henry et al., 1994). The first magnetostratigraphic work in the area was by Margaret Testarmata (1978; Testarmata and

Gose, 1979). She sampled rocks in the Vieja area (Fig. 1) between the Buckshot Ignimbrite and the Mitchell Mesa Rhyolite, which included the Chambers and Capote Mountain tuffs. However, Testarmata did not sample the late Uintan sections of Colmena Tuff, or many of the fossil localities of Duchesnean and Chadronian age in the area. Her sites were spaced about 1.5 m apart stratigraphically, but only a single sample was taken at each site, so no site statistics were calculated. Rock magnetic analyses (Testarmata, 1978, pp. 15-18) showed that the remanence was carried by fine-grained magnetite partially oxidized to hematite. Thermal demagnetization at 400-500°C gave the best results in her study.

Testarmata and Gose (1979, fig. 7) summarized their magnetic stratigraphy, which was extremely noisy, with many single-site normal "zones." Most of these "zones" are probably due to unremoved overprinting, which is hard to detect when only one sample per site is analyzed. It is possible, however, that some of these short polarity events might represent true "tiny wiggles" of the magnetic polarity time scale, which have been recently documented in the early Oligocene (Hartl et al., 1993). This "tiny wiggle" interpretation might be more likely in the cases of short polarity zones 2-3 sites in thickness. However, since the short zones cannot be correlated over distance, and do not seem to match any version of the magnetic polarity time scale, they are ignored in my correlations.

Once the shortest polarity events are discounted, Testarmata and Gose (1979) found a relatively long zone of reversed polarity running from just below the Bracks Rhyolite to above the Ford siltstone, which they correlated with Chron C12r (Fig. 2). They based these correlations on the K-Ar dates then known from the volcanic units, and using the magnetic time scale of LaBrecque et al. (1977). At that time, Chron C12r was thought to range from 33-35 Ma, and the K-Ar dates on the Bracks Rhyolite were 36.5 and 36.8 Ma. Unfortunately, a key

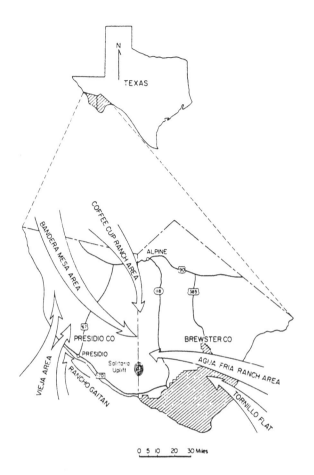

N.A. LAND MAMMAL AGE		VIEJA-OJINAGA AREA				AGUA FRIA-GREEN VALLEY AREA		
		Lithologic unit	K Ar	Local fauna Mexico	Local fauna Texas	Lithologic unit	K Ar	Local fauna or locality
CHADRONIAN		Mitchell Mesa Rhyolite	32.28		Ash Spring (Vieja Group undif.) Airstrip	Mitchell Mesa Rhyolite	32.28 33.04 basalt	
		Capote Mt. Tuff						Coffee Cup
		Bracks Rhyolite	37.4 37.7	Rancho Gaitan	Little Egypt	Bandera Mesa Member		
DUCHESNEAN	Late	Chambers Tuff			Porvenir			Montgomery Bonebed
	Early	Buckshot Ignimbrite	37.3 37.0			Middle member / Skyline channels	42.70 tuff	Cotter Channels Skyline
UINTAN	C	Colmena Tuff			Candelaria	Titanothere channels		Serendipity
	B					Lower member / Basal Tert. congl	43.90 biotite 49.70, 46.90 tuff	Whistler Squat

Figure 1. Index map and summary stratigraphy of the Trans-Pecos Texas region (modified from Wilson, 1986).

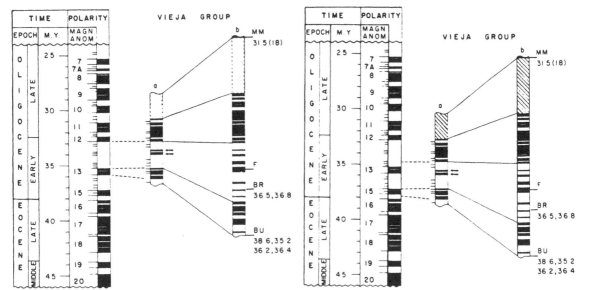

Figure 2. Interpretations of the magnetic polarity results of Testarmata and Gose (1979) on the left, and as reinterpreted by Prothero et al. (1982, 1983) on the right; from Wilson (1986).

interval just below the Mitchell Mesa Rhyolite did not produce a stable polarity signal, and thus was considered of indeterminate polarity.

As a result of magnetic correlations with Flagstaff Rim, Wyoming, Prothero et al. (1982, 1983) reinterpreted Testarmata and Gose's (1979) Vieja correlations (Fig. 2). Prothero et al. (1982, 1983) had found what they believed to be Chron C12r in middle Chadronian rocks at Flagstaff Rim between K-Ar dates of 32 and 36 Ma. Consequently, they thought that the Duchesnean-early Chadronian reversed interval in the Vieja Group (identified as Chron C12r by Testarmata and Gose) was probably correlative with the early Chadronian C13r. This was also in better agreement with the known K-Ar dates on the Bracks Rhyolite and the LaBrecque et al. (1977) time scale.

Another paleomagnetic study in the region was undertaken by Walton (1986, 1992), who sampled the lower and middle members of the Devil's Graveyard Formation in the Agua Fria area (Fig. 1; see also Stevens et al., 1984). Although multiple samples were typically collected per site, no site statistics were calculated. Rock magnetic analyses showed that most samples in her study contained detrital magnetite partially altered to hematite or goethite. Most normal overprints were apparently contained in goethite, as they were typically removed by about 250°C.

Walton (1992, fig. 3.8) correlated her results with Chrons C20n to C20r of the Berggren et al. (1992) time scale, based on the limited K-Ar dates that had been done previously. This placed late Uintan faunas (such as Purple Bench and Serendipity) in Chron C20n, supported by an overlying K-Ar date of 42.7 ± 1.6 Ma on a tuff high in the middle member. This correlation was

accepted by Prothero and Swisher (1992) in tying the late Uintan faunas of the Uinta Basin to Chron C20n (see Prothero, this volume, Chapter 1).

Since that time, much better radiometric dating (especially by $^{40}Ar/^{39}Ar$ methods) and major changes in the magnetic polarity time scale (Cande and Kent, 1993; Berggren et al., 1995) have made all the previous correlations obsolete. The revised correlation is discussed below.

MAGNETIC ANALYSIS

For this study, over 100 sites (each containing a minimum of three samples) were collected from a number of key localities between 1990 and 1993. In most sections, sites were spaced at about 10 m intervals. All samples were taken with simple hand tools, then trimmed on a band saw with a tungsten carbide blade, and analyzed at the paleomagnetics laboratory at the California Institute of Technology. Each sample was first measured at NRM, and then demagnetized in alternating fields (AF) of 25-200 Gauss to remove any multidomain remanence. Then each sample was thermally demagnetized between 200° and 600°C in multiple steps.

As can be seen in the vector demagnetization ("Zijderveld") plots (Fig. 3), the magnetic intensity declined rapidly under AF treatment, indicating that a significant component of the remanence (especially in the overprinting) was carried by magnetite. A stable reversed component was apparent at temperatures of 200-300°C. In some cases, this reversed component persisted up to 600°C. This indicates that some of the remanence is carried by hematite, since the Curie point of magnetite is 580°C. Testarmata and Gose (1979, p.

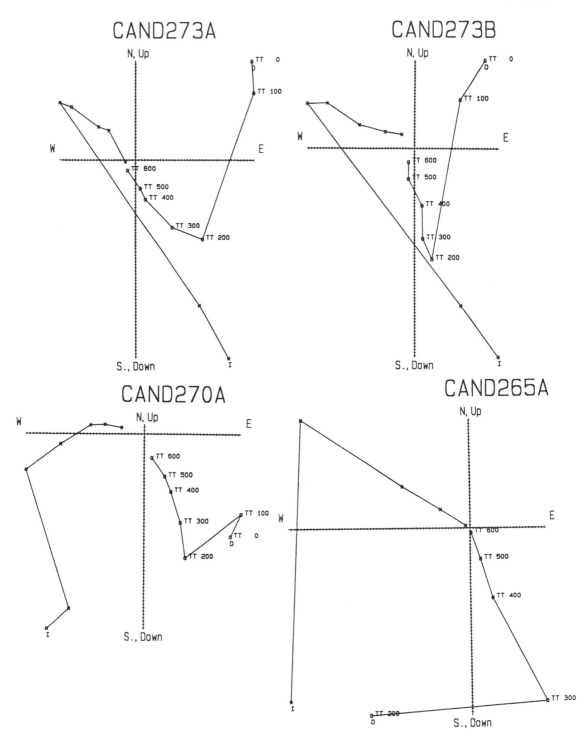

Figure 3. Vector demagnetization plots of thermal results on typical samples from Trans-Pecos Texas. All four samples are reversed, with a component of normal overprinting removed at temperatures of 200-400°C. From 300-600°C, most samples reveal a single stable component of magnetization. Since there is frequently some remanence left at 600°C, some of the magnetization must be due to a high Curie point mineral such as hematite. Open circles indicate the horizontal component, stars show the vertical component; I is the NRM value. Each increment equals 10^{-5} emu.

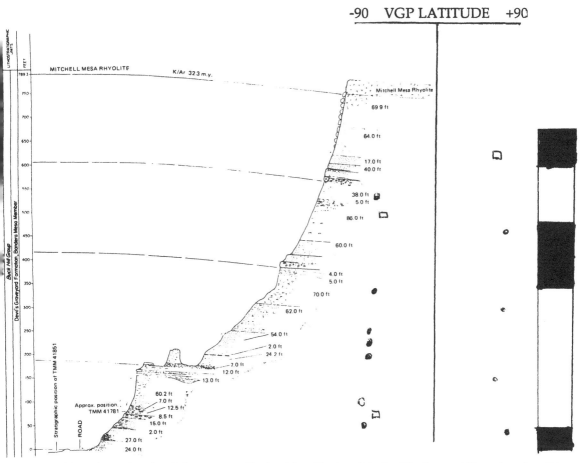

Figure 4. Magnetic stratigraphy of the stratotype Bandera Mesa Member of the Devil's Graveyard Formation, Agua Fria area, Texas (Stevens et al., 1984, fig. 10). The Chadronian Red Hill fauna (TMM 41781) is found at the base, and the Mitchell Mesa Ignimbrite occurs at the top. Positive virtual geomagnetic pole (VGP) latitudes indicate normal polarity; negative VGP latitudes indicate reversed polarity. Solid circles: Class I sites of Opdyke et al. (1977), which were significantly clustered at the 95% confidence level; open squares: Class II sites, in which one sample was lost or crumbled; open circles: Class III sites, in which two directions out of three showed a clear polarity preference. Stratigraphy from Wilson (1978, fig. 6).

57) reported that the polished sections of these rocks examined under reflected light showed magnetite which was partially oxidized to hematite, and this is in good agreement with our results.

Once the overprint was removed, thermal demagnetization at 400-500°C seemed to produce the most stable results, and these directions were used for further analysis. The directions showed good clustering around normal and reversed poles, and produced a positive reversal test, since the mean for the normal sites [n = 24, D = 342.8, I = 57.5, k = 7.3, α95 = 11.8] was antipodal to the mean for the reversed sites [n = 26, D = 163.9, I = -50.8, k = 6.4, α95 = 12.1].

A number of important sections were sampled. One section spanned about 800 feet of the stratotype Bandera Mesa Member, or upper member of the Devil's Graveyard Formation (Stevens et al., 1984, fig. 10). This

section supplemented the research of Walton (1992), who studied the lower and middle members of this formation. As can be seen in Figure 4, the section was mostly of reversed polarity, with a normal polarity zone between 400 and 550 feet on the measured section, and another normal polarity zone from 650 feet to the Mitchell Mesa Rhyolite at the top.

A second long section spanned over 500 feet of the Colmena Tuff (Wilson, 1978, fig. 6), the primary locality for the early Duchesnean Candelaria local fauna. As shown in Figure 5, the first 100 feet of this section were reversed, followed by normal rocks from 100 to 375 feet, and reversed rocks to the top.

Sixty feet of the section in the Reeves Bonebed (Wilson, 1978, fig. 9), the primary collecting locality of the early Chadronian Little Egypt local fauna, were entirely of normal polarity.

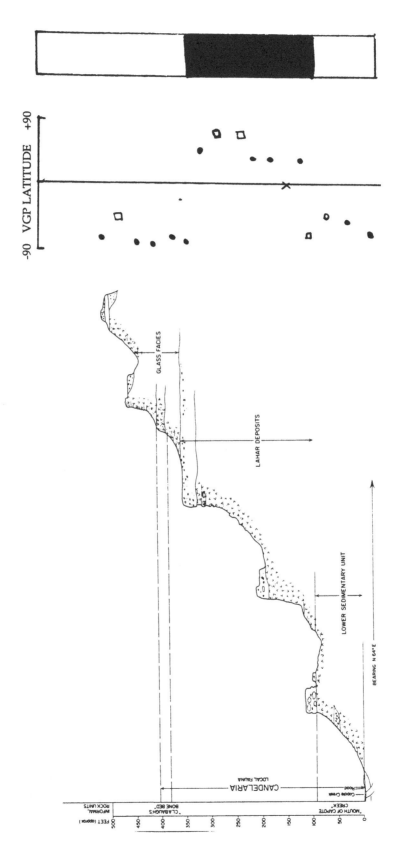

Figure 5. Preliminary magnetic polarity of the Colmena Tuff, near the mouth of Capote Creek, which contains the early Duchesnean Candelaria l.f. Magnetic conventions as in Figure 4.

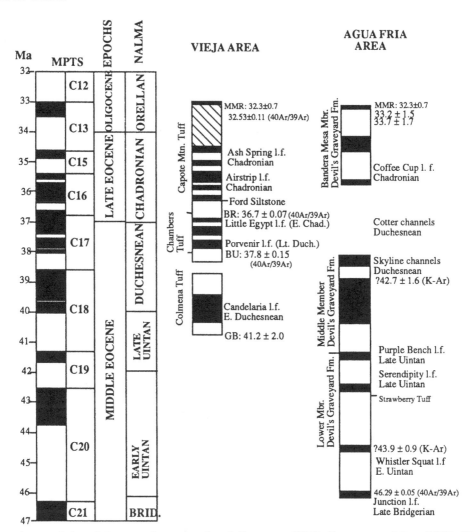

Figure 6. Tentative correlation of the magnetic stratigraphy of Testarmata (1978), Testarmata and Gose (1979), Walton (1992), and this study with the new time scale of Berggren et al. (1995), Cande and Kent (1993) and Prothero and Swisher (1992). Position of important mammalian faunas and radioisotopic dates are shown. All dates are K-Ar (from Henry et al., 1986), unless indicated otherwise. Diagonal cross-hachured pattern denotes section of indeterminate polarity. Abbreviations: BRID. = Bridgerian; BR = Bracks Rhyolite; BU = Buckshot Ignimbrite; GB = Gill Breccia; l.f. = local fauna; MMR = Mitchell Mesa Rhyolite; MPTS = magnetic polarity time scale; NALMA = North American land mammal "age."

The middle Chadronian Airstrip local fauna locality (Wilson, 1978, p. 22), spanning about 80 feet, was of normal polarity for the lower 60 feet; the top of the section was reversed in polarity.

CORRELATIONS

Figure 6 shows summary of the correlations of the sections analyzed in this study, plus those originally reported by Testarmata (1978), Testarmata and Gose (1979) and Walton (1986, 1992). They can be divided into two main areas, the Vieja area north of Candelaria (Wilson, 1978), and the Agua Fria-Green Valley area, just north of the west entrance to Big Bend National Park (Stevens et al., 1984; Wilson, 1986).

Agua Fria-Green Valley area

The Bandera Mesa section (Fig. 4) is constrained by the K-Ar date of 32.3 ± 0.7 Ma on the overlying Mitchell Mesa Rhyolite (Henry et al., 1986). Henry et al. (1994) also obtained $^{40}Ar/^{39}Ar$ dates of 32.53 ± 0.11 and 32.59 ± 0.11 Ma, so the two methods are in good agreement and the age of this unit is well established. Stevens et al. (1984) reported K-Ar dates of 33.2 ± 1.5 on a basalt and 33.7 ± 1.7 on a tuff from the upper part of the section (Fig. 6). The lowest part of the section contains the Chadronian Red Hill local fauna, part of the Coffee Cup fauna (Stevens et al., 1984; Wilson, 1986). Given these constraints, the simplest interpretation is that the lower part of the section probably correlates with C15r, the middle normal zone with C15n, and the upper reversed part of the section with C13r (Fig. 6).

Walton (1992, fig. 3.8) reported magnetic results for the lower and middle members of the Devil's Graveyard Formation. She correlated her section with Chrons C19n-C20r. Since her paper, however, the new magnetic polarity time scale (Cande and Kent, 1993; Berggren et al., 1995) have suggested a reinterpretation. The late Uintan in the Uinta Basin is now thought to begin in Chron C20n, so this moves the late Uintan Serendipity local fauna from C20r (as Walton suggested) to C19r-C19n (in agreement with other late Uintan localities in North America). This revised correlation, in turn, suggests that the long normal zone in the middle of the middle member is probably C18n. The bulk of the reversed rock in the lower member probably represents C19r to C20r, and the late Bridgerian faunas at the base of the section appear to come from rocks of normal magnetic polarity representing Chron C21n. These new interpretations are in agreement with the K-Ar date of 43.9 ± 0.9 Ma on the dated tuff just above the early Uintan Whistler Squat local fauna, with K-Ar dates of 46 Ma on the Alamo Creek Basalt near the base of the section (Henry et al., 1986), and with $^{40}Ar/^{39}Ar$ dates of 46.0 ± 1.0 Ma and 46.29 ± 0.05 Ma on a tuff just below the Alamo Creek basalt (Henry et al., 1994). Only the K-Ar date of 42.7 ± 1.6 from just below the Duchesnean Skyline channels is discordant; it should be reanalyzed by $^{40}Ar/^{39}Ar$ methods. According to Margaret Stevens (personal communication), this 42.7 Ma date came from very tiny flakes of biotite, so it is probably not very reliable. The Duchesnean begins around 39-40 Ma (Prothero and Swisher, 1992), so the 42.7 date seems much too old.

Between the two sections is a 200-foot interval represented by numerous stacked river channels, and little or no continuous section (Stevens et al., 1984). The correlations shown in Figure 6 suggest a 3 million-year gap in the Devil's Graveyard Formation section between the upper C16n rocks at the base of the Bandera Mesa Member section (35.5 Ma) and the C18n rocks just below the Skyline channels (38.5 Ma). These channel sequences produce early Duchesnean mammals from the Skyline channels, and late Duchesnean mammals from the overlying Cotter channels (Wilson, 1986; Wilson and Stevens, 1986). Because this section is discontinuous, the polarity of the individual channel deposits is not very informative, and so they were not sampled.

Vieja area

The thick section of Colmena Tuff (Fig. 5), which produces the Candelaria local fauna, overlies the Gill Breccia (K-Ar dated at 41.2 ± 2.0 Ma), and underlies the Buckshot Ignimbrite ($^{40}Ar/^{39}Ar$ dated at 37.8 ± 0.15, according to Swisher and Prothero, 1990, and Prothero

and Swisher, 1992; 37.54 ± 0.11 Ma by Henry et al., 1994). The Candelaria l.f. was originally interpreted at late Uintan by Wilson (1978). Krishtalka et al. (1987, p. 104) compare it to the late Uintan Randlett and Halfway faunas of the Duchesne River Formation. Prothero and Swisher (1992) and Prothero (this volume, Chapter 1) found that the lowest or Brennan Basin Member of the Duchesne River Formation (containing the Randlett horizon) correlates with Chrons C19r-C18r. Consequently, the long normal zone within the Colmena Tuff appears to correlate with Chron C19n, with parts of C19r and C18r also represented (Fig. 6).

The magnetic results of Testarmata (1978) and Testarmata and Gose (1979) have long been controversial (Prothero et al., 1982, 1983; Prothero, 1985; Wilson, 1986; Swisher and Prothero, 1990; Prothero and Swisher, 1992). As discussed above, these rocks produced a very noisy pattern, with many short zones of normal and reversed polarity. By resampling and reanalyzing these data, and rejecting normal polarity zones less than 3 sites thick (typically prone to unremoved normal overprinting), some clarification is possible (Fig. 6). There are three undoubted short zones of normal polarity in the Chambers Tuff between the Buckshot Ignimbrite and the Bracks Rhyolite ($^{40}Ar/^{39}Ar$ dated at 36.7 ± 0.07 Ma by Prothero and Swisher, 1992, and 36.73 ± 0.13 by Henry et al., 1994). These dates, and the Cande and Kent (1993) or Berggren et al. (1995) magnetic polarity time scale, suggest that these normal intervals are the multiple events of Chron C17n. The late Duchesnean Porvenir local fauna occurs in the lower part of the Chambers Tuff, and is probably correlative with early Chron C17n. The early Chadronian Little Egypt local fauna (based on the Reeves Bonebed section, which was entirely of normal polarity) probably correlates with the upper part of C17n.

Although the magnetic results in the Capote Mountain Tuff were particularly noisy and difficult to interpret, there are some constraints on its age. The section is bracketed by dates on the Bracks Rhyolite at the base, and the Mitchell Mesa Rhyolite at the top. The lower third of the section (between the Bracks Rhyolite and Ford Siltstone) was entirely reversed, and probably corresponds to C16r. Above the Ford Siltstone, there were two longer zones of normal polarity, spanning 80-130 m on Testarmata's (1978, fig. 15) section "D"; these plus other short normals probably correlate to the multiple events of Chron C16n. Above 130 meters, the section is mostly of reversed polarity, and probably represents C15r. The upper third of the Chambers Tuff section in the Adobe Spring area was unstably magnetized, and so is labeled "indeterminate" by Testarmata and Gose (1979), and also here (Fig. 6). This interval is probably equivalent to Chrons C15r to C13n, based on the constraining date (see above) of the normally polarized Mitchell Mesa Rhyolite at the top.

On this basis, the normal-reversed pattern in the Airstrip section (which produced middle Chadronian mammals) probably correlates to the events within Chron C16n, based on similarities with the faunas below Ash A at Flagstaff Rim, Wyoming (see Emry, 1992). The Ash Spring section (which produced slightly younger Chadronian mammals) probably correlates with C13r, based on the presence of late Chadronian taxa such as *Merycoidodon culbertsoni* (see Stevens and Stevens, this volume, Chapter 25).

ACKNOWLEDGMENTS

Prothero acknowledges the support of NSF grant EAR87-08221 and EAR91-17819 during this research. Drs. Jim and Margaret Stevens, Christopher Henry, and John A. Wilson were extremely helpful in guiding me through the stratigraphy and geochronology. Dr. Melissa Winans provided important locality data from the University of Texas archives. Dr. Joseph Kirschvink graciously allowed Prothero to use the California Institute of Technology paleomagnetics laboratory to run the samples. I thank Boyd and Johnnie Chambers and Clegg Fowlkes for access to their property. The field sampling would never have been possible without the hard work, enthusiasm, and good spirits of several undergraduate field crews. In 1990, they included Erin Campbell, Dani Crosby, and Tim Tierney; in 1993, they included Joby Campbell, Walter Lohr, and Erin Wilson. I thank Margaret Stevens, Anne Walton, and John A. Wilson for reviewing the paper.

LITERATURE CITED

Berggren, W. A., D. V. Kent, J. D. Obradovich, and C. C. Swisher, III. 1992. Toward a revised Paleogene geochronology; pp. 29-45 in D. R. Prothero and W. A. Berggren (eds.), Eocene-Oligocene Climatic and Biotic Evolution. Princeton University Press, Princeton, N. J.

Berggren, W. A., D. V. Kent, M.-P. Aubry, C. C. Swisher III, and K. G. Miller. 1995. A revised Paleogene geochronology and chronostratigraphy. SEPM Special Publication 54:129-212.

Cande, S. C., and D. V. Kent. 1993. A new geomagnetic polarity time scale for the late Cretaceous and Cenozoic. Journal of Geophysical Research 97(B10):13917-13951.

Emry, R. 1992. Mammalian range zones in the Chadronian White River Formation at Flagstaff Rim, Wyoming; pp. 106-115 in D. R. Prothero and W. A. Berggren (eds.), Eocene-Oligocene Climatic and Biotic Evolution. Princeton University Press, Princeton, N. J.

Hartl, P., L. Tauxe, and C. Constable. 1993. Early Oligocene geomagnetic field behavior from Deep Sea Drilling Project Site 522. Journal of Geophysical Research 98 (B11):19649-19665.

Henry, C. D., and F. W. McDowell. 1986. Geochronology of magmatism in the Tertiary volcanic field, Trans-Pecos Texas. Bureau of Economic Geology of the University of Texas Guidebook 23:99-122.

Henry, C. D., F. W. McDowell, J. G. Price, and R. C. Smyth. 1986. Compilation of potassium-argon ages of Tertiary igneous rocks, Trans-Pecos Texas. Bureau of Economic Geology of the University of Texas Circular 86-2:1-34.

Henry, C. D., M. Kunk, and W. McIntosh. 1994. $^{40}Ar/^{39}Ar$ chronology and volcanology of silicic volcanism in the Davis Mountains, Trans-Pecos Texas. Geological Society of America Bulletin 106:1359-1376.

Krishtalka, L., R. K. Stucky, R. M. West, M. C. McKenna, C. C. Black, T. M. Bown, M. R. Dawson, D. J. Golz, J. J. Flynn, J. A. Lillegraven, and W. D. Turnbull. 1987. Eocene (Wasatchian through Duchesnean) biochronology of North America; pp. 77-117 in M. O. Woodburne (ed.), Cenozoic Mammals of North America, Geochronology and Biostratigraphy, University of California Press, Berkeley.

LaBrecque, J. L., D. V. Kent, and S. C. Cande. 1977. Revised magnetic polarity time scale for the late Cretaceous and Cenozoic. Geology 5:330-335.

McDowell, F. W. 1979. Potassium-argon dating in the Trans-Pecos Texas volcanic field. Bureau of Economic Geology of the University of Texas Guidebook 19:10-18.

Opdyke, N. D., E. H. Lindsay, N .M. Johnson, and T. Downs. 1977. The paleomagnetism and magnetic polarity stratigraphy of the mammal-bearing section of Anza-Borrego State Park, California. Journal of Quaternary Research 7:316-329.

Prothero, D. R. 1985. Chadronian (early Oligocene) magnetostratigraphy of eastern Wyoming: implications for the Eocene-Oligocene boundary. Journal of Geology 93:555-565.

Prothero, D. R., C. R. Denham, and H. G. Farmer. 1982. Oligocene calibration of the magnetic polarity time scale. Geology, 10:650-653.

Prothero, D. R., C. R. Denham, and H. G. Farmer. 1983. Magnetostratigraphy of the White River Group and its implications for Oligocene geochronology. Palaeogeography, Palaeoclimatology, Palaeoecology 42:151-166.

Prothero, D. R., and C. C. Swisher, III. 1992. Magnetostratigraphy and geochronology of the terrestrial Eocene-Oligocene transition in North America; pp. 46-74 in D. R. Prothero and W. A. Berggren (eds.), Eocene-Oligocene Climatic and Biotic Evolution, Princeton University Press, Princeton, N. J.

Stevens, J. B., M. S. Stevens, and J. A. Wilson. 1984. Devil's Graveyard Formation (new), Eocene and Oligocene age, Trans-Pecos Texas. Texas Memorial Museum Bulletin 32:1-21.

Swisher, C. C., III, and D. R. Prothero. 1990. Single-crystal $^{40}Ar/^{39}Ar$ dating of the Eocene-Oligocene transition in North America. Science 249:760-762.

Testarmata, M. M. 1978. Magnetostratigraphy of the Eocene-Oligocene Vieja Group, Trans-Pecos Texas. M.A. Thesis, University of Texas, Austin. 56 pp.

Testarmata, M. M., and W. A. Gose. 1979. Magnetostratigraphy of the Eocene-Oligocene Vieja Group, Trans-Pecos Texas. Bureau of Economic Geology of the University of Texas Guidebook 19:55-66.

Walton, A. H. 1986. Magnetostratigraphy and the ages of Bridgerian and Uintan faunas in the lower and middle members of the Devil's Graveyard Formation, Trans-Pecos Texas. M.A. Thesis, University of Texas, Austin. 135 pp.

Walton, A. H. 1992. Magnetostratigraphy and the ages of Bridgerian and Uintan faunas in the lower and middle members of the Devil's Graveyard Formation, Trans-

Pecos Texas; pp. 74-87 *in* D. R. Prothero and W. A. Berggren (eds.), Eocene-Oligocene Climatic and Biotic Evolution. Princeton University Press, Princeton, N. J.

Wilson, J. A. 1978. Stratigraphic occurrence and correlation of early Tertiary vertebrate faunas, Trans-Pecos Texas, Part 1: Vieja area. Texas Memorial Museum Bulletin 25:1-42.

Wilson, J. A. 1980. Geochronology of the Trans-Pecos volcanic field. New Mexico Geological Society Guidebook, 31:205-211.

Wilson, J. A. 1984. Vertebrate fossil faunas 49 to 36 million years ago and additions to the species of *Leptoreodon* found in Texas. Journal of Vertebrate Paleontology 4:199-207.

Wilson, J. A. 1986. Stratigraphic occurrence and correlation of early Tertiary vertebrate faunas, Trans-Pecos Texas: Agua Fria-Green Valley areas. Journal of Vertebrate Paleontology 6:350-373.

Wilson, J. A., and M. S. Stevens. 1986. Fossil vertebrates from the latest Eocene Skyline Channels, Trans-Pecos Texas. Contributions to Geology of the University of Wyoming, Special Paper 3:221-235.

Wilson, J. A., P. C. Twiss, R. K. DeFord, and S. E. Clabaugh. 1968. Stratigraphic succession, potassium-argon dates, and vertebrate faunas, Rim Rock Country, Trans-Pecos Texas. American Journal of Science 266:590-604.

10. Magnetic Stratigraphy of the Duchesnean Part of the Galisteo Formation, New Mexico

DONALD R. PROTHERO AND SPENCER G. LUCAS

ABSTRACT

Two sections of the Galisteo Formation in central New Mexico, which contain the Duchesnean Tonque local fauna, were sampled for magnetic polarity stratigraphy. The section at Arroyo del Tuerto is of mixed polarity, and probably correlates with Chrons C17r1 to C17n3 (37.5-38.0 Ma), based on similarities with other well-dated Duchesnean faunas. The section of the type Galisteo Formation in the Cerrillos area is mostly of reversed polarity, and probably correlates with Chron C17r (38.0-38.5 Ma) based on comparisons of its faunas to other late Duchesnean faunas.

INTRODUCTION

In the southwestern United States, the sharpest pulse of the Laramide orogeny began during the middle Eocene. In northern New Mexico and Colorado, the northeastwardly moving Colorado Plateau impinged upon the basement buttresses of the Rocky Mountain foreland, producing significant tectonism. This formed a series of en-echelon, asymmetric downwarps, and tilted and subsided blocks that Chapin and Cather (1981) referred to as Echo Park-type basins, and Dickinson et al. (1988) referred to as axial basins.

One of these basins was the Galisteo-El Rito basin of north-central New Mexico. Its principal basin fill, the Galisteo Formation, is as much as 1300 m thick and consists of fluvial sandstone, mudstone, and conglomerate (Gorham and Ingersoll, 1979) (Fig. 1). Two Eocene fossil mammal local faunas (l.f.) are known from the Galisteo Formation. The lower part of the unit produces the Cerrillos local fauna of Wasatchian age, and the upper part produces the Tonque local fauna of Duchesnean age (Fig. 1; see also Lucas, 1982). The Tonque local fauna is one of the more diverse and better known Duchesnean mammal faunas from western North America (Lucas, 1992). Here, we establish the magnetic-polarity stratigraphy of the Duchesnean interval of the Galisteo Formation to correlate these strata and their fossil mammals to other Duchesnean strata outside of New Mexico.

MAGNETIC STRATIGRAPHY

In the spring of 1988 and the summer of 1990, magnetic sampling was conducted in the two main sections of the Galisteo Formation which contain Duchesnean mammals. In the Cerrillos area (Fig. 1), samples of the type Galisteo were taken, starting with Lucas' (1982, fig. 4) unit 73 (magnetic site 241, Fig. 4) and concluding with the base of unit 82 (magnetic site 259), for a total of 200 m of section. Fossil localities C5 and C6 of the Duchesnean Tonque l.f. are located in units 74 and 82, respectively. Sampling was also conducted in the upper Galisteo exposures in the Arroyo del Tuerto (Fig. 5), beginning with the top of Lucas's (1982, fig. 7) unit 1 (magnetic site 230) and concluding with the base of the Espinaso Formation (magnetic site 240), a total of 70 m of section. Note that this is the base of the Espinaso Formation as defined originally by Stearns (1943) and used by Lucas (1982), which is the most readily mappable contact between two lithostratigraphic units. Disbrow and Stoll (1957), Kautz et al. (1981), and Smith et al. (1991) placed the Galisteo-Espinaso contact at the lowest occurrence of volcanic detritus in the Arroyo del Tuerto section, which is at the base of unit 3 of Lucas (1982, fig. 7), a bentonitic mudstone, but this is not so readily mappable as a contact between the two formations.

Horizontally oriented blocks of rock were sampled with simple hand tools. A minimum of three samples was taken at each site. In the type Galisteo section, sites were spaced approximately 10 m apart stratigraphically; in the shorter Arroyo del Tuerto section, sites were spaced about 5-6 m apart. Samples were trimmed on a band saw equipped with a tungsten carbide blade, and analyzed at the California Institute of Technology paleomagnetic laboratory. Poorly indurated specimens were hardened with sodium-silicate solution to preserve them during thermal demagnetization.

After measurement of the NRM (natural remanent magnetization), a suite of samples was subjected to AF (alternating field) demagnetization. Many samples, such as 240A (Fig. 2A), showed a steady decline in

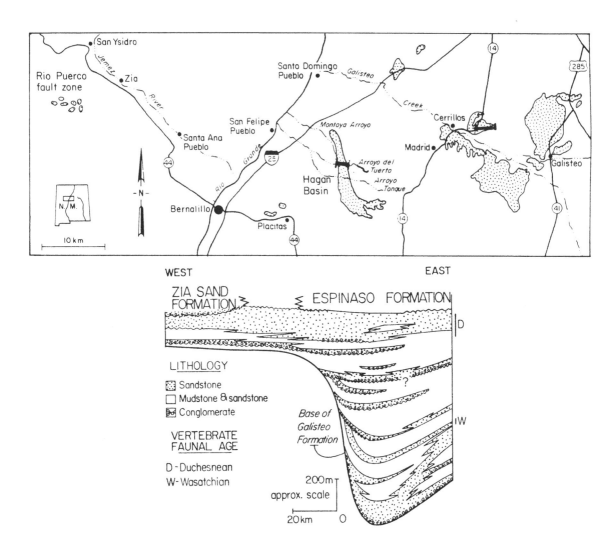

Figure 1. Index map showing location of Cerrillos and Arroyo del Tuerto sections discussed in text (from Lucas, 1982, fig. 1), and diagrammatic west-east cross-section shown stratigraphic relations within Galisteo Formation (from Lucas and Williamson, 1993).

intensity during AF demagnetization. This suggests that a low-coercivity mineral such as magnetite is the main carrier of remanence. Under thermal demagnetization (Fig. 2B), most samples had lost nearly all their remanence by 600°C, which is above the Curie point of magnetite. However, some samples (Fig. 2C) still had significant remanence left by 600°C, suggesting that significant hematite was present. A few reddish samples (Fig. 2D) showed no drop in intensity, even at 600°C; most of their remanence is apparently carried in hematite.

Analyses of IRM (isothermal remanent magnetization) acquisition further confirmed that both magnetite and hematite were present in many samples. Some rocks (Fig. 3A) reached saturation IRM values at 100-300 mT (millitesla); the remanence in these rocks is carried mostly by magnetite. Other specimens (Fig. 3B) showed no evidence of IRM saturation, even at fields of 1300 mT; these samples clearly contained much hematite. A modified Lowrie-Fuller ARM (anhysteretic remanent magnetization) test (e.g., Johnson et al., 1975) was also conducted during the IRM analysis (see Pluhar et al., 1991, for details). This test compares the resistance of AF demagnetization of both an IRM acquired in a 100 mT peak field, and an ARM gained in a 100 mT oscillating field. In almost all samples, the ARM (black squares) demagnetizes at higher peak fields than does the IRM (open squares), indicating that the remanence is carried by single-domain or pseudo-single-domain grains.

Finally, thin sections of representative samples were examined under reflected light. In most slides, euhedral magnetite grains with oxidized rims of hematite could be seen. In addition, several samples had a strong red

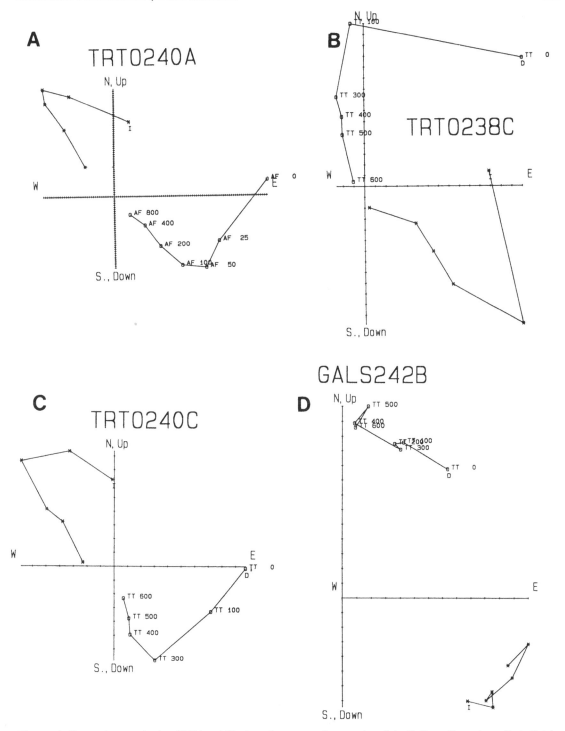

Figure 2. Vector demagnetization ("Zijderveld") plots of representative samples of the Galisteo Formation. Each division = 10^{-6} emu. Horizontal component shown with open circles; vertical component shown with asterisks; NRM direction of vertical component indicated by "I". A) AF demagnetization of sample 240A from Arroyo del Tuerto, showing the overprint removed in fields from NRM ("AF 0") to 50 Gauss, then a stable reversed component at higher fields. The rapid drop in intensity suggests that a low-coercivity mineral such as magnetite is the major source of the remanence. B) Thermal demagnetization plot, showing thermal steps ("TT") in degrees Centigrade. After removal of overprints, a stable normal component was revealed at temperatures between 100-600°C. C) Thermal demagnetization of a reversed sample, with a stable reversed component apparent between 300-600°C. D) Reddish sample with much chemical overprinting from hematite. Note that there was no decline in intensity, even at 600°C, which is close the Curie point of hematite.

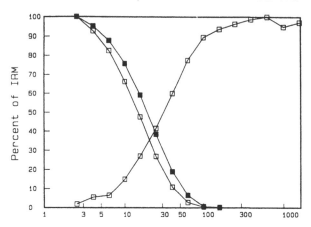

GALISTEO FM., ARROYO DEL TUERTO

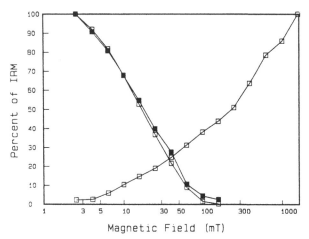

RED SS., GALISTEO FM., CERILLOS

Magnetic Field (mT)

Figure 3. IRM acquisition and Lowrie-Fuller tests for representative samples (see Pluhar et al., 1991, for details). Open boxes are IRM intensities, and solid boxes are ARM values. The IRM acquisition curves (ascending curves on right) show saturation in some samples (top), suggesting that magnetite is the primary carrier of remanence. In red sandstones (bottom), there is no IRM saturation, indicating that hematite is dominant. Both samples showed ARM demagnetization curves which were more resistant to AF demagnetization than the IRM curves (descending curves on left), suggesting that a fine-grained, single-domain or pseudo-single domain mineral is the carrier of the remance.

hematite or goethite stain, which may account for the behavior in some samples (e.g, Fig. 2D).

Based on these analyses, it is clear that a complex magnetic signal is present in the Galisteo samples. All remaining samples were subjected to stepwise thermal demagnetization at 100°C steps. This procedure should remove overprinting by a high-coercivity iron hydroxide (such as goethite), which is not affected by AF demagnetization. As can be seen from Figure 2B and 2C, an

overprinted component was removed by 100-300°C, and a stable characteristic reversed component was apparent at temperatures from 300-600°C. The directions of the component between 300 and 500°C were used for further statistical analyses.

The stable sample directions were then clustered by site, and statistically analyzed by the methods of Fisher (1953; see Butler, 1992). Class I sites of Opdyke et al. (1977) showed a clustering that differed significantly from random at the 95% confidence level. In Class II sites, one sample was lost or crumbled, but the remaining samples gave a clear polarity indication. In Class III sites of Opdyke et al. (1977), two samples showed a clear polarity preference, but the third sample was divergent because of insufficient removal of overprinting. A few samples were considered indeterminate if their magnetic signature was unstable, or their direction uninterpretable.

The beds of the Cerrillos area range in dip from 30° to vertical, so it was possible to conduct a modified fold test for stability. Before the dip correction, the cleaned mean declination (D) was $1.4°$, and inclination (I) was $338.9°$; the precision parameter (k) was 1.2 and the ellipse of confidence (α_{95}) was $89.9°$. After dip correction, the directions were much less scattered [D = 350.9, I = 44.6, k = 31.8, α_{95} = 16.6], showing that the remanence was acquired before tilting. In addition, the cleaned but uncorrected normal direction above is not parallel to the modern normal magnetic field, nor is the cleaned but uncorrected reversed mean [D = 178.5, I = 50.9, k = 2.7, α_{95} = 18.4]. This suggests that the magnetization was acquired prior to tilting.

Reversal tests for stability were conducted on the samples from both areas. The mean directions of all cleaned normal sites in the Cerrillos area shown above were antipodal to the mean direction of all reversed samples [D = 159.3, I = -36.8, k = 11.5, α_{95} = 12.8], indicating that the magnetization was probably acquired during deposition of the beds. The mean of normal samples in Arroyo del Tuerto [D = 325.3, I = 38.8, k = 13.3, α_{95} = 19.1] is also antipodal to the mean of reversed samples [D = 189.0, I = -48.2, k = 9.1, α_{95} = 19.4].

CORRELATION

The magnetic stratigraphy of the two sections is shown in Figures 4 and 5. The Arroyo del Tuerto section (Fig. 5) was mostly of normal polarity, except for reversed sites (upper unit 2) just below the brontothere quarry (locality T4) and in the uppermost site in the Espinaso Formation (unit 7). The Cerrillos section (Fig. 4) is mostly of reversed polarity, except for the two lowest sites just below locality C5 near Ambush Rock (units 73 and 74), and the top three sites (units 81 and 82), which include localities C6 and C8.

Without radioisotopic dates, these magnetic patterns are not diagnostic by themselves. Kautz et al. (1981,

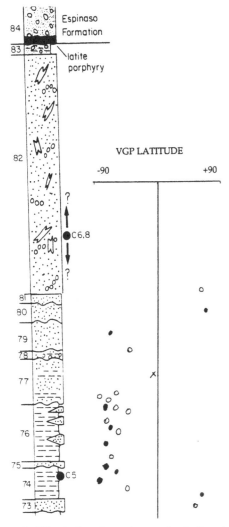

Figure 4. Lithostratigraphy and magnetic stratigraphy of the Duchesnean section in the Cerrillos area, based on Lucas (1982, fig. 4). Solid circles are Class I (significant) sites of Opdyke et al. (1977); open squares are Class II (only two samples) sites; open circles are Class III (one sample divergent) sites. Vertebrate fossil localities C5 and C6 and C8 are shown.

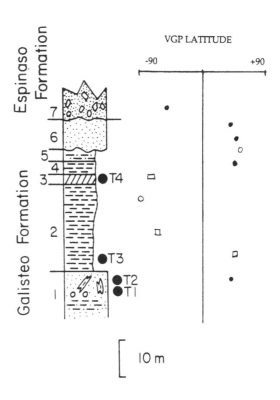

Figure 5. Lithostratigraphy and magnetic stratigraphy of the Arroyo del Tuerto section, after Lucas (1982, fig. 7). All conventions as in Figure 4. Vertebrate fossil localities T1–T4 are shown.

table 1) reported a K-Ar age of 34.3 ± 0.8 Ma for a calcic-quartz-latite clast from conglomeratic sandstone 10 m above the base of the Espinaso Formation in Arroyo del Tuerto, stratigraphically above our highest magnetic sample. This age seems much too young (see following discussion) unless there is a significant unconformity in the Espinaso Formation between our highest magnetic site and the radiometrically dated horizon. Indeed, Kautz et al. (1981, table 1) also reported a K-Ar age of 34.6 ± 0.7 Ma for a calcic-quartz-latite clast in a debris flow unit 200 m above the base of the Espinaso Formation. This supports the conclusion that their stratigraphically lower numerical age is anomalously young.

By comparing the Tonque l.f. with other Duchesnean faunas, it it possible to narrow down the possible correlations. Figure 6 shows the current interpretation of the age of several important Duchesnean faunas. In California, the early Duchesnean Pearson Ranch l.f. occurs in Chron C18n, and the late Duchesnean Simi Valley Landfill l.f. occurs in Chron C17r (Kelly, 1990; Kelly et al., 1991; Kelly and Whistler, 1994; Prothero et al., this volume, Chapter 8). In Trans-Pecos Texas, the late Duchesnean Skyline Channels l.f. probably occurs in Chron C17n3, and the late Duchesnean Porvenir l.f. occurs in Chron C17n2 (Prothero, this volume, Chapter 9). The Porvenir l.f. was recovered just above the Buckshot Ignimbrite, which has an $^{40}Ar/^{39}Ar$ date of 37.8 ± 0.15 Ma (Prothero and Swisher, 1992).

The Lapoint Ash in the type Duchesne River Formation of Utah yields a $^{40}Ar/^{39}Ar$ date of 39.74 ± 0.74 Ma (Prothero and Swisher, 1992). Much of the sparse Lapoint l.f. has no locality information, and so occurs an unknown distance above this dated ash (Emry, 1981). The best known locality, the Carnegie Museum *Teleodus* (now *Duchesneodus*—Lucas and Schoch, 1982, 1989) quarry, occurs approximately 50 m (150 feet) above it (Kay, 1934, Plate XLVI). No magnetic stratigraphy was possible in this part of the Duchesne

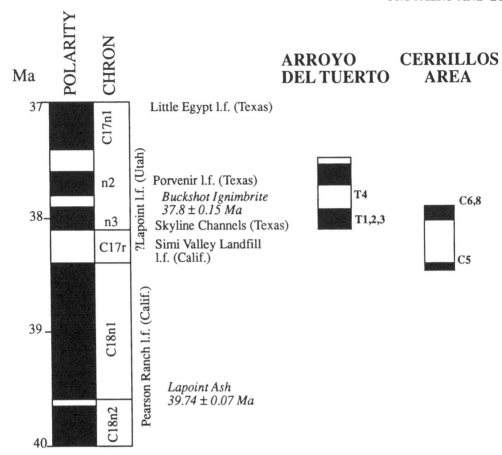

Figure 6. Suggested correlation of the magnetic sections in the Galisteo Formation. Time scale after Berggren et al. (1995). Other correlations and age assignments discussed in text.

River Formation (Prothero and Swisher, 1992; Prothero, this volume, Chapter 1). Some authors (Wilson, 1986; Kelly, 1990; Lucas, 1992) consider the Lapoint l.f. to be late Duchesnean, comparable to the Porvenir l.f. in age, and younger than the Pearson Ranch l.f.

Lucas (1992) suggested that the Tonque l.f. was also late Duchesnean, and most similar to the Porvenir and Lapoint local faunas. If this is true, then the long reversed polarity interval in the Cerrillos section is most likely Chron C17r, the only relatively long reversed interval in the late Duchesnean. The section probably spans the interval between C17n3 and C18n1, or about 38.0-38.5 Ma (Fig. 6). The two long normal polarity intervals of the Arroyo del Tuerto section, on the other hand, suggest a correlation with Chrons C17n2 and C17n3. If the section were younger, it would make the top of the section C17n1, which is correlatable with the Chadronian Little Egypt l.f. in Texas (Prothero, this volume, Chapter 9), a possibility we reject. If the section were older, then the lower normal interval would be C18n1, which is early Duchesnean. Thus, a correlation with Chrons C17r1 to C17n3, or about 37.5-38.0 Ma, appears to be the most reasonable interpretation.

ACKNOWLEDGMENTS

Prothero acknowledges the support of NSF grant EAR87-08221 during this research. Dr. Joseph Kirschvink graciously allowed the use of the California Institute of Technology paleomagnetic laboratory to run the samples. The field sampling would never have been possible without the hard work, enthusiasm, and good spirits of several undergraduate field crews. In 1988, they included Jeff Amato, Greg Brown, and Jennifer Chean; in 1990, they included Erin Campbell, Dani Crosby, and Tim Tierney. Ray Ingersoll and Tom Kelly provided thoughtful reviews of the manuscript.

LITERATURE CITED

Berggren, W. A., D. V. Kent, M.-P. Aubry, C. C. Swisher III, and K. G. Miller. 1995. A revised Paleogene geochronology and chronostratigraphy. SEPM Special Publication 54:129-212.

Butler, R. F. 1992. Paleomagnetism. Blackwell, London.

Chapin, C. E., and S. M. Cather. 1981. Eocene tectonics and sedimentation in the Colorado Plateau-Rocky Mountain area. Arizona Geological Society Digest 14:173-198.

Dickinson, W. R., M. A. Klute, M. J. Hayes, S. U. Janecke, E. R. Lundin, M. A. McKittrick, and M. D. Olivares. 1988. Paleogeographic and plate tectonic setting of the Laramide sedimentary basins in the central Rocky

Mountain region. Geological Society of America Bulletin 100:1023-1039.

Disbrow, A. E., and W. C. Stoll. 1957. Geology of the Cerrillos area, Santa Fe County, New Mexico. New Mexico Bureau of Mines and Mineral Resources Bulletin 48:1-73.

Emry, R. J. 1981. Additions to the mammalian fauna of the type Duchesnean, with comments on the status of the Duchesnean "age." Journal of Paleontology 55:563-570.

Fisher, R. A. 1953. Dispersion on a sphere. Proceedings of the Royal Society A217:295-305.

Gorham, T. W. and R. V. Ingersoll. 1989. Evolution of the Eocene Galisteo basin, north-central New Mexico. New Mexico Geological Society Guidebook 30:219-224.

Johnson, H. P., W. Lowrie, and D. V. Kent. 1975. Stability of anhysteretic remanent magnetization in fine and coarse magnetite and maghemite particles. Geophysical Journal of the Royal Astronomical Society 41:1-10.

Kautz, P. F., R. V. Ingersoll, W. S. Baldridge, P. E. Damon, and M. Shafiqullah. 1981. Geology of the Espinaso Formation (Oligocene), north-central New Mexico. Geological Society of America Bulletin 92:Part I: 980-983; Part II: 2318-2400.

Kay, J. L. 1934. The Tertiary formations of the Uinta Basin, Utah. Annals of the Carnegie Museum 23:357-372.

Kelly, T. S. 1990. Biostratigraphy of Uintan and Duchesnean land mammal assemblages from the middle member of the Sespe Formation, Simi Valley, California. Los Angeles County Museum Contributions in Science 419:1-42.

Kelly, T. S., and D. P. Whistler, 1994. Additional Uintan and Duchesnean (middle and late Eocene) mammals from the Sespe Formation, Simi Valley, California. Contributions to Science of the Los Angeles County Museum 439:1-29.

Kelly, T. S., E. B. Lander, D. P. Whistler, M. A. Roeder, and R. E. Reynolds. 1991. Preliminary report on a paleontologic investigation of the lower and middle members, Sespe Formation, Simi Valley Landfill, Ventura County, California. PaleoBios 13(50):1-13.

Lucas, S. G. 1982. Vertebrate paleontology, stratigraphy, and biostratigraphy of Eocene Galisteo Formation, north-central New Mexico. New Mexico Bureau of Mines and Mineral Resources Circular 186:1-34.

Lucas, S. G. 1992. Redefinition of the Duchesnean land

mammal "age," late Eocene of western North America; pp. 88-105 in D. R. Prothero and W. A. Berggren (eds.), Eocene-Oligocene Climatic and Biotic Evolution, Princeton University Press, Princeton, N. J.

Lucas, S. G., and R. M. Schoch. 1982. Duchesneodus, a new name for some titanotheres (Perissodactyla, Brontotheriidae) from the late Eocene of western North America. Journal of Paleontology 56:1018-1023.

Lucas, S. G., and R. M. Schoch. 1989. Taxonomy of Duchesneodus (Brontotheriidae) from the late Eocene of North America: pp. 490-503 in D. R. Prothero and R. M. Schoch (eds.), The Evolution of Perissodactyls, Oxford University Press, New York.

Lucas, S. G., and T. E. Williamson. 1993. Eocene vertebrates and late Laramide stratigraphy of New Mexico. New Mexico Museum of Natural History and Science Bulletin 2: 145-158.

Opdyke, N. D., E. H. Lindsay, N. M. Johnson, and T. Downs. 1977. The paleomagnetism and magnetic polarity stratigraphy of the mammal-bearing section of Anza-Borrego State Park, California. Journal of Quaternary Research 7:316-329.

Pluhar, C. J., J. L. Kirschvink, and R. W. Adams. 1991. Magnetostratigraphy and clockwise rotation of the Plio-Pleistocene Mojave River Formation, central Mojave Desert, California. San Bernardino County Museum Association Quarterly 38(2):31-42.

Prothero, D. R. and C. C. Swisher III. 1992. Magnetostratigraphy and geochronology of the terrestrial Eocene-Oligocene transition in North America; pp. 46-74 in D. R. Prothero and W. A. Berggren (eds.), Eocene-Oligocene Climatic and Biotic Evolution, Princeton University Press, Princeton, N. J.

Smith, G. A., D. Larsen, S. S. Harlan, W. C. McIntosh, D. W. Erskine, and S. Taylor. 1991. A tale of two volcaniclastic aprons: field guide to the sedimentology and physical volcanology of the Oligocene Espinaso Formation and Miocene Peralta Tuff, north-central New Mexico. New Mexico Bureau of Mines and Mineral Resources Bulletin 137:87-103.

Stearns, C. E. 1943. The Galisteo Formation of north-central New Mexico. Journal of Geology 51:301-319.

Wilson, J. A. 1986. Stratigraphic occurrence and correlation of early Tertiary vertebrate faunas, Trans-Pecos Texas: Agua Fria-Green Valley areas. Journal of Vertebrate Paleontology 6:350-373.

11. Stratigraphy and Vertebrate Faunas of the Bridgerian-Duchesnean Clarno Formation, North-Central Oregon

C. BRUCE HANSON

ABSTRACT

Within its type area, the Clarno Formation (Eocene, north-central Oregon) includes Oregon's two oldest Tertiary terrestrial mammal localities, numerous fossil plant localities, and many radiometrically dated rock units. Detailed lithostratigraphic mapping has now clarified the complex geologic history and stratigraphic framework which relate these localities and dates to each other and to the geologic record of adjacent areas. Of the five allostratigraphic units recognized within the local Clarno section, the second oldest includes the Nut Beds locality with vertebrates characteristic of the Bridgerian land mammal "age," and the youngest includes the Hancock Quarry locality, the source of a large assemblage of Duchesnean vertebrates. Fossil plants are also preserved in both of these localities and in the middle unit. Those dates which are consistent with the local superpositional sequence also accord with dates on biostratigraphically correlated faunal assemblages elsewhere in North America

The Nut Beds local fauna, preserved within a lacustrine delta complex, includes species of *Patriofelis*, *Orohippus*, *Hyrachyus*, and *Telmatherium* which indicate a late Bridgerian North American Land Mammal "Age" (NALMA). This accords with an ^{40}Ar/^{39}Ar date of 48.32 Ma, but not with other dates near 44 Ma. A remarkably diverse floral record documents a paratropical to tropical rain forest environment, but, at about 14 million years older than the presently recognized Eocene/Oligocene boundary, it provides little evidence bearing on the rate of climatic cooling near the end of the Eocene.

Hancock Quarry has yielded more than 2000 specimens of large mammals. The fossil-bearing strata lie about 5 to 10 m concordantly below a widespread ignimbrite at the base of member A of the John Day Formation A recently published ^{40}Ar/^{39}Ar date of 39.2 Ma on the ignimbrite is probably more accurate than older K-Ar and fission track dates near 37 Ma. The assemblage represents 14 vertebrate taxa (11 of mammals) consistent with assignment to the Duchesnean NALMA. Two genera (*Epihippus* and *Diplobunops*) are late records of North American lineages. Many of the rest of the mammals appear related to subcoastal Asian taxa (*Procadurcodon* is known only from Hancock Quarry and eastern Asia), and some taxa (*Hemipsalodon*, Nimravidae, *Haplohippus*, *Protitanops*, *Teletaceras* and *Heptacodon*) also appear in Duchesnean or later North American localities. *Protapirus* occurs in Uintan through Whitneyan North American localities and in the late Eocene of Asia and Europe.

INTRODUCTION

In contrast with its rich post-Eocene record, the Pacific Northwest has yielded a sparse record of older terrestrial vertebrate faunas. To date, the only documented Eocene mammalian faunal assemblages of the region, and many fossil plant localities, occur within the small area known as Hancock Canyon, near the John Day River in north-central Oregon (Fig. 1). Here, the Clarno Formation (early? to late Eocene) with its scattered, localized fossil sites, directly underlies the John Day Formation (late Eocene to mid-Miocene) which has yielded vertebrate fossils from hundreds of localities in its post-Eocene portion. The two Clarno vertebrate localities extend the mammalian faunal record of the region earlier by about 10 and 20 million years. The geographic concentration of fossil plant and vertebrate localities, interspersed with datable volcanic rocks, provides an unusual opportunity to interrelate and cross-check independently derived chronologies of Eocene floral, faunal, climatic, and tectonic evolution of the Pacific Northwest.

Hancock Canyon lies about 130 km east of the present Cascade Range crest in Wheeler County, Oregon. Its western rim follows the northwest flank of the Blue Mountains Anticline, a regional post-Clarno uplift that extends into northeastern Oregon (Walker and Robinson, 1990). Centered 3 km east of the former site of Clarno's Ferry, the canyon also encompasses most of the type area of the Clarno Formation, described by J. C. Merriam (1901a, 1901b) after his pioneering geologic and paleontologic exploration of the region. Because of its importance to the earth sciences and the scenic value of its cliffs and pinnacles, most of the Hancock Canyon vicinity is now included in the John Day Fossil Beds National Monument. The term

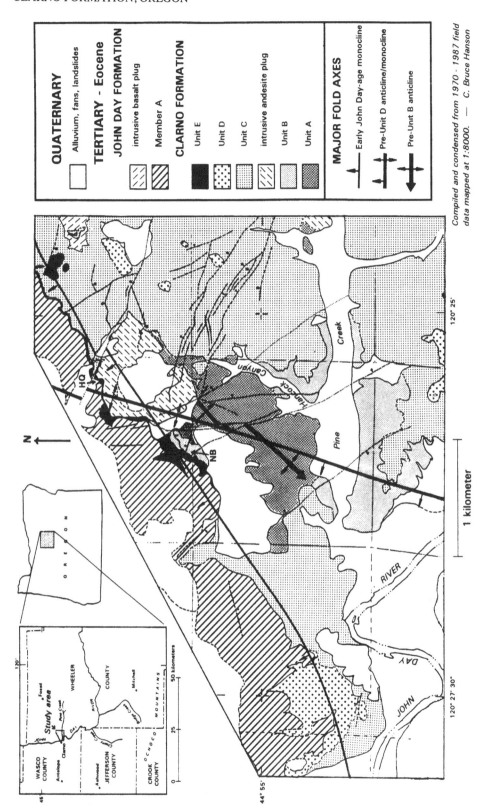

Figure 1. Geologic sketch map of Hancock Canyon and vicinity. HQ = Hancock Quarry, UCMP loc. no's V75203, V75204; NB = Nut Beds locality, UCMP loc. no's V4203, V65513, V78127, PA33. Geology in the vicinity of the localities is complicated by the near convergence of three successive fold axes, two fault zones, an intrusive plug, and a late Clarno paleovalley.

"Hancock Canyon" is used here in its broad sense, encompassing the triangular area delineated by prominent divides to the east (separating it from Indian Canyon) and northwest (a ridge formed by the resistant basal John Day Formation ignimbrite) and by Pine Creek and the John Day River to the south. This area includes the drainage areas of several small tributaries to Pine Creek and the John Day River as well as the three-km-long main ephemeral drainage which enters Pine Creek in the south-central part of Sec. 34, T7S R19E.

Deposited on a low-lying marginal arc with scattered volcanic centers (White and Robinson, 1992), the Clarno Formation is a stratigraphically and structurally complex sequence of predominantly calc-alkaline volcanic, intrusive, and reworked volcanogenic rocks. Metamorphic and reworked sedimentary clasts also appear in the older parts of the formation (Noblett, 1981). Andesitic to basaltic lava flows and volcanic debris flow (lahar) deposits predominate and are interspersed with generally tuffaceous fluvio-lacustrine sediments. Although the formation is exposed over a large area of north-central Oregon, many of the included rock units extend no more than a few kilometers and exhibit rapid lateral facies changes. Intraformational unconformities interrupt most local sequences (Swanson and Robinson, 1968; Oles and Enlows, 1971; Noblett, 1981; Walker and Robinson, 1990; White and Robinson, 1992). These characteristics reflect the moderate to high relief maintained in the region during most of Clarno time, a result of the proximity of eruptive centers as well as ongoing tectonism.

The intent of this paper is to summarize those aspects of the physical stratigraphy of the type area which relate to interpretation of the biostratigraphic and chronometric data now available, and to outline the taxonomic composition and biochronologic implications of the vertebrate faunas.

ABBREVIATIONS

FMNH, Field Museum of Natural History, Chicago; LACM, Los Angeles County Museum of Natural History; OMSI, Oregon Museum of Science and Industry, Portland; TMM, Texas Memorial Museum, Austin; UCMP, University of California Museum of Paleontology, Berkeley; UO, University of Oregon, Condon Museum of Natural History, Eugene; UW, University of Wyoming Museum, Laramie: UWBM, University of Washington, Burke Memorial Museum, Seattle

PHYSICAL STRATIGRAPHY

The existing large-scale geologic maps including the Clarno Formation of the Hancock Canyon area (Taylor, 1960; Swanson, 1969; Robinson, 1975), which were available when this study began, did not clarify the stratigraphic relationships among the many fossil localities and dated rock units in the immediate vicinity. I therefore mapped details of bedrock lithology on an airphoto base at a scale of 1:8000 in the Hancock Canyon–lower Pine Creek area, and on 1:24,000 topographic maps in key areas nearby. Preliminary conclusions, now recognized as oversimplified, were presented by Hanson (1979).

The more complete stratigraphic sections available in areas somewhat removed from the principal fossil localities have provided a basis for reconstruction of a spatial and temporal context which frames details of significant but localized geologic events (Fig. 2). Only within this context can the abbreviated and isolated sections which include the two vertebrate-bearing localities be tied into the regional picture.

Stratigraphic units

Several unconformities have been found to interrupt the stratigraphic record of the Clarno Formation in the Hancock Canyon area. I have used these unconformities as boundaries to define five local subunits of the formation. Each of these is lithologically varied, and some rock types occur in more than one subunit. Such unconformity-bounded bodies of rock are termed *allostratigraphic units* by the North American Stratigraphic Code (North American Commission on Stratigraphic Nomenclature, 1983). Although the Code permits formal naming of such units, formalization in this case must await broader regional study of the extent and utility of the units. In the interim, I assign to the subunits informal letter designations intended to apply only in the vicinity of Hancock Canyon, lower Pine Creek, and the adjacent reach of the John Day River Canyon. In this area, sedimentary facies correspond in general to G. A. Smith's (1991) syneruptive and intereruptive sequences on alluvial aprons adjacent to volcanic centers. In addition to the volcanic controls on sedimentation inherent to this model, tectonically induced terrain features of local to regional scale also control the character and geometry of Clarno Formation depositional subunits.

Within the general vicinity of Hancock Canyon and lower Pine Creek, the allostratigraphic units and their boundaries are characterized as follows:

(*Top of Clarno Formation concordant with basal airfall tuff and ignimbrite of John Day Formation*)

Unit E—Localized intereruptive colluvial and fluvial valley-fill sediments, mostly conglomerates, breccias, and mudstones ranging in color from drab gray and yellow-brown to brick red and dark to light green. Reworked volcanogenic clasts with little indication of contemporaneous volcanism and no debris flows. Many paleosols within unit. Channeled into Unit D flows and older Clarno terraces beyond flows. Maximum thickness ca. 30 m.

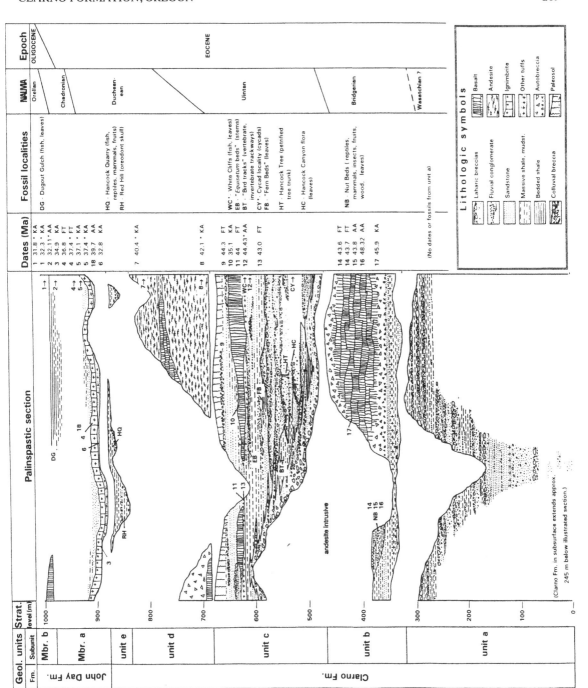

Figure 2. Physical stratigraphic framework of the type Clarno Formation and stratigraphic distribution of major fossil localities and dated samples. Date and locality references with arrow are outside Hancock Canyon area. References for dates in Table 1. Fossil locality partial references: DG - Dugout Gulch (OMSI loc's 240, 241; Bridge Creek flora equivalent; Manchester and Meyer, 1987); HQ - Hancock Quarry (="Clarno Mammal Quarry," UCMP loc's V75203, 75204; McKenna in Evernden et al., 1964; this paper); RH - Red Hill (Hypsodont oreodont skull; M.C. McKenna, pers. comm., 1989, 1993); WC - White Cliffs (Manchester, in press); EB - "*Equisetum* Beds" (unpublished); BT - "Bird tracks" (JDNM loc. 66; unpublished); FB - "Fern Beds" (UCMP loc. 832; = "Leaf beds" dated by Vance, 1988); HC - Hancock Canyon flora (UCMP loc. 842; Knowlton, 1902; ?= "uppermost assemblage of the Clarno Formation near Iron Mountain"; Wolfe, 1994); HT - Hancock Tree (Retallack, 1991); CY - Cycad locality (UCMP loc. PA653; Chaney, 1937. Stratigraphic level within unit C uncertain); NB - Nut Beds (UCMP loc's. V4203, V65513, V78127, PA33; UWBM Loc. C0144; Scott, 1954; Chandler, 1964; Manchester, 1981, 1988, 1994; this paper; ="Upper Clarno flora" of Wolfe and Hopkins, 1967, and Wolfe, 1971).

(High-relief disconformity where underlain by Unit D; angular unconformity, subplanar to channeled, often with thin, brightly colored paleosol where underlain by hills and terraces developed on older Clarno rocks)

Unit D—Andesitic to basaltic lava flows and flow breccias with local maroon to dark red interflow sediments. Lacks lahar deposits. Undeformed through remainder of Clarno time. Maximum thickness greater than 170 m south of Pine Creek, but originally thin or absent within Hancock Canyon.

(Nearly planar angular unconformity marked by deep paleosol)

Unit C—Diverse syneruptive to intereruptive sequence dominated by lahar and associated hyperconcentrated to normal stream channel deposits with lesser lacustrine and paludal sediments (generally tuffaceous). Volcanic facies in lower half of unit intertongue with transgressive basal colluvial facies of large-boulder breccias to silty mudstones. Unit also includes basaltic and andesitic lava flows and an ash-flow tuff ("Rajneesh tuff" of Vance, 1988; "Muddy Ranch tuff" of Bestland et al., 1994a). Gently folded, many minor WNW-trending strike-slip faults. Includes palisade cliff exposures figured by Merriam (1901b). 0 to 220 m.

(High-relief angular unconformity on faulted sediments, lavas, and intrusive plug)

Unit B—Localized unit including a lower (30 m) intereruptive sedimentary subunit of drab green-gray to yellow-brown tuffaceous fluvial and lacustrine mudstone, sandstone, and conglomerate, with thin bentonite beds, anomalous boulders, and occasional metamorphic clasts, overlain in some areas by a thick (<90 m) domelike brecciated andesitic basalt flow. Less folded than Unit A, but cut by large-displacement NNW-trending left lateral faults.

(Irregular angular unconformity developed on anticline)

Unit A—Folded and faulted altered syneruptive sequence: boulder lahar deposits interbedded with sandy, tuffaceous fluvial sediments including metamorphic and igneous clasts. Maximum exposed thickness 305 m; total thickness estimated from boreholes approximately 550 m.

(Base of formation not exposed. Well data indicates angular unconformity on Cretaceous marine sediments and older metamorphic rocks; Taylor, 1960)

The total of maximum thicknesses of the units exceeds 1100 m, although actual sections (including subsurface rocks) probably reach no more than about 950 m near Hancock Canyon and about 350 m on the pre-John Day structural high near the center of the canyon. Vance (1988) estimated a thickness of 300 m for the Clarno Formation in this area and Hammond (1993) indicated it is at least 450 m thick.

Between the two vertebrate localities is a kilometer-wide intrusive plug of distinctive light gray to yellowish-tan heavily altered porphyritic hornblende andesite (Taylor, 1960; Hammond 1993) or dacite (Bestland, Retallack, and Fremd, 1994a), emplaced during or after deposition of Unit B. Where not in intrusive contact, it is unconformably overlain by Unit C or E.

After this manuscript had been submitted for review, Bestland, Retallack, Fremd, and Swisher (1994) and Bestland, Retallack, and Fremd (1994a, 1994b) published brief descriptions of the geology of approximately the same area. Although the measured Clarno Formation sequences depicted in Bestland, Retallack, and Fremd (1994a) generally correspond to the section described here, their composite section and proposed correlations of major fossil localities and dated units to this sequence differ significantly from those presented below. The differing conclusions appear to arise from the conceptually different (lithostratigraphic vs. allostratigraphic) approaches to subdividing the formation and to an underestimation on the part of those authors of the complexity of ongoing structural, topographic, and consequent stratigraphic evolution of the area during Clarno time, especially within the immediate vicinity of the Nut Beds and Hancock Quarry. Their geologic sketch map depicts no faults and only a single fold axis of unspecified age. The conclusions presented here summarize the results of detailed structural and stratigraphic mapping, section measurements, and repeated field testing of a succession of interim hypotheses spanning nearly 20 years, and have not required modification in light of information presented in these recent papers. Time and space constraints during final revision have allowed only partial presentation of available data crucial to the resolution of some of these differences of interpretation.

Stratigraphy, paleoenvironments, and physical correlation of major localities

Units A through E are exposed in clear stratigraphic superposition near the confluence of Hancock Canyon and Pine Creek. However, within the canyon, each of the units is at least locally absent due to exclusion by pre-existing topographic highs, original limits of lava flows, intrusive displacement, and/or uplift and erosion during Clarno time. Both of the major vertebrate fossil localities lie within a portion of Hancock Canyon which has experienced the combined effects of all of these factors, further complicated by topographic inversions, faulting, and subrecent mass movement. As a result, lateral relationships among outcrop areas are obscure and complex. However, several independent lines of evidence (lithologically unique beds and sequences, and consistency of structural features, paleotopography, depositional environments, and sedimentary provenance) allow confident physical correlation of the sections at the vertebrate localities to the more complete sections elsewhere in the canyon. Most of the other plant-bearing localities and a track locality occur in less complex, well-exposed sections which include a widely recognizable marker horizon, the "Rajneesh tuff" (Vance, 1988) or "Muddy Ranch Tuff" (Bestland et al., 1994a) and associated vesicular basalt flow.

Nut Beds locality

The Nut Beds locality, long known for its diverse and uniquely preserved fossil flora, has also yielded a small assemblage (about 100 specimens) of fossil vertebrates. A dated pumice lapilli tuff is directly associated with some of the vertebrate remains. Accurate resolution of its time-stratigraphic position is particularly important because of controversies concerning its radiometric age (cf. Evernden and James, 1964; Vance, 1988; Manchester, 1990; Bestland, Retallack, Fremd, and Swisher, 1994), relationship to the rest of the formation (Hanson, 1973, 1979; Bestland, Retallack, and Fremd, 1994a, b), and correlation to the chronologies of mammalian evolution (Stirton, 1944; Hanson, 1973; Bestland, Retallack, and Fremd, 1994b) and floral and climatic change (Scott, 1954; Wolfe, 1971, 1981; Manchester, 1981, 1990, 1994; Bestland, Retallack, and Fremd, 1994a, b).

Through local usage and published accounts over more than 50 years, the term "Nut Beds" has come to apply to a small area on the west side of Hancock Canyon characterized by conspicuous light-colored cliff outcrops of varied, highly indurated sediments locally bearing abundant fossils. The term is used here in its sense as a locality name, capitalized and without quotes, and not as the name of a formal geologic unit.

Paleoenvironmental indicators and geometry of stratigraphic units within the Nut Beds sedimentary complex are consistent with facies expected in and near a lacustrine delta as suggested by Scott (1954), probably in a small, shallow lake with water level varying in response to climatic and/or tectonic changes (cf. Dunbar and Rogers, 1957, fig. 21B). An alternative interpretation of stream channel and levee deposition (Retallack, 1981, 1991) and lahar runout (Bestland, Retallack, and Fremd, 1994b) has been proposed, but supporting evidence has not yet been offered. Features favoring the lacustrine model include the following: (1) clast size generally increases upward, a typical trend in prograding deltas but rare in fluvial sequences; coarse, well-sorted conglomerates (probable proximal distributary channel deposits) appear only in the uppermost part of the section in shallow channels lacking sharp incisure; (2) channel fills including fine conglomerate and coarse sandstone lower in the section are poorly sorted and most are matrix-supported, unlike open-channel sediments, and probably represent density-current deposits in sublacustrine prodelta and delta-front channels; (3) stratigraphic subdivisions demarcated by thin (5 to 25 cm) but laterally extensive sandy tuff beds (which, themselves, would not be preserved in a channel/levee setting) exhibit classic lacustrine geometry; each subdivision is gently concave upward and thins in both directions from a maximum near the middle of the outcrop area, where the overlying conglomeratic deposits are also thickest and coarsest (Janes, 1971; Hanson, unpubublished data).

Lenses of dark, sandy siltstone in the upper levels (interpreted as interdistributary pond deposits) include monospecific clusters of fruits and seeds (T. J. Bones, personal communication, 1971), apparently buried where they fell. Slender, vertical *Equisetum* stems as long as 60 cm are buried in thin-bedded siliceous clayey siltstone within the lower half of the Nut Beds section. The fast-growing *Equisetum* and sparse ferns (not associated with any woody plants in life position) may have grown during a brief lowstand, but the unbent stems and the enclosing siltstone, locally exhibiting cyclic, steeply draped bedding, indicate burial in standing water. Identification of this level as a paleosol (Retallack, 1981) does not imply that it was subaerially exposed for more than a single season. A basal unit of laminated leaf-bearing siltstone (Manchester, 1981), laterally traceable over the length of the outcrop area, represents lake bed facies.

Pervasive primary and secondary mineralization (e.g., chalcedony, opal, calcite, gilsonite) of the Nut Beds sediments and fossils probably resulted from hydrothermal springs induced by the intrusion just to the north of (and probably extending below) the Nut Beds. Chalcedony veins similar to those cutting Nut Beds sediments also cut the intrusion. Mineralization of decomposable organic structures as fragile as uncrushed insect exoskeletons and internal seed structures must have required high concentrations of dissolved silica and minimal flushing by through-going streams at the time of deposition.

Near the middle of the Nut Beds section, a laterally extensive and very indurated bed, 15 to 40 cm thick, consists of very poorly sorted pebbly mudstone with abundant pumice lapilli and plant remains and a high concentration of large-vertebrate bones and teeth (UCMP locality V78127). This deposit overlies well-sorted, plane-bedded sandstone and siltstone (probable toeset sediments) and is overlain, at a very contorted but unchanneled contact, by poorly sorted conglomerate, interpreted as lower foreset facies. The bone-bearing unit most probably represents multiple episodes of gravity-flow slurry deposition at the foreset-toeset slope break, a bone-concentrating mechanism which has been reproduced in laboratory model experiments (Hanson, unpublished data). Large-scale foresets are ephemeral features of deltas of lakes with fluctuating levels (Dunbar and Rogers, 1957) and are not apparent in the Nut Beds deposit.

All of the vertebrate specimens were found in the coarser-grained upper half of the Nut Beds section. Though most were recovered from blocks of matrix jackhammered from the rich horizon described above, a number of specimens (including that described by Stirton, 1944) were collected from the higher distributary channel and interdistributary pond deposits (UCMP locality V62403), mostly by the late Thomas J. Bones and his wife Lorene during their 40-year investigation of the Nut Beds flora (Bones, personal communication, 1971).

Most of the vertebrate specimens are isolated teeth, bones, and portions of skulls and jaws with teeth, but one *Hyrachyus* specimen (Fig. 7) from a high channel conglomerate apparently had been a complete skull with articulated jaws. The basal foreset level yielded a nearly articulated pair of juvenile brontothere maxillae with a slightly displaced molar. These associated remains argue against the possibility of reworking from older deposits. None of the specimens exhibits rounding or other evidence of extensive stream transport, and the represented taxa undoubtedly coexisted within the nearby riparian and delta-surface environments. The volcanic ash associated with the bones at the lower level may have caused the death of some or all of the individuals represented there.

The lithologic subunit which characterizes the Nut Beds ranges from 7.5 m to 12 m in exposed thickness and extends about 240 m along the western side of Hancock Canyon. The subunit is cut by several strike-slip faults with low-angle slickenside striations plunging toward the hill (westward), and there is no evidence that offset of blocks is due to slumping as indicated by Retallack (1991).

Overlying the faulted Nut Beds subunit with prominent unconformity, and extending to a concordant contact with the basal tuff and ignimbrite of member A of the John Day Formation, is a 30 m unfaulted section of extremely clay-rich, stratified, gray-green and brick-red channel and overbank sediments ("mudstones and siltstones with some muddy sandstones and gravels"; G. S. Smith, 1988) and heavily altered igneous and sedimentary clasts ranging in size to 30-cm boulders (S. Manchester, personal communication, 1973). The faults which cut the Nut Beds subunit and produce stratigraphic offsets ranging to 10 m are capped by the overlying nonsiliceous sediments, and the Nut Beds subunit dips more steeply westward (19° to 21°) than do the overlying sediments (about 8°). Forming a steep slope locally known as Red Hill, these sediments include stacked paleosols discussed by Retallack (1981, 1991) and described in detail by G. S. Smith (1988).

Structural, paleotopographic, and lithologic evidence (e.g., the near-absence of primary volcanic clasts) support assignment of the Red Hill sedimentary sequence to Unit E, the localized, terminal Clarno valley-fill complex. This lenticular deposit is channeled into the deeply altered subplanar surface that had developed on Unit C and older rocks in an area that had not been buried by the lavas of Unit D. Much of the Red Hill parent sediment must have been derived from the adjacent saprolitic terrace developed on this surface and would have included abundant pre-existing pedogenic minerals. With the near-filling of the paleovalley, and covering of the terrace by new floodplains, the parent material of the Red Hill paleosols would include a smaller percent of these reworked minerals. This may alternatively account for the rather abrupt changes in color (red to tan) and paleosol chemistry (e.g., reduction in the ratio of kaolinite to smectite) reported by Smith (1988) and Retallack (1991) near the top of the section, changes they cite as evidence for pronounced paleoclimatic change. Because of the unconformity between the Nut Beds subunit and the Red Hill sediments (shown below to represent a temporal gap of about 10 million years), Retallack's (1981, 1991) assumption that the paleosols of Red Hill were formed under climatic conditions comparable to those of Nut Beds time is not justified.

The Nut Beds subunit overlies about 30 m of drab, well-bedded fluvio-lacustrine clays to fine gravels (including common metamorphic clasts) with several cream-colored bentonitic beds and scattered andesite porphyry boulders. This unusual lithology closely matches that of the lower part of Unit B which was temporarily exposed in 1976 within a more complete section below the Palisades Cliffs, just east of the mouth of Hancock Canyon. Although structurally concordant with the Nut Beds subunit, slightly altered sediments below the sharp contact suggest a possible disconformity, probably representing only a brief hiatus in deposition during Unit B time. There is no colluvial breccia at this contact, and it is clearly not the same prominent angular unconformity as that at the base of Unit C.

These Unit B sediments, in turn, overlie a heavily altered and faulted sequence of boulder conglomerates and mudstones with lenses of fine, subangular to well-rounded metamorphic and volcanic pebbles. Except for the degree of alteration, consistent with development of a deep paleosol, these sediments are identical to those of Unit A exposed elsewhere.

Superpositional relationships alone, then, broadly constrain the time of deposition of the Nut Beds subunit to the interval after deposition of lower Unit B and before the valley filling of Unit E. Bestland, Retallack, and Fremd (1994a), in their composite section, place the Nut Beds within 10 m of the top of their "Conglomerates of Hancock Canyon" (equivalent to the upper half of Unit C as defined here), 25 m above the amygdaloidal basalt, and above the closely associated "Muddy Ranch tuff." The tuff below the Nut Beds, however, is unwelded, much more altered than the "Muddy Ranch tuff" in other exposures, and closely associated with other thin bentonite beds and abundant metamorphic clasts. These characteristics indicate a much more probable correlation to the older unnamed tuff of Unit B. Careful tracing of the Unit C basalt and tuff horizon throughout the study area demonstrates that the Nut Beds lie very near the crest of the anticline affecting Unit C and older rocks (Fig. 1), and the nearest exposures of the basalt, about 400 m northeast and 800 m southwest of the Nut Beds, are already within 5 m of the top of the Clarno Formation. Rocks of Unit C at and above this level were eroded away from the Nut Beds area before the time of deposition of Unit D. The Nut Beds subunit must also predate the older portions of Unit C and the time of surface exposure of the intrusive plug just 60 m to the north. The fluvio-deltaic deposits of the Nut Beds and all underlying units completely lack clasts derived from the intrusive, although such clasts are virtually the only components of thick exposures of basal Unit C just to the north and south. The peri-lacustrine (or alternatively, channel and levee) deposi-tional environment of the Nut Beds subunit is incom-patible with the colluvial and erosional hillslope regimes which demonstrably prevailed in the immediate vicinity of the plug during early Unit C time. Polycy-clic metamorphic clasts have been found only in the fluvial facies of the upper part of Unit A, the lower part of Unit B, and in the Nut Beds conglomerates, indicat-ing that these units shared a source provenance which was not yet buried by volcanic debris. Only volcanic and intrusive sources contributed to Unit C and younger units.

Collectively, the physical evidence indicates that the Nut Beds subunit falls within the boundaries of Unit B and that the angular unconformity directly above the Nut Beds subunit then corresponds to a convergence of three unconformities recognized elsewhere in the canyon (those at the bases of units C, D, and E).

Climatic interpretations of the Nut Beds flora (Wolfe and Hopkins, 1967; Wolfe, 1971; Scott, 1954; Chandler, 1964; Manchester, 1981, 1990, 1994) have been quite consistent. Leaf-margin analysis, foliar physiognomy, wood structure, liane or epiphytic habit of many taxa, and taxonomic diversity (Manchester, 1994, reports 145 genera and 173 species of fruits and seeds, and about 65 genera of leaves) in comparison to modern floras all indicate a moist, equable subtropical to tropical rain forest environment.

Localities of Unit C

Folding, faulting, intrusion, and erosion subsequent to the deposition of Unit B had radically altered the local topography by the time deposition of Unit C began. The resulting high-relief terrain continued to evolve with the sporadic addition of thick boulder-bearing debris flows to valley floors, blocking side drainages and occasionally shunting main stream courses to valley margins. Fossil localities within Unit C reflect the resulting topographic diversity. Standing tree trunks, logs, limbs, and intact forest floor leaf assemblages appear at and near the bases of some of the debris flows. Pond delta deposits preserve ferns and leaves, and at one locality, pond-bottom sediments record the passage of turtles, birds, and microscopic worms. Higher in Unit C, airfall tuffs supplemented overbank deposits to bury stands of *Equisetum*. All of these localities are in the lower part of the unit, stratigraphically below the vesicular basalt flow.

Some of the plant localities in this unit have been known since the turn of the century. Knowlton (1902) commented on a few of the taxa, and Chaney (1937) described a fossil cycad from a nearby locality, but no systematic floral descriptions have yet been published. Laminated pond-bottom shales on the east side of the canyon have yielded well-preserved leaves and a fission-track date of 43 Ma (Vance, 1988, and personal commu-nication).

Localities in the northern and eastern parts of the canyon mostly represent forest floor accumulations and standing tree trunks buried by debris flows. Assem-blages from these localities collectively constitute the Hancock Canyon flora (Manchester, 1990). This flora was mentioned in Armentrout (1981), where it is referred to as the "Iron Mountain flora" and related to a radiometric date from the basalt flow higher in the section, but still within Unit C. This is apparently the "dominantly broad-leaved evergreen" flora cited by Wolfe (1992) as "the uppermost assemblage of the Clarno Formation near Iron Mountain . . . from a stratigraphic interval radiometrically dated at ~35 Ma; this interval also contains probable Chadronian mammals." The claimed stratigraphic position, faunal association, biostratigraphic correlation, and radiometric age have all now proven incorrect, and the floral assem-blage has only indirect bearing on investigations of climatic change at the end of the Eocene. Earlier references had cited the still older Nut Beds locality as the source of the "upper Clarno flora" (Wolfe and Hopkins, 1967; Wolfe, 1971, 1981), and erroneously stated that this flora "is associated with Chadronian mammals" (Wolfe, 1981).

The only known Hancock Canyon area vertebrate locality in Unit C (JDNM 66), which I found in 1987 at a stratigraphic level comparable to that of the Hancock Canyon flora, has yielded a small assemblage of bird and turtle tracks. The ichnofossils have not been identified below ordinal level, and probably will not yield diagnostic biochronologic data. The correlated deposits near Cherry Creek (see below) include several lacustrine shale localities which have produced an unstudied fish fauna as well as the White Cliffs floral assemblage (Manchester, 1990).

Despite the paucity of published taxonomic data, the paleoclimate of Unit C time has been inferred from modern analogues and diversity level. Retallack (1981), citing Manchester (personal communication, 1980) for taxonomic identifications, concluded that the flora preserved in the debris flows grew under cooler conditions than that inferred for the Nut Beds, and assuming them contemporaneous, hypothesized that the cooler assemblage had been transported with the lahar from an upland forest. It is difficult to imagine, however, how leaves, many of them quite large, could have survived intact the high energy and coarse clasts in this transport environment. More recently, Bestland, Retallack, and Fremd (1994a) hypothesized an ecotone separating the Nut Beds "lowland tropical" forest from the "upland paratropical" forest represented by the Hancock Canyon flora, but offer no explanation as to how the necessary elevation difference could be attained in little more than 1 km within a single aggrading basin. Recognition of the unconformity between the two localities permits a more probable interpretation of the differences as a reflection of climatic change over time rather than elevation.

A cooler climate during Unit C time is consistent with correlation of Unit C, via the "Rajneesh tuff," to Wolfe's "Middle Clarno" (John Day Gulch) flora 20 km to the south (see below), which Wolfe (1971) concludes grew under much more temperate conditions than the Nut Beds flora. However Wolfe (1981) correlates the John Day Gulch flora with the late Ravenian megafloral stage, supposedly earlier than the "late Kummerian" "Upper Clarno" (Nut Beds) flora. The conflict in correlation is discussed below.

Hancock Quarry locality

Hancock Quarry lies within a very restricted Clarno sedimentary section just 1 km northeast of the Nut Beds. About 10 m of conglomerate, mudstone, and minor sandstone, lacking primary volcanic clasts, rest unconformably on the porphyritic andesite intrusive, and immediately underlie the basal John Day ignimbrite with no apparent unconformity, although a number of paleosols occur within this section (Pratt, 1988a, b). More complete sections nearby demonstrate the unconformable superposition of the quarry unit above Unit C. The lava flows characterizing Unit D did not extend to this area, but angular fragments of platy andesite derived from Unit D occur at the basal contact of sediments northeast of and equivalent to those of Hancock Quarry. To the southwest, very similar sediments lie very near the top of the Red Hill sequence described above and assigned to Unit E. These relationships support the correlation of the sediments of Hancock Quarry to Unit E, the localized alluvial and colluvial deposits which immediately preceded the explosive onset of John Day deposition. The paucity of primary volcanic clasts in all

observed outcrops of Unit E suggests that this unit postdates the volcanism which produced Unit D and that Unit E deposition was not "in response to the eruption of the andesite of Horse Mountain" [= Unit D] (Bestland, Retallack, and Fremd, 1994a).

The diverse paleotopography of the region at the time the Hancock Quarry fauna was living can be reconstructed from existing outcrop geometry of the basal ignimbrite of member A of the John Day Formation. This densely welded tuff concordantly caps the quarry deposit about 5 m above the highest specimens, and originally blanketed the composite high-relief terminal Clarno terrain as far as 60 km south of Hancock Quarry (Robinson, 1975). Within this outcrop region, isopachs between the originally horizontal, subplanar unconformity below Unit D and the irregular base of the ignimbrite reveal features of a topography developed on the north flank of a large lava flow and dome complex and on the adjacent, relatively flat terrain developed on older Clarno rocks. Many trough-like features preserved by the ignimbrite, at least some of which have been mapped as synclines (Taylor, 1960; Swanson, 1969; Robinson, 1975; Hammond, 1993), can be shown to be paleovalleys which had been partially backfilled by sediments here assigned to Unit E. The area now occupied by Hancock Canyon lay between two distal lava flow lobes of Unit D, along a major north-flowing drainage flanked by low terraces developed on the pre-Unit D surface. Near the center of the 2-km-wide valley between the lava lobes was a cluster of small hills, inselbergs of the andesite plug. The sediments of Hancock Quarry accumulated along the southeast flank of one of these hills.

Within the main locality (Hancock Quarry 1, UCMP locality V75203), the bones, teeth, and occasional carbonized plant remains are distributed through 5.3 stratigraphic m of conglomerate (at lowest levels) and massive silty to sandy mudstones, with the majority of specimens from the lower 2 m of the deposit. Excavations in an area about 10 by 25 m have produced more than 2000 specimens, although the lateral limits of the bone-bearing deposit have not yet been reached. The most numerous vertebrate specimens are isolated bones and teeth, but many complete skulls and jaws are present, and several clusters of disarticulated elements from single individuals have been found. Small fragments of large bones are common, but nearly all the animals represented exceeded about 10 kg in body mass. A minimum of 90 individual mammals has been calculated from 273 teeth or tooth-bearing specimens prepared to date and in collections at UO, UCMP, and OMSI. This does not include hundreds of unprepared specimens at UO and UCMP, unknown numbers of specimens privately collected between 1958 and 1968, or those previously eroded from or still remaining at the site. Individuals actually represented in the original deposit most likely numbered many hundreds.

Carbonized seeds associated with the vertebrate remains have been described by McKee (1970). The assemblage includes taxa also present in the Nut Beds flora, but reflects a much lower taxonomic diversity.

Features of the stratigraphy and geometry of rock units indicate initial deposition in a pond setting, possibly a sag pond or abandoned channel segment, subjected to occasional floods. Pratt (1988a, b) offers a somewhat different interpretation of active channel and point bar deposition. A detailed taphonomic and paleoecologic study of the site is in preparation.

Physical correlations to Horse Heaven Mine–Cherry Creek area

Another paleontologically important area of Clarno Formation exposures lies about 20 km south of Hancock Canyon, in the vicinity of Cherry Creek, Horse Heaven Mine, and John Day Gulch. Links between the geochemistry of the Cherry Creek volcanics and regional tectonic events have been proposed by Noblett (1981). Two stratigraphic horizons in this area—a welded ash-flow tuff and a major saprolite-mantled unconformity—offer ties to the Hancock Canyon section.

The distinctive tuff has been noted in both areas and in intervening locations west of Muddy Ranch, briefly known as Rajneeshpuram. White welded tuffs are otherwise unrecorded in the Clarno Formation. Taylor (1960) noted these occurrences of the "Clarno lower welded tuff." Vance (1988) informally dubbed it the "Rajneesh tuff," and the same unit is called the "Muddy Ranch Tuff" by Bestland, Retallack, and Fremd (1994a). Taylor (1960) describes hand specimens "resembling porous chalk" and provides detailed petrographic data and modal analyses. Though discontinuous, it reaches thicknesses in excess of 10 m in the Cherry Creek area at the White Cliffs locality (Manchester, 1990), overlying and interbedded with lacustrine shales which have yielded well-preserved leaves and articulated fish remains. The tuff occurs at or near the top of Noblett's (1981) "lower sedimentary unit." Noblett's (1979) map and my own reconnaissance show that these shales are stratigraphically equivalent to the John Day Gulch ("middle Clarno") flora (Wolfe, 1971, 1981). Both Taylor (1960) and Vance (1988) have indicated that this is the same tuff as that exposed west of the mouth of Hancock Canyon. In an obscure exposure on the John Day River, the tuff is seen to directly underlie the vesicular basalt flow near the middle of Unit C. The basalt is traceable through most of the canyon.

The unconformity between units C and D in the Hancock Canyon area exhibits several features in common with a prominent unconformity in the Ashwood and Cherry Creek–Horse Heaven Mine areas (Waters, et al, 1951; Peck, 1964; Noblett, 1979, 1981).

In all areas, a series of varied lava flows, domes, and minor tuffs—notably lacking debris flow deposits and horizontally bedded except where affected by post-Clarno deformation—rests on a deep paleosol, developed with low-relief angular discordance on eroded older Clarno rocks. In the Horse Heaven Mine area, Waters et al. (1951) referred to the extrusive complex above the paleosol as the "post-Clarno" rocks, but Swanson and Robinson (1967) and subsequent workers have included these in the Clarno Formation. Although the relationships suggest a general temporal correspondence of the unconformity throughout the larger area, the length of the hiatus differs among subareas so that, for example, Noblett's (1981) thick middle and upper sedimentary units at Cherry Creek (below the unconformity there) have no correlatives near Hancock Canyon.

CHRONOSTRATIGRAPHY

A growing number of isotopic and fission-track dates on rocks in the Hancock Canyon area and in other areas of Clarno and lower John Day Formation exposures are forcing a revision of chronostratigraphic correlations of these formations (Table 1, Fig. 1). The emerging picture agrees with the physical stratigraphic conclusions presented above, and necessitates reconsideration of the dates reported in older publications.

Within Hancock Canyon, all published dates are from rocks here assigned to units B and C and to the basal John Day Formation. Dates from other areas on rocks physically correlated to the Hancock Canyon section accord with newer dates from Hancock Canyon.

Radiometric dates from Unit B include six (ostensibly) from the Nut Beds locality and one on the lava flow elsewhere in Unit B. Evernden and James's (1964) widely cited "Nut Beds" date of 34.9 Ma (corrected for new constants) has proven anomalously young in comparison to the others in Unit B and to many additional dates now known to be from stratigraphically higher rocks. Archival samples of the crystal separates used to produce the date were associated with a rock sample (C. Swisher, personal communication, 1989) lithologically unlike any Nut Beds rocks but closely resembling the ignimbrite in the lowermost John Day Formation, strongly suggesting that the dated "Nut Beds" sample had been displaced from the much younger stratigraphic level. Two fission track dates with 10% error margins (Vance, 1988, erroneously cited as K-Ar dates in Bestland, Retallack, and Fremd, 1994b), and one $^{40}Ar/^{39}Ar$ date (43.8±0.31 Ma, obtained by B. Turrin, reported in Manchester, 1990) from within the Nut Beds locality fall near 44 Ma. A reported $^{40}Ar/^{39}Ar$ date of "approximately 44 my" (Bestland, Retallack, Fremd, and Swisher, 1994) and "single crystal Ar/Ar age determinations on plagioclase crystals by C. Swisher of 43.8 Ma" (Bestland, Retallack, and Fremd, 1994) from the Nut Beds locality could

Table 1. Summary of radiometric and fission track dates within and physically correlated (dates with *) to the Clarno and lower John Day Formations in the Hancock Canyon area. Stratigraphic levels shown indicate approximate distance from base of locally exposed Clarno Fm. section to cumulative height of sample locality when structurally compensated allostratigraphic units are stacked as shown in Fig. 2. Abbreviations: @ = approximate; AA = ^{40}Ar/^{39}Ar single-crystal laser fusion; FT = fission track; JD = John Day Formation; JDa = member a of John Day Formation; KA = potassium-argon; a = apatite; h = hornblende; s = sanidine; p = plagioclase; wr = whole rock; z = zircon; fl = floral locality; fn = faunal locality. References: 1—Evernden, Savage, Curtis, and James (1964; recalculated); 2, 12, 15—Manchester (1990); 3, 6—Evernden and James (1964; recalculated); 4, 7, 9, 11, 13, 14—Vance (1988); 5, 8—Swanson and Robinson (1968; recalculated); 10, 17—R. Drake, pers. comm. (1987) (See also Armentrout, 1981); 16—Swisher (1992); 18—Bestland, Retallack, and Fremd (1994b).

Stratigr. level(m)	Date	±	Method	Unit	Locality	Ref
1000 @	31.8 *	?	KA, s	JD	Bridge Creek fl	1
1000 @	32.3 *	?	KA, wr	JD	Below Bridge Creek fl	1
1000 @	32.11*	0.69	AA, p	JD	Fossil School fl.	2
885 ?	34.9	?	KA, p	uncert.	uncertain	3
885	36.8	3.7	FT, z	JDa tuff	Above Hancock Q. fn	4
885	37.4	3.7	FT, z	JDa tuff	Above Hancock Q. fn	4
885	37.1 *	1.0	KA	JDa tuff	Above Hancock Q. fn	5
885	37.4 *	1.1	KA	JDa tuff	Above Hancock Q. fn	5
885	39.2	?	AA, p	JDa tuff	Above Hancock Q. fn	18
885	39.7 *	?	AA, p	JDa tuff	Painted Hills area	18
870	32.8	?	KA, s	JDa tuff	Above Hancock Q. fn	6
823 @	40.4 *	1.6	KA, h	Clarno d	Wagner Mountain	7
700 @	42.1 *	0.8	KA, s	Clarno d	Horse Heaven area	8
670	44.3	4.4	FT, z	Clarno c	Above unit c fls	9
640	35.1	0.9	KA, wr	Clarno c	Above unit c fls	10
625	44	4.4	FT, z	Clarno c	Above unit c fls	11
625	44.43*	0.17	AA, p	Clarno c	White Cliffs fl, fn	12
580	43.0	4.3	FT, z	Clarno c	"Fern beds" fl	13
425	43.6	4.4	FT, z	Clarno b	Nut Beds fl, fn	14
425	43.7	4.4	FT, z	Clarno b	Nut Beds fl, fn	14
425	43.8	0.31	AA, p	Clarno b	Nut Beds fl, fn	15
425	48.32	0.11	AA, p	Clarno b	Nut Beds fl, fn	16
365-455	45.9	0.9	KA, wr	Clarno b	Approx. Nut Beds	17

not be confirmed by Swisher, after a review of his field and laboratory records, as having been obtained by him (Swisher, personal communication, July 5, 1995). These reports apparently repeat Turrin's determination. In the same communication, Swisher reaffirmed an ^{40}Ar/^{39}Ar date of 48.32±0.11 Ma (Swisher, 1992) which is probably the most accurate date available for the Nut Beds. Individual plagioclase crystals from pumice lapilli directly associated with the vertebrate fossils near the middle of the Nut Beds sequence yielded a much tighter cluster of individual dates than did those used to produce the weighted mean date obtained by B. Turrin (Turrin, personal communication, 1990). A whole-rock K-Ar date of 45.9±0.9 Ma (K-Ar No. 2963, R. Drake, personal communication, 1987) was obtained from a sample I collected from the lava flow in upper Unit B, an approximate correlative of the Nut Beds.

One of four published dates from Unit C of Hancock Canyon appears erroneous. This is a whole-rock K-Ar date (K-Ar No. 2962, 35.1± 0.9 Ma, corrected from that appearing in Armentrout, 1981), obtained by R. Drake on a sample I collected from the vesicular basalt in the

SW 1/4 Sec. 26, T7S R19E, Wheeler Co., Oregon. It is far out of sequence compared to dates from superposed, associated, and correlated rocks and is not supported by biostratigraphic correlations. The basalt flow which yielded this date directly overlies the "Rajneesh tuff," whose Hancock Canyon fission-track dates averaging 44 Ma (Vance, 1988) agree well with ^{40}Ar/^{39}Ar dates on the correlated tuff near Cherry Creek (obtained by B. Turrin, reported in Manchester, 1990). Two other fission-track dates on other samples in Unit C within Hancock Canyon (Vance, 1988 and written communication, 1990) differ insignificantly from 44 Ma.

Robinson and McKee (1983, correcting a date in Swanson and Robinson, 1968) report a K-Ar date of 42.1 ± 0.8 Ma near Horse Heaven Mine on a sample which falls between the same pair of unconformities that define Unit D near Hancock Canyon.

The densely welded tuff which defines the base of the John Day Formation in this region (Peck, 1964; Swanson and Robinson, 1968; Robinson, 1975; Robinson, Walker, and McKee, 1990) provides a regionally recognizable time horizon slightly younger than the Hancock Quarry locality. The date first reported for this unit (32.8 Ma; Evernden and James, 1964, corrected for new constants) falls well outside the range of newer dates. Four K-Ar (Swanson and Robinson, 1968) and fission track (Vance, 1988) dates cluster near 37 to 38 Ma. Recently published $^{40}Ar/^{39}Ar$ dates by C. Swisher of 39.2 Ma (Retallack et al., 1993) on this unit in the Hancock Canyon area and 39.7 Ma (Bestland, Retallack, and Fremd, 1994b) on a correlated tuff in the Painted Hills area are probably more accurate.

An $^{40}Ar/^{39}Ar$ date from the lower part of the John Day Formation closely associated with the Fossil School locality, which is florally and lithologically similar to that at Bridge Creek in the Painted Hills area (Manchester, 1990; Manchester and Meyer, 1987), closely matches the Bridge Creek dates near 32 Ma reported by Evernden and James (1964).

VERTEBRATE BIOSTRATIGRAPHY
Nut Beds local fauna

Despite the locally high concentrations of bones and teeth, their fragile nature, extremely indurated matrix, and vertical exposure have limited the size of the recovered and prepared assemblage to about 80 identifiable specimens. These represent six taxa of large vertebrates, and it is not clear whether taphonomic factors or collecting biases imposed by the camouflaging matrix are primarily responsible for the apparent exclusion of small vertebrates.

Three UCMP locality numbers have been assigned to various parts of the Nut Beds exposures. Locality V62403 includes the uppermost strata, mostly exposed on the dip slope of the indurated distributary complex. This is also the source area for most of the plant specimens collected by the late Thomas J. Bones and his wife, whose collections are conserved at the U.S. National Museum, UCMP, the University of Florida Museum, and John Day Fossil Beds National Monument, Oregon (Bones, personal communication, 1971). Locality V78127 designates the rich bone-bearing stratum about midway down the east-facing cliff exposure. A small number of specimens for which specific sites of origin are uncertain have been assigned a general Nut Beds locality number, V65513.

SYSTEMATIC PALEONTOLOGY
Class REPTILIA Laurenti, 1768
Order TESTUDINES Batsch, 1788
Family TESTUDINIDAE Gray, 1825

Hadrianus sp.
Figure 3

Specimen—Peripheral, UCMP 12799, locality V78127.

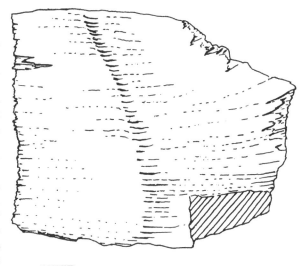

Figure 3. *Hadrianus* sp. from Nut Beds l.f. 11th peripheral plate, UCMP 12799.

Description—The single peripheral plate, probably the 11th peripheral, measures 78 mm between adjacent peripheral and pygal sutures, 67 mm from the costal suture to the margin, and 16 mm in maximum thickness. A prominent, oblique sulcus crosses the dorsal surface.

Discussion—The large size and deep sulcus are characters known only in the genus *Hadrianus* among North American Tertiary tortoises (J. H. Hutchison, personal communication, 1993). The material is inadequate for specific identification. *Hadrianus* occurs in Bridgerian and Uintan localities in the Rocky Mountains and southern California.

Order CROCODYLIA (Gmelin, 1789)
Family CROCODYLIDAE (Cuvier, 1807)

Pristichampsus sp.
Figure 4a,b

Specimens—Isolated tooth crowns, UCMP 141374 and 141375, UCMP locality V65513. Edentulous L maxilla fragment, UCMP 121798, UCMP locality V78127.

Description—Teeth laterally compressed with fine serrations (6.2 per mm) on distal and most of proximolingual ridges. Posterior tooth (UCMP 141375) slightly bulbous with D-shaped outline. Caniniform tooth (UCMP 141374) lanceolate. Maxilla fragment includes oval alveoli, one of which measures 4.9 mm labio-lingually by 9.1 mm. Ten depressions for accommodation of lower teeth lie lingual to the dental ridge and are elongate mesio-distally.

Discussion—Compressed teeth with serrate ridges are known to occur only in the genus *Pristichampsus*

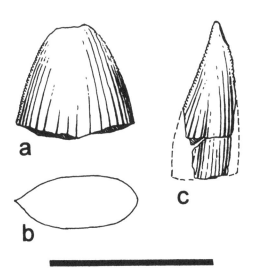

Figure 4. *Pristichampsus* sp. from Nut Beds l.f.; a. caniniform tooth, UCMP 141374; b. lateral tooth, UCMP 141375. Both labial views.

among crocodilians, but the specimens are inadequate to provide specific identification. *Pristichampsus* occurs in Bridgerian and Uintan localities in Wyoming and southern California (Hutchison, 1992 and personal communication).

Pristichampsus is incorrectly listed as a member of the Hancock Quarry ("mammal quarry") assemblage by Retallack (1991) and Bestland, Retallack, and Fremd (1994b).

Class MAMMALIA Linnaeus, 1758
Family OXYAENIDAE Cope, 1877

Patriofelis ferox (Marsh, 1872)
Figure 5a, b

Specimen—L dentary with P_3-M_2, UCMP 121800, UCMP locality V78127.

Description—P_3 length, 17.1 mm, width, 12.8 mm; P_4 length, 23.0 mm, width, 14.0 mm; M_1 length, 18.7 mm; M_2 length, 24.3 mm, width, 11.6 mm. Length P_2-M_2, 82.5 mm.

Discussion—All of the characters and measurements of this specimen fall within or very close to those reported for the small but variable (Matthew, 1909) sample of *Patriofelis ferox*, the larger of the two known species of *Patriofelis*. All previously reported specimens of *P. ferox* have been collected from late Bridgerian localities in Wyoming; the Black's Fork (upper) Member of the Bridger Formation (Gazin, 1976), Adobe Town Member of the Washakie Formation (see McCarroll et al., this volume, Chapter 2), and the upper Aycross Formation (Bown, 1982; Flynn, 1986).

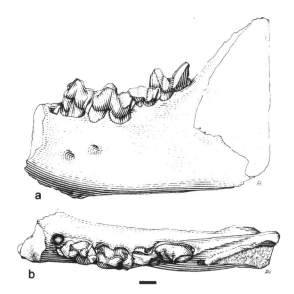

Figure 5. *Patriofelis ferox* from Nut Beds l.f. Left dentary with P_3 - M_2, UCMP 121800, a. Lateral view; b. occlusal view.

Order PERISSODACTYLA Owen, 1848
Family EQUIDAE Gray, 1821

Orohippus major Marsh, 1874
Figure 6a–e, Table 2

Specimens—L maxilla with P^3-M^1, M^3, UWBM 56991, UWBM locality C0144 = UCMP locality V4203; LM^2, UCMP 121818; R dentary with P_2-M_3, UCMP 121819; R dentary with P_3-M_3, UCMP 121820; RM_3, UCMP 121816; M_x fragment, UCMP 121817, all from locality V78127.

Description—P^3 lacking paraconule; metaconule very small, with low ridges connecting it to metacone and hypocone. P^4 submolariform with weak metaconule. Molars bear distinct conules, but metaconules much smaller than paraconules. Mesostyle distinct on M^2 (UCMP 121818) but barely discernible on M^1 and M^3 (UWBM 56991). P_2 very narrow and simple with single trigonid cusp bearing only faint lingual groove demarcating paraconid and metaconid; talonid with low hypoconid and no entoconid. P_3 similar but wider. P_4 molariform. Morphologic details and dental measurements within or close to range of small sample of *O. major* (Kitts, 1957) except slightly smaller M_1.

Discussion—Of the five species of *Orohippus* recognized by Kitts (1957), measurements of the Nut Beds specimens approach only those of *O. major* and *O. agilis*, but the latter is distinguished by its much shorter M_3, more quadrate P^4, and more prominent mesostyles. The isolated M^2 (UCMP 121818) matches Kitts's description of the M^2 of *O. major*.

The maxillary specimen (UWBM 56991) exhibits

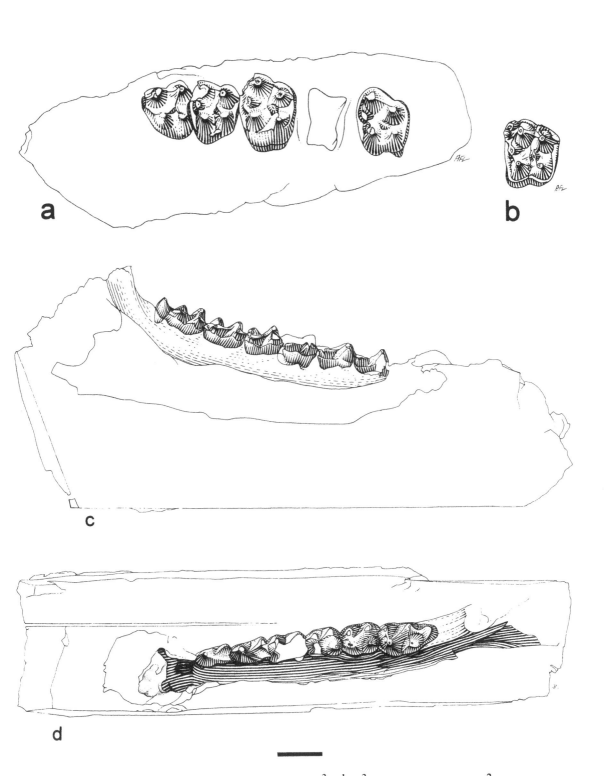

Figure 6. *Orohippus major* from Nut Beds l.f. a. Left maxilla with P^3-M^1, M^3, UWBM 56991.; b. left M^2, UCMP 121818, occlusal view; c. right dentary with P_2-M_3, UCMP 121819; d. same, occlusal view.

Table 2. Dental measurements of *Orohippus major* from Nut Beds l.f. and recognized species from Bridger Basin, Wyoming (Kitts, 1957). AP = anteroposterior dimension along midline; CHI = crown height index (HPac/Tr); C. V. = coefficient of variation; hcld = length of hypoconulid.; HPac = paracone height; S. Dev. = standard deviation; Tr = transverse dimension of single-lobed cheek teeth; TrA = transverse dimension of anterior lobe; TrP = transverse dimension of posterior lobe. Measurements in millimeters.

| | Nut Beds l.f | | Bridger | | Basin * | | |
| | | | *O. major* | *O. pumilus* | *O. progressus* | *O. sylvaticus* | *O. agilis* |
	N	OR	OR	OR	OR	OR	OR
P^3 AP	1	7.6		7.1-7.2	6.1-7.0	6.8-7.0	7.4-8.0
TrA	1	7.3		6.3-8.4	6.1-7.2	6.3-6.7	7.8-8.7
TrP	1	8.3					
P^4 AP	1	7.8		6.2-6.3	6.5-6.9	6.7-7.1	7.5-8.0
TrA	1	9.5		7.2-8.0	6.9-7.2	7.3-7.9	8.5-9.1
TrP	1	7.9					
M^1 AP	1	8.2		6.2-7.9	7.0	7.0-7.6	7.8-8.7
TrA	1	10.3		7.4-9.5	8.7	8.4-8.8	9.5-10.5
TrP	1	8.7					
M^2 AP	1	9.0	9.3	6.2-8.0			8.1-9.2
TrA	1	10.8	10.5	8.4-10.0			10.4-11.0
TrP	1	9.3					
M^3 AP	1	8.4	8.5	6.6-8.0			7.8-7.9
TrA	1	10.0	9.8	8.0-9.0			9.8-10.5
TrP	1	8.3					
P_2 AP	1	8.3				5.9	
TrA	1	3.8				3.0	
TrP	2	4.5					
P_3 AP	2	7.5-8.5		6.2-7.6		6.3-6.9	7.5
TrA	2	4.1-5.0		4.0-4.7		3.8-4.6	5.2
TrP	1	5.7					
P_4 AP	2	7.7-8.5		6.3-7.3		6.3-6.9	7.6
TrA	2	4.8-6.1		4.3-5.6		4.5-5.2	6.2
TrP	2	5.1-6.7					
M_1 AP	2	8.3-8.9	9.5	6.5-8.0		6.9-7.6	8.3
TrA	2	5.9-6.2	6.8	4.2-5.7		4.9-5.7	6.1
TrP	2	5.8-6.5					
M_2 AP	2	8.4-9.0	8.4-9.5	6.7-7.9	7.0-7.0	7.5-7.7	8.4
TrA	1	6.6	6.3-7.1	4.7-5.9	4.7-5.4	5.2-5.6	6.3
TrP	2	6.0-6.6					
M_3 AP	3	12.0-13.0	12.1-13.7	9.3-11.3	9.2-9.3	10.0-11.3	10.4
TrA	3	6.2-6.5	6.0-6.8	4.4-5.9	4.4-5.1	5.0-5.2	5.4
TrP	3	5.9-5.9					
M_{1-3} AP	2	29.0-30.5					

* Anterior and posterior widths not differentiated by Kitts (1957). Probable maximum widths listed as TrA for Bridger Basin specimens.

differences from all of the other *Orohippus* species figured by Kitts (1957); the premolars and molars are less quadratic in outline and have less prominent labial cingula, and P^3 nearly lacks conules. Neither upper nor lower premolars of *O. major* are preserved in other specimens, but the premolar morphology exhibited by the Nut Beds sample is consistent with the generally

primitive character of the molars which Kitts noted are very similar to those of *Hyracotherium craspedotum*.

Kitts (1957) notes the presence of *O. major* in both members of the Bridger Formation, accepting Granger's (1908) referrals of two lower jaws. However, Gazin (1976) records it only in the Black's Fork (lower) member.

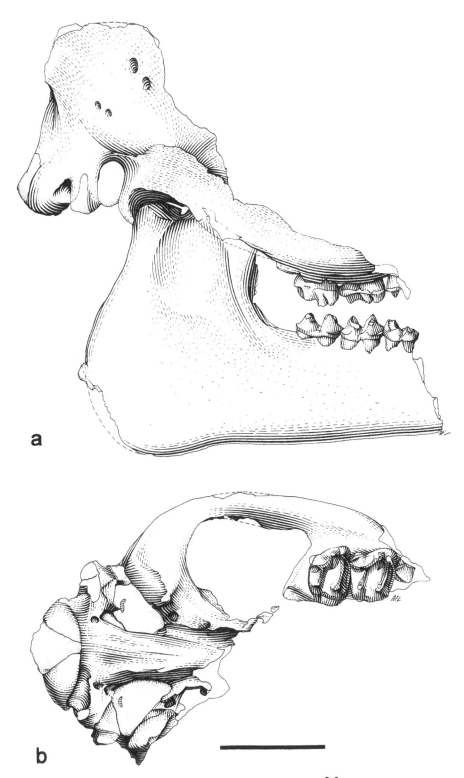

Figure 7. *Hyrachyus eximius* from Nut Beds l.f. a. Posterior skull with right M^{2-3}, right dentary with M_{1-3}, UCMP 121815, loc. V4203, lateral view; b. Same skull, ventral view.

Table 3. Dental measurements of *Hyrachyus eximius* from Nut Beds l.f. Measurements in millimeters. Abbreviations given in Table 2.

	N	Mean	OR	
P^4 AP	1	14.7		
Tr	1	19.0		
M^2 AP	1	22.0		
Tr	1	24.0		
M^3 AP	1	19.4		
Tr	2	22.1	19.7 - 24.5	
HPac	1	10.3		
CHI	1	0.43		
P_2 AP	2	10.2	10.2 - 10.2	
Tr	1	7.5		
P_3 AP	1	13.8		
Tr	1	9.2		
M_1 AP	1	18.0		
Tr P	1	13.0		
M_2 AP	1	20.2		
TrA	1	14.5		
TrP	1	14.2		
M_3 AP	2	20.0	19.0 - 21.0	
TrA	2	13.7	13.0 - 14.3	
TrP	2	13.6	13.3 - 13.9	

Superfamily RHINOCEROTOIDEA Owen, 1845
Hyrachyus eximius Leidy, 1871
Figure 7a, b, Table 3

Specimens—Posterior skull with RM^{2-3}, R dentary with M_{1-3}, UCMP 121815, locality V4203; premaxilla with I^1 and I^3, UCMP 121812, locality V78127; LP^2, UCMP 121811, locality V4203; RP^4, UCMP 121814, locality V78127; RM^2 fragment, UCMP 121813, locality V78127; LM^2 fragment, UCMP 121809, locality V4203; LM^x fragment, UCMP 121810, locality V4203; LM^x fragment, UCMP 121806, locality V4203; RP_2, UCMP 37078, locality V4203; RP_2, UCMP 121807, locality V4203; LP4, UCMP121805, locality V65513; RM_1 or 2, UCMP 121808, locality V4203.

Description—Moderately large, robust, brachydont (crown height index 0.43). Molars with rounded median valleys lacking sharp creases. Incisors spatulate, I^3 crown expanded mesio-distally.

Discussion—Most dental measurements fall within the range of those included by Radinsky (1967) in *Hyrachyus eximius*, the larger of the two Bridgerian species he recognized. *H. eximius* has been recorded only from late Bridgerian localities (Radinsky, 1967; Gazin, 1976). The specimens are also close in size to the one known Uintan specimen of the genus, UW 1937 from Washakie B, an early Uintan correlative, referred by Hanson (1989) to *H. douglassi*, but the teeth are much more brachydont than in that species (crown height index 0.43 vs. 0.54), and the Nut Beds skull is somewhat more robust.

The specimens do not resemble those of the nearest published record of *Hyrachyus*, a much smaller species from the Elderberry Canyon local fauna, Nevada (Emry, 1990).

Family BRONTOTHERIIDAE Marsh, 1873
Telmatherium Marsh, 1872
Telmatherium n. sp.

Specimens—RM^1, UCMP 121844; incomplete RM^2, UCMP 121845; incomplete RM^2, UCMP 121846; RM^3, UCMP 121847, incomplete M^x, UCMP 121852; RM_2, UCMP 121848; RM_3 hypoconulid, UCMP 121849; M_x fragment, UCMP 121850; RM_2 or 3, UCMP 121851; all from locality V65513.

Discussion—This small brontothere is the most abundantly represented vertebrate in the Nut Beds assemblage. It closely resembles specimens of *Telmatherium* and *Manteoceras* (probably a junior synonym of *Telmatherium*) in characters of, e.g., nasal incision, malar shape, and cheek tooth outline, but is smaller than described species of either. A detailed description of this new species will be published elsewhere.

Telmatherium (including *Manteoceras*) ranges through the Bridgerian and Uintan of Wyoming (Osborn, 1929; Eaton, 1985) where there is a trend toward size increase through time. Small size is probably a primitive character for this genus.

Biostratigraphic correlation and discussion

All subsequent vertebrate finds have reinforced Stirton's (1944) initial conclusion, based on a single *Hyrachyus* premolar (the first vertebrate specimen found in the Clarno Formation), that the Nut Beds are of Bridgerian age. All of the vertebrate genera now known from the Nut Beds also appear in the type Bridger Formation (south-central Wyoming). Three of the species (*Orohippus major*, *Hyrachyus eximius*, and *Patriofelis ferox*) are unknown from older or younger deposits, and *P. ferox* occurs only in the Twin Buttes (upper) of the two members of the Bridger Formation (Gazin, 1976). *P. ferox* and *H. eximius* also occur in the late Bridgerian upper Aycross Formation, the nearest late Bridgerian assemblage, about 700 km from Hancock Canyon in northwestern Wyoming (Bown, 1982; Flynn, 1986). The two reptilian genera as well as *Hyrachyus* (Eaton, 1985; Hanson, 1989) and *Telmatherium*, are also known from Uintan localities, but the two mammalian genera in those faunas are represented by more derived species.

Paleobotanists have reached disparate conclusions regarding the age of the Nut Beds flora. Wolfe (1971, 1981) correlated the Nut Beds to the Kummerian megafloral stage which he placed in the early Oligocene (now recognized as latest Eocene), although the biostratigraphic basis for this assignment was not stated.

Without proposing that they were similar in age, Scott (1954) and Chandler (1964) nevertheless noted that many of the fruits and seeds are congeneric with those of the London Clay flora, now correlated with the middle early Eocene (mid-Wasatchian; Lucas, 1989). Manchester (1981, 1990, 1994), in papers dealing with Nut Beds plant fossils, accepted Stirton's (1944) and Hanson's (1973) mammal-based correlations to the Bridgerian (early middle Eocene).

Hancock Quarry local fauna

Less than 1 km northeast of the Nut Beds is the rich bone-bearing deposit which I have named Hancock Quarry in recognition of the late Alonzo W. Hancock, whose enthusiasm for the natural history of the area led to the discovery of the locality in 1954. Since then, excavations have been undertaken by field crews under the auspices of the University of Oregon Museum of Natural History (UO) and the Oregon Museum of Science and Industry (OMSI), the latter with grants from the National Science Foundation. Most of the collections made by OMSI are now on permanent loan to the University of California Museum of Paleontology (UCMP), where they were prepared and curated.

Although all of the mammalian genera are known from other localities, most are represented in the Hancock Quarry assemblage by more complete and/or more numerous specimens than have been found elsewhere. However, only three of the taxa have been described (Radinsky, 1963; Mellett, 1969; Hanson, 1989), and two have been discussed in other publications (Schoch, 1989; Mader, 1989). Brief discussions of the taxa are included here, with emphasis on descriptions of species referred to known taxa and on documentation of generic identifications of new species. More complete descriptions and taxonomic reviews of new taxa are in preparation and will be published elsewhere.

All of the specimens described here are from the main locality, UCMP V75203. Many of the same taxa (but no additional ones) are also represented in the smaller channel locality (UCMP V75204), but most of those specimens are poorly preserved and have not yet been prepared or curated.

SYSTEMATIC PALEONTOLOGY
Class OSTEICHTHYES
(Not studied)

Class REPTILIA Laurenti, 1768
Order TESTUDINES Batsch, 1788
Family CHELYDRIDAE Gray, 1870
Genus and species not determined
Figure 8

Specimen—Epiplastron, UCMP 141370

Description—Small elongate epiplastron: maximum length 38 mm; length along midline suture 4.8

Figure 8. Chelydridae epiplastron, UCMP 141370, from Hancock Quarry l.f., UCMP loc. V75203. Ventral view.

mm. Broadest about midway along length, tapering posterad and with a shallow emargination along most of anterolateral border. Surface striations subparallel to entoplastral suture.

Discussion—The available material is inadequate to diagnose a new taxon, but the occurrence is significant as it constitutes the only North American record of the Chelydridae during the 10-million-year interval between Uintan and Whitneyan NALMAs (Hutchison, 1992).

Order CROCODILIA (Gmelin, 1789)
Family ALLIGATORIDAE (Cuvier, 1807)
Genus and species not determined
(Not studied)

Class MAMMALIA Linnaeus, 1758
Order DELTATHERIDEA
Family HYAENODONTIDAE

Hemipsalodon grandis Cope, 1885H

Referred specimen—RC_1, UCMP 141371.

Description of referred specimen—Total preserved length, 120 mm (root tip missing) enamel crown measures 59.5 mm along mesio-ventral border, 25.9 mm transverse diameter. Enamel worn from strip along lateral (posterior) side and a small portion of the tip. Maximum root diameter 33.8 mm mesio-distally, 24.8 mm labio-lingually.

Discussion—A skull of *Hemipsalodon grandis* with articulated jaws, and a maxilla of a second individual from Hancock Quarry were described, figured, and discussed by Mellett (1969). His conclusions remain unmodified in light of subsequent discoveries.

H. grandis is based on a type from the Cypress Hills Formation, Saskatchewan, and has been recognized from the early Chadronian Calf Creek l.f., Saskatchewan (J. Storer, personal communication, 1995) and Yoder l.f., Wyoming (Schlaikjer, 1935; Kihm, 1987). A smaller species occurs in the Porvenir l.f. (Duchesnean, Trans-Pecos, Texas) and Emry (1992) reports *Hemipsalodon* sp. from the lowermost (Duchesnean?) portion of the Flagstaff Rim sequence in Wyoming. The genus also appears in the Duchesnean Tonque l.f. of New Mexico (Lucas, 1982, 1992; this volume, Chapter 10).

Figure 9.   Nimravidae right C^1, UCMP 141372, from Hancock Quarry l.f., UCMP loc. V75203. Lateral view.

Order CARNIVORA
Family NIMRAVIDAE Cope, 1880
Genus and species indeterminate
Figure 9

Specimen—RC^1 tip, UCMP 141372

Description—C^1 with straight, finely serrate distal margin, weak, serrate proximo-lingual ridge not extending to crown tip.   Serrations formed of "hourglass-shaped beads" (Bryant, 1991) which are slightly irregular in size and spacing.   Cross-section teardrop-shaped; crown only slightly compressed laterally.   Lingual face slightly convex in anterior view. Very finely rugose pattern on unworn surfaces.

Specimen measures 21.8 mm from broken base to crown tip, 10.7 mm in greatest preserved mesio-lateral dimension, and 7.6 mm in maximum preserved labio-lingual dimension, but does not include base of enamel.

Discussion—Bryant (1991) indicates that the type of serration present in the Hancock Quarry specimen is restricted to the Nimravidae.

This is the earliest record of the family which is otherwise known from North American local faunas of Chadronian to Clarendonian (late Miocene) ages.

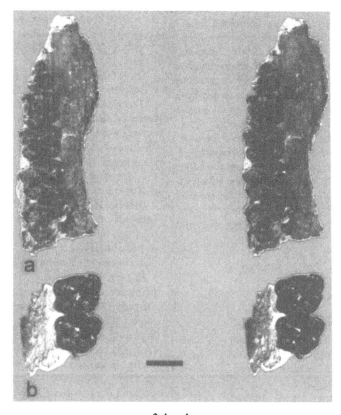

Figure 10.   *Haplohippus texanus*; a. right maxilla with P$^{2-4}$, M^1, UCMP 114475, occlusal view; b. right M$^{2-3}$, UCMP 114476, occlusal view. Both from Hancock Quarry l.f., UCMP loc. V75203.

Table 4. Dental measurements of *Haplohippus texanus* from the Hancock Quarry l.f. and Porvenir l.f. (Type). Measurements in millimeters. Abbreviations given in Table 2.

		Hancock Quarry				Type
	N	Mean	OR	S.D.	C.V.	
P^2 AP	1	8.7				
TrA	1	4.6				
TrP	1	7.7				
P^3 AP	2	9.5	9.1 - 9.8			
TrA	2	10.8	10.6 - 10.9			
TrP	3	10.4	10.2 - 10.7	0.21	2.0	
P^4 AP	3	10.4	10.0 - 10.7	0.29	2.8	
TrA	2	11.4	11.1 - 11.7			
TrP	2	10.8	10.6 - 10.9			
M^1 AP	3	10.5	10.4 - 10.6	0.08	0.8	
TrA	2	13.1	13.0 - 13.1			
TrP	2	12.1	11.9 - 12.2			
M^2 AP	6	10.6	10.4 - 10.7	0.11	1.1	
TrA	4	13.7	13.5 - 13.9	0.17	1.2	
TrP	4	12.5	11.7 - 13.3	0.57	4.6	
M^3 AP	3	10.6	10.0 - 11.5	0.65	6.1	
TrA	3	13.1	13.1 - 13.2	0.05	0.4	
TrP	2	11.7	11.4 - 12.0			
P_2 AP	2	8.7	8.6 - 8.7			7.9
TrA	2	4.1	3.9 - 4.2			4.0
TrP	2	4.4	4.2 - 4.5			4.0
P_3 AP	3	9.2	9.0 - 9.5	0.24	2.6	8.5
TrA	3	5.2	4.9 - 5.4	0.22	4.2	5.8
TrP	3	5.7	5.5 - 5.8	0.14	2.5	6.3
P_4 AP	3	4.0	8.7 - 9.4	0.29	3.3	8.7
TrA	3	6.2	6.0 - 6.6	0.26	4.2	6.0
TrP	3	6.8	6.7 - 6.9	0.09	1.4	6.6
M_1 AP	4	9.4	8.2 - 10.3	0.83	8.9	8.9
TrA	3	6.9	6.5 - 7.5	0.45	6.5	6.0
TrP	3	6.8	6.6 - 7.0	0.17	2.5	6.5
M_2 AP	5	10.3	9.5 - 11.5	0.80	7.8	9.1
TrA	3	7.6	7.1 - 7.9	0.36	4.7	7.0
TrP	3	7.2	7.1 - 7.3	0.09	1.3	6.3
M_3 AP	5	13.5	13@ - 14.0	0.40	2.9	
TrA	4	6.9	6.0 - 7.9	0.68	9.9	
TrP	5	6.9	6.4 - 7.4	0.37	5.4	

Order PERISSODACTYLA Owen, 1848
Family EQUIDAE Gray, 1821

Haplohippus texanus McGrew, 1953
Figure 10a,b, Table 4

Holotype—FMNH PM 17, nearly complete lower jaws lacking incisors, right P_1, P_4, and both coronoid processes and condyles.

Type locality—Porvenir l.f., lower Chambers Tuff, Presidio County, Texas.

Known distribution—Late Eocene, North America: Type locality and Hancock Quarry l.f., Clarno Formation, Wheeler County, Oregon.

Revised diagnosis—$I^?_3$, C^1_1, P^4_4, M^3_3. Symphysis and postcanine diastema very long; cheek teeth relatively low-crowned (as in *Epihippus*) with distinct cusps; P^2_2 non-molariform; protocone, conules, and mesostyle absent on P^2, single anterior cusp and weakly distinguished posterior cusps on P_2; P^3_3 sub-molariform, mesostyle lacking on P^3; P_3 protoconid and metaconid appressed; P_4 and lower molars with abbreviated paralophids, low parastylids, and distinct entoconids; P^4- M^3 with distinct conules weakly connected to adjacent cusps, prominent, bulbous parastyles and mesostyles, strong, rounded labial cingula with sharp inflections anterior to mesostyles; anterior lacerate foramen separate from rotund foramen; frontal not in contact with squamosal.

Referred specimens—Skull, L P^4-M^3, P^2 frag., R M^{1-3}, UO 27626; L dentary frag., M_{2-3}, UO 27628; L dentary, P_2- M_3, UO 27630; L dentary frag., M_{1-3}, UO 27631; R M^3, UO 27632; L P_{2-3}, UO 27635; L M_2 frag., M_3, UO 27636; R M_2, UO 27637; R M^3, UO 27640; L M_2 frag, UO 27642; L M_3, UO 27643;

R dentary, P_4-M_3, P_{2-3} roots, UO 28310; R maxilla, P^{2-4}, M^1, UCMP 114475; R M^{2-3}, UCMP 114476; L dentary, P_3-M_3, UCMP 130341; L dentary, P_4-M_3, UCMP 130342; R P^3, UCMP 130343; L M^2 crown, UCMP 130344; R M^2 crown, UCMP 130348; L dentary, P_{3-4}, incompl. M_{1-2}, P_{1-2} alveoli, UCMP 130349; incompl. R P_4, UCMP 130350; R M_1 or $_2$, UCMP 130351; R M_x frag, UCMP 130352.

Discussion—Lower teeth and jaws of the more common of the two equoids in the assemblage exhibit only minor differences from the type dentary of *Haplohippus texanus* McGrew, 1953 from the Porvenir l.f., Presidio County, Texas, the only other specimen recorded for the genus. In that local fauna, *H. texanus* is associated with *Mesohippus texanus* McGrew, 1971, a species which is very close to *Epihippus intermedius* and may be referable to that genus. Primitive skull characters, most notably a broad contact between the parietal and orbitosphenoid (rather than the derived frontal-squamosal contact), suggest derivation from Old World Eocene equoids rather than from the more derived clade which appears to include all other North American horses. A detailed description of specimens referred to this species is in manuscript and will be published elsewhere.

Epihippus gracilis (Marsh, 1871)
Figure 11, Table 5

Anchitherium gracilis Marsh, 1871
Epihippus gracilis (Marsh), 1871
Orohippus uintensis Marsh, 1875
Epihippus uintensis (Marsh), 1877
Epihippus parvus Granger, 1908
Epihippus gracilis (Marsh), (MacFadden, 1980)
Epihippus uintensis (Marsh), (in part; MacFadden, 1980)

Holotype—YPM 11271, lower jaws with left DP_2 - M_2, right DP_3 - M_1.

Type locality—White River, Uinta Basin, Utah.

Known distribution—Early Uintan to Duchesnean (middle to late Eocene), western North America.

Referred specimens—L dentary with P_3–M_3, UCMP 130341; L dentary with P_4–M_3, UCMP 130342.

Description—P_3 essentially molariform, although the part of the paralophid immediately anterior to the protoconid is more anteriorly directed than in the molars. Remainder of paralophid angles sharply linguad, nearly parallel to protolophid. Moderately high but constricted protolophid extends from protoconid to midpoint between distinct metaconid and metastylid. From this junction, but lower on the metastylid, the metalophid descends to a relatively low hypoconid. Narrow cingulid along paralophid and between principal labial cusps.

P_4 paralophid directed anteromediad; shorter in

	N	OR	
P_3 AP	1	8.1 -	8.1
TrA	1	5.0 -	5.0
P_4 AP	2	7.6 -	8.1
TrA	2	5.1 -	5.5
TrP	2	5.6 -	6.2
M_1 AP	2	8.7 -	8.9
TrA	2	5.9 -	5.9
TrP	2	5.8 -	6.1
M_2 AP	2	8.4 -	8.9
TrA	2	5.7 -	6.0
TrP	2	5.6 -	5.9
M_3 AP	2	10.8 -	12@
TrA	2	5.3 -	5.7
TrP	2	5.1 -	5.5
M_{1-3} AP	2	30.0 -	31.0
P_2-M_3	1	50@ -	50@

Table 5. Dental measurements of *Epihippus gracilis* from the Hancock Quarry l.f. Measurements in millimeters. Abbreviations given in Table 2.

UCMP 130342 than in UCMP 130341. UCMP 130342 smaller, principal cusps more crowded toward midline. Both bear distinct entoconids. Hypolophid less developed than protolophid. Hypostylid present. Anterolabial and medial segments of labial cingulid faintly connected around protoconid of smaller specimen only.

M_1-M_3 with distinct metaconids and metastylids demarcated by faint lingual furrows near apices. Paralophids extend obliquely from protoconids to point near midline of anterior margin, then sharply linguad, terminating low on anterolingual side of metaconid. Protolophids approached in height and strength by hypolophids. M_1 and M_2 of both specimens bear hypostylids. M_3 hypoconulid basin-shaped. UCMP 130341 with small, prominent cingular stylid in cleft between protoconid and hypoconid of each molar. Similar but smaller cusp only on M_3 between hypoconid and hypoconulid in UCMP 130342.

Discussion—The two specimens clearly differ from those referred here to *Haplohippus texanus*. In addition to their smaller size, the teeth have longer, more anteriorly directed paralophids, higher metalophids, and stronger transverse lophids. The specimens closely resemble those referred by Forsten (1971) to *Epihippus* cf. *E. gracilis* from the late Uintan Candelaria l.f., Trans-Pecos area, Texas.

MacFadden (1980, p. 136) discusses species distinction in *Epihippus*, concluding that premolar cuspid and lophid morphology are of limited taxonomic value and that "size differences are most useful in allocating specimens of *Epihippus* to a particular species." However, of his tabulated length and width measurements of the five posterior lower cheek teeth, only those of P_4 widths suggest a bimodal distribution, but

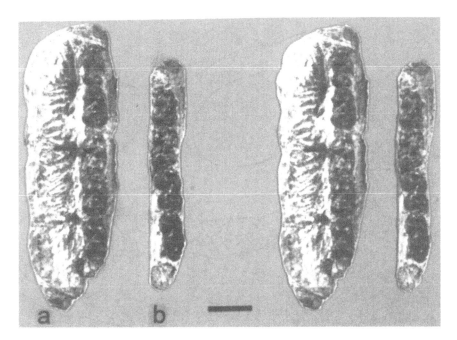

Figure 11. *Epihippus gracilis*; a. left dentary with P$_3$-M$_3$. UCMP 130341, stereo occlusal view; b. left dentary with P$_4$-M$_3$. UCMP 130342, stereo occlusal view. Both from Hancock Quarry l.f., UCMP locality V75203. Scale bar = 1 cm.

the broken right P$_4$ of the *E. gracilis* type, excluded from his bivariate plots, has a transverse dimension about halfway between the two apparent clusters. The apparent P$_4$ bimodality is probably an accident of sampling and the available specimens reveal no size trends with time. Allocation of the known specimens of *Epihippus* to two or more species based on size of cheek teeth would therefore depend on arbitrary cutoff points which could require different species designations for different teeth of single individuals. In the absence of diagnostic differences among the samples which include the type specimens of *E. gracilis*, *E. uintensis*, and *E. parvus*, only the first-named species, *E. gracilis*, remains valid for this group.

The type dentary of *E. intermedius* (CM 11845) does differ from other *Epihippus* specimens in its much more hypsodont cheek teeth with better developed crosslophids which lack anterior grooves demarcating the primary cusps. The paralophids are longer, straighter, and more obliquely directed, nearly parallelling the metalophids. In lateral view, the axes of the labial molar cusps align in an oblique orientation compared with near-vertical cusps in other *Epihippus* specimens. *E. intermedius* therefore warrants recognition as a distinct species, but should not include all large *Epihippus*. The type is from the Dry Gulch Creek Member of the Duchesne River Formation The fauna from this unit is included in the Duchesnean NALMA by Lucas (1992).

The characters which distinguish *E. intermedius* from other *Epihippus* specimens approach the degree of expression seen in *Mesohippus texanus* McGrew, 1971, from the Porvenir l.f., Texas, and these species could prove synonymous. The premolar cusp development in *M. texanus* appears more similar to that of *Epihippus*, especially *E. intermedius*, than to other species of *Mesohippus*.

The specimens from the Hancock Quarry and Candelaria local faunas bear slightly higher and stronger transverse lophids than do *E. gracilis* specimens from the Uinta and Tepee Trail formations. Nevertheless, the molar lophids do retain the medial indentations between the principal cusps and the erect labial cusps characterizing *E. gracilis*. Dental measurements of specimens from both localities fall well within the clusters defined by other specimens here referred to *E. gracilis*.

The Hancock Quarry specimens bear closer resemblance to two of the Candelaria specimens (TMM 40498-4 and 40276-20) than does a third Candelaria specimen (TMM 40497-3), which has longer, more anteriorly-directed paralophids, less distinct metastylids, and much more prominent labial cingulids. Given the small samples and demonstrated intrapopulation variability of *Epihippus* elsewhere, it must be assumed for the present that all of these specimens are conspecific, though future additions may justify subspecific distinction of the Hancock Quarry-Candelaria group from other *E. gracilis* population groups.

The record of *Epihippus* in the Hancock Quarry l.f. documents the last known occurrence of the genus unless *M. texanus* also proves referable to *Epihippus*.

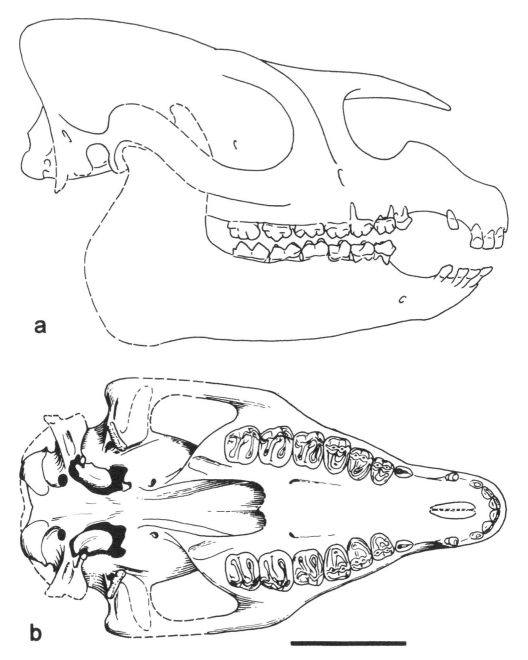

Figure 12.   Reconstruction of skull of *Protapirus hancocki* based on: UO 20377 (holotype—skull lacking postorbital portion); UO 27539, a laterally crushed skull lacking postorbital portion; UO 28318, skull; UCMP 125545, skull lacking nasals and premaxillae; UCMP 125555, left dentary. a. lateral view of skull and jaw; b. ventral view of skull. Hancock Quarry l.f., UCMP loc. V75203.

Family TAPIRIDAE Burnett, 1830
Protapirus Filhol, 1877

Type species—*P. priscus* Filhol, 1877.
Revised generic diagnosis—Dental formula; I^3_3, C^1_1, P^4_3, M^3_3. Long, sigmoidal facial groove incorporating infraorbital foramen and extending above orbit. Maxilla not constricted in advance of premolars. Paired dorsal flanges near maxillary-premaxillary suture well-separated from midline; no ossified mesethmoid. Incisors non-cingulate. Molar postparacrista straight, aligned with axis of metacone, separated by prominent notch. Premolars nonmolariform to molariform. M_3 with or without hypoconulid.

Table 6. Dental statistics of *Protapirus hancocki* from the Hancock Quarry l.f. Measurements in millimeters. Abbreviations given in Table 2.

	N	Mean	OR			S.D.	C.V.
P^1 AP	1	7.0	7.0	-	7.0	-	-
Tr	1	4.5	4.5	-	4.5	-	-
P^2 AP	5	9.2	8.2	-	9.8	0.78	8.5
Tr	5	10.6	10.0	-	11.3	0.57	5.4
P^3 AP	4	10.2	9.8	-	10.7	0.39	3.8
Tr	4	14.1	13.3	-	15.3	0.88	6.3
P^4 AP	9	10.8	10.0	-	11.4	0.44	4.1
Tr	8	15.3	14.5	-	16.0	0.54	3.5
M^1 AP	10	13.1	12.3	-	14.2	0.55	4.2
Tr	8	15.8	14.5	-	16.4	0.67	4.2
M^2 AP	7	14.8	13.6	-	15.8	0.97	6.6
Tr	7	17.2	16.5	-	17.9	0.49	2.8
M^3 AP	6	15.3	14.2	-	16.9	1.00	6.5
Tr	8	16.8	16.2	-	19.0	0.94	5.6
P^{2-4}	4	30.1	28.7	-	32.0	1.53	5.1
M^{1-3}	3	43.2	41.5	-	46.0	2.50	5.8
P^1-M^3	1	83.7	83.7	-	83.7	-	-
P_2 AP	3	8.3	7.8	-	9.1	0.72	8.7
W	3	6.1	5.6	-	6.5	0.47	7.7
P_3 AP	5	10.3	10.0	-	10.5	0.25	2.4
TrA	6	7.5	7.1	-	8.2	0.43	5.7
TrP	5	8.2	7.9	-	8.6	0.26	3.1
P_4 AP	9	10.6	10.0	-	11.5	0.46	4.4
TrA	9	8.5	7.7	-	9.2	0.48	5.6
TrP	9	9.5	9.1	-	10.0	0.38	3.9
M_1 AP	5	13.2	12.7	-	14.0	0.52	3.9
TrA	6	10.3	9.6	-	10.9	0.54	5.2
TrP	6	9.7	9.0	-	10.7	0.69	7.1
M_2 AP	8	14.8	14.4	-	15.7	0.43	2.9
TrA	9	11.0	10.2	-	11.7	0.49	4.5
TrP	7	10.4	9.8	-	11.7	0.73	7.0
M_3 AP	8	17.5	15.4	-	18.5	0.94	5.4
TrA	8	10.9	10.3	-	11.5	0.37	3.4
TrP	9	10.3	9.7	-	11.1	0.44	4.3
P_{2-4} AP	3	28.8	28.0	-	29.3	0.68	2.4
M_{1-3} AP	7	45.7	42.4	-	47.5	2.17	4.8
P_2-M_3	3	74.8	73.0	-	76@	1.61	2.2
M_3 hcld	9	2.8	1.2	-	3.7	0.82	29.6

Protapirus hancocki (Radinsky, 1963) new combination
Figure 12a-b, Table 6
Colodon? *hancocki* Radinsky, 1963:67.
Plesiocolopirus hancocki Schoch, 1989:306.

Holotype—UO 20377, anterior part of skull with left P^1-M^3, roots of canines and incisors.

Type locality—UCMP Loc. V75203, Hancock Quarry 1, Clarno Formation, Wheeler County, Oregon.

Known distribution—Late Eocene, North America: Hancock Quarry l.f., uppermost Clarno Formation, Wheeler Co., Oregon; Candelaria l.f., Presidio Co., Texas.

Hypodigm—Type and UO Nos. 20376, L dentary, P_4-M_3; 20378, L dentary, P_{2-3}, M_{1-3}; 20379, L maxilla, P^2- M^2; 20380, right P^4; 20381, R P_4; 20500, L M^3; 20547, R M^3; 21389, L M_3; 21392, L M^3; 21436, L P^4, incompl. M^1; 27539, laterally crushed skull with teeth missing or distorted, lacking postorbital portion; 27625, R P^4-M^1; 27713, L P_4; 27714, L P_4, roots of M_{1-3}; 27715, L P^4-M^3; 28328, R dentary frag., anter. M_2, poster. M_3; 27731, R dentary frag., M_2, incompl. M_1, M_3; 28318, skull, R I^2, L I^3; 28323, R M^1; 28332, L M^1; 28345, L P^3. UCMP Nos. 125545, skull lacking nasals, premaxillae, R and L P^1; 125546, L M^1; 125547, R P^4; 125548, L P^4 frag.; 125549, R P^2; 125550, R M^3; 125551, R M^2; 125552, R dentary, P_3-M_3; 125553, L M_2? frag.; 125554, L dentary, P_3-M_2, incompl. P_2, roots of M_3;

125555, L dentary, P_2-M_3; 125556, R dentary frag., M_{2-3}. UWBM Nos. 56951, R and L dentaries, R P_2, R and L P_3-M_3, L I_3, R C_1; 56955, L maxilla, M^{1-2}, roots of P^{1-4}.

Revised diagnosis—Small-sized tapirids; mean length M^{1-3}, 43.2 mm. P^1 small, narrow, lacking lingual cusps. P^2 with small, distinct anterolingual cusp. P^{3-4} with at most a faint anterolingual groove demarcating protocone and hypocone; metaloph slightly less prominent than protoloph. Molar paracones pyramidal with posterior ridge extending straight (in occlusal view) toward anterior side of prominent, labially convex metacones. Less hypsodont than *Protapirus priscus*. Smaller than *P. obliquidens*.

Discussion—The species listed in Table 2 as *Protapirus hancocki* was referred with a query to *Colodon* in Radinsky's (1963) original description. Schoch (1989), apparently unaware that the hypodigm subsequently had been expanded by a factor of six, used Radinsky's data to erect a new genus, *Plesiocolopirus*, for the Hancock Quarry tapir. The new material includes three skulls and numerous upper and lower dentitions whose characters and intrapopulation variants effectively eliminate arguments (Radinsky, 1963; Schoch, 1989) against inclusion of the species in the genus *Protapirus*. The M_3 hypoconulid, whose presence has been historically cited as a family-level diagnostic character, has been found to be extremely variable in this population, and completely absent in one well-preserved molar series. The dentition is virtually identical in size to that of the type of *P. priscus*, the holotype of the type species of the genus from Quercy, France, and differs from it only in features related to crown height; *P. hancocki* is more brachydont, and its labial molar cingula slightly extended, but its molar metacones are not lingually displaced relative to the other principal cusps. Details of the metacone and labial cingulum are quite variable in the Hancock Quarry sample, hence are of doubtful taxonomic significance. Several derived skull characters (e.g., long, sigmoidal facial groove incorporating the infraorbital foramen and extending above the orbit) unite *Protapirus obliquidens* and *P. hancocki* and distinguish them from *Colodon*. The nasals are not shortened as indicated by Schoch (1989), but extend forward to the level of the canines (UO 27539) and are deeply undercut by the nasal incision. *Plesiocolopirus* is therefore regarded as a junior synonym of *Protapirus*.

Collections from the Vieja Group, Trans-Pecos, Texas, include several specimens which are very similar to the Hancock Quarry sample of *Protapirus hancocki*. Three specimens from the Candelaria l.f. (TMM 31281-5, 31281-10, and 31281-12), including upper molars and lower cheek teeth, are virtually indistinguishable from the Hancock Quarry sample except that M^2 of TMM 31281-10 bears an enlarged crest on the labial

cingulum adjacent to the metacone. The M^2 cingulum in the Hancock Quarry suite is prominent and somewhat variable, but none attains the height seen in the Texas specimen. This difference, however, is no greater than might be expected between geographically and probably temporally separated populations of the same species.

Family AMYNODONTIDAE
Scott and Osborn, 1883
Procadurcodon Gromova, 1960

Type species—*P. orientalis* Gromova, 1960
Type locality—Artëm Coal Field, Maritime Province, Russia.
Known distribution—Late Eocene of eastern Asia and western North America.
Revised generic diagnosis—$I^3$$_{1?-3}$, C^1_1, P^3_2, M^3_3. Very large amynodontids (basilar length of skull < 760 mm, M^{1-3} < 195 mm). Tips of nasals retracted about to level of P^3. "Inner cingulum" on P^2 and P^3. Differs from *Zaisanamynodon* in more posterior position of orbit (above M^2), smaller incisors, less curved lower canines, absence of prominent vertical groove on labial faces of lower molars, molar crosslophs more perpendicular to ectoloph, and less hypsodont cheek teeth. Differs from *Metamynodon* in its more prominent facial fossae which are broadly open dorsally and excavated behind level of anterior margin of orbit, postorbital constriction narrower than muzzle, narrower braincase, more brachydont cheek teeth, and broader third metacarpal.

Procadurcodon n. sp.
Figure 13
Discussion—In his revision of the Amynodontidae, Wall (1989) expressed the opinion that *Procadurcodon* was inadequately diagnosed, but left the genus standing as Amynodontidae, *incertae sedis*. The type material of the type species of the genus, *P. orientalis* Gromova (1960), from a coal field locality near Artëm, Russia, includes most of the upper and lower teeth. Illustrations in the original description show significant characters also seen in the very large Hancock Quarry amynodontid. Most notable are taxonomically restricted characters of the upper premolars: the second and third each bear an "inner cingulum," a slender accessory lophule parallel to the posterolingual cingulum and lying just inside it, and connecting the protocone and metacone. This character is otherwise seen only in *Zaisanamynodon* which Beliaeva (1971) described and distinguished from *Procadurcodon*. The Artëm and Hancock Quarry specimens also share a complex P^2 in which the lingual end of the slender metaloph branches, sending small spurs antero- and posteromedially. (Although premolar details have been proven quite variable in many perissodactyls, the consistency of these

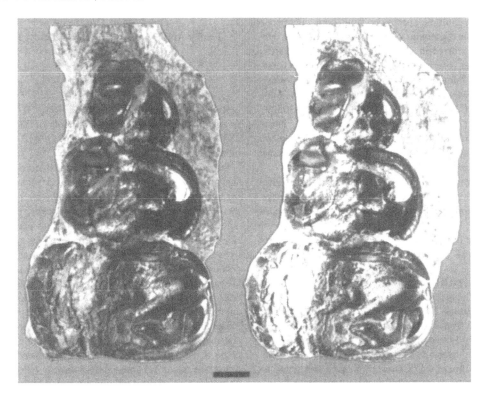

Figure 13. *Procadurcodon* n. sp.; right P$^{2-4}$, UO 27729, occlusal view. Hancock Quarry l.f., UCMP loc. V75203.

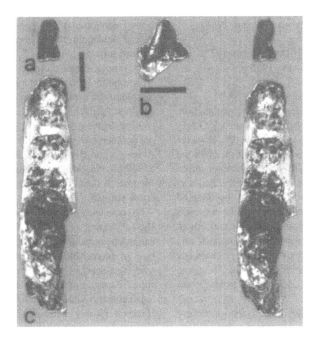

Figure 14. *Heptacodon* sp.; a. incomplete left P4, UCMP 141373, stereo occlusal view; b. same specimen, lateral view; c. right dentary fragment with incomplete M$_3$, UO 27721, occlusal view. Hancock Quarry l.f., UCMP loc. V75203. Scale bars = 1 cm.

characters in multiple specimens from both continents supports their taxonomic value in this case.) These two samples also agree in nearly all other dental features, including the absence of a labial groove separating the trigonid and talonid, and crown height, more brachydont than *Metamynodon*, *Megalamynodon*, or *Paramynodon*.

Casts of C^1 and M^2 (UCMP 136521 and 136522, respectively) from the Ube coal field, Honshu, Japan closely resemble corresponding specimens from Artëm and Hancock Quarry, and probably represent the same genus.

"*Metamynodon*" *mckinneyi* from the Porvenir l.f., Texas (Wilson and Schiebout, 1981) closely resembles *Zaisanamynodon*, the probable sister taxon to *Procadurcodon*.

A detailed description of this new species is in manuscript and will be published elsewhere.

Family RHINOCEROTIDAE Owen, 1845
Teletaceras radinskyi Hanson, 1989

Discussion — The large sample of this primitive rhinocerotid from Hancock Quarry, its type locality, has been described by Hanson (1989). Its taxonomic relationships are discussed by Prothero et al. (1986), where it is referred to as the "Clarno rhino" and the genus is shown to be the sister taxon to all other known rhinocerotids. No new material from this locality has come to light.

Other species of *Teletaceras* appear in the early Chadronian Titus Canyon l.f. near Death Valley, California, USA, and in the Sharamurunian (?) Artëm l.f., Maritime Province, eastern Russia (Hanson, 1989).

Family BRONTOTHERIIDAE Marsh, 1873
Protitanops Stock, 1936

Protitanops n. sp.

Discussion — Without reference or explanation, Mader (1989) expressed doubt that the Hancock Quarry brontothere should be referred to *Protitanops*. However direct comparison with the type of that genus, from the Titus Canyon l.f., California, reveals only minor differences but numerous unusual shared skull characters (e.g., large, rounded nasals, broad, flat postglenoid processes, thickened preorbital bar, widely separated foramen ovale and posterior alisphenoid canal, sharp medial notch in occipital crest) which support this generic assignment. At least some of these characters are also shared with *Notiotitanops*, which may be found synonymous, and suggest derivation from *Protitanotherium*.

A detailed description of the large Hancock Quarry sample of this brontothere is in manuscript for publication elsewhere.

Order ARTIODACTYLA
Family ANTHRACOTHERIIDAE Leidy, 1869
Heptacodon Marsh, 1894

Heptacodon sp.
Figure 14

Referred specimens — UO 27721, R dentary fragment with incomplete M_3 and roots of M_{1-2}; UO 28346, R P_1 crown and most of root; UCMP 141373, incomplete L P_4.

Description — M_3 protoconid with distinct but not sharp paracristid descending anterolinguad from the cusp tip toward middle of the anterior margin of the tooth, then sharply posterolinguad to meet a weak crest at the anterior base of the metaconid about 1 mm above anterior cingulid. Protoconid and metaconid also connected by crests descending from their tips to a point slightly posterior to the midpoint between them, forming low, U-shaped protolophid. Metaconid bears a third low crest along its posterolingual side, but lacks fourth, anterolabial crest seen in *Bothriodon*, *Elomeryx*, and most other anthracotheres. Posterior faces of protoconid and metaconid nearly flat and form a single oblique plane. Cristid obliqua descends from hypoconid to meet this plane about 1.5 mm. almost directly below saddle of protolophid. Cristid obliqua not joined by crest from entoconid as in *Bothriodon* and *Elomeryx*, and postcristid less prominent. Incipient hypolophid on lingual face of hypoconid. Subdued labial cingulid extends from protoconid-hypoconid valley back to broken posterior half of hypoconulid. Anterior cingulid extends from anterolabial side of protoconid to anterolingual side of metaconid. Very finely rugose pattern on most unworn surfaces of crown.

P_1 simple, single-rooted. Crown triangular in profile with minute metaconid at posterior margin of crown, merging with straight postprotocristid. Labial face of crown strongly biconvex. Lingual face approximating conical segment; shallow grooves accentuate pre- and postprotocristids. Crown not laterally compressed as in *Elomeryx* or *Bothiodon*. Resembles P_1 of SDSM 53309 (*Heptacodon* sp.; Macdonald, 1956) from Yoder Member of the Chadron Formation (Kihm, 1987), eastern Wyoming.

P_4 double-rooted with small, distinct paraconid and short, sloping anterior cingulid. Lingual cingulid extends posterad to shallow vertical groove at middle of base of protoconid. Protoconid slightly inflated with oval cross-section, bearing ridges down mesial and distal sides. Posterior cusp(s) broken off specimen.

Discussion — Despite the fragmentary nature of the Hancock Quarry specimens, enough of the nearly unworn M_3 (UO 27721) remains to permit reasonably certain generic assignment. The tooth is relatively low-crowned and bears "rounded and non-angular cusps" typical of *Heptacodon* (Macdonald, 1956). Cristids are less prominent and less numerous than in other anthracothere genera.

Comparisons with published figures, casts, and specimens of anthracotheres from North America, Asia, Europe, and Africa led to the conclusion that the Hancock Quarry specimens are best referred to *Heptacodon*, a genus whose reported distribution extends from the Duchesnean through the Whitneyan of western North America (Scott, 1940; Macdonald, 1956; Storer, 1983, 1987; Stucky, 1992). The genus is nowhere common, and temporal and geographic range extensions would not be unexpected.

Several species of *Heptacodon* have been erected, most of them based on upper cheek teeth or skulls lacking associated lower dentitions. Nearly all of the known specimens represent species which are substantially larger than the Hancock Quarry species. Directly comparable specimens from the Chadron Formation possess lower molar series ranging from 22% to 26% larger than a reasonable estimate for the Hancock Quarry jaw specimen. A specifically unassigned lower jaw, SDSM 53309, from the Yoder Member of the Chadron Formation (early Chadronian of eastern Wyoming; Macdonald, 1956; Kihm, 1987) is slightly smaller than the other Chadron and Brule Formation specimens, and possesses a P_1 whose antero-posterior dimension is close to that of the isolated P_1 from Hancock Quarry. However, M_1 and M_2 of the Yoder specimen are much closer in size to those of the Brule and Chadron specimens than to reasonable estimates, based on the preserved roots, of the same teeth of UO 27721.

Heptacodon pellionis Storer, 1983, from the early or middle Duchesnean Lac Pelletier Lower Fauna (Storer, 1987), Cypress Hills Formation, Saskatchewan, is the only species of *Heptacodon* whose type is a dentary with cheek teeth. The basic dental measurements are very close to those known or inferred for UO 27721, but morphological differences appear much greater than expectable within a species, and, in fact, call into question the generic assignment of *pellionis*. The M_3 of this species bears a cristid obliqua which extends to the metaconid rather than the middle of the protolophid. The paracristid is much straighter and terminates near the anterolingual corner of the crown. There is no anterior cristid on the metaconid such as that which, in UO 27721, meets the more medially placed termination of the paracristid. The hypoconid and especially the protoconid of *H. pellionis* appear pinched anteroposteriorly, creating more strongly convex labial surfaces than in the Hancock Quarry specimen.

The Hancock Quarry specimens thus appear to represent an undescribed species of *Heptacodon*, but the presently available material is not adequate to justify establishment of a new species.

With the possible exception of *H. pellionis*, the sample from Hancock Quarry appears to represent the earliest record of the genus, though not the earliest North American record of the family. Macdonald (1956) noted that the *Heptacodon* specimen from the Yoder l.f. (which is probably slightly younger than the Hancock Quarry local fauna) represented the earliest then known North American anthracothere. More recently, however, Coombs and Coombs's (1977) recognition of *Apriculus* (known only from a single maxillary fragment with P^4-M^3 from the Hendry Ranch Member of the Tepee Trail Formation, Wyoming) as a primitive anthracothere extends the North American record of the family back into the Uintan.

It remains unclear whether *Heptacodon* originated within North America from the *Apriculus* lineage or whether it immigrated independently. The former hypothesis would require independent derivation of, at least, parastyles and mesostyles in parallel with similar changes within exclusively Old World lineages. On the other hand, derivation of *Heptacodon* from the known Old World stock would necessitate loss of some characters which appear to be synapomorphic for all other anthracotheres (except possibly *Apriculus*), such as the anterolabial metaconid cristid, and generally more prominent crests on all molars.

This occurrence of *Heptacodon* in central Oregon is the only North American record of the family west of the Rocky Mountains.

Family AGRIOCHOERIDAE Leidy, 1869
Diplobunops Peterson, 1919

Diplobunops n. sp.

Discussion — The only artiodactyl commonly represented in the quarry sample is an undescribed agriochoerid with very large C^1 and P_1 and broadly expanded premaxillae. The snout morphology reflects, to an exaggerated degree, the characterizing feature of Uintan species of *Diplobunops*. Although the present specimens are larger and have a variably present post-P_1 diastema, they appear referable to that genus. The cheek teeth are much smaller than those of *"Agriochoerus" antiquus* known from other localities of similar age (Emry, 1981).

Diplobunops is an uncommon agriochoerid reported from Uintan localities in Arkansas (Westgate and Emry, 1985), Wyoming, and Utah. Utah localities include those in the Brennan Basin ("Randlett") Member of the Duchesne River Formation, but not the younger Lapoint Member whose faunal assemblage typifies the Duchesnean NALMA.

A full description of this new species is in preparation and will be published elsewhere.

Order RODENTIA Bowdich, 1821F
Family not determined

Specimens—UO 28304, 28307.

Discussion—Rodents are represented only by two fragmentary incisors, one of which is that of a juvenile. Identification of these specimens has not been attempted.

Biostratigraphic correlation and discussion

The unique Hancock Quarry local fauna is temporally and geographically transitional. No other local fauna includes more than two of the genera associated here, and the temporal ranges of some of the genera were not formerly known to overlap. *Epihippus* and *Diplobunops* are otherwise known from the Uintan (*Epihippus* extends into the Duchesnean) whereas two other genera and their families (Nimravidae and Rhinocerotidae) first appear elsewhere in North America in the Chadronian. *Heptacodon*, known from Chadronian through Whitneyan localities (Macdonald, 1956; Stucky, 1992) has also been reported from the Duchesnean Lac Pelletier Lower l.f (Storer, 1983, 1988). Only four of the Hancock Quarry species have been recognized elsewhere, in local faunas ascribed to the late Uintan (*Protapirus hancocki* and *Epihippus* cf. *E. gracilis* from the Candelaria l.f.) and to the late Duchesnean and/or early Chadronian (*Hemipsalodon grandis* from Yoder and Calf Creek l.f.—Schlaikjer, 1935; Russell, 1938—and *Haplohippus texanus* from Porvenir l.f.—McGrew, 1953). The other species include an advanced member of the typically Uintan genus *Diplobunops*, primitive species of genera known from early Chadronian faunas (*Teletaceras*, *Protitanops*), and a new species of *Procadurcodon*, a genus otherwise unknown in North America.

The relative weight generally accorded to first (as opposed to last) appearances in biostratigraphic correlation must be reassessed in the case of the late Eocene North American terrestrial record. Because of the much smaller numbers of localities, specimens, and recorded taxa of Duchesnean age compared with the Chadronian, taxa whose first *recorded* appearance is in the Chadronian may well have arrived or evolved earlier but escaped discovery. This is especially true for the rarer taxa, such as the Nimravidae and *Heptacodon* which Lucas (1992) lists as Chadronian first appearances. Furthermore, locally common taxa such as rhinocerotids may have been hindered in their intracontinental dispersal by the same factors which were responsible for the high degree of endemism documented for the Duchesnean (see references in Stucky, 1992, p. 471). Under these circumstances, the argument that the Hancock Quarry and Lac Pelletier Lower local faunas are Chadronian because "*Heptacodon* is a Chadronian first appearance" (Lucas, 1992) becomes somewhat circular.

Although the Hancock Quarry l.f shares no genera with the Lapoint l.f. (the assemblage which typifies the Duchesnean NALMA), the combined range zone and stage-of-evolution evidence indicates correlation with the Duchesnean, probably bracketed in time by the Candelaria and the younger Porvenir local faunas of Texas.

One of the most striking and significant aspects of the fauna is its high percentage of taxa with close relatives in the late Eocene subcoastal faunas of eastern Asia. Geographic ranges of at least three and probably four of the ten Hancock Quarry large-mammal genera extended to Japan, Russia, and/or North Korea. *Procadurcodon* was first described from the Artëm locality in Maritime Province, Russia (Gromova, 1960), and additional specimens apparently referable to the genus have been collected in the Ube coal field, Honshu, Japan. *Protapirus* is represented in Hokkaido, Japan ("*Colodon?*" *kushiroensis*, Tomida, 1983), Ube, and Hosan, North Korea ("*Desmatotherium*" spp., Takai, 1939, 1944; Tomida, 1983), as well as the younger type locality near Quercy, France, and other European localities (Cerdeno and Ginsburg, 1988). Hanson (1989) included specimens from the Artëm locality in the genus *Teletaceras*. Teeth comprising a nearly complete dental series from Hosan (*Protitanotherium koreanicum*, Takai, 1939) closely resemble those of *Protitanops* n. sp. from Hancock Quarry, and appear to represent a congeneric species. Three other taxa (*Hemipsalodon*, *Haplohippus*, and Nimravidae) represent first or very early North American occurrences of clades probably rooted in Eurasia. North American ancestry and exclusive distribution can be clearly documented for only two genera, *Epihippus* and *Diplobunops*.

The Asian links not only document a high rate of intercontinental migration during the Duchesnean, but also offer potential improvements in intercontinental biostratigraphic correlation. Lucas (1992) has commented on the affinities of the Hancock Quarry local fauna to the later Sharamurunian LMA of Li and Ting (1983), although his suggestion of a separate faunal province encompassing northwestern North America and eastern Asia during this interval may be premature in view of the similar numbers of taxa shared with other North American localities as well as those in Asia, and the degree of provinciality apparent among Duchesnean sites within North America (Storer, 1988, 1992; Golz and Lillegraven, 1977).

Little evidence seems to support the contention that this assemblage is "adapted to cooler and drier conditions" (Bestland, Retallack, and Fremd, 1994a) than that of the Nut Beds assemblage. All of the Hancock Quarry herbivores are relatively brachydont browsers and each of the Nut Beds taxa has a related or ecologically equivalent counterpart in the Hancock Quarry assemblage (except the Chelonia, represented by a terrestrial tortoise in the Nut Beds and an aquatic predator at Hancock Quarry). High-crowned, selenodont artiodactyls, common and diverse in most Uintan through Chadronian local faunas, are conspicuously absent here.

RESOLUTION OF CORRELATIONS AND DATES

In the foregoing, I have attempted to develop three independent lines of evidence—physical stratigraphic superposition, radiometric dates, and biostratigraphic correlations—bearing on the chronology of the Hancock Canyon stratigraphic section. After elimination of the K-Ar dates which conflict with more recent analyses, the remaining dates were found generally consistent with observed stratigraphic relationships. (References for these and dates mentioned below may be found in the "Chronostratigraphy" section above.) It remains to test the concordance of Clarno Formation dates with other dates on biostratigraphically correlated rocks elsewhere in western North America.

In this regard, Unit B, including the Nut Beds locality, presents some discrepancies. Mammalian correlation of the Nut Beds to dated Bridgerian localities in the Rocky Mountains region suggests either that the actual age is close to the oldest of the five Unit B dates or that the Bridgerian/Uintan boundary is diachronous. Stucky (1992) places this boundary at 48 Ma. Data summarized in Krishtalka et al. (1987) suggest a boundary age near 48.4 Ma, which would separate all but one of a large number of published Rocky Mountains K-Ar dates on Bridgerian and older assemblages from those on Uintan and younger assemblages. However, Prothero and Swisher (1992) indicate that some of these dates may have been influenced by inclusion of detrital biotite and "may be some 1.5 million years too old," possibly shifting the Bridgerian/ Uintan boundary to about 47 Ma. In view of the proximity of the Nut Beds to a younger intrusive plug, the evidence for metasomatic alteration of the Nut Beds tuff and the broad spread of its individual ^{40}Ar/^{39}Ar dates (B. Turrin, personal communication, 1990), large (±10%) error margins associated with the fission track dates, and the angular unconformity separating Unit B from higher rocks dated at about 44 Ma, I am of the opinion that the age of the Nut Beds is near the oldest of the available dates, and that the Nut Beds l.f. was essentially contemporaneous with the late Bridgerian faunas of the intermontane region, approximately 47 to 49 Ma. None of the published dates would be inconsistent with paleobotanical correlation (Wolfe, 1981) to the early type Kummerian megafloral stage, which Turner et al. (1983) have dated at older (perhaps substantially older) than 43.4 ± 0.95 Ma.

Fission-track, ^{40}Ar/^{39}Ar, and (with one exception) K-Ar dates from units C and D in Hancock Canyon and from correlated rocks to the south all fall near 41 to 44 Ma. This is in accord with the stratigraphic position of units C and D between the Bridgerian Nut Beds l.f. and the Duchesnean Hancock Quarry l.f.

Revised interpretation of the Hancock Canyon section, placing the Nut Beds below Unit C, and correlation of Unit C to the John Day Gulch flora via the Rajneesh tuff, invert the relative stratigraphic positions of the "Middle" and "Upper" Clarno floras of Wolfe (1981) and eliminate the evidence for the "terminal Eocene event" that the floras and dates of the area formerly appeared to provide. The revisions support the interpretation that the climatic deterioration started much earlier (between about 48 and 44 Ma vs. 34 to 32 Ma) and proceeded at a slower rate than formerly thought.

Unit E, including Hancock Quarry, contains no datable minerals, but concordantly underlies an ignimbrite dated by ^{40}Ar/^{39}Ar, K-Ar, and fission-track methods. The ^{40}Ar/^{39}Ar dates near 39.5 Ma are probably more accurate than dates by the other methods, all of which fall near 37.5 Ma. The fossil locality disconformably overlies rocks of Unit D, with dates as young as 41 Ma. As discussed above, biochronologic data suggest correlation of the Hancock Quarry local fauna with the interval approximately bounded by the Candelaria l.f. (late Uintan; Wilson, 1984) and the Porvenir l.f. (late Duchesnean and/or earliest Chadronian; Wilson, 1984, figs. 1 and 2 respectively) from the dated sequence in Texas. The Porvenir l.f. is bracketed by two volcanic units, recently redated at 36.5 Ma, and 37.5 Ma (Prothero and Swisher, 1992). The Candelaria correlates with nearby local faunas dated between 43.9 Ma and 42.7 Ma (Wilson, 1984). The ^{40}Ar/^{39}Ar ignimbrite dates also predate the 37.2 ± 0.7 Ma K-Ar date on a tuff near the top of the Ahearn Member of the Chadron Formation, South Dakota, but approach in age the base of the type Duchesnean Lapoint member, Duchesne River Formation, Utah, dated at 40.3 ± 0.8 Ma (McDowell et al., 1973, corrected for revised decay constants).

In light of revisions of North American Paleogene chronology by Swisher and Prothero (1990), the consistent dates near 32 Ma on the Bridge Creek flora of the lower John Day Formation indicate that the nearly unfossiliferous portion of the John Day Formation below the Bridge Creek flora corresponds to the interval spanned by the latest Duchesnean, Chadronian, and Orellan NALMA's.

CONCLUSIONS

Establishment of a physical stratigraphic framework for the numerous dates and fossil localities of the type Clarno Formation facilitates the resolution of long-standing controversies in interpretation of the area's abundant faunal, floral, climatic, and chronometric data. Most of the controversies arose from a lack of recognition of the local geological complexities, especially the presence of several unconformities within the fossiliferous Clarno sequence. Of the five recognized unconformity-bounded Clarno subunits (allostratigraphic units), the second lowest and the highest include both faunal and floral assemblages, and the middle unit has produced a large undescribed flora.

The older local fauna, from the Nut Beds locality, is known from only six vertebrate taxa, but all are consistent with biostratigraphic correlation to the late Bridgerian NALMA. None of the genera and only one species is endemic to the western coastal states. Nominal values of most of the published dates from this unit are slightly younger than those associated with Bridgerian faunas in Wyoming, probably a result of both post-depositional alteration of the Clarno samples and contamination of the Wyoming samples by older mineral grains (Prothero and Swisher, 1992). An $^{40}Ar/^{39}Ar$ date of 48.32 Ma (Swisher, 1992) is probably the most accurate on technological grounds, and is the most consistent with physical stratigraphy and mammalian biochronology,

The faunal assemblage from the younger Hancock Quarry locality correlates both biostratigraphically and chronostratigraphically with Duchesnean faunas elsewhere in western North America, and documents both endemism and a high rate of intermigration between Asia and North America at this time.

Stratigraphic and chronometric considerations dictate reinterpretation of the Clarno-John Day floral sequence and its implications for Paleogene climatic change in western North America.

ACKNOWLEDGMENTS

This study would not have begun without the help and encouragement of Dr. D. E. Savage who made me aware of the large, unstudied Hancock Quarry collection at the University of Oregon. Dr. J. A. Shotwell, former director of the Condon Museum at the University of Oregon, loaned a large part of this invaluable collection for this study. Dr. J. Armentrout, former director of Camp Hancock (now Hancock Field Station), was instrumental in arranging programs involving student volunteers to collect additional material from the same locality in 1969 through 1972, and in providing an ideal base of field operations for the research reported here. Subsequent directors, especially Joseph Jones, and the Oregon Museum of Science and Industry, operator of Hancock Field Station, continued to provide similar support. I gratefully (and belatedly) thank the following student volunteers who spent long, arduous days of their summer vacations extracting fragile specimens from hard rock under the immediate direction of Dave Taylor, Annalisa Berta, and Ron Wolff: Dave Anderson, Scott Ballew, Scott Bruce, Casey Burns, Nanci Courtney, Graham Deacon, John Faulhaber, Sheryl Fleming, Jackie Freeman, Chris Galen, "Happy" Heiberg, Stewart Janes, Greg Lippert, Herb Meyer, John Norwood, Greg Paul, Tom McKee, Karen Sweetman, Tom Thompson, Brian Warrington, Dan Wagner, and Steve Zann.

I am grateful to the University of California Museum of Paleontology for financial assistance for part of the field work and for countless hours of preparation time. Financial support for one field season was provided by the Geological Society of America. Wes Wehr relocated a long-lost collection of Clarno specimens collected by A. W. "Lon" Hancock, and generously donated important specimens to UCMP. I thank Dr. John Rensberger of UWBM for the loan of additional Hancock Quarry specimens from the same collection.

I am also indebted to the organizers of the 1989 Rapid City Penrose Conference for their efforts. I especially thank Dr. Donald R. Prothero for his patience and encouragement, without which this would not have been completed. I thank Ted Fremd, Spencer Lucas, Greg Retallack, and John Storer for reviewing an earlier version of this contribution.

Dr. Howard Schorn provided help and advice in producing the photographs included here. Jaime Lufkin drew the line illustrations in Figures 5, 6, and 7.

Finally, I would like to express my gratitude to my wife, Vicki, for her boundless patience, understanding, and support throughout the course of this research.

LITERATURE CITED

Armentrout, J. A. 1981. Correlation and ages of Cenozoic chronostratigraphic units in Oregon and Washington. Geological Society of America Special Paper 184:137-148.

Beliaeva, E. I. 1971. Novye dannye po aminodontam SSSR [New data on the amynodonts of the USSR]. Trudy Paleontologicheskogo Instituto, Akademii nauk SSSR 130:39-61.

Bestland, E. A., G. J. Retallack, T. Fremd, and C. C. Swisher, III. 1994. Geology and age assessment of late Eocene fossil localities in the Clarno unit, John Day Fossil Beds National Monument, central Oregon. Geological Society of America Abstracts with Programs 26(6):4.

Bestland, E. A., G. J. Retallack, and T. Fremd. 1994a. Geology of late Eocene Clarno unit, John Day Fossil Beds National Monument, central Oregon. National Park Service Paleontological Research Symposium, Geological Society of America.

Bestland, E. A., G. J. Retallack, and T. Fremd. 1994b. Sequence stratigraphy of the Eocene-Oligocene transition: examples from the non-marine volcanically influenced John Day Basin; pp. 11-19 in D. A. Swanson and R. A. Haugerud (eds.), Geologic Field Trips in the Pacific Northwest, University of Washington, Seattle.

Bown, T. M. 1982. Geology, paleontology, and correlation of Eocene volcaniclastic rocks, southwest Absaroka Range, Hot Springs County, Wyoming. U. S. Geological Survey Professional Paper 1201-A:A1-A75.

Bryant, H. N. 1991. Phylogenetic relationships and systematics of the Nimravidae (Carnivora). Journal of Mammalogy 72(1):56-78.

Cerdeno, E., and L. Ginsburg. 1988. Les Tapiridae (Perissodactyla, Mammalia) de l'Oligocene et du Miocene inferieur Europeens. Annales de Paleontologie, Paris 74:71-96.

Chandler, M. E. J. 1964. The Lower Tertiary Floras of Southern England, IV, London, British Museum (Natural History), 151 pp.

Chaney, R. W. 1937. Cycads from the upper Eocene of Oregon. Proceedings of the Geological Society of America for 1936, p. 397 (abstract).

Coombs, W. P., Jr., and M. C. Coombs. 1977. The origin of the anthracotheres. Neues Jahrbuch für Geologie und Paläontologie, Monatshefte 1977:84-599.

Dunbar, C. O., and J. Rogers. 1957. Principles of Stratigraphy. Wiley, New York, 356 pp.

Eaton, J. G. 1985. Paleontology and correlation of the Eocene Tepee Trail and Wiggins Formations in the North Fork of Owl Creek Area, southeastern Absaroka Range, Hot Springs County, Wyoming. Journal of Vertebrate Paleontology 5:345-370.

Emry, R. 1981. Additions to the mammalian fauna of the type Duchesnean, with comments on the status of the Duchesnean "age." Journal of Paleontology 55:563-570.

Emry, R. 1990. Mammals of the Bridgerian (middle Eocene) Elderberry Canyon Local Fauna of eastern Nevada. Geological Society of America Special Paper 243:187-210.

Emry, R. 1992. Mammalian range zones in the Chadronian White River Formation at Flagstaff Rim, Wyoming; pp. 106-115 in D. R. Prothero and W. A. Berggren (eds.), Eocene-Oligocene Climatic and Biotic Evolution. Princeton University Press, Princeton, N. J.

Evernden, J. F., and G. T. James. 1964. Potassium-argon dates and the Tertiary floras of North America. American Journal of Science 262:945-974.

Evernden, J. F., D. E. Savage, G. H. Curtis, and G. T. James. 1964. Potassium-argon dates and the Cenozoic mammalian chronology of North America. American Journal of Science 262:145-198.

Fiebelkorn, R. B., G. W. Walker, N. S. MacLeod, E. H. McKee, and J. G. Smith. 1983. Index to K-Ar determinations for the State of Oregon. Isochron West 37:3-60.

Flynn, J. J. 1986. Correlation and geochronology of Middle Eocene strata from the western United States. Palaeogeography, Palaeoclimatology, Palaeoecology 55:335-406.

Forsten, A. 1971. Early Tertiary vertebrate faunas, Vieja Group, Trans-Pecos, Texas: Part 1: Epihippus from the Vieja Group, Trans-Pecos Texas. Texas Memorial Museum Pearce-Sellards Series 18:1-5.

Gazin, C. L. 1976. Mammalian faunal zones of the Bridger Middle Eocene. Smithsonian Contributions in Paleobiology 2625.

Golz, D. J., and J. A. Lillegraven. 1977. Summary of known occurrences of terrestrial vertebrates from Eocene strata of southern California. University of Wyoming Contributions in Geology 15:43-65.

Granger, W. 1908. A revision of the American Eocene horses. Bulletin of the American Museum of Natural History 24:221-264.

Gromova, V. I. 1960. Pervaia nakhodkha v sovetskom soiuze aminodonta (novyi rod Procadurcodon) [First find in the Soviet Union of an amynodont (new genus Procadurcodon)]. Trudy Paleontologicheskogo Instituto, Akademii nauk SSSR 77:128-151.

Hammond, P. E. 1993. Preliminary geologic map of the Clarno Unit (Camp Hancock area), John Day Fossil Beds National Monument, Oregon. (Abstract) Proceedings of the Oregon Academy of Sciences 1993:34.

Hanson, C. B. 1973. Geology and vertebrate faunas in the type area of the Clarno Formation, Oregon. Geological Society of America, Cordilleran Section, Abstracts with Programs 1973:50.

Hanson, C. B. 1979. Geologic map of Hancock Canyon and vicinity; in B. Hansen (ed.), An Introduction to the Natural History of Hancock Field Station and the Clarno Basin, North Central Oregon, 3rd Edition, The Oregon Museum of Science and Industry Research Center, Portland.

Hanson, C. B. 1980. Fluvial taphonomic processes: Models and experiments; pp. 156-181 in A. K. Behrensmeyer and A. P. Hill (eds.), Fossils in the Making. University of Chicago Press, Chicago.

Hanson, C. B. 1989. Teletaceras radinskyi, a new primitive rhinocerotid from the late Eocene Clarno Formation, Oregon; pp. 379-398 in D. R. Prothero and R. M. Schoch (eds.), The Evolution of Perissodactyls. Oxford University Press, New York.

Hutchison, J. H. 1992. Western North American reptile and amphibian record across the Eocene/Oligocene boundary and its climatic implications; pp. 451-463 in D. R. Prothero and W. A. Berggren (eds.), Eocene-Oligocene Climatic and Biotic Evolution. Princeton University Press, Princeton, N. J.

Janes, S. 1971. Stratigraphy and mapping of the Clarno Nut Beds. Oregon Museum of Science and Industry Student Research Reports 11:35-40.

Kihm, A. J. 1987. Mammalian paleontology and geology of the Yoder Member, Chadron Formation, east-central Wyoming; pp. 28-45 in J. E. Martin and G. E. Ostrander (eds.), Papers in Vertebrate Paleontology in Honor of Morton Green. Dakoterra 3.

Kitts, D. B. 1957. A revision of the genus Orohippus (Perissodactyla, Equidae). American Museum Novitates 1864:1-40.

Knowlton, F. H. 1902. Fossil flora of the John Day Basin, Oregon. U. S. Geological Survey Bulletin 204:9-153.

Krishtalka, L., R. K. Stucky, R. M. West, M. C. McKenna, C. C. Black, T. M. Bown, M. R. Dawson, D. J. Golz, J. J. Flynn, J. A. Lillegraven, and W. D. Turnbull. 1987. Eocene (Wasatchian through Duchesnean) biochronology of North America; pp. 77-117 in M. O. Woodburne (ed.), Cenozoic Mammals of North America, Geochronology and Biostratigraphy, University of California Press, Berkeley.

Li, C., and S. Ting. 1983. The Paleogene mammals of China. Bulletin of the Carnegie Museum of Natural History 21:1-93.

Lucas, S. G. 1982. Vertebrate paleontology, stratigraphy and biostratigraphy of the Eocene Galisteo Formation, north-central New Mexico. New Mexico Bureau of Mines and Mineral Resources Circular 186:1-34.

Lucas, S. G . 1989. Fossil mammals and Paleocene-Eocene boundary in Europe, North America and Asia. 28th International Geological Congress (Washington, D.C.) Abstracts 2:335.

Lucas, S. G. 1992. Redefinition of the Duchesnean land mammal "age," late Eocene of western North America; pp. 88-105 in D. R. Prothero and W. A. Berggren (eds.), Eocene-Oligocene Climatic and Biotic Evolution. Princeton University Press, Princeton, N. J.

Macdonald, J. R. 1956. The North American anthracotheres. Journal of Paleontology 30:615-645.

MacFadden, B . J. 1980. Eocene perissodactyls from the type section of the Tepee Trail Formation of northwestern Wyoming. Contributions to Geology, University of Wyoming 18(2):135-143.

Mader, B. J. 1989. The Brontotheriidae: a systematic revision and preliminary phylogeny of North American genera; pp. 458-484 in D. R. Prothero and R. M. Schoch (eds.), The Evolution of Perissodactyls. Oxford University Press, New York.

Manchester, S. R. 1981. Fossil plants of the Eocene Clarno Nut Beds. Oregon Geology 43:75-81.

Manchester, S. R. 1990. Eocene to Oligocene floristic changes recorded in the Clarno and John Day Formations, Oregon, USA; pp. 183-187 in E. Knobloch and Z. Kvacek (eds.), Symposium Proceedings, Paleofloristic and paleoclimatic changes in the Cretaceous and Tertiary, Geological Survey Press, Prague, Czechoslovakia.

Manchester, S. R. 1994. Fruits and seeds of the middle Eocene Nut Beds flora, Clarno Formation, Oregon. Paleontographica Americana 58:1-205.

Manchester, S. R., and H. W. Meyer. 1987. Oligocene fossil plants of the John Day Formation, Fossil, Oregon. Oregon Geology 49:115-127.

Matthew, W. D. 1909. The Carnivora and Insectivora of the Bridger Basin, middle Eocene. Memoirs of the American Museum of Natural History 9:289-567.

McDowell, F. W., J. A. Wilson, and J. Clark. 1973. K-Ar dates for biotite from two paleontologically significant localities: Duchesne River Formation, Utah and Chadron Formation, South Dakota. Isochron West 7:11-12.

McGrew, P. O. 1953. A new and primitive early Oligocene horse from Trans-Pecos, Texas. Fieldiana, Geology 10:167-171.

McKee, T. M. 1970. Preliminary report on fossil fruits and seeds from the Mammal Quarry of the Clarno Formation, Oregon. Ore Bin 32:117-132.

Mellett, J. S. 1969. A skull of Hemipsalodon (Mammalia, Deltatheridia) from the Clarno Formation of Oregon. American Museum Novitates 2387:1-19.

Merriam, J. C. 1901a. A geological section through the John Day Basin. Journal of Geology 9:71-72.

Merriam, J. C. 1901b. A contribution to the geology of the John Day Basin. University of of California Publications in Geological Sciences 2:269-314.

Noblett, J. B. 1979. Volcanic petrology of the Eocene Clarno Formation on the John Day River near Cherry Creek, Oregon. Ph.D. dissertation, Stanford University, Stanford, CA. 162 pp.

Noblett, J. B. 1981. Subduction-related origin of the volcanic rocks of the Eocene Clarno Formation near Cherry Creek, Oregon. Oregon Geology 43:91-99.

North American Commission on Stratigraphic Nomenclature. 1983. North American Stratigraphic Code. Bulletin of the American Association of Petroleum Geologists 67:841-875.

Oles, K. F., and H. E. Enlows. 1971. Bedrock geology of the Mitchell Quadrangle, Wheeler County, Oregon. Oregon Department of Geology and Mineral Industries Bulletin 72:1-62.

Osborn, H. F. 1929. The titanotheres of ancient Wyoming, Dakota, and Nebraska. U.S. Geological Survey Monograph 55:1-953.

Peck, D. L. 1964. Geologic reconnaissance of the Antelope-Ashwood area, north-central Oregon, with emphasis on the John Day Formation of late Oligocene and early Miocene age. U. S. Geological Survey Bulletin 1161-D:D1- D26.

Pratt, J. A. 1988a. Paleoenvironment of the Eocene/Oligocene Hancock Mammal Quarry, upper Clarno Formation, Oregon. Geological Society of America, Cordilleran Section, Abstracts with Programs, p. 222.

Pratt, J. A. 1988b. Paleoenvironment of the Eocene/Oligocene Hancock Mammal Quarry, upper Clarno Formation, Oregon. M.S. Thesis, University of Oregon, Eugene, OR. 104 pp.

Prothero, D. R., E. M. Manning, and C. B. Hanson. 1986. The phylogeny of the Rhinocerotoidea (Mammalia, Perissodactyla). Zoological Journal of the Linnean Society 87:341-366.

Prothero, D. R., and C. C. Swisher, III. 1992. Magnetostratigraphy and geochronology of the terrestrial Eocene-Oligocene transition in North America; pp. 74-87 in D. R. Prothero and W. A. Berggren (eds.), Eocene-Oligocene Climatic and Biotic Evolution. Princeton University Press, Princeton, N. J.

Radinsky, L. B. 1963. Origin and early evolution of North American Tapiroidea. Peabody Museum of Natural History Bulletin 17:1-106.

Radinsky, L. B. 1967. Hyrachyus, Chasmotherium, and the early evolution of helaletid tapiroids. American Museum Novitates 2313:1-23.

Retallack, G. 1981. Preliminary observations on fossil soils in the Clarno Formation (Eocene to early Oligocene) near Clarno, Oregon. Oregon Geology 43:147-150.

Retallack, G. 1991. A field guide to mid-Tertiary paleosols and paleoclimatic changes in the high desert of central Oregon—Part 1. Oregon Geology 53:51-59.

Retallack, G., E. A. Bestland, and T. Fremd. 1993. Reassessment of the age of fossil localities in the Clarno Formation, Hancock Field Station, Wheeler County, Oregon. Oregon Academy of Science Annual Meeting.

Robinson, P. T. 1975. Reconnaissance geologic map of the John Day Formation in the southwestern part of the Blue Mountains and adjacent areas, north-central Oregon. U. S. Geological Survey Miscellaneous Investigations Map I-872.

Robinson, P. T., and E. H. McKee. 1983. Index to K-Ar determinations for the state of Oregon. Isochron West 37:3-60.

Robinson, P. T., G. W. Walker, and E. H. McKee. 1990. Eocene(?), Oligocene and lower Miocene rocks of the Blue Mountains region. U.S. Geological Survey Professional Paper 1437:29-62.

Russell, L. S. 1938. The skull of Hemipsalodon grandis, a giant Oligocene creodont. Transactions of the Royal Society of Canada, Series 3, 32:61-66.

Schlaikjer, E. M. 1935. Contributions to the stratigraphy and paleontology of the Goshen Hole area, Wyoming. III: A new basal Oligocene formation. Bulletin of the Museum of Comparative Zoology 76:71-93.

Schoch, R. M. 1989. A review of the tapiroids; pp. 298-320 in D. R. Prothero and R. M. Schoch (eds.), The Evolution of Perissodactyls. Oxford University Press, New York.

Scott, R. A. 1954. Fossil fruits and seeds from the Eocene Clarno Formation of Oregon. Stuttgart, Palaeontographica, Abteilung B 96:66-97.

Scott, W. B. 1940. The mammalian fauna of the White River Oligocene. Part IV—Artiodactyla. American Philosophical Society Transactions 28:363-746.

Smith, G. A. 1991. Facies sequences and geometries in continental volcaniclastic sediments; pp. 109-121 in R. V. Fisher and G. A. Smith (eds.), Sedimentation in volcanic settings. SEPM Special Publication 45.

Smith, G. S. 1988. Paleoenvironmental reconstruction of Eocene fossil soils from the Clarno Formation in eastern Oregon. M.S. thesis, University of Oregon, Eugene, OR, 167 pp.

Stirton, R. A. 1944. A rhinoceros tooth from the Clarno Eocene of Oregon. Journal of Paleontology 18:265-267.

Stock, C. 1936. Titanotheres from the Titus Canyon Formation, California. Proceedings of the National Academy of Sciences 22(11):656-661.

Storer, J. E. 1983. A new species of the artiodactyl *Heptacodon* from the Cypress Hills Formation, Lac Pelletier, Saskatchewan. Canadian Journal of Earth Sciences 20:1344-1347.

Storer, J. E. 1987. Dental evolution and radiation of Eocene and Early Oligocene Eomyidae (Mammalia, Rodentia) of North America, with new material from the Duchesnean of Saskatchewan. Dakoterra 3:108-117.

Storer, J. E. 1988. The rodents of the Lac Pelletier Lower fauna, late Eocene (Duchesnean) of Saskatchewan. Journal of Vertebrate Paleontology 8(1):84-101.

Stucky, R. K. 1992. Mammalian faunas in North America of Bridgerian to early Arikareean "ages" (Eocene and Oligocene); pp. 463-493 *in* D. R. Prothero and W. A. Berggren (eds.), Eocene-Oligocene Climatic and Biotic Evolution. Princeton University Press, Princeton, N. J.

Swanson, D. R. 1969. Reconnaissance map of the east half of Bend quadrangle, Crook, Wheeler, Jefferson, Wasco, and Deschutes counties, Oregon. U.S. Geological Survey Miscellaneous Investigations Map I-568. Scale 1:250,000.

Swanson, D. R. and P. T. Robinson. 1968. Base of the John Day Formation in and near the Horse Heaven mining district, north-central Oregon. U.S. Geological Survey Professional Paper 600-D:D154-D161.

Swisher, C. C. III. 1992. ^{40}Ar/^{39}Ar dating and its application to the calibration of the North American Land Mammal ages. Ph.D. dissertation, University of California, Berkeley.

Swisher, C. C. III, and D. R. Prothero. 1990. Single-crystal ^{40}Ar/^{39}Ar dating of the Eocene-Oligocene transition in North America. Science 249:760-762.

Takai, F. 1939. Eocene mammals found from the Hosan coal-field, Tyosen. Journal of the Faculty of Science, Imperial University, Tokyo 5:199-217.

Takai, F. 1944. Eocene mammals found in the Ube and Hosan coal-fields in Nippon. Proceedings of the Imperial Academy 20:736-741.

Taylor, E. M. 1960. Geology of the Clarno Basin, Mitchell Quadrangle, Oregon. M.S. thesis, Oregon State College, Corvallis. 173 pp.

Taylor, E. M. 1981. A mafic dike system in the vicinity of Mitchell, Oregon, and its bearing on the timing of Clarno-John Day volcanism and early Oligocene deformation in central Oregon. Oregon Geology 43(8):107-112.

Tomida, Y. 1983. A new helaletid tapiroid (Perissodactyla, Mammalia) from the Paleogene of Hokkaido, Japan, and the age of the Urahoro Group. Bulletin of the National Science Museum, Tokyo, Series C 9:151-163.

Turner, D. L., V. A. Frizzell, D. M. Triplehorn, and C. W. Naeser. 1983. Radiometric dating of ash partings in the Eocene Puget Group, Washington: Implications for paleobotanical stages. Geology 44:527-531.

Vance, J. A. 1988. New fission track and K-Ar ages from the Clarno Formation, Challis age volcanic rocks in north central Oregon. Geological Society of America Rocky Mountain Section Abstracts with Program 20(6):473.

Walker, G. W., and P. T. Robinson. 1990. Paleocene(?), Eocene, and Oligocene(?) rocks of the Blue Mountains region. U.S. Geological Survey Professional Paper 1437:13-27.

Wall, W. P. 1989. The phylogenetic history and adaptive radiation of the Amynodontidae; pp. 341-354 *in* D. R. Prothero and R. M. Schoch (eds.), The Evolution of Perissodactyls. Oxford University Press, New York.

Waters, A. C., R. E. Brown, R. R. Compton, L. W. Staples, G. W. Walker, and H. Williams. 1951. Quicksilver deposits of the Horse Heaven mining district, Oregon. U.S. Geological Survey Bulletin 969-E:E105-E149.

Westgate, J. W., and R. J. Emry. 1985. Land mammals of the Crow Creek local fauna, late Eocene, Jackson Group, St. Francis County, Arkansas. Journal of Paleontology 59:242-248.

White, J. D. L., and P. T. Robinson. 1992. Intra-arc sedimentation in a low-lying marginal arc, Eocene Clarno Formation, central Oregon. Sedimentary Geology 80:89-114.

Wilson, J. A. 1977. Stratigraphic occurrence and correlation of early Tertiary vertebrate faunas, Trans-Pecos, Texas; Part 1: Vieja area. Texas Memorial Museum Bulletin 25:1-42.

Wilson, J. A. 1984. Vertebrate faunas 49 to 36 million years ago and additions to the species of Leptoreodon (Mammalia, Artiodactyla) found in Texas. Journal of Vertebrate Paleontology 4:199-207.

Wilson, J. A. 1984. Early Tertiary vertebrate faunas, Trans-Pecos Texas: Ceratomorpha less Amynodontidae. Texas Memorial Museum Pearce-Sellards Series 39:1-47.

Wilson, J. A. and J. A. Schiebout. 1981. Early Tertiary vertebrate faunas, Trans-Pecos Texas: Amynodontidae. Texas Memorial Museum Pearce-Sellards Series 33:1-62.

Wolfe, J. A. 1971. Tertiary climatic fluctuations and methods of analysis of Tertiary floras. Palaeogeography, Palaeoclimatology, Palaeoecology 9:27-57.

Wolfe, J. A. 1981. A chronologic framework for Cenozoic megafossil floras of northwestern North America and its relation to marine geochronology. Geological Society of America Special Paper 184:39-47.

Wolfe, J. A. 1992. Climatic, floristic, and vegetational changes near the Eocene/Oligocene boundary in North America; pp. 421-436 *in* D. R. Prothero and W. A. Berggren (eds.), Eocene-Oligocene Climatic and Biotic Evolution. Princeton University Press, Princeton, N. J.

Wolfe, J.A. and D. M. Hopkins. 1967. Climatic changes recorded by Tertiary land floras in northwestern North America; pp. 67-76 *in* K. Hatai (ed.), Tertiary Correlations and Climatic Changes in the Pacific. Sasaki Printing and Publishing Co., Sendai, Japan.

12. Eocene-Oligocene Faunas of the Cypress Hills Formation, Saskatchewan

JOHN E. STORER

ABSTRACT

The Cypress Hills Formation of southwestern Saskatchewan preserves fossil mammals of Uintan (middle Eocene) to Hemingfordian (early Miocene) age. Deposition was nearly continuous during this interval, and the widely held belief that a limited period of Uintan sedimentation on the Swift Current Plateau was followed by Chadronian sedimentation throughout the area, the two separated by a period of non-deposition, is incorrect. Eocene assemblages from the Cypress Hills Formation have aided in identifying the profound middle Eocene faunal turnover as taxonomic displacement, and in documenting the greatly increased faunal provinciality of the Duchesnean. Late Eocene and Oligocene assemblages, many of them newly discovered, provide a rare opportunity to trace the evolution and eventual replacement of the White River Chronofauna in the northern Great Plains. The classic concept of the "Cypress Hills Oligocene" is based on Chadronian to Hemingfordian specimens collected from a large geographical area, and must be abandoned in favor of a model involving many successive local faunas. The local fauna most similar to the "Cypress Hills Oligocene" of the literature is the middle Chadronian (late Eocene) Calf Creek Local Fauna.

INTRODUCTION

The Cypress Hills Formation of the Swift Current and Cypress Hills Plateaus (Figs. 1, 2), southwestern Saskatchewan, preserves a series of mammalian paleofaunas representing middle Eocene to early Miocene time. Uintan to Chadronian local faunas in the Swift Current Plateau and Chadronian to Hemingfordian assemblages in the Cypress Hills Plateau furnish a unique opportunity to trace the evolution of mammals in the northern Great Plains through the period of profound middle Eocene faunal replacement, much of the history of the White River Chronofauna, and into the beginnings of the Miocene Chronofauna. In this chapter I will discuss the Eocene and Oligocene local faunas of the Cypress Hills Formation, those belonging to the Uintan through Arikareean land mammal "ages."

The Cypress Hills Formation reaches nearly 76 meters in thickness (Vonhof, 1965). It is composed of fluviatile sands and gravels, in places cemented into sandstones and conglomerates, and debris flow deposits (Leckie and Cheel, 1989), alternating with mudstones and sandy mudstones. These deposits were formed over a long period of time (at least 28 million years). This is an unexpectedly long aggradational phase, 15-20 million years longer than recognized in previous interpretations (Taylor et al., 1964).

Recent analysis indicates that the source of the Cypress Hills deposits was "a northeasterly-flowing braided-river system on a braidplain that headed in the Sweetgrass Hills, Bearpaw Mountains, and possibly Highwood Mountains" of Montana (Leckie and Cheel, 1989, p. 1930). Following conclusions based on study of the herpetofauna (Holman, 1969, 1972) and citing the presence of silcrete nodules within the fluviatile deposits, Leckie and Cheel (1989, p. 1927) considered that the area was semi-arid, with "subtropical to tropical xeric climate with periodic or seasonal rainfall, similar to the climate of modern Mexican coastal lowlands."

Cypress Hills sands and gravels and debris flows are fossiliferous at many localities. Although a few well preserved fossils have been found in the mudstones, specimens from mudstone layers are rare. A detailed stratigraphy of the Cypress Hills Formation can probably be constructed using occurrence of mudstone beds combined with less extreme changes of facies as lithologic markers, correlated by fossil mammals as a biostratigraphic control. This study is just beginning, and in building up a picture of faunal succession it will be necessary to examine assemblages recovered from single localities. Study of Cypress Hills biostratigraphy in the Eastend and Swift Current areas offers two advantages that may make it possible to construct a regional interpretation: (1) many localities yield abundant small fossil vertebrates, so that extensive local faunas can be assembled for study and correlation; and (2) superposition of local faunas is present in several sections, with as many as four assemblages of different ages documented in successive beds. With further study it may be possible to identify beds that can be dated

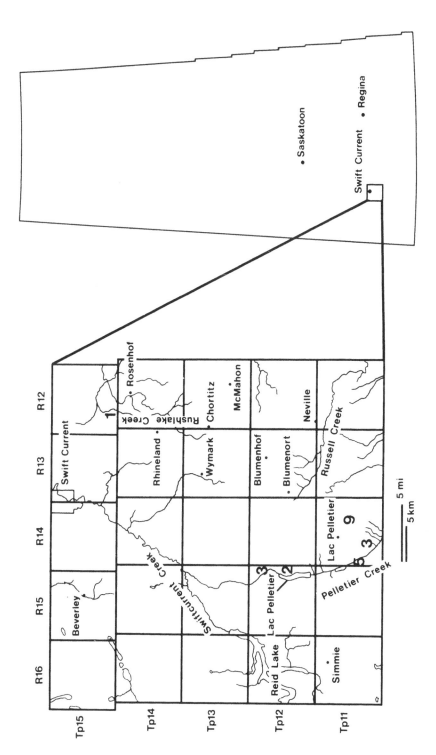

Figure 1. Outline drawing of part of Swift Current Plateau south of Swift Current, showing place names and sources of Eocene-Oligocene mammalian paleofaunas. All ranges west of the Third Meridian. 1, Swift Current Creek Local Fauna; 2, Lac Pelletier Lower Fauna; 3, Lac Pelletier Upper Fauna; 5, Simmie Local Fauna; 9, Blumenort Local Fauna.

radiometrically, although no ash beds or other suitable deposits are now known.

This view of the Cypress Hills Formation has not always been the accepted one. Until recently the formation was regarded as preserving faunas of two ages: Uintan (middle Eocene) in the Swift Current beds of the Swift Current Plateau, and Chadronian (late Eocene) in the rest of the formation on both plateaus. The concept of restricted periods of deposition has been surprisingly persistent, and the source of much confusion among paleontologists.

The Tertiary paleofaunas of Saskatchewan should be part of a broadly distributed Great Plains fauna, because there have been no prominent physiographic barriers during any part of the Cenozoic to hinder the spread of animals throughout the region. Various local faunas throughout the Great Plains region should differ only in representation of various "ecological subzones" (Emry et al., 1987, p. 121). Contrary to this expectation, however, the "Cypress Hills Oligocene" fauna of the Eastend area seemed to present many features that were strongly discordant with other Plains faunas, and included taxa that appeared to be anomalously advanced. The "Cypress Hills Oligocene" eventually proved to be a mixed fauna, containing mostly Chadronian taxa but also including material from younger strata, with a few specimens derived from beds as young as Hemingfordian (Storer and Bryant, 1993).

The "Cypress Hills Oligocene"

The study of the Eocene to Miocene faunas of western Canada began in 1883 when R. G. McConnell discovered fossil vertebrates in the Cypress Hills (McConnell, 1886). This study, and the classical "Cypress Hills Oligocene" concept, involved some of North America's foremost paleontologists, including E. D. Cope, L. M. Lambe, and L. S. Russell. Cypress Hills local faunas became important reference points for study of the evolution of mammals in the Great Plains, and the Wood Committee designated the Cypress Hills fauna and the Pipestone Springs fauna of southwestern Montana as "principal correlatives" of the Chadronian Provincial Age (Wood et al., 1941, p. 11). Our present view has finally come to include an extensive series of faunas of the Cypress Hills Formation ranging in age from middle Eocene through early Miocene.

McConnell (1886) first applied the term "Cypress Hills" to the deposits capping the Cypress Hills Plateau, and the unit has variously been referred to as the Cypress Hills beds (Fraser et al., 1935) and Cypress Hills Formation (Williams and Dyer, 1930). Early collections were studied by Cope (1885, 1891) and Lambe (1905a, b, c, 1908). Their work was revised and extended by Russell (1934).

The concept of the "Cypress Hills Oligocene" has been important in the development of North American mammalian biochronology. Cope's early papers (1885, 1891) described fossils from Anxiety Butte and Bone Coulee (Fig. 2), both in the Eastend area of the Cypress Hills. Chadronian fossil mammals had previously been reported from the Big Badlands of South Dakota and Nebraska (summary in Scott and Jepsen, 1936, p. 4-5) and from northeastern Colorado (summary in Galbreath, 1953, p. 8), but the Cypress Hills fauna was described significantly earlier than either the Pipestone Springs Local Fauna of Montana (Douglass, 1901; Matthew, 1903) or others from the northern Great Plains, Wyoming, and Texas. Even the later work on the Cypress Hills fauna by Lambe (1905a, b, c, 1908) came during the relatively early stages of study of the Chadronian fauna of North America. Thus the "Cypress Hills Oligocene" and the many species based on its fossils assumed major importance as paleontologists attempted to establish the succession of Chadronian paleofaunas.

The Pipestone Springs and Thompson Creek assemblages of southwestern Montana (Douglass, 1901, 1903; Matthew, 1903) were the first North American Chadronian faunas described that would now be termed local faunas. Pipestone Springs and Thompson Creek were the first Chadronian assemblages described from limited stratigraphic intervals, and from localities situated within limited areas, so that the integrity of the samples was assured. Attempts to treat the "Cypress Hills Oligocene" as a comparable local fauna immediately resulted in problems. Nonetheless, the single-fauna concept was not easily abandoned; perhaps it was not realized that a thick sequence of deposits (up to 76 meters) was represented, in a formation covering a very large area (the entire Swift Current and Cypress Hills plateaus).

One early worker, L. M. Lambe, recognized the diachroneity of the Cypress Hills deposits. Lambe described five new species of equids (1905b) and in another paper of the same year (Lambe, 1905a) discussed new specimens clarifying the nature of *Mesohippus westoni* (Cope, 1889). Lambe recognized that the teeth he was describing, collected from several localities at Anxiety Butte and Bone Coulee (Fig. 2), could represent several different stages of evolution and that some of the species were relatively advanced. As a result, he postulated that several faunal levels were present in the Eastend area, representing ages that we would now call Chadronian, Orellan, and early and late Whitneyan.

Although I do not agree with Lambe on the ages of the Cypress Hills equid species (Storer and Bryant, 1993), I am convinced that he was nearer to the truth than any other worker for a long time to come. It is unfortunate that Lambe did not attempt to verify his ideas by making stratigraphically controlled collections.

The first unmixed, entirely Chadronian local fauna

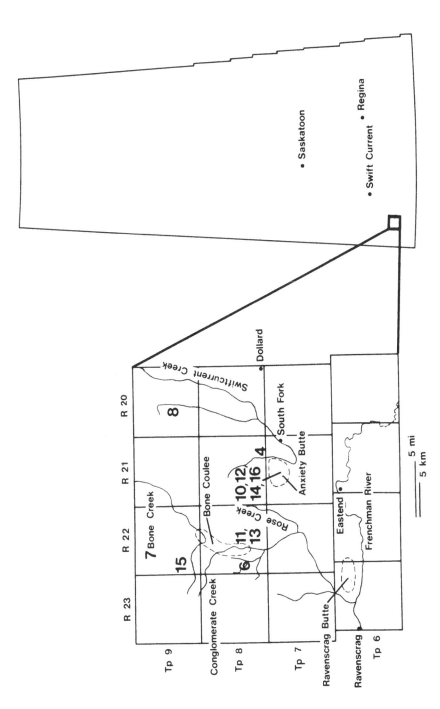

Figure 2. Outline drawing of Eastend area, Cypress Hills of southwestern Saskatchewan, showing place names and sources of Eocene-Oligocene mammalian paleofaunas. All ranges west of the Third Meridian. 4, Southfork Local Fauna, Chadronian; 6, Calf Creek Local Fauna, Chadronian; 7, Carnagh Local Fauna, Chadronian; 8, Irish Spring Local Fauna, Chadronian; 10, Anxiety Butte, Chadronian; 11, KSW Local Fauna, Chadronian; 12, Fossil Bush Local Fauna, Orellan; 13, Anxiety Butte, Orellan; 14, Rodent Hill Local Fauna, Whitneyan; 15, Anxiety Butte, Whitneyan; 16, Kealey Springs Local Fauna, early Arikarecan; 17, Anxiety Butte, late Arikarecan.

in the Cypress Hills was discovered in 1936 and 1937 by Fenley and Hazel Hunter for the National Museum of Canada, now Canadian Museum of Nature (Russell, 1972, p. 3). Hunter Quarry has yielded many skulls of brontotheres, rhinoceroses, entelodonts, and other mammals. It and the microvertebrate locality at the same stratigraphic level on the other side of Calf Creek valley are the source of the middle Chadronian Calf Creek Local Fauna (Fig. 2:6). This local fauna has produced so many "Cypress Hills Oligocene" species that it has come to typify the fossil fauna of the area.

Stratigraphically documented collections from a much wider area were first made in 1963 by B. A. McCorquodale, of the Saskatchewan Museum of Natural History, now Royal Saskatchewan Museum. Unfortunately, his sections and fossils did not become available for study until 1990, when he deposited the material in the Royal Saskatchewan Museum. The belief persisted that the Cypress Hills Formation in the Eastend area was almost entirely Chadronian in age (Emry et al., 1987), even though doubts about the fauna's integrity had continued to be expressed (e.g., Clark and Beerbower, 1967, p. 59).

Broadening the time range: the Swift Current Eocene

Description of the Swift Current beds and their middle Eocene (late Uintan) mammalian fauna by Russell and Wickenden (1933) added a significantly older level to Saskatchewan's Eocene-Oligocene faunas. The Swift Current beds, later renamed the Swift Current Creek beds by Russell (1950), were considered to lie beneath the Cypress Hills beds in the Swift Current Plateau (Russell and Wickenden, 1933; Fraser et al., 1935). Although both were mapped together as Cypress Hills Formation on the classic Regina Sheet (Geological Survey of Canada Map 267A, 1935), the idea gradually developed that the Swift Current beds were separated from the Cypress Hills beds by a significant unconformity (Wood et al., 1941, Plate 1; Taylor et al., 1964, p. 193).

J. A. Vonhof, in an unpublished Ph.D. thesis (1969), included the Swift Current Creek beds within the Cypress Hills Formation on lithological grounds. Recent work on the Swift Current Creek Local Fauna (Fig. 1:1) (Storer, 1984b) and geological mapping (Whitaker and Pearson, 1972; Macdonald and Broughton, 1980) have followed this assignment.

Hemingfordian and earlier Chadronian: a first look at the complexity of Cypress Hills deposits

The late Hemingfordian Topham Local Fauna (Storer, 1975; Skwara, 1986, 1988) was the first assemblage to challenge seriously the assumption that Cypress Hills deposition was restricted to Uintan and Chadronian time. Discovery of this assemblage was both surprising and fortuitous: D. B. Schowalter, then an undergraduate at the University of Alberta, asked me for advice in identifying a group of specimens collected in 1967 by L. A. Lindoe. I recognized that the material was Miocene, an age completely unexpected in the Cypress Hills Formation, and later was able to relocate the sites and make collections that were adequate for analysis. Skwara (1986, 1988) concluded that the fauna was late Hemingfordian, the youngest assemblage known to date from the Cypress Hills Formation.

The Southfork Locality was discovered by Albert Swanston of the Saskatchewan Museum of Natural History about 1962 (Russell, 1980b, p. 2). Elements of the Southfork Local Fauna (Fig. 2:4) were originally described by Mellett (1977) and Russell (1980b, 1982). Storer (1984a) later recognized that this assemblage was older than the Calf Creek Local Fauna (Fig. 2:6), adding yet another faunal level to an increasingly complex biostratigraphic framework.

Although it was never anticipated that the Cypress Hills Formation would yield faunas of so many distinct ages, discovery of the Topham and Southfork local faunas paved the way for future investigations by suggesting the potential time range of the deposits. With the recognition of at least four distinct local faunas in the formation, three of them from the Eastend area, it became evident that detailed stratigraphic and geographic data were vital for faunal studies. It also became evident that the local faunas must be studied on a very restricted basis, essentially one locality at a time.

EOCENE-OLIGOCENE MAMMALIAN FAUNAS OF THE CYPRESS HILLS FORMATION

In this chapter I do not refer in detail to the pertinent geological sections, but list the mammals known from each local fauna and discuss the conclusions that have been drawn from the available fossil evidence. Local faunas are listed in Table 1, and are located by number on Figures 1 and 2.

Swift Current Creek Local Fauna, late Uintan, Swift Current Plateau

The Swift Current Creek Local Fauna (1), discovered by R. T. D. Wickenden in 1930, is the oldest of the known assemblages from the Cypress Hills Formation (Russell and Wickenden, 1933; Russell, 1965; Storer, 1984b). A few fossils recovered near the base of the formation may be older, but this scanty material cannot be identified precisely.

Russell (1965, p. 18) judged that "the closest resemblances appear to be with the fauna of the Upper Uinta or Myton Member." This correlation has been accepted by all subsequent authors and was further strengthened by Storer (1984b). Despite a well established temporal correlation with Uinta C assemblages,

Table 1. Eocene-Oligocene mammalian paleofaunas of Saskatchewan.

Epoch	Stage or North American Land Mammal "Age"	Locality or fauna
OLIGOCENE	Late Arikareean	Anxiety Butte (17)
	Early Arikareean	Kealey Springs Local Fauna (16)
	Whitneyan	Anxiety Butte (15)
	Whitneyan	Rodent Hill Local Fauna (14)
	Orellan	Anxiety Butte (13)
	Orellan	Fossil Bush Local Fauna (12)
EOCENE	Chadronian	KSW Local Fauna (11)
	Chadronian	Anxiety Butte (10)
	Chadronian	Blumenort Local Fauna (9)
	Chadronian	Irish Spring Local Fauna (8)
	Chadronian	Carnagh Local Fauna (7)
	Chadronian	Calf Creek Local Fauna (6)
	Chadronian	Simmie Local Fauna (5)
	Chadronian	Southfork Local Fauna (4)
	Duchesnean	Lac Pelletier Upper Fauna (3)
	Duchesnean	Lac Pelletier Lower Fauna (2)
	Uintan	Swift Current Creek Local Fauna (1)

the Swift Current Creek Local Fauna shows only 69% generic similarity to the approximately contemporaneous faunas of the Rocky Mountain region, and only about half the species occur in both faunal provinces. This is comparable to the level of endemism that Golz and Lillegraven (1977) found when comparing late Uintan assemblages of southern California with those of the Rocky Mountains region, and is strongly suggestive of increasing faunal provinciality throughout western North America as compared to earlier in the Uintan.

The Swift Current Creek Local Fauna is the only known Uintan assemblage from the Great Plains. In addition to being at the base of the known faunal succession of the Cypress Hills Formation, the Swift Current Creek Local Fauna also gives us our earliest look at a post-Wasatchian Great Plains fauna. The following list is based principally on Storer (1984b), with a few corrections that have been made since, both published (Storer and Tokaryk, 1992) and unpublished.

In this and other faunal lists in this chapter, species based on Cypress Hills fossils are indicated with an asterisk. Names of taxa are followed by citations of their type descriptions, as listed in Bibliography of Fossil Vertebrates and the preceding bibliographies of O. P. Hay. The only type descriptions listed among the references for this chapter are those published since latest coverage by Bibliography of Fossil Vertebrates, about 1990.

Mammals of the Swift Current Creek Local Fauna

Order Multituberculata
  Family Neoplagiaulacidae
    Ectypodus lovei (Sloan, 1966)
Order Peradectida
  Family Peradectidae
    Peradectes californicus (Stock, 1936A)
    Peradectes cf. *P. minutus* (McGrew, 1937A)
Order Didelphimorphia
  Family Didelphidae
    Herpetotherium marsupium Troxell, 1923F
    Herpetotherium knighti (McGrew, 1959)
    Herpetotherium innominatum (Simpson, 1928G)
Order Leptictida
  Family Leptictidae
    * *Palaeictops borealis* (Russell, 1965D)
    Leptictis sp.
Order Cimolestida
  Family Palaeoryctidae
    * *Didelphodus serus* Storer, 84-1249
Order Apatotheria
  Family Apatemyidae
    Apatemys hendryi Robinson, 1966C
Order incertae sedis
  Family Plagiomenidae
    * *Thylacaelurus campester*

Storer, 84-1249

Order Insectivora
  Family Dormaaliidae
    Sespedectes sp.
  Family Erinaceidae
    Erinaceidae, gen. et sp. indet.
  Family Nyctitheriidae
    Nyctitherium sp.
  Family Geolabididae
    Centetodon aztecus Lillegraven et al., 82-778
  Family Proscalopidae
    * *Wallia scalopidens* Storer, 84-1249
  Family Soricidae
    Domnina sp.
    Soricidae, sp. A Krishtalka and Setoguchi, 77-3102
  Family Apternodontidae
    Oligoryctes sp
  Family *incertae sedis*
    Insectivora, *incertae sedis*
    Insectivora, gen. et sp. nov.
Order Chiroptera
  Family *incertae sedis*
    Chiroptera, sp. 1 Storer, 84-1249
    Chiroptera, sp. 2 Storer, 84-1249
Order Primates
  Family Omomyidae
    Omomys sp.
    Omomyidae, sp. 2 Storer, 84-1249
Order Carnivora
  Family Viverravidae
    Viverravus sp.
  Family Miacidae
    Miocyon cf. *M. scotti* (Wortman and Matthew, 1899A)
  Family incertae sedis
    Caniform, sp. 1 Storer, 84-1249
    Caniform, sp. 2 Storer, 84-1249
    Caniform, gen. et sp. indet. (for *Alloeodectes mcgrewi* (Russell, 84-1094); see Bryant, 1991:2064)
Order Mesonychia
  Family Mesonychidae
    Harpagolestes sp.
Order Condylarthra
  Family Hyopsodontidae
    * *Hyopsodus fastigatus* Russell *in* Russell and Wickenden, 1933
Order Perissodactyla
  Family Equidae
    Epihippus cf. *E. gracilis* Marsh, 1871D
  Family Brontotheriidae
    Brontotheriidae, gen. et sp. indet.
  Family Helaletidae
    Colodon cf. *C. kayi* (Hough, 1955)

Dilophodon sp.
Order Artiodactyla
  Family Homacodontidae
    Mytonomeryx scotti Gazin, 1955
  Family Antiacodontidae
    * *Auxontodon processus* Storer, 84-1249
  Family Leptochoeridae
    * *Ibarus ignotus* Storer, 84-1249
  Family Agriochoeridae
    Protoreodon pumilus (Marsh, 1875B)
  Family Oromerycidae
    Oromerycidae, gen. et sp. indet. Storer, 84-8249
  Family Protoceratidae
    Leptoreodon marshi Wortman, 1898A
Order Rodentia
  Family Ischyromyidae
    Leptotomus nr. *L. guildayi* Black, 1971A
    Leptotomus sp.
    Pseudotomus cf. *P. eugenei* (Burke, 1935B)
    Microparamys dubius (Wood, 1949B)
    * *Microparamys solidus* Storer, 84-1249
    Anonymus sp.
  Family Cylindrodontidae
    . * *Pseudocylindrodon citofluminis* Storer, 84-1249
  Family Aplodontidae
    Prosciurinae, gen. et sp. indet.
  Family Eutypomyidae
    * *Janimus mirus* Storer, 84-1249
  Family Eomyidae
    * *"Namatomys" fugitivus* Storer, 84-1249
    * *Protadjidaumo altilophus* Storer, 84-1249
Order Lagomorpha
  Family Leporidae
    * *Procaprolagus vusillus* Storer, 84-1249
    Leporidae, gen. et sp. indet.

Lac Pelletier Lower Fauna, Duchesnean, Swift Current Plateau

The Lac Pelletier Lower Fauna (2) occurs in a series of exposures on the east side of Lac Pelletier, an artificial lake south of Swift Current. The fossils are preserved in sands and gravels low in the formation which are overlain by the lowest of a thick series of mudstones that are interbedded with channel deposits. This is the earliest of four local faunas in the Lac Pelletier area.

Storer (1987) characterized this local fauna as Duchesnean because it includes archaic taxa otherwise known from middle Eocene strata, together with early representatives of groups characteristic of the Chadronian and younger White River Chronofauna. The assemblage includes several first appearances of families and genera otherwise known from Chadronian faunas, leading some authors to believe that the Lac

Pelletier Lower Fauna is a mixed assemblage with Chadronian fossils and a Uintan or Duchesnean fauna incorporated into the same deposits. The geological setting from which the fauna is derived precludes such reworking, however. Above the channel deposits that preserve the Lac Pelletier Lower Fauna are massive mudstones. Above these mudstones are sands and gravels that contain another transitional assemblage, the Lac Pelletier Upper Fauna (3). Above the Lac Pelletier Upper Fauna are more mudstones, which are in turn overlain by channel deposits which yield the Simmie Local Fauna (5), an early Chadronian assemblage containing "*Leptomeryx*" cf. "*L.*" *blacki*. Thus it appears impossible for mixing to have occurred with Chadronian fossils from strata of later origin four levels higher in the section. First appearances of White River taxa in the Lac Pelletier Lower Fauna should be regarded as actual early records rather than as reworked fossils. First appearances not previously recognized by Lucas (1992) include *Sinclairella, Hesperocyon, Heptacodon, Pseudoprotoceras, Trigenicus, Eumys, Adjidaumo*, and *Heliscomys*. Of these, the records of *Sinclairella* and *Trigenicus* are unpublished and consist of fragmentary material. *Pseudoprotoceras* (Emry and Storer, 1981) and *Heliscomys* (Storer, 1988a) are represented by few specimens, and the Duchesnean forms are not obviously at different stages of evolution than Chadronian species. Lac Pelletier *Hesperocyon* (Bryant, 1992), *Heptacodon* (Storer, 1983), *Eumys* (Storer, 1988a), and *Adjidaumo* (Storer, 1987) are distinctly more primitive forms than are known from Chadronian faunas, and make entirely believable early records.

The Lac Pelletier Lower Fauna shows only 51% generic similarity with reported Duchesnean local faunas of the Rocky Mountains region, suggesting very strong faunal provinciality during the Duchesnean. Faunal provinciality is expressed to a greater extent among some groups of mammals than among others, and a few genera, e.g., *Duchesneodus* (Lucas, 1992), are known from mammalian assemblages throughout much of North America.

A figure of 51% generic similarity between the Great Plains and Rocky Mountains during the Duchesnean cannot be regarded as highly reliable, however, because much of the Duchesnean fauna of the Rocky Mountains region has not been studied. Few species in the Lac Pelletier Lower Fauna are known from other regions, and the assemblage is consequently very difficult to correlate accurately. Nevertheless, increases in both faunal provinciality and diversity (Storer, 1988a,b, 1989) are consistent with the idea that the middle Eocene faunal turnover was a taxonomic displacement (Maas et al., 1988, p. 412), a rapid faunal turnover driven by a wave of originations rather than by extinctions and featuring newly evolved forms in each paleobiogeographic province replacing "archaic" forms

over an interval of nearly four million years (Storer, 1988a). The middle Eocene faunal turnover does not show characteristics consistent with a mass extinction (catastrophic or otherwise) with subsequent filling of niches by survivors and immigrants; in the Lac Pelletier Lower Fauna newly evolved and "archaic" taxa are found together in the same strata.

Stucky (1992, p. 466) listed this as a late Duchesnean local fauna, but I feel that much more research will be necessary before we can assign most of the North American Duchesnean assemblages to the early or late part of the land-mammal "age." The Lac Pelletier Lower Fauna is certainly older than the latest Duchesnean Porvenir Local Fauna (Wilson, 1986), but I am hesitant to correlate it more precisely.

The Duchesnean assemblages of Lac Pelletier are nearly unique for the Great Plains. The only other Duchesnean local fauna from the Great Plains is the Antelope Creek Local Fauna from Slim Buttes, northwestern South Dakota (Bjork, 1967). The Antelope Creek assemblage is significant but not taxonomically diverse; like the Lac Pelletier local faunas, it has so far proven difficult to correlate precisely.

A preliminary faunal list of the Lac Pelletier Lower Fauna was given by Storer (1987). This has been substantially updated by Storer (1988a, 1990, 1992, 1993), Storer and Tokaryk (1992), and Bryant (1992). Further changes dictated by research in progress are also included in the following faunal list.

Mammals of the Lac Pelletier Lower Fauna
Order Multituberculata
  Family Neoplagiaulacidae
    Ectypodus lovei (Sloan, 1966)
    ?Neoplagiaulacidae, gen. et sp. indet.
Order Peradectida
  Family Peradectidae
    Peradectes californicus (Stock, 1936A)
Order Didelphimorphia
  Family Didelphidae
    Herpetotherium marsupium Troxell, 1923F
    Herpetotherium knighti (McGrew, 1959)
    Herpetotherium innominatum (Simpson, 1928G)
Order Leptictida
  Family Leptictidae
    * *Palaeictops* cf. *P. borealis* (Russell, 1965D)
    Leptictis sp.
Order Cimolestida
  Family Palaeoryctidae
    Didelphodus, sp. nov.
Order Pantolestida
  Family Pantolestidae
    Pantolestidae, gen. et sp. indet.

Order Apatotheria
  Family Apatemyidae
    Apatemys hendryi Robinson, 1966C
    Apatemys sp.
    Sinclairella sp.
Order *incertae sedis*
  Family Plagiomenidae
    * *Thylacaelurus campester* Storer, 84-
      1249
Order Insectivora
  Family Dormaaliidae
    Talpavus, sp. nov.
    Sespedectes sp.
  Family Nyctitheriidae
    Nyctitherium sp.
  Family Geolabididae
    Centetodon aztecus Lillegraven et al., 82-
      778
    Centetodon magnus (Clark, 1936)
  Family Proscalopidae
    * *Wallia scalopidens* Storer, 84-1249
  Family Soricidae
    Domnina cf. *D. gradata* Cope, 1873T
    Soricidae, sp. A Krishtalka and
      Setoguchi, 77-3102
  Family Apternodontidae
    Apternodus sp.
    Oligoryctes sp.
  Family incertae sedis
    Insectivora, gen. et sp. nov.
Order Chiroptera
  Family *incertae sedis*
    Chiroptera, 2 spp.
Order Primates
  Family Microsyopidae
    Uintasorex sp.
  Family Omomyidae
    * *Trogolemur leonardi* (Storer, 90-490)
    Macrotarsius cf. *M. montanus* Clark,
      1941B
    Omomys sp.
Order Carnivora
  Family Miacidae
    * *Miocyon magnus* Bryant, 1992
  Family Amphicyonidae
    Daphoenus cf. *D. demilo* Dawson, 80-0437
  Family Canidae
    Hesperocyon sp.
  Family incertae sedis
    Carnivora, gen. et sp. indet.
Order Perissodactyla
  Family Equidae
    Equidae, gen. et sp. indet.
  Family Brontotheriidae
    Brontotheriidae, gen. et sp. indet.

  Family Helaletidae
    Colodon sp.
  Family Hyracodontidae
    Hyracodon sp.
  Family Rhinocerotidae
    Rhinocerotidae, gen. et sp. indet.
Order Artiodactyla
  Family Leptochoeridae
    Leptochoeridae, gen. et sp. indet.
  Family Anthracotheriidae
    * *Heptacodon pellionis* Storer, 83-857
  Family Agriochoeridae
    Protoreodon sp.
    Diplobunops sp.
  Family Oromerycidae
    Oromerycidae, cf. Swift Current form
      Storer, 84-1249
  Family Protoceratidae
    Pseudoprotoceras sp.
    Trigenicus sp.
Order Rodentia
  Family Ischyromyidae
    Leptotomus cf. *L. guildayi* Black, 1971A
    Leptotomus sp. Storer, 88-519
    * *Microparamys nimius* Storer, 88-519
    * *Anonymus baroni* Storer, 88-519
    * *Pseudotomus timmys* Storer, 88-519
    Ischyromys, sp. nov. Storer, 88-519
    Ischyromyidae, sp. 1 Storer, 88-519
    Ischyromyidae, sp. 2 Storer, 88-519
  Family Cylindrodontidae
    * *Pseudocylindrodon lateriviae* Storer,
      88-519
  Family Aplodontidae
    Prosciurinae, sp. 1 Storer, 88-519
    Prosciurinae, sp. 2 Storer, 88-519
  Family Eutypomyidae
    * *Janimus dawsonae* Storer, 88-519
    * *Microeutypomys tilliei* (Storer, 88-519)
    * *Eutypomys acares* Storer, 88-519
    * *Eutypomys obliquidens* Storer, 88-519
  Family Cricetidae
    Eumys sp. Storer, 88-519
  Family Eomyidae
    Yoderimys sp. (doubtful, acc. Emry and
      Korth, 1993, p. 1047)
    * *"Namatomys" lacus* Storer, 87-898
    * *Protadjidaumo pauli* Storer, 87-898
    * *Adjidaumo craigi* Storer, 87-898
  Family Heliscomyidae
    Heliscomys sp. Storer, 88-519
  Family incertae sedis
    Rodentia, gen. et sp. indet. Storer, 88-519
Order Lagomorpha
  Family Leporidae
    * *Tachylagus gawneae* Storer, 1992

Lac Pelletier Upper Fauna, Duchesnean, Swift Current Plateau

The Lac Pelletier Upper Fauna (3) has been collected from two localities, one southeast of Lac Pelletier (south of Lac Pelletier village) and the other northeast of the lake. Although the two localities are separated by more than 15 km, I have tentatively combined their assemblages into a single local fauna because both are in sediments immediately overlying the lowest sequence of massive mudstones in the area; these lowest mudstones directly overlie the sands and gravels that are the source of the Lac Pelletier Lower Fauna.

The only species so far described from the Lac Pelletier Upper Fauna is the anthracothere *Heptacodon pellionis* (Storer, 1983). The other records listed here should be considered preliminary, but even this preliminary faunal list conveys the impression of a transitional assemblage slightly more advanced than that of the Lac Pelletier Lower Fauna (*Hendryomeryx*, ?*Aulolithomys*, and ?*Paradjidaumo* present), but still Duchesnean in age. The fact that this local fauna is overlain by a transitional Duchesnean-Chadronian assemblage (Simmie Local Fauna) also restricts it to the Duchesnean. The Lac Pelletier Upper Fauna will probably prove to be as difficult to correlate with other Duchesnean assemblages as the Lac Pelletier Lower Fauna, and for the same reasons: absence of well documented Duchesnean local faunas with which to compare it, and lack of close similarity to coeval faunas because of strong faunal provinciality.

Mammals of the Lac Pelletier Upper Fauna
Order Multituberculata
  Family Neoplagiaulacidae
    Ectypodus lovei (Sloan, 1966)
Order Didelphimorphia
  Family Didelphidae
    Herpetotherium sp.
Order Leptictida
  Family Leptictidae
    Leptictis sp.
Order Cimolestida
  Family Palaeoryctidae
    Didelphodus sp.
Order *incertae sedis*
  Family Plagiomenidae
    Thylacaelurus sp.
Order Insectivora
  Family Dormaaliidae
    Talpavus sp.
  Family Nyctitheriidae
    Nyctitherium sp.
  Family Geolabididae
    Centetodon magnus (Clark, 1936)
  Family Soricidae
    Soricidae, gen. et sp. indet.

Order Primates
  Family Microsyopidae
    Uintasorex sp.
  Family Omomyidae
    ?*Trogolemur* sp.
    Omomys sp.
Order Perissodactyla
  Family Equidae
    Equidae, cf. *Orohippus* (not
      Haplohippus)
  Family Brontotheriidae
    Brontotheriidae, gen. et sp. indet.
  Family Helaletidae
    Colodon sp.
Order Artiodactyla
  Family Anthracotheriidae
    * *Heptacodon pellionis* Storer, 83-857
  Family Protoceratidae
    Leptoreodon sp.
  Family Leptomerycidae
    Hendryomeryx sp.
Order Rodentia
  Family Ischyromyidae
    Ischyromyidae, gen. et sp. indet.
  Family Cylindrodontidae
    Pseudocylindrodon sp.
  Family Aplodontidae
    ?*Prosciurus* sp.
  Family Eutypomyidae
    Janimus sp.
    Eutypomys sp.
  Family Eomyidae
    "*Namatomys*" sp.
    ?*Aulolithomys* sp.
    ?*Paradjidaumo* sp.
    Adjidaumo sp.
Order Lagomorpha
  Family Leporidae
    ?*Tachylagus* sp.

Southfork Local Fauna, early Chadronian, Cypress Hills Plateau

The Southfork Local Fauna (4) derives from a locality low in the Cypress Hills Formation. The Southfork locality (Russell, 1980b) is about 4 km northeast of Anxiety Butte, northeast of Eastend. The following faunal list is slightly modified after Storer (1984a).

The Southfork Local Fauna is one of a group that include *Brachyhyops*, "*Leptomeryx*" cf. "*L.*" *blacki*, and "*Leptomeryx*" *yoderi*. Comparable assemblages include the Yoder Local Fauna of Wyoming (Kihm, 1987), the Titus Canyon Local Fauna of California (Stock, 1949), the McCarty's Mountain Local Fauna of Montana (Douglass, 1906), and the fauna of Dry Hole Quarry, below Ash A at Flagstaff Rim, Wyoming (Emry and

Gawne, 1986), although most of these assemblages do not include all three taxa.

These local faunas have been variously regarded as early Chadronian (Storer, 1984a; Kihm, 1987) or latest Duchesnean (Storer, 1989) in age. It is in these assemblages and in the late Duchesnean Porvenir Local Fauna of West Texas that a well developed phase of the relatively stable White River Chronofauna (Emry et al., 1987) is first established. These local faunas share many widely distributed taxa, indicating greatly diminished faunal provinciality (Storer, 1988a, b).

Emry et al. (1987) have suggested a three-fold division of the Chadronian based on range zones of mammals in the Flagstaff Rim section, and the presence of "*Leptomeryx*" *yoderi* at Southfork seems to put the local fauna in the interval designated CH1 by Stucky (1992). Many of the key taxa listed by Emry et al. (1987) and Emry (1992) are not present in the Southfork Local Fauna, however.

Taxa belonging to the "*Leptomeryx*" *speciosus* and "*Leptomeryx*" *mammifer* lineages are cited with quotation marks in this chapter because they are almost certainly distinct at the generic level from true *Leptomeryx*, typified by the *L. evansi* lineage. Work in progress by other authors may furnish new generic names to clarify the situation.

Mammals of the Southfork Local Fauna
Order Creodonta
    Family Hyaenodontidae
        Hyaenodon horridus Leidy, 1853D
Order Perissodactyla
    Family Equidae
        * *Mesohippus westoni* (Cope, 1889I)
    Family Brontotheriidae
        Brontotheriidae, gen. et sp. indet.
    Family Helaletidae
        Colodon occidentalis (Leidy, 1863I)
    Family Hyracodontidae
        Hyracodon petersoni Wood, 1927B
Order Artiodactyla
    Family En](telodontidae
        * *Brachyhyops viensis* Russell, 80-1556
    Family Leptomerycidae
        "*Leptomeryx*" cf. "*L.*" *blacki* Stock, 1949
        "*Leptomeryx*" cf. "*L.*" *yoderi* Schlaikjer,
            1935B

Simmie Local Fauna, early Chadronian, Swift Current Plateau
The Simmie local fauna (5) was collected from the Robitaille Farm south of the town of Simmie, southwest of Lac Pelletier. Beds that yield this assemblage are higher in the section than the Lac Pelletier Upper Fauna, and occur above the second massive mudstone layer, which overlies the Lac Pelletier Upper Fauna.

Unfortunately, this part of the section is not widely exposed. The presence of "*Leptomeryx*" cf. "*L.*" *blacki* suggests correlation with the Southfork Local Fauna. Thus, the Simmie Local Fauna is placed in the early Chadronian. This is the earliest local fauna of the Cypress Hills Formation that contains the rodent *Cylindrodon*, a genus that has often been cited as being limited to the Chadronian, but is listed by Stucky (1992, table 24.2) as occurring in the late Duchesnean and in the Orellan.

Mammals of the Simmie Local Fauna
Order Perissodactyla
    Family Equidae
        Mesohippus sp.
    Family Brontotheriidae
        Brontotheriidae, gen. et sp. indet.
    Family Helaletidae
        Colodon sp.
Order Artiodactyla
    Family Leptomerycidae
        "*Leptomeryx*" cf. "*L.*" *blacki* Stock, 1949
Order Rodentia
    Family Cylindrodontidae
        Cylindrodon sp.
    Family Eutypomyidae
        Eutypomys sp.
    Family Eomyidae
        Adjidaumo sp.
Order Lagomorpha
    Family Leporidae
        Leporidae, gen. et sp. indet.

Calf Creek Local Fauna, middle Chadronian, Cypress Hills Plateau
The Calf Creek Local Fauna (6), originally named by Holman (1972), includes about 75 mammalian species that are represented by a total of several thousand specimens. The local fauna comes from two localities in Calf Creek valley, northwest of Eastend: the Calf Creek screening locality (also Canadian Museum of Nature Locality 117 and Royal Ontario Museum Locality V-38) and the Hunter Quarry. This local fauna contains most of the elements of the classic "Cypress Hills Oligocene," and is easily the best candidate to replace this widely cited but unsatisfactory "fauna" which was described from scattered localities from a wide range of stratigraphic levels. The Calf Creek Local Fauna most closely approximates the principal correlative of the Chadronian Provincial Age designated by the Wood Committee (Wood et al., 1941).

Fossil mammals of the Calf Creek Local Fauna have been described by Russell (1938, 1972, 1975, 1978, 1980a, 1982, 1984), Bryant (1991, 1993), Emry and Storer (1981), Krishtalka et al. (1982), Storer

(1978, 1981a, b), and Storer and Bryant (1993). The presence of ?*Apatemys, Hemipsalodon, Parictis parvus, Hendryomeryx, Leptotomus, Ischyromys junctus*, a primitive form of *Palaeolagus temnodon*, the relatively primitive *Megalagus* nr. *M. brachyodon*, and possibly *Toxotherium* indicates that the fauna is older than the Pipestone Springs Local Fauna of Montana. The Calf Creek Local Fauna is younger than the early Chadronian Southfork, Yoder, Titus Canyon, and McCarty's Mountain local faunas and the fauna of the Dry Hole Quarry below Ash A at Flagstaff Rim. All of these assemblages are in the *Brachyhyops-"Leptomeryx" blacki-"Leptomeryx" yoderi* range zone, but the Calf Creek Local Fauna includes the more advanced forms *Archaeotherium, "Leptomeryx" speciosus*, and *"Leptomeryx" mammifer*.

Whether to call the Calf Creek Local Fauna early or middle Chadronian is a matter of some debate. The assemblage contains *Yoderimys stewarti, Hemipsalodon, Parictis parvus* (Bryant, 1993, p. 1045), *Toxotherium*, and *Pseudoprotoceras semicinctus*, all of which occur low in the Flagstaff Rim section, and *Hendryomeryx*, which does not occur at Flagstaff Rim but is preserved in Duchesnean and early Chadronian beds elsewhere. Emry (1992, p. 111) has suggested that the disappearance of *Toxotherium* may mark the end of the early Chadronian. I doubt the validity of a *Toxotherium* datum, and later examples of *Toxotherium* have been identified tentatively from Cypress Hills deposits. The Calf Creek Local Fauna closely resembles assemblages near and below Ash B at Flagstaff Rim, and may best be assigned to the later part of the early Chadronian rather than to the middle Chadronian CH2 interval as proposed by Stucky (1992). For the present, however, I follow other authors in assigning the local fauna to the middle Chadronian. If this assignment is correct, the Calf Creek Local Fauna includes several latest occurrences of taxa previously known from early Chadronian assemblages.

The following faunal list is based on published work, work in press, and research in progress. Most, but not all, of the records have been documented in the literature.

The cylindrodontid *Ardynomys saskatchewensis* is maintained in the position proposed by Storer (1978), despite Wood's (1980, p. 24) opinion to the contrary. I think that Lambe's original illustration of the specimen (1908, Plate 8, nos. 16-17) stands almost by itself in showing the cylindrodontid affinities of this species. Detailed examination of the holotype shows that the species is a large cylindrodontid, very similar to (and perhaps the senior synonym of) *Ardynomys occidentalis*. The protoceratid *"Leptotragulus" profectus* Matthew (1903) is referred to *Trigenicus* Douglass (1903). It appears obvious to me that *Trigenicus socialis* Douglass (1903, p. 162) is a junior synonym of *Leptotragulus profectus* Matthew (1903, p. 224), because the two species are similar in dental morphology, including the distinctive buccolingual expansion of the anterior ends of P2-3, are of similar size, and are derived from faunas that are not widely separated in age or geographical location (Thompson Creek and Pipestone Springs, Montana). Matthew's paper of April, 1903 appears to have been published slightly before Douglass's September, 1903, work (despite an internal date of February 25, 1903, at the end of Douglass's paper), so Matthew's species *Leptotragulus profectus* must take priority, but the species shows many points of dissimilarity from true *Leptotragulus*. Douglass's *Trigenicus* is the senior available generic name to which the species can be referred.

Mammals of the Calf Creek Local Fauna
Order Multituberculata
  Family Neoplagiaulacidae
    Ectypodus lovei (Sloan, 1966)
Order Peradectida
  Family Peradectidae
    Didelphidectes pumilus Hough, 1961
Order Didelphimorphia
  Family Didelphidae
    * *Herpetotherium valens* (Lambe, 1908A)
    Copedelphys titanelix (Matthew, 1903B)
Order Leptictida
  Family Leptictidae
    Leptictis cf. *L. acutidens* (Douglass, 1901B)
Order Apatotheria
  Family Apatemyidae
    ?*Apatemys* sp.
    Sinclairella sp.
Order Insectivora
  Family Dormaaliidae
    ?*Ankylodon* sp.
  Family Geolabididae
    Centetodon cf. *C. magnus* (Clark, 1936)
  Family Proscalopidae
    Wallia sp.
  Family Soricidae
    Domnina sp.
  Family Apternodontidae
    Apternodus sp.
    Oligoryctes sp.
  Family Micropternodontidae
    Micropternodus sp.
  Family *incertae sedis*
    Cryptoryctes sp.
Order Palaeanodonta
  Family Epoicotheriidae
    Epoicotherium sp.
Order Creodonta
  Family Hyaenodontidae

  * *Hemipsalodon grandis* Cope, 1885H
  Hyaenodon horridus Leidy, 1853D
  Hyaenodon microdon Mellett, 1977
  Hyaenodon spp. "probably two" (Bryant,
      1993)
Order Carnivora
  Family Nimravidae
    ?*Dinictis* sp.
    ?*Hoplophoneus* sp.
  Family Amphicyonidae
    Daphoenus sp.
    Daphoenocyon dodgei (Scott, 1898B)
    Daphoeninae, gen. et sp. indet.
  Family Canidae
    Hesperocyon gregarius (including
        Alloeodectes mcgrewi (Russell, 84-
        1094); acc. Bryant, 1991)
    Canidae, gen. et sp. indet.
  Family Amphicynodontidae
    Parictis parvus Clark and Beerbower,
        1967
    Parictis sp.
  Family incertae sedis
    Carnivora, gen. et sp. indet.
Order Perissodactyla
  Family Equidae
    * *Mesohippus westoni* (Cope, 1889I)
    * *Mesohippus propinquus* Lambe, 1905C
    Miohippus grandis (Clark and
        Beerbower, 1967)
  Family Brontotheriidae
    Megacerops sp.
    Menodus sp.
  Family Helaletidae
    Colodon occidentalis (Leidy, 1863I)
  Family Hyracodontidae
    * *Hyracodon priscidens* Lambe, 1905B
  Family Lophiodontidae
    * *Toxotherium hunteri* Wood, 1961
  Family Rhinocerotidae
    Trigonias osborni Lucas, 1900D
    * *Penetrigonias sagittatus* (Russell, 82-
        1256)
Order Artiodactyla
  Family Leptochoeridae
    Stibarus montanus Matthew, 1903B
  Family Entelodontidae
    * *Archaeotherium coarctatum* (Cope,
        1889I)
  Family Anthracotheriidae
    * *Bothriodon advena* Russell, 78-1042
  Family Agriochoeridae
    Agriochoerus sp.
  Family Merycoidodontidae
    Merycoidodon macrorhinus (Douglass,
        1903A)

  Bathygenys sp.
Family Oromerycidae
    Eotylopus cf. *E. reedi* Matthew, 1910A
    Malaquiferus sp.
Family Camelidae
    Poebrotherium sp.
Family Protoceratidae
    * *Pseudoprotoceras semicinctus* (Cope,
        1889I)
    Trigenicus profectus (Matthew, 1903B)
Family Leptomerycidae
    * *Hendryomeryx esulcatus* (Cope, 1889I)
    * *"Leptomeryx" speciosus* Lambe, 1908A
    * *"Leptomeryx" mammifer* Cope, 1886U
Order Rodentia
  Family Ischyromyidae
    Leptotomus guildayi Black, 1971A
    * *Ischyromys junctus* Russell, 1972B
    Ischyromyidae, gen. et sp. indet. Storer,
        78-1150
  Family Cylindrodontidae
    Pseudocylindrodon neglectus Burke,
        1935A
    * *Cylindrodon collinus* Russell, 1972B
    * *Ardynomys saskatchewensis* (Lambe,
        1908A)
  Family Aplodontidae
    Prosciurus vetustus Matthew, 1903B
  Family Sciuridae
    ?*Protosciurus jeffersoni* (Douglass, 1901B)
  Family Eutypomyidae
    * *Eutypomys parvus* Lambe, 1908A
  Family ?Castoridae
    ?Castoridae, gen. et sp. indet. Storer, 78-
        1158
  Family Eomyidae
    * *Yoderimys stewarti* (Russell, 1972B)
    * *Cupressimus barbarae* Storer, 78-1158
    * *Paradjidaumo hansonorum* (Russell,
        1972B)
    Aulolithomys bounites Black, 1965A
    Adjidaumo minimus (Cope, 1873T)
    Eomyidae, gen. et sp. indet. Storer, 78-
        1158
  Family Heliscomyidae
    Heliscomys ostranderi Korth et al., 1991
Order Lagomorpha
  Family Leporidae
    Palaeolagus temnodon Douglass, 1901B
    Megalagus nr. *M. brachyodon* (Matthew,
        1903B)

**Carnagh Local Fauna, Chadronian, Cypress
Hills Plateau**

The Carnagh Local Fauna (7), recovered from the
northernmost fossiliferous exposure of the Cypress

Hills Formation in the Eastend area, is Chadronian in age but younger than the Calf Creek Local Fauna because it contains *Ischyromys veterior,* a more advanced species than the Calf Creek *Ischyromys junctus.* The Carnagh Local Fauna probably is middle Chadronian, CH2 of Stucky (1992).

I tentatively include specimens from the nearby Horse Locality in the Carnagh Local Fauna; graduate research in progress should determine whether the faunas of the two localities are contemporaneous. The faunal list below is very preliminary and will increase several-fold as the abundant collections are studied.

Mammals of the Carnagh Local Fauna
Order Multituberculata
  Family Neoplagiaulacidae
    Ectypodus lovei (Sloan, 1966)
Order Didelphimorphia
  Family Didelphidae
    Herpetotherium sp.
Order Leptictida
  Family Leptictidae
    Leptictis sp.
Order Carnivora
  Family Amphicyonidae
    Amphicyonidae, gen. et sp. indet.
  Family Canidae
    Hesperocyon sp.
Order Perissodactyla
  Family Equidae
    * *Mesohippus westoni* (Cope, 1889I)
    * *Mesohippus propinquus* Lambe, 1905C
  Family Brontotheriidae
    Brontotheriidae, gen. et sp. indet.
  Family Rhinocerotidae
    Rhinocerotidae, gen. et sp. indet.
Order Artiodactyla
  Family Leptochoeridae
    Stibarus montanus Matthew, 1903B
  Family Entelodontidae
    Entelodontidae, gen. et sp. indet.
  Family Anthracotheriidae
    Bothriodon sp.
  Family Agriochoeridae
    Agriochoerus sp.
  Family Merycoidodontidae
    Merycoidodontidae, gen. et sp. indet.
  Family Protoceratidae
    Pseudoprotoceras sp.
    Trigenicus profectus (Matthew, 1903B)
  Family Leptomerycidae
    * *"Leptomeryx" speciosus* Lambe, 1908A
Order Rodentia
  Family Ischyromyidae
    Ischyromys veterior (Matthew, 1903B)
  Family Cylindrodontidae

Pseudocylindrodon neglectus Burke, 1935A
    Cylindrodon sp.
  Family Aplodontidae
    Prosciurus sp.
  Family Eutypomyidae
    Eutypomys sp.
  Family Eomyidae
    Namatomys sp.
    Meliakrouniomys sp.
Order Lagomorpha
  Family Leporidae
    Palaeolagus temnodon Douglass, 1901B
    Megalagus sp.

Irish Spring Local Fauna, Chadronian, Cypress Hills Plateau
The Irish Spring Local Fauna (8) is far removed from the other known Cypress Hills assemblages, and it is not possible to trace the section from one area to the other. The local fauna is Chadronian in age, but research is in a very preliminary stage, so the list below should grow considerably in the future.

Mammals of the Irish Spring Local Fauna
Order Didelphimorphia
  Family Didelphidae
    Herpetotherium sp.
Order Insectivora
  Family Geolabididae
    Centetodon sp.
  Family Proscalopidae
    Wallia sp.
Order Carnivora
  Family Canidae
    Canidae, gen. et sp. indet.
Order Perissodactyla
  Family Equidae
    Mesohippus sp.
  Family Brontotheriidae
    Brontotheriidae, gen. et sp. indet.
  Family Lophiodontidae
    * *Toxotherium hunteri* Wood, 1961
  Family Rhinocerotidae
    * *Penetrigonias sagittatus* (Russell, 82-1256)
Order Artiodactyla
  Family Entelodontidae
    Archaeotherium sp.
  Family Merycoidodontidae
    Merycoidodon sp.
  Family Protoceratidae
    Pseudoprotoceras sp.
    Trigenicus profectus (Matthew, 1903B)
  Family Leptomerycidae
    * *"Leptomeryx" speciosus* Lambe, 1908A

Order Rodentia
  Family Cylindrodontidae
    Pseudocylindrodon sp.
    Cylindrodon sp.
  Family Aplodontidae
    Prosciurus sp.
  Family Sciuridae
    ?*Protosciurus jeffersoni* (Douglass, 1901B)
  Family Cricetidae
    Eumys sp.
  Family Eomyidae
    Namatomys sp.
    Paradjidaumo sp.
    Aulolithomys sp.
  Family Heliscomyidae
    Heliscomys ostranderi Korth et al., 1991
Order Lagomorpha
  Family Leporidae
    Palaeolagus cf. *P. temnodon* Douglass,
      1901B
    Megalagus brachyodon (Matthew, 1903B)

Blumenort Local Fauna, Chadronian, Swift Current Plateau

The Blumenort Local Fauna (9) is preserved in beds at the top of the section southeast of Lac Pelletier. It is evidently Chadronian in age because it contains brontotheres and occurs in beds overlying those that contain the early Chadronian Simmie Local Fauna.

Mammals of the Blumenort Local Fauna
Order Perissodactyla
  Family Brontotheriidae
    Brontotheriidae, gen. et sp. indet.
  Family Helaletidae
    Colodon occidentalis (Leidy, 1863I)
Order Artiodactyla
  Family Leptomerycidae
    "*Leptomeryx*" sp.
Order Lagomorpha
  Family Leporidae
    Procaprolagus sp.

Chadronian fauna of Anxiety Butte, Cypress Hills Plateau

The lower beds on Anxiety Butte (10) produce Chadronian fossils. The section has not been collected intensively. The stratigraphically controlled collections, many of them made by B. A. McCorquodale in the early 1960s, were derived from several beds, and more than one faunal level is probably represented. The following faunal list is based on the stratigraphically controlled SMNH collection.

Mammals from Chadronian beds on Anxiety Butte
Order Perissodactyla
  Family Equidae
    * *Mesohippus westoni* (Cope, 1889I)
  Family Brontotheriidae
    Brontotheriidae, gen. et sp. indet.
  Family Hyracodontidae
    * *Hyracodon priscidens* Lambe, 1905B
Order Artiodactyla
  Family Entelodontidae
    * *Archaeotherium coarctatum* (Cope,
      1889I)
  Family Anthracotheriidae
    Anthracotheriidae, gen. et sp. indet.
  Family Agriochoeridae
    Agriochoerus cf. *A. antiquus* Leidy, 1850C
  Family Camelidae
    Poebrotherium sp.
  Family Protoceratidae
    * *Pseudoprotoceras semicinctus* (Cope,
      1889I)
  Family Leptomerycidae
    * "*Leptomeryx*" *speciosus* Lambe, 1908A

KSW Local Fauna, late Chadronian, Cypress Hills Plateau

The KSW Local Fauna (11) is based on a small collection from the bottom of the section that also contains the Kealey Springs Local Fauna (16), early Arikareean. This local fauna contains several species that have been cited as characteristic of the early part of the Orellan land mammal "age." Several other taxa in the KSW assemblage are known from both the Chadronian and Orellan. Presence of *Pseudocylindrodon neglectus*, however, is strong evidence that the assemblage is pre-Orellan.

The assemblage is probably latest Chadronian in age (Storer, 1994). It is the northernmost local fauna representing a time near the Eocene-Oligocene boundary in North America.

Mammals of the KSW Local Fauna
Order Didelphimorphia
  Family Didelphidae
    * *Herpetotherium valens* (Lambe, 1908A)
    Herpetotherium fugax Cope, 1873T
Order Insectivora
  Family Soricidae
    Domnina cf. *D. gradata* Cope, 1873T
Order Perissodactyla
  Family Equidae
    Mesohippus cf. *M. bairdi* (Leidy, 1850C)
  Family Rhinocerotidae
    Rhinocerotidae, gen. et sp. indet.
Order Rodentia

Family Aplodontidae
   Prosciurus sp.
Family Cylindrodontidae
   Pseudocylindrodon neglectus Burke,
    1935A
Family Sciuridae
   ?*Protosciurus* sp.
Family Cricetidae
   Eumys elegans Leidy, 1856I
   Cricetidae, gen. et sp. indet.
Family Eomyidae
   Namatomys cf. *N. lloydi* Black, 1965A
   Centimanomys sp.
   Paradjidaumo sp.
   Adjidaumo maximus Korth, 90-878
Family Heliscomyidae
   Heliscomys hatcheri Wood, 1935C
Order Lagomorpha
Family Leporidae
   Palaeolagus cf. *P. hemirhizis* Korth and
    Hageman, 88-1539
   Megalagus cf. *M. turgidus* (Cope, 1873T)

Fossil Bush Local Fauna, Orellan, Cypress Hills Plateau

The Fossil Bush Local Fauna (12) is one of a series of four superposed assemblages in a section northwest of Eastend. Near the base of this section is a quarry from which a brontothere skeleton (*Megacerops* sp.) was collected in the 1970s. Above the Fossil Bush assemblage is the late Whitneyan Rodent Hill Local Fauna (14), and at the top of the section is an early Hemingfordian assemblage not discussed here.

The Fossil Bush Local Fauna appears to correlate with faunas of Orella C or D age in Nebraska (Korth, 1989b), based on the joint occurrence of *Eumys elegans*, *Palaeolagus burkei*, and *Palaeolagus haydeni*.

Specimens from this locality were regarded as Chadronian in age by Russell (1972), but with more detailed stratigraphic data and additional collections now available we know the assemblage to be Orellan in age. Russell (1972) cited "Rodent Hill," ROM V-37, as the locality for several specimens, apparently putting them in what is now regarded as a late Whitneyan fauna. I have consulted Royal Ontario locality records, with the kind assistance of Kevin Seymour, and have examined collections associated with the cited specimens, and it appears certain that the material in question actually derived from the Fossil Bush locality.

This assemblage will be discussed in more detail than the other local faunas, because several seemingly anomalous taxa can now be excluded from the "Cypress Hills Oligocene," the Calf Creek Local Fauna discussed earlier in this chapter.

The marsupials listed below were discussed by Russell (1972): specimens ROM 6234 and 6237, lower molars, were referred to *Peratherium valens* but not illustrated, and ROM 6242, LM3, was figured and identified as cf. *Nanodelphys minutus*. I have changed these identifications to *Herpetotherium fugax* and *Peradectes* cf. *P. minutus*.

A single specimen, ROM 6339, RM2, was identified as *Hesperocyon gregarius* (Russell, 1972, pp. 54-55), and I have retained the identification.

Among the rodents, ROM 6283, RM1, referred to *Prosciurus vetustus* by Russell (1972, p. 24, fig. 7A), is almost certainly *Prosciurus relictus*, and Russell described the specimen as having a single metaconule. ROM 6282, RM3, referred to *Prosciurus saskatchewensis* by Russell (1972, pp. 21-22, fig. 6C) is probably an aplodontid, but ROM 6284, RP4, referred to *Prosciurus major* (Russell, 1972, p. 25, fig. 7C) and ROM 6286, RM1 or M2, referred to *Prosciurus altidens* (Russell, 1972, p. 26, fig. 7E) are indeterminate, pending study of further specimens. ROM 6306 (Russell, 1972, p. 31) was identified as LM1 or M2 of *Cylindrodon* but not illustrated. I have not examined this specimen.

ROM 6315, Rp4, was one of the specimens referred to *Eutypomys* cf. *magnus* by Russell (1972, p. 36, fig. 10A). I list it below under the same name, though more work is needed. Interestingly, neither large *Eutypomys* specimen listed by Russell (1972) is from the Calf Creek Local Fauna.

ROM 6318, L maxillary fragment with M2, referred to *Adjidaumo stewarti* (Russell, 1972, pp. 37-38) was not figured and has not been studied. ROM 6320, Lm3 and ROM 6321, Rp4, both identified as *Adjidaumo hansonorum* (Russell, 1972, pp. 38-39, figs. 10G-H) are listed here as *Paradjidaumo* sp.

Three specimens that Russell referred to *Eumys pristinus* are from the Fossil Bush locality, including the type RM1, ROM 6324 (Russell, 1972, p. 41, fig. 10J). Russell (1972, p. 43) noted similarities between the material he studied and teeth of *Eumys brachyodus*, but I consider *Eumys pristinus* to be a junior synonym of *E. elegans*. I identify the holotype of *E. pristinus* and ROM 6327, Rm2 (Russell, 1972, pp. 41-42, fig. 11B) as *E. elegans*. The third specimen referred by Russell to *E. pristinus*, ROM 6329, Lm2, was not figured.

Deletion of these specimens and taxa from the "Cypress Hills Oligocene" removes several sources of confusion, and may facilitate understanding of the Chadronian fauna as well as that of the Fossil Bush Orellan assemblage.

Mammals of the Fossil Bush Local Fauna
Order Peradectida
   Family Peradectidae
     Peradectes cf. *P. minutus* (McGrew,
      1937A)

Order Didelphimorphia
  Family Didelphidae
    Herpetotherium fugax Cope, 1873T
Order Leptictida
  Family Leptictidae
    Leptictis sp.
Order Carnivora
  Family Viverravidae
    Palaeogale sp.
  Family Amphicyonidae
    Daphoenus sp.
  Family Canidae
    Hesperocyon gregarius (Cope, 1873T)
Order Perissodactyla
  Family Equidae
    Mesohippus bairdi (Leidy, 1850C)
    Miohippus cf. *M. grandis* (Clark and
      Beerbower, 1967)
  Family Rhinocerotidae
    Subhyracodon sp.
Order Artiodactyla
  Family Leptochoeridae
    Leptochoerus sp.
  Family Anthracotheriidae
    Elomeryx armatus (Marsh, 1894F)
  Family Agriochoeridae
    Agriochoerus sp.
  Family Protoceratidae
    Protoceras celer Marsh, 1891A
  Family Leptomerycidae
    Leptomeryx evansi Leidy, 1853D
Order Rodentia
  Family Aplodontidae
    Prosciurus relictus (Cope, 1873CC)
  Family Eutypomyidae
    Eutypomys cf. *E. magnus* Wood, 1937D
  Family Cricetidae
    Eumys elegans Leidy, 1856I
  Family Eomyidae
    Paradjidaumo sp.
    Adjidaumo sp.
Order Lagomorpha
  Family Leporidae
    Palaeolagus burkei Wood, 1940
    Palaeolagus haydeni Leidy, 1856I
    Megalagus turgidus (Cope, 1873T)

Orellan fauna of Anxiety Butte, Cypress Hills Plateau

The Orellan beds on Anxiety Butte (13) have not been studied intensively. The faunal list below is derived from four localities and several beds, and it could represent a mixed fauna.

Mammals from Orellan beds on Anxiety Butte

Order Perissodactyla
  Family Equidae
    Mesohippus bairdi (Leidy, 1850C)
    Miohippus sp.
  Family Helaletidae
    Colodon sp.
  Family Lophiodontidae
    ?*Toxotherium* sp.
  Family Rhinocerotidae
    Diceratherium sp.
Order Artiodactyla
  Family Entelodontidae
    Archaeotherium mortoni Leidy, 1850A
  Family Anthracotheriidae
    Elomeryx sp.
  Family Camelidae
    Poebrotherium sp.
  Family Protoceratidae
    Protoceras sp.
  Family Leptomerycidae
    Leptomeryx evansi Leidy, 1853D
Order Rodentia
  Family Cricetidae
    Eumys elegans Leidy, 1856I
Order Lagomorpha
  Family Leporidae
    Palaeolagus haydeni Leidy, 1856I

Rodent Hill Local Fauna, Whitneyan, Cypress Hills Plateau

The Rodent Hill Locality (14) is part of the same section as the Fossil Bush Locality (12) discussed above. It is assigned a late Whitneyan age because of the occurrence of the variant of *Palaeolagus* cf. *P. burkei* discussed by Dawson (1958, pp. 22-23), with an anteroexternal fold on the trigonid of p3.

Mammals of the Rodent Hill Local Fauna

Order Didelphimorphia
  Family Didelphidae
    Herpetotherium sp.
Order Leptictida
  Family Leptictidae
    Leptictis sp.
Order Insectivora
  Family Proscalopidae
    Proscalopidae, gen. et sp. indet.
  Family Soricidae
    Soricidae, gen. et sp. indet.
Order Carnivora
  Family Nimravidae
    Dinictis sp.
  Family Mustelidae
    Mustelidae, gen. et sp. indet.

Family Amphicyonidae
  Daphoenus sp.
Family Canidae
  ?*Hesperocyon* sp.
Order Perissodactyla
Family Equidae
  Miohippus nr. *M. equiceps* (Cope, 1879D)
Family Rhinocerotidae
  Diceratherium sp.
Order Artiodactyla
Family Leptochoeridae
  Leptochoerus sp.
Family Anthracotheriidae
  Elomeryx sp.
Family Agriochoeridae
  Agriochoerus sp.
Family Merycoidodontidae
  Merycoidodontidae, gen. et sp. indet.
Family Protoceratidae
  Protoceras sp.
Family Leptomerycidae
  Leptomeryx sp.
  Pronodens sp.
Order Rodentia
Family Aplodontidae
  Prosciurus relictus (Cope, 1877CC)
  Aplodontidae, gen. et sp. indet.
Family Cricetidae
  Eumys brachyodus Wood, 1937D
Family Eomyidae
  Namatomys sp.
  Paradjidaumo sp.
Family Heteromyidae
  Proheteromys sp.
Order Lagomorpha
Family Leporidae
  Palaeolagus cf. *P. burkei* Wood, 1940
  Megalagus primitivus Dawson, 1958

Whitneyan fauna of Anxiety Butte, Cypress Hills Plateau

Whitneyan beds on Anxiety Butte (15) have been described by Storer and Bryant (1993) to document the occurrence of *Miohippus assiniboiensis* at this stratigraphic level. The following faunal list is a composite of the assemblages from three localities. Although all appear to be late Whitneyan (preserving the variant *Palaeolagus* cf. *P. burkei* described by Dawson, 1958, pp. 22-23), the exact relationships of the assemblages are not known.

Mammals from Whitneyan beds of Anxiety Butte

Order Carnivora
Family Canidae
  ?*Hesperocyon* sp.

Order Perissodactyla
Family Equidae
  * *Miohippus assiniboiensis* (Lambe, 1905C)
  Miohippus cf. *M. intermedius* (Osborn and Wortman 1895A)
  Miohippus sp.
Family Rhinocerotidae
  Diceratherium armatum Marsh, 1875A
  Subhyracodon sp.
Order Artiodactyla
Family Entelodontidae
  Archaeotherium cf. *A. zygomaticum* Troxell, 1920B
Family Anthracotheriidae
  Elomeryx armatus (Marsh, 1894F)
Family Agriochoeridae
  Agriochoerus sp.
Family Merycoidodontidae
  Merycoidodon (*Anomerycoidodon*) sp.
Family Camelidae
  Poebrotherium sp.
Family Protoceratidae
  Protoceras celer Marsh, 1891A
Family Leptomerycidae
  Leptomeryx sp.
Order Lagomorpha
Family Leporidae
  Palaeolagus cf. *P. burkei* Wood, 1940

Kealey Springs Local Fauna, early Arikareean, Cypress Hills Plateau

The Kealey Springs Local Fauna (16) is preserved on a high hilltop northwest of Eastend. Several hundred specimens have been collected but research is in a very preliminary stage.

Early Arikareean beds may also be present on Anxiety Butte, but no fossils certainly diagnostic of an early Arikareean age have been recovered there.

Mammals of the Kealey Springs Local Fauna

Order Didelphimorphia
Family Didelphidae
  Herpetotherium sp.
Order Insectivora
Family Proscalopidae
  Proscalops sp.
Order Carnivora
Family Nimravidae
  Nimravidae, gen. et sp. indet.
Family Canidae
  Canidae, gen. et sp. indet.
Family Mustelidae
  Mustelidae, gen. et sp. indet.
Order Perissodactyla
Family Equidae

Miohippus sp.
Order Artiodactyla
  Family Tayassuidae
    Tayassuidae, gen. et sp. indet.
  Family Anthracotheriidae
    Elomeryx sp.
  Family Merycoidodontidae
    Promerycochoerus sp.
    Sespia sp.
  Family Camelidae
    Camelidae, gen. et sp. indet.
  Family Leptomerycidae
    Leptomeryx sp.
Order Rodentia
  Family Aplodontidae
    Crucimys sp.
    Allomys sp.
  Family Mylagaulidae
    Mylagaulidae, gen. et sp. indet.
  Family Eutypomyidae
    Eutypomys sp.
  Family Castoridae
    Palaeocastor sp.
  Family Eomyidae
    Eomyidae, gen. et sp. indet.
  Family Geomyidae
    Pleurolicus sp.
  Family Heteromyidae
    Heteromyidae, gen. et sp. indet.
  Family Zapodidae
    Zapodidae, gen. et sp. indet.
  Family Cricetidae
    Leidymys sp.
    Cricetidae, gen. et sp. indet.
Order Lagomorpha
  Family Leporidae
    Palaeolagus sp.

Late Arikareean fauna of Anxiety Butte, Cypress Hills Plateau

Late Arikareean fossils appear to be present at four localities at Anxiety Butte (17). More exploration will be needed to confirm that the specimens are indeed of this age.

Mammals from late Arikareean beds of Anxiety Butte

Order Perissodactyla
  Family Equidae
    Miohippus cf. *M. gemmarosae* Osborn, 1918A
    "*Parahippus*" cf. "*P.*" *nebrascensis* Peterson, 1906C
Order Artiodactyla
  Family Anthracotheriidae
    Arretotherium sp.

Family Camelidae
  Stenomylus sp.

CONCLUSIONS

Eocene-Oligocene mammalian paleofaunas of the Cypress Hills Formation span most of the interval from the late Uintan through Arikareean Land Mammal "Ages." The late Uintan Swift Current Creek Local Fauna, most of the Duchesnean Lac Pelletier Lower Fauna, the early Chadronian Southfork Local Fauna, most of the middle Chadronian Calf Creek Local Fauna, and the latest Chadronian KSW Local Fauna have been studied in detail, but much work remains to be done on the many additional Cypress Hills assemblages.

At present only the middle (Uintan-Duchesnean) and late Eocene (Chadronian) faunal assemblages have been firmly documented. These have been used to elucidate part of the middle to late Eocene faunal turnover on the Great Plains. Studies of the Saskatchewan Uintan-Duchesnean evidence in conjunction with studies of assemblages from other paleobiogeographic provinces have made it clear that faunal turnover resulted from taxonomic displacement, in which newly evolved species displaced "archaic" groups over an interval of at least four million years in a regime characterized by increasing climatic seasonality, increasing faunal diversity, and increasing biogeographic provinciality. The late Duchesnean and early Chadronian local faunas of western North America document the return to a much more stable, less diverse, and more widely distributed fauna, the White River Chronofauna.

The recognition that the "Cypress Hills Oligocene" of the literature is actually a mixed fauna ranging in age from Chadronian to Hemingfordian has been a major advance in mammalian biochronology. This classic fauna had been regarded as Chadronian (late Eocene) in age because most of the specimens assigned to it were clearly of that age. But other specimens attributed to the "Cypress Hills Oligocene" (e.g., equid species, Storer and Bryant, 1993; Fossil Bush Local Fauna, Orellan, reported in this chapter) include some that are as young as Hemingfordian (early Miocene).

Future research on the Oligocene and Miocene assemblages of the Cypress Hills Formation will establish in detail the content and correlation of the local faunas. This research should also reduce the number of extraneous taxa formerly included in the "Cypress Hills Oligocene." Continuing research on the Calf Creek Local Fauna will clarify the nature of the Chadronian fauna of the Cypress Hills Formation.

Establishment of a detailed biostratigraphy for the Cypress Hills Formation will result from continued work on correlation of the local faunas coupled with our rapidly increasing knowledge of the lithostratigraphy of the formation.

ACKNOWLEDGMENTS

I thank Dr. Loris S. Russell for introducing me to the Tertiary of Saskatchewan. Dr. Russell and Dr. C. S. Churcher supervised my graduate research at the University of Toronto; both showed exemplary patience and encouraged independent thinking. They have been valued colleagues and friends for more than 25 years.

Mr. Bruce McCorquodale made many collections from the Cypress Hills Formation in the 1950s and 1960s. With field assistant Clive Elliott, he measured sections and made the first stratigraphically controlled collections in 1963. I am grateful to him for many valuable discussions of Saskatchewan paleontology over the years.

I thank fellow collectors and field companions David Baron, Dr. Harold Bryant, James MacEachern, Ron Tillie, Tim Tokaryk, and many others for their contributions to Cypress Hills studies. I also thank the many scientists who have discussed with me the many aspects of the mammalian fossil record.

Reviewers Alan Tabrum and Spencer Lucas made many helpful comments on the manuscript, and helped strengthen it greatly. Any errors in fact or interpretation are my own, however.

This work would not have been possible without the constant encouragement of my family.

LITERATURE CITED

Bjork, P. R. 1967. Latest Eocene vertebrates from northwestern South Dakota. Journal of Paleontology 41:227-236.

Bryant, H. N. 1991. Reidentification of the Chadronian supposed didelphid marsupial *Alloeodectes mcgrewi* as part of the deciduous dentition of the canid *Hesperocyon*. Canadian Journal of Earth Sciences 28:2062-2065.

Bryant, H. N. 1992. The Carnivora of the Lac Pelletier Lower Fauna (Eocene: Duchesnean), Cypress Hills Formation, Saskatchewan. Journal of Paleontology 66:847-855.

Bryant, H. N. 1993. Carnivora and Creodonta of the Calf Creek Local Fauna (Late Eocene, Chadronian), Cypress Hills Formation, Saskatchewan. Journal of Paleontology 67:1032-1046.

Clark, J. and J. R. Beerbower. 1967. Geology, paleoecology, and paleoclimatology of the Chadron Formation; pp. 27-74, *in* J. Clark, J. R. Beerbower, and K. Kietzke, Oligocene sedimentation, stratigraphy, paleoecology and paleoclimatology in the Big Badlands of South Dakota. Fieldiana, Geology Memoirs 5.

Cope, E. D. 1885. The White River beds of Swift Current River, Northwest Territory. American Naturalist 19(2):163.

Cope, E. D. 1889. The Vertebrata of the Swift Current River. 2. American Naturalist 23:151-155.

Cope, E. D. 1891. On Vertebrata from the Tertiary and Cretaceous rocks of the North West Territory. I. The species from the Oligocene or Lower Miocene beds of the Cypress Hills. Contributions to Canadian Palaeontology 3(1):1-25.

Dawson, M. R. 1958. Later Tertiary Leporidae of North America. Paleontological Contributions, University of Kansas, Article 6:1-75.

Douglass, E. 1901. Fossil Mammalia of the White River beds of Montana. Transactions, American Philosophical Society, n.s. 20:237-279.

Douglass, E. 1903. New vertebrates from the Montana Tertiary. Annals, Carnegie Museum 2:145-199

Douglass, E. 1906. The Tertiary of Montana. Memoirs, Carnegie Museum 2:203-223.

Emry, R. J. 1992. Mammalian range zones in the Chadronian White River Formation at Flagstaff Rim, Wyoming; pp. 106-115 *in* D. R. Prothero and W. A. Berggren (eds.), Eocene-Oligocene Climatic and Biotic Evolution, Princeton University Press, Princeton, N. J.

Emry, R. J., P. R. Bjork, and L. S. Russell. 1987. The Chadronian, Orellan, and Whitneyan North American Land Mammal Ages; pp. 118-152 *in* M. O. Woodburne (ed.), Cenozoic Mammals of North America. Geochronology and Biostratigraphy. University of California Press, Berkeley.

Emry, R. J., and C. E. Gawne. 1986. A primitive, early Oligocene species of *Palaeolagus* (Mammalia, Lagomorpha) from the Flagstaff Rim area of Wyoming. Journal of Vertebrate Paleontology 6:271-280.

Emry, R. J., and W. W. Korth. 1993. Evolution in Yoderimyinae (Eomyidae: Rodentia), with new material from the White River Formation (Chadronian) at Flagstaff Rim, Wyoming. Journal of Vertebrate Paleontology 67:1047-1057.

Emry, R. J., and J. E. Storer. 1981. The hornless protoceratid *Pseudoprotoceras* (Tylopoda: Artiodactyla) in the early Oligocene of Saskatchewan and Wyoming. Journal of Vertebrate Paleontology 1:101-110.

Fraser, F. J., F. H. McLearn, L. S. Russell, P. S. Warren, and R. T. D. Wickenden. 1935. Geology of southern Saskatchewan. Memoirs, Geological Survey of Canada 176:1-137.

Galbreath, E. C. 1953. A contribution to the Tertiary geology and paleontology of northeastern Colorado. Paleontological Contributions, University of Kansas, Article 4:1-120.

Golz, D. J. and J. A. Lillegraven. 1977. Summary of known occurrences of terrestrial vertebrates from Eocene strata of southern California. Contributions to Geology, University of Wyoming 15:43-65.

Holman, J. A. 1969. Lower Oligocene amphibians from Saskatchewan. Quarterly Journal, Florida Academy of Sciences 31:273-289.

Holman, J. A. 1972. Herpetofauna of the Calf Creek Local Fauna (Lower Oligocene: Cypress Hills Formation) of Saskatchewan. Canadian Journal of Earth Sciences 9:1612-1631.

Kihm, A. J. 1987. Mammalian paleontology and geology of the Yoder Member, Chadron Formation, east-central Wyoming. Dakoterra 3:28-45.

Korth, W. W. 1989a. Geomyoid rodents (Mammalia) from the Orellan (middle Oligocene) of Nebraska; pp. 31-46, *in* C. C. Black and M. R. Dawson (eds.), Papers on Fossil Rodents in Honor of Albert Elmer Wood. Science Series, Natural History Museum of Los Angeles County 33.

Korth, W. W. 1989b. Stratigraphic occurrence of rodents and lagomorphs in the Orella Member, Brule Formation (Oligocene), northwestern Nebraska. Contributions to Geology, University of Wyoming 27:15-20.

Korth, W. W., J. H. Wahlert, and R. J. Emry. 1991. A new species of *Heliscomys* and recognition of the Family Heliscomyidae (Geomyoidea: Rodentia). Journal of Vertebrate Paleontology 11:247-256.

Krishtalka, L., R. J. Emry, J. E. Storer, and J. F. Sutton.

1982. Oligocene multituberculates (Mammalia: Allotheria): youngest known record. Journal of Paleontology 56:791-794.

Lambe, L. M. 1905a. A new species of *Hyracodon* (*H. priscidens*) from the Oligocene of the Cypress Hills, Assiniboia. Transactions, Royal Society of Canada, Series 2, 11(Section 4):37-42.

Lambe, L. M. 1905b. Fossil horses from the Oligocene of the Cypress Hills, Assiniboia. Transactions, Royal Society of Canada, Series 2, 11(Section 4):43-52.

Lambe, L. M. 1905c. On the tooth structure of *Mesohippus westoni* (Cope). American Geologist 35:243-245.

Lambe, L. M. 1908. The Vertebrata of the Oligocene of the Cypress Hills, Saskatchewan. Contributions to Canadian Palaeontology 3(4):5-64.

Leckie, D. A., and R. J. Cheel. 1989. The Cypress Hills Formation (Upper Eocene to Miocene): A semi-arid braidplain deposit resulting from intrusive uplift. Canadian Journal of Earth Sciences 26:1918-1931.

Lucas, S. G. 1992. Redefinition of the Duchesnean Land Mammal "Age," late Eocene of western North America; pp. 88-105 *in* D. R. Prothero and W. A. Berggren (eds.), Eocene-Oligocene Climatic and Biotic Evolution. Princeton University Press, Princeton, N. J.

Maas, M. C., D. W. Krause, and S. G. Strait. 1988. The decline and extinction of Plesiadapiformes (Mammalia: ?Primates) in North America: Displacement or replacement? Paleobiology 14:410-431.

Macdonald, R., and P. Broughton. 1980. Geological map of Saskatchewan, provisional edition. Saskatchewan Department of Mineral Resources, Saskatchewan Geological Survey.

Matthew, W. D. 1903. The fauna of the Titanotherium beds at Pipestone Springs, Montana. Bulletin, American Museum of Natural History 19:197-226.

McConnell, R. G. 1886. Report on the Cypress Hills, Wood Mountain and adjacent country. Report, Geological Survey of Canada, New Series 1(Part C):1-85.

Mellett, J. S. 1977. Paleobiology of North American *Hyaenodon* (Mammalia, Creodonta). Contributions to Vertebrate Evolution 1:1-134.

Russell, L. S. 1934. Revision of the Lower Oligocene vertebrate fauna of the Cypress Hills, Saskatchewan. Transactions, Royal Canadian Institute 29(1), no. 43:49-67.

Russell, L. S. 1938. The skull of *Hemipsalodon grandis,* a giant Oligocene creodont. Transactions, Royal Society of Canada, Series 3, 32(Section 4):61-66.

Russell, L. S. 1950. The Tertiary gravels of Saskatchewan. Transactions, Royal Society of Canada, Series 3, 44(Sec. 4):51-59.

Russell, L. S. 1965. Tertiary mammals of Saskatchewan. Part I. The Eocene fauna. Life Sciences Contributions, Royal Ontario Museum 67:1-33.

Russell, L. S. 1972. Tertiary mammals of Saskatchewan. Part. II: The Oligocene fauna, non-ungulate orders. Life Sciences Contributions, Royal Ontario Museum 84:1-97.

Russell, L. S. 1975. Revision of the fossil horses from the Cypress Hills Formation (Lower Oligocene) of Saskatchewan. Canadian Journal of Earth Sciences 12:636-648.

Russell, L. S. 1978. Tertiary mammals of Saskatchewan. Part IV: The Oligocene anthracotheres. Life Sciences Contributions, Royal Ontario Museum 115:1-16.

Russell, L. S. 1980a. Tertiary mammals of Saskatchewan. Part V: The Oligocene entelodonts. Life Sciences Contributions, Royal Ontario Museum 122:1-42.

Russell, L. S. 1980b. A new species of *Brachyhyops*? (Mammalia, Artiodactyla) from the Oligocene Cypress Hills Formation of Saskatchewan. Life Sciences Occasional Papers, Royal Ontario Museum 33:1-5.

Russell, L. S. 1982. Tertiary mammals of Saskatchewan. Part VI: The Oligocene rhinoceroses. Life Sciences Contributions, Royal Ontario Museum 133:1-58.

Russell, L. S. 1984. Tertiary mammals of Saskatchewan. Part VII: Oligocene marsupials. Life Sciences Contributions, Royal Ontario Museum 139:1-13.

Russell, L. S., and R. T. D. Wickenden. 1933. An Upper Eocene vertebrate fauna from Saskatchewan. Transactions, Royal Society of Canada, Series 3, 27(Section 4):53-65.

Scott, W. B., and G. L. Jepsen. 1936. The mammalian fauna of the White River Oligocene. Part I. Insectivora and Carnivora. Transactions, American Philosophical Society, New Series 28:1-153.

Skwara, T. 1986. A new "flying squirrel" from the Early Miocene of southwestern Saskatchewan. Journal of Vertebrate Paleontology 6:290-294.

Skwara, T. 1988. Mammals of the Topham local fauna: Early Miocene (Hemingfordian), Cypress Hills Formation, Saskatchewan. Natural History Contributions, Saskatchewan Museum of Natural History 9:1-169.

Stock, C. 1949. Mammalian fauna from the Titus Canyon Formation, California. Publications, Carnegie Institution of Washington 584:229-244.

Storer, J. E. 1975. Middle Miocene mammals from the Cypress Hills, Canada. Canadian Journal of Earth Sciences 12:520-522.

Storer, J. E. 1978. Rodents of the Calf Creek Local Fauna (Cypress Hills Formation, Oligocene, Chadronian), Saskatchewan. Natural History Contributions, Saskatchewan Museum of Natural History 1:1-54.

Storer, J. E. 1981a. Leptomerycid Artiodactyla of the Calf Creek Local Fauna (Cypress Hills Formation, Oligocene, Chadronian), Saskatchewan. Natural History Contributions, Saskatchewan Museum of Natural History 3:1-32.

Storer, J. E. 1981b. Lagomorphs of the Calf Creek Local Fauna (Cypress Hills Formation, Oligocene, Chadronian), Saskatchewan. Natural History Contributions, Saskatchewan Museum of Natural History 4:1-14.

Storer, J. E. 1983. A new species of the artiodactyl *Heptacodon* from the Cypress Hills Formation, Lac Pelletier, Saskatchewan. Canadian Journal of Earth Sciences 20:1344-1347.

Storer, J. E. 1984a. Fossil mammals of the Southfork Local Fauna (Early Chadronian) of Saskatchewan. Canadian Journal of Earth Sciences 21:1400-1405.

Storer, J. E. 1984b. Mammals of the Swift Current Creek Local Fauna (Eocene: Uintan), Saskatchewan. Natural History Contributions, Saskatchewan Museum of Natural History 7:1-158.

Storer, J. E. 1987. Dental evolution and radiation of Eocene and Early Oligocene Eomyidae (Mammalia, Rodentia) of North America, with new material from the Duchesnean of Saskatchewan. Dakoterra 3:108-117.

Storer, J. E. 1988a. The rodents of the Lac Pelletier Lower Fauna, Late Eocene (Duchesnean) of Saskatchewan. Journal of Vertebrate Paleontology 8:84-101.

Storer, J. E. 1988b. Mammalian faunal provinces, Eocene-Oligocene of North America. Abstracts with Programs, 41st Annual Meeting, Rocky Mountain Section, Geological Society of America 20(6):471.

Storer, J. E. 1989. Rodent faunal provinces, Paleocene-Miocene of North America; pp. 17-29 *in* C. C. Black and

M. R. Dawson (eds.), Papers on Fossil Rodents in Honor of Albert Elmer Wood. Science Series, Natural History Museum of Los Angeles County 33:17-29.

Storer, J. E. 1990. Primates of the Lac Pelletier Lower Fauna (Eocene: Duchesnean), Saskatchewan. Canadian Journal of Earth Sciences 27:520-524.

Storer, J. E. 1992. *Tachylagus*, a new lagomorph from the Lac Pelletier Lower Fauna (Eocene: Duchesnean) of Saskatchewan. Journal of Vertebrate Paleontology 12:230-235.

Storer, J. E. 1993. Multituberculates of the Lac Pelletier Lower Fauna, Late Eocene (Duchesnean), of Saskatchewan. Canadian Journal of Earth Sciences 30:1613-1617.

Storer, J. E. 1994. A latest Chadronian (late Eocene) mammalian fauna from the Cypress Hills, Saskatchewan. Canadian Journal of Earth Sciences 31:1335-1341.

Storer, J. E., and H. N. Bryant. 1993. Biostratigraphy of the Cypress Hills Formation (Eocene to Miocene), Saskatchewan: equid types (Mammalia: Perissodactyla) and associated faunal assemblages. Journal of Paleontology 67:660-669.

Storer, J. E., and T. T. Tokaryk. 1992. Catalogue of type and figured fossils, Saskatchewan Museum of Natural History, Regina. Natural History Contributions, Saskatchewan Museum of Natural History 10:1-67.

Stucky, R. K. 1992. Mammalian faunas in North America of Bridgerian to early Arikareean "ages" (Eocene and Oligocene); pp. 464-493 in D. R. Prothero and W. A. Berggren (eds.), Eocene-Oligocene Climatic and Biotic Evolution. Princeton University Press, Princeton, N. J.

Taylor, R. S., W. H. Mathews, and W. O. Kupsch. 1964. Tertiary; pp. 190-194 in R. G. McCrossan and R. P. Glaister (eds.), Geological History of Western Canada. Alberta Society of Petroleum Geologists, Calgary, Alberta.

Vonhof, J. A. 1965. Tertiary sands and gravels in southern Saskatchewan. M.Sc. thesis, University of Saskatchewan, Saskatoon, 99 pp.

Vonhof, J. A. 1969. Tertiary gravels and sands in the Canadian Great Plains. Ph.D. thesis, University of Saskatchewan, Saskatoon, 287 pp.

Whitaker, S. H., and D. E. Pearson. 1972. Geological map of Saskatchewan. Mineral Resources Branch, Saskatchewan Department of Mineral Resources, and Geology Division, Saskatchewan Research Council.

Williams, M. Y., and W. S. Dyer. 1930. Geology of southern Alberta and southwestern Saskatchewan. Memoirs of the Geological Survey of Canada 163:1-160.

Wilson, J. A. 1986. Stratigraphic occurrence and correlation of early Tertiary vertebrate faunas, Trans-Pecos Texas: Agua Fria-Green Valley areas. Journal of Vertebrate Paleontology 6:350-373.

Wood, A. E. 1980. The Oligocene rodents of North America. Transactions of the American Philosophical Society 70(5):1-68.

Wood, H. E., J. Clark, E. H. Colbert, G. L. Jepsen, J. B. Reeside, Jr., and C. Stock. 1941. Nomenclature and correlation of the North American continental Tertiary. Bulletin, Geological Society of America 52:1-48.

13. Magnetic Stratigraphy of the White River Group in the High Plains

ABSTRACT

Magnetostratigraphic sampling was conducted on key fossiliferous exposures of the upper Eocene-lower Oligocene (Chadronian-Whitneyan) White River Group in North Dakota, the Pine Ridge area of Nebraska, the North Platte River-Wildcat Ridge area of Nebraska and Wyoming, and the Cedar Creek area of northeastern Colorado. Together with results from other White River outcrops reported elsewhere, a general magnetostratigraphic pattern can be seen. Most sections are of predominantly reversed polarity, but there are two short zones of normal polarity in the middle and late Chadronian, another in the early Orellan, and another in the late Whitneyan. Based on ^{40}Ar/^{39}Ar dates and correlation to the Berggren et al. (1995) time scale, the middle Chadronian normal magnetozone correlates with Chron C16n (35.4-35.6 Ma), the late Chadronian normal magnetozone with Chron C15n (35.7-35.9 Ma), the early Orellan normal magnetozone with Chron C13n (33.0-33.5 Ma), and the late Whitneyan with Chron C12n (30.5-30.9 Ma). The Chadronian spans the interval from about 37 Ma to about 33.8 Ma; the Orellan from 33.8-32 Ma; the Whitneyan from 32 to about 30 Ma.

INTRODUCTION

The richly fossiliferous and spectacularly scenic badlands of the White River Group in the High Plains (Fig. 1) have been a magnet to fossil collectors and paleontologists ever since the first fossils were collected and described in 1846. In the century and a half since then, tens of thousands of fossils of extraordinary quality have been recovered from White River rocks, making them the most productive mammal-bearing sequence in North America. When the Wood Committee (Wood et al., 1941) set up a provincial time scale for North America, they based the Chadronian through Whitneyan land mammal "ages" (then thought to be early-late Oligocene, but now considered to be late Eocene-early Oligocene) on the White River Group and its mammals. Many detailed lithostratigraphic studies have been published over the years (Sinclair, 1921; Wanless, 1923; Schultz and Stout, 1955; Clark et al., 1967; Singler and Picard, 1980; Retallack, 1983; Seeland, 1985, among others; reviewed by Emry et al., 1987). Despite the amazing quality and quantity of fossils, biostratigraphic studies have been much less

detailed (see Emry et al., 1987), but the updated systematic reviews of the critical fossil mammal groups in this volume finally allow us to construct a detailed range-zone biostratigraphy (see Prothero and Whittlesey, in press, and the summary chapter for this volume).

Since 1976, magnetic stratigraphy has also figured prominently in improving correlation within the White River Group, and with the global magnetic time scale. The initial studies in 1976-1977 by Charles Denham and Harlow Farmer (then of the Woods Hole Oceanographic Institute) were a first attempt to determine if a stable magnetic polarity stratigraphy could be obtained from White River outcrops. In 1979-1980 I conducted my initial sampling program as part of my doctoral dissertation (Prothero, 1982). Further sampling was conducted in the summers of 1983, 1986, and 1987. Brief summaries of this research have been published over the years (Prothero et al., 1982, 1983; Prothero, 1985a, b; Swisher and Prothero, 1990; Evanoff et al., 1992; Prothero and Swisher, 1992), but space limitations prevented a detailed discussion of the polarity stratigraphy of each section. This chapter, and another paper now in press (Prothero and Whittlesey, in press), finally present the details of the past 20 years of magnetostratigraphic research in the White River Group.

METHODS

As previously described in Prothero et al. (1983) and Prothero (1985a), all samples were collected with simple hand tools as horizontally oriented blocks of rock. Three samples were collected per site, and most sites were spaced 5.5 feet (1.7 m) stratigraphically. In the laboratory, the samples were trimmed into oriented cubes approximately 2.5 cm in length.

Different laboratories were used for the samples as they were collected over the years, and my procedures have evolved as well. For the sections that were done in the field summers of 1979-1980, most samples were run on the ScT cryogenic magnetometer at Woods Hole Oceanographic Institute. Most samples were treated

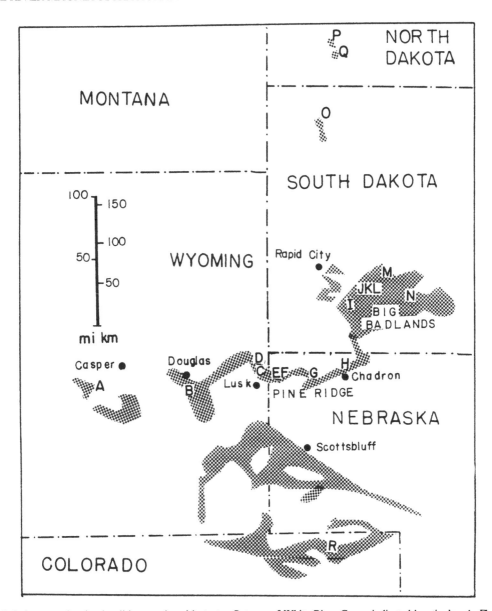

Figure 1. Index map showing localities mentioned in text. Outcrop of White River Group indicated by stipple. A, Flagstaff Rim, Natrona County, Wyoming (see Emry, 1992; Prothero and Swisher, 1992); B, Douglas-Dilts Ranch area, Converse County, Wyoming (see Evanoff et al., 1992); C, Sherrill Hills and Thompson Ranch, and D, Boner Ranch, in the Seaman Hills, Niobrara County, Wyoming (see Prothero and Whittlesey, in press); E, Geike Ranch; F, Munson Ranch; G, Toadstool Park and Raben Ranch area (E-G in Sioux County, Nebraska); H, Trunk Butte and Morris-Bartlett Ranches, Dawes County, Nebraska; I-N, sections in Big Badlands of South Dakota (see Prothero and Swisher, 1992, and Prothero and Whittlesey, in press); O, Slim Buttes, Harding County, South Dakota (see Prothero, 1985b); P, Little Badlands, and Q, Fitterer Ranch, Stark County, North Dakota; R, Flat Top and Chimney Canyons, Logan County, Colorado.

primarily with AF demagnetization; thermal demagnetization was undertaken later at Lamont-Doherty Geological Observatory when it became apparent that AF demagnetization could not remove overprints due to iron hydroxides. These 1979-1980 sections include most of the Pine Ridge sequences: Geike and Munson Ranches, and Toadstool Park in Sioux County, Nebraska, and Trunk Butte and Bartlett-Morris Ranches in Dawes County, Nebraska. In addition, the 1980 sampling included the White River strata in Logan County, Colorado; Slim Buttes, Harding County, South Dakota; and the Little Badlands and Fitterer Ranch, Stark County, North Dakota. I also sampled the Red Shirt Table section in the Big Badlands of South Dakota to supplement Denham's sections elsewhere in the Big Badlands (Denham, 1984). A more detailed discussion of the geology and stratigraphy of most of these regions follows; for further details, see Prothero (1982). Of

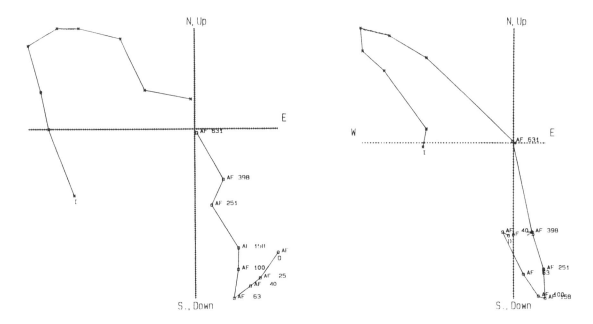

Figure 2. Typical orthogonal demagnetization plots ("Zijderveld" plots) of alternating field (AF) demagnetization of representative samples. Declination is shown by open boxes (with the AF field in Gauss indicated); inclination by asterisks. I = NRM direction of inclination. Each division = 10^{-6} emu. The intensity drops rapidly through AF demagnetization, suggesting that magnetite is the primary carrier of the remanence.

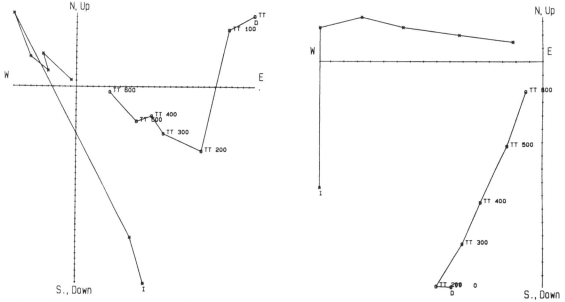

Figure 3. Typical orthogonal demagnetization plots ("Zijderveld" plots) of thermal demagnetization results. Conventions as in Figure 2. "TT" indicates thermal demagnetization temperature. In each case, a normal overprint is removed by 200°C, and a stable reversed component is apparent between 300-500°C. This component was used in further analysis.

these 1979-1980 sections, only the Toadstool Park section (Prothero et al., 1983, fig. 5; Prothero, 1985b, fig. 2; Prothero and Swisher, 1992, fig. 2.5) and Slim Buttes section (Prothero, 1985b, fig. 3) have been previously published.

In 1983, the original data base was expanded with

sampling at Scottsbluff (Prothero and Swisher, 1992, fig. 2.6), and Castle Rock in the North Platte Valley, Scottsbluff County, Nebraska, and the Harvard Fossil Reserve near Torrington, Goshen County, Wyoming. In addition, samples were taken in Cottonwood Pass and in the Indian Creek drainage west of Sheep Mountain

Table in the Big Badlands, site of Clark's (1937) Ahearn, Crazy Johnson, and Peanut Peak members of the Chadron Formation. Sections were taken at Ledge Creek and Flagstaff Rim, Natrona County, Wyoming (Prothero, 1985a); except for these and the Scottsbluff section, the rest of these sections have not yet been published. These samples were run on the ScT cryogenic magnetometer at the laboratory of the South Dakota School of Mines in Rapid City, using mostly thermal demagnetization, with AF demagnetization of pilot samples undertaken to determine coercivity behavior.

In the early 1980s, it became apparent that the original magnetic analysis by Denham and Farmer was inadequate, because they used only one sample per site (preventing calculation of site statistics), and used almost no thermal demagnetization to remove overprinting that might be due to high-coercivity iron hydroxides. The resampling and thermal demagnetization analysis of samples from the Flagstaff Rim section (Prothero, 1985a; Prothero and Swisher, 1992, fig. 2.4) radically changed the pattern originally obtained by Denham (in Prothero et al., 1982, fig. 1; Prothero et al., 1983, fig. 6). In addition, the Big Badlands sections had to be resampled, since none of the results reported by Denham (1984) used thermal demagnetization. In 1986, several areas were sampled: Raben Ranch, Sioux County Nebraska; the upper part of the Roundtop-Toadstool section (Prothero and Swisher, 1992, fig. 2.5), and a dense resampling program in the Big Badlands, following the measured sections and field notes of the Frick Laboratory (particularly those of Morris Skinner). The Cedar Pass, Pinnacles, and Wolf Table-Wanblee sections were sampled that summer. In 1987, we continued the Big Badlands sampling with a section at Sheep Mountain Table. These Badlands magnetic sections were summarized by Prothero and Swisher (1992, fig. 2.7). In 1994, Karen Whittlesey supplemented the Badlands coverage with sections at Sage Creek and the type Scenic area; these are described elsewhere (Prothero and Whittlesey, in press).

The samples taken in 1979-1980 and 1983 were treated with thermal demagnetization at 300-500°C, based on detailed stepwise thermal and AF demagnetization of a pilot suite (e.g., Prothero et al., 1983, fig. 4). All of the samples run since 1986 have been analyzed on a 2G cryogenic magnetometer using extensive thermal and AF demagnetization at the paleomagnetics laboratory at the California Institute of Technology. AF demagnetization (Fig. 2) showed that most samples declined in intensity rapidly, showing that the primary carrier of remanence was a low-coercivity mineral such as magnetite. Thermal demagnetization (Fig. 3) showed that overprints were removed between 200-300°C, and results obtained between 300-400°C were used in the analysis.

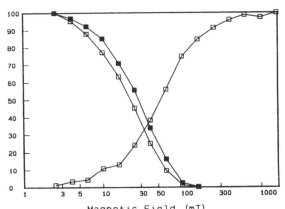

Figure 4. IRM (isothermal remanent magnetization) acquisition (ascending curve on the right) and Lowrie-Fuller test (two descending curves on left) of a typical sample. Open boxes indicate IRM; solid boxes indicate ARM (anhysteretic remanent magnetization). The IRM saturates by 300 mT (millitesla), showing that magnetite is the dominant magnetic mineral phase. In the Lowrie-Fuller test, the ARM (black squares) is more resistant to AF demagnetization than the IRM (open squares) suggesting that single-domain or pseudo-single-domain grains predominate (see Pluhar et al., 1991, for details of the methods).

IRM (isothermal remanent magnetization) acquisition studies (Fig. 4) clearly showed saturation between 300 and 1000 mT (millitesla), indicating that the carrier of remanence is mostly magnetite. A modified Lowrie-Fuller ARM (anhysteretic remanent magnetization) test (e.g., Johnson et al., 1975) was also conducted during the IRM analysis (see Pluhar et al., 1991, for details). This test compares the resistance of AF demagnetization of both an IRM acquired in a 100 mT peak field, and an ARM gained in a 100 mT oscillating field. In almost all samples, the ARM (black squares) demagnetizes at higher peak fields than does the IRM (open squares), indicating that the remanence is carried by single-domain or pseudo-single-domain grains.

Based on these results, the vectors obtained between 300-500°C were averaged using the methods of Fisher (1953; see Butler, 1992). Class I sites of Opdyke et al. (1977) showed a clustering that differed significantly from random at the 95% confidence level. In Class II sites, one sample was lost or crumbled, but the remaining samples gave a clear polarity indication. In Class III sites of Opdyke et al. (1977), two samples showed a clear polarity preference, but the third sample was divergent because of insufficient removal of overprinting. A few samples were considered indeterminate if their magnetic signature was unstable, or their direction uninterpretable.

The means for the normal and reversed sites at each locality were also averaged using the methods of Fisher (1953). The mean of all 143 Class I reversed sites (D = 155.8, I = -40.8, k = 7.8, α_{95} = 4.5) is antipodal to the mean of all 272 Class I normal sites (D = 351.2, I =

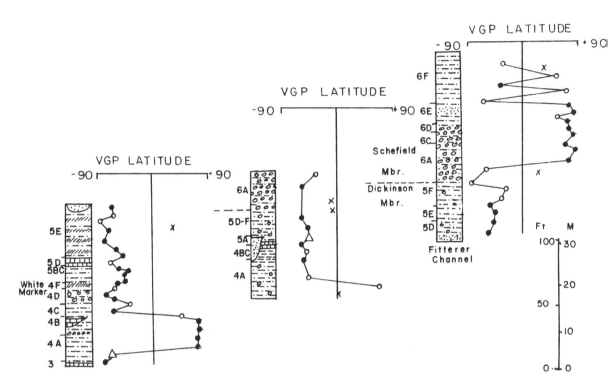

Figure 5. Magnetic stratigraphy of of the "Little Badlands" (left) and "Fitterer Ranch" (center and right) sections in Stark County, North Dakota. Exact location of sections given in text. Stratigraphic terminology follows Skinner (1951) and Stone (1972). Solid circles are Class I sites of Opdyke et al. (1977), which are statistically separated from a random distribution at the 95% confidence level. Triangles are Class II sites, which could not be statistically analyzed because only two samples remained; the third was lost or crumbled. Open circles are Class III sites, in which two samples showed a clear polarity preference, but the third sample was divergent. "x" indicates site of indeterminate polarity.

67.4, k = 25.7, $\alpha95$ = 1.7) within the $\alpha95$ error estimate. This positive reversal test suggests that the magnetization is primary and not due to secondary overprinting. Most of the strata are horizontal, so no fold test could be conducted.

In some sections, it was not possible to completely remove the overprinting on every site, so that there are isolated single-site "polarity events." In most cases (such as single normal sites within a long reversed interval), these are most likely due to unremoved normal overprinting. Consequently, the correlations discussed below are based only on magnetozones that are at least two or more sites thick; single-site "polarity events" are not correlated between regions, since they are not consistently found in every section. However, the possibility that these short "polarity events" are real cannot be ruled out, since detailed analysis of Oligocene deep-sea cores (Hartl et al., 1993) demonstrates that there were a number of brief polarity events during the Eocene and Oligocene.

RESULTS

The magnetic polarity patterns of each of these regions are discussed below. Details of the Douglas area sections have already been published elsewhere

(Evanoff et al., 1992). The Big Badlands sections, and the sections in the Lusk area, Niobrara County, Wyoming, are described by Prothero and Whittlesey (in press). The Slim Buttes section was described by Prothero (1985b, fig. 3), and the Scottsbluff section was detailed in Prothero and Swisher (1992, fig. 2.6). From north to south, the rest of the sections are described below.

Southwestern North Dakota

Isolated remnants of the White River "blanket" (Trimble, 1980) occur as scattered buttes and badlands in several places in southwestern North Dakota, eastern Montana, and northwest South Dakota (Denson and Gill, 1965; Murphy et al., 1993). The most important of these are the "Little Badlands" and related exposures in Stark County, North Dakota. First described by Cope (1883), Douglass (1909), and Leonard (1922), these outcrops and their fossils have been collected and studied by a number of field parties and institutions over the years. In addition, exploration for uranium led to much detailed geological research (summarized by Murphy et al., 1993).

The first detailed report on the stratigraphy of the mammal-bearing beds was by Skinner (1951), based on

Frick Laboratory collecting in the area starting in 1944. Skinner presented a detailed lithostratigraphy for the classic collecting areas in the Little Badlands, and for the newly discovered Frick localities at Fitterer Ranch. Stone (1972) presented a more detailed lithostratigraphy of the White River outcrops, coining several new names for the members he recognized. However, those names have proven to be of limited utility (Hoganson, 1986; Murphy et al., 1993). Since the 1980s, the White River outcrops of North Dakota have been extensively studied and published by personnel of the North Dakota Geological Survey (Hoganson, 1986; Hoganson and Lammers, 1985, 1992). The White River Group in the region was monographed in detail by Murphy et al. (1993).

In 1980 I had only Skinner (1951) and Skinner's unpublished field notes on which to base a magnetostratigraphy of the important Frick collections from the area. A magnetic section was taken in the main area of the Little Badlands ("7 miles south of South Heart" in the Frick Laboratory terminology; "Privratsky Ranch" of Murphy et al., 1993, p. 35). The section (Fig. 5) began in SE NW NW Sec. 23, and then concluded in SW SW NE Sec. 23, T138N R98W, Belfield SE 7.5' quadrangle, Stark County, North Dakota. It spans about 120 feet of Orellan strata (units 3-5E of Skinner, 1951, p. 57) and produces an early Orellan fauna. Units 4A-4B of Skinner (1951) are of normal polarity, and probably correlate with the early Orellan normal magnetozone found throughout the White River Group. The rest of the section is of reversed polarity.

Important collections were also made by the Frick Laboratory at Fitterer Ranch. The measured section began in NW SW Sec. 7, T137N R98W, New England NW 7.5' quadrangle, Stark County, North Dakota (Fig. 5); the upper part of the section above Fitterer Ranch channel was taken in NW SW NW Sec. 17. The sections and stratigraphic terminology shown in Figure 5 follow Skinner (1951); the section is also described by Murphy et al. (1993, p. 37). In these sections, unit 4A of Skinner (1951) is of normal polarity, as it is in the Little Badlands. However, unit 4B is of reversed polarity, unlike the Little Badlands section. The rest of the lower half of the Fitterer Ranch section is of reversed polarity. The upper half of the section began at the top of the highly fossiliferous Fitterer Ranch channel sandstone, and units 5D through lower unit 6A of Skinner (1951) are of reversed polarity. The rest of units 6A through 6E of Skinner (1951) are of normal polarity. In the Frick Laboratory notes, these strata are thought to be Whitneyan, which if true would make this normal magnetozone correlative with the upper Whitneyan normal magnetozone (Prothero and Swisher, 1992). Murphy et al. (1993, p. 106) questioned the age assignment of these strata, since few diagnostic Whitneyan mammals have been described from them so far. However, the majority of the Frick collections

from these beds have not yet been studied. They may eventually substantiate the suggestions of Morris Skinner and the others in the Frick collecting parties.

Pine Ridge area, Sioux County, Nebraska

Outside the Big Badlands and the Lusk-Douglas sections along the Pine Ridge in Wyoming, the most fossiliferous collecting areas occur along the Pine Ridge in Nebraska. The thickest and most complete of these sections is the Toadstool Park-Roundtop section, described by Schultz and Stout (1955, fig. 3, sections 8-9; also shown in Schultz and Stout, 1961), Harvey (1960), and Singler and Picard (1979; 1980, Section 1). It has been used as the basis for zonation of the University of Nebraska State Museum fossil collections (e.g., Schultz and Falkenbach, 1968; Korth, 1989). The revised magnetostratigraphy of the Toadstool Park-Roundtop section is shown by Prothero and Swisher (1992, fig. 2.5).

Within the Toadstool Park area (Fig. 6), however, a thick fluvial channel-fill sequence cuts down from a level near the top of Orella B (Schultz and Stout, 1955, fig. 3, Section 4; also shown in Schultz and Stout, 1961). In the main Toadstool Park section, this channeled disconformity at the Orella B/C boundary also occurs at a transition from normal to reversed polarity. In 1980 a magnetostratigraphic section was taken through the main part of the channel sequence (S 1/2 Sec. 5, T33N, R53W, Roundtop 7.5' quadrangle, Sioux County, Nebraska). From the base of the channel to about 100 feet upsection (near the top of Orella C), the channel fill is of normal polarity (Fig. 6A). In the channel sequence, the normal-reversed transition occurs just above the upper nodules in Orella C, slightly higher than it appears in the main Toadstool-Roundtop section. The remaining parts of Orella C and D are reversed in polarity.

Further west along the Pine Ridge are numerous important collecting localities of both the University of Nebraska State Museum and the Frick Laboratory. Most of these are discussed in the unpublished dissertation of Harvey (1960) and in the field notebooks of Morris Skinner in the Archives of the Department of Vertebrate Paleontology of the American Museum of Natural History. For example, the Albert Meng ranch (SE Sec. 2 T33N R54W, Sioux County, Nebraska) is described in the Skinner notebooks (1944, Vol. 5, pp. 20-23) and is also known as University of Nebraska locality Sx-14. In this area, Skinner measured 60 feet of Chadron below the PWL ("Purplish-White Layer" of Schultz and Stout, 1955, 1961; known as the "Persistent White Layer" in the Frick Laboratory), and over 115 feet of the Orella Member (through Orella C). The stratigraphy is so similar to that of the Toadstool Park area (just 2 miles to the east) that no magnetostratigraphic section was taken.

The next important collecting area along the Pine

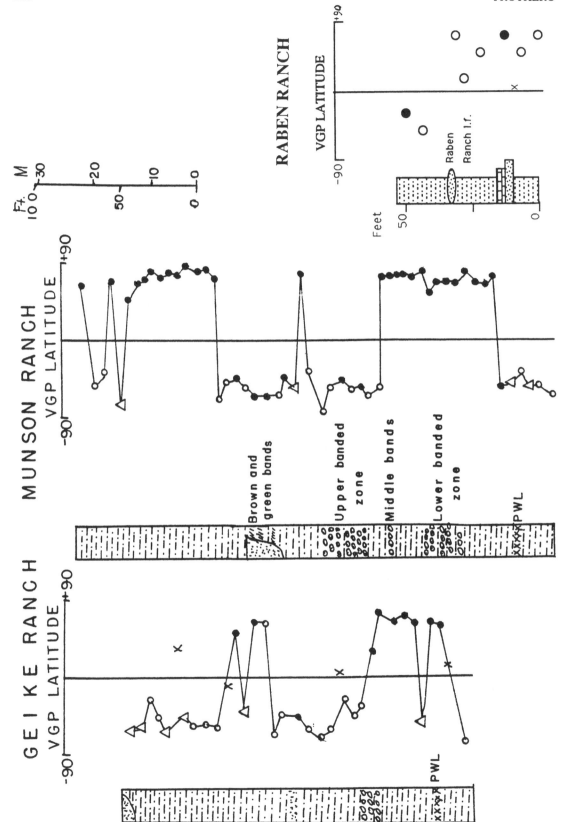

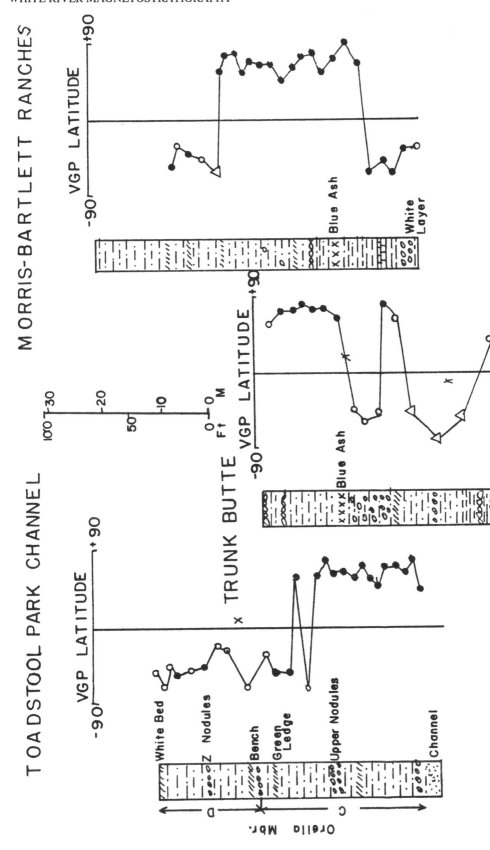

Figure 6. Magnetic stratigraphy of sections in northern Sioux and Dawes counties, northwestern Nebraska. Exact locations of sections described in text. Lithostratigraphy of most sections follows Skinner (unpublished field notes, on file in the Archives of the Department of Vertebrate Paleontology, American Museum of Natural History). Toadstool Park channel section after Schultz and Stout (1951). Raben Ranch section after Ostrander (1985). All other symbols as in Figure 5.

Ridge occurs in the Prairie Dog Creek drainage, just east of the road north from Harrison as it descends through Monroe Creek Canyon. Known in the Frick Collection as "Plunkett-Parsons Ranch" (after the ranch owners in the 1960s when Morris Skinner and Bob Emry measured these sections; Skinner section books, Vol. 6, pp. 88-89), it has also been called "Munson Ranch" (after the landowners in the 1970s), "10 miles north of Harrison," "Hall Ranch," and "Warbonnet Creek" in the Frick Collection; localities Sx 1-6 and Sx 37-38 of the University of Nebraska State Museum cover the same area. The 1980 edition of the Warbonnet Ranch 7.5' quadrangle calls it "Shalimar Ranch." The main part of the measured section occurs in NW NW NW Sec. 31 (lower part) and then continues up the southeast-trending ravine from center to SW SE Sec. 36, T33N R56W, Bodarc 7.5' quadrangle, Sioux County. A nearby section is shown by Singler and Picard (1980, fig. 4, section 4). This collecting area is about 15 miles west of Toadstool Park, and the detailed stratigraphy differs considerably from the units recognized by Schultz and Stout (1955, 1961) and Harvey (1960). Although the PWL is clearly present (Fig. 6B), it is not at all clear whether the Orella A-D units can be recognized (*contra* Harvey, 1960, and Singler and Picard, 1980, fig. 4, section 4). In the Frick Laboratory notes, the section was zoned by "lower," "middle," and "upper banded zones" which were extraordinarily rich in fossils (especially small mammals). Based on the magnetostratigraphic pattern (Fig. 6), the early Orellan normal magnetozone occurs only 15 feet above the PWL (so Orella "A" is extremely short), and the top of the early Orellan normal zone occurs in the "middle bands" (corresponding to the "Upper Nodules" at Toadstool Park, which mark the division between Orella B and C). Another long normal zone occurs between 190 and 240 feet in the section, and appears to correspond to the middle Whitneyan normal zone, which brackets the Upper Whitney Ash (upper Whitney B and lower Whitney C) at Toadstool Park. However, neither Whitneyan mammals nor the Lower or Upper Whitney Ashes have been recognized at Munson Ranch.

On the west side of the road north out of Harrison is another, smaller collecting area known to the Frick Laboratory as Geike Ranch. The main section, as measured by Morris Skinner in October, 1955 (Skinner section books, Vol. 5, pp. 24-25), begins in NE SW SW Sec. 17 and moves up the ravine to the top of the butte (BM 4384) in SE SE SW Sec. 17, T33N R56W, Warbonnet Ranch 7.5' quadrangle, Sioux County. In the University of Nebraska catalogues, it corresponds to locality Sx-36. The section differs so much from the Munson Ranch section (just a mile to the east) that Skinner used a completely different stratigraphic terminology (Fig. 6). Based on the magnetic polarity pattern, the early Orellan normal magnetozone is much shorter (from 15-50 feet on the measured section), and a

short normal magnetozone, between 110 and 130 feet in the section, may correspond the late Whitneyan normal magnetozone (unfortunately, no Whitneyan fossils have yet been reported from Geike Ranch). The bulk of the collections, however, is from the lower (Orellan) part of the section.

North of the Pine Ridge proper in the low flatlands of Chadron Formation is the Raben Ranch locality, described by Ostrander (1985). It contains an important middle Chadronian mammalian fauna, the most diverse Chadronian fauna reported from Sioux County. The exposures occur in a small eroded ravine running from NE SE SW Sec. 30 to SW NW NE Sec. 31, T34N R53W, Orella 7.5' quadrangle, Sioux County, Nebraska. Although Ostrander (1985, pp. 226-227) measured 65.6 feet of section, in 1986 we could measure and sample only the top 50 feet in one continuous exposure (Fig. 6). All but the top two sites (the upper 15 feet) of the section is of normal polarity.

Chadron area, Dawes County, Nebraska

East of the Toadstool Park area are additional exposures that have proven to be important for White River stratigraphy and faunas. Clustered around the town of Chadron, in Dawes County, Nebraska, these areas have been collected and studied by the Frick Laboratory, the University of Nebraska State Museum, and many others, but the only published stratigraphic descriptions were by Vondra (1960) and Gustafson (1986). The latter paper summarizes the stratigraphy and faunas as currently known.

One of the most important sections was measured and sampled in 1980 on a prominent landmark known as Trunk Butte (northeast face, in NW SW NW Sec. 31, T33N R49W, Trunk Butte 7.5' quadrangle, Dawes County, Nebraska). First studied by Morris Skinner in 1951 (Skinner section books, Vol. 4, pp. 14-15), it was also measured and described by Gustafson (1986, section 1). The Trunk Butte section includes 45 feet of typical bentonitic Chadron clays at the base, overlain by 50 feet of white ashy Chadronian (based on brontothere tooth fragments) deposits below the PWL (Fig. 6). Skinner (field notes, dated August 2, 1953) planned to name these volcaniclastic-rich strata the "Trunk Butte Member" of the Chadron Formation. They closely resembled the white, ashy Chadronian deposits he had studied in the Lusk and Douglas areas of Wyoming, as well as other exposures around Chadron, Nebraska, and the Chadronian exposures at Slim Buttes, Harding County, South Dakota (Lillegraven, 1970). Over 40 years later, the same concept is now being resurrected, but Terry et al. (1995) call this volcaniclastic-rich Chadronian unit the "Big Cottonwood Creek Member" of the Chadron Formation.

Above this Chadronian sequence at Trunk Butte is the PWL (called the "white clay with brown spots" by Skinner, but referred to the PWL or "UPW" by

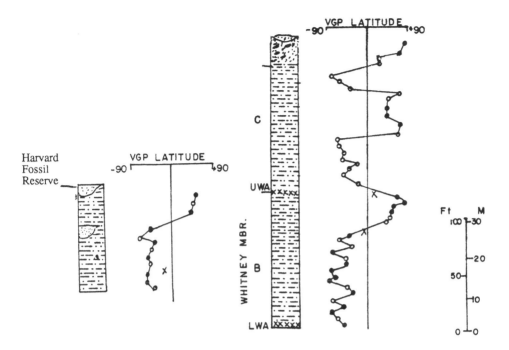

Figure 7. Magnetic stratigraphy of the Torrington (left) and Castle Rock (right) sections along the North Platte River Valley, Scottsbluff County, Nebraska, and Goshen County, Wyoming. Exact location of sections described in text. Castle Rock section after Schultz and Stout (1955); Torrington section after Schlaikjer (1935). All symbols as in Fig. 5.

Gustafson, 1986), overlain by 60 feet of typical Orella Member (including the "Blue Ash" marker bed of Skinner). The upper 50 feet of section (starting just below the Blue Ash) produces an early Orellan fauna (Gustafson, 1986) and is of normal polarity, so it probably correlates with the early Orellan normal magnetozone found throughout the White River Group. The lower 85 feet of section is of reversed polarity, except for two sites near the PWL (Fig. 6).

Another important collecting area for the Frick Laboratory is about 2.5 miles north of Chadron, in low exposures known as the Morris and Bartlett ranches (NW Sec. 5 and W1/2 NE Sec. 6, T33N R48W, Chadron West 7.5' quadrangle, Dawes County, Nebraska). The 1980 sampling and sections were based on Skinner's notes (Vol. 4, pp. 10-12; Vol. 5, pp. 104-105); this section also corresponds to Gustafson's (1986, p. 11) section 4. The very base of the exposures appears to include the PWL; above this layer are about 190 feet of Orella Member with an early Orellan fauna (Fig. 6). The section above the Blue Ash was sampled at Bartlett Ranch (NW Sec. 5) and the section below this marker was sampled at Morris Ranch (W1/2 NE Sec. 6). As in the Trunk Butte section, the early Orellan normal magnetozone begins just below the Blue Ash and extends upward for about 90 feet (as indicated by Gustafson, 1986, fig. 3, column 5). The sections below and above this zone were of reversed polarity, although the upper 40 feet of section could not be sampled due to poor exposures.

North Platte River Valley, Wyoming and Nebraska

In addition to the Pine Ridge escarpment, important White River outcrops occur along the North Platte River Valley in the Wildcat Ridge area, Scottsbluff County, Nebraska, and on into Goshen County, Wyoming (Schlaikjer, 1935; Schultz and Stout, 1955, fig. 10; Schultz and Stout, 1961; Swinehart et al., 1985). The thickest and most complete section is found at Scottsbluff National Monument (published in Prothero and Swisher, 1992, fig. 2.6). About eight miles to the east, near the town of McGrew, Nebraska, is an isolated butte known as Castle Rock (SW NW Sec. 6, T20N R53W, McGrew 7.5' quadrangle, Scottsbluff County, Nebraska). This section was first described by Schultz and Stout (1955, fig. 10, section 6). In 1983, we sampled on the southwest face of the bluff from the Lower Whitney Ash (Fig. 7) through Whitney C and the brown siltstone beds (Swinehart et al., 1985). As in the section at Scottsbluff and elsewhere, a normal magnetozone occurs below and above the Upper Whitney Ash. Based on the ^{40}Ar/^{39}Ar date of 30.58 ± 0.18 Ma on the UWA at Scottsbluff (Swisher and Prothero, 1992), this magnetozone correlates with Chron C12n. Above this level are two other normal magnetozones which may correlate with Chrons C11n and C10n (see Tedford et al., this volume, Chapter 15).

Schlaikjer (1935) and Schultz and Stout (1955, fig. 10, section 15-18) described a short section at Harvard Fossil Reserve, just south of Torrington, Wyoming

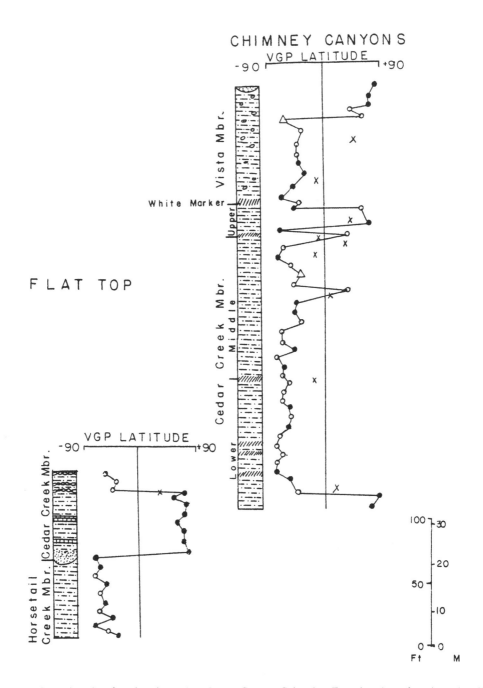

Figure 8. Magnetic stratigraphy of sections in northern Logan County, Colorado. Exact location of sections described in text. Lithostratigraphy after Galbreath (1953). All symbols as in Fig. 5.

(NW NW Sec. 32, T24N R61W, Torrington 7.5' quadrangle, Goshen County, Wyoming). Approximately 80 feet of Orella Member were sampled in 1983, starting at the lowest exposures in the ravine and ending with the limestone that caps the escarpment. The lower 60 feet of the section (Fig. 7) are of reversed polarity, and the upper 20 feet (containing the "Harvard Fossil Reserve Quarry" of Schlaikjer, 1935, or "Torrington Quarry" of Schultz and Stout, 1955) is of normal polarity. Based on the late early Orellan nature of the fossil assemblage (based on taxa such as *Mesohippus barbouri*), this normal magnetozone probably correlates with the early Orellan normal magnetozone found elsewhere in the Brule Formation.

Northeastern Colorado

Galbreath (1953) described a thick sequence of White River exposures along the southern rim of the High Plains in Logan and Weld counties, northeastern Colorado (Fig. 1). These areas had been important for fossil mammals since Marsh first collected there in 1870, and Cope (1874) first described Colorado fossil mammals; the first detailed study was published by Matthew (1901). Since that time, large collections have been made by the American Museum of Natural History, the Frick Laboratory, the Denver Museum of Natural History, the University of California, the University of Kansas Museum of Natural History, and Southern Illinois University. However, only the last two collections (made largely by Galbreath) have adequately detailed biostratigraphic data.

Although the stratigraphic units are similar to other White River deposits, they are different enough (with a much higher ash content, and few nodular beds or river channel sandstones) that they have their own stratigraphic terminology. Galbreath (1953) called the Chadronian beds at the base of the section the Horsetail Creek Member, the Orellan exposures the Cedar Creek Member, and the high Whitneyan cliffs the Vista Member of the White River Formation. The mammalian faunas allow some biostratigraphic correlation with other areas, but none of the marker beds found in Nebraska or Wyoming (such as the PWL or the Lower or Upper Whitney Ashes) was thought to occur in these sections. However, Evanoff (personal communication) has sampled the White Marker bed at the base of the Vista Member, and believes it can be geochemically correlated with the Lower Whitney Ash.

At Galbreath's suggestion, two main sections were sampled. The Chadronian Horsetail Creek Member was sampled along a low hill known as "Flat Top Butte" (Galbreath, 1953, plate 1C, figs. 6-7, section IX). The section (Fig. 8) was located in S1/2 Sec. 1, T10N R54W, North Sterling Reservoir 7.5' quadrangle, Logan County, Colorado. The lower 60 feet of the section include the maximum thickness of Horsetail Creek Member, and are entirely of reversed polarity. Disconformably incised into this sequence is the late early Orellan lower Cedar Creek Member; the lower 50 feet of this section at Flat Top are of normal polarity, and the top 15 feet are of reversed polarity. Based on the presence of late early Orellan fossils, this Cedar Creek normal magnetozone probably correlates with the early Orellan normal magnetozone found elsewhere in the White River group.

The second section (Fig. 8) spans about 250 feet of the Cedar Creek Member and about 90 feet of the Vista Member in Chimney Canyons (Galbreath, 1953, plate 1A-B, figures 6-7, section VII). The lower part of the section was taken in the ravine in NW NE Sec. 10, and continued up the canyon between "West Chimney" and "East Chimney" in SW SE Sec. 3, T11N R54W,

Chimney Canyons 7.5' quadrangle, Logan County, Colorado. The upper part of the section went up the south face of "East Chimney" (SE SE SE Sec. 3) and then the top of the section was taken in the canyon behind "West Chimney" (SE NW Sec. 3). At the base of the lower Cedar Creek Member is the early Orellan normal magnetozone, but the rest of the lower and middle parts of the Cedar Creek Member is of reversed polarity. A short normal magnetozone occurs in the thin upper part of the Cedar Creek Member. The lower 60 feet of the Vista Member are again of reversed polarity, but the upper 30 feet are of normal polarity.

CALIBRATION AND CORRELATION

As discussed by Swisher and Prothero (1990) and Prothero and Swisher (1992), the White River magnetic stratigraphy can be correlated to the magnetic polarity time scale of Berggren et al. (1995) by numerous $^{40}Ar/^{39}Ar$ dates (Fig. 9). The oldest exposures sampled in this study produce the middle Chadronian Raben Ranch fauna. Based on Ostrander's correlation of these mammals with those found between Ashes B and F at Flagstaff Rim, Wyoming (Emry, 1992; Prothero and Swisher, 1992, fig. 2.4), the Raben Ranch section probably correlates with the normal and reversed magnetozones just above Ash B (but below Ash D), which would correlate with Chrons C16n.1 and C15r (Prothero and Swisher, 1992, fig. 2.3).

The widespread early Orellan normal magnetozone consistently lies above the PWL. This layer has long been correlated with the "5 tuff" of Evanoff et al. (1992) near Douglas, Wyoming, which has been $^{40}Ar/^{39}Ar$ dated at 33.91 ± 0.06 Ma (Prothero and Swisher, 1992). Obradovich et al. (1995) report a $^{40}Ar/^{39}Ar$ date of 33.59 ± 0.14 Ma for the same tuff. However, Larson and Evanoff (in press) have shown that the Douglas 5 tuff is geochemically distinct from the PWL in Lusk or to the east in Nebraska. Instead, the PWL seems to geochemically match Ash J at Flagstaff Rim, Wyoming, which has been $^{40}Ar/^{39}Ar$ dated at 34.7 ± 0.04 Ma (Swisher and Prothero, 1992) or 34.36 ± 0.11 Ma (Obradovich et al., 1995). According to Larson and Evanoff (in press), the PWL and Ash J also appear to correlate with the "4 tuff" at Douglas (Evanoff et al., 1992), rather than the 5 tuff. Ash J at Flagstaff Rim and the 4 and 5 tuffs at Douglas all occur in Chron C13r (Prothero and Swisher, 1992; Evanoff et al., 1992). Despite the changes in ash correlations, these dates would place the early Orellan normal magnetozone in Chron C13n (33.0-33.4 Ma in the time scale of Berggren et al., 1995).

The long reversed interval that spans most of the later Orellan and early Whitneyan thus correlates with Chron C12r (30.9-33.0 Ma). This is corroborated by the $^{40}Ar/^{39}Ar$ date of 31.8 ± 0.007 on the Lower Whitney Ash (Swisher and Prothero, 1990; Prothero and Swisher, 1992). The late Whitneyan normal magneto

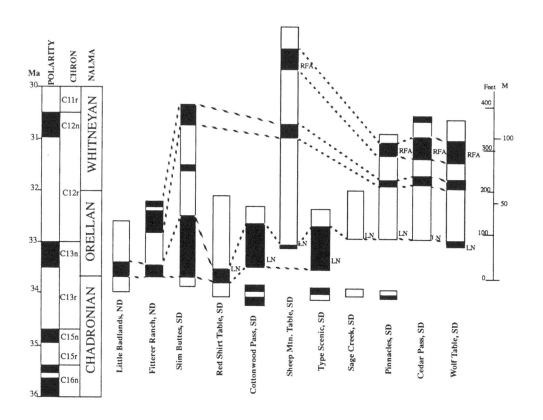

Figure 9. Magnetic correlation of sections within the White River Group, based on sections presented in this paper, Prothero and Whittlesey (in press), Swisher and Prothero (1992), and Prothero (1985b). Datum for sections is the Chadron-Brule contact or the PWL. Correlation to the time scale of Berggren et al. (1995) based on dates presented in Prothero and Swisher (1992).

zone correlates with Chron C12n (30.5-30.9 Ma), which is calibrated by the Upper Whitney Ash within this zone, $^{40}Ar/^{39}Ar$ dated at 30.58 ± 0.18 Ma (Swisher and Prothero, 1990; Prothero and Swisher, 1992). The Whitneyan/Arikareean boundary is a bit more difficult to define, but based on the criteria discussed by Tedford et al. (this volume, Chapter 15), it occurs in Chron C11n, about 30 Ma. The contact between the Whitney Member and "brown siltstone" beds of the Brule Formation appears to lie within Chron C11r (about 30.0 Ma), and the oldest rocks of the Arikaree Group occur in Chron C9r (about 28.3 Ma).

Based on the information presented above and in Prothero and Swisher (1992), the Chadronian appears to span the interval from about 37 Ma to about 33.8 Ma. The Orellan runs from 33.8-32 Ma, and the Whitneyan from 32 Ma to about 30 Ma. This is slightly different from the calibrations reported by Prothero and Swisher (1992), mostly due to the changes in the Berggren et al. (1995) time scale and the new dates reported by Obradovich et al. (1995).

ACKNOWLEDGMENTS

Over the past 15 years, this research has been supported by NSF grant EAR87-08221, a grant from the Donors of the Petroleum Research Fund of the American Chemical Society, a grant in aid of research from Sigma Xi, and by field funds from the Department of Geological Sciences of Columbia University. I thank Charles Denham, Dennis Kent, Neil Opdyke, William Roggenthen, and Joseph Kirschvink for access to their paleomagnetics laboratories. I thank Malcolm C. McKenna, Robert J. Emry, Philip R. Bjork, James E. Martin, Richard H. Tedford, Michael R. Voorhies, James B. Swinehart, Emmett Evanoff, the late Gregg Ostrander, the late Edwin C. Galbreath, and the late Morris F. Skinner for all their help with White River stratigraphy over the years. The field sampling would never have been possible without the hard work and cheerful good spirits of several field crews: in 1979, Karen Gonzalez and Heidi Shlosar; in 1980, Priscilla Duskin, Jon Frenzel, and Heidi Shlosar; in 1983, Allison Kozak, Rob Lander, and Annie Walton; in

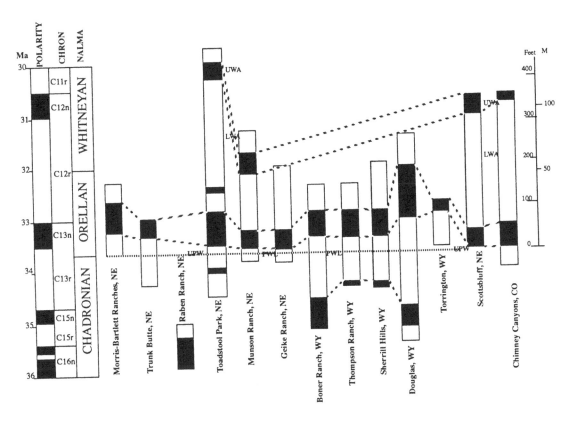

1986, Allison Kozak, Dana Gilchrist, and Kecia Harris; in 1987, John Foster, Steve King, and Susan Briggs. Karen Gonzalez (1979), Priscilla Duskin (1980), Rob Lander (1983), and Steve King (1987) also helped with the processing of samples and laboratory analysis. I thank Bob Emry, Bill Korth, Margaret Stevens, Jim Swinehart, and Richard Tedford for helpful reviews of the manuscript. This chapter is dedicated to the memory of Morris F. Skinner, who did more to advance our understanding of White River mammals and stratigraphy than any person who has ever lived.

LITERATURE CITED

Berggren, W. A., D. V. Kent, M.-P. Aubry, C. C. Swisher III, and K. G. Miller. 1995. A revised Paleogene geochronology and chronostratigraphy. SEPM Special Publication 54:129-212.

Butler, R. F. 1992. Paleomagnetism. Blackwell, Boston.

Clark, J. 1937. The stratigraphy and paleontology of the Chadron Formation in the Big Badlands of South Dakota. Annals of the Carnegie Museum 25:261-350.

Clark, J., J. R. Beerbower, and K. K. Kietzke. 1967. Oligocene sedimentation, stratigraphy, paleoecology and paleoclimatology of the Big Badlands of South Dakota. Fieldiana: Geology Memoir 5:1-158.

Cope, E. D. 1874. Report on the stratigraphy and Pliocene vertebrate palaeontology of northern Colorado. U.S. Geological and Geographic Survey of the Territories, Bulletin 1, ser. 1, Vol. 1, pp. 9-28.

Cope, E. D. 1883. A letter to the secretary. Proceedings of the American Philosophical Society, 21:216-227.

Denham, C. R. 1984. Statistical sedimentation and magnetic polarity stratigraphy; pp. 101-112 in W. A. Berggren and J. A. Van Couvering (eds.), Catastrophes and Earth History, The New Uniformitarianism. Princeton University Press, Princeton, N. J.

Denson, N. M., and J. R. Gill. 1965. Uranium-bearing lignite and carbonaceous shale in the southwestern part of the Williston Basin—a regional study. U.S. Geological Survey Professional Paper 463:1-75.

Douglass, E. 1909. A geological reconnaissance in North Dakota, Montana, and Idaho, with notes on Mesozoic and Cenozoic geology. Carnegie Museum Annals, 5:211-288.

Emry, R. J. 1992, Mammalian range zones in the Chadronian White River Formation at Flagstaff Rim, Wyoming; pp. 106-115 in D. R. Prothero and W. A. Berggren (eds.), Eocene-Oligocene Climatic and Biotic Evolution. Princeton University Press, Princeton, N. J.

Emry, R. J., P. R. Bjork, and L. S. Russell. 1987. The Chadronian. Orellan, and Whitneyan land mammal ages; pp. 118-152 in M. O. Woodburne (ed.), Cenozoic Mammals of North America, Geochronology and Biostratigraphy. University of California Press, Berkeley.

Evanoff, E., D. R. Prothero, and R. H. Lander. 1992. Eocene-Oligocene climatic change in North America: the White River Formation near Douglas, east-central Wyoming; pp. 116-130 in D. R. Prothero and W. A. Berggren (eds.), Eocene-Oligocene Climatic and Biotic Evolution. Princeton University Press, Princeton, N. J.

Fisher, R. A. 1953. Dispersion on a sphere. Proceedings of the Royal Astronomical Society A217:295-305.

Galbreath, E. C. 1953. A contribution to the Tertiary

geology and paleontology of northeastern Colorado. University of Kansas Paleontological Contributions, Vertebrata 4.

Gustafson, E. P. 1986. Preliminary biostratigraphy of the White River Group (Oligocene, Chadron and Brule Formations) in the vicinity of Chadron, Nebraska. Transactions of the Nebraska Academy of Sciences, XIV:7-19.

Hartl, P., L. Tauxe, and C. Constable. 1993. Early Oligocene geomagnetic field behavior from Deep Sea Drilling Project Site 522. Journal of Geophysical Research 98 (B11):19649-19665.

Harvey, C. 1960. Stratigraphy, sedimentation, and environment of the White River Group of the Oligocene of northern Sioux County, Nebraska. Ph. D. Dissertation, University of Nebraska, Lincoln, 151 pp.

Hoganson, J. W. 1986. Oligocene stratigraphy of North Dakota; pp. 36-40 in E. Clausen, and A. J. Kihm (eds.), Tertiary and Upper Cretaceous of south-central and western North Dakota. North Dakota Geological Society Field Trip, 1986.

Hoganson, J. W., and G. E. Lammers. 1985. The vertebrate fauna and paleoecology of the Dickinson Member, Brule Formation (Oligocene) in Stark County, North Dakota. Proceedings of the North Dakota Academy of Sciences 39:1-15.

Hoganson, J. W., and G. E. Lammers. 1992. Vertebrate fossil record, age, and depositional environments of the Brule Formation (Oligocene) in North Dakota; pp. 243-257 in J. M. Erickson and J. W. Hoganson (eds.), Frank D. Holland Jr. Memorial Symposium. North Dakota Geological Survey Miscellaneous Series 76:243-257.

Johnson, H. P., W. Lowrie, and D. V. Kent. 1975. Stability of anhysteretic remanent magnetization in fine and coarse magnetite and maghemite particles. Geophysical Journal of the Royal Astronomical Society 41:1-10.

Korth, W. W. 1989. Stratigraphic occurrence of rodents and lagomorphs in the Orella Member, Brule Formation (Oligocene), northwestern Nebraska. Contributions to Geology, University of Wyoming 27(1):15-20.

Leonard, A. G. 1922. The White River Formation in North Dakota. North Dakota Quarterly Journal 12:27-114.

Lillegraven, J. A. 1970. Stratigraphy, structure, and vertebrate fossils of the Oligocene Brule Formation, Slim Buttes, northwestern South Dakota. Bulletin of the Geological Society of America 81:831-850.

Matthew, W. D. 1901. Fossil mammals of the Teritary of northeastern Colorado. American Museum of Natural History Memoirs, Vol. 1, Part 7, pp. 355-447.

Murphy, E. C., J. W. Hoganson, and N. F. Forsman. 1993. The Chadron, Brule, and Arikaree Formations in North Dakota, the buttes of southwestern North Dakota. North Dakota Geological Survey, Reports of Investigations 96:1-144.

Obradovich, J. D., E. Evanoff, and E. E. Larson. 1995. Revised single-crystal laser-fusion ^{40}Ar/^{39}Ar ages of Chadronian tuffs in the White River Formation of Wyoming. Geological Society of America, Abstracts with Programs 27(3):77-78.

Opdyke, N. D., E. H. Lindsay, N. M. Johnson, and T. Downs. 1977. The paleomagnetism and magnetic polarity stratigraphy of the mammal-bearing section of Anza-Borrego State Park, California. Journal of Quaternary Research 7:316-329.

Ostrander, G. E. 1985. Correlation of the early Oligocene (Chadronian) in northwestern Nebraska; pp. 205-231 in J. E. Martin (ed.), Fossiliferous Cenozoic deposits of western South Dakota and northwestern Nebraska. Dakoterra, Museum of Geology, South Dakota School of Mines 2.

Pluhar, C. J., J. L. Kirschvink, and R. W. Adams. 1991. Magnetostratigraphy and clockwise rotation of the Plio-Pleistocene Mojave River Formation, central Mojave Desert, California. San Bernardino County Museum Association Quarterly 38 (2):31-42.

Prothero, D. R. 1982. Medial Oligocene magnetostratigraphy and mammalian biostratigraphy: testing the isochroneity of mammalian biostratigraphic events. Ph.D. Dissertation, Columbia University, New York.

Prothero, D. R. 1985a. Chadronian (early Oligocene) magnetostratigraphy of eastern Wyoming: implications for the Eocene-Oligocene boundary. Journal of Geology 93:555-565.

Prothero, D. R. 1985b. Correlation of the White River Group by magnetostratigraphy; pp. 265-276 in J. E. Martin (ed.), Fossiliferous Cenozoic deposits of western South Dakota and northwestern Nebraska. Dakoterra, Museum of Geology, South Dakota School of Mines 2.

Prothero, D. R. 1996. Biostratigraphic zonation and chronostratigraphy of the Orellan and Whitneyan land mammal "ages." Geological Society of America Special Paper (in press).

Prothero, D. R., C. R. Denham, and H. G. Farmer. 1982. Oligocene calibration of the magnetic polarity time scale. Geology 10:650-653.

Prothero, D. R., C. R. Denham, and H. G. Farmer. 1983. Magnetostratigraphy of the White River Group and its implications for Oligocene geochronology. Palaeogeography, Palaeoclimatology, Palaeoecology 42:151-166.

Prothero, D. R., and C. C. Swisher III. 1992. Magnetostratigraphy and geochronology of the terrestrial Eocene-Oligocene transition in North America; pp. 46-74 in D. R. Prothero and W. A. Berggren (eds.), Eocene-Oligocene Climatic and Biotic Evolution, Princeton University Press, Princeton, N.J.

Prothero, D. R., and K. E. Whittlesey. In press. Magnetostratigraphy and biostratigraphy of the Orellan and Whitneyan land mammal "ages" in the White River Group. Geological Society of America Special Paper (in press).

Retallack, G. 1983. Late Eocene and Oligocene fossil Paleosols from Badlands National Park, South Dakota. Geological Society of America Special Paper 193.

Schlaikjer, E. M. 1935. Contributions to the stratigraphy and paleontology of the Goshen Hole area, Wyoming. IV. New vertebrates and the stratigraphy of the Oligocene and early Miocene. Bulletin of the Museum of Comparative Zoology, Harvard University 76:97-189.

Schultz, C. B., and C. H. Falkenbach. 1968, The phylogeny of the oreodonts, Parts 1 and 2. Bulletin of the American Museum of Natural History 139:1-498.

Schultz, C. B., and T. M. Stout. 1955. Classification of the Oligocene sediments in Nebraska. Bulletin of the University of Nebraska State Museum 4:17-52.

Schultz, C. B., and T. M. Stout. 1961. Field conference on the Tertiary and Pleistocene of western Nebraska. Special Publication of the University of Nebraska State Museum 2:1-54.

Seeland, D. 1985. Oligocene paleogeography of the northern Great Plains and adjacent mountains; pp. 187-205 in R. M. Flores and S. Kaplan (eds.), Cenozoic Paleogeography of the west central United States. Special Publication of the Rocky Mountain Section, SEPM.

Sinclair, W. J. 1921. The "Turtle-Oreodon" layer or "Red Layer," a contribution to the stratigraphy of the White River Oligocene. Proceedings of the American Philosophical Society, 60:457-466.

Singler, C. R., and M. D. Picard. 1979, Petrography of the White River Group (Oligocene) in northwestern

Nebraska and adjacent Wyoming. Contributions to Geology, University of Wyoming 18(1):51-67.

Singler, C. R., and M. D. Picard. 1980. Stratigraphic review of the Oligocene beds in the northern Great Plains. Wyoming Geological Association and Earth Science Bulletin 13(1):1-18.

Skinner, M. F. 1951. The Oligocene of western North Dakota; pp. 51-58 *in* J. D. Bump (ed.), Society of Vertebrate Paleontology Guidebook, 5th Annual Field Conference, Western South Dakota, August-September 1951.

Stone, W. J. 1972. Middle Cenozoic stratigraphy of North Dakota; pp. 123-132 *in* F. T. C. Ting (ed.), Depositional environments of the lignite-bearing strata in western North Dakota. North Dakota Geological Survey Miscellaneous Series 50.

Swinehart, J. B., V. L. Souders, H. M. Degraw, and R. F. Diffendal, Jr. 1985. Cenozoic paleogeography of western Nebraska; pp. 209-229 in R. M. Flores and S. Kaplan (eds.), Cenozoic Paleogeography of the west central United States. Special Publication of the Rocky Mountain Section, SEPM.

Swisher, C. C., III, and D. R. Prothero. 1990. Single-crystal ^{40}Ar/^{39}Ar dating of the Eocene-Oligocene transition in North America. Science 249:760-762.

Terry, D. O., H. LaGarry, and W. B. Wells. 1995. The White River Group revisited: vertebrate trackways, ecosystems, and stratigraphic revision, reinterpretation, and redescription. Nebraska Conservation and Survey Division Guidebook 10:43-57.

Trimble, D. E. 1980. The geologic story of the Great Plains. U.S. Geological Survey Bulletin 1493:1-55.

Vondra, C. F. 1960. Stratigraphy of the Chadron Formation in northwestern Nebraska. Compass of Sigma Epsilon 37(2):73-90.

Wanless, H. R. 1923. The stratigraphy of the White River beds of South Dakota. Proceedings of the American Philosophical Society 62:190-269.

Wood, H. E., R. W. Chaney, J. Clark, E. H. Colbert, G. L. Jepsen, J. B. Reeside, Jr., and C. Stock. 1941. Nomenclature and correlation of the North American continental Tertiary. Geological Society of America Bulletin 52:1-48.

14. Magnetostratigraphy and Biostratigraphy of the Eocene-Oligocene Transition, Southwestern Montana

ALAN R. TABRUM, DONALD R. PROTHERO, AND DANIEL GARCIA

ABSTRACT

Detailed magnetostratigraphic and biostratigraphic studies of Eocene-Oligocene deposits in southwestern Montana have greatly refined the dating of the fossil mammal assemblages known from the region. In the Jefferson Basin, the middle Chadronian Pipestone Springs l.f. correlates with Chrons C15r to C16r1 (35.0-35.5 Ma). The assemblages known from the middle to late(?) Chadronian Little Pipestone Creek localities correlate at least in part with Chrons C15r to C16n1 (35.0-35.4 Ma). The late Chadronian West Easter Lily l.f. and the succeeding early Orellan Easter Lily l.f. span much of Chron C13r (33.5-34.3 Ma). In the Beaverhead Basin, the early Chadronian McCarty's Mountain fauna correlates with Chrons C16r1 to C17n1 (35.5-36.7 Ma). The late Duchesnean Diamond O Ranch l.f. may correlate with part of the long episode of normal polarity during Chron C17n. In the Sage Creek Basin, the late Uintan Dell beds have produced the stratigraphically successive Douglass Draw and Hough Draw local faunas, both of which appear to fall in Chron C18r (40.2-41.0 Ma). The Matador Ranch and Cook Ranch local faunas from the stratigraphically higher Cook Ranch Formation are correlated with the Orellan part of Chron C12r (32.0-33.0 Ma).

The mammalian local faunas of southwestern Montana strongly reflect the significant faunal provincialism that characterized western North America during the Eocene-Oligocene transition. A varying percentage (generally 20-40%) of the mammalian species known from the late Uintan through late Orellan local faunas of the region appear to have been endemic to southwestern Montana. Severeal taxa, including talpids, *Ocajila*, *Megalagus*, *Palaeolagus*, *Palaeogale*, and *Leptomeryx sensu stricto*, appear to occur significantly earlier in southwestern Montana than they do in the Great Plains region, while others, including apternodontids and cylindrodontids, persist later.

Sedimentation was highly episodic and locally very rapid throughout the Eocene-Oligocene interval in southwestern Montana. The timing of sedimentation in each basin or group of closely related basins appears to have been largely dependent on local basin tectonics. Significant, but not regionally synchronous, gaps occur in the sedimentary sequence preserved in each of the basins. Most of the exposed sequences, some of which are very thick, span less than one million years.

INTRODUCTION

Eocene and Oligocene sediments and associated volcanics are sporadically but locally well exposed in the small intermontane basins of southwestern Montana. Moderately extensive remnants of these deposits are also present at high elevation in some of the adjacent mountain ranges. Although exposures are generally limited and discontinuous, some of the sediments are richly fossiliferous and a temporally extensive suite of mammalian local faunas has been developed within the region. These preserve one of the most extensive and important records of the Eocene-Oligocene transition in continental rocks of western North America outside the Great Plains region (Fig. 1).

The Cenozoic geologic history of southwestern Montana is complex, characterized by episodic sedimentation and multiple episodes of faulting (Fields et al., 1985; Hanneman and Wideman, 1991). Tertiary basins initially developed and were locally receiving sediments by early middle Eocene (Bridgerian) time. By late Eocene (Chadronian) time all of the basins, with the possible exception of some of the small peripheral basins, were the site of significant sediment accumulation.

Although the largest of the intermontane basins of southwestern Montana is only about 60 miles long by 30 miles wide, the basins locally contain very thick sequences of Cenozoic continental deposits—approximately 16,000 feet in the relatively undissected Big Hole Basin and more than 10,000 feet in the Deer Lodge Basin (Fields et al., 1985). The present basins of southwestern Montana were largely outlined during an interval of intense Basin-and-Range style faulting in the middle and late Miocene and have been further fragmented by latest Tertiary and Quaternary faulting. The presence of Eocene and Oligocene sediments in some of the adjacent mountain ranges indicates that at least some of the Paleogene basins were more extensive than the present small basins of the region. Vertebrate

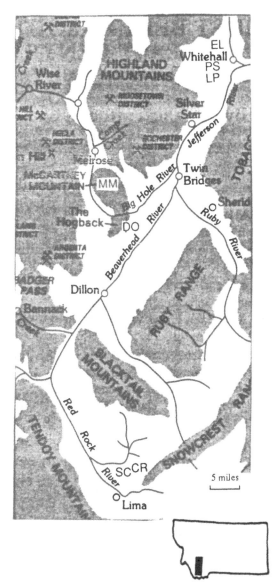

Figure 1. Index map showing location of sites mentioned in text. Localities: CR, Cook Ranch; DO, Diamond O Ranch; EL, Easter Lily; LP, Little Pipestone; MM, McCarty's Mountain; PS, Pipestone Springs; SC, Sage Creek.

paleontological field work in southwestern Montana was largely pioneered by Earl Douglass. Between 1894 and 1905, Douglass conducted extensive field investigations in the region, working in most of the intermontane basins of southwestern Montana and amassing a large collection of Eocene, Oligocene, and Miocene vertebrates. During this period Douglass discovered many of the most significant Tertiary localities known from the region, including the Pipestone Springs, Little Pipestone Creek, and McCarty's Mountain localities discussed below.

Douglass described some of the Paleogene vertebrates he had collected in a series of papers published between 1901 and 1908 (Douglass, 1901, 1903, 1905, 1907, 1908a,b).

Douglass was hired by the Carnegie Museum in 1902, and his large collection of fossil vertebrates from southwestern Montana became one of the first key acquisitions made by the museum. Douglass'ss field notes were not, however, acquired at the time and for many years were unavailable and presumed lost (see Wood et al., 1941, p. 25). In 1976 Gawin Douglass donated his father's extensive surviving papers to the Marriott Library of the University of Utah; these contain a wealth of useful information, including much more precise locality data for many of the specimens Douglass collected than had been retained in the Carnegie Museum records.

In 1937 J. LeRoy Kay renewed collecting in southwestern Montana for the Carnegie Museum. Kay collected in the region for more than twenty years, discovering several important new localities and greatly increasing the sample available from many of the localities previously worked by Douglass. The collection amassed by Douglass and Kay and housed at the Carnegie Museum is both the largest and historically most important collection of fossil vertebrates known from southwestern Montana.

Major progress in unraveling the complicated Cenozoic history of southwestern Montana was made by Robert W. Fields and his students at the University of Montana. In the late 1950s, Fields initiated a long-term project to investigate the stratigraphy and vertebrate paleontology of the intermontane basins of the region. Theses by his students describing the Cenozoic geology and vertebrate paleontology of the Beaverhead (Riel, 1963; Hoffman, 1972; Petkewich, 1972), Flint Creek (Rasmussen, 1969), Jefferson (Kuenzi, 1966), Muddy Creek (Dunlap, 1982), North Boulder (Lofgren, 1985), Smith River (Runkel, 1986), and Upper Ruby River (Monroe, 1976) basins, combined with data collected from most of the other intermontane basins of southwestern Montana, have greatly clarified the stratigraphic and temporal relationships of the Cenozoic deposits known from the region. The information contained in these studies, as well as that from many other published and unpublished sources, was summarized in Fields et al. (1985). During the course of their work, Fields and his students greatly increased the available sample of fossil vertebrates known from southwestern Montana. In particular, the University of Montana collection retains the precise locality information that is often lacking in the older collections.

In their review paper, Fields et al. (1985, fig. 4 and Appendix A) attempted to assign relatively precise ages to many of the mammalian local faunas known from southwestern Montana. These were based in large part on the then unpublished, but widely circulated, bichronological studies of Emry et al. (1987) and Tedford et al. (1987), combined with the preliminary results of some of Prothero's magnetostratigraphic studies in the region. While we generally endorse the placement of the southwestern Montana local faunas within the North American Land Mammal "Ages" advocated by Fields et al. (1985), significant advances in calibration of the late Paleogene global time scale (Berggren et al., 1992, 1995) coupled with refinements in both the dating and correlation of the North American Land Mammal "Ages" (Woodburne, 1987; Prothero and Swisher, 1992) have necessarily modified some of the numerical age estimates they provided. For example, Fields et al. (1985, fig. 2) placed the Eocene/Oligocene boundary at 36.6 Ma, within the Chadronian Land Mammal "Age" which they illustrated as spanning the interval from 32.4-38.7 Ma; more recent estimates, however, place the Eocene/Oligocene boundary at about 34 Ma (Berggren et al., 1992, 1995) and the age range for the Chadronian as approximately 33.7-37.0 Ma (Prothero and Swisher, 1992).

The late Uintan through late Orellan mammalian local faunas from southwestern Montana, discussed below, differ significantly from contemporaneous assemblages known from other areas of North America. The moderate to high level of provincialism exhibited by these local faunas has tended to inhibit precise correlation with assemblages known from other parts of the continent. High-resolution chronostratigraphy, combining detailed biostratigraphic and magnetostratigraphic analysis, is, however, now available for several of the key sections in the region and allows us to correlate many of the local faunas from southwestern Montana with a degree of precision hitherto unattainable. In the following pages we report the results of our analyses of key localities in the Beaverhead, Jefferson, and Sage Creek basins. These span much of the Eocene-Oligocene interval in southwestern Montana and include many of the most fossiliferous and historically most important localities known from the region. We should, however, add the cautionary caveats that some of our faunal analyses are preliminary and that much work remains to be done in the region.

MAGNETIC METHODS

Magnetic sampling in southwestern Montana was conducted over several field seasons. In 1980, the Easter Lily section, and the Cook Ranch Formation were sampled; both were briefly summarized in Prothero (1982). The main section at Pipestone Springs was sampled in 1983, and described the following year (Prothero, 1984). The McCarty's Mountain section was sampled in 1986, and the Dell beds, Little Pipestone

Creek, and Diamond O Ranch were sampled in 1987. A minimum of three oriented block samples was taken at each site with simple hand tools. In most sections, sites were spaced 1.5-3 m (5.5-11 feet) stratigraphically, depending upon exposures. Samples were then trimmed on a band saw with a tungsten carbide blade; crumbly samples were hardened with sodium silicate solution.

The 1980 samples were run at the paleomagnetics laboratory at Woods Hole Oceanographic Institute, and the 1983 samples were analyzed at the South Dakota School of Mines and Technology. All other samples were analyzed at the paleomagnetics laboratory of the California Institute of Technology. After measurement of NRM (natural remanent magnetization), a suite of samples was subjected to stepwise demagnetization to determine the origin of magnetization. AF (alternating field) demagnetization (Fig. 2A) showed a steady and rapid decline in intensity, suggesting that the primary carrier of the remanence is a low-coercivity mineral such as magnetite. Under thermal demagnetiza-tion (Fig. 2B-D), almost all remanence had vanished by temperatures of 600°C (above the Curie point of magnetite), indicating that very little remanence is carried by hematite.

This interpretation was corroborated by IRM (isothermal remanent magnetization) studies of representative lithologies, as described by Pluhar et al. (1991). Most samples (Fig. 3) reached saturation IRM values between 100 and 300 mT (millitesla) and did not gain any further magnetization even at fields of 1300 mT. This suggests that fine-grained magnetite is the primary carrier of the remanence. The same IRM samples were also subjected to a modified Lowrie-Fuller ARM test (Johnson et al., 1975; Pluhar et al., 1991). This test compares the resistance to AF demagnetization of both the IRM produced in a 100-mT peak field, and the ARM (anhysteretic remanent magnetization) gained in a 100 mT oscillating field. As can be seen in Figure 3, the ARM (solid squares) demagnetizes at higher fields than does the IRM (open squares), showing that the magnetic remanence is carried by fine-grained (<10 micron) particles of magnetite in the single-domain or pseudo-single-domain size range.

Because AF demagnetization does not remove overprints caused by chemical remanence from iron hydroxides, all remaining samples were analyzed by thermal demagnetization. Reversely magnetized samples had slight overprints that were typically removed at temperatures of 200°C, and a stable reversed magnetic component was isolated at temperatures of 300-500°C (Fig. 2C-D). For statistical averaging purposes, the component between 300-400°C was typically used.

Most samples came from flat-lying beds, so no fold test for stability could be conducted. However, in the McCarty's Mountain area, the dip varied between 30° and 50° from the top to the bottom of the section, and a

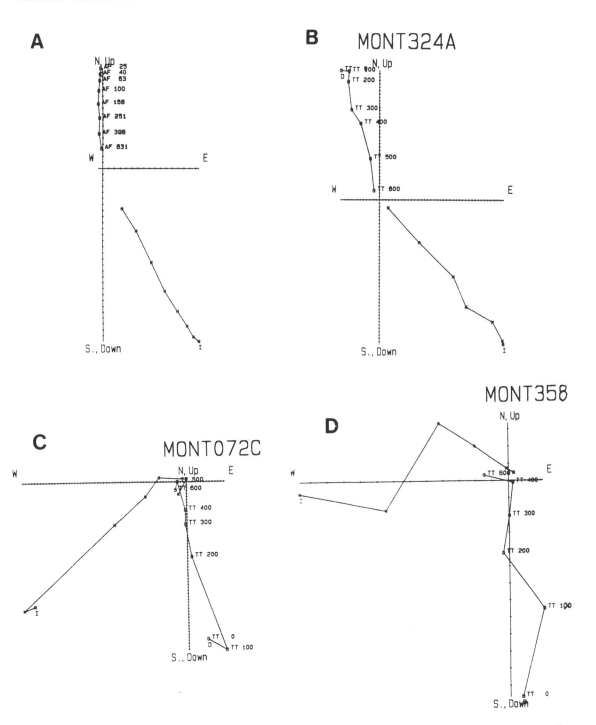

Figure 2. Orthogonal demagnetization ("Zijderveld") plots of representative samples in this study. Each increment is 10^{-7} emu. Horizontal components are indicated by open circles and vertical components by asterisks. "I" indicates NRM vector in the vertical component. A. AF demagnetization behavior, showing rapid decline in intensity from 25 to 631 Gauss. This indicates that a low coercivity mineral such as magnetite is the primary carrier of the remanence. B-D. Thermal demagnetization plots, showing vectors at temperatures ranging from NRM (TT0) to 600°C. Sample 324A (Fig. 2B) is a normally magnetized sample; sample 72C (Fig. 2C) is a reversed sample that finally reached the upper hemisphere at 400-500°C; sample 358B (Fig. 2D) is a reversed sample that reached the upper hemisphere at 200-400°C.

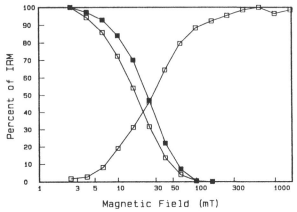

Figure 3. IRM acquisition and Lowrie-Fuller ARM test of a representative sample (see Pluhar et al., 1991, for experimental details). Open boxes indicate IRM intensity; solid boxes represent ARM intensity. The IRM acquisition curve (ascending curve on right) saturates at about 300 mT, showing that magnetite is the primary carrier of the remanence. Demagnetization of the IRM (descending curve on left with open boxes) and ARM (solid boxes) shows the ARM more resistant to AF demagnetization than the IRM. This suggests that the carrier is a fine-grained (<10 microns), single-domain or pseudo-single-domain magnetite.

modified fold test could be conducted. After demagnetization, the uncorrected magnetic vectors (D = 23.0°, I = -4.9°, k = 2.9, α_{95} = 32.4) are clearly not overprinted by a post-tilting magnetic field, because they deviate significantly from the present magnetic field. In addition, the uncorrected vectors were more scattered than the same directions after the dip correction (D = 8.2, I = 62.4, k = 5.7, α_{95} = 20.9), indicating that the magnetization was acquired prior to tilting.

Reversal tests for stability were also conducted. At McCarty's Mountain, the mean for reversed sites (D = 198.6, I = -47.7, k = 6.2, α_{95} = 13.0) was antipodal to the mean for normal sites given above. Similar results were obtained for the Dell beds (normal sites, D = 7.9, I = 59.0, k = 3.3, α_{95} = 49.7; reversed sites, D = 176.4, I = -49.8, k = 9.1, α_{95} = 12.1). This suggests that the magnetic component isolated between 300-500°C is the characteristic component of the remanence.

The stable characteristic directions for all samples at a given site were then averaged using the methods of Fisher (1953; see Butler, 1992). Sites whose samples clustered well enough that they could be distinguished from random scatter at the 95% confidence interval were designated Class I sites, following Opdyke et al. (1977). If a sample was lost or crumbled, then the mean direction of the two remaining samples was used; these were Class II sites of Opdyke et al. (1977). If two samples showed a clear polarity preference, but the third sample was divergent, then the site was designated Class III (Opdyke et al., 1977). A few sites gave highly

scattered or unstable results, or showed no clear polarity preference, and were labeled indeterminate.

JEFFERSON BASIN

In 1899, Douglass discovered rich concentrations of fossil vertebrates, including abundant small mammals, at exposures near Pipestone Springs and south of Little Pipestone Creek along the western margin of the Jefferson Basin. His subsequent description (Douglass, 1901) of a variety of new mammals from the "White River Beds" of Montana, including eight new species from the Pipestone Springs area, spurred the American Museum of Natural History to send an expedition to western Montana in the summer of 1902 to further investigate these deposits. Matthew (1903) described the large collection obtained by the American Museum from the Pipestone Springs locality, naming a dozen additional new species of mammals, and first noted the presence of a younger fauna in beds exposed north of the railroad tracks. The Pipestone Springs area has since been worked by many institutions, with the largest collections being those made by Kay for the Carnegie Museum and by Fields and his students for the University of Montana.

The Cenozoic geology of the Jefferson Basin was thoroughly reviewed by Kuenzi (1966) in his Ph.D. dissertation and summarized in an influential paper published by Kuenzi and Fields (1971). Based on homotaxis and strong lithologic similarity to the Climbing Arrow and Dunbar Creek formations mapped by Robinson (1963, 1967) in the Three Forks and Toston quadrangles, Kuenzi and Fields (1971) divided the Paleogene sediments exposed in the western part of the Jefferson Basin into the Climbing Arrow and Dunbar Creek members of their newly named Renova Formation. Sediments referred to the Climbing Arrow Member are discontinuously exposed between the Pipestone Springs localities and the Little Pipestone Creek area about five miles to the south. Sediments assigned to the Dunbar Creek Member are largely restricted to the area north of the Burlington Northern Railroad tracks, although exposures in the Colbert Creek area (parts of sections 6 and 7, T1N R5W), originally included in the Climbing Arrow Member by Kuenzi and Fields (1971), have since been reassigned to the Dunbar Creek Member by Garcia (1992).

Exposures of Tertiary sediments in the western part of the Jefferson Basin are highly discontinuous. Most of the exposures are small and several have produced only very limited vertebrate assemblages. Attempts to correlate between some of these localities have proven difficult. Garcia (1992) has consequently divided the Pipestone Creek region into eight separate "collecting areas" and has recognized nine geographically and/or temporally distinct local faunas. Each of Garcia's "collecting areas" consists of either a single exposure or a few closely related exposures: these generally coincide with the principal fossil localities illustrated on the map provided by Kuenzi and Fields (1971, fig. 3).

The most richly fossiliferous outcrops in the Jefferson Basin are the small pockets of Climbing Arrow sediments exposed in section 29, T2N R5W, near Pipestone Springs, and the small area of exposures south of Little Pipestone Creek in the SW section 8, and NW section 17, T1N R5W. Except for the few specimens reported by Matthew (1903) from exposures north of the railroad tracks, all of the specimens obtained by the early collectors appear to have been derived from these two areas.

The quality of available locality data for specimens collected from the Pipestone Springs area is highly variable. Many specimens, especially in the older collections, are simply listed as having been collected from "Pipestone Creek," "Pipestone Springs," or the "Pipestone beds" without any more specific locality information (Garcia, 1992). Field parties from the University of Montana have since divided the exposures in the region into multiple geographically and stratigraphically restricted localities, and the locality information retained in the University of Montana records is generally the most precise available for any of the larger collections of vertebrates from the area. Kay also seems to have consistently distinguished between the major localities in the large collection he made for the Carnegie Museum.

Some mixing between localities appears to be demonstrable. Many specimens that have been attributed to Pipestone Springs were actually collected from other localities. For example, information contained in Douglass's surviving field notes indicates that he made a sizable collection from the Little Pipestone Creek area, yet only a few specimens that he collected are attributed to Little Pipestone Creek in the Carnegie Museum records. The bulk of the material that Douglass obtained from Little Pipestone Creek appears to have become hopelessly mixed with his collection from the Pipestone Springs locality. Similarly, some of the specimens (e.g., "?*Limnenetes* sp.") attributed to the Pipestone Springs locality by Matthew (1903) also seem likely to have been collected from other localities, and may not even have been derived from the Jefferson Basin. Some of the other collections from the Pipestone Springs area are probably also mixed. Most, but not all, of the apparent mixing seems to involve the assignment of specimens collected from the Little Pipestone Creek area to the Pipestone Springs l.f. Because the assemblages known from the Pipestone Springs and Little Pipestone Creek localities do appear to differ somewhat in composition, and possibly in age, the many specimens from the Pipestone Springs area that have only vague or questionable locality data must be treated cautiously.

Garcia (1992) has recently comprehensively reviewed the mammalian paleontology of the Pipestone Creeks region. In the following sections we briefly review those assemblages for which we also have magneto-stratigraphic data: the Pipestone Springs, Little Pipestone Creek, and Easter Lily areas. Several other nearby localities have also produced significant vertebrate faunas (Kuenzi and Fields, 1971; Garcia, 1992) but are only mentioned in passing.

Pipestone Springs Local Fauna

Specimens assigned to the Pipestone Springs l.f. are derived from three small pockets of richly fossiliferous Climbing Arrow sediments exposed in section 29, T2N R5W, about 1.5 miles west of Pipestone Hot Springs. Kay termed the most fossiliferous of these the Main Pocket and Fence Pocket localities. Collecting parties from the University of Montana have since divided Kay's Main Pocket locality into multiple geographically and/or stratigraphically restricted localities (Kuenzi and Fields, 1971; Tabrum and Fields, 1980; Lillegraven and Tabrum, 1983; Garcia, 1992). The prominent dip-slope near the northern end of the Main Pocket exposures (Orr, 1958, Photo 1), University of Montana locality MV 5811, is the richest of the Pipestone Springs localities and has probably produced the majority of the exquisite small mammal specimens known from the Pipestone Springs l.f.

A second isolated pocket of Climbing Arrow sediments occurs about 250 yards south of the Main Pocket localities and has produced the moderately diverse assemblage assigned to University of Montana locality MV 5903. Garcia (1992) called this area Pipestone Springs South, although the name South Pocket might be preferable. Lithologically and faunally these beds are indistinguishable from those at the Main Pocket exposures. Some of the specimens obtained by Kay and other collectors and attributed to Pipestone Springs Main Pocket were probably derived from the South Pocket exposures. Specimens collected from this area have traditionally been included with those from the Main Pocket (Kuenzi and Fields, 1971; Tabrum and Fields, 1980), a procedure that we follow, but do not necessarily endorse, here.

The Fence Pocket locality, University of Montana locality MV 6103, is about 0.5 miles southeast of the Main Pocket localities and is separated from them by an eastwardly projecting spur of granitic rocks of the Boulder Batholith. The thin section (only 12 m, approximately 39 feet) of sediments exposed at the Fence Pocket locality is lithologically similar to, and seems to correlate with, the most productive horizon at the Main Pocket exposures (Garcia, 1992). However, because the Fence Pocket locality is somewhat removed from the other Pipestone Springs localities we have elected to list the fauna separately in Table 1.

Table 1. Pipestone Springs local fauna, Jefferson Basin: Localities in Sec. 29, T2N R5W, Jefferson County, Montana. Climbing Arrow Member, Renova Formation. Middle Chadronian. PSMP, Pipestone Springs Main Pocket. PSFP, Pipestone Springs Fence Pocket. (* = type locality.)

	PSMP	PSFP
PERADECTIA		
Peradectidae		
Didelphidectes pumilus Hough*	X	
DIDELPHIMORPHIA		
Didelphidae		
Herpetotherium valens (Lambe)	X	X
Herpetotherium fugax Cope	X	X
Copedelphys titanelix (Matthew)*	X	X
PROTEUTHERIA		
Leptictidae		
Leptictis acutidens (Douglass)*	X	X
Leptictis thomsoni (Matthew)*	X	X
INSECTIVORA		
Geolabididae		
Centetodon kuenzii Lillegraven and Tabrum*	X	X
Centetodon magnus (Clark)	X	
Micropternodontidae		
Micropternodus borealis Matthew*	X	X
Apternodontidae		
Oligoryctes altitalonidus (Clark)	X	X
Oligoryctes cameronensis Hough	X	
Apternodus mediaevus Matthew*	X	X
Soricidae		
Domnina thompsoni Simpson*	X	X
Proscalopidae		
Oligoscalops? new species	X	
Insectivora, *incertae sedis*		
Cryptoryctes kayi Reed*	X	
PHOLIDOTA		
Manidae		
Patriomanis americanus Emry		
LAGOMORPHA		
Leporidae		
Megalagus brachyodon (Matthew)*	X	X
Palaeolagus temnodon Douglass*	X	X
Chadrolagus emryi Gawne	X	
RODENTIA		
Ischyromyidae		
Ischyromys veterior Matthew*	X	X
Ischyromys new species	X	X
Cylindrodontidae		
Pseudocylindrodon neglectus Burke*	X	
Cylindrodon fontis Douglass*	X	X
Aplodontidae		
Prosciurus vetustus Matthew*	X	
Sciuridae		
Protosciurus jeffersoni (Douglass)*	X	
Eutypomyidae		
Eutypomys parvus Lambe	X	
Eomyidae		
Zemiodontomys burkei (Black)*	X	
Namatomys lloydi Black*	X	X
Montanamus bjorki Ostrander*	X	
Adjidaumo minimus (Matthew)*	X	
Paradjidaumo trilophus (Cope)	X	X
Aulolithomys bounites Black*	X	
New genus and species	X	

	PSMP	PSFP
Heliscomyidae		
Heliscomys ostranderi Korth et al.	X	
Rodentia, *incertae sedis*		
Pipestoneomys bisulcatus Donohoe*	X	
CREODONTA		
Hyaenodontidae		
Hyaenodon crucians Leidy		
Hyaenodon microdon Mellett*		
CARNIVORA		
Amphicyonidae		
Brachyrhynchocyon dodgei (Scott)	X	
Brachyrhynchocyon new species	X	
Daphoenictis sp.	X	
Canidae		
Hesperocyon gregarius (Cope)	X	X
Procyonidae		
Mustelavus priscus Clark	X	
Ursidae		
Parictis montanus Clark and Guensburg*	X	
Nimravidae		
Dinictis sp.	X	
Carnivora, *incertae sedis*		
Palaeogale sectoria (Gervais)	X	
PERISSODACTYLA		
Equidae		
Mesohippus westoni (Cope)	X	
Mesohippus portentus Douglass*		
Miohippus grandis (Clark and Beerbower)	X	X
Brontotheriidae		
Brontotheriid sp.	X	
Helaletidae		
Colodon sp., cf. *C. kayi* (Hough)	X	
Hyracodontidae		
Hyracodon sp.	X	?
Rhinocerotidae		
Rhinocerotid sp.	X	
ARTIODACTYLA		
Leptochoeridae		
Stibarus montanus Matthew*	X	
Leptochoerus sp.		
Agriochoeridae		
Agriochoerus sp.	X	
Merycoidodontidae		
Bathygenys alpha Douglass*	X	X
Merycoidodon sp.	X	X
Camelidae		
Poebrotherium sp.	X	
Protoceratidae		
"*Leptotragulus*" *profectus* (Matthew)*	X	X
Leptomerycidae		
Hendryomeryx new species	X	
Leptomeryx new species	X	
"*Leptomeryx*" *mammifer* Cope	X	X
"*Leptomeryx*" *speciosus* Lambe	X	X
"*Leptomeryx*" new species I	X	X
"*Leptomeryx*" new species M	X	

Kuenzi and Fields (1971) cited a thickness of 176 feet for the Climbing Arrow sediments exposed at the Pipestone Springs Main Pocket localities, and Prothero (1984) reported a comparable thickness of about 60 m (197 feet) in his magnetostratigraphic study of the Pipestone Springs locality. Garcia (1992) has recently measured additional stratigraphic sections in the area and reports a thickness of as much as 100 m (328 feet) for the deposits.

The Pipestone Springs l.f. is easily the most diversified and best studied mammalian fauna known from any of the intermontane basins of southwestern Montana. Pipestone Springs is particularly well known for superb specimens of small mammals and provided the first clear glimpse of a Chadronian small-mammal fauna (Matthew, 1903). If the number of type specimens collected from a locality is a measure of its significance, then the Main Pocket at Pipestone Springs is arguably the single most important locality of Chadronian age known. The type specimens of thirty currently recognized species of vertebrates are attributed to the Pipestone Springs l.f., including the type species of the lizard *Helodermoides*, and the mammals *Apternodus, Aulolithomys, Bathygenys, Copedelphys, Cryptoryctes, Cylindrodon, Didelphidectes, Micropternodus, Montanamus, Namatomys, Pipestoneomys, Prosciurus, Pseudocylindrodon*, and *Zemiodontomys*. Several additional new taxa remain to be described.

Tabrum and Fields (1980) presented a revised faunal list for the Pipestone Springs l.f. and briefly summarized the extensive literature on Pipestone Springs mammals. Since publication of their paper, additional taxa have been described from the Pipestone Springs localities by Emry and Hunt (1980), Korth (1994), Korth et al. (1991), Lillegraven et al. (1981), Lillegraven and Tabrum (1983), Ostrander (1983), and Storer (1984a). Taxonomic name changes accepted here have been introduced by de Bonis (1981), Emry and Korth (1993), Heaton (1993 and this volume, Chapter 27), Hunt (this volume, Chapter 23), Korth (1980, 1994), Storer (1981), Storer and Bryant (1993), and Wang (1994). Further study by Garcia (1992) has significantly increased the Pipestone Springs faunal list and has necessitated additional taxonomic name changes. Most of these have involved only minor taxonomic shifts; however, specimens thought to pertain to an oromerycid, and identified as "?*Eotylopus* new sp." by Tabrum and Fields (1980) are now instead believed to represent a new species of the leptomerycid *Hendryomeryx* comparable in size to the large "*Leptomeryx*" *mammifer*.

Some of the specimens that have been attributed to Pipestone Springs are now known to have been collected from other localities. Many others, including some type materials (e.g., the holotype and referred specimens of *Pipestoneomys bisulcatus* described by Donohoe, 1956), are suspected to have been derived

from other localities. Two of the four species of *Hyaenodon* attributed to Pipestone Springs by Mellett (1977) were based on specimens demonstrably not collected from the Pipestone Springs localities. The single specimen of *H. horridus* (F:AM 95716) reported by Mellett was actually collected from younger beds exposed about two miles to the northeast in the Easter Lily area (Garcia, 1992), and Mellett's reference to *H. montanus* in the Pipestone Springs assemblage was based only on the type specimen, collected from the Toston area approximately 50 miles to the northeast. Matthew (1903, p. 222), in his description of the American Museum collection from Pipestone Springs, provisionally assigned "a number of lower jaws and parts of jaws" to ?*Limnenetes* sp. Matthew suggested that more than one species was represented but did not cite any specimen numbers. Schultz and Falkenbach (1956, p. 459) noted this but reported that "a tray of material from the Pipestone Springs area . . . may represent the specimens in question." This sample consisted of ten partial lower jaws of *Oreonetes anceps*. *Oreonetes anceps* is not, however, represented in any of the other collections that have been made from the Pipestone Springs area. The 1902 American Museum expedition also collected at various places in the Three Forks Basin and North Boulder Valley, at least in the general area from which specimens of *O. anceps* have been recovered, and it seems most likely that the material reported by Schultz and Falkenbach was actually collected from one of these two areas.

In the faunal list presented in Table 1 we attribute 65 mammalian taxa to the Pipestone Springs l.f. Accurate locality information is available for at least some of the specimens of all of the taxa listed in Table 1 except *Patriomanis americanus, Hyaenodon crucians, H. microdon, Mesohippus portentus*, and *Leptochoerus* sp. Of these we believe that the available specimens of *P. americanus, H. crucians*, and *H. microdon* were derived from the Main Pocket exposures. The specimens assigned to *Mesohippus portentus* and *Leptochoerus* sp., however, were collected in 1899 by Douglass and have not been duplicated in later collections from the area; they may have been collected from one of the Pipestone Springs localities or may instead have been collected from other localities in the general area.

The Pipestone Springs l.f. was designated as one of the principal correlatives of the Chadronian provincial age by the Wood Committee, who regarded the locality as early Chadronian (Wood et al., 1941, p. 28). Later, Clark and Beerbower (in Clark et al., 1967), based on relatively weak evidence, believed that Pipestone Springs was late Chadronian in age, about equivalent to the Peanut Peak Member of the Chadron Formation in South Dakota. The significance of this evidence was strongly disputed by Emry (1973). Based on his biostratigraphic study of the Chadronian deposits at

Flagstaff Rim, Wyoming, Emry (1973) concluded that the Pipestone Springs l.f. was not late Chadronian in age, but as old as middle Chadronian.

The age of the Pipestone Springs l.f. has since been firmly established as middle Chadronian (Prothero, 1984; Ostrander, 1985; Emry et al., 1987; Emry, 1992). Based in part on the then unpublished work of Emry et al. (1987) coupled with his own magnetostratigraphic study of the Pipestone Springs exposures, Prothero (1984) assigned an early middle Chadronian age to the Pipestone Springs l.f. and was followed in this age assignment by Fields et al. (1985). Ostrander (1985) regarded the very diverse assemblage known from Pipestone Springs as effectively the type assemblage for the middle Chadronian. Emry et al. (1987) noted that

the taxonomically diverse fauna occurring from about 250 to 400 feet above the base of the generalized zonation section of the White River Formation in the Flagstaff Rim area of Wyoming . . . essentially duplicates the Pipestone Springs l.f., predominantly at the species level

and defined the middle Chadronian based principally on these two assemblages. The strong similarities (and some significant differences) between the assemblages known from Pipestone Springs and that part of the Flagstaff Rim section between the Ash B and Ash G levels is emphasized in the biostratigraphic range zones compiled by Emry (1992) for the Flagstaff Rim section.

Prothero (1984) sampled and described the magnetic stratigraphy of the main section at Pipestone Springs, running from NW NW SW section 29 to SE SW NW of the same section, T2N R5W Delmoe Lake 7.5' quadrangle, Jefferson County, Montana. The 130 feet of section in the Climbing Arrow Member of the Renova Formation produced reversed sites at base, followed by about 80 feet of rocks of normal polarity; the remaining section was reversed (Prothero, 1984, fig. 3). Although the polarity pattern remains unchanged from that illustrated by Prothero (1984), its interpretation (Prothero, 1984, fig. 4) has changed due to major revisions in the correlation of the Chadronian with the magnetic polarity time scale (Prothero and Swisher, 1992). Based on the latest interpretation of the Flagstaff Rim sequence, the Pipestone Springs local fauna probably correlates with Chrons C15r-C16r1, or about 35.0-35.5 Ma (Fig. 7).

Little Pipestone Creek Local Fauna

Fossil vertebrates assigned to the Little Pipestone Creek local fauna were collected from a small area of good exposures south of Little Pipestone Creek in the SW section 8, and NW section 17, T1N R5W, immediately east of Montana Highway 41. These exposures are about three miles south of the Pipestone

Springs localities and include University of Montana localities MV 5905, MV 6001, and MV 8603. The small sample from Honeymoon Quarry, part of MV 5905, is one of the few quarry samples known from the Cenozoic deposits of southwestern Montana.

Kuenzi and Fields (1971) referred the sediments exposed at the Little Pipestone Creek localities to the Climbing Arrow Member of the Renova Formation. Three partially overlapping sections measured by Garcia (1992) total about 185 feet, close to the thickness of 178 feet reported for the Little Pipestone Creek section by Kuenzi (1966). The lower part of the section is poorly exposed and very sparingly fossiliferous. All of the vertebrates listed in Table 2 are believed to have been collected from the upper 75 feet of the Little Pipestone Creek section.

Kay et al. (1958) and Kuenzi and Fields (1971) provided faunal lists of the taxa known by them to have been collected from the Little Pipestone Creek area, but fossil vertebrates from Little Pipestone Creek have been much less intensively investigated than those from the main Pipestone Springs exposures. The localities are fairly rich, but of the many specimens described from the Pipestone Creek region, only the holotype of *Agriochoerus maximus* Douglass (1901) was reported to have been collected from the Little Pipestone Creek area. Some specimens that have been attributed to Pipestone Springs (e.g., CM 9287, the most complete lower jaw from the Pipestone Springs area referred to *Daphoenocyon dodgei* by Clark and Beerbower *in* Clark et al., 1967) were, however, actually collected from Little Pipestone Creek.

The Little Pipestone Creek local fauna includes at least 41 species of mammals. Of the 37 taxa that appear to be specifically determinate, all except the very rare *Agriochoerus*? *maximus* have also been recorded from the Pipestone Springs local fauna. Faunal composition, however, appears to differ markedly. "*Leptomeryx*" *speciosus*, the most common artiodactyl at the Pipestone Springs localities, is not represented at Little Pipestone Creek (Garcia, 1992) and appears to have been replaced by Garcia's "*Leptomeryx*" n. sp. M. *Cylindrodon fontis*, abundant at Pipestone Springs, is represented by only a single specimen from Little Pipestone Creek, and *Pseudocylindrodon neglectus* and *Heliscomys ostranderi*, both fairly common at Pipestone Springs, have not been recorded from the Little Pipestone Creek local fauna. The enigmatic rodent *Pipestoneomys bisulcatus*, however, is vastly more common at Little Pipestone Creek than it is at Pipestone Springs, and *Adjidaumo minimus* also appears to be relatively more common.

There also appear to be potentially significant differences in the ischyromyids and equids. *Ischyromys veterior* is much more common than *Ischyromys* n. sp. at Pipestone Springs but much less common at Little

Table 2. Little Pipestone Creek local fauna, Jefferson Basin: Localities in SW Sec. 8, and NW Sec. 17, T1N R5W, Jefferson County, Montana. Climbing Arrow Member, Renova Formation. Middle to late Chadronian. (* = type locality.)

DIDELPHOMORPHIA
  Didelphidae
    Herpetotherium valens (Lambe)
    Herpetotherium fugax Cope
    Copedelphys titanelix (Matthew)
PROTEUTHERIA
  Leptictidae
    Leptictis acutidens (Douglass)
INSECTIVORA
  Geolabididae
    Centetodon kuenzii Lillegraven and Tabrum
  Micropternodontidae
    Micropternodus borealis Matthew
  Apternodontidae
    Oligoryctes altitalonidus (Clark)
    Oligoryctes cameronensis Hough
    Apternodus mediaevus Matthew
  Soricidae
    Domnina thompsoni Simpson
  Proscalopidae
    Oligoscalops? new species
  Insectivora, *incertae sedis*
    Cryptoryctes kayi Reed
LAGOMORPHA
  Leporidae
    Megalagus brachyodon (Matthew)
    Palaeolagus temnodon Douglass
    Chadrolagus emryi Gawne
RODENTIA
  Ischyromyidae
    Ischyromys veterior Matthew
    Ischyromys new species
  Cylindrodontidae
    Cylindrodon fontis Douglass
  Aplodontidae
    Prosciurus vetustus Matthew
  Eomyidae
    Adjidaumo minimus (Matthew)
    Paradjidaumo trilophus (Cope)

    Aulolithomys bounites Black
  Rodentia, *incertae sedis*
    Pipestoneomys bisulcatus Donohoe
CREODONTA
  Hyaenodontidae
    Hyaenodon crucians Leidy
CARNIVORA
  Amphicyonidae
    Brachyrhynchocyon dodgei (Scott)
  Canidae
    Hesperocyon gregarius (Cope)
PERISSODACTYLA
  Equidae
    Mesohippus westoni (Cope)
    Miohippus grandis (Clark and Beerbower)
  Brontotheriidae
    Brontotheriid sp.
  Rhinocerotidae
    Rhinocerotid sp.
ARTIODACTYLA
  Leptochoeridae
    Stibarus montanus Matthew
  Agriochoeridae
    Agriochoerus? maximus Douglass*
  Merycoidodontidae
    Bathygenys alpha Douglass
    Merycoidodon sp.
  Camelidae
    Poebrotherium sp.
  Protoceratidae
    "Leptotragulus" profectus (Matthew)
  Leptomerycidae
    Hendryomeryx new species
    Leptomeryx new species
    "Leptomeryx" mammifer Cope
    "Leptomeryx" new species I
    "Leptomeryx" new species M

Pipestone Creek and may in fact be replaced by *Ischyromys* n. sp. in the stratigraphically highest parts of the Little Pipestone Creek section. The relative abundance of *Mesohippus westoni* and *Miohippus grandis* may also be reversed, with *Mesohippus westoni* the more common equid at Pipestone Springs and *Miohippus grandis* more common at Little Pipestone Creek.

Black (1965) suggested that sediments exposed in the Little Pipestone Creek area might, at least in part, be younger than those exposed at the Pipestone Springs localities. Tabrum and Fields (1980) also noted that a difference in age might be involved. Further collecting in the area subsequently led Fields et al. (1985, p. 34) to assign a "Middle to Late (?) Chadronian" age to the Climbing Arrow sediments exposed south of Little Pipestone Creek.

Garcia (1992), however, later concluded that the entire Little Pipestone Creek sequence was middle Chadronian

in age and that the Little Pipestone Creek l.f. was about equivalent in age to the Pipestone Springs l.f. Garcia's conclusions were based on the strong similarities between the stratigraphic sections exposed in the two areas, coupled with the fact that all of the species recorded from the Little Pipestone Creek localities, except *Agriochoerus? maximus*, have also been reported from the Pipestone Springs l.f. He correlated the Little Pipestone Creek stratigraphic section even more closely with that preserved in the Delmoe Ditch area about one mile to the east in Sec. 9, T1N R5W, suggesting that the strata preserved in the Delmoe Ditch section were the same as those preserved in the upper part of the Little Pipestone Creek section. The very limited fauna known from the Delmoe Ditch area (University of Montana locality MV 6106) includes a specimen assigned to *Eutypomys thompsoni* by Kuenzi and Fields (1971). This specimen appears to be correctly identified and represents a species that is clearly distinct

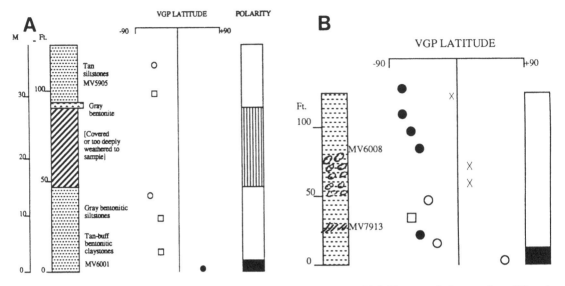

Figure 4. Magnetic stratigraphy of sections in the Jefferson River Basin. A. The Little Pipestone Springs section of Kuenzi and Fields (1971). Open boxes indicate Class II sites of Opdyke et al. (1977). B. Easter Lily section. Solid circles are Class I sites of Opdyke et al. (1977); open circles are Class III sites of Opdyke et al. (1977). "x" indicates an indeterminate site.

from the single specimen of *E. parvus* known from the Pipestone Springs Fence Pocket locality.

Emry et al. (1987, p. 138) listed the first appearance of "*Eutypomys* near *E. thompsoni*" as indicative of a late Chadronian age, and on this basis at least part of the Delmoe Ditch section is late Chadronian in age. If Garcia's (1992) correlation based on lithostratigraphic criteria is correct, then part of the Little Pipestone Creek section would also appear to be late Chadronian in age, which may, in the absence of any obvious differences in ecology, help to account for some of the clear differences in faunal composition between the Pipestone Springs and Little Pipestone Creek localities.

At least part of the Little Pipestone Creek section is, however, demonstrably of middle Chadronian age. The middle Chadronian index taxon "*Leptomeryx*" *mammifer* has been recovered from the stratigraphically lowest Little Pipestone Creek locality, MV 6001, and possibly from the somewhat higher MV 5905, but not from the stratigraphically highest locality, MV 8603. *Ischyromys veterior* has also not been recovered from MV 8603, although *Ischyromys* n. sp. is fairly common. University of Montana locality MV 6001 is middle Chadronian in age, MV 8603 possibly late Chadronian, and MV 5905 apparently intermediate in age between the two. All may be somewhat younger than the Pipestone Springs local fauna.

About 120 feet of section in the Climbing Arrow Member of the Renova Formation were sampled at Little Pipestone Creek (Fig. 4). The stratigraphic section followed the transect shown in Kuenzi and Fields (1971, fig. 3), and began in the lowest exposures in NE NW NW section 17 and followed a northeasterly

traverse to the uppermost continuous exposures in SE SE SW section 8, T1N R5W Vendome 7.5' quadrangle, Jefferson County, Montana. Although the middle part of the section was too deeply weathered or too covered for sampling, the remaining section (except for the lowest site) was entirely of reversed polarity. Based on the middle to late Chadronian age of the fauna and the similarity to the Pipestone Springs local fauna, the Little Pipestone Creek l.f. probably correlates with Chrons C13r-C15r, or about 34.5-35.0 Ma (Fig. 7).

Easter Lily section

Sediments assigned to the Dunbar Creek Member of the Renova Formation by Kuenzi and Fields (1971) are moderately well exposed north of the Burlington Northern Railroad tracks, about two miles northeast of the Pipestone Springs exposures. Matthew (1903) equated these beds with the Oreodon Beds of South Dakota and Colorado and first reported the presence of *Eumys* and other taxa indicative of an age younger than that of the Pipestone Springs l.f.

Two principal areas of good exposures are present: the Easter Lily (or Easter Lily Mine) section in sections 16 and 21, T2N R5W, and the apparently stratigraphically higher Palisades (also known as the Palisades Cliff or Palisade Cliffs) section about one mile to the east in sections 15 and 22. Kuenzi and Fields (1971) reported a thickness of 275 feet for the Dunbar Creek sediments discontinuously exposed in the Easter Lily section and 320 feet for the stratigraphically higher Palisades section.

Kuenzi and Fields (1971) assigned an Orellan age to the fossil vertebrates collected from both the Easter Lily

Table 3. West Easter Lily local fauna, Jefferson Basin: Localities in NE and SW Sec. 16, and NW Sec. 21, T2N R5W, Jefferson County, Montana. Dunbar Creek Member, Renova Formation. Late Chadronian.

DIDELPHIMORPHIA
    Didelphidae
        Herpetotherium valens (Lambe)
INSECTIVORA
    Geolabididae
        Centetodon kuenzii Lillegraven and Tabrum
    Aptenodontidae
        Apternodus mediaevus Matthew
LAGOMORPHA
    Leporidae
        Megalagus brachyodon (Matthew)
        Palaeolagus temnodon Douglass
        Palaeolagus haydeni Leidy
RODENTIA
    Ischyromyidae
        Ischyromys veterior Matthew
        Ischyromys sp.
    Eomyidae
        Paradjidaumo trilophus (Cope)

CARNIVORA
    Canidae
        Hesperocyon? sp.
PERISSODACTYLA
    Equidae
        Mesohippus bairdi Leidy
        Miohippus obliquidens (Osborn)
    Brontotheriidae
        Brontotheriid sp.
    Hyracodontidae
        Hyracodon sp.
    Rhinocerotidae
        Trigonias sp., cf. *T. osborni* Lucas
ARTIODACTYLA
    Merycoidodontidae
        Merycoidodon sp., cf. *M. culbertsoni* Leidy
    Leptomerycidae
        "*Leptomeryx*" new species I
        "*Leptomeryx*" new species M

and Palisades sections. However, further work has since established that most of the Easter Lily section is Chadronian in age (Lillegraven and Tabrum, 1983; French, 1988; Garcia, 1992). Kuenzi (1966) and Kuenzi and Fields (1971) assigned all of the vertebrates collected from the Easter Lily section, regardless of stratigraphic position, to the single University of Montana locality MV 6008, and their published faunal list (Kuenzi and Fields, 1971, table 2) consequently appears to consist of a mixture of Chadronian and Orellan taxa.

Field work initiated in 1978 by Tabrum, C. C. Swisher III, R. Nichols, and later by L. B. French, resulted in the establishment of six new geographically and stratigraphically restricted localities in the Easter Lily exposures (French, 1988) and the recognition of two stratigraphically separated and biostratigraphically distinct local faunas. The name Easter Lily l.f. is retained for the Orellan assemblage recovered from the small area of exposures east of the access road to the Easter Lily Mine in the SW SE section 16. The late Chadronian West Easter Lily l.f. was derived from stratigraphically lower exposures in areas to the west and north (see French, 1988; Garcia, 1992).

The Easter Lily section is not richly fossiliferous, and unfortunately none of the specimens collected prior to 1978 retains sufficiently precise locality information to allow their unequivocal assignment to either the Easter Lily or West Easter Lily l.f. In addition to the taxa listed in Tables 3 and 4, specimens of *Cryptoryctes kayi, Cylindrodon fontis, Hyaenodon horridus, H. crucians, H. microdon, Agriochoerus* sp., and "*Leptotragulus*" *profectus* have been recovered from the Easter Lily section (Garcia, 1992). Although most of these appear to be Chadronian forms, available locality information is at present inadequate to allow assignment

of any of these specimens to either of the local faunas currently recognized from the Easter Lily section. The following discussions of the West Easter Lily and Easter Lily local faunas are based entirely on specimens for which we have precise geographic and stratigraphic data.

West Easter Lily Local Fauna

Fossils recovered from University of Montana localities MV 7913, MV 7915, and MV 8105, in the SE SW section 16, and the NE NW section 21, T2N R5W, west of the unimproved dirt road that runs through the area, are assigned to the West Easter Lily l.f. The most fossiliferous and stratigraphically lowest locality, MV 7913, is approximately 44 m (144 feet) above the base of the Easter Lily section; MV 8105 is approximately 55 m (180 feet), and MV 7915 approximately 57 m (187 feet) above the base of the Easter Lily section (French, 1988). A fourth locality, MV 8111, from a separate area of exposures in the SW NE section 16, has produced a very sparse fauna that does not appear to differ from that known from the other localities and is also tentatively included in the West Easter Lily l.f.

As currently known, the West Easter Lily l.f. includes at least eighteen mammalian taxa (Table 3). Most of the available specimens were collected from MV 7913, the stratigraphically lowest locality. However, the only specimens of *Centetodon kuenzii* and *Mesohippus bairdi* known from the West Easter Lily l.f. were collected from MV 8105, and the only material of "*Leptomeryx*" n. sp. M was derived from MV 8111.

Although most of the species in the rather limited assemblage known from the West Easter Lily l.f. also occur in the much more diverse Pipestone Springs and Little Pipestone Creek local faunas, advances in the

Table 4. Easter Lily local fauna, Jefferson Basin: Localities in SE Sec. 16, and NE Sec. 21, T2N R5W, Jefferson County, Montana. Dunbar Creek Member, Renova Formation. Early Orellan.

DIDELPHIMORPHIA
   Didelphidae
     Herpetotherium valens (Lambe)
     Herpetotherium fugax Cope
     Copedelphys titanelix (Matthew)
INSECTIVORA
   Geolabididae
     Centetodon kuenzii Lillegraven and Tabrum
LAGOMORPHA
   Leporidae
     Palaeolagus haydeni Leidy
     "Palaeolagus" intermedius Matthew
     "Palaeolagus" burkei Wood

RODENTIA
   Aplodontidae
     Prosciurus relictus (Cope)
   Cricetidae
     Eumys sp., aff. *E. obliquidens* Wood
     Eumys sp., cf. *E. cricetodontoides* White
CARNIVORA
   Canidae
     Hesperocyon? sp.
PERISSODACTYLA
   Equidae
     Mesohippus bairdi Leidy
     Miohippus obliquidens (Osborn)
ARTIODACTYLA
   Leptomerycidae
     Leptomeryx sp., cf. *L. evansi* Leidy

equids and possibly in the leporids suggest that the West Easter Lily l.f. is significantly younger than either Pipestone Springs or Little Pipestone Creek. The West Easter Lily l.f. is best regarded as late Chadronian in age. The most significant difference seems to be in the equids, where *Mesohippus bairdi* and *Miohippus obliquidens* have replaced the characteristic Pipestone Springs and Little Pipestone Creek species *Mesohippus westoni* and *Miohippus grandis* (Garcia, 1992). Both of these species first appear in the late Chadronian (Prothero and Shubin, 1989; see summary chapter to this volume).

Garcia (1992) has also identified specimens of both *Palaeolagus temnodon* and *P. haydeni* in the small sample of lagomorphs (mostly isolated teeth) known from MV 7913. This sample may be comparable to the large sample from the Orella A beds of Nebraska referred to *P. hemirhizis* by Korth and Hageman (1988), who suggested that some late Chadronian samples might also represent this species. Prothero and Whittlesey (in press) have suggested that *P. hemirhizis* is an artificial composite that includes specimens of both *P. temnodon* and *P. haydeni* and that these two species overlap slightly in range.

A magnetic section through the main sequence of the Easter Lily section was collected in 1980, starting at the lowest exposures in the wash in SE SE SW SW section 16, and ending at the top of the hill in the NE SW SW SE section 16. About 125 feet of section were sampled (Fig. 4B), and except for the lowest site, the entire section was of reversed polarity. Given the late Chadronian-early Orellan fauna, this section most likely correlates with Chron C13r, or about 33.7-34.3 Ma (Fig. 7).

Easter Lily Local Fauna

Fossil vertebrates assigned to the Easter Lily l.f. were collected from three closely related localities in a small area of fairly continuous exposures in the SW SW section 16, T2N R5W, east of the road that leads to the Easter Lily Mine. All specimens were derived from a single thick unit of massive tuffaceous mudstone with occassional calcareous horizons. The stratigraphically lowest specimens assigned to the Easter Lily l.f. were collected approximately 67 m (220 feet) above the base of the Easter Lily section (French, 1988).

The Easter Lily l.f. is a very limited assemblage that as presently known consists of only fourteen species of mammals (Table 4). About half of the species represented are shared with either the West Easter Lily l.f. or with earlier local faunas from the Pipestone Creeks area. Biochronologically important taxa that first appear in the Easter Lily l.f. include *"Palaeolagus" intermedius*, *"P." burkei*, *Prosciurus relictus*, *Eumys*, and a species of *Leptomeryx* that is closely comparable to *L. evansi*. The Easter Lily l.f. also includes the last known occurrence of *Centetodon* in western Montana.

The presence of *Palaeolagus haydeni*, *"P." intermedius*, *"P." burkei*, *Prosciurus relictus*, possibly two species of *Eumys*, and *Leptomeryx* sp. cf. *L. evansi* all indicate an Orellan age for the Easter Lily l.f. The vertebrates from this stratigraphic level were assigned an early Orellan age by Lillegraven and Tabrum (1983) and Fields et al. (1985). The unstated reasons for this age assignment were: (1) the relatively short stratigraphic separation, without an obvious hiatus, between these localities and localities of demonstrable late Chadronian age; and (2) the presence of a morphologically primitive species of *Eumys* similar to, but larger than, *E. obliquidens*, then regarded, following Galbreath (1953), as probably indicative of a relatively early Orellan age. An early Orellan age assignment also seemed to be supported by the presence of *Eumys cricetodontoides* (or *E. elegans* if one accepts the synonymy of Martin, 1980) in the stratigraphically higher Palisades l.f. from exposures about one mile to the east.

An early Orellan age assignment for the Easter Lily l.f. also seems to be supported by the magnetic polarity stratigraphy of this part of the Easter Lily section (if we are correct in our belief that no major, unrecognized hiatus is present in the Easter Lily section). The sites sampled by Prothero in this part of the Easter Lily section were all of reversed polarity and appear to represent a continuation of the interval of reversed polarity sampled in the late Chadronian part of the section, interpreted here as Chron C13r, and would suggest an age of about 33.5-33.7 Ma for the Orellan part of the Easter Lily section.

If this interpretation is correct, then both *Eumys* and *"Palaeolagus" burkei* occur significantly earlier in southwestern Montana than they do in the central Great Plains region. In western Nebraska and eastern Wyoming, *Eumys* first appears late in Chron C13n in deposits of Orella B age, and *"P." burkei* early in Chron C12r in deposits of Orella C age (Korth, 1989; Prothero and Emry, this volume, Summary chapter). Storer (1994) has recently reported the presence of an *Eumys obliquidens*-like cricetid in the "latest Chadronian" Kealey Springs West l.f. of Saskatchewan. Although this l.f. is likely of Orellan age, rather than Chadronian as Storer suggested, the occurrence of *Eumys* seems certainly to predate the first record of the genus in Nebraska and Wyoming. Available evidence thus seems to suggest that the first appearance of *Eumys* is diachronous in different parts of the Great Plains region, and an unusually early record of *Eumys* in southwestern Montana does not seem unreasonable.

The presence of *"Palaeolagus" burkei* in the Easter Lily l.f. is more problematic. However, an obvious ancestor for *"P." burkei* is lacking in the central Great Plains region, and the species appears to have been an immigrant to the Wyoming-Nebraska region early in Chron C12r. The association of *"P." burkei* with a "primitive" species of *Eumys* in the Easter Lily l.f., coupled with the interpretation that this part of the Easter Lily section was deposited during C13r, suggests that *"P." burkei* occurs significantly earlier in southwestern Montana than it does in the central Great Plains region.

BEAVERHEAD BASIN

Paleogene sediments referable to the Renova Formation are sporadically but locally well exposed in the Beaverhead Basin and have produced significant vertebrate assemblages from several localities. The two most diverse assemblages currently known are the early Chadronian fauna from the McCarty's Mountain locality, initially reported by Douglass (1905, 1908), and the previously almost entirely unreported late Duchesnean Diamond O Ranch l.f.

McCarty's Mountain Local Fauna

Fossil vertebrates were first recovered from the thick section of Tertiary sediments exposed at the McCarty's Mountain locality by Earl Douglass, collecting for the Carnegie Museum, on July 4 or 5, 1903. The fossiliferous sediments are moderately well exposed over an area of about one quarter square mile in section 28, T4S R8W, Madison County, on the north side of the Big Hole River about five miles south-southeast of McCarty's (now McCartney) Mountain.

Douglass worked the McCarty's Mountain locality, and other localities in the general area, for about four weeks, making a sizable collection. The Carnegie Museum collection was greatly enhanced by subsequent work conducted by Kay, who spent parts of several field seasons between 1937 and 1960 collecting at the locality. Several other institutions have also worked the McCarty's Mountain locality, with by far the largest and most important of these later collections those made by S. J. Riel for the University of Montana and by J. M. Rensberger and M. Asnake for the Burke Museum of the University of Washington.

Douglass (1905) briefly described the McCarty's Mountain deposits and variously referred to them as the *"Titanotherium* Beds" and "Lower White River Beds." Riel (1963), in his comprehensive review of the McCarty's Mountain locality, referred to the deposits only as "Lower Oligocene beds." Hoffman (1972) assigned the McCarty's Mountain beds to the Renova Formation, and Asnake (1984) later referred them to the Climbing Arrow Member. The beds exposed at the McCarty's Mountain locality do not, however, particularly closely resemble the type Climbing Arrow Formation in the Three Forks Basin, or sediments referred to the Climbing Arrow Formation or to the Climbing Arrow Member of the Renova Formation in other basins of southwestern Montana, and are probably best regarded as undifferentiated Renova Formation.

The McCarty's Mountain sediments dip 30–45° to the southwest and are discontinuously exposed in small fault blocks, complicating measurement of an accurate stratigraphic section. Douglass (1905) reported that the beds were more than 700 feet thick, and Riel (1963) cited a thickness of at least 1200 feet for the deposits. The stratigraphic section measured by Prothero (Fig. 5) followed Riel's (1963, plate 2) locality map and extended from the stratigraphically lowest deposits exposed near the center of section 28 (from a short distance northeast of the 5232' hill illustrated on the Block Mountain 7.5' Quadrangle) to the uppermost beds exposed in the NE SW section 28. Approximately 850 feet of continuous section were measured. This stratigraphic section appears to span nearly all of the McCarty's Mountain sequence.

Precise locality information is available for many of the specimens collected from the McCarty's Mountain section, and it should eventually prove possible to biostratigraphically zone the beds, although this is beyond the scope of the present chapter. In his 1903 field notes, Douglass divided the McCarty's Mountain beds into stratigraphically successive levels Q (lowest) through Z (highest). Most of the specimens that Douglass collected still retain their original field labels and, based on a comprehensive field catalog and two cross-sectional sketches in his field notes, can be assigned to their approximate position in the McCarty's Mountain stratigraphic sequence (Tabrum, 1994). Riel (1963, plate 2) plotted the precise location of the specimens he collected for the University of Montana on a greatly enlarged (scale: 1 inch = 234 feet) copy of the then preliminary USGS topographic map of the area. Most of Riel's specimens can be directly tied to the magnetostratigraphic section measured by Prothero. Detailed locality and stratigraphic data are also available for the University of Washington collection (Asnake, 1984).

Wood et al. (1941, p. 25) questioned both the unity and age of the McCarty's Mountain fauna, noting that

McCarty's Mountain, so far as available knowledge goes, is merely a locality term for Oligocene exposures on its slopes or at its base . . . Douglass divided the exposures into several successive fossiliferous levels, but his unpublished notes have not been located; much or all of the Oligocene may be represented.

There was some justification for their concern, in that Douglass (1905) had indicated that he had collected specimens from more than one locality. These potential problems were not clarified by Kay et al. (1958), who included in their McCarty's Mountain faunal list some taxa that were based on specimens demonstrably collected from localities other than McCarty's Mountain. The concerns expressed by the Wood Committee were finally resolved in Riel's (1963) careful study of the McCarty's Mountain locality and by the subsequent recovery of Douglass's field notes. We restrict, as have most authors, the term "McCarty's Mountain fauna" to include only specimens derived from exposures in section 28, T4S R8W.

The known vertebrate fauna from the McCarty's Mountain locality includes turtles, lizards, and a moderately diversified assemblage of mammals. However, less than half the taxa in the McCarty's Mountain fauna have ever been formally described. For several of those that have, much larger samples are now available than are reflected in the moderately extensive but widely scattered literature. Good biostratigraphic data are available for many of the specimens, and further study of the McCarty's Mountain fauna should prove extremely profitable. Only the following elements of

the fauna have thus far been described: the lizard *Helodermoides* (Douglass, 1908; Gilmore, 1928; Sullivan, 1989); lepticids (Douglass, 1905); *Centetodon* (Lillegraven et al., 1981); *Epoicotherium* (Douglass, 1905; Simpson, 1927); *Ischyromys* (Black, 1968; Wood, 1976); *Ardynomys* (Burke, 1936; Wood, 1970); *Pseudocylindrodon* (Burke, 1938; Black, 1974); *Colodon*? (Radinsky, 1963); *Triplopides* (Radinsky, 1967); *Protoreodon* (Wilson, 1971); the oreodonts *Limnenetes* and *Oreonetes* (Loomis, 1924; Thorpe, 1937; Scott, 1940; Schultz and Falkenbach, 1956); and *Montanatylopus* (Prothero, 1986). Wood (1980) attributed *Cylindrodon fontis* to the McCarty's Mountain fauna but did not list any referred specimens, and Black (1978) indicated that two specimens in the Carnegie Museum collection represented a new species of *Hendryomeryx* but did not describe the material. Storer (1984b) briefly discussed the McCarty's Mountain leptomerycids and concluded that *Hendryomeryx* was not present, a view endorsed here. In addition, Novacek (1976) has provided useful commentary on the lepticids.

The small oreodont *Oreonetes anceps* is probably the most common mammal in available collections, but *Colodon*? *cingulatus, Ischyromys douglassi,* and one or more species of *Pseudocylindrodon* are also represented by fairly large samples. Many of the small mammals, especially the marsupials, insectivores, and some of the smaller rodents, common in most of the other Paleogene local faunas of southwestern Montana, are not well represented in existing collections from McCarty's Mountain. Several of the small mammal species that were probably very common are known from only one or two specimens. A revised list of the McCarty's Mountain fauna is presented in Table 5.

Although Douglass (1905, 1908b) and subsequent authors regarded the McCarty's Mountain beds as "Lower Oligocene," Wood et al. (1941) questioned the age of the McCarty's Mountain fauna, and Kay et al. (1958), based in part on specimens that were not collected from the McCarty's Mountain locality, cited an age range of Chadronian to Whitneyan for the deposits. Riel (1963) recovered fragmentary brontothere remains from the uppermost exposures and firmly established the upper age limit of the McCarty's Mountain locality as Chadronian. However, the precise placement of the McCarty's Mountain fauna within the Chadronian has been disputed.

Based on the presence of *Oreonetes anceps* and *Limnenetes platyceps,* Schultz and Falkenbach (1956) assigned the McCarty's Mountain deposits to "the middle part of the Chadron formation," although neither of these taxa has ever been reported from the Chadron Formation in the Great Plains region. Wood (1974) suggested, based on the presence of *Ardynomys occidentalis* in both and the similar stage of evolution

Table 5. McCarty's Mountain local fauna, Beaverhead Basin: Localities in Sec. 28, T4S R8W, Madison County, Montana. Renova Formation undifferentiated. Early Chadronian. Carnivores identified by H. N. Bryant. (* = type locality.)

DIDELPHIMORPHIA
 Didelphidae
  Herpetotherium valens (Lambe)
  Copedelphys sp., cf. *C. titanelix* (Matthew)
PROTEUTHERIA
 Leptictidae
  Leptictus montanus (Douglass)*
  Leptictis major (Douglass)*
INSECTIVORA
 Geolabididae
  Centetodon magnus (Clark)
 Micropternodontidae
  Micropternodus sp.
 Apternodontidae
  Apternodus sp., cf. *A. mediaevus* Matthew
PALAEANODONTA
 Epoicotheriidae
  Epoicotherium unicum (Douglass)*
LAGOMORPHA
 Leporidae
  Megalagus? new species
  Palaeolagus sp.
RODENTIA
 Ischyromyidae
  Ischyromys douglassi Black*
 Cylindrodontidae
  Pseudocylindrodon medius Burke*
  Ardynomys occidentalis Burke*
  Cylindrodon sp., cf. *C. fontis* Douglass
 Eomyidae
  Paradjidaumo sp., cf. *P. trilophus* (Cope)
  Paradjidaumo new species, aff. *P. trilophus* (Cope)
 Heliscomyidae
  Heliscomys sp., cf. *H. ostranderi* Korth, Wahlert, and
   Emry

CREODONTA
 Hyaenodontidae
  Hyaenodon sp.
CARNIVORA
 Miacidae
  "*Miacis*" new species A
 Amphicyonidae
  Daphoenine (cf. *Daphoenictis*) new species
 Canidae
  Hesperocyon sp.
PERISSODACTYLA
 Equidae
  Mesohippus sp.
 Brontotheriidae
  Brontotheriid sp.
 Helaletidae
  Colodon? *cingulatus* Douglass
 Hyracodontidae
  Triplopides rieli Radinsky*
  Hyracodon sp.
 Rhinocerotidae
  Rhinocerotid sp.
ARTIODACTYLA
 Agriochoeridae
  Protoreodon minimus (Douglass)
 Merycoidodontidae
  Bathygenys sp., cf. *B. alpha* Douglass
  Oreonetes anceps (Douglass)
  Limnenetes platyceps Douglass
  Merycoidodontid large sp.
 Oromerycidae
  Montanatylopus matthewi Prothero*
 Camelidae
  Paratylopus? sp.
 Leptomerycidae
  "*Leptomeryx*" new species, aff. "*L.*" *speciosus* Lambe
  "*Leptomeryx*" new species

exhibited by *Ischyromys douglassi* and *I. blacki*, that the McCarty's Mountain fauna was most nearly equivalent in age to, but slightly younger than, the Porvenir fauna of West Texas, now considered to be late Duchesnean in age. Wood (1980, table 1) later revised this correlation slightly and listed the McCarty's Mountain fauna as about the same age as the early Chadronian Little Egypt l.f. Ostrander (1985) believed, based on the presence of *Cylindrodon fontis* and *Oreonetes*, that McCarty's Mountain was middle Chadronian in age. Emry et al. (1987) only peripherally discussed the McCarty's Mountain fauna, but noted similarities to the Airstrip l.f., which they assigned a middle Chadronian age. On their correlation chart (Emry et al., 1987, fig. 5.3) the McCarty's Mountain and Airstrip localities are illustrated as about equivalent in age, with both somewhat older than the Pipestone Springs and Ash Springs local faunas. More recently, Emry (1992) has suggested that the Airstrip l.f. is early Chadronian in age. Storer (1984b) strongly argued for

an early Chadronian age for the McCarty's Mountain fauna based on the leptomerycids and suggested that McCarty's Mountain correlated in at least a general way with the lower part of the Flagstaff Rim section and with the Yoder, Southfork, and Titus Canyon local faunas. Storer (1989) later assigned a late Duchesnean age to these assemblages, but his definition of the Duchesnean/Chadronian boundary (Storer, 1990) approximates the early/middle Chadronian boundary of most other authors, and he has since abandoned this age assignment (Storer, this volume, Chapter 12).

Based principally on the occurrence of *Ischyromys douglassi*, *Pseudocylindrodon medius*, *Ardynomys occidentalis*, *Colodon*? *cingulatus*, *Triplopides rieli*, *Protoreodon minimus*, *Oreonetes anceps*, *Limnenetes platyceps*, *Montanatylopus matthewi*, and the small leptomerycids here referred to "*Leptomeryx*" n. sp. aff. "*L.*" *speciosus* and "*Leptomeryx*" n. sp., the McCarty's Mountain fauna is considered to be early Chadronian in age. None of these species, which include some of the most common mammals in the McCarty's Mountain fauna, is present in the middle Chadronian Pipestone Springs l.f. McCarty's Mountain and Pipestone Springs share only a few relatively long-ranging species.

The McCarty's Mountain fauna correlates most closely with the early Chadronian part of the composite Thompson Creek fauna of authors, derived from the upper part of the Climbing Arrow Formation and possibly the lower part of the Dunbar Creek Formation in the Three Forks Basin. The described fauna from Thompson Creek is not very diverse, but several key taxa are shared with McCarty's Mountain. *Colodon? cingulatus, Protoreodon minimus, Oreonetes anceps,* and *Limnenetes platyceps* are present in both the McCarty's Mountain and Thompson Creek faunas but are not represented in the much more diverse Pipestone Springs l.f. At University of Montana locality MV 6403 (= loc. f158 of Robinson, 1963) in the upper part of the Climbing Arrow Formation, a small lepto-merycid conspecific with "*Leptomeryx*" n. sp. aff. "*L.*" *speciosus* from McCarty's Mountain occurs with a larger leptomerycid that appears to be "*Leptomeryx*" *yoderi* and supports correlation of the McCarty's Mountain fauna with the lower part of the Flagstaff Rim section and the Yoder and Southfork local faunas.

The McCarty's Mountain fauna shares relatively few key taxa with early Chadronian localities outside western Montana. However, as Emry et al. (1987) noted, McCarty's Mountain does show a fairly strong resemblence to the very limited assemblage known from the Airstrip l.f. of West Texas. *Ardynomys, Bathygenys,* and *Limnenetes* occur in both, and the large oreodont that Wilson (1971) referred to ?*Prodesmatochoerus meekae* (*Merycoidodon presidioensis* of Stevens and Stevens, this volume, Chapter 25) may be the same taxon as the large unidentified oreodont from McCarty's Mountain.

The McCarty's Mountain section is very thick, and there is mounting evidence that some faunal turnover occurred during the course of deposition of the McCarty's Mountain beds. In the most detailed analysis of any group of mammals from the McCarty's Mountain fauna, Asnake (1984) divided the *Pseudocylindrodon medius* lineage into five stratigraphically superposed species. Some of his samples were small, and although we are presently unable to evaluate his work, at least some of the species Asnake recognized are probably valid. The McCarty's Mountain section also appears to document the transition between *Paradjidaumo* n. sp. aff. *P. trilophus* and *Paradjidaumo trilophus*. There also may be some discernible change in the oreodonts *Oreonetes* and *Limnenetes*. Schultz and Falkenbach (1956) named *Oreonetes anceps douglassi* for a smaller specimen with "lighter" premolars than other specimens of *O. anceps*, and ?*Limnenetes*, species undetermined, for a larger, more robust specimen than

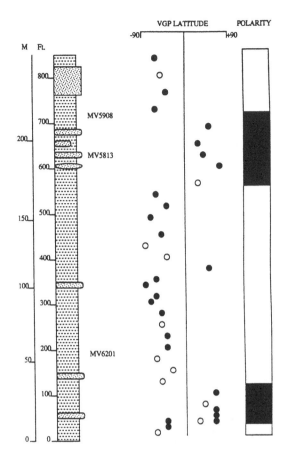

Figure 5. Magnetic stratigraphy of the main section at McCarty's Mountain. Conventions as in Figure 4.

the holotype of *L. platyceps*; both specimens come from the stratigraphically highest part of the McCarty's Mountain sequence (Level Z of Douglass).

Although our present knowledge of the biostratigraphy of the McCarty's Mountain locality is inadequate, the entire section appears to be early Chadronian in age. The holotype and most of the referred specimens of *Ischyromys douglassi* were collected from the stratigraphically highest part of the sequence (Level Z of Douglass), as were referred specimens of *Oreonetes anceps* and *Limnenetes platyceps*. The stratigraphically lowest part of the McCarty's Mountain sequence is more difficult to date, but *Oreonetes* has been recovered from near the base of the beds (Level R of Douglass) and "*Leptomeryx*" n. sp. aff. "*L.*" *speciosus* from only slightly higher in the section (Level S).

The magnetic stratigraphy and location of the main fossil localities are shown in Figure 5. Five polarity zones were recognized, with normal polarity occurring in rocks between 40-120 feet, and 550-720 feet on the local section; all of the remaining section was of reversed magnetic polarity. Based on the early

Table 6. Diamond O Ranch local fauna, Beaverhead Basin: Localities in Sec. 20, T5S R7W, Beaverhead County, Montana. Climbing Arrow Member, Renova Formation. Late Duchesnean. Carnivores identified by H.N. Bryant.

DIDELPHIMORPHIA
 Didelphidae
  Herpetotherium valens (Lambe)
  Copedelphys sp., cf. *C. titanelix* (Matthew)
PROTEUTHERIA
 Leptictidae
  Leptictis sp.
INSECTIVORA
 Geolabididae
  Centetodon sp., cf. *C. kuenzii* Lillegraven and Tabrum
 Apternodontidae
  Oligoryctes sp., cf. *O. cameronensis* Hough
  Apternodus sp., cf. *A. mediaevus* Matthew
 Soricidae
  Domnina sp., cf. *D. thompsoni* Simpson
 Proscalopidae
  Proscalopid sp.
 Insectivora, *incertae sedis*
  Cryptoryctes kayi Reed
  New genus and species
PALAEANODONTA
 Epoicotheriidae
  Epoicotherium sp.
LAGOMORPHA
 Leporidae
  Megalagus? new species
  Palaeolagus sp.
RODENTIA
 Ischyromyidae
  Ischyromys sp., cf. *I. douglassi* Black

Cylindrodontidae
 Pareumys? sp.
 Pseudocylindrodon sp.
 Ardynomys occidentalis Burke
Eutypomyidae
 Eutypomys sp.
Eomyidae
 Adjidaumo new species, aff. *A. minimus* (Matthew)
 Paradjidaumo new species, aff. *P. trilophus* (Cope)
CREODONTA
 Hyaenodontidae
  Hyaenodon sp.
CARNIVORA
 Miacidae
  "*Miacis*" new species B
 Canidae
  Hesperocyon? sp.
PERISSODACTYLA
 Equidae
  Mesohippus sp.
 Brontotheriidae
  Brontotheriid sp.
 Helaletidae
  Colodon? *cingulatus* Douglass
 Hyracodontidae
  Hyracodon? sp.
 Rhinocerotidae
  Rhinocerotid sp.
ARTIODACTYLA
 Agriochoeridae
  Protoreodon sp.
 Leptomerycidae
  Hendryomeryx sp., cf. *H. defordi* (Wilson)

Chadronian character of the mammalian fauna, the McCarty's Mountain section is correlated with Chrons C16r1 to C17n1, which would indicate a time span of 35.5.-36.7 Ma (Fig. 7).

Diamond O Ranch Local Fauna

The Diamond O Ranch l.f. is a previously unpublished assemblage derived from five closely related localities (University of Montana localities MV 6726 to MV 6730) on the north side of the Beaverhead River in section 20, T5S R7W, Beaverhead County, about seven miles southeast of the McCarty's Mountain localities and about two miles west of Beaverhead Rock. The Diamond O Ranch localities were first discovered by D. S. Hoffman in 1967 and the fauna was described by him in his unpublished Ph.D. dissertation (Hoffman, 1972). In 1984 Tabrum began reworking the localities, and sampling of a short magnetostratigraphic section was attempted by Prothero in 1987.

Hoffman (1972) assigned the relatively thin sequence (<100 feet) of gray bentonitic mudstones and interbedded sandstones exposed at the Diamond O Ranch localities to the Renova Formation, named by Kuenzi (1966) in his Ph.D. dissertation. Lithologically similar, but somewhat younger sediments about two miles to the east were subsequently assigned to the Climbing Arrow Member of the Renova Formation by Petkewich (1972). The Diamond O Ranch deposits are lithologically similar and approximately equivalent in age to part of the type Climbing Arrow Formation in the Three Forks Basin and to sediments referred to the Climbing Arrow Member of the Renova Formation in several of the other intermontane basins of southwestern Montana.

The Diamond O Ranch localities have produced a moderately diversified mammalian fauna, although only a single lower jaw of *Herpetotherium valens*, briefly mentioned by Korth (1994, p. 380), has previously been reported in the published literature. A small blowout near the base of the sequence, University of Montana locality MV 6726, is particularly noteworthy as the source of abundant well-preserved small-mammal specimens.

A list of the mammals from the Diamond O Ranch l.f. is presented in Table 6. A new species of *Megalagus*, comparable in size to species of *Palaeolagus*, is by far the most common mammal in the available collections, although its apparent abundance may in part be due to the ease with which isolated lagomorph teeth can be seen and collected. Other common small mammals include *Paradjidaumo* n. sp., *Apternodus* sp. cf. *A. mediaevus*, and *Herpetotherium valens*. The most common large mammals are *Hendryomeryx* sp. cf. *H. defordi*, followed by *Colodon?* *cingulatus* and *Mesohippus* sp.

Hoffman (1972) assigned a Chadronian age to the Diamond O Ranch l.f. Based principally on the presence of *Ardynomys occidentalis* and *Colodon? cingulatus* in both, as well as some similarities in other groups of mammals, Hoffman suggested that the Diamond O Ranch and McCarty's Mountain faunas were approximately contemporaneous. Further study, however, has indicated that the faunal similarities are not so great as Hoffman believed and that Diamond O Ranch is significantly older than McCarty's Mountain. The Diamond O Ranch l.f. appears to fall very near the Duchesnean/Chadronian boundary and is tentatively assigned a late Duchesnean age.

The Diamond O Ranch l.f. lacks the moderately diverse assemblage of oreodonts that dominates the McCarty's Mountain fauna and also lacks other taxa that are here considered indicative of an early Chadronian age. The leptomerycid at Diamond O Ranch is a species of *Hendryomeryx* that is probably conspecific with *H. defordi* from the late Duchesnean Porvenir fauna of West Texas and differs significantly from the small species of "*Leptomeryx*" present in McCarty's Mountain.

Only a few of the taxa present in the Diamond O Ranch l.f. are not known from Chadronian or younger faunas. *Hendryomeryx defordi* is otherwise known only from the late Duchesnean Porvenir fauna, although the apparently closely related *H. esulcatus* occurs in the middle Chadronian Calf Creek l.f. of Saskatchewan (Storer, 1981). An as yet unnamed new genus and species of insectivore (Adapisoricidae, gen. et sp. indet., in part, of Storer, 1984a) is also known from the late Uintan Swift Current Creek l.f. and the Duchesnean Lac Pelletier Lower Fauna of Saskatchewan and from the early Duchesnean Badwater locality 20 of Wyoming. *Pareumys* has also not been recorded from localities younger than Duchesnean in age.

Several specimens in the Diamond O Ranch l.f. seem to pertain to the leporid *Palaeolagus*, which has generally been regarded as first appearing in the Chadronian Mammal Age (e.g., Wood et al., 1941; Emry et al., 1987; Lucas, 1992). *Palaeolagus* first appears in the central Great Plains in the early Chadronian (Emry and Gawne, 1986) and is the earliest lagomorph recorded from the region. However, *Palaeolagus* appears to occur earlier in southwestern Montana than in the Great Plains region and may well occur much earlier if undescribed specimens from the Mantle Ranch l.f. of early or middle Duchesnean age are correctly identified.

The Diamond O Ranch l.f. contains the earliest known records of a number of lineages of small mammals that continue into or through the Chadronian of southwestern Montana, and form an important component of indigenous Chadronian small-mammal faunas. Species conspecific with or closely related to *Herpetotherium valens*, *Copedelphys titanelix*, *Apterno-dus mediaevus*, *Domnina thompsoni*, *Cryptoryctes kayi*, *Adjidaumo minimus*, and *Paradjidaumo trilophus* are all first recorded from the Diamond O Ranch l.f. The records of *Herpetotherium*, *Copedelphys*, and *Crypto-ryctes* are at present the earliest known for these genera.

Although about 100 feet of Renova Formation were exposed at the Diamond O Ranch localities, the bulk of the section proved too coarse-grained and poorly indurated for magnetic sampling. Only a single site near the base of the section at the main fossil locality was suitable for magnetic analysis, and it was of normal polarity. Such limited polarity information does not produce an unambiguous correlation, but based on the magnetic pattern of such late Duchesnean sequences as the Chambers Tuff which produces the Porvenir local fauna (Prothero, this volume, Chapter 9), the Diamond O ranch strata probably correlate with some part of the long episode of normal polarity during Chron C17n (Fig. 7).

SAGE CREEK BASIN

The Sage Creek Basin has produced the temporally most extensive suite of Cenozoic vertebrate localities currently known from any of the intermontane basins of southwestern Montana. Horizons of Bridgerian, Uintan, Chadronian, Orellan, Whitneyan, Arikareean, Hemingfordian, and Barstovian ages are all represented, with at least two biostratigraphically distinct horizons present in the Bridgerian, Uintan, Chadronian, and Orellan parts of the sequence. Bridgerian through Orellan sediments and associated volcanics are locally well exposed along the crest and eastern flank of the Red Rock Hills, north and west of Sage Creek. The most extensive, most fossiliferous, and historically most important of these deposits are discontinuously exposed over an area of approximately five square miles on the southeast flank of the Red Rock Hills, northwest of the "Big Bend" of Sage Creek.

Douglass (1909, p. 281) early noted that the Tertiary sediments exposed in the Sage Creek Basin appeared to be "partly of Eocene, partly of Oligocene, and partly of Miocene age." The complicated Cenozoic stratigraphy of the Sage Creek Basin has, for a variety of reasons, proven exceptionally difficult to unravel, principally because exposures are discontinuous, distinctions between some of the stratigraphic units are subtle, and the area has been extensively faulted. Interpretation of that part of the Tertiary sequence in the Sage Creek

Basin traditionally regarded as Eocene in age has been particularly controversial, with conflicting or partially conflicting views presented by Wood (1934), Hough (1955, 1958), Scholten et al. (1955), and Fields et al. (1985).

Much of the interest in the Cenozoic deposits of the Sage Creek Basin stems from the early description by Douglass (1903) of four fragmentary vertebrates of Eocene age that he collected in 1897 from deposits he called the Sage Creek Beds. Douglass's four fossils— here identified as *Helaletes nanus, Hyrachyus douglassi, Colodon* sp. cf. *C. woodi,* and *Amynodon* sp. cf. *A. advenus*—were derived from two localities "about a half mile apart" (Douglass, 1903, p. 158). Douglass was somewhat equivocal about assigning an age to the Sage Creek Beds and did not suggest that his four fossils were all necessarily of the same age. Subsequent authors (Matthew, *in* Osborn and Matthew, 1909; Wood, 1934; Hough, 1955, 1958; Kay et al., 1958), however, all concluded that only a single horizon was represented, although earlier in his paper Wood (1934) had suggested the possibilty that both Bridgerian and Uintan horizons might be present. Because considerable taxonomic manipulation is required to interpret Douglass's four fossils as all of the same age, each has had a checkered taxonomic history.

In a largely unsuccessful attempt to resolve the apparent discrepancy posed by Douglass's four fossils, H. E. Wood briefly investigated the Sage Creek area in the summers of 1931 and 1933. The principal result of his first trip was the recovery of a small collection of "middle Oligocene" vertebrates from east-facing exposures opposite the Cook Sheep Company Home Ranch (Rock Island Ranch on the USGS 7.5' Quadrangle, now the Matador Ranch). Wood (1934) formally proposed the name Cook Ranch Formation for these beds.

Douglass (1903, p. 146) had described the locality that produced the specimens here identified as *Helaletes nanus* and *Hyrachyus douglassi* as "composed of stratified material, and it contains quartz geodes, tubes lined with crystals both of calcite and quartz, and calcified trunks and twigs of trees." On his second trip to the Sage Creek Basin, Wood succeeded in relocating this site. Wood (1934, p. 255) failed to find any additional identifiable Eocene fossils but eventually concluded that all four of Douglass's fossils were of "Lower Uinta age" and formally defined the Sage Creek Formation as consisting of "regularly bedded, fine-grained, greenish-gray sandstones, with interspersed, coarser, cross-bedded channel sandstones, ranging into conglomerates in some places." Wood noted that the contact between the Sage Creek Formation and the immediately overlying beds, which he referred to the Cook Ranch Formation, was strongly unconformable and marked by a unit of recemented debris derived from

the Sage Creek Formation. The hiatus between the Sage Creek Formation and the overlying sediments referred to the Cook Ranch Formation was interpreted as representing the interval between "Lower Uinta" and "Middle Oligocene" time.

Shortly after publication of Wood's paper, Kay began collecting in the Sage Creek Basin for the Carnegie Museum. In 1937 he secured a few specimens of Eocene age from beds overlying the type Sage Creek Formation of Wood (1934). In 1939 Kay made a slightly larger collection of Eocene vertebrates from exposures about one half mile east of the Sage Creek type locality in the area between Draws 2 and 3 of Hough (1955, fig. 1— Hough and Kay Draws of this chapter). At least part, perhaps all, of Kay's 1937 collection also appears to have been derived from this area, and in 1940 a much larger collection was made from this locality. Kay devoted much of the 1940 field season to collecting in the Sage Creek Basin, also developing a late Chadronian quarry assemblage from exposures of the Cook Ranch Formation on the north side of Little Spring Gulch, about six miles north of the Cook Ranch type exposures. Kay continued to work intermittently at these and other localities in the area until his retirement from the Carnegie Museum in 1957.

In the early 1950s Hough became interested in the fossil vertebrates from the Sage Creek Basin, particularly the Eocene vertebrates, and conducted limited field investigations in the area during the 1950 through 1953 field seasons. Hough (1955) described some of the material that Kay and she had collected from Eocene beds overlying the type Sage Creek Formation of Wood (1934). Douglass (1903, p. 156) had stated (possibly erroneously) that the holotype of his *Hyrachyus? priscus* (= *H. douglassi*) was collected from "a breccia formed by the breaking up and recementing of the sandstone" and that "it was found a few feet below the specimen of *Heptodon*? [= *Helaletes nanus*]." This statement by Douglass, combined with the failure of later collecting parties to secure any additional specimens, or even bone scrap, from the greenish-gray sandstones of Wood's type Sage Creek Formation and the subsequent recovery of a moderately diverse and demonstrably Eocene fauna from the overlying beds, led Hough (1955, 1958) to conclude that all four of Douglass's Eocene specimens had been derived from the beds overlying the type Sage Creek Formation of Wood. Hough (1955, p. 25) suggested that the name Sage Creek beds more properly applied to these higher beds, noting that "the Sage Creek beds are obviously those beds from which Douglass obtained his fossils, i.e. the pinkish gray, fossiliferous tuffs above the disconformity at the type locality." Hough (1955, 1958) ultimately concluded that only a single Paleogene unit of formational rank was present in the Sage Creek area and synonymized the Cook Ranch Formation with the Sage Creek Formation, interpreting the Sage Creek Formation of Wood as a stream-channel facies of her redefined and expanded unit.

Extensive further investigations of the Cenozoic geology and vertebrate paleontology of the Sage Creek Basin initiated by Tabrum in 1977 have greatly increased the collections available from the area and have largely resolved the principal problems noted by the earlier investigators. Paleogene strata exposed in the area west of Sage Creek are divisible into the Bridgerian Sage Creek Formation, the Uintan Dell beds, and the Chadronian-Orellan Cook Ranch Formation (Fields et al., 1985). The Hall Spring basalt (= Sage Creek basalt of Scholten et al., 1955) locally intervenes between the Sage Creek Formation and the Dell beds, and was illlustrated in Fields et al. (1985, fig. 4) as a unit intermediate in age between the two. Further study indicates that the Hall Spring basalt is probably best treated as an upper member of the Sage Creek Formation.

During the course of his investigations, Tabrum was able to recover specimens from most of the key stratigraphic horizons in the area, including additional material of two of the four taxa reported by Douglass (1903). Several specimens of *Helaletes nanus*, two of which duplicate Douglass's specimen of "*Heptodon?*" were collected from the type Sage Creek Formation of Wood (1934), stratigraphically below the breccia that marks the contact between the Sage Creek Formation and the overlying Dell beds, which produced the "Upper Eocene" fossils reported by Hough (1955). These and a few additional specimens from the type locality establish the age of the type Sage Creek Formation as middle or late Bridgerian. Other localities in the Sage Creek Formation have produced specimens of early Bridgerian age. The presence of early Bridgerian specimens now identified as *Eotitanops* sp. and *Palaeosyops fontinalis* in the Sage Creek Formation was briefly noted by Wallace (1980) and Stucky (1984).

An amynodont jaw collected from exposures of the Dell beds about 0.75 miles northeast of the type locality of the Sage Creek Formation is conspecific with the specimen reported by Douglass from his second Sage Creek locality. The specimens from Douglass's second locality, *Amynodon* sp. cf. *A. advenus* and *Colodon* sp. cf. *C. woodi,* now seem certainly to have been collected from the Dell beds, probably from exposures on the east side of Douglass Draw (= Draw no. 6 of Hough, 1955, fig. 1) in the NW Sec. 33, T12S R8W. Wood (1934, p. 255) also suggested that this area was the probable site of Douglass's second Sage Creek locality, a conclusion further supported by information contained in Douglass's 1897 field notes.

Tabrum's field work in the Sage Creek Basin was significantly supplemented by the measurement of five magnetostratigraphic sections by Prothero. Sections were measured through the "lower," "middle," and "upper" parts of the Cook Ranch Formation in 1980 (Prothero, 1982). In 1987, Prothero and his field crew sampled two stratigraphic sections in the Dell beds. In the following sections we briefly summarize the biostratigraphic and magnetostratigraphic data currently available for the Dell beds and Cook Ranch Formation in the classic part of the Sage Creek area worked by Wood (1934) and Hough (1955, 1958).

Dell Beds

Based on Tabrum's work in the Sage Creek Basin, the name Dell beds was introduced in Fields et al. (1985) for an informal unit of formational rank that unconformably overlies the Sage Creek Formation (of Wood, 1934) and "Hall Spring basalt" (= Sage Creek basalt of Scholten et al., 1955). The Dell beds are in turn disconformably overlain by the Cook Ranch Formation. The formation is widely exposed in the area worked by Wood (1934) and Hough (1955, 1958), especially in Secs. 20, 21, 28, 29, and 33, T12S R8W, and is sporadically exposed along the crest and eastern flank of the Red Rock Hills as far north as the north side of Little Spring Gulch (Secs. 31 and 32, T12S R8W).

The Dell beds consist largely of poorly sorted tuffaceous mudstone and pebbly to cobbly tuffaceous mudstones interbedded with sandstone and conglomerate. Most of the conglomerates occur as lenses or steep-sided channel fills and locally contain abundant debris reworked from the Sage Creek Formation and Hall Spring basalt. Brightly colored cobbles and boulders of rhyolitic ash-flow tuff with sanidine phenocrysts occur as matrix-supported clasts in some of the tuffaceous mudstone units, in some of the conglomeratic units dominated by other rock types, and in monolithologic channel fills. Cobbles of this type have not been observed in any other part of the Cenozoic stratigraphic sequence in the Sage Creek Basin and appear to be restricted to the Dell beds.

The uppermost unit of the Dell beds is an approximately 60-foot thick unit of massive, well indurated tuffaceous mudstone which has produced most of the fossils known from the formation. Vertebrates recovered from this unit are referred to the Hough Draw l.f. Specimens collected from stratigraphically lower horizons in the Dell beds appear to represent a faunally distinct assemblage and are assigned to the Douglass Draw l.f.

Discontinuous exposures and extensive faulting make it difficult to reliably estimate the thickness of the Dell beds. Two apparently non-overlapping sections measured by Tabrum totaled 278 feet, but did not cover the entire formation. The total thickness of the Dell beds is estimated to be at least 300 to 400 feet in the

Table 7. Douglass Draw local fauna, Sage Creek Basin: Localities in Secs. 27, 28, 29, and 33, T12S R8W, Beaverhead County, Montana. Lower part of "Dell beds" (see Fields, et al., 1985). Late Uintan.

RODENTIA
 Cylindrodontidae
  Cylindrodontid sp.
PERISSODACTYLA
 Brontotheriidae
  Brontotheriid sp.
 Helaletidae
  Dilophodon leotanus (Peterson)
  Colodon kayi (Hough)
  Colodon sp., cf. *C. woodi* (Gazin)
 Amynodontidae
  Amynodon sp., cf. *A. advenus* (Marsh)

Hyracodontidae
 Triplopus sp. cf. *T. rhinocerorhinus* (Wood)
 Hyracodontid sp.
ARTIODACTYLA
 Homacodontidae
  Homacodontid sp. A
 Agriochoeridae
  Protoreodon or *Diplobunops* sp.
 Protoceratidae
  Leptoreodon marshi Wortman

Table 8. Hough Draw local fauna, Sage Creek Basin: Localities in SE Sec. 28, and NE Sec. 33, T12S R8W, Beaverhead County, Montana. Upper part of "Dell beds" (see Fields et al.,1985). Late Uintan. (* = type locality.)

DIDELPHIMORPHIA
 Didelphidae
  Peratherium sp., cf. *P. knighti* McGrew
PALAEANODONTA
 Epoicotheriidae
  Epoicotherium sp.
LAGOMORPHA
 Leporidae
  Mytonolagus sp., cf. *M. petersoni* Burke
RODENTIA
 Ischyromyidae
  Ischyromys? sp.
 Cylindrodontidae
  Pareumys? new species
  Pseudocylindrodon new species
 Sciuravidae
  Sciuravid sp.
CARNIVORA
 Miacidae
  Tapocyon sp., nr. *T. robustus* (Peterson)

PERISSODACTYLA
 Equidae
  Epihippus sp., cf. *E. uintensis* (Marsh)
 Brontotheriidae
  Brontotheriid sp.
 Lophiodontidae
  Schizotheriodes parvus Hough*
 Helaletidae
  Dilophodon leotanus (Peterson)
  Colodon kayi (Hough)*
 Hyracodontidae
  Hyracodontid sp., cf. *Mesamynodon medius* Peterson
ARTIODACTYLA
 Homacodontidae
  Homacodontid sp. B
 Agriochoeridae
  "*Protoreodon pearcei*" Gazin
  Protoreodon sp., cf. *P. pumilus* (Marsh)
  Protoreodon small sp.
 Protoceratidae
  Leptotraguline sp.

area northwest of the "Big Bend" of Sage Creek. The Dell beds appear to thicken to the north, and may be in excess of 500 feet thick on the north side of Little Spring Gulch.

Douglass Draw Local Fauna

Fossil vertebrates collected from widely scattered localities in Secs. 28, 29, and 33, T12S R8W, have been grouped together, possibly artificially, as the Douglass Draw l.f. The available specimens were collected from various horizons stratigraphically below the massive mudstone in the uppermost part of the Dell beds that has produced the specimens assigned below to the Hough Draw l.f. Localities established by Tabrum in this part of the Dell beds include University of Montana localities MV 7757, MV 8112, MV 8114, and MV 8115. The specimens described by Douglass (1903) from his second Sage Creek locality—here assigned to *Colodon* sp. cf. *C. woodi* and *Amynodon* sp. cf. *A.*

advenus—appear to have been derived from this part of the section, almost certainly from exposures on the east side of Douglass Draw. A few specimens collected by Kay for the Carnegie Museum retain locality data which indicate that they also were collected from this part of the Dell beds.

The stratigraphically lower part of the Dell beds is poorly fossiliferous, and at present only a very limited fauna is known (Table 7). The only relatively commonly encountered fossils are the fragmentary remains of tortoises. Identifiable mammals are rare, with only *Leptoreodon marshi*, *Amynodon* sp. cf. *A. advenus*, and one of the species of *Colodon* currently represented by more than a single specimen.

The available material seems adequate to establish the age of the Douglass Draw l.f. as late Uintan. *Dilophodon leotanus*, *Colodon woodi*, and *C. kayi* are first recorded from late Uintan deposits, and the remainder of the known assemblage is also consistent with a late

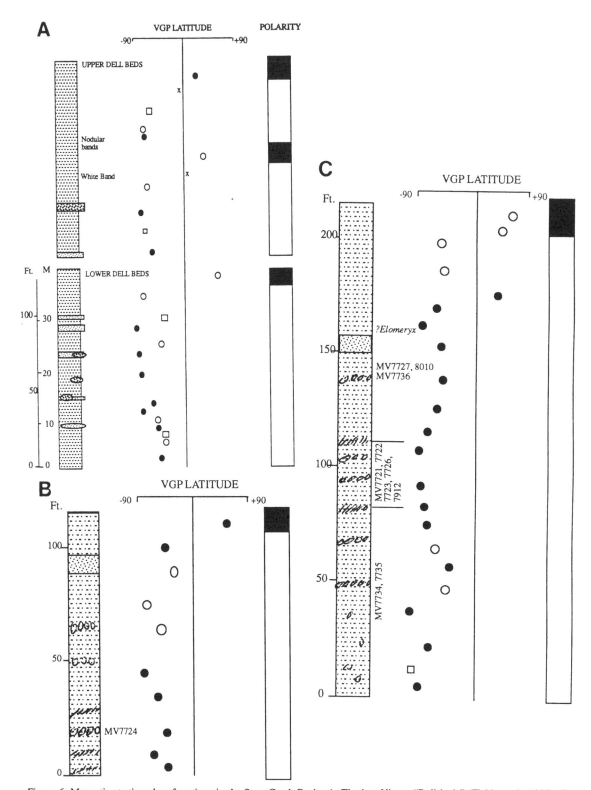

Figure 6. Magnetic stratigraphy of sections in the Sage Creek Basin. A. The late Uintan "Dell beds" (Fields et al., 1985). B. The middle-late Orellan Matador Ranch l.f. C. The late Orellan Cook Ranch l.f. All conventions as in Figs. 4 and 5.

Uintan age assignment. The presence of both *Colodon kayi* and *Colodon* sp. cf. *C. woodi* (if *C. woodi* is valid and Douglass's specimen is correctly identified as closer to this species than to *C. kayi*) in the Douglass Draw l.f. may indicate that more than one level is represented. It may also indicate that part or all of the Douglass Draw assemblage is equivalent in age to the Badwater late Uintan localities of central Wyoming.

In 1987 Prothero measured a magnetic section through the exposures on the east side of Douglass Draw in the NE NW NW Sec. 33, T12S R8W, Rock Island Ranch 7.5' Quadrangle, Beaverhead County, Montana. As noted above, the beds exposed at this locality now seem almost certainly to have been the source of Douglass's (1903) specimens here identified as *Amynodon* sp. cf. *A. advenus* and *Colodon* sp. cf. *C. woodi*. Prothero's section extended from the base of the exposures at an elevation of about 6440 feet eastward to the knob exposed at an elevation of approximately 6580 feet. The measured section totaled 133 feet, and except for the uppermost site, was entirely of reversed polarity (Fig. 6). Based on the late Uintan age of the fauna, the Douglass Draw section probably correlates with the early part of Chron C18r, about 40.7-41.0 Ma.

Hough Draw Local Fauna

Specimens assigned to the Hough Draw l.f. were collected from exposures of the Dell beds on the east side of Kay Draw and at various places along the short western branch of Hough Draw in the NE Sec. 33, and the southernmost part of the SE Sec. 28, T12S R8W (the area of localities C, D, and X of Hough, 1955, fig. 1). Specimens were derived from the stratigraphically highest part of the Dell beds, from an approximately 20-m-thick unit of massive tuffaceous mudstone exposed just below the contact of the Dell beds with the Cook Ranch Formation. This is easily the most fossiliferous part of the Dell beds, and most of the specimens collected by Kay and Hough were derived from this stratigraphic level. Four University of Montana localities (MV 6803, MV 7629, MV 7729, and MV 7730) were established; all four are stratigraphically equivalent but refer to slightly different collecting areas.

As presently known the Hough Draw l.f. includes at least nineteen mammalian taxa (Table 8). The fauna is dominated by "*Protoreodon pearcei*" and *Protoreodon* sp. cf. *P. pumilus*. *Dilophodon leotanus* and *Colodon kayi* are also reasonably common. *Mytonolagus* sp. cf. *M. petersoni* is probably the most common small mammal, but the available material consists largely of isolated teeth.

Hough (1955) assigned a ?Duchesnean age to the fossil vertebrates she reported from the Sage Creek area, correlating the fauna with those then known from the

Badwater localities and from the Randlett Horizon of the Duchesne River Formation. Hough's correlation still seems to be essentially correct, although these localities are now regarded as late Uintan in age (Gazin, 1955, 1956). The Hough Draw l.f. is, however, probably somewhat advanced over that known from the late Uintan Badwater localities. *Colodon kayi* appears to be a later species than the Badwater *C. woodi* and persists into Chadronian time (Bjork, 1968). The Hough Draw specimens attributed to *Dilophodon leotanus* compare very closely to the type materials from the Randlett Horizon of the Duchesne River Formation; both are probably specifically distinct from the material from the Badwater localities referred to *D. leotanus* by Black (1979). The possible presence of *Ischyromys* in the Hough Draw l.f. also suggests an age somewhat younger than that of the Badwater late Uintan localities. The Hough Draw l.f. appears to be among the youngest Uintan local faunas known and may be of latest Uintan age. It is probably of about the same age as the Randlett Horizon of the Duchesne River Formation.

A magnetic section measured by Prothero in 1987 sampled the Dell beds exposed on the east side of Kay Draw (= Draw no. 3 of Hough, 1955, fig. 1) in the NE NW NE Sec. 33, T12S R8W, Rock Island Ranch 7.5' Quadrangle, Beaverhead County, Montana. The section extended from near the bottom of the wash at approximately the 6400' contour eastward to the uppermost beds exposed slightly below the crest of the ridge (elevation 6537 feet) that separates Kay Draw from the short, western branch of Hough Draw. Approximately 130 feet of continuous section were measured. The upper 61 feet of this section sampled a massive mudstone that has produced the Hough Draw local fauna. Except for two sites, the entire section was of reversed polarity (Fig. 6). The uppermost site, and site near the base of the fossiliferous massive mudstone unit, were of normal polarity. Based on the late (possibly latest) Uintan age of the Hough Draw local fauna and the superpositional relationship of these beds to those that produced the Douglass Draw local fauna, this part of the section appears to correlate with the later part of Chron C18r, about 40.2-40.7 Ma (Fig. 7).

Cook Ranch Formation

The Cook Ranch Formation was named by Wood (1934, p. 254) for east-facing exposures in the NW Sec. 34, T12S R8W, that produced a small assemblage of "middle Oligocene" vertebrates. As originally defined by Wood, the Cook Ranch Formation included nearly all of the "buff" colored Tertiary beds exposed above the type Sage Creek Formation. As discussed above, the stratigraphically lower part of these beds has since been excluded from the Cook Ranch Formation by Tabrum and is now assigned to the lithologically distinct Dell

Beds of late Uintan age (Fields et al., 1985).

The Cook Ranch Formation is discontinuosly exposed along the eastern flank of the Red Rock Hills in a narrow belt generally less than one mile wide, extending from the type exposures in Sec. 34, T12S R8W, northerly to Sec. 33, T11S R8W. The Cook Ranch exposures have been truncated on the east by late Cenozoic faulting, and for a distance of about three miles, from slightly north of the type exposures to the northern part of Sec. 15, the Cook Ranch Formation has been faulted out of the sequence.

In the type area, the Cook Ranch Formation rests disconformably on the Dell Beds. The contact is marked by a well developed paleosol, well exposed on the east side of Kay Draw and along the scarp at the northern end of the short western branch of Hough Draw (NE Sec. 33, and SE Sec. 28, T12S R8W) Farther to the north, in the Little Spring Gulch area, the Cook Ranch Formation has been downfaulted relative to the Dell Beds.

At the type locality, the Cook Ranch Formation is unconformably overlain by approximately 90 feet of upper Tertiary conglomerate. Exposures to the north of the type locality, as well as those immediately south of it, are faulted into contact with the Little Sage Creek Beds of Fields et al. (1985) of late Hemingfordian/early Barstovian age. Stratigraphic relationships in the Sage Creek Basin are complex, and the Little Sage Creek Beds are also locally overlain by the conglomerate that unconformably overlies the type Cook Ranch Formation.

The Cook Ranch Formation consists largely of light-colored (typically very pale orange 10YR8/2, dry) tuffaceous mudstones with abundant calcareous nodular horizons interbedded with lesser amounts of conglomerate, sandstone, and a few beds of relatively pure, but partially devitrified volcanic ash. Nodular horizons are most common and best developed in the middle and upper parts of the Cook Ranch Formation. Several channel fills in the Orellan part of the section consist largely or entirely of cobbles and boulders of penecontemporaneous basalt, possibly reworked, at least in part, from an autoclastic debris-flow breccia exposed within the Cook Ranch Formation in the southern part of Sec. 27, T12S R8W.

Pervasive faulting and discontinuous exposures make it difficult to estimate the thickness of the Cook Ranch Formation. Four apparently non-overlapping sections measured by Tabrum in the type area totaled 484 feet, but probably did not include the entire formation. A better estimate for the thickness of the Cook Ranch Formation would appear to be approximately 600 to 700 feet.

The Cook Ranch Formation ranges in age from late Chadronian to late Orellan. The small assemblage reported by A.E. Wood (1933) and H.E. Wood (1934)

from the type exposures was designated as a "principal correlative" of the Orellan Provincial Age by Wood et al. (1941). Further collecting in this area has since produced a large and diversified fauna of late Orellan age. Other localities in the Cook Ranch Formation have produced significantly older assemblages. A quarry assemblage developed by Kay from exposures on the north side of Little Spring Gulch, stratigraphically very low in the Cook Ranch Formation, includes abundant specimens of *"Leptotragulus" profectus* and *"Leptomeryx" speciosus*, but lacks the characteristic middle Chadronian species *"Leptomeryx" mammifer*, and appears to be of late Chadronian age. The other mammals known from the Little Spring Gulch l.f. are also consistent with a late Chadronian age assignment. Clark and Beerbower (*in* Clark et al., 1967) reported a specimen of *Daphoenocyon* (=*Brachyrhynchocyon*) *dodgei* from this locality, and it appears that it is the assemblage from Little Spring Gulch which led Clark and Beerbower to assign "Cook Ranch" to the late Chadronian on their correlation chart (Clark and Beerbower *in* Clark et al., 1967, fig. 24).

East Hough Draw Localities

The stratigraphically lowest part of the Cook Ranch Formation in the type area is very sparingly fossiliferous. Exposures just above the contact with the Dell beds in the SE SW SE Sec. 28, T12S R8W (University of Montana locality MV 7733, North Hough Draw no. 1) have produced only a few isolated teeth of *Palaeolagus temnodon* and a single cylindrodontid incisor. Approximately equivalent exposures on the east side of Hough Draw in the NE NE Sec. 33, and the adjacent part of Sec. 34, have provided a slightly more diverse, but still very limited, assemblage. The stratigraphically lower of the two East Hough Draw localities, MV 7732, has produced a fragmentary lower jaw of *Herpetotherium*? sp., and isolated teeth of an unidentified ischyromyid rodent, *Palaeolagus temnodon*, and a second leporid, possibly *Chadrolagus emryi*. The second East Hough Draw locality, MV 7731, at the eastern margin of these exposures in the NW NW NW Sec. 34, has produced isolated teeth of *"Leptomeryx" speciosus*, *Pardajidaumo trilophus*, *Ischyromys* sp., and a lagomorph that appears to be either *Chadrolagus emryi* or a related species. The sparse fauna from this area is probably late Chadronian in age.

Prothero measured a stratigraphic section through the East Hough Draw exposures in 1980. The section (Fig. 6) began at the 6480-foot contour in the SW NE NE NE Sec. 33, and continued up the spine of a ridge to the 6600-foot contour in the SE SE SE SE Sec. 28. Approximately 170 feet of section were sampled. Nearly the entire section was of reversed polarity. Based on the probable late Chadronian age for the vertebrates collected from this part of the Cook Ranch Formation,

Table 9. Matador Ranch local fauna, Sage Creek Basin: Localities in NE Sec. 33, NW nd SW Sec. 34, T12S R8W, Beaverhead County, Montana. Middle part of Cook Ranch Formation. Middle and/or late Orellan.

DIDELPHIMORPHIA
  Didelphidae
    Herpetotherium fugax Cope
    Copedelphys sp., cf. *C. stevensoni* (Cope)
INSECTIVORA
  Soricidae?
    Soricid? sp.
  Proscalopidae
    Oligoscalops? sp.
LAGOMORPHA
  Leporidae
    "Palaeolagus" burkei Wood
    Palaeolagus? sp.
RODENTIA
  Aplodontidae
    Prosciurus sp.
  Eomyidae
    Adjidaumo sp., cf. *A. minimus* (Matthew)

  Heliscomyidae
    Heliscomys sp.
  Cricetidae
    Eumys sp., cf. *E. elegans* Leidy
    Eumys new species
CARNIVORA
  Canidae
    Hesperocyon gregarius (Cope)
PERISSODACTYLA
  Equidae
    Mesohippus sp.
  Hyracodontidae
    Hyracodon? sp.
ARTIODACTYLA
  Camelidae
    Poebrotherium sp.
  Leptomerycidae
    Leptomeryx sp., aff. *L. evansi* Leidy

the section most likely correlates with Chron C13r, or about 33.7-34.3 Ma (Fig. 7).

Matador Ranch Local Fauna

Fossils collected from the "middle part" of the Cook Ranch Formation are assigned to the Matador Ranch l.f. The available specimens were collected from south-facing exposures in the SE NE, Sec. 33, and west-facing exposures in the NW Sec. 34, T12S R8W. Four localities were established (University of Montana localities MV 7724, MV 7725, MV 7728, and MV 8116). The western localities, MV 7725 and MV 8116, appear to be stratigraphically lower than the eastern localities, MV 7724 and MV 7728. The intervening area is largely covered, but based on interpretation of photolinears it appears to be complexly faulted.

The Matador Ranch l.f. (Table 9) has produced a rather limited assemblage of Orellan age. The assemblage is dominated, especially at MV 7724, by abundant specimens of *"Palaeolagus" burkei*. *Eumys* is also fairly common, but the remaining taxa are represented by very small samples.

The stratigraphically lower localities, MV 7725 and MV 8116, have produced only a few specimens of *"Palaeolagus" burkei*, *Eumys*, and *Leptomeryx*, and may differ in age from the stratigraphically higher localities, MV 7724 and MV 7728. Specimens of *Eumys* from MV 7725 and MV 8116 compare closely with *E. elegans* and are probably referable to this species. Specimens from MV 7724 and MV 7728, however, appear to represent a more derived species of *Eumys* that parallels species of *Wilsoneumys* in the presence of a slender, delicate lower incisor and higher crowned cheek teeth with a more planar occlusal surface.

The Matador Ranch localities appear to be of middle or late Orellan age. *"Palaeolagus" burkei* is first recorded

from beds of Orella C age, mid Chron C12r, in the Central Great Plains region (Korth, 1989a; Prothero and Emry, this volume, Summary chapter), but as noted above may occur earlier in southwestern Montana. *Eumys elegans* is a long-ranging, but typically middle to late Orellan, species. The presence of a relatively derived species of *Eumys* in the stratigraphically higher localities MV 7724 and MV 7728 may indicate a late Orellan age for this part of the Matador Ranch section.

Only the stratigraphically higher part of the Matador Ranch section has been paleomagnetically sampled. A section measured by Prothero in 1980 in the SW SE NW NW Sec. 34, T12S R8W, began at the base of the exposures at the 6420-foot contour and extended northeasterly to the 6560-foot contour. The richest of the Matador Ranch localities, MV 7724, occurs in the lower part of this section. A total of 130 feet of the "middle part" of the Cook Ranch Formation was sampled, and all except the uppermost sample were of reversed polarity. The section appears to correlate with the early part of Chron C12r, suggesting an age of about 32.5-33.0 Ma for the Matador Ranch l.f.

Cook Ranch Local Fauna

Fossil vertebrates collected from the east-facing type exposures of the Cook Ranch Formation in the NW Sec. 34, T12S R8W, are assigned to the Cook Ranch l.f. At the type locality approximately 207 feet (63.0 meters) of continuous section are exposed. The Cook Ranch type section is dominated by tuffaceous mudstone and pebbly tuffaceous mudstone with numerous nodular horizons, interbedded with lesser amounts of sandstone, conglomerate, and intraformational mudstone-pebble conglomerate. Most of the available specimens were collected from the middle part of the section, from an approximately 60-foot thick

Table 10. Cook Ranch local fauna, Sage Creek Basin: Localities in NW Sec. 34, T12S R8W, Beaverhead County, Montana. Upper part of Cook Ranch Formation (= type Cook Ranch Formation of Wood, 1934). Late Orellan. (* = type locality.)

PERADECTIA
  Peradectidae
   Nanodelphys hunti (Cope)
DIDELPHIMORPHIA
  Didelphidae
   Herpetotherium fugax Cope
   Copedelphys sp., cf. *C. stevensoni* (Cope)
INSECTIVORA
  Erinaceidae
   Ocajila sp., cf. *O. makpiyahe* MacDonald
  Apternodontidae
   Oligoryctes sp., cf. *O. altitalonidus* (Clark)
  Soricidae
   Domnina new species, aff. *D. gradata* Cope
  Proscalopidae
   Oligoscalops new species
  Talpidae
   Talpid new species A
   Talpid? new species B
CHIROPTERA
  Vespertilionidae
   Vespertilionid sp.
LAGOMORPHA
  Leporidae
   Megalagus sp., cf. *M. turgidus* (Cope)
   Palaeolagus sp., cf. *P. haydeni* Leidy
   "*Palaeolagus*" *burkei* Wood
RODENTIA
  Ischyromyidae
   Ischyromys new species?
  Cylindrodontidae
   Pseudocylindrodon new species
  Aplodontidae
   Pelycomys sp.
   Prosciurus sp., cf. *P. parvus* Korth
   Prosciurus? new species
   Sespemys? new species
   Prosciurine, new genus and species
  Castoridae
   Agnotocastor sp.

Eomyidae
  Adjidaumo minimus (Matthew)
  Metadjidaumo new species
  Paradjidaumo new species
Heliscomyidae
  Heliscomys gregoryi Wood*
  Heliscomys sp., cf. *H. mcgrewi* Korth
Cricetidae
  Scottimus viduus Korth
  Scottimus sp.
  Eumys sp., cf. *E. parvidens* Wood
  Eumys sp., cf. *E. elegans* Leidy
    and/or *E. cricetodontoides* White
  Eumys new species, aff. *E. brachyodus* Wood
  Wilsoneumys new species
CARNIVORA
  Canidae
   Hesperocyon gregarius (Cope)
  Carnivora, *incertae sedis*
   Palaeogale sectoria (Gervais)
PERISSODACTYLA
  Equidae
   Mesohippus sp.
  Hyracodontidae
   Hyracodon? sp.
  Rhinocerotidae
   Subhyracodon sp.
ARTIODACTYLA
  Anthracotheriidae
   Elomeryx? sp.
  Merycoidodontidae
   Merycoidodon sp., cf. *M. culbertsoni* Leidy
   Miniochoerus? sp.
  Camelidae
   Poebrotherium sp.
  Leptomerycidae
   Hendryomeryx? sp.
   Leptomeryx sp., aff. *L. evansi* Leidy
   Leptomeryx new species

interval 80 to 140 feet above the base. The lower half of this interval is the most richly fossiliferous part of the sequence. Identifiable vertebrates have not thus far been recovered from the upper part of the type exposures, and only a few specimens are known from the lower part of the section.

The type Cook Ranch Formation is locally faulted and slumped, and some of the exposures are separated by covered intervals. Largely for this reason, Tabrum has divided the type Cook Ranch exposures into fourteen geographically and/or stratigraphically restricted localities. The most productive of these are University of Montana localities MV 7721, MV 7722, MV 7723, and MV 7726, which are stratigraphically equivalent, but geographically separated, localities in the middle part of the Cook Ranch type section. Localities MV 7734 and MV 7735 sample the stratigraphically lower part of the section, but each has produced only a few

specimens.

A. E. Wood (1933) and H. E. Wood (1934) briefly reported on the small collections of fossil vertebrates made by their parties from the type Cook Ranch Formation during the 1931 and 1933 field seasons. Additional small collections were made by Kay and Hough during the course of their field work in the Sage Creek Basin. Further work by Tabrum has since greatly increased the available collections, and a large sample is now available from the Cook Ranch l.f.

After Pipestone Springs, the Cook Ranch l.f. (Table 10) is the most diverse mammalian local fauna currently known from any of the intermontane basins of south-western Montana, but it is almost entirely undescribed. Only three specimens have thus far been formally reported in the published literature. A. E. Wood (1933) described the holotype of *Heliscomys gregoryi*, and Hough (1961) attributed two specimens to her new genus and species of marsupial, *Didelphidectes pumilus*. Both of Hough's specimens appear, however, to pertain to the common Cook Ranch didelphid *Herpetotherium fugax*.

The Cook Ranch l.f. is predominantly a small-mammal fauna and has produced a diverse assemblage of marsupials, insectivores, chiropterans, lagomorphs, and rodents. Large mammals are present, but are relatively uncommon. "*Palaeolagus*" *burkei* is the most common mammal in the available, largely surface-collected sample, as it is in the stratigraphically lower Matador Ranch l.f. *Wilsoneumys* n. sp., *Megalagus* sp. cf. *M. turgidus*, and *Herpetotherium fugax* are also represented by large samples. Other common small mammals include *Prosciurus*? n. sp., *Sespemys*? n. sp., and *Adjidaumo minimus*. *Leptomeryx* n. sp. is the only moderately common large mammal. Rare mammals include late records of an apternodontid (*Oligoryctes* sp. cf. *O. altitalonidus*) and a cylindrodontid (*Pseudocylindrodon*? n. sp.) and early occurrences of talpids and a species of *Ocajila* close to *O. makpiyahe*.

The Cook Ranch l.f. is highly endemic, which hinders attempts at correlation with approximately contemporaneous assemblages from the Great Plains region. Of the 32 Cook Ranch taxa that are here regarded as specifically determinate, only 13 appear to be conspecific with forms known from contemporaneous deposits of the Great Plains region, and most of these are relatively long-ranging forms. This figure will probably increase somewhat when some of the large mammals are more precisely identified but will probably not significantly enhance correlation.

Wood et al. (1941, p. 11) designated "Cook Ranch and other scattered Montana deposits" as the principal correlatives of the Orellan Provincial Age, even though only a small assemblage was then known from the Cook Ranch Formation, and only from the east-facing type exposures (= Cook Ranch l.f.). The large sample now known from the Cook Ranch l.f. seems to be indicative of a late, probably latest, Orellan age, equivalent to, or perhaps slightly later than, the Orella D beds of Nebraska. Because of the high degree of endemism exhibited by the Cook Ranch l.f. only a very few biostratigraphically useful species are shared with approximately contemporaneous assemblages from the Great Plains region. *Heliscomys* sp. cf. *H. mcgrewi* from the Cook Ranch l.f. (Korth and Tabrum, MS) is, however, probably conspecific with *H. mcgrewi*, otherwise known only from beds of Orella D age in Nebraska and Wyoming (Korth, 1989a). *Heliscomys* sp. cf. *H. mcgrewi* is one of the few taxa known from both the lower and middle parts of the type section of the Cook Ranch Formation; the presence of *H.* sp. cf. *H. mcgrewi* in both suggests that the entire fossiliferous part of the type section is of late Orellan age.

Several taxa from the Cook Ranch l.f. appear to be somewhat advanced over their counterparts in the most nearly contemporaneous assemblages from the Great Plains region. The cricetid *Wilsoneumys planidens* was reported by Korth (1989a) to be restricted to beds of Orella D age in the Great Plains region. *Wilsoneumys* n. sp. from Cook Ranch is a larger, dentally more derived species, which may indicate that the Cook Ranch l.f. is of very late Orella D age. Some of the Cook Ranch aplodontids also seem advanced compared to those described by Korth (1989b) from the Orellan of Nebraska: *Prosciurus*? n. sp. is comparable in size to *Prosciurus relictus* but possesses features suggestive of relationship to the Whitneyan *Haplomys liolophus*; *Sespemys*? n. sp. is a small species possibly related to the Whitneyan *Sespemys thurstoni*; and an as yet unnamed new genus and species of prosciurine exhibits some characters usually associated with allomyines. These, and other, differences between mammals from the Cook Ranch l.f. and those from approximately contemporaneous assemblages of the Great Plains region appear largely to be related to faunal endemism but possibly also reflect a slight difference in age.

The Cook Ranch l.f. most strongly resembles the late Orellan Cedar Ridge l.f. of the Badwater Creek area, central Wyoming (Setoguchi, 1978; Korth, 1989a, 1994), although only a few species are actually shared between the two. The eomyid rodent *Metadjidaumo* is known only from the Cedar Ridge and Cook Ranch local faunas; *Metadjidaumo* n. sp. from Cook Ranch is very close to the Cedar Ridge *M. hendryi*, differing principally in having slightly wider cheek teeth (Korth and Tabrum, MS). Specimens from the Cedar Ridge l.f. referred to *Paradjidaumo hypsodus* by Setoguchi (1978) appear to represent at least three distinct species of eomyid, and two small, atypical teeth may instead pertain to the new species of *Paradjidaumo* known from the Cook Ranch l.f. (Korth and Tabrum, MS). Both local faunas also possess a small species of *Adjidaumo*, which, although not conspecific, differ significantly from the contemporaneous *A. minutus* of the Great Plains region. In addition, the assemblage of cricetid rodents in both the Cedar Ridge and Cook Ranch local faunas is dominated by *Wilsoneumys*. The Cook Ranch *Wilsoneumys* is a more derived species than *W. planidens* from Cedar Ridge, which may suggest that the Cook Ranch l.f. is slightly later in age. Of the few species that are present in both the Cedar Ridge and Cook Ranch local faunas, *Heliscomys mcgrewi* is probably the most biochronologically useful and, as noted above, is restricted to beds of late Orellan (Orella D) age (Korth, 1989a).

Prothero sampled the type Cook Ranch Formation for paleomagnetic analysis in 1980. Approximately 230 feet of continuous section were measured (Fig. 6). The section ascended the east face of the prominent ridge exposed in the SW SW NE NW Sec. 34, T12S R8W

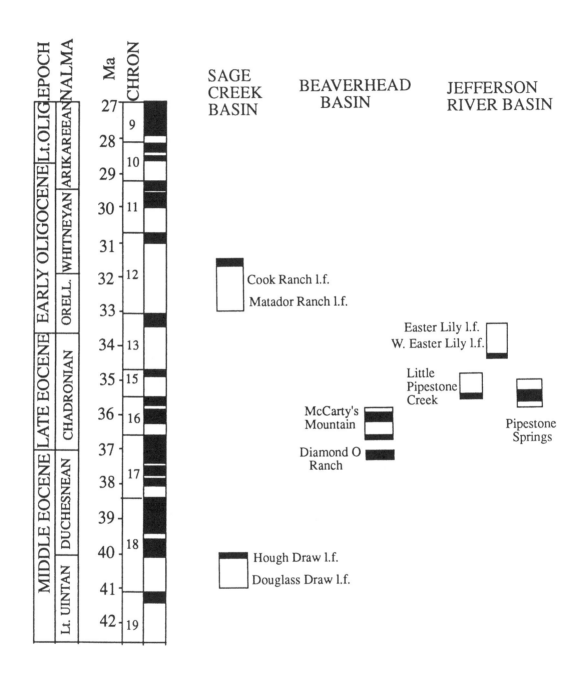

Figure 7. Temporal correlation of the sections in this study. Time scale after Berggren et al. (1995).

Rock Island Ranch 7.5' Quadrangle, Beaverhead County, Montana, beginning at the base of the exposures at the 6400-foot contour and extending due west to the uppermost exposures at the 6620-foot contour. Except for two sites, the entire section was of reversed polarity. Based on the late Orellan age of the Cook Ranch l.f., the type Cook Ranch Formation appears to correlate with the early part of Chron C12r, about 32.0-32.5 Ma (Fig. 7). The Cook Ranch l.f. seems to fall very near the Orellan/Whitneyan boundary and thus appears to be approximately 32.0 Ma in age.

DISCUSSION

The mammalian local faunas of southwestern Montana strongly reflect the significant biogeographic provincialism that characterized western North America during the Eocene-Oligocene transition. The local faunas discussed here contain many species that appear to have been endemic to the region. The largest and most diversified of these local faunas (Pipestone Springs, Little Pipestone Creek, McCarty's Mountain, Diamond O Ranch, Hough Draw, Cook Ranch) exhibit moderate to high levels of faunal endemism. The late Uintan Hough Draw local fauna and the middle Chadronian Pipestone Springs and Little Pipestone Creek local faunas each contain about 20-25% species that seem to have been endemic to southwestern Montana. Slightly more than 40% of the species in the McCarty's Mountain, Cook Ranch, and the Diamond O Ranch local faunas have not been recorded from contemporaneous localities in other parts of North America. Peak endemism thus occurs in the early Chadronian, late Orellan, and possibly in the late Duchesnean part of the Montana sequence.

Several biochronologically important mammalian taxa, including talpids, *Ocajila*, *Megalagus*, *Palaeolagus*, *Palaeogale*, and *Leptomeryx sensu stricto*, appear to occur significantly earlier in Montana that they do in the High Plains, which others, including apternodontids and cylindrodontids, persist later. Of the taxa that have earlier records in southwestern Montana than elsewhere in North America, talpids, *Palaeogale*, and *Ocajila* are either demonstrably or probably immigrants from Eurasia, which may indicate that southwestern Montana provided an environmentally more favorable dispersal route for these forms than was available in the High Plains. This is also supported by the presence in southwestern Montana of taxa closely related to those that gave rise to Asian species of *Ardynomys* and *Eomys*, both groups immigrating to Asia from North America.

The principal reasons for the significant differences between the Eocene-Oligocene mammalian faunas of southwestern Montana and the generally better-studied coeval faunas of the High Plains appear straightforward. Lillegraven and Tabrum (1983) discussed the major differences in paleoenvironmental setting between the relatively open environment of the High Plains region and that of the small intermontane basins of southwestern Montana. They concluded that the "overall large-scale habitats" represented in the two regions were demonstrably quite different (Lillegraven and Tabrum, 1983, p. 69). The generally high level of endemism exhibited in the local faunas of southwestern Montana appears to be due to the higher elevation of the sites in Montana, the much greater topographic complexity of the region, and close proximity of many of the sites to areas of significant topographic relief, coupled with differences in climatic regime.

The available paleobotanical evidence, reviewed by Wing (1987), also clearly supports an upland setting for the Paleogene deposits of southwestern Montana. The "Beaverhead Basin" floras (from the contiguous Medicine Lodge, Horse Prairie, and Grasshopper basins), of probable Chadronian age (Fields et al., 1985), and the apparently somewhat younger floras from the Upper Ruby River Basin, represent mixed coniferous and deciduous broad-leaved forests that are similar to the Florissant and Red Rock Ranch floras of Colorado and New Mexico. Wing (1987, p. 763) suggested that these floras all indicate growth at an "intermediate elevation" under a seasonally dry climate. Wing also noted that the greater diversity of conifers and mesic taxa in the "Beaverhead Basin" floras than in the Florissant and Red Rock Ranch floras might indicate higher rainfall in the northern part of the Rocky Moutains than in parts of Colorado and New Mexico at comparable elevation. The apparently younger (possibly Orellan to Whitneyan) Upper Ruby River Basin floras, though generally similar to the "Beaverhead Basin" floras, appear to indiate some increase in seasonal aridity.

CONCLUSION

Although the Paleogene record in southwestern Montana is patchy, significant local faunas ranging in age from Bridgerian to Arikareean age are preserved in the numerous small intermontane basins of the region. Magnetic stratigraphy has provided a powerful adjunct to biochronology in our continuing quest to generate progressively more precise age assignments for the local faunas known from the region. Our results confirm and strengthen the age assignments suggested by Fields et al. (1985) and add a level of precision hitherto unattainable. Our suggested correlations of the magnetic sections is summarized in Figure 7. These results show the limited temporal span of most of these intermontane sequences. Although some are thousands of feet thick, few appear to span more than a million years, and most are much shorter in duration. This demonstrates just how rapidly these fault-bounded basins were subsiding and filling up during the Paleogene.

ACKNOWLEDGMENTS

We thank Tony Barnosky, Bob Emry, John Rensberger, and John Storer for their helpful reviews of this chapter. Prothero was supported by Columbia University Department of Geology field funds during the 1980 season, and by the Donors of the Petroleum Research Fund of the American Chemical Society during the 1983 and 1986 field seasons. The 1987 field work was supported by NSF grant EAR87-08221. The 1980 research was undertaken under the supervision of Dr. Malcolm C. McKenna. Prothero thanks Chuck Denham, Bill Roggenthen, and Joe Kirschvink for graciously allowing access to their paleomagnetics laboratories. The magnetic sampling and analysis would never have been possible without the hard work and cheerful good spirits of several field crews. In 1980, they included Priscilla Duskin, Jon Frenzel, and Heidi Shlosar. In 1983, they included Rob Lander and Annie Walton. The 1986 samples were collected by Dana Gilchrist, Kecia Harris, and Allison Kozak. The 1987 samples were collected by Jill Bush, John Foster, and Steve King.

Tabrum's field work in southwestern Montana was supported by grants from Amoco Production Company, the Penrose Fund of the Geological Society of America, the STATEMAP program, and the O'Neil (to Dr. Mary Dawson) and M. Graham Netting research funds of the Carnegie Museum. Copies of Earl Douglass's field notes were also acquired via a grant from the Netting Research Fund. Heartfelt thanks are owed to many people. Dennis Dunlap provided lodging, many meals, and his great wit during the 1981 field season. Malcolm McKenna generously provided financial and logistical support for field work conducted during the 1984, 1985, and 1987 field seasons. Landowners Norman and Gay Ashcraft, and Thomas and Ann Dooling graciously allowed unlimited access to localities on their land. During the course of five extended and ten shorter field seasons devoted to the study of the Cenozoic rocks and faunas of southwestern Montana, many people assisted in the collection of materials. Principal among these were Carol Goozey, Dennis Dunlap, Andy Wyss, Ralph Nichols, Carl Swisher, Larry French, Kim Bideganeta, Barbara Pitman, and Bill Leistner. Special thanks are extended to Debbie Hanneman and Chuck Wideman for their warm hospitality and enthusiastic support of Tabrum's field work, and to Susan Vuke for providing a base of operations for the 1994 field season. The continuing support of Mary Dawson of the Carnegie Museum for field endeavors in southwestern Montana is deeply appreciated.

Garcia's field work was funded by the Museum of Paleontology, University of California, and by a grant from the Society of Sigma Xi. Landowners in the Little Pipestone Creek area, particularly the members of the Alley family, generously permitted access to the localities on their land. Assistance in the field was provided by Mark Garcia, Don Lofgren, and John Rittel. Debbie Hanneman and Chuck Wideman provided key logistical support.

Tabrum and Garcia acknowledge a tremendous debt to the late Dr. Robert W. Fields (September 17, 1920–August 23, 1995), Professor of Geology at the University of Montana, mentor, and friend. He introduced them to the wonders of the Tertiary of southwestern Montana, and unceasingly supported them during their years as graduate students at the University of Montana. We dedicate this paper to his memory.

LITERATURE CITED

Asnake, M. 1984. Biostratigraphic and evolutionary relationships of cylindrodontid rodents of the Chadronian (Early Oligocene) of Montana. M.S. Thesis, University of Washington, Seattle, 97 p.

Berggren, W. A., D. V. Kent, J. D. Obradovich, and C. C. Swisher III. 1992. Toward a revised Paleogene geochronology; pp. 29-45 in D. R. Prothero and W. A. Berggren (eds.), Eocene-Oligocene Climatic and Biotic Evolution, Princeton University Press, Princeton, N. J.

Berggren, W. A., D. V. Kent, M.-P. Aubry, C. C. Swisher III, and K. G. Miller. 1995. A revised Paleogene geochronology and chronostratigraphy. SEPM Special Publication 54:129-212.

Bjork, P. R. 1968. New records of helaletid tapiroids from the Oligocene of South Dakota. Papers of the Michigan Academy of Science, Arts, and Letters 53:73-78.

Black, C. C. 1965. Fossil mammals from Montana. Pt. 2. Rodents from the early Oligocene Pipestone Springs local fauna. Annals of Carnegie Museum 38:1-48.

Black, C. C. 1968. The Oligocene rodent genus *Ischyromys* and discussion of the family Ischyromyidae. Annals of Carnegie Museum 39:273-305.

Black, C. C. 1974. Paleontology and geology of the Badwater Creek area, central Wyoming. Part 9. Additions to the cylindrodont rodents from the late Eocene. Annals of Carnegie Museum 45:151-160.

Black, C. C. 1978. Paleontology and geology of the Badwater Creek area, central Wyoming. Part 14. The artiodactyls. Annals of Carnegie Museum 47:223-259.

Black, C. C. 1979. Paleontology and geology of the Badwater Creek area, central Wyoming. Part 19. Perissodactyla. Annals of Carnegie Museum 48:391-401.

Burke, J. J. 1936. *Ardynomys* and *Desmatolagus* in the North American Oligocene. Annals of Carnegie Museum 25:135-154.

Burke, J. J. 1938. A new cylindrodont rodent from the Oligocene of Montana. Annals of Carnegie Museum 27:255-274.

Butler, R. F. 1992. Paleomagnetism. Blackwell, New York.

Clark, J., J. R. Beerbower, and K. K. Kietzke. 1967. Oligocene sedimentation, stratigraphy, paleoecology, and paleoclimatology in the Big Badlands of South Dakota. Fieldiana: Geology Memoirs 5:1-158.

de Bonis, L. 1981. Contribution a l'etude du genre *Palaeogale* Meyer (Mammalia, Carnivora). Annales de Paleotologie, Vertebres, 67:37-56.

Donohoe, J. C. 1956. New aplodontid rodent from Montana Oligocene. Journal of Mammalogy 37:264-268.

Douglass, E. 1901. Fossil Mammalia of the White River Beds of Montana. Transactions of the American Philosophical Society n.s., 20:237-279.

Douglass, E. 1903. New vertebrates from the Montana Tertiary. Annals of Carnegie Museum 2:145-199.

Douglass, E. 1905. The Tertiary of Montana. Memoirs of the Carnegie Museum 2:146-199.

Douglass, E. 1907. Some new merycoidodonts. Annals of the Carnegie Museum 4:99-109.

Douglass, E. 1908a. Fossil horses from North Dakota and Montana. Annals of the Carnegie Museum 4:267-277.

Douglass, E. 1908b. Some Oligocene lizards. Annals of Carnegie Museum 4:278-285.

Douglass, E. 1909. A geological reconnaissance in North Dakota, Montana, and Idaho; with notes on Mesozoic and Cenozoic geology. Annals of Carnegie Museum 5:211-288.

Dunlap, D. G. 1982. Tertiary geology of the Muddy Creek Basin, Beaverhead County, Montana. M.S. thesis, University of Montana, Missoula, 133 pp.

Emry, R. J. 1973. Stratigraphy and preliminary biostratigraphy of the Flagstaff Rim area, Natrona County, Wyoming. Smithsonian Contributions to Paleobiology 18.

Emry, R. J. 1992, Mammalian range zones in the Chadronian White River Formation at Flagstaff Rim, Wyoming; pp. 106-115 in D. R. Prothero and W. A. Berggren (eds.), Eocene-Oligocene Climatic and Biotic Evolution. Princeton University Press, Princeton, N. J.

Emry, R. J., P. R. Bjork, and L. S. Russell. 1987. The Chadronian. Orellan, and Whitneyan land mammal ages; pp. 118-152 in M. O. Woodburne (ed.), Cenozoic Mammals of North America, Geochronology and Biostratigraphy. University of California Press, Berkeley.

Emry, R. J., and C. E. Gawne. 1986. A primitive, early Oligocene species of Palaeolagus (Mammalia, Lagomorpha) from the Flagstaff Rim area of Wyoming. Journal of Vertebrate Paleontology 6: 271-280.

Emry, R. J., and R. M. Hunt, Jr. 1980. Maxillary dentition and new records of Daphoenictis, an Oligocene amphicyonid carnivore. Journal of Mammalogy 61(4):720-723.

Emry, R. J., and W. W. Korth. 1993. Evolution in Yoderimyinae (Eomyidae: Rodentia), with new material from the White River Formation (Chadronian) at Flagstaff Rim, Wyoming. Journal of Paleontology 67: 1047-1057.

Fields, R. W., D. L. Rasmussen, A. R. Tabrum, and R. Nichols. 1985. Cenozoic rocks of the intermontane basins of western Montana and eastern Idaho: a summary; pp. 9-36, in R. M. Flores and S. S. Kaplan (eds.), Cenozoic Paleogeography of the West-central United States. Rocky Mountain Paleogeography Symposium 3. Rocky Mountain Section, SEPM.

Fisher, R. A. 1953. Dispersion on a sphere. Proceedings of the Royal Astronomical Society A217: 295-305.

French, L. B. 1988. The Lower Oligocene (Chadronian) - Middle Oligocene (Orellan) boundary in the Easter Lily Mine section (Renova Formation) near Whitehall, Jefferson County, Montana. Northwest Geology 17:51-56.

Galbreath, E. C. 1953. A contribution to the Tertiary geology and paleontology of northeastern Colorado. University of Kansas Paleontological Contributions, Vertebrata 4:1-120.

Garcia, D. 1992. Fossil mammals from the Pipestone Creeks region, Late Eocene and Oligocene (Chadronian and Orellan), Jefferson County, Montana. Ph.D. Dissertation, University of California, Berkeley, 215 p.

Gazin, C. L. 1955. A review of the Upper Eocene Artiodactyla of North America. Smithsonian Miscellaneous Collections 128(8):1-96.

Gazin, C. L. 1956. The geology and vertebrate paleontology of Upper Eocene strata in the northeastern part of the Wind River Basin, Wyoming. Part 2. The mammalian fauna of the Badwater area. Smithsonian Miscellaneous Collections 131(8):1-35.

Gilmore, C. W. 1928. Fossil lizards of North America. Memoirs of the National Academy of Sciences, 22.

Hanneman, D. L., and C. J. Wideman. 1991. Sequence stratigraphy of Cenozoic continental rocks, southwestern Montana. Geological Society of America Bulletin 103:1335-1345.

Heaton, T. H. 1993. The Oligocene rodent Ischyromys of the Great Plains: replacement mistaken for anagenesis. Journal of Paleontology 67:297-308.

Hoffman, D. S. 1972. Tertiary stratigraphy, vertebrtae paleontology, and paleoecology of a portion of the lower Beaverhead River Basin, Madison and Beaverhead counties, Montana. Ph.D. Dissertation, University of Montana, Missoula, 174 pp.

Hough, J. R. 1955. An Upper Eocene fauna from the Sage Creek area, Beaverhead County, Montana. Journal of Paleontology 29:22-36.

Hough, J. R. 1958. Tertiary beds of the Sage Creek area, Beaverhead County, Montana. Society of Vertebrate Paleontology, 8th Annual Field Conference Guidebook, p. 41-45.

Hough, J. R. 1961. Review of Oligocene didelphid marsupials. Journal of Paleontology 35:218-228.

Johnson, H. P., W. Lowrie, and D. V. Kent. 1975. Stability of anhysteretic remanent magnetization in fine and coarse magnetite and maghemite particles. Geophysical Journal of the Royal Astronomical Society 41:1-10.

Kay, J. L., R. W. Fields, and J. B. Orr. 1958. Faunal lists of Tertiary vertebrates from western and southwestern Montana. Society of Vertebrate Paleontology, 8th Annual Field Conference Guidebook, pp. 33-39.

Korth, W. W. 1980. Paradjidaumo (Eomyidae, Rodentia) from the Brule Formation, Nebraska. Journal of Paleontology 54:933-941.

Korth, W. W. 1989a. Stratigraphic occurrence of rodents and lagomorphs in the Orella Member, Brule Formation (Oligocene), northwestern Nebraska. Contributions to Geology, University of Wyoming 27:15-20.

Korth, W. W. 1989b. Aplodontid rodents (Mammalia) from the Oligocene (Orellan and Whitneyan) Brule Formation, Nebraska. Journal of Vertebrate Paleontology 9:400-414.

Korth, W. W. 1994. Middle Tertiary marsupials (Mammalia) from North America. Journal of Paleontology 68:376-397.

Korth, W. W., and J. Hageman. 1988. Lagomorphs (Mammalia) from the Oligocene (Orellan and Whitneyan) Brule Formation, Nebraska. Transactions of the Nebraska Academy of Sciences 16:141-152.

Korth, W. W., and A. R. Tabrum. (In press). Eomyid and heliscomyid rodents from the late Orellan Cook Ranch local fauna of southwestern Montana.

Korth, W. W., J. H. Wahlert, and R. J. Emry. 1991. A new species of Heliscomys and recognition of the family Heliscomyidae (Geomyoidea: Rodentia). Journal of Vertebrate Paleontology 11:247-256.

Kuenzi, W. D. 1966. Tertiary stratigraphy in the Jefferson River Basin, Montana. Ph.D. Dissertation, University of Montana, Missoula, 293 pp.

Kuenzi, W. D., and R. W. Fields. 1971. Tertiary stratigraphy, structure, and geologic history, Jefferson Basin, Montana. Bulletin of the Geological Society of America 82:3373-3394.

Lillegraven, J. A. 1979. A biogeographical problem involving comparisons of later Eocene terrestrial vertebrate faunas of western North America; pp. 333-347 in J. Gray and A. J. Boucot (eds.), Historical biogeography, plate tectonics, and the changing environment. Oregon State University Press, Corvallis.

Lillegraven, J. A., M. C. McKenna, and L. Krishtalka. 1981. Evolutionary relationships of Middle Eocene and younger species of Centetodon (Mammalia, Insectivora, Geolabididae) with a description of the dentition of Ankylodon. University of Wyoming Publications 45:1-115.

Lillegraven, J. A., and A. R. Tabrum. 1983. A new species of Centetodon (Mammalia, Insectivora, Geolabididae) from southwestern Montana and its biogeographical implications. Contributions to Geology, University of Wyoming 22:57-73.

Lofgren, D. L. 1985. Tertiary vertebrate paleontology, stratigraphy, and structure, north Boulder River Basin, Jefferson County, Montana. M. S. thesis, University of Montana, Missoula, 113 pp.

Loomis, F. B. 1924. The oreodonts of the Lower Oligocene. Annals of Carnegie Museum 15: 368-378.

Lucas, S. G. 1992. Redefinition of the Duchesnean land mammal "age," late Eocene of western North America; pp. 88-105 in D. R. Prothero and W. A. Berggren (eds.), Eocene-Oligocene Climatic and Biotic Evolution. Princeton University Press, Princeton, N. J.

Martin, L. D. 1980. The early evolution of the Cricetidae in North America. University of Kansas Paleontological Contributions 102:1-42.

Matthew, W. D. 1903. The fauna of the Titanotherium Beds at Pipestone Springs, Montana. Bulletin of the American Museum of Natural History 19:197-226.

Mellett, J. S. 1977. Paleobiology of North American Hyaenodon (Mammalia, Creodonta). Contributions to Vertebrate Evolution 1:1-134.

Monroe, J. S. 1976. Vertebrate paleontology, stratigraphy, and sedimentation of the Upper Ruby River Basin, Madison County, Montana. Ph.D. Dissertation, University of Montana, Missoula, 301 pp.

Novacek, M. J. 1976. Early Tertiary vertebrate faunas, Vieja Group, Trans-Pecos Texas: Insectivora. Texas Memorial Museum, Pearce-Sellards Series 23:1-18.

Opdyke, N. D., E. H. Lindsay, N. M. Johnson, and T. Downs. 1977. The paleomagnetism and magnetic polarity stratigraphy of the mammal-bearing section of Anza-Borrego State Park, California. Journal of Quaternary Research 7:316-329.

Orr, J. B. 1958. The Tertiary of Western Montana. Society of Vertebrate Paleontology, 8th Annual Field Conference Guidebook, pp. 25-33.

Osborn, H. F., and W. D. Matthew. 1909. Cenozoic mammal horizons of western North America: United States Geological Survey Bulletin 361:1-138.

Ostrander, G. E. 1983. A new genus of eomyid (Mammalia, Rodentia) from the Early Oligocene (Chadronian), Pipestone Springs, Montana. Journal of Paleontology 57:140-144.

Ostrander, G. E. 1985. Correlation of the Early Oligocene (Chadronian) in northwestern Nebraska; pp. 205-231 in J. E. Martin (ed.), Fossiliferous Cenozoic Deposits of Western South Dakota and Northwestern Nebraska. Dakoterra 2(2).

Petkewich, R. M. 1972. Tertiary geology and paleontology of the Beaverhead East area, southwetern Montana. Ph.D. Dissertation, University of Montana, Missoula, 365 pp.

Pluhar, C. J., J. L. Kirschvink, and R. W. Adams. 1991. Magnetostratigraphy and clockwise rotation of the Plio-Pleistocene Mojave River Formation, central Mojave Desert, California. San Bernardino County Museum Association Quarterly 38(2):31-42.

Prothero, D. R. 1982. Middle Oligocene magnetostratigraphy and mammalian biostratigraphy: testing the isochroneity of mammalian biostratigraphic events. Ph.D. Dissertation, Columbia University, New York.

Prothero, D. R. 1984. Magnetostratigraphy of the Early Oligocene Pipestone Springs locality, Jefferson County, Montana. Contributions to Geology, University of Wyoming 23 (1): 33-36.

Prothero, D. R. 1985a. Chadronian (early Oligocene) magnetostratigraphy of eastern Wyoming: implica-tions for the Eocene-Oligocene boundary. Journal of Geology 93:555-565.

Prothero, D. R. 1986. A new oromerycid (Mammalia, Artiodactyla) from the early Oligocene of Montana. Journal of Paleontology 60:458-465.

Prothero, D. R., and N. Shubin. 1989. The evolution of Oligocene horses; pp. 142-175 in D. R. Prothero and R. M. Schoch (eds.), The Evolution of Perissodactyls. Oxford University Press, New York.

Prothero, D. R., and C. C. Swisher III. 1992. Magnetostratigraphy and geochronology of the terrestrial Eocene-Oligocene transition in North America; pp. 46-74 in D. R. Prothero and W. A. Berggren (eds.), Eocene-Oligocene Climatic and Biotic Evolution, Princeton University Press, Princeton, N. J.

Prothero, D. R., and K. E. Whittlesey. 1996. Magnetostratigraphy and biostratigraphy of the Chadronian, Orellan, and Whitneyan. Geological Society of America Special Paper (in press).

Radinsky, L. B. 1963. Origin and early evolution of North American Tapiroidea. Bulletin of the Peabody Museum of Natural History 17:1-106.

Radinsky, L. B. 1967. A review of the rhinocerotoid family Hyracodontidae (Perissodactyla). Bulletin of the American Museum of Natural History 136:1-45.

Rasmussen, D. L. 1969. Late Cenozoic geology of the Cabbage Patch area, Granite and Powell counties, Montana. M.A. thesis, University of Montana, Missoula, 188 pp.

Riel, S. J. 1966. A basal Oligocene local fauna from McCarty's Mountain, southwestern Montana. M.S. thesis, University of Montana, Missoula, 74 p.

Robinson, G. D. 1963. Geology of the Three Forks Quadrangle, Montana. U.S. Geological Survey Professional Paper 370:1-143.

Robinson, G. D. 1967. Geologic map of the Toston Quadrangle, southwestern Montana. U.S. Geological Survey Miscellaneous Geological Investigations Map I-486.

Runkel, A. C.1986. Geology and vertebrate paleontology of the Smith River Basin, Montana. M. S. thesis, University of Montana, Missoula, 80 pp.

Ruppel, E. T., and D. A. Lopez. 1984. The thrust belt in southwest Montana and east-central Idaho. U. S.

Geological Survey Professional Paper 1278:1-41.

Scholten, R., K. A. Keenmon, and W. O. Kupsch. 1955. Geology of the Lima region, southwestern Montana and adjacent Idaho. Bulletin of the Geological Society of America 66:345-404.

Schultz, C. B., and C. H. Falkenbach. 1956. Miniochoerinae and Oreonetinae, two new subfamilies of oreodonts. Bulletin of the American Museum of Natural History 109:373-482.

Scott, W. B. 1940. The mammalian fauna of the White River Oligocene. Part IV. Artiodactyla. Transactions of the American Philosophical Society 28:363-746.

Setoguchi, T. 1978. Paleontology and geology of the Badwater Creek area, central Wyoming. Part 16. The Cedar Ridge local fauna (Late Oligocene). Bulletin of Carnegie Museum 9:1-61.

Simpson, G. G. 1927. A North American Oligocene edentate. Annals of Carnegie Museum 17:283-298.

Storer, J. E. 1981. Leptomerycid Artiodactyla of the Calf Creek l.f. (Cypress Hills Formation, Oligocene, Chadronian), Saskatchewan. Saskatchewan Museum of Natural History, Natural History Contributions 3:1-32.

Storer, J. E. 1984a. Mammals of the Swift Current Creek local fauna (Eocene): Uintan, Saskatchewan. Saskatchewan Museum of Natural History, Natural History Contributions 7:1-158.

Storer, J. E. 1984b. Fossil mammals of the Southfork l.f. (early Chadronian) of Saskatchewan. Canadian Journal of Earth Sciences 21:1400-1405.

Storer, J. E. 1989. Rodent faunal provinces, Paleocene-Miocene of North America; pp. 17-29 in C. C. Black and M. R. Dawson (eds.), Papers on Fossil Rodents in Honor of Albert Elmer Wood. Science Series, Natural History Museum of Los Angeles County 33:1-192.

Storer, J. E. 1990. Primates of the Lac Pelletier Lower Fauna (Eocene: Duchesnean), Saskatchewan. Canadian Journal of Earth Sciences 27:520-524.

Storer, J. E. 1994. A latest Chadronian (Late Eocene) mammalian fauna from the Cypress Hills, Saskatchewan. Canadian Journal of Earth Sciences 31:1335-1341.

Storer, J. E., and H. N. Bryant. 1993. Biostratigraphy of the Cypress Hills Formation (Eocene to Miocene), Saskatchewan: Equid types (Mammalia, Perissodactyla) and associated faunal assemblages. Journal of Paleontology 67:660-669.

Stucky, R. K. 1984. The Wasatchian-Bridgerian Land Mammal Age boundary (Early to Middle Eocene) in western North America. Annals of Carnegie Museum 53:347-382.

Sullivan, R. M. 1979. Revision of the Paleogene genus Glyptosaurus (Reptilia, Anguidae). Bulletin of the American Museum of Natural History 163:1-72.

Tabrum, A. R. 1994. Biostratigraphic data from the 1903 McCarty's Mountain field notes of Earl Douglass. Journal of Vertebrate Paleontology 14:49A.

Tabrum, A. R., and R. W. Fields. 1980. Revised mammalian faunal list for the Pipestone Springs local fauna (Chadronian, Early Oligocene), Jefferson County, Montana. Northwest Geology 9:45-51.

Tedford, R. H., T. Galusha, M. F. Skinner, B. E. Taylor, R. W. Fields, J. R. Macdonald, J. M. Rensberger, S. D. Webb, and D. P. Whistler. 1987. Faunal succession and biochronology of the Arikareean through Hemphillian interval (late Oligocene through earliest Pliocene epochs) in North America; pp. 153-210 in M. O. Woodburne (ed)., Cenozoic Mammals of North America, Geochronology and Biostratigraphy. University of California Press, Berkeley.

Thorpe, M. R. 1937. The Merycoidodontidae, an extinct group of ruminant mammals. Memoirs of the Peabody Museum of Natural History, 3:1-428.

Wallace, S. M. 1980. A revision of North American Early Eocene Brontotheriidae (Mammalia, Perissodactyla). M.S. thesis, University of Colorado, Boulder, 154 p.

Wang, X. 1994. Phylogenetic systematics of the Hesperocyoninae (Carnivora: Canidae). Bulletin of the American Museum of Natural History 221:1-207.

Wilson, J. A. 1971. Early Tertiary vertebrate faunas, Vieja Group, Trans-Pecos Texas: Agriochoeridae and Merycoidodontidae. Texas Memorial Museum Bulletin 18:1-83.

Wing, S. L. 1987. Eocene and Oligocene floras and vegetation of the Rocky Mountains. Annals of the Missouri Botanical Garden 274:748-784.

Wood, A. E. 1933. A new heteromyid rodent from the Oligocene of Montana. Journal of Mammalogy 14:134-141.

Wood, A. E. 1970. The early Oligocene rodent Ardynomys (Family Cylindrodontidae) from Mongolia and Montana. American Museum Novitates 2366:1-8.

Wood, A. E. 1974. Early Tertiary vertebrate faunas, Vieja Group, Trans-Pecos Texas: Rodentia. Texas Memorial Museum Bulletin 21:1-112.

Wood, A. E. 1976. The Oligocene rodents Ischyromys and Titanotheriomys and the content of the Family Ischyromyidae; pp. 244-277 in C. S. Churcher (ed.), Athlon: Essays on Palaeontology in Honour of Loris Shano Russell. Royal Ontario Museum, Life Sciences Miscellaneous Publications.

Wood, A. E. 1980. The Oligocene rodents of North America. Transactions of the American Philosophical Society 70:1-68.

Wood, H. E. 1934. Revision of the Hyrachyidae. Bulletin of the American Museum of Natural History 67:181-295.

Wood, H. E., R. W. Chaney, J. Clark, E. H. Colbert, G. L. Jepsen, J. B. Reeside Jr., and C. Stock. 1941. Nomenclature and correlation of the North American continental Tertiary. Bulletin of the Geological Society of America 52:1-48.

Woodburne, M. O. (ed.) 1987. Cenozoic Mammals of North America, Geochronology and Biostratigraphy. University of California Press, Berkeley.

15. The Whitneyan-Arikareean Transition in the High Plains

RICHARD H. TEDFORD, JAMES B. SWINEHART, CARL C. SWISHER III,
DONALD R. PROTHERO, STEVEN A. KING, AND TIMOTHY E. TIERNEY

ABSTRACT

We bring together lithostratigraphic, biostratigraphic, and magnetostratigraphic data from Nebraska and South Dakota to detail faunal change between 28-30 Ma in medial Oligocene time. This span records the transition from the White River chronofauna to the new assemblages that characterize the younger part of the Arikareean "age." Although a regional disconformity of approximately a half-million year duration breaks the biostratigraphic sequence, the fossil record is reasonably continuous and mostly confined to the eolian facies. Between 28-30 Ma the White River chronofauna experienced significant enrichment in autochthonous clades especially hesperocyonine canids, oreodonts, camels, hypertragulids, and burrowing castoid and geomyoid rodents. Few allochthonous taxa are encountered so that the chronofauna was enriched without marked immigration or extinction. At approximately 28 Ma most of the White River genera leave the record, thus terminating the chronofauna. The fauna that emerges contains representatives of autochthonous lineages, some of which appeared during the enrichment phase of the White River chronofauna. In addition there are allochthonous genera that represent taxa new to mid-continental North America. The better resolved and calibrated fossil record allows re-examination of the definition and characterization of the beginning of the Arikareean mammal "age." We propose that the initiation of the Arikareean Mammal "age" is signaled by the first appearance of taxa that enrich the White River chronofauna in latest Chron C11r and earliest Chron C11n (about 30 Ma). The "age" is defined by the first appearance of the allochthone *Plesiosminthus* and characterized by the autochthonous *Palaeolagus hypsodus* and *P. philoi*, *Palaeocastor nebrascensis*, *Shunkahetanka geringensis*, *Nanotragulus loomisi*, *Sespia nitida*, and ?*Mesoreodon minor*.

INTRODUCTION

When the North American Land Mammal "ages" were proposed by Wood and others in 1941, the Arikareean "age" was defined primarily as the geochron of the Arikaree Group. The better documented faunal succession in the later part of that span was used to characterize the "age." The nature of the faunal succession at the beginning of the "age," its relationship to the preceding Whitneyan "age," and the lithostratigraphic relationships between the White River and Arikaree groups were poorly documented. Major rock units, the Sharps Formation of South Dakota (Harksen et al., 1961) and the informal Brown Siltstone beds (Swinehart et al., 1985) of Nebraska, which lie in the upper White River and lower Arikaree groups, were confused with other rock bodies and unrecognized until late in this century. As the faunal content of these units was not known in 1941, its nature and significance to the definition and characterization of the Whitneyan-Arikareean boundary has only been recently established (Tedford et al., 1987) and is still not fully deciphered. This contribution reviews the Whitneyan-Arikareean transition in light of previous work (Tedford et al., 1985; Tedford et al., 1987), and explores, with improved biostratigraphic evidence, the nature of faunal change in this interval.

The attempt of Tedford et al. (1985) to correlate rocks and faunas across the boundary between the White River and Arikaree groups raised many questions that created a focus for further study. These included detailed geologic mapping of North Platte Valley exposures (Swinehart and Diffendal, 1995), and the gathering of additional radiometric and paleomagnetic data at previously studied localities, as well as at newly studied exposures. The most significant advance in our ability to construct tighter correlations came with ^{40}Ar/^{39}Ar single-crystal laser fusion dates (Swisher and Prothero, 1990). Ashes that had not yielded reliable ages (eg., the Lower Ash of the Whitney) or had large errors in fission-track or standard K-Ar dates were redated and helped tie the paleomagnetic, stratigraphic, and paleontologic results into a tighter framework. We report four new ^{40}Ar/^{39}Ar dates (Table 1) from the Wildcat Ridge and North Platte Valley mapping projects of Swinehart and Diffendal (1995). In addition, although no significant new fossil collections have been made, two of us (Swisher and Tedford) have reviewed the pertinent existing fauna in light of the new information. Pertinent

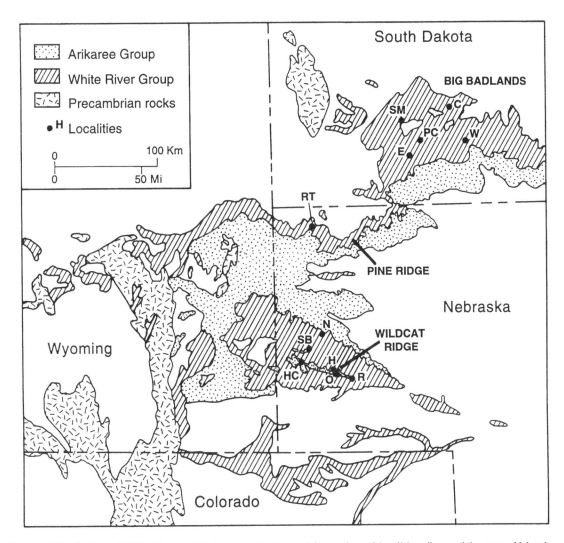

Figure 1. Distribution of White River and Arikaree strata, Precambrian rocks and localities discussed in text. Nebraska geology modified from Swinehart et al. (1985); Wyoming, South Dakota and Colorado modified from Bennison and Chenoweth (1984). Locality abbreviations as follows: C, Cedar Pass; E, Evergreen; H, Horn Ranch; HC, Helvas Canyon; N. Nuss Ranch; O, Olsen Ranch; PC, Palmer Creek; R, Roundhouse Rock; RT, Roundtop/Toadstool; SM, Sheep Mountain Table; W, Wolff Table.

fossil localities of the Nebraska State Museum are indicated by the acronym UNSM, those of the South Dakota School of Mines by SDSM.

LITHOSTRATIGRAPHY

Historically, the region encompassed by this study (Fig. 1) has served as the foundation for development of concepts of Cenozoic stratigraphy of the High Plains for nearly 150 years. The excellent outcrops provided by the dissection of the Cenozoic sedimentary blanket by major tributaries of the Missouri River in western Nebraska and southwestern South Dakota are still the focus of lithostratigraphic study.

Nebraska Panhandle

Recently the extensive drilling program of the University of Nebraska-Lincoln Conservation and Survey Division, coupled with analysis of petroleum and water wells in the region, has extended the stratigraphic record beneath covered terrain for a better appreciation of the regional relationships of the rock bodies recognized from outcrops (Swinehart et al., 1985). The following summary draws on all these sources of data.

Whitney Member, Brule Formation, White River Group

Typically the Whitney Member is a massive, brown, volcaniclastic, eolian siltstone as much as 90 meters thick with local, fine- to coarse-grained fluvial sandstones and mudstone. The type section in the "Round Top to Adelia" traverse (Schultz and Stout, 1955) in the Pine Ridge (Fig. 1) contains essentially no fluvial sediments. The oldest of two regionally extensive ash beds, the Lower Ash (LW) has been ^{40}Ar/^{39}Ar dated at 31.85 ± 0.02 Ma and the youngest, the Upper Ash (UW), has a date of 30.58 ± 0.61 Ma (Prothero and Swisher, 1992). The volcaniclastic siltstones themselves are composed of about 50% glass shards, a value close to the average for the Brule Formation as a whole.

Brown Siltstone Beds, Brule Formation, White River Group

Swinehart et al. (1985) significantly modified the regional lithostratigraphy of the upper part of the White River Group as described by Darton (1899) and Schultz and Stout (1955). They demonstrated that the Whitney Member of the Brule Formation, considered to be the uppermost part of the White River Group, was overlain (locally disconformably) by a slightly coarser grained, volcaniclastic siltstone sequence, as much as 135 m thick. These beds were informally named the "Brown Siltstone." Deposited primarily as a volcaniclastic eolian blanket (like the Whitney), it typically contains glass shard percentages from 50 to 60%. Swinehart et al. (1985, fig. 12) described an interval 10 to 20 m thick near the base of the Brown Siltstone that contains three ash beds (Nonpareil Ash Zone, NPZ) traceable over much of western Nebraska. At Roundtop in the Nebraska Pine Ridge (Fig. 1) Swisher and Prothero (1990) obtained a ^{40}Ar/^{39}Ar date of 30.05 ± 0.19 Ma on a NPZ correlated with NP_3. Fluvial mudstones and fine- to medium-grained sandstones also occur in paleovalleys that locally cut out the Nonpareil Ash Zone (Fig. 5) and the Upper Ash of the Whitney (Tedford et al., 1985, fig. 2; Souders, 1981, 1986). Geologic mapping by Swinehart and Diffendal (1995) has shown that the Brown Siltstone lies in a complex paleovalley system cropping out in the Wildcat Ridge (Fig. 5), the deposits of which had been included in the Gering Formation (Vondra et al., 1969). In establishing a type section for Darton's Gering Formation at Helvas Canyon (Figs. 1 and 4), Vondra et al. (1969) included within the Gering about 20 m of the Brown Siltstone beds consisting of a local, multiple cut-and-fill fluvial sequence. In addition, they were apparently misled, as was Darton, to include in the basal Gering another local outcrop of a 30-m thick, stacked fluvial sequence that actually lies within the Whitney.

Gering Formation, Lower Arikaree Group

The oldest Arikaree Group strata represent a regional interruption in the aggradation of fine-grained, eolian deposits by widespread fluvial erosion and subsequent deposition of coarser clastics. Two major paleovalleys were developed in the north half of the Nebraska panhandle and a minor one in the south half (Swinehart et al., 1985, fig. 14). There is uncertainty as to how close in time these basal Arikaree fluvial strata were deposited.

Darton (1899) named the Gering Formation for a series of fine- to coarse-grained, predominantly fluvial, volcaniclastic sandstones and conglomerates in the Wildcat Ridge. These rocks contain only 20-30% glass shards reflecting the importance of fluviatile processes during deposition of the Gering. The Gering Formation is separated by an erosional unconformity from the Brule and overlain conformably by rocks of the middle Arikaree Group. Although Darton and later workers were aware that fluvial sequences occurred within the Brule, a series of fluvial cuts and fills that occurred stacked together were often included in the Gering Formation. This concept led to the inclusion of the Brown Siltstone and some Whitney rocks in the Gering type section as discussed above. Believing the Brown Siltstone beds to be Gering, the often conformable relationship with the Whitney was a source of confusion for Darton and other workers. The Gering Formation, as mapped by Swinehart and Diffendal (1995), is redefined to include fluvial pumice-bearing siltstones, sandstones, and conglomerates at its base overlain (and possibly coeval in part) by thin-bedded, horizontally stratified sandstones and mudstones. It contains zones of calcite pseudomorphs after gypsum and vertebrate tracks (Loope, 1986; Swinehart and Loope, 1987) well exposed on Scotts Bluff National Monument (Fig. 4). This widely scattered but distinctive facies probably represents ephemeral stream, playa, and eolian sandsheet deposits.

New ^{40}Ar/^{39}Ar dates from the Gering and a basal post-Gering Arikaree ash (Table 1), combined with a consistent paleomagnetic stratigraphy (Fig. 4), suggests that the redefined Gering Formation spans only a few hundred thousand years. This conclusion is supported by dates on sanidine from pumice pebbles in the Twin Sisters Pumice Conglomerate low in the Gering at 28.31 ± 0.03, the Chimney Rock perrierite ash at the base of the thin horizontally bedded sequence at 28.26 ± 0.05 Ma, and the Roundhouse Rock Pisolitic Ash (RRP, Fig. 4) occurring near the top of the pumice-bearing sandstone at 28.11 ± 0.18 Ma. The post-Gering ash (Olsen's third ash, OT, Figs. 4, 5) dates at 27.79 ± 0.08 Ma. The revised Gering attains a maximum thickness of about 25 m (compared with 60 m in Darton, 1899).

Along the Pine Ridge, north of the North Platte Valley, the 100-m thick sequence of coarse- to fine-grained fluvial strata of the basal Arikaree (Swinehart et al., 1985) appears, on faunal criteria discussed below, to be coeval with the type Gering, as assumed by most workers since Hatcher (1902). To date no pumice has been found in these rocks and no reliable dates have been obtained on the several ashes present.

Middle Arikaree Group, Undivided

Conformably overlying the Gering Formation, the undivided Arikaree Group includes gray and grayish brown, fine-grained, poorly indurated, volcaniclastic, eolian sandstone and silty sandstone with carbonate-cemented "pipy concretions." Stratification is typically lacking or indistinct, but zones of eolian, low-angle cross-stratification occur locally (Swinehart and Loope, 1987). Along the Pine Ridge, between Harrison and Crawford, Nebraska, these rocks have been divided into two units, the Monroe Creek and Harrison formations (Hatcher, 1902; Hunt, 1985), but in the North Platte Valley no consistent lithologic criteria or marker beds have been discerned with which to differentiate these units (Swinehart et al., 1985). There are no dated ashes from the lower part of the Middle Arikaree in the Pine Ridge. The oldest dated ash is the 21.9 Ma Agate Ash (corrected from Evernden et al., 1964) in the upper part of the Harrison Formation.

Big Badlands of South Dakota
White River Group

In South Dakota, the White River Group was historically subdivided using biostratigraphic criteria, and later lithogenetic units were recognized (Darton, 1899; Bump, 1956). Bump (1956) subdivided the Brule Formation into the Scenic Member, thin-bedded claystone, siltstones, and channel-form sandstones with prominent beds of nodules cemented by groundwater, and the conformably overlying, massive-appearing siltstones and claystones of the Poleslide Member that include channel-form sandstones ("*Protoceras* channels") but are otherwise rather uniform in lithology and contain scattered calcareous nodules. Traditionally, the Poleslide Member was considered the top of the Brule Formation but the discovery by Harksen et al. (1961) of an additional unit, the Sharps Formation, previously mistaken for Poleslide strata, significantly extended the White River Group lithology. A widespread ash bed, the Rockyford Ash, was used as the base of the Sharps, but the Sharps is coarser grained than the Poleslide with nodular sandy siltstones, silty, very fine sandstones of massive appearance and lenses of marl being the most common lithologies. The unit is about 75 m thick.

Arikaree Group

In the highest outcrops north of the White River, such as at Cedar Pass and the Pinnacles in the Badlands National Monument, the Sharps siltstones are capped by sand and silt-filled stream channels that have incised through the Sharps and Rockyford Ash and deeply into the underlying Poleslide (Parris and Green, 1969, Harksen, 1974; complete section from M. F. Skinner, American Museum of Natural History, MS). Harksen (1974) suggested that the same disconformity is present south of the White River near the mouth of Porcupine Creek, where the Godsell Ranch Channel of Macdonald (1963) likewise cuts through the Sharps and deeply into the Poleslide. Skinner (MS) traced a similar disconformity eastward along the south side of the White River to the Craven Basin. Harksen and Macdonald (1969, p. 19) postulated that this disconformity marks a regional event that corresponds to the White River-Arikaree disconformity in Nebraska. South of the White River, the regional paleoslope carries this disconformity beneath deposits designated by Harksen et al. (1961) as the upper part of the Sharps Formation. The upper Sharps consists of nearly 50 m of friable (but locally silica-cemented) massive concretionary siltstone and fine sandstone. It is gradationally overlain by more than 25 m of indurated massive siltstones and very fine sandstones correlated with the Monroe Creek Formation of adjacent Nebraska largely on the basis of their bold, cliff-forming outcrops.

SINGLE-CRYSTAL LASER-FUSION $^{40}AR/^{39}AR$ DATING

As mentioned in the introduction we utilize a number of new single-crystal laser fusion $^{40}Ar/^{39}Ar$ dates in this study (Table 1). We do not use the 26.3 Ma (corrected) K-Ar date reported by Evernden et al. (1964) for an ash at the base of the Gering at Scotts Bluff National Monument (Fig. 4). Our correlations with the magnetic polarity time scale of Cande and Kent (1992) would place ashes in this part of the Gering close to 28.3 Ma. We report a new $^{40}Ar/^{39}Ar$ date of 28.26 ±0.05 Ma on such an ash (Chimney Rock perrierite ash) from the base of the thin-bedded horizontally stratified facies of the Gering. Naeser et al. (1980) reported a fission-track age of 27.2 ± 0.9 Ma for this ash. We also do not use the 27.0 ± 0.7 Ma K-Ar date on the Carter Canyon Ash (CC) obtained by Obradovich et al. (1973) from Helvas Canyon (Fig. 4) on the grounds that stratigraphic and paleomagnetic data argue that this ash should be between 29 and 29.5 Ma old. We use our new $^{40}Ar/^{39}Ar$ date on the Twin Sisters Pumice of 28.31 ± 0.03 Ma in place of the K-Ar date of 27.0 ± 0.6 Ma reported by Obradovich et al. (1973).

The sanidine and plagioclase separated from the Nebraska ashes were each irradiated twice in the hydraulic rabbit facility core of the Omega West research reactor at Los Alamos National Laboratory, with a fast neutron fluence of 5.7 x 10^{13} neutrons/cm². In both irradiations, the minerals were loaded into wells of an aluminum (Al) sample disk along with a centrally located monitor mineral and wrapped in Al foil. Samples 4812 and 4818 were irradiated for 28 hours, and samples 2561 and 2562 were irradiated for 24 hours.

Following irradiation, individual grains were placed in separate wells of a copper sample disk, enclosed within the sample chamber, bolted onto the extraction system, and baked at 200°C for eight hours. Total fusion of the minerals was accomplished with a 6W Coherent Ar ion laser. The released gases were then purified by two Zr-

Table 1. $^{40}Ar/^{39}Ar$ laser fusion ash dates from Morrill County and Wildcat Ridge, western Nebraska. All dates by Carl Swisher. * = radiogenic; SD = 1 standard deviation; (SE) = standard error of mean. $\lambda\varepsilon + \lambda\varepsilon' = 0.581 \times 10^{-10}$ yr; $\lambda\beta = 4.9662 \times 10^{-10}$ yr; $^{40}K/^{40}Ktotal = 1.167 \times 10^{-4}$

Lab number	Mineral	$^{37}Ar/^{39}Ar$	$^{36}Ar/^{39}Ar$	$^{40}Ar*/^{39}Ar$	$\%^{40}Ar*$	Age(Ma)	SD(Ma)
Olsen's third ash. Arikaree Group, undivided. Center NW NW sec. 13, T19N R53W							
4812-01	Sanidine	0.01066	0.00008	0.70232	96.4	27.94	0.81
4812-02	Sanidine	0.01284	0.00013	0.68249	94.7	27.20	0.21
4812-03	Sanidine	0.01323	0.00008	0.69524	96.5	27.66	0.62
4812-04	Sanidine	0.01067	0.00010	0.70066	95.8	27.88	0.15
4812-06	Sanidine	0.01369	0.00007	0.70465	96.9	28.03	0.40
4812-07	Sanidine	0.01416	0.00007	0.70081	96.8	27.88	0.16
					Mean =	27.76	0.30
				Weighted mean =		27.79	0.08 (SE)
Roundhouse Rock pisolitic ash. Gering Formation. NE SW sec. 21, T19N R51W							
4818-06	Sanidine	0.04588	0.00018	0.70142	93.2	27.90	0.11
4818-02	Plagioclase	1.14853	0.00052	0.70409	91.7	28.01	0.37
4818-07	Sanidine	0.00908	0.00010	0.70958	95.7	28.23	0.13
4818-03	Sanidine	0.00758	0.00012	0.71102	95.1	28.28	0.18
					Mean =	28.11	0.18
				Weighted mean =		28.07	0.07 (SE)
4818-08	Sanidine	0.01748	0.00005	0.72968	97.9	29.02	0.09
4818-05	Plagioclase	3.33706	0.02903	80.52205	90.6	1850.70	5.89
Chimney Rock perrierite ash. Gering Formation. NW NW sec. 20, T20N R52W							
2561-02	Sanidine	0.02938	0.00010	0.89260	97.0	28.23	0.11
2561-03	Sanidine	0.04364	0.00023	0.90022	93.0	28.47	0.12
2461-04	Sanidine	0.03485	0.00067	0.89125	81.8	28.19	0.20
2561-06	Sanidine	0.03125	0.00009	0.89243	97.0	28.23	0.12
2561-10	Sanidine	0.02590	0.00026	0.89233	92.1	28.23	0.10
2561-13	Sanidine	2686	0.00013	0.89217	95.9	28.22	0.10
					Mean =	28.26	0.10
				Weighted mean =		28.26	0.05 (SE)
Twin Sisters pumice conglomerate. Gering Formation. NE NE sec. 6, T20N R55W							
2562-02	Sanidine	0.01856	0.00044	0.88987	87.1	28.15	0.11
2562-03	Sanidine	0.03409	0.00019	0.88760	94.2	28.08	0.09
2562-04	Sanidine	0.08135	0.00008	0.89176	97.3	28.21	0.09
2562-05	Sanidine	0.02098	0.00010	0.88812	96.9	28.09	0.20
2562-10	Sanidine	9.11327	0.00007	0.89907	98.5	28.44	0.10
2562-11	Sanidine	0.01300	0.00004	0.89640	98.7	28.35	0.08
2562-12	Sanidine	0.01580	0.00003	0.89953	99.0	28.45	0.08
2562-13	Sanidine	0.01686	0.00003	0.89756	98.9	28.39	0.08
2562-15	Sanidine	3611	0.00009	0.89706	98.0	28.37	0.09
					Mean =	28.28	0.15
				Weighted mean =		28.31	0.03 (SE)
2562-14	Sanidine	0.54865	0.00033	0.919685	94.4	29.08	0.33
2562-01	Sanidine	0.01811	0.00037	0.877881	88.9	27.77	0.36

Fe-V getters operated at approximately 150°C. Argon was measured with an on-line Mass Analyzer Product 215 noble-gas mass spectrometer, operated in the static mode, using automated data-collection techniques. Sample fusion, gas purification, and mass spectrometry were completely automated following computer-programmed schedules.

Ages were calculated using a J value of 2.222 ± 0.0024 x 10^{-2} for the 28-hour irradiation and 1.77 ±

0.002 x 10^{-2} for the 24-hour irradiation, as calculated from replicate analyses of individual grains of the co-irradiated monitor mineral Fish Canyon Tuff sanidine with a reference age of 27.84 Ma (modified from Cebula et al., 1986), intercalibrated in-house with Minnesota hornblende MMhb-1 at 520.4 Ma (Samson and Alexander, 1987). Ca and K corrections used during this study were determined from laboratory salts: $(36/37)Ca = 2.53 \times 10^{-4} \pm 5.0 \times 10^{-6}$, $(39/37)Ca =$

$6.38 \times 10^{-4} \pm 1.0 \times 10^{-5}$ and $(40/39)K = 2.4 \times 10^{-3} \pm 7.0 \times 10^{-4}$. Mass discrimination as determined from replicate analyses from an on-line pipette system during this study yielded a mean value of 1.005 ± 0.0005. Weighted means were derived using the method of Taylor (1982). Age determinations not used in calculation of mean age were considered to be contaminant grains.

MAGNETIC STRATIGRAPHY

Magnetic sampling was conducted by Prothero and a variety of field crews over the last decade. Roundtop-Toadstool, Olsen Ranch, Nuss Ranch, and the Sharps sections at Cedar Pass and Wolff Table were sampled in 1986. The Sharps sections at Sheep Mountain Table, Evergreen, and Palmer Creek were sampled in 1987. The lower part of Roundhouse Rock was sampled in 1989, and the remaining Wildcat Ridge sections (Helvas Canyon, Horn Ranch, and the upper parts of Roundhouse Rock and Scottsbluff) were sampled in 1990. Steve King was responsible for the laboratory analysis and interpretation of the Sharps Formation type sections (Evergreen and Palmer Creek). Tim Tierney analyzed and interpreted the 1990 Wildcat Ridge sections.

In all of these field seasons, sampling was conducted with simple hand tools, recovering three oriented blocks from a single horizon at each site. Sites were located about 5.5 ft. (1.6 m) apart stratigraphically, except where precluded by poor exposures. This resulted in more than 700 magnetic sites and more than 2,100 individual samples. All samples were trimmed into 2-cm cubes on a tungsten-carbide-bladed band saw and analyzed at the paleomagnetics laboratory of the California Institute of Technology.

After measurement of natural remanent magnetization (NRM), a suite of samples was demagnetized using both alternating field (AF) and thermal demagnetization. AF demagnetization (Fig. 2) showed that most samples rapidly declined in intensity at higher AF fields, suggesting that a low-coercivity mineral such as magnetite was the primary carrier of the remanence. This was borne out by stepwise thermal demagnetization (Fig. 2) of most samples. Almost all remanence had disappeared by 580-600°C, which is above the Curie point of magnetite.

Analyses of IRM (isothermal remanent magnetization) acquisition further confirmed that both magnetite and hematite were present in many samples. Most rocks (Fig. 3) reached saturation IRM values at 800 mT (millitesla); the remanence in these rocks is carried mostly by magnetite. A modified Lowrie-Fuller ARM (anhysteretic remanent magnetization) test (e.g., Johnson et al., 1975) was also conducted during the IRM analysis (see Pluhar et al., 1991, for details). This test compares the resistance of AF demagnetization of both an IRM acquired in a 100 mT peak field, and an ARM gained in a 100 mT oscillating field. In almost all samples, the ARM (black squares) demagnetizes at higher peak fields than does the IRM (open squares), indicating that the remanence is carried by single-domain or pseudo-single-domain grains.

Since AF demagnetization does not remove high-coercivity overprints due to iron hydroxides, all remaining samples were subjected to stepwise thermal demagnetization. Overprinted components of magnetization were typically removed by 200°C (Fig. 2), and a characteristic component was isolated between 300-500°C. This component was used for statistical purposes, as described by Fisher (1953) and Butler (1992). After statistical analysis, sites were classified according to the scheme of Opdyke et al. (1977). Class I sites were significantly clustered at the 95% confidence level. Class II sites could not be analyzed because one sample was lost or crumbled. In Class III sites, two samples showed a clear directional preference, but the third site was divergent. Sites which gave highly scattered magnetic results, or ambiguous polarity directions, were classed as indeterminate. The results for each site can be seen by the symbols in Figures 4 and 6.

In almost all cases, the strata were nearly horizontal, so it was impossible to conduct a fold test for stability. However, several reversal tests were conducted, and the reversed sites were antipodal to the normal sites (within the margin of error of the analysis). For example, in the Sharps Formation in Evergreen and Palmer Creek, Class I normal sites had a mean declination (D) of 3.6°, and a mean inclination (I) of 61.1°. The precision parameter (k) was 55.5 and the ellipse of confidence had a radius (α_{95}) of 5.4. Class I reversed sites at these same localities had a mean D = 194.2°, I = -50.3° (k = 48.6, and α_{95} = 5.3). The 1990 Wildcat Ridge normal samples produced a mean D = 354.3°, I = 59.1° (k = 5.9, α_{95} = 7.7), and reversed samples had a mean D = 169.5°, I = -42.3° (k = 10.0, α_{95} = 6.3); these are nearly antipodal. Thus, the component isolated between 300-500°C yields a positive reversal test, and is probably the characteristic component of the remanence.

Stratigraphic and paleomagnetic analysis of all the sections studied in this work shows that the Nonpareil Ash Zone (NPZ) consists of three distinct ash beds covering a 20-m interval, that the lowest ash lies at the top of Chron C11r, and that the remaining lie at the base and top of Chron C11n (Fig. 7). The Rockyford Ash (RF) lies low in Chron C11n and may be correlated with the middle ash of the NPZ (NP_2). Correlation of the lower part of the Sharps Formation with the Brown Siltstone shows that both lie in Chrons 11n-10r. The Wolff Table section (Prothero and Swisher, 1992) indicates that the "Second White Layer" (2W) of the Sharps is still within Chron C10r (Fig. 7). The Nebraska sections show a hiatus in Chron C10r with Chron C10n missing. There is weak evidence of part of Chron C10n in the Nuss Ranch section (Figs. 4 and 7).

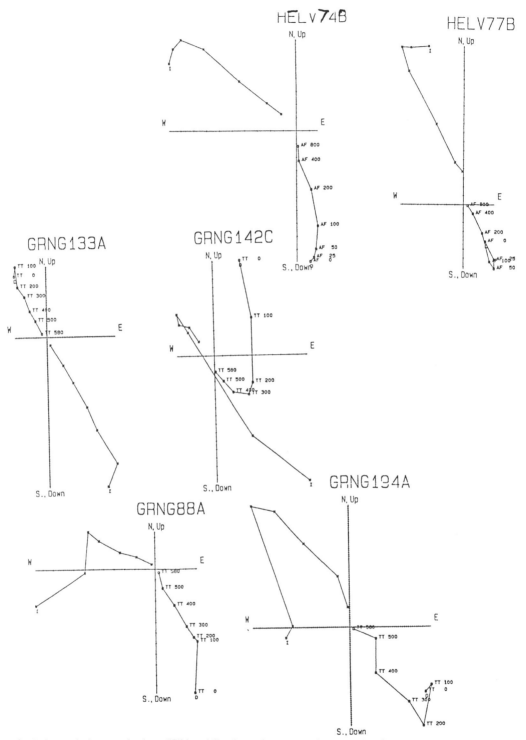

Figure 2. Orthogonal demagnetization ("Zijderveld") plots of representative samples from the Whitneyan-Arikareean transition. Circles show the horizontal component, and asterisks show the vertical component in each plot. Each increment = 10^{-5} emu. In samples HELV74B and 77B, AF demagnetization to 800 Gauss is shown. Note that both samples demagnetized rapidly, indicating that the remanence is carried by a low-coercivity mineral. In samples GRNG133A, 132C, 88A, and 194A, thermal demagnetization to 580°C is shown. Note that after removal of an overprinted component at temperatures up to 200°C, a stable component was isolated between 300-500°C. In most samples, nearly all the remanence was gone at the Curie point of magnetite, suggesting that very little was carried by hematite.

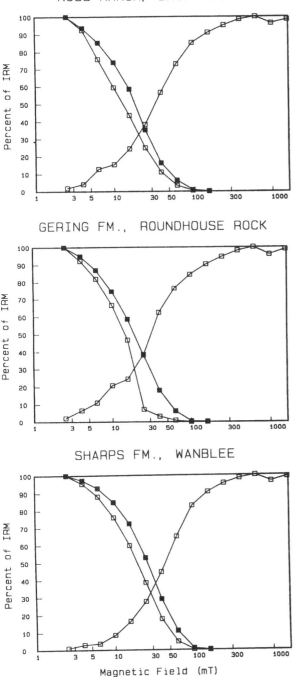

NUSS RANCH, BROWN SILTST.

GERING FM., ROUNDHOUSE ROCK

SHARPS FM., WANBLEE

Magnetic Field (mT)

Figure 3. IRM acquisition analysis and modified Lowrie-Fuller test of three representative samples. IRM indicated by open squares, ARM by solid squares. Note that all three samples reached magnetic saturation (ascending curves on right) by about 800 mT, suggesting that the primary carrier of remanence is magnetite. In addition, the ARM in all three samples (descending curves on left) was more resistant to AF demagnetization than the IRM, indicating that the remanence was carried by single-domain or pseudo-single-domain grains. (See Pluhar et al., 1991, for further details).

However, given the generally poor quality of the paleomagnetic data in the upper half of this section, we do not have much confidence in the presence of Chron C10n here. The paleovalley fluviatile sequence in the Brown Siltstone of the Roundhouse Rock section in the eastern Wildcat Ridge (Swisher's 1982 Unit A; Figs. 4 and 8b) contains a reversal boundary which must be equated with Chron C11n-Chron C10r if our biostratigraphic correlations are correct. Unfortunately, hiatuses separate depositional units in this area (Fig. 5), preventing assembly of a complete section. At the moment, radiometric dating is of no service either because of the large error ($\sigma = 0.96$ Ma) for the Roundhouse Rock Ash (RR) at the top of the postulated Chron C11n at this site.

Correlation of the Gering Formation (Swisher's 1982 Units B-D; Fig. 8b) with the upper Sharps is not so readily tested paleomagnetically for lack of data. Only the Evergreen (Slope) segment of the Sharps type section has a relevant magnetostratigraphy (Fig. 6). The regional unconformity breaking the Sharps Formation occurs in Chron C10n or the top of Chron C10r as normally polarized strata occur in the channels capping the Cedar Pass section (Prothero and Swisher, 1992) and, in truncated form, at the base of unit 14c (Harksen et al., 1961) in the upper 50 m of the Evergreen (Slope) subsection of the Sharps. Reversely polarized strata comprise most of the remainder of the Sharps (the sampled traverse did not reach the top) that must represent Chron C9r. Pumice-bearing Gering Formation reaches upward into the base of Chron C9n in the North Platte Valley (Figs. 4 and 7). The biostratigraphic correlation of the upper Sharps and Gering is supported by these data.

The younger Arikaree strata in the North Platte Valley begin within Chron C9n and our biostratigraphic correlations suggest that the "Monroe Creek" in South Dakota should do the same. The youngest fossil site mentioned below from Scotts Bluff National Monument occurs in a reversed interval that may represent Chron 8r (Figs. 4 and 7).

BIOSTRATIGRAPHY

The Cenozoic rocks of the High Plains contain a rich fossil record that forms the foundation of a biochronologic scheme for the North American Oligocene and Miocene that not only characterizes faunal change in this region but can be extended over the mid-latitudes of the continent (Emry et al., 1987; Tedford et al., 1987). In a previous report, Tedford et al. (1985) assembled biostratigraphic data from the upper White River and lower Arikaree groups of the Niobrara and White River valleys of northwestern Nebraska together with similar information from the Big Badlands of South Dakota. These data were compared and correlations proposed between Nebraska and South Dakota. Very limited paleomagnetic data were available to test the biostrati

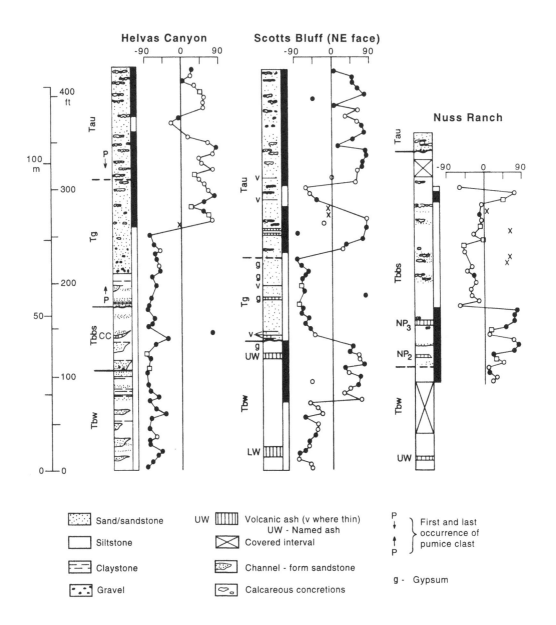

graphic correlation. New information from these areas and from the North Platte Valley in Nebraska now makes it possible to extend the biostratigraphic analysis over most of western Nebraska and southwestern South Dakota. The resulting correlation network can in turn be tested against the more comprehensive paleomagnetic and radiometric data summarized in previous sections of this report.

It has long been recognized that lithologic facies of the Brule Formation, White River Group, contain contrasting faunas (Matthew, 1901; Clark et al. 1967; Wilson, 1975) apparently representing coeval biotopes. The most striking contrasts are between the faunas of the shallow stream channels and their proximal over-

bank facies and those of the interfluve sheet-flood and eolian facies distal to the stream tracts. This phenomenon is best exemplified by the well-known contrast in faunas between the "*Protoceras* channels" and the "*Leptauchenia* clays" within the Poleslide Member of the Brule Formation of South Dakota. Stream channels are rare in the correlative Whitney Member of the Brule Formation and in the lower part of the Sharps Formation and correlative Brown Siltstone beds of the Brule Formation, all of which are dominated by massive to crudely bedded eolian silts. The faunas contained in these rocks contrast with those of the channels in having lower taxonomic diversity. This generalization may be tempered somewhat by the difficulty of

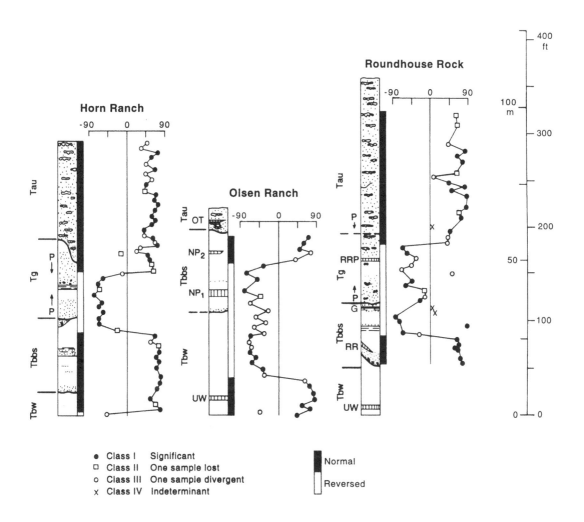

Figure 4. Magnetic polarity stratigraphy of Wildcat Ridge localities, Nebraska. Location of sections as follows: Helvas Canyon (Type section Gering Formation of Vondra and others, 1969)— NW sec. 6, T.20N., R56W.; Scotts Bluff-- E1/2 sec. 33, T.22N., R.55W.; Nuss Ranch—NW SW sec. 18, T.23N., R.53W.; Horn Ranch—NW NE sec. 34, T.20N., R.53W.; Olsen Ranch—SW sec. 21, T.19N., R.51W; Roundhouse Rock—NE SW sec. 21, T. 19N., R. 51W. Abbreviations: Tbw, Brule Formation, Whitney Member; Tbbs, Brule Formation, Brown Siltstone beds; Tg, Gering Formation; Tau, Arikaree Group, undivided; LW, Lower Ash of Whitney Member; UW, Upper Ash of Whitney Member; NP1-3, Nonpareil Ash (1, 2, or 3); RR, Roundhouse Rock Ash; CC, Carter Canyon Ash; G, Grey Ash bed; RRP, Roundhouse Rock Pisolitic Ash; OT, Olson's Third Ash. The pre-Gering portion of the Scotts Bluff section modified from Prothero and Swisher (1992). Reversals shown when indicated by two or more sites.

assessing the diversity of small mammals scattered in the eolian facies in comparison with those in channels where hydraulic processes concentrate the remains.

Rocks of the succeeding Arikaree Group were laid down initially in deeply incised stream valleys, the deposits of which are an intimate mixture of fluviatile and eolian facies. These facies still maintain their faunal characteristics, but are interbedded as depositional tracts shifted across the valley axes (Bart, 1976). Most investigators have tended to collect from bulk units and to ignore the details of the internal stratigraphy of these valley fills so that presently available assemblages are

composites of these faunal facies. The upper part of the Sharps Formation and the conformably succeeding "Monroe Creek Formation" in South Dakota may be an exception since the initial paleovalleys, although deeply incised, were narrow and aggradation soon formed a broad depositional plain in which fluviatile systems were a minor component. The fauna obtained from the massive nodular siltstones of the upper part of the Sharps Formation came from sediments largely deposited by eolian processes, a continuation of the lithotope common in the lower part of the unit. In the North Platte valley of Nebraska, fluviatile systems are impor

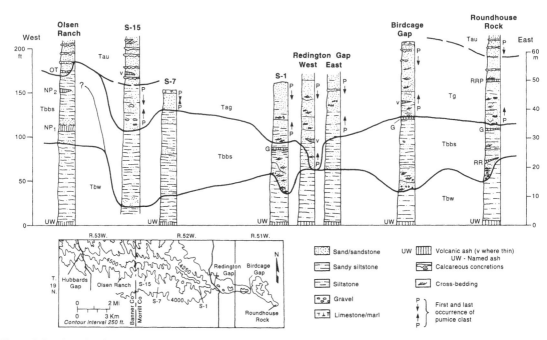

Figure 5. Stratigraphy of Brule and Arikaree strata exposed along southeastern Wildcat Ridge (See Figure 1 for location of Wildcat Ridge). The Brown Siltstone (Tbbs) includes a paleochannel-fill (Unit A of Swisher, 1982) cutting from an undetermined level in the upper part of the Tbbs through the Nonpariel Ash beds (NP_1, and NP_2) and enclosing older Brown Siltstone strata and deeply into the Whitney Member between the Olsen Ranch and S-15 outcrops. Likewise the pumice-bearing Gering (Unit B of Swisher, 1982) fills a paleovalley that locally removed all of the Brown Siltstone in the Redington Gap area. Both paleovalleys are truncated by undivided Arikaree strata in the Olsen Ranch section. Upper Whitney Ash used as a datum for the measured sections. Location of sections not given in Figure 4: S-15, Center NE sec. 18, T. 19N., R. 52W.; S-7, SE SW sec. 17, T. 19N., R. 52W.; S-1, SE NE sec. 22, T. 19N, R. 52W.; Redington Gap, Center N 1/2 SW sec. 14, T. 19N., R. 52W.; Birdcage Gap, NE NE sec. 15, T. 19N., R. 51W. Abbreviations for rock units and ash beds follow those of Figure 4.

tant in the correlative Gering Formation, as they are in the basal Arikaree of the White River valley. Higher in the Arikaree, eolian deposits again become dominant.

Facies localization of taxa and stratigraphic changes in the environment of deposition pose limitations on determination of the range zones of given taxa (Hunt, 1985, p. 188). The ideal would be a long stratigraphic sequence in a single lithotope where hiatuses would be the only complexity to determination of stratigraphic range. The best approximation of the ideal is the Poleslide-Sharps-"Monroe Creek" succession in South Dakota, which lies in the eolian facies for the most part and is broken by a single important hiatus.

In the discussions that follow, the biostratigraphic succession within the White River valley of northwestern Nebraska and the Big Badlands of South Dakota presented in 1985 is re-examined in light of additional evidence. The succession in the North Platte River valley of western Nebraska is added, and a biostratigraphic synthesis of all data for the upper White River and lower Arikaree groups is advanced. As oreodonts are an important part of this record, we follow the taxonomic revisions of CoBabe (this volume, Chapter 26) for leptauchenines and Stevens and Stevens (this

volume, Chapter 25) for merycoidodontines and Stevens (MS) for promerycochoerines.

White River Valley, Nebraska

The biostratigraphy of this area lies within the eolian facies and was based largely on the type section of the Whitney Member of the Brule Formation in the "Round Top to Adelia" traverse in Sioux County (UNSM Loc. Sx-22). In this section the widespread Lower (LW) and Upper (UW) Whitney ash beds have been identified, as has the Nonpareil Ash Zone (NPZ). In outcrop the NPZ is represented by a single ashfall that can be correlated magnetically with the third ash (NP_3) of the thick NPZ of the North Platte River valley. These ash beds thus serve as lithostratigraphic markers to anchor the biostratigraphy and to test correlations based on the latter.

In Tedford et al. (1985), the Whitney and succeeding Brown Siltstone interval was partially subdivided faunally in order to discuss the biostratigraphy as a set of exclusive intervals called Fauna I through III that traced the succession of certain taxa into the base of the Brown Siltstone. Fauna I from the top of the UW to the upper ashy zone of the Whitney (UA), possibly part

Sharps Formation Type Sections

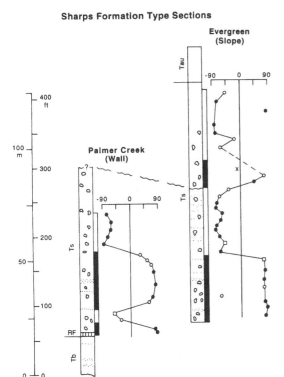

Figure 6. Magnetic polarity stratigraphy of the Sharps Formation type subsections. Lithology modified from Harksen et al. (1961). Lithographic correlation of the subsections proposed by Harksen et al. (1961) is corroborated magnetically. The top of the Palmer Creek subsection approximates the position of the regional disconformity separating the Sharps Formation into two units (see discussion pp. 325-326). This disconformity is identified in the Evergreen subsection by the truncated normally polarized interval postulated to represent Chron C10n (see Fig. 7). Location as follows: Palmer Creek— SW 1/4 NW 1/4 Sec. 31, T. 41N, R. 42W.; Evergreen—NW 1/4 sec. 30 and NE 1/4 sec. 20, T. 39 N., R. 43W.

of NPZ, perhaps NP_1, contains *Hyracodon* and *Leptauchenia decora* plus *Merycoidodon major*. Fauna II, which was obtained from the UA to 4.6 m below the base of what was taken as the NPZ (here correlated with the NP_3 ash of the Wildcat Ridge, Fig. 4), contains the lowest local occurrence of the large *Leptauchenia major* associated with *L. decora* and *Hyracodon*. About 4.6 m below the NPZ, at the base of the interval characterized by Fauna III, *Sespia nitida* has its first local occurrence associated with *Leptauchenia major*. Along the Beaver Wall, 65 km northeast of Roundtop in Dawes County, *?Mesoreodon minor* has its first appearance at the same position below the NPZ if this ash equates with NP_3. These taxa extend for an additional 18.3 m above the NPZ. In the Niobrara River valley, 45 km south of Roundtop in Box Butte County, scattered remains occur in the Brown Siltstone (correlated with the Monroe Creek Formation by Yatkola, 1978) over 61 m above the NPZ. These data show the upward continuation of

the range zones of *Sespia*, *Leptauchenia major* and *?Mesoreodon minor*.

A regional disconformity limits the Brown Siltstone in northwestern Nebraska and the succeeding Arikaree Group fluviatile and eolian deposits were laid down in deep valleys excavated up to 91 m into the White River Group (Swinehart et al., 1985). The scattered faunas in these deposits are much more diverse than those of the upper White River because of the fluviatile component in the valley fill.

The massive silty sandstones representing the eolian facies within this Arikaree valley fill contain the oreodonts *Leptauchenia decora* and *L. major*. The thin and lenticular fine to medium fluviatile sands with lenses of lithic gravels bear a large mammal fauna lacking oreodonts, but containing *Miohippus*, *Diceratherium*, *Arretotherium*, and *Pseudolabis*. These friable channel sands also contain a diverse microfauna exemplified by that described by Martin (1973) from UNSM locality Dw-108, a site low in the Arikaree paleovalley sequence south of Chadron. Among the taxa he recorded from this site are the insectivores *Domnina dakotensis* and *Ocajila makpiyahe*; the rabbits *Palaeolagus philoi* and *?Megalagus primitivus*; and the rodents *Kirkomys schlaikjeri*, *Sanctimus stuartae*, *Leidymys blacki* and *Geringia mcgregori*. Another nearby site (UNSM Dw-121), recently discovered by Bruce Bailey (personal communication, 1995), is stratigraphically lower than Dw-108, in the base of the paleovalley, produces a more diverse fauna, but contains the same taxa including *Heliscomys woodi*. In addition, a species of the allochthonous talpine mole *Scalopoides* is present representing the lowest stratigraphic occurrence of this taxon demonstrated to date. Except for *Scalopoides* and *Heliscomys woodi*, both recorded from younger Arikaree rocks in South Dakota, most of the remainder of the taxa listed have local range zones that are either restricted to or include the upper part of the Sharps Formation (Sharps Fauna C of Tedford et al., 1985, and this chapter; see below) of nearby South Dakota.

Big Badlands, Southwestern South Dakota
(*local range zones of selected taxa shared with Nebraska are shown in Fig. 8a*).

Wilson (1975) contrasted the faunas obtained from the classic "*Leptauchenia* clays" and "*Protoceras* channels" in the Poleslide Member of the Brule Formation in the Palmer Creek Unit of Badlands National Park. In addition to the common occurrence of the oreodonts *Leptauchenia decora* and the large *Merycoidodon major*, the thick-bedded siltstones and claystones distal to the "*Protoceras* channels" contain the rabbit *Palaeolagus*, a round-incisor beaver, *Agnotocastor*, a canid, and the small rhino *Hyracodon* (only *Agnotocastor* was shared between the facies: Wilson 1975, pp. 80-81). Larger collections obtained by Morris Skinner for the American Museum of Natural History from the same

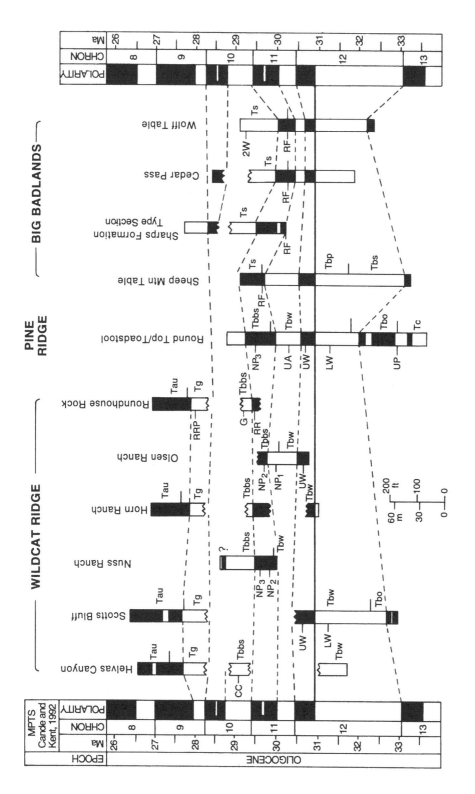

Figure 7. Correlation of Oligocene strata in western Nebraska and southwestern South Dakota with the magnetic polarity time scale (MPTS) of Cande and Kent (1992). The sections are adjusted to the base of Chron 12n which contains the Upper Ash of the Whitney (UW). Round Top, Sheep Mountain Table, Wolff Table, and pre-Arikaree of Scotts Bluff from Prothero and Swisher (1992). Abbreviations not on Figure 4: Tc, Chadron Formation; Tbo, Brule Formation; Orella Member; Tbs, Brule Formation, Scenic Member; Tbp, Brule Formation, Poleslide member; Ts, Sharps Formation; UA, upper ashy-zone of the Whitney; UP, Upper Purplish White bed; RF, Rockyford Ash; 2W, second White Layer.

lithofacies and interval at Cedar Pass in the National Monument were also dominated by *Leptauchenia decora, L. major* and large merycoidodontine oreodonts with rarer *Hyracodon, Miohippus obliquidens,* a carnivore, and peccary. Recent intensive collecting of the Poleslide and lower Sharps by parties from the South Dakota School of Mines has demonstrated the presence of *Leidymys blacki* 15.2 m beneath the Rockyford Ash (P. Bjork, personal communication, 1993) raising the possibility that other Sharps microfauna may also be extended downward in range.

The lithology of the lower part of the Sharps Formation closely resembles the Poleslide Member, but it carries a somewhat more diverse fauna. Tedford et al. (1985) grouped as Sharps Fauna A the material collected by Skinner's and Macdonald's (1963, 1970) parties from the lowest 30.5 m of the Sharps Formation (including material within the basal Rockyford Ash) from outcrops on Sheep Mountain, Cedar Pass and south of the White River within and east of the National Monument (Wolff Ranch area and eastward). These levels contain the last local occurrence of *Hyaenodon* (*H. brevirostris*) and *Hyracodon* (*H. nebraskensis, H. leidyanus*), in addition to the continuation of *Leptauchenia decora* and *L. major*; and the first local occurrences of *Nanotragulus loomisi,* ?*Mesoreodon minor, Diceratherium* (Green, 1958), *Eusmilus olsontau, Shunkahetanka geringensis, Capatanka cankpeopi, Capacikala gradatus* (Parris and Green, 1969), *Palaeocastor nebraskensis, Palaeolagus hypsodus, Palaeolagus philoi* and *Sanctimus stuartae* (P. Bjork, personal communication, 1993).

A second fauna, Sharps Fauna B, was obtained by Morris Skinner and party from the deposits that begin at the ash bed ("Second White layer," 2W) 50 m above the RF and extend through the next 21 m. This assemblage includes the first local occurrence of *Sespia nitida,* associated with *Leptauchenia decora, L. major,* ?*Mesoreodon minor,* the flat-incisor beavers, and *Palaeolagus* species recorded in Fauna A. Tedford et al. (1985) correlated this assemblage with Faunal Unit III of northwestern Nebraska largely on the first occurrence of *Sespia,* but this correlation has been rejected paleomagnetically. *Sespia* is rare in these rocks, making it difficult to determine its range zone. At Cedar Pass, Parris and Green (1969) report a fauna containing the giant entelodont *Dinohyus* sp., and a microfauna with *Domnina, Tamias, Proheteromys, Hitonkala andersontau,* and *Plesiosminthus* from the sand- and silt-filled channel that cuts from 53 m above the Rockyford Ash through the ash and 14 m into the underlying Poleslide (Parris and Green, 1969; Harksen, 1974; complete section from M. F. Skinner, MS). Collections made by M. F. Skinner add *Eusmilus olsontau, Leptomeryx, Palaeocastor* cf. *nebraskensis,* and *Leptauchenia* to the fauna from this channel. The section at Cedar Pass is capped by a final channel-form sandstone only 5 m thick that contains a similar fauna, including *Palaeocastor*

nebraskensis, Cormocyon cf. *copei* and *Leptauchenia.* To the southwest, near the mouth of Porcupine Creek, the Godsell Ranch Channel of Macdonald (1963) has a similar stratigraphic position and likewise cuts through the lower part of the Sharps Formation and deeply into the Poleslide. Macdonald (1963) found a small mammal fauna (SDSM Loc. V5413) from the Godsell Ranch Channel that is similar to that recorded at Cedar Pass, including *Domnina greeni, Tamias, Heliscomys, Proheteromys bumpi, Hitonkala andersontau,* and *Scottimus.* Lateral equivalents of this channel (SDSM Loc. V5410) also yielded additional taxa including *Palaeocastor nebrascensis, Capacikala gradatus, Leidymys blacki, Cormocyon copei, Oxydactylus* cf. *wyomingensis, Leptomeryx, Nanotragulus loomisi,* and *Sespia nitida.* All of these taxa range upward into the upper part of the Sharps Formation.

The local faunas from these channels represent the earliest assemblages from the base of the younger part of the Sharps Formation. At the stratotype of the Sharps Formation this level lies at the top of the first subsection (the "Wall"; Palmer Creek magnetic section, Fig. 6). The widespread topographic bench above the "Wall" escarpment from Porcupine Creek to the east denotes the base of the siltstones and fine sandstones of the middle and upper Sharps Formation that yielded most of the fossils used by Macdonald (1963, 1970) to typify his "Wounded Knee-Sharps Fauna." The biostratigraphic data provided by Macdonald (1963, fig. 2, and descriptions of fossil sites, pp. 153-162) indicate that the middle and upper sites lie above the lower subsection 75 m or more above the Rockyford Ash. In the second subsection of the stratotype (the "Slope"; Evergreen magnetostratigraphic Section, Fig. 6), correlative sites occur in the upper third of the section. A comparison of the local faunas of the middle Sharps with the upper Sharps as listed by Macdonald (1963; new locality data were not stratigraphically localized in Macdonald, 1970) reveals more than 70% similarity at the species level. Thus the term "Wounded Knee-Sharps Fauna" should be applied to the collective assemblage that occurs in the Sharps Formation above the disconformity discussed above. This is equivalent in concept to the Sharps Fauna C of Tedford et al. (1985).

Macdonald (1963, 1970) and his associates assiduously collected smaller mammal remains from the upper part of the Sharps. As a result there are a number of first-occurring taxa at this level, and a major enrichment of the fauna occurs when compared to the lower Sharps. Significantly, these include the appearance of burrowing insectivores (*Arctoryctes* and moles) and burrowing rodents, such as the flat-incisor beavers and diverse geomyoid rodents, the latter outnumbering the cricetids. First occurrences of rodent taxa also found outside South Dakota, and hence valuable in correlation, are: *Paciculus woodi, Geringia mcgregori,* and *Kirkomys*

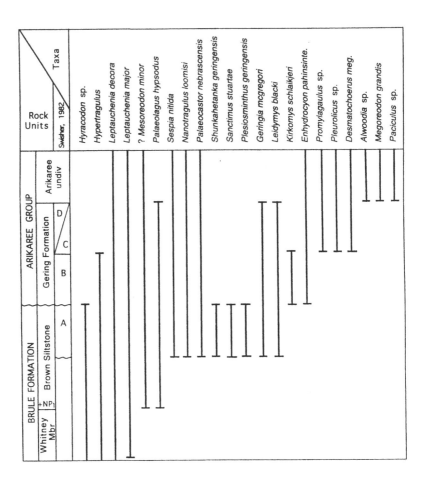

Figures 8a, b. Biostratigraphy of the upper White River through lower Arikaree groups, a) in South Dakota; b) in Nebraska. Stratigraphic ranges of selected taxa common to both South Dakota and Nebraska generalized to the limit of data available to this study. For ease in comparison, lithostratigraphic units are dimensionless and reflect succession only, rather than stratigraphic thickness.

schlaikjeri, which are limited to the upper part of the Sharps. A number of aplodontoid rodents also appear toward the top of the upper Sharps, including species of *Promylagaulus, Meniscomys,* and *Niglarodon.* Among the larger mammals, Fauna C shows the continuation of most of the Sharps Fauna B large taxa, with the addition of *Enhydrocyon pahinsintewakpa, Megoreodon grandis,* and *Oxydactylus* cf. *wyomingensis* near the top of the Sharps. A number of taxa common in the Brule and lower Sharps formations do not occur above the top of the Sharps, such as the nimravid "cats" *Dinictis, Eusmilus,* and possibly *Nimravus* (but see discussion of *N. secator* in Macdonald, 1970), *Palaeolagus hypsodus, Megalagus, Palaeocastor nebraskensis, Capatanka cankpeopi, Capacikala gradatus, Shunkahetanka, Chaenohyus, Elomeryx, Leptochoerus, Leptomeryx,* and *Agriochoerus.*

Gradationally overlying the Sharps Formation are massive siltstones and very fine sandstones correlated

with the Monroe Creek Formation of Nebraska. This unit also bears a reasonably diverse mammal fauna (J. R. Macdonald, 1963, 1970; L. J. Macdonald, 1972), containing many of the same elements seen in the Sharps, reflecting a similar depositional environment. Local first occurrences in this unit of the genera *Amphecinus, Archaeolagus,* ?*Gripholagomys* (Green, 1972), ?*Desmatolagus, Alwoodia* (*A. harkseni,* Korth 1992), *Pleurolicus* (Rensberger, 1973), *Gregorymys, Promartes,* and *Oreodontoides* accompany last generic records within this unit for *Palaeolagus, Proscalops, Geolabis, Meniscomys, Eutypomys, Miohippus, Megoreodon* and possibly leptauchenine oreodonts. In addition, L. J. MacDonald (1972), on the basis of anthill collections below cliff outcrops of the upper part of the "Monroe Creek," recorded *Brachyerix, Metechinus, Parvericus, Trimylus, Scalopoides, Pseudotheridomys* and *Plesiosminthus.* Except for *Scalopoides* and *Plesiosminthus,* the remaining taxa had not previously

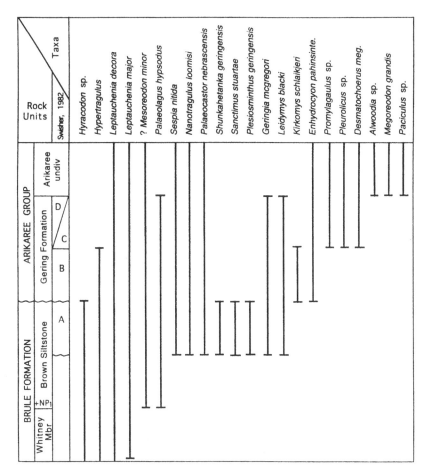

been recorded in the lower Arikaree. All are based on single teeth and could represent float from younger rocks. However, Korth (1992) has reported most of these genera (except *Brachyerix* and *Trimylus*) in the microfaunal sites at McCann Canyon, north-central Nebraska. There they are associated with a fauna otherwise closely comparable to that obtained from the "Monroe Creek" anthills. All of these changes within the "Monroe Creek" coupled with limitations of range of other taxa at the top of the Sharps, indicate that a major faunal turnover took place during the span of this unit (Fig. 8a).

North Platte Valley, Western Nebraska
(*local range zones of selected taxa shared with South Dakota are shown in Fig. 8b*).

Although initially thought to be poorly fossiliferous (Darton, 1899), the Arikaree sequence in the Platte Valley has been intensively prospected by parties from the University of Nebraska over the past 50 years with significant results. Martin's dissertation (1973) contains the first comprehensive review of the faunal succession obtained from the "lower Miocene Gering Formation" in the Wildcat Ridge area on the south side of the North Platte River valley in Scottsbluff, Banner, and Morrill

counties, Nebraska. Vondra's study (1963, and Vondra et al., 1969) of the lithostratigraphy of the upper White River and lower Arikaree groups in this area had been completed as Martin began his work, and this stratigraphic study formed the basis of Martin's biostratigraphy. This lithostratigraphy was too generalized to provide an adequate basis for depicting the biological history of the region at the scale now required by our studies. In 1982, Swisher completed a master's thesis on the lithostratigraphy and biostratigraphy of the eastern Wildcat Ridge in Morrill County where the most fossiliferous Brule and Arikaree sequence is found. Careful mapping by Swisher (1982), and further work by Swinehart and Diffendal (1995) reported herein, have provided a more detailed lithostratigraphic basis for the biostratigraphy.

In this region the eolian facies of the Brown Siltstone beds are typically gradational with the Whitney Member of the Brule Formation and show a consistent trend to slightly coarser grain size. The base can be locally defined in the field and in the subsurface as the first occurrence of silty sand beds above the UW ash. The Nonpareil Ash Zone occurs near the base of the Brown Siltstone and usually is a single ash bed, but at some localities it includes two ash beds separated by as much

as 18 m. In the eastern Wildcat Ridge, the Brown Siltstone beds also contain the fill of a paleovalley (Unit A
of Swisher, 1982) cutting obliquely southeastward
across the line of outcrop (Fig. 5). We infer that the
cutting of this paleovalley has removed the lower 15 to
24 m of the Brown Siltstone (including the NPZ) so
that the biostratigraphic sequence contained in Unit A
begins above the entire ash zone. The fauna of this
paleovalley fill contains the last local occurrence of
Eumys, Hyaenodon, and *Hyracodon*; part of the local
range zones of *Leptauchenia decora* and *L. major*; the
lower part of the local range zones of *Palaeocastor
nebraskensis, Sespia nitida,* ?*Mesoreodon minor,
Shunkahetanka geringensis, Miotylopus gibbi,* and
Nanotragulus loomisi. In addition, there are a number
of lagomorphs and rodents making their first local
appearance at this level: *Palaeolagus hypsodus* and *P.
philoi, Leidymys blacki, Geringia mcgregori,* and the
?primate or ?dermopteran (see McKenna, 1990) *Ekgmo-
wechashala.* *Shunkahetanka geringensis, Sanctimus
stuartae,* and *Plesiosminthus geringensis* appear to be
confined to this interval. *Plesiosminthus* is an
immigrant whose earliest North American record is
based on two specimens from the Durnal Ranch Quarry
assigned by Swisher to his Unit A.

Since disconformities have truncated the Brown Siltstone above the NPZ in the eastern Wildcat Ridge, we
must turn to the north side of the North Platte valley to
find an unbroken succession of massive siltstones above
this ash interval. Outcrops from north of Lake Mina-
tare on the Nuss Ranch in Scottsbluff County northwest to the Spottedtail Creek drainage in Sioux County
show about 70 m of Brown Siltstone beds deposited
conformably on the Whitney (Fig. 4). The NPZ is
usually present near the base, but may be removed by
the stream courses within the paleovalley.

The Sioux County outcrops of the Brown Siltstone
about 30 km west of Nuss Ranch have been the most
consistently collected. The lower part of this section,
including taxa from within the NPZ, contain *Hoplophoneus occidentalis, Hypisodus, Heliscomys, Eumys,*
and *Leptauchenia decora,* also known in older White
River strata, along with *Diceratherium* cf. *armatum.*
Higher in the sequence *Leptauchenia major* occurs with
Palaeolagus hypsodus and ?*Mesoreodon minor*
(including the holotype of a synonym, *Desmatochoerus
sanfordi*). These rocks are disconformably overlain by
20 m of cross-bedded pebbly sands and thin-bedded silty
sandstones that give way upward to massive silty fine
sand containing the fauna from the old Ledingham
Ranch area mentioned by Skinner et al. (1968). This
assemblage contains large leptauchenines, *Desmatochoerus megalodon,* ?*Mesoreodon minor, Megoreodon
grandis,* and *Enhydrocyon pahinsintewakpa* (Wang,
1994, identified as *E. crassidens* in Skinner et al., 1968)
which, along with the lithostratigraphy, indicate correla-

tion with the Gering and post-Gering Arikaree strata of
the Wildcat Ridge to the south.

The Gering Formation is incised as much as 15 to 20
m into the Brown Siltstone along the Wildcat Ridge
(Fig. 5). Its paleochannel axis roughly parallels the
east-to-west trend of the Ridge (Swinehart et al., 1985,
fig. 14). The lower, pumice-bearing part of this unit
(Unit B of Swisher, 1982) contains a locally abundant
fauna, which for the most part represents an upward
continuation of the assemblage contained in the Brown
Siltstone with the addition of *Kirkomys schlaikjeri* and
Enhydrocyon cf. *pahinsintewakpa.* Some taxa, mostly
surviving White River forms, have their last local
appearance in these beds: *Palaeolagus philoi, Heliscomys, Chaenohyus, Agriochoerus, Hypertragulus,* and
Hypisodus.

In the upper part of the Gering Formation (Unit C of
Swisher, 1982) a number of small mammals appear,
which along with the last appearances of taxa from Unit
B, marks a faunal turnover within the Gering. The
newcomers include the mylagaulid *Promylagaulus,* the
beaver *Capatanka,* and the geomyoid *Pleurolicus* cf.
sulcifrons. *Geringia, Palaeolagus hypsodus,* and
Leidymys blacki have their last local appearances in this
unit.

Little is known of the fauna from the thin-bedded fine
sands and silts of Unit D, which seem to lie in a broad,
low-relief paleovalley within the upper part of Unit C.
These beds probably represent a sequence of fluvial
sheet-flood, eolian sand-sheet and local playa deposits.
The entelodont *Dinohyus* occurs at the base of Unit D
and scattered large mammal remains have been obtained
higher in the section from the massive, fine sands that
cap the cycles of thin-bedded to laminated fine sands
filling the paleovalley. *Cormocyon copei, Nimravus
brachyops, Desmatochoerus megalodon,* and *Leptomeryx* sp. were obtained from these rocks in Black
Hank's Canyon (UNSM Loc. Mo-109), northwest of
Redington Gap. Most of these species have ranges that
descend into the Brown Siltstone beds and the Whitney
Member of the Brule Formation, although the occurrence of the large oreodont *D. megalodon* may represent
the initiation of its local range zone in this sequence.

Units C and D grade upward to predominantly
massive, silty fine sandstones of the undivided Arikaree
Group. Swisher found micromammals in anthills on
the lower part of the latter deposits that indicated the
upward continuation of *Proscalops, Promylagaulus,
Palaeocastor nebraskensis, Proheteromys,* and *Pleurolicus* from the Gering and Brown Siltstone. The
occurrence of *Paciculus nebraskensis* marks the first
appearance of the genus in the local section. References
to this genus in older rocks (Martin, 1973, 1980) are
based on species within or close to *Geringia* (Swisher,
1982). Scattered large mammal remains occur in these
rocks, especially in the western part of the Wildcat
Ridge.

At Scotts Bluff National Monument, 27 m of thin-bedded Unit D lies directly on the Whitney Member of the Brule Formation and is overlain by thick-bedded or massive undivided Arikaree (Swinehart and Loope, 1987). Fossil mammal remains occur in eolian deposits 18 to 20 m above the base of the undivided Arikaree, including *Promylagaulus*, *Alwoodia*, *Sespia*, and *Nanotragulus*. At Roubadeau Pass, west of the National Monument, large mammal remains occur in the undivided Arikaree at a similar stratigraphic position, including *Sespia*, *Megoreodon* and *Diceratherium*.

It is clear from the data presented above that the Big Badlands and North Platte valley successions are the longest and most completely documented sequences. The White River valley sequence is shorter and less complete, but informative regarding the ranges of some taxa. It is used in an ancillary fashion in determining the total range zones for some species.

If we inspect the South Dakota and Nebraska sections (Fig. 8a-b), a biostratigraphic correlation readily emerges (Fig. 9). The lower part of the Sharps and Brown Siltstone intervals are correlative. An important faunal enrichment occurs in both the eolian and fluviatile facies. A few Whitneyan taxa have their last appearances at these levels but the striking effect is of augmented diversity, magnified by the visibility of the microfauna in the stream channel deposits (Unit A of Swisher) that disconformably overlie the massive siltstones in the North Platte valley section. The upper part of the Sharps and the Gering formations (units B-D of Swisher) are also correlative; the former shows the upper parts of the ranges of *Geringia*, *Leidymys*, and *Kirkomys* and the base of the range of *Promylagaulus*. In the White River valley of Nebraska the disconformity between the Arikaree and White River strata corresponds to an episode of erosion that removed a considerable part of the Brown Siltstone (all of it in the Arikaree paleovalley axis). Arikaree aggradation in this paleovalley began during the span of deposition of the upper part of the Sharps Formation judging from the co-existence of rodent taxa collected from the lower Arikaree Group in the White River valley.

An important turnover takes place in the uppermost Sharps and "Monroe Creek" and the equivalent upper Gering and overlying undivided Arikaree in Nebraska. In South Dakota most of the long-ranging Sharps taxa, with the exception of the oreodonts, terminate at the top of the Sharps. In Nebraska the correlative level records a similar phenomenon. Evidence from both areas shows that some genera ranging up from the Whitney also reach no higher than the top of the Sharps (e.g., most nimravids, *Leptochoerus*, *Chaenohyus*, *Elomeryx*, *Hypertragulus*, *Leptomeryx*, *Hypisodus*, *Agriochoerus*). This turnover marks the termination of the White River Chronofauna (Emry et al., 1987) and a natural biochronological boundary.

BIOCHRONOLOGY

Having tested our biostratigraphic correlations against other data and found no conflict, we can assemble the local range zones of some taxa shared between Nebraska and South Dakota into a magnetically calibrated composite chronostratigraphic chart (Fig. 9). These data place into temporal context the major aspects of the succession detailed above. A phase of enrichment of the White River chronofauna (Emry et al., 1987) is seen early in the geochron of the Brown Siltstone-latest Poleslide interval; i.e., during Chron C11n beginning at 30 Ma. However, true faunal turnover, with extinction and origination components, takes place later in the sequence, and interestingly, does not correspond to the regional stratigraphic hiatus beneath the Gering and upper Sharps. This turnover event, beginning approximately at 28 Ma, effectively ends the White River Chronofauna through extinction of many of its characteristic elements, including some taxa that appeared during the enrichment episode. The new fauna that emerges in the overlying Arikaree Group includes *Megoreodon* and *Oreodontoides* and a suite of oreodonts known in the older deposits, plus more derived species of genera that are recorded earlier (e.g., *Enhydrocyon crassidens*) and a suite of small mammals, including genera which appeared in the immediately underlying strata (*Promylagaulus* and *Paciculus*) as well as genera that represent new lineages (*Archaeolagus*, *Alwoodia*, *Pleurolicus*, and *Gregorymys*), and others such as *Amphechinus*, *Promartes*, *Pseudotheridomys* and *Gripholagomys*, that appear to be immigrants to the Great Plains. The fauna of the McCann Canyon sites reported by Korth (1992) seems to be closely comparable to that of the "Monroe Creek" so the occurrence there of the European aplodontid *Parallomys* may be one of the expected additions to the list of immigrant taxa for this interval.

When Wood et al. (1941) first proposed the North American "provincial ages" to subdivide the mammalian record of the continent, the Arikareean faunas were poorly known, particularly those from the lower part of the Arikaree Group. Thus faunal typification of this mammal age was based largely on the later part of the interval. The fauna of the Gering was included largely because the Arikareean was envisioned as the geochron of the Arikaree Group, rather than a faunal sequence (for discussion see Tedford et al., 1987). With the data now in hand we can re-evaluate the definition of the lower limit of the Arikareean and the faunal content of the early Arikareean.

Even though our data identify an important faunal turnover event lying at the top of the Gering Formation and extending into the immediately overlying strata, typological considerations persuade us to retain the Gering faunas in the Arikareean as originally suggested. Faunal changes leading to the enrichment in Chron C11n could constitute the characterizing fauna of the

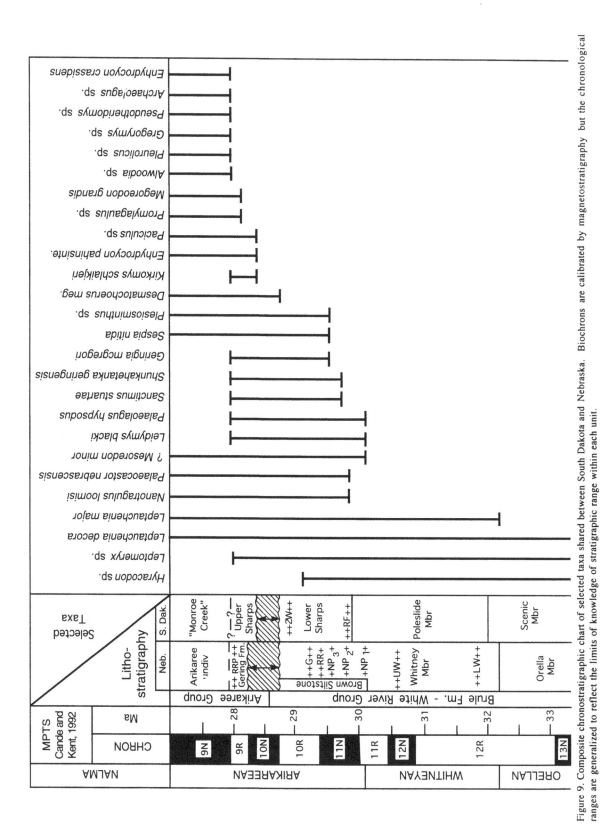

Figure 9. Composite chronostratigraphic chart of selected taxa shared between South Dakota and Nebraska. Biochrons are calibrated by magnetostratigraphy but the chronological ranges are generalized to reflect the limits of knowledge of stratigraphic range within each unit.

earliest Arikareean, with *Plesiosminthus* being the lone immigrant defining the beginning of the age. In addition, the first appearance of a number of autocthonous species, *Palaeolagus hypsodus, P. philoi, Palaeocastor nebrascensis, Shunkahetanka geringensis, Nanotragulus loomisi, Sespia nitida,* and *?Mesoreodon minor,* could be used to characterize the age. The coexistence of these new taxa and surviving Whitneyan forms could serve as a biostratigraphic typification of the earliest part of Arikareean time in the Great Plains (Fig. 9). This characterization would be particularly useful in the Great Plains where eolian lithofacies are widespread and account for a significant percentage of the lithologic column pertaining to this interval (Swinehart et al., 1985).

In their review of the Arikareean, Tedford et al. (1987) designated a number of defining taxa for the beginning of the age. The present study indicates that of these taxa, the galericine hedgehog *Ocajila,* the talpine *Scalopoides,* the sicistine zapodid *Plesiosminthus,* and the aplodontid *Allomys* (here recognized as *Alwoodia*), have different first appearances with regard to the biostratigraphy presented herein. Of the four taxa cited, only *Plesiosminthus* appears in Chron C11n, during the enrichment phase of faunal change recommended here to mark the beginning of the Arikareean. *Ocajila* and *Scalopoides* appear in the Gering and upper Sharps and *Alwoodia* in the "Monroe Creek" as part of the turnover fauna at that level. The characterizing fauna for the early Arikareean listed by Tedford et al. (1987, pp. 184-185, 193) can also be modified with the data provided in this analysis. Of first occurrences listed, most apply to the turnover fauna. Only *Diceratherium, Nanotragulus, Sespia* ("hypsodont leptauchenine oreodonts"), and "*Mesoreodon*" remain with demonstrated first appearances in Chron C11n.

Tedford et al. (1987) also noted that the early Arikareean could be subdivided and that the later part could be defined by the first appearance of the hedgehog *Parvericius,* the eomyid *Pseudotheridomys,* the mustelid *Promartes,* and the chalicothere *Moropus.* These taxa, along with the autochthonous lineages mentioned above represent the turnover fauna distinguished here and noted by Korth (1992) in his analysis of the correlative fauna of the McCann Canyon sites in north-central Nebraska.

CONCLUSIONS

In this study we have brought together lithostratigraphic, biostratigraphic, and magnetostratigraphic data, to elucidate the details of faunal change through an interval of medial Oligocene time that records the transition from the White River Chronofauna (Emry et al., 1987) to new assemblages that are part of the Arikareean "age." These events are most completely recorded in the upper White River and lower Arikaree group rocks exposed in western Nebraska and adjacent South Dakota. An ^{40}Ar/^{39}Ar calibrated magnetostrati-

graphy indicates that the succession is broken by a single regional disconformity of less than 0.5 m.y. duration that removed most of the evidence for Chron C10n. In South Dakota this disconformity breaks the Sharps Formation of Harksen et al. (1961) into two parts, the lower of which (including its basal Rockyford Ash) is continuous with the Poleslide Member of the Brule Formation beneath it. Correlative rocks in Nebraska are the recently recognized (Swinehart et al., 1985), informally designated, Brown Siltstone beds. Both the lower Sharps and Brown Siltstone are lithogenetically parts of the Brule Formation of the White River Group. Above the regional disconformity the upper part of the Sharps Formation of South Dakota grades upward into Arikaree Group strata locally referred to the Monroe Creek Formation of Nebraska. Correlative rocks in the North Platte Valley of Nebraska are the Gering Formation and the conformably overlying undivided Arikaree Group.

Although the lower Arikaree Group deposits are on average coarser-grained than the upper White River sediments, both have a high glass-shard content and are dominated by eolian facies (loess and sand-sheets) that accumulated on broad, low-relief interfluves under a seasonal climatic regime. Fossil remains are usually scattered in eolian deposits and only the common mammalian taxa are well represented, often by partial skeletons. Many mammals entombed in these rocks show special adaptations to dry environments such as burrowing habit for smaller mammals (insectivores and rodents such as castorids and geomyoids). Large mammals show low taxic diversity as contrasted with the faunas obtained from contemporary fluviatile deposits. Strong lithofacies localization of taxa makes it difficult to determine biostratigraphic range zones if contrasting sedimentary environments infrequently alternate in the stratigraphic column. Nevertheless the dominance of eolian environments in the mid-continent Oligocene makes it possible to reconstruct the biostratigraphy of the eolian lithofacies, a situation most completely realized in the South Dakota sequence.

Analysis of the biostratigraphic record for the interval studied reveals that the fauna of the highest part of the White River Group (29 Ma) not only shows continuation of most genera typical of the White River Chronofauna (often as derived species) but includes the appearance of new taxa in autochthonous clades (hesperocyonine canids, oreodonts, camels, hypertragulids, burrowing castorid and geomyoid rodents) that are inserted into the fauna enriching taxic diversity. This enriched fauna persisted across the regional depositional hiatus into the base of the Arikaree Group. Faunal turnover (extinction and origination) takes place approximately at the conformable contact of the Gering and Sharps formations with the overlying Arikaree Group strata, at approximately 28 Ma. At that time most of the White River genera leave the record thus terminating the White River Chronofauna. Some genera that appeared in the enrichment phase of the White River Chronofauna continue into the younger Arikareean (especially tragulids, oreodonts, and burrowing beavers) but the fauna that emerges contains new clades within autochthonous lineages (hesperocyonine canids, large oreodonts, anthracotheres, giant entelodonts, mylagaulid, geomyoid and cricetid rodents, and rabbits), plus allochthonous lagomorphs, insectivores, aplodontid and eomyid rodents, arctoid carnivores, and chalicotheres that invaded mid-continental North America from sources outside the region contributing to the fossil record.

The biological changes summarized above, and in more detail in the preceding text, indicate that an important phase of reorganization of the mammal fauna of mid-latitude interior North America was taking place in mid-Oligocene time. These changes were spread over a 2-m.y. span (28-30 Ma) almost exactly coincident with the last half of the longest glacial episode of the Cenozoic (Miller et al., 1991). According to the data assembled by Prothero (1994), other Oligocene biotas lack evidence of a comparable response, even though eustatic regression should have widely exposed continental shelves and increased the isolation of continental interiors. Gradients of the coastal segments of rivers must have increased, but it is moot whether this would translate base level change nearly two thousand kilometers into the North American continent to account for the White River-Arikaree disconformity. There is clear evidence from the Great Plains that coarser epiclastic debris was contributed to the initial aggradation in the Arikaree paleovalleys, and that the cores of the Front Ranges and their volcanic edifices were being tapped by the headward erosion of Arikaree streams. White River and Arikaree sedimentary blankets ramp up onto (or can be projected onto) the Front Range often with little or no tectonic offset (Evanoff, 1990; Leonard and Langford, 1994) suggesting little differential uplift during the Oligocene and thus a lack of tectonic influence on fluviatile systems (Gregory and Chase, 1994). These observations seem to focus attention on a climate-driven shift to drier and more seasonal regimes, evident from Whitneyan to earliest Arikareean time (Retallack, 1983). A decrease of plant cover may have allowed deep erosion of the higher gradient alluvial apron near the mountain front. Unfortunately the region of interest lacks a detailed botanical record of the critical period that could be used to test a climate-driven model of Oligocene degradation and aggradation.

In 1941 when Wood and others proposed the "provincial" North American Land Mammal "ages," the base of the Arikareean was set at the lower limit of the Arikaree Group, essentially the base of the Gering Formation in the Wildcat Ridge of western Nebraska. Meager knowledge of the fauna of this level led to a biologic characterization of the Arikareean based on the better known assemblages from the upper part of the Arikaree Group. This situation prevailed for nearly fifty years until it was re-examined by Tedford et al. (1987) using the data on "Gering" mammals provided by Martin (1973), the detailed biostratigraphy of Swisher (1982), and the data published by Macdonald (1963, 1970) on the Sharps Formation biostratigraphy. A number of allochthonous insectivores and rodents were proposed to define the beginning of the Arikareean. The record detailed here shows that the basal Arikaree Group retains an augmented White River Chronofauna and that the defining taxa of Tedford et al. (1987) have first appearances that are scattered from these levels to those higher in the Arikaree. Only *Plesiosminthus* accompanies the earliest records of the suite of autochthonous taxa that appear in the Brown Siltstone and its correlate the lower Sharps Formation. It is proposed here that the initiation of the Arikareean Mammal "age" is signaled by the first appearance of these taxa that enrich the White River Chronofauna in the latest part of Chron C11r and early part of Chron C11n, at approximately 30 Ma. The "age" is defined by the first appearance of *Plesiosminthus* and characterized by the appearances of *Palaeolagus hypsodus*, *P. philoi*, *Palaeocastor nebraskensis*, *Shunkahetanka geringensis*, *Nanotragulus loomisi*, *Sespia nitida*, and *?Mesoreodon minor*. The coexistence of these taxa, and those surviving the Whitneyan, serve as a biochronological typification of earliest Arikareean time in the Great Plains. The turnover event that terminates the White River Chronofauna could be used to define and characterize the beginning of the medial Arikareean "age."

ACKNOWLEDGMENTS

Prothero was supported by grants from the donors of the Petroleum Research Fund of the American Chemical Society, National Science Foundation grant EAR87-08221, and funds from the University of Nebraska-Lincoln Conservation and Survey Division, during this research. We thank Dr. J. L. Kirschvink of the California Institute of Technology for access to his paleomagnetics laboratory. The Nuss Ranch and Olsen Ranch samples were analyzed by Christopher Brown and Ka Wing Kam. Without the help of the following field crews, the paleomagnetic sampling would never have been possible: 1986—Allison Kozak, Dana Gilchrist, and Kecia Harris; 1987—John Foster, Susan Briggs, and Jill Bush; 1990—Erin Campbell and Dani Crosby.

We thank R. F. Diffendal and V. L. Souders for their help in the field and suggesting important revisions in the manuscript. E. Evanoff, R. M. Hunt Jr., L. D. Martin, and S. D. Webb critically reviewed the manuscript and we are grateful for their insightful comments.

LITERATURE CITED

Bart, H. A. 1976. Sedimentology of cross-stratified sandstones in Arikaree Group, Miocene, southeastern Wyoming. Sedimentary Geology 19:165-184.

Bennison, A. P., and P. A. Chenoweth. 1984. Geological Highway Map—Northern Great Plains Region. American Association of Petroleum Geologists, Tulsa, OK.

Bump, J. D. 1956. Geographic names for members of the Brule Formation of the Big Badlands of South Dakota. American Journal of Science 254:429-432.

Butler, R. F. 1992. Paleomagnetism. Blackwell, Boston.

Cande, S. C. and D. V. Kent. 1993. A new geomagnetic time scale for the late Cretaceous and Cenozoic. Journal of Geophysical Research 97:13917-13951.

Cebula, G. T., M. J. Kunk, H. H. Mehnert, C. W. Naeser, J. D. Obradovich, and J. F. Sutter. 1986. The Fish Canyon Tuff, a potential standard for the ^{40}Ar/^{39}Ar and fission-track methods. Terra Cognita 6:139-140.

Clark, J., J. R. Beerbower, and K. K. Kietzke. 1967. Oligocene sedimentology, stratigraphy, paleoecology and paleoclimatology in the Big Badlands of South Dakota. Fieldiana: Geology Memoirs 5:1-158.

Darton, N. H. 1899. Preliminary report on the geology and the water resources of Nebraska west of the one hundredth and third meridian. United States Geological Survey, 19th Annual Report 1897-1898, Part 4, pp. 719-814.

Emry, R. J., P. R. Bjork, and L. S. Russell. 1987. The Chadronian, Orellan and Whitneyan North American Land Mammal Ages; pp. 118-152 in M. O. Woodburne (ed.), Cenozoic Mammals of North America: Geochronology and Biostratigraphy. University of California Press, Berkeley.

Evanoff, E. 1990. Early Oligocene paleovalleys in southern and central Wyoming: evidence of high local relief on the late Eocene unconformity. Geology 18:443-446.

Evernden, J. F., D. E. Savage, G. H. Curtis, and G. T. James. 1964. Potassium-argon dates and the Cenozoic mammalian chronology of North America. American Journal of Science 262:145-198.

Fisher, R. A. 1953. Dispersion on a sphere. Proceedings of the Royal Society A217:295-305.

Green, M. 1958. Arikareean rhinoceroses from South Dakota. Journal of Paleontology 32:587-594.

Green, M. 1972. Lagomorpha from the Rosebud Formation, South Dakota. Journal of Paleontology 46:377-385.

Gregory, K. M., and C. G. Chase. 1994. Tectonic significance of a late Eocene low-relief, high-level geomorphic surface, Colorado. Journal of Geophysical Research 99:20141-20160.

Harksen, J. C., J. R. Macdonald, and W. D. Sevon. 1961. New Miocene formation in South Dakota. South Dakota Geological Survey, Report of Investigation No. 3, 11 pp.

Harksen, J. C. and J. R. Macdonald. 1969. Guidebook to the major Cenozoic deposits of Southwestern South Dakota. South Dakota Geological Survey, Guidebook 2, 103 pp.

Harksen, J. C. 1974. Miocene channels in the Cedar Pass area, Jackson County, South Dakota. South Dakota Geological Survey, Report of Investigations, no. 111, 10 pp.

Hatcher, J. B, 1902. Origin of the Oligocene and Miocene deposits of the Great Plains. Proceedings of the American Philosophical Society 41:113-131.

Hunt, R. M. Jr. 1985. Faunal succession, lithofacies, and depositional environments in Arikaree rocks (lower Miocene) of the Hartville Table, Nebraska and Wyoming. Dakoterra 2:155-204.

Johnson, H. P., W. Lowrie, and D. V. Kent. 1975. Stability of anhysteretic remanent magnetization in fine and coarse magnetite and maghemite particles. Geophysical Journal of the Royal Astronomical Society 41:1-10

Korth, W. W. 1992. Fossil small mammals from the Harrison Formation (late Arikareean: earliest Miocene), Cherry County, Nebraska. Annals of the Carnegie Museum 61:69-131.

Leonard, E. M., and R. P. Langford. 1994. Post-Laramide deformation along the eastern margin of the Colorado Front Range—a case against significant faulting. The Mountain Geologist 31:45-52.

Loope, D. B. 1986. Recognizing and utilizing vertebrate tracks in cross section: Cenozoic hoofprints from Nebraska. Palaios 1:141-151.

Macdonald, J. R. 1963. The Miocene faunas from the Wounded Knee area of western South Dakota. Bulletin of the American Museum of Natural History 125:139-238.

Macdonald, J. R. 1970. Review of the Miocene Wounded Knee faunas of southwestern South Dakota. Bulletin of the Los Angeles County Museum of Natural History 8:1-82.

Macdonald, L. J. 1972. Monroe Creek (Early Miocene) microfossils from the Wounded Knee area, South Dakota. South Dakota Geological Survey, Report of Investigations 105:1-43.

Martin, L. D. 1973. The mammalian fauna of the lower Miocene Gering Formation of Western Nebraska and the early evolution of the North American Cricetidae. Ph.D. dissertation, University of Kansas, 219 pp.

Martin, L. D. 1980. The early evolution of the Cricetidae in North America. University of Kansas, Paleontological Contributions 102:1-42.

Matthew, W. D. 1901. Fossil mammals of the Tertiary of northeastern Colorado. Memoirs of the American Museum of Natural History 1:355-447.

McKenna, M. C. 1990. Plagiomenids (Mammalia: ?Dermoptera) from the Oligocene of Oregon, Montana, and South Dakota, and the middle Eocene of northwestern Wyoming. Geological Society of America Special Paper 243:211-234.

Miller, K. B., J. D. Wright, and R. G. Fairbanks. 1991. Unlocking the Ice House: Oligocene-Miocene oxygen isotopes, eustasy, and margin erosion. Journal of Geophysical Research 96:6829-6848.

Naeser, C. W., G. A. Izett, and J. D. Obradovich. 1980. Fission-track and K-Ar ages of natural glasses. United States Geological Survey Bulletin 1489:1-31.

Obradovich, J. D., G. A. Izett, and C. W. Naeser. 1973. Radiometric ages of volcanic ash and pumice beds in the Gering sandstone (earliest Miocene) of the Arikaree Group, southwestern Nebraska. Geological Society of America Abstracts with Programs 5:499-500.

Opdyke, N. D., E. H. Lindsay, N. M. Johnson, and T. Downs. 1977. The paleomagnetism and magnetic polarity stratigraphy of the mammal-bearing section of Anza-Borrego State Park, California. Quaternary Research 7:316-329.

Parris, D. C. and M. Green. 1969. Dinohyus (Mammalia: Entelodontidae) in the Sharps Formation, South Dakota Journal of Paleontology 43:1277-1279.

Pluhar, C. J., J. L. Kirschvink, and R. W. Adams. 1991. Magnetostratigraphy and clockwise rotation of the Plio-Pleistocene Mojave River Formation, central Mojave Desert, California: San Bernardino County Museum Association Quarterly 38(2):31-42.

Prothero, D. R. 1994. The Eocene-Oligocene Transition: Paradise Lost. Columbia University Press, New York.

Prothero, D. R., and Swisher, C. C. III. 1992. Magnetostratigraphy and geochronology of the terrestrial Eocene-Oligocene transition in North America; pp. 46-73 in D. R. Prothero and W. A. Berggren (eds). Eocene-Oligocene Climatic and Biotic Evolution, Princeton University Press, Princeton, N.J.

Rensberger, J. M. 1973. Pleurolicine rodents (Geomyoidea) of the John Day Formation (Oregon) and their relationships to taxa from the early and middle Miocene, South Dakota. University of California Publications in Geological Sciences 102:1-95.

Retallack, G. J. 1983. Late Eocene and Oligocene paleosols from Badlands National Park, South Dakota. Geological Society of America, Special Paper 193:1-82.

Samson, S. D. and E. C. Alexander, Jr. 1987. Calibration of the interlaboratory ^{40}Ar/^{39}Ar standard, MMhb-1, Chemical Geology (Isotope Geoscience Section) 66:27-34.

Schultz, C. B. and T. M. Stout. 1955. Classification of Oligocene sediments in Nebraska. Bulletin of the Nebraska State Museum 4:17-52.

Skinner, M. F. (MS). American Museum of Natural History, Department of Vertebrate Paleontology, Archives, section book, Volume 3.

Skinner, M. F., S. M. Skinner, and R. J. Gooris. 1968. Cenozoic rocks and faunas of Turtle Butte, southcentral South Dakota. Bulletin of the American Museum of Natural History 138:379-436.

Souders, V. L. 1981. Geology and groundwater supplies of southern Dawes and northern Sheridan counties. University of Nebraska Conservation and Survey Division Open File Report, 125 pp.

Souders, V. L. 1986. Geologic sections, groundwater maps, and logs of test holes, Morril County, Nebraska. University of Nebraska Conservation and Survey Division Open File Report, 90 pp.

Stevens, M. S. (MS) Re-evaluation of taxonomy and phylogeny of some oreodonts (Arctiodactyla, Merycoidodontidae): Part II, Promerycochoerinae and Merycochoerinae.

Swinehart, J. B., and R. F. Diffendal. 1995. Geologic map of Morrill County, Nebraska. United States Geological Survey, Miscellaneous Investigations Series Map I-2496.

Swinehart, J. B., and D. B. Loope. 1987. Late Cenozoic geology along the summit to museum hiking trail, Scotts Bluff National Monument, western Nebraska. Geological Society of America Centennial Field Guide-North Central Section, pp. 13-18.

Swinehart, J. B., V. L. Souders, H. M. DeGraw, and R. F. Diffendal. 1985. Cenozoic paleogeography of western Nebraska; pp. 209-229 in R. M. Flores and S. S. Kaplan (eds.), Cenozoic Paleogeography of West-Central United States. Rocky Mountain Section-SEPM, Denver, CO.

Swisher, C. C. 1982. Stratigraphy and biostratigraphy of the eastern portion of the Wildcat Ridge, western Nebraska. M. S. thesis, University of Nebraska, Lincoln, 172 pp.

Swisher, C. C., III and D. R. Prothero. 1990. Single crystal ^{40}Ar/^{39}Ar dating of the Eocene-Oligocene transition in North America. Science 249:760-762.

Taylor, J. R. 1982. An Introduction to Error Analysis. University Science Books, Mill Valley, CA.

Tedford, R. H., J. B. Swinehart, R. M. Hunt, Jr., and M. R. Voorhies. 1985. Uppermost White River and lowermost Arikaree rocks and faunas, White River Valley, northwestern Nebraska and their correlation with South Dakota. Dakoterra 2:335-352.

Tedford, R. H., T. Galusha, M. F. Skinner, B. E. Taylor, R. W. Fields, J. R. Macdonald, J. M. Rensberger, S. D. Webb, and D. P. Whistler. 1987. Faunal succession and biochronology of the Arikareean through Hemphillian interval (late Oligocene through earliest Pliocene epochs) in North America; pp. 153-210 in M. O. Woodburne (ed)., Cenozoic Mammals of North America, Geochronology and Biostratigraphy. University of California Press, Berkeley.

Vondra, C. F. 1963. The stratigraphy of the Gering Formation in the Wildcat Ridge in Western Nebraska. Ph.D. dissertation, University of Nebraska, Lincoln, 155 pp.

Vondra, C. F., C. B. Schultz, and T. M. Stout. 1969. New members of the Gering Formation (Miocene) in western Nebraska. Nebraska Geological Survey Paper 18:1-18.

Wang, X.-M. 1994. Phylogenetic systematics of the Hesperocyoninae (Carnivora: Canidae), Bulletin of the American Museum of Natural History 221:1-207.

Wilson, R. W. 1975. The National Geographic Society-South Dakota School of Mines and Technology Expedition into the Big Badlands of South Dakota, 1940. National Geographic Society Research Reports, 1890-1954 Projects, pp. 79-85.

Wood, H. E., R. W. Chaney, J. Clark, E. H. Colbert, G. L. Jepsen, J. B. Reeside, Jr., and C. Stock. 1941. Nomenclature and correlation of the North American continental Tertiary. Bulletin of the Geological Society of America 52:1-48.

Yatkola, D. A. 1978. Tertiary stratigraphy of the Niobrara River Valley, Marsland Quadrangle, western Nebraska. Nebraska Geological Survey Paper 19:1-66.

PART II

COMMON VERTEBRATES OF THE WHITE RIVER CHRONOFAUNA

16. Testudines

J. HOWARD HUTCHISON

ABSTRACT

Of the eleven turtle taxa recognized from the White River Group, eleven occur in the Chadronian, five in the Orellan and three in the Whitneyan. *Chrysemys inornata* Loomis 1904 and *Graptemys cordifera* Clark 1937 are synonymized and placed in *Pseudograptemys*, new genus. *Trachemys*? *antiqua* Clark 1937 is transferred to *Chrysemys*. *Testudo praeextens* Lambe 1913, *T. quadratus* Cope 1885, *T. thomsoni* Hay 1908 are included in *Gopherus* (*Oligopherus*, new subgenus) *laticuneus* (Cope 1873). Despite the addition of records of aquatic turtles to the Orellan and Whitneyan, the general pattern of reduction of the aquatic turtle fauna between the Chadronian and Orellan (Eocene-Oligocene) transition remains marked.

INTRODUCTION

The White River Group occupies an historical and geographic center for studies of fossil turtles in North America. The first described North American nonmarine turtle (*Emys nebrascensis* Leidy 1851) was based on fossils from the White River Group. The group also transits a critical time in the record of North American turtles, recording the demise of the great Eocene turtle diversity in the continental interior (Hay, 1908; Hutchison, 1992). The major change occurs across the Chadronian/Orellan (Eocene/Oligocene) boundary interval (Hutchison, 1992; Prothero and Swisher, 1992) and sets the stage for the evolution of the modern turtle fauna.

The scope of this paper is centered on the White River Group in the strict sense, that is, the Chadron and Brule formations and correlatives in North and South Dakota, Nebraska, Wyoming, and Colorado (Emry et al., 1987). These units contain the most complete and richest collection of turtles during the represented temporal interval (Chadronian-Whitneyan) in North America. Temporal calibrations follow Prothero and Swisher (1992). The following systematic discussion is based primarily upon a survey of the literature and a few selected specimens seen or readily at hand. Despite the long history of study, much work remains to be done on the description, systematics, and stratigraphic and geographic distribution of the turtles from the White River Group. Many unseen and undescribed specimens now exist in collections that will modify, expand, or correct interpretations used below.

ABBREVIATIONS

Institutions: ACM = Pratt Museum of Natural History, Amherst College, Amherst; AMNH and F:AM = Department of Vertebrate Paleontology and Frick collection, American Museum of Natural History, New York; CM = Carnegie Museum, Pittsburgh; DMNH = Denver Museum of Natural History; FMNH and FMNH PM and PR = Field Museum of Natural History, Chicago; NMC = Canadian Museum of Nature, Ottawa; SDSM = Museum of Geology, South Dakota School of Mines and Technology, Rapid City; UCM = University of Colorado Museum, Boulder; UCMP = University of California Museum of Paleontology, Berkeley; USNM = Department of Paleobiology, United States National Museum, Washington; YPM and YPM (PU) = Yale Peabody Museum (Princeton collection), New Haven. Elements: Bones: C = costal (e.g. C1 = first costal), ENT = entoplastron, EPI = epiplastron, HYO = hyoplastron, HYPO = hypoplastron, N = neural, P = peripheral, XIPH = xiphiplastron. Scales: ABD = abdominal, FEM = femoral, GUL = gular, HUM = humeral, M = marginal, PEC = pectoral, PL = pleural, V = vertebral.

MATERIALS AND METHODS

Diagnoses of the genera and species are meant only to separate each taxon from other closely related taxa within the White River Group and may or may not be more generally applied. The author of a taxon is not separated from the Latin taxon name by a comma but subsequent cited authors are. Terminology follows Zangerl (1969) for bones and scales and Tinkle (1962) for the carapacial scale sulci (seam) fomulae. For neural formulae, the number of contacts and direction of the parenthesis indicates the wide end of the neural [e.g., "6)" indicates that the neural is hexagonal with the wide end posteriorly].

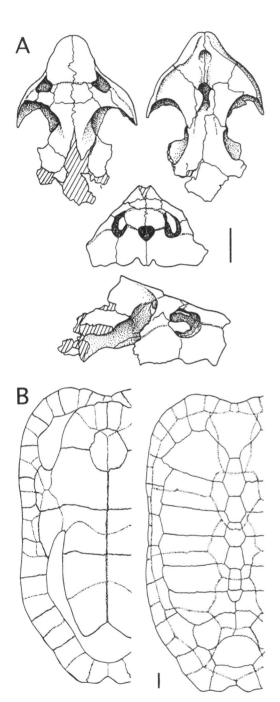

SYSTEMATIC PALEONTOLOGY
TESTUDINES
CRYPTODIRA
Family KINOSTERNIDAE Baur 1893
Genus *Xenochelys* Hay 1906

Diagnosis (from Hutchison, 1991)—Plastron lobes moderately broad; plastron length 67–78% of carapace length; ENT present; intergular present; no anterior or posterior lobe kinesis; HUM undivided; scar for the origin of the acromial ligament restricted to the area of the EPI-ENT suture; EPI with anteriolaterally, anteriodorsally directed projection; HUM-FEM sulcus lies on anterior moiety of HYPO; moderate to well-developed caudal notch between XIPH; C2 not contacting P3; carapace distinctly tricarinate to smooth; C1 rib enters middle or posterior moiety of P3 and terminates in anterior P4 or P3-P4 suture; six neurals; perimeter length of nuchal longer to shorter than any peripheral; distinct cusp to no cusp at posteriolateral margin of FEM; posterior buttress terminates in anterior two-thirds of P7; GUL-HUM sulcus on anterior moiety of ENT; M10 distinctly elevated.

Xenochelys formosa Hay 1906
Figure 1

Xenochelys formosa Hay 1906:29, figs. 2-3.
Type—AMNH 1097, complete shell (Hay, 1906, figs. 2-3; 1908, figs. 355-356; Hutchison, 1991, fig. 1), Quinn Draw, Shannon County (previously part of Washington County), South Dakota, Chadron Formation, Chadronian.
Diagnosis (from Hutchison, 1991)—Carapace length to 208 mm; plastron about 82% of carapace length; plastron and carapace relatively thin to moderately robust; carapace carinae weak or absent; EPI wider than long; suture between the XIPH and HYPO vertical and with relatively shallow lateral interlocking teeth; height of M9 averages about 62% of M10; M10 and M12 about equally high; V1-PL1 sulcus contacts M1 on P1 or very near the nuchal-P1 suture; N4 contacts C5; anterior outline of carapace in dorsal view with cephalic concavity; perimeter length of nuchal not longer than any peripheral; no cusp formed at posteriolateral margin of FEM; intergular separates gulars; well-developed caudal notch between XIPH.
Referred material—South Dakota, Chadron Formation, Harding County—FMNH (PR) 1404, XIPH fragment; Pennington County—CM 27435, crushed shell; NE of Hutenmacher Table, Indian Creek drainage, Crazy Johnson Member, YPM (PU) 13686, partial skull (Williams, 1952, plate 2; Gaffney, 1979, fig. 169). Wyoming, Yoder Formation, Goshen County—SDSM loc. V5369: SDSM 8662, partial shell.
Comment—Hutchison (1991) recently revised the genus and included species on the basis of the shell, noting that the type specimen is anomalous in having seven rather than eight costal pairs. Williams (1952a, p. 2; Gaffney, 1979: fig. 169) described and figured the only known skull.

Figure 1. *Xenochelys formosa.* A. AMNH 1097 (type), shell, dorsal and ventral views (after Hay, 1908). B. YPM (PU) 13686, partial skull, dorsal, ventral, anterior and right lateral views (after Williams, 1952). Scale lines equal 1 cm.

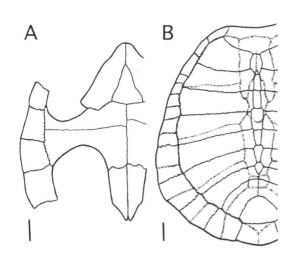

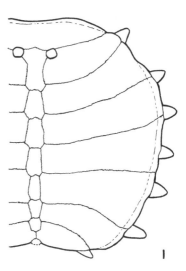

Figure 2. Anosteirinae. A. *Anosteira ornata*, AMNH 6132 ventral view of plastron and peripherals (after Hay, 1908). B. CM 11808, *Pseudanosteira pulchra*, type, modified from Clark (1937), with restorations from *A. ornata*. Scale lines equal 1 cm.

Figure 3. *Apalone leucopotamica*, AMNH 6045, dorsal view carapace (after Hay, 1908, fig. 704). Scale lines equal 1 cm.

Family CARETTOCHELYIDAE Boulenger 1887
Subfamily ANOSTEIRINAE Lydekker 1889

Diagnosis (shell)—Carapace retaining scales; plastron cruciform with narrow bridges; one or more neurals supporting a keel or spike.

Anosteirinae, genus indet.
Figure 2

Anosteirine, Clark et al., 1967:26.
Pseudanosteira or *Anosteira*, Clark et al., 1967:26.
Referred material—South Dakota: Chadron Formation: FMNH PM 16302, peripheral, and 16303, two peripherals, from Crazy Johnson Member, (Clark et al., 1967); FMNH uncatalogued "scrap," from Ahearn Member (Clark et al., 1967).
Discussion—The two North American genera of this subfamily, *Pseudanosteira* and *Anosteira*, are distinguished only by the arrangement of the vertebral scales and neurals and shape of the dorsal spines (Clark, 1932, p. 162), none of which have been recorded from the White River Group. *Anosteira* is only known from the Bridger Formation (Bridgerian) but the more derived *Pseudanosteira* is known from the Uintan (Uinta Formation), and it is likely, but not demonstrable, that this is the genus represented in the Chadron Formation and correlative units. These records represent the last known occurrences of the family in North America. References to *Anosteira ornata?* Leidy 1871 from the Cypress Hills Formation of Saskatchewan (Lambe, 1908, p. 17, pl. 1, figs. 17-19; Holman, 1976) are generically and specifically suspect.

Family TRIONYCHIDAE Bell 1828
Subfamily TRIONYCHINAE Bell 1828
Tribe TRIONYCHINI Bell 1828
Subtribe APALONINA Meylan 1987

Diagnosis (shell after Meylan, 1987, p. 91)—C8 reduced or absent.

Genus *Apalone* Rafinesque 1832

Diagnosis (shell after Meylan, 1987, p. 91)—Location of posteriormost neural reversal is highly variable but occurs at or anterior to N6; marked size sexual dimorphism.
Discussion—The genus *Trionyx* was revised by Meylan (1987) which resulted in the resurrection of the genus *Apalone* to encompass the Recent North American trionychids. Meylan also recognized two subgenera, *Apalone* and *Platypeltis* but listed only non-carapacial characters as diagnostic.

Apalone leucopotamica (Cope 1891)
Figure 3

Trionyx leucopotamica Cope 1891:5, plate I, figs. 8-9.
Trionyx punctiger Cope 1891:5, footnote 1.
Platypeltis leucopotamica (Cope 1891), Hay, 1908:546, plate 113, figs. 1-3, text-fig. 704.
Amyda sp., Clark et al., 1967:26.
Aspideretes sp., Harksen and Macdonald, 1969:15.
Types—NMC 6232, fragment of costal plate, Cypress Hills, Saskatchewan, Canada, Cypress Hills Formation, Chadronian. Costal fragments [type of *T. punctiger*], White Buttes, Billings County, North Dakota, Chadron Formation?, Chadronian?
Diagnosis—Carapace length to 325 mm; fontanel

on each side of N1; neural reversal at N5; nuchal length to nuchal width ratio 5.6; nuchal width to carapace width ratio about 0.6; carapace length to width ratio about 0.9; costal margins tapered; carapace truncate posteriorly; sculpture without welts.

Referred material—Nebraska, Chadron Formation: Hat Creek Basin, Chadronian, AMNH 6045, nearly complete carapace (Hay, 1908, text-fig. 704, plate 113, figs. 1-3). South Dakota, Chadron Formation: Custer County—Phinney Breaks, FMNH PR 663, partial carapace, Crazy Johnson Member, Chadronian; Pennington County—NW 1/8, NW 1/4, Sec. 13, T5S, R12E, FMNH PR 588, posterior part of carapace and C fragments, Ahearn Member (middle), Chadronian (Clark et al., 1967); from SW, NE, sec. 1, T42N, R46W, Battle Creek Draw, SDSM 5822, partial nuchal, Crazy Mountain Member, Chadronian. North Dakota, Brule Formation: Stark County—Fitterer Channel, Fitterer Ranch, AMNH 14245, two costal fragments, Orellan; Slope County—White Buttes, F:AM 12418, costal, Orellan.

Discussion—The types of this species are not specifically or generically diagnostic and any meaningful characterization of this taxon rests on Hay's (1908) reference of a complete carapace (AMNH 6045) from Nebraska which he compared to fragments (Cope, 1891) from the White Buttes of North Dakota. Although the proximity of geography and age provide a probabalistic justification for these assignments in the absence of convincing evidence of more than one taxon, better topotypic material or selection of AMNH 6045 as a neotype is needed to stabilize the name. Holman (1972, p. 1615) referred additional material from the Cypress Hills Formation to this species. The "*Trionyx* sp." listed by Lammers and Hoganson (1988) from the Brule (Orellan) of North Dakota are also included here.

Attribution of *Apalone leucopotamica* to the subtribe Apalonina is indicated by the reduction (loss) of the C8. Reference to *Apalone* is indicated by the neural reversal at N6. The fossil species of the genus need revision and the shell in the genus is known to be variable (Webb, 1962). Meylan provided no carapacial features of the subgenera that would allow assignment of *Apalone leucopotamica* to subgenus. The ratio of nuchal length to width is 5.6 (Hay, 1908, p. 547), greater than that cited by Meylan for Recent *Apalone (Apalone)* sp. The ratio of the carapace length to width is 0.92, within only the observed range of that in *A. (A.) mutica* (Le Sueur 1827) and below that of *A. (A.) spinifera* (Le Sueur 1827) and *A. (P.) ferox* (Schneider 1827).

In his key to the fossil "*Platypeltis*" [=*Apalone*] of North America, Hay (1908, p. 537) used age as his first-level division, thus obviating the need for character differences because *Apalone leucopotamica* was the only "Oligocene" [Chadronian] species. Nevertheless, he listed the carapace as "thin, broader than long; nuchal 0.6 the width of the carapace; a fontanel each side of first neural." Of the other ten extinct species Hay discussed, *A. antiqua* (Hay 1907), *A serialis* (Cope 1877), and *A. trepida* (Hay 1907) are juveniles and probably undiagnosable. *A. amnicola* (Hay 1907) is only represented by the HYO-HYPO. *A. trionychoides* (Cope 1872) is smaller, has a distinctly narrower

carapace, has longitudinal welts, lacks fontanels, and has only very slight lateral rib extensions. *A. extensa* (Hay 1908) has a distinctly narrower nuchal, relatively larger C8, and no rib extensions. *A. heteroglypta* (Cope 1873) lacks fontanels and has a distinctly narrow nuchal. *A. postera* (Hay 1908) lacks fontanels, has a more polished and less dense sculpture, and thicker carapace with abruptly beveled or truncated free margins. Which, if any, of the above features will actually serve to separate these species considering the marked sexual and morphological variability of the carapace in this genus (Webb, 1964; Meylan, 1987) awaits a thorough revision.

Family BATAGURIDAE McDowell 1964
Echmatemys Hay 1906

Diagnosis (of shell)—Neurals normally hexagonal, HYO and HYPO buttresses large, HYPO buttress sutured to costals 5-6, marginal 12 onlaps suprapygal 2, shell unsculptured except for broad keels in early species, well-developed EPI lips and dorsal GUL overlap.

Discussion—The genus was extensively monographed in Hay (1908) and has not been extensively revised since. Roberts (1962) revised some of the Uintan species. The genus appears abruptly in the earliest Wasatchian and is last recorded, excluding the record below, in the Uintan. It is one of the most common turtles in the Eocene of North America.

cf. *Echmatemys* sp.
Figure 4

Referred material—South Dakota, Chadron Formation, Chadronian: Pennington County—SDSM loc. V828, SDSM 27730, posterior right P; SDSM 27729, right C1. Wyoming, Yoder Formation, Chadronian: Goshen County—SDSM loc. V742, SDSM 8663, left C4-7, P8?, left HYO.

Description—The most complete specimen (SDSM 8663) is represented by a series of costals, a HYPO and one peripheral. These show that there were two suprapygals, that the neurals were normally hexagonal (wide end anterior), that the HYPO buttress extended about two-thirds the way up the C5-6 suture, that the dorsocentral area of the pleurals formed irregular raised bosses surrounded by distinct growth annuli (7-8) but lacked other sculpture or keels, and that the costals do not thicken notably toward the dorsal ends. The base or the dorsal free rib ends are small, circular to subovoid, and set back slightly from the neural suture. The associated peripheral (P8?) exhibits deep marginal covered areas and also shows that the central marginal areas were irregularly raised. The HYPO is relatively thin and unsculptured and supports a vertical and long HYPO buttress. The FEM laps onto the dorsal surface but not as extensively as in Chadron emydids. The apparently unfinished margins of the costals suggest that the specimen was not an adult.

Discussion—The fragments of an akinetic batagurid turtle in the Yoder Formation (SDSM 8663) resemble those of *Echmatemys* in the presence of a

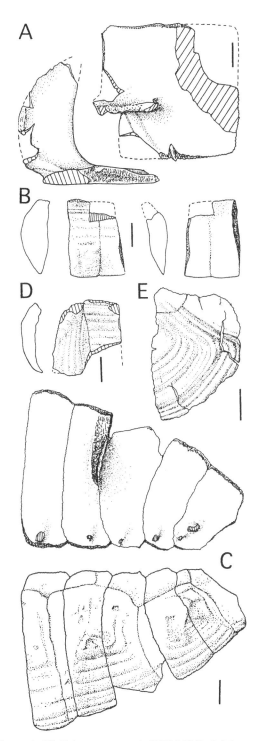

Figure 4. Cf. *Echmatemys* sp. A. SDSM 8663, left hypo-plastron, dorsal and posterior views. B. SDSM 8663, left P 8?, external and anterior and posterior sutural views. C. SDSM 8663, left costals 4-8, dorsal and ventral views. D. SDSM 27730, posterior right peripheral, posterior sutural and external views. E. SDSM 27729, right costal 1, external view. Scale lines equal 1 cm.

vertical and tall HYPO buttress extending up the C5-6 suture, lack of keels, and normal N series. These differ from *Echmatemys* in lacking the flattened dorsal rib ends and shorter space between the rib bases and the N sutures. *Echmatemys* also usually lacks the surficial growth rings, and has distinct dorsal thickening of the costals and thicker plastral elements. These differences may only be a reflection of the immaturity of the specimen but juvenile *Echmatemys* tend to resemble the adults in most of these features. The present material is not sufficient to clearly separate it from some Old World batagurids, but *Echmatemys* is one of the most common Eocene turtles and is the only known batagurid in North America with such well-developed buttresses. If this assignment is correct, then these specimens are the last known specimens of the genus. The specimens from the Chadron Formation also exhibit strong growth corrugations, but SDSM 27729 has a relatively greater number of rings despite its relatively small size. Direct comparison of this material with the type and referred material of *Testudo exornata* (see below) may result in synonymy.

Tribe PTYCHOGASTRINI De Stefano, 1917

Diagnosis (modified from Mlynarski, 1976)— Moderate-sized turtles; shell usually co-ossified in adults; carapace strongly arched; plastron broad; anterior lobe akinetic with HYO firmly anchored to C1 by ascending anterior buttress; posterior lobe of plastron kinetic along the anterior and lateral HYPO sutures; large cervical scale; V5 encroaching onto pygal; neural complex containing reversed hexagonal, octagonal, and quadratic shapes; HUM-PEC sulcus crosses the ENT; V1 contacts M1 on the nuchal; axillary, when present, on HYO buttress and peripherals.

Undescribed genus and species

Diagnosis—HYPO buttress extending well up onto C5-6? and loosely articulating with a large basined scar; ABD-PEC sulcus distinctly anterior to the HYO-HYPO suture; dorsal gular areas inflated (convex) dorsally and not distinctly truncate anteriorly; inguinal and axillary scales present; distinct anal notch.

Material—South Dakota, Chadron Formation, Chadronian: Custer County—SDSM loc. V653, SDSM 10052, anterior half of carapace, nearly complete plastron, right and left scapula-coracoid, parts of left and right pubis, partial ilium, cervical vertebrae 5 to 8; Pennington County—SDSM Loc. V828, SDSM 11714, partial right HYPO; SDSM 28132, right P7. Nebraska, Chadron Formation (middle): Sioux County—SDSM loc. V772, SDSM 28133, partial posterior lobe of plastron.

Discussion—This genus and species, to be described elsewhere, is the first member of this group of batagurids to be found in North America. The Ptychogastrini are otherwise only known from the middle Eocene to Miocene of Europe (Mlynarski, 1976). Because there is no obvious indication of their presence in North America earlier and a direct route to

Europe was blocked by the opening of the North Atlantic, I suggest that the group exists unrecorded in Asia in the later Eocene.

Family EMYDIDAE?
Pseudograptemys, new genus

?Graptemys Agassiz 1857, Hay, 1908:358.
Graptemys (?) Agassiz 1857, Clark, 1937:294.
Graptemys Agassiz 1857 (in part), Dobie, 1981:97.

Type species—*Pseudograptemys inornata* (Loomis 1904)

Diagnosis—Carapace unsculptured except for traces of a low rounded median keel, posterior peripherals notched at marginal sulcus, marginal 12 onlapping suprapygal 2, hypoplastron buttress well developed and articulating with costal 5 but at costal 5-6 suture, humeral-pectoral sulcus posterior to entoplastron, anal notch well defined by distinctly pointed xiphiplastron ends, cervical scale about as broad as long; small inguinal and axillary musk duct (Rathke's gland) foramina.

Pseudograptemys inornata
(Loomis 1904), new combination
Figure 5

Chrysemys inornata Loomis 1904:429, figs. 10-11.
Graptemys? inornata (Loomis 1904), Hay 1908: 358, figs. 455-456.
Graptemys (?) cordifera Clark 1937:294, figs. 11-12.
Graptemys inornata (Loomis 1904), Dobie 1981:97.
Graptemys cordifera Clark 1937, Dobie 1981:97.

Types—ACM 3607, carapace and plastron, head of Bear Creek in Spring Draw Basin, 10 miles east of Creston, Meade County, South Dakota. Chadron Formation [*Titanotherium* beds], Chadronian. YPM (PU) 13838 [type of *G.* (?) *cordifera*], carapace and plastron, West side of Quinn Draw five miles from the mouth of Quinn Draw, Washington County, South Dakota, middle Chadron Formation, channel sandstone.

Diagnosis —as for genus.

Referred material—South Dakota, Chadron Formation, Chadronian: Custer County—SDSM loc. V653, SDSM 10053, carapace and plastron; Pennington County—SDSM loc. V828, SDSM 27727, partial right P9, SDSM 28130, right P11.

Description—SDSM 10053 is well enough preserved to distinguish the presence of the inguinal and axillary scales which were not or incompletely determined in the types (types not seen). The axillary scale is small and well up on the HYO buttress anterior to the posterior point of the axillary notch and onlaps P3 and P4. In contrast, the inguinal is large and covers most of the anterior margin of the inguinal notch and the ventral part of P7 with broad contact with M6-7. Small musk duct foramina pierce the axillary and inguinal scales on the P3 and P7 respectively. The cervical scale underlap is shorter than the overlap and the scale expands to wider than long ventrally. The HYPO buttress articulates directly only with C5 but

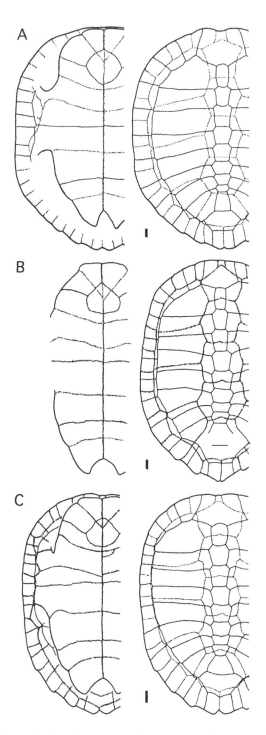

Figure 5. *Pseudograptemys inornata*, ventral views of plastron and dorsal views of carapace. A. ACM 3607, type (after Hay, 1908). B. YPM(PU) 13838, reconstruction type of *G.? cordifera* (after Clark, 1937). C. SDSM 10053, reconstruction of carapace and plastron. Scale lines equal 1 cm.

Table 1. Types of Chadron *Pseudograptemys* ["*?Graptemys*"] compared to new specimen.

Character	*?G. inornata*	*?G. cordifera*	SDSM 10053
Carapace length	303 mm	267	192
Suprapygal 1	weakly rectangular	distinctly rhomboid	distinctly rhomboid
Entoplastron	roughly quadrangular	heart-shaped	intermediate
N7/N8 ratio(%)	166	67	76
Anterior lobe sides	parallel	divergent, parallel	divergent
V1 lateral borders	concave	convex	sinusoidal
V1-Pl1 sulcus to M1	on nuchal	off nuchal	at nuchal suture

this suture lies at the extreme posterior margin adjacent to and buttressed by a descending flange of C6.

Discussion—The two nominal late Eocene species from the Chadron Formation previously included in *?Graptemys* differ from all the modern species of *Graptemys* by posterior articulation of the HYPO buttress on C5 rather than near middle of C5 (Dobie, 1981, fig. 8A), presence of musk duct foramina (small), at least on SDSM 10053, and the apparently consistent onlap of M12 onto the suprapygal. Although the latter character seems relatively minor, it is a rare variant in emydids and was noted by McDowell (1964) in his diagnosis of the Emydinae [Emydidae *s.s.*]. Onlap of M12 may be primitive because it is the standard condition in the Bataguridae (at least in part, Gaffney and Meylan, 1988), the presumptive outgroup of the Emydidae. Hay (1908, p. 358) and Clark (1937, p. 294) questioned assignment of these species to *Graptemys* but placed them questionably under *?Graptemys* on the basis of the dorsal keel and (for *?G. inornata*) the elongated first suprapygal. The former is very weak but best expressed on the posterior slope and also occurs in the *Pseudemys* complex; the latter is variable even in extant *Graptemys* and only known in the genotypic specimen. Both authors as well as Loomis (1904) suggested relationship to the *Chrysemys* complex as also likely. Although Wood (1977) suggested that *Graptemys* and *Malaclemys* are congeneric, Dobie (1981, p. 97) noted that these species did not belong with *Malaclemys* but did not question the assignment to *Graptemys*. Whatever the relationships of these species, they do not appear to belong within the concept of the extant *Chrysemys* or *Graptemys* crown groups and should be separated from them generically. *Pseudograptemys* may yet prove to be a sister taxon of one or the other when more material is known.

Two species of *?Graptemys*, each based on a single shell, were named from the Chadron Formation of South Dakota. No significant new material has been referred to either species since their descriptions. A comparison of Clark's (1937) and Hay's (1908) descriptions yields about eight points of difference that may be used to distinguish the two species, but it is difficult to distinguish the significance of these features when they vary within modern species and the states represented for each of the distinguishing features may only represent extremes in normal distribution. Although it is not unlikely that they may be sympatric

species, as is common in extant *Pseudemys*, *Trachemys* and *Graptemys*, the addition of one more complete shell (SDSM 10053) appears to bridge the gap between these extremes, and argues for regarding them all as a single normally variable species (Table 1).

Assignment of this taxon to the Emydidae is suspect and rests primarily on geography, shift of the HYPO buttress onto C5, the presence of notched posterior periperhals, and posterior position of the HUM-PEC sulcus (both occur in some batagurids). The presence of normal hexagonal neurals, musk ducts (although small and only observed in SDSM 10053, types not seen), HYPO buttress near C5-6 suture, and onlap of M12 on suprapygal 2 are all regarded as plesiomorphous (e.g., *Echmatemys*). Additional evidence (e.g., skull) is needed to secure its placement here versus the Bataguridae.

Family EMYDIDAE Gray 1825
Chrysemys Gray 1844

Diagnosis—Shell low domed; musk duct foramina absent; anal notch weak or absent; no posterior peripheral notching; no dorsal keel; V1 with only slight nuchal overlap and little or no anterior constriction.

Chrysemys antiqua (Clark 1937)
Figure 6

Trachemys? antiqua Clark 1937:292, fig. 10.

Type—YPM (PU) 13839, plastron and right posterior border of carapace and right bridge, two miles southeast of "Cedar Butte," Indian Creek drainage basin, Pennington County, South Dakota, middle Chadron Formation, channel sandstone.

Referred material—Colorado, White River Formation, Horsetail Creek member, Chadronian: Weld County—UCM loc. 69002, UCM 24708, partial carapace and anterior lobe of plastron. Nebraska, Brule Formation, Dawes County—UCM loc. 83194, UCM 48599, juvenile shell, probably Orellan. North Dakota, Chadron Formation, Chadronian: Stark County - Little Badlands, CM 2053 (in part), EPI and ENT fragment, anteriolateral part of carapace, left P2. South Dakota, Chadron Formation, Chadronian: Pennington County—Indian Creek, CM 27434, plastron and partial carapace; SDSM loc. V828, SDSM 27726, left P2; SDSM 27728, right HYPO. Brule Formation: Shannon County—SDSM loc. V363, SDSM 3632, carapace

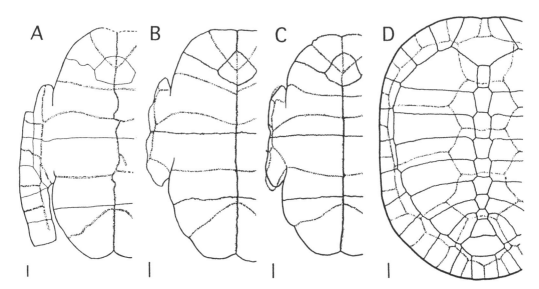

Figure 6. *Chrysemys* plastra and carapace. A. *Chrysemys antiqua* [type], YPM (PU) 13839, ventral view of plastron and P5-9 (from Clark, 1937). B. *Chrysemys antiqua*, SDSM 2754, reconstruction of ventral view of plastron. C. *Chrysemys picta belli*, UCMP 131092, ventral view of plastron. D. *Chrysemys antiqua*, SDSM 406, 2754 and 3632, dorsal view of reconstructed carapace. Scale lines equal 1 cm.

lacking margins of P4-6, Scenic Member, Metamynodon channels, Orellan; SDSM loc. V401, SDSM 406, complete plastron and nearly complete carapace lacking P4-6, Poleslide Member, *Leptauchenia* nodules, Whitneyan; SDSM loc. V271, SDSM 2754, compositely complete plastron and carapace lacking the bridge peripherals and the margins of P3-4, Poleslide Member, *Protoceras* channels, Whitneyan; SDSM loc. V402, SDSM 4062, carapace fragments and plastron lacking bridge and most of XIPH, Poleslide Member, *Protoceras* channels, Whitneyan; SDSM loc. V6026, SDSM 60161, distal costal, Scenic Member, lower nodular zone, Orellan.

Description—The type of *Chrysemys antiqua* consists of a plastron and a few posterior peripherals. The referred material provides additional evidence of the carapace that warrants description. The carapace is of moderate size (to 190 mm), moderately domed, with or without sculpture, lacks keels, has a sharp but moderately incised sulci. The HYPO buttress contacts the ventral part of the C5 only. The scale numbers and relative positions are the same as in emydids. The anterior nuchal margin is straight to moderately concave. The cervical scale is narrow and dorsally about twice as long as wide and broadening or not posteriorly, underlap is also longer than wide but shorter than the overlap. The pleural-marginal sulci formula (after Tinkle, 1962: left of colon = pleural sulcus, right of colon = marginal scale number the sulcus contacts, < = anterior, > = posterior, and s = at the anterior sulcus of the numbered marginal): V1-PL1:1> to 2<, PL1-2: 5< to 1/2, PL2-3:7<, PL3-4: 9< to 1/2, PL4-V5: s11-11<. V1 is slightly longer than wide, broadening anteriorly with slight constriction or not near the anterior margin. Juncture of the V1-PL1-M varies from just off, just on,

or coincident with nuchal-P1 suture. V2-4 are about equally long as wide to wider than long, subrectangular to hexagonal. V5 is equal or narrower than V4 and about as wide as long, broadening posteriorly. The pleural-marginal sulcus is relatively straight except for the elevated M9. The anterior peripherals are not flared and posterior peripherals only slightly so.

The referred plastra closely resemble the holotype of *C. antiqua* but preserve additional morphology not mentioned in the latter, the axillary scale is restricted to the HYO buttress and adjacent peripherals and does not extend posterior to axillary notch. The inguinal scale is large with broad contact with M7 and just touching M8 ventrally. No musk duct foramina were observed. The entoplastra are small to moderate sized with convexly arched or semi-rectangular posterior sutures. The presence of fine lineal sculpture is most evident in the areas posterior to the HUM-PEC and ABD-FEM sulci with only faint indications of a vaguely oriented sculpture on the other parts of the plastron.

Discussion—The normal hexagonal neurals, absence of musk duct foramina, and contact of the posterior buttress with C5 are consistent with assignment to the Emydidae. The lack of kinesis, and posterior position of the HUM-PEC sulcus behind the ENT exclude it from *Clemmys*, *Terrapene* and *Emydoidea*. One or more of the following excludes *Deirochelys*, *Graptemys*, *Pseudemys,* and *Trachemys*: absence of dorsal keels, shallow and weakly defined anal notch, absence or nature of carapacial sculpture, little or no overlap of pleurals on the nuchal, slight or no anterior constriction of the V1, weak contact of the buttresses with the costals, normal proximal rib ends.

Aside from the sculpture, the Oligocene species falls within the definition of the shell of *Chrysemys* by Hay

(1908, p. 345). Excluding Old World species referred to the broad concept of *Chrysemys* of McDowell (1964), the new material is referable to the genus in the restricted sense (Hay, 1908; Ward, 1984). The plastra of the Poleslide material are practically indistinguishable from the type of *Chrysemys antiqua*, but these specimens differ from the referred Chadron material and the few peripherals of the type in the lack or reduction of carapacial sculpture and growth rings. The latter are best developed in an immature Poleslide specimen (SDSM 406). The shallow doming of the carapace and reduction of costal and neural sculpture in later samples, shape and arrangement of the nuchal and associated scales, lack of peripheral notching, and reduction of the anal notch are more consistent with reference to *Chrysemys* than *Trachemys*.

The non-growth ring sculpture of the carapace and fine lineal sculpture on the plastron is a resemblance to some *Pseudemys* (e.g., Rose and Weaver, 1966, fig. 4) and *Deirochelys* but sculpture is probably primitive for Emydidae because it more closely resembles that in probable outgroups including the *Pseudochrysemys* from the Paleocene of Asia (Sukhanov and Narmandakh (1976, plate 7) and the Eocene emydid from Ellesmere Island, Canada (Estes and Hutchison, 1980, fig. 4). The sculpturing of the carapace is well and consistently developed in the Chadronian adult specimens but absent in adult specimens from the Whitneyan. The reduction in sculpture between these two samples may represent species or subspecies level distinctions and may document the reduction in sculpture presaging its absence in the later species. At the present time, species distinctions, other than sculpture, are not supported by the samples. Only two North American species comprise *Chrysemys sensu strictu*. The extant *C. picta* (Schneider 1783) are known as fossils into the Barstovian (Holman and Sullivan, 1981) and *C. timida* Hay 1908 from the later Cenozoic. *C. timida* has not been diagnosed except for the mention by Hay of somewhat narrower neurals than in *C. picta*. *C. antiqua* differs from *C. timida* Hay 1908 in having only two rather than three suprapygals, having a faint plastral or carapacial sculpture, and PL1-V1 sulcus contacting posterior M1 rather than anterior M2 (both states occur in *C. picta*). *C. antiqua* differs from *C. picta* in lack of the fine EPI and nuchal "teeth" on the free margins, presence of a fine sculpture, and somewhat narrow neurals as in *C. timida*.

It is notable that the only emydid represented in the Brule Formation is *Chrysemys*, a genus today exhibiting the largest mid-continental range. The single extant species, *C. picta*, is the most drought- and temperature-tolerant North American emydid.

Family TESTUDINIDAE Gray 1825

The Testudinidae are the most common turtles in the White River Group (Hay, 1908; Hutchison, 1992) and are found throughout the Group. The tortoises from the White River Group have been assigned classically to *Testudo*, *Stylemys*, *Gopherus*, and *Geochelone*. *Testudo* and *Geochelone* have been restricted more recently to

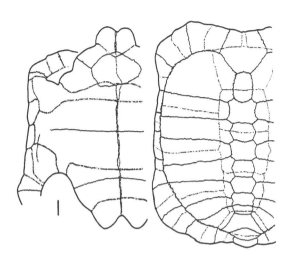

Figure 7. *Hesperotestudo brontops*, YPM "Box 27, 1985," type, dorsal and ventral (reversed) views of shell (modified after Hay, 1908). Scale line equals 5 cm.

non-North America taxa (Williams, 1952b; Preston, 1979).

Genus *Hesperotestudo* (Williams 1950)

Testudo (Hesperotestudo) Williams 1950:25

Diagnosis (modified from Auffenberg, 1964, 1974; Bramble, 1974, 1982)—Premaxillary ridge reduced or absent; no symphyseal dentary groove; os transiliens absent; sacculus of inner ear not containing a small otolithic mass; costals moderately to distinctly alternately narrower and wider laterally; a moderate to short tail and caudal vertebrae with interpostzygopophyseal notches; posterior EPI excavation moderate to deep; free proximal portion of the ribs short; armor on forelimbs moderate to well-developed and with patch of sutured dermal ossicles on thigh and perhaps tail in adults; nuchal scale longer than wide; inguinal scale enlarged anteriomedially and may broadly contact FEM; normal neural shape formula = 4(8)4(8)4(6(6(6; phalangeal formula 2-2-2-2-2.

Hesperotestudo brontops (Marsh 1890)
Figure 7

Testudo brontops Marsh 1890:179, plate 8.
Testudo (Hesperotestudo) brontops Williams, 1950:20.

Type—YPM [Box 27, 1885], shell. Indian Creek, SE corner of Pennington County, South Dakota. Chadron Formation (*Titanotherium* beds), Chadronian.

Diagnosis—PEC expanded medially and contacting ENT; inguinal scale large but not contacting FEM.

Referred material—YPM (PU) unnumbered (Clark, 1937), plastron, South Dakota, Indian Creek drainage basin, middle Chadron Formation.

Discussion—Williams (1950, p. 20) placed *H. brontops* in and near the base *Hesperotestudo*. Auffenberg (1974, p. 145) listed it in *Geochelone*

(?*Caudochelys*) although no direct evidence is given for this placement other than that his *Caudochelys* is morphologically plesiomorphous. Crumly (1984, p. 119) also placed it in *Geochelone*. Bramble (1971, unpublished) placed this taxon in *Stylemys* but noted that this species is transitional to *Hesperotestudo*. Derived features shared with *Hesperotestudo* include the derived neural formula 4(8)4(8)4, large inguinal scale, and strong gular projection. The character states indicate that *H. brontops* is a member of the Hesperotestudo clade where it is here retained.

Hesperotestudo sp.

Material—South Dakota, Brule Formation, Poleslide member, Whitneyan: Shannon County—UCMP loc.V71115, UCMP 65650, right inguinal region of shell and anterior part of anterior lobe of plastron, UCMP 65649, anterior part of anterior lobe of plastron; SDSM Godsell Ranch, SDSM 55159, isolated partial skull.

Discussion—The fragments of a shell and partial skull attest to the presence of a primitive *Hesperotestudo* in the Whitneyan. UCMP 65650 exhibits a large inguinal scale approaching but not contacting the FEM and a distinctly excavated EPI with some overhang by the posterior surface of the gular area. The skull retains a jugal-pterygoid contact and partial posterior premaxillary ridge but has the strong medial maxillary triturating ridge typical of *Hesperotestudo*. Bramble (1971, unpublished) placed it in *Stylemys* but noted that it is intermediate between *Stylemys* and Miocene *Hesperotestudo*. If correctly identified, the presence of a partial premaxillary ridge indicates that this feature is shared by *Stylemys*, *Gopherus*, *Xerobates*, and primitive *Hesperotestudo*, and lost in later *Hesperotestudo* and supports a close relationship of these genera.

Stylemys Leidy 1851

Diagnosis (modified from Auffenberg, 1964, 1974; Bramble, 1974, 1982)—Premaxillary ridge present; a symphyseal dentary groove; sacculus of inner ear not containing a small otolithic mass; os transiliens absent; costals only slightly alternately narrower and wider laterally; a moderately, long, unspecialized tail with some of the caudal vertebrae lacking interpostzygopophyseal notches; posterior EPI excavation shallow or absent; free proximal portion of the ribs long; minimum of armor on forelimbs but with small patch of sutured dermal ossicles on thigh in some adults; nuchal scale longer than wide; inguinal scale moderate to small, never contacts FEM; normal neural shape formula = 4(6(6(6(6(6(6(6 or 4(8)4(6(6(6(6(6; phalangeal formula 2-2-2-2-2.

Discussion—Auffenberg (1964, 1974) provided the most recent redefinition and list of included species of the genus respectively but reference of several of these species to this genus needs reexamination. Bramble (1982, p. 854) added three additional species previously included in *Gopherus* but provided no stated justifications.

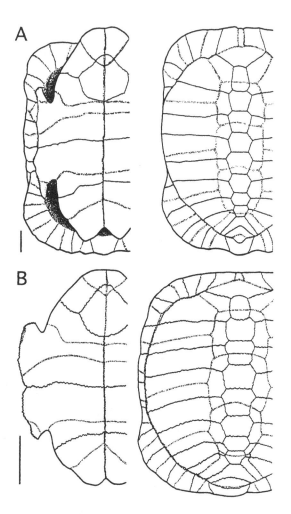

Figure 8. *Stylemys*, dorsal and ventral views of shell. A. *S. amphithorax*, UCM 20575 (after Auffenberg, 1962). B. *S. nebrascensis*, AMNH 1433 (after Hay, 1908). Scale lines equal 5 cm.

Stylemys amphithorax (Cope 1873)
Figure 8A

Testudo amphithorax Cope 1873a:6.
Testudo ligonius Cope 1873a, Cope, 1884:762, 766, pl. 61, figs. 2-3.
Testudo (Hesperotestudo) ligonia (Cope 1873a), Williams, 1950:25.
Stylemys amphithorax (Cope 1873), Auffenberg, 1962:1.

Types—AMNH 1145, cotypes, parts of two plastra, head of Horse Tail Creek, Weld County, Colorado. White River Formation, Horsetail Creek Member, Oreodon beds, Chadronian. AMNH 1148, [type of ?*Stylemys ligonia* (Cope 1873a)], part of nuchal bone, some peripherals and parts of plastron, same locality as above.

Referred material—Colorado, White River Formation: Weld County—AMNH 1139, parts of

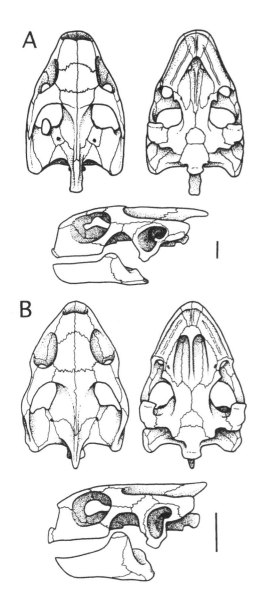

Figure 9. Testudinid skulls, dorsal, ventral and left lateral views and lower mandibles, left lateral view. A. *Gopherus (Oligopherus) laticuneus*, USNM 15874 (modified from Gilmore, 1946, and UCM 57043), os transiliens from UCMP 94707 (after Bramble, 1982). B. *Stylemys nebrascensis*, UMMP 9318, reconstruction from CM 1571, UMMP 9318 and SDSM 2891 (after Hay, 1908; Case, 1925). Scale lines equal 1 cm.

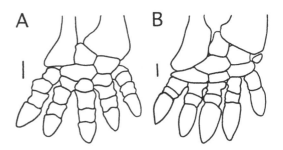

Figure 10. Testudinid pes, dorsal views. A. *Gopherus (Oligogopherus) laticuneus*, composite left adult pes from UCMP 94708 and USNM 15854) (modified from Bramble, 1982). B. *Stylemys nebrascensis*, UMMP 17600, (modified from Case, 1936, and Auffenberg, 1964). Scale lines equal 1 cm

plastron and carapace (Hay, 1908, p. 408), and AMNH 1147, left HYPO and XIPH (Hay, 1908, p. fig.531), same area as types; UCM loc. 77293, UCM 20575, male shell and most of the postcranial skeleton (Auffenberg, 1963), Horse Tail Creek Member.

Diagnosis (after Auffenberg, 1962)—Lacks peripheral pits for reception of costal ribs; proportionately thinner and more elongate shell; narrow and elongate cervical scale; distinct lateral notch on XIPH; anterior lobe of plastron longer than wide; length of posterior lobe nearly equal or greater than bridge length.

Discussion—Auffenberg (1962) and Hay (1908, p. 407) provide the most complete descriptions of the shell and skeleton including the girdles, caudal vertebrae, femur, humerus, tibia, partial pes and fibula, and thigh armor. Auffenberg (1962, 1964, 1974) regarded *S. ligonia* as probably conspecific with *S. amphithorax*. Williams (1950, fig. 2) placed this species in *Hesperotestudo* without comment but it is retained in *Stylemys* by Auffenberg (1962) and here.

Stylemys nebrascensis Leidy 1851
Figures 8B, 9B, 10B

Stylemys nebrascensis Leidy 1851:173.
Emys hemispherica Leidy 1851:173.
Testudo lata Leidy 1851:173.
Emys oweni Leidy 1851:327.
Emys culbertsonii Leidy 1852a:34.

Holotype—USNM 97, most of carapace and plastron. Brule Formation, Scenic Member, South Dakota.

Diagnosis (inverse after Auffenberg, 1962)—Carapace to 530 mm or more, peripheral pits for reception of costal ribs; proportionately thicker and more rounded shell; broad and relatively short cervical scale; weak lateral notch on XIPH; anterior lobe of plastron wider than long; posterior lobe length distinctly less than bridge length.

Referred material (hundreds of shells of this species reside in museums and only a few are listed here as vouchers)—Colorado, White River Formation, Orellan: Logan County—KU 8239, shell (Galbreath, 1953), middle Cedar Creek Member; UCMP loc. V71042, UCMP 95648, shell. Weld County—DMNH 13469, partial shell; DMNH loc, 960, DMNH 14174, shell with partial skull and jaws. Nebraska, Brule Formation, Orellan: Dawes County—UCM loc.

80078, UCM 48582, 48584, 48585, 48587, 48590, 48591, 48595, 48607, 53353, shells including three skulls and limb material; Sioux County—Hat Creek Basin near Prairie Dog Creek, CM 1571, skull and jaws (Hay, 1908). South Dakota, Chadron Formation, Chadronian: Pennington County—SDSM loc. V8940, SDSM 27732, shell. Brule Formation, Orellan: Pennington County—UCMP loc. V80035, UCMP 124374, shell; Big Badlands, UMMP 17600, skeleton lacking head. Wyoming, Brule Formation, Orellan: Niobrara County—Hat Creek, Tributary of Old Woman Creek, UMMP 9310, skull, jaws, and partial skeleton (Case, 1925, plates 1-2); Seaman's old ranch, Sage Creek (Lambe, 1913, p. 57, plates 6-7), .

Discussion—Aside from being the first fossil tortoise named from North America, this species is the most common, most completely known and most studied of the White River Group turtles. Descriptions and figures of the skeleton include: skull (Hay, 1907, 1908), lower mandible (Hay, 1907, 1908), cervical vertebrae (Case, 1919), carapace (Hay, 1908; Case, 1936; Auffenberg, 1964), plastron (Hay, 1908; Case, 1919; Auffenberg, 1964), shoulder girdle (Case, 1925, 1936), humerus (Hay, 1908; Case, 1936), forearm bones (Hay, 1908; Case, 1936; Auffenberg, 1961, 1964), manus (Hay, 1908; Case, 1936; Auffenberg, 1961, 1964), pelvis (Hay, 1908; Case, 1919, 1925, 1936), femur (Hay, 1908; Case, 1925, 1936), lower leg (Hay, 1908; Case, 1936; Auffenberg, 1961, 1964), pes (Case, 1936; Auffenberg, 1961, 1964), tail (Case, 1936), and egg (Hay, 1908). The interpretation by Case (1936, p. 71) (reiterated by Williams, 1952b, p. 557) that *S. nebrascensis* has an emydine phalangeal formula is refuted by Auffenberg (1961, 1964). Auffenberg (1964) discussed the ontogenetic development of the shell. Despite these works, there is much to learn about the ontogenetic and individual variation of this common species.

Stylemys sp.

Discussion—Galbreath (1953, p. 38) suggests that the genus may range into the Whitneyan. *Stylemys* is known from the Arikareean of Oregon, California, Nebraska, and South Dakota but has not been positively identified from the Whitneyan of the White River Group, although it was assuredly present.

Gopherus Rafinesque 1832

Bysmachelys Johnston 1937
Diagnosis—Premaxillary ridge present; a symphyseal dentary groove; os transiliens present; sacculus of inner ear containing an otolithic mass; costals distinctly alternately narrower and wider laterally; a moderately long to short unspecialized tail and caudal vertebrae with interpostzygopophyseal notches; free proximal portion of the ribs moderate to short; minimum of armor on forelimbs but without patch of sutured dermal ossicles on thigh in adults; nuchal scale as wide or wider than long; inguinal scale moderate to small but contacts FEM; humerus and coracoid expanded, phalangeal formula 2-2-2-2-1 or less; [and from *Xerobates*]

incipient to well-developed differentiation of the radial and ulnar facets of the humerus; enlarged and distally displaced insertion of the biceps on the ulna; medial maxillary triturating ridge occasionally or normally joining the median premaxillary ridge.

Discussion—The gopher turtles of the White River Group have classically been assigned to *Testudo* (now restricted to an Old World group) or *Gopherus*. Bramble (1982) divided *Gopherus* into two groups, *Gopherus* and *Scaptochelys* [= *Xerobates*].

Oligopherus, new subgenus

Type species of the subgenus—*Gopherus (Oligopherus) laticuneus* (Cope 1873).
Diagnosis—Posterior epiplastron excavation shallow; normal neural formula 6)6)4(6(6(6(6(6 or 4(8)4(6(6(6(6(6; pes phalangeal formula 2-2-2-2-1; incipient differentiation of the radial and ulnar facets of the humerus; insertion of the biceps on the ulna proximal to midpoint; medial maxillary triturating ridge occasionally or normally not joining the median premaxillary ridge; sacculus of inner ear containing a small otolithic mass; cranium relatively dolicocephalic; inner ear chambers not inflated; cervical vertebrae not appreciably shortened, pre- and postzygopophyses not enlarged, widely separated; cervical 8 without elongated postzygopophyses; first dorsal vertebra with small zygopophyses and neural arch suturally united with neural 1; manus with two subradial bones; ungual phalanges not greatly enlarged or spatulate; mesocarpal joint well-developed; manus digitigrade.

Gopherus (Oligopherus) laticuneus (Cope 1873)
Figures 9A, 10A, 11

Testudo laticunea Cope 1873a:6.
Testudo quadratus Cope 1884:762, plate 61.
Testudo thomsoni Hay 1908:400, plate 66, figs. 1-5.
Testudo laticunea Cope 1873, Hay, 1908:402, plate 67, figs, 1-2; text-fig. 509-515.
Testudo praeextans Lambe 1913:57, plate 4-5.
Gopherus laticunea (Cope 1873), Williams, 1950:25.
Geochelone thompsoni (Hay 1908), Auffenberg, 1963:94 [error].
Scaptochelys laticunea (Cope 1873), Bramble, 1982:854.

Types—AMNH 1160 (Hay, 1908, figs 509-510), almost complete carapace, pelvis, partial humerus, femur, pelvis, head of Horse Tail Creek, Weld County, Colorado, Horsetail Creek Member, White River Formation. Orellan; NMC 8401, (Type of *Testudo praeextans* Lambe 1913) male shell (*contra* Gilmore, 1946), Sage Creek, Niobrara County, Wyoming, Orellan; AMNH 1149 (type of *T. quadrata*), nuchal fragment and EPI beak, head of Horse Tail Creek, Weld County, Colorado, Chadronian; AMNH 3940 (type of *T. thomsoni*), skull and anterior lobe of plastron, cervical vertebrae and parts of foreleg. Corral Draw, Ziebach County, South Dakota, Lower Oreodon beds, Orellan.
Referred material—Colorado, White River Formation, Cedar Creek Member, Orellan: Weld

County—UCMP loc. V70106, UCMP 94707, UCMP loc. V71048, UCMP 94708, UCMP V71042, UCMP 94709, partial shells, limb material, partial skulls (Bramble, 1974). Nebraska, Brule Formation, Orellan: Dawes County—UCM loc. 80019, UCM 47043, male skeleton with skull; UCM 48600, female shell and limb bones; UCM 48601, partial skeleton; UCM 48892, male shell; UCMP loc. V6902, UCMP 86301, shell; Sioux County—Warbonnet Creek, CM 238, shell (Hay, 1908, p. 405); UMMP 61138, male shell and limbs and girdles. Chadron Formation, Chadronian: Dawes County—UCM loc. 83195, UCM 48602, partial shell and limb bones. South Dakota, Brule Formation, Orellan: Ziebach County—Corral Draw, skull (Hay, 1908, p. 401); Chadron Formation, Chadronian: YPM "2051, Box 1," partial shell (Hay, 1908, p. 404). Wyoming, Brule Formation, Orellan: Niobrara County—Anderson Ranch on south side of Young Woman Creek, USNM 15874, most of skeleton including the skull (Gilmore, 1946, p. 294, Text-figs. 20-23, plate 38, left); east side of Little Indian Creek, USNM 15878, most of postcranial skeleton (Gilmore, 1946, p. 294, plate 38, right; plate 40, right; plate 44, fig. 2) and USNM 16732, shell and partial postcranial skeleton (Gilmore, 1946, p. 294, plate 38, left; plate 41, left); about one mile north of Whitman Post office, USNM 16728, shell and partial postcranial skeleton (Gilmore, 1946, p. 294, plate 41, right), USNM 16731, complete shell (Gilmore, 1946, plate 42), and USNM 16737, EPI beak (Gilmore, 1946, p. 309, p;. 44, fig. 1); Thomas Ranch, USNM 15854, nearly complete shell (Gilmore, 1946, plate 43). Crumly (1994) concluded that *Xerobates* is paraphyletic and should not be recognized. He regarded *G. laticuneus* as the sister taxon of all *Gopherus* but did not include or discuss the characters used by Bramble to relate *G. laticuneus* to *Gopherus s.s.*

Discussion—Hay (1908, p. 403) noted the resemblance of the shell of *Testudo laticunea* to *G. polyphemus*. Williams (1950, p. 25) placed it in *Gopherus* as did subsequent authors until Bramble (1982, p. 853) referred it to *Xerobates* [*Scaptochelys*] without comment other than to note that it is generally primitive. Bramble (1971, unpublished and *contra* 1982), nevertheless, noted that *G. laticuneus* exhibited derived features indicative of *Gopherus* that precluded it from direct ancestry to later *Xerobates*. Derived character states shared with *Gopherus s.s.* include incipient differentiation of the radial and ulnar facets of the humerus, enlarged and distally placed insertion of the biceps on the ulna (Bramble, 1982, fig. 11), expansion of the distal end of the humerus and coracoid, medial maxillary triturating ridge occasionally joining the median premaxillary ridge (SDSM 3143, UCM 57043), and generally low dome of the adult shell. As he noted, the many similarities stated or implied by Bramble (1982) between *Xerobates* and *G. laticuneus* are plesiomorphous. Because *G. laticuneus* shares no derived characters with *Xerobates* but at least three or more with *Gopherus*, it should be retained in *Gopherus*. The monophyly of *Gopherus* excluding *G. laticuneus* is exceptionally well defined (Bramble, 1982)

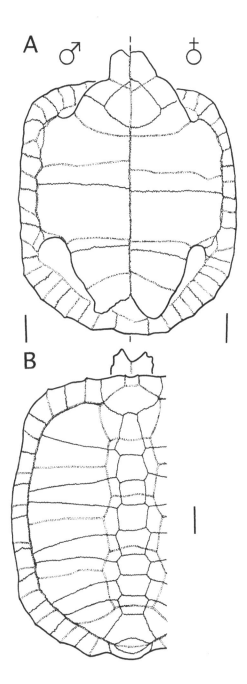

Figure 11. *Gopherus (Oligopherus) laticuneus* (after Gilmore, 1946). A. Composite ventral view of shell, male on left (USNM 15878), female on right (USNM 16731) . B. Dorsal view of male carapace (USNM 16728 and 16732). Scale lines equal 5 cm.

and the inclusion of *G. laticuneus* dilutes his diagnosis of the genus. The subgenus *Oligopherus* is here created to receive the plesiomorphous species while *G. (Gopherus)* is retained for *Gopherus sensu* Bramble (1982).

G. laticuneus is one of the better known fossil testudinids. Descriptions and figures of the skeleton include: skull (Hay, 1908; Gilmore, 1946; Bramble, 1974), lower mandible (Gilmore, 1946), carapace (Cope, 1873a; Hay, 1908; Lambe, 1913; Gilmore, 1946), plastron (Cope, 1873a; Hay, 1902, 1908; Lambe, 1913; Gilmore, 1946), humerus (Hay, 1908), forearm bones (Gilmore, 1946; Bramble, 1982), manus (Gilmore, 1946; Bramble, 1982), pelvis (Hay, 1908; Gilmore, 1946), and femur (Hay, 1908).

Lambe (1913) and Gilmore (1946, p. 306) noted the close similarity of the EPI beak of *Testudo praeextans* and *Testudo thomsoni*. With a large sample of *T. praeextans* at hand, Gilmore (1946) noted that Lambe would not be justified in establishing a new species (*Testudo praeextans*) on the plastral parts alone and Gilmore would "unhesitatingly regard *Testudo praeextans* to be a synonym of *T. thomsoni*." Nevertheless, Gilmore retained them as separate taxa because "different proportion of the elements forming the skull roofs and different widths of the channels on the triturating surfaces, strongly suggest that the discovery of more complete materials of *T. thomsoni* may disclose other and more important distinctive characters." The latter statement could just as well be reversed because each species was only known from a single skull, incomplete in *T. thomsoni*. Comparison of the illustrated skulls and lower mandibles of the two species reveals differences most notably in the constriction of the orbital contribution of the frontal and narrower inner triturating shelf in *T. thomsoni*. Both these character states vary comparably within *Xerobates agassizi* Cooper 1863.

Auffenberg (1974, p. 185) stated that "*Gopherus praeextans* and *G. neglectus* [*G. neglectus* Brattstrom 1961, Arikareean of California] may be synonyms of *G. laticuneus*." Further on, Auffenberg (1974, p. 187) states that *Gopherus praeextans* is "probably only a chronologic race of *G. laticuneus*." Bramble (1982) listed *Gopherus praeextans* but not *G. neglectus* as a synonym of *S. laticuneus*. (Bramble, 1971 unpublished, regarded *G. neglectus* as a *Stylemys*). With regard to *T. praeextans*, Gilmore (1946, p. 306) considered the possibility that the differences in the development of the EPI lip and "more truncate carapace border" of *T. laticunea* "might be attributed to the female sex" of *T. praeextans*. He rejected this hypothesis, however, because of the consistencies of these characters in the sample he assigned to *T. praeextans*. The pronounced EPI projections, concave rather than flat plastra, and larger size (to 540 mm) of his *T. praeextans* indicate that they are indeed males of *G. laticuneus*, thus supporting the synonymy with *G. laticuneus* by Bramble (1982).

The last fragmentary taxon that may be included here is *T. quadrata*, represented by an EPI beak and half nuchal, although it is not entirely demonstrable that these are from a single individual. Hay (1908, p. 411) noted that *Testudo quadrata* differs in having the HUM-GUL sulcus anterior to the ENT and departing from the midline at a right angle. Gilmore (1946, p. 310) and Williams (1950, p. 28) suggested that the transverse sulcus is probably an individual variation and, in any

event, is not far from the high angle seen in male *G. laticuneus* (Gilmore, 1946, plates 40-41). Williams (1950) regarded the type as undiagnostic. The flattened and parallel sided EPI beak, raised sulci on the nuchal, and relatively wide cervical scale are consistent with *Gopherus* and *Xerobates*. Despite the meagerness of this material, Gilmore (1946, p. 309, plate 44, fig. 1) managed to refer another EPI beak to this taxon, although it has a normal angle of the HUM-GUL sulci. Regardless of the identification of the latter, the type EPI cannot be positively excluded as that of a somewhat truncate male EPI beak of *G. laticuneus* and is here also considered a synonym of *G. laticuneus*.

cf. *Gopherus (Oligopherus)* sp.

Referred material—Wyoming, Brule Formation, Whitneyan?: Goshen County—UCMP loc. V5246, UCMP 42830, EPI and anterior ENT.

Subgenus *Gopherus* Rafinesque 1832

Diagnosis (emended from Bramble, 1982)— posterior EPI excavation well-developed; normal neural formula 4(8)4(8)4(6(6(6; pes phalangeal formula 2-2-2-2-0; well-developed differentiation of the radial and ulnar facets of the humerus; insertion of the biceps on the ulna at mid point; medial maxillary triturating ridge normally or not joining the median premaxillary ridge; sacculus of inner ear containing a massive otolithic mass; cranium relatively brachycephalic; inner ear chambers hypertrophied; cervical vertebrae short, pre- and postzygopophyses enlarged and drawn together; cervical 8 with elongated postzygopophyses; first dorsal vertebra with expanded zygopophyses and neural arch extended anteriorly and suturally joined to strut on nuchal; manus with three or four subradial bones; ungual phalanges enlarged and spatulate; mesocarpal joint restricted or eliminated; manus sub- or fully unguligrade.

Discussion—The diagnosis above applies to *Gopherus* of Bramble (1982) (type species: *Gopherus polyphemus* Daudin 1802). No *Gopherus (Gopherus)* are reported prior to the Hemingfordian.

"*Testudo*" *cultrata* Cope 1873, *nomen dubium*

Testudo cultratus Cope 1873a:6, syntype.
Testudo cultrata Cope 1873, Hay 1902b:451.

Lectotype—Syntype of two individuals (now lost) gular part of left EPI and pygal and right P10-11 and gular beak of right EPI (Cope, 1884, plate 68, figs. 1-3), Head of Horse Tail Creek, Weld County, Colorado, Oreodon beds, middle White River. Orellan?

Discussion—The original syntypes of this taxon have been lost and the EPI fragment that has functioned as the lectotype (Auffenberg, 1974) is apparently undiagnostic (Williams, 1950, p. 29). It is likely that this material was referable to either *H. brontops* or *G. laticunea* (male) but is here regarded as a *nomen dubium*.

Table 2. Named taxa and present status of White River Group turtles (*new combination or synonym).

Original name	This paper
Chrysemys inornata Loomis 1904	*Pseudograptemys inornata* (Loomis 1904)*
Emys culbertsonii Leidy 1851	*Stylemys nebrascensis* Leidy 1851
Emys hemisphericus Leidy 1851	*Stylemys nebrascensis* Leidy 1851
Emys oweni Leidy 1851	*Stylemys nebrascensis* Leidy 1851
?Graptemys cordifera Clark 1937	*Pseudograptemys inornata* (Loomis 1904)*
Stylemys nebrascensis Leidy 1851	unchanged
Testudo amphithorax Cope 1873	*Stylemys amphithorax* (Cope 1873)
Testudo brontops Marsh 1890	*Hesperotestudo brontops* (Marsh 1890)
Testudo cultrata Cope 1873	nomen dubium
Testudo lata Leidy 1851	*Stylemys nebrascensis* Leidy 1851
Testudo laticunea Cope 1873	*Gopherus laticuneus* (Cope 1873)
Testudo ligonia Cope 1873	*Stylemys amphithorax* (Cope 1873)
Testudo praeextans Lambe 1913	*Gopherus laticuneus* (Cope 1873)*
Testudo quadratus Cope 1885	*Gopherus laticuneus* (Cope 1873)*
Testudo thomsoni Hay 1908	*Gopherus laticuneus* (Cope 1873)*
Trachemys? antiqua Clark 1937	*Chrysemys antiqua* (Clark 1937)
Trionyx leucopotamica Cope 1891	*Apalone leucopotamica* (Cope 1891)*
Xenochelys formosa Hay 1908	unchanged

Table 3. Distribution of turtles in White River Group by North American Land Mammal "Age"*

Taxon	Chadronian	Orellan	Whitneyan
Xenochelys formosa	x		
Anosteirinae unident.	x		
Apalone leucopotamica	x	x	
cf. *Echmatemys* sp.	x		
Ptychogastrini undescribed	x		
Pseudograptemys inornata	x		
Chrysemys antiqua	x	x	x
Stylemys amphithorax	x		
Stylemys nebrascensis	x	x	
Stylemys sp.			x
Hesperotestudo brontops	x		
Hesperotestudo sp.		x	
Gopherus laticuneus	x	x	
Gopherus sp.			x

* Unconfirmed taxa listed in Harksen and Macdonald (1969) include *Graptemys* sp. from the Orellan and *S. nebrascensis* and *G. laticuneus* from the Whitneyan of the Brule Formation of South Dakota.

Testudo exornata Lambe 1906, *nomen dubium*

Testudo exornata Lambe 1906:187, plate 3, figs 1-3.

Type—NMC 6229, distal half of juvenile left C5, Bone Coulee, Cypress Hills, southwest Saskatchewan, Chadronian.

Hypodigm—NMC 6230, 6231, costal fragments from the type locality (Lambe, 1908, p. 18).

Diagnosis—Estimated length of carapace to 200 mm (Hay, 1908, p. 402), carapace strongly sculptured with growth corrugations, costals distinctly alternately wedged, inguinal articulates with costal 5-6.

Discussion—Lambe (1906) erected this taxon based on a few carapace fragments (Lambe, 1906, p. 187, plate 3, figs 1-3; 1908, p. 18, plate 1, figs. 20-22; Hay, 1908, plate 66, figs. 6-7, text fig. 506-508). Hay (1908, p. 402) noted that the specimen appears to represent a juvenile, and Williams (1950, p. 27) stated that the species "may never be determinate." Auffenberg (1974, p. 161), while noting that the species is very poorly defined, suggested that the species is close to the *Hesperotestudo turgida* (Cope 1870) lineage. The articulation of the hypoplastral buttress with C5-6 is a distinctive feature in testudinids as this usually contacts only C6 in other tortoises (Hay, 1908), but is common in batagurids. Although I agree with Williams (1950) that the types are indeterminate, I take a less pessimistic view that eventual accumulation of topotypic material may permit an adequate diagnosis of the species. Although, this taxon has not been identified from the White River Group, it is not clear that this taxon is a testudinid and comparison with the material referred to cf. *Echmatemys* sp. above may result in synonymy.

SUMMARY

Of the 18 taxa previously described from the White River Group, ten of these are synonyms or indeterminate (Table 2). I recognize eleven taxa from the White River Group, but six of these are only from the Chadronian. Of the eleven taxa known from the Chadronian, these drop to five in the Orellan and three in the Whitneyan. A cursory survey of a few collections expanded the record of aquatic turtles into the Orellan and Whitneyan (Table 3) although they appear to be rare. Further collecting of channel facies may add a few more. Nevertheless, the general pattern of reduction of the aquatic turtle fauna between the Chadronian and Orellan (Eocene-Oligocene) transition is still marked (Hutchison, 1992). Despite the long history of collecting in the White River Group, detailed studies of the turtle faunas and taxa are rare and additional research is needed to resolve persistent taxonomic, morphological, distributional, and stratigraphic uncertainties. I recognized no stratigraphic species distinctions within or between the Orellan and Whitneyan but this is likely a reflection of the poor study and/or small size of the samples and collecting bias.

ACKNOWLEDGMENTS

Contribution No. 1630 of the University of California Museum of Paleontology. Institutional support was provided by the Annie M. Alexander Endowment to the Museum of Paleontology. Philip Bjork and Janet Whitmore (SDSM), and Mary R. Dawson (CM) generously provided access to specimens and arranged loans. Kieran Shepherd (NMC), Peter Robinson (UCM), Bob Purdy (USNM), Jim Martin (SDSM), and Linda L. Thomas (ACM) provided data on types. I thank Peter Meylan and Donald Brinkman for their reviews and helpful criticisms.

LITERATURE CITED

Auffenberg, W. A. 1961. A correction regarding the phalangeal formula of the turtle *Stylemys nebrascensis* Leidy. Copeia 1961:496-498.

Auffenberg, W. A. 1962. *Testudo amphithorax* Cope referred to *Stylemys*. American Museum Novitates 2120:1-10.

Auffenberg, W. A. 1963. The fossil testudinine turtles of Florida, genera *Geochelone* and *Floridemys*. Bulletin Florida State Museum 7(2):53-97.

Auffenberg, W. A. 1964. A redefinition of the fossil tortoise genus *Stylemys* Leidy. Journal of Paleontology 38:316-324.

Auffenberg, W. A. 1974. Checklist of fossil land tortoises (Testudines). Bulletin Florida State Museum, Biological Sciences 18:121-251.

Bramble, D. M. 1971. Functional morphology, evolution, and paleoecology of gopher tortoises. Ph.D. Dissertation, University of California, Berkeley, 341 pp.

Bramble, D. M. 1974. Occurrence and significance of the Os Transiliens in gopher tortoises. Copeia 1974:102-109.

Bramble, D. M. 1982. *Scaptochelys*: generic revision and evolution of gopher tortoises. Copeia, 1982:852-867.

Case, E. C. 1919. Notes on a specimen of *Stylemys nebrascensis* (Leidy). American Journal of Science Series 4, 47: 435-438.

Case, E. C. 1925. A specimen of *Stylemys nebrascensis* Leidy, with the skull preserved. Contributions of the Museum Geology University of Michigan 2:87-73.

Case, E. C. 1936. A specimen of *Stylemys nebrascensis* Leidy, with the the bones of the feet and limbs. Contributions of the Museum of Geology, University of Michigan 5:69-73.

Clark, J. 1932. A new anosteirid from the Uinta Eocene. Annals of the Carnegie Museum 21:161-170.

Clark, J. 1937. The stratigraphy and paleontology of the Chadron formation in the Big Badlands of South Dakota. Annals of the Carnegie Museum, 25:261-350.

Clark, J., J. R. Beerbower, and K. K. Kietske. 1967. Oligocene sedimentation, stratigraphy, paleoecology and paleoclimatology in the Big Badlands of South Dakota. Fieldiana: Geology Memoirs 5:5-158.

Cope, E. D. 1873a. Second notice of extinct Vertebrata from the Tertiary of the Plains. Paleontological Bulletin 15:1-6.

Cope, E. D. 1873b. Synopsis of new Vertebrata from the Tertiary of Colorado. Washington, Government Printer Office 1-21.

Cope, E. D. 1884. The Vertebrata of the Tertiary formations of the West. Reports of the United States Geological Survey of the Territories 3:1-1781.

Cope, E. D. 1891. On Vertebrata from the Tertiary and Cretaceous rocks of the North West Territory. Contributions to Canadian Paleontology 3:1-25.

Crumly, C.R. 1984. A hypothesis for the relationships of land tortoise genera (family Testudinidae). Studia Geológica Salamanticensia, spec. vol. I:115-124

Crumly, C. R. 1994. Phylogenetic systematics of the North American tortoises (genus *Gopherus*): evidence for their classification. Fish and Wildlife Research 13:1-32.

Dobie, J. L. 1981. The taxonomic relationship between *Malaclemys* Gray, 1844 and *Graptemys* Agassiz, 1857 (Testudines: Emydidae). Tulane Studies in Zoology and Botany 23:85-102.

Emry, R. J., P. R. Bjork, and L. S. Russell. 1987. The Chadronian, Orellan, and Whitneyan North American Land Mammal Ages; pp. 118-152 *in* M. O. Woodburne (ed.), Cenozoic Mammals of North America. University of California Press, Berkeley.

Estes, R. and Hutchison, J. H. 1980. Eocene lower vertebrates from Ellesmere Island, Canadian Arctic Archipelago. Palaeogeography, Palaeoclimatology, Palaeoecology 30:325-347.

Gaffney, E. S. 1979. Comparative cranial morphology of recent and fossil turtles. Bulletin of the American Museum of Natural History 164:65-376.

Gaffney, E. S., and P. A. Meylan. 1988. A phylogeny of turtles. Systematics Association Spec. Vol. 35A:157-519.

Galbreath, E. C. 1953. A contribution to the Tertiary geology and paleontology of north-central Colorado. University of Kansas Paleontological Contributions 4:1-119.

Gilmore, C. W. 1946. The osteology of the fossil turtle *Testudo praeextans* Lambe with notes on other species of *Testudo* from the Oligocene of Wyoming. Proceedings of the United States National Museum 96:293-310.

Harksen, J. C., and J. R. Macdonald. 1969. Type sections for the Chadron and Brule Formations of the White River Oligocene in the Big Badlands, South Dakota. South Dakota Geological Survey Report of Investigations 99: 1-23.

Hay, O. P. 1902. Bibliography and catalogue of fossil vertebrates of North America. Bulletin of the United

States Geological Survey 179:1-449.

Hay, O. P. 1906. Descriptions of two new genera (*Echmatemys* and *Xenochelys*) and two new species (*Xenochelys formosa* and *Terrapene putnami*) of fossil turtles. Bulletin of the American Museum of Natural History 22:27-31.

Hay, O. P. 1907. Descriptions of new turtles of the genus *Testudo*, collected from the Miocene by Carnegie Museum: together with a description of the skull of *Stylemys*. Annals of the Carnegie Museum (1906) 4: 5-20.

Hay, O. P. 1908. The fossil turtles of North America. Carnegie Institute of Washington Publication 75:1-568.

Holman, J. A. 1972. Herpetofauna of the Calf Creek local fauna (lower Oligocene): Cypress Hills Formation of Saskatchewan. Canadian Journal of Earth Sciences 9:1612-1631.

Holman, J. A. 1976. Cenozoic herpetofaunas of Saskatchewan; pp. 80-92 *in* C. S. Churcher (ed.), Athlon, Essays on Paleontology in Honour of Loris Shano Russell. Royal Ontario Museum Life Sciences Miscellaneous Publications.

Holman, J. A., and R. M. Sullivan. 1981. A small herpetofauna from the type section of the Valentine Formation (Miocene, Barstovian), Cherry County, Nebraska. Journal of Paleontology 55:138-144.

Hutchison, J. H. 1991. Early Kinosterninae (Reptilia: Testudines) and their phylogenetic significance. Journal of Vertebrate Paleontology 11:145-167.

Hutchison, J. H. 1992. Western North American reptile and amphibian record across the Eocene/Oligocene boundary and its climatic implications; pp. 451-467 *in* D. R. Prothero and W. A. Berggren (eds.), Eocene-Oligocene Climatic and Biotic Evolution. Princeton University Press, Princeton, N. J.

Lambe, L. M. 1906. Description of a new species of *Testudo* and *Baena* wtih remarks on some Cretaceous forms. Ottawa Naturalist 19:187-196.

Lambe, L. M. 1908. Vertebrata of the Oligocene of the Cypress Hills, Saskatchewan. Contributions to Canadian Paleontology 3(4):1-65.

Lambe, L. M. 1913. Description of a new species of *Testudo* and of a remarkable specimen of *Stylemys nebrascensis* from the Oligocene of Wyoming. Ottawa Naturalist 27:57-63.

Lammers, G. E., and J. W. Hoganson. 1988. Oligocene fauna additions and new localities from North Dakota. Proceeding of the North Dakota Academy of Science 42:14.

Leidy, J. 1851. [On a new species of fossil tortoise]. Proceedings of the Academy of Natural Sciences, Philadelphia 5:172-173.

Loomis, F. B. 1904. Two new reptiles from the Titanotherium beds. American Journal of Science, Ser. 4, 18:427-432.

Marsh, O. C. 1890. Notice of some extinct Testudinata. American Journal of Science, Ser. 3, 40:177-179.

McDowell, S. B. 1964. Partition of the genus *Clemmys* and

related problems in the taxonomy of the aquatic Testudinidae. Proceedings of the Zoological Society of London 143:239-279.

Meylan, P. A. 1987. The phylogenetic relationships of soft-shelled turtles (Family Trionychidae). Bulletin of the American Museum of Natural History 186:1-101.

Mlynarski, M. 1976. Testudines. Encyclopaedia of Paleoherpetology 7:i-vi, 1-130.

Preston, R. E. 1979. Late Pleistocene cold-blooded vertebrate faunas from the mid-continental United States. University of Michigan Papers on Paleon-tology, 19:1-53.

Prothero, D. R., and C. C. Swisher III. 1992. Magnetostra-tigraphy and geochronology of the terrestrial Eocene-Oligocene transition in North America; pp. 46-53 *in* D. R. Prothero and W. A. Berggren (eds.) Eocene-Oligocene Climatic and Biotic Evolution. Princeton University Press, Princeton, N. J.

Roberts, D. C. 1962. A study of *Echmatemys callopyge* from the Uinta Eocene of Utah, and its redefinition as a subspecies of *E. septaria*. Bulletin of the Museum of Comparative Zoology, Harvard University 127: 375-400.

Rose, F. L., and W. G. Weaver, Jr. 1966. Two new species of *Chrysemys* (= *Pseudemys*) from the Florida Pliocene. Tulane Studies in Geology 5:41-48.

Sukhanov, V. B., and P. Narmandakh, 1976. Paleotol-senovyye cherepakhi Mongolii [Paleocene turtles from Mongolia.]; in N. N. Kramarenko (ed.), Paleontologiya i Biostratigrafiya Mongolii, 3, Sovmestnaya Sovetsko-Mongol'skaya Paleontolo-giche-skaya Ekspeditsiya. Trudy, Moscow.

Tinkle, D. W. 1962. Variation in shell morphology of North American turtles. I. The carapacial seam arrange-ments. Tulane Studies in Zoology 9:331-349.

Ward, J. P. 1984. Relationships of chrysemyid turtles of North America. Texas Tech University Museum Special Publications 21:1-50.

Webb, R. G. 1962. North American Recent soft-shelled turtles (Family Trionychidae). University of Kansas Publications Museum of Natural History 13:429-611.

Williams, E. E. 1950. *Testudo cubensis* and the evolution of western hemisphere tortoises. Bulletin of the American Museum of Natural History 95:1-36.

Williams, E. E. 1952a. A staurotypine skull from the Oligocene of South Dakota (Testudinata, Chelydridae). Breviora 2:1-16.

Williams, E. E. 1952b. A new fossil tortoise from Mona Island, West Indies, and a tentative arrangement of the tortoises of the World. Bulletin of the American Museum of Natural History 99:541-560.

Wood, R. C. 1977. Evolution of the emydine turtles *Graptemys* and *Malaclemys* (Reptilia, Testudines, Emydidae). Journal of Herpetology 11:415-421.

Zangerl, R. 1969. The turtle shell; pp. 311-339 *in* C. Gans, A. Bellairs, and T. S. Parsons. Biology of the Reptilia. I. Academic Press, New York.

17. Squamata

ROBERT M. SULLIVAN AND J. ALAN HOLMAN

ABSTRACT

The central North American squamatofauna is best preserved in the Orellan-aged strata, where it has its greatest diversity. By contrast the Chadronian and Whitneyan intervals are much less diverse with respect to genera present which may, in part, reflect preservation biases. Unlike the European squamatofauna associated with the "Grande Coupure," the North American squamatofauna has some taxa in common on both sides of the Eocene/Oligocene boundary, and persistence of some genera into the Miocene, and later Cenozoic, suggest that the transition of this part of the central North American herpetofauna was stepwise in nature.

A re-evaluation of the systematics of the Eocene-Oligocene Squamata from central North America (Great Plains and adjacent eastern Rocky Mountain regions) suggests a squamatofauna that is more primitive than previously thought. Prior taxonomic assignments of some fossil lizards have been based largely on primitive characters that permitted assignment to a few late Neogene-Recent taxa. Some other lizard (and snake) taxa, which included numerous species, are largely now demonstrably monotypic.

A synopsis of our study concludes that *Aciprion* is a sister taxon to the iguanines and is a monotypic taxon (*A. formosum*). *Paraphrynosoma* (new genus) is established for the species *P. greeni*, which was originally diagnosed largely by primitive features and is considered a sister taxon to, and differs from, Recent *Phrynosoma*. *Aciprion formosum*, *Crotaphytus*? *oligocenicus*, *Cypressaurus hypsodontus*, *Paraphrynosoma greeni*, cannot be assigned to any modern day families of Iguania owing to their plesiomorphic nature, and are designated Iguania *incertae sedis*.

A new unnamed species of scincomorph (previously referred to *Eumeces*) is known from two skulls and numerous isolated dentaries and maxillae. *Palaeoxantusia* cf. *P. borealis* is noted from the first skull material recovered from the late Eocene of Flagstaff Rim. Additional material of *Parophisaurus pawneensis* and *Helodermoides tuberculatus* add to our knowledge of these fossil anguids. *Peltosaurus granulosus* is a monotypic taxon; *P. abbotti* and *P. floridanus* are subjective junior synonyms. *Lowesaurus matthewi* is an extremely rare taxon known from only three specimens, all from different intervals within the Oligocene. Two vertebrae previously referred to *Saniwa* sp. are reassigned to *Helodermoides*, consequently *Saniwa* is known with certainty from only early and middle Eocene strata of North America.

Only one species of rhineurid amphisbaenian is valid: *Rhineura hatcheri*. *Gilmoreia attenuatus*, *Pseudorhineura minuta*, *R. amblyceps*, *R. attenuatus*, *R. hibbardi*, *R. minutus*, *R. sternbergii*, and *R. wilsoni* are all subjective junior synonyms of *R. hatcheri*. Thus, *Gilmoreia* and *Pseudorhineura* are synonyms of *Rhineura*. The genus *Lestophis* and the species *L. crassus* and *L. anceps* are *nomina dubia*. *Hyporhina* is known by two species: *H. tertia* from the late Eocene and *H. antiqua* from the early-middle Oligocene. *H. galbreathi* is placed into synonymy with *H. antiqua*. The holotypes of the following problematic lizard taxa have been lost and are *nomina dubia*: *Cremastosaurus carinicollis*, "*Cremastosaurus*" *rhambastes*, and *Diacium quinquepedale*.

Valid snake taxa include: *Coprophis dakotaensis*, *Boavus* cf. *B. occidentalis*, *Calamagras angulatus*, *Calamagras murivorus*, *Calamagras weigeli*, *Helagras orellanensis*, *Ogmophis compactus*, *Geringophis vetus*, and *Texasophis galbreathi*. *Calamagras talpivorus* is synonymized with *C. murivorus*. Snake taxa are less diverse in the late Eocene than in the early Oligocene, which saw the first appearance of the boid genus *Coprophis*, the distinctive boid genus *Geringophis*, and the earliest North American colubrid, *Texasophis*.

INTRODUCTION

The Tertiary sediments of the northern plains of North America that embrace the Eocene/Oligocene boundary have yielded numerous salamanders, frogs, lizards, amphisbaenians, snakes and turtles. These lower vertebrates are an extremely important part of the transitional (from archaic to modern) terrestrial vertebrate fauna, important not only from an evolutionary standpoint, but from a biostratigraphic one as well. Here we review and document the occurrence of Squamata (see usage below) from the Great Plains and adjacent eastern Rocky Mountain regions of the northern portion of the United States and south-central Saskatchewan, Canada, presenting a revised biostratigraphy for this part of the paleoherpetofauna, as well as a revision of some taxa. Because a number of studies concerning specific taxa within this squamatofauna are currently in preparation (e.g., a revision of *Peltosaurus* and additions to *Parophisaurus*, by Sullivan), the present study is intended to provide a synopsis of current understanding of central North American squamates at the Eocene/

Oligocene boundary, noting new material and commenting on the taxonomic status of all known species from these deposits. Although we acknowledge that some of the taxa discussed herein are not diagnosable by apomorphies (Gauthier, personal communication to RMS, 1994) we recognize that these fossil taxa can be distinguished (characterized) by a combination of primitive and derived features. We do not attempt to address the validity of higher taxonomic categories to which many of the fossil taxa have been previously placed. In this way, the present chapter is intended only to update and revise the accounts of relevant taxa presented by Holman (1979), Estes (1983), Rage (1984a), and Hutchison (1992) in light of recent geochronological studies by Prothero (1985a, b, and c), Prothero et al. (1982, 1983), and Prothero and Swisher (1992). This contribution is dedicated to the memory of our friend and colleague, the late Richard Estes.

ABBREVIATIONS

AMNH, American Museum of Natural History, New York; BHI, Black Hills Institute, Hill City; CM, Carnegie Museum of Natural History, Pittsburgh; F:AM, Frick Collection, American Museum of Natural History, New York; FHSM, Fort Hays Sternberg Museum, Hays; KU, University of Kansas, Museum of Natural History, Lawrence; LACM, Natural History Museum of Los Angeles County, Los Angeles; MSUVP, Michigan State University Museum, East Lansing; NMC, National Museum of Canada, Ottawa; YPM-PU, Princeton University Collection, Yale Peabody Museum, New Haven; SDSM, South Dakota School of Mines, Rapid City; SMNH, Saskatchewan Museum of Natural History, Regina; UMMP, University of Michigan Museum of Paleontology, Ann Arbor; UNSM, University of Nebraska State Museum, Lincoln; USNM, United States National Museum, Washington, D.C.

SYSTEMATIC PALEONTOLOGY

We follow the recent taxonomy proposed by Estes et al. (1988), that refers to Amphisbaenia and Serpentes as Scleroglossa *incertae sedis*, and that of Frost and Etheridge (1989) for more inclusive taxonomic categories within Iguania.

SQUAMATA Merrem, 1820
IGUANIA Cope, 1864
IGUANIA *incertae sedis*
Aciprion Cope, 1873a
Aciprion formosum Cope, 1873a

Aciprion formosum Cope, 1873a, p. 17.
Aciprion majus Gilmore, 1928, p. 20, fig. 7; Plate 20, fig. 11.
Holotype—AMNH 1609, dentary with seven teeth.
Type Locality—Cedar Creek, Logan Co., Colorado.
Previously Referred Material—See Estes (1983, p. 24).
New Referred Material—LACM 4893, skull and

lower jaws, Brule Fm. (Orella Mbr.), Converse Co., Wyoming; MSUVP (unnumbered) numerous dentary and maxillary fragments, Brule Fm., Logan Co., Colorado.
Stratigraphic Distribution—Brule Fm. (Orella Mbr.).
Age Distribution—Early Oligocene (Orellan).
Remarks—Estes (1983) provided the most recent historical overview of this enigmatic lizard. This monotypic taxon was recently considered in a study by David Hauser (unpublished masters thesis, San Diego State University), who reports (personal communication to RMS, 1993) that *Aciprion* is the sister taxon to the iguanines (*Brachylophus* + *Dipsosaurus* + iguanas; Frost and Etheridge, 1989). Holman (1989) referred some incomplete dentaries and maxillary fragments to cf. *Aciprion* sp., noting that they departed from known *Aciprion formosum* in having more widely spaced teeth. Despite this difference, *Aciprion* is still considered to be a monotypic taxon.

In an abstract, Rossman (1993) recently established the subfamily Messelosaurinae for *Aciprion, Caduciguana, Cypressaurus, Geiseltaliellus* and "*Holmanisaurus*" (= *Crotaphytus*? *oligocenicus*, see below) based on undisclosed "close relationships" to extant "basiliscine" lizards. We cannot comment on the placement of all these North American and European taxa within this group at this time. Furthermore, we are unable to corroborate the monophyly of the Messelosaurinae as defined by Rossman (1993). Given the recent assessment of the Iguania by Frost and Etheridge (1989) and the work of Hauser (personal communication to RMS), it seems doubtful that *Aciprion* has "basiliscine" affinities.

Unnamed new genus
Crotaphytus? *oligocenicus* (Holman, 1972)

Crotaphytus oligocenicus Holman, 1972, p. 1618, fig. 3.
Crotaphytus? *oligocenicus* (Holman, 1972) Estes, 1983, p. 30.
Holotype—SMNH 1444, right dentary.
Type Locality—Calf Creek screening locality, Calf Creek Valley on NE facing slope, Saskatchewan, Canada (see Storer and Tokaryk, 1992).
Previously Referred Material—SMNH 1445, three left dentaries and two right dentaries.
New Referred Material—None.
Stratigraphic Distribution—Cypress Hills Fm.
Age Distribution—Late Eocene (Chadronian).
Remarks—Recently, Rossman (1993), in an abstract, published the taxon "*Holmanisaurus*" without a definition or diagnosis. It has been determined that he established this taxon in an unpublished thesis (Rossman, 1992) for the reception of the species *Crotaphytus*? *oligocenicus*. Consequently, "*Holmanisaurus*" is not a valid taxon and is presently a *nomen nudum*. However, we recognized that it may eventually become a valid taxon when, and if, proper publication is realized. We also note here that the holotype and

paratypes of *Crotaphytus*? *oligocenicus* (and other Cypress Hills squamate type specimens described by Holman, 1972) were reported lost by Storer and Tokaryk (1992), but all of this missing material has recently been relocated.

Estes (1983) noted that *C.*? *oligocenicus* was more primitive than extant species of the genus, and that it lacked characters that would unequivocally permit referral to *Crotaphytus*. Estes (1983) also suggested that it might be more allied to the tropodurine (e.g., "*Tropidurus*" group of Frost and Etheridge, 1989) iguanines rather than *Crotaphytus* based on the primitive nature of its tooth implantation and the Meckelian sulcus opening. Furthermore, Estes (1983, p. 30) stated that the lack of definitive characters made it impossible to refer the Cypress Hills taxon to any extant (iguanine) taxon and that it was not adequate to diagnose a new taxon. Because we recognize that this fossil taxon can be distinguished from modern *Crotaphytus*, and all other fossil Iguania based on distinct features, some of which are plesiomorphic, we believe that reference of this species to modern *Crotaphytus* is not defensible.

Cypressaurus Holman, 1972
Cypressaurus hypsodontus Holman, 1972

Cypressaurus hypsodontus Holman, 1972, p. 1617, fig. 2.
Holotype—SMNH 1442, left dentary.
Type Locality—Calf Creek screening locality, Calf Creek Valley on NE facing slope, Saskatchewan, Canada (see Storer and Tokaryk, 1992).
Previously Referred Material—SMNH 1443, left dentary (paratype).
Stratigraphic Distribution—Cypress Hills Fm.
Age Distribution—Late Eocene (Chadronian).
Remarks—Estes (1983) believed this species to be a primitive form of *Sceloporus*. As with *Crotaphytus*? *oligocenicus*, *Cypressaurus hypsodontus* is distinctive, yet undiagnosable, but can be differentiated from other Iguania.

Paraphrynosoma new genus

Type Species—*Paraphrynosoma greeni* (Holman, 1989, p. 18, fig. 4).
Diagnosis—Same as for species.
Remarks—See below.

Paraphrynosoma greeni (Holman, 1989)

Phrynosoma greeni Holman, 1989, p. 18, fig 4.
Holotype—MSUVP 1135, right dentary.
Type Locality—Flats Locality, W1/2 Sec. 7, T11N, R33W, Logan Co., Colorado.
Previously Referred Material—None.
New Referred Material—None.
Stratigraphic Distribution—Brule Fm. (Scenic Mbr., levels 11-12).
Age Distribution—Early Oligocene (Orellan).
Diagnosis—See Holman (1989, p. 18).

Remarks—The holotype (MSUVP 1135) is the sole specimen of this species. Holman (1989) cited five characters that serve to diagnose this species and separate it from the Pliocene species *Phrynosoma holmani*. These characters include: 1) lack of strong sculpturing on the open posterolabial expanse of the dentary; 2) absence of the bevelled posterior part of the ventrolateral part of the dentary; 3) dorsolateral surface of the dentary flattened below the tooth row; 4) the groove located below the Meckelian canal more lateral in position (compared to *Phrynosoma holmani*); and 5) teeth slightly lower crowned and incipiently tricuspid. Because these characters are plesiomorphic compared to those seen in the Pliocene *Phrynosoma holmani* and Recent species of *Phrynosoma*, we believe that removal from *Phrynosoma* is warranted. The oldest putative record of *Phrynosoma* is from the late Miocene (see Estes, 1983). As with so many fragmentary Late Cretaceous, Paleogene, and early Neogene taxa (see Estes, 1983; Good, 1988; Pregill, 1992; see below), a number of previous taxonomic referrals to Recent lizard taxa have been made based on the presence of primitive characters which have no phylogenetic value. Because of their nature, these and other referrals constitute form taxa. *Paraphrynosoma greeni* lacks characters that would unambiguously allow reference to the Phrynosomatidae as characterized by Frost and Etheridge (1989). Therefore we place this species in Iguania *incertae sedis*.

SCLEROGLOSSA
Estes, de Queiroz, and Gauthier, 1988
SCINCOIDEA Oppel, 1811
?SCINCIDAE Gray, 1825
Unnamed new taxon

Holotype—F:AM 42916, nearly complete skull and lower jaws.
Type Locality—Southwest corner of White Buttes, south of Amidon, Slope Co., North Dakota.
Referred Material—CM 428, incomplete skull and lower jaws, from Prairie Dog Creek, Nebraska.
Stratigraphic Distribution—Brule Fm. (Dickinson Mbr.) for F:AM 42916; Brule Fm. (Orella Mbr.) for CM 428.
Age Distribution—Early Oligocene (Orellan).
Remarks—Estes (1983) cited two specimens of *Eumeces* sp. (CM 428 and F:AM 42916) from the "middle Oligocene" of Nebraska and North Dakota, respectively, along with other specimens from various strata of the later Cenozoic of North America. He concluded that they were all too fragmentary to determine their specific status. However, the two Oligocene specimens (both from the Brule Fm.) appear to be diagnosable and will be described elsewhere (Sullivan, in preparation). They are tentatively referred to *Eumeces* pending further study, but as in the case with the aforementioned iguanines, reference to the Recent taxon (i.e., *Eumeces*) is suspect.

XANTUSIIDAE Baird, 1859
Palaeoxantusia Hecht, 1956

Type Species—*Palaeoxantusia fera*, Hecht, 1956, p. 4, fig. 1.

Remarks—The genus *Palaeoxantusia* is known from the Paleocene (Torrejonian) through Eocene (Chadronian) strata of North America (Estes, 1983; this chapter, see below).

Palaeoxantusia borealis Holman, 1972

Palaeoxantusia borealis Holman, 1972, p. 1621, fig. 4.

Holotype—SMNH 1435, right spleniodentary.

Type Locality—Calf Creek screening locality, Calf Creek Valley on NE facing slope, Saskatchewan, Canada (see Storer and Tokaryk, 1992).

Referred Material—SMNH 1436 five left and four right spleniodentaries (paratypes).

Stratigraphic Distribution—Cypress Hills Fm. (see *Palaeoxantusia* sp. indet. below for further comments).

Age Distribution—Late Eocene (Chadronian).

Remarks—Holman (1972) diagnosed this species from *Palaeoxantusia fera* as "1) somewhat smaller, 2) groove between teeth and lingual shelf deeper, 3) coronoid incision deeper and with its borders stronger, 4) nutritive foramina larger, irregularly spaced, [and] 5) groove for M. adductor mandibulae externis extending more posteriad." Cladistically, *Palaeoxantusia*, *P. fera* and *P. borealis* cannot be diagnosed using apomorphies (Gauthier, personal communication to Sullivan, 1994) but can be differentiated.

Palaeoxantusia cf. *P. borealis*

New Referred Material—USNM 214716, left mandible and unidentified element; USNM 214717, incomplete skull consisting of left mandible, right spleniodentary, both nasals, articulated thoracic vertebrae, and an unidentified bone fragment; USNM 214719, crushed skull consisting of paired nasals, frontals, parietals, portions of both postfrontals, both squamosals, incomplete right jugal, incomplete right maxilla, both mandibles, and ventral part of the basioccipital; USNM 214720, right mandible and unidentified skull fragments; USNM 214721, incomplete skull consisting of portions of the right mandible and maxilla, left spleniodentary and unidentified skull fragments; USNM 214728, left frontal and medial part of left spleniodentary; and USNM (unnumbered), left spleniodentary.

Stratigraphic Distribution—White River Fm., Natrona Co., Wyoming.

Age Distribution—Late Eocene (Chadronian).

Remarks—This material (all of which was collected from Al's Pocket, south fork of Lone Tree Gulch, Chadron Formation, approximately 18.2 meters below ash F, Flagstaff Rim, Wyoming) greatly supplements our knowledge of late Eocene *Palaeoxantusia*. The Flagstaff Rim specimens (particularly USNM 214719) will

allow critical comparison between *P. kyrentos* from the middle Eocene Mission Valley Formation of San Diego, *P. allisoni* from the same formation (Schatzinger, 1980), and the type species, *P. fera*, from various Paleocene and early-mid Eocene localities in the Rocky Mountain region (Estes 1976, Hecht 1956; and Sullivan, 1982). Estes (1983) citing Gauthier (personal communication) noted some derived features shared between *P. kyrentos* and modern *Lepidophyma* that led him to suggest that the former may represent a new genus of xantusiid lizard. The USNM specimens will be described elsewhere (Gauthier, in preparation).

ANGUIMORPHA Fürbringer, 1900
XENOSAURIDAE Cope, 1886
Exostinus Cope, 1873a

Type Species—*Exostinus serratus* Cope, 1873a, p. 16.

Remarks—Two species of *Exostinus* are currently recognized: *E. serratus*, from the Oligocene of Colorado and Wyoming and *E. lancensis*, from the Late Cretaceous and Paleocene (Torrejonian) of Montana (Estes, 1983).

Exostinus serratus Cope, 1873a

Exostinus serratus Cope, 1873a, p. 16.

Holotype—AMNH 1608, frontal, left jugal, and dentary fragment.

Type Locality—Cedar Creek, Logan Co., Colorado.

Referred Material—none.

Stratigraphic Distribution—Brule Fm. ("Oreodon beds").

Age Distribution—Early Oligocene (Orellan).

Remarks—*Exostinus serratus* is known only from the type specimen, so its stratigraphic occurrence is restricted. Estes (1983) noted that comparison of *E. serratus* to *E. lancensis* is difficult because the type specimens of each species consist largely of different elements, making comparison difficult. However, the latter species differs from the former in having a more rudimentary intramandibular septum and by the presence of a sulcus dentalis.

ANGUIDAE Gray, 1825
ANGUINAE (Gray, 1825)
Parophisaurus Sullivan, 1987

?*Xestops* Gilmore, 1928, p. 150 (in part).
Pancelosaurus Meszoely, 1970, p. 117 (in part).
Machaerosaurus Meszoely et al., 1978, p. 157 (in part).
Parophisaurus Sullivan, 1987, p. 118, fig. 2.

Type Species—*Parophisaurus pawneensis* (Gilmore, 1928, p. 150).

Remarks—Sullivan (1987) established this genus for material previously referred to *Machaerosaurus, Pancelosaurus* (= *Odaxosaurus*, in part: see Meszoely, Estes, and Haubold, 1978, p. 157) and ?*Xestops*.

Parophisaurus pawneensis (Gilmore, 1928)

?*Xestops pawneensis* Gilmore, 1928, p. 150.
Pancelosaurus pawneensis Meszoely, 1970, p. 117.
Machaerosaurus pawneensis Meszoely et al., 1978, p. 157.
Parophisaurus pawneensis Sullivan, 1987, p. 118, fig. 2.

Holotype—KU 1281, incomplete left maxilla, medial section of right mandible, posterior fragment of right frontal with overlying dermal armor, two unidentifiable bone fragments (see Sullivan, 1987).

Type Locality—48.2 kilometers north of Sterling, Sec. 28, T11N, R53W, Logan Co., Colorado.

Previously Referred Material—See Sullivan (1987) for list of previously referred material.

New Referred Material—BHI 1262, nearly complete skull and lower jaws, Brule Fm. (Orella Mbr.), Sioux Co., Nebraska; CM 60405, nearly complete skull and lower jaws, Camel horizon, White River Fm. (Cedar Creek Mbr.), Pawnee Buttes, Weld Co., Colorado; KU 7662, anterior part of right mandible (including dentary, coronoid process and surangular), anterior part of left dentary, medial fragment of left maxilla, nearly complete premaxilla with fragment of left maxilla and distal portions of both vomers and anterior part of right maxilla, associated with unidentified fragments and a glyptosaurine osteoderm, ?Brule Fm. ?Casement Ranch, ?Logan Co., ?Colorado; KU 7668, anterior part of left dentary, right dentary with anterior parts of right surangular and right coronoid, left pterygoid-palatine-vomer and one vertebra, ?Brule Fm., ?Casement Ranch, ?Logan Co., ?Colorado; KU 125307, anterior part of frontal, right prefrontal, proximal part of left prefrontal, posterior fragment of right maxilla and parts of both vomers and palatines, Brule Fm. (Orella Mbr.), W1/2 Sec. 7, T11N, R53E, Logan Co., Colorado; LACM 5103, medial part of right dentary, CIT 272, Brule Fm. (Orella Mbr.), Grover, Weld, Co., Colorado; LACM 81640, medial part of right dentary, Brule Fm. (Orella Mbr.), 6459/Orin, Converse Co., Wyoming; MSUVP (unnumbered) numerous isolated dentary and maxillary fragments, SW1/4 Sec 12, T11N, R54W, Brule Fm. (?Orella Mbr.), Logan Co., Colorado; SDSM 2716, maxilla fragment, humerus, numerous vertebrae, osteoderms and unidentified bone fragments in a bone conglomerate, Brule Fm. (Scenic Mbr.), Cottonwood Pass, Shannon Co., South Dakota; SDSM 20183, incomplete parietal, fragment of left dentary, posterior part of right maxilla, posterior part of right mandible, proximal end of humerus, osteoderms and miscellaneous bone fragments, Brule Fm. (Scenic Mbr.), Pass Creek, Jackson Co., South Dakota; SDSM 20203, incomplete parietal table, Brule Fm. (Scenic Mbr., Zone D of Lillegraven, 1970), Reva Gap, Harding Co., South Dakota; and SDSM 63130, a minimum of two individuals represented by: two frontals, two parietals; two left dentaries; two maxillary fragments, numerous vertebrae and bone fragments (all are heavily concreted with limonite), Brule Fm. (Zone D of Lillegraven, 1970), V6320, Reva Gap, Slim Butte locality # 11, Harding Co., South Dakota.

Stratigraphic Distribution—Brule and White River fms. (Cedar Creek, Orella and Scenic mbrs.).

Age Distribution—Early Oligocene (Orellan).

Remarks—This new material is a significant addition to previously referred material reported by Sullivan (1987) and will be described elsewhere (Sullivan, in preparation). *Parophisaurus* is known only from the early Oligocene (Orellan) of North America and is thought to be the sister taxon of extant North American *Ophisaurus* (*sensu stricto*).

GLYPTOSAURINAE Marsh, 1872
"MELANOSAURINI" Sullivan, 1979
Peltosaurus Cope, 1873b

Type Species—*Peltosaurus granulosus* Cope, 1873b, p. 5.

Remarks—Five species of *Peltosaurus* have been named: only two, *P. granulosus* and *P. abbotti*, were recognized as valid by Estes (1983) although he suggested that the latter species differed little from the former, thereby implying synonymy with *P. granulosus* (see remarks below).

Peltosaurus granulosus Cope, 1873b

Peltosaurus granulosus Cope, 1873b, p. 5.
Peltosaurus abbotti Gilmore, 1928, p. 135, fig. 83; Plate 22, figs. 1, 2.
Peltosaurus abbottii (*lapsus calami*) Meszoely, 1970, p. 130.
Peltosaurus floridanus Vanzolini, 1952, p. 457, plate 56, figs. 7, 8.

Holotype—AMNH 1610, medial parts of right and left mandibles, anterior part of left dentary, posterior part of left maxilla, fragment of right dentary, incomplete right jugal, anterior part of frontal, anterior part of parietal table, premaxilla, vertebra, osteoderms and miscellaneous skeletal fragments, all of which are not part of the holotype.

Type Locality—Cedar Creek, Logan Co., Colorado.

Previously Referred Material—See Gilmore (1928), Meszoely (1970), and Estes (1983).

New Referred Material—A number of newly discovered specimens of *Peltosaurus* are now known (too numerous to cite here) and will be the subject of a forthcoming paper (Sullivan, in preparation).

Stratigraphic Distribution—Brule, Gering and White River formations.

Age Distribution—Oligocene (Orellan through early Arikareean).

Remarks—*Peltosaurus* is the most common of all Oligocene "melanosaur" lizards and is known from numerous specimens, many of which have not been previously reported. A study of the new and previously referred material by one of us (RMS) suggests that *P. abbotti* is actually an aberrant *P. granulosus*, rather than a distinct species. A more formal assessment of this taxon is currently being conducted (Sullivan, in prep.).

Estes (1963) synonymized *P. floridanus* with *P. granulosus*, recognizing that the Florida specimen was indistinguishable from *Peltosaurus granulosus* and that the preservation of the "Florida" holotype is identical to the preservation of typical "White River" fossils, an observation with which we concur. Numerous pre-Oligocene lizards have been referred to the taxon *Peltosaurus*, but all of these are misidentifications and represent mostly undescribed basal glyptosaurines.

In a recent paper, Albright (1994) referred some early Miocene (Arikareean) osteoderms to ?*Peltosaurus* sp. The figures of his osteoderms (Albright, 1994, p. 1137, figs. 3.3 and 3.4) are similar to those of *Peltosaurus* and reference to the Glyptosaurinae is certain. However, inspection of casts of these osteoderms suggests that they might have been reworked from older sediments (Oligocene strata) reported in the area (Albright, 1994), as the sides and sculptural surfaces are smoothly abraded. The youngest definitive *Peltosaurus* is a specimen (UNSM 81001) from the Gering Formation which is considered early Arikareean (late Oligocene) in age by Tedford et al. (1987). Putative Miocene occurrences of *Peltosaurus* have been reported (see Estes, 1983), but like the putative pre-Oligocene specimens, these Miocene specimens are also misidentified. Therefore, *Peltosaurus* is restricted with certainty to the Oligocene deposits of North America, although an early Miocene occurrence is possible.

GLYPTOSAURINI Sullivan, 1979
Helodermoides Douglass, 1903
Helodermoides tuberculatus Douglass, 1903

Helodermoides tuberculatus Douglass, 1903, p. 160, figs. 4, 5.
Glyptosaurus montanus Douglass, 1908, p. 278, fig. 1.
Glyptosaurus giganteus Gilmore, 1928, p. 119, Plate 14, fig 1.
Holotype — CM 707, incomplete paired frontals, nasals and partial left dentary and other skull fragments (see Gilmore, 1928; Sullivan, 1979).
Type Locality — Pipestone Springs, Jefferson Co., Montana.
Previously Referred Material — See Sullivan (1979) and Estes (1983).
New Referred Material — YPM-PU 16304, posterior part of left frontal with cephalic osteoderms; SDSM 5317, nearly complete right frontal, incomplete basicranium, parietal with distal part of squamosal in articulation, cephalic and body osteoderms, vertebrae and unidentified bone fragments (associated with three articulated snake vertebrae), V5311, Yoder Quarry 2, Chadron Fm. (Yoder Mbr.), Goshen Co., Wyoming; SDSM 5320, numerous body osteoderms, V5310, Yoder Quarry 1, Goshen Co., Wyoming; SDSM 6339, incomplete trunk vertebra and numerous body osteoderms (associated with numerous unidentifiable bone fragments), V5369 General Yoder Locality, Chadron Fm. (Yoder Mbr.), Goshen Co., Wyoming; SDSM 6349, anterior part of left maxilla with two teeth, V5369 General Yoder Locality, Chadron Fm.

(Yoder Mbr.), Goshen Co., Wyoming; SDSM 53303, incomplete right pterygoid with tooth patch, V5311, Yoder Quarry 2, Chadron Fm. (Yoder Mbr.), Goshen Co., Wyoming; SDSM 53465, anterior part of left maxilla with five complete teeth, three broken teeth and three incomplete trunk vertebrae (associated with numerous unidentified bone fragments and three incomplete snake vertebrae), Yoder Quarry 1, Chadron Fm. (Yoder Mbr.), Goshen Co., Wyoming; SDSM 53522, left scapulocoracoid and five incomplete body osteoderms (associated with snake vertebrae, unidentifiable ribs, and other incomplete material not referable to this taxon), and incomplete juvenile skull, Quarry 1, Chadron Fm. (Yoder Mbr.), Goshen Co., Wyoming; USNM 466053, juvenile individual consisting of both frontals and incomplete anterior parts of both maxillae, left and right ectopterygoids, anterior parts of both pterygoids, and unidentified bone fragments, White River Group (undifferentiated), pocket 13.4 m below Ash B, Little Lone Tree Gulch, Natrona Co., Wyoming.

cf. *Helodermoides tuberculatus* Douglass, 1903

New Referred Material — SDSM 5335, two anterior caudal vertebrae (associated with snake vertebrae and mammalian elements, Quarry 1, Chadron Fm. (Yoder Mbr.), Goshen Co., Wyoming; SDSM 6347, distal end of right humerus, no specific locality, Chadron Fm.; SDSM 6348, proximal end of right humerus, type locality, Chadron Fm. (Yoder Mbr.), Goshen Co., Wyoming; SDSM 53297, two dentary fragments, Yoder Quarry 2?, Chadron Fm. (Yoder Mbr.), Goshen Co., Wyoming; SDSM 53303, right ectopterygoid and pterygoid with tooth patch, Yoder Quarry 2?, Chadron Fm. (Yoder Mbr.), Goshen Co., Wyoming; and SDSM 53520, trunk vertebra, distal end of fibula, and proximal end of tibia, Yoder Quarry 1, Chadron Fm. (Yoder Mbr.), Goshen Co., Wyoming; USNM 10968, nearly complete caudal vertebra, White River Group, Hat Creek Basin, Sioux Co., Nebraska (removed from *Saniwa*, see below); and USNM 241330, incomplete vertebra, White River Fm. (undifferentiated), south fork of Lone Tree Gulch, Natrona Co., Wyoming (removed from *Saniwa*, see below).
Stratigraphic Distribution — Brule, Chadron, Renova and White River formations (undifferentiated).
Age Distribution — Late Eocene (Chadronian) to early Oligocene (Orellan).
Remarks — *Helodermoides tuberculatus* was first described by Douglass (1903) based on paired frontals and portions of the left dentary and other lesser elements from the Renova Formation (= "Pipestone Creek beds") of Jefferson Co., Montana. Gilmore (1928) synonymized *Helodermoides* with *Glyptosaurus*. Sullivan (1979) resurrected the genus *Helodermoides*, recognizing numerous characters that served to separate it from *Glyptosaurus* and other glyptosaurs (see also Sullivan, 1989). *Helodermoides* is known from both Chadronian and Orellan age deposits.

VARANOIDEA Camp, 1923
HELODERMATIDAE Gray, 1837
Lowesaurus Pregill, Gauthier, and Greene, 1986

Heloderma Gilmore, 1928, p. 89, fig. 59; Plate 11, figs. 1, 1a (in part).
Type Species—*Lowesaurus matthewi* Pregill, Gauthier, and Greene, 1986, p. 181, fig. 4.
Remarks—See below.

Lowesaurus matthewi (Gilmore, 1928)

Heloderma matthewi Gilmore, 1928, p. 89, fig. 59; Plate 11, figs. 1, 1a.
Lowesaurus matthewi Pregill, Gauthier and Greene, 1986, p. 181, fig. 4.
Holotype—AMNH 990A, posterior part of left maxilla with three teeth.
Type Locality—Lewis Creek, Logan Co., Colorado.
Previously Referred Material—See Pregill et al. (1986, p. 183).
Stratigraphic Distribution—Brule Fm. (?Orella Mbr.), Logan Co., Colorado; Brule Fm. (Whitney Mbr.), Morrill Co., Nebraska; and Gering Fm., Morrill Co., Nebraska.
Age Distribution—Oligocene (Orellan, Whitneyan, and Arikareean).
Remarks—Pregill et al. (1986) reassessed fossils referred to the Recent genus *Heloderma* and assigned these specimens to a new genus, *Lowesaurus matthewi* (Gilmore). Only three specimens are known; one from each of the first three Oligocene land-mammal "ages." Because previous correlations put the Arikareean within the Miocene (rather than the Oligocene), this lizard was thought to have a greater biostratigraphic range by earlier workers. The new stratigraphic calibration restricts *Lowesaurus* to the Oligocene of North America.

VARANIDAE Gray, 1827
Saniwa Leidy, 1870

Type Species—*Saniwa ensidens* Leidy, 1870, p. 124.
Remarks—The fossil varanid lizard *Saniwa* is known by eight named species, and only one of them (*S. ensidens*) is adequately known (Estes, 1983). *Saniwa* may occur as early as the Paleocene (Torrejonian) of North America (Sullivan, 1982), and its post-Eocene occurrences are no longer accepted (see below).

Saniwa sp. indet.

Previously Referred Material—USNM 10968, caudal vertebra and USNM 241330, vertebra.
New referred material—None.
Age Distribution—Early-middle Eocene (Wasatchian through ?Uintan).
Remarks—It is worth noting here that there is often confusion regarding the identification of vertebrae belonging to *Helodermoides* and those of *Saniwa*. In *Helodermoides*, the dorsal vertebral condyle (as seen in ventral view) is smaller and the centrum is more triangular. In ventral view, the dorsal vertebrae of *Saniwa* are distinguished by an incipient posterior constriction (in front of the condyle) resulting in a subparallel form of the centrum. Typically, the dorsal vertebrae of *Saniwa* can be further distinguished by a smooth ventral surface of the centrum. In *Helodermoides* this region is slightly raised and extends from the cotyle to the condyle.

Two isolated vertebrae were cited by Estes (1983) as being Oligocene in age. Regarding the first, Estes (1983, p. 187) remarked that "the undescribed Lower Oligocene specimen (USNM 241330) is well preserved and distinctively varanid" and tentatively referred it to *Saniwa*. The specimen is an incomplete dorsal vertebra consisting solely of the posterior ventral portion of the centrum and condyle. Sullivan (1979) referred USNM 241330 to cf. Anguidae and noted that the specimen consisted of two vertebrae. Estes (1983) only noted one vertebra, and presently there is only one vertebra assigned this number. This apparent discrepancy was a mistake on Sullivan's part, but has no bearing on our taxonomic conclusion. Contrary to Estes's observation, this specimen conforms to the condition in *Helodermoides*, where there is present a distinct demarcation (depression) between the ventroanteriad part of the condyle and the ventral surface of the centrum. The second specimen, USNM 10968, is a distal caudal vertebra. Estes assigned this specimen to *Saniwa* based on the fact that the haemal arches articulate directly to the ventral portion of the centrum, but this is a condition also seen in *Helodermoides*. This particular specimen compares readily to other distal caudal vertebrae belonging to *Helodermoides*. A fracture plane is present, albeit faint, and the autotomic septum can be traced laterally on both sides of the vertebra, reaching anteriorly to coincide with the ventral margin of the cotyle. Varanids, to which *Saniwa* belongs, lack autotomic septa on their caudal vertebrae (Etheridge, 1967). Both specimens are reassigned to cf. *Helodermoides tuberculatus* (see above) and are therefore removed from *Saniwa*. *Saniwa* is restricted to the early and middle Eocene (Wasatchian-Bridgerian and questionably Uintan) of North America.

SCLEROGLOSSA *incertae sedis*
AMPHISBAENIA Gray, 1844
RHINEURIDAE Vanzolini, 1951
Rhineura Cope, 1861

Lepidosternon (Baird, 1859).
Type Species—*Rhineura floridana* (Baird, 1859, p. 255).
Remarks—If the fossil genus is correctly referred, the amphisbaenian *Rhineura* has a relatively long stratigraphic range (early Oligocene to Recent). It is the only Recent squamate genus represented in the early Oligocene squamatofauna.

Rhineura coloradoensis (Cope) 1873a

Platyrhachis coloradoensis Cope, 1873a, p. 19.
Rhineura coloradoensis Gilmore, 1928, p. 42.
Platyrhachis coloradensis (lapsus calami) Estes 1983, p. 199.
Rhineura coloradensis (lapsus calami) Estes 1983, p. 199.

Holotype—AMNH 1607, three vertebrae.
Type Locality—Horse Tail Creek, Logan Co., Colorado.
Referred Material—None.
Stratigraphic Distribution—Chadron Fm.
Age Distribution—Late Eocene (Chadronian).
Remarks—Estes (1983) offered little comment on the validity of "Rhineura (= Platyrachis) coloradensis" [sic] other than the fact that Gilmore (1928) had suggested earlier that this taxon may in fact be conspecific with Rhineura hatcherii. The holotype has been lost, and the species is not diagnosable. Rhineura coloradoensis is a nomen dubium.

Rhineura hatcheri Baur, 1893

Rhineura hatcherii Baur, 1893, p. 998.
Rhineura sternbergii Walker, 1932, p. 225.
Rhineura minutus Gilmore, 1938a, p. 12, fig. 1.
Gilmoreia attenuatus Taylor, 1951, p. 527, fig. 1; Plate 58, figs. 3, 4, 5; text fig. 1 a, b, c.
Rhineura hibbardi Taylor, 1951, p. 539.
Rhineura amblyceps Taylor, 1951, p. 543, fig 5; Plate 59, figs. 1-5; text figs. 5a-c, 6a-c.
Rhineura wilsoni Taylor, 1951, p. 548.
Pseudorhineura minuta Vanzolini, 1951, p. 116.
Rhineura attenuatus Estes, 1983, p. 198.
Rhineura hatcheri emendation Estes, 1983, p. 199.

Holotype—YPM-PU 11389, skull and jaws.
Type Locality—Battle Draw Spring, Washington Co., South Dakota.
Previous Referred Material—See Estes (1983).
New Referred Material—MSUVP (unnumbered) two right dentaries, posterior part of right dentary, anterior part of right dentary, anterior fragment of left dentary, left maxilla from the Brule Fm., Logan Co., Colorado; MSUVP (unnumbered) right maxillary fragment, Brule Fm., Logan Co., Colorado; SDSM 498, cranium (posterior part of skull) from the Brule Fm., Logan Co., Colorado; and SDSM 20198, nearly complete skull, from the Brule Fm., Reva Gap, Slim Buttes, Harding Co., South Dakota (see Estes, 1983) for previously referred material.
Stratigraphic Distribution—Brule and White River formations (Orella, Cedar Creek, ?Poleslide and Scenic members).
Age Distribution—Early through middle Oligocene (Orellan-Whitneyan).
Remarks—Previous to this study, a number of early Oligocene amphisbaenians were recognized, including six species of Rhineura, one species of Pseudorhineura, and two species of Hyporhineura (Estes, 1983). Estes (1983) stated that a revision of the fossil

rhineurids was needed and alluded to the synonymy of a few species. Although an exhaustive study of the Amphisbaenia is beyond the scope of the present chapter, we feel that formal synonymy can be made for the early Oligocene species of Rhineura.

Berman (1973) noted that the type specimen of Gilmoreia attenuatus was not only incomplete, but badly weathered, and that the differences between Gilmoreia and Rhineura, cited by Taylor (1951), could be attributed to the poor condition of the holotype. Estes (1983) synonymized Gilmoreia with Rhineura but retained the species R. attenuatus, believing it to be distinct. Although this skull is relatively small for Rhineura (approximately 7.5 mm in length), osteologically it compares readily to Rhineura hatcheri, and elements of the braincase appear to be fused. However, despite its small size it is our opinion that R. attenuatus is not distinct from R. hatcheri and is here placed in synonymy with this species.

Another small rhineurid, Pseudorhineura minuta (= Rhineura minutus), which has a skull length of approximately 7.8 mm, shares similar features with the holotype of Gilmoreia attenuatus (= R. attenuatus, see above), including small cranial facial angle and lack of sculpturing on the anterodorsal surface of the skull. Differences noted by Estes (1983) include lack of sagittal crest (for P. minuta) and a wide, deflected facial region. Some of these features are no doubt an artifact of its preservation: the holotype appears eroded, obliterating features such as the sagittal crest. Gilmore (1938a) believed the features seen in P. minuta were significant because he thought the holotype represented a mature specimen based on the co-ossification of the skull. However, Estes (1983) noted that ossification occurs early in Rhineura, suggesting that the differences in this species were ontogenetic. It is interesting to note that the holotypes of P. minuta and R. attenuatus share the same type locality with R. sternbergii. R. sternbergii is another small, yet slightly larger species (approximately 8.4 mm in length), and, contrary to Estes (1983), differs only in its having a more highly arched skull and blunter snout (compared to P. minuta). Estes (1983), however, thought it was similar to other Rhineura species, an observation with which we concur, and that these differences are also ontogenetic in nature. The fact that the three species come from the same type locality is consistent with our suspicions of conspecificity. However, we recognize that identical stratigraphic and geographic occurrence are not sufficient criteria for suggesting synonymy, or for that matter, distinct taxa.

Rhineura wilsoni is intermediate in size (skull length of 14.1 mm) between the largest Oligocene rhineurids (R. amblyceps and R. hibbardi) and the smaller species just discussed (above). While noting differences in this species (i.e., shallow braincase, elongated frontals, and seven maxillary teeth), Estes (1983) believed R. wilsoni to be similar to R. hatcheri. An increase in maxillary tooth count from six (in smaller individuals) to seven in larger individuals cannot be ruled out. Gilmore (1928) reported six maxillary teeth in the holotype of R. hatcheri, whereas both the holotypes of *R. hibbardi* and *R. wilsoni* are known to have seven maxillary teeth and are among the larger Oligocene rhineurids. Extant *Rhineura* has five maxillary teeth (Gans, 1967). The maxillary tooth count for *R. amblyceps* is unknown, owing to the fact that the maxillary is not preserved in the holotype. Because tooth counts tend to be variable among some members of the same species of some Sauria (see various species accounts in Estes, 1983), we believe that an additional tooth is taxonomically insignificant in the Amphisbaenia and that an increased number of maxillary teeth (from six to seven) in *Rhineura* may be solely an ontogenetic feature.

Rhineura amblyceps and *R. hibbardi*, which are the largest Oligocene species (18 mm+ and 20 mm, respectively), were recognized by Estes (1983) as being inadequately diagnosed. Both are from the Oligocene deposits of Logan Co., Colorado. Brattstrom (1958) was the first to suggest that *R. amblyceps* was a synonym of *R. hatcheri*, a fact reiterated by Estes (1983). Features that set these taxa apart are largely size related along with minor characteristics of the dentition (for *R. hibbardi*), which we believe are insignificant for species level taxonomy, and the degree of separation between the orbitosphenoid and the squamosal (for *R. amblyceps*). The differences in size among the sample of *Rhineura* skulls we believe represent specimens of various growth stages. Size differences, representing ontogenetic stages, have been reported in saurian squamates such as *Helodermoides* and *Parophisaurus* (Sullivan, 1979, 1987).

In summary, the various characters used by previous workers to diagnose species of *Rhineura* and distinct genera (*Gilmoreia* and *Pseudorhineura*) may not be valid for recognizing distinct taxa. Oligocene *Rhineura* is known solely by one species: *R. hatcheri*.

Lestophis Marsh 1885

Lestophis Marsh, 1885, p. 169 [proposed as replacement name for *Limnophis* Marsh, 1871, p. 326; preoccupied by *Limnophis* Macquart 1834].
    Type Species—*Lestophis crassus* (Marsh, 1871) p. 326.
    Holotype—YPM 531, cervical vertebra.
    Type Locality—Uinta Co., Wyoming.
    Remarks—See below.

Lestophis sp.

    Referred Material—AMNH 3824 and 3825 (vertebrae); SMNH 1439, forty-three vertebrae.
    Stratigraphic Distribution—Bridger Fm.
    Age Distribution—Middle Eocene (Bridgerian).
    Remarks—Estes (1983) reviewed the material referred to *Lestophis* and concluded that neither the type species (*Lestophis crassus*) nor *Lestophis anceps* were diagnostic, hence they were designated *nomina dubia*. Moreover, the genus itself was considered a *nomen dubium* because "rhineurid vertebrae are not diagnostic to genus" (Estes, 1983, p. 202). Estes noted that

Holman (1972) referred Cypress Hills material to this taxon without justification while recognizing that vertebrae assigned to *Lestophis* were of rhineurid affinities, implying that they could belong to *Rhineura*.

Spathorhynchus Berman, 1973

    Type Species—*Spathorhynchus fossorium* Berman, 1973, p. 705, figs. 1, 2c-e, 3a, b.
    Remarks—See below.

Spathorhynchus natronicus Berman, 1977

Spathorhynchus natronicus Berman, 1977, p. 986, text-fig. 1, text- fig. 2a, d.
    Holotype—AMNH 8677, skull and mandibles.
    Type Locality—South fork of Lone Tree Gulch, Flagstaff Rim region, Natrona Co., Wyoming.
    Referred Material—AMNH 8678, skull and mandibles.
    Stratigraphic Distribution—White River Group (undifferentiated).
    Age Distribution—Late Eocene (Chadronian).
    Remarks—*Spathorhynchus* is a robust rhineurid known by two species: *S. fossorium* (from the middle Eocene Bridger Fm.) and *S. natronicus* (Berman 1973, 1977). Berman (1977) cited a number of characters which serve to separate the latter species from the former. *Spathorhynchus* is distinct from *Rhineura* and is considered to be a primitive member of the Rhineuridae (Berman, 1973; Estes, 1983). *Spathorhynchus* is restricted to the middle and late Eocene of North America.

HYPORHINIDAE Baur, 1893
Hyporhina Baur, 1893
Hyporhina antiqua Baur, 1893

Hyporhina antiqua Baur, 1893, p. 998.
Hypsorhina antiqua (*lapsus calami*) Cope, 1900, p. 684.
Hyporhina antiqua Gilmore, 1928, p. 47, figs. 24-26; Plate 1, figs 1, 1a, 1b.
Hyporhina galbreathi Taylor, 1951, p. 532.
    Holotype—YPM-PU 11390, skull and jaws.
    Type Locality—Battle Spring Draw, Washington Co., South Dakota.
    New Referred Material—KU 8221, nearly complete skull. Clyde Ward Ranch, Logan Co., Colorado.
    Previously Referred Material—KU 8219 and 8222, partial skulls, both from SW1/4 of Sec. 12, T11N, R54W, Logan Co., Colorado.
    Stratigraphic Distribution—White River Fm. (?Cedar Creek Mbr.) and Brule Fm. (?Poleslide Mbr).
    Age Distribution—Early–middle Oligocene (Orellan-Whitneyan).
    Remarks—Three species of *Hyporhina* have been named on the basis of a meager sample of five specimens which include the holotypes of the named species. A reassessment of these taxa suggests that *Hyporhina galbreathi* is a subjective junior synonym of *H. antiqua* based on having identical sutural patterns of the premax-

illa, maxillae, prefrontals, frontals and nasals. We believe that this sutural configuration is the critical characteristic in determining species among *Hyporhina*. The primitive suture pattern seen in *H. tertia* (see below) complements its older biostratigraphic occurrence. Furthermore, we suggest here that the holotype of *H. galbreathi* is a juvenile of *H. antiqua* based on the fact that the latter is: 1) smaller than the former and that 2) the larger form (holotype of *H. antiqua*) is more rugose, suggesting an older individual. The other characters previously cited to separate these two (i.e., six maxillary teeth in *H. galbreathi*, four in *H. antiqua*; reduced prefrontal excluded from orbit in *H. antigua*; spine-like lateral processes on the anterior parietal in *H. antiqua*; thin lateral process in *H. galbreathi*; smaller prefrontal in *H. antiqua*; larger prefrontal in *H. galbreathi*) may be ontogenetic differences, rather than features having phylogenetic significance. Contrary to Berman (1972) and Estes (1983), the apparent morphological series of prefrontal reduction (largest to smallest: *H. galbreathi* - *H. tertia* - *H. antiqua*) cannot be substantiated. When viewed at the same scale, the prefrontals of the holotypes of *H. tertia* and *H. galbreathi* are nearly identical in size. We believe that the apparent prefrontal reduction between the holotypes of *H. galbreathi* and *H. antiqua* may also be an ontogenetic feature.

Hyporhina tertia Berman, 1972

Hyporhina tertia Berman, 1972, p. 3, fig. 1.
   Holotype—CM 17179, snout region of skull.
   Type Locality—Near Cameron Springs, Fremont Co., Wyoming.
   Referred Material—None.
   Stratigraphic Distribution—Chadron Fm.
   Age Distribution—Late Eocene (Chadronian).
   Remarks—*Hyporhina tertia* is the earliest and most primitive member of the Hyporhinidae. Estes (1983) noted that this amphisbaenian family shares the same diagnosis as the genus *Hyporhina*. From a phylogenetic systematic perspective, we question whether a family ranking is warranted for only two species of *Hyporhina*. *H. tertia* is the sister taxon to *H. antiqua* based on the primitive arrangement and morphology of the anterior skull elements (see above).

SQUAMATA *INCERTAE SEDIS*
Cremastosaurus Cope, 1873a
Cremastosaurus carinicollis Cope, 1873a

Cremastosaurus carinicollis Cope, 1873a, p. 18 (in part).
   Holotype—AMNH 1604, articulated vertebrae.
   Type Locality—Horse Tail Creek, Logan Co., Colorado.
   Referred Material—None.
   Stratigraphic Distribution—Chadron Fm.
   Age Distribution—Late Eocene (Chadronian).
   Remarks—Estes (1983) noted similarities to cervical vertebrae associated with specimens of the taxon *Aciprion*, but believed the holotype to be undiagnostic and designated it a *nomen dubium*. The holotype has since been lost.

"Cremastosaurus" rhambastes Cope, 1873a

Cremastosaurus carniocollis Cope, 1873a, p. 516 (in part).
Platyrhachis rhambastes Cope, 1884 p. 779, pl. 60, figs. 18a, b, c.
Cremastosaurus rhambastes Gilmore, 1928, p. 152, Plate 25, figs. 8a-c.
   Holotype—AMNH 1606, seven trunk vertebrae.
   Type Locality—Horse Tail Creek, Logan Co., Colorado.
   Referred Material—None.
   Stratigraphic Distribution—Chadron Fm.
   Age Distribution—Late Eocene (Chadronian).
   Remarks—Estes (1983) reported that the holotype of this species came from the "middle Oligocene White River Formation" (i.e., Orellan age). However, the holotype came from the Chadron Fm. of Horse Tail Creek (see Galbreath, 1953; Gilmore, 1928). Anguid affinities were suggested by Estes (1983), who believed that the holotype might be referable to *Parophisaurus* (= *Machaerosaurus* [in part]) *pawneensis*, which is known from only the Orellan (see above), but designated it a *nomen dubium*. The holotype of this species has been lost.

Diacium Cope, 1873a
Diacium quinquepedale (Cope, 1873a)

Diacium quinquepedalis Cope, 1873a, p. 17.
Diacium quinquepedale Cope, 1884, p. 777 [emendation of *D. quinquepedalis*]
Diacium sesquipedale (*lapsus calami*) Cope, 1884, Plate 60, fig. 20.
   Holotype—AMNH 1602, vertebra.
   Type Locality—No specific locality, Logan Co., Colorado.
   Referred Material—None.
   Stratigraphic Distribution—?Brule Fm.
   Age Distribution—?Early Oligocene (Orellan).
   Remarks—Estes (1983) stated that this holotype was not diagnostic and therefore *Diacium quinquepedale* is a *nomen dubium*.

SERPENTES Linnaeus, 1758

In this section we follow the taxonomy of Rage (1984a) for fossil snakes.
   The snake fauna of the Eocene/Oligocene boundary is diminished, and consists mainly of a few taxa of boids, mostly small in size, and a single colubrid. Three of the boids, *Calamagras*, *Helagras*, and *Ogmophis*, may be rather closely related to one another, whereas *Boavus* and *Geringophis* are distinct. The main event within the Serpentes during this time period was the appearance of the distinct boid genus *Geringophis* and the first North American colubrid, *Texasophis*, in the early Oligocene.

SERPENTES Linnaeus, 1758
ALETHINOPHIDEA Nopsca, 1925
BOOIDEA Gray, 1825
BOOIDEA *INCERTAE SEDIS*
Coprophis Parris and Holman, 1978
Coprophis dakotaensis Parris and Holman, 1978

Coprophis dakotaensis Parris and Holman, 1978, p. 259, fig. 1.

Holotype—YPM-PU 20732(a), vertebra.

Type Locality—From the "Big Badlands," precise locality unknown.

Original Referred Material—One trunk and two cervical vertebrae, YPM-PU 20732(b,c, and d), vertebrae.

New Referred Material—None.

Stratigraphic Distribution—Brule Fm.

Age Distribution—Early Oligocene (Orellan).

Remarks—The vertebrae of *Coprophis* were discovered during a comprehensive study by D. C. Parris of mammalian coprolites from the Brule Formation. These vertebrae are somewhat eroded because of their partially digested state. They were assigned to the new genus *Coprophis* based on a combination of characters (Parris and Holman, 1978, p. 259). Rage (1984a) suggested that this snake may belong to either the Booidea or the Anilioidea. But based on the narrow, well-developed hemal keel (Parris and Holman, 1978, fig. 1b), we suggest that this taxon belongs to the Booidea rather than to the Anilioidea, which have wider, much more indistinct hemal keels.

BOIDAE Gray 1825
BOINAE Gray, 1825
Boavus Marsh, 1871

Boavus Marsh, 1871, p. 322.
Protagras Cope, 1872, p. 471.

Type Species—*Boavus occidentalis* Marsh, 1871, p. 323.

Remarks—This genus has previously been reported only from the early middle Eocene (Bridgerian) through the late middle Eocene (Duchesnean), but the Duchesnean record of the genus has been questioned (Rage, 1984a). Newly available material has provided the first records of *Boavus* from both the late Eocene (Chadronian) and the early Oligocene (Orellan).

Boavus cf. *Boavus occidentalis* Marsh, 1871

Locality—Reva Gap, Slim Buttes, Harding Co., South Dakota.

New Referred Material—SDSM 53465, three fragmentary vertebrae.

Stratigraphic Distribution—Brule Fm. "D."

Age Distribution—Early Oligocene (Orellan).

Remarks—These large boid vertebrae are consistent with *Boavus* and one vertebra with a complete ventral centrum especially resembles *Boavus occidentalis*, in its sharp hemal keel. This is the first record of this genus from the Orellan.

Boavus sp. indet.

Locality—Flats locality (top of low silts) of Galbreath (1953). Logan Co., Colorado (MSUVP 1357). Shable Butte locality of Galbreath, Weld Co., Colorado (MSUVP 1358).

New Referred Material—MSUVP 1357, fragmentary vertebra; MSUVP 1358, fragmentary vertebra.

Stratigraphic Distribution—Chadron Fm.

Age Distribution—Late Eocene (Chadronian).

Remarks—These large boid vertebrae also have characters that are consistent with *Boavus*; but MSUVP 1357 has a hemal keel that is not as sharp as in *Boavus occidentalis* and may represent one of the other species.

ERYCINAE Bonaparte, 1831
Calamagras Cope, 1873a

Calamagras Cope, 1873a, p. 15.
Aphelophis Cope, 1873a, p. 16.

Type Species—*Calamagras murivorus* Cope, 1873a, p. 15.

Remarks—The genus *Calamagras*, and the following genera, *Helagras* and *Ogmophis*, may be closely related. In fact, Rage (1984a) thought that the latter two genera belong to the same taxon, *Calamagras*, which has priority. The vertebrae of these snakes are mainly separated on the basis that *Ogmophis* has a relatively long neural spine, while *Calamagras* a short one, and *Helagras* a short, tubular one.

Calamagras angulatus Cope, 1873a

Calamagras angulatus Cope, 1873a, p. 16.
Ogmophis angulatus Cope, 1874, p. 783.

Holotype—AMNH 1654, an incomplete trunk vertebra.

Type Locality—Cedar Creek, Colorado.

Original Referred Material—None.

New Referred Material—SDSM 20189, trunk vertebra of a juvenile and SDSM 20197, trunk vertebra. Brule Fm. (Zone D), Reva Gap, Slim Buttes, Harding Co., South Dakota.

Stratigraphic Distribution—Brule, Gering, Harrison, Monroe Creek, and White River fms.

Age Distribution—Early Oligocene (Orellan)– early Miocene (Arikareean, Harrison Fm.).

Remarks—Of the three Brule Formation species named by Cope in 1873, *Calamagras angulatus* differs from *Calamagras murivorus* Cope in having a much more depressed vertebral form (see Gilmore, 1938b, fig. 16) and appears to be a distinct species.

Calamagras murivorus Cope, 1873a

Calamagras murivorus Cope, 1873a, p. 15.
Aphelophis talpivorus Cope, 1873a, p. 16.
Calamagras truxalis Cope, 1873a, p. 15.
Calamagras talpivorus Cope, 1873a, p. 16.

Holotype—AMNH 1603, six articulated vertebrae.

Type Locality—Cedar Creek, Logan Co., Colorado.

Original Referred Material—From the type locality AMNH 1657, four articulated vertebrae and AMNH 1658, 3 vertebrae. From the Oligocene of Sioux Co., Nebraska, USNM 13824, three trunk vertebrae collected by J. B. Hatcher.

New Referred Material—AMNH 1598, three articulated vertebrae (holotype of *Calamagras talpivorus*), White River Fm. (Orellan), Cedar Creek, Logan Co., Colorado and YPM-PU 19517, three disarticulated and two articulated vertebrae from the "uppermost White River Beds" = Brule Fm. (Zone D), Battle Springs Draw, Washington Co., South Dakota.

Stratigraphic Distribution—Brule and White River fms.

Age Distribution—Early Oligocene (Orellan).

Remarks—*Calamagras murivorus* differs from *Calamagras angulatus* in having a more vaulted neural arch and a less depressed vertebral shape (see Gilmore, 1938b, figs 14a, 15a, 16b, and 17a). Re-evaluation of the type material of *Calamagras angulatus*, *Calamagras murivorus*, and *Calamagras talpivorus*, suggests that *Calamagras angulatus*, with its depressed vertebral form, is probably a distinct species. Characters used to separate *Calamagras murivorus* from *Calamagras talpivorus* are individually variable ones (e.g., Gilmore, 1938b, pp. 42-43, 45-46). Therefore, *Calamagras talpivorus* is a subjective junior synonym of *Calamagras murivorus*. All of Cope's type species are from the White River Formation (Orellan) of Cedar Creek, Logan Co., Colorado.

Calamagras weigeli Holman, 1972

Calamagras weigeli Holman, 1972, p. 1626, fig. 7.

Holotype—SMNH 1437, trunk vertebra.

Type Locality—North branch of Calf Creek, legal subdivision 4, Sec. 8, T8, R22, W 3rd mer., Saskatchewan.

Referred Material—SMNH 1438, trunk vertebra.

Stratigraphic Distribution—Cypress Hills Fm.

Age Distribution—Late Eocene (Chadronian).

Remarks—This late Eocene form differs from the above early Oligocene species in having a stronger, wider hemal keel, and in having more strongly developed subcentral ridges (Holman, 1972).

Helagras Cope, 1883

Type Species—*Helagras prisciformis* Cope, 1883, p. 545.

Remarks—As cited above, Rage (1984a) has stated that the differences between *Helagras*, *Calamagras*, and *Ogmophis* are not great and that only the discovery of more complete specimens will clear up the status of *Helagras*. Nevertheless, the very short, tubular neural spine appears to be a distinct derived character in this genus (Holman, 1983, figs. 1 and 2). The distribution of this genus is puzzling. Thus far *Helagras* (i.e, *H. prisciformis* Cope, 1883) is only known from the Paleocene

(Puercan and Torrejonian) of New Mexico and the Paleocene (Torrejonian) of Wyoming (Sullivan, 1980) and from the early Oligocene of Kansas (*Helagras orellanensis*, see below).

Helagras orellanensis Holman, 1983

Helagras orellanensis Holman, 1983, p. 417, fig 1.

Holotype—KU 49127, a trunk vertebra.

Type Locality—Near Toadstool Park, Sioux Co., Nebraska.

Referred Material—KU 49128, two tiny fused vertebrae.

Stratigraphic Distribution—Brule Fm.

Age Distribution—Early Oligocene (Orellan).

Remarks—*Helagras orellanensis* is separated from *Helagras prisciformis* of the Paleocene (Holman, 1983, p. 417) mainly on the basis of the enlargement of the tubular neural spine.

Ogmophis Cope, 1884

Type Species—*Ogmophis oregonensis* Cope, 1884, p. 783.

Remarks—This genus is known on the basis of several species that range in time from the upper middle Eocene (Duchesnean) of Georgia (Holman, 1977) to the Pliocene (Blancan) of Texas (Rogers, 1976). Only one species, *Ogmophis compactus*, is known from the late Eocene-early Oligocene transition.

Ogmophis compactus Lambe, 1908

Ogmophis compactus Lambe, 1908, p. 20, Plate 1, figs. 26-30.

Holotype—NMC 6237, a trunk vertebra.

Type Locality—The type locality in Gilmore (1938b) is stated "4 miles above east end of Post Office, Cypress Hills, Saskatchewan, Can." This makes no sense and the phrase must actually refer to the post office in the town of Eastend, Saskatchewan. Thus, the type locality probably is four miles north of the post office in Eastend, Saskatchewan, Canada.

Previously Referred Material—Paratypes: apparently three vertebrae, NMC 6238, 6239, and 6240. Additional material: NMC 4241, one thoracic vertebra from the town of Eastend, Saskatchewan; SMNH 1433, one cervical vertebra and 27 trunk vertebrae from the Calf Creek local fauna, north branch of Calf Creek, legal subdivision 4, Sec. 8, T 8, R 22 W. 3rd mer., Saskatchewan (Holman, 1972); USNM 13675, two provisionally referred lumbar vertebrae from the Oligocene (presumably Orellan) of Sioux Co., Nebraska (Gilmore, 1938b).

Stratigraphic Distribution—Cypress Hills Fm.

Age Distribution—Late Eocene (Chadronian) through presumably early Oligocene (Orellan).

Remarks—A very important feature of *Ogmophis compactus* compared to the other erycinine boid snakes in North America is its large size. In fact it is about twice as large as any other species in the genus. Based on vertebral projections on modern boid snakes, Holman (1972) estimated that *Ogmophis compactus* from the Chadronian Calf Creek local fauna of Saskatchewan was about 1.4 to 1.5 m in length.

Relative to the presumably early Oligocene record of *Ogmophis compactus* from Sioux County, Nebraska, Gilmore (1938b) states "Two articulated vertebrae, No. 13675 U.S.N.M., from the Oligocene of Sioux County, Nebraska, except for their smaller size, have the closest resemblance to *O. compactus* and are provisionally referred to it." If correct in this identification it marks the first record of this species outside of Canada. It seems parsimonious to suggest that Oligocene *Ogmophis* vertebrae from Sioux Co., Nebraska were from the Brule Formation.

?ERYCINAE
Geringophis Holman, 1976

Type Species—*Geringophis depressus* Holman, 1976, p. 90.

Remarks—The genus *Geringophis* is very distinct from the other small boid genera of the late Eocene-early Oligocene transition in that it has a very flattened shape and a long, high neural spine (Holman, 1982, figs. 1-3). This genus occurs from the early Oligocene (Orellan) to the late Miocene (late Barstovian) and consists of three species. The earliest of these is *Geringophis vetus*.

Geringophis vetus Holman, 1982

Geringophis vetus Holman, 1982, p. 490, fig. 1.
   Holotype—KU 49126, a trunk vertebra.
   Original Referred Material—None.
   New Referred Material—MSUVP 1123, five vertebrae from the Flats locality (levels 11-12) of E. C. Galbreath, Logan Co., Colorado.
   Type Locality—KU Site KU-NEBR-22, near Orella, Sioux Co., Nebraska.
   Stratigraphic Distribution—Brule Fm.
   Age Distribution—Early Oligocene (Orellan).
   Remarks—This is the earliest fossil record of this genus. The flattened neural arch and high neural spine is an unusual combination of characters in the Boidæ that is known only in *Cadurcoboa* of the late Eocene of France and in some living Tropidopheinae (Rage, 1984a). Whether this genus arose from *Cadurcoboa* and immigrated to North America from the Old World, or originated from a erycinine boid with a flattened vertebral form such as *Calamagras angulatus*, is unknown.

COLUBROIDEA Oppel, 1811
COLUBRIDAE Oppel, 1811
Texasophis Holman, 1977

Type Species—*Texasophis fossilis* Holman, 1977, p. 397.
Remarks—The genus *Texasophis* is a small colubrid snake that ranges from the early Oligocene (Orellan or Whitneyan) of North America to the late Miocene

(Clarendonian). It also occurs in the Miocene (Astaracian) of France. Four species of *Texasophis* have been named (Holman, 1989). *Texasophis galbreathi* represents the earliest record of a colubrid snake in the New World.

Texasophis galbreathi Holman, 1984a

Texasophis galbreathi Holman, 1984a, p. 223, fig. 1.
   Holotype—MSUVP 1038, a trunk vertebra.
   Type Locality—A discussion of the type locality of this specimen is important. The original "type locality and horizon" for this specimen in Holman (1984a) was modified (Holman, 1984b) after a letter from Edwin C. Galbreath, who collected the specimen in 1958. The emended type locality and horizon is "Flats, E. 1/2 Sec. 12, T11N, R54W, Logan Co., Colorado, Scenic Member of the Brule Formation. White River Group: early Orellan to Whitneyan age: middle to late Oligocene." (verbatim from Galbreath's letter).

The Scenic Member of the Brule Formation is now considered to be Orellan (early Oligocene) in age (Emry et al., 1987; Prothero and Swisher, 1992). This has important implications in the following remarks section.
   Referred Material—None.
   Stratigraphic Distribution—Brule Fm. (Scenic Mbr.).
   Age Distribution—Early Oligocene (Orellan).
   Remarks—The advanced snake family Colubridae may be the most successful of all living reptile groups and has over 1,500 living species. It has recently been shown that the earliest Colubridae are from the late Eocene of Thailand (Rage, 1992), and it was postulated that the family spread first to Europe and then to North America from Asia. This is based on the fact that in 1992 the next oldest colubrid snakes known were from the early Oligocene of western Europe and from the Arabian Peninsula, and it was believed that the North American *Texasophis galbreathi* was from the middle Oligocene. Now that the Orellan is considered early Oligocene, it would have to be reasoned that the Colubridae invaded Western Europe and North America from Asia at about the same time.

DISCUSSION

The appearance of Recent squamate taxa in the fossil record is of interest to paleoherpetologists and neoherpetologists from both a taxonomic and an evolutionary standpoint. The ability to recognize Recent genera and species (and the higher taxonomic rankings to which they belong, such as families), in the fossil record provides information regarding the timing of origin and evolutionary rates.

Estes (1970) was the first to provide a comprehensive review of the origin of the Recent North American herpetofauna (frogs, salamanders, turtles, crocodilians, lizards and amphisbaenians; snakes were not included) and ichthyofauna. The Late Cretaceous and Paleogene North American lower vertebrate faunas are based largely on the fossil record from the Rocky Mountain and Great Plains regions. Some previous interpretations concerning the first appearances of fossil squamates in the North American Cretaceous-Paleogene record have been made based on very fragmentary material. These taxonomic assignments have later proven to be erroneous resulting in inaccurate geochronological occurrences and extended ranges for taxa.

The erroneous conclusions reached by earlier workers resulted from assigning fragmentary specimens to modern genera (and families) based on what is now generally regarded to be insufficient material (see, for example, Good, 1988). These mistakes were made primarily because of the inability of workers to recognize phylogenetic relationships, based on the presence of apomorphies, that may be diagnostic of a particular taxon. It is worth noting that Estes's earlier studies (i.e., Estes, 1964, 1970) of North American paleoherpetofaunas employed this methodology, a practice that preceded the advent of the rigorous cladistic analysis that he later championed.

In his first comprehensive overview of North American herpetofaunas, Estes (1970) concluded that, unlike the North American mammalian fauna, many of the Recent lower vertebrate genera had evolved by Oligocene time, and that modern "families" to which many of the taxa belong, had an antiquity that could be traced back to Late Cretaceous times. This is unlike the picture for the Cenozoic mammalian genera, where Recent mammal taxa (genera) have a Pleistocene or Pliocene origination at best (Estes, 1970). Recent paleoherpetological studies have forced us to re-evaluate putative first occurrences of modern taxa and challenge the view that rates of evolution among lower vertebrates are demonstrably slower than that of mammals.

Many of the aforementioned earlier assignments were made based on primitive features (plesiomorphies), mere look-a-likes. At this point, a review of some recently published taxonomic reassessments is in order to emphasize revised interpretations of evolution and antiquity of Recent taxa.

Good (1988) studied 166 fragmentary fossil specimens previously referred to the anguid subfamily Gerrhonotinae and concluded that only two specimens had the diagnostic characters that would permit reference to this group. Previously, it had long been held that gerrhonotine lizards were present as far back as the Late Cretaceous (Estes, 1964; Meszoely, 1970) but Good's study forcefully demonstrated that the earliest gerrhonotine lizard recorded in the fossil record is of early Pliocene age and that there is no evidence for an older (Late Cretaceous to Miocene) occurrence.

Pregill (1992), in his recent review of *Leiocephalus*, demonstrated that previous reports of this taxon from the Oligocene and Miocene of North America are erroneous based on undiagnostic features present in referred fossil material. The earliest *bona fide Leiocephalus* is from the late Pleistocene of the West Indies.

Recently, Augé (1992) named a new species of "*Ophisaurus*" (*O. roqueprunensis*) from the Oligocene Phosphorites du Quercy (France). This taxon (genus) as employed by Augé is paraphyletic because *Ophisaurus*

(*sensu stricto*) is restricted to North America (Sullivan, 1987; Augé, 1992). The earliest record of *Ophisaurus* (*sensu stricto*) is late Miocene (for *O. canadensis* and *O. ventralis*; see Estes, 1983). When asked why he assigned the new species to *Ophisaurus* rather than *Ophisauriscus* or *Anguis* he replied (in letter to RMS, 1993) that the species could not be referred to either *Ophisauriscus* or *Anguis*. He further stated that "*Ophisaurus*" *roqueprunensis* and *Dopasia* ("*Ophisaurus*") *harti* belong to the same (unnamed) genus (which we note here might be *Dopasia* and not a new taxon). *Ophisaurus* (*sensu stricto*) is not known from the Old World; specimens assigned to this taxon have been based on similar morphology (i.e., limblessness, frontal and parietal scutellation pattern, etc.). Previously, *Ophisaurus* was thought to occur in strata as early as the middle Eocene but these serpent-like lizards have been returned to the taxon *Ophisauriscus quadrupes* (Sullivan, 1987; Sullivan et al., in preparation).

As a consequence of these earlier and other studies, the taxonomic range of some genera (and in some cases, families) have been erroneously extended back in time, thus giving the impression that these squamates evolved early in the Late Mesozoic, or Early Cenozoic, and have changed little since. Moreover, it would appear that, based on these studies, lower vertebrates evolved slowly compared to mammals. Therefore, misidentification of taxa can directly lead to inaccurate biostratigraphic ranges and obscure the time of origination and rates of evolution in some squamate taxa.

Numerous additional biostratigraphic errors that pertain to taxa discussed in this chapter were recently introduced by Hutchison (1992), and they need to be corrected here. *Aciprion* is restricted to the Orellan; its occurrence in Chadronian age strata has not been documented. *Glyptosaurus* is not known from the Chadronian but it is restricted to the Bridgerian; two fragmentary specimens (*Glyptosaurus* sp. indet.) were noted from Uintan age strata (Sullivan, 1979, p. 19), but owing to their fragmentary nature it is impossible to extend the stratigraphic range of *Glyptosaurus* to the Uintan with any degree of confidence. None of the material assigned to *Helodermoides tuberculatus*, or *H.* sp., is known from the late Uintan; *Helodermoides* is known from only Chadronian and Orellan age strata (Sullivan, 1979). *Peltosaurus* is not known from the Chadronian interval, only from the Orellan, Whitneyan, and Arikareean. *Parophisaurus* is not known from the Chadronian, only the Orellan (Sullivan, 1987; and this chapter).

Leiocephalus cannot be confirmed to occur in the Oligocene (or Miocene) strata of North America (Pregill, 1992; and see above). The earliest record of *Sceloporus* is from the early Miocene (late Arikareean), not early Arikareean, as indicated by Hutchison (1992). The occurrence of *Saniwa* in late Eocene (Chadronian) and early Oligocene (Orellan) strata are erroneous (see above); *Saniwa* is restricted to the middle Eocene (Bridgerian).

Other occurrences can be summarized as follows: an unnamed new taxon (referred to *Eumeces* by Estes,

Table 1. Stratigraphic distribution of definitive Squamata, by genus, across the Eocene-Oligocene (Chadronian-Orellan) boundary, central North America. Taxa considered *nomina dubia* (see text) are not listed. Abbreviations: C = Chadronian; O = Orellan; W = Whitneyan; A = Arikareean.

EPOCH	EOCENE --- OLIGOCENE			
NALMA	C	O	W	A
Taxon				
Aciprion (Iguania *incertae sedis*)		x		
Crotaphytus? (Iguania *incertae sedis*)	x			
Cypressaurus (Iguania *incertae sedis*)	x			
Paraphrynosoma (Iguania *incertae sedis*)		x		
unnamed new taxon (Scincidae)		x		
Palaeoxantusia (Xantusiidae)	x			
Exostinus (Xenosauridae)		x		
Parophisaurus (Anguidae)		x		
Peltosaurus (Anguidae)		x	x	x
Helodermoides (Anguidae)	x	x		
Lowesaurus (Helodermatidae)		x	x	x
Rhineura (Rhineuridae)		x		
Spathorhynchus (Rhineuridae)	x			
Hyporhina (Hyporhinidae)		x	x	
Coprophis (Henophidia *incertae sedis*)		x		
Boavus (Boidae)	x	x		
Calamagras (Boidae)	x	x	x	x
Helagras (Boidae)		x		
Ogmophis (Boidae)	x	x		
Geringophis (?Erycinae)		x		
Texasophis (Colubridae)		x		

1983) is present in the Orellan. *Palaeoxantusia* occurs in the Chadronian and is not present in Orellan age strata. The amphisbaenian *Spathorhynchus* is known from the Chadronian; *Rhineura* and *Hyporhina* are known only from the Orellan.

Snake genera that appear for the first time in the early Oligocene (Orellan) include the boid *Coprophis*, the distinctive boid *Geringophis*, and the first North American colubrid, *Texasophis*. The boid genera *Boavus* and *Ogmophis* are present in both Chadronian and Orellan. The early Paleogene taxon *Helagras* "re-appears" in the early Oligocene and may reflect a paleogeographic range shift for this taxon. *Calamagras* is present in Chadronian through Arikareean. A summary of the stratigraphic ranges of valid squamate taxa (genera) on both sides of the Eocene/Oligocene boundary is summarized in Table 1.

The European Eocene/Oligocene boundary squamates have also been the subject of scrutiny (Rage, 1984a, 1986). This boundary in western Europe, more specifically France, has been dubbed the "Grande Coupure," a term used to described an apparent major faunal turnover at the Eocene-Oligocene transition. What is known of the late Eocene lizards, amphisbaenians, and snakes in Europe comes largely from various localities of the Phosphorites du Quercy (France), with the Bretou site being the richest in terms of diversity of taxa (Rage, 1988a, b). Rage (1986) summarized the "Grande Coupure" herpetofaunal event as a major change from a diverse and comparatively rich herpetofauna present during late Eocene time, followed by a decrease in taxa

(75%). The Eocene/Oligocene boundary herpetofauna (i.e., lowermost Oligocene) included a few Eocene holdovers that survived the transition along with the appearance of a few Asian "immigrant" taxa. The lower Oligocene herpetofauna of Europe is poor in comparison to its late Eocene counterpart. Rage (1986) speculated that this extinction event was prominent and resulted from a number of factors such as climatic deterioration, invasion of Asian immigrant taxa (which competed with the indigenous taxa), and other factors (Rage, 1986). A summary of European Eocene/ Oligocene boundary squamatofauna is presented in Table 2.

How then does the European squamatofaunal "Grande Coupure" compare to the squamatofaunal turnover of central North America at the Eocene/Oligocene boundary? First, the fossil record for lizards, amphisbaenians, and snakes is just the opposite of that for Europe (France). Where a diverse squamatofauna is present in the late Eocene strata of Europe (see Table 2), a rather depauperate squamatofauna exists in central North America during this time interval (see Table 1). Conversely, a rich Oligocene squamatofauna is known from the early Oligocene (Orellan) in central North America, while its faunal counterpart in Europe is diminished (Rage, 1986; Szyndlar, 1994). The post-early Oligocene trends are also opposite for the two continents. The European fossil record indicates an apparent increase in species diversity for the later Oligocene, whereas the North American is less clear, and less diverse, probably owing to changes in depositional environments rather than the result of any

Table 2. Stratigraphic distribution of definitive Squamata, by genus, across the Eocene-Oligocene (Headonian-Suevian) boundary, Europe (Phosphorites du Quercy). Only definitive squamate taxa are considered and only genera are listed. Rage (1984b) listed a number of boid taxa that may represent distinct species (rather than genera) which are listed here (Boidae C-H; J-M). *Enigmatosaurus bottii* was determined to be a frog (Estes, 1983) and is therefore not considered. Stratigraphic levels are not equivalent to NALMA's. Abbreviations: E = Escamps (Montmartre); H = Hoogbutsel; M = Montalban; C = Les Chapelins; A = Antoingt. (Compiled from Crochet et al., 1981; Rage, 1984a; and 1988a).

EPOCH	EOCENE --- OLIGOCENE				
STRATIGRAPHIC LEVELS	E	H	M	C	A
Taxon					
Cadurcogekko (Gekkonidae)	x				
Plesiolacerta (Lacertidae)			x		
Pseudolacerta (Cordylidae)	x				
"*?Ophisaurus*"[1] (Anguinae)	x	x	x	x	x
Placosaurus (Anguidae)	x				
Eurheloderma (Helodermatidae)	x				
Necrosaurus (Varanidae)	x				
Coniophis (Aniliidae)	x				
Eoanilius (Aniliidae)	x				
Dunnophis (Boidae)	x				
Palaeopython (Boidae)	x				
Bransateryx (Boidae)					x
Cadurceryx (Boidae)	x				
Cadurocoboa (Boidae)	x				
Platyspondylia[2] (Boidae)			x	x	x
Boidae (C)	x	x			
Boidae (D)	x				
Boidae (E)		x	x		
Boidae (F)			x		
Boidae (G)			x		
Boidae (H)			x	x	x
Boidae (J)	x				
Boidae (K)	x				
Boidae (L)	x				
Boidae (M)	x				
Coluber (Colubridae)			x	x	x

[1] This record of "*Ophisaurus*" (*sensu stricto*) is not accepted as a valid occurrence for this taxon. It probably is a different serpentine-like anguine (see Discussion).

[2] This genus is also known from later Oligocene horizons (Rage, 1984a).

natural extinction event.

Some general comparisons can be made between the central North American and western European squamatofaunas at the Eocene/Oligocene boundary. In the North American squamatofauna there are nine lizard genera representing seven families; three amphisbaenians representing two families; and eight snake genera representing three families. The European squamatofauna has seven lizard genera representing six families; amphisbaenians are notably absent; and snakes are represented by nine named genera distributed among three families and nine boids of uncertain relationships. The snakes are clearly the most diverse and dominant part of this Old World faunal assemblage, but this fauna is rivaled by the central North American squamatofauna which is diverse in its own right. Diversity of the two lizard faunal components are nearly the same in both squamatofaunas in terms of number of genera and families represented. They differ solely by the lack of genera in common (i.e., none).

On the face of it, it would appear that these two squamatofaunas are out of phase with each other. The peak of diversity in the central North American squamatofauna is in the early Oligocene, which suggests a radiation of new taxa or sudden appearance of immigrant taxa. In contrast, the western European squamatofauna has its peak diversity in the late Eocene. Preservation and sampling biases must be considered along with correlation of the Eocene-Oligocene terrestrial boundary between North America and Europe.

Six squamate taxa cross the Eocene/Oligocene boundary in North America. These include the lizards *Exostinus* and *Helodermoides* and the snakes *Boavus*, *Calamagras*, *Helagras*, and *Ogmophis*. By contrast, none of the European squamate taxa are found in strata from both sides of the boundary (Escamps-Hoogbutsel stratigraphic intervals) (Rage, 1984b, 1986).

The "Grande Coupure" for squamates of central North

America would appear to be at the Orellan-Whitneyan boundary, rather than the Eocene/Oligocene boundary, based on the disappearance of taxa known from early Oligocene strata. However, a few taxa continue to the late Miocene (i.e., *Geringophis* and *Texasophis*); others presumably share common ancestry with Recent taxa (*Lowesaurus* with *Heloderma*; *Parophisaurus* with *Ophisaurus* [*sensu stricto*]); and one (*Rhineura*) is known from the Recent. The squamatofaunal turnover at the Eocene/Oligocene boundary in central North America appears to be stepwise.

ACKNOWLEDGMENTS

We thank the following individuals for the loan of specimens surveyed in this study: Eugene Gaffney, Mark Norell, and Charlotte Holton (American Museum of Natural History); Peter Larson (Black Hills Institute); Mary Dawson and Elizabeth Hill (Carnegie Museum of Natural History); Richard J. Zakrzewski (Fort Hays Sternberg Memorial Museum); Gerald Smith (Museum of Natural History, University of Kansas); Philip Gingerich (Museum of Paleontology, University of Michigan); Philip Bjork and Janet Whitmore (South Dakota School of Mines); Robert J. Emry, Nicholas Hotton, and Robert Purdy (United States National Museum); and Mary Ann Turner (Peabody Museum of Natural History, Yale University).

Sullivan thanks L. Barry Albright (University of California, Riverside) for the privilege of examining casts of osteoderms from the Toledo Bend site; Richard Stucky (Denver Museum of Natural History) for bringing to his attention new material in the Carnegie Museum collection; Robert J. Emry (United States National Museum) for additional stratigraphic and locality data for the *Palaeoxantusia* specimens; David Hauser for information on *Aciprion*; Mary Ann Turner for additional stratigraphic data concerning former Princeton University specimens; Alan Tabrum (Carnegie Museum of Natural History) and Donald R. Prothero (Occidental College) for discussions regarding White River stratigraphy; Brent Briethaupt (University of Wyoming) for relinquishing the holotypes of *Calamagras* on loan to him; and especially Jacques A. Gauthier (California Academy of Sciences) for discussions regarding phylogenetic systematics of Squamata, and his critique of our manuscript. David S Berman (Carnegie Museum of Natural History) and Spencer G. Lucas (New Mexico Museum of Natural History and Science) also reviewed the manuscript and we are grateful for their comments and suggestions.

R. M. Sullivan was responsible for the fossil lizard and amphisbaenian sections and J. A. Holman was responsible for the snake section in this report. The joint authorship does not indicate mutual agreement on all of the taxonomic decisions made.

LITERATURE CITED

Albright, L. B. 1994. Lower vertebrates from an Arikareean (earliest Miocene) fauna near the Toledo Bend Dam, Newton County, Texas. Journal of Paleontology 68(5):1131-1145.

Augé, M. L. 1992. Une espèce nouvelle d'*Ophisaurus* (Lacertilia, Anguidae) de l'Oligocene des Phosphorites du Quercy. Révision de la sous-famille des Anguinae. Paläontologische Zeitschrift 66(1/2):159-175.

Baird, S. F. 1859. Description of new genera and species of North American lizards in the museum of the Smithsonian Institution. Proceedings of the Academy of Natural Sciences, Philadelphia 11:253-256.

Baur, G. 1893. The discovery of Miocene amphisbaenians. American Naturalist 27:998-999.

Berman, D. S 1972. *Hyporhina tertia*, new species (Reptilia: Amphisbaenia), from the early Oligocene (Chadronian) White River Formation of Wyoming. Annals of Carnegie Museum 44:1-10.

Berman, D. S 1973. *Spathorhynchus fossorium*, a middle Eocene amphisbaenian (Reptilia) from Wyoming. Copeia 1973(4):704-721.

Berman, D. 1977. *Spathorhynchus natronicus*, a new species of rhineurid amphisbaenian (Reptilia) from the early Oligocene of Wyoming. Journal of Paleontology 51(5):986-991.

Brattstrom, B. 1958. Two Oligocene lizards. Herpetologica 14:43-44.

Cope, E. D. 1872. Third account of new Vertebrata from the Bridger Eocene of Wyoming Territory. Palaeontological Bulletin 3:1-4.

Cope, E. D. 1873a. Synopsis of new Vertebrata from Tertiary of Colorado, obtained during the summer of 1873. Washington, United States Government Printing Office, October 1873, pp. 1-19.

Cope, E. D. 1873b. Second notice of extinct Vertebrata from the Tertiary of the plains. Palaeontological Bulletin 15:1-6.

Cope, E. D. 1874. Report on the vertebrate paleontology of Colorado. Annual Report of the Geological and Geographical Survey Territories for 1873, Hayden 1874:427-533.

Cope, E. D. 1883. First addition to the fauna of the Puerco Eocene. Proceedings of the American Philosophical Society 20:545-546.

Cope, E. D. 1884. The Vertebrata of the Tertiary formations of the West. Book 1. Report of the United States Geological Survey of the Territories, Hayden, 1884; 1009 pp.

Cope, E. D. 1900. The crocodilians, lizards and snakes of North America. Annual Report of the Smithsonian Institution, 1898, United States National Museum, Part 2:153-1270.

Crochet, J.-Y., J.-L. Hartenberger, J.-C. Rage, J. A. Rémy, B. Sigé, B. Sudre, and M. Vianey-Liaud. 1981. Les nouvelles faunes de vertébrés antérieures à la "Grande Coupure" découvertes dans les Phosphorites du Quercy. Bulletin de la National d'Histoire Museum Naturelle, Paris, Series 4, 3(3):245-265.

Douglass, E. 1903. New vertebrates from the Montana Tertiary. Annals of Carnegie Museum 2:145-199.

Douglass, E. 1908. Some Oligocene lizards. Annals of Carnegie Museum 4:278-285.

Emry, R. J., L. S. Russell, and P. R. Bjork. 1987. The Chadronian, Orellan, and Whitneyan North American Land Mammal Ages; pp. 118-152, *in* M. O. Woodburne (ed.), Cenozoic Mammals of North America, Geochronology and Biostratigraphy. University of California Press, Berkeley.

Estes, R. 1963. Early Miocene salamanders and lizards from Florida. Quarterly Journal of the Florida Academy of Sciences 26:234-256.

Estes, R. 1964. Fossil vertebrates from the Late Cretaceous Lance Formation, eastern Wyoming. University of California Publications in the Geological Sciences 49:1-180.

Estes, R. 1970. Origin of the Recent North American lower vertebrate fauna: An inquiry into the fossil record. Forma et Functio 3:139-163.

Estes, R. 1976. Middle Paleocene lower vertebrates from the Tongue River Formation, southeastern Montana. Journal of Paleontology 50:500-520.

Estes, R. 1983. Sauria terrestria, Amphisbaenia. Part 10A. Handbuch der Paläoherpetologie. Gustav Fischer Verlag, Stuttgart.

Estes, R., K. de Queiroz, and J. Gauthier. 1988. Phylogenetic relationships within Squamata; pp. 119-281, in R. Estes and G. Pregill (eds.), Phylogenetic Relationships of the Lizard Families. Essays Commemorating Charles L. Camp. Stanford University Press, Stanford, CA.

Etheridge, R. 1967. Lizard caudal vertebrae. Copeia 4(1967):699-721.

Frost, D. R., and R. Etheridge. 1989. A phylogenetic analysis and taxonomy of iguanian lizards (Reptilia: Squamata). University of Kansas Museum of Natural History, Miscellaneous Publication 81:1-65.

Galbreath, E. C. 1953. A contribution to the Tertiary geology and paleontology of northeastern Colorado. University of Kansas Paleontological Contributions, Vertebrata, Article 4:1-120.

Gans, C. 1967. Rhineura. Catalogue of American Amphibians and Reptiles, p. 42.1-42.2.

Gilmore, C. W. 1928. Fossil lizards of North America. Memoirs, National Academy of Sciences 22:1-197.

Gilmore, C. W. 1938a. Descriptions of new and little known lizards of North America. Proceedings of the United States National Museum 86:11-26.

Gilmore, C. W. 1938b. Fossil snakes of North America. Geological Society of America, Special Paper 9(1):1-96.

Good, D. A. 1988. The phylogenetic position of fossils assigned to the Gerrhonotinae (Squamata: Anguidae). Journal of Vertebrate Paleontology 8(2):188-195.

Hecht, M. K. 1956. A new xantusiid lizard from the Eocene of Wyoming. American Museum Novitates 1174:1-8.

Holman, J. A. 1972. Herpetofauna of the Calf Creek local fauna (lower Oligocene: Cypress Hills Formation) of Saskatchewan. Canadian Journal of Earth Sciences 9(12):1612-1631.

Holman, J. A. 1976. Cenozoic herpetofaunas of Saskatchewan; pp. 80-92, in C. S. Churcher, (ed.), Athlon, Essays on Palaeontology in honour of Loris Shano Russell. Royal Ontario Museum Life Sciences Miscellaneous Publications.

Holman, J. A. 1977. Amphibians and reptiles from the Gulf Coast Miocene of Texas. Herpetologica 33(4):391-403.

Holman, J. A. 1979. A review of North American Tertiary snakes. Publications of the Museum, Michigan State University, Paleontological Series 1(6):200-260.

Holman, J. A. 1981. A herpetofauna from an eastern extension of the Harrison Formation (early Miocene: Arikareean), Cherry County, Nebraska. Journal of Vertebrate Paleontology 1(1):49-56.

Holman, J. A. 1982. Geringophis (Serpentes: Boidae) from the middle Oligocene of Nebraska. Herpetologica 38(4):489-492.

Holman, J. A. 1983. A new species of Helagras (Serpentes) from the middle Oligocene of Nebraska. Journal of Herpetology 17(4):417-419.

Holman, J. A. 1984a. Texasophis galbreathi, new species, the earliest New World colubrid snake. Journal of Vertebrate Paleontology 3(4):223-225.

Holman, J. A. 1984b. Correction to Holman, J. A., Texasophis galbreathi, new species, the earliest New World colubrid snake. [Journal of Vertebrate Paleontology 3(4):223-225, March 1984]. Journal of Vertebrate Paleontology 4(1):168.

Holman, J. A. 1989. Some amphibians and reptiles from the Oligocene of northeastern Colorado. Dakoterra 3:16-21.

Hutchison, J. H. 1992. Western North American reptile and amphibian record across the Eocene/Oligocene boundary and its climatic implications; pp. 451-463, in D. R. Prothero and W. A. Breggren (eds.), Eocene-Oligocene Climatic and Biotic Evolution. Princeton University Press, Princeton, N. J.

Lambe, L. M. 1908. The Vertebrata of the Oligocene of the Cypress Hills, Saskatchewan. Canada Department of Mines, Geological Survey Branch of Canada, Contributions to Canadian Palaeontology, Vol. 3, Part 4, pp. 1-65.

Leidy, J. 1870. [Descriptions of Emys jeansei, E. haydeni, Bäena arenosa and Saniwa ensidens]. Proceedings of the Academy of Natural Sciences, Philadelphia 1870:122.

Lillegraven, J. A. 1970. Stratigraphy, structure, and vertebrate fossils of the Oligocene Brule Fm., Slim Buttes, northwestern South Dakota. Geological Society of America Bulletin 81:831-850.

Marsh, O. C. 1871. Description of some new fossil serpents, from the Tertiary deposits of Wyoming. American Journal of Science, 101:322-329.

Marsh, O. C. 1885. Names of extinct reptiles. American Journal of Science 29:169.

Meszoely, C. A. M. 1970. North American fossil anguid lizards. Bulletin of the Museum of Comparative Zoology 139(2):87-149.

Meszoely, C. A. M., R. Estes, and H. Haubold. 1978. Eocene anguid lizards from Europe and a revision of the genus Xestops. Herpetologica 34:156-166.

Parris, D. C., and J. A. Holman. 1978. An Oligocene snake from a coprolite. Herpetologica 34(3):258-264.

Pregill, G. K. 1992. Systematics of the West Indian lizard genus Leiocephalus (Squamata: Iguania: Tropiduridae). The University of Kansas Museum of Natural History Miscellaneous Publication 84:1-69.

Pregill, G. K., J. A. Gauthier, and H. W. Greene. 1986. The evolution of helodermatid squamates, with a description of a new taxon and an overview of Varanoidea. Transactions of the San Diego Society of Natural History 21(11):167-202.

Prothero, D. R. 1985a. Chadronian (early Oligocene) magnetostratigraphy of eastern Wyoming: implications for the age of the Eocene-Oligocene boundary. Journal of Geology 93:555-565.

Prothero, D. R. 1985b. Correlation of the White River Group by magnetostratigraphy; pp. 265-276, in J. E. Martin (ed.), Fossiliferous Cenozoic Deposits of Western South Dakota and Northwestern Nebraska. Dakoterra 2(2).

Prothero, D. R. 1985c. North American mammalian diversity and Eocene-Oligocene extinctions. Paleobiology 11(4):389-405.

Prothero, D. R., C. R. Denham, and H. G. Farmer. 1982. Oligocene calibration of the magnetic polarity time scale. Geology 10:650-653.

Prothero, D. R., C. R. Denham, and H. G. Farmer. 1983.

Magnetostratigraphy of the White River Group and its implications for Oligocene geochronology. Palaeogeography, Palaeoclimatology, Palaeoecology 42(1983): 151-166.

Prothero, D. R., and C. C. Swisher III. 1992. Magnetostratigraphy and geochronology of the terrestrial Eocene-Oligocene transition in North America; pp. 46-73, in D. R. Prothero and W. A. Breggren (eds.), Eocene-Oligocene Climatic and Biotic Evolution. Princeton University Press, Princeton, N. J.

Rage, J.-C. 1984a. Serpentes. Part 11. Handbuch der Paläoherpetologie. Gustav Fischer Verlag, Stuttgart.

Rage, J.-C. 1984b. La "Grande Coupure" éocène/oligocene et les herpetofaunes (amphibiens et reptiles): problèmes du synchronisme des événements paléobiogeographiques. Bulletin de la Société Géologique de France, Nouvelle Serie, 26(6):1251- 1257.

Rage, J.-C. 1986. The amphibians and reptiles at the Eocene-Oligocene transition in western Europe: An outline of the faunal alterations; pp. 309-310, in C. Pomerol and I. Premoli-Silva (eds.), Terminal Eocene Events. Elsevier Science Publishers B. V., Amsterdam.

Rage, J.-C. 1988a. Le gisement du Bretou (Phosphorites du Quercy, Tarn-et-Garonne, France) et sa faune de vertébrés de Eocène supérieur. I. Amphibiens et reptiles. Palaeontographica. Beiträge zur Naturgeschichte der Vorzeit. Abteilung A., 205(1-6):3-27.

Rage, J.-C. 1988b. Le gisement du Bretou (Phosphorites du Quercy, Tarn-et-Garonne, France) et sa faune de vertébrés de Eocène supérieur. X. Conclusions generales. Palaeontographica Abt. A., 205(1-6):183-189.

Rage, J.-C. 1992. A colubrid snake in the late Eocene of Thailand: the oldest known Colubridae (Reptilia, Serpentes). Comptes Rendus de l'Académie des Sciences, Paris 314:1085-1089.

Rogers, K. 1976. Herpetofauna of the Beck Ranch local fauna (upper Pliocene: Blancan) of Texas. Publications of the Museum, Michigan State University, Paleontological Series 1(5):167-200.

Rossman, T. 1992. Vollständig erhattene "Iguanider", cf. Geiseltaliellus longicaudus Kuhn, 1944 ("Reptilia": Squamata) aus dem mitteleozän der Grube Messel bei Darmstadt. Diplomarbeit, Am Fachbereich Biologie der Technischen Huchschle Darmstadt in Zusammenarbeit mit der Abteilung für Geologie, Mineralogie und Paläontologie des Hessischen Landesmuseum in Darmstadt. July 1992 (unpublished).

Rossman, T. 1993. "Iguanids" from the middle Eocene (Lower Lutetian) of 'Grube Messel', Germany. Journal of Vertebrate Paleontology, Abstract of Papers, 13(3), Supplement to Number 3, 55A. (abstract).

Schatzinger, R. A. 1980. New species of Palaeoxantusia (Reptilia: Sauria) from the Uintan (Eocene) of San Diego Co., California. Journal of Paleontology 54(2): 460-471.

Storer, J. E., and T. T. Tokaryk. 1992. Catalogue of type and figured fossils, Saskatchewan Museum of Natural History, Regina. Natural History Contributions, Saskatchewan Museum of Natural History, No. 10, 67 pp.

Sullivan, R. M. 1979. Revision of the Paleogene genus Glyptosaurus (Reptilia, Anguidae). Bulletin of the American Museum of Natural History 163(1):1-72.

Sullivan, R. M. 1980. Lower vertebrates from Swain Quarry "Fort Union Formation.," middle Paleocene (Torrejonian), Carbon County, Wyoming. Ph.D. dissertation, Michigan State University, 56p.

Sullivan, R. M. 1982. Fossil lizards from Swain Quarry "Fort Union Formation," middle Paleocene (Torrejonian), Carbon County, Wyoming. Journal of Paleontology 56(4):996-1010.

Sullivan, R. M. 1987. Parophisaurus pawneensis (Gilmore, 1928), new genus of anguid lizard from the middle Oligocene of North America. Journal of Herpetology 21(2):115-133.

Sullivan, R. M. 1989. Proglyptosaurus huerfanensis, new genus, new species: Glyptosaurine lizard (Squamata: Anguidae) from the early Eocene of Colorado. American Museum Novitates 2949:1-8.

Szyndlar, Z. 1994. Oligocene snakes of southern Germany. Journal of Vertebrate Paleontology 14(1):24-37.

Taylor, E. 1951. Concerning Oligocene amphisbaenid reptiles. Bulletin of the University of Kansas 34:521-558.

Tedford, R. H., M. F. Skinner, R. W. Fields, J. M. Rensberger, D. P. Whistler, T. Galusha, B. E. Taylor, J. R. Macdonald, and S. D. Webb. 1987. Faunal succession and biochronology of the Arikareean through Hemphillian interval (late Oligocene through earliest Pliocene epochs) in North America; pp. 153-210 in M. O. Woodburne (ed.), Cenozoic Mammals of North America. University of California Press, Berkeley and Los Angeles.

Vanzolini, P. E. 1951. A systematic arrangement of the family Amphisbaenidae (Sauria). Herpetologica 7:113-123.

Vanzolini, P. E. 1952. Fossil snakes and lizards from the lower Miocene of Florida. Journal of Paleontology 26(3):452-457.

Walker, M. 1932. A new burrowing lizard from the Oligocene of central Wyoming. Transactions of the Kansas Academy of Sciences 35:224-231.

18. Ischyromyidae

TIMOTHY H. HEATON

ABSTRACT

Ischyromys is known from the early Duchesnean to the early Whitneyan of the Rocky Mountains and Great Plains of North America. Early species are morphologically diverse and are known from small and sometimes fragmentary samples, so the Duchesnean history of the group is hard to unravel. A diverse radiation of species with a derived but variable jaw musculature dominates the Chadronian of the Rocky Mountains. These species have been assigned to *Ischyromys* by some authors and *Titanotheriomys* by others. The name *Titanotheriomys* is recognized herein as a subgenus of *Ischyromys*. The Orellan of the Great Plains is dominated by two species of *Ischyromys* that lack the *Titanotheriomys* specializations. Both these species are also known from the Chadronian of the Great Plains based on a few fragmentary remains. *Ischyromys* became less abundant in the Rocky Mountains and more abundant in the Great Plains at the Chadronian/Orellan (Eocene/Oligocene) boundary, but no new species originated and no existing species underwent measurable change during the transition.

INTRODUCTION

Ischyromys is one of the most common rodents of the Eocene-Oligocene transition in North America, with a history of discovery going back to the early F. V. Hayden expeditions of the 1850s. Clark and Kietzke (1967), in their systematic paleoecological study of the South Dakota badlands, found *Ischyromys* to be the most common rodent in every environment and to be most abundant in dry plains environments lacking trees. More than five thousand jaws of *Ischyromys* exist in public museum collections and have been examined for this study.

In the Great Plains, *Ischyromys* is dominant during the Orellan but is also occasionally found in sediments of Chadronian and Whitneyan age. Farther west, particularly in Montana, *Ischyromys* is common in the Chadronian but rare during the Duchesnean and Orellan. The most geographically remote populations of *Ischyromys* come from Duchesnean and Chadronian deposits of Saskatchewan and Texas. The history of *Ischyromys* is complicated by the fact that Chadronian skulls typically have a more derived musculature than do their later Orellan counterparts (Wood, 1937, 1980). Several authors have used the name *Titanotheriomys* for

Table 1. Subjective scores for wear stage assigned to each cheek tooth.

Wear Stage	Description
0	Tooth unerupted but visible
1	Tooth partially erupted
2	Tooth fully erupted but unworn
3	Cusp peaks worn but not down to ridge saddles
4	Ridge saddles worn but valley cirques not migrated
5	Valley cirques migrated but all valleys continuous
6	Posterior lingual valley forms an island
7	Posterior lingual valley worn away
8	All valleys worn away but tooth margins distinct
9	Tooth worn down to roots

Table 2. Subjective size scores assigned to each of the four possible accessory cusps on each cheek tooth.

Cusp Size	Description
0	Not present
1	Barely visible
2	Small but very distinct and clearly developed
3	Large and well developed
4	Unusually large or forming a ridge across valley

Chadronian specimens with this derived condition, as discussed below.

METHODS

This study of *Ischyromys* has focused on lower dentitions because dentaries are by far the most common elements recovered. Each dentary or partial dentary containing teeth was placed on a special mount and photographed in occlusal and lingual view. These images were projected onto a digitizing tablet where a periphery/area/centroid measurement and 25 point coordinates were made on each available tooth (Fig. 1). These data were used to calculate measurements for length and width of the teeth, positions of cusps and valleys, and various angles in order to quantify tooth size and shape (Fig. 2). Three point coordinates were also established along the ventral margin of each jaw in

Occlusal View of Lower Molar

Lingual View of Lower Molar

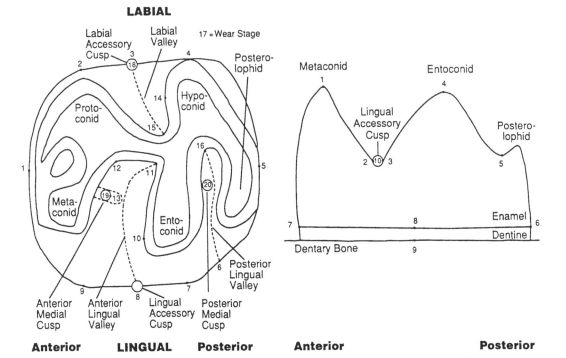

Figure 1. Diagram of an idealized lower molar of *Ischyromys* in occlusal and lingual view showing points digitized and positions of accessory cusps (after Heaton, 1988).

Occlusal View of Lower Molar

Lingual View of Lower Molar

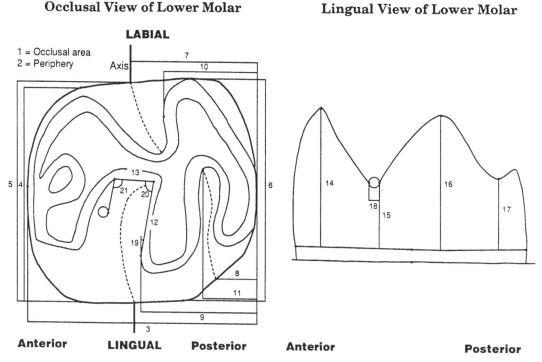

Figure 2. Diagram of an idealized lower molar of *Ischyromys* in occlusal and lingual view showing measurements generated for this study (after Heaton, 1988).

Table 3. List of measurements made on each molar (Figure 2) and dentary (Figure 3).

#	View	Description
Measurements on each tooth (p4–m3)		
1	Occlusal	Square root of tooth area
2	Occlusal	Tooth Periphery
3	Occlusal	Tooth length
4	Occlusal	Tooth width (anterior)
5	Occlusal	Tooth width (medial)
6	Occlusal	Tooth width (posterior)
7	Occlusal	Anterior lingual valley mouth to end of tooth
8	Occlusal	Posterior lingual valley mouth to end of tooth
9	Occlusal	Anterior lingual valley posterior wall to end of tooth
10	Occlusal	Labial valley posterior wall to end of tooth
11	Occlusal	Posterior lingual valley head to end of tooth
12	Occlusal	Length of anterior lingual valley
13	Occlusal	Width of anterior lingual valley
14	Lingual	Height of anterior lingual cusp (metaconid)
15	Lingual	Height of anterior lingual valley bottom
16	Lingual	Height of posterior lingual cusp (entoconid)
17	Lingual	Height of posterior lingual valley bottom
18	Lingual	Width of lingual accessory cusp
19	Occlusal	Anterior lingual valley posterior wall angle
20	Occlusal	Anterior lingual valley posterior head angle
21	Occlusal	Anterior lingual valley anterior head angle
22	Occlusal	Wear stage (0–9)
23	Occlusal	Labial accessory cusp size (0–4)
24	Occlusal	Anterior medial accessory cusp size (0–4)
25	Occlusal	Posterior medial accessory cusp size (0–4)
26	Lingual	Lingual accessory cusp size (0–4)

Measurements on each dentary		
1	Lingual	Depth of anterior jaw from M_{1-2} base line
2	Lingual	Depth of medial jaw from M_{1-2} base line
3	Lingual	Depth of posterior jaw from M_{1-2} base line
4	Lingual	Depth of posterior jaw from M_3 base
5	Lingual	Distance between two anterior jaw margin measurements
6	Lingual	Distance between two posterior jaw margin measurements
7	Lingual	Angle between two anterior jaw margin measurements
8	Lingual	Angle between two posterior jaw margin measurements

lingual view, and jaw depth and shape were determined therefrom (Fig. 3). Subjective scores were given for the wear stage of each tooth (Table 1) and for the size of the four accessory cusps (Table 2). Table 3 lists all measurements used in this study.

By far the largest collections of *Ischyromys* come from Orellan deposits of the Great Plains states, and these show remarkable morphologic uniformity. For this reason they will be discussed first, after which the problem of the Chadronian "*Titanotheriomys*" complex will be addressed.

ISCHYROMYS OF THE CENTRAL GREAT PLAINS

The following is a list of ischyromyid genera and species named from the Great Plains of the United States, together with the locality of each type specimen:

Ischyromys typus Leidy, 1856, 1869
  SW South Dakota
Colotaxis cristatus Cope, 1873a
  NE Colorado

Gymnoptychus chrysodon Cope, 1873b
  NE Colorado
Ischyromys parvidens Miller and Gidley, 1920
  SW South Dakota
Ischyromys pliacus Troxell, 1922
  SE Wyoming (?)
Ischyromys typus nanus Troxell, 1922
  NW Nebraska
Ischyromys typus lloydi Troxell, 1922
  NW Nebraska
Ischyromys troxelli Wood, 1937
  SW South Dakota
Ischyromys sp. (large) Howe, 1966
  NW Nebraska

The number of valid species has been disputed, though all were assigned to the genus *Ischyromys* at an early date. Wood (1937, 1980) accepted four species: *I. typus*, *I. parvidens*, *I. pliacus*, and *I. troxelli*. Howe (1956, 1966) proposed that *I. parvidens*, *I. typus*, *I. pliacus*, and *I.* sp. (large) represent chronospecies of a lineage that increased in size during the Orellan and early Whitneyan.

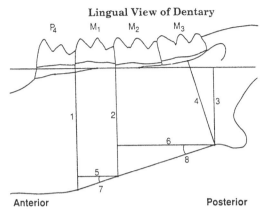

Figure 3. Diagram of an idealized dentary of *Ischyromys* in lingual view showing points digitized and measurements generated for this study (after Heaton, 1988).

Black (1968), on the other hand, accepted only two valid species, one large (*I. typus*) and one small (either the small Montana species, *I. veterior*, or *I. parvidens*). In an unpublished thesis that included much statistical analysis, O'Neill (1979) went so far as to consider all central Great Plains *Ischyromys* to be a single species.

My recent study of Great Plains *Ischyromys* has been the most comprehensive to date (Heaton, 1993a). A reevaluation was made of Howe's (1956, 1966) claim of anagenesis using a much larger sample, and it was concluded that two distinct and rather static species coexisted during the late Chadronian and Orellan in the Pine Ridge area (northwestern Nebraska and adjacent Wyoming). The first line of evidence for this conclusion came from the bimodal, or highly skewed, size distributions found in each stratigraphic interval. Figure 4 shows size histograms for the Chadron Formation, the lower portion of the Orella Member of the Brule Formation, and the middle Orella to Whitney interval of the Brule Formation. All of these populations differ significantly from a normal distribution, especially the lower Orella sample which has the smallest mean and is extremely right skewed. Another striking feature of the three populations in Figure 4 is that they all have about the same overall size range (same minimum and maximum) despite having very different means, medians, and modes. Although the mean size at each stratigraphic level is different, as Howe noted, it is not a case of a shifting normal distribution as would be expected for anagenetic change in a single lineage.

My alternative hypothesis, that two species of different sizes coexisted but varied in proportion, fits the data nicely. Separating a mixed distribution into its constitu-ent normal distributions can be done mathematically using various assumptions, and a tabular approach was presented in Heaton (1993a). Using this method, the two mean values derived from the three stratigraphic populations are remarkably similar. Only the ratio of the small to large species

differs, being about 1:1 in the Chadron, 10:1 in the lower Orella, and 1:20 in the middle Orella to Whitney. Based on the priority of type specimens of appropriate size from the Big Badlands of South Dakota, Heaton (1993a) used the names *Ischyromys typus* and *I. parvidens* for the large and small species, respectively. Estimated values for the mean and standard deviation of these two Pine Ridge species are given in Figure 4.

The second line of evidence suggesting the presence of two coexisting species is the absence of the smaller form outside the Pine Ridge area, even in lower Orellan strata. Bivariate plots showing size vs. stratigraphic level were presented in Heaton (1993a) for southwestern North Dakota, northwestern South Dakota, and northeastern Colorado, and corresponding histograms (all levels mixed) are shown in Figure 5. Only slight increases in mean size occurred over time, and for each locality the population closely fits a normal distribution. Mean sizes of these populations match well with the calculated mean for *Ischyromys typus* at Pine Ridge, so they clearly belong to that species. Therefore *I. typus* had a much wider geographic range than its smaller contemporary.

My previous report only briefly considered the Big Badlands of southwestern South Dakota despite the enormous collections made there. This was in part because most Big Badlands specimens are not zoned, and most of those that are come from the narrow stratigraphic interval known as the "Lower Nodular Zone." Contrary to my initial impression (Heaton, 1993a), the Big Badlands does not seem to contain the small species of *Ischyromys* found at Pine Ridge despite its close proximity, which brings into question the validity of the name *I. parvidens* (Heaton, 1993b).

Figure 6 contains histograms of all Big Badlands specimens and those known to come from the "Lower Nodular Zone." Figure 7 is a bivariate plot of the smaller zoned sample. Unfortunately the stratigraphic level for most of these specimens is not precise, and correlation across the Big Badlands is questionable (Prothero, 1985). But since plots made from separate localities look virtually identical, combining them seems justified and provides a reasonable density for the sparser intervals. A small increase in mean size over time is indicated, much like that in the northern Dakotas localities (Heaton, 1993a), but more notable is the near uniformity in size at all levels. Given this uniformity, it is ironic that Clark (1967) used the geographic uniformity of *Ischyromys typus* in the "Lower Nodular Zone" as evidence for the rapid and geographically synchronous deposition of the unit.

The mean size of Big Badlands *Ischyromys* is similar to that of *I. typus* at Pine Ridge and other Great Plains localities, and the sample very closely approximates a normal distribution. No evidence has been found for a small Pine Ridge species. Fortunately the type specimen of *I. parvidens* is a dentary with teeth; its size is marked on Figure 6. This appears to be a classic case

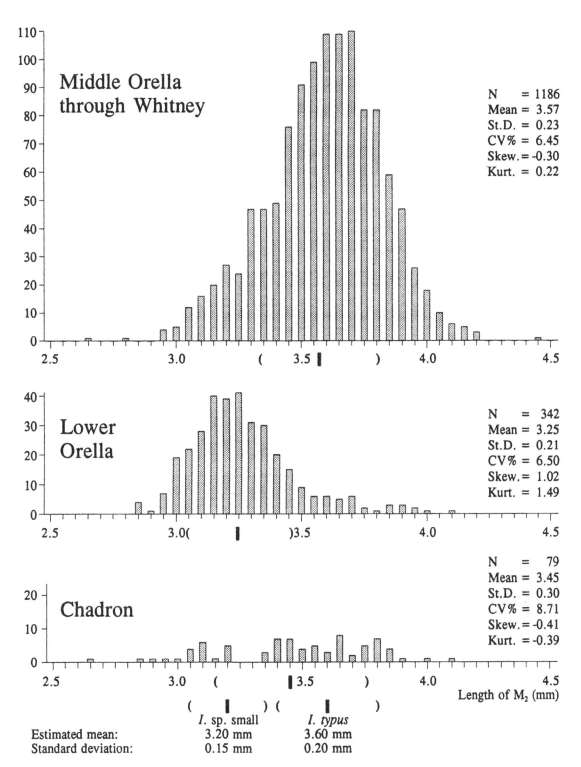

Figure 4. Histograms of M$_2$ length values for three stratigraphic samples of *Ischyromys* at Pine Ridge, Nebraska. The upper two histograms include all appropriate specimens from the Toadstool Park and Munson Ranch localities (see Heaton, 1993a). The lower histogram also includes Chadronian specimens from all Pine Ridge and Douglas localities (see map on page 263). The bar and parentheses below each histogram show graphically the mean and standard deviation of each sample. Below the histograms are the estimated statistics for the two species that are thought to make up these three samples.

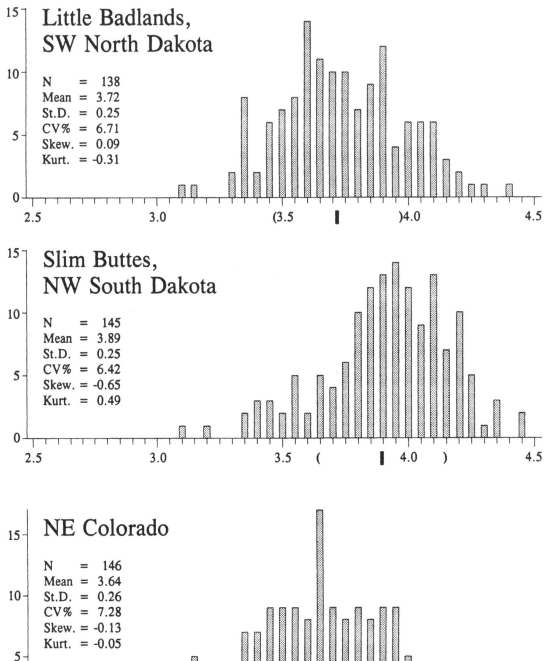

Figure 5. Histograms of M₂ length values for three geographic populations of *Ischyromys*: southwestern North Dakota, northwestern South Dakota, and northeastern Colorado (see map on page 263). Each of these populations closely approximates a normal distribution whose mean and standard deviation closely matches the larger species from Pine Ridge (*I. typus*) shown in Figure 4. Note the paucity of specimens in the 3.0-3.3 mm range where the small Pine Ridge species has its mode. The skewness of the Slim Buttes sample is a result of stratigraphic mixing and anagenesis, not coexisting species (Heaton, 1993a).

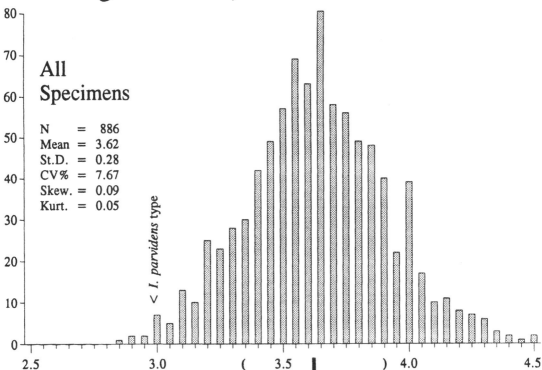

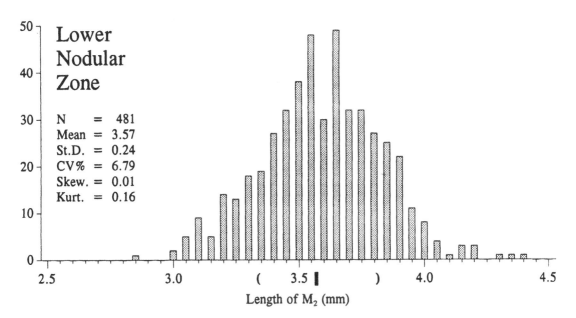

Figure 6. Histograms of M₂ length values for *Ischyromys* from the Big Badlands, southwestern South Dakota (see map on page 263). The lower histogram contains only specimens known to have come from the "Lower Nodular Zone" (the most fossiliferous stratigraphic interval). Like the localities in Figure 5, this population closely fits a normal distribution and is therefore interpreted to contain only *I. typus*. The type specimen of *I. parvidens* falls in the extreme left tail of the size distribution.

Big Badlands, South Dakota

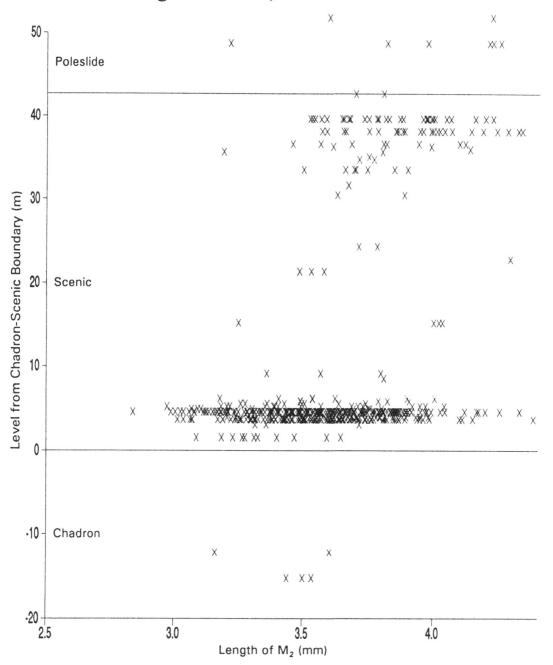

Figure 7. Bivariate plot of 605 zoned *Ischyromys* m2s from the Big Badlands of South Dakota, a subset of the sample shown in the upper histogram of Figure 6. Stratigraphic levels are shown with reference to the Chadron/Brule boundary. Most specimens are from the "Lower Nodular Zone."

in which a specimen from the tail end of a normal distribution was set apart as a distinct species when in fact it is not. This problem is discussed below.

Wood (1976), Howe (1966), and others have commented on the uniform pattern of *Ischyromys* skulls from the Great Plains. Attempts to divide the sample into multiple species using my large multivariate dataset on lower dentitions all failed. Even the small and large species from Pine Ridge seem indistinguishable except for their difference in size. The only other possible distinction found was the incidence of accessory cusps on the teeth, but upon further investigation this turned out to be an allometric correlate of size and not a species difference (Heaton, 1993a). Therefore there is no justifica-tion for recognizing multiple species of *Ischyromys* in the central Great Plains beyond *I. typus* and the smaller Pine Ridge species, and long-used names such as *I. pliacus* and *I. troxelli* are synonyms of *I. typus*.

ISCHYROMYS (TITANOTHERIOMYS) OF THE ROCKY MOUNTAINS

Ischyromyid fossils have been found in the Rocky Mountains and other western localities from Saskatchewan to Texas with a majority coming from basinal deposits of Montana and Wyoming. Nearly all of these predate the more abundant Orellan *Ischyromys* of the Great Plains, and most collections are small. By far the largest sample is from the Pipestone Springs Local Fauna in southwestern Montana. Pipestone Springs is the type locality for the first species of *Ischyromys* named from outside the Great Plains and the species most commonly compared to Great Plains *Ischyromys*.

Matthew (1903) named the Pipestone Springs material *Ischyromys veterior* and later gave this species the subgeneric name *Titanotheriomys* (Matthew, 1910). Miller and Gidley (1920), in their paper describing *I. parvidens*, considered *Titanotheriomys* to be of full genus rank, and they noted the similar small size of their new species and *T. veterior*. But Black (1968) dismissed the differences between *Titanotheriomys* and *Ischyromys* and considered *I. parvidens* to be a junior synonym of *I. veterior*. This synonymy began a dispute that will be a major focus of this paper.

Wood (1937) was the first to do a thorough taxonomic analysis of the ischyromyids, and his influence on the taxonomy of the group has been long and important (Wood, 1976, 1980). Wood (1937) considered *Ischyromys* and *Titanotheriomys* to be distinct genera, but he ignored most of the differences noted by Matthew (1903, 1910) and Miller and Gidley (1920) and instead drew up his own list of distinctions based mainly on skull structure. *Titanotheriomys* was described as having a broader, flatter skull than *Ischyromys* with a shorter pre-orbital region and more slender snout, zygoma, and braincase. The temporal crests of *Titanotheriomys*, as noted by Matthew (1910), do not meet to form a sagittal crest as they do in

Ischyromys. Wood (1937) also stated that in *Titanotheriomys* the masseter extends farther onto the zygoma than in *Ischyromys* with a thin strip that extends farther onto the snout, thus approaching the advanced sciuromorph pattern found in some other rodent families. The infraorbital foramen is visible dorsally in *Titanotheriomys* but not in *Ischyromys*. Wood also stated that the mandible and postcranial elements, insofar as known, tend to be more slender in *Titanotheriomys*.

Black (1965) briefly mentioned *Ischyromys* and the problems of its taxonomy in his review of the rodents of the Pipestone Springs Local Fauna, and in a later paper he undertook his major revision of the genus. Black (1968) claimed that the distinctive characters used by Wood (1937) to separate *Titanotheriomys* from *Ischyromys* were not real and were due only to crushing and distortion of the two Chadronian skulls available to Wood. With better material from Pipestone Springs, Black stated that the skulls were indistinguishable and, therefore, that *Titanotheriomys* was a junior synonym of *Ischyromys*. He did recognize that some skulls have a sagittal crest while others do not, but on finding several intermediate configurations in skulls from Pipestone Springs he attributed this difference to individual or sexual variation. Wood (1969) initially conceded that Black's synonymy was correct, but in later publications he persuasively argued for the validity of *Titanotheriomys* (Wood, 1974, 1976, 1980).

Early descriptions of *Titanotheriomys* mentioned a number of distinctive characters of the lower cheek teeth. Both Matthew (1903) and Wood (1937) claimed that the molars of *T. veterior* were relatively narrow compared to those of Great Plains *Ischyromys*, and Wood (1937) even erected a new species, *T. wyomingensis*, based on a skull similar to that of *T. veterior* but with "lower teeth [that] are no longer than they are broad." Wood (1937) also asserted that *T. veterior* was unique in possessing a partial barrier across the median valley of the lower molars, even though a similar feature was seen in some *Ischyromys* of the Great Plains.

Black (1968) erected the new species *Ischyromys douglassi* for specimens from the early Chadronian McCarty's Mountain Local Fauna. He described *I. douglassi* as being similar in size to *I. typus* but having teeth that are wider in relation to their length. He also listed many minor dental differences.

Russell (1972) reported ten isolated lower cheek teeth from a Chadronian locality in Saskatchewan and erected a new species, *Ischyromys junctus*, which he described as being intermediate in size between *I. typus* and *I. parvidens*. He stated that its unique features are the joining of the anterolophid and metalophid to the metaconid, forming a triangular valley, and the curving of the posterolophid around the posterior margin of the crown nearly to the entoconid. Wood (1974) erected the most recently named new species, *Ischyromys blacki*, for a specimen from Texas that he described as having a skull similar to that of *I. typus* but with teeth that have many primitive characters including prominent metaconules and incomplete metalophs, comparable to the condition seen in *T. douglassi*.

PIPESTONE SPRINGS, MONTANA

This study includes a much larger sample of Pipestone Springs ischyromyids than has previously been used, and multivariate analysis of the many characters measured can be used to evaluate the various claims of earlier authors. First, however, appropriate samples must be chosen for comparison.

As was the case for the Orellan sample from Pine Ridge, the ischyromyid sample from Pipestone Springs seems to be a mixture of two species. Wood (1937) and Black (1965) both recognized the presence of a second species larger than *Titanotheriomys veterior* which they provisionally assigned to the same species as the largest Great Plains specimens, *Ischyromys pliacus*. However, both authors ignored these larger specimens in their later publications.

Figure 8 shows two histograms of Pipestone Springs specimens. Note the strong skewness to the right caused by the small number of large specimens, especially apparent in the occlusal area measurements. Since the deviation from a normal distribution is highly signifi-cant, the presence of two species is clearly indicated, and their means can be estimated mathematically. The estimated mean for *T. veterior*, the smaller species, is similar to that of the small Pine Ridge species that dominates the lower Orella. These species of similar size make for a useful comparison because they exemplify the *Ischyromys* and *Titanotheriomys* conditions and are available in roughly equal numbers. Because each of these populations is contaminated by specimens of a larger species that overlaps it in size but is not readily distinguishable, pure samples cannot be used. My solution was to eliminate all specimens for which the square root of occlusal area of M_2 was 3.4 mm or greater in an attempt to reduce the level of contamination to a minimum.

Discriminant analysis is the ideal statistical technique for determining the degree to which two populations can be distinguished based on a suite of measured characters. Figures 1 through 3 show an idealized *Ischyromys* tooth and jaw along with the measurements made thereon, and the measurements are listed in Table 3. Heaton (1988, 1993c) reported results on five step-wise discriminant analyses attempting to distinguish Pipestone Springs *Titanotheriomys* from Great Plains *Ischyromys*: one analysis for each lower cheek tooth and one using only measurements of the dentary. Table 4 shows the success rates of these analyses and the number of variables selected by the step-wise routine in each case. Analyses of the last two molars were the most successful.

Table 5 lists several statistics for M_2 and M_3 measurements for the two populations. Note that the

Table 4. Success rates of discriminant analyses at distinguishing *Ischyromys* from Pipestone Springs, Montana, and from the lower Orella, Pine Ridge, Nebraska. The first four are stepwise discriminant analyses done with BMDP (Heaton, 1988). The last one was done with SYSTAT and includes previously unmeasured Pipestone Springs material.

Element used	Jaw	P_4	M_1	M_2	M_3	M_{2-3}
Success rate	84%	87%	82%	90%	91%	98%
Number of characters	2	7	9	10	9	38

Table 5. Selected measurements on M_2 and M_3 for populations of *Ischyromys* from the lower Orella, Pine Ridge, Nebraska and Pipestone Springs, Montana. L. = length.

	M_2			M_3		
	SQRT AREA	TOTAL L.	POST. END L.	SQRT AREA	TOTAL L.	POST. END L.
Pipestone Springs, Jefferson Co., Montana						
N	323	332	322	230	251	231
Max.	3.94	4.30	2.20	3.77	4.40	2.31
Mean	3.15	3.33	1.61	2.96	3.32	1.56
Min.	2.68	2.65	1.19	2.50	2.68	1.15
CV%	6.46	7.19	9.68	7.18	9.01	13.87
Lower Orella, Sioux Co., Nebraska						
N	372	380	362	195	238	202
Max.	3.78	4.06	2.18	3.77	4.19	2.33
Mean	3.09	3.29	1.66	3.09	3.51	1.80
Min.	2.65	2.81	1.28	2.62	2.98	1.37
CV%	7.43	7.09	9.66	6.92	7.42	10.06

mean tooth length is larger for M_2 than for M_3 in the Pipestone Springs sample, but that the opposite is true in the Pine Ridge sample. This led me to believe that combining characters from these two teeth would improve the success of the discriminant analysis even further, and this turned out to be the case (Table 4). Figure 9 diagrams the result of a discriminant analysis on 38 characters, 19 from M_2 and 19 from M_3. Table 6 lists the variables used and shows the factor loadings provided by the routine. The success rate was 98% with only five Pipestone Springs specimens and one Pine Ridge specimen being misclassified.

The magnitude of the factor loadings shows how much they contributed to the discrimination (Table 6). For the factor loadings on whole tooth measurements (1-6), the positive values for M_2 and negative values for M_3 show that, on average, M_2 is the larger tooth at Pipestone while M_3 is larger at Pine Ridge (thus

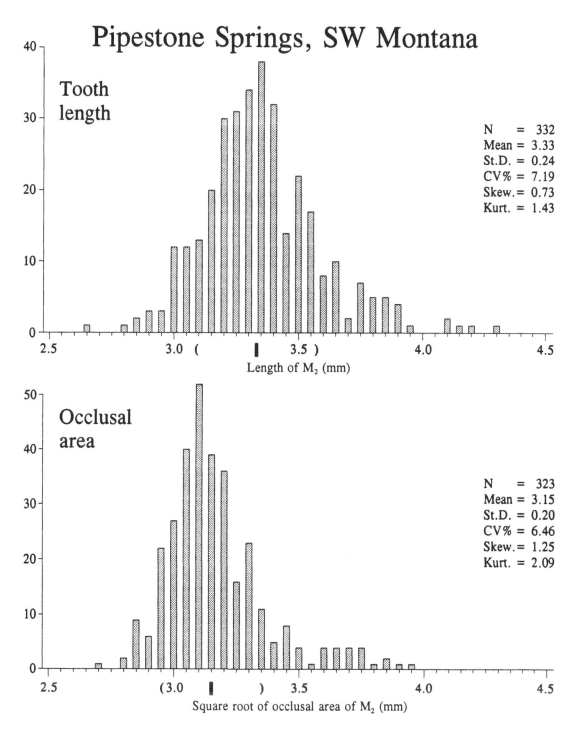

Figure 8. Histograms of M_2 length and square root of occlusal area values for *Ischyromys* (*Titanotheriomys*) from Pipestone Springs, Montana. Like the lower Orella sample in Figure 4, this population is clearly a mixed sample containing a small number of specimens from a large species as well as a large sample of *I.* (*T.*) *veterior*.

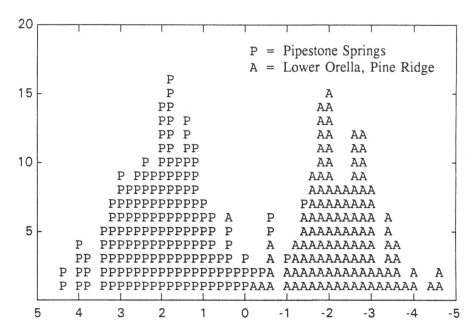

Figure 9. Histogram showing results of discriminant analysis on Pine Ridge lower Orella and Pipestone Springs populations based on 38 characters of M_2 and M_3. Only dentaries for which both molars are present and fall in wear classes 2 through 5 are included (see Table 1). Table 6 lists the characters used and their corresponding factor loadings. Only five of the Pipestone Springs and one of the lower Orella specimens were misclassified, giving the analysis a 98% classification success rate.

reinforcing the indications in Table 5). Not only does the last tooth tend to be shorter in *Titanotheriomys*, but the posterior end of each tooth is also relatively shortened, especially in M_3, as seen in the strong negative loadings on the factors that measure posterior features (7-11).

Figure 10 shows the results of an attempt to distinguish the two populations using only two characters, total length of M_2 and posterior shortening of M_3. One length measurement was divided by the other to create these histograms, thus factoring out differences in overall size. Although these two characters alone are not diagnostic, the modes of the Pipestone Springs and Pine Ridge distributions are very different. Note that the Pipestone Springs sample has a wider range and is left skewed, but the Pine Ridge sample is not right skewed. Note also in Table 5 the low mean but high coefficient of variation for the posterior end of both M_2 and M_3 in the Pipestone Springs sample. This higher variability in *Titanotheriomys*, along with a tendency to mimic the *Ischyromys* condition, strongly suggests that the posterior shortening seen in the tooth row of the Pipestone Springs species is a derived condition.

It is interesting that the only character other than size that Matthew (1903) used to distinguish his new Montana species was that the last lower molar "has always a narrow heel with the last crest imperfect internally, while in all the Colorado specimens the heel is as wide as the rest of the tooth, and the third (last)

crest perfectly developed." Wood (1937) claimed to be "entirely unable to find [this] difference," but it now stands out as an important feature in distinguishing *Titanotheriomys veterior* from Great Plains *Ischyromys*.

Table 6 shows that *Titanotheriomys veterior* of Pipestone Springs also tends to have lower-crowned molars than *Ischyromys* of Pine Ridge as suggested by the consistent negative loadings on the tooth height measurements (14-17), a difference suggested by Wood (1937). Several features of the lingual valleys also show consistent differences on both M_2 and M_3, though the magnitude of the loadings is low (Table 6). On average the anterior lingual valley is smaller in *T. veterior*, and it runs at a slightly greater angle from the main tooth axis (measurements 12-13, 18-19). Both the anterior and posterior lingual valleys sometimes contain an accessory cusp that in extreme cases can dam the valleys, though this feature is not consistent in *Titanotheriomys* and sometimes occurs in *Ischyromys* as well (Heaton, 1993a; Wood, 1937).

After examining a great many specimens I came to recognize subtle character differences that, unfortunately, do not show up in any of the measurements. The lingual valleys, especially the anterior part of the tooth, commonly make a bend in *Titanotheriomys veterior* that is never seen in *Ischyromys*. When this bend is present, a sharp ridge may extend posterolingually off the metaconid which appears as a bump in lingual view. Figure 11 illustrates this condition compared with a typical Great Plains *Ischyromys*. This condition, like

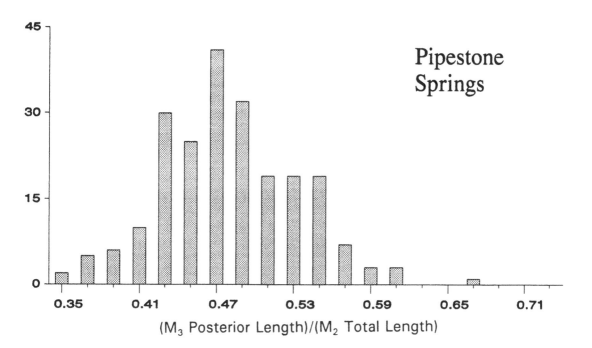

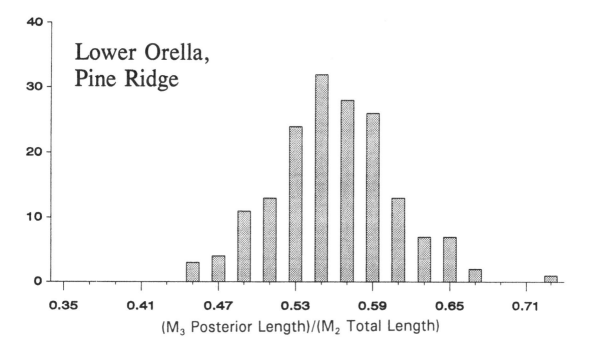

Figure 10. Histograms for Pipestone Springs and Pine Ridge lower Orella populations showing values generated by dividing the length of the posterior end of M_3 by the total length of M_2. This calculated variable combines the effects of the shortening of M_3 relative to M_2 and the relative shortening of the posterior end of M_3 in *Ischyromys* (*Titanotheriomys*) *veterior*. Note that there is still considerable overlap, mainly because the posterior shortening in *I. veterior* is a highly variable character (see Table 5).

Table 6. Factor loadings from discriminant analysis on *Ischyromys* using 38 characters on M_2 and M_3. Positive values indicate features that tend to be larger on Pipestone Springs specimens; negative values indicate features that tend to be larger on lower Orella specimens from Pine Ridge.

		M_2	M_3
Whole Tooth Measurements			
1	Square root of area	0.148	-0.201
2	Periphery	0.109	-0.258
3	Total length	0.101	-0.181
4	Anterior width	0.029	-0.042
5	Medial width	0.036	-0.084
6	Posterior width	-0.005	-0.186
Features to End of Tooth			
7	Ant. lingual valley mouth	-0.109	-0.384
8	Post. lingual valley mouth	-0.125	-0.364
9	Ant. lingual valley wall	0.184	-0.143
10	Labial valley wall	0.107	-0.172
11	Post. lingual valley head	0.115	-0.130
Tooth Valley Measurements			
12	Ant. lingual valley length	-0.131	-0.101
13	Ant. lingual valley width	-0.027	-0.009
Tooth Height Measurements			
14	Anterior cusp height	-0.186	-0.319
15	Anterior valley height	-0.087	-0.189
16	Posterior cusp height	-0.039	-0.089
17	Posterior valley height	-0.112	-0.311
Valley Shape Angles			
18	Ant. lingual valley wall	0.082	0.046
19	Ant. lingual valley head	0.019	0.038

the posterior shortening of the molars, is quite variable in *T. veterior*, so it cannot be used to distinguish the two forms with perfect accuracy.

My success in distinguishing the tooth rows of the small Pipestone Springs species from the small Pine Ridge species is ironic in that the only thing that Black (1968) and Wood (1980) agreed on was that the teeth of the two populations were indistinguishable. This might seem a strong argument in favor of generically separating *Titanotheriomys* from *Ischyromys*, but one further consideration must be taken into account. The Pipestone Springs and Pine Ridge populations are separated by 750 km and are significantly different in age. Several smaller samples have been collected from the intervening parts of Wyoming. Most of the localities involved are geographically closer to Pine Ridge but closer in age to Pipestone Springs (i.e., Chadronian). The question to be asked is whether the fossils from the intervening localities match *Ischyromys*, match *Titanotheriomys*, or fall somewhere between the two.

The teeth will be considered first. Table 7 lists the pertinent localities arranged in order from northwest to southeast. Table 7 also shows the classification of the small ischyromyid teeth (M_{2-3}) provided by the discriminant factors of Table 6. By far the largest sample comes from Flagstaff Rim (discussed below). The Pine Ridge group consists of small dentaries from

the Chadron Formation underlying the lower Orella sample. The discriminant analysis tends to classify specimens from the northwestern localities with the Pipestone Springs species and those from the southeastern localities with the Pine Ridge species, but no distinct boundary exists. Figure 12 diagrams the results of the classification on the same scale as Figure 9, with specimens from each locality coded separately. Taken together these specimens form a unimodal distribution centered just to the right of the dividing line between the Montana and Nebraska samples. This clearly demonstrates that the intervening localities contain populations of intermediate character and not simply a mixture of two distinct species. The latter case would produce a broad bimodal curve similar to that in Figure 9.

FLAGSTAFF RIM, WYOMING

Figure 13 is a bivariate plot showing length of M_2 and stratigraphic position of ischyromyids from Flagstaff Rim. This section represents most of the Chadronian, and the collections are well zoned thanks to the work of Emry (1973, 1992). Flynn (1977) studied both skulls and dentitions of ischyromyids from Flagstaff Rim and concluded that the sample represented coexisting lineages of *Ischyromys* and *Titanotheriomys* that were not distinguishable by size or tooth morphology. He based this conclusion on the fact that each of the 16 available skulls could clearly be classified as either *Ischyromys* (4) or *Titanotheriomys* (12) based on the characters outlined by Wood (1976). See Table 8 for a list of Wood's characters and Table 9 for Flynn's classifications. Flynn (1977) also suggested that a small number of larger specimens (seen in the upper portion of Figure 13) might represent a second species of *Titanotheriomys* at Flagstaff Rim. These larger specimens are not included in Table 7 or Figure 12. Kron (1978) also reported coexisting *Ischyromys* and *Titanotheriomys* from the latest Chadronian of the nearby Douglas locality based on skull morphology.

Wood (1976) discussed whether the differences in skull structure listed in Table 8 could be found as individual variation within a single population. No such variation exists in the abundant Orellan *Ischyromys* of the Great Plains, but the skull structure of *Titanotheriomys* is much more variable as Wood's character descriptions suggest (Table 8). The discriminant classification shown in Figure 12 strongly suggests that Flagstaff Rim and other Wyoming localities contain a population of intermediate character. Measurements on lower dentitions from Flagstaff Rim were scrutinized using statistical techniques to search for any sign of mixing. The population shows exceptionally low coefficients of variation for most characters, and a principal components analysis showed a good multivariate normal distribution with no evidence of bimodality or skewness on any principal component (Heaton, 1988). The conclusion seems inescapable

Table 7. Classification ischyromyids based on the discriminant factor loadings listed in Table 6. In addition to the Pipestone Springs and lower Orella samples on which the discriminant analysis is based, Chadronian specimens from Wyoming localities and Pine Ridge are also classified. The mean value of the canonical variable for each population is also given (see scale in Figures 9 and 12).

| | NW Montana | | Wyoming | | | | | Nebraska | | SE | |
	Pipestone Springs	Beaver Divide	Cameron Springs	Flagstaff Low Q	Flagstaff Mid Q	Flagstaff High	Ledge Creek	Harshman Quarry	Pine Ridge	Lower Orella	Total
Mean	1.8	1.0	-0.3	-0.9	-2.1	-0.3	-2.0	-0.8	-2.0	-2.4	
Pipestone	157	3	5	1	0	17	0	0	2	1	186
Orella	5	0	5	3	6	16	1	1	11	122	170
Total	162	3	10	4	6	33	1	1	13	123	356

Table 8. List of skull characters used by Wood (1976) to distinguish *Titanotheriomys* of Pipestone Springs from *Ischyromys* of the Great Plains.

Feature	*Ischyromys*	*Titanotheriomys*
Dorsal profile of skull	Always flat	Often arched
Anterior face of zygoma	Steep for an ischyromyid	Gently inclined
Origin of M. masseter lateralis	Ends anteriorly at strong crest curving medially on ventral side of zygoma, only slightly anterior to front of P^3	Extends lateral and forward of infraorbital foramen, at least as far as premaxillary-maxillary suture
Angle between central fibers of M. masseter lateralis and of M. masseter superficialis	About 60°	About 20°
M. masseter medialis penetrating infraorbital foramen	Never	Possible sometimes
Possible subdivision of infraorbital foramen	Never	Sometimes
Descriptive name of jaw musculature	Protrogomorphous	Sciuromorphous and/or myomorphous
Sagittal crest	Always well developed in adult	Usually absent; present in one specimen from Pipestone Springs
Temporal crests	Unite above eye in adult to form sagittal crest; in juvenile may not unite but are parallel	Usually curved with lyrate space between them; may unite near front of interparietal
Origin of pars maxillaris anterior of M. buccinator	A fossa, usually shallow, along lateral edge of palate well in front of P^3	A fossa, usually deep, sometimes closed ventrally, along lateral edge of palate in front of P^3
Bucco-naso-labialis	Scar for origin weak or absent	Scar for origin extends across entire premaxilla, almost continuous with that for masseter lateralis

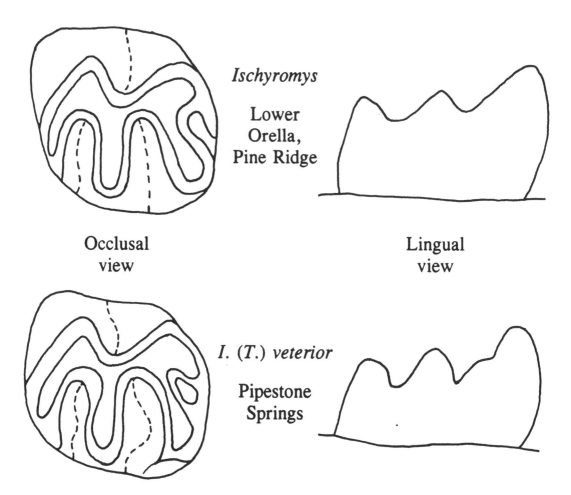

Figure 11. Idealized diagram showing differences between *Ischyromys* (*Ischyromys*) from Pine Ridge, Nebraska, and *I.* (*Titanotheriomys*) *veterior* from Pipestone Springs, Montana.

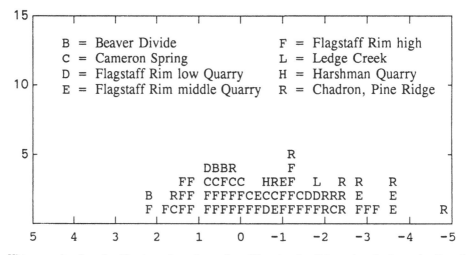

Figure 12. Histogram showing classification of specimens from Wyoming localities using the factor loadings from the discriminant analysis shown in Figure 9. The Wyoming specimens do not match either the Pipestone Springs or the Pine Ridge samples but form a single population of intermediate character.

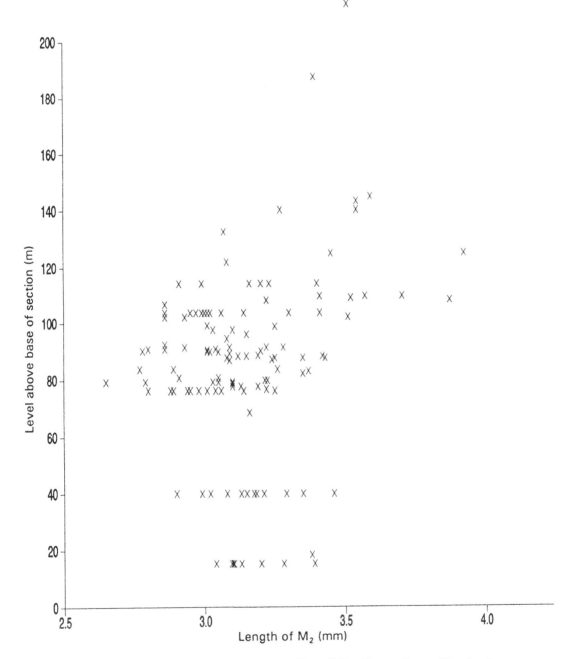

Flagstaff Rim, Wyoming

Figure 13. Bivariate plot of 120 zoned *Ischyromys* m2s from Flagstaff Rim, Natrona County, Wyoming (see map on page 263). The entire section is Chadronian in age.

Table 9. Classification by Flynn (1977) of ischyromyid skulls from Flagstaff Rim, Wyoming, using selected characters from Table 8. "I" = *Ischyromys*, "T" = *Titanotheriomys*, "-" = not observable. USNM = U.S. National Museum, Smithsonian Institution, Washington, D.C.; AMNH = Frick Collection, American Museum of Natural History, New York. Specimens (columns) coded as follows: A = USNM 215394; B = USNM 215404; C = USNM 215396; D = USNM 215403; E = AMNH 42993; F = AMNH 79313; G = AMNH 101453; H = AMNH 79312; I = AMNH 97315; J = AMNH 101447; K = AMNH 101454; L= USNM 242513; M = AMNH 79314; N = USNM 242514; O = AMNH 101449; P = AMNH 79318.

CHARACTER

	A	B	C	D	E	F	G	H	I	J	K	L	M	N	O	P
Dorsal profile of skull	T	I?	-	-	T	-	-	-	-	T	-	-	-	-	-	-
Anterior face of zygoma	T	I	I?	I	T	-	-	-	-	T	-	-	-	-	T	-
Origin of M. masseter lateralis	T	I	I	I	T	T	-	T	T	T	T?	T	T	-	T	T
Sagittal crest	T	I	I	I?	T	-	T	-	-	T	T?	-	T?	I	T?	-
Temporal crests	T	I	I	I?	T	-	T	T	T	T	T?	-	T?	I	T	-
Bucco-naso-labialis	T	I	-	I	T	T	-	T	-	T	T	T	T	-	T	T

that the small Flagstaff Rim ischyromyids represent a single species, contra Flynn (1977).

There is one additional piece of evidence to suggest that only one lineage of small ischyromyid exists at Flagstaff Rim. The first four skulls listed in Table 9 (USNM 215394, 215396, 215403, 215404), plus another partial skull (USNM 215402), came from a fossiliferous burrow filling 0.3 m in diameter and 1.2 m long (Robert J. Emry, personal communication). All five skulls represent juveniles at roughly the same stage of development (M^2 fully erupted, M^3 unerupted). The burrow filling also contained 5 left and 4 right juvenile dentaries and part of a juvenile articulated skeleton as well as two dentaries of older individuals. The excellent preservation and similar developmental stage make it highly likely that these juveniles are siblings that lived in the burrow rather than scavenged remains. However, using Wood's skull characters (Table 8), three of the five match *Ischyromys*, one matches *Titanotheriomys*, and one is not complete enough to determine (Table 9). If both skull types are present among siblings, then certainly both can exist in a single species.

THE CHADRONIAN-ORELLAN "TRANSITION"

The information presented here might seem to suggest that the small ischyromyids from Montana and Nebraska are merely end-members of a cline and that the Wyoming specimens represent an intermediate condition because of their central location within the cline. However, despite the fact that certain populations exhibit unusual variation in skull shape and musculature, the fact still remains that the abundant *Ischyromys* of the Great Plains shows no such variation but consistently retains the primitive condition. It therefore seems highly unlikely that Great Plains *Ischyromys* descended from the variable ischyromyids of Flagstaff Rim despite their close fall proximity. As explained previously and illustrated in Figure 4, Chadronian *Ischyromys* of Pine Ridge seems to into the same two

size classes as Orellan *Ischyromys*, so they are presumably continuous lineages (Heaton, 1993a). Unfortunately, Chadronian ischyromyids are rare in Nebraska, and no skulls are available for comparison. Wood (1976) and Kron (1978) reported several incomplete skulls from the Douglas Chadronian, only one of which shows affinity to *Titanotheriomys*.

Figure 14 is a dendrogram from cluster analysis based on mean values from 31 populations of ischyromyids (after Heaton, 1993a). It shows a clear distinction between Chadronian and Orellan populations except that Chadronian teeth from Pine Ridge and the nearby Douglas locality cluster with their Orellan counterparts.

There are many striking differences between ischyromyids of the Great Plains and those of the Rocky Mountains. The vast majority of ischyromyid specimens collected from the Great Plains belong to *Ischyromys typus*, a species that is widespread, long-lived, shows remarkable morphologic uniformity, and consistently retains the primitive protrogomorphous skull condition. The smaller Great Plains species, though more restricted geographically, is similar to *I. typus* in its morphologic uniformity and primitive skull structure. Both species were abundant in the Orellan but have a documented Chadronian ancestry. Rocky Mountain ischyromyids, on the other hand, are virtually all Chadronian or older, are generally represented by small samples, exhibit much greater variation within populations and between localities, and show innovations in the skull that are not present in the Great Plains species. Their diversity may result from the more diverse topography and the likely presence of habitat islands in the more mountainous regions.

The ischyromyid populations of the Great Plains and Rocky Mountains are separated by time, space, and regional physiography. It is therefore difficult to assess which of these factors is most important in characterizing their differences. If time alone were considered, it would appear that a major transition occurred at the

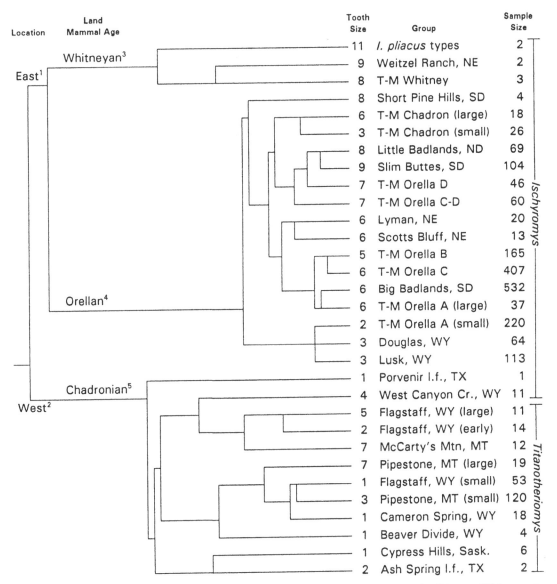

[1] All groups in this cluster occur at about 105° longitude and eastward, and the most distant localities are 650 km apart.

[2] All groups in this cluster occur at about 105° longitude and westward, and the most distant localities are 2300 km apart.

[3] The age of the *Ischyromys pliacus* type specimens is unknown but is believed to be late Orellan; the type locality is in northeastern Colorado, southeastern Wyoming, or southwestern Nebraska.

[4] All groups in this cluster are Orellan except the two "Toadstool-Munson Chadron" groups which are Chadronian and are ancestral to all Orellan and Whitneyan groups.

[5] All groups in this cluster are Chadronian except the Porvenir and West Canyon Creek groups which are late Duchesnean.

Figure 14. Dendrogram from cluster analysis on 31 populations using the Ward method (after Heaton, 1993a). Analysis is based on mean population values for 83 characters: 25 on each molar and 8 on the jaw. Overall size has been factored out so that similarities are based on shape alone. Size is given for comparison, with values representing tenths of a mm over 3.0 mm for mean length of M_2. "T-M" groups are combined samples from Toadstool Park and Munson Ranch. "T-M Chadron" groups also include specimens from other Pine Ridge and Douglas localities. "Orella A" refers to the lower Orella, and the other lettered units refer to higher intervals in the section. The "T-M Orella C-D" group comes from a very productive level at the Orella C-D boundary. Groups cluster primarily by location and age with Chadronian populations forming a more diverse group than Orellan populations. The separation of Whitneyan populations from Orellan populations is an artifact as explained in Heaton (1993a).

Chadronian/Orellan boundary from a diverse and progressive family of rodents to a pair of primitive and stable species. There may be an element of truth to this, but at least part of the apparent transition is an artifact of the location, age, and productivity of fossiliferous sediments. Ischyromyid fossils have been found in the Chadronian of Nebraska and South Dakota and in the Orellan of Montana, but nearly all are isolated teeth, and they are too few in number for adequate statistical evaluation. Because it is highly unlikely that the primitive and morphologically uniform *Ischyromys typus* descended from a more derived and variable species like *Titanotheriomys veterior*, geographic factors must be considered to explain the differences between Chadronian and Orellan populations. Although many changes occurred in land mammal communities at the Eocene/Oligocene boundary as documented in this volume, there may not be such a clear demarcation for the ischyromyid rodents.

A purely geographic explanation for the ischyromyid dichotomy would hold that *Ischyromys typus* (and its smaller sister species) persisted throughout the Chadronian and Orellan of the Great Plains while an adaptive radiation of more variable and shorter-lived ischyromyid species occurred in the Rocky Mountains. The greater environmental uniformity and higher population density in the Great Plains might help explain such long-term stasis. The similarity of late Chadronian and Orellan ischyromyids from Pine Ridge shown in Table 7 and Figures 4, 12, and 14 supports this model, but the paucity of fossiliferous deposits of early Chadronian age in the Great Plains makes a full evaluation impossible. Because a likely ancestor for Great Plains *Ischyromys* has not been found in the extensive Chadronian deposits of the Rocky Mountains, the simplest explanation is that such an ancestor lived in the Great Plains.

Stasis in the small ischyromyid species at Flagstaff Rim also supports a model of long-term regional stability. In Figures 12 and 13, specimens from two quarries low in the section exhibit no significant difference from those in the upper part of the section, although the quarry specimens group with those from some earlier localities in the cluster analysis (Fig. 14). Although I consider the small ischyromyid from Flagstaff Rim to be conspecific with *Titanotheriomys veterior* from Pipestone Springs, the Flagstaff Rim population of central Wyoming is morphologically much more similar to *Ischyromys* of the Great Plains than is its Montana counterpart, as is clearly shown by the discriminant analysis in Table 7 and Figure 12. It should also be noted that the Flagstaff Rim section spans a much longer time interval than the Pipestone Springs beds (Emry, 1992; Emry et al., 1987), yet specimens from Flagstaff Rim that both predate and postdate the Pipestone Springs Local Fauna were classified similarly by the discriminant analysis comparing Montana and Nebraska ischyromyids (Table 7, Fig. 12).

Although geographic factors seem to hold greater weight in explaining ischyromyid evolution and diversity, time considerations are also important. The relative abundance of ischyromyids compared to other rodents was highest in the Chadronian of the Rocky Mountains but in the Orellan of the Great Plains. Rocky Mountain ischyromyids barely survived the Chadronian/Orellan boundary and are extremely uncommon in Orellan deposits; their Great Plains counterparts barely survived the Orellan/Whitneyan boundary before going extinct.

ISCHYROMYID TAXONOMY

Pending the completion of a detailed study of the skulls and upper dentitions of ischyromyids, I am not prepared to make a final taxonomic analysis or to erect any new species. This will be done in the near future. I will, however, present my tentative conclusions based on my extensive study of lower dentitions and on studies of skulls conducted by Wood (1976) and others.

Population studies such as those I have conducted with *Ischyromys* provide satisfactory bases for determining the limits of fossil species so that inappropriate lumping or splitting of taxa can be avoided. Such bases are needed in the science of paleontology for resolving taxonomic disputes at the species level. However, defining the statistical means and limits of a species does not always ease the problem of assessing the validity of old type specimens or, for that matter, of assigning any individual specimen to its appropriate species. Even after doing elaborate statistical analyses of large populations, it remains frustratingly difficult to evaluate samples that are small or fragmentary. Taxonomy has been the most problematic and unrewarding aspect of this lengthy paleobiological study.

THE STATUS OF *TITANOTHERIOMYS*

The fact that *Titanotheriomys veterior* and related species have specializations in their skulls that are generally considered to be characters of subordinal rank among rodents would suggest that genus recognition is appropriate (Wood, 1976). The multivariate statistical studies presented in Heaton (1993a) and in this chapter suggest that the *Ischyromys/Titanotheriomys* distinction, at least among some populations, can also be made with dentaries. As can be seen in Figure 14, Chadronian populations with *Titanotheriomys*-type skulls cluster, along with two Duchesnean populations having *Ischyromys*-type skulls, into a coherent group that diversified and subsequently went extinct (or became exceedingly rare) in the Chadronian. Orellan *Ischyromys* of the Great Plains did not evolve from this group but retains the primitive skull condition of its Duchesnean ancestors.

Titanotheriomys appears to be a monophyletic group with skull specializations that are shared and derived, though rather variable and inconsistent, characters. If *Titanotheriomys* is granted generic status, then

Ischyromys becomes a paraphyletic group sharing the primitive protrogomorphous skull condition.

The status of *Titanotheriomys* is subjective, and utility should be a prime consideration. There is no single character that consistently separates any element of *Ischyromys* from that of *Titanotheriomys*. In both skulls and lower dentitions, some members of the *Titanotheriomys* group are indistinguishable from *Ischyromys*. Because of this, and because the *Ischyromys/Titanotheriomys* group as a whole is so easily distinguished from all other rodents, I prefer Matthew's (1910) original designation of *Titanotheriomys* as a subgenus of *Ischyromys*. Such a designation allows the recognition of the species of *Titanotheriomys* as a coherent adaptive radiation of rodents, but it also allows ambiguous specimens (or even species) to be labeled "*Ischyromys* sp." rather than "*Ischyromys* or *Titanotheriomys*."

The subgenus *Ischyromys* will apply to *I. typus* and other species that retain the primitive skull condition of the genus. The subgenus *Titanotheriomys* will apply to *I. (T.) veterior* and its relatives which have a more derived skull musculature.

ISCHYROMYS (ISCHYROMYS) TYPUS OF THE GREAT PLAINS

Range: Late Chadronian (or earlier) to early Whitneyan.

Ischyromys typus is the latest surviving ischyromyid species, one of the most wide ranging geographically, and by far the most common in museum collections. It is unusually uniform in the structure of its skull and teeth, although it increased slightly in mean size between the Chadronian and early Whitneyan (Heaton, 1993a). *Ischyromys typus* is most abundant at Orellan localities from Colorado to North Dakota.

As the type species of the genus, the type specimen of *Ischyromys typus* (ANSP 11015) from the Big Badlands of South Dakota has priority. Fortunately the type specimen is a fairly complete skull with the typical protrogomorphous condition. As shown earlier in this chapter, only one species of *Ischyromys* is indicated by the large sample of lower dentitions from South Dakota, so this species is *I. typus*.

The following taxa are synonyms of *Ischyromys typus* Leidy (1856, 1869): *Colotaxis cristatus* (Cope, 1873a), *Gymnoptychus chrysodon* (Cope, 1873b), *I. pliacus* (Troxell, 1922), *I. typus lloydi* (Troxell, 1922), and *I. troxelli* (Wood, 1937). The types of *I. parvidens* (Miller and Gidley, 1920) and *I. typus nanus* (Troxell, 1922) are small jaws that may belong to *I. typus* or may instead belong to the smaller ischyromyid of the Great Plains discussed below.

ISCHYROMYS (ISCHYROMYS) (SMALL) OF THE GREAT PLAINS

Range: Late Chadronian (or earlier) and Orellan.

As originally proposed by Miller and Gidley (1920) and demonstrated by Heaton (1993a), a second, smaller species of *Ischyromys* existed in the central Great Plains. This species has usually been called *I. parvidens*. As shown above, it is distinct from *I. veterior* of Pipestone Springs and does not belong to that species, contra Black (1968). Also, it did not give rise to *I. typus* during the Orellan as proposed by Stout (1937) and Howe (1956, 1966). Both species coexisted in the Orellan and for an undetermined portion of the Chadronian (Heaton, 1993a).

The decision as to the proper name for this small species has been one of the biggest dilemmas faced in this study. The type of *Ischyromys parvidens* is a small, unzoned dentary (USNM 9134) from the Big Badlands of South Dakota (Miller and Gidley, 1920). As can be seen in Figure 6, the Big Badlands appears to have only a single species of which this specimen is but a small member at the tail end of the size distribution. As such *I. parvidens* would be a junior synonym of *I. typus*. However, this specimen is smaller than the mean size of the small species at Pine Ridge just 100 km to the southwest, and it is conceivable that the small Pine Ridge species could have ranged north into the Big Badlands. All that can be stated with certainty is that the small species does not appear in the Big Badlands in statistically significant numbers, and therefore the probability that such a specimen belongs to the small Pine Ridge species is exceedingly low.

Second in priority to *Ischyromys parvidens* is *I. typus nanus* named by Troxell (1922) for a dentary from Pine Ridge north of Harrison, Nebraska. The type specimen (YPM 12519) is slightly larger than the type of *I. parvidens* but is still smaller than the mean of the small Pine Ridge species. Although there are clearly two coexisting species at Pine Ridge, only the very largest and smallest specimens can be assigned to one species or the other with any degree of confidence because the only features that distinguish the two are size and characters allometrically correlated with size (Heaton, 1993a). Most specimens from the ranches north of Harrison are *I. typus* from the middle and late Orellan, so the type of *I. typus nanus* may be nothing more than a small specimen of *I. typus* as proposed by Troxell (1922).

The resolution of this problem must await a careful study of the skulls and upper dentitions. If a new type must be designated, there is a nearly complete skeleton with skull and dentaries (UNSM 68129) from the early Orellan which almost certainly belongs to this species.

ISCHYROMYS (TITANOTHERIOMYS) VETERIOR OF THE ROCKY MOUNTAINS

Range: Early to late (but not latest) Chadronian.

This species was named by Matthew (1903) who mentioned "the anterior part of a skull and some forty jaws or parts of jaws, upper and lower" from Pipestone Springs but provided no illustrations or museum catalog numbers. Records at the American Museum of

Natural History list a skull (AMNH 9647) as Matthew's intended type specimen, but Wood (1937) designated a dentary (AMNH 9658) as the lectoholotype. The material available to Matthew adequately characterizes this small Pipestone Springs species, so *Ischyromys (Titanotheriomys) veterior* is the valid name. Wood (1937) named the species *T. wyomingensis* for a skull and associated dentaries (AMNH 14579) from Beaver Divide, central Wyoming, but this has since been considered a synonym of *I. veterior* (Black, 1968; Wood, 1980).

As shown in Figures 9 and 10 and Tables 5 and 6, *Ischyromys veterior* from Pipestone Springs is distinct from the small Orellan species of Pine Ridge in having a short M_3 relative to the other teeth and in having a shortening of the posterior region of all the lower cheek teeth. This appears to be a derived character that is most distinctive in the Montana specimens; teeth referred to *I. veterior* from Wyoming localities are more similar to those from Pine Ridge (Figure 12, Table 7). Although the Montana and Wyoming populations show measurable differences, the overlap is so great that the two should be regarded as the same species. The Montana population could be considered a more derived subspecies or variety.

The study of *Ischyromys veterior* from Wyoming provides some of the clearest insights into ischyromyid evolution. Figure 13 shows that this species underwent a minor decrease in size during the Chadronian. The cluster analysis in Figure 14 grouped the specimens from the upper part of the Flagstaff Rim section with the Pipestone Springs population and with other populations of similar age from Wyoming, but it grouped the quarry samples from low in the section with several early Chadronian populations, one of which (West Canyon Creek) has the *Ischyromys*-type skull. This suggests that *I. veterior* might have been the stem species that first developed the *Titanotheriomys*-type skull and subsequently gave rise to the other, more distinctive members of the subgenus (Heaton, 1988). The fact that Flagstaff Rim *I. veterior* dentaries, regardless of stratigraphic level, lack the specializations of the Pipestone Springs population (Figures 11 and 12, Table 6) suggests that *Titanotheriomys* may have originated in a region closer to Wyoming than to Montana.

The probable descendants of *Ischyromys veterior* are discussed below, together with some populations that may prove to be conspecific. As can be seen in Figure 13, *I. veterior* may not have survived to the end of the Chadronian or even survived as long as the larger Flagstaff Rim species, although the Flagstaff Rim section becomes less fossiliferous toward the top and makes the available evidence inconclusive.

ISCHYROMYS (TITANOTHERIOMYS) (LARGE) OF FLAGSTAFF RIM, WYOMING
Range: Middle to late Chadronian.

A second, larger species of *Ischyromys (Titanotheriomys)* first appears near the middle of the Flagstaff Rim section as noted by Flynn (1977) and Emry (1992) and shown in Figure 13. Its teeth are nearly identical in shape to those from the lower quarries at Flagstaff Rim, as indicated by the cluster analysis in Figure 14. The appearance of this species may represent a case of sympatric speciation in *I. veterior*, although such an assertion is impossible to prove. The only distinguishing feature of this species is its large size, but this alone, following cladogenesis, warrants its recognition as a new species.

There is a striking similarity in evolutionary pattern between the ischyromyids and the leptomerycids at Flagstaff Rim (Emry, 1973; Heaton and Emry, this volume, Chapter 27). In both cases an ancestral species seems to have split into two: one that became slightly smaller and underwent a change in shape and another that became much larger but retained the ancestral morphology.

ISCHYROMYS (TITANOTHERIOMYS) (LARGE) OF PIPESTONE SPRINGS, MONTANA
Range: Middle Chadronian.

As discussed above and illustrated in Figure 8, a large ischyromyid occurs in small numbers at Pipestone Springs along with *Ischyromys veterior*. This species was erroneously identified as *I. pliacus* by Wood (1937) and Black (1965) but was later recognized as a new species of *Titanotheriomys* based on skull characters (Wood, 1976, 1980). This species clusters with the middle Chadronian populations of *I. veterior* (Figure 14) and probably arose from that species later than did the large Flagstaff Rim species. It differs from the later species and from *I. veterior* in having dental characters that are convergent with *I. typus* of the Great Plains (Heaton, 1988). It is a distinct but as yet unnamed species.

ISCHYROMYS (TITANOTHERIOMYS) DOUGLASSI OF McCARTY'S MOUNTAIN, MONTANA
Range: Early Chadronian.

Ischyromys douglassi is sufficiently distinct to be considered a valid species as proposed by Black (1968). It has the snout musculature of *Titanotheriomys* but has a sagittal crest in all known specimens (Wood, 1976, 1980; see Table 8). It is unique in being the largest species of *Ischyromys* with low cusps and shallow valleys, and the anterior lingual valleys of the teeth have a distinctive shape (Heaton, 1988). The M_3 of *I. douglassi* has not undergone the posterior shortening seen in *I. veterior*; in fact it has the longest mean posterior M_3 length of all the populations studied. This species is known only from McCarty's Mountain, Montana; it has not been found in sediments of similar age in Wyoming.

ISCHYROMYS (TITANOTHERIOMYS) JUNCTUS OF CYPRESS HILLS, SASKATCHEWAN
Range: Early and/or middle Chadronian.

Russell (1972) named *Ischyromys junctus* for a small number of isolated teeth from the Cypress Hills in western Saskatchewan. The cluster analysis of Figure 14 groups it with *Titanotheriomys*, but the skull is unknown. When mean values for various populations are compared, *I. junctus* has the shortest length for M_3 of any ischyromyid (Heaton, 1988). This suggests that it may have taken the relative shortening of M_3 to an even greater extreme than did *I. veterior* of Pipestone Springs, but with the present small sample of isolated teeth it is impossible to prove this conjecture. The cusps of the teeth are very low and rounded and the valleys shallow, so they are distinct from *I. veterior* despite their similar size. Storer (1978) mentioned this and listed several other distinguishing characters. *Ischyromys junctus* appears to be a valid species, but without more complete material it is difficult to establish its relationship to other ischyromyids.

ISCHYROMYS (TITANOTHERIOMYS) OF ASH SPRING, TEXAS
Range: Middle Chadronian.

Dentaries of two small individuals of *Ischyromys* were found in the Big Bend region of Texas (Harris, 1967) and were considered by Wood (1974) to belong to *Ischyromys (Titanotheriomys) veterior*. The cluster analysis of Figure 14 groups this population with *I. junctus* from the Cypress Hills despite the 2200 km that separates the two localities. The Ash Spring species, however, lacks the low, rounded cusps of *I. junctus*. In fact the cusps would be unusually high for *I. veterior*, so the Ash Spring form may represent a distinct species (Heaton, 1988). Morphological differences of small magnitude in such a small sample do not warrant species recognition, but the distant geographic locality may make such a distinction useful.

ISCHYROMYS (ISCHYROMYS) BLACKI OF PORVENIR, TEXAS
Range: Latest Duchesnean and/or early Chadronian

The Porvenir Local Fauna of Texas is the southern-most *Ischyromys* locality as well as one of the earliest. Wood (1974) erected the new species *I. blacki* for a skull and dentary from the Porvenir Local Fauna. Like the other early species discussed below, *I. blacki* lacks the skull specializations of *Titanotheriomys* (Wood, 1976).

The single known dentary of *Ischyromys blacki* has an M_3 that is exceedingly long compared to M_2, a trend opposite that seen in *I. veterior*. The cluster analysis of Figure 14 groups this species with other early populations (mostly *Titanotheriomys*), though as a very distinct member of that group. Both its early age and distinctive suite of characters make it a valid species.

ISCHYROMYS (ISCHYROMYS) OF WEST CANYON CREEK, WYOMING
Range: Late Duchesnean

A channel fill in central Wyoming has produced a late Duchesnean fauna that is currently under study by Robert J. Emry. Several well-preserved skulls and jaws of *Ischyromys* have been found there which, like *I. blacki*, lack the skull specializations of *Titanotheriomys*. The West Canyon Creek specimens comprise a new species that is typical in most respects but which has a unique dental feature that separates it from all other species of *Ischyromys* and gives the molars a distinctive appearance. This unique feature is the exceptionally high incidence of medial and lingual accessory cusps (Figures 1 and 2, Tables 2 and 3), higher than in any other *Ischyromys* population on both M_2 and M_3. These accessory cusps are so strongly developed that they completely dam the lingual valleys in most cases. In addition, the posterior lingual valley extends unusually far anteriorly (Heaton, 1988).

It is difficult to determine whether this species represents a likely ancestor for later species of *Ischyromys* or whether it is a unique, terminal lineage. The species is of medium size and is quite average in most respects. Like *I. blacki*, it tends to cluster with the *Titanotheriomys* populations (Figure 14) in spite of having an *Ischyromys*-type skull. However, the West Canyon Creek form lacks the elongate M_3 of *I. blacki*. In these respects it makes a likely ancestor for all later species. The high incidence of accessory cusps might most easily be explained as a derived character, but an alternative explanation is available. Since these cusps show up sporadically in all later species of *Ischyromys* for which large samples are available, they may represent atavisms of a former, more consistent character trait. The M^3 on the type skull of *I. blacki* has the peculiar look of the West Canyon Creek molars, but the other teeth lack it. Of these two known late Duchesnean species, it is difficult to say which, if either, would make a good ancestor for Chadronian and Orellan *Ischyromys*.

ISCHYROMYS OF BADWATER CREEK, WYOMING
Range: Early Duchesnean

Black (1971), using screening techniques, recovered four upper and one lower cheek teeth from the Hendry Ranch Member of the Wagonbed Formation in central Wyoming which he called ?*Ischyromys* sp. These appear to be the oldest known *Ischyromys* specimens. Despite the close proximity of this site to West Canyon Creek, none of these teeth has accessory cusps. Even the major cusps appear weakly developed and the valleys shallow as in *I. junctus*. Without better specimens it is difficult to evaluate this material, but it may represent a new species.

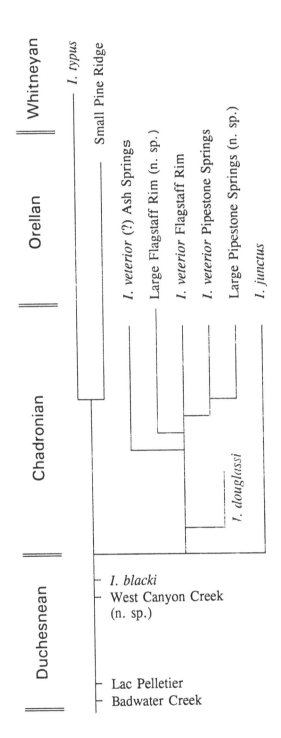

Figure 15. Proposed evolutionary tree of the species of *Ischyromys*. The species in the Chadronian radiation on the right belong to the subgenus *Titanotheriomys*.

ISCHYROMYS OF LAC PELLETIER, SASKATCHEWAN

Range: Early or middle Duchesnean.

Storer (1988) recovered five upper and four lower isolated cheek teeth of a small form of *Ischyromys* from the Lac Pelletier Lower Fauna which he believes to represent a new species. Some of these teeth have very low and indistinct cusps like those of *I. junctus*; others are similar to *I. veterior* of the lower Flagstaff Rim section.

As with the Badwater Creek specimens, the lack of large samples, associated teeth, and skull remains from the Lac Pelletier Lower Fauna makes evaluation of this material impossible using the methods employed in this study. At present it is not even possible to determine how many species might be represented.

CONCLUSIONS

Working out the early evolution of *Ischyromys* is difficult because most of the available populations are small and seem to represent local species. Early Duchesnean samples consist of only a few isolated teeth, whereas late Duchesnean samples are difficult to interpret because they possess unique characters. All known Duchesnean skulls are of the *Ischyromys* type and lack the specializations seen later in the *Titanotheriomys* radiation (Table 8). *Ischyromys douglassi* of the early Chadronian is intermediate in this respect.

The Chadronian of the Rocky Mountains hosted an adaptive radiation of species that possess the skull specializations of *Titanotheriomys*. *Ischyromys (T.) veterior* of the lower Flagstaff Rim section is the most generalized member of the group and may be ancestral to the others. Populations from the upper part of the Flagstaff Rim section, Pipestone Springs and other Montana localities, and at Ash Spring, Texas, all have more derived characters. Figure 15 shows possible relationships among *Ischyromys* species and among geographic populations of *I. veterior*. The *Titanotheriomys* radiation dominates the Chadronian, but this subgenus is not known from the Orellan, with the possible exception of a few isolated teeth from Montana.

Orellan *Ischyromys* of the central Great Plains is not derived from *Titanotheriomys* but had a separate, as yet unknown ancestry from some Duchesnean species that lacked the *Titanotheriomys* skull specializations. *Ischyromys typus* is by far the most common ischyromyid fossil and is known from the late Chadronian to early Whitneyan at localities from Colorado to North Dakota. Its smaller sister species is known only from the late Chadronian and Orellan of Pine Ridge, Nebraska, to Douglas, Wyoming, where it dominated over *I. typus* during the early Orellan at all localities in its range (Heaton, 1993a).

The Rocky Mountains hosted many localized species and populations of *Ischyromys* throughout the Duchesnean and Chadronian. The Great Plains, in contrast, hosted only two very conservative, similar, static species, one of which had a wide geographic range. Although this difference may be due in part to the later (Orellan) age of most Great Plains specimens, it seems largely related to geographic differences between the Rocky Mountains and Great Plains physiographic provinces. For instance, the isolated basins of the Rocky Mountains might have promoted the formation of isolated populations capable of speciation while the environmental uniformity of the Great Plains might have inhibited this. Some of the apparent differences between Chadronian and Orellan ischyromyids may, therefore, be artifacts of the location of fossiliferous sediments. However, the Chadronian-Orellan transition is marked by the decline and extinction (or extreme rarity) of *Titanotheriomys* in the Rocky Mountains and by the increasing dominance of *Ischyromys* in the Great Plains.

ACKNOWLEDGMENTS

This project began as a doctoral dissertation under the direction of S. J. Gould at Harvard University. It was proposed by T. Mylan Stout, who also introduced me to the Nebraska Oligocene. R. J. Emry and D. R. Prothero provided extensive help and encouragement throughout the duration of the project. A. R. Tabrum and J. E. Storer read the manuscript and made many helpful comments. To these individuals, and many others who helped in numerous ways, I offer my warmest thanks.

This study included nearly every *Ischyromys* jaw available in a public museum. I therefore owe the greatest debt to the hundreds of collectors and curators over the last 150 years who made a study such as this possible by preserving these fossils, along with pertinent data, and making them available for study.

I especially want to thank my wife, Julia S. Heaton, and our daughters, Christy, Amy, and Holly, for their support throughout the duration of this project.

This research was supported by fellowships and grants from Harvard University, the University of South Dakota, the Geological Society of America, the Smithsonian Institution, and the American Museum of Natural History.

LITERATURE CITED

Black, C. C. 1965. Fossil mammals from Montana. Part 2. Rodents from the early Oligocene Pipestone Springs local fauna. Annals of Carnegie Museum 38(1):1-48.

Black, C. C. 1968. The Oligocene rodent *Ischyromys* and discussion of the Family Ischyromyidae. Annals of Carnegie Museum 39(18):273-305.

Black, C. C. 1971. Paleontology and geology of the Badwater Creek area, central Wyoming. Part 7. Rodents of the family Ischyromyidae. Annals of Carnegie Museum 43(6):179-217.

Clark, J. 1967. Paleoecology of the Scenic Member of the Brule Formation; in J. Clark, J. R. Beerbower, and K. K. Kietzke (eds.), Oligocene Sedimentation, Stratigraphy,

Paleoecology and Paleoclimatology of the Big Badlands of South Dakota. Fieldiana: Geology Memoirs, 5:75-110.

Clark, J., and K. K. Kietzke. 1967. Paleoecology of the Lower Nodular Zone, Brule Formation, in the Big Badlands of South Dakota; in J. Clark, J. R. Beer-bower, and K. K. Kietzke (eds.), Oligocene Sedimentation, Stratigraphy, Paleoecology and Paleoclimatology of the Big Badlands of South Dakota. Fieldiana: Geology Memoirs, 5:111-137.

Cope, E. D. 1873a. Second notice of extinct vertebrates from the Tertiary of the plains. Paleontological Bulletin, 15:1-8.

Cope, E. D. 1873b. Third notice of extinct vertebrates from the Tertiary of the plains. Paleontological Bulletin, 16:1-8.

Emry, R. J. 1973. Stratigraphy and preliminary biostratigraphy of the Flagstaff Rim area, Natrona County, Wyoming. Smithsonian Contributions to Paleobiology, 18:1-43.

Emry, R. J. 1992. Mammalian range zones in the Chadronian White River Formation at Flagstaff Rim, Wyoming: pp. 106-115 in D. R. Prothero and W. A. Berggren (eds.), Eocene-Oligocene Climatic and Biotic Evolution. Princeton University Press, Princeton, N. J.

Emry, R. J., P. R. Bjork, and L. S. Russell. 1987. The Chadronian, Orellan, and Whitneyan Land Mammal Ages: pp. 119-152 in M. O. Woodburne (ed.), Cenozoic Mammals of North America: Geochronology and Biostratigraphy. University of California Press, Berkeley.

Flynn, J. J. 1977. Morphological variation in the Oligocene Ischyromyidae (Rodentia) from Flagstaff Rim, Wyoming. Yale University, Department of Geology and Geophysics, Senior Paper Series, 46 pp.

Harris, J. M. 1967. Oligocene vertebrates from western Jeff Davis County, Trans-Pecos Texas. M.A. thesis, University of Texas, Austin, 165 pp.

Heaton, T. H. 1988. Patterns of evolution in *Ischyromys* and *Titanotheriomys* (Rodentia: Ischyromyidae) from Oligocene deposits of western North America. Ph.D. dissertation, Harvard University, Cambridge, 165 pp.

Heaton, T. H. 1993a. The Oligocene rodent *Ischyromys* of the Great Plains: replacement mistaken for anagenesis. Journal of Paleontology 67(2):297-308.

Heaton, T. H. 1993b. The fossil rodent *Ischyromys* in Badlands National Park, South Dakota (Abstract). Proceedings of the South Dakota Academy of Sciences 72:332-333.

Heaton, T. H. 1993c. Dentaries of *Ischyromys* and *Titanotheriomys* (Oligocene, Rodentia) can be distinguished by discriminant analysis, cluster analysis, and multidimensional scaling (Abstract). Journal of Vertebrate Paleontology 13(3):41A.

Howe, J. A. 1956. The Oligocene rodent *Ischyromys* in relationship to the paleosols of the Brule Formation. M.S. thesis, University of Nebraska, Lincoln, 89 pp.

Howe, J. A. 1966. The Oligocene rodent *Ischyromys* in Nebraska. Journal of Paleontology 40(5):1200-1210.

Kron, D. G. 1978. Oligocene vertebrate paleontology of the Dilts Ranch area, Converse County, Wyoming. M.S. thesis, University of Wyoming, Laramie, 185 pp.

Leidy, J. 1856. Notices of remains of extinct Mammalia discovered by Dr. F. V. Hayden in Nebraska Territory. Proceedings of the Academy of Natural Sciences of Philadelphia 8:88-90.

Leidy, J. 1869. The extinct mammalian fauna of Dakota and Nebraska, including an account of some allied forms from other localities, together with a synopsis of the mammalian remains of North America. Journal of the Academy of Natural Sciences of Philadelphia Series 2, 7:23-472.

Matthew, W. D. 1903. The fauna of the *Titanotherium* beds at Pipestone Springs, Montana. Bulletin of the American Museum of Natural History 19:197-226.

Matthew, W. D. 1910. On the osteology and relationships of *Paramys*, and the affinities of the Ischyromyidae. Bulletin of the American Museum of Natural History 28:43-71.

Miller, G. S., Jr., and J. W. Gidley. 1920. A new fossil rodent from the Oligocene of South Dakota. Journal of Mammalogy 1(2):73-74.

O'Neill, K. E. 1979. Taxonomic study of middle Oligocene specimens of the genus *Ischyromys*: multivariate analysis of the cheek tooth dentition. M.A. thesis, University of Wisconsin, 358 pp.

Prothero, D. R. 1985. Correlation of the White River Group by magnetostratigraphy. Dakoterra, Museum of Geology, South Dakota School of Mines 2(2):265-276.

Russell, L. S. 1972. Tertiary mammals of Saskatchewan. Part II: The Oligocene fauna, non-ungulate orders. Life Science Contributions of the Royal Ontario Museum 84:1-63.

Storer, J. E. 1978. Rodents of the Calf Creek Local Fauna (Cypress Hills Formation, Oligocene, Chadronian), Saskatchewan. Contributions of the Saskatchewan Museum of Natural History 1:1-54.

Storer, J. E. 1988. The rodents of the Lac Pelletier Lower Fauna, late Eocene (Duchesnean) of Saskatchewan. Journal of Vertebrate Paleontology 8(1):84-101.

Stout, T. M. 1937. A stratigraphic study of the Oligocene rodents in the Nebraska State Museum. M.S. thesis, University of Nebraska, Lincoln, 83 pp.

Troxell, E. L. 1922. Oligocene rodents of the genus *Ischyromys*. American Journal of Science, Series 5, 3(8):123-130.

Wood, A. E. 1937. The mammalian fauna of the White River Oligocene. Part II. Rodentia. Transactions of the American Philosophical Society, New Series, 28(2):155-269.

Wood, A. E. 1969. Rodents and lagomorphs from the 'Chadronia Pocket,' Early Oligocene of Nebraska. American Museum Novitates 2366:1-8.

Wood, A. E. 1974. Early Tertiary vertebrate faunas, Vieja Group, Trans-Pecos Texas: Rodentia. Bulletin of the Texas Memorial Museum 21:1-112.

Wood, A. E. 1976. The Oligocene rodents *Ischyromys* and *Titanotheriomys* and the content of the Family Ischyromyidae; pp. 244-277 *in* C. S. Churcher (ed.), Athlon: Essays on Paleontology in Honor of Loris Shano Russell. Royal Ontario Museum Life Science Miscellaneous Publication, University of Toronto Press.

Wood, A. E. 1980. The Oligocene rodents of North America. Transactions of the American Philosophical Society 70(5):1-68.

19. Cylindrodontidae

ROBERT J. EMRY AND WILLIAM W. KORTH

ABSTRACT

The rodent family Cylindrodontidae reached its greatest diversity in North America in Chadronian time, and cylindrodonts are among the most common rodents in White River faunas of Chadronian age. This chapter provides a general systematic summary of all Chadronian cylindrodont rodents, but focuses on those from the White River Group. We describe two new species, *Cylindrodon natronensis* and *C. solarborus*, from the early to middle Chadronian of the Flagstaff Rim area, Wyoming. Previously undescribed material affords much new information on the cranial and mandibular morphology of *Cylindrodon* and prompts a reinterpretation of its zygomasseteric structure.

INTRODUCTION

The earliest record of cylindrodontid rodents is *Dawsonomys woodi* from the Wasatchian of Wyoming (Korth, 1984). The family reached its greatest diversity in the Chadronian, where four genera and 13 species have been recognized, but before the end of Chadronian time the family became extinct in North America. Two subfamilies, the Cylindrodontinae and Jaywilsonomyinae, are recognized on the basis of dental characters (Wood, 1974), and can be traced back to the Bridgerian and Uintan, respectively; the subfamilial allocation of the Wasatchian *Dawsonomys* is uncertain, but is probably to Cylindrodontinae.

Outside North America, cylindrodontids are rare. The genus *Ardynomys* (originally referred to the Ischyromyidae) was first described from the latest Eocene or earliest Oligocene Ardyn Obo Formation of Mongolia by Matthew and Granger (1925). From the same deposits, Shrevyreva (1972) later named *Morosomys silentiumis*, which she classified as a cylindrodont; we consider *Morosomys* a synonym of *Tsaganomys*, and not a cylindrodont. Subsequent to its discovery in Asia, *Ardynomys* was recognized by Burke (1936) in the Chadronian of North America. The only other cylindrodont presently known from Asia is *Anomoemys* Wang (1986), from the late Eocene or early Oligocene of Mongolia and China.

Wood (1974) included the Asian Oligocene protrogo-morphous-hystricognathous Tsaganomyinae (Matthew and Granger, 1923) as a subfamily of the Cylindrodontidae. Similarities between the tsaganomyines and *Ardynomys* had been observed as early as 1935 by Burke, although Matthew and Granger (1923) originally recognized the Tsaganomyinae as a subfamily of the African Bathyergidae because of the hystricognathous mandible, protrogomorphous skull, procumbent incisors, and posteroventrally sloping occipital region of the skull. All of these features are shared with bathyergids, and most are not characteristic of cylindrodonts. Moreover, the microstructure of the incisor enamel in cylindrodonts is uniserial (Wahlert, 1968) and that of tsaganomyines is multiserial, typical of bathyergids and hystricomorphous-hystricognathous rodents. The tsaganomyines appear not to be cylindrodonts, and may be part of an early radiation of the Old World hystricomorphous rodents.

Dawson (1968) described *Hulgana*, an ischyromyid with a massive skull and cheek teeth, from the Oligocene of Mongolia. Flynn et al. (1986) included this genus in the Cylindrodontidae. However, details of the dentition of *Hulgana* are better interpreted as those of a simplified ischyromyid than of a cylindrodontid. The similarity of *Hulgana* to the cylindrodonts is reduced to the relative massiveness of the skull and dentition.

The systematic position of the Cylindrodontidae is still not definitely resolved. The family is usually placed within the Ischyromyoidea as a separate family (Miller and Gidley, 1918; Wood, 1937), but has also been considered a subfamily of the Ischyromyidae (Wilson, 1949; Wahlert, 1974). At the subordinal level, cylindrodonts have been included in the Sciuromorpha (Simpson, 1945; Wilson, 1949; Wood, 1955) and in the Protrogomorpha (Wood, 1937, 1958, 1965). In all cases, the cylindrodonts have been considered a primitive group of rodents, not far derived from the protrogomorphous/sciurognathous rodent stock.

More recently, Wood (1980, 1981, 1984) included the Cylindrodontidae in his new infraorder Franimorpha, as

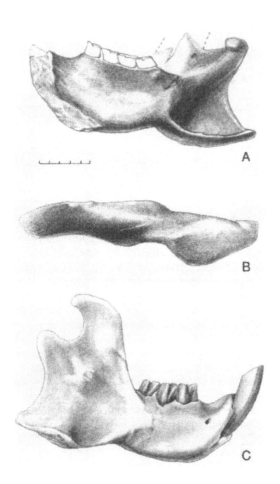

Figure 1. Right mandible of *Cylindrodon solarborus*. A-B, USNM 475973, in internal view (A) to show the deep pocketing of the petrygoid fossa, and in ventral view (B) to show the profile of the prominent flange for insertion of the superficial masseter muscle. C, F:AM 79109, showing extent of masseteric fossa with its laterally flared ventral border, and position of base of incisor, indicated by swelling anterior to condyle. Approximately X3, scale in millimeters.

part of the suborder Hystricognathi. Wood regarded the Franimorpha as the stock from which the later hystricognathous-hystricomorphous New World Caviomorpha were derived. Wood's placement of Cylindrodontidae here was supported by his interpretation of the mandible of *Cylindrodon* as "subhystricognathous" (Wood, 1984), an interpretation based on mandibles with imperfectly preserved angular processes. In the better preserved material examined for this study, none of the mandibles of *Cylindrodon* appears to have the angular process originating lateral to the plane of the incisor, or lateral to the body of the horizontal ramus (the definition of hystricognathy). In *Cylindrodon* the

ventral surface of the angular process has a very broad ventral border (Fig. 1), due in part to the deep pocketing of the pterygoid fossa internally, and accentuated by the development of a laterally projecting flange along the lower edge of the masseteric fossa for attachment of the superficial masseter muscle. In the specimens available to Wood, the middle edge of this broad angular process had been broken, making it appear that the angular process originated lateral to the body of the horizontal ramus. We interpret the mandible of *Cylindrodon* as sciurognathous, differing little from those of Eocene ischyromyids and sciuravids.

Some authors (Korth, 1984; Luckett and Hartenberger, 1985; Wilson, 1986) have argued that Franimorpha is not a natural group, and that "subhystricognathy" is but a minor variation of the primitive condition in rodents, derived independently in several groups of rodents. We find no cranial and dental characters of the Cylindrodontidae to support the inclusion of this family in the Hystricomorpha (*sensu lato*). We regard it as protrogomorphous, with its position within the Rodentia not confidently established.

ABBREVIATIONS

Acronyms as part of specimen numbers refer to collections in the following museums: F:AM, Frick Collection, The American Museum of Natural History; USNM, National Museum of Natural History. When used with measurements, AP means anteroposterior and TR means transverse. In dental notation, upper case letters indicate upper (maxillary and premaxillary) teeth and lower case letters indicate lower (dentary) teeth; for example, M1 is the upper first molar, p4 is the lower fourth premolar.

SYSTEMATIC PALEONTOLOGY
Order RODENTIA Bowdich, 1821
Family CYLINDRODONTIDAE
Miller and Gidley, 1918
Subfamily JAYWILSONOMYINAE
Wood, 1974
Jaywilsonomys Ferrusquia and Wood, 1969

Discussion—*Jaywilsonomys* is the only known Chadronian genus of this subfamily, which ranges as far back as Uintan (Wood, 1974). *Jaywilsonomys* is definitely known only from the early Chadronian of Trans-Pecos Texas and adjacent Mexico (Ferrusquia and Wood, 1969; Wood, 1974, 1980). Ostrander (1985, p. 228) included "*Jaywilsonomys* n. sp." in a faunal list of the middle Chadronian Raben Ranch Local Fauna of northwestern Nebraska, but he neither described nor cited any specimens; if this identification were substantiated, it would be the only record of the genus outside of Texas and northernmost Chihuahua, Mexico.

Wood (1974) diagnosed this genus (and the subfamily) as having upper cheek teeth that were shorter (anteroposteriorly) relative to their width than in cylindrodontines and having only three complete transverse lophs on the cheek teeth rather than the four of cylindrodontines.

The skull of *Jaywilsonomys* is not known, and the angular process of the jaw is not preserved in any known specimen.

Jaywilsonomys ojinagaensis
Ferrusquia and Wood, 1969

Discussion—*Jaywilsonomys ojinagaensis*, the type species, known only from the early Chadronian of Chihuahua, Mexico, is the largest cylindrodontid known in North America.

Jaywilsonomys pintoensis
Ferrusquia and Wood, 1969

Discussion—Specimens of *J. pintoensis* have been recovered from the early Chadronian Airstrip Local Fauna, from the Capote Mountain Tuff Formation, Texas, as well as from its type locality in Chihuahua (Wood, 1980). This species is smaller than *J. ojinagaensis*.

Subfamily CYLINDRODONTINAE
Miller and Gidley, 1918
Pseudocylindrodon Burke, 1935

Discussion—Species of *Pseudocylindrodon* are known from the Uintan and Duchesnean of Utah, Wyoming, and Saskatchewan, and from the Chadronian of Montana, Texas, and Saskatchewan (Black, 1965, 1970a, b, 1974; Burke, 1935, 1938; Galbreath, 1969; Russell, 1954; Storer, 1978, 1984, 1988; Wood 1974, 1980).

The occlusal morphology of the cheek teeth of *Pseudocylindrodon* is generally similar to that of *Cylindrodon*, with four distinct transverse lophs, but in contrast to *Cylindrodon*, the cheek teeth of *Pseudocylindrodon* are much lower crowned, ranging from brachydont to mesodont. The occlusal pattern of the cheek teeth in *Pseudocylindrodon* is maintained until very late stages of wear. The skull of *Pseudocylindrodon* is very similar to that of *Cylindrodon* (see below description of skull of *C. solarborus* for comparison) and is neither so large nor so robust as that of *Ardynomys*. The premaxillary-maxillary suture across the palate intersects the incisive foramina at their center in *Pseudocylindrodon*. In other cylindrodonts this suture intersects the incisive foramina at their posterior margins.

The angular process is not preserved in any known

mandible of *Pseudocylindrodon*. However, some are sufficiently complete to suggest a deep pterygoid pocket on the medial side of the angle (Burke, 1936), much as in *Cylindrodon*, although it cannot be determined whether this feature is as well developed as in the latter. All mandibles of *Pseudocylindrodon* appear sciurognathous in ventral view. The position of the mental foramen in *Pseudocylindrodon* is farther ventral and posterior (below p4) than in other Chadronian cylindrodontids where this foramen is anterior to the tooth row.

The only change in this genus through its history is the slight increase in crown height and lophodonty; the Uintan and Duchesnean species of the genus are generally lower crowned than the Chadronian species.

Pseudocylindrodon neglectus Burke, 1935

Cylindrodon fontis Douglass, Matthew, 1903 (in part)
Pseudocylindrodon sylvaticus Russell, 1954

Discussion—*Pseudocylindrodon neglectus*, the type species, was originally described from the middle Chadronian Pipestone Springs fauna (Burke, 1935; Black, 1965) but has also been reported from the early Chadronian of Texas (Wood, 1974), from the early to middle Chadronian of Saskatchewan, and from beds of uncertain age in southeastern British Columbia (Storer, 1978).

This species differs from other species in having slightly higher crowned cheek teeth and lacking a hypolophid on p4. In size it is near *P. medius*, and smaller than *P. texanus*.

Pseudocylindrodon medius Burke, 1938

Discussion—This species is known only from the early Chadronian of McCarty's Mountain, Montana. It has a more gracile skull than that of *P. texanus*, and has lower crowned cheek teeth than *P. neglectus*.

Pseudocylindrodon texanus Wood, 1974

Discussion—*Pseudocylindrodon texanus* was described from the early Chadronian of Texas (Wood, 1974). It is distinct from the other species of the genus mainly in its much larger size and larger proportional size of M3.

Wood (1974, 1980) viewed *P. texanus* as morphologically intermediate between other species of *Pseudocylindrodon* and *Ardynomys occidentalis*, and therefore considered that it may have also been phylogenetically intermediate.

Pseudocylindrodon sp. Galbreath, 1969

Discussion—Galbreath (1969) described two

specimens of *Pseudocylindrodon* from the late Chadronian of northeastern Colorado that he believed should constitute a distinct species. These specimens differed from other species of the genus in the more anterior position of the mental foramen and size of the capsule for the base of the incisor on the ascending ramus of the mandible, and the relatively narrow lower incisor. Galbreath (1969, p. 94) refrained from naming a new species because neither of the two available specimens was "fit to be a type." Judging from Galbreath's figures, the crown height of these specimens is similar to that of *P. neglectus*, higher crowned than the other Chadronian species of the genus.

Ardynomys Matthew and Granger, 1925

Discussion—In North America, *Ardynomys* is known from the early Chadronian of Montana and Texas, and from the middle Chadronian of Saskatchewan. The Asian species of this genus are from the latest Eocene or early Oligocene of Mongolia (Matthew and Granger, 1925; Wood, 1970) and from deposits of about the same age in Kazakhstan (Vinogradov and Gambarian, 1952; Wood, 1970).

The skull and mandible of *Ardynomys* are shorter and broader than in other cylindrodontids. One result of the shortening of the rostrum is that the capsule of bone that houses the base of the upper incisor is exposed in the anterior wall of the orbit; this condition is not seen in other cylindrodonts except *Pseudocylindrodon texanus*. *Ardynomys* has a more pronounced "chin" process on the mandible than is seen in other cylindrodontids. The angular process of the mandible is not preserved in any specimen of *Ardynomys* from North America. One specimen of the Mongolian *A. olseni* (Wood, 1970, fig. 2A) has enough of the angle preserved to suggest a deep pocket for attachment of the pterygoid muscles as in *Cylindrodon* and *Pseudocylindrodon*.

The dentition of *Ardynomys* is relatively more robust than in other cylindrodontids; the incisors are broader anteriorly, and the cheek teeth are broader buccolingually. The cheek teeth are brachydont to mesodont, similar in crown height to those of *Pseudo-cylindrodon*. The Asian species of *Ardynomys* are markedly larger than those from North America (Wood, 1970).

Ardynomys occidentalis Burke, 1936

Discussion—*A. occidentalis* is limited to the early Chadronian of Montana and Texas (Burke, 1936, Wood, 1970, 1974). An isolated upper molar of *Ardynomys* from the middle Chadronian of Nebraska may prove to be referable to *A. occidentalis* (Korth, 1992); if correctly referred, this is the latest occurrence of the genus in North America.

Ardynomys saskatchewaensis (Lambe, 1908)

Sciurus? saskatchewaensis Lambe, 1908
Prosciurus saskatchewaensis (Lambe), Russell, 1934
Ardynomys saskatchewaensis (Lambe), Storer, 1978
?Prosciurus saskatchewaensis (Lambe), Wood, 1980

Discussion—The only specimen definitely pertaining to this species is the holotype, a single upper molar from the early or middle Chadronian of the Cypress Hills Formation, Saskatchewan. Russell (1972) referred several isolated teeth to this species, but these were later shown to be referable to other rodent taxa (Storer, 1978; Wood, 1980). Storer (1978) recognized the cylindrodontid affinities of the holotype and referred a worn upper molar and an edentulous maxilla to this species. Wood (1980) argued that the holotype belonged to a prosciurine and the maxilla referred by Storer (1978) differed in a number of characters from the maxilla of *Ardynomys occidentalis*. The holotype of "*S.*" *saskatchewaensis* is clearly referable to *Ardynomys* (Russell, 1972, fig. 5D); its overall proportions, unilateral hypsodonty, continuous cingulum around the entire tooth, and morphology of the protoloph and metaloph are characters of *Ardynomys* and not of the aplodontid *Prosciurus*.

The holotype of *A. saskatchewaensis* differs from upper molars of *A. occidentalis* in a number of features mentioned by Storer (1978, p. 16), and also in having a short loph connecting the metaconule to the posteroloph, a feature not seen in any of the referred specimens of *A. occidentalis*.

Cylindrodon Douglass, 1902

Discussion—*Cylindrodon* is limited to the Chadronian of North America and is among the more common taxa in many faunas of this age. It is distinguished from all other genera of the family by the hypsodonty of its cheek teeth. The greatest crown height attained in other genera could best be termed mesodonty, with some showing a slight shift toward unilateral hypsodonty on the upper cheek teeth. The basic occlusal pattern of the cheek teeth, both upper and lower, in *Cylindrodon* consists of four transverse lophs that enclose three basins of varying depth. However, most of this occlusal pattern disappears with moderate wear, leaving a single central enamel fossette (-id). This feature also eventually disappears with wear, leaving the tooth as a cylinder of dentine surrounded by a band of enamel during late stages of wear. Because the cheek teeth of *Cylindrodon* are tapered toward the base, the shape of the occlusal surface changes during wear, from longer (anteroposteriorly) than wide (transversely) when unworn to wider than long in later wear stages.

Table 1. Mean values of measurements, in mm, of five species of *Cylindrodon* (values for *C. fontis* from Black, 1965, pp. 18-19).

Measurement	C. fontis	C. natron- ensis	C. solar- borus	C. collinus	C. nebrask- ensis
Alveolar length, P4-M3	7.10	6.85	6.95	—	—
Alveolar length, p4-m3	7.70	7.79	8.05	—	8.33
I1-P4 diastema	—	6.32	7.05	—	—
i1-p4 diastema	3.40	2.60	3.20	—	—
Width palate at P4	—	3.30	3.16	—	—
Width palate at M3	—	2.37	2.73	—	—
Depth ramus at m1	6.10	5.59	6.47	—	6.75
I1, AP	1.75	1.85	1.92	—	—
I1, TR	1.80	1.95	2.04	—	—
dP4, AP	1.77	—	1.91	—	—
dP4, TR	1.70	—	1.67	—	—
P4, AP	1.73	1.75	1.76	2.23	—
P4, TR	2.03	1.73	1.86	2.41	—
M1, AP	1.86	1.57	1.71	1.95*	2.18*
M1, TR	1.99	1.93	1.97	2.04*	2.25*
M2, AP	1.78	1.70	1.73	—	—
M2, TR	1.78	1.70	1.79	—	—
M3, AP	1.56	1.46	1.51	1.73	1.76
M3, TR	1.46	1.34	1.43	1.60	1.66
i1, AP	2.03	1.85	2.05	—	2.54
i1, TR	1.87	1.70	1.87	—	2.06
dp4, AP	1.87	1.91	2.05	2.09	—
dp4, TR	1.53	1.38	1.34	1.63	—
p4, AP	1.77	1.80	1.88	2.09	2.25
p4, TR	1.71	1.58	1.69	1.97	1.99
m1, AP	1.87	1.81	1.93	2.08*	2.39*
m1, TR	1.97	1.74	1.96	2.04*	2.23*
m2, AP	1.82	1.80	1.92	—	—
m2, TR	1.95	1.71	1.93	—	—
m3, AP	1.64	1.59	1.64	1.90	1.72
m3, TR	1.45	1.32	1.41	1.67	1.67

*M1/m1 or M2/m2

In no species of *Cylindrodon* is dP3 replaced by a permanent P3, so that the adult dentition has only the single permanent upper premolar, P4; in rare adult individuals dP3 is retained. All other cylindrodontids have a P3 in the adult dentition.

Except for changes in minor dental details, and a slight increase in size, there is little evidence of evolutionary trends within *Cylindrodon*. The earliest known species from the early Chadronian (new species from Flagstaff Rim described below) differs only in these small details from the latest known species.

Cylindrodon fontis Douglass, 1902

Discussion — *Cylindrodon fontis*, the type species, was originally described from the middle Chadronian Pipestone Springs fauna of Montana (Douglass, 1902). A more thorough description and diagnosis of this species were provided by Black (1965). All specimens of *C. fontis* reported from other faunas (Galbreath,

1969; Wood, 1974) have subsequently been referred to other species (see discussions below); as presently known, *C. fontis* is restricted to the Pipestone Springs fauna.

One feature of *C. fontis* mentioned by Black (1965) appears not to characterize all specimens. This is the morphology of the anterior cusps of p4. Black (1965) described the metaconid and protoconid as fusing very near their apices, appearing as a single cusp after only minimal wear. However, in some topotypic specimens available, evidence that this column is composed of two cusps remains until much later stages of wear.

Cylindrodon fontis is distinguished from all other species of the genus by the lack of a short, antero-posteriorly oriented loph on the upper molars that divides the basin between the metaloph and the posteroloph. This short loph is at least variably present on P4-M2 of all other known species. In size, *C. fontis* is intermediate, being larger than the new, small species of *Cylindrodon* from the Flagstaff Rim area of

Wyoming (described below), and statistically smaller than *C. collinus* and *C. nebraskensis* (see Table 1).

Cylindrodon nebraskensis Hough and Alf, 1956

Cylindrodon fontis Douglass, Galbreath, 1969 (in part)
Cylindrodon fontis Douglass, Wood, 1980 (in part)
Cylindrodon galbreathi Ostrander, 1983
Discussion—*Cylindrodon nebraskensis* is known from the middle to late Chadronian of Nebraska and Colorado (Korth, 1992). It is distinguished by its larger size (mean measurements greater than any other species, see Table 1) and the unique "spur" that runs posteriorly from the metalophid of the lower cheek teeth and from the protoloph of the upper cheek teeth.

Cylindrodon collinus Russell, 1972

Prosciurus altidens Russell, 1972
?*Plesispermophilus altidens* (Russell), Wood, 1980
Discussion—*Cylindrodon collinus* is known only from the early to middle Chadronian of the Cypress Hills Formation of Saskatchewan, and is represented only by isolated cheek teeth. This species is statistically larger than all other species of the genus except *C. nebraskensis* (see Table 1). However, the size range of specimens of *C. collinus* overlaps at least part of all of the species of the genus. Its size and the variable nature of the loph on the upper molars that connects the metaconule to the posteroloph distinguish it from all other species.

Storer (1978) synonymized *Prosciurus altidens* Russell (1972) with *C. collinus*. Wood (1980), however, resurrected Russell's species under the otherwise European aplodontid genus *Plesispermophilus*. The descriptions and figures of *P. altidens* (Russell, 1972, fig. 7D, E) clearly show that the two cheek teeth on which the species is based belong to *Cylindrodon* (teeth hypsodont) and are not those of a prosciurine. Wood's (1980) referral of these teeth to *Plesispermophilus* cannot be substantiated, based on dental morphology.

Cylindrodon natronensis n. sp.
Figure 2

Type specimen—F:AM 79153, palate and parts of attached premaxillaries, containing LP4-M3 and RM1-M3, incisors broken just inside alveoli.
Referred specimens—F:AM 79154 to 79284, including partial skulls, palates, dentaries. Also numerous uncatalogued specimens in the USNM collections, including partial skulls, palates, and dentaries.
Horizon and locality—Type and all referred specimens are from the Flagstaff Rim area of Natrona

County, Wyoming. The type is from the head of Little Lone Tree Gulch at 35 feet (10.7 m) below ash D, or at 250 feet (76.2 m) on the generalized zonation section (Emry, 1973, fig. 16), White River Formation. Referred specimens are from the Little Lone Tree Gulch, Lone Tree Gulch, and Blue Gulch drainages, from 70 feet (21.3 m) to 320 feet (97.5 m) on the same generalized section.

Age—Early to middle Chadronian.
Diagnosis—Statistically smallest species of the genus, but showing some overlap in size with all species; anterior and posterior basins of upper cheek teeth closed buccally (anterior basin open on P4 of *C. fontis* and *C. collinus*; posterior basin of M3 open in *C. fontis*); posterior basin of P4-M2 divided by short lophs from metaconule to posteroloph as in *C. solarborus* and *C. nebraskensis* (absent in *C. fontis* and variably present in *C. collinus*); dp4 with complete hypolophid and metalophid (hypolophid incomplete in *C. fontis*); trigonid basin of p4 closed posteriorly and anteriorly after moderate wear; p4 with complete hypolophid (incomplete on *C. solarborus*).
Etymology—Named for Natrona County, Wyoming, in which the type locality occurs.
Description and comparisons—No complete skull of *C. natronensis* is known, but the many partial skulls show no obvious differences in cranial morphology from the skulls of *C. solarborus* described below.

The upper incisors are oval in cross section with enamel only on the rounded anterior faces.

Several specimens representing young individuals have a small single-cusped dP3. The dP4 is much lower crowned and more nearly triangular in outline than P4 or the molars, but has a similar arrangement of lophs. Both dP3 and dP4 are replaced by P4. One specimen, F:AM 79184, has a small peg-like tooth anterior to the left P4, apparently a dP3 that was not shed with the eruption of P4.

When unworn, the occlusal outline of the P4 is somewhat triangular, narrower anteriorly, with the anteroposterior dimension about half again the transverse dimension. This differs from *C. fontis* and *C. collinus* in which P4 is more nearly square and equidimensional in occlusal outline. The lingual border of P4 is a smooth curve that continues anterolaterally as an anterior cingulum that meets the anterior base of the paracone below the level of the protoloph. The protoloph is complete from the paracone to the lingual border. Together, the anterior cingulum and the protoloph enclose a basin that develops with wear into an elliptical enamel fossette. In comparison, P4 of *C. fontis* lacks an anterior basin, protoloph, and anterior cingulum (Black, 1965). P4 of *C. collinus* has an anterior basin, but it is shallower and narrower than in *C. natronensis*. In *C. natronensis* the posteroloph and

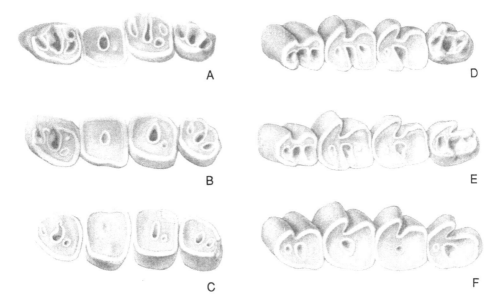

Figure 2. Upper (A-C) and lower (D-F) cheek teeth of *Cylindrodon natronensis*, new species, selected to show appearance of occlusal pattern at progressive wear stages. A, F:AM 79153, holotype, left P4-M3. B, F:AM 79155, left P4-M3. C, F:AM 79154, left P4-M3. D, F:AM 79156, left p4-m3. E, F:AM 79157, left p4-m3. F, F:AM 79158, left p4-m3. Buccal is toward top in all. All approximately X7.5, scale in millimeters.

metaloph are complete. The latter has a swelling medially that apparently represents the metaconule; a short anteroposteriorly directed crest (in a few individuals, two parallel crests) connects the metaconule to the posteroloph, so that with slight wear two (in the few individuals, three) small enamel fossettes are formed between the metaloph and posteroloph. In later wear stages, P4 retains its subtriangular-triangular occlusal outline but becomes more nearly equidimensional with more rounded sides.

M1 and M2 are so nearly identical that one description applies to both. Each has four transverse lophs enclosing three basins. The central basin is much deeper than the others and opens labially, via a shallow notch between paracone and metacone, only during the very earliest stage of wear. The anterior basin is shallowest and is usually lost by the time P4 and M3 begin to wear. The swollen central part of the metaloph (metaconule) is connected to the posteroloph by a short anteroposteriorly directed loph. This short loph is slightly lower than the metaloph and posteroloph; with slight wear the posterior basin becomes two small enamel fossettes, the lingual of which is the deeper.

The unworn outlines of M1 and M2 are more nearly rectangular than in *C. fontis*, due primarily to a longer anteroloph. The upper molars of *C. collinus* (Storer, 1978) and *C. nebraskensis* (Ostrander, 1983; Korth, 1992) resemble those of *C. natronensis*. The short loph from the metaconule to the posteroloph is not present in molars of *C. fontis* and is only variably present in specimens of *C. collinus*; it is present in all known

upper molars of *C. nebraskensis* and *C. solarborus* (described below). The crowns of M1 and M2 taper toward the base so that with wear the anteroposterior dimension decreases and the transverse dimension increases.

M3 of *C. natronensis* is the smallest of the permanent upper cheek teeth. Its occlusal outline is elliptical, with the long axis oblique to the tooth row. Like M1 and M2, it has four lophs. The internal border is more arcuate than in the other molars, with the anteroloph and posteroloph being part of the same curving crest. This tooth is similar to that of other species of *Cylindrodon*, with the posterior basin being completely enclosed on unworn teeth. In *C. fontis* this basin opens posteriorly due to the lack of a complete posteroloph (Black, 1965).

The mandible of *C. natronensis* is generally much like that of the other species, but the horizontal ramus is somewhat more slender and the diastema relatively shorter. Details of the posterior part of the jaw and the ascending ramus are not known for *C. natronensis*. The anterior limit of the masseteric fossa is beneath the anterior edge of m2. The anterior edge of the ascending ramus passes the alveolar border at the posterior edge of m2. The mental foramen is usually below the anterior edge of p4. In addition, there are usually numerous small vascular foramina on the outer surface of the jaw just below the diastema and most specimens also have a small circular foramen about half way down the outer surface of the jaw below the posterior part of p4 or the anterior part of m1.

Table 2. Statistical summary of measurements, in millimeters, of *Cylindrodon natronensis*, sp. nov.

Measurement	N	M	OR	s	V
Alveolar length, P4-M3	33	6.85	6.15-7.60	.359	5.24
Alveolar length, p4-m3	39	7.79	7.00-8.92	.487	6.25
Length I1-P4 diastema	30	6.32	5.85-7.05	.276	4.38
Length i1-p4 diastema	29	2.60	2.05-4.20	.541	20.77
Inside alveolar width					
of palate at P4	23	3.30	3.00-3.65	.170	5.14
Inside alveolar width					
of palate at M3	15	2.37	2.10-2.80	.216	9.12
Depth of ramus below m1	49	5.59	4.00-6.60	.554	9.91
I1, AP	22	1.85	1.70-1.95	.078	4.26
I1, TR	22	1.95	1.75-2.10	.110	5.56
dP4, AP	—	—	—	—	—
dP4, TR	—	—	—	—	—
P4, AP	30	1.75	1.50-1.90	.082	4.71
P4, TR	25	1.73	1.35-2.25	.236	13.63
M1, AP	35	1.57	1.35-1.70	.097	6.19
M1, TR	18	1.93	1.55-2.45	.285	14.81
M2, AP	35	1.70	1.40-1.85	.100	5.90
M2, TR	26	1.70	1.45-2.25	.195	11.51
M3, AP	23	1.46	1.25-1.70	.112	7.71
M3, TR	19	1.34	1.05-1.65	.114	10.76
i1, AP	43	1.85	1.30-2.30	.199	10.78
i1, TR	43	1.70	1.20-2.00	.147	8.63
dp4, AP	5	1.91	1.70-2.10	—	—
dp4, TR	5	1.38	1.20-1.55	—	—
p4, AP	49	1.80	1.45-2.20	.164	9.10
p4, TR	48	1.58	1.30-1.90	.144	9.12
m1, AP	77	1.81	1.40-2.20	.161	8.90
m1, TR	63	1.74	1.35-2.15	.169	9.69
m2, AP	79	1.80	1.30-2.10	.126	6.99
m2, TR	74	1.71	1.20-2.00	.162	9.45
m3, AP	48	1.59	1.30-1.80	.104	6.54
m3, TR	40	1.32	1.00-1.50	.104	7.87

The cross section of the lower incisors approximates an equilateral triangle, with enamel only on the gently rounded anterior face. The pulp cavity is narrow and short.

Dp4 is longer and narrower than any of the permanent lower teeth. The protoconid and metaconid are separated by a narrow valley except posteriorly. The hypolophid is complete, connecting the hypoconid and entoconid as in *C. collinus*; this differs from *C. fontis*, in which the hypolophid does not reach the entoconid (Black, 1965). The posterior cingulid is in the form of an arc from the hypoconid to the base of the entoconid.

The permanent lower premolar is broader and shorter than dp4. The metaconid is the highest cusp on an unworn tooth. It is connected to the smaller protoconid posteriorly by a short metalophid and anteriorly by a short, low anterior cingulid, enclosing a shallow trigonid basin. The hypolophid and posterolpohid are both complete, the former passing directly across the tooth from hypoconid to entoconid and the latter curving to the rear to connect these two cusps. These two lophids enclose a basin which, after slight wear, becomes an enamel fossetid. The central basin of p4 is the deepest and becomes an enamel fossetttid after early

wear stages, before which it opens lingually. Black (1965, fig. 3e) figured a p4 of *C. fontis* that differs from p4s of *C. natronensis* and all other species of *Cylindrodon* by having the metaconid and protoconid fused near their apices, becoming indistinct from one another after only slight to moderate wear with no trigonid basin. However, in all of the topotypic specimens of *C. fontis* in the USNM collections, the metaconid and protoconid of p4 are distinct and separated by a shallow groove, as in all other *Cylindrodon*

The first two lower molars of *C. natronensis* are longer than wide in early wear (Fig. 2D) due to the anterior and posterior expansion of the anterolophid and posterolophid, respectively. When unworn, the teeth have three basins. The anterior basin is shallowest and is completely enclosed by the anterolophid and metalophid. The central basin is deepest and, in unworn teeth, opens lingually through a shallow notch between the metaconid and entoconid. The posterior basin remains as an enamel fossettid after the anterior basin has disappeared. The posterior basin opens lingually in unworn teeth because the posterior cingulid is separated by a shallow notch from the entoconid. With wear, the posterior border of the teeth becomes rounded. The hypoconid is expanded labially and anteriorly so that the posterior half of the tooth is wider than the anterior half. The hypoconulid is only barely defined on unworn teeth.

The last lower molar is the smallest of the lower cheek teeth. The anterior part of this tooth is wider than the posterior part. The central basin is deepest and the anterior basin shallowest. The morphology is similar to the other two molars except that the hypolophid and posterolophid are shorter and the hypoconid is not expanded as much labially.

In p4-m2, the deep central basin often has a fold of enamel projecting from the anterointernal side toward the center of the basin. This fold does not extend the full depth of the central basin and consequently is seen only during certain wear stages; it is not seen at the occlusal surface during earliest and latest wear stages of individual teeth. It is not certain that this fold is present in all individuals. A similar fold was reported by Galbreath (1969) in specimens from Colorado later referred to C. nebraskensis (Ostrander, 1983; Korth, 1992).

Size—Statistical summaries of dental measurements of C. natronensis are given in Table 2.

Discussion—Cylindrodon natronensis is statistically the smallest species of the genus. It differs from C. fontis, as do all other species, by the possession of the loph connecting the metaconule to the posteroloph on P4-M2. C. natronensis differs from C. solarborus (described below) in the proportions of p4 and m3, the lack of the accessory loph on M3, and having a complete hypolophid on p4 (incomplete on C. solarborus). C. natronensis also occurs stratigraphically lower in the section at Flagstaff Rim than the larger C. solarborus.

A large collection of specimens of Cylindrodon from Cameron Springs, Wyoming, was examined as a part of this study. These specimens appear to have the morphological characters of C. natronensis and should probably be referred to this species.

Cylindrodon solarborus n. sp.
Figures 3-4

Type specimen—F:AM 79100, articulated skull and mandible with all teeth present, part of skull roof and occipital surface missing.

Referred specimens—F:AM 79101 to 79152, includes partial skulls, some with mandibles associated, palates, dentaries. Also referred are many uncatalogued specimens in the USNM collections, including partial skulls, palates, dentaries.

Horizon and locality—Type and referred specimens all from the Flagstaff Rim area, Natrona County, Wyoming. Type is from the South Fork of Lone Tree Gulch, at 35 feet (10.7 m) below ash G or

400 feet (121.9 m) on the generalized zonation section, White River Formation (Emry, 1973, fig. 16). All referred specimens are from Little Lone Tree Gulch, Lone Tree Gulch, and Blue Gulch drainages, ranging from 370 feet (112.8 m) to 620 feet (189 m) on the same generalized zonation section.

Age—Middle Chadronian.

Diagnosis—Near size of C. fontis, statistically larger than C. natronensis and smaller than C. collinus and C. nebraskensis (see Table 1); anterior and posterior basins of all upper cheek teeth enclosed as in C. natronensis; posterior basin of P4-M3 enclosed and divided by a short loph from metaconule to posterior cingulum (loph lacking in M3 of C. natronensis, variably present in all upper molars of C. collinus, and lacking in all molars of C. fontis); dp4 with complete hypolophid and metalophid as in C. natronensis and C. collinus; trigonid basin of p4 enclosed anteriorly and posteriorly; hypolophid of p4 incomplete (complete in all other species).

Etymology—From Latin solus, alone, lone, single, and arbor, tree; alluding to Lone Tree Gulch, the type locality.

Description and comparisons—The skull of Cylindrodon was described in part by Wood (1937), Black (1965), and Wahlert (1974). The more nearly complete skulls of C. solarborus allow a more detailed description. Since Wahlert (1974) has described the cranial foramina of Cylindrodon, comments on the cranial foramina are made only where our observations differ from those of Wahlert. The skulls of some other cylindrodontids have been described in some detail: Pseudocylindrodon—Burke (1938) and Wood (1974); Ardynomys—Burke (1936), Wood (1974), and Wahlert (1974). Comparisons made below with these other cylindrodonts are based on the above descriptions.

In general shape, the skull is much like that of Pseudocylindrodon, and does not have the shorter rostrum and greater overall breadth of Ardynomys. In lateral view (Fig. 3A), the dorsal profile of the skull is evenly convex. The profile of the diastema is convex, nearly paralleling the dorsal profile of the snout except posteriorly, where it curves downward. The tooth row is essentially horizontal. The auditory bullae are inflated well below the level of the pterygoids, but not so much as in Pseudocylindrodon.

In ventral view (Fig. 3B), the rostrum is quite broad, the sides diverging slightly toward the rear but yet more nearly parallel than in Pseudocylindrodon, similar to the condition in Ardynomys. The nasal bones are narrower and more nearly parallel-sided than those of Pseudocylindrodon and Ardynomys, which are broader anteriorly. The nasals do not extend anterior to the premaxillaries and are not inflated as they are in Pseudocylindrodon and Ardynomys. The nasal-frontal and premaxillary-frontal sutures are even with the

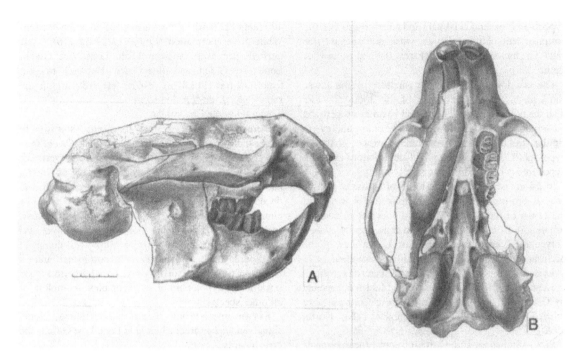

Figure 3. Skull and right dentary of F:AM 79100, holotype of *Cylindrodon solarborus*, new species, in lateral (A) and ventral (B) views. Approximately X3, scale in millimeters.

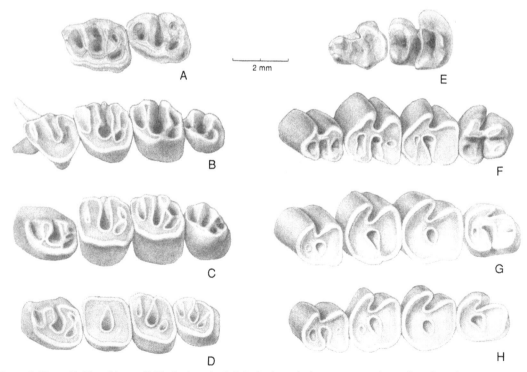

Figure 4. Upper (A-D) and lower (E-H) cheek teeth of *Cylindrodon solarborus*, new species, selected to show appearance of occlusal pattern at progressive wear stages. A, F:AM 79289, left M1-M2. B, F:AM 79102, left dP3, dP4, M1-M3. C, F:AM 79100, holotype, left P4-M3. D, F:AM 79107, left P4-M3. E, F:AM 79298, left dp4-m1. F, F:AM 79100, holotype, left p4-m3. G, F:AM 79115, left p4-m3. H, F:AM 79123, left p4-m3. Buccal is toward top in all. All approximately X7.5, scale in millimeters.

anterior edges of the orbits, slightly more posterior than in *Pseudocylindrodon*, *Ardynomys*, and *C. fontis*. In most specimens the nasal bones extend slightly more posteriad than the premaxillaries.

The postorbital constriction is of about the same degree as in *P. medius* and less constricted than in *Ardynomys* and *P. texanus*. The braincase expands laterally immediately posterior to the postorbital constriction. The braincase of *C. solarborus* is narrower than that of *P. medius* or *Ardynomys*. The very thin bone of the dorsal and occipital surfaces of the braincase is missing on most specimens. In a few skull fragments that have the parietal region at least partially preserved, a single low but distinct sagittal crest is indicated.

The occipital surface is oriented vertically but is gently convex posteriorly. The occipital crests (nuchal crests) are most prominent laterally. The squamoso-mastoid foramen pierces the mastoid bone within a shallow depression of the mastoid dorsal to the bulla, along the mastoid-squamosal suture. The condyles border only the lower half of the foramen magnum and are broad and quite flat. The paraoccipital processes are short and blunt, pressed against the posterior side of the auditory bullae. The auditory bullae are slightly drawn out into a rounded point anteromedially. The meatus has a recurved lip. The basioccipital has a prominent median ventral keel between parallel concave sulci. The edges of the basioccipital turn downward slightly where they contact the bullae. The basisphenoid is flat and in the same plane as the basioccipital.

The squamosals apparently played little part in the formation of the skull roof, being restricted to the sides of the braincase as in other cylindrodonts and *Ischyromys*.

Anterior to the anteromedial end of the bulla is a relatively large rounded recess bordered by the pterygoid flanges (Fig. 3A). Wahlert (1974) noted that the foramen ovale and transverse canal open into this recess; in addition, apparently in all cylindrodonts, the median lacerate foramen is present at the posterolateral limit of this recess, within which is an opening for the eustachian canal. On the middle side of the median lacerate foramen a shallow depression has an additional opening, apparently the anterior carotid canal.

The pterygoid is a thin plate with the ventral border drawn out posteriorly into a slender hamular process that extends toward the bullae but ends in a blunt point that does not contact the bulla; this is similar to the condition in *Ardynomys* and in contrast to that in *Pseudocylindrodon* where the process contacts the bulla. Lateral to the pterygoid plate is the external palatine plate, which extends laterally and caudally, terminating in a free process. The pterygoid fossa is a fairly deep and broad pocket between these two plates.

The posterior palatine border is carried anteriorly to a point opposite the posterior part of M2. The postnareal trench, between the pterygoid processes, is narrower than in *Pseudocylindrodon* and has more nearly parallel sides. The maxillary-premaxillary suture is about a third of the distance from P4 to the incisor and intersects the posterior limit of the incisive foramina as it does in *Ardynomys* and more posteriorly than in *Pseudocylindrodon* where the suture crosses the incisive foramina near their centers. The palate is narrow with a low median ridge. It is narrowest in the region of M3, as in *P. texanus* and *Ardynomys* (in *P. medius* it is narrowest in the area of M1 and M2).

The zygomatic arch is relatively stouter than that of *Pseudocylindrodon*. The zygomatic process of the maxilla extends backward to a point opposite the middle of M1 (Fig. 3A). The jugal is more robust and thicker dorsoventrally than in *Pseudocylindrodon*, similar in size to that of *Ardynomys*. It extends forward and upward above the maxillary to form the ventral and most of the anterior rim of the orbit. Posteriorly, the jugal extends beneath the squamosal to a point lateral and ventral to the glenoid fossa and together with the squamosal forms the most posterior part of the arch. Midway along the dorsal border of the jugal the edge is drawn up slightly into a low postorbital process. The well defined masseteric muscle scar is limited to the ventral part of the zygoma. The anterior edge of this scar is only slightly anterior to P4. The anterior root of the zygoma is tilted dorsolaterally so that this muscle scar is inclined about 45° from the frontal plane.

The infraorbital foramen is of moderate to small size and is usually oval, higher than wide. In some specimens of *C. solarborus* this foramen is divided by a septum of bone so that two foramina show anteriorly, one dorsal to the other. This condition is seen on both sides of the type of *C. solarborus* (F:AM 79100) and also in two other specimens (F:AM 79104 and 79113). The foramen is divided on only one side in four other specimens (F:AM 79101, 79103, 79108, 79109). This condition has not been observed in other cylindrodontids.

On the middle orbital wall, the contribution of the palatine bone is obscured by the dorsal swelling of the maxilla due to the hypsodonty of the cheek teeth. The foramina of the middle orbital wall (ethmoid, sphenopalatine, dorsal palatine) have been displaced more dorsally than in the other cylindrodontids with lower crowned cheek teeth. One foramen not observed by Wahlert (1974) is a minute frontal foramen, within the frontal bone dorsal to M1. This foramen may occur just within the dorsal margin of the orbit or just outside the orbit on the skull roof.

The upper incisors of *C. solarborus* are much like those of the other species of the genus, being nearly oval in cross section, the transverse diameter greatest. Enamel is limited to the anterior surface. The pulp

Table 3. Statistical summary of measurements, in millimeters, of *Cylindrodon solarborus*, sp. nov.

Measurement	N	M	OR	s	V
Alveolar length, P4-M3	33	6.85	6.15-7.60	.359	5.24
Alveolar length, p4-m3	39	7.79	7.00-8.92	.487	6.25
Length I1-P4 diastema	30	6.32	5.85-7.05	.276	4.38
Length i1-p4 diastema	29	2.60	2.05-4.20	.541	20.77
Inside alveolar width of palate at P4	23	3.30	3.00-3.65	.170	5.14
Inside alveolar width of palate at M3	15	2.37	2.10-2.80	.216	9.12
Depth of ramus below m1	49	5.59	4.00-6.60	.554	9.91
I1, AP	22	1.85	1.70-1.95	.078	4.26
I1, TR	22	1.95	1.75-2.10	.110	5.65
dP4, AP	—	—	—	—	—
dP4, TR	—	—	—	—	—
P4, AP	30	1.75	1.50-1.90	.082	4.71
P4, TR	25	1.73	1.35-2.25	.236	13.63
M1, AP	35	1.57	1.35-1.70	.097	6.19
M1, TR	18	1.93	1.55-2.45	.285	14.81
M2, AP	35	1.70	1.40-1.85	.100	5.90
M2, TR	26	1.70	1.45-2.25	.195	11.51
M3, AP	23	1.46	1.25-1.70	.112	7.71
M3, TR	19	1.34	1.05-1.65	.144	10.76
i1, AP	43	1.85	1.30-2.30	.199	10.78
i1, TR	43	1.70	1.20-2.00	.147	8.63
dp4, AP	5	1.91	1.70-2.10	—	—
dp4, TR	5	1.38	1.20-1.55	—	—
p4, AP	49	1.80	1.45-2.20	.164	9.10
p4, TR	48	1.58	1.30-1.90	.144	9.12
m1, AP	77	1.81	1.40-2.20	.161	8.90
m1, TR	63	1.74	1.35-2.15	.169	9.69
m2, AP	79	1.80	1.30-2.10	.126	6.99
m2, TR	74	1.71	1.20-2.00	.162	9.45
m3, AP	48	1.59	1.30-1.80	.104	6.54
m3, TR	40	1.32	1.00-1.50	.104	7.87

cavity is oval.

Both upper and lower permanent cheek teeth are hypsodont. DP3 is present in young individuals but is not replaced by a permanent P3; DP3 and DP4 are apparently lost at the same time and replaced only by P4, as in other species of the genus. Three specimens of adult individuals, otherwise normal and indistinguishable from other specimens of *C. solarborus*, have a small simple conical tooth anterior to P4. In F:AM 79103 and 79106, this small tooth is present on both sides, and in F:AM 79151 it is present only on the left side. These small teeth are morphologically like dP3s and are believed to be retained dP3s. DP4 is much like that of the other species of the genus. The occlusal outline is triangular; posterior and anterior basins are enclosed. The central basin is blocked buccally by a small mesostyle.

In unworn teeth P4 is more elongate anteroposteriorly than in other species. The paracone bulges into the central basin but is strongly connected to the lingual border, forming a distinct protoloph. The anteroloph curves around to join the anterobuccal edge of the paracone; this connection is below the level of the protoloph but nevertheless closes off a deep anterior basin, which with wear becomes an elliptical enamel fossette. The buccal side of P4 is usually flatter than in the other species and often has a rounded crest at the anterobuccal and posterobuccal edges of the tooth. The metaloph is usually complete, but on some specimens a very narrow but relatively deep notch separates the metaconule from the lingual border. Either one or two short crests connect the metaloph and posteroloph so that either two or three basins are developed between these two lophs. The lingual of these two or three basins is the deepest, as in *C. natronensis*.

Unworn M1 and M2 are so nearly identical that they cannot be separated as isolated teeth. However, some differences can be seen in specimens where the teeth are in place: in most cases M1 has a small mesostyle and M2 does not; in a few specimens neither M1 or M2 has a mesostyle, and even more rarely both have it; the case has not been observed in which M2 has a mesostyle and M1 does not. M1 and M2 are a little broader anteriorly than in *C. fontis* or *C. natronensis*, due to expansion of the anterolingual part of the anteroloph, giving them a more rectangular occlusal outline. M1 and M2 of *C. solarborus*, like those of *C. natronensis* and *C. nebraskensis*, have a loph connecting the metaconule to the posteroloph so that two small basins are formed between the metaloph and posteroloph. This loph is lacking in *C. fontis* and only variably present in *C. collinus*. As in *C. natronensis*, the lingual of these two basins formed is deepest. Some teeth have a fold of enamel extending from the paracone into the central basin. This fold does not extend the entire depth of the basin and consequently is seen on the occlusal surface only during certain wear stages; it is not seen in early and late stages of wear.

M3 of *C. solarborus* differs from that of *C. fontis* and *C. natronensis* and is similar to that of *C. collinus* in the greater development of the posterior part. The tooth is altogether more rectangular than that of *C. fontis* and *C. natronensis* and the metaloph and posteroloph enclose two basins. This is elsewhere only variably present in *C. collinus* (M3 is not known for *C. nebraskensis*). A crest, at the same level as the metaloph, extends from the center of the metaloph to the posteroloph; buccal to this junction, the posteroloph is lower and joins the metacone below the level of the metaloph (Fig. 4B-C). This part of the posteroloph might more properly be termed a posterior cingulum, but it nevertheless encloses a small basin that becomes a circular enamel fossette with wear. In the holotype, F:AM 79100, this small basin is even further divided by another short loph so that there are three small basins posterior to the metaloph.

Wood (1984) described the posterior parts of two mandibles that he referred to *C. fontis*. The first (and more nearly complete) of the mandibles was AMNH 94501 from Cameron Springs, Wyoming, the second (F:AM 105370) from Pipestone Springs, Montana. The first of these specimens is probably referable to *C. natronensis*, based on observations of additional material of *Cylindrodon* from Cameron Springs in the collections of the USNM. Although Wood (1984) reconstructed the entire mandible, he noted that the material on which he based the reconstruction was not complete. Better preserved material of *C. solarborus* in the Flagstaff Rim samples suggests that Wood's reconstructions and interpretations of "*C. fontis*" are inaccurate.

The mandible of *C. solarborus* is relatively deeper dorsoventrally than in *C. fontis* and *C. natronensis* but not so deep as in *C. nebraskensis*. The condyle is globular, elevated above the tooth row. The coronoid process extends above the condyle (Figs. 1, 3B). Wood (1981, 1984) interpreted the angle of the mandible as being laterally displaced from the horizontal ramus, a condition he termed "subhystricognathous." However, in completely preserved specimens (Fig. 1), it appears that the angle does not originate lateral to the body of the ramus but in line with it. The ventral surface of the angular process is quite broad, due in part to the development of a deep "pocket" internally for the

insertion of the pterygoid musculature, forming a wide ventral shelf on the internal side of the mandible. This is combined with a laterally deflected flange along the ventral edge of the angular process for attachment of the masseter superficialis (this flange in completely preserved specimens is larger than figured by Wood, 1984, fig. 1A). In the specimens on which Wood (1984) based his observations, the internal flange (that which forms the lower surface of the pterygoid pocket) is broken away, making it appear that the angular process is lateral to the plane of the incisor. In fact the angular process originates in line with the incisor, and posterior to its origination is expanded both inward and outward.

Near the center of the ascending ramus, on the lateral surface, is a small circular swelling that houses the base of the incisor (Figs. 1C, 3A). Posterior to this swelling, and ventral to the condyle is a shallow depression.

The anterior edge of the masseteric fossa is beneath the anterior edge of m2. None of the specimens of *Cylindrodon* examined during this study has the distinct scar for the masseter medialis indicated by Wood (1984, fig. 1C). The symphysis is quite heavy and rugose, with the posterior edge beneath the anterior part of m1. There is a small pocket by the posterodorsal edge of the symphysis under the anterior part of p4. The diastema is relatively longer than that of *C. fontis* and *C. natronensis*. The mental foramen is almost directly below the anterior edge of p4, sometimes slightly ahead of this, and well down on the side of the jaw. It is often divided into two closely adjacent small foramina or, in some specimens, two foramina opening into a small pocket.

The lower incisor is triangular in cross section, with enamel only on the anterior face. The pulp cavity is narrow and short.

As in other species, dp4 is narrower and longer than any of the permanent lower teeth. Only one specimen of *C. solarborus* has a relatively unworn dp4 and in this specimen the hypolophid is incomplete, in contrast to *C. natronensis*. Dp4 of *C. solarborus* is relatively longer and narrower than in the other species in which it is known.

The permanent lower premolar differs from that of the other species in several details; the anterior part of the tooth is relatively broader and the hypolophid is incomplete. The protoconid and metaconid are distinct from one another but joined by a short, low, anterolophid and by a complete metalophid; with slight wear an elliptical enamel fossettid (trigonid basin) is formed between the two cusps. The central basin opens lingually in unworn teeth, but with slight wear it becomes an enclosed enamel fossettid. A loph extends from the entoconid toward the hypoconid, but rather than joining the hypoconid and forming a hypolophid, it turns backward to join the center of the posterolophid and encloses a small basin in the posterolingual part of the tooth.

The first two lower molars are nearly identical. As in the other species, the crown is longer than wide when unworn, due to anterior and posterior expansion of the anterolophid and posterolophid. The occlusal surface becomes more nearly equidimensional in later wear stages. When unworn, each of the three lower molars has three basins, the anterior being shallowest, the central basin being deepest. In unworn molars the central basin opens lingually through a shallow notch between the metaconid and entoconid. The lower molars of *C. solarborus* differ from those of *C. fontis* and *C. natronensis* in the relatively greater width of the anterior half of m1 and m2 and the relatively greater width of the posterior half of m3. The proportions of the lower molars of *C. collinus* and *C. nebraskensis* are similar to those of *C. solarborus*.

Size—Statistical summaries of the dental measurements of *C. solarborus* are given in Table 3.

Discussion—The unique features of the dentition of *C. solarborus* are the partial hypolophid on p4, the proportions of p4 and m3, and presence of the accessory loph from the metaconule on M3. Of these features, the loph on M3 is known elsewhere only in *C. collinus*, where it is variably present; the proportions of p4 and m3 are similar in some specimens of *C. collinus* and *C. nebraskensis*; no other species is known to have the incomplete hypolophid of p4.

The occurrence of *C. natronensis* and *C. solarborus* in lower and higher parts, respectively, of the same stratigraphic succession, suggests in situ evolution of the latter from the former. We see no characters of the skull or dentition of these species that would rule out such an interpretation.

DENTAL MEASUREMENTS OF *CYLINDRODON*

Statistics of measurements of the new species, *Cylindrodon natronensis* and *C. solarborus*, are summarized in 13 tables. Table 1 also includes comparative measurements of the other species of *Cylindrodon*

Tables 2 and 3 include all of the F:AM specimens referred to these two new species. It can be seen that the coefficient of variation (V) for some of the measurements is quite high, particularly with nondental measurements such as lengths of diastemata, depth of mandibular ramus, and width of palate. And among the dental measurements, it can be noted that the coefficients of variation of the transverse measurements of cheek teeth, particularly the upper teeth, are considerably greater than those of anteroposterior measurements. This is clearly due to the change in proportion of the occlusal surface of these teeth during wear. In unworn teeth, the anteroposterior dimension is

greater than the transverse, but as the teeth wear down, the anteroposterior dimension decreases slightly, and the transverse dimension increases greatly, so that in late wear stages the transverse dimension is greater.

So that measurements of single specimens from other localities, or of other species, can be more meaningfully compared, each tooth was assigned to one of five different wear stages when measured. The statistics for these measurements are summarized in Tables 4 to 13. Tables 2 and 3, above, including specimens of all wear stages, are presented in spite of the variation introduced by proportional changes during wear; these two tables allow comparisons with other samples that are large enough to treat statistically as a whole, but not large enough to provide statistically valid samples of individual wear stages.

Particularly in *C. solarborus*, samples of some wear stages of some teeth were not large enough to be statistically meaningful. It can be seen however, that in nearly all cases the coefficient of variation (V) of any one wear stage of any tooth (Tables 4 to 13) is considerably smaller than the coefficient of variation of the same tooth in all wear stages (Tables 2 and 3).

The wear stages of *Cylindrodon* cheek teeth can be conveniently and objectively divided into five wear stages, as follows:

Wear Stage I—unworn, or worn so slightly that dentine is exposed only at the points of the main cusps.

Wear Stage II—dentine exposed on occlusal surface of lophs, but three enamel lakes or basins still present.

Wear State III—anterior basin eliminated, only two basins remain.

Wear Stage IV—posterior basin eliminated, only central basin remains.

Wear State V—central basin eliminated, occlusal surface is dentine surrounded by a band of enamel, but no internal enamel pattern.

Although all individual teeth can be objectively assigned to one of these wear stages, it seems not possible to assign individual animals to wear stage groups. The sequence of tooth eruption is not sufficiently uniform, and neither is the rate of wear of all teeth. In some individuals, P4/p4 begin to wear before M3/m3, and in other individuals the reverse is true. In Figure 4G, p4 is in wear stage II, m2 is in wear stage IV, and m3 is in wear stage III, whereas in Figure 4H, p4 is more heavily worn (wear stage IV) whereas m2 and m3 are less heavily worn (stages III and II, respectively).

Table 4. Statistical summary of measurements of teeth of *Cylindrodon natronensis* in Wear Stage I (measurements in mm, AP = anteroposterior, TR = transverse).

Measure	N	M	OR	s	V
P4, AP	8	1.76	1.70-1.85	.054	3.08
P4, TR	8	1.62	1.35-1.80	.185	11.43
M1, AP	—	—	—	—	—
M1, TR	—	—	—	—	—
M2, AP	2	1.63	1.60-1.65	—	—
M2, TR	1	1.50	1.50	—	—
M3, AP	3	1.45	1.35-1.50	—	—
M3, TR	3	1.22	1.05-1.50	—	—
p4, AP	13	1.66	1.45-1.87	.109	6.56
p4, TR	13	1.46	1.30-1.55	.092	6.29
m1, AP	9	1.97	1.87-2.10	.083	4.19
m1, TR	8	1.57	1.35-1.97	.233	14.88
m2, AP	7	1.83	1.70-2.10	.138	7.54
m2, TR	7	1.49	1.20-1.85	.266	17.88
m3, AP	7	1.48	1.30-1.70	.133	8.97
m3, TR	5	1.20	1.00-1.30	.115	9.65

Table 5. Statistical summary of measurements of teeth of *Cylindrodon natronensis* in Wear Stage II (measurements in mm; AP = anteroposterior, TR = transverse).

Measurement	N	M	OR	s	V
P4, AP	5	1.78	1.70-1.90	.083	4.67
P4, TR	5	1.67	1.48-1.85	.166	9.96
M1, AP	—	—	—	—	—
M1, TR	—	—	—	—	—
M2, AP	3	1.78	1.75-1.85	—	—
M2, TR	2	1.54	1.50-1.57	—	—
M3, AP	6	1.44	1.35-1.48	.092	6.35
M3, TR	6	1.31	1.20-1.46	.122	9.35
p4, AP	7	1.72	1.60-1.85	.094	5.49
p4, TR	6	1.52	1.30-1.60	.121	7.98
m1, AP	10	1.97	1.75-2.20	.126	6.41
m1, TR	7	1.79	1.50-2.00	.199	11.17
m2, AP	20	1.89	1.70-2.10	.113	5.98
m2, TR	20	1.69	1.50-2.00	.128	7.55
m3, AP	15	1.59	1.40-1.80	.109	6.82
m3, TR	14	1.30	1.15-1.50	.106	8.11

Table 6. Statistical summary of measurements of teeth of *Cylindrodon natronensis* in Wear Stage III (measurements in mm; AP = anteroposterior, TR = transverse).

Measurement	N	M	OR	s	V
P4, AP	7	1.75	1.65-1.85	.070	3.99
P4, TR	5	1.71	1.40-2.12	—	—
M1, AP	3	1.62	1.60-1.65	—	—
M1, TR	—	—	—	—	—
M2, AP	6	1.78	1.65-1.85	.082	4.63
M2, TR	4	1.57	1.45-1.65	—	—
M3, AP	6	1.48	1.40-1.55	.069	4.67
M3, TR	5	1.35	1.25-1.45	—	—
p4, AP	3	1.81	1.80-1.83	—	—
p4, TR	3	1.56	1.55-1.58	—	—
m1, AP	5	1.87	1.75-2.00	—	—
m1, TR	5	1.76	1.60-1.85	—	—
m2, AP	8	1.85	1.75-1.93	.058	3.14
m2, TR	8	1.73	1.50-1.90	.141	8.16
m3, AP	5	1.62	1.55-1.70	—	—
m3, TR	4	1.35	1.25-1.50	—	—

Table 7. Statistical summary of measurements of teeth of *Cylindrodon natronensis* in Wear Stage IV (measurements in mm; AP = anteroposterior, TR = transverse).

Measurement	N	M	OR	s	V
P4, AP	6	1.74	1.65-1.80	.055	3.16
P4, TR	4	1.80	1.65-2.00	—	—
M1, AP	12	1.61	1.50-1.70	.072	4.51
M1, TR	6	1.78	1.55-2.10	.221	12.37
M2, AP	18	1.68	1.60-1.75	.067	4.00
M2, TR	15	1.69	1.52-1.95	.133	7.88
M3, AP	4	1.45	1.40-1.60	—	—
M3, TR	2	1.38	1.35-1.40	—	—
p4, AP	18	1.88	1.55-2.20	.148	7.87
p4, TR	18	1.63	1.50-1.82	.087	5.36
m1, AP	33	1.82	1.63-2.10	.116	6.37
m1, TR	29	1.75	1.55-2.15	.144	8.24
m2, AP	32	1.78	1.63-1.90	.083	4.72
m2, TR	29	1.74	1.55-2.00	.111	6.40
m3, AP	19	1.61	1.50-1.75	.066	4.08
m3, TR	15	1.35	1.20-1.50	.078	5.78

Table 8. Statistical summary of measurements of teeth of *Cylindrodon natronensis* in Wear Stage V (measurements in mm; AP = anteroposterior, TR = transverse).

Measurement	N	M	OR	s	V
P4, AP	2	1.55	1.50-1.60	—	—
P4, TR	2	2.00	2.00	—	—
M1, AP	18	1.53	1.35-1.70	.102	6.68
M1, TR	10	1.92	1.65-2.15	.228	11.79
M2, AP	4	1.63	1.40-1.85	—	—
M2, TR	3	1.93	1.75-2.15	—	—
M3, AP	2	1.28	1.25-1.30	—	—
M3, TR	2	1.38	1.35-1.40	—	—
p4, AP	7	1.94	1.70-2.15	.160	8.26
p4, TR	7	1.79	1.65-1.90	.098	5.44
m1, AP	19	1.65	1.40-1.85	.124	7.52
m1, TR	13	1.81	1.65-2.10	.126	6.98
m2, AP	11	1.67	1.30-1.85	.152	9.09
m2, TR	9	1.80	1.50-1.95	.162	8.98
m3, AP	1	1.80	1.80	—	—
m3, TR	1	1.35	1.35	—	—

Table 9. Statistical summary of measurements of teeth of *Cylindrodon solarborus* in Wear State I (measurements in mm; AR = anteroposterior, TR = transverse).

Measurement	N	M	OR	s	V
P4, AP	7	1.77	1.72-1.90	.062	3.50
P4, TR	7	1.40	1.30-1.57	.113	8.04
M1, AP	5	2.03	1.95-2.15	—	—
M1, TR	5	1.63	1.55-1.70	—	—
M2, AP	7	1.95	1.90-2.00	.041	2.09
M2, TR	7	1.51	1.40-1.60	.080	5.29
M3, AP	4	1.51	1.40-1.60	—	—
M3, TR	4	1.24	1.20-1.30	—	—
p4, AP	5	1.76	1.65-2.00	—	—
p4, TR	5	1.58	1.45-1.75	—	—
m1, AP	2	2.13	2.00-2.25	—	—
m1, TR	2	1.55	1.45-1.65	—	—
m2, AP	1	1.85	1.85	—	—
m2, TR	1	1.40	1.40	—	—
m3, AP	2	1.55	1.50-1.60	—	—
m3, TR	2	1.28	1.25-1.30	—	—

Table 10. Statistical summary of measurements of the teeth of *Cylindrodon solarborus* in Wear Stage II (measurements in mm; AP = anteroposterior, TR = transverse).

Measurement	N	M	OR	s	V
P4, AP	2	1.90	1.90	—	—
P4, TR	2	2.00	2.00	—	—
M1, AP	2	1.93	1.85-2.00	—	—
M1, TR	2	1.75	1.75	—	—
M2, AP	6	1.74	1.58-1.93	.140	8.05
M2, TR	6	1.71	1.65-1.75	.038	2.20
M3, AP	4	1.50	1.45-1.55	—	—
M3, TR	4	1.38	1.25-1.55	—	—
p4, AP	4	1.82	1.75-1.90	—	—
p4, TR	4	1.68	1.50-1.80	—	—
m1, AP	5	2.05	2.00-2.10	—	—
m1, TR	5	1.91	1.75-2.00	—	—
m2, AP	6	2.01	1.85-2.15	.107	5.32
m2, TR	6	1.86	1.75-2.05	.107	5.75
m3, AP	3	1.60	1.50-1.70	—	—
m3, TR	3	1.40	1.35-1.50	—	—

Table 11. Statistical summary of measurements of teeth of *Cylindrodon solarborus* in Wear Stage III (measurements in mm; AP = anteroposterior, TR = transverse).

Measurement	N	M	OR	s	V
P4, AP	—	—	—	—	—
P4, TR	—	—	—	—	—
M1, AP	—	—	—	—	—
M1, TR	—	—	—	—	—
M2, AP	—	—	—	—	—
M2, TR	—	—	—	—	—
M3, AP	3	1.62	1.50-1.85	—	—
M3, TR	3	1.55	1.41-1.60	—	—
p4, AP	1	1.90	1.90	—	—
p4, TR	1	1.80	1.80	—	—
m1, AP	1	2.03	2.03	—	—
m1, TR	1	1.95	1.95	—	—
m2, AP	5	1.96	1.85-2.00	—	—
m2, TR	5	1.94	1.70-2.10	—	—
m3, AP	4	1.71	1.63-1.80	—	—
m3, TR	4	1.47	1.35-1.60	—	—

Table 12. Statistical summary of measurements of teeth of *Cylindrodon solarborus* in Wear Stage IV (measurements in mm; AP = anteroposterior, TR = transverse).

Measurement	N	M	OR	s	V
P4, AP	10	1.76	1.65-1.95	.101	5.77
P4, TR	10	2.11	1.90-2.30	.159	7.53
M1, AP	7	1.78	1.73-1.85	.066	3.69
M1, TR	7	1.93	1.72-2.10	.118	6.10
M2, AP	11	1.67	1.60-1.85	.081	4.87
M2, TR	9	1.92	1.75-2.25	.201	10.48
M3, AP	6	1.47	1.35-1.60	.108	7.36
M3, TR	5	1.45	1.30-1.65	—	—
p4, AP	3	1.88	1.80-1.95	—	—
p4, TR	3	1.74	1.65-1.80	—	—
m1, AP	10	1.97	1.75-2.15	.119	6.05
m1, TR	10	2.05	1.95-2.15	.071	3.49
m2, AP	6	1.90	1.82-1.95	.052	2.77
m2, TR	6	2.05	1.90-2.15	.090	4.39
m3, AP	2	1.62	1.58-1.65	—	—
m3, TR	2	1.43	1.40-1.45	—	—

Table 13. Statistical summary of measurements of teeth of *Cylindrodon solarborus* in Wear Stage V (measurements in mm; AP = anteroposterior, TR = transverse).

Measurement	N	M	OR	s	V
P4, AP	4	1.70	1.65-1.80	—	—
P4, TR	3	1.98	1.90-2.10	—	—
M1, AP	14	1.53	1.40-1.75	.123	8.06
M1, TR	12	2.18	2.00-2.40	.127	5.84
M2, AP	6	1.58	1.50-1.65	.052	3.26
M2, TR	4	2.13	2.00-2.25	—	—
M3, AP	3	1.47	1.45-1.50	—	—
M3, TR	2	1.73	1.65-1.80	—	—
p4, AP	4	2.08	1.95-2.20	—	—
p4, TR	3	1.80	1.65-1.90	—	—
m1, AP	6	1.70	1.50-1.83	.129	7.58
m1, TR	4	2.01	1.80-2.20	—	—
m2, AP	4	1.79	1.70-2.05	—	—
m2, TR	3	1.98	1.90-2.10	—	—
m3, AP	4	1.68	1.60-1.70	—	—
m3, TR	4	1.44	1.30-1.60	—	—

SUMMARY

In the White River Formation at Flagstaff Rim, Wyoming, species of *Cylindrodon* are easily the most common rodents found. They occur from near the base of the section to at least as high as Ash H (Emry, 1973, fig. 16), which is near the top of the local section. This range encompasses early and middle Chadronian, and possibly extends into late Chadronian time. Similarly, at Cameron Springs in central Wyoming, and at Pipestone Springs in western Montana, both localities of middle Chadronian age, species of *Cylindrodon* are among the most common rodents. The fact that *Cylindrodon* is less common in the Chadron Formation of western Nebraska and southwestern South Dakota might suggest that *Cylindrodon* preferred intermontane over plains habitats. However it seems more likely that this lesser abundance simply reflects the fact that in the Chadron Formation small mammals in general are poorly represented. Where special efforts

have been made to recover small mammals from the Chadron Formation (e.g., the Raben Ranch local fauna—Ostrander, 1983, 1985), *Cylindrodon* is in fact relatively common. Moreover, *Cylindrodon* is also well represented in the Cypress Hills area of southern Saskatchewan, far from the mountains.

Little is known about the postcranial anatomy of cylindrodonts. The Chadronian cylindrodonts are generally about the size of ground squirrels such as *Spermophilus*, and some features of the skull suggest that they may have been semi-fossorial. A fossorial habit might also account for the fact that they are so abundantly preserved.

There are no substantiated occurrences of cylindrodonts beyond the end of Chadronian time (except possibly in Montana; see Chapter 14 of this volume). Their extinction coincides approximately with that of the largest land mammals of the time, the brontotheres, and with some other smaller mammals such as the fossorial palaeanodont *Xenocranium*, which last occurs at the Chadronian-Orellan transition. The reason for the extinction of cylindrodonts is not clear. It may be more than coincidence that just about at the time cylindrodonts disappear, the eumyine cricetids first appear in the North American record, as immigrants from Asia. Shortly after the beginning of Orellan time the eumyine cricetids become the dominant rodents and continue so through much of Orellan time. The eumyines are also generally of about the same size range as cylindrodonts. The disappearance of cylindrodonts might be due in part to competition with newly arrived immigrants for the same resources.

ACKNOWLEDGMENTS

This report is one of the results of many years of field work in central Wyoming by the first author, during which he was ably assisted by numerous volunteers, students, and, in later years, support staff at the National Museum of Natural History; to all those who collected cylindrodont rodents at Flagstaff Rim, our thanks for your efforts. The illustrations are by Jennifer Emry. For improvements in the manuscript as a result of their careful critical reviews, we thank Craig Black and John Storer.

LITERATURE CITED

Black, C. C. 1965. Fossil mammals from Montana. Part 2. Rodents from the early Oligocene Pipestone Springs Local Fauna. Annals of Carnegie Museum 38: 1-48.

Black, C. C. 1970a. A new *Pareumys* (Rodentia: Cylindrodontidae) from the Duchesne River Formation, Utah. Fieldiana, Geology 16: 453-459.

Black, C. C. 1970b. Paleontology and geology of the Badwater Creek area, central Wyoming. Part 5. The cylindrodont rodents. Annals of Carnegie Museum 41: 201-214.

Black, C. C. 1974. Paleontology and geology of the Badwater Creek area, central Wyoming. Part 9. Additions to the cylindrodont rodents from the late Eocene. Annals of Carnegie Museum 45: 151-160.

Bowdich, T. E. 1821. An Analysis of the Natural Classifications of Mammalia for the Use of Students and Travellers. J. Smith, Paris, 115 pp.

Burke, J. J. 1935. *Pseudocylindrodon*, a new rodent genus from the Pipestone Springs Oligocene of Montana. Annals of Carnegie Museum 25: 1-4.

Burke, J. J. 1936. *Ardynomys* and *Desmatolagus* in the North American Oligocene. Annals of Carnegie Museum 25: 135-154.

Burke, J. J. 1938. A new cylindrodont rodent from the Oligocene of Montana. Annals of Carnegie Museum 27: 255-274.

Dawson, M. R. 1968. Oligocene rodents (Mammalia) from East Mesa, Inner Mongolia. American Museum Novitates 2324: 1-12.

Douglass, E. 1902. Fossil Mammalia of the White River beds of Montana. Transactions of the American Philosophical Society 20: 237-279.

Emry, R. J. 1973. Stratigraphy and preliminary biostratigraphy of the Flagstaff Rim area, Natrona County, Wyoming. Smithsonian Contributions to Paleobiology 18: 1-43.

Ferrusquia-Villafranca, I., and A. E. Wood. 1969. New fossil rodents from the early Oligocene Rancho Gaitan local fauna, northeastern Chihuahua, Mexico. Pearce-Sellards Series 16: 1-13.

Flynn, L. J., L. L. Jacobs, and I. U. Cheema. 1986. Baluchimyinae, a new ctenodactyloid rodent subfamily from the Miocene of Baluchistan. American Museum Novitates 2841: 1-25.

Galbreath, E. C. 1969. Cylindrodont rodents from the lower Oligocene of northeastern Colorado. Transactions of the Illinois State Academy of Science 62: 94-97.

Hough, J. R., and R. Alf. 1956. A Chadronian mammalian fauna from Nebraska. Journal of Paleontology 30: 132-140.

Korth, W. W. 1984. Earliest Tertiary evolution and radiation of rodents in North America. Bulletin of Carnegie Museum of Natural History 24: 1-71.

Korth, W. W. 1992. Cylindrodonts (Cylindrodontidae, Rodentia) and a new genus of eomyid, *Paranamatomys* (Eomyidae, Rodentia) from the Chadronian of Sioux County, Nebraska. Transactions of the Nebraska Academy of Sciences 19: 75-82.

Lambe, L. M. 1908. The Vertebrata of the Oligocene of the Cypress Hills, Saskatchewan. Geological Survey of Canada, Contributions to Paleontology 3: 1-65.

Luckett, W. P., and J.-L. Hartenberger. 1985. Evolutionary relationships among rodents: comments and conclusions; pp. 685-712 *in* W. P. Luckett and J. L. Hartenberger (eds.), Evolutionary Relationships among Rodents: A Multidisciplinary Analysis. Plenum Press, New York.

Matthew, W. D. 1903. The fauna of the *Titanotherium* beds at Pipestone Springs, Montana. Bulletin of the American Museum of Natural History 19: 197-226.

Matthew, W. D., and W. Granger. 1923. New Bathyergidae from the Oligocene of Mongolia. American Museum Novitates 101: 1-5.

Matthew, W. D., and W. Granger. 1925. New creodonts and rodents from the Ardyn Obo Formation of Mongolia. American Museum Novitates 193: 1-7.

Miller, G. S., Jr., and J. W. Gidley. 1918. Synopsis of the supergeneric groups of rodents. Journal of the Washington Adacemy of Sciences 8: 431-448.

Ostrander, G. E. 1983. New early Oligocene (Chadronian) mammals from the Raben Ranch Local Fauna, northwest Nebraska. Journal of Paleontology 57: 128-139.

Ostrander, G. E. 1985. Correlation of the early Oligocene (Chadronian) in northwestern Nebraska; pp. 205-231 *in* J. E. Martin (ed.), Fossiliferous Cenozoic Deposits of Western South Dakota and Northwestern Nebraska. Dakoterra 2.

Russell, L. S. 1934. Revision of the Lower Oligocene vertebrate fauna□of the Cypress Hills, Saskatchewan. Transactions of the Royal Canadian Institute 20: 49-67.

Russell, L. S. 1954. Mammalian fauna of the Kishenehn Formation, southeastern British Columbia. Bulletin of the National Museum of Canada 132: 92-111.

Russell, L. S. 1972. Tertiary mammals of Saskatchewan. Part II: The Oligocene fauna, non-ungulate orders. Life Sciences Contributions of the Royal Ontario Museum 84: 1-63.

Shevyreva, N. S. 1972. New rodents from the Paleogene of Mongolia. Paleontologicheskii Zhurnal (Akademia Nauk SSSR). 1972 (3): 134-145.

Simpson, G. G. 1945. The principles of classification and a classification of mammals. Bulletin of the American Museum of Natural History 85: 1-350.

Storer, J. E. 1978. Rodents of the Calf Creek Local Fauna (Cypress Hills Formation, Oligocene, Chadronian) Saskatchewan. Natural History Contributions, Saskatchewan Museum of Natural History 1: 1-54.

Storer, J. E. 1984. Mammals of the Swift Current Creek Local Fauna (Eocene: Uintan), Saskatchewan. Natural History Contributions, Saskatchewan Museum of Natural History 7: 1-158.

Storer, J. E. 1988. The rodents of the Lac Pelletier lower fauna, late Eocene (Duchesnean) of Saskatchewan. Journal of Vertebrate Paleontology 8: 84-101.

Vinogradov, B. C., and P. P. Gambarian. 1952. Oligocene cylindrodonts of Mongolia and Kazakhstan (Cylindrodontidae, Glires, Mammalia). Trudy Paleontologicheskogo Instituta, Akademiya Nauk SSSR. Moscow, USSR 41: 13-42. [Russian]

Wahlert, J. H. 1968. Variability of rodent incisor enamel as viewed in thin section, and the microstructure of the enamel in fossil and Recent rodent groups. Breviora 309: 1-18.

Wahlert, J. H. 1974. The cranial foramina of protrogomorphous rodents; an anatomical and phylogenetic study. Bulletin of the Museum of Comparative Zoology 146: 363-410.

Wang, B. 1986. The systematic position of *Prosciurus lohiculus*. Vertebrata PalAsiatica 24(4): 284-294.

Wilson, R. W. 1949. Early Tertiary rodents of North America. Carnegie Institute of Washington Publication 584: 59-83.

Wilson, R. W. 1986. The Paleogene record of the rodents: fact and interpretation; pp. 163-176 *in* K. M. Flanagan and J. A. Lillegraven (eds.), Vertebrates, Phylogeny, and

Philosophy. Contributions to Geology, University of Wyoming Special Paper 3.

Wood, A. E. 1937. The mammalian fauna of the White River Oligocene. Part II. Rodentia. Transactions of the American Philosophical Society 28: 155-269.

Wood, A. E. 1955. A revised classificaion of the rodents. Journal of Mammalogy 36: 165-187.

Wood, A. E. 1958. Are there rodent suborders? Systematic Zoology 7: 169-173.

Wood, A. E. 1965. Grades and clades among rodents. Evolution 19: 115-130.

Wood, A. E. 1970. The early Oligocene rodent *Ardynomys* (Family Cylindrodontidae) from Mongolia and Montana. American Museum Novitates 2418: 1-18.

Wood, A. E. 1974. Early Tertiary vertebrate faunas, Vieja Group, Trans-Pecos Texas: Rodentia. Bulletin of the Texas Memorial Museum 21: 1-112.

Wood, A. E. 1980. The Oligocene rodents of North America. Transactions of the American Philosophical Society 70: 1-68.

Wood, A. E. 1981. The origin of the caviomorph rodents from a source in Middle America: a clue to the area of origin of the platyrhine primates; pp. 79-91 *in* R. L. Ciochon and A. B. Chiarelli (eds.), Evolutionary Biology of the New World Monkeys and Continental Drift. Plenum Publishing, New York.

Wood, A. E. 1984. Hystricognathy in the North American Oligocene rodent *Cylindrodon* and the origin of the Caviomorpha; pp. 151-160 *in* R. M. Mengel (ed.), Papers in Vertebrate Paleontology Honoring Robert Warren Wilson. Special Publication of the Carnegie Museum of Natural History 9.

20. Castoridae

XIAOFENG XU

ABSTRACT

North American Eocene-Oligocene beavers compise 18 species in 7 genera, including one new genus (*Nannasfiber*) and 5 new species (*Capacikala parvus, Capatanka minor, Euhapsis luskensis, Nannasfiber ostellatus*, and *Nannasfiber osmagnus*). They can be divided into three groups. The first includes only *Agnotocastor*; the second group consists of *Palaeocastor* and *Capacikala*; the third group includes *Capatanka, Euhapsis, Nannasfiber*, and *Fossorcastor*. Differences among the three groups are recognized from skull, mandible, and dental patterns. During their evolution in the Chadronian through the Arikareean, dental pattern becomes simpler, and skull shape becomes diversified.

The only beaver represented in the Chadronian and Orellan is *Agnotocastor*, which survived into the earliest Arikareean. *Palaeocastor* first appeared in the Whitneyan, but was most abundant in the early Arikareean. A radiation of beavers took place in the early Arikareean when 4 castorid genera (*Capacikala, Capatanka, Euhapsis*, and *Fossorcastor*) first appeared. The rapid evolution and radiation of the Arikareean beavers make them excellent biochronological correlation tools. *Palaeocastor* and *Capatanka* are significant references for the early early Arikareean; *Capacikala* comes from the early Arikareean; *Nannasfiber* implies an early late Arikareean age; *Euhapsis* and *Fossorcastor* are known from the late early through late late Arikareean.

INTRODUCTION

The North American Eocene-Oligocene beavers range from the Chadronian through the Arikareean. Previously described genera include *Agnotocastor, Palaeocastor, Capacikala, Capatanka, Euhapsis, Pseudopalaeocastor*, and *Fossorcastor* (Stirton, 1935; Wilson, 1949a, b; Emry, 1972; J. Macdonald, 1963, 1970; L. Macdonald, 1972; Martin, 1987). *Agnotocastor* is known from the Chadronian (Emry, 1972), Orellan (Wilson, 1949a), and Whitneyan (Stirton, 1935; Galbreath, 1953). *Palaeocastor* and *Capatanka* are found in the Sharps, Monroe Creek, and Harrison Formations of Arikareean age (Stirton, 1935; J. R. Macdonald, 1963, 1970; L. Macdonald, 1972; Martin, 1987). *Capacikala* is found in the Sharps, Harrison and John Day Formations (J. Macdonald, 1963, 1970; L. Macdonald, 1972; Martin, 1987). However, the generic definitions and relationships of North American Eocene-Oligocene beavers are not well understood. This survey is aimed at (1) redefinition of

each Eocene-Oligocene species based on the comparison of skull, mandible, and dental characters, (2) producing a phylogeny for all North American Eocene-Oligocene beavers and giving a perspective on their later evolutionary history, and (3) constructing a revised biochronology of beavers based on the new phyletic revision in the context of the North American land mammal "ages" (Prothero and Swisher, 1992; Tedford et al., 1987).

METHODS

Skull and mandible morphology and dental pattern analyses are used in this study. Morphometric analysis of beaver skulls was performed using Rohlf's Thin-Plate Spline program (TPS). TPS reveals local deformations in skull shape between pairs of skulls by determining bending energy (the larger the bending energy, the more strongly deformed an object is) and distorted square grids (the more distorted the grid is, the stronger the deformation in a local area) (Bookstein, 1991). TPS excludes uniform transformations caused by non-biological factors, for example, uniform deformation caused by compressive stress. TPS provides a useful tool for tracing skull shape change through time or through evolutionary lineages.

Skull and Mandible Terminology

All skull measurements are performed as in Figure 1, and tooth comparisons are based on maximum length and breadth of occlusal surface.

Terms for skull and mandible morphology and dental pattern are defined in or redefined from Stirton (1935).

Masseter Ridge—A vertical ridge on lateral sides of skull. It may be convex, "C" shaped, or straight; the ridge consists of two parts: the origin of masseter lateralis in the upper portion and masseter superficialis in the lower portion.

Position of Infraorbital Foramen—A foramen located at the medial side of masseter ridge. It can be low, at the lower end of masseter ridge; intermediate, at the middle point of the ridge; or high, about a third from the top end of the ridge.

Jugal-Lacrimal Connection—Jugal may contact lacrimal strongly, weakly, or not at all.

Posterior view of mandible—Positional relationship among coronoid, condyloid, and angular processes can

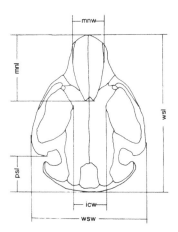

Figure 1. Skull measurements: wsl, whole skull length; wsw, whole skull width; mnl, maximum nasal length; mnw, maximum nasal width; psl, postzygomatic skull length; icw, interorbital constriction width.

be seen in posterior view of mandible. They may be aligned or alternate. The TPS orientation of mandible posterior view is determined by putting symphyseal and anterior end of angle on the same horizontal plane and letting incisor sit between coronoid and condyloid processes in posterior view (for alternate pattern), or letting incisor be aligned with coronoid and condyloid processes (for aligned pattern) if the incisor is there.

Dental Terminology

Incisors—Incisor frontal face (in cross section) convex; incisor frontal face flat.

Fossette or Fossettid—An isolated lake-like structure formed by enamel of upper and lower cheek teeth, respectively, at occlusal surface of tooth.

Flexus or Flexid—Similar to fossette or fossettid, but with a labial or lingual opening in occlusal view.

Stria or Striid—The structure of flexus or flexid in the crown from a labial or lingual view.

Body Size

All genera concerned here are small beavers with a skull length less than 85 mm. Their body size is not larger than 50 cm, exclusive of tail, estimated on the basis of relationship between skull length and body length in modern beavers (Xu, 1995; Jenkins and Busher, 1979).

ABBREVIATIONS

AMNH, American Museum of Natural History, New York; BHI, Black Hills Institute of Geology, Hill City, South Dakota; CIT, California Institute of Technology (specimens now in Los Angeles County Museum of Natural History); CM, Carnegie Museum, Vertebrate Paleontology, Pittsburgh; F:AM, Frick Collection, American Museum of Natural History, New York; KUVP, University of Kansas, Museum of Natural History, Lawrence, Kansas; LACM, Los Angeles County Museum of Natural History, Los Angeles; FMNH PM, Field Museum of Natural History,

Chicago; SDSM, South Dakota School of Mines and Technology, Museum of Natural History, Rapid City, South Dakota; UC & UM, Fossil collection now in Field Museum of Natural History, Chicago; UCMP, University of California, Museum of Paleontology, Berkeley, California; UCM, University of Colorado Museum, Denver, Colorado.

SYSTEMATIC PALEONTOLOGY
Family CASTORIDAE Gray, 1821
Agnotocastor Stirton, 1935

Emended Diagnosis—Narrowest skull in beavers; skull length/width ratio around 1.5; maximum nasal length/width ratio of greater than 2.5; postzygomatic skull portion 0.20-0.25 of skull length. Single or parallel double sagittal crests extending from the interorbital constriction. Masseter ridge straight and short; position of infraorbital foramen low; jugal-lacrimal connection strong. Bulla kidney-shaped in ventral view. Internal nares (Macdonald, 1963) terminated at level of anterior edge of M3. Digastric eminence strong (Emry, 1972). Posterior view of mandible aligned. Upper tooth rows parallel; convex incisors; P3 present occasionally, and only 0.17-0.25 of P4 length if present; occlusal outline of P4 nearly square; width of upper molars greater than length. Two or three fossettes (fossettids) anterior and posterior to elongated mesofossette (-fossettid) in cheek teeth, and sometimes only one or two fossettes (fossetids) in the first and second molars in late wear stages.

Distribution—Colorado, Wyoming, South Dakota; Chadronian to early early Arikareean.

Agnotocastor praetereadens Stirton, 1935

Steneofiber nebrascensis Matthew, 1902
Type—AMNH 1428, crushed skull, lacking zygomatic arches and incisors.
Referred Specimens—SDSM 6420, slightly crushed skull and mandibles; BHI 1218, skull missing only left M3; SDSM 40123, skull with P4s; SDSM 40167 and 40168, right mandibles missing only m3.
Type Locality and Horizon—Locality unknown, but in South Dakota. *Protoceras* Beds on Cheyenne River, White River Formation, Whitneyan.
Other Localities and Horizons—*Protoceras* channel sandstone and *Leptauchenia* nodules, near Rockyford, South Dakota, Whitneyan.
Emended Diagnosis—Upper cheek teeth elongate transversely, with length/width ratio equal to or less than 0.70 in m1-3, but P4 nearly square in occlusal outline. Length of P3 as small as 0.17 of P4 length. Width gradually smaller from P4 to M3. Hypostria longer than mesostria. Hypoflexus equal to, or longer than mesoflexus or mesofossette. Isolated tiny circular parafossette and metafossette in M1-2; P4 and M3 complex: two fossettes anterior to mesofossette and one or two fossettes posterior to mesofossette. Outline of lower cheek teeth nearly square, and lower dental pattern simpler than upper. Hypoflexid longest in lower cheek teeth. Parafossettid, mesofossettid, and metafossettid of p4 equal in transverse length. Parafossettid and

metafossettid from m1 to m3 gradually smaller transversely, but metafossettid longer than parafossettid.

Agnotocastor coloradensis Wilson, 1949

Type—UCM 19809, fragmentary left ramus missing only m3.
Referred Specimens—SDSM 6441, fragmentary skull missing only left incisor; SDSM 6526 and UCMP 1631, fragmentary skulls.
Type Locality and Horizon—Near boundary of Sec. 16 and Sec. 21, T. 11 N., R. 53 W., Logan County, Colorado; Cedar Creek, Brule Formation, Orellan.
Other Localities and Horizons—SDSM locality No. v.6221, Sharps Formation, South Dakota; General v.827, John Day Formation, Grant County, Oregon. Orellan to early Arikareean.
Emended Diagnosis—Size 15% larger than *A. praetereadens, A. galushai,* and *A. aubekerovi.* Upper cheek teeth elongated transversely, with length/width ratio 0.75-0.82 in m1-2; length/width ratio of m3 close to 1. Width of P4-M3 smaller posteriorly. Mesostria almost as developed as hypostria. Two fossettes anterior to mesoflexus and two or three fossettes posterior to mesoflexus. Length of lower cheek teeth slightly greater than their width. Lower dental pattern more simple than upper. Length of m3 greater than width.

Agnotocastor galushai Emry, 1972

Type—F:AM 79310, left and right mandibles with p4-m3 and broken incisors, lacking the angular and coronoid processes.
Type Locality and Horizon—South Fork of Lone Tree Gulch, Sec. 27, T31N, R83W, Natrona County, Wyoming; Chadronian.
Emended Diagnosis—Based on lower teeth. Cheek teeth 15% smaller than *A. coloradensis* and slightly smaller than in *A. praetereadens.* Length/width ratio of m1-3 greater than 0.80. Molar structure simpler (4-5 of fossettids and flexids) than in *A. coloradensis* (6-7 of fossettids and flexids), but more complicated than in *A. praetereadens* (3-4 of fossettes and flexi). Occlusal structure of p4 complicated, having more than ten circular fossettids (Emry, 1972, fig. 2A, 2C).
Discussion—The phyletic position of *Agnotocastor* is controversial. The incisors are similar to *Eucastor* (Wilson, 1949a; Lytshev, 1978; Martin, 1987); based on cheek teeth, it is similar to the group including *Castor* (Stirton, 1935; Crusafont-Pairo et al., 1948). *Agnotocastor* has primitive skull characters similar to *Eutypomys,* for instance, narrow skull and nasal, complicated enamel structure. On the other hand, *Agnotocastor* shares derived characters with all beavers, for example, five flexi (flexids) dental pattern, relatively smaller interparietal than in *Eutypomys,* digastric eminence, and the angle extending up posteriorly.
Agnotocastor does not have any autapomorphic characters. Furthermore, it shares with all beavers some important derived characters mentioned above, but it does not share any derived characters uniquely found in any other beaver genus. *Agnotocastor* shares convex incisors with *Steneofiber,* castoroidine beavers, and other sciuromorphic rodents (for example, *Marmota*). This character is primitive because it is universal in rodents except some beavers.

Agnotocastor is the most primitive known beaver. Of the four species reported in the genus, three are from North America and one from Kazakhstan, Asia (Stirton, 1935; Wilson, 1949a; Emry, 1972; Lytshev, 1978). *A. aubekerovi* from Kazakhstan is more similar to *A. galushai* than to other species (Lytshev, 1978) in p4 and molar structures and size. Also, *A. aubekerovi* from the lower Oligocene is probably slightly younger than *A. galushai* of the Chadronian of North America (Emry, 1972; Lytshev, 1978) because elongate fossettids are relatively derived character in *Agnotocastor,* and the Chadronian is considered to be late Eocene, beginning at 37 Ma (Prothero and Swisher, 1992). *A. coloradensis* is known from the Orellan to early Arikareean, and *A. praetereadens* is known from the Whitneyan. *A. galushai* and *A. aubekerovi* have the most complicated p4; *A. coloradensis* has an intermediate complication of p4. The p4 of *A. praetereadens* is unknown, but its P4 is simpler than *A. coloradensis.* Premolar complexity apparently decreases through time.

Palaeocastor Leidy, 1869

Emended Diagnosis—Skull shorter and broader than in *Agnotocastor.* Skull length-breadth ratio: 1.2-1.3. Single or double well-developed sagittal crests starting at interorbital constriction. Postzygomatic skull length is 0.20-0.25 of skull length. Masseter ridge straight line but longer than in *Agnotocastor;* jugal-lacrimal connection strong; position of infraorbital foramen low. Bullae slightly kidney-shaped or rounded but not extending ventrally. Internal nares terminated at the line connecting the anterior edge of the M3. Three processes of mandible aligned in posterior view. Incisor face flat. P4 complicated in early wear stage, but simpler in late wear stage. Two elongated fossettes and one circular fossette on upper molars in late wear stage.
Distribution—South Dakota, North Dakota, Oregon; Whitneyan to earliest Arikareean.

Palaeocastor nebrascensis (Leidy, 1856)

Steneofiber nebrascensis Leidy, 1856
?*Capacikala sciuroides* Macdonald, 1963
Type—Fragmentary cranium and mandibles (Leidy, 1869, Plate 26, figures 7-8) not numbered, at Academy of Natural Sciences of Philadelphia.
Referred Specimens—AMNH 12901 and 55545, fragmentary skulls; F:AM 64097, F:AM 65023, and SDSM 53346, fragmentary skulls; UCMP 114635, skull and fragmentary mandibles; UCMP 114795, fragmentary cranium; LACM 29985, skull without right zygomatic arch.
Type Locality and Horizon—"from the Mauvaises Terres of White River" (Leidy, 1869);

probably Sharps Formation, earliest Arikareean.

Other Localities and Horizons—AMNH Loc. 5 of Hough, Stark County, North Dakota; Basal *Leptauchenia* beds, 15-20 feet below the First White Layer, Interior, South Dakota; Lusk 276-2252, Horse Creek; John Day, Loc. 3109/Picture Gorge Quadrangle, Wheeler, Oregon; SDSM Loc. v. 5351 and v. 544, South Dakota; Whitneyan to earliest Arikareean.

Emended Diagnosis—Skull length 53–58 mm. Well-developed parallel sagittal crests extending from relatively broad interorbital constriction. Bullae slightly kidney-shaped in ventral view, but not so obviously as in *Agnotocastor*. Seven or more fossettes on upper fourth premolar in early wear stage, 4 or 5 in late wear stage. Two elongated fossettes and one or two circular fossettes on upper molars. Cheek teeth gradually smaller from P4 to M3, but no obvious difference between P4 and M1. Hypoflexid developed and hypostriid close to crown base. Para-, meso-, and metafossettids well developed and elongate. Labial end of hypoflexid opposite to metafossettid.

Palaeocastor peninsulatus (Cope, 1881)

Steneofiber peninsulatus, Cope, 1881

Type—AMNH 8669, skull with complete upper teeth.

Referred Specimens—BHI 961, skull with complete upper cheek teeth; CIT 890, skull without left zygomatic arch; F:AM 64026, skull; SDSM 53347, skull; LACM 21658, fragmentary skull with complete cheek teeth; LACM 9319, fragmentary skull; LACM 9431, skull without zygomatic arches; LACM 16005 and 33215, skulls without left zygomatic arch; LACM 30891, skull missing only right M3; SDSM 40123, skull with P4s; SDSM 53344, skull fragment; SDSM 55108, skull fragment with complete cheek teeth.

Type Locality and Age—*Diceratherium* Beds, John Day Basin, Oregon; early (probably earliest) Arikareean.

Other Localities and Horizons—LACM Loc. 1959, 1975, and 1982, Wounded Knee Area, Shannon County, South Dakota; SDSM Loc. v.5341, v.5352 (RB#14), v.5353, v.5361, Sharps Formation, South Dakota; above the First White Layer, Washabaugh County, South Dakota; the John Day Formation, CIT22, Wheeler and Grant Counties, Oregon; earliest Arikareean.

Emended Diagnosis—Relatively large *Palaeocastor*; skull length 50-70 mm. Nasals relatively narrower than in *P. nebrascensis*. Incisor face about 0.2 mm broader than *P. nebrascensis*. Upper cheek teeth one fossette less than in *P. nebrascensis*. Width of occlusal face of molars slightly greater than length. Elongate hypofossette and mesofossette oppositely convex.

Discussion—*Palaeocastor* has been a "waste basket" taxon for polyphyletic groups of species. The revised genus comprises only two species. *Palaeocastor* is characterized by the following features: (1) Dental pattern simpler than in *Agnotocastor*, but more complicated than in other beavers; (2) straight masseter ridge (longer than in *Agnotocastor*); (3) low position of infraorbital foramen; (4) aligned three mandibular

processes in posterior view; (5) nasals and skull relatively shorter or broader than in *Agnotocastor*; (6) sagittal crest pattern similar to that of *Agnotocastor*. Characters 1, 2, and 5 are derived compared with those of *Agnotocastor*, and can separate *Palaeocastor* from *Agnotocastor*.

P. nebrascensis first appeared in the Whitneyan, and another species first appeared in the earliest Arikareean. This is the only beaver genus that shared a Whitneyan occurrence with *Agnotocastor*. No *Palaeocastor* has yet been found in deposits younger than the early Arikareean.

Capacikala Macdonald, 1963

Emended Diagnosis—Smallest castorid genus. Skull length 43-55 mm. Masseter ridge straight and longer than in *Palaeocastor*; position of infraorbital foramen low; jugal-lacrimal connection strong. Sagittal crest lyrate. Maximum width of nasals nearly equal to maximum width of snout. Postzygomatic skull length is 0.2-0.25 of skull length. Internal nares terminating at the line connecting M2's. Three processes of mandible aligned in posterior view. Incisor face flat. Mesoflexus (-flexid) and hypoflexus (-flexid) strongest in dental structures; other dental structures variable.

Distribution—Oregon, South Dakota, and Wyoming; early Arikareean.

Capacikala gradatus (Cope), 1879

Steneofiber gradatus Cope, 1879
Palaeocastor gradatus Stirton, 1935

Type—AMNH 7008, a cranium without left M3.

Referred Specimens—F:AM 64115-A, skull without right M1-2 and right jugal; F:AM 64152, complete skull and mandibles; F:AM 64114, juvenile skull, P4 just erupted; F:AM 64156-A, skull with broken nasals and incisors. F:AM 64167, fragmentary skull; F:AM 64220, fragmentary skull with complete dentition.

Type Locality and Horizon—*Diceratherium* beds, John Day Formation, John Day region, Oregon; early Arikareean.

Other Localities and Horizons—Lower and middle Little Muddy Creek, Niobrara County, Wyoming; early Arikareean.

Emended Diagnosis—Largest species of *Capacikala*, skull length 50-55 mm. Maximum width of lyrate sagittal crest around 0.5 snout width. Snout broader than interorbital constriction. Nasal length/width ratio around 1.5. Angle between anteroventral edge of zygomatic arch and the skull roof approximately 45 degrees in lateral view. One circular and two elongate fossettes on upper cheek teeth in late wear stage; complex fossettes on upper cheek teeth in early wear stage.

Capacikala parvus sp. nov
Figures 2A-2C

Type—F:AM 64552, a complete skull and mandibles.

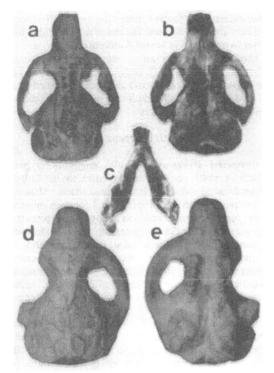

Figure 2. *Capacikala.* A-C, *Capacikala parvus* sp. nov., holotype (F:AM 64552), 4/3 natural size. A, dorsal view; B, ventral view; C, mandible occlusal view. D-E, *Capatanka minor* sp. nov., holotype (SDSM 5489). D, dorsal view; E, ventral view.

Referred Specimens—AMNH 10281, skull fragment without right P4; UCMP 114819, palate without left M3.

Type Locality and Horizon—High level of middle brown sand, Little Muddy Creek, Niobrara, Wyoming; early Arikareean.

Other Locality and Horizon—Sharps loc. V76047, Sharps Formation, Shannon County, South Dakota; early Arikareean.

Etymology—Latin *parvus* for small, referring to this species having the smallest body size of all beavers.

Diagnosis—One of the smallest known species of beaver; skull length of type specimen = 43 mm; maximum lyrate sagittal crest width nearly equal to maximum snout width and interorbital constriction width; ratio (maximum nasal length/width) around 2. Angle between anteroventral edge of zygomatic arch and the skull roof in lateral view about 60°. Enamel pattern of adult simple, consisting of three elongate fossettes (fossettids) on premolars and two elongate fossettes (fossettids) on molars.

Discussion—*Capacikala* was proposed by Macdonald (1963). He used the following characters to define the genus, (1) the smallest beaver, (2) strongly developed lyrate sagittal crests, (3) cheek teeth gradually smaller in transverse diameter from P4 to M3, (4) internal nares at the line connecting M2s, and (5) bullae large, rounded in ventral view, and expanded ventrally.

This is a well-defined genus. The emendations made here are elimination of character 5 and adding nasal and mandible characters. Characters 2 and 4 are derived characters (or autapomorphic characters of the genus).

?*C. sciuroides* (AMNH 12901) proposed by Macdonald (1963) belongs to *Palaeocastor nebrascensis.* Macdonald also referred *Paleocastor milleri* (UC1582) to *Capacikala*; however, this specimen is better referred to a new genus because it has characters that exclude it from *Capacikala* and *Palaeocastor.* (1) maximum width of nasals equal to 0.5 snout width, (2) masseter ridge C-shaped, position of infraorbital foramen high, and jugal-lacrimal connection weak, (3) internal nares at the line connecting the M3s, and (4) three processes of mandible alternate in posterior view. Therefore, there are only two species in this genus, *C. gradatus* and *C. parvus.* The new species is distinguished from *C. gradatus* by its smaller size, broader lyrate sagittal crest, relatively simple dental pattern, steeper anterior zygomatic edge, relatively wide interorbital constriction, and long nasals.

The type of *C. gradatus* is from the *Diceratherium* beds of the John Day Formation in Oregon. Other specimens are from the Wounded Knee-Sharps fauna in South Dakota (Macdonald, 1963) and lower and middle Little Muddy Creek, Wyoming. These localities are probably equivalent to the earliest Arikareean, approximately 25-29 Ma (Tedford et al., 1987; Prothero and Swisher, 1992). The type of *C. parvus* is from a high level of Little Muddy Creek, Wyoming. It is probably late early Arikareean, approximately 23-25 Ma (Tedford et al., 1987).

Capatanka Macdonald, 1963

Emended Diagnosis—Relatively broad and short massive skull, with supraorbital ridges that join to form sagittal crest at least midway between interorbital constriction and lambdoidal crest. Snout relatively broad. Postzygomatic skull length around 0.2 of skull length. Masseter ridge convex; position of infraorbital foramen intermediate; jugal-lacrimal connection strong. Well-developed lambdoidal crest and glenoid fossa; large bullae elongated anteroposteriorly. Internal nares terminated at line connecting M3s. Three processes of mandible alternate in posterior view. First and second molars with maximum transverse diameters nearly as great as premolars; M3 slightly smaller. Incisor face flattened or slightly convex.

Distribution—Sharps Formation, South Dakota; earliest Arikareean.

Capatanka cankpeopi Macdonald, 1963

Euhapsis gaulodon Matthew, 1907

Type—SDSM 53421, a skull with complete upper teeth.

Referred Specimens—AMNH 12897, partial skull without zygomatic arches; F:AM 64251, a skull without right zygomatic arch; F:AM 64030, skull and mandibles; LACM 9429, fragmentary skull without left incisor and M3; LACM 9450, fragmentary skull with M2-3; LACM 17436, skull without right zygomatic arch; LACM 17692, fragmentary skull missing zygomatic arches; LACM 22443 and SDSM 5672, skulls.

Type Locality and Horizon—SDSM Loc. v.5354, Sharps Formation, Shannon County, South Dakota; early early Arikareean.

Other Localities and Horizons—Lower Rosebud, Porcupine Creek, Pine Ridge; 8 miles north on east side of Porcupine Creek; LACM Loc. 1819/ St. John's; LACM Loc. 1955 and 6457, Wounded Knee Area; LACM Loc. 1820, Horncloud; LACM Loc. 1994, Wolff Ranch Badlands; SDSM Loc. V5672. All localities in the Sharps Formation, Shannon County, South Dakota; earliest Arikareean.

Emended Diagnosis—Larger *Capatanka* species; skull length 64-75.4 mm. Masseter ridge approaching premaxilla-maxilla suture. Up to 8 fossettes on P4 in early wear stage, similar to that of *Agnotocastor*; two elongate fossettes or fossettids on molars and three elongate fossettes or fossettid on premolars in late wear stage. Incisor face slightly convex.

Capatanka minor sp. nov.
Figures 2D-2E

Type—F:AM 64221, skull without right zygomatic arch.

Referred Specimens—SDSM 5489, skull without left zygomatic arch; LACM 17435, fragmentary skull without zygomatic arches and M3s; SDSM 62427, fragmentary skull without right incisor and M3s.

Type Locality and Horizon—SDSM Loc. v.5359, Lower Rosebud, Sharps Formation, South Dakota; earliest Arikareean.

Other Localities and Horizons—Lusk 378-3541, Bear Mountain; LACM Loc. 1820, Horncloud and SDSM Loc. V6211, Sharps Formation, South Dakota; earliest Arikareean.

Etymology—Latin *minor* for smaller, referring to the smallest size in the genus.

Diagnosis—Smallest *Capatanka* species, skull length 50-55 mm. Post-constriction part of supraorbital ridges laterally convex, not straight as in *C. cankpeopi*. Premaxilla-maxilla suture more anterior to masseter ridge than in *C. cankpeopi*. Incisor face slightly more convex than the larger species.

Discussion—*Capatanka* was proposed by Macdonald in 1963, and is a well-defined genus. This revision adds skull and mandible characters. Two species, *C. cankpeopi* and *C. brachyceps* (Matthew), were recognized by Macdonald (1963); however, the type of *C. brachyceps* (AMNH 12902) has been assigned to *Fossorcastor* (Martin, 1987), and another referred specimen (AMNH 12897) of *C. brachyceps* belongs to *C. cankpeopi*. *C. brachyceps* is not a valid species in this genus because the only two specimens referred to it pertain to other species.

Two skull shapes are recognized in *C. cankpeopi*: one (LACM 9429 and F:AM 64030) is shallower; another (other specimens) is deeper. Shallow skulls are somewhat dorsoventrally crushed, but their original shape should be shallower than the deeper skulls. No other difference is found between the two skull shapes, and they are from the same age and area. Therefore, it is better to treat them as sexual dimorphism.

The new small species, *C. minor*, is found in the Sharps Formation with *C. cankpeopi*. This species has the *Capatanka* skull characters and is distinguished from *C. cankpeopi* by smaller skull size and laterally convex posterior supraorbital ridges.

Euhapsis Peterson, 1905

Emended Diagnosis—Skull length 48-65 mm. Width of skull equal or nearly equal to its length. Parietals broad and short. Interparietal absent. Masseter ridge convex; position of infraorbital foramen intermediate; jugal-lacrimal connection strong. Three processes in mandible alternate in posterior view. Occipital region inclined. Masseter ridge slightly posterior to premaxilla-maxilla suture. Masseter ridge and maxilla-premaxilla suture located at the anterior base of maxillary zygomatic process. Zygomatic arches extended laterally making skull width equal to length, and jugal depth 0.50-0.67 of skull depth. Postzygomatic skull length 0.17 of skull length. Mastoid processes directed laterally, instead of ventrally. Auditory tubes reach edge of occiput. Incisor face flat. Dental pattern simple: two elongated fossettes or fossettids on molars; three elongated fossettes or fossettids, and another circular fossette or fossettid occasionally on premolars.

Distribution—Nebraska, Wyoming; Monroe Creek and Harrison Formations, late early to late Arikareean.

Euhapsis platyceps Peterson, 1905

Type—CM 1220, skull missing cheek teeth; fragmentary left mandible without m3.

Type Locality and Horizon—Upper Monroe Creek beds, near the head of Warbonnet Creek, Sioux, Nebraska; immediately underlies the Harrison beds, late early Arikareean.

Emended Diagnosis—Skull length about 58 mm. Skull length equal to its breadth; zygomatic arches strongly expanded laterally. Nasals broad anteriorly and suddenly narrowed at posterior part. Maxilla snout width equal to minimum interorbital constriction width. Parietals relatively broad and short. Occiput straight in the lower half and inclined anteriorly in upper. Occipital slope around 45° from vertical direction. Jugal depth equal to at least 0.67 of skull depth. Masseter ridge slightly posterior to premaxilla-maxilla suture, and both about 5 mm anterior to anterior base of maxillary jugal process. Incisor face flat. The m1-2 anteroposteriorly smaller than p4, but tranversely equal.

Euhapsis ellicottae Martin, 1987

Type—KUVP 48015, skull missing zygomatic arches.

Referred Specimens—KUVP 48016, partial skull; KUVP 48017, fragmentary skull; UM 1617, snout with incisors.

Type Locality and Horizon—KUVP Coll. Loc.

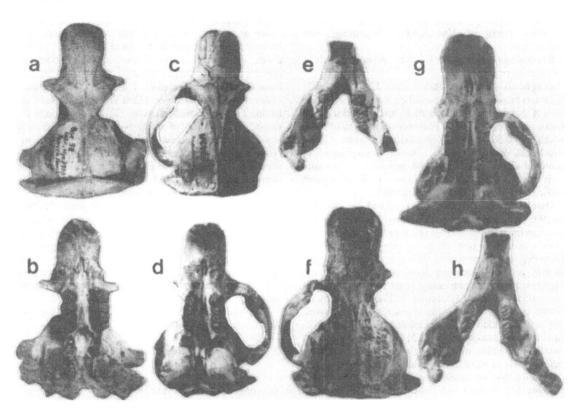

Figure 3. *Euhapsis* and *Nannasfiber* (natural size). A-B, *Euhapsis luskensis* sp. nov., holotype (F:AM 64595). A, dorsal view; B, ventral view. C-E, *Nannasfiber ostellatus* gen. et sp. nov., holotype and type species (UC 1461). C, dorsal view; D, ventral view; E, mandible occlusal view. F-H, *Nannasfiber osmagnus* sp. nov., holotype (F:AM 64606). F, dorsal view; g, ventral view; H, mandible occlusal view.

Wy-115, Niobrara County, Wyoming; Harrison Formation, late Arikareean.

Other Locality and Horizon—Near Rawhide Buttes, Goshen County, Wyoming; lower Harrison Formation; late Arikareean.

Emended Diagnosis—Skull breadth equal to length. Nasals heart-shaped (broader anteriorly) and roughened. Snout broader than interorbital constriction. Parietals narrow and squamosal broad. Orbit small. Skull deeper than in *E. platyceps*; occiput slope approximately 45°–60° from vertical. Masseter ridge close to premaxilla-maxilla suture; masseter ridge and maxilla-premaxilla suture at anterior base of maxillary jugal process. Bullae more spherical in ventral view than in *E. platyceps*. P4 much larger than upper molars. Two elongate fossettes on molars and four elongate fossettes on premolars.

Euhapsis breugerorum Martin, 1987

Type—KUVP 28376, skull with left complete mandible, missing zygomatic arches.

Referred Specimen—KUVP 28373, fragmentary skull missing snout, left zygomatic arch, and right cheek teeth.

Type Locality and Horizon—KUVP Coll. Loc. Neb.-31 Lower Harrison; early late Arikareean. KUVP

28373 is from the same locality and horizon as the type specimen.

Emended Diagnosis—Skull slightly smaller than in *Euhapsis ellicottae*. Skull length approximately 53 mm (*Euhapsis ellicottae* around 60 mm). Parietals broader than in *Euhapsis ellicottae*. Snout slightly broader than interorbital constriction. Distinct notch between external auditory meatus and posterior zygomatic root, and auditory tubes reaching edge of occiput (Martin, 1987). Occiput vertical in lower half and inclined anteriorly in upper. Bullae relatively smaller than in *E. platyceps* and *E. ellicottae*. Mandible stout. Relative distance between angular-condyloid processes smaller than in other beavers. Two elongate fossettes or fossettids on molars and three elongated fossettes or fossettids on premolars.

Euhapsis luskensis sp. nov.
Figures 3A-3B

Type—F:AM 64595, skull with complete upper teeth but missing zygomatic arches.

Referred Specimen—F:AM 64599, skull with complete cheek teeth but missing left zygomatic arch and snout.

Type Locality and Horizon—Patrick Ranch, near Lusk, Wyoming; late Arikareean.

Other Locality and Horizon—Lusk 280-2292, 3 miles North of Van Tassel, Wyoming; late Arikareean.

Etymology—Named for Lusk, Wyoming, where this species occurs.

Diagnosis—Skull length less than 55 mm. Nasals large and broader than interorbital constriction. Anterior half of nasals broad and parallel, and posterior nasal width narrowed suddenly. Posterior (or minimum) nasal width only 0.5 of anterior (or maximum) width. Skull width less than its length. Bullae hemispherical in posterior view. Maxillary-jugal process of zygomatic arch broader dorsoventrally. Lower 0.67 of occiput vertical and upper 0.33 inclined anteriorly.

Discussion—*Euhapsis* was proposed by Peterson in 1905. The type species is *Euhapsis platyceps*. Only one specimen is referred to the type species. Martin (1987) described two other species in the genus. Including the present species, four species are recognized in this genus. Characters defining *Euhapsis* are: (1) broad skull (skull width nearly equal to its length), (2) interparietal absent, (3) occiput inclined anteriorly in dorsal part or totally, (4) masseter ridge convex, position of infraorbital foramen intermediate, and jugal-lacrimal connection strong, (5) masseter ridge and maxilla-premaxilla suture slightly anterior to or at the anterior base of maxillary zygomatic process.

Euhapsis species are easily distinguished. *E. platyceps* has the broadest and flattest skull, the broadest parietals, and the deepest zygomatic arches. *E. ellicottae* is characterized by (1) rough and heart-shaped nasals, (2) masseter ridge and maxilla-premaxilla suture located at anterior base of maxillary-jugal process. *E. breugerorum* is characterized by heart-shaped nasals, smaller bullae, and relative distance between angular-condyloid processes smaller than in other beavers. *E. luskensis* has the narrowest skull in this genus, and only upper one third of occiput is inclined anteriorly.

This genus is restricted to the Arikareean of Nebraska and Wyoming. *E. platyceps*, *E. breugerorum*, and *E. ellicottae* were found in the lower Harrison Formation (early late Arikareean). *E. luskensis* is from the late late Arikareean.

Nannasfiber gen. nov.

Type Species—*Nannasfiber ostellatus* sp. nov.

Generic Diagnosis—Skull length 43-62 mm. Nasals long and narrow, making up only half of snout breadth. Supraorbital ridges joining and forming single sagittal crest somewhere between lambdoidal crest and interorbital constriction. Interparietal tiny. Postzygomatic skull length approximately 0.13-0.14 of skull length. Masseter ridge C-shaped; jugal-lacrimal connection weak; position of infraorbital foramen high. Three mandible processes alternate in posterior view. Anterior edge of ascending ramus vertically raised below p4.

Distribution—Wyoming and Nebraska; Harrison Formation, early late Arikareean.

Etymology—Latin, *nann-* for small; *nas-* for nose; *fiber* for beaver. *Nannasfiber* means "small-nosed beaver."

Discussion—Martin (1987) named *Pseudopalaeocastor* as a new genus in a new tribe Capacikalini. However, the diagnosis proposed by him cannot distinguish his genus from *Capatanka*. I consider the specimen to be closely related to *Capatanka*, but far from *Capacikala*. Therefore, I name it a new genus, *Nannasfiber*. *Nannasfiber* differs from *Capacikala* by having (1) much smaller nasals, (2) masseter ridge C-shaped, (3) position of infraorbital foramen high, (4) jugal-lacrimal connection weak, (5) sagittal crests non-lyrate, (6) skull broader, and (7) processes of mandible alternate. It is similar to *Euhapsis* in skull shape but three characters (1, 2, and small interparietal) can be used to distinguish *Nannasfiber* from *Euhapsis*.

Four characters can be used to distinguish the two species in this genus: (1) as their names imply, relative snout breadths are different; (2) the large elongate mesofossettid on p4 of *N. osmagnus* is almost perpendicular to the small elongate parafossettid; (3) elongate hypofossette on P4 of *N. osmagnus* is almost parallel to the sagittal plane; (4) snout is relatively longer in *N. osmagnus*.

This genus has been found only in the early late Arikareean deposits of Nebraska and Wyoming.

Nannasfiber ostellatus sp. nov.
Figures 3C-3E

Steneofiber barbouri Peterson, 1905
Palaeocastor barbouri Stirton, 1935
Palaeocastor milleri Olson, 1940
Pseudopalaeocastor barbouri Martin, 1987

Type—UC 1461.

Referred Specimens—FMNH PM3916, posterior skull fragment; F:AM 64594, skull without occiput and left zygomatic arch; F:AM 64609, skull without left zygomatic arch; CM 1210, right mandible and a skull without zygomatic arches; UC 1582, skull without left zygomatic arch and occiput.

Type Locality and Horizon—33 Ranch, Niobrara River, Sioux County, Nebraska; lower Harrison Formation, early late Arikareean.

Other Localities and Horizons—Lusk, Van Tassel, Niobrara River, Wyoming; early late Arikareean.

Diagnosis—One of the smallest castorids. Skull length 43-53 mm. Breadth of snout less than or equal to maximum breadth of cheek tooth rows. Snout/skull length ratio 0.30 and snout shorter than in *Nannasfiber osmagnus*. Parafossettid and metafossettid equally weak; hypoflexid best developed; mesofossettid intermediate. Transversely elongate parafossettid on p4.

Etymology—Latin *ostellatus* means "small mouth," because this species has a premaxillary breadth less than or equal to the breadth of cheek tooth rows.

Nannasfiber osmagnus sp. nov.
Figures 3F-3H

Type—F:AM 64606, skull missing right zygomatic arch; left and right mandibles.

Referred Specimens—UC 1575, skull without zygomatic arches, left mandible without m3.

Type Locality and Horizon—Lusk, near Van Tassel, Wyoming; early late Arikareean.

Other Locality and Horizon—Near Rawhide Buttes, Goshen County, Wyoming; lower Harrison Formation, early late Arikareean.

Diagnosis—Largest *Nannasfiber*. Skull length 57-62 mm. Breadth of snout greater than cheek tooth rows. Snout/skull length ratio 0.36 and snout relatively longer than in *Nannasfiber ostellatus*. Elongate hypofossette almost parallel to the sagittal plane; large elongate mesofossettid on p4 nearly perpendicular to small elongated parafossettid.

Etymology—Latin *osmagnus* means "large mouth," referring to this species having a premaxillary width greater than the breadth of cheek tooth rows.

Fossorcastor Martin, 1987

Emended Diagnosis—Masseter ridge C-shaped; position of infraorbital foramen intermediate; jugal-lacrimal connection extremely weak. Skull relatively narrower than in *Nannasfiber* and *Euhapsis*, but wider than in *Capacikala* and *Palaeocastor*, similar to *Capatanka*. Nasal maximum breadth less than one half maximum length, about two thirds of snout breadth, and maximum nasal breadth gradually smaller posteriorly. Interparietal small, and parietals narrow. Length of postzygomatic skull portion approximately 0.167 of skull length. Internal nares terminated posterior to line connecting M3s. Posterior view of mandible processes alternate. Ascending ramus extends near vertically from below p4.

Distribution—Nebraska, South Dakota, Wyoming; Harrison Formation, late early to late Arikareean.

Fossorcastor greeni Martin, 1987

Type—KUVP 80845, skull and mandibles.

Type Locality and Horizon—KUVP Coll. Loc. Wyo.-91, lower part of Harrison Formation; early late Arikareean.

Emended Diagnosis—Small *Fossorcastor* species. Skull length about 50 mm. Sagittal and lambdoidal crests undeveloped. Occiput slightly inclined. Maxillary zygomatic process relatively narrow in lateral view, compared to *Fossorcastor fossor*. Maxilla-premaxilla suture closer to the base of maxillary zygomatic process, compared to *F. fossor* and *F. brachyceps*. P4 length greater than upper molars.

Fossorcastor brachyceps Martin, 1987

Type—AMNH 12902; skull, mandible, and fragments of the postcranial skeleton.

Type Locality and Horizon—Porcupine Creek, Shannon County, South Dakota; Monroe Creek Formation (Lower Rosebud), late early Arikareean.

Emended Diagnosis—Skull broad and *Euhapsis*-like, and slightly larger than in *Fossorcastor greeni*. Skull length around 54 mm and breadth about 52 mm. Nasals relatively wider than in *Fossorcastor fossor*. Skull deeper than in *Fossorcastor fossor*. Maxillary

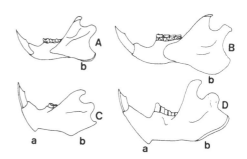

Figure 4. Mandible comparisons among beavers, *Marmota*, and *Eutypomys*. A, *Eutypomys* (1/2); B, *Marmota* (1/2); C, *Fossorcastor* (1/2); D, *Castor* (1/4). a, digastric eminence; b, angle.

zygomatic process relatively narrow in lateral view, compared to *Fossorcastor fossor*. Occiput more vertical than in *Fossorcastor greeni*. P4 length greater than upper molars, and tooth size gradually reduced from p4 to m3.

Fossorcastor fossor (Peterson), 1905

Steneofiber fossor Peterson, 1905.
Steneofiber simplicidens Matthew, 1907.
Palaeocastor major Romer and McCormack, 1928.
Palaeocastor simplicidens Stirton, 1935.
Palaeocastor fossor Stirton, 1935.
Palaeocastor (Capatanka) magnus Martin, 1987

Holotype—CM 1217, complete skull and mandibles.

Referred Specimens—AMNH 12900, F:AM 64217-A, UC 1581, and UM 1618, skulls without zygomatic arches; AMNH 83388, fragmentary skull without zygomatic arches; F:AM 65022 and KUVP 28372, skulls without right zygomatic arch; KUVP 28380 and KUVP 28386, skulls without left zygomatic arch; KUVP 28383 and KUVP 28388, skulls and mandibles; KUVP 28384, fragmentary anterior skull; KUVP 28389, fragmentary skull; UC 1408, skull and fragmentary mandibles; UC 1580, fragmentary posterior skull.

Type Locality and Horizon—Near Niobrara River, Wyoming; "*Daemonelix*" Beds in lower Harrison Formation, late Arikareean.

Other Localities and Horizons—Lusk 442-4053; Lusk 252; Porcupine Creek, Shannon County, South Dakota; Ellicott Ranch, KUVP Coll. Loc. Neb.-26; Breuger Ranch, KUVP Coll. Loc. Wy.-91; Hanson Ranch, Rawhide Buttes, Goshen County, Wyoming. Lower Rosebud (equivalent to Monroe Creek), Harrison Formation, late Arikareean.

Emended Diagnosis—Relatively larger *Fossorcastor*. Skull length 62-80 mm. Nasals oval in shape. Sagittal and lambdoidal crests well developed in adults, but weak in juveniles. Maxillary zygomatic process relatively wider in lateral view than in *F. greeni* and *F. brachyceps*. Extremely weak connection between lacrimal and jugal. Upper and lower cheek teeth similar in size.

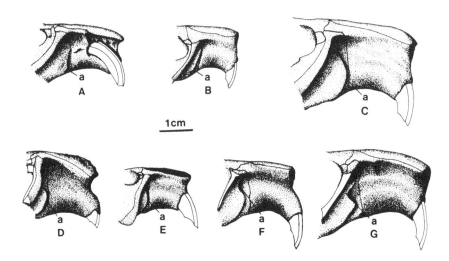

Figure 5. Morphological comparisons of infraorbital foramen and masseter ridge on skulls among North American Oligocene beavers. a, infraorbital foramen. A, *Palaeocastor peninsulatus* (LACM 16005); B, *Capacikala gradatus* (F:AM 64152); C, *Capatanka cankpeopi* (SDSM 53421); D, *Euhapsis ellicottae* (UM 1617); E, *Euhapsis platyceps* (CM 1220); F, *Nannasfiber ostellatus* (UC 1461); G, *Fossorcastor fossor* (UM 1618).

Discussion—This genus, proposed by Martin (1987), includes two species: *F. greeni* and *F. brachyceps*. Specimens here referred to *F. fossor* were originally placed in *Palaeocastor fossor* (Peterson, 1905; Stirton, 1935) or *Palaeocastor (Capatanka) magnus* (Martin, 1987). However, they do not pertain to *Palaeocastor* because they have C-shaped masseter ridge, intermediate position of infraorbital foramen, extremely weak jugal-lacrimal connection, and alternate processes of mandible. *Fossorcastor greeni* and *F. brachyceps* also have those characters. Therefore, specimens previously called *P. fossor* and *P. (C.) magnus* should be referred to *Fossorcastor*.

The difference originally recognized between *P. fossor* and *P. magnus* is that the former has stronger sagittal and lambdoidal crests than the latter. However, the difference is ontogenetic variation rather than species-level difference because weak crests are always accompanied by unfused sutures, and closed cranial sutures occur with strong crests. *Palaeocastor simplicidens* (type specimen: AMNH 12900) from the Monroe Creek Formation is a synonym of *Fossorcastor fossor* because it has all characters found only in *F. fossor*, though the characters are not strongly developed.

This genus is an important part of the Harrison fauna, but only one specimen (AMNH 12900) is known from the Monroe Creek Formation. *F. greeni* and *F. brachyceps* are known from one specimen each, but they provide important information for their relationship to *F. fossor*. *F. fossor* is one of the most common beaver fossils from the late Arikareean.

ORIGIN, GENERIC PHYLOGENY, AND
RADIATION OF NORTH AMERICAN EOCENE-
OLIGOCENE BEAVERS

Similarities of dental pattern and skull foramina between *Eutypomys* and *Agnotocastor* seem to encourage researchers to relate the two genera (Wahlert, 1977;

Wood, 1965, 1980; Wilson, 1949b). Wood (1980) proposed that *Eutypomys* belongs to Infraorder Castorimorpha. However, in addition to the dental pattern and foramina, beavers have derived mandible characters that *Eutypomys* and other sciuromorphous rodents do not share. These characters include a digastric eminence and the angle extending up posteriorly (Fig. 4). *Eutypomys* and beavers have a sciuromorphous masseter arrangement that relates them to sciuromorphous rodents (Wood, 1965). Beavers are here defined as the rodents that have sciuromorphous masseter arrangement on the skull and a derived mandible here termed the beaver-pattern mandible. *Eutypomys* is a primitive sister-group of beavers. *Propalaeocastor kumbulakensis* from the Kazahkstan Oligocene (Lytschev, 1970) is not a beaver because it does not have the derived beaver-pattern mandible. It is probably an *Eutypomys* with a simplified dental pattern.

Seven North American Eocene-Oligocene beaver genera can be divided into three groups. *Agnotocastor* is a primitive group; *Palaeocastor* and *Capacikala* belong to one group; *Capatanka*, *Euhapsis*, *Nannasfiber*, and *Fossorcastor* are in another. *Agnotocastor* is characterized by relatively long and narrow skull and complicated dental pattern. *Palaeocastor* and *Capacikala* are represented by relatively short and broad skulls, simplified dental pattern, and longer straight masseter ridge. The third group is characterized by convex or C-shaped masseter ridge, alternate processes of mandible in posterior view, higher or intermediate position of infraorbital foramen. Figure 5 displays the character comparisons of position of infraorbital foramen and masseter ridge among North American Eocene-Oligocene beavers. Character distributions and a cladogram of North American Eocene-Oligocene beavers are shown in Table 1, Table 2, and Figure 6.

Convex-faced incisors in *Agnotocastor* are plesiomorphic because they are universal in rodents. Flat-faced

Table 1. List of characters used in North American Eocene-Oligocene beaver phyletic analysis (Fig. 10).

C1[1] Masseter arrangement: 0^2, Non-sciuromorphic masseter arrangement; 1, Sciuromorphic masseter arrangement
C2 Dentition: 0, Non-beaver pattern; 1, Five-crest pattern; 2, Four-crest pattern; 3. Three-crest pattern in molars; 4, Three-crest pattern in premolars and molars
C3 Mandible digastric eminence: 0, No digastric eminence; 1, Digastric eminence
C4 Mandible angle: 0, Mandible angle extending horizontally or ventrally; 1, Mandible angle extending up posteriorly
C5 Skull shape: 0, Narrow (length/width approximately 1.5); 1, Intermediate (length/width approximately 1.3); 2, Broad (length/width approximately 1.2); 3, Broadest (length/width approximately 1.0)
C6 Masseter ridge: 0, Short and straight; 1, Long and straight; 2, Convex; 3, "C" shape
C7 Infraorbital foramen position: 0, Low; 1, intermediate; 2, High
C8 Nasal Shape: 0, Maximum length/width greater than 2.5; 1, Maximum length/width smaller than 2.0
C9 Ratio of maximum width of nasals vs maximum width of snout: 0, Greater than 0.85; 1, Between 0.80-0.65; 2, Smaller than 0.55
C10 Interparietal: 0, Large; 1, Small; 2, Absent
C11 Connection between jugal and lacrimal: 0, Strong; 1, Weak; 2, Weakest
C12 Mandible posterior view: 0, Alignment; 1, Alternate
C13 Sagittal crest: 0, Sagittal crests meet at interorbital constriction; 1, Sagittal crests meet at post-interorbital constriction; 2, Lyrate sagittal crests
C14 Mandible ramus: 0, Ramus frontal margin raising from m3; 1, Ramus frontal margin raising from m2; 2, Ramus frontal margin raising from m1 or p4
C15 Palatal Termination: 0, At level of M3s: 1, At level of M2s; 2, At level of post M3s
C16 Postzygomatic skull length vs skull length: 0, 0.20-0.25; 1, 0.17-0.20; 2, 0.14-0.17; 3, 0.12-0.14
C17 Development of crest at posterior sagittal-lamdboidal area: 0, Undeveloped; 1, Well developed
C18 Relative distance of maxilla-premaxilla suture to masseter ridge: 0, Close; 1, Anterior
C19 Incisor frontal face: 0, Convex; 1, Flat
C20 Fossette number of premolars: 0, Non-beaver pattern; 1, Equal or more than 5; 2, Equal or less than 4

Table 2. Data matrix of character states for North American Eocene-Oligocene beavers

Character	1	2	3	4	5	6	7	8	9	10	11	12	13	14	15	16	17	18	19	20
Marmota	1	0	0	0	0	0	0	0	0	0	0	0	0	0	0	0	0	0	0	0
Eutypomys	1	1	0	0	0	0	0	0	0	0	0	0	0	0	0	0	0	0	0	1
Agnotocastor	1	1	1	1	0	0	0	0	1	0	0	0	0	1	0	1	0	0	0	1
Palaeocastor	1	2	1	1	1	1	0	0	1	0	0	0	0	1	1	1	0	0	1	1
Capacikala	1	2	1	1	1	1	0	0	1	0	0	0	2	1	1	1	0	0	1	2
Capatanka	1	3	1	1	3	2	1	1	1	1	1	1	1	2	2	2	0	0	1	1
Euhapsis	1	3	1	1	3	2	1	1	2	1	1	1	1	2	2	3	0	1	1	2
Nannasfiber	1	4	1	1	2	3	2	1	1	2	2	1	1	2	2	3	0	1	1	2
Fossorcastor	1	4	1	1	2	3	1	1	1	1	2	1	1	2	2	3	1	1	1	2

Table 3. Skull size comparison among North American Eocene-Oligocene beavers and post-Oligocene beavers

Taxon	Sample size	Skull length(mm)	Mean	S.D.	Skull width (mm)	Mean	S.D.
Agnotocastor	5	59.5-65.4	62.06	2.35	38.0-43.1	40.70	2.16
Palaeocastor	13	50.0-70.0	60.82	6.58	39.8-54.2	46.62	5.26
Capacikala	6	43.0-55.0	49.82	4.60	35.6-47.5	40.15	4.21
Capatanka	12	50.0-75.4	63.14	8.25	42.3-69.6	56.82	9.03
Euhapsis	5	53.0-58.0	55.6	1.95	52.7-58.0	55.48	2.12
Nannasfiber	6	43.0-56.5	49.78	4.60	41.4-45.6	42.08	1.59
Fossorcastor	15	60.7-80.0	71.80	5.80	51.2-65.8	59.78	4.70

C1 means character 1.
Numbers represent character states. 0 represents primitive state and non-zero numbers are derived states compared with outgroups.

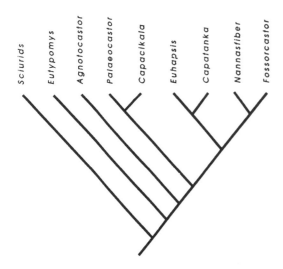

Figure 6. A cladogram of the North American Eocene-Oligocene beavers from PAUP analysis. Refer to Table 1 and Table 2 for characters and data matrix. Tree length = 38 and CI = 0.974.

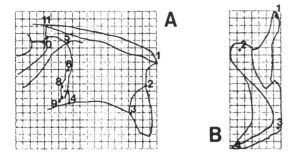

Figure 7. Landmarks chosen for TPS analysis. A, lateral view of anterior skull; landmark 7 is for infraorbital foramen, landmark 11 is posterior edge of nasal, and landmarks 6, 8, and 9 are used for localizing masseter ridge. B, posterior view of mandible; landmarks are for processes.

incisors (character 19 in Table 1) are apomorphic and are known in *Palaeocastor, Capacikala, Castor, Euhapsis, Fossorcastor, Capatanka,* and *Nannasfiber.* The incisor face of *Capatanka* is slightly convex, and it may be secondarily derived.

Eocene-Oligocene beavers are small compared to Miocene and Pliocene beavers, and their size did not greatly change during the Eocene-Oligocene (Table 3). In the *Palaeocastor* and *Capacikala* group, masseter ridge remains straight and progressively becomes longer, which means that area of masseter origins increases during the evolution. In the third group, position of infraorbital foramen is intermediate or high (*Nannasfiber*); mandibular processes in posterior view becomes alternating, which may imply enhanced chewing muscles or chewing function; masseter ridge appears convex or C-shaped; jugal-lacrimal connection becomes progressively weaker. TPS analysis result of mandible and skull characters is shown in Figures 7, 8, and 9.

TPS results positively support the phylogenetic relationships (Table 1, Table 2, and Fig. 6) from PAUP (v.3.1) analysis. Eutypomyid rodents are outgroup for castorids, and polarity and states for each character are shown in Table 1 and Table 2. Non-affine deformation (not proportional) takes place in the anterior skull portion. Bending energy in A through D and E through H in Figure 8 increases progressively. It displays progressively evolutionary trends because all deformed grids are of similar shape. The similarity between A-D and E-H indirectly shows the close relationship between *Palaeocastor* and *Capacikala.* In other words, if any two taxa are sister-groups, they should show similar trends in TPS series comparisons when they are compared with a monophyletic group respectively. The compared monophyletic group should be arranged in the same order for comparisons. Therefore, TPS is able to be used for testing phyloge-netic relationships. Local deformation concentrates at the infraorbital foramen, masseteric ridge, and nasals, which is coincident with those characters chosen in this study. Mandibular characters display affine deformation, and bending energy is near zero (Fig. 9). However, both non-affine and affine deformation are consistent with two evolutionary lineages. One includes *Palaeo-castor* and *Capacikala;* another comprises *Capatanka, Euhapsis, Fossorcastor,* and *Nannasfiber.*

Based on skull and mandibular characters analyzed above, the North American Eocene-Oligocene mandible and skull patterns can be defined as follows:

Skull Patterns

Agnotocastor Skull Pattern—The longest and narrow-est skull with the long snouts and nasals. Masseter ridge straight and shortest, position of infraorbital foramen low, and jugal-lacrimal connection strong. This pattern includes only *Agnotocastor.*

Palaeocastor Skull Pattern—Shorter and broader skull than in *Agnotocastor.* Masseter ridge straight and longer than in *Agnotocastor;* position of infraorbital foramen low, and jugal-lacrimal connection strong. This pattern comprises *Palaeocastor, Capacikala, Steneofiber,* and *Castor.*

Euhapsis Skull Pattern—Broadest, shallowest skull in beavers. Masseter ridge convex, position of infraorbital foramen intermediate, and jugal-lacrimal connection weak. Masseter ridge close to maxillary-premaxillary suture. The pattern is found in *Euhapsis* and *Capatanka.*

Fossorcastor Skull Pattern—Skull narrower than in *Euhapsis,* but broader than in *Palaeocastor.* Masseter ridge C-shaped, position of infraorbital foramen intermediate or high, jugal-lacrimal connection extremely weak. *Nannasfiber* and *Fossorcastor* have this pattern.

Mandible Patterns

Palaeocastor Mandible Pattern—In posterior view, coronoid, condyloid, and angular processes aligned vertically. *Agnotocastor, Palaeocastor, Capacikala, Steneofiber,* and *Castor* have this mandible pattern.

Fossorcastor Mandible Pattern—In posterior view, coronoid, condyloid, and angular processes alternating.

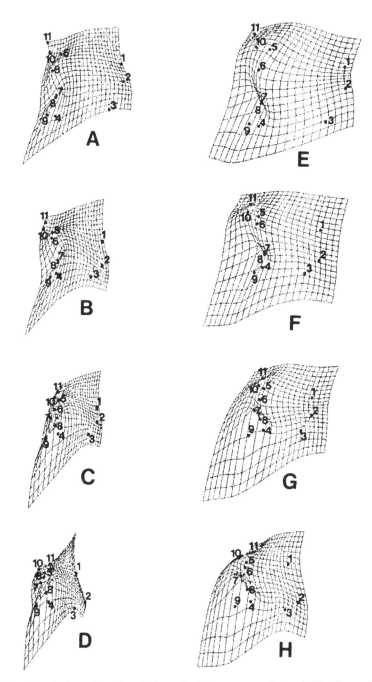

Figure 8. TPS result from lateral view of anterior skull. A-D, *Palaeocastor* vs *Euhapsis* (A), *Capatanka* (B), *Fossorcastor* (C), and *Nannasfiber* (D) (bending energy = 2.16515, 4.40926, 8.80411, and 11.22376, respectively); E-H, *Capacikala* vs *Euhapsis* (E), *Capatanka* (F), *Fossorcastor* (G), and *Nannasfiber* (H) (bending energy = 2.32746, 4.66661, 9.06990, and 10.83305, respectively). The deformed grids result in the corresponding displacement of each point on squared grids with those landmarks of *Palaeocastor* or *Capacikala* moving to overlap each corresponding landmark of compared specimen.

Coronoid and angular processes extending laterally, and condyloid process extending medially. This pattern includes *Capatanka*, *Euhapsis*, *Nannasfiber*, *Fossorcastor*, and giant beaver group (*Eucastor*, *Dipoides*, *Procastoroides*, and *Castoroides*).

The first occurrence of beavers is *Agnotocastor* in the North American Chadronian, late Eocene (Emry, 1972; Prothero and Swisher, 1992), and earlier than the Kazahkstanian early Oligocene *Agnotocastor* (Lytschev, 1978). Three North American *Agnotocastor* species are distributed from the Chadronian through the earliest Arikareean. *A. galushai* is from the Chadronian; *A.*

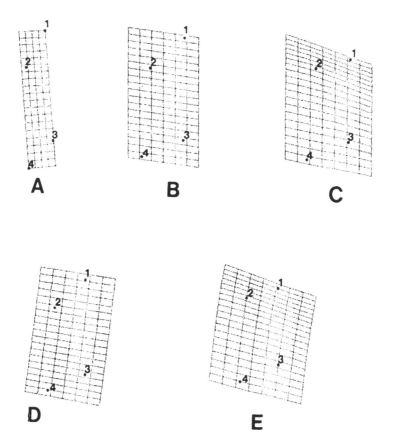

Figure 9. TPS result from posterior view of mandible characters. A, *Palaeocastor* vs *Capacikala* (bending energy = 0.00057); B-C, *Palaeocastor* vs *Nannasfiber* (B) and *Fossorcastor* (C) (bending energy = 0.00000 and 0.01823, respectively); D-E, *Capacikala* vs *Nannasfiber* (D) and *Fossorcastor* (E) (bending energy = 0.00077 and 0.01346, respectively).

coloradensis is found in the Orellan and the earliest Arikareean; *A. praetereadens* is known in the Whitneyan. *Palaeocastor* was probably derived from *Agnotocastor*, but only one species, *P. nebrascensis*, is known from the Whitneyan.

The earliest beaver radiation occurred in the early Arikareean. Five genera first appear at that time, four in the early Arikareean and one in the late Arikareean. As a result, the rapid diversification of beavers makes an excellent tool for biochronological correlations in the North American Arikareean (Fig. 10).

Modern beavers have a *Palaeocastor* skull pattern and *Palaeocastor* mandible pattern, and they are, via *Steneofiber*, closely related to the *Palaeocastor* and *Capacikala* group. The significant differences are that modern beavers have better developed stria (-iids) and flexuses (-ids), extremely strong masseter ridges that can be seen in skull dorsal view, and greater body size. The North American giant beaver group, including *Eucastor*, *Dipoides*, *Procastoroides*, and *Castoroides* has the *Fossorcastor* mandible pattern and the skull pattern similar to *Fossorcastor* except that the lacrimal in the giant beaver group does not contact the jugal. *Fossorcastor* is closely related to the giant beaver group.

CONCLUSIONS

1. Eighteen species in seven genera of North American Eocene-Oligocene beavers (the Chadronian through Arikareean), including one new genus (*Nannasfiber*) and five new species (*Capacikala parvus*, *Capatanka minor*, *Euhapsis luskensis*, *Nannasfiber ostellatus*, and *N. osmagnus*), are recognized in the present study.

2. Four skull patterns and two mandible patterns of North American Eocene-Oligocene beavers are defined, and they are important criteria for beaver classification and phylogenetic analysis.

3. The rapid evolution and great radiation of the Arikareean beavers make them excellent for biochronological correlation: *Agnotocastor* for Chadronian throught the earliest Arikareean, *Palaeocastor* and *Capatanka* for the early early Arikareean, *Capacikala* for the early Arikareean, *Nannasfiber* for the early late Arikareean, and *Euhapsis* and *Fossorcastor* for the late early through late late Arikareean.

4. Modern beavers are, via *Steneofiber*, closely related to *Palaeocastor* and *Capacikala*, and the giant beaver group has a close relationship to the *Fossorcastor* group.

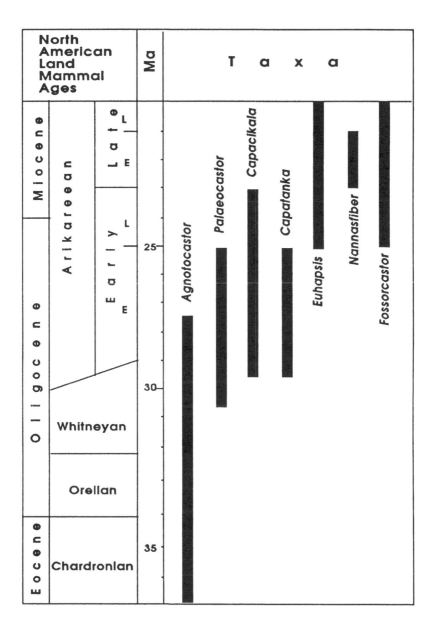

Figure 10. Generic distributions of North American Eocene-Oligocene beavers in North American land mammal "ages." After Prothero and Swisher (1992) from the Chadronian through Whitneyan, and after Tedford et al. (1987) for the Arikareean subdivisions.

ACKNOWLEDGMENTS

I thank L. Jacobs and D. Winkler for comments and expertise. For access to fossil material, I am indebted to R. Tedford (American Museum of Natural History), J. Whitmore (South Dakota School of Mines and Technology), L. Barnes (Los Angeles County Museum of Natural History), H. Hutchison (University of California, Berkeley), J. Flynn and W. Simpson (Field Museum of Natural History), L. Martin (University of Kansas, Lawrence), R. Hunt and T. M. Stout (University of Nebraska, at Lincoln). L. Martin and R. Emry (National Museum of Natural History, Washington, D.C.) reviewed the manuscript and made helpful comments. Also, thanks go to X. Wang (American Museum of Natural History) for stratigraphic discussion.

This research was supported by a Seed Grant at Institute for Study of Earth and Man at SMU, Research Funds of Graduate Student Assembly in SMU, AMNH Short-Term Visiting Support, and Short-Term Visiting Support of Smithsonian Institution.

LITERATURE CITED

Bookstein, F. L. 1991. Morphometric Tools for Landmark Data: Geometry and Biology. Cambridge University Press, New York.

Cope, E. D. 1879a. On some characters of the Miocene fauna of Oregon. Proceedings of the American Philosophical Society 18:63-78.

Cope, E. D. 1879b. Observations on the faunae of the Miocene Tertiaries of Oregon. Bulletin of the United States Geological and Geographical Survey 5: 55-69.

Cope, E. D. 1881. Review of the Rodentia of the Miocene period of North America. Bulletin of the United States Geological and Geographic Survey 6: 361-386.

Crusafont-Pairo, M., J. F. Villalta, and J. R. Bataller. 1948. Los castores fossils de España; Boletin del Instituto Geológico y Minero 61:323-448.

Emry, R. J. 1972. A new species of *Agnotocastor* (Rodentia, Castoridae) from the early Oligocene of Wyoming. American Museum Novitates No. 2485:1-7.

Galbreath, E. G. 1953. A contribution to the Tertiary geology and paleontology of northeastern Colorado. University of Kansas Paleontological Contributions, Vertebrata, Article 4:1-120.

Jenkins, S. H. and P. E. Busher. 1979. *Castor canadensis*. Mammalian Species 120:1-8.

Leidy, J. 1856. Notices of remains of extinct mammalia, discovered by Dr. F. V. Hayden in Nebraska Territory. Proceedings of Academy of Natural Sciences of Philadelphia 8:88-90.

Leidy, J. 1869. The extinct mammalian fauna of Dakota and Nebraska. Journal of the Academy of Natural Sciences of Philadelphia 7:1-472.

Lytschev, G. F. 1970. A new Oligocene beaver species from the northern Aral Region. Paleontological Journal 1970(2):229-233.

Lytschev, G. F. 1978. A new early Oligocene beaver of the genus *Agnotocastor* from Kazakhstan. Paleontological Journal 1978(4):542-544.

Macdonald, J. R. 1963. The Miocene faunas from the Wounded Knee area of western South Dakota. Bulletin of the American Museum of Natural History 125(3):139-238.

Macdonald, J. R. 1970. Review of the Miocene Wounded Knee faunas of southwestern South Dakota. Bulletin of the Los Angeles County Museum of Natural History, Science 8:1-82.

Macdonald, L. J. 1972. Monroe Creek (early Miocene) microfossils from the Wounded Knee area, South Dakota. South Dakota Geological Survey-1972, Report of Investigations 105:1-43.

Martin, L. D. 1987. Beavers from the Harrison Formation (early Miocene) with a revision of *Euhapsis*. Papers in Vertebrate Paleontology in Honor of Morton Green.

Dakoterra 3:73-91.

Matthew, W. D. 1902. A horned rodent from the Colorado Miocene. With a revision of the mylaguali, beavers, and hares of the American Tertiary. Bulletin of the American Museum of Natural History 16(22):291-310.

Matthew, W. D. 1907. A Lower Miocene fauna from South Dakota. Bulletin of the American Museum of Natural History 23:205-210.

Olson, E. C. 1940. Cranial foramina of North American beavers. Journal of Paleontology 14(5):495-501.

Peterson, O. A. 1905. Description of new rodents and discussion of the origin of *Daemonelix*. Memoirs of the Carnegie Museum 2(4):139-203.

Prothero, D. R., and C. C. Swisher III. 1992. Magnetostratigraphy and geochronology of the terrestrial Eocene-Oligocene transition in North America; pp. 46-73 *in* D. R. Prothero and W. A. Berggren (eds.), Eocene-Oligocene Climatic and Biotic Evolution. Princeton University Press, Princeton, N. J.

Romer, A. S., and J. T. McCormack. 1928. A large *Palaeocastor* from the Lower Miocene. American Journal of Science 15:58-60.

Stirton, R. A. 1935. A review of the Tertiary beavers. Bulletin of the Department of Geological Sciences, University of California Publications 23(13):391-485.

Tedford, R. H., M. F. Skinner, R. W. Fields, J. M. Rensberger, D. P. Whistler, T. Galusha, B. E. Taylor, J. R. Macdonald, and S. D. Webb. 1987. Faunal succession and biochronology of the Arikareean through Hemphillian interval (late Oligocene through earliest Pliocene epochs); pp. 153-210 *in* M. O. Woodburne (ed.), Cenozoic Mammals of North America: Geochronology and Biostratigraphy. University of California Press, Berkeley.

Wahlert, J. H, 1977. Cranial foramina and relationships of *Eutypomys* (Rodentia, Eutypomyidae). American Museum Novitates No. 2626:1-8.

Wilson, R. W. 1949a. On some White River fossil rodents. Some Tertiary mammals and birds from North America. Contributions to Paleontology, Carnegie Institution of Washington Publication 584: 27-50.

Wilson, R. W. 1949b. Early Tertiary rodents of North America. Contributions to Paleontology, Carnegie Institution of Washington Publication 584: 69-164.

Wood, A. E. 1965. Grades and clades among rodents. Evolution 19:115-130.

Wood, A. E. 1980. The Oligocene rodents of North America. Transactions of the American Philosophical Society 70(5):1-68.

Xu, X. 1995. Phylogeny of beavers (Family Castoridae): Applications to faunal dynamics and biostratigraphy since the Eocene. Ph.D. dissertation, Southern Methodist University, Dallas, TX.

21. Canidae

XIAOMING WANG AND RICHARD H. TEDFORD

ABSTRACT

The White River Group of the late Eocene and Oligocene age contains some of the earliest and most primitive canids in North America. These archaic canids, mainly *Hesperocyon* and its close relatives, gave rise to several clades that contain ancestral forms for all three subfamilies of the Canidae: the paraphyletic Hesperocyoninae which became the dominate canids during the Arikareean, the monophyletic Borophaginae which flourished during the Miocene, and the monophyletic Caninae which did not diversify until Pliocene through Recent time. The middle-late Eocene (Duchesnean to Chadronian) saw the emergence of the first canids with a fully ossified entotympanic bulla and canid dentition (3 1 4 2/3 1 4 3), as represented by *Prohesperocyon wilsoni* and *H. gregarius*, which probably evolved from primitive cynoids such as *Procynodictis*. The White River canids began an initial diversification in the beginning of Oligocene (Orellan), giving rise to the earliest members of the borophagine (*Cormocyon*) and canine (*Leptocyon*) clades. At least six species can be recognized at this time: *H. gregarius*, "*H.*" *coloradensis*, "*Mesocyon*" *temnodon*, *Osbornodon renjiei*, *Cormocyon pavidus*, and *Leptocyon* sp. In Whitneyan time, the majority of the hesperocyonine clades began to be readily distinguishable, and the borophagines began to develop small, hypocarnivorous taxa. The number of species in the Whitneyan had increased to nine: *H. gregarius*, *Paraenhydrocyon josephi*, "*M.*" *temnodon*, *Cynodesmus thooides*, *Osbornodon renjiei*, *O. sesnoni*, *Ectopocynus antiquus*, *Cormocyon pavidus*, and *Oxetocyon cuspidatus*. Near the top of the White River sequence, an early precursor of the hypercarnivorous *Enhydrocyon* clade, *Sunkahetanka geringensis*, had emerged, and most clades of the Hesperocyoninae began independently developing meso- to hypercarnivorous forms.

INTRODUCTION

Early members of the dog family (Canidae) left an abundant fossil record in the White River Group of North America with a fairly continuous presence throughout the strata. This is especially true of *Hesperocyon*, an ubiquitous, small fox-sized animal frequently found in the White River Group, leading some authors to speculate that it was responsible for keeping the numbers of its contemporary rodents, lagomorphs, and small artiodactyls in check (Cope, 1874; Clark et al., 1967). Less conspicuous than their contemporary large predators such as hyaenodontids and nimravids, these early canids include the ancestral stocks

that went on to become the dominant carnivores of later times.

The late Eocene and Oligocene *Hesperocyon* has attracted the attention of vertebrate paleontologists, not only because of the well-preserved skeletal materials commonly found in the White River strata, but also because it is an important *bona fide* canid which possesses the right morphology and occurs at the right time to be ancestral to or close to the ancestry of all later canids. In addition to representing the basal group for the first canid radiation, the subfamily Hesperocyoninae, *H. gregarius* is also the closest sister-group of the presumed common ancestor of the Borophaginae and Caninae, two subfamilies which flourished in the late Tertiary and Quaternary-Recent times (Tedford, 1978).

Since the classic treatment of the White River *Hesperocyon* by Scott and Jepsen (1936), no systematic effort has been devoted to re-evaluating the taxonomy of these early canids despite a number of attempts (e.g., Galbreath, 1953; Hough and Alf, 1956; Elliott, 1980). Meanwhile, our knowledge of fossil canids during this crucial period of time has increased considerably, thanks mainly to the Frick Collection at the American Museum of Natural History, which contains a large number of undescribed taxa and more complete materials of known taxa from the White River Group. We are now in possession of an unprecedented data base to better evaluate morphological variations and to more accurately reconstruct phylogenies. Such a wealth of information has, in recent years, stimulated a number of studies that establish a new outlook the evolutionary relationships of primitive canids (Tedford, 1978; Wang, 1990, 1993, 1994; Wang and Tedford, 1992, 1994; Tedford and Taylor, MS). It is thus opportune to update, in this volume, the current systematics of the White River canids.

The main purpose of the present review is an evaluation of the known taxonomic diversity and phylogeny of canids from the White River Group of the northern Great Plains in light of the new calibration of Eocene-Oligocene transition as recently proposed by Swisher and Prothero (1990). The nature and scope of this chapter do not allow exhaustive treatment of all White River canids, and undescribed taxa from the Frick Collection, mostly small borophagines, will not be discussed.

ABBREVIATIONS

AMNH, American Museum of Natural History, New York; F:AM, Frick Collection, American Museum of Natural History, New York; CMNH, Carnegie Museum of Natural History, Pittsburgh; KUVP, Museum of Natural History, University of Kansas, Lawrence; LACM, Los Angeles County Museum, Los Angeles; SDSM, South Dakota School of Mines and Technology, Rapid City; TMM, Texas Memorial Museum, Austin; UNSM, State Museum, University of Nebraska, Lincoln; USNM, United States National Museum of Natural History, Smithsonian Institution, Washington, D.C.; YPM-PU, Yale Peabody Museum (Princeton University Collection), New Haven.

TAXONOMIC OVERVIEW OF THE EARLY CANIDAE

Detailed systematic treatments of the hesperocyonines, borophagines, and canines are the subjects of separate studies (Wang, 1994; Tedford and Taylor, MS), and for the present purpose, only a synopsis is presented to summarize the taxonomy of early canids in the White River and equivalent strata. Since only published taxa are included, the White River canid diversity presented here is incomplete. This is especially the case for the Borophaginae, which has more undescribed taxa than has been documented. Complete synonym lists for each species of the Hesperocyoninae can be found in Wang (1994).

Higher level taxonomic nomenclature (e.g., Cynoidea) follows Wang and Tedford (1994). The archaic family Miacidae is most likely paraphyletic because of its presumed ancestral relationship to more derived families of caniform carnivorans (cynoids and arctoids). Unless stated otherwise, the present usage of miacids refers to the nominal genus *Miacis* (type species *M. parvivorus*). As is the case with many primitive fossil groups, paraphyly cannot be easily avoided without naming of numerous new genera. As a result, several paraphyletic genera, *Hesperocyon*, *Mesocyon*, *Cormocyon*, and *Leptocyon*, are still in use in this overview. Those that are clearly paraphyletic in view of the phylogeny presented here (Fig. 10) are enclosed in quotation marks.

Order CARNIVORA Bowditch, 1821
Suborder CANIFORMIA Kretzoi, 1943
Infraorder CYNOIDEA Flower, 1869
Family CANIDAE Gray, 1821

Diagnosis—Canids can be characterized by the following derived basicranial and dental features: an inflated entotympanic bulla that is composed of a rostral entotympanic and a caudal entotympanic; medial expansion of the petrosal which is in full contact with the basioccipital and basisphenoid; fully ossified tegmen tympani forming the roof of the facial nerve canal; extrabullar position of the internal carotid artery (and thus loss of the stapedial artery) which lies between the entotympanic and the petrosal; a small suprameatal fossa anterior to the mastoid process (in primitive

canids only); presence of posterior accessory cusps on the upper and lower third premolars; reduction of the M1 parastyle; and absence of M3 (Wang and Tedford, 1994). In addition, all canids, except the most basal taxon *Prohesperocyon*, have a low septum on the internal surface of the bulla at the suture of the ecto- and entotympanic.

Geologic and Geographic Range—Late Eocene to Recent of North America, Late Miocene to Recent of Europe, Pliocene to Recent of South America, Asia, and Africa.

Comments—The derived features in the above diagnosis easily distinguish canids from "miacids," but some of them may have been shared with arctoid carnivorans (e.g., suprameatal fossa; see this volume, Chapter 24). These basicranial and dental characteristics, conservative once acquired, permit easy recognition of canids among carnivorans.

The family Canidae underwent at least two major radiations in the Oligocene and Miocene that were exclusively confined to North America: the paraphyletic Hesperocyoninae and the monophyletic Borophaginae (Tedford, 1978). A third radiation, the monophyletic Caninae, originated in North America but did not reach the Palaearctic until some time in the Late Miocene. All three subfamilies appear in the White River deposits with animals smaller than a living red fox and diversified into larger, wolf-sized carnivorans. There are striking ecological parallels among the three subfamilies, each with its own small predators, fruit eaters, and large hypercarnivorous predators (Tedford, 1978; Martin, 1989; Van Valkenburgh, 1991).

The ancestry of all three subfamilies can be traced back to *Hesperocyon*, traditionally considered the most basal canid because of its fully ossified entotympanic bulla and canid dental formula (3 1 4 2/3 1 4 3). Such easily recognized features have served well as a practical and convenient diagnosis of the Canidae. Discoveries of more archaic taxa, however, blur these distinctions, and the increased knowledge about the phylogeny of caniform "miacids" makes possible the identification of ancestral cynoids (e.g., *Procynodictis vulpiceps*) lacking the above traditional diagnostic features, thus pushing cynoid origins into an early phase of caniform phylogeny (Wang and Tedford, 1994).

CANIDAE *incertae sedis*

Prohesperocyon wilsoni (Gustafson, 1986)
Figure 1B

Holotype—TMM 40504-126, partial skull and mandible with left C1, P3-M2, right C1-M2, left p4-m2, and right p2-m3.

Type Locality—Airstrip local fauna, in Capote Mountain Tuff, 174 m above the Bracks Rhyolite, in southern part of the Sierra Vieja area, Presidio County, Texas (Wilson et al., 1968). Middle Chadronian.

Diagnosis—*Prohesperocyon wilsoni* is characterized by a mixture of derived canid and primitive "miacid" features. Derived features include: an inflated and fully ossified entotympanic bulla; an accessory cusp on upper and lower third premolars; a rather posteriorly positioned internal cingulum of M1; loss of a notch between the parastyle and paracone of M1; and lack of M3. Among primitive features it retains: a large, anteriorly located protocone of P4; a still large parastyle of M1; a short shearing blade of m1; a narrow talonid relative to trigonid of m1; a reduced metaconid of m2; and a posteriorly reclined ascending ramus of mandible. In addition, *P. wilsoni* has autapomorphies that further differentiate it from *Hesperocyon gregarius* and preclude it from the direct ancestry of the later canids: elongated rostrum, anteriorly expanded bulla that exceeds in lateral view the posterior wall of the postglenoid fossa, slender premolars, and reduced metacone of M2.

Geologic and Geographic Range—From the type locality only.

Comments—Although the presence of a fully ossified entotympanic bulla (and the essentially canid middle ear region), a posteriorly positioned internal cingulum of M1, the lack of parastyle notch on M1, and the loss of M3 in *Prohesperocyon wilsoni* signal canid affinity, its dental morphology, for the most part, is primitive. It had, in fact, been assigned to *Miacis gracilis* by Wilson et al. (1968). Gustafson (1986) noted this primitive similarity between *P. wilsoni* and *Miacis* but nevertheless placed *P. wilsoni* in the Canidae because of its complete bulla. Recent study of the basicranial anatomy of *P. wilsoni* reveals that it lacks a low septum on the internal surface of the bulla in contrast to the rest of canids, which have the structure (Wang and Tedford, 1994).

A monotypic genus, *Prohesperocyon*, was erected by Wang (1994) to preserve monophyly of the taxon and to mark its largely primitive dentition distinct from *Hesperocyon* of the northern Great Plains. The intermediate nature of *P. wilsoni* narrows the morphologic gap between canids and "miacids," and reduces the number of characteristics of the Canidae previously regarded as first appearing together in *Hesperocyon gregarius*. Stratigraphically, however, *P. wilsoni* (middle Chadronian) occurs later than the earliest *Hesperocyon* (Duchesnean) from the Cypress Hills Formation of southwestern Saskatchewan, Canada (Bryant, 1992). This late occurrence of *P. wilsoni* and its possession of autapomorphies (above) preclude it from being in the direct ancestry of the more derived canids.

Subfamily HESPEROCYONINAE Tedford, 1978

Included Genera—*Hesperocyon* Scott, 1890; *Mesocyon* Scott, 1890; *Cynodesmus* Scott, 1893; *Sunkahetanka* Macdonald, 1963; *Philotrox* Merriam, 1906; *Enhydrocyon* Cope, 1879; *Osbornodon* Wang, 1994; *Paraenhydrocyon* Wang, 1994; *Caedocyon* Wang, 1994; and *Ectopocynus* Wang, 1994.

Geologic and Geographic Range—Duchesnean (late Eocene) to Early Barstovian (middle Miocene) of North America.

Comments—A cladistic diagnosis of the Hespero-cyoninae is not possible because of its paraphyletic nature. A trenchant talonid of lower molars (or much smaller entoconid than hypoconid), a primitive character, is all there is to distinguish it from the two more derived subfamilies Borophaginae and Caninae, which share the synapomorphy of basined (bicuspid) talonid. Associated with this single-cusped talonid is the absence or reduced presence of a metaconule in a transversely elongated M1, in contrast to the well-developed metaconule on a quadrate M1 often present in boro-phagines and canines. Even this single character does not hold true for all hesperocyonines. One lineage of the hesperocyonine, *Osbornodon*, has acquired a basined talonid and quadrate upper molars. Despite the difficulty in diagnosing the Hesperocyoninae, it is nevertheless a useful term to encompass the basal stocks of canids that are not related to later subfamilies.

The subfamily Hesperocyoninae includes four mono-phyletic clades whose ancestry can be traced to a *Hesperocyon*-like form: *Mesocyon–Enhydrocyon* clade, *Osbornodon* clade, *Paraenhydrocyon* clade, and *Ectopocynus* clade. These clades represent part of the initial radiation of the Canidae. They parallel one another or later canids in some morphological features, and tend to evolve into medium- to large-sized species that take the roles of powerful hypercarnivorous predators. Through-out most of their history, the hesperocyonines coexisted with borophagines and canines, although Oligocene members of the latter two subfamilies were mainly represented by a few small forms.

Hesperocyon Scott, 1890

Type Species—*Hesperocyon gregarius* (Cope, 1873a).

Included Species—*Hesperocyon gregarius* (Cope, 1873a), and "*H.*" *coloradensis* Wang, 1994.

Diagnosis—Compared to *Prohesperocyon*, *Hesperocyon* has the following derived characters (Fig. 10, node 2): a reduced protocone of P4; a further reduced parastyle of M1; an elongated shearing blade of m1; nearly equal metaconid and protoconid of m2; a more upright ascending mandibular ramus, and a low septum within the bulla. Compared to more derived hespero-cyonines and borophagines/canines, *Hesperocyon* is still primitive in having a posteriorly directed paroccipital process which is not fused with the entotympanic (except in "*H.*" *coloradensis*), an unreduced metaconid of m1, and presence of a low entoconid shelf of m1.

Geologic and Geographic Range—Cypress Hills Formation, southwestern Saskatchewan, Canada; Chadron, Brule, and their correlative formations in the White River Group of southwestern Montana, eastern Wyoming, western South Dakota, southwestern North Dakota, northeastern Colorado, and northwestern Nebraska. Duchesnean to Whitneyan of North America (late Eocene to early Oligocene).

Comments—*Hesperocyon* is important, morpho-logically and stratigraphically, in the understanding of the early canid radiation. It is one of the best known early canids (Cope, 1884; Scott, 1898; Matthew, 1901;

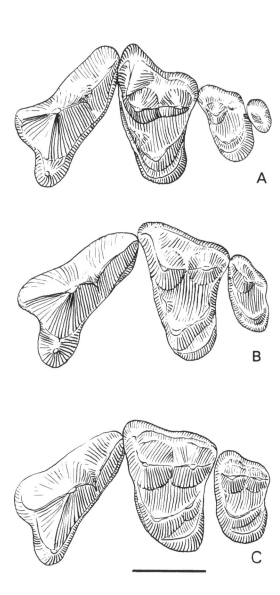

Figure 1. Upper P4–M2 (or M3) of A, *Procynodictis vulpiceps*, CMNH 12063; B, *Prohesperocyon wilsoni*, TMM 40504-126; C, *Hesperocyon gregarius*, F:AM 63930. Partially adopted from Wang and Tedford (1994, fig. 7). Scale = 5 mm.

Scott and Jepsen, 1936). Excellently preserved basicrania have been the subject of detailed studies of this small canid (Hough, 1948; Mitchell and Tedford, 1973; Tedford, 1976; Flynn et al., 1988; Wang and Tedford, 1994). The basicranium of *H. gregarius* has undoubted canid characteristics, a testimony to the considerable stability of the canid basicranium throughout its long history. *Hesperocyon* is clearly paraphyletic because of its ancestral position to some

hesperocyonine clades (e.g., *Paraenhydrocyon*) and because it lacks a synapomorphy shared by its two species.

A recent study by Bryant (1992) identified isolated teeth of *Hesperocyon gregarius* in the lower part of Cypress Hills Formation in Saskatchewan, Canada, and thus extended its record into the Duchesnean.

Hesperocyon gregarius (Cope, 1873)
Figures 1C, 2

Lectotype—AMNH 5297a, fragment of left ramus with p2–m1 and alveolus for p1.

Type Locality—White River Formation of northeastern Colorado, ?Orellan.

Diagnosis—Distinctions among species of *Hesperocyon* are quite subtle, and no single character could be used as the sole criterion of identification. *H. gregarius* tends to have a more posteriorly enlarged internal cingulum (hypocone) of M1 which often lacks the anterior segment of the cingulum. The parastyle of M1, although more reduced than in "miacids," is relatively large compared with that of the borophagine *Cormocyon pavidus*. The metaconid of m2 is smaller than the protoconid; the reverse is usually true for *C. pavidus*. Compared to the larger "*H.*" *coloradensis*, *H. gregarius* lacks a ventrally directed paroccipital process.

Geologic and Geographic Range—Cypress Hills Formation, Saskatchewan, Canada; Renova Formation of western Montana; Chadron, Brule, and their correlative formations in Colorado, Nebraska, North Dakota, South Dakota, and Wyoming; Duchesnean to Whitneyan.

Comments—The rich materials of *Hesperocyon gregarius* from the White River Group of the northern Great Plains is unsurpassed by any other group of canids from the Tertiary of North America. The species is the earliest discovered Tertiary canid and its morphology the best known among early canids.

Hesperocyon gregarius plays a key role in understanding the phylogeny of ancestral canids, and its central position in the early canid radiation has been recognized for more than a century (e.g., Cope, 1883). It has acquired such canid characters as a low septum inside the bulla, reduced parastyle of M1, and lost M3 (a few individuals, approximately 7% of all samples, still have an M3), but is still primitive enough to be close to the ancestry of many later canids. Therefore, except for a few minor peculiarities (such as a posteriorly positioned internal cingulum of M1), *H. gregarius* possesses no diagnostic autapomorphies and serves well as a primitive morphotype of the family Canidae. However, also because of its generalized morphologies, *H. gregarius* betrays no special similarity that can link it with any particular clade of canids.

"*Hesperocyon*" *coloradensis* Wang, 1994

Holotype—KUVP 85067, partial skull and mandible with left I3–C1, P2–M1, right I2–C1, P2–M1, left c1–m2, right i3–m2, and partial postcranial skeleton.

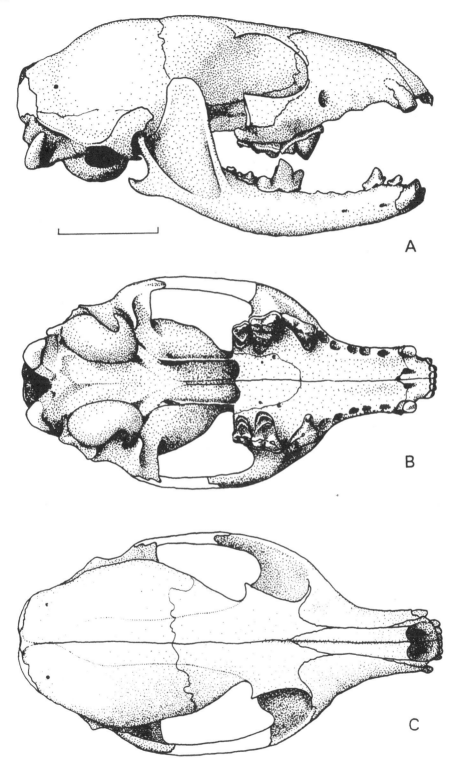

Figure 2. Skull and mandible of *Hesperocyon gregarius*, USNM 437888, from Wulff Ranch, southeast of Douglas, Converse County, Wyoming, in the Orella Member of the Brule Formation, Orellan. A, lateral view; B, ventral view; and C, dorsal view. Scale equals 20 mm.

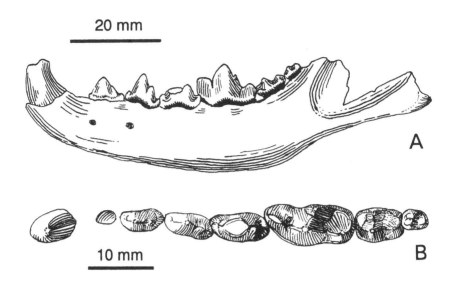

Figure 3. Mandible and teeth of *"Mesocyon" temnodon*, AMNH 8753, lectotype, from Castle Rock, Logan County, Colorado, in the Vista Member of the White River Formation, Whitneyan. A, lateral view of left mandible; and B, occlusal view of lower teeth.

Type Locality—Colorado locality number 46 of University of Kansas, Cedar Creek Member of White River Formation, Logan County, Colorado. Orellan.

Diagnosis—Signifying its more advanced status than *Hesperocyon gregarius*, *"H." coloradensis* has a ventrally directed paroccipital process, although it is still primitive in its lack of fusion of the process with the entotympanic (as in *Mesocyon*). *"H." coloradensis* is also 18% larger than *H. gregarius* and has a more robust mandible. It tends to have a less posteriorly positioned internal cingulum than in *H. gregarius*.

Geologic and Geographic Range—Cedar Creek Member of the White River Formation, northeastern Colorado. Orellan.

Comments—Cope (1873b, p. 9) recognized a species from the Tertiary of northeast Colorado, *Canis lippincottianus*, which was frequently used to include the larger-sized individuals that could not fit comfortably into *H. gregarius* (Matthew, 1901; Scott and Jepsen, 1936; Elliott, 1980; Galbreath, 1953). The type of *C. lippincottianus* (AMNH 5327a, a ramus fragment with an isolated m2), however, may be a small *Daphoenus* because of the wide talonid of its m2 with a well-developed entoconid (see Wang, 1994). Nonetheless, Cope's concept of a form larger than *H. gregarius* is confirmed.

Intermediate in size between *Hesperocyon gregarius* and *"Mesocyon" temnodon*, *"H." coloradensis* is at the base of the *Mesocyon-Enhydrocyon-Osbornodon* clade because of its initial development of a ventrally directed paroccipital process (node 7 of Fig. 10 and Table 1). Its membership in this latter clade further renders *Hesperocyon* a paraphyletic taxon, a fact already apparent before the inclusion of *"H." coloradensis* (see comments under the genus).

Mesocyon Scott, 1890

Type Species—*Mesocyon coryphaeus* (Cope, 1879).

Included Species—*Mesocyon coryphaeus* (Cope, 1879), *"M." temnodon* (Wortman and Matthew, 1899), and *M. brachyops* Merriam, 1906.

Diagnosis—Compared to *Hesperocyon* and *Paraenhydrocyon*, *Mesocyon* is more derived in its robust and long paroccipital process which is directed ventrally. The two derived species, *M. coryphaeus* and *M. brachyops*, have further acquired a small fossa on the supraoccipital above foramen magnum. *Mesocyon* is still primitive in its lack of robust premolars and broadened M2 found in *Cynodesmus*. In addition, the paroccipital process of *Mesocyon* is not thickened at the base as seen in more derived (geologically later) individuals of *Cynodesmus*. *Mesocyon* is paraphyletic and has no autapomorphy of its own (see Comments below).

Geologic and Geographic Range—John Day Formation of central Oregon; Otay and Sespe formations of southern California; Brule-Gering and correlative formations of the northern Great Plains. Orellan to Arikareean.

Comments—*Mesocyon* includes primitive members of the *Mesocyon-Enhydrocyon* clade (*M. coryphaeus* and *M. brachyops*). Its most primitive species *"M." temnodon*, however, is probably a basal taxon of the larger *Mesocyon-Enhydrocyon-Osbornodon* clade (Fig. 10), and the genus *Mesocyon* is thus paraphyletic. Only *"M." temnodon* is summarized below, which has fossil records in the White River Group. The Arikareean species, *M. coryphaeus* and *M. brachyops*, are not considered in this paper.

"Mesocyon" temnodon
(Wortman and Matthew, 1899)
Figure 3

Lectotype—AMNH 8753, partial left ramus with c1, p2–m3.

Type Locality—Horizon C of Castle Rock "in either R53 or R54, T11N, Logan County" (Galbreath, 1953, p. 50), Vista Member of White River Formation, northeastern Colorado. Whitneyan.

Diagnosis—Compared to "Hesperocyon" coloradensis, the paroccipital process of "Mesocyon" temnodon is fully ventrally directed so that its base touches and fuses with the posterior surface of the entotympanic, a derived character shared by all subsequent taxa of the Mesocyon-Enhydrocyon-Osbornodon clade. In contrast to M. coryphaeus and M. brachyops, "M." temnodon possesses primitive features such as a well-developed anterior cingulum of P4, the absence of a pair of round fossae above the foramen magnum, and the lack of strong elongation of the paroccipital process.

Geologic and Geographic Range—Whitneyan of northeastern Colorado and western Nebraska; Orellan to early Arikareean of southwestern South Dakota.

Comments—Since its first establishment by Wortman and Matthew (1899), "Mesocyon" temnodon has largely remained in obscurity because of the lack of adequate description and clear assignment of a type. "M." temnodon is now known from larger numbers of specimens and more complete material, and has a long geologic history spanning the Orellan to early Arikareean. This species is at the base of a clade which includes such meso- and hypercarnivorous taxa as Mesocyon, Cynodesmus, Sunkahetanka, Philotrox, Enhydrocyon, and Osbornodon (Fig. 10, node 8).

Cynodesmus Scott, 1893

Type Species—Cynodesmus thooides Scott, 1893.

Included Species—Cynodesmus thooides Scott, 1893; and C. martini Wang, 1994.

Diagnosis—Compared to Mesocyon, Cynodesmus has more robust and closely spaced premolars and a larger protocone of P4. Furthermore, Cynodesmus has two autapomorphies: a strong paroccipital process with a posterior keel and a transversely broadened M2.

Geologic and Geographic Range—Whitneyan to early Arikareean of Nebraska, South Dakota, and Montana.

Comments—Considerable confusion exists in the taxonomy of Cynodesmus and 13 species had been assigned to this genus (see Wang, 1994). It is now restricted to two species. Only the type species, C. thooides, occurs in the White River Group and is summarized below.

Cynodesmus thooides Scott, 1893
Figure 4

Holotype—YPM-PU 10412, partial skull with nearly complete upper dentitions and partial mandible with left i1–m3 and right i1–p3.

Type Locality—Fort Logan Formation of western Montana, early Arikareean.

Diagnosis—Cynodesmus thooides has derived characters that distinguish it from Mesocyon: stronger premolars, heavy paroccipital process with a keel, and broad M2. However, it has many primitive features compared to its sister-species C. martini: smaller size, less robust skull and teeth, slender rostrum, lower sagittal crest, and presence of a P1.

Geologic and Geographic Range —Whitneyan to early Arikareean of western Nebraska, South Dakota, and western Montana.

Comments—Cynodesmus thooides is of the same size as Mesocyon coryphaeus west of the continental divide. Morphological differences between them are rather subtle, and this species pair probably played the same ecological roles. C. thooides is a primitive species of the Mesocyon-Enhydrocyon clade, and is at the basal stem of a series of progressively more hypercarnivorous taxa in the Arikareean such as Sunkahetanka and Philotrox, terminating in Enhydrocyon.

Sunkahetanka geringensis
(Barbour and Schultz, 1935)

Holotype—UNSM 1092, nearly complete skull, mandible, and postcranial skeleton.

Type Locality—Redington Gap, west of Bridgeport, Morrill County, Nebraska. Brown Siltstone beds, uppermost Brule Formation, earliest early Arikareean.

Diagnosis—Sunkahetanka geringensis has two synapomorphies that distinguish it from Cynodesmus: more massive premolars that tend to be imbricated and reduced metaconid on lower molars. In addition, S. geringensis possesses some subtle features that further differentiate it from Cynodesmus and Mesocyon: larger postorbital process; posteriorly extended nuchal crest that does not overhang the occipital condyle; deep and laterally arched zygomatic arch; bulla narrowed anteriorly; slightly antero-posteriorly constricted M1 trigon; crowded lower incisors; and wider m1.

Geologic and Geographic Range—Uppermost part of the Brule Formation in western Nebraska, and lower Sharps Formation of western South Dakota. Early Arikareean.

Comments—Following the study by Tedford et al. (1985, 1987), the sandy rocks containing the type of Sunkahetanka geringensis are now regarded as the channel facies of the Brown Siltstone in the uppermost part of the Brule Formation, instead of the traditional Gering Formation. S. geringensis is thus properly discussed here, even though it is already in the Arikareean. Corresponding to this young age, S. geringensis is the largest and most hypercarnivorous among all White River canids, and shows indications of robustness more fully developed only in Enhydrocyon. It is otherwise transitional between Cynodesmus and the Philotrox-Enhydrocyon clade in cranial and dental morphology.

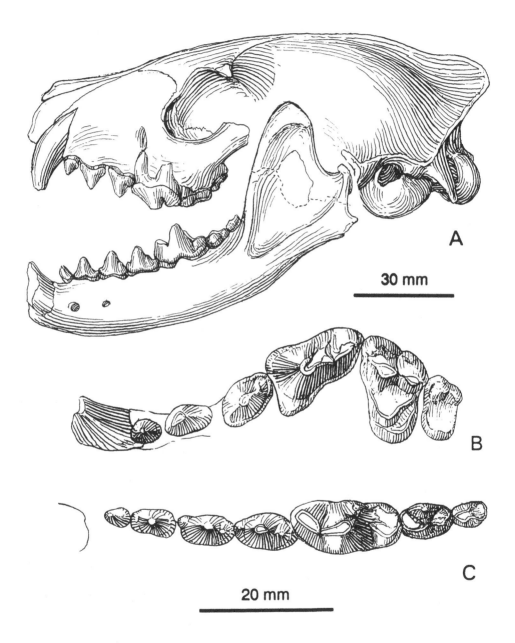

Figure 4. Skull, mandible, and dentitions of *Cynodesmus thooides*, F:AM 63382, from west side of Potato Creek, 3 miles east of Morgan Ranch near the head of east branch of Red Water Creek, Washabaugh County, South Dakota, in the Brule Formation, Whitneyan. A, lateral view of skull and mandible; B, occlusal view of upper left dentition, P3–M2 drawn from right side; and C, occlusal view of lower left dentition (reversed from right side).

Osbornodon Wang, 1994

Type Species—*Osbornodon fricki* Wang, 1994.
Included Species—*Osbornodon fricki* Wang, 1994; *O. renjiei* Wang, 1994; *O. sesnoni* (Macdonald, 1967); *O. iamonensis* (Sellards, 1916); and *O. brachypus* (Cope, 1881).
Diagnosis—*Osbornodon* is mainly distinguished from other hesperocyonines by 2 synapomorphies: antero-posteriorly elongated (thus more quadrate) upper molars and enlargement of M2 and m2; and basined (bicuspid) talonid of m1-2. Later species of *Osbornodon* progressively acquire additional derived characters that set the genus further apart from other lineages of hesperocyonines: long rostrum; large frontal sinus; posteriorly expanded paroccipital process; short, medio-laterally flattened angular process; and enlarged p4 relative to p3.
Geologic and Geographic Range—Orellan–Whitneyan of the White River Group in northern Great Plains, late Arikareean–early Barstovian of Nebraska, Wyoming, and Hemingfordian–early Barstovian of Florida, New Mexico, and California.
Comments—*Osbornodon* represents one clade of hesperocyonines that has independently acquired many skull and dental features commonly found in the subfamilies Borophaginae and Caninae. It is superficially closer to the Caninae with such features as: slender rostrum, elongated premolars, quadrate M1s, and basined talonids. However, the talonid basin, a key synapomorphy of the sister-taxa Borophaginae and Caninae, is fundamentally different in *Osbornodon* from that of the two later subfamilies. The bicuspid talonid in *Osbornodon* lacks a transverse ridge between the hypoconid and entoconid, present in most borophagines and canines, and the talonid basin is surrounded by crest-like entoconid on the lingual side rather than a cusp-shaped entoconid as in borophagines and canines.

Osbornodon renjiei Wang, 1994
Figure 5

Holotype—F:AM 63316, maxillary and mandible fragments with right P4–M1, left c1–p3, and right p2, p4–m2.
Type Locality—Leo Fitterer Ranch locality, 13 miles south and 8 miles west of Dickinson, Stark County, North Dakota, late Orellan.
Diagnosis—*O. renjiei* is the most primitive species of *Osbornodon* and has no additional synapomorphy beside those for the genus (quadrate, enlarged M1–2 and basined talonid). In addition to its small size, *O. renjiei* is distinct from *O. sesnoni*, the next more derived species in *Osbornodon* clade, by its relatively broad premolars (primitive). Compared to *"Hesperocyon" coloradensis*, *O. renjiei* has a completely ventrally directed paroccipital process, a derived character shared with the *Mesocyon-Enhydrocyon* clade.
Geologic and Geographic Range—Orellan of southwestern North Dakota; Whitneyan of southwestern South Dakota and western Nebraska.

Comments—Smallest of all *Osbornodon*, *O. renjiei* is also the earliest and most primitive species of the genus. Besides the above diagnostic characters, it is little different from such primitive hesperocyonines as *"Hesperocyon" coloradensis* and *"Mesocyon" temnodon*, which are also of approximately the same size as *O. renjiei*.

Osbornodon sesnoni (Macdonald, 1967)

Holotype—LACM 17039, partial broken skull and mandible with left P1–M2, right I3–C1, P2–M2, left c1, p2–m3, and right c1, p3–m1; postcranial fragments.
Type Locality—Wolff Ranch, Shannon County, South Dakota. Poleslide Member of the Brule Formation, Whitneyan.
Diagnosis—Larger than *O. renjiei*, *Osbornodon sesnoni* has one derived character distinguishing it from the former: slender upper and lower premolars (high length/width ratio). The overall skull proportion of *O. sesnoni* is also slightly more robust than in *O. renjiei*. Compared to later species of *Osbornodon* (*O. iamonensis*, *O. brachypus*, and *O. fricki*), *O. sesnoni* shows the following primitive characteristics: relatively short rostrum; low sagittal crest; lack of an inflated frontal sinus; a round infraorbital foramen; unreduced bulla; paroccipital process not posteriorly expanded; mastoid process not reduced; p4 not enlarged; and a M2 less enlarged relative to M1.
Geologic and Geographic Range—*Osbornodon sesnoni* is known only in the Poleslide Member of Brule Formation, Whitneyan of southwestern South Dakota.
Comments—Beside its larger size and slender premolars, *Osbornodon sesnoni* is little different from the basal species *O. renjiei*. It is a transitional species that gave rise to the successively larger *O. iamonensis*, *O. brachypus*, and *O. fricki* in the late Arikareean to early Barstovian. There is a long hiatus of fossil record between *O. sesnoni* and the first occurrence of *O. iamonensis* in the Marsland Formation of western Nebraska. This is also reflected in the large morphological gap between these two species (see Diagnosis).

Paraenhydrocyon Wang, 1994

Type Species—*Paraenhydrocyon wallovianus* (Cope, 1881).
Included Species—*Paraenhydrocyon wallovianus* (Cope, 1881); *P. josephi* (Cope, 1881); *P. robustus* (Matthew, 1907).
Diagnosis—Members of *Paraenhydrocyon* can be distinguished from other genera of hesperocyonines in having three synapomorphies: a deeply pocketed angular process, a long, slender mandible with a deep masseteric fossa, and a dorso-ventrally compressed paroccipital process. The two more advanced species, *P. robustus* and *P. wallovianus*, are further derived in their narrow premolars with reduced/absent accessory cusps and in their reduced/absent metaconids on m1–2. *Paraenhydrocyon* retains a posteriorly directed paroccipital process, a

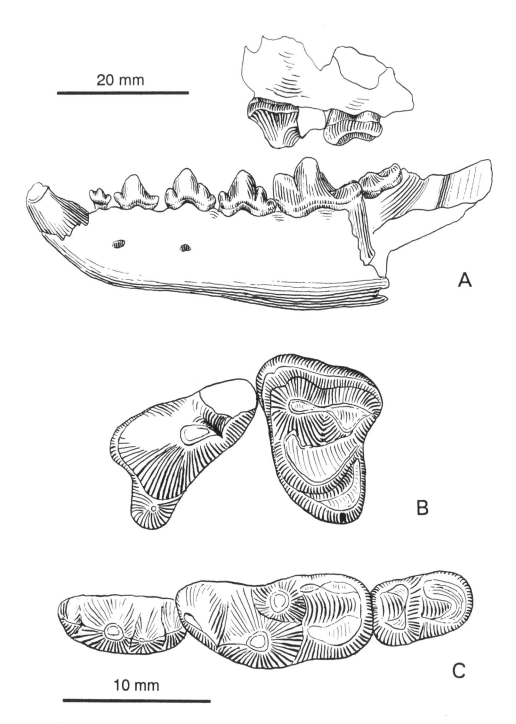

Figure 5. Mandible and teeth of *Osbornodon renjiei*, F:AM 63316, holotype, from Leo Fitterer Ranch, 13 miles south and 8 miles west of Dickinson, Stark County, North Dakota, Orellan. A, lateral view of upper P4–M1 and mandible (the anterior part between c1–p3 is reconstructed from that of left side); B, occlusal view of the upper teeth; and C, occlusal view of the lower teeth.

primitive character in contrast to ventrally oriented processes in most clades of hesperocyonines.

Geologic and Geographic Range—Brule Formation of southwestern South Dakota; Gering-Marsland formations of eastern Wyoming, western Nebraska, and western South Dakota; John Day Formation of central Oregon; Delaho Formation of Texas. Whitneyan to late Arikareean.

Comments—*Paraenhydrocyon* has striking parallels to *Enhydrocyon*, particularly in the development of hypercarnivorous carnassial teeth. The reduction or loss of the metaconid in lower first and second molars and the semicircular internal cingulum of the upper first molar makes isolated molars of the two genera difficult to differentiate. This dental similarity was noted by Loomis (1936), who thus placed *Temnocyon venator* (= *P. wallovianus*) under *Enhydrocyon*. However, there are some fundamental differences in the shape of the premolars and the proportions of the skulls between *Paraenhydrocyon* and *Enhydrocyon* that suggest independent evolutionary lineages. Only *P. josephi*, which has White River occurrence, is summarized below; the two more derived species of *Paraenhydrocyon* (*P. robustus* and *P. wallovianus*) occur only in the Arikareean, and will not be further discussed.

Paraenhydrocyon josephi (Cope, 1881)
Figure 6

Holotype—AMNH 6878, anterior half of skull with complete upper teeth except broken canines and incisors.

Type Locality—John Day Formation of Oregon, Arikareean.

Diagnosis—*P. josephi* is at the base of the *Paraenhydrocyon* clade and can be distinguished from other hesperocyonines by the synapomorphies of the genus: a deeply excavated masseteric fossa that expands downward so that the ventral margin of the fossa is closer to the ventral border of the mandible; a dorso-ventrally flattened paroccipital process; and an angular process of mandible with a deeply pocketed fossa for the superior ramus of the medial pterygoid muscle. *P. josephi* is distinct from the more advanced *P. robustus* and *P. wallovianus* mostly in such primitive dental structures as: a distinct anterior and posterior cingular cusp and a posterior accessory cusp in most of the premolars; presence of a small cingulum-like parastyle on P4; M1 trigon basin not deeply excavated; and relatively large metaconid on m1–2.

Geologic and Geographic Range—Brule Formation of southwestern South Dakota; Arikaree Group of eastern Wyoming and western Nebraska; John Day Formation of central Oregon. Whitneyan to Late Arikareean.

Comments—Except for its larger size and the derived characters listed above, *P. josephi* is quite similar to *Hesperocyon*. Furthermore, *P. josephi* retains a primitively posteriorly oriented paroccipital process as opposed to ventrally directed ones in all other derived clades of hesperocyonines. *P. josephi* is

postulated to be directly derived from *Hesperocyon gregarius* (Fig. 10). In fact, isolated individuals of *H. gregarius* may exhibit tendencies toward a pocketed angular process, a synapomorphy of *Paraenhydrocyon*. *P. josephi* gave rise to *P. robustus*, an intermediate species that eventually leads to the hypercarnivorous *P. wallovianus*.

Ectopocynus Wang, 1994

Type Species—*Ectopocynus simplicidens* Wang, 1994.

Included Species—*Ectopocynus simplicidens* Wang, 1994; *E. antiquus* Wang, 1994; *E. intermedius* Wang, 1994.

Diagnosis—*Ectopocynus* is distinguished on the basis of its short, blunt, and robust lower premolars with extremely reduced or absent accessory and cingular cusps, derived characters unique to the genus. Such simple cusp patterns of the premolars are distinct from those of *Enhydrocyon* which has similarly robust premolars but has well-developed accessory and cingular cusps. Except for one referred individual of *Ec. antiquus* from the Whitneyan, all have lost their p1s (derived). The metaconids of m1–2 are also significantly reduced even in the basal species *Ec. antiquus*, in contrast to other lineages of hesperocyonines which reduce their metaconids only in larger-sized, highly derived forms.

Geologic and Geographic Range—Poleslide Member of Brule Formation, western South Dakota; Lower Arikaree Group of eastern Wyoming; and Runningwater Formation of western Nebraska. Whitneyan to Middle Hemingfordian.

Comments—This rare lineage is known only from mandibles and a few upper teeth (all in the Frick Collection of the AMNH). The limited materials available do not permit a precise assessment of the phylogenetic position of *Ectopocynus* to other hesperocyonines. No synapomorphy unites it with any other clades, although a few characteristics associated with the robust mandible and teeth (e.g., loss of metaconids, loss of p1 and m3, etc.) are also found in derived species of *Enhydrocyon*. These characters are most likely to be independently derived, not only because of the present recognition of the primitive species *Ectopocynus antiquus* that does not possess these derived characters, but also because of the detailed anatomical differences between *Ectopocynus* and *Enhydrocyon* (see Diagnosis).

Only one specimen (F:AM 63376) is referred to *Ectopocynus antiquus* from the Brule Formation of South Dakota, which will be discussed below. The two later (Arikareean to Hemingfordian) and more advanced species, *E. intermedius* and *E. simplicidens*, are not further discussed.

Ectopocynus antiquus Wang, 1994

Holotype—F:AM 54090, partial right ramus with i3–c1, p2, p4–m3.

Type Locality—From Little Muddy Creek locality, in Lower Arikaree Group, southeast of Lusk, 20

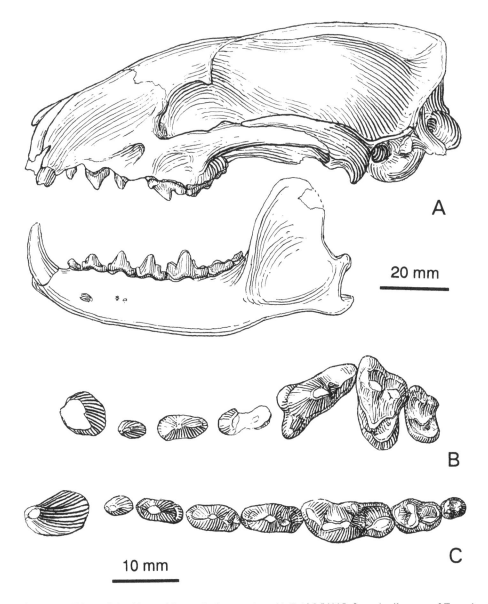

Figure 6. Skull, mandible, and dentitions of *Paraenhydrocyon josephi*, F:AM 54115, from 4 miles east of Tremain, Goshen County, Wyoming, in lower Arikaree Group, early Arikareean. A, lateral view of skull and mandible; B, occlusal view of upper left teeth; and C, occlusal view of lower left teeth, P4 restored from that of right side.

miles south of Keeline and west of Spanish Diggings, Niobrara County, Wyoming. Early Arikareean.

Diagnosis—Approximately the size of *"Hesperocyon" coloradensis*, *Ectopocynus antiquus* can be distinguished from the former by its loss of p1, its short but wide (in occlusal view) lower premolars, its reduced metaconid of m1, and its reduced m2–3, characters that are synapomorphies for the genus *Ectopocynus* (except the reduced metaconid, most of these characters are seen in holotype but not in F:AM 63376, see Comments below). In addition to its much smaller (21%) size than *E. intermedius*, *E. antiquus* also

has relatively smaller and narrower lower premolars and molars. *E. antiquus* retains primitive characters that can be used to differentiate it from *E. simplicidens*: the presence of a small metaconid on m1 and an m3.

Geologic and Geographic Range—Brule Formation of western South Dakota and Lower Arikaree Group of eastern Wyoming. Whitneyan to early Arikareean.

Comments—The presence of this small, *Hesperocyon*-sized canid in the Whitneyan of the White River Group is indicated by a single specimen, F:AM 63376, a partial right ramus from Hay Creek of Washabaugh

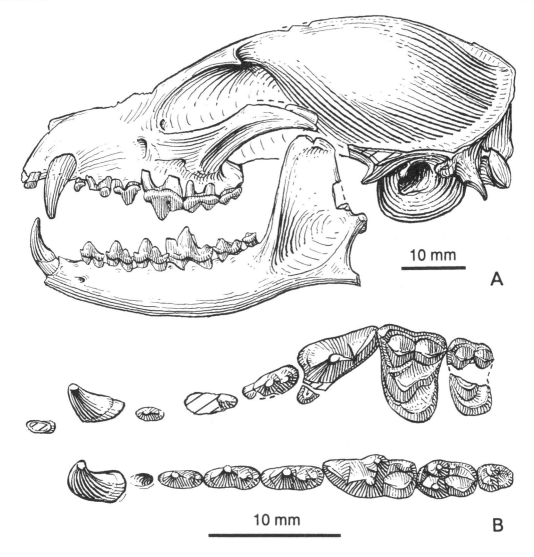

Figure 7. Skull, mandible, and dentitions of *Cormocyon pavidus*, F:AM 63970, from east side of the Roundhouse Rock, 5 miles southwest of Bridgeport, Morrill County, Nebraska, in upper part of the Whitney Member of the Brule Formation, Whitneyan. A, lateral view of skull and mandible; B, occlusal view of upper and lower left teeth.

County, South Dakota. The reference of F:AM 63376 to *Ectopocynus antiquus* is mainly based on its reduced metaconids on m1–2. Although reduction of metaconid has been independently developed in other clades of hesperocyonines (e.g., *Enhydrocyon* and *Paraenhydrocyon*), most lineages that acquired it had done so in association with a number of hypercarnivorous features including large size, robust premolars, etc. The reduction of the metaconid in *Ec. antiquus* is thus unique in its lack of association with other hypercarnivorous characters, and allows the recognition of F:AM 63367 as belonging to the *Ectopocynus* clade. On the other hand, the possession of many primitive characters in F:AM 63367 (e.g., presence of a p1 and unshortened premolars) helps to link the *Ectopocynus* clade directly to *Hesperocyon*.

Subfamily BOROPHAGINAE Simpson, 1945

Included White River Genera—*Cormocyon* Wang and Tedford, 1992, and *Oxetocyon* Green, 1954.

Geologic and Geographic Range of White River Taxa—Orellan through Arikareean of North America.

Diagnosis—The subfamily Borophaginae shares with the Caninae the following derived characteristics that distinguish them from the primitive hesperocyonines: a bicuspid talonid with a salient entoconid on the lower first molars, reduced labial cingulum and parastyle of upper molars, and metaconid larger than protoconid in m2. Borophagines after the Arikareean developed more synapomorphies, such as broadened and elongated posterior process of the premaxillary that meets the frontal, lateral accessory cusps on the upper third incisors, well-developed posterior accessory cusps on premolars, and transverse ridges on the talonid cusps of lower first molars.

Comments—The earliest Borophaginae can be recognized in the Orellan strata of the White River Group, although specimens of the borophagines are far less abundant than those of the hesperocyonines. Two White River species are presently known in the literature (summarized below), but additional undescribed taxa are present in the Frick Collection and the University of Nebraska.

Cormocyon pavidus (Stock, 1933)
Figure 7

Holotype—LACM 466, crushed anterior half skull and mandible with left P2–M1, right P3–M2, left and right p2–m2, and alveoli for p1 and m3.

Type Locality—Kew Quarry Local Fauna, Las Posas Hills, Ventura County, California. Sespe Formation, early Arikareean.

Diagnosis—The smallest of all White River canids, *Cormocyon pavidus* can be distinguished from species of *Hesperocyon* on the basis of a number of subtle but significant features, which lead to the recognition of the earliest Borophaginae: reduced labial cingulum and parastyle of M1, relatively large M2, a salient entoconid on m1 enclosing a talonid basin, and m2 metaconid slightly larger than protoconid. *C. pavidus* is still primitive compared to the more derived taxa in the Arikareean of John Day and Great Plains (e.g., *C. oregonensis*, *C. copei*, *C. leptodus*) in the following plesiomorphic conditions: unexpanded braincase, posteriorly directed paroccipital process, and rather transversely elongated upper molars which lack a well-developed metaconule.

Geologic and Geographic Range—Sespe Formation, California; Cedar Creek Member of White River Formation, Colorado; Whitney Member of Brule Formation, Nebraska; Poleslide Member of Brule Formation, South Dakota. Orellan to Whitneyan.

Comments—Since its initial description by Stock (1933), *Cormocyon pavidus* has remained an enigma avoided by most students of early canids. The type specimen's geographic location far from the main occurrences of *Hesperocyon* in the northern Great Plains, and its poor condition of preservation posed special difficulties in linking *C. pavidus* with taxa from the Great Plains.

Galbreath (1953) recognized a small-sized *Hesperocyon* from the Tertiary of northeastern Colorado and informally called it "*Pseudocynodictis* sp. (small form)." The Colorado specimens are here referred to *Cormocyon pavidus*. In addition, a few Whitneyan specimens from Nebraska and South Dakota are also referable to this species. In particular, F:AM 63970, a slightly crushed skull and mandible from the Whitneyan of Nebraska, represents the most complete individual available (Fig. 7). Little different from that of *Hesperocyon*, the skull of *Cormocyon pavidus* remains mostly

primitive in its posteriorly oriented paroccipital process, its unexpanded braincase, and its possession of a shallow suprameatal fossa.

The present reference of the Great Plains materials to *C. pavidus* greatly increases its geological and geographical distribution away from its formerly isolated occurrence in California. Our recognition of *C. pavidus* in the Plains establishes it as the earliest and most primitive borophagine, and thus provides an ideal model for the primitive morphotype of the Borophaginae (Wang, 1990, 1994). From this basal position, a number of primitive borophagines are derivable through slight dental modification and size increase, such as *C. leptodus* from the Arikareean of the Great Plains, and *C. oregonensis* and *C. copei* from the ?Whitneyan/Arikareean of John Day, Oregon. Also derivable from *C. pavidus*, is a small clade of hypocarnivorous taxa represented by *Oxetocyon* described below.

Oxetocyon cuspidatus Green, 1954
Figure 8

Holotype—SDSM 2980, left maxillary fragment with M1 and alveola of P4.

Type Locality—"*Protoceras* channels, 7 miles east of Rockyford, Shannon County, South Dakota" (Green, 1954).

Diagnosis—*Oxetocyon cuspidatus* is highly apomorphic among White River canids. It is characterized by a hypocarnivorous dentition with prominent developments of a metaconule and a "hypocone" created by a transverse notch on the internal cingulum of M1. The M2 is enlarged and similarly constructed as the M1. The P4 has a well-developed lingual cingulum although no hypocone is developed. Talonid cusps of the lower molars are conical in contrast to primitively crest-like cusps in *Hesperocyon* and *Cormocyon*. The protoconid and metaconid of the m2 are similarly cuspidate.

Geologic and Geographic Range—Whitneyan of South Dakota and Nebraska.

Comments—The peculiar M1 in *Oxetocyon* is unique among White River canids. The transverse groove on the internal cingulum is continuous with notches between the paracone/metacone and protocone/metaconule pairs so that the upper molars are evenly divided into anterior and posterior halves. Such a divided appearance of M1 allows unambiguous recognition of *Oxetocyon*, even based on the single M1 of the holotype.

Specimens collected subsequent to the first publication of the holotype have greatly increased our knowledge of this rare taxon. The morphology of the P4-M2 and part of the skull has been revealed (UNSM 2665, see Tanner, 1973), and the lower teeth and the horizontal ramus (UNSM 25698, partial right ramus with p4-m3 and alveola of p3, see Fig. 8) are presently referred to *Oxetocyon cuspidatus* based on agreement in size of the ramus with the upper jaws. Its p3 is very similar to that of *Cormocyon pavidus* with a distinct posterior accessory cusp and small anterior and posterior cingular cusps. The lower molars have the same proportion as

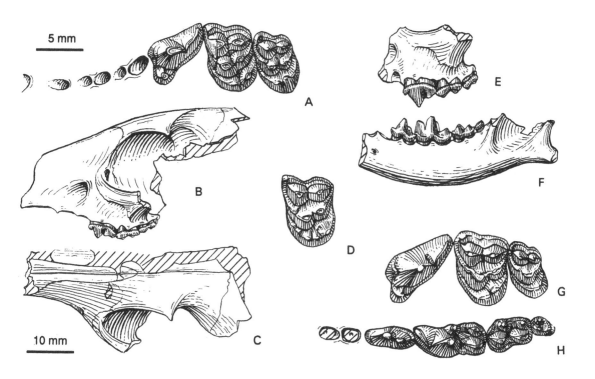

Figure 8. Skull and teeth of *Oxetocyon cuspidatus*. A–C, UNSM 2665, from Roundhouse Rock, Morrill County, Nebraska, in the Whitney Member of the Brule Formation; D, SDSM 2980, holotype, from 7 miles east of Rockyford, Shannon County, South Dakota, in the Poleslide Member of the Brule Formation; E and G, UNSM 25381, from 7 miles southeast of Broadwater, Morrill County, Nebraska, in the Whitney Member of the Brule Formation; F and H, UNSM 25698. The scale in A also applies to D, G, and H, and that in C to B, E, and F.

those of *C. pavidus*. However, the talonid cusps of m1 and all cusps of m2 are conical in form, instead of crest-like as in *Cormocyon* and all hesperocyonines. Such cuspidate (bunodont) cusp morphology is typically found in hypocarnivorous carnivorans, and forms the basis of our reference of UNSM 25698 to *Oxetocyon*, which is the only candidate for a hypocarnivorous canid in the right size range. The trigonid of the m1 is still relatively high for a hypocarnivorous taxon.

Subfamily CANINAE Gray, 1821
Leptocyon sp.
Figure 9

Comments—The earliest known records of the subfamily Caninae are represented by a series of small species of *Leptocyon* in the Arikareean and later deposits (Tedford and Taylor, MS). Although no published record exists for the presence of canines in pre-Arikareean deposits, our preliminary survey suggests potential candidates in the White River deposits. An Orellan specimen (UNSM 25354, left ramus fragment with p3–m1, from Sioux County, Nebraska, Fig. 9) shows such *Leptocyon*-like features as: slender, shallow ramus, slender premolars, complete loss of p3 posterior accessory cusp (that on p4 is still

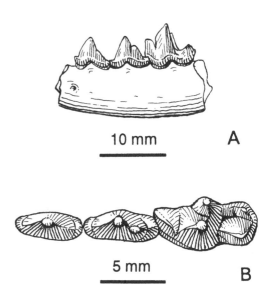

Figure 9. Mandible fragment of *Leptocyon* sp., UNSM 25354, from Sioux County, Nebraska, in the Orella Member of the Brule Formation. A, lateral view of ramus; B, occlusal view of lower teeth.

well developed), and well-developed entoconid crest on m1 nearly equal in height to that of hypoconid and occupying nearly half of the talonid basin. UNSM 25354 primitively has a short (closed) trigonid of the m1. Such an Orellan occurrence of the Caninae is consistent with a sister-group relationship between the Borophaginae (earliest record Orellan) and Caninae (Tedford, 1978).

PHYLOGENY OF EARLY CANIDS

The phylogeny shown in Figure 10 represents a synthesis of our cladistic analyses of the basal cynoids (Wang and Tedford, 1994), the hesperocyonines (Wang, 1994), the canines (Tedford and Taylor, MS), and our on-going research on primitive borophagines. Taxa are included if they have records in the White River Group (only published taxa are presented). Major characters that support this arrangement are listed in Table 1.

Cope (1877, 1883) proposed that the origin of canids was to be found among forms then placed in the Creodonta. More specifically, the "miacids" were postulated as the probable ancestral stock for canids (Cope, 1880, p. 81), and this broad statement has since gained nearly universal acceptance by almost every student of carnivorans (e.g., Matthew, 1930; Clark, 1939; Tedford, 1978; Gustafson, 1986), although considerable confusion exists concerning the precise phylogenetic relationships among the "miacids," and how they in turn relate to the living families of Carnivora. Wortman and Matthew (1899) erected the genus *Procynodictis* in the Uintan and indicated an ancestor-descendant relationship between *Procynodictis* and *Hesperocyon* (then known as *Cynodictis*). Clark (1939) more explicitly allied his late Uintan species *Miacis gracilis* (here synonymized with *Procynodictis vulpiceps*) to *Hesperocyon*. The main evidence cited by Clark was the canid-like postcranial skeleton of *P. vulpiceps*, e.g., slender and straight limb bones, reduction of deltoid ridge in humerus. However, as acknowledged by Clark (1939), the dentition of *P. vulpiceps* (Fig. 1A) is little advanced toward canids except its possession of a posterior accessory cusp in the P3 and p3. As in most "miacids," the P4 protocone of *P. vulpiceps* is still large and isolated from the base of paracone, in contrast to a reduced protocone more appressed to the paracone in *H. gregarius* and other canids. *P. vulpiceps* has a primitive M1 with a large parastyle and an internal cingulum surrounding the protocone, unlike that of *H. gregarius* whose parastyle is reduced to a narrow cingulum and whose internal cingulum is incomplete at the anterior segment. The m1 of *P. vulpiceps* is equally primitive with a wide trigonid relative to the talonid. Our recent cladistic analysis (Wang and Tedford, 1994) is in essential agreement with the conclusions by Wortman and Matthew (1899) and particularly Clark (1939), and

placed *P. vulpiceps* ("*Miacis*" *gracilis* in Wang and Tedford, 1994) as the closest sister-taxon of the Canidae.

The above difference in dental morphology between *Procynodictis* and *Hesperocyon* is partially bridged by the recent discovery of *Prohesperocyon wilsoni* from the Vieja Group of southwestern Texas (Gustafson, 1986). Teeth of *P. wilsoni* are slightly advanced toward the condition present in *Hesperocyon* in having a more reduced anterior portion of the internal cingulum of M1, reduction of M1 parastyle and loss of the parastyle notch, absence of M3 (Fig. 1B), and a slightly more elongated shearing blade of m1. The remaining dental morphology of *P. wilsoni* is still primitive and similar to those of "miacid" carnivorans: a large protocone on P4, a large parastyle on M1 (relative to that in *Hesperocyon*), a wide trigonid and a narrow talonid on m1, and an m2 with the protoconid larger than the metaconid. Our study of the basicranium further demonstrates that the middle ear region of *P. wilsoni* is essentially canid-like except for its lack of a low septum within the well ossified entotympanic bulla (Wang and Tedford, 1994; Table 1, node 1).

The advanced status of the White River and more derived canids over *Prohesperocyon wilsoni* is well supported by a number of basicranial and dental characters: a partial bony septum on the internal surface of the bulla, a reduced protocone of P4, a M1 with a reduced parastyle, an elongated shearing blade (trigonid) of m1, and nearly equal size of metaconid and protoconid on m2 (node 2). Resolutions of phylogenetic relationships among the White River and later canids, however, are poor (multichotomy at node 2) despite well-defined autapomorphies for subsequent clades. *Hesperocyon*, as presently defined, is clearly paraphyletic. Paraphyly would not be avoidable even for the single type species *H. gregarius* if the ancestor-descendant relationships between it and other hesperocyonines, as well as the borophagines and canines, are correctly reconstructed (Fig. 10). The Hesperocyoninae is also paraphyletic for lack of a synapomorphy uniting all of its genera.

The clade represented by such familiar genera as *Mesocyon* and *Enhydrocyon* has the best documented fossil records (node 6). Indeed, this clade is the most enduring part of canid phylogenies proposed by many authors (Cope, 1883; Matthew, 1930; Macdonald, 1963; Tedford, 1978). It involves a progressive increase in hypercarnivory from a small ancestral "*H.*" *coloradensis*, through median-sized *Mesocyon* (paraphyletic) and *Cynodesmus*, to the transitional *Sunkahetanka* and *Philotrox*, and ending with the large *Enhydrocyon*.

The recently recognized clade *Osbornodon* is characterized by quadrate and enlarged upper molars, basined talonids on lower molars, narrowed premolars, as well as a suite of synapomorphies (e.g., long rostrum,

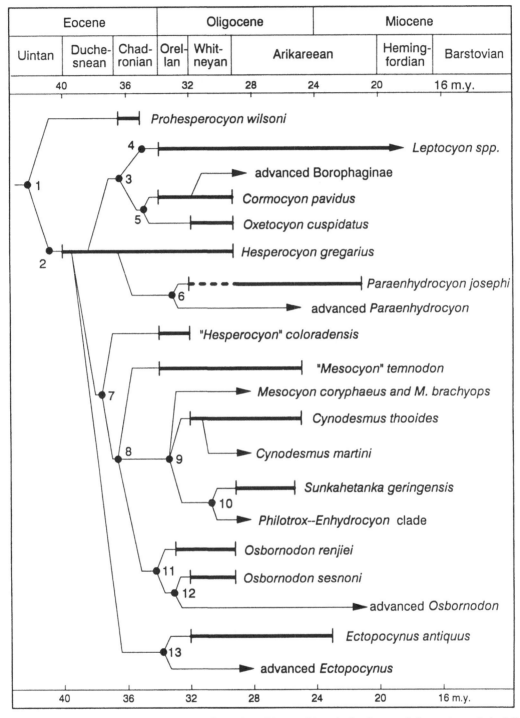

Figure 10. Phylogeny and geologic distribution of the White River canids and related taxa; phylogenetic analysis follows Tedford and Taylor (MS), Wang (1994), and Wang and Tedford (1994). Numbers beside the nodes (black dots) correspond to those in Table 1. Arrows lead to non-White River canids not reviewed in this paper, and indicate the time of origin (at base of the arrow) of these taxa. In some cases (*Paraenhydrocyon* and *Cynodesmus*), ancestral-descendant relationships are suggested by stratigraphic and/or morphologic continuity. Stratigraphic correlations are based on Emry et al. (1987) and Tedford et al. (1987). Calibrations by radiometric dates are based on Swisher and Prothero (1990) and Prothero and Swisher (1992).

Table 1. Shared-derived characters for early canids as identified at major nodes in Fig. 10. Character analysis for hesperocyonine part of the cladogram is based on Wang (1994).

Node 1, Family Canidae
Ossified entotympanic bulla; extrabullar position of internal carotid artery and loss of stapedial artery; presence of a small suprameatal fossa; reduction of M1 parastyle and loss of a parastyle notch; reduction of anterior segment of internal cingulum of M1; absent or extremely reduced M3; presence of posterior accessory cusps on P3 and p3.

Node 2, Subfamily Hesperocyoninae (minus Node 3, Borophaginae–Caninae clade)
Low septum inside bulla; reduced protocone of P4; further reduction of parastyle of M1; elongated shearing blade of m1; metaconid and protoconid of m2 subequal.

Node 3, Borophaginae–Caninae clade
Bicuspid talonid of m1 with a salient entoconid; reduction of parastyle in upper molars; internal cingulum of M1 expanded anteriorly to surround the protocone.

Node 4, Subfamily Caninae (as defined by primitive *Leptocyon*)
Narrow premolars with reduced posterior accessory cusps; presence of diastemata between premolars; elongated shearing blade of m1; slender horizontal ramus.

Node 5, Subfamily Borophaginae
Primitive borophagines (e.g., *Cormocyon pavidus*) have no synapomorphies distinct from the Caninae clade, although slightly more derived borophagines progressively acquired characters such as: strong posterior premaxillary process fused with the frontal process, and presence of lateral accessory cusps on upper third incisors.

Node 6, *Paraenhydrocyon* clade
Dorso-ventrally compressed paroccipital process; deep masseteric fossa which forms a pocket at the ventral border of the fossa; medially pocketed angular process of mandible.

Node 7, *"Hesperocyon" coloradensis–Osbornodon* clade
Paroccipital process beginning to rotate ventrally and to touch the posterior base entotympanic.

Node 8, *"Mesocyon" temnodon–Osbornodon* clade
Paroccipital process completely ventrally directed and extensively fused with posterior surface of entotympanic.

Node 9, *Mesocyon–Enhydrocyon* clade
Fossa on supraoccipital above foramen magnum; paroccipital elongated and robust.

Node 10, *Sunkahetanka–Enhydrocyon* clade
Massive premolars that tend to be imbricated; reduced metaconid on lower molars.

Node 11, *Osbornodon* clade
Enlarged entoconid to encircle a basined talonid of m1; quadrate and enlarged upper molars.

Node 12, *Osbornodon sesnoni* and advanced *Osbornodon*
Narrowed premolars.

Node 13, *Ectopocynus* clade
Short, blunt premolars lacking accessory and cingular cusps; reduced metaconid of lower molars; loss of p1.

frontal sinus, and enlarged p4) occurring only in non-White River species (*O. iamonensis*, *O. brachypus*, and *O. fricki*). *Osbornodon* is united with the *Mesocyon-Enhydrocyon* clade by a ventrally directed paroccipital process, a character known to be homoplastic in canids.

The *Paraenhydrocyon* clade appears to be directly descended from *Hesperocyon* (as speculated in Fig. 10). In fact, dental morphology of its primitive species *P. josephi* is little different from that of *H. gregarius*. In addition, the basicranium of *Paraenhydrocyon* is conservative in having a posteriorly oriented paroccipital process (primitive), which turns ventrally in all other canids when a body size larger than *Hesperocyon* is acquired.

The mostly Arikareean *Ectopocynus* is linked to *Hesperocyon* through a referred specimen (F:AM 63376) of *E. antiquus* in the Whitneyan which, except for its reduced metaconid on m1, is very similar to *Hesperocyon*. Other than such primitive similarity, however, there is no shared-derived character in *Ectopocynus* to indicate special relationship with other clades, although *Ectopocynus* does share a few hypercarnivorous features, here deemed convergences, with *Enhydrocyon*.

The relationship between *Hesperocyon gregarius* and the borophagine clade is bridged by *Cormocyon pavidus*, one of the smallest and most primitive borophagines recognized in the White River Group. Little different from *Hesperocyon*, *C. pavidus* serves as an ideal morphotype for the Borophaginae. Its apomorphic characters thus distinguish the borophagine clade: bicuspid talonid of m1, reduced labial cingulum and parastyle of M1, enlarged M2, and m2 metaconid slightly larger than paraconid. Besides *Cormocyon pavidus*, the Borophaginae of the White River Group had evolved a distinct hypocarnivorous clade as represented by *Oxetocyon cuspidatus*. Teeth of the *Oxetocyon* are highly derived and suddenly appear in the Whitneyan without primitive precursors.

A sister-group relationship between the Borophaginae and Caninae is supported by basined talonids on the m1, reduced parastyle of upper molars, and enlarged metaconids of m2. The canine clade is further distinguished from the borophagine clade by dental synapomorphies present in primitive *Leptocyon*: narrow premolars, reduction of posterior accessory cusps in premolars, presence of diastema between premolars, elongated shearing blade (open trigonid) of lower carnassial, and slender horizontal ramus.

CONCLUSIONS

If the reconstructed pattern of relationship (Fig. 10) is reasonably accurate, the timespan of the White River Group corresponded to a spectacular period of adaptive radiation in the Canidae. Sister-group relationships and the chronological appearance of taxa imply that the major diversification shown in Figure 10 had taken place by the beginning of Oligocene time. The Duchesnean and Chadronian record lacks the predicted diversity, which must have resulted from speciation events peripheral to the mid-continent where the geologic record is preserved. The recorded diversity was assembled in Orellan time corresponding to marked climatic change toward drier and more seasonal climates and presumably greater diversity of habitats. It might be hypothesized that canid cladogenesis took place in locations where such diversity existed in the late Eocene possibly in more northerly or higher elevation sites in habitats that became widespread at lower latitudes and elevations in the early Oligocene.

ACKNOWLEDGMENTS

Part of this study was initiated as a Ph.D. research for one of the authors (XW), who would like to express his sincere gratitude to members of the dissertation committee: Drs. Larry D. Martin, Robert S. Hoffmann, Robert W. Wilson, and Hans-Peter Schultze. An earlier draft of this manuscript was read by the above listed committee members. We would like to thank the anonymous reviewers for their valuable comments and suggestions that greatly improved this paper. We would also like to acknowledge the prodigious field work carried out over a 40-year span by the Frick Laboratory parties, especially by Charles Falkenbach, Ted Galusha, and Morris Skinner, that brought together the remarkable sample of canid remains that formed the subject of this report. Figures 4 and 6 were prepared in the Frick Laboratory of the AMNH before the 1960s; the artists are not known. Figures 7-9 are the work of Ray Gooris. George Corner of the University of Nebraska provided stratigraphic information about the Orellan *Leptocyon*.

LITERATURE CITED

Barbour, E. H., and C. B. Schultz. 1935. A new Miocene dog, *Mesocyon geringensis*, sp. nov. Bulletin of Nebraska State Museum 1:407–418.

Bryant, H. N. 1992. The Carnivora of the Lac Pelletier Lower Fauna (Eocene: Duchesnean), Cypress Hills Formation, Saskatchewan. Journal of Paleontology 66:847–855.

Clark, J. 1939. *Miacis gracilis*, a new carnivore from the Unita Eocene (Utah). Annals of the Carnegie Museum 27:349–370.

Clark, J., J. R. Beerbower, and K. K. Kietzke. 1967.

Oligocene sedimentation, stratigraphy, paleoecology and paleoclimatology in the Big Badlands of South Dakota. Fieldiana Geology Memoir 5:1–158.

Cope, E. D. 1873a. Third notice of extinct Vertebrata from the Tertiary of the plains. Palaeontological Bulletin 16:1–8.

Cope, E. D. 1873b. Synopsis of new Vertebrata from the Tertiary of Colorado obtained during the summer of 1873. Seventh Annual Report of the United States Geological Survey of the Territories, pp. 1–19.

Cope, E. D. 1874. Report on the vertebrate paleontology of Colorado. Annual Report Geological and Geographical Survey of the Territories for 1873, F. V. Hayden, U. S. Geologist, Washington, D.C., pp. 427–533.

Cope, E. D. 1877. Report upon the extinct Vertebrata obtained in New Mexico by parties of the expedition of 1874. Report upon United States Geological Surveys west of the one hundredth meridian, Part II 4:1–324.

Cope, E. D. 1879. Observations on the faunae of the Miocene Tertiaries of Oregon. Bulletin of the United States Geological and Geographical Survey of the Territories 1880 5:55–69.

Cope, E. D. 1880. On the genera of the Creodonta. Proceedings of American Philosophical Society of Philadelphia 19:76–82.

Cope, E. D. 1881. On the Nimravidae and Canidae of the Miocene Period. Bulletin of the United States Geological and Geographical Survey of the Territories 6:165–181.

Cope, E. D. 1883. On the extinct dogs of North America. American Naturalist 17:235–249.

Cope, E. D. 1884. The Vertebrata of the Tertiary formations of the west, book I. Report of the U. S. Geological Survey of the Territories 3:1–1009.

Elliott, R. J., 1980. A population study of the systematics and stratigraphic variation of *Hesperocyon* (Mammalia, Canidae). M.S. Thesis, South Dakota School of Mines and Technology, Rapid City, 178 pp.

Emry, R. J., P. R. Bjork, and L. S. Russell. 1987. The Chadronian, Orellan, and Whitneyan North American land mammal ages; pp. 118–152 *in* M. O. Woodburne (ed.), Cenozoic Mammals of North America, Geochronology and Biostratigraphy. University of California Press.

Flynn, J. J., N. A. Neff, and R. H. Tedford. 1988. Phylogeny of the Carnivora; pp. 73–116 *in* M. J. Benton (ed.), The Phylogeny and Classification of the Tetrapods, Vol. 2: Mammals. Systematics Association Special Volume No. 35B, Clarendon Press, Oxford.

Galbreath, E. C. 1953. A contribution to the Tertiary geology and paleontology of northeastern Colorado. University of Kansas Paleontological Contributions 4:1–120.

Green, M. 1954. A cynarctine from the Upper Oligocene of South Dakota. Transactions of the Kansas Academy of Science 57:218–220.

Gustafson, E. P. 1986. Carnivorous mammals of the Late Eocene and Early Oligocene of Trans-Pecos Texas. Texas Memorial Museum Bulletin 33:1–66.

Hough, J. R. 1948. The auditory region in some members of the Procyonidae, Canidae, and Ursidae, its significance in the phylogeny of the carnivora. Bulletin of the American Museum of Natural History 92:71–118.

Hough, J. R., and R. Alf. 1956. A Chadron mammalian fauna from Nebraska. Journal of Paleontology 30:132–140.

Loomis, F. B. 1936. Three new Miocene dogs and their phylogeny. Journal of Paleontology 10:44–52.

Macdonald, J. R. 1963. The Miocene faunas from the Wounded Knee area of western South Dakota. Bulletin of the American Museum of Natural History 125:141–238.

Macdonald, J. R. 1967. A new species of Late Oligocene dog, *Brachyrhynchocyon sesnoni*, from South Dakota. Los Angeles County Museum Contribution in Sciences 126:1–5.

Martin, L. D. 1989. Fossil history of the terrestrial Carnivora; pp. 536–568 in J. L. Gittleman (ed.), Carnivoran Behavior, Ecology, and Evolution, Cornell University Press, Ithaca, N. Y.

Matthew, W. D. 1901. Fossil mammals of the Tertiary of northeastern Colorado. Memoirs of the American Museum of Natural History 1:355–448.

Matthew, W. D. 1903. The fauna of the *Titanotherium* beds at Pipestone Springs, Montana. Bulletin of the American Museum of Natural History 19:197–226.

Matthew, W. D. 1907. A lower Miocene fauna from South Dakota. Fossil Vertebrates in the American Museum of Natural History 3:169–219.

Matthew, W. D. 1930. The phylogeny of dogs. Journal of Mammology 11:117–138.

Merriam, J. C. 1906. Carnivora from the Tertiary formations of the John Day region. University of California Bulletin of Department of Geological Science 5:1–64.

Mitchell, E., and R. H. Tedford. 1973. The Enaliarctinae, a new group of extinct aquatic Carnivora and a consideration of the origin of the Otariidae. Bulletin of the American Museum of Natural History 151:203–284.

Prothero, D. R., and C. C. Swisher III. 1992. Magnetostratigraphy and geochronology of the terrestrial Eocene–Oligocene Transition in North America; pp. 46–73 in D. R. Prothero and W. A. Berggren (eds.), Eocene-Oligocene Climatic and Biotic Evolution. Princeton University Press, Princeton, N. J.

Scott, W. B. 1890. The dogs of the American Miocene. Princeton College Bulletin 2:37–39.

Scott, W. B. 1893. The mammals of the Deep River Beds. American Naturalist 27:659–662.

Scott, W. B. 1898. Notes on the Canidae of the White River Oligocene. Transactions of the American Philosophical Society Philadelphia 19:325–415.

Scott, W. B., and G. L. Jepsen. 1936. The mammalian fauna of the White River Oligocene, Part I. Insectivora and Carnivora. Transactions of the American Philosophical Society Philadelphia 28:1–980.

Sellards, E. H. 1916. Fossil vertebrates from Florida: a new Miocene fauna; new Pliocene species; the Pleistocene fauna. The Eighth Annual Report of the Florida State Geological Survey 77–160.

Stock, C. 1933. Carnivora from the Sespe of the Las Posas Hills, California. Carnegie Institute of Washington Publication, Contributions to Palaeontology IV 440:29–41.

Swisher, C. C., and D. R. Prothero. 1990. Single-crystal ^{40}Ar/^{39}Ar dating of the Eocene-Oligocene transition in North America. Science 249: 760–762.

Tanner, L. G. 1973. Notes regarding skull characteristics of *Oxetocyon cuspidatus* Green (Mammalia, Canidae).

Transactions of the Nebraska Academy of Sciences 2:66–69.

Tedford, R. H. 1976. Relationship of pinnipeds to other carnivorans (Mammalia). Systematic Zoology 25:363–374.

Tedford, R. H. 1978. History of dogs and cats: a view from the fossil record; pp. 1–10 *In* Nutrition and Management of Dogs and Cats, Ralston Purina Company, St. Louis.

Tedford, R. H., T. Galusha, M. S. Skinner, B. E. Taylor, R. W. Fields, J. R. Macdonald, M. F. Rensberger, S. D. Webb, and D. P. Whistler. 1987. Faunal succession and biochronology of the Arikareean through Hemphillian interval (Late Oligocene through Earliest Miocene epochs) in North America; pp. 153–210 in M. O. Woodburne (ed.), Cenozoic Mammals of North America, Geochronology and Biostratigraphy. University of California Press, Berkeley.

Tedford, R. H., J. B. Swinehart, R. M. Hunt Jr., and M. R. Voorhies. 1985. Uppermost White River and lowermost Arikaree rocks and faunas, White River Valley, northwestern Nebraska, and their correlation with South Dakota; in J. E. Martin (ed.), Fossiliferous Cenozoic Deposits of Western South Dakota and Northwestern Nebraska, A Guidebook for the 45th Annual Meeting of the Society of Vertebrate Paleontology. Dakoterra 2:335–352.

Tedford, R. H., and B. E. Taylor. MS. North American fossil Canidae (Mammalia: Carnivora): the tribe Canini (Caninae).

Van Valkenburgh, B. 1991. Iterative evolution of hypercarnivory in canids (Mammalia: Carnivora): evolutionary interactions among sympatric predators. Paleobiology 17:340–362.

Wang, X. 1990. Systematics, functional morphology, and evolution of primitive Canidae (Mammalia: Carnivora). Ph.D. dissertation, University of Kansas, Lawrence, 222 pp.

Wang, X. 1993. Transformation from plantigrady to digitigrady: functional morphology of locomotion in *Hesperocyon* (Canidae: Carnivora). American Museum Novitates 3069:1–23.

Wang, X. 1994. Phylogenetic systematics of the Hesperocyoninae (Carnivora: Canidae). Bulletin of the American Museum of Natural History 221:1–207.

Wang, X., and R. H. Tedford. 1992. The status of genus *Nothocyon* Matthew, 1899 (Carnivora): an arctoid not a canid. Journal of Vertebrate Paleontology 12:223–229.

Wang, X., and R. H. Tedford. 1994. Basicranial anatomy and phylogeny of primitive canids and closely related miacids (Carnivora: Mammalia). American Museum Novitates 3092:1–34.

Wilson, J. A., P. C. Twiss, R. K. DeFord, and S. E. Clabauch. 1968. Stratigraphic succession, potassium-argon dates, and vertebrate faunas, Vieja Group, Rim Rock country, Trans-Pecos Texas. American Journal of Science 266:590–604.

Wortman, J. L., and W. D. Matthew. 1899. The ancestry of certain members of the Canidae, the Viverridae, and Procyonidae. Bulletin of the American Museum of Natural History 12:109–139.

22. Nimravidae

HAROLD N. BRYANT

ABSTRACT

The Eocene to Oligocene Nimravinae was the first radiation of cat-like carnivorans. This radiation was centered in North America. Generic and specific taxonomy and temporal ranges are discussed. Eleven North American species, in six genera, are recognized as valid. Cladistic analysis based on a preliminary character analysis provides strong support for clades consisting of *Nimravus* and *Dinaelurus* (Nimravini) and the species referred to *Hoplophoneus* and *Eusmilus* (Hoplophoneini). *Pogonodon* and the Hoplophoneini are sister taxa, and *Dinictis* and *Hoplophoneus* are probably paraphyletic. *Dinictis*, *Pogonodon* and the Hoplophoneini display moderate to extreme development of sabertooth morphologies, whereas the Nimravini lacks these cranial and mandibular features and *Dinaelurus crassus* has conical teeth. Species diversity increased from the Chadronian to the Whitneyan and declined in the early Arikareean. Although the postcranial skeleton of the nimravines most resembles that of modern carnivorans that inhabit closed forest habitats, the diversification of the group coincided with the initial stages of the development of grassland habitats in central North America during the Eocene to Oligocene transition. Nonetheless, the extinction of the clade at the end of the early Arikareean might be associated with the widespread establishment of grassland ecosystems in the late Oligocene.

INTRODUCTION

The Nimravidae are cat-like, predominately saber-toothed, carnivorans of late Eocene to late Miocene age that are known from North America, Europe, Asia, and Africa. Nimravids superficially resemble the true cats (Felidae) in their cranial morphology, hypercarnivorous dentition, and retractile claws. Scott and Jepsen (1936), in the last comprehensive review of the Carnivora of the White River deposits, followed a long tradition of referral of all cat-like carnivorans to the Felidae and the subdivision of the family primarily on the basis of dental characters (e.g., Matthew, 1910). Recently, our understanding of the phylogenetic relationships of cat-like carnivorans has undergone a major conceptual change. The North American Eocene to Oligocene "cats" are now referred to a separate clade, the Nimravidae, and are no longer considered most closely related to the Felidae (Tedford, 1978; Baskin, 1981; Hunt, 1987; Flynn et al., 1988; Bryant, 1991a). The Nimravidae includes two subclades, the late Eocene to Oligocene Nimravinae and the middle to late Miocene Barbourofelinae (Bryant, 1991a).

This chapter is a synopsis and review of the generic and specific taxonomy and the biostratigraphy of the North American members of the Nimravinae. Eleven North American species in six genera are recognized as valid; four additional species of uncertain or indeterminate status are also discussed. A preliminary phylogenetic analysis of valid species is presented in the appendix (Fig. 1). The higher-level taxonomy of the Nimravidae was discussed by Bryant (1991a); only abbreviated diagnoses of the Nimravidae and Nimravinae that emphasize features pertinent to relationships within the Nimravinae and generic and specific diagnoses are included here. *Nimravus* and *Dinaelurus* constitute a clade (Nimravini) which displays little or no development of sabertooth features, more rounded cheek teeth without serrated ridges, and a relatively gracile skeleton. *Dinictis* and *Pogonodon* have a more hypsodont C1 and some sabertooth cranial modifications, more trenchant, fully serrated cheek teeth, and a lightly built skeleton. *Hoplophoneus* and *Eusmilus* constitute a clade (Hoplophoneini) which has marked to extreme sabertooth features, higher crowned carnassials, conical incisors and reduced premolars, and a robust, heavily built skeleton. This last clade is the most speciose and morphologically diverse. Cladistic analysis (Fig. 1) suggests that *Dinictis* and *Hoplophoneus* are paraphyletic.

The systematic paleontology of genera and species relies heavily on previous taxonomic revisions of *Nimravus* (Toohey, 1959), *Hoplophoneus* (Simpson, 1941; Hough, 1949; Morea, 1975), and *Eusmilus* (Morea, 1975), and includes a preliminary reconsideration of the taxonomy of *Dinictis* and *Pogonodon*, which have not been revised since Scott and Jepsen (1936). The generic-level taxonomy follows recent usage and is provisional pending more detailed revision at the species level and cladistic analysis of valid species. In the absence of more detailed study of hypodigms, generic and specific diagnoses are necessarily somewhat

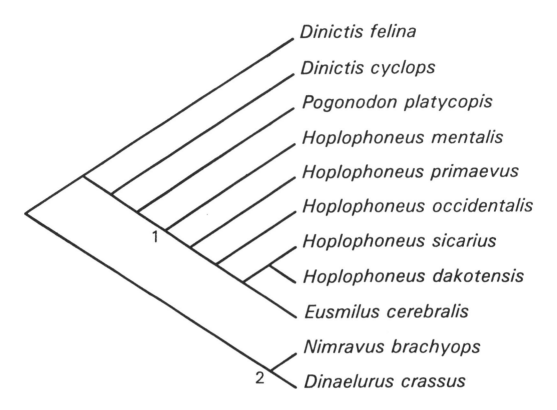

Figure 1. Suggested phylogenetic relationships within the North American Nimravinae. Most parsimonious tree (43 steps; CI = 0.814; RI = 0.884) based on a branch and bound analysis of a preliminary data matrix of 27 characters (Table 1). Multistate characters were unordered. Labeled nodes: 1, Hoplophoneini; 2, Nimravini.

imprecise. Because some genera are probably not monophyletic (Fig. 1) diagnoses do not always comply completely with cladistic principles. Diagnoses may include apomorphies, plesiomorphies (as indicated by the phylogenetic analysis), autapomorphies (unique apomorphies not included in the phylogenetic analysis), and additional descriptive features (undetermined phylogenetic status), as appropriate. Skull lengths (SL) are condylobasal. Humeroradial and femorotibial indices equal the ratio of the length of the distal element to that of the proximal element expressed as a percentage and are taken from Gonyea (1976).

Preliminary temporal ranges of recognized species are presented (Fig. 2) but determination of more precise ranges must await further biostratigraphic study of museum collections, especially the Frick Collection at the American Museum of Natural History. Nimravids are well represented in the highly fossiliferous Chadronian to Whitneyan White River deposits in North and South Dakota, Wyoming, Nebraska, and Colorado, and are known from contemporary deposits in Montana and Saskatchewan. The relatively large, stratigraphically well-constrained samples from the White River deposits, together with magnetostratigraphy and the geochronology provided by the dating of

numerous ash beds, provide the potential for the development of precise biostratigraphic ranges of species.

In the early Arikareean the geographic distribution of the group broadened to include the California Coast Ranges and Oregon, but the record from the Great Plains is more limited. This geographic range extension may be an artifact of the broader distribution of early Arikareean deposits. The youngest portions of the temporal range of the early Arikareean nimravines are recorded in the John Day Formation in Oregon. The early collections lack a precise stratigraphic context and determination of the exact younger limit of temporal ranges must await detailed biostratigraphic study of these taxa. Merriam and Sinclair (1907) indicated that nimravines occur in the middle and upper John Day Formation. Fisher and Rensberger (1972) renamed the middle unit (Turtle Cove Member) and divided the upper unit into the Kimberly and Haystack Valley members; the latter is Hemingfordian and does not include nimravines. Although color is important for distinguishing between the Turtle Cove and Kimberly members, it is not completely reliable and the boundary is temporally transgressive (Fisher and Rensberger, 1972). Biostratigraphic (e.g., Rensberger, 1971, 1973)

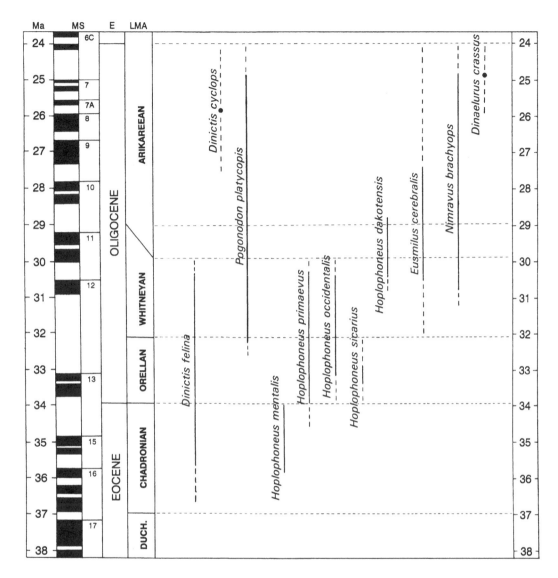

Figure 2. Temporal ranges of nimravine species. Minimum ranges are indicated by solid lines; possible, but unconfirmed, extensions to those ranges are indicated by dashed lines. Dots with dashed lines indicate uncertainty regarding the level of origin of specimens from the John Day Formation, Oregon. Correlations among the numerical time scale, magneto-stratigraphy, epochs and land mammal ages are after Prothero and Swisher (1992) (>27 Ma) and Berggren et al. (1985) (<27 Ma). Correlations among the first three differ slightly from those of Cande and Kent (1992). Abbreviations: E, epoch; DUCH., Duchesnean Land Mammal "Age"; LMA, North American Land Mammal "Age"; Ma, million years ago; MS, magnetostratigraphy.

and paleomagnetic studies (Prothero and Rensberger, 1985) are providing an improved temporal context for the John Day Formation.

ABBREVIATIONS

AMNH, American Museum of Natural History, New York; ANSP, Academy of Natural Sciences, Philadelphia; F:AM, Frick American Mammals, AMNH, New York; KU, University of Kansas,

Lawrence; LACM, Los Angeles County Museum of Natural History, Los Angeles; PU, Princeton University Collection, Yale University, New Haven; SDSM, South Dakota School of Mines, Rapid City; SMNH, Royal Saskatchewan Museum, Regina; UCM, University of Colorado Museum, Boulder; UCMP, University of California Museum of Paleontology, Berkeley; UNSM, University of Nebraska, Lincoln; USNM, National Museum of Natural History, Smithsonian

Institution, Washington, D.C.; YPM, Peabody Museum, Yale University, New Haven.

SYSTEMATIC PALEONTOLOGY
CARNIVORA Bowdich, 1821
FELIFORMIA Kretzoi, 1945
NIMRAVIDAE Cope, 1880a

Definition—The most recent common ancestor of Nimravinae and Barbourofelinae, and all of its descendants.

Partial Diagnosis—Synapomorphies: skull with short rostrum and mesocranium (cat-like); palate short, not extending posteriad of the distal end of the tooth row; C1 significantly larger than c1; loss or reduction of the anterior premolars and posterior molars so that P1/p1 usually absent, P2/p2 small, M1 and m2 small, and M2, M3 and m3 usually absent; dC1 moderately to extremely enlarged; hourglass-shaped serrations on at least the distal margin of the canines (see Bryant, 1991a, for complete diagnosis).

NIMRAVINAE Cope, 1880a

Definition—The most recent common ancestor of Hoplophoneini and Nimravini, and all of its descendants.

Included Genera—*Dinictis, Pogonodon, Hoplophoneus, Eusmilus, Nimravus, Dinaelurus.*

Distribution—?Late Duchesnean or earliest Chadronian to end of the early Arikareean in North America; early and middle Oligocene of Europe; Oligocene of Asia.

Partial Diagnosis—Synapomorphies: various attributes of the basicranium and auditory bulla, including a thick entotympanic that does not fuse with the ectotympanic and a bulla that is cartilaginous ventrally (Neff, 1983; Hunt, 1987); premolars without complete distal cingulum; P4 without cingular parastyle (see Bryant, 1991a, for complete diagnosis).

Plesiomorphies relevant to relationships within the Nimravinae (see Appendix: character analysis): sutural contact between the lacrimal and jugal; masseteric fossa on the lateral surface of the maxilla and jugal shallow or absent; no fossa on the ventral surface of the zygomatic arch below the postorbital process; anterior opening of palatine canal level with P3; postglenoid foramen large; paroccipital process large and projecting posteriorly; little or no development of features characteristic of sabertoothed carnivores (see Emerson and Radinsky, 1980); C1 hypsodont and blade-like (moderately laterally compressed); anteroposterior length of C1 less than that of P4; P3 size typical of that of most carnivorans; P4 with distinct protocone; P4 crown height typical of carnivorans; no lingual rotation of P4 as it wears; M1 with strong parastylar wing, distinct protocone and reduced metacone; m1 with small metaconid; m1 talonid with a single trenchant cusp; atlas with relatively short dorsal arch and atlantal foramen in medial position; femorotibial index similar to that of open woodland- to savanna-dwelling felids; no articulation between the calcaneum and navicular.

Discussion—Although the earliest nimravine material of well-constrained age that can be referred to a particular species comes from early in the middle Chadronian (*Hoplophoneus mentalis*; just under 36 Ma), nimravines are known from the early Chadronian and possibly from the Duchesnean. A nimravid canine tooth was recovered from the early Chadronian 60-foot level of the section at Flagstaff Rim, Wyoming (Emry, 1992, fig. 5.2–"paleofelid"). Fragmentary remains of nimravines, possibly representing both *Dinictis* and *Hoplophoneus*, occur in the Calf Creek Local Fauna, Cypress Hills Formation, Saskatchewan (Bryant, 1993); although this fauna is usually considered of medial Chadronian age, it may date from late in the early Chadronian (Storer, 1978, 1981, this volume, Chapter 12). If the Hancock Quarry, Clarno Formation, Oregon, is Duchesnean in age, the occurrence of a nimravid C1 from that fauna (B. Hanson, personal communication; this volume, Chapter 11) provides the first pre-Chadronian record of nimravids. The "felid or nimravid, genus and species indeterminate," from the Little Egypt Local Fauna, Trans-Pecos, Texas (Gustafson, 1986) may also represent an early nimravine record. New ^{40}Ar/^{39}Ar dates on the Buckshot Ignimbrite suggest an age close to the Duchesnean-Chadronian boundary (Prothero and Swisher, 1992). *Eosictis avinoffi* from the Dry Gulch Creek Member of the Duchesne River Formation (Emry, 1981), which was originally identified as a felid (Scott, 1945), is not a felid or a nimravid. Gustafson (1986) recognized the similarity of the C1 of this specimen to the amphicyonid *Daphoenocyon* and the meager additional morphological information is consistent with that referral.

Dinictis Leidy, 1854

Dinictis Leidy, 1854:127
Daptophilus Cope, 1873a:2

Type Species—*Dinictis felina* Leidy, 1854.
Referred Species—*Dinictis cyclops.*
North American Distribution—Chadronian of the northern Great Plains and Rocky Mountains; Orellan and Whitneyan of the northern Great Plains; early Arikareean of the northern Rocky Mountains and Great Basin.

Diagnosis—*Dinictis* is paraphyletic and diagnosed by its synapomorphies with *Pogonodon* and the Hoplophoneini, together with the absence of the synapomorphies of *Pogonodon* and Hoplophoneini. Synapomorphies: sabertooth morphologies including the glenoid fossa lowered on a pedicle below the level of the basicranium, mandible with geneal flange, and coronoid process reduced in size; serrated ridges on the incisors and the post-canine teeth. Plesiomorphies (relative to the morphology of *Pogonodon* and the Hoplophoneini): anterior opening of the palatine canal level with P3; anteroposterior length of C1 less than that of P4; size of P3 typical of that of most carnivorans; P4 with distinct protocone; m2 present. Descriptive features: mastoid process small; incisors spatulate and pointed.

Discussion—Leidy's (1869) concept of *Dinictis* entailed sabertoothed features similar to, but somewhat less developed than, those of *Drepanodon* (= *Hoplophoneus*) *primaevus*, together with the molar complement of certain mustelines. Cope (1884) dismissed the possibility of musteline affinities and considered *Dinictis* to be the most generalized of the nimravine sabertooths based on its less-reduced dental formula and the lesser development of sabertooth features. *Dinictis* has a unique combination of dental features but cannot be diagnosed using only apomorphies; for most authors the genus has represented a low to moderate grade of sabertooth development within the Nimravidae.

Scott and Jepsen (1936) referred six species from the White River deposits to *Dinictis*. Scott and Jepsen stated that the species-level taxonomy of *Dinictis* was in a state of confusion but did not attempt to synonymize any of the named species; they concluded that proper revision would require more material than was then available. *Dinictis felina* and *D. squalidens* were considered well defined whereas the others were considered of doubtful status. *Pogonodon cismontanus* Thorpe was referred to *Dinictis*. Further collecting, especially by the Frick lab, has greatly alleviated the problem of sample sizes but no revision has yet been published. The present suggestions regarding revision are preliminary and are based primarily on evaluation of the validity of the features used to characterize the named species. Suggested synonymies must be confirmed by further detailed qualitative and quantitative study.

Dinictis felina Leidy, 1854
Figure 3A

Dinictis felina Leidy, 1854:127
Daptophilis squalidens Cope, 1873a:2
Dinictis squalidens Cope, 1879a:170
Dinictis fortis Adams, 1895:574
Dinictis bombifrons Adams, 1895:577
Dinictis paucidens Riggs, 1896a:237

Type—AMNH 455, skull.

Hypodigm—Pending more detailed revision of this genus, the types of the above species and referred specimens that match the diagnosis below are tentatively referred to *Dinictis felina*.

Distribution—Chadronian of Montana, Nebraska, North Dakota, Saskatchewan, South Dakota, Wyoming; Orellan of Colorado, Nebraska, North Dakota, South Dakota, Wyoming; Whitneyan of Nebraska, Saskatchewan, South Dakota.

Diagnosis—Descriptive features: moderate size (mean SL of 150 mm [n = 9]; Van Valkenburgh, 1987); dentition trenchant; short diastema between canines and premolars; P3, p3, and p4 well developed and of subequal size; limbs relatively gracile with humeroradial and femorotibial indices above 90 (94.5 and 97.2, respectively; Gonyea, 1976).

Discussion—Preliminary review of the type material of the five species listed above, and referred materials in various collections suggests that these specimens may represent a single species. The range of morphological and mensural variation seems consistent with that view and closely matches that of *Hoplophoneus primaevus*.

Leidy (1854) based *Dinictis felina* on a poorly preserved skull and mandible from the White River of the Dakota Territory. Leidy (1869) noted the similarity in many cranial features of this species to *Drepanodon* (= *Hoplophoneus*) *primaevus*. Cope (1873a) established *Daptophilus squalidens* on a mandible with m1 and deciduous premolars, and a fragmentary upper canine. Cope recognized its similarity to *Dinictis* but established a new genus because of the apparent absence of m2. Subsequent identification of this tooth led to referral of the species to *Dinictis* (Cope, 1879a, 1880b). Cope noted the peculiar morphology of the upper canine on the type specimen and suggested (correctly) that it might be deciduous. Cope (1884) diagnosed *D. squalidens* by a single-rooted p2 and a short, rounded mandibular flange. Subsequent recognition of *D. squalidens* was based primarily on size (e.g., Sinclair, 1924). Scott and Jepsen (1936) noted the intergradation in size between the specimens referred to the two species and separated the two on cranial characters. *Dinictis squalidens* was said to have a more horizontal braincase and smaller glenoid and mastoid processes. These features show considerable variation within the hypodigm and are probably at least partly age- and size-related. Adams (1895) established two new species, *D. fortis* and *D. bombifrons*, on material considered significantly larger than that of *D. felina*. *Dinictis bombifrons* was subsequently placed in synonymy with *D. fortis* (Adams, 1896a,b). In addition to larger size, *D. fortis* was differentiated from *D. felina* by its shorter, broader muzzle, the absence of the anterior accessory cusp on p3, and the larger C1 (Adams, 1895). Larger specimens that might be referred to *D. fortis* do not appear to have consistently shorter, broader rostra, and the large C1 is probably an allometric feature associated with larger size. Most larger specimens, including PU 10502 which was referred to *D. fortis* by Adams (1896b), have a small accessory cusp on p3. Riggs

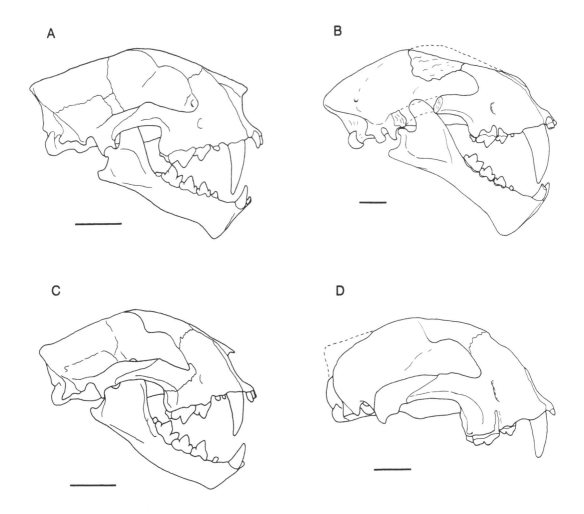

Figure 3. Skulls and lower jaws of *Dinictis*, *Pogonodon*, *Nimravus*, and *Dinaelurus* in right lateral view. A, *Dinictis felina* (PU 13587, after Scott and Jepsen, 1936). B, *Pogonodon platycopis* (AMNH 6938 [type], after Cope, 1880a). C, *Nimravus brachyops* (SDSM 348, after Scott and Jepsen, 1936; reversed). D, *Dinaelurus crassus* (YPM 10518 [type]; reversed). Scale bars = 30mm.

(1896a) established *D. paucidens* based on a specimen that lacked i1 and m2, and had a differently shaped P4. The more slender base and concave outer border of P4 are subtle features that occur on specimens that are typically referred to other species of *Dinictis*, and the absence of i1 may be a preservational artifact. *Pogonodon* also lacks m2 but the type specimen of *D. paucidens* lacks the other apomorphies of this genus. Although more detailed study of the hypodigm may identify additional species, the characters that have been used to diagnose the proposed synonyms of *D. felina* appear to be invalid or insufficient.

The lowest record of *Dinictis* at Flagstaff Rim is at the 250-foot level (early middle Chadronian; Emry, 1992); if this material is referred to *D. felina*, the earliest record of the species is early in Chron C15r at just over 35.5 Ma (Prothero and Swisher, 1992). One definite, and a second probable, record of *Dinictis* may predate this occurrence. *Dinictis* probably occurs in the Pipestone Springs Local Fauna, Jefferson County, Montana; because the bottom of this section probably spans Chron C16n, this occurrence of *Dinictis* could predate that at Flagstaff Rim. *Dinictis* is also known from the Cypress Hills Formation of Saskatchewan (Russell, 1972). Given the revised temporal framework of the formation (Storer and Bryant, 1993; Storer, this volume, Chapter 12), the age of many of the specimens in earlier collections is uncertain. However, the Calf Creek Local Fauna, which may include *Dinictis* (Bryant, 1993), may be of early Chadronian age. Precise determination of the younger end of the temporal range of *D. felina* must await a detailed revision of the genus and allocation of specimens between *D. felina* and *Pogonodon platycopis*. Smaller specimens that probably represent *D. felina* are known from the Whitneyan *Leptauchenia* beds in South Dakota and equivalent deposits elsewhere. This suggests that *D. felina* survived at least into Chron C11r (less than 30.4 Ma; Prothero and Swisher, 1992). This temporal range of at least 5 million years must be considered provisional pending detailed revision of the genus. Some of the Chadronian specimens appear to be plesiomorphic in various dental features (e.g., m1 with more transverse trigonid blade and larger metaconid). Nonetheless, this temporal range is similar to that of *Nimravus brachyops* and shorter than that of *P. platycopis*.

Dinictis cyclops Cope, 1879a

Dinictis cyclops Cope, 1879a:176

Type—AMNH 6937, skull, anterior portions of mandible.
Referred Specimen—?YPM 10047, skull; no other specimens are unequivocally referred to this species.
Distribution—Early Arikareean of Oregon.
Diagnosis—Synapomorphy with *Pogonodon* and Hoplophoneini: p3 reduced in size relative to that in *Dinictis felina*. Descriptive features: SL of 132 mm (Eaton, 1922); skull high with convex dorsal profile; face shorter than that of *D. felina*; premaxillaries almost in contact with frontals; infraorbital foramen larger than that of *D. felina*; C1 more laterally compressed than in *D. felina*; P2 reduced in size.
Discussion—AMNH 6937 demonstrates various differences in cranial and dental morphology from specimens of *Dinictis felina* that suggest that it represents a distinct species. It is considerably smaller than *Pogonodon* and lacks synapomorphies of *Pogonodon* and the Hoplophoneini. Possible assignment of additional specimens (e.g., YPM 10047) to this species must await detailed revision of the genus; Toohey (1959) noted that USNM 16558, which was identified as *D. cyclops* (Hough, 1953, fig. 9), is a specimen of *Nimravus brachyops*. Merriam and Sinclair (1907) did not assign the type specimen to a particular subdivision of the John Day Formation and the stratigraphic level of its origin within the extent of the Turtle Cove and Kimberly members must be considered indeterminate. There is similar uncertainty concerning the origin of YPM 10047.

Pogonodon Cope 1880b

Hoplophoneus Cope 1879b:798b
Pogonodon Cope, 1880b:142
Dinictis Adams, 1896b:431

Type Species—*Hoplophoneus platycopis* Cope, 1879b.
Referred Species—?"*Dinictis*" *eileenae*.
Diagnosis—As for *Pogonodon platycopis*.

Discussion—Cope (1880b) based *Pogonodon* on *Hoplophoneus platycopis* because the type specimen retains p2, which is absent in *Hoplophoneus*, but lacks m2 which occurs in most specimens of *Dinictis*. Adams (1896b) considered *Pogonodon* a synonym of *Dinictis* because of the similarity of the dentition and the absence of m2 in some specimens of the latter; this taxonomy was followed by some authors (e.g., Scott and Jepsen, 1936). Others (e.g., Merriam, 1906; Thorpe, 1920; Eaton, 1922) emphasized the mixture of features characteristic of *Dinictis* and *Hoplophoneus* in the species referred to *Pogonodon* and argued that the taxon should at least be considered a separate subgenus of *Dinictis*, thus emphasizing the intermediate morphology of *Pogonodon*. Although *Pogonodon* resembles *Dinictis* in its cranial and postcranial proportions and in the detailed morphology of the dentition, these features appear to be plesiomorphic. The occurrence of apomorphies of *Pogonodon* (absence of the metaconid on m1; absence of m2) in some specimens of *D. felina* may reflect the paraphyly of *Dinictis*. *Pogonodon* and the Hoplophoneini have the following five synapomorphies: anterior opening of the palatine canal anterior to P3, anteroposterior length of C1 equal to that of P4, P3 reduced in size, P4 protocone reduced or absent, and m2 absent. The considerable support for a sister-group relationship between *P. platycopis* and the Hoplophoneini, together with the large number of apomorphies that characterize the latter and are absent in *P. platycopis*, support the recognition of *Pogonodon* as a genus distinct from both *Dinictis* and *Hoplophoneus*.

Pogonodon platycopis (Cope, 1879b)
Figure 3B

Hoplophoneus platycopis Cope, 1879b:798b
Pogonodon platycopis Cope, 1880b:142
Dinictis platycopis Adams, 1896b:431
Pogonodon davisi Merriam, 1906:53
Pogonodon cismontanus Thorpe, 1920:222
Pogonodon serrulidens Eaton, 1922:429

Type—AMNH 6938, skull and mandible (Fig. 3B).
Referred Specimens—AMNH 1399, 1403; F:AM 62004, 62019, 62020, 62024, 62026, 62042, 62043, 62059, 69369, 102155 125664; SDSM 2865; UCMP 789; UNSM 2509-59; YPM 10053, 10520, 14386.
Distribution—Orellan of South Dakota; Whitneyan of Nebraska, South Dakota; early Arikareean of Montana, Oregon.
Diagnosis—Apomorphies: m1 without metaconid; articulation between the calcaneum and the navicular. Descriptive features: large (estimated adult SL of approximately 200 mm); most dental features as in

Dinictis but teeth more robust and C1 relatively larger; longer diastema between canines and premolars; second and third premolars reduced in size.

Discussion—Cope (1879b) based *Hoplophoneus platycopis* on a large skull and lower jaws from the John Day Formation of Oregon; he initially placed this species in *Hoplophoneus* probably because of the absence of m2, and differentiated it from other species in the genus by its larger premolars. Merriam (1906) named a second species of *Pogonodon*, *P. davisi*, based on a skull from the upper John Day Formation. It was differentiated from *P. platycopis* by its smaller size, its narrow M1, and the proportions of the posterior portion of the cranium. Thorpe (1920) made the first referral of a specimen from the White River deposits to *Pogonodon* and used it as the basis for *P. cismontanus*. He stated that the specimen, a complete left and partial right dentary, was slightly smaller than that of *P. platycopis* but did not list additional specific characters. *Pogonodon serrulidens* is based on a poorly preserved skull, fragmentary right dentary and some postcrania of a juvenile with erupting permanent canines (Eaton, 1922). Eaton (1922) considered its closest affinity to be with *P. davisi* and differentiated it by its smaller size, lower sagittal crest and other cranial proportions, and more compressed upper canines. Preliminary study of these specimens and referred material in the Frick Collection suggests that none of these features adequately divides the hypodigm into separate species. The range in size is not extreme and is partially age-related; other features are subject to considerable individual variation. Eaton's (1922) diagnostic features of *P. serrulidens* are all juvenile characters. All specimens are provisionally referred to a single species, *P. platycopis*.

Most of the specimens of *Pogonodon platycopis* are of Whitneyan or Arikareean age. However, several specimens of late Orellan age have been referred to this species. At least some of these referrals are probably valid. References to the "middle Oreodon beds" do not provide a precise age constraint but at least one specimen (F:AM 125664) comes from below the upper nodules in the Scenic Member of the Brule Formation, South Dakota (slightly older than 32 Ma). The youngest specimens of this species come from the John Day Formation. Merriam and Sinclair (1907) stated that the type of *P. platycopis* came from the Middle John Day Formation whereas the type of *P. davisi* came from the Upper John Day Formation. The stratigraphic range of this species is considered to extend into the Kimberly Member of the formation (?25 Ma) but assessment of whether its temporal range extends to the end of the early Arikareean must await the determination of the detailed biostratigraphy of this species in the John Day Formation. As delimited, the species has an extremely long temporal range of over 7 million years.

HOPLOPHONEINI Kretzoi, 1929

Definition—The most recent common ancestor of *Hoplophoneus mentalis*, *H. primaevus*, and *Eusmilus bidentatus*, and all of its descendants.

Included Genera—*Hoplophoneus*, *Eusmilus*.

Diagnosis—Synapomorphies: no sutural contact between the lacrimal and the jugal; more pronounced sabertooth morphologies than in *Dinictis* and *Pogonodon* including C1 extremely hypsodont and laterally compressed, incisors hypsodont and conical, glenoid fossa lowered on pedicle well below the basicranium, and mandible with large geneal flange, much reduced coronoid process and ventrally placed condyle; p2 absent; parastyle on P4; P4 hypsodont; P4 rotates lingually as it wears (unconfirmed in *Hoplophoneus mentalis*, *H. occidentalis* and *H. sicarius*; see Bryant and Russell, 1995); M1 with protocone and parastylar region reduced; m1 talonid reduced in size; femorotibial index similar to that of forest-dwelling felids.

Discussion—The present analysis suggests that there is strong support for this clade (bootstrap percentage of over 99) but not for particular relationships within the clade. The diagnoses of individual species list proposed apomorphies within the clade.

Hoplophoneus Cope, 1874

Machairodus Leidy, 1851:329
Drepanodon Leidy, 1857:176
Hoplophoneus Cope, 1874:509
Dinotomius Williston, 1895:170
Eusmilus Hatcher, 1895:1091
Eusmilus Sinclair and Jepsen, 1927:391
Drepanodon Scott and Jepsen, 1936:125

Type Species—*Machaerodus oreodontis* Cope, 1873b.

Included Species—*Hoplophoneus mentalis*, *H. primaevus*, *H. occidentalis*, *H. sicarius*, *H. dakotensis*.

North American Distribution—Chadronian to earliest Arikareean of the northern Great Plains.

Diagnosis—*Hoplophoneus* is paraphyletic and is diagnosed by the synapomorphies of the Hoplophoneini (see above), together with the absence of the diagnostic features of *Eusmilus* (see below). Descriptive features: mastoid process large and projecting ventrally.

Discussion—Cope's (1884) concept of *Hoplophoneus* entailed the marked development of sabertooth features, together with the absence of p2 (present in *Dinictis* and *Pogonodon*), the absence of m2 (present in most *Dinictis*), and the presence of p3 (absent in *Eusmilus*). By 1936, 14 nominal species had been referred to this genus; many from contemporary deposits in the Great Plains. Scott and Jepsen (1936) attempted only minor revision by synonymizing *H. robustus* and *H. insolens* with *H. occidentalis*. Simpson (1941) reviewed the Chadronian to Whitneyan species and concluded that four, and possibly five, species were valid: *H. primaevus*, *H. occidentalis*, *H. molossus*, *H. mentalis*, and possibly *H. oharrai*. Hough (1949) accentuated the lumping of taxa by suggesting that all post-Chadronian specimens represented a single species which was divided into a number of geographic subspecies.

Morea (1975) dramatically changed the conceptualization of *Hoplophoneus* by separating it from *Eusmilus* on cranial morphology rather than dental formula and degree of sabertooth specialization. Morea's primary criterion for separating *Hoplophoneus* and *Eusmilus* involved the relative orientation between the dorsal surfaces of the face and the cranium. In *H. primaevus* the dorsal surface of the braincase slopes posteroventrad from just behind the orbits; this line forms an oblique angle with that drawn along the dorsal surface of the face. In *E. bidentatus* the dorsal surface of the braincase is nearly in line with that of the face, resulting in a craniofacial angle of nearly 180°. On this criterion Morea (1975) referred *Eusmilus dakotensis* Hatcher and *Eusmilus sicarius* Scott and Jepsen to *Hoplophoneus*. Morea's (1975) revision had a phylogenetic rationale and was an advance over previous gradistic conceptualizations of these genera. He argued that *H. dakotensis* and *H. sicarius* were derived from *H. mentalis* and therefore these species were more closely related to *H. mentalis* than to *E. bidentatus*. Morea's conclusions were not based on an explicit phylogenetic analysis and do not agree with the preliminary phylogeny presented here. Based on the present phylogeny, both concepts of *Eusmilus* render *Hoplophoneus* paraphyletic. Cladistic analyses based on more detailed character analyses may support further reallocation of species between the two genera or may suggest that *Eusmilus* should be synonymized with *Hoplophoneus*.

Hoplophoneus mentalis Sinclair, 1921

Hoplophoneus mentalis Sinclair, 1921:96
Hoplophoneus oharrai Jepsen, 1926:1

Type—PU 12515, left dentary.

Referred Specimens—AMNH 27798, 32668, 82911; PU 13635; SDSM 2417; UCM 19163; various specimens in the F:AM and USNM collections from Flagstaff Rim and adjacent areas, Natrona County, Wyoming.

Distribution—Chadronian of ?Nebraska, ?Saskatchewan, South Dakota, Wyoming.

Diagnosis—Apomorphy within Hoplophoneini: occurrence of m2 variable. Descriptive features: relatively large (SL of SDSM 2417 = 183 mm; Morea, 1975); sabertooth features more strongly developed than in *Hoplophoneus primaevus*, including a more hypsodont C1, longer geneal flange, and more reduced

coronoid process; zygomatic arches not bowed out laterally.

Discussion—Sinclair (1921) named *Hoplophoneus mentalis* because of the larger flange and more reduced premolars on PU 12515 as compared to other known specimens of the genus. It was the first specimen of *Hoplophoneus* from the Chadron Formation and came from the uppermost part of the formation (Peanut Peak Member: Clark et al., 1967). Jepsen (1926) referred a skull and lower jaws (SDSM 2417), also with strongly developed sabertooth features, from lower in the formation (Middle [= Crazy Johnson] Member: Clark, 1937) to a second species, *H. oharrai*. He separated it from *H. mentalis* because of its larger flange and the presence of i1, which was missing on PU 12515. Simpson (1941) suggested that the two species may be synonymous and Hough (1949) formalized that arrangement. Morea (1975) judged that the difference in the size of the flange between the types of *H. mentalis* and *H. oharrai* was probably within the range expected in a normal population.

The Chadronian samples of *Hoplophoneus* need to be studied in more detail to determine whether all specimens demonstrate the more highly developed sabertooth features characteristic of *H. mentalis*. The sample from Flagstaff Rim, Natrona County, Wyoming, will be especially important in this regard. Specimens may have been referred to *H. mentalis* or *H. oharrai* solely because of their stratigraphic position. Hough (1949) concluded that with larger samples from the Chadronian *H. mentalis* might be found to be synonymous with the post-Chadronian taxon, *H. primaevus*.

The lowest record of *Hoplophoneus mentalis* at Flagstaff Rim is at the 200-foot level (Emry, 1992); this places the earliest record of the species toward the end of Chron C16n, between 35 and 36 Ma (Prothero and Swisher, 1992). This is early in the middle Chadronian (*sensu* Emry et al., 1987; Emry, 1992). The type specimen came from the uppermost part of the Chadron Formation, Pennington County, South Dakota, which suggests that the temporal range of the species extended to the end of the Chadronian. No post-Chadronian specimens have been referred to this species.

Hoplophoneus primaevus (Leidy, 1851)
Figure 4A

Machairodus primaevus Leidy, 1851:329
Drepanodon primaevus Leidy, 1857:176
Machaerodus oreodontis Cope, 1873b:9
Hoplophoneus oreodontis Cope, 1874:509
Hoplophoneus primaevus Cope, 1880a:850
Hoplophoneus insolens Adams, 1896a:48
Hoplophoneus robustus Adams, 1896a:49
Hoplophoneus marshi Thorpe, 1920:211

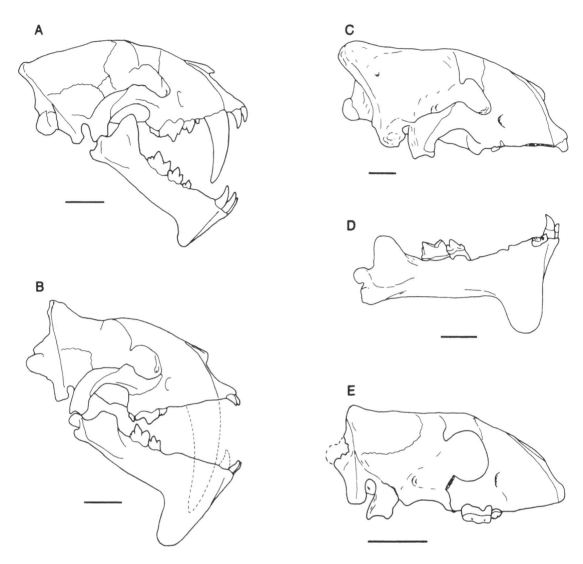

Figure 4. Skulls and lower jaws of members of the Hoplophoneini in right lateral view. A, *Hoplophoneus primaevus* (PU 10540, after Scott and Jepsen, 1936). B, *Hoplophoneus sicarius* (PU 12953 [type], after Scott and Jepsen, 1936). C, *Hoplophoneus dakotensis* (SDSM 2815, after Jepsen, 1933; reversed). D, *Hoplophoneus dakotensis* (PU 11079 [type]). E, *Eusmilus cerebralis* (AMNH 6941 [type]). Scale bars = 30mm.

Hoplophoneus latidens Thorpe, 1920:214
Hoplophoneus molossus Thorpe, 1920:220
Hoplophoneus primaevus primaevus Hough, 1949:553
Hoplophoneus primaevus latidens Hough, 1949:554

Type—USNM 99, skull and partial mandible.

Hypodigm—The types of the above species and referred specimens that match the diagnosis below; referral of Chadronian specimens is uncertain.

Distribution—?Chadronian of Nebraska, South Dakota, Wyoming; Orellan of Colorado, Nebraska, North Dakota, South Dakota, Wyoming; Whitneyan of Nebraska, South Dakota.

Diagnosis—Apomorphies within Hoplophoneini: paroccipital process reduced in size but projecting posteriorly; length of C1 less than that of P4.

Descriptive features: moderate size (mean SL of 156 mm (n = 28); Van Valkenburgh, 1987); zygomatic arches somewhat bowed out laterally; limbs robust with humeroradial and femorotibial indices below 85 (Gonyea, 1976).

Discussion—Simpson (1941) demonstrated that specimens referred to *Hoplophoneus oreodontis*, *H. primaevus*, *H. robustus*, and *H. insolens* form a nearly continuous series of increasing size, and argued that the sample represented a single species with the smaller specimens being females and the larger specimens being males. Thorpe's (1920) *H. latidens* and *H. marshi* were also considered synonymous with *H. primaevus*, whereas *H. molossus* was judged to be distinct. Hough (1949) formalized Simpson's conclusions and also synonymized *H. molossus* with *H. primaevus*; two subspecies were named, *H. primaevus primaevus* from South Dakota and *H. primaevus latidens* from Nebraska, Wyoming, and Colorado. Morea (1975) followed Hough's species-level taxonomy but did not recognize her subspecies of *H. primaevus*.

Scott and Jepsen (1936) restricted the range of *Hoplophoneus primaevus* and its synonyms to the Brule Formation. Clark (1937) was the first to refer a specimen (PU 13635) from the Chadron Formation to a synonym of *H. primaevus* (*H. robustus*). Hough (1949) implicitly rejected that referral by continuing to limit *H. primaevus* to the Brule Formation, and Morea (1975) explicitly referred the specimen to *H. mentalis*. Evaluation of possible pre-Orellan records of this species must await more detailed study of the Chadronian material. Preliminary study of the sample from Flagstaff Rim, Natrona County, Wyoming, suggests that *H. primaevus* may occur in the Chadronian. *Hoplophoneus primaevus* is the most common nimravid in most Orellan deposits but becomes extremely rare in the Whitneyan. The exact upper limit of its range is uncertain but specimens have come from the basal *Leptauchenia* beds of the Poleslide Member in South Dakota and just below the upper Whitney ash in Nebraska. Thus, this species survived through most of the Whitneyan, probably into Chron C11r or approximately 30 to 30.5 Ma.

Hoplophoneus occidentalis (Leidy, 1869)

Drepanodon occidentalis Leidy, 1869:63
Hoplophoneus occidentalis Cope, 1880a:850
Dinotomius atrox Williston, 1895:170

Type—ANSP 11074, left dentary fragment with p4, roots of p3 and m1.

Referred Specimens—AMNH 1001, 1401, 1407; ANSP 11073; F:AM 62025, 62060, 69390, uncatalogued specimens; KU 2540, 2651; SDSM 40126; USNM 16812; YPM 12756.

Distribution—Orellan of Nebraska, North Dakota, South Dakota, Wyoming; Whitneyan of South Dakota.

Diagnosis—Apomorphies within Hoplophoneini: postglenoid foramen small; paroccipital process reduced in size but projecting posteriorly; P2 absent. Descriptive features: linear dimensions significantly larger than those of *Hoplophoneus primaevus* (mean SL of 236 mm [n = 2]; Van Valkenburgh, 1987); cranial and mandibular morphology as in *H. primaevus* but with changes in proportions associated with larger size (e.g., C1 relatively larger and more robust; face proportionally longer).

Discussion—Leidy (1869) differentiated *Hoplophoneus occidentalis* from *H. primaevus* on the basis of its larger size. Hough (1949) suggested that *H.*

occidentalis might represent a large subspecies of *H. primaevus*, but this proposal has not been accepted by subsequent authors. Both Simpson (1941) and Morea (1975) concluded that the larger size of the specimens of *H. occidentalis* was indicative of a separate, rarer species. Morea (1975) noted that *H. occidentalis* is only slightly smaller than *H. dakotensis*. He also discussed the morphological similarity of the lower jaw in the two species and argued that the only available criterion for differentiating the two (presence of p3) may be unreliable.

Hoplophoneus occidentalis occurs throughout most of the Orellan and Whitneyan. The earliest stratigraphically constrained specimens come from the lower nodules of the Scenic Member of the Brule Formation in South Dakota; these nodules are early (33 to 33.5 Ma), but not earliest, Orellan (Prothero and Swisher, 1992). Specimens from the *Protoceras* channels of the Poleslide Member of the Brule Formation extend the range of the species into the late Whitneyan (approximately 30.5 Ma).

Hoplophoneus sicarius (Sinclair and Jepsen, 1927)
Figures 4B, 5A

Eusmilus sicarius Sinclair and Jepsen, 1927:391
Hoplophoneus sicarius Morea, 1975:28

Type—PU 12953, skull, right dentary (Fig. 4B).
Referred Specimens—USNM 12820, 18214 (Fig. 5A), skulls.
Distribution—Orellan of South Dakota, Wyoming.

Diagnosis—Apomorphies within Hoplophoneini: posterior margin of the palate level with the anterior surface of M1; postglenoid foramen small; paroccipital process extremely reduced; anteroposterior length of C1 greater than that of P4; P2 reduced or absent; p3 absent; atlas with relatively long dorsal arch and atlantal foramen in lateral position. Autapomorphies: upward rotation of the face relative to the basicranium exceeds 20 degrees; rostrum extremely long with large C1 to P3 diastema. Descriptive features: moderately large (SL of 187 to 211 mm; Morea, 1975); C1 extremely hypsodont, anteroposteriorly elongated and laterally compressed (length to width ratio of approximately 3:1).

Discussion—The original referral of this species to *Eusmilus* was based on similarities in dental formula and the morphology of the lower jaw to those of *E. bidentatus* (Sinclair and Jepsen, 1927). Morea (1975) recognized significant differences in cranial morphology between *E. sicarius* and *E. bidentatus* and referred the former to *Hoplophoneus*. Nonetheless, *H. sicarius* displays significant differences in cranial morphology from other species of *Hoplophoneus* that in traditional evolutionary systematics would be considered of generic

A

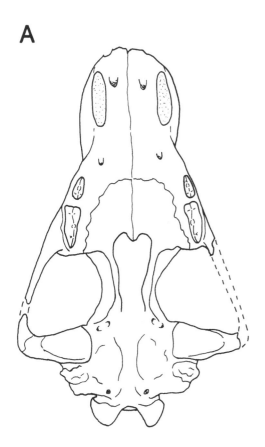

B

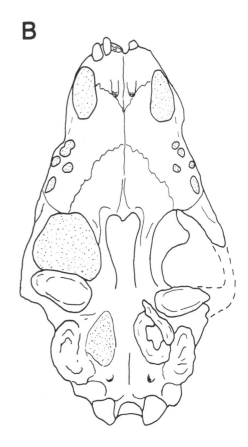

Figure 5. Ventral view of skulls of: A, *Hoplophoneus sicarius* (USNM 18214), and B, *Hoplophoneus dakotensis* (F:AM 102387) drawn to a standard length. Note the differences in length of the C1-P3 diastema, the shape of C1, the shape of the palate, the length of the zygomatic arch, and the size of the mastoid. The upward facial rotation in *H. sicarius* causes the basicranium to appear overly short in this view.

value. The most important features include the marked upward rotation of the face and the extremely long rostrum. Emerson and Radinsky (1980) demonstrated that in sabertoothed carnivorans, the face is usually rotated upward relative to the cranium such that the palate is angled upward relative to the basicranium. In *H. sicarius* the palate is rotated upward more than 20°; in other nimravid and felid sabertooths this angle never exceeds 13°. *Hoplophoneus sicarius* also has an extremely long rostrum. In Radinsky's (1982) factor analysis of cranial morphology in fossil carnivorans USNM 18214 was an extreme outlier to the nimravid-felid cluster and placed within the polygon delimited by extant canids (plot of the first two rotated axes of a 14 variable principal components analysis). Radinsky (1982) attributed this result to the specimen's short temporal and masseteric fossae and long jaw and upper tooth row. The latter reflects the elongated rostrum. This elongation is not simply the result of the large upper canine; the distance between the saber and P3 is much greater than that in either *E. bidentatus* or other species of *Hoplophoneus* (Fig. 5). Given the phylogeny

in Figure 1, referral of *H. sicarius* to a new genus would contribute to the paraphyly of *Hoplophoneus*.

The type specimen came from just above a red layer that is considered equivalent to the lower nodule zone in the Scenic Member, Brule Formation, Pennington County, South Dakota (early, but not earliest, Orellan). Both referred skulls came from sections of the Brule Formation in Niobrara County, Wyoming that are of Orellan age but precise stratigraphic data appears to be lacking. Thus, a more precise range for this species within the Orellan cannot be determined at present.

Hoplophoneus dakotensis (Hatcher, 1895)
Figures 4C,D, 5B

Eusmilus dakotensis Hatcher, 1895:1091
Hoplophoneus? *dakotensis* Kretzoi, 1929:1304
Hoplophoneus dakotensis Morea, 1975:36
    Type—PU 11079, right dentary (Fig. 4D).
    Referred Specimens—F:AM 102387 (Fig. 5B); SDSM 2815 (Fig. 4C), 2830; UNSM 6-12-7-95; complete or partial skulls.

Distribution—Whitneyan of South Dakota; early Arikareean of Nebraska.

Diagnosis—Apomorphies within Hoplophoneini: postglenoid foramen small; paroccipital process extremely reduced; anteroposterior length of C1 greater than that of P4; P2 absent; p3 absent; m1 metaconid reduced to a ridge. Descriptive features: skull very large (SL of 238 mm [SDSM 2830] and 251 mm [SDSM 2815]; Morea, 1975); palate broad, especially anteriorly; zygomatic arch extremely short; jugal extends ventrally on posterior glenoid pedicle almost to level of the mandibular fossa; mastoid process extremely large and rugose.

Discussion—None of the referred skulls has an associated lower jaw and thus the referral of these specimens to this species remains provisional. Jepsen (1933) presumably referred the first three skulls to *Hoplophoneus dakotensis* because of their large size and stratigraphic context similar to that of the type jaw. The traditional referral of this species to *Eusmilus* was based solely on the dental formula of the lower jaw and the extreme development of sabertooth features. Whereas Kretzoi (1929) questionably referred this species to *Hoplophoneus*, Jepsen (1933) elaborated on the specific differences between this species and "*Eusmilus*" *sicarius* and suggested features that separated both from *Hoplophoneus*. As argued by Morea (1975), the cranial and mandibular morphology of *H. dakotensis* differs considerably from that of *E. bidentatus* and more closely resembles that of *H. primaevus* and most other species of *Hoplophoneus*; however, these similarities are probably plesiomorphic within the Hoplophoneini.

Morea (1975) argued that aside from dental formula the lower jaws of *Hoplophoneus dakotensis* and *H. occidentalis* are essentially indistinguishable. If the presence of p3 is variable in these species as suggested by Morea (1975), *H. dakotensis* may eventually be synonymized with *H. occidentalis*. Nonetheless, the skulls referred to *H. dakotensis* are morphologically distinct from those referred to other species of *Hoplophoneus*. The skulls of *H. dakotensis* are more robust than those of the similarly sized *H. occidentalis*. The mastoid process is extremely large and robust (Fig. 4C). *Hoplophoneus dakotensis* may also be unique in the greater ventral extension of the posterior end of the jugal (Fig. 4C; jugal in *H. occidentalis* unstudied by the author but Riggs, 1896b, Plate 1, shows morphology as in *H. primaevus*; Fig. 4A). Suggestions that *H. sicarius* may be ancestral to (Jepsen, 1933; Morea, 1975), or even conspecific with (Morea, 1975), *H. dakotensis* are untenable. *Hoplophoneus dakotensis* lacks the extreme facial rotation and elongated rostrum that are autapomorphies of *H. sicarius* (Fig. 5). The shape of the skull differs considerably in the two species (Figs. 4B,C, 5) with that of *H. dakotensis* more closely resembling that of *H. primaevus*.

Jepsen's (1933) hypodigm of *Hoplophoneus dakotensis* was restricted to the Whitneyan. The type specimen and the two SDSM skulls come from the *Protoceras* beds of the Poleslide Member of the Brule Formation in South Dakota; Jepsen (1933) inferred a similar origin for the UNSM specimen. This probably constrains the appearance of this species to the later Whitneyan (during Chron C12n or C11r, approximately 30.5 Ma). The referral of F:AM 102387, which comes from a channel of the Gering Formation that is cut into the Brule Formation (Sioux County, Nebraska), to this species extends its temporal range into the earliest Arikareean. Thus, the known temporal range of this rare species is relatively short at probably no more than 1.5 million years.

Eusmilus Gervais, 1876

Machaerodus Filhol, 1873:205
Eusmilus Gervais, 1876:53
Machaerodus Cope, 1880b:143
Hoplophoneus Cope, 1880a:850
Ekgmoiteptecela Macdonald, 1963:221

Type species—*Machaerodus bidentatus* Filhol, 1873.

Referred Species—*Eusmilus cerebralis*.

Diagnosis—Apomorphies within Hoplophoneini: paroccipital process extremely reduced; P2 absent; p3 absent; atlas with relatively long dorsal arch and atlantal foramen in lateral position; femorotibial index intermediate, within the range of values of both forest- and open woodland-dwelling felids. Possible autapomorphies: angle between the dorsal surfaces of the braincase and face approaches 180°; occiput oval. Descriptive features: small size; extreme development of sabertooth morphology, including C1 extremely hypsodont and somewhat laterally compressed, glenoid fossa lowered on pedicle well below the basicranium, and mandible with extremely large flange, reduced coronoid process, and ventrally placed condyle; mastoid process large and projecting ventrally; i1 extremely reduced or absent; postcranial skeleton robust (partly modified from Morea, 1975).

Discussion—Initial referrals of North American species to the European genus *Eusmilus* were based primarily on the purported occurrence of only two lower incisors, the absence of p3, the extremely large geneal flange, and, by extrapolation, the extremely hypsodont saber (*E. dakotensis*, Hatcher, 1895; *E. sicarius*, Sinclair and Jepsen, 1927). Scott and Jepsen (1936) followed this taxonomy but suggested that information regarding the cranial anatomy of *E. bidentatus* (they were unaware of the cranial material described by Piveteau, 1931) might necessitate the referral of the North American species to a new genus. Morea (1975) argued that the purported closest relationship of these

North American forms to *E. bidentatus* was based on the parallel development of highly derived sabertooth features in the lower jaw. Based on comparison of the skull of *E. bidentatus* with those of "*E.*" *dakotensis*, "*E.*" *sicarius*, and *Hoplophoneus*, Morea argued that the latter two species should be removed from *Eusmilus* and referred to *Hoplophoneus*.

Toohey (1959) suggested that the small sabertooths *Hoplophoneus cerebralis* Cope and *H. belli* Stock should be referred to *Eusmilus*. Macdonald (1963) recognized the relationship between these two forms and his *Ekgmoiteptecela olsontau* from the Sharps Formation of South Dakota. Schultz et al. (1970) referred *H. belli* to *Ekgmoiteptecela*. Morea (1975) concluded that the published North American records of *Eusmilus* were restricted to *H. cerebralis* Cope, *H. belli* Stock, and *Ekgmoiteptecela olsontau* Macdonald. These three forms share features with *Eusmilus bidentatus* that suggest that they are more closely related to this species than to other nimravids. Morea's (1975) allocation of species to *Hoplophoneus* and *Eusmilus* was followed by Martin (1980) and Bryant (1984).

The horizontally oriented dorsal surface of the braincase, Morea's (1975) primary feature for differentiating *Eusmilus* from *Hoplophoneus*, may be associated with small size; some juvenile specimens of *Hoplophoneus* share this feature. Morea (1975) also argued that the occiput in *Eusmilus* is oval whereas that of *Hoplophoneus* is triangular. The apparent shape of the occiput varies considerably in *Hoplophoneus* depending on the size of the nuchal crests and the size and orientation of the mastoid process; no consistent difference from *Eusmilus* is evident. Although Morea (1975) argued that resemblances among *Eusmilus*, *H. sicarius*, and *H. dakotensis* are homoplasious features associated with the extreme development of sabertooth morphology, these three taxa form a clade in Figure 1. On this basis, the referral of these two larger North American species to *Eusmilus* does not violate cladistic principles. However, Figure 1 suggests that both Morea's (1975) and traditional conceptualizations of *Eusmilus* result in the paraphyly of *Hoplophoneus*.

Eusmilus cerebralis (Cope, 1880b)
Figure 4E

Machaerodus cerebralis Cope, 1880b:143
Hoplophoneus cerebralis Cope, 1880a:850
Hoplophoneus belli Stock, 1933:37
Ekgmoiteptecela olsontau Macdonald, 1963:221

Type—AMNH 6941, skull (Fig. 4E).

Referred Specimens—F:AM 69377, 69409, 98189, 98769, 98770; LACM 463, 5465; SDSM 54247; UCMP 123180, 123181.

Distribution—Whitneyan of South Dakota; early Arikareean of California, Oregon, South Dakota, Wyoming.

Diagnosis—Descriptive features: smaller (SL of 97 mm [F:AM 98189] and approximately 110 mm [AMNH 6941]) than *Eusmilus bidentatus*; caliber of infraorbital canal smaller than that of *E. bidentatus*; P3 with anterior accessory cusp; P4 with large parastyle (modified from Morea, 1975).

Discussion—Morea (1975) synonymized *Hoplophoneus belli* (Stock, 1933) with *Eusmilus cerebralis*. He attributed the differences between *H. belli* and *Eusmilus cerebralis* to individual variation and the juvenile age of the former. Morea (1975) suggested that *Ekgmoiteptecela olsontau* may also be conspecific with *Eusmilus cerebralis* but did not synonymize these species because the absence of cranial material of the former precluded direct comparison with the latter. Bryant (1984) described two partial nimravine skeletons that included associated fragmentary cranial and mandibular material, and concluded that the cranial and mandibular elements compared well with AMNH 6941 (skull of *Eusmilus cerebralis*), and SDSM 54247 (type jaw of *Ekgmoiteptecela olsontau*), respectively. As a result, Bryant (1984) considered *Ekgmoiteptecela olsontau* a junior synonym of *Eusmilus cerebralis*. This taxonomy is provisionally accepted but cursory study of the hypodigm suggests considerable variation in detailed cranial and mandibular morphology.

Of Morea's (1975) diagnostic characters for *Eusmilus cerebralis*, only the anterior accessory cusp on P3 appears to clearly differentiate this species from *E. bidentatus*. The size difference between species is small and the parastyle on P4 in *E. cerebralis* does not appear to be consistently larger. On the other hand, the caliber of the infraorbital canal of *E. bidentatus* may be significantly larger than that of *E. cerebralis*. Morea (1975) suggested that the morphological differences may be insufficient for specific separation. The revised correlation between the North American and European records (Swisher and Prothero, 1990; Prothero and Swisher, 1992) has also removed much of the previously proposed discrepancy in age.

Eusmilus cerebralis is Whitneyan to early Arikareean in age but the precise older and younger limits of its temporal range cannot be determined at present. Several referred specimens in the Frick Collection came from deposits of Whitneyan age in South Dakota but it is uncertain whether any of them come from below the *Leptauchenia* nodules (approximately 30.5 Ma). Stock's (1933) *Hoplophoneus belli* from the Sespe Formation of California and Macdonald's (1963, 1970) *Ekgmoiteptecela olsontau* from the Sharps Formation of South Dakota are of earliest Arikareean age. The stratigraphic position of the type (AMNH 6941) within the John Day Formation is indeterminate but its age is probably younger (<28 Ma) than that of any of the referred specimens.

NIMRAVINI Cope, 1880a

Definition—The most recent common ancestor of *Nimravus brachyops, N. intermedius, N. edwardsi,* and *Dinaelurus crassus,* and all of its descendants.

Included Genera—*Nimravus, Dinaelurus.*

Diagnosis—Synapomorphies: deep masseteric fossa with a distinct dorsal margin on the lateral surface of the maxilla and jugal; fossa on the ventral surface of the zygomatic arch below the postorbital process; posterior margin of the palate level with or anterior to the anterior surface of M1; P4 protocone reduced or absent; M1 smaller than that of *Dinictis felina.*

Discussion—This clade occurred on over 90% of the trees generated using bootstrap resampling (300 replicates). This clade is plesiomorphic in the absence of sabertooth features in the cranium and mandible and serrated ridges on the incisors and post-canine teeth that occur in *Dinictis, Pogonodon,* and the Hoplophoneini.

Nimravus Cope, 1879a

Machairodus Cope, 1878:72
Hoplophoneus Cope, 1879c:56
Nimravus Cope, 1879a:170
Archaelurus Cope, 1879b:798a
Pogonodon Cope, 1880a:849
Dinictis Adams, 1896b:431

Type Species—*Machairodus brachyops* Cope, 1878.

Referred North American Species—?*Nimravus sector.*

Diagnosis—Apomorphies within Nimravini: m1 without metaconid; articulation between the calcaneum and navicular (these features are unknown in *Dinaelurus*). Descriptive features: alisphenoid canal short; accessory ossicle associated with the external auditory meatus; buccal surface of mandible with alveolar torus below m1; third premolars high crowned with P3 equal in height to the paracone on P4, and p3 often higher than p4; P4 metastyle low; occurrence of m2 and m3 variable; postcranial skeleton relatively gracile (partly modified from Toohey, 1959).

Discussion—Cope (1880a) characterized *Nimravus* and *Archaelurus* as the "false sabretooths" and considered this group transitional between the nimravid sabretooths and the primitive unspecialized cats (e.g., *Pseudaelurus*). Scott and Jepsen (1936) held to this view and placed *Nimravus* and *Archaelurus* in a separate subfamily within the Felidae. This intermediate status for *Nimravus* and its allies became untenable with the rejection of any ancestral-descendant relationship between nimravids and felids (Piveteau, 1931; de Beaumont, 1964; Tedford, 1978; Baskin, 1981). If the absence of sabertooth morphology is plesiomorphic for the Nimravinae, the morphology of *Nimravus* may approximate that of the ancestral nimravine.

Nimravus brachyops (Cope, 1878)
Figure 3C

Machairodus brachyops Cope, 1878:72
Hoplophoneus brachyops Cope, 1879c:56
Nimravus brachyops Cope, 1879a:170
Archaelurus debilis Cope, 1879b:798a
Nimravus gomphodus Cope, 1880a:844
Nimravus confertus Cope, 1880a:844
Pogonodon brachyops Cope, 1880a:849
Dinictis brachyops Adams, 1896b:431
Dinictis major Lucas, 1898:399
Archaelurus debilis major Merriam, 1906:43
Archaelurus debilis merriami Hay, 1930:543
Nimravus meridianus Stock, 1933:39
Nimravus bumpensis Scott and Jepsen, 1936:147
Nimravus altidens Macdonald, 1950:601

Lectotype—AMNH 6935, partial right dentary.

Referred Specimens—See list in Toohey (1959); additional specimens in the Frick Collection, AMNH; SMNH P2270.1.

Distribution—Whitneyan of Nebraska, South Dakota; late Whitneyan or early Arikareean of Saskatchewan; early Arikareean of California, Nebraska, Oregon, Wyoming.

Diagnosis—Descriptive features: moderately large size (mean SL of John Day sample of 188 mm [n = 7]; Toohey, 1959); P4 protocone extremely reduced or absent (larger in *Nimravus intermedius*).

Discussion—Toohey (1959) provided a detailed account of the taxonomic history of North American *Nimravus.* He demonstrated that the variation among the North American specimens of *Nimravus,* which had been the basis for the named species listed above, could be explained as individual variation and sexual dimorphism within a single species. Toohey argued for a single ancestor-descendant series of populations that, given additional specimens, might be divisible into spatiotemporally separated subspecies.

Toohey (1959) argued that *Nimravus brachyops* and *N. intermedius* of Europe were distinguishable only on the basis of geographic distribution and relative age (the European *N. edwardsi* is considerably smaller). The revised correlation between the European and North American records (Swisher and Prothero, 1990; Prothero and Swisher, 1992) has removed the apparent age discrepancy. *N. intermedius* may have a relatively larger protocone on P4.

Nimravus brachyops is characteristic of the late Whitneyan and early Arikareean. Macdonald (1950) stated that the type specimen of *N. altidens* may be the oldest North American specimen. Certainly no known specimen unequivocally predates its provenience in the *Protoceras* channel sandstones of the Poleslide Member of the Brule Formation, Shannon County, South Dakota (30 to 31 Ma). A partial m1 of *Nimravus* from the Cypress Hills Formation of Saskatchewan of probably late Whitneyan or early Arikareean age extends the geographic range of this species into Canada. *Nimravus* occurs in both the Turtle Cove and Kimberly members of the John Day Formation (Toohey, 1959), but the lack of precise stratigraphic data precludes the accurate determination of the upper limit of its temporal range (probably <25 Ma). The temporal range of this species is probably at least 5 million years.

Dinaelurus Eaton, 1922

Dinaelurus Eaton, 1922:437
   Type Species—*Dinaelurus crassus* Eaton, 1922.

Dinaelurus crassus Eaton, 1922
Figure 3D

Dinaelurus crassus Eaton, 1922:437
   Type—YPM 10518, skull (Fig. 3D).
   Referred Specimens—none.
   Distribution—Early Arikareean of Oregon.
   Diagnosis—Apomorphies within Nimravini: C1 shorter and more conically shaped than in other nimravids; P2 absent. Descriptive features: skull extremely broad for its length (SL = 174 mm; Eaton, 1922); facial region short; anterior zygomatic pedicle extremely deep; cheek teeth robust.
   Discussion—The known morphology of *Dinaelurus crassus* is unique among nimravids in lacking any features associated with the sabertooth morphotype. Unlike other nimravids, *D. crassus* is best categorized as a conical-toothed cat (*sensu* Martin, 1980). C1 is slightly laterally compressed and has at least a distal ridge but wear and poor preservation precludes the determination of the presence or absence of serrations. The morphology of the other teeth most resemble those of *Nimravus*; as in that genus, the other teeth are not serrated.
   The precise stratigraphic origin of YPM 10518 from the John Day Formation is unknown. Eaton (1922) assigned it to the upper John Day because of the light gray color of the skull and associated matrix. Fisher and Rensberger (1972) indicated, however, that color is often the result of diagenesis or facies differences and cannot be used to confidently assign specimens to a particular member of the formation.

SPECIES OF UNCERTAIN OR INDETERMINATE STATUS

Dinictis eileenae Macdonald, 1970

Dinictis eileenae Macdonald, 1970:67
   Type—LACM 9195, right dentary missing condyle, angular process, and much of masseteric fossa.
   Distribution—Early Arikareean of South Dakota.

Diagnosis—"Long diastema between canine and p2; p2 single rooted; well developed p3; m1 without metaconid; m2 missing" (Macdonald, 1970:68).
   Discussion—The diagnosis of this species includes several features that characterize *Pogonodon platycopis*. The specimen is also of comparable size but pending direct study of the specimen this species is not placed in synonymy. The association of *D. eileenae* with the canid *Enhydrocyon geringensis* (Macdonald, 1970) suggests a relatively early age within the temporal range recorded by the Sharps Formation (Tedford et al., 1985, fig. 3). As a result, the age of this specimen may be earliest Arikareean.

Hoplophoneus strigidens (Cope, 1878)

Machaerodus strigidens Cope, 1878:71
Hoplophoneus strigidens Cope, 1880a:851
   Type—AMNH 6942, fragmentary upper canine.
   Distribution—Early Arikareean of Oregon.
   Diagnosis—Cope (1878, 1884) characterized this species by the central, basal grooves that occur on both faces of this tooth; its extreme hypsodonty and lateral compression were also considered distinctive.
   Discussion—Cope (1884) suggested that this tooth represented an animal of about the size of "*Hoplophoneus*" *cerebralis*. Adams (1896a) noted that the specimen lacked any identifiable features that allowed referral to *Hoplophoneus* rather than any other genus. The basal grooving might suggest a deciduous upper canine but this feature normally characterizes only the lingual face of the crown. The serrations are indicative of the Nimravidae but the specimen is otherwise indeterminate at present. The stratigraphic level of origin within the John Day Formation is unknown.

Nanosmilus kurteni Martin, 1992

Nanosmilus kurteni Martin, 1992:343
   Type—UNSM 25505, skull and mandible of a young individual (C1 not erupted).
   Distribution—Orellan of Nebraska.
   Diagnosis—Descriptive features: size of a small bobcat; skull narrow; sagittal crest separating into a "V" above the glenoid fossa; frontals less broadened than in *Hoplophoneus* and *Eusmilus*; premaxillaries end posteriorly above the posterior edge of C1; glenoid fossa level with the "gum line" and the ventral border of the mastoid process; optic foramen and orbital fissure separate (simplified from Martin's [1992] description).
   Discussion—Martin (1992) argued for the origin of the *Eusmilus* lineage from a different *Dinictis*-like ancestor than that of *Hoplophoneus*. UNSM 25505 was considered an early stage in the lineage leading to *Eusmilus* and plesiomorphic relative to *Hoplophoneus* in its 1) weak dependent flange, 2) strongly inclined

occiput, and 3) relatively well-developed premolars. The first two features are characteristic of juvenile specimens of *Hoplophoneus* and the latter occurs in some *Hoplophoneus*. Martin argued that the following synapomorphies separate *Nanosmilus* and *Eusmilus* from *Hoplophoneus*: 1) a parastyle on P4, 2) reduction of the talonid on m1 to a basal projection, 3) lack of elevation of the occiput, 4) shortened nasals, 5) extreme anterior placement of the anterior palatine foramen, and 6) small size. Character 1 occurs in all *Hoplophoneus*, characters 2, 4, and 5 occur in some *Hoplophoneus*, and character 3 is a juvenile character that also appears to be associated with small size. UNSM 25505 is only slightly smaller than the smallest individuals of *H. primaevus* of similar ontogenetic age. UNSM 25505 does have features, such as the absence of the metaconid on m1, that differ from those of *H. primaevus*, but the overall morphology of this specimen suggests that it is best considered a somewhat unusual individual of this species. However, pending further direct study of the specimen, this species is not placed in synonymy.

Nimravus sectator Matthew, 1907

Nimravus sectator Matthew, 1907:204

Type—AMNH 12882, anterior portion of left dentary.

Distribution—Early Arikareean of South Dakota.

Diagnosis—Descriptive features: large (length of m1 = 28.5 mm); short c1-p3 diastema; p2 absent; p4 broad and robust; m1 without metaconid; no evidence of alveolar torus; m2 absent.

Discussion—Matthew (1907) differentiated this mandible from previously named species of *Nimravus* and *Dinictis* by its larger size and robustness, and the absence of the geneal flange. Matthew included this specimen in the "Lower Rosebud" fauna but Macdonald (1963, 1970) concluded that it probably came from the Monroe Creek Formation. Matthew considered its referral to *Nimravus* as provisional, and its mixture of features considered typical of *Nimravus* and *Pogonodon* leaves its generic allocation in doubt. Toohey (1959) did not consider the specimen in his comprehensive review of *Nimravus*, suggesting that he excluded it from the genus. It resembles *Nimravus* in the short c1-p3 diastema, the absence of a geneal flange, and the large size of p3. It resembles *Pogonodon platycopis* in its extremely large size, the robust p4, and lack of evidence for the alveolar torus. It differs from both *Nimravus* and *Pogonodon* in the apparent absence of p2.

THE EVOLUTIONARY HISTORY OF THE NIMRAVINAE

The Nimravinae was an important component of the North American carnivore guild from the latest Eocene to the end of the Oligocene. The composition of this guild was remarkably stable, especially throughout the Chadronian to Whitneyan interval, and also included hesperocyonine and borophagine canids, daphoenine amphicyonids, and hyaenodontids. Changes in this guild, both at the Eocene-Oligocene boundary and throughout much of the Oligocene, were limited primarily to shifts in relative diversity (Van Valkenburgh, 1994). Trophic diversity, at least in the Orellan, resembled that of modern faunas (Van Valkenburgh, 1988). Being hypercarnivores without advanced cursorial adaptations, the Nimravinae probably fulfilled an ecological role similar to that of felids in modern carnivore guilds.

The evolutionary history of the Nimravinae in North America occurred during a period of climatic and floristic change that included a decline in mean annual temperature and an increase in mean annual range of temperature about one million years into the Oligocene (Wolfe, 1992), and a continuous decline in precipitation from early in the late Eocene (40 Ma) to the end of the Oligocene, at least in the northern Great Plains (Retallack, 1992). The paleosols of Badlands National Park suggest a sequential change from late Eocene forest, through latest Eocene dry woodland and early Oligocene wooded grassland to gallery woodland, to middle Oligocene open grassland with few trees along watercourses (Retallack, 1992). Changes in the large herbivore guild were concentrated at the Eocene-Oligocene boundary. However, despite the extinction of various groups of archaic herbivores during this transition, Orellan faunas demonstrate no major ecological differences from those of the Chadronian (Prothero, 1985).

Recent assignment of the Chadronian to the Eocene (Prothero and Swisher, 1992) suggests that the earliest known nimravids are North American, but the continent of origin is uncertain. The absence of obvious close relatives in North American faunas has been the basis for arguments that nimravids were immigrants, probably from Asia (Martin, 1989). The "miacoid" *Chailicyon* from the Eocene of China has been suggested as a plausible close plesiomorphic relative (Flynn et al., 1988). Although nimravids are an important component of Oligocene faunas in Eurasia, they were more diverse in North America. At least two species (*Dinictis felina*, *Hoplophoneus mentalis*) were present in the Chadronian. Both genera survived the extinction event at the end of the Chadronian (Prothero, 1985) but the disappearance of *H. mentalis* at this time may be a true extinction. Diversification within the Nimravinae was the first major radiation of cat-like carnivorans (Fig. 6) which resulted to a large degree from *in situ* cladogenesis within the North American Hoplophoneini during the Orellan and Whitneyan.

Because of differences in limb proportions *Hoplophoneus* and *Dinictis* have been considered ambush and

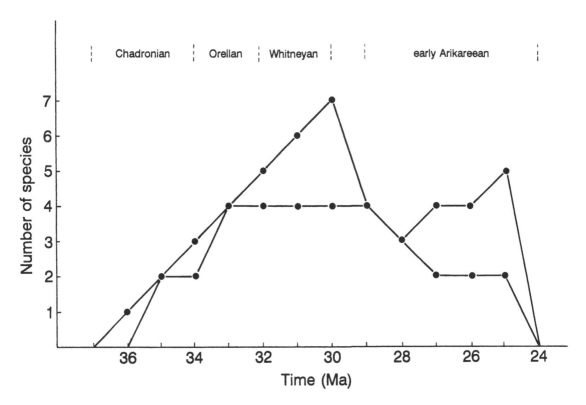

Figure 6. Species diversity of the Nimravinae in North America from the beginning of the Chadronian to the end of the early Arikareean. The curves are based on the temporal ranges on Figure 2; the lower curve is based on the minimum ranges (solid lines, Fig. 2) and the upper curve includes possible, but unconfirmed, extended ranges (dashed lines, Fig. 2).

pursuit predators, respectively (Gonyea, 1976; Martin, 1980). The increase in species diversity within the Hoplophoneini in the early Oligocene coincides with the opening up of forest habitats, a correlation that seems somewhat at odds with these interpretations of *Hoplophoneus* as a closed habitat, ambush predator. Predictions of locomotory, predatory, and other behaviors by analogy with extant carnivorans are often problematic. Although a discriminant function analysis based on correlations between morphologic indices and locomotor behavior in extant carnivorans classified both *Dinictis* and *Hoplophoneus* as arboreal, the size of these animals and the paleoenvironmental and faunal evidence indicates that strictly arboreal habits were unlikely (Van Valkenburgh, 1987). So-called arboreal characters (curved claws, long olecranon, high femur/metatarsal ratio) may be associated with killing behavior (Van Valkenburgh, 1987) or may be plesiomorphic features inherited from ancestral carnivorans. In addition, in a second multivariate discriminant function analysis using different morphological features, both *Dinictis* and *Hoplophoneus* fell outside the morphospace occupied by extant pursuit, ambush, and generalist predators (Janis

and Wilhelm, 1993). Probable nimravine prey such as oreodonts were less cursorial than extant grassland ungulates (Stanley et al., 1983); consequently, the locomotor requirements for predatory success in modern grassland ecosystems did not apply to the same degree in the early Oligocene. In addition, the early Oligocene wooded grassland to gallery woodland would have provided some cover. Despite arguments that *Dinictis* was not a pursuit predator, at least in the modern sense (Janis and Wilhelm, 1993), differences in postcranial morphology between *Dinictis* and *Hoplophoneus* are consistent with some partitioning of resource use in local habitats with the former probably being more adapted to open country.

Together with the speciation within the Hoplophoneini, the appearance of *Nimravus* resulted in maximum diversity within the Nimravinae in the late Whitneyan (Fig. 6). The Whitneyan appearances of *Eusmilus cerebralis* and *Nimravus brachyops* have usually been attributed to immigration from Eurasia. However, with the revised temporal correlation between the European and North American terrestrial records (Prothero and Swisher, 1992), the Old and New World records of these two genera are more or less contemporary and provide no evidence regarding the direction of dispersal. The absence of the Nimravini from the earlier, highly fossiliferous portion of the North American record, together with the antiquity of this clade implied by Figure 1, suggests that *Nimravus* was an immigrant, probably from Asia. Because North America was the major center of the radiation within the Hoplophoneini, dispersal in *Eusmilus* may have been from the New to the Old World.

Nimravine diversity drops slightly in the earliest Arikareean due to the extinction of the *Hoplophoneus* species. The apparent rarity of the previously common *H. primaevus* in the Whitneyan and the extinction of most of the Hoplophoneini by the early Arikareean may be associated with the increasing predominance of open habitats. Bakker (1983) suggested that a considerable gap in locomotor grade existed between nimravines and their prey; such a gap might have become increasingly detrimental with the continued development of grasslands and may have impacted earlier on *Hoplophoneus* than on the more gracile, longer-limbed *Dinictis* and *Nimravus*. *Dinictis*, *Pogonodon* and the Nimravini survive until late in the early Arikareean but the precise timing of these extinctions is uncertain at present. The extinction of these taxa seems less correlated with environmental and other faunal changes but these nimravines may have been less well adapted to prevailing conditions than the contemporary canids and amphicyonids. The extinction of the Nimravinae resulted in the absence of cat-like carnivorans in North America until the immigration of pseudaelurine felids in the Hemingfordian.

ACKNOWLEDGMENTS

I thank Don Prothero for his central role in, and invitation to participate in, the 1989 Penrose Conference on the Eocene-Oligocene transition, his efforts toward the publication of the proceedings, and his refusal to allow me not to contribute to this book. I thank the following for the opportunity to study or borrow materials in their care, discussion and information regarding nimravids, or support and encouragement of various kinds: J. Alexander, P. W. Bjork, C. S. Churcher, M. Dawson, T. Deaschler, R. J. Emry, J. J. Flynn, R. M. Hunt Jr., L. D. Martin, N. A. Neff, M. J. Novacek, A. P. Russell, K. Seymour, J. E. Storer, R. H. Tedford, M. A. Turner, and B. Van Valkenburgh. Ben Lepage provided access to a Macintosh LCIII and PAUP 3.1.1. Blaire Van Valkenburgh and two anonymous referees reviewed the manuscript. Colleen Bryant helped with editing. Financial support came primarily from Natural Sciences and Engineering Research of Canada grants to C. S. Churcher (University of Toronto) and to the author and J. E. Storer (University of Regina).

LITERATURE CITED

Adams, G. I. 1895. Two new species of *Dinictis* from the White River beds. American Naturalist 30:573-578.

Adams, G. I. 1896a. On the species of *Hoplophoneus*. American Naturalist 30:46-51.

Adams, G. I. 1896b. The extinct Felidae of North America. American Journal of Science, 4th series 1:419-444.

Bakker, R. T. 1983. The deer flees, the wolf pursues: incongruences in predator-prey coevolution; pp. 350-382 *in* D. J. Futuyma and M. Slatkin (eds.), Coevolution. Sinauer Associates Inc., Sunderland, MA.

Baskin, J. A. 1981. *Barbourofelis* (Nimravidae) and *Nimravides* (Felidae) with a description of two new species from the late Miocene of Florida. Journal of Mammalogy 62:122-139.

Berggren, W. A., D. V. Kent, and J. J. Flynn. 1985. Jurassic to Paleogene: Part 2. Paleogene geochronology and chronostratigraphy; pp. 141-195 *in* N. J. Snelling (ed.), The Chronology of the Geological Record. The Geological Society, Memoir 10, Blackwell Scientific Publications.

Bowdich, T. E. 1821. An analysis of the natural classifications of Mammalia for the use of students and travellers. J. Smith, Paris, 115 pp.

Bryant, H. N. 1991a. Phylogenetic relationships and systematics of the Nimravidae (Carnivora). Journal of Mammalogy 72:56-78.

Bryant, H. N. 1991b. The polarization of character transformations in phylogenetic systematics: role of axiomatic and auxiliary assumptions. Systematic Zoology 40:433-445.

Bryant, H. N. 1993. Carnivora and Creodonta of the Calf Creek Local Fauna (late Eocene, Chadronian), Cypress Hills Formation, Saskatchewan. Journal of Paleontology 67:1032-1046.

Bryant, H. N., and A. P. Russell. 1995. Carnassial functioning in nimravid and felid sabertooths: theoretical basis and robustness of inferences; pp. 116-135 *in* J. J. Thomason (ed.), Functional Morphology in Vertebrate Paleontology. Cambridge University Press, Cambridge.

Bryant, L. J. 1984. Skeletons of the diminutive sabertooth *Eusmilus* from the Arikareean of South Dakota; pp. 161-170 *in* R. M. Mengel (ed.), Papers in Vertebrate Paleontology Honoring Robert Warren Wilson. Carnegie Museum of Natural History Special Publication 9.

Cande, S. C., and D. V. Kent. 1992. A new geomagnetic polarity time scale of the late Cretaceous and Cenozoic. Journal of Geophysical Research 97:13,917-13,951.

Clark, J. 1937. The stratigraphy and paleontology of the Chadron Formation in the Big Badlands of South Dakota. Annals of the Carnegie Museum 25:261-350.

Clark, J., J. R. Beerbower, and K. K. Kietzke. 1967. Oligocene sedimentation, stratigraphy, paleoecology, and paleoclimatology in the Big Badlands of South Dakota. Fieldiana Geology Memoirs 5:1-158.

Cope, E. D. 1873a. Third notice of extinct Vertebrata from the Tertiary of the plains. Palaeontological Bulletin 16:1-8.

Cope, E. D. 1873b. Synopsis of new Vertebrata from the Tertiary of Colorado, obtained during the summer of 1873. Annual Report of the Hayden Geological and Geographical Survey of the Territories (published separately) 7:1-19.

Cope, E. D. 1874. Report on the vertebrate paleontology of Colorado. Annual Report of the Hayden Geological and Geographical Survey of the Territories for 1873:427-533.

Cope, E. D. 1878. On some characters of the Miocene fauna of Oregon. Proceedings of the American Philosophical Society 18:63-78.

Cope, E. D. 1879a. On the genera of Felidae and Canidae. Proceedings of the Academy of Natural Sciences of Philadelphia 1879:168-194.

Cope, E. D. 1879b. Scientific news. American Naturalist 13:798a-798b.

Cope, E. D. 1879c. Observations on the faunae of the Miocene Tertiaries of Oregon. Bulletin United States Geological and Geographical Survey of the Territories, 1880 5:55-69.

Cope, E. D. 1880a. On the extinct cats of America. American Naturalist 14:833-858.

Cope, E. D. 1880b. Notes on sabre-tooths. American Naturalist 14:142-143.

Cope, E. D. 1884. The Vertebrata of the Tertiary Formations of the west. United States Geological Survey of the Territories for 1884, Vol. 3.

de Beaumont, G. 1964. Remarques sur la classification des Felidae. Eclogae Geologica Helvetica 57:837-845.

Eaton, G. F. 1922. John Day Felidae in the Marsh Collection. American Journal of Science, 5th series 4:425-452.

Emerson, S. B., and L. B. Radinsky. 1980. Functional analysis of sabertooth cranial morphology. Paleobiology 6:295-312.

Emry, R. J. 1981. Additions to the mammalian fauna of the type Duchesnean, with comments on the status of the Duchesnean "age." Journal of Paleontology 55:563-570.

Emry, R. J. 1992. Mammalian range zones in the Chadronian White River Formation at Flagstaff Rim, Wyoming; pp. 106-115 in D. R. Prothero and W. A. Berggren (eds.), Eocene-Oligocene Climatic and Biotic Evolution. Princeton University Press, Princeton, N. J.

Emry, R. J., P. R. Bjork, and L. S. Russell. 1987. The Chadronian, Orellan, and Whitneyan North American Land Mammal Ages; pp. 118-152 in M. O. Woodburne (ed.), Cenozoic Mammals of North America. University of California Press, Berkeley.

Filhol, H. 1873. Note relative à la découverte dans les gisements de phosphate de chaux du Lot d'un mammifère fossile nouveau. Bulletin Société des Sciences physical et naturale Toulouse 1:205-208.

Fisher, R. V., and J. M. Rensberger. 1972. Physical stratigraphy of the John Day Formation, central Oregon. University of California Publications in Geological Sciences 101:1-33.

Flynn, J. J., N. A. Neff, and R. H. Tedford. 1988. Phylogeny of the Carnivora; pp. 73-115 in M. J. Benton (ed.), The Phylogeny and Classification of the Tetrapods. Vol. 2: Mammals. Systematics Association Special Volume 35B, Clarendon Press, Oxford.

Gervais, P. 1876. Zoologie et paléontologie générales. Nouvelles recherches sur les animaux vertébrés vivants et fossiles. 4 Paris, 2e ser. 72 pp.

Gonyea, W. J. 1976. Behavioral implications of saber-toothed felid morphology. Paleobiology 2:332-342.

Gustafson, E. P. 1986. Carnivorous mammals of the late Eocene and early Oligocene of Trans-Pecos Texas. Texas Memorial Museum Bulletin 33:1-66.

Hatcher, J. B. 1895. Discovery, in the Oligocene of South Dakota, of Eusmilus, a genus of saber-toothed cat new to North America. American Naturalist 29:1091-1093.

Hay, O. P. 1930. Second bibliography and catalog of the fossil Vertebrata of North America. Carnegie Institute of Washington Publication 390:1-916.

Hough, J. 1949. The subspecies of Hoplophoneus: a statistical study. Journal of Paleontology 23:536-555.

Hough, J. 1953. Auditory region in North American fossil Felidae: its significance in phylogeny. U.S. Geological Survey Professional Paper 243-G:95-115.

Hunt, R. M., Jr. 1987. Evolution of the aeluroid Carnivora: significance of auditory structure in the nimravid cat Dinictis. American Museum Novitates 2886:1-74.

Janis, C. M., and P. B. Wilhelm. 1993. Were there mammalian pursuit predators in the Tertiary? Dances with wolf avatars. Journal of Mammalian Evolution 1:103-125.

Jepsen, G. L. 1926. The oldest known cat, Hoplophoneus oharrai. Black Hills Engineer 14:87-92.

Jepsen, G. L. 1933. American eusmiloid sabre-tooth cats of the Oligocene epoch. Proceedings of the American Philosophical Society 72:355-369.

Kretzoi, N. 1929. Materialien zur phylogenetischen Klassifikation der Aeluroïdeen. Congrès international de Zoologie 10, Budapest, 1927, Section 8:1293-1355.

Kretzoi, N. 1945. Bemerkungen über das Raubtiersystem. Annales Historico Naturales Musei Nationalis Hungarici 38:59-83.

Leidy, J. 1851. [No Title] Proceedings of the Academy of Natural Sciences of Philadelphia 5:329-330.

Leidy, J. 1854. Remarks on a new species of mammal from Nebraska, Dinictis felina. Proceedings of the Academy of Natural Sciences Philadelphia 7:127.

Leidy, J. 1857. Rectification of the references of certain of the extinct mammalian genera of Nebraska. Proceedings of the Academy of Natural Sciences of Philadelphia 9:175-176.

Leidy, J. 1869. The extinct mammalian fauna of Dakota and Nebraska. Journal of the Academy of Natural Sciences of Philadelphia 7:1-472.

Lucas, F. A. 1898. Contributions to paleontology 2. A new species of Dinictis (D. major). American Journal of Science, series 4, 6:399-400.

Macdonald, J. R. 1950. A new species of Nimravus from the upper Oligocene of South Dakota. Journal of Paleontology 24:601-603.

Macdonald, J. R. 1963. The Miocene faunas from the Wounded Knee area of western South Dakota. Bulletin of the American Museum of Natural History 125:139-238.

Macdonald, J. R. 1970. Review of the Miocene Wounded Knee faunas of southwestern South Dakota. Bulletin of the Los Angeles County Museum of Natural History 8:1-82.

Martin, L. D. 1980. Functional morphology and the evolution of cats. Transactions of the Nebraska Academy of Sciences 8:141-154.

Martin, L. D. 1989. Fossil history of the terrestrial Carnivora; pp. 536-568 in J. L. Gittleman (ed.), Carnivore Behavior, Ecology, and Evolution. Cornell University Press, Ithaca, NY.

Martin, L. D. 1992. A new miniature saber-toothed nimravid from the Oligocene of Nebraska. Annales Zoologici Fennici 28:341-348.

Matthew, W. D. 1907. A lower Miocene fauna from South Dakota. Bulletin of the American Museum of Natural History 23:169-219.

Matthew, W. D. 1910. The phylogeny of the Felidae. Bulletin of the American Museum of Natural History 28:289-316.

Merriam, J. C. 1906. Carnivora from the Tertiary formations of the John Day region. University of California Publications, Bulletin of the Department of Geology 5:1-64.

Merriam, J. C., and W. J. Sinclair. 1907. Tertiary faunas of the John Day region. University of California Publications, Bulletin of the Department of Geology 5:171-205.

Morea, M. F. 1975. On the species of *Hoplophoneus* and *Eusmilus* (Carnivora, Felidae). M.Sc. Thesis, South Dakota School of Mines and Technology, Rapid City, 74 pp.

Neff, N. A. 1983. The basicranial anatomy of the Nimravidae (Mammalia: Carnivora): character analysis and phylogenetic inferences. Ph.D. dissertation, City University of New York, 642 pp.

Piveteau, J. 1931. Les chats des Phosphorites du Quercy. Annales de Paléontologie 20:105-163.

Prothero, D. R. 1985. North American mammalian diversity and Eocene-Oligocene extinctions. Paleobiology 11:389-405.

Prothero, D. R., and J. M. Rensberger. 1985. Preliminary magnetostratigraphy of the John Day Formation, Oregon, and the North American Oligocene-Miocene boundary. Newsletter in Stratigraphy 15:59-70.

Prothero, D. R., and C. C. Swisher. 1992. Magnetostratigraphy and geochronology of the terrestrial Eocene-Oligocene transition in North America; pp. 46-73 *in* D. R. Prothero and W. A. Berggren (eds.), Eocene-Oligocene Climatic and Biotic Evolution. Princeton University Press, Princeton, N. J.

Radinsky, L. B. 1982. Evolution of skull shape in carnivores. 3. The origin and early radiation of the modern carnivore families. Paleobiology 8:177-195.

Rensberger, J. M. 1971. Entoptychine pocket gophers (Mammalia, Geomyoidea) of the early Miocene John Day Formation, Oregon. University of California Publications in Geological Sciences 90:1-163.

Rensberger, J. M. 1973. Pleurolicine rodents (Geomyoidea) of the John Day Formation, Oregon, and their relationships to taxa from the early and middle Miocene, South Dakota. University of California Publications in Geological Sciences 102:1-95.

Retallack, G. J. 1992. Paleosols and changes in climate and vegetation across the Eocene/Oligocene boundary; pp. 382-398 *in* D. R. Prothero and W. A. Berggren (eds.), Eocene-Oligocene Climatic and Biotic Evolution. Princeton University Press, Princeton, N. J.

Riggs, E. S. 1896a. A new species of *Dinictis* from the White River Miocene of Wyoming. Kansas University Quarterly 4:237-241.

Riggs, E. S. 1896b. *Hoplophoneus occidentalis*. Kansas University Quarterly 5:37-52.

Russell, L. S. 1972. Tertiary mammals of Saskatchewan. Part II: the Oligocene fauna, non-ungulate orders. Life Sciences Contributions, Royal Ontario Museum 84:1-97.

Schultz, C. B., M. R. Schultz, and L. D. Martin. 1970. A new tribe of saber-toothed cats (Barbourofelini) from the Pliocene of North America. Bulletin of the University of Nebraska State Museum 9:1-31.

Scott, W. B. 1945. The mammalian fauna of the Duchesne River Oligocene. Transactions of the American Philosophical Society, N.S. 34:209-253.

Scott, W. B., and G. L. Jepsen. 1936. The mammalian fauna of the White River Oligocene. Part 1. Insectivora and Carnivora. Transactions of the American Philosophical Society 28:1-153.

Simpson, G. G. 1941. The species of *Hoplophoneus*. American Museum Novitates 1123:1-21.

Sinclair, W. J. 1921. A new *Hoplophoneus* from the *Titanotherium* beds. Proceedings of the American Philosophical Society 60:96-98.

Sinclair, W. J. 1924. The faunas of the concretionary zones of the Oreodon beds, White River Oligocene. Proceedings of the American Philosophical Society 63:94-133.

Sinclair, W. J., and G. L. Jepsen. 1927. The skull of *Eusmilus*. Proceedings of the American Philosophical Society 66:391-407.

Stanley, S. M., B. Van Valkenburgh, and R. S. Steneck. 1983. Coevolution and the fossil record; pp. 328-349 *in* D. J. Futuyma and M. Slatkin (eds.), Coevolution. Sinauer Associates Inc., Sunderland, MA.

Stock, C. 1933. Carnivora from the Sespe of the Las Posas Hills, California. Carnegie Institution of Washington, Publication 440:29-41.

Storer, J. E. 1978. Rodents of the Calf Creek Local Fauna (Cypress Hills Formation, Oligocene, Chadronian) Saskatchewan. Natural History Contributions, Saskatchewan Museum of Natural History 1:1-54.

Storer, J. E. 1981. Lagomorphs of the Calf Creek Local Fauna (Cypress Hills Formation, Oligocene, Chadronian) Saskatchewan. Natural History Contributions, Saskatchewan Museum of Natural History 4:1-14.

Storer, J. E., and H. N. Bryant. 1993. Biostratigraphy of the Cypress Hills Formation (Eocene to Miocene), Saskatchewan: Equid types (Mammalia: Perissodactyla) and associated faunal assemblages. Journal of Paleontology 67:660-669.

Swisher, C. C., and D. R. Prothero. 1990. Single-crystal $^{40}Ar/^{39}Ar$ dating of the Eocene-Oligocene transition in North America. Science 249:760-762.

Swofford, D. L. 1993. PAUP: phylogenetic analysis using parsimony, version 3.1.1. Computer program distributed by the Illinois Natural History Survey, Champaign, IL.

Tedford, R. H. 1978. History of dogs and cats: a view from the fossil record; Chapter M23 *in* Nutrition and Management of Dogs and Cats. Ralston Purina Co., St. Louis.

Tedford, R. H., J. B. Swinehart, R. M. Hunt Jr., and M. R. Voorhies. 1985. Uppermost White River and lowermost Arikaree rocks and faunas, White River Valley, northwestern Nebraska, and their correlation with South Dakota; pp. 335-352 *in* J. E. Martin (ed.), Fossiliferous Cenozoic deposits of western South Dakota and northwestern Nebraska. Dakoterra 2, South Dakota School of Mines and Technology, Rapid City.

Thorpe, M. R. 1920. New species of Oligocene (White River) Felidae. American Journal of Science 50:207-224.

Toohey, L. 1959. The species of *Nimravus* (Carnivora, Felidae). Bulletin of the American Museum of Natural History 118:71-112.

Van Valkenburgh, B. 1987. Skeletal indicators of locomotor behavior in living and extinct carnivores. Journal of Vertebrate Paleontology 7:162-182.

Van Valkenburgh, B. 1988. Trophic diversity in past and present guilds of large predatory mammals. Paleobiology 14:155-173.

Van Valkenburgh, B. 1994. Extinction and replacement among predatory mammals in the North American late Eocene and Oligocene: tracking a paleoguild over twelve million years. Historical Biology 8: 129-150.

Williston, S. W. 1895. New or little known extinct vertebrates. Kansas University Quarterly 3:165-176.

Wolfe, J. A. 1992. Climatic, floristic, and vegetational changes near the Eocene/Oligocene boundary in North America; pp. 421-436 *in* D. R. Prothero and W. A. Berggren (eds.), Eocene-Oligocene Climatic and Biotic Evolution. Princeton University Press, Princeton, N. J.

Table 1. Character state distribution among eleven nimravine species and a hypothetical ancestor for the 27 characters listed in the character analysis. (? = missing datum; a = taxon polymorphic, treated as uncertain; n = character not applicable.

ancestor	000?0	000?0	00000	0000?	0?000	?0
Dinictis felina	00000	00101	00000	00000	00010	10
Dinictis cyclops	00000	00101	00001	000?0	?001?	??
Pogonodon platycopis	000a0	00101	10011	0100a	10210	11
Hoplophoneus mentalis	1001?	?0111	10111	111?2	01110	00
Hoplophoneus primaevus	10010	01111	00111	11112	01210	00
Hoplophoneus occidentalis	100??	11111	?2111	111?2	0121?	00
Hoplophoneus sicarius	?0011	12111	21112	111?2	01211	??
Hoplophoneus dakotensis	10010	12111	22112	11112	2121?	??
Eusmilus cerebralis	100a0	02111	12112	11112	01211	20
Nimravus brachyops	01101	000n1	00000	01001	10000	11
Dinaelurus crassus	?1101	000n0	02?0?	01001	???0?	??

APPENDIX: PHYLOGENETIC ANALYSIS

Phylogenetic relationships among the eleven nimravine species recognized in this study were analyzed based on a preliminary data matrix of 27 characters. The data matrix includes the cranial, mandibular, and dental features that have been used to differentiate genera and species, and additional characters that were recognized as potentially informative for resolving relationships at this level during taxonomic revision of the Nimravidae (Bryant, 1991a). Results must be considered preliminary pending more detailed revision of the alpha taxonomy and subsequent cladistic analysis of valid species. The data matrix includes eight multistate characters; missing and non-applicable data constitute 11.1% of the matrix. Multistate characters were unordered in initial analyses; the ordering of particular characters in subsequent searches produced no major differences in pattern. Characters were polarized using outgroup comparison (Barbourofelinae, Aeluroidea, and Caniformia as increasingly more distant outgroups) or the paleontological method, as appropriate (Bryant, 1991b). The resulting hypothetical ancestor was included in the cladistic analysis and used to root the tree. Branch and bound analysis (PAUP 3.1.1; Swofford, 1993) using a Macintosh LCIII produced a single most parsimonious tree (Fig. 1) with 43 steps, a consistency index of 0.814, and a retention index of 0.884. Character states were optimized using the MINF algorithm. In bootstrap analyses (300 replications) the Hoplophoneini and Nimravini occurred on over 90% of the trees based on resampled matrices; support for other clades was under 80% and these clades are unnamed.

Character Analysis

The characters and character states used in the phylogenetic analysis are listed below. The coding of the character states is given in parentheses and matches that in the data matrix (Table 1). Plesiomorphic character states of polarized characters are denoted by "0". Five characters were not polarized prior to cladistic analysis (see Table 1).

1. Sutural contact between the lacrimal and jugal: present (0); absent (1).
2. Masseteric fossa on the lateral surface of the maxilla and jugal: shallow or absent (0); deep with distinct dorsal margin (1).
3. Marked fossa on the ventral surface of the zygomatic arch below the postorbital process: absent (0); present (1).
4. Anterior opening of the palatine canal (Bryant, 1991a, fig. 5): level with P3 (0); anterior to P3 (1).
5. Posterior margin of the palate (Fig. 5): posterior to the anterior surface of M1 (0); anterior to or level with the anterior surface of M1 (1).
6. Postglenoid foramen: large (0); reduced in size or absent (1).
7. Paroccipital process: large and projecting strongly posteriorly (0); reduced in size and projecting posteriorly (1); extremely reduced in size, not projecting posteriorly (2).
8. Cranial and mandibular features characteristic of sabertooth carnivores (as outlined by Emerson and Radinsky, 1980): absent (0); present (1).
9. Development of sabertooth features (as in character 8): moderate (0); marked (1).
10. Shape of C1: conical (0); blade-like (1) (modified from the categories of Martin, 1980).
11. Anteroposterior length of C1: less than that of P4 (0); equal to that of P4 (1); greater than that of P4 (2).
12. P2: present (0); reduced to nubbin, sometimes absent (1); absent (2).
13. p2: present (0); absent (1).
14. Size of P3: typical of carnivorans (0); reduced (1).
15. p3: present (0); size reduced relative to p4 (1); absent (2).

16. Parastyle on P4: absent (0); present (1).

17. Protocone on P4: distinct and of moderate size (0); reduced in size or absent (1).

18. Crown height of P4: typical of carnivorans (0); hypsodont (1).

19. Lingual rotation of the crown of P4 as it wears due to bending of roots (Bryant and Russell, 1995): absent (0); present (1).

20. M1: with strong parastylar wing, distinct protocone and reduced to absent metacone (0); morphology similar but size reduced (1); protocone and parastylar regions reduced (2).

21. Metaconid on m1: present (0); absent (1); reduced to a ridge (2).

22. Talonid on m1: single trenchant cusp (0); size reduced (1).

23. m2: present (0); occurrence variable (1); absent (2).

24. Serration of ridges on incisors and post-canine teeth (Bryant, 1991a, fig. 8): absent (0); present (1).

25. Atlas: dorsal arch relatively short anteroposteriorly, atlantal foramen in medial position (0); length of dorsal arch increased, atlantal foramen in lateral position (1).

26. Femorotibial index: similar to those of forest dwelling felids (0); similar to those of open woodland to savanna dwelling felids (1); intermediate, within range of values in both felid groups (2) (based on data in Gonyea, 1976).

27. Articulation between the calcaneum and navicular absent (0); present (1).

23. Amphicyonidae

ROBERT M. HUNT, JR.

ABSTRACT

Beardogs (Mammalia, Amphicyonidae) are mid-Cenozoic arctoid carnivorans ranging in size from less than 5 to over 200 kg found as fossils on the northern continents and Africa. The earliest North American species are generally small animals of late Eocene and Oligocene age, placed in the genera *Daphoenus, Daphoenictis, Paradaphoenus,* and *Brachyrhynchocyon*. The most complete skeletons of these carnivorans come from the White River beds of the North American mid-continent, but more fragmentary remains found throughout the United States and southern Canada indicate that the family was widely dispersed in North America at the time of the Eocene-Oligocene transition.

INTRODUCTION

Amphicyonid carnivorans at the time of the Eocene-Oligocene transition are the oldest known representatives of the family, poised at the threshhold of a Holarctic radiation of marked diversity centered in the Oligocene and Miocene epochs. Amphicyonids, commonly known as "beardogs," are the large carnivorans of the mid-Cenozoic that, together with the hemicyonine ursids, dominate the carnivore faunas of the northern continents in the Miocene. By the beginning of the Pliocene the amphicyonid-hemicyonine ursid interval has ended, and the dominant large terrestrial carnivores belong to the modern groups familiar to us today, the felids, ursine ursids, and canids.

Although the earliest amphicyonids are found in the late Eocene of both Eurasia and North America, the fossil record is predominantly North American and European. The oldest European (*Cynodictis*) and North American (*Daphoenus*) amphicyonids share a derived basicranial anatomy that proclaims a not-too-distant common ancestry in the Eocene. At the time of their first appearance in the late Eocene these ancestral genera are distinct in dentition, and hence identifiable as separate lineages. From the European *Cynodictis* springs an Old World radiation of amphicyonine beardogs, and similarly from North American *Daphoenus* arises a New World radiation of daphoenines (Hunt, in press).

Old World amphicyonines evolved rapidly, as evidenced by the diversity of beardogs found in the Quercy fissure fillings (France) and in Oligocene sediments of western Europe (Teilhard, 1915; Ginsburg, 1966; Remy et al., 1987). The daphoenines of the New World appear to be less diverse on present evidence, yielding only four genera (*Daphoenus, Paradaphoenus, Daphoenictis,* and *Brachyrhynchocyon*) in the Duchesnean-Chadronian-Orellan interval that brackets the Eocene-Oligocene transition in North America. These genera comprise the greater part of the daphoenine radiation that begins at about 40 Ma and terminates with the extinction of the last North American daphoenine by ~17 Ma in the early Miocene (Fig. 1). Daphoenines are replaced in the middle and late Miocene by amphicyonine beardogs from the Old World that migrate into the New World at intervals to become key members of the North American mid-Cenozoic carnivoran fauna (Hunt, in press).

ABBREVIATIONS

ACM, Pratt Museum, Amherst College, Amherst, MA; AMNH, American Museum of Natural History, New York, NY; CM, Carnegie Museum of Natural History, Pittsburgh, PA; FAD, first appearance datum; F:AM, Frick Collection, American Museum of Natural History, New York, NY; FMNH, Field Museum of Natural History, Chicago, IL; LAD, last appearance datum; PU, Princeton University (now housed at Yale University, New Haven, CT); TMM, Texas Memorial Museum, University of Texas, Austin, TX; UCR, University of California, Riverside, CA; UNSM, University of Nebraska State Museum, Lincoln, NE.

GEOGRAPHIC AND GEOLOGIC DISTRIBUTION

Amphicyonids of Chadronian-Orellan age in North America are almost entirely preserved in fine-grained volcaniclastic sediments of the White River Group. These deposits are regionally extensive, thick, lithologically uniform units developed in western Nebraska, southwest South Dakota, Wyoming, Colorado, and Montana, with important correlative outliers in Saskatchewan and North Dakota (Emry et al., 1987). White River tuffaceous silts and clays buried a rich mammalian fauna, often beautifully preserved as three-dimensional skeletons in association with intact crania. Fine-grained sediment slurries surrounded articulated

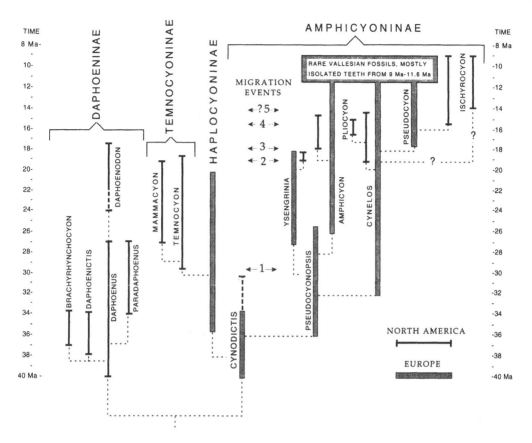

Figure 1. Temporal ranges, migration events, and suggested phylogenetic relationships of North American and European Cenozoic Amphicyonidae (1, 29.5 Ma, haplocyonine; 2, 19.2 Ma, amphicyonines *Cynelos*, *Ysengrinia*; 3, 18-18.5 Ma, *Amphicyon*; 4, 15.5 Ma, *Pseudocyon*; 5, ?14-15 Ma, *Ischyrocyon*). Four subfamilies constitute the Amphicyonidae: Daphoeninae, Temnocyoninae, Haplocyoninae, and Amphicyoninae (from Hunt, in press).

bones and penetrated cavities of skulls at the time of sediment burial to create undistorted fossils that retain highly detailed anatomical features. As a result, our knowledge of the White River genera of amphicyonids is most complete, and the species from the North American mid-continent are adequately known.

Sediments of Chadronian-Orellan age elsewhere in North America have produced far fewer beardogs than the sample from the White River beds. Important specimens come from the Vieja Group of west Texas (Gustafson, 1986b), the John Day sequence of Oregon (Hunt, in press), and the I-75 locality of north Florida (Patton, 1969, considered Whitneyan in age). These fossils belong to the same genera as those occurring in the White River rocks of the mid-continent, indicating a regional homogeneity, but suggest a degree of geographic variation at the level of species or subspecies.

Duchesnean amphicyonids are the oldest in North America and are very rare. Only a few localities can be reliably attributed to this as yet controversial and newly redefined age (Lucas, 1992).

Here I review the principal genera of Duchesnean, Chadronian, Orellan, and Whitneyan amphicyonids and

describe their basic skeletal and dental characteristics, time ranges, and geographic distribution.

DUCHESNEAN AND EARLIEST CHADRONIAN AMPHICYONIDS

The few amphicyonids attributable to the Duchesnean/earliest Chadronian indicate a wide geographic distribution for the family at its appearance in North America. These remains belong to *Daphoenus* and occur in Wyoming, Texas, and Saskatchewan (Figs. 1 and 2). These are small carnivorans with lower carnassial lengths of 10-12 mm and body weights of less than 5 kg. Because these are the oldest North American amphicyonids, the dental measurements useful in defining this sample are presented in Table 1. These measurements identify a small species of *Daphoenus* that occurs at a limited number of localities in the North American mid-continent: the Cypress Hills, Saskatchewan; Beaver Divide (West Canyon Creek), Wyoming; Shirley Basin, Wyoming; Badwater Creek area, Wyoming; and the Porvenir local fauna, Texas.

Two name-bearing holotypes fall in this hypodigm,

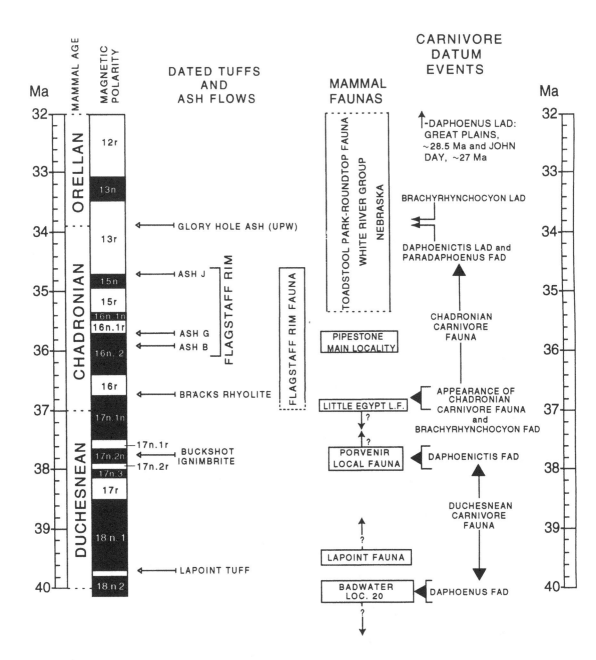

Figure 2. Chronology of the Eocene-Oligocene transition relative to first (FAD) and last (LAD) appearances of amphicyonid genera in North America. Geomagnetic polarity time scale after Cande and Kent (1992) and dated tuffs and ash flows from Prothero and Swisher (1992). Placement of mammal faunas and carnivore datum events is discussed in the text.

Daphoenus lambei Russell 1934 and *D. demilo* Dawson 1980. Based on the very limited range of dental variation observed in this hypodigm (considered in relation to the known range of dental variation in the much larger Orellan samples of *Daphoenus vetus* and *D. hartshornianus*), there is sufficient justification to place these oldest *Daphoenus* in a single geographically wide-ranging morphospecies defined by the dental dimensions

of Table 1. The species *lambei* has priority for this hypodigm, thus *D. lambei* is the oldest North American amphicyonid, ranging from southern Canada to Texas.

In addition to size, various aspects of tooth form characterize the *Daphoenus lambei* sample: (a) m1-2 are relatively wide robust teeth, more so than in European *Cynodictis*; (b) M2/m2 are large developed teeth and M3/m3 are smaller but present (in *Cynodictis* the

Table 1.   Dental measurements of the amphicyonid *Daphoenus lambei* Russell 1934 from the Duchesnean-early Chadronian of North America (in mm).   [ ] = estimated

	Length			Width	
p4	8.6- 9.5	(N=4)	3.8- 4.5	(N=4)
m1	10.9-11.9	(N=4)	6.2- 6.6	(N=4)
m2	7.0- 7.8	(N=4)	4.4- 5.4	(N=4)
P4	[10.8]-11.8	(N=4)	7.5- 8.0	(N=4)
M1	7.9-8.4	(N=3)	12.1-12.6	(N=3)
M2	5.6-[6.7]	(N=3)	9.2-[10]	(N=3)

M2/m2 are often reduced in size, and M3/m3 are vestigial to absent); (c) in *Daphoenus* m1 retains a "closed" trigonid in which the paraconid is not anteriorly displaced nor the metaconid retracted (in *Cynodictis* the m1 paraconid is more anteriorly placed, hence the trigonid is more "open"), and the broad basined talonid is encircled by well-defined hypoconid and entoconid ridges; (d) sharply defined cingula occur on the carnassial teeth (the holotype of *D. lambei*, an isolated P4, has this distinct cingulum); (e) M1 retains a plesiomorphic buccal shelf and developed parastylar region as well as distinct para- and metaconules (also in *Cynodictis*). Both basicranial and dental traits demonstrate a close relationship between *Daphoenus* and *Cynodictis* but in the former the M2-3/m2-3 are retained and enlarged, a hallmark of the *Daphoenus* lineage. Although the m1 of *Daphoenus* is similar in form to the m1 of the small amphicyonodont ursid *Parictis*, it differs from the latter in its larger size and in the form of the premolars in the two genera (for example, p4 is elongate with a prominent posterior accessory cusp in *Daphoenus* but in *Parictis* it is a short conical tooth with only a small accessory cusp).

The age of all known specimens of *D. lambei* falls in the Duchesnean and/or earliest Chadronian. Current discussions (Emry, 1981; Emry et al., 1987; Krishtalka et al., 1987; Storer, 1990; Lucas, 1992) of the Duchesnean and Chadronian mammal ages generally agree on faunal sequence but differ in placement of biochronologic boundaries. I employ the redefinition of Lucas (1992) for clarity and simplicity in discussing amphicyonid distribution in Duchesnean and earliest Chadronian faunas.

Daphoenus lambei occurs in relationship to radiometrically dated tuffs at two localities, Badwater Locality 20, Wyoming, and the Porvenir local fauna, Texas (Fig. 2). The Porvenir assemblages found between the Bracks Rhyolite (~36.7 Ma) and the Buckshot Ignimbrite (~37.8 Ma) have been recently redated employing the ^{40}Ar/^{39}Ar method (Prothero and Swisher, 1992, p. 69). The Porvenir sites occur in the lower third of the Chambers Tuff above the ignimbrite and thus can be no older than ~37.8 Ma. Badwater Locality 20 was long considered validly dated at 42.3 ± 1.4 Ma by a biotite-bearing bed above Locality 20 in the Hendry Ranch Member of the Wagonbed Formation, central Wyoming

(Black, 1969). However, recent attempts to redate this biotite by the ^{40}Ar/^{39}Ar method suggest the crystals are detrital, derived from a nonvolcanic source, and as a consequence the potassium-argon date has been rejected as a Duchesnean calibration point (Prothero and Swisher, 1992, p. 49). Lucas (1992) nevertheless believes that the Locality 20 assemblage with its first appearances of amphicyonids and the protoceratid *Poabromylus* may be among the oldest Duchesnean sites known. The remaining sites producing *D. lambei* in the Cypress Hills, Shirley Basin, and Beaver Divide either lack reliable radiometric control or have not been dated.

Employing the Duchesnean concept of Lucas (1992) and Wilson (1984, 1986) in which Badwater Locality 20 is considered one of the oldest faunas and Porvenir one of the youngest, one can speculate that *Daphoenus* appears at least by ~38 Ma and possibly extends in the Badwater sites to ~40 Ma. The date of ~40 Ma is derived from the occurrence of the Duchesnean Lapoint fauna of eastern Utah above the Lapoint Tuff, recently dated at 39.7 Ma by both K-Ar and ^{40}Ar/^{39}Ar methods (Prothero and Swisher, 1992), and the belief that the Badwater Locality 20 mammals are older than the Lapoint fauna (Fig. 2).

The only remains of *Daphoenus lambei* found in association with postcranial elements are the dental remains from the Shirley Basin, Wyoming, collected by E. N. Harshman in 1962 (USNM 22970). His fossils were collected from the lower 75 feet of the White River Formation in section 15, T.29N, R.80W, Natrona County, Wyoming (Harshman, 1972). These rocks are very likely early Chadronian (Emry, written communication, September 1990; Hunt, field observations, July 1994) and may be equivalent in age to the lowermost part of the Flagstaff Rim section. Several of the fossils retain a reddish patina suggestive of the lower part of the White River Formation in the Shirley Basin, Flagstaff Rim, and Yoder sections.   Two partial mandibles with m1 and partial p4, two isolated P4s, complete upper and lower canines, an incisor, and fragments of other teeth were found together with two large proximal caudal vertebrae, two calcanea, 5 distal metapodials, and several proximal and intermediate phalanges. The metapodials and phalanges all appear to belong to the hindfoot, suggesting that more of this skeleton may have been available at the time of its discovery.   The anatomical features of the calcanea, vertebrae, and hind foot bones are identical to these elements in associated postcranial skeletons of the Orellan species of *Daphoenus*, indicating that these carnivorans share and maintain their generalized postcranial skeleton during the early history of the group. The ratio of calcaneal length/m1 length for the Shirley Basin *Daphoenus* is 2.94, falling in the range of the smaller Orellan *D. hartshornianus* (2.75-3.21, n = 5) relative to the range of the large Orellan *D. vetus* (3.10-

3.34, n = 4). *Daphoenus lambei* may be ancestral to both of the Orellan species. Anatomy of calcaneum, metapodials, and phalanges indicates that these earliest New World amphicyonids were paraxonic digitigrade to subdigitigrade cursors with ability to evert and invert the hindfoot. An increase in body size occurs in *Daphoenus* from the earliest species of the Duchesnean-early Chadronian into the Orellan, continuing into the Whitneyan and early Arikareean.

In addition to *Daphoenus*, Duchesnean localities (*sensu* Lucas, 1992) have also produced fragmentary amphicyonid remains of uncertain generic assignment. In the Porvenir fauna, Gustafson (1986b) described a partial mandible (TMM 40688-72) as an indeterminate daphoenine amphicyonid. The damaged teeth (partial p4-m2) preserve features reminiscent of *Daphoenictis* as noted by Gustafson. Although he hesitated to place this specimen in the genus, I believe this is the earliest example of *Daphoenictis* in North America. Hence, the genus ranges from the later Duchesnean to the end of the Chadronian (Hunt, 1974; Emry and Hunt, 1980). The temporal range in North America is established using the ^{40}Ar/^{39}Ar dates for the Buckshot Ignimbrite (~37.8 Ma) immediately below the Porvenir local fauna, and the Glory Hole Ash (~33.9 Ma) in the Douglas area, Wyoming, that approximates the end of the Chadronian (Prothero and Swisher, 1992), thus a span of about 4 million years (Fig. 2). At its earliest appearance in the Porvenir fauna, *Daphoenictis* is a carnivore about the size of *Daphoenus lambei* and may have evolved from the *Daphoenus* lineage.

From the Badwater localities in the Wagonbed Formation of central Wyoming, Dawson (1980) also mentioned and figured carnivore teeth from Locality 6 that could not be assigned to family. The elongate p4 relative to m1 (CM 16053) eliminates amphicynodont ursids such as *Parictis* with a short conical p4, and there is nothing to support an amphicyonid assignment for this very small species. More complete material is necessary to determine this taxon.

In summary, Duchesnean and earliest Chadronian faunas have the oldest North American amphicyonids: a small species of *Daphoenus* and a possible early record of the catlike *Daphoenictis*. No certain evidence of the short-jawed *Brachyrhynchocyon* Loomis 1931 (a senior synonym for *Daphoenocyon* Hough 1948) occurs in these localities to date.

EARLY CHADRONIAN AMPHICYONIDS AND THE ADVENT OF THE WHITE RIVER CARNIVORE FAUNA

The Flagstaff Rim area of central Wyoming preserves a thick sequence of White River Formation sediments of Chadronian age that has been studied by Emry (1973, 1992). Emry (1992) has established biostratigraphic ranges for numerous mammals in the Flagstaff Rim

section, including many carnivorans. In the lowest part of the section which he considers early Chadronian, the suite of carnivorans comprises the creodonts *Hyaenodon* and *Hemipsalodon*, the amphicynodont *Parictis*, the early canid *Hesperocyon*, a possible nimravid cat, and the amphicyonids *Daphoenictis* and *Brachyrhynchocyon* (= *Daphoenocyon*). This assemblage constitutes a typical Chadronian carnivoran suite, made up primarily of new appearances and lacking in archaic species. The carnivorans and other mammals of early Chadronian age at Flagstaff Rim in the lower 100 feet of Emry's generalized section were largely collected from quarries. Within the lowermost 85 feet of the generalized section of Emry (1992, p. 107), the 60-85 foot interval that has been allocated to Chron 16n using Prothero and Swisher's (1992) magnetostratigraphy also includes the early Chadronian carnivoran fauna (Emry 1992, p. 109). The Chron 16n magnetozone has been dated at ~38.2-38.5 Ma using some current estimates (McGowran et al., 1992; Miller, 1992) but was placed by Prothero and Swisher (1992) at ~36.3-36.4. However, the new geomagnetic polarity time scale of Cande and Kent (1992) provides a further modification of these interpretations which I follow here: Ash B at Flagstaff Rim dated at ~35.9 Ma would fall in Cande and Kent's Chron 16n.2, suggesting that the reversed zone beneath it at Flagstaff Rim is Chron 16r. Thus the normal interval at the base of the Flagstaff Rim section that includes the fossil sites (Dry Hole and Low Red quarries) from 60-85' would fall in the upper part of Chron 17n.1, giving an age for this interval of ~36.6-37 Ma. Hence the White River carnivore fauna characteristic of the Chadronian age probably makes its appearance by ~36.6-37 Ma (Fig. 2).

The Cande and Kent interpretation also is in good agreement with the ^{40}Ar/^{39}Ar radiometry associated with the Porvenir and Little Egypt faunas of Texas. Since the Porvenir local fauna occurs close above the Buckshot Ignimbrite dated at ~37.8 Ma, the age of this latest Duchesnean fauna arguably falls at perhaps ~37-37.8. The Little Egypt local fauna, which has been considered an earliest Chadronian assemblage (Lucas, 1992), lies about 60 feet below the 36.7 Ma Bracks Rhyolite and may fall within the 36.8-37 Ma interval, and thus can be considered a near-correlative of the lower 85 feet of the Flagstaff Rim section, using the polarity time scale of Cande and Kent. Consequently, the Duchesnean may span an interval from ~40 Ma to at least 37 Ma and possibly extends to the 36.6-36.8 Ma interval. The Chadronian carnivore fauna appears, then, during the 36.6-37 Ma interval and continues to ~34 Ma. It is typified by the carnivore genera indicated in the lower 200 feet of the Flagstaff Rim section by Emry (1992).

The amphicyonid *Brachyrhynchocyon* does not seem to occur prior to the advent of the Chadronian chrono-

Table 2. Dental measurements of medial-late Chadronian *Daphoenus* from North America (in mm).

	Length		**Width**	
p4	[10.2]-[12.5]	(N=9)	4.0- 5.4	(N=7)
m1	[12.6]- 15.4	(N=10)	5.9- 7.6	(N=9)
m2	7.3 - 10.5	(N=8)	4.7- 7.0	(N=7)
P4	[11.9]- 15.7	(N=4)	9.0-10.0	(N=3)
M1	9.9 - 12.5	(N=3)	13.3-17.0	(N=4)
M2	6.0 - 8.5	(N=4)	[8.6]-13	(N=4)

Table 3. Dental measurements of Orellan *Daphoenus* from North America (in mm).
HARTS = *D. hartshornianus*; VETUS = *D. vetus*

		Length		**Width**	
p4	HARTS	10.2 - 12.0	(N=24)	4.0 - 5.3	(N=21)
	VETUS	10.9 - 13.0	(N=24)	4.4 - 6.0	(N=20)
m1	HARTS	13.0 - 14.9	(N=24)	5.7 - 7.4	(N=23)
	VETUS	15.0 - 16.9	(N=23)	6.4 - 8.0	(N=20)
m2	HARTS	7.0 - 9.2	(N=25)	4.9 - 6.0	(N=22)
	VETUS	8.9 - 10.9			

fauna. It is reported in the Little Egypt local fauna in Texas and the Flagstaff Rim section below Ash A (Gustafson, 1986b; Emry, 1992). A mandible is known from 140 feet below Ash B, thus about 35 feet above the base of the generalized section and is the stratigraphically lowest occurrence at Flagstaff Rim. It also occurs in the Yoder fauna of southeast Wyoming which appears to be a correlative (Kihm, 1987). I regard these three occurrences as the oldest evidence of the genus in North America. A fourth occurrence in the upper third of the Ahearn Member of the Chadron Formation, Indian Creek, South Dakota, must also be an early record. The suggestion that this taxon might be present in the Lapoint fauna of Duchesnean age (Emry, 1981), based on an edentulous mandible, is difficult to evaluate but appears improbable based on the large size of the mandible. The early occurrences of the genus are not large carnivores and it is more likely that the Lapoint mandible belongs to the large miacid *Miocyon magnus* (Bryant, 1992).

In summary, the three oldest amphicyonid genera in North America appear at staggered intervals if the preceding analysis is correct. *Daphoenus* first occurs in the Locality 20 fauna at Badwater in the early part of the Duchesnean and may serve as an indicator of the beginning of this age. *Daphoenictis* probably first arrives in the Porvenir fauna of the later Duchesnean and is certainly present at the early Chadronian Dry Hole Quarry, 70-80 feet above the base of the White River Formation at Flagstaff Rim. *Brachyrhynchocyon* (= *Daphoenocyon*) is first observed at three early Chadronian sites in Wyoming and Texas that are probably penecontemporaneous, and at present is an index fossil for the Chadronian.

MEDIAL AND LATE CHADRONIAN AMPHICYONIDS

The medial and late Chadronian are characterized by the common ocurrence of the amphicyonid genera *Daphoenus*, *Daphoenictis*, and *Brachyrhynchocyon* (Fig. 1). Although none of these is represented by large samples in Chadronian rocks, *Brachyrhynchocyon* is most common, followed by somewhat scarcer remains of *Daphoenus*. *Daphoenictis* remains an extremely rare carnivoran in any setting.

I believe that dental and mandibular measurements demonstrate that all three genera experience some measure of size increase during the Chadronian. Measurements of the teeth of the North American sample of *Daphoenictis* appear to increase in size over time, using length of the lower carnassial as an index (Hunt, 1974; Emry and Hunt, 1980): (1) the probable specimen from the Porvenir fauna (Gustafson, 1986b, TMM 40688-72), 12.5 mm, late Duchesnean; (2) the Flagstaff Rim (10 feet above Ash E) and Pipestone Main Locality carnassials, both ~14.2 mm, medial Chadronian; (3) carnassials, isolated or in partial mandibles, from northwest Nebraska (Norman Ranch, Brecht Ranch) and the Cypress Hills (Loc. 133), 16 to 16.6 mm, late Chadronian.

Medial to late Chadronian *Daphoenus* is best known from the White River badlands of northwest Nebraska and southwest South Dakota where the genus is represented by about 13-14 individuals: these fossils are on average larger than those from the Duchesnean-early Chadronian (compare Tables 1 and 2). In some dimensions there is no range overlap (p4 length, m1 length, P4 and M1 length and width), and sample size is sufficiently large in several of these to suggest a significant difference.

In contrast, there is a much greater overlap in the ranges of dental measurements of Orellan *Daphoenus* (allocated to two species in the White River beds of Nebraska, South Dakota, Colorado, and Wyoming, Table 3), when compared to measurements of medial-late Chadronian animals. These data (compare Tables 2 and 3) suggest, as does the form of the teeth, that the roots of the two Orellan species lie in the clearly variable medial and late Chadronian sample. It is not possible to segregate all medial and late Chadronian *Daphoenus* specimens into two discrete species because of small sample sizes but the two morphs presaging the Orellan species are clearly present. By employing the much larger samples from the Orellan age the two species, one small and one larger, can be defined and prove to be markedly divergent in some dental measurements (Table 3, m1 and m2 lengths).

Consequently, the Duchesnean to Orellan record in North America seems to document not only a size increase in *Daphoenus* during that interval but also a phyletic branching event in which two daughter species

(*D. hartshornianus*, *D. vetus*) evolve from a parent species (*D. lambei*).

In addition to *Daphoenus* and *Daphoenictis*, the medial to late Chadronian faunas include the short-faced beardog *Brachyrhynchocyon* as a common element. The holotype of Loomis (1931) is a well-preserved cranium and mandibles of a small individual (ACM 2752) from the Seaman Hills of Wyoming, undoubtedly referable to the hypodigm.

Brachyrhynchocyon can be distinguished from both *Daphoenus* and *Daphoenictis* by its short robust premolars, an abbreviated snout, and thick massive mandible. These features are already apparent in the oldest known fossils from the early Chadronian (Yoder local fauna, Wyoming; Little Egypt local fauna, Texas; 140 feet below Ash B, Flagstaff Rim, Wyoming; Ahearn Member, Indian Creek drainage, South Dakota). Marked size differences in mandibles, teeth, and crania from the medial and late Chadronian caused earlier authors to segregate these fossils into several species (Clark et al., 1967). However the large sample currently available, including the occurrence at a single stratigraphic horizon of dimorphic crania and mandibles, indicates that the explanation of dental variation probably lies in pronounced sexual dimorphism. The key specimens were collected in 1982-1983 at the Bartlett West #1 locality, Brecht Ranch local fauna, north of Chadron, Nebraska, by Gustafson (1986a, p. 15). Gustafson's careful work documented the occurrence of large robust male and gracile female crania in the same stratigraphic horizon. Although there was probably a small degree of size increase in the teeth during the Chadronian, the creation of a small species, *minor* (Clark et al., 1967, p. 32), for the small form is probably not valid; many or all of these smaller animals are very likely females. Thus, except for a single mandible (CM 8851) from Montana, all Chadronian *Brachyrhynchocyon* are referred here to the species *dodgei*.

An extremely small mandible with teeth (CM 8851) has been collected in the Pipestone Springs Main Locality together with mandibles of large medial Chadronian *Brachyrhynchocyon dodgei*. This rare individual is a young adult much too small to be referred to the larger species and is named *B. montanus*, new species, a designation intended for CM 8851 by the late John Clark but never published. It is the only known individual of the species. Table 4 indicates that the dimensions of its p4-m2 fall entirely outside the range of the larger *B. dodgei*. The teeth of *B. montanus* and *B. dodgei* are so similar in form that the two species are distinguished entirely on size.

Teeth measurements of *Brachyrhynchocyon dodgei* greatly exceed similar dimensions of early Chadronian *Daphoenus* (Table 1) but are comparable to these dimensions in medial-late Chadronian *Daphoenus* (Table

Table 4.   Dental measurements of the amphicyonid *Brachyrhynchocyon* from the Chadronian of North America (in mm).

	Length		**Width**	
B. dodgei				
p 4	10.0 - 12.7	(N=30)	5.0 - 6.3	(N=2)
m 1	13.4 - 17.0	(N=30)	6.1 - 7.7	(N=24)
m 2	7.0 - [9.8]	(N=26)	5.1 - 6.5	(N=18)
B. montanus (N=1)				
p 4	8.6		4.1	
m 1	11.4		5.2	
m 2	6.7		4.8	

2). However, in p4 width and m1 length, *Brachyrhynchocyon dodgei* averages larger than *Daphoenus* of Table 2. Moreover, the short broad premolars and shortened premolar row (with anterior premolars often *en echelon*), the distinctive stabbing canines, and the broad short-snouted skull with characteristic basicranial structure allows the genus to be distinguished from *Daphoenus*.

ORELLAN AMPHICYONIDS

The existence of Orellan *Brachyrhynchocyon* has long been doubted and the genus is usually believed to be limited to the Chadronian. However, three mandibles found in 1959 on the Amer Ranch, Sioux County, Nebraska (UCR 11057-11058, Locality V5704) are indicated as Orellan in field records. Such a designation commonly is given to fossils collected from the Orella Member of the Brule Formation that occurs above the "Upper Purple White" Ash marking the boundary between the Chadron Formation and the Orella Member in this area. Both Chadronian and Orellan fossils have been collected from White River sediments on the Amer Ranch. These mandibles and teeth are not larger or more dentally advanced relative to the large sample of *Brachyrhynchocyon* from the Chadronian and fall within the measurements of the Chadronian hypodigm. Since there are other lithically similar "purple-white" ashes in the area, one might question whether the field collectors accurately placed these mandibles relative to the Chadron-Orella boundary on the Amer Ranch. However, in 1963 a maxilla (UNSM 2564-63, Locality Sx-29) of *Brachyrhynchocyon* was also found on the Amer Ranch and was recorded in field notes as collected from 10' above the "Upper Purple-White" in the Orella Member. We do not know whether this specimen was collected in place or if care was exercised in the designation of the collecting horizon, but these combined occurrences suggest that the genus may occur a short distance above the Chadron-Orella boundary in the Sand Creek-Toadstool badlands of northwest Nebraska.

Such an occurrence stratigraphically low in the Orella Member would not be surprising. In the Sand Creek-Toadstool Park area the lithologies of both the upper Chadron Formation and the lower Orella Member are alike–gray-green claystones that show no significant lithologic difference or evidence of a depositional hiatus above or below the "Upper Purple-White" ash. Moreover, specimens of *Daphoenus* from the upper Chadron and lower Orella units differ little and would seem to indicate that minimal time in the geologic sense was involved in the passage from one unit to the other. Elsewhere in North America *Brachyrhynchocyon* is not known in sediments of Orellan or Whitneyan age.

The common beardogs of the Orellan are two species of *Daphoenus* represented by abundant fossils from the White River badlands of South Dakota, Nebraska, and Wyoming. Collected initially by John Evans in 1849, the larger *Daphoenus vetus* was one of the first mammals described from the White River beds by Joseph Leidy in 1854. The smaller species, *Daphoenus hartshornianus*, was first collected from the White River beds of Cedar Creek, Colorado, by E. D. Cope in 1873. Since that time, these amphicyonids have been common elements in numerous White River collections. As a consequence, the remains of these species include crania and mandibles in association with postcranial skeletons. The skeletons of both species are remarkably similar except for size, and the skeleton of a male *D. vetus* was described by J. B. Hatcher (1902) based on an individual with associated baculum collected by O. A. Peterson in Sioux County, Nebraska, in 1901. Peterson's skeleton was remarkable in allowing Hatcher to describe the enormous exostoses developed bilaterally on the inner edge of the distal radii in this animal. The much larger sample of these carnivorans that has accumulated in the last 90 years since Hatcher's description has allowed further clarification of the nature and distribution of these growths. They are not pathologic but represent bilateral development of normal bone. In *Daphoenus vetus* they are restricted to males and increase in size with age (males are identified by larger canines and/or the presence of a baculum). In some specimens they become truly enormous; one would think they could have easily fractured or even impeded locomotion. In the smaller *Daphoenus hartshornianus* no exostoses have been observed, but we do not know as yet if the species developed them because a distal radius is associated with a dentition in only three cases. These are either confirmed or probable females based on canine size (PU 11424, Prairie Dog Creek, Sioux County, Nebraska; F:AM 63332, 1 mile east of Babby Butte, Shannon County, S.D.; FMNH UM-488, Warbonnet Creek, Sioux County, Nebraska), thus the situation in a male remains uncertain. Collecting associated skeletons of these carnivores has not received sufficient attention in the past and should be pursued.

Dental measurements of Orellan *Daphoenus* in Table 3 demonstrate the size differences in lower teeth of the two species using the large samples available from the Orella Member. Generally there is some overlap in range except for the length of the lower carnassial which can be a useful indicator in a first attempt to identify the species. Very small or large specimens are easily placed whereas the individuals that fall in the zone of overlapping dental dimensions are difficult. Canine size and other evidence of gender can be used since the overlap involves large males of the small species and small females of the large species. The two species are so morphologically similar that they must have partitioned food resources chiefly on the basis of size.

The only other amphicyonid encountered in Orellan faunas is the very small "*Daphoenus*" *minimus* first described by Hough in 1948. She based this taxon on a small cranium (AMNH 39099) from the Orellan near Scenic, South Dakota. As Hough noted, the basicranial anatomy is distinct and allows discrimination from canids. The auditory bulla was undoubtedly a simple bony ectotympanic crescent, and not the inflated bulla of early canids formed by fusion of ectotympanic with entotympanic elements. The upper teeth (there is no associated mandible) are configured as in amphicyonids. However, Hough and subsequent workers were not aware that "*Daphoenus*" *minimus* ·is the oldest North American species in the lineage of *Paradaphoenus*, a small amphicyonid genus based on a cranium and associated mandibles from the John Day beds of Oregon. *Paradaphoenus minimus* is known from the holotype skull and a few fragmentary lower jaws from the Orellan and Whitneyan of the Great Plains. It is continued in the early Arikareean of the John Day region of Oregon and in lower Arikaree rocks of South Dakota and Nebraska by more dentally derived lineages of slightly larger size (Hunt, in press). Although the cranium has resemblances to Old World *Cynodictis*, it seems more likely that the genus is a North American endemic that may arise from "*Miacis*" *cognitus* of the early Chadronian Little Egypt fauna of Texas (Gustafson, 1986b, fig. 26). "*Miacis*" *cognitus* differs from many members of *Miacis* in its upper molars symmetrical about a transverse axis and in the reduced buccal (stylar) shelves on M1-2, features also observed in the upper molars of *Paradaphoenus*. Skull form and basicranial structure also seem comparable. However, because the ancestry of *Paradaphoenus* within or near "*M.*" *cognitus* is uncertain, the FAD of *Paradaphoenus* is placed at ~34 Ma in Figures 1 and 2.

WHITNEYAN AMPHICYONIDS

The only amphicyonid lineages certainly present in the Whitneyan are *Daphoenus* and *Paradaphoenus*. Based on recent $^{40}Ar/^{39}Ar$ dating of Whitneyan and Arikareean localities, the oldest temnocyonine amphicyonids (Fig. 1) appear to be early Arikareean in age and are not found in Whitneyan faunas (Hunt, in press).

Whitneyan *Daphoenus* is known from rare dental remains in the Great Plains that average larger in size than Orellan species.

In the John Day beds of Oregon, two species of *Daphoenus* also occur, and are similar in size to the two Orellan species of South Dakota and Nebraska. However, minor differences in the form of the teeth, particularly the carnassials and molars, suggest that the two John Day species are distinct and constitute geographically disjunct species relative to the Great Plains Orellan lineages. It is possible that the John Day species are of Whitneyan age and may survive as relict forms into the early Arikareean of the Pacific Northwest based on radiometric dating of a few of the more recently discovered fossils. The holotypes and many of the best preserved John Day *Daphoenus* specimens cannot be allocated to a limited stratigraphic interval or dated radiometrically because no record of the horizon, or in some cases even the geographic locality, was noted by early field collectors.

SUMMARY

The Eocene-Oligocene transition, as currently identified in North America, witnesses the appearance of amphicyonids and the development of a moderate diversity of species. Several extinctions and one orgination event characterize the Chadronian-Orellan boundary. *Daphoenictis* and possibly *Brachyrhynchocyon* terminate at the end of the Chadronian, and *Paradaphoenus* appears in the Orellan. *Daphoenus*, however, continues from the early part of the Duchesnean into the Chadronian and Orellan, with sparse fossils testifying to its persistence in Whitneyan and early Arikareean faunas. However, the passage from the archaic carnivore groups of the Uintan to the transitional Duchesnean assemblage heralding the Chadronian carnivore fauna is a faunal transformation of consequence in North America. The Uintan carnivores, to the extent these rare species are known, are vastly different from the White River carnivore fauna that appears at the beginning of the Chadronian at about 36.6-37 Ma. This event unfolds within Eocene time and appears to be related to the global climatic changes now identified as the hallmark of this period in Earth history.

LITERATURE CITED

Black, C. C. 1969. Fossil vertebrates from the late Eocene and Oligocene, Badwater Creek area, Wyoming, and some regional correlations; pp. 43-47 *in* J. A. Barlow (ed.), Wyoming Geological Association 21st Field Conference Guidebook, 1969 Symposium on Tertiary Rocks of Wyoming. Wyoming Geological Association, Casper.

Bryant, H. N. 1992. The Carnivora of the Lac Pelletier Lower Fauna (Eocene: Duchesnean), Cypress Hills Formation, Saskatchewan. Journal of Paleontology 66:847-855.

Cande, S. C., and D. V. Kent. 1992. A new geomagnetic polarity time scale for the late Cretaceous and Cenozoic. Journal of Geophysical Research 97(B10):13917-13951.

Clark, J., J. R. Beerbower, and K. K. Kietzke. 1967. Oligocene sedimentation, stratigraphy, paleoecology and paleoclimatology in the Big Badlands of South Dakota. Fieldiana: Geology Memoirs 5:1-158.

Dawson, M. R. 1980. Paleontology and geology of the Badwater Creek area, central Wyoming. Part 20: The late Eocene Creodonta and Carnivora. Annals of Carnegie Museum 49(4):79-91.

Emry R. J. 1973. Stratigraphy and preliminary biostratigraphy of the Flagstaff Rim area, Natrona County, Wyoming. Smithsonian Contributions to Paleobiology 18:1-43.

Emry, R. J. 1981. Additions to the mammalian fauna of the type Duchesnean, with comments on the status of the Duchesnean "age." Journal of Paleontology 55:563-570.

Emry, R. J. 1992. Mammalian range zones in the Chadronian White River Formation at Flagstaff Rim, Wyoming; pp. 106-115 *in* D. R. Prothero and W. A. Berggren (eds.), Eocene-Oligocene Climatic and Biotic Evolution. Princeton University Press, Princeton, N. J.

Emry, R. J., and R. M. Hunt, Jr. 1980. Maxillary dentition and new records of *Daphoenictis*, an Oligocene amphicyonid carnivore. Journal of Mammalogy 61(4):720-723.

Emry, R. J., P. R. Bjork, and L. S. Russell. 1987. The Chadronian, Orellan, and Whitneyan North American land mammal ages; pp. 118-152 *in* M. O. Woodburne (ed.), Cenozoic Mammals of North America: Geochronology and Biostratigraphy. University of California Press, Berkeley.

Ginsburg, L. 1961. La faune des carnivores Miocènes de Sansan. Mémoires du Muséum National d'Histoire Naturelle (n.s.), Serie C, Sciences de la Terre 9:1-190.

Ginsburg, L. 1966. Les Amphicyons des Phosphorites du Quercy. Annales de Paléontologie 52:23-64.

Gustafson, E. P. 1986a. Preliminary biostratigraphy of the White River Group (Oligocene, Chadron and Brule Formations) in the vicinity of Chadron, Nebraska. Transactions of the Nebraska Academy of Sciences 14:7-19.

Gustafson, E. P. 1986b. Carnivorous mammals of the late Eocene and early Oligocene of Trans-Pecos Texas. Texas Memorial Museum Bulletin 33:1-66.

Harshman, E. N. 1972. Geology and uranium deposits, Shirley Basin area, Wyoming. U.S. Geological Survey Professional Paper 745:1-82.

Hatcher, J. B. 1902. Oligocene Canidae. Memoirs of the Carnegie Museum 1(2):65-108.

Hough. J. R. 1948. A systematic revision of *Daphoenus* and some allied genera. Journal of Paleontology 22:573-600.

Hunt, R. M., Jr. 1974. *Daphoenictis*, a cat-like carnivore (Mammalia, Amphicyonidae) from the Oligocene of North America. Journal of Paleontology 48:1030-1047.

Hunt, R. M., Jr. In press. North American Cenozoic Amphicyonidae; *in* C. Janis, K. Scott, and L. Jacobs (eds.), Tertiary Mammals of North America. Cambridge University Press, Cambridge.

Kihm, A. J. 1987. Mammalian paleontology and geology of the Yoder Member, Chadron Formation, east-central Wyoming. Dakoterra 3:28-45.

Krishtalka, L., R. K. Stucky, R. M. West, M. C. McKenna, C. C. Black, T. M. Bown, M. R. Dawson, D. J. Golz, J. J. Flynn, J. A. Lillegraven, and W. D. Turnbull. 1987.

Eocene (Wasatchian through Duchesnean) biochronology of North America; pp. 77-117 *in* M. O. Woodburne (ed.), Cenozoic Mammals of North America, Geochronology and Biostratigraphy. University of California Press, Berkeley.

Leidy, J. 1854. [Untitled communication]. Proceedings of the Academy of Natural Sciences of Philadelphia 6 (1852-53):392-394.

Loomis, F. B. 1931. A new Oligocene dog. American Journal of Science 22:100-102.

Lucas, S. G. 1992. Redefinition of the Duchesnean land mammal "age," late Eocene of western North America; pp. 88-105 *in* D. R. Prothero and W. A. Berggren (eds.), Eocene-Oligocene Climatic and Biotic Evolution. Princeton University Press, Princeton, N. J.

McGowran, B., G. Moss, and A. Beecroft. 1992. Late Eocene and early Oligocene in southern Australia: local neritic signals of global oceanic changes; pp. 178-201 *in* D. R. Prothero and W. A. Berggren (eds.), Eocene-Oligocene Climatic and Biotic Evolution. Princeton University Press, Princeton, N. J.

Miller, K. G. 1992. Middle Eocene to Oligocene stable isotopes, climate and deep-water history: the terminal Eocene event? pp. 160-177 *in* D. R. Prothero and W. A. Berggren (eds.), Eocene-Oligocene Climatic and Biotic Evolution. Princeton University Press, Princeton, N. J.

Patton, T. H. 1969. An Oligocene land vertebrate fauna from Florida. Journal of Paleontology 43:543-546.

Prothero, D. R., and C. C. Swisher. 1992. Magnetostratigraphy and geochronology of the terrestrial Eocene-Oligocene transition in North America; pp. 46-73 *in* D. R. Prothero and W. A. Berggren (eds.), Eocene-Oligocene Climatic and Biotic Evolution. Princeton University Press, Princeton, N. J.

Remy, J. A., J.-Y. Crochet, B. Sigé, J. Sudre, L. de Bonis, M. Vianey-Liaud, M. Godinot, J.-L. Hartenberger, B. Lange-Badré and B. Comte. 1987. Biochronologie des phosphorites du Quercy: Mise à jour des listes fauniques et nouveaux gisements de mammifères fossiles. Abhandlungen der Münchner geowissenschaften (A) 10:169-188.

Russell, L. S. 1934. Revision of the Lower Oligocene vertebrate fauna of the Cypress Hills, Saskatchewan. Transactions of the Royal Canadian Institute 20(1):49-67.

Storer, J. E. 1990. Primates of the Lac Pelletier Lower Fauna (Eocene: Duchesnean), Saskatchewan. Canadian Journal of Earth Sciences 27:520-524.

Teilhard de Chardin, P. 1915. Les Carnassiers des Phosphorites du Quercy. Annales de Paléontologie 9:103-191.

Wilson, J. A. 1984. Vertebrate faunas 49 to 36 million years ago and additions to the species of *Leptoreodon* (Mammalia: Artiodactyla) found in Texas. Journal of Vertebrate Paleontology 4:199-207.

Wilson, J. A. 1986. Stratigraphic occurrence and correlation of early Tertiary vertebrate faunas, Trans-Pecos Texas: Agua Fria-Green Valley areas. Journal of Vertebrate Paleontology 6:350-373.

24. Small Arctoid and Feliform Carnivorans

JON A. BASKIN AND RICHARD H. TEDFORD

ABSTRACT

The Subparictidae (new family) are an endemic North American radiation of Chadronian (*Subparictis*) to Arikareean ursoids. The Amphicynodontinae are the earliest known members of the Family Ursidae and first occur in the latest Eocene (middle and late Chadronian) of North America, where they are represented by the middle Chadronian *Campylocynodon personi* and the late Chadronian ?*Amphicynodon major*. *Drassonax harpagops* is an Orellan member of this subfamily. The Musteloidea (*Mustelavus priscus*) and the feliform carnivoran *Palaeogale* also have their earliest records in North America.

INTRODUCTION

This chapter examines a variety of rare, small, mainly Chadronian and Orellan carnivorans, which occur in rocks of the White River Group or equivalents in North and South Dakota, Wyoming, Nebraska, Colorado, Montana, and Saskatchewan. These include the genera *Subparictis*, *Campylocynodon*, *Drassonax*, ?*Amphicynodon*, *Mustelavus*, and *Palaeogale*. The first four are assigned to the Ursoidea, the superfamily containing the extant Ursidae and Pinnipedia. *Subparictis* is a member of the Subparictidae, a new family which also includes *Nothocyon* and probably *Parictis*. *Campylocynodon*, *Drassonax*, and ?*Amphicynodon major* are amphicynodontines, a subfamily that is much better represented in the Oligocene of Eurasia (Teilhard, 1915; Cirot and de Bonis, 1992). *Mustelavus* is a very primitive North American member of the Musteloidea, the superfamily which includes the Procyonidae and Mustelidae. The affinities of *Palaeogale* have not been resolved, but it is probably a primitive feliform. Amphicynodontines, musteloids, and *Palaeogale* all have a Holarctic distribution in the Oligocene, but their earliest records are in the late Eocene of North America.

PREVIOUS WORK

Clark (1936, *in* Scott and Jepsen, 1936) was the first to describe musteloid (*Mustelavus priscus*) and ursoid ("*Parictis*" *dakotensis*) carnivorans from the White River

Group. Clark (1937) compared *Mustelavus* with the European Oligocene and early Miocene mustelid *Plesictis*. Simpson (1946) tentatively synonymized the two. *Parictis*, named for a specimen from the John Day Formation of Oregon (Scott, 1893), was recognized to be closely related to Old World amphicynodontines (Hall, 1931; Clark, 1937). The two additional White River ursoid genera are the Chadronian *Campylocynodon* (Chaffee, 1954) and the Orellan *Drassonax* (Galbreath, 1953). Clark and Guensburg (1972) reviewed the systematics of North American late Eocene and early Oligocene ursoids. These and more recent interpretations are discussed below.

Bunaelurus lagophagus was described by Cope (1873) from the Orellan of Colorado. Scott and Jepsen (1936) illustrated a skull from the Orellan of South Dakota. Matthew (1903) described a second species, *B. infelix*, from the Chadronian of Montana. Simpson (1946) synonymized both species of *Bunaelurus* and placed them in the Old World genus *Palaeogale*. The species of *Palaeogale* have been reviewed by de Bonis (1981). He synonymized *P. lagophaga* and *P. infelix* with the Old World *P. sectoria*.

ABBREVIATIONS

AMNH, American Museum of Natural History, New York; BM(NH), British Museum (Natural History), London; CMNH, Carnegie Museum of Natural History, Pittsburgh; CU, University of Colorado, Boulder; FMNH, Field Museum of Natural History, Chicago; KUVP, University of Kansas, Lawrence; MNNH, Muséum National d'Histoire naturelle, Paris; PUMN, Princeton University Natural History Museum (now in the collections of the Yale Peabody Museum, New Haven); SDSM, South Dakota School of Mines, Rapid City; UMMP, University of Michigan, Museum of Paleontology, Ann Arbor; UNSM, University of Nebraska State Museum, Lincoln; MP, European Mammal Paleogene zone; NALMA, North American Land Mammal "Age"; Ma, megaannum.

SYSTEMATIC PALEONTOLOGY
Order CARNIVORA Bowdich, 1821
Infraorder ARCTOIDEA Flower, 1869
Parvorder URSIDA Tedford, 1976
Superfamily URSOIDEA Gray, 1825

Diagnosis of Early Ursoids—Premolars relatively simple but with tendency to develop complete cingula; P4 with protocone broader and more posteriorly placed than in Amphicyonidae, metastylar blade shortened, and internal cingulum present but poorly developed; M1 parastyle small or absent, metaconule posteroexternal to protocone, broadening the tooth relative to canids and amphicyonids; M2 posteriorly expanded and three-rooted; M3 absent; m1 trigonid relatively low, paraconid blade angled internally, and talonid with prominent entoconid and hypoconid; m2 paraconid reduced to absent, and if present connected to protoconid by a convex-anterior preprotocristid, metaconid subequal in size to protoconid; m3 small and round. Type A (Hunt, 1974b) auditory bulla with somewhat inflated ectotympanic, alisphenoid canal and shallow suprameatal fossa present.

Discussion—The superfamily Ursoidea includes the extant Ursidae, Otariidae, Odobenidae, and Phocidae. The above diagnosis characterizes late Eocene and early Oligocene ursoids (Subparictidae and Amphicynodontinae).

Family SUBPARICTIDAE new

Type Genus—*Subparictis* Clark and Guensburg, 1972.

Included Genera—The type genus; *Nothocyon*, Matthew, 1899; and, tentatively, *Parictis* Scott, 1893.

Diagnosis—Small to medium-sized ursoid carnivores. Primitive in having P4 protocone low, conical, and anteriorly situated. Derived in having M2 expanded posteroexternally and m2 with medially displaced protoconid and prominent anteroexternal cingulum.

Discussion—Clark and Guensburg (1972) established *Subparictis* as a subgenus for four of the seven species they included in *Parictis*. Wang and Tedford (1992) elevated *Subparictis* from subgeneric to generic rank, because of the paucity of diagnostic characteristics in the holotype of *Parictis*. *Subparictis* is an endemic genus that evolved in North America and, although not abundant, is present in many Chadronian faunas. The relationship of *Nothocyon* to *Subparictis* is discussed in Wang and Tedford (1992). Although the systematic position of *Parictis* is questionable because of the poor quality of the genotypic specimen, it appears to be most closely related to *Subparictis* and *Nothocyon*.

Parictis Scott, 1893

Type Species—*Parictis primaevus* Scott, 1893.

Included Species—Type species only.
Diagnosis—Same as for *P. primaevus*.

Parictis primaevus Scott, 1893

Type Specimen—PUMN 10583, left mandible with p2-p3.
Type Locality—Camp Creek, John Day Formation, Oregon, early Arikareean NALMA.
Diagnosis—p2-p3 are simple conical teeth with crenulated enamel, completely encircled by cingula; p2 only slightly larger than p3; p2 and p3 broad anteriorly and with subrectangular occlusal outlines; p4 only slightly longer than p3.
Discussion—*Parictis primaevus* was initially described as a mustelid (Scott, 1893). Hall (1931) recognized that it was most closely related to the European *Pachycynodon* and *Amphicynodon*, which were included in the Canidae at that time. Subsequently Clark (1937) and Clark et al. (1967, 1972) referred six additional species to this genus. Clark and Guensburg (1972) recognized three subgenera of *Parictis*: *Parictis*, *Subparictis*, and *Campylocynodon*.

The type species, *Parictis primaevus*, is known from a partial lower jaw with p2-p3 (Clark and Guensburg, 1972, fig. 14). The specimen was described by Scott (1893) as coming from the John Day Formation at Silver Wells, Oregon. Hall (1931) reported the locality as Camp Creek, following the information on the specimen label. The exact locality is unknown, but fossiliferous sediments in the Camp Creek area on the Crooked River are from the middle and upper part of the John Day and are early Arikareean in age. Wang and Tedford (1992) recognized *Parictis* and *Subparictis* as separate genera, because of the absence of generically diagnostic characters on the type specimen of *Parictis*, and cited the disparity in geologic ages (*Subparictis* is restricted to the White River Chadronian) as another reason to question generic equivalence. Since material of *Parictis* with stratigraphic provenance has yet to be recovered from the John Day, the age assignment is uncertain, although most likely middle Oligocene. The type specimen of *P. primaevus* is certainly closely related to the taxa assigned to *Subparictis*, as indicated by the subequal p2 and p3. However, the material is much too limited to provide the diagnostic generic characters and for that reason *Parictis* should be considered a *nomen dubium*. Material referred to *P. primaevus* by Clark and Guensburg (1972) from the Orellan of South Dakota and Nebraska can be assigned to other taxa, as noted by Tedford (cited in Clark and Guensburg, 1972, p. 33). The maxillary fragment from Nebraska is probably assignable to *Drassonax*. The mandible from South Dakota is referable to *S*. cf. *dakotensis*.

Subparictis   Clark and Guensburg, 1972

Parictis (*Subparictis*) Clark and Guensburg, 1972:11
Subparictis Wang and Tedford, 1992:228

Type Species—*Subparictis gilpini* (Clark and Guensburg, 1972).

Included Species—*S. gilpini*, *S. montanus*, *S. dakotensis*, and *S. parvus*.

Emended Diagnosis—Ursoids with crowded and relatively simple premolars that have oval to subquadrate occlusal outlines and well-developed, nearly encircling cingula; P4 protocone slightly to moderately well separated from paracone and situated at a level not greatly posterior to parastyle, metastylar blade short, and internal cingulum poorly developed; M1 metaconule posteroexternal to protocone; M2 not reduced, wider externally than internally, and with metaconule posteroexternally situated and not connected to protocone; c1 with bulbous base; p2 slightly shorter than p3 (approximately 90% of p3 length); m1 paraconid low and angled internally, metaconid large and moderately well separated from protoconid, and talonid basined and bordered externally and internally by ridgelike hypoconid and entoconid; m2 trigonid compressed anteroposteriorly, paraconid reduced, protoconid medially displaced, anteroexternal cingulum present, and talonid relatively elongate.

Comparisons and Discussion—For further information on specimens, localities, descriptions, measurements, and illustrations of the species of *Subparictis* consult Clark and Guensburg (1972). Table 1 gives selected measurements of *Subparictis* and related taxa. Clark (1937) distinguished *Subparictis dakotensis* from *Amphicynodon* and *Pachycynodon* by differences in the m2 and in having teeth that are usually broader, heavier, and more blunt, and a more massive mandible. The M1 internal cingulum is more posteriorly expanded in *Subparictis*. The m2 pattern of *Subparictis* (Fig. 1) is very distinctive; unfortunately it is not readily apparent in previous illustrations (Clark and Guensburg, 1972). Galbreath (1953) and Wang and Tedford (1992) precisely described the m2 morphology in *Subparictis*. Wang and Tedford (1992) illustrated the m2 in the related (but more derived) genus *Nothocyon*.

Subparictis gilpini   Clark and Guensburg, 1972

Parictis (*Subparictis*) *gilpini* Clark and Guensburg, 1972:14
Subparictis gilpini Wang and Tedford, 1992:228

Type Specimen—FMNH 22405, anterior portion of skull with lower jaws.

Type Locality—Middle of Ahearn Member, Chadron Formation, Pennington County, South Dakota, early Chadronian NALMA.

Distribution—Ahearn Member to near top of Chadron Formation, Big Badlands, South Dakota;

Table 1.   Length of selected teeth of Chadronian and Orellan Ursoidea.   Measurements are in millimeters. Parentheses indicate approximate measurements.

	P4	M1	m1	m2
Subparictis parvus				
PUMN 16265			6.4	3.4
Subparictis gilpini				
FMNH 22405	7.4	6.2	8.1	4.5
FMNH 729			8.5	4.7
AMNH 50241			8.0	5.0
UNSM 19930			8.3	4.5
UNSM 19932			8.2	
SDSM 2567			8.2	
AMNH 63933			8.3	
AMNH 76196			8.4	5.2
Subparictis montanus				
CMNH 9571	7.3	5.8	7.6	4.0
CMNH 9068			7.6	3.3
FMNH 3843			7.9	
Subparictis dakotensis				
SDSM 2476			9.0	4.5
AMNH 50240			8.5	4.3
AMNH 12244			9.4	4.8
UNSM 25154	8.9	(6.5)		
Subparictis cf. *dakotensis*				
CU 22749			8.4	
?*Amphicynodon major*				
FMNH 22404			10.8	
Campylocynodon personi				
PUMN 17795	7.0	6.0	7.6	4.0
Drassonax harpagops				
KUVP 121			8.2	3.6
FMNH 27157	8.0	6.1		
Parictis primaevus				
PUMN 10583			(7.1)	(4.1)

Figure 1. Occlusal view of *Subparictis dakotensis*, UMMP 14579, right m2; length = 4.7 mm

Chadron Formation, Niobrara County, Wyoming; and Chadron Formation, Sioux County, Nebraska; Chadronian NALMA.

Emended Diagnosis—P4 with small, very low, narrow, and elongate protocone situated at anterointernal base of paracone; M1 narrow internally with poorly developed paraconule and metaconule, preparacrista extends to small parastyle.

Description—The P3 is not as expanded internally as illustrated by Clark and Guensburg (1972, fig. 4). On P4, the internal cingulum extends to the posterior edge of the metastyle, the apex of the paracone is situated medially, and the metastylar blade is smaller than the paracone. The M1 has a subtriangular occlusal outline, the internal cingulum is only slightly enlarged posteriorly, and the paracone is larger than the metacone. The M2 is suboval in occlusal outline, narrower internally, with a broad posterior shelf and a posteroexternally situated metaconule that is not connected to the postprotocrista. The lower premolars have weak cingula. In the p2 and p3, the external cingulum is absent. The p3 is slightly larger than the p2. The p4 is subquadrate in occlusal outline, with the external cingulum poorly developed, and is much larger than the p3. The m1 has an anteroexternal cingulum, a prominent metaconid, and the entoconid smaller than hypoconid. The m2 is ovate in occlusal outline, with the trigonid taller than the talonid and anteroposteriorly compressed, with an internally displaced protoconid; the protoconid and metaconid are subequal in size; the metaconid is posterior to the protoconid; the paraconid is extremely small; the talonid is elongate and basined with a low hypoconid and entoconid.

Comparisons—The lower dentition of *S. gilpini* (Clark and Guensburg, 1972, fig. 6) differs from that of *S. montanus* in its slightly larger size and relatively larger m2. It is questionable whether the two species can be distinguished on the basis of lower dentitions alone. The upper dentition of *S. gilpini* (Clark and Guensburg, 1972, figs. 4-5) is less derived than in other subparictids. The protocone of the upper carnassial is similar in morphology to that of the late Eocene European amphicyonid *Cynodictis*. *Cynodictis* has a longer metastylar blade.

Subparictis montanus Clark and Guensburg, 1972

Parictis (Subparictis) montanus Clark and Guensburg, 1972:16

Type Specimen—CMNH 9571, right and left mandibles, right maxilla.

Type Locality—Pipestone Springs Formation, Jefferson County, Montana, middle Chadronian NALMA.

Referred Distribution—Douglas Creek valley, Powell County, Montana, middle Chadronian NALMA.

Emended Diagnosis—Small species of *Subparictis*; P4 with strong anterior indentation, external and internal cingula, small parastyle, low and conical protocone anteriorly situated and separated from paracone; m2 trigonid very compressed, paraconid very small and close to metaconid, and metaconid only slightly posterior to protoconid.

Description—The P4 internal margin is relatively straight and the protocone is low and conical. The M1 has pre- and postprotocristae, the paracone slightly larger than the metacone, the preprotocrista connected to a small parastyle, the metaconule small and connected

to the postprotocrista, low ridges connecting the paracone and metacone with the anterior and posterior arms of the protocone respectively, and the internal cingulum broad, particularly posteriorly, giving the tooth a subquadrate occlusal outline. The M2 is very worn and broken posteriorly, but appears to be expanded posterior to the metacone. The lower premolars have weak cingula. The p2 is blunt anteriorly. The p3 and p4 have subquadrate occlusal outlines.

Comparisons—Clark and Guensburg (1972, figs. 7-9) illustrated the upper and lower dentition of the type specimen of *S. montanus*. *Subparictis montanus* is smaller than *S. gilpini*, but the lower dentition is otherwise very similar. It differs further from *S. gilpini* in having P4 with a larger and rounder protocone that is more anteromedially separated from the paracone, and an external cingulum, The M1 has a much broader anterointernal cingulum and a broader, more posteriorly expanded internal cingulum. The m2 is relatively smaller with a more anteroposteriorly compressed trigonid.

Subparictis dakotensis (Clark, 1936)

Parictis dakotensis Clark, *in* Scott and Jepsen, 1936:106

Parictis dakotensis Clark, 1937:312

Parictis (Parictis) dakotensis Clark and Beerbower, 1967:28

Parictis (Subparictis) dakotensis Clark and Guensburg, 1972:21

Subparictis dakotensis Wang and Tedford, 1992:228

Type Specimen—SDSM 2476, right mandible with p2-m2.

Type Locality—Questionably from the Peanut Peak Member, Chadron Formation, Shannon County, South Dakota, late Chadronian NALMA.

Distribution—Chadron Formation of South Dakota and Nebraska; Horsetail Creek Member, White River Formation, Logan County, Colorado; Chadron Member, White River Formation, Niobrara County, Wyoming; late Chadronian NALMA.

Emended Diagnosis—Large species of *Subparictis* with deep horizontal ramus; anterior lower premolars quadrate and with well-developed external and internal cingula; and m1 with prominent anterointernal and posteroexternal cingula.

Description—The p2-p4 have subquadrate occlusal outlines and prominent external and internal cingula. The m2 has a broad anteroexternal cingulum. The m2 on AMNH 12244 has the paraconid absent, the apex of protoconid subcentral, the metaconid slightly posterior to the protoconid, and the entoconid larger and possibly taller than the metaconid.

UNSM 25154, an undescribed P4-M2 from the Chadron Formation of Sioux County, Nebraska, has a

P4 with a short metastylar blade, a minute parastyle, a conical protocone slightly anterior and well separated from the paracone. The internal cingulum is broad opposite the paracone and extends medial and anterior to the protocone, and there is a well developed external cingulum. The M1 is subquadrate, wider than long, and has the paracone low and larger than the metacone. It has a wide lingual cingular shelf bearing a parastyle at the anteroexternal corner. The protocone is low, with a beaded crista connecting small para-□andmetaconules. The internal cingulum forms a broad shelf posterior to the protocone. The M2 is expanded posteroexternally, with the metaconule on the posterior margin approximately halfway between the protocone and metacone.

Comparisons—Clark and Guensburg (1972, fig. 11) illustrated the type specimen of *Subparictis dakotensis*. They considered *S. dakotensis* a probable senior synonym of *S. gilpini*, but the two are distinct. The p1-m2 length of the holotype of *S. dakotensis* is 8% greater than that of *S. gilpini* and 17% greater than that of *S. montanus*. The premolars are more quadrate, have better developed cingula, and have the p2 with a flattened rather than rounded anterior end. The m1 has an anterointernal cingulum and the talonid is broadened by the posteroexternal cingulum. The m2 on the holotype is very worn and the protoconid appears close to the external margin. In less worn specimens in the AMNH collections, the apex of the protoconid is close to the midline, although the base of the protoconid extends further externally.

Subparictis cf. *dakotensis* (Clark, 1936)

Referred Specimen—CU 22749.
Distribution—Lower Nodule Zone, Scenic Member, Brule Formation, Shannon County, South Dakota; Orellan NALMA.
Discussion—This specimen was questionably referred to *Parictis primaevus* by Clark and Guensburg (1972). The specimen is distinct in that the p2 is longer than p3. One specimen referred to *S. gilpini* (ANNH 50241) has p2 almost equal in length to p3, indicating a relatively large p2 is not diagnostic only of *Parictis*. The external cingulum on m1 is better developed than in *S. dakotensis* and the main cusp of the premolars is more centrally located. Otherwise this specimen is similar in size and morphology to *S. dakotensis*.

Subparictis parvus (Clark and Beerbower, 1967)

Parictis (Campylocynodon) parvus Clark and
  Beerbower, 1967:28
  Type Specimen—PUMN 16265, right mandible.
  Type Locality—Red layer near base of Ahearn Member, Chadron Formation, Shannon County, South Dakota, early Chadronian NALMA.
  Referred Distribution—Yoder Local Fauna,

Chadron Formation, Goshen County, Wyoming, early Chadronian NALMA; Flagstaff Rim Fauna below ash A, White River Formation, Natrona County, Wyoming, early Chadronian NALMA; Calf Creek Local Fauna, Cypress Hills Formation, southwestern Saskatchewan, early or middle Chadronian NALMA.

Diagnosis—Smallest species of *Subparictis*; premolars increase in size posteriorly; p3 and p4 narrow anteriorly; p4 noticeably larger than p3; m1 with very short talonid; m2 trigonid but little compressed, ectostylid on posteroexternal base of protoconid, talonid not elongate.

Discussion—The generic assignment is tentative. Clark and Guensburg (1972) referred *S. parvus* to *Campylocynodon* and suggested that it might be a junior synonym of *C. personi*. The slenderness of the mandible (Clark and Guensburg, 1972, fig. 3), considered diagnostic of *Campylocynodon*, may be a function of individual age or a primitive characteristic. The m2 of *S. parvus* is much more primitive than other species. The trigonid is not so compressed anteroposteriorly and has a definite "V" shape. However, the protoconid is slightly displaced medially and there is an anteroexternal cingulum, two derived characteristics of *Subparictis*.

Bryant (1993) described the additional material of *S. parvus* from Saskatchewan and Wyoming. The upper molars that he illustrated and tentatively referred to this species are much more primitive than those of other species of *Subparictis*.

Subparictis sp.

Distribution—Flagstaff Rim Fauna between ashes A and G, Wyoming, White River Formation, Natrona County, Wyoming, middle Chadronian NALMA; Calf Creek Local Fauna, Cypress Hills Formation, southwestern Saskatchewan, early or middle Chadronian NALMA.
Comment—Bryant (1993) noted these occurrences of *Subparictis*.

Family URSIDAE Gray, 1825
Subfamily AMPHICYNODONTINAE
Simpson, 1945

Diagnosis—Small to medium-sized, ursoid carnivores; infraorbital foramen relatively large; and infraorbital canal short. Characters shared with ursids include P4 with protocone well separated from paracone, posteriorly displaced, and anteroposteriorly elongate; M2 expanded posterointernally; and m2 paraconid absent.
Discussion—Amphicynodontines are well known from the Quercy deposits of France (Teilhard, 1915), but are rare elsewhere. The European and Asian genus *Amphicynodon* has recently been revised (Cirot and de Bonis, 1992). *Amphicynodon*, as defined by Cirot and de Bonis (1992), encompasses a wide range of mor-

phologies and the more primitive species probably should be placed in a distinct genus. *Drassonax* and *Campylocynodon* share characteristics with derived species of *Amphicynodon*. Tedford et al. (1994) have further explored amphicynodontine phylogeny at the generic level and demonstrate that the group is also the paraphyletic, primitive sister taxon of the Pinnipedimorpha.

Amphicynodon Filhol, 1881
?Amphicynodon major
(Clark and Guensburg, 1972)

Parictis (Subparictis) major Clark and Guensburg, 1972:25

Type Specimen—FMNH 22404, left mandible with c-m1.

Type Locality—Peanut Peak Member, Chadron Formation, Shannon County, South Dakota, late Chadronian NALMA.

Diagnosis and Description—A large species with a deep mandible; massive and long crowned lower canine; premolars with strong external and internal cingula; p2 and p3 subround anteriorly, subequal in size, and lower crowned than p4; p4 blunt anteriorly and with a small posterior accessory cusp; m1 protoconid taller than the paraconid and metaconid, metaconid reduced and posterior to the protoconid, hypoconid with a low external cingulum, talonid with hypoconid, hypoconulid, and entoconid equal in height, and a low anteroexternal cingulum.

Comparisons—*?Amphicynodon major* is represented only by the type mandible (Clark and Guensburg, 1972, fig. 12) that lacks the m2, which makes generic identification difficult, if not impossible. Clark and Guensburg (1972) compared this species with *Parictis* (including *Subparictis* and *Campylocynodon*) and *Drassonax*, and concluded that it was most closely related to *S. dakotensis*. It differs from *Subparictis* in its larger size, more massive canine, and in having m1 with talonid closed posteriorly by a hypoconulid and a smaller, more posteriorly situated metaconid. Additionally it differs from *S. dakotensis* in having somewhat less quadrate premolars. It differs from *Campylocynodon* in its larger size, much more massive mandible and canine, having p2 only slightly shorter than p3, p3 noticeably shorter than p4, and m1 with a more posteriorly situated metaconid. It differs from *Pachycynodon* in having less inflated cusps and m1 relatively long and taller than p4, and with a reduced metaconid. It resembles *Drassonax*, certain species of *Amphicynodon*, and the hemicyonine ursid *Cephalogale* in having an m1 with a reduced, posteriorly displaced metaconid. It differs from *Drassonax* in its larger size, having p2 and p3 subequal in size, much better developed encircling cingula on the premolars, and the talonid of m1 closed

posteriorly by a hypoconulid. *Cephalogale* differs in having m1 with a more open trigonid, entoconid a very low ridge, and more laterally compressed premolars, although in side view, the premolar morphology is similar. *Cephalogale* is similar in having the m1 talonid closed posteriorly, but by the posteroexternal continuation of the entoconid, not by a hypoconulid. *?Amphicynodon major* is similar in size to the type species of *Amphicynodon*, *A. velaunus*, which also has a very large canine (Cirot and de Bonis, 1992, fig. 1), a posteriorly displaced metaconid, and an enclosed basined talonid. The large canine could be a sexually dimorphic feature. It also resembles the smaller *A. teilhardi* and *A. gracilis*, which have a reduced, posteriorly situated metaconid on m1.

Campylocynodon Chaffee, 1954

Type Species—*Campylocynodon personi* Chaffee, 1954.
Included Species—Type species only.
Diagnosis—Same as for *C. personi*.

Campylocynodon personi Chaffee, 1954

Type Specimen—PUMN 17795, skull and jaws, badly weathered.

Type Locality—White River Formation, Beaver Divide, Fremont County, Wyoming, Chadronian NALMA.

Emended Diagnosis—Small amphicynodontine, with short face, short and wide basicranium, narrow basioccipital, and basisphenoid region narrow between inflated auditory bullae; mandible slender; lower premolars narrower anteriorly than in *Subparictis*; p2 smaller than p3; p3 and p4 subequal in length and height; m1 metaconid small and appressed to protoconid; P4 protocone low, conical, slightly anteroposteriorly elongate, and widely separated from and slightly anterior to paracone; M1 and M2 subquadrate in occlusal outline, wider than long, and slightly narrower internally.

Description—The skull is short and broad, with a postorbital process, a large infraorbital foramen, and a short infraorbital canal. There is a shallow suprameatal fossa excavated into the mastoid. The palate does not extend behind the M2. The canines are relatively small. The P3 has a small but prominent protocone. The P4 has a concave internal margin posterior to the protocone; posteriorly the protocone is connected broadly to the paracone, with a slight basin on the shelf, a low internal cingulum along the internal margin of the low metastylar blade, and an external cingulum. The M2 is apparently similar, but smaller, than the M1 and has a more convex posterior margin and is slightly wider internally than externally. The m1 metaconid is lower than the paraconid. The m2 trigonid appears to be unreduced, occupying approximately half the length of the tooth.

Comparisons—Unfortunately the preservation of the type specimen is so poor, it is almost impossible to discern the details of the dentition. One of us (RHT) believes that the specimen may be a natural cast. *Campylocynodon* differs from *Subparictis* in having a more slender mandible; smaller canines; P4 with a more elongate metastylar blade; and the M2 apparently lacking significant posteroexternal expansion. The skull of *Campylocynodon* (Chaffee, 1954), has a relatively shorter muzzle, a narrower basicranium between the more inflated auditory bullae, and apparently less enlarged paroccipital and especially mastoid processes than *Amphicynodon* (Teilhard, 1915; Cirot and de Bonis, 1992). The morphology of the m2 may have been more like *A. chardini*, which has the least reduced trigonid of the species of *Amphicynodon* (Cirot and de Bonis, 1992).

Discussion—Van Houten (1964) indicated that this specimen was collected from 50 to 200 feet above the base of the White River Formation. Emry (1975) stated that the White River Formation of the western Beaver Divide area is Chadronian in age and Emry et al. (1987, fig. 5.3) suggested a middle Chadronian age for fossils from above the Beaver Divide Conglomerate.

Drassonax Galbreath, 1953

Type Species—*Drassonax harpagops* Galbreath, 1953.

Included Species—Known from type species only.

Diagnosis—Same as for *D. harpagops*.

Drassonax harpagops Galbreath, 1953

Holotype—KUVP 121, left mandible with p1-m2.

Type Locality—Cedar Creek Member of the White River Formation, Logan County, Colorado; Orellan NALMA.

Referred Specimens—FMNH 27157, left maxillary with P4-M1, base of the Orella Member, Brule Formation, Dawes County, Nebraska; F:AM 50358, left ramus with p3 and m1, and associated right ramus with m2, Scenic Member, Brule Formation, Washington County, South Dakota, Orellan NALMA.

Emended Diagnosis—Enamel faintly rugose; premolars narrow anteriorly and simple, with no posterior accessory cusp, external and internal cingula well developed; premolars increase in size from p1 to p4; m1 with small metaconid closely appressed to protoconid and short talonid, open posteriorly, and bordered by ridgelike entoconid and hypoconid; m2 with very short trigonid and short talonid, paraconid absent, protoconid on the external margin, and metaconid subequal in size and anterior to protoconid.

Description—The p1 is smaller than in *Subparictis*; the p2 is noticeably smaller than the p3, narrower anteriorly, with an anteriorly situated main cusp, and with internal and external cingula; the p3 is narrow anteriorly, external cingulum narrow, but well developed, internal cingulum less well developed, subcentral main cusp; the p4 has complete cingula and a tiny anterior cingular cusp; and m1 with talonid short and narrower than trigonid, a short anterointernal cingulum bridging the bases of the paraconid and metaconid, and an anteroexternal cingulum present.

An undescribed associated upper and lower dentition in the University of Kansas collections indicates that the upper dentition referred to *Parictis primaevus* by Clark and Guensburg (1972, fig. 13) is assignable instead to *Drassonax*. The P4 has strong internal, external, and anterior cingula; the parastyle is small; the protocone is anteroposteriorly elongate, conical, and anterior to and separate from the paracone; the hypocone is a slight rise on the internal cingulum; and the metastylar blade is very short. The M1 is trapezoidal in occlusal outline, the interior margin is slightly narrower than the external margin; the paracone is the tallest cusp; the protocone is situated just posterolabial to the paracone and is connected to the paracone by a low preprotocrista that lacks a paraconule; a stronger crest passes from the protocone to the anterior cingulum; a low postprotocrista extends from the metacone to the posterior edge of the protocone; the metaconule is situated almost directly posterior to the protocone on the posterior margin of the tooth but retains its connection to the protocone through a strong branch of the postprotocrista; there is a low, narrow cusp situated internal to the metaconule on the posterior margin of the internal cingulum; the external cingulum is narrow; the preparacrista extends to a small parastyle; the postparacrista extends to the metacone; a weak posterior cingulum extends from the metaconule to the external margin; and there is a strong external cingulum and cingular shelf external to the principal cusps.

Discussion—*Drassonax* was initially described as a mustelid (Galbreath, 1953). Clark and Guensburg (1972) recognized that it was an amphicynodontine, closely related to *Subparictis*. *Drassonax* differs from *Subparictis* in having premolars that are relatively slender, especially anteriorly. The lower premolars are graded, increasing uniformly in length and height from p1 to p4. The p4 lacks a posterior accessory cusp. In *Subparictis* the large p2 is only slightly shorter and subequal in height to p3. The larger p4 has a posterior accessory cusp. In *Drassonax*, the m1 metaconid is much smaller and closely appressed to the protoconid and the m2 is relatively small, with an external protoconid and no paraconid or anterior cingulum. If the association of FMNH 27157 with *Drassonax* is correct, then *Drassonax* also differs from *Subparictis* in having a much smaller (particularly less wide) M2, as predicted by the size of the corresponding m2. It differs from

Campylocynodon in having P4 with a shorter metasty-lar blade and a complete internal cingulum and a smaller M2. Both *Campylocynodon* and *Drassonax* have slender premolars. The p3 is noticeably larger than p2 in both genera. *Campylocynodon* has a more slender mandible and p3 subequal in size to p4.

Although there are strong resemblances between *Drassonax* and *Amphicynodon velaunus*, the type species of the genus, and *A. mongoliensis* (Cirot and de Bonis, 1992), all of which have the m1 with somewhat reduced metaconid and relatively open trigonid and the m2 reduced, there are sufficient differences, e.g., the p4 lacks a posterior accessory cusp, the m2 trigonid with cusps as described above, and the M1 with salient metaconule and preprotocrista anteriorly directed, that *Drassonax* stands apart from *Amphicynodon*.

SYSTEMATIC RELATIONSHIPS OF CHADRONIAN AND ORELLAN URSOIDEA

Figure 2 is a cladogram showing the relationships of the primitive ursoids discussed in this chapter. Arctoidea include the Amphicyonidae, Ursoidea (Subparictidae and Ursidae), and Musteloidea (Procyonidae and Mustelidae). Characters uniting this infraorder are given in Flynn et al. (1988). Clark and Guensburg (1972) included the subparictids in the Amphicynodontinae. The Subparictidae is herein recognized as a family distinct from, but closely related to, the Ursidae (including the Amphicynodontinae). Amphicynodontinae has been used to group together a variety of primitive arctoids. Amphicynodontines were initially classified within the Canidae as the Cynodontinae (Schlosser, 1911). Teilhard (1915) noted the relationships of *Amphicynodon* and *Pachycynodon* to arctoid carnivores. Pilgrim (1931, pp. 5, 21), in brief remarks, recognized the Cynodontinae as a subfamily of the Ursidae. Simpson (1945) emended the subfamily name to Amphicynodontinae and returned the amphi-cynodontines to the Canidae, because he considered they had few derived arctoid dental characters. Nonetheless, he considered this group ancestral to the ursids and perhaps the procyonids. Ginsburg (1966) classified the amphicynodontines (restricted to the genera *Amphi-cynodon, Pachycynodon,* and *Parictis*) as a subfamily of the Ursidae, because of the ursoid basicranium of *Pachycynodon*. De Bonis (1969) stated that amphi-cynodontines were ancestral to ursids, mustelids, and procyonids. Tedford (1976, and in Flynn et al., 1988) listed dental and basicranial characters that placed the Amphicynodontinae as primitive members of the Ursidae. Tedford et al. (1994) list synapomorphies for the amphicynodontines that make them a paraphyletic group of ursids that are the sister taxa of the Pinnipedi-morpha. Primitive members of the Amphicyonidae and Musteloidea constitute outgroups used in this analysis of the Subparictidae and Amphicynodontinae.

1. ARCTOIDEA—Characters are cited in Flynn et al. (1988). These include a: Type A auditory bulla with a tubular external auditory meatus; b: mastoid process of petrosal prominent, salient; c: shallow, but evident, suprameatal fossa (but probably primitive for the Caniformia, since a suprameatal fossa is also present in the primitive canid *Hesperocyon*); d: hypotympanic space is expanded into the base of the auditory tube; e: internal carotid artery with an "ursid loop" (Hunt, 1977); and f: scapula with a postscapular fossa.

2. URSIDA—a: M3 lost; b: upper molar internal cingulum is not symmetrical because of a posteriad expansion; c: m2 paraconid reduced, metaconid equal to or larger than protoconid; and d: bulla fully ossified.

3. URSOIDEA—a: P4 with a shortened metastylar blade; b: upper molars with reduced parastyles; c: M1 with metaconule and postprotocrista more posteriorly directed opening up the talon valley and making the tooth more quadrate; d: M2 posteriorly expanded with metaconule on posterointernal margin; e: premolars simple, but with a tendency to develop complete cingula; f: m1 with a relatively low trigonid; and g: wrinkled enamel.

4. SUBPARICTIDAE—a: M2 expanded posteroex-ternally; and b: m2 with protoconid displaced medially and a prominent anteroexternal cingulum.

5. Unnamed Group—a: same as 4b, but more pronounced.

6. Unnamed Group—a: P4 with protocone enlarged and more separate from the paracone; b: P4 with external cingulum; c: M1 with the talon valley more open posteriad; d: M1 with ridges connecting the paracone with the anterior arm of the protocone and the metacone with the metaconule; and e: M1 more quadrate because of further posteriad expansion of internal cingulum.

7. *Subparictis dakotensis*—a: premolars with more prominent cingula; b: anterior lower premolars more quadrate; c: m1 with an anterointernal cingulum; and d: m1 with a wide posteroexternal cingulum.

8. *Nothocyon geismarianus*—a: M1 with a posteri-orly shifted metaconule and a large hypocone contribut-ing to rectangular occlusal outline; b: m1 with hypoconulid and entoconulid cusps; and c: more robust premolars.

9. URSIDAE—a: P4 with protocone well separated from the paracone and posteriorly displaced; b: P4 protocone anteroposteriorly elongate; c: M2 expanded posterointernally; and d: m2 paraconid absent.

10. URSINAE (including hemicyonines and ursines) a: M1 with enlarged metaconule; b: M2 with a posteri-orly projecting platform; c: enlarged m2 and m3.

11. AMPHICYNODONTINAE—Tedford et al. (1994) should be consulted for a detailed discussion of the relationships of the amphicynodontines to the ursines and the Pinnipedimorpha. Synapomorphies listed by

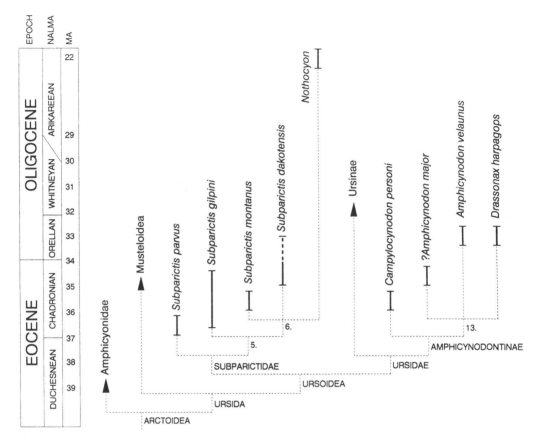

Figure 2. Hypothesis of relationships among North American Chadronian and Orellan ursoids. Characters are discussed in the text.

Tedford et al. (1994) include a: short infraorbital canal; b: large infraorbital foramen; and c: premolars robust with encircling cingula.

12. *Campylocynodon personi*—a: facial region and basicranium short anteroposteriorly; and b: basicranial region narrow between inflated auditory bullae.

13. Unnamed Group—m1 metaconid posteriorly displaced; and b: m2 talonid short.

14. *?Amphicynodon major*—a: m1 basined talonid (possible synapomorphy with 15) closed by a hypoconulid; and b: m1 metaconid reduced (possible synapomorphy with 16).

15. *Amphicynodon velaunus*—a: m1 basined talonid.

16. *Drassonax harpagops*—a: M2 and m2 reduced; and b: m1 metaconid reduced.

Superfamily MUSTELOIDEA Swainson, 1835

Comment—Musteloids are derived relative to ursoids in having enlargement and inflation of the caudal entotympanic, increased separation of the posterior carotid foramen from the posterior lacerate foramen, and loss of m3. Schmidt-Kittler (1981) placed early Miocene and older musteloids in an undifferentiated stem-group that gave rise to the Mustelidae and Procyonidae. However, most of these primitive musteloid taxa can be allocated either to the Procyonidae or Mustelidae (Baskin, 1987, and in press). A more restricted assemblage of late Eocene to middle Oligocene taxa (including *Mustelavus* and the European *Mustelictis*), probably represents the musteloid stem-group. The alisphenoid canal (present in primitive musteloids and *Ailurus*, the red panda) was independently lost in procyonids and mustelids, as it was in the ursid *Ailuropoda* (the giant panda) and phocids (seals).

?Family MUSTELIDAE Swainson, 1835
Mustelavus Clark,
in Scott and Jepsen, 1936

Type Species—*Mustelavus priscus* Clark, in Scott and Jepsen, 1936.

Included Species—Known from type species only.

Diagnosis—Same as for *M. priscus*.

Mustelavus priscus Clark,
in Scott and Jepsen, 1936

Mustelavus priscus Clark, in Scott and Jepsen,
    1936:107
Mustelavus priscus Clark, 1937:317
Plesictis priscus Simpson, 1946:13
Mustelavus priscus Hough, 1948:90
Mustelavus priscus Clark and Beerbower, 1967:32
Plesictis priscus Macdonald 1970:64
Mustelavus priscus Schmidt-Kittler, 1981:785
    Type Specimen—PUMN 13775, skull and lower
jaws.
    Type Locality—Peanut Peak Member, Chadron
Formation, South Dakota; late Chadronian NALMA.
    Referred Distribution—*Mustelavus* sp., Scenic
Member, Brule Formation, South Dakota; Orellan
NALMA.
    Description—The alisphenoid canal is possibly
present, but "damage precludes certain determination"
(Clark and Guensburg, 1972, p. 45). The suprameatal
fossa is shallow. The auditory bulla is moderately
inflated and in contact with the base of a posteriorly
projecting paroccipital process, which has a broad
ventral connection with mastoid process. The dentition
is secant. There are four upper and lower premolars.
The P4 has an anteriorly placed, rounded, conical
protocone and a deep carnassial notch. The M1 is
subtriangular in occlusal outline, with a moderate-sized
parastyle, a small paraconule, a very reduced postprotocrista which does not connect to the very small
metaconule, and narrow anterointernal and wider
posterointernal cingula. The M2 is small and three-
rooted. The p4 has weak external and internal cingula.
The m1 has a tall trigonid forming an isosceles triangle
in occlusal view, the protoconid taller and larger than
the subequal anterior paraconid and posterior metaconid,
the metaconid is prominent and separated from the
protoconid, the paraconid blade extends anterointernally,
there is a small accessory cusp at the internal base of
the paraconid, a low basined talonid, a small hypoconid
and smaller entoconid, and a notch separates the
protoconid and hypoconid. The m2 is small, double
rooted, with a basined and elongate talonid, a compressed trigonid with a narrow anterior cingulum
connected to a very reduced paraconid, and the protoconid is opposite the smaller metaconid. The m3 is
absent.
    Discussion—*Mustelavus* has been placed in the
Mustelidae, Procyonidae, and most recently has been
considered closely related to the amphicyonid *Cynodictis*
(Schmidt-Kittler, 1981). Ursida synapomorphies of
Mustelavus include M3 absent and m2 with a low
trigonid, reduced paraconid, and enlarged metaconid.
The m2 metaconid is more anterior than in most
amphicyonids and canids. Procyonids have a postpro-

tocrista and a metaconule on M1, but these features are
primitive for musteloids. The reduction of these
features in *Mustelavus*, as well as the loss of m3,
reduction of M2/m2, and the presence of a strong
parastyle on M1 suggest this genus is a primitive
mustelid. Simpson (1946) tentatively synonymized
Mustelavus with *Plesictis*, although he noted that the
more primitive species of *Plesictis* (including *M.
priscus*) probably deserved generic distinction. *Mustelavus* differs from *Plesictis sensu stricto* in characters of
the dentition and the auditory region (Schmidt-Kittler,
1981). *Mustelavus* is closely related to the Old World
Mustelictis, the very primitive musteloid genus that
incorporates Oligocene species previously included in
Plesictis (Schmidt-Kittler, 1981). The dentition of
Mustelavus is more primitive than that of *Mustelictis*.

Order CARNIVORA *incertae sedis*
Palaeogale von Meyer, 1846

    Diagnosis—A very small carnivoran with auditory
bullae complete, inflated, and undivided; alisphenoid
canal absent; P4 with parastyle and carnassial notch;
M1 internal cingulum absent; M2 vestigial to absent;
M3 absent; p1 vestigial to absent; m1 trigonid open and
bladelike, metaconid absent, and talonid trenchant; m2
very reduced but ovate and two-rooted, cusps linear, and
talonid absent; and m3 absent.
    Distribution—Stampian to Orleanean, Europe;
Chadronian to Hemingfordian, North America; early
Oligocene, Mongolia.
    Comment—*Palaeogale* has long been considered a
primitive mustelid. However, according to recent
interpretations (Hunt, 1974a, 1989; Flynn and Galiano,
1982), it is closely related to the early Tertiary family
Viverravidae (i.e., "Miacoidea"). Flynn et al. (1988)
conclude that *Palaeogale* is best considered *incertae sedis*
within the Carnivora, not having any unambiguous
synapomorphies with either the Feliformia or Caniformia.

Palaeogale sectoria (Gervais, 1848-1852)

Mustela sectoria Gervais, 1848-1852, plate 28
Bunaelurus lagophagus Cope, 1873:8
Palaeogale sectoria Schlosser, 1888
Bunaelurus infelix Matthew, 1903:210
Bunaelurus parvulus Matthew and Granger, 1924:8
Bunaelurus ulysses Matthew and Granger, 1924:8
Palaeogale lagophaga Simpson, 1946:12
Palaeogale sectoria de Bonis, 1981:50 (see for a more
    complete synonymy).
    Type specimen—BM(NH) 27816, left mandible.
    Type Locality—Antoingt, France, late Stampian.
    Diagnosis—Lower premolars slender, lacking an
anterior cingular cuspule, p1 small to absent, p4 with

posterior accessory cusp; M2 present to absent.

North American Distribution—Cedar Creek Member, White River Formation, Colorado, Orellan NALMA (type locality of *P. lagophaga*); Pipestone Springs Local Fauna, Renova Formation, Montana, middle Chadronian NALMA (type locality of *P. infelix*); Cook Ranch Local Fauna, Cook Ranch Formation, Montana, Orellan NALMA; and Scenic Member, Brule Formation, South Dakota, Orellan NALMA.

Discussion—*Palaeogale sectoria* is known from the Oligocene of Europe and Mongolia and the late Eocene and early Oligocene of North America (de Bonis, 1981). The type locality at Antoingt is in MP25, approximately 29 Ma. The earliest well-correlated occurrence in Eurasia is from MP22, approximately 32 Ma. The Pipestone Springs Local Fauna is dated at 35-36 Ma, which marks the first known occurrence of this genus.

CONCLUSIONS

In North America, the dominant small carnivorans of the late Eocene and early Oligocene are the Canidae, particularly the genus *Hesperocyon*. Additional small carnivorans include daphoenine amphicyonids, eusmiline nimravids, and the genera discussed in this chapter. During this time period, canids and ursoids occur for the most part allopatrically; canids are found exclusively in North America and ursoids, with the exception of the genera discussed in this chapter, are predominantly Eurasian (Tedford, 1976). The Amphicyonidae and Nimravidae have Holarctic distributions during this interval.

Amphicyonids and canids are the earliest known caniform carnivores. In North America, daphoenines and canids appear in the Duchesnean NALMA and ursoids (*Subparictis*) and musteloids (*Mustelavus*) appear in the Chadronian NALMA. *Subparictis* is first known from Ahearn Member of the Chadron Formation in South Dakota and the Yoder Local Fauna of Wyoming, both early Chadronian (Late Eocene) from about 36-37 Ma (Emry et al., 1987, fig. 5.3). *Mustelavus* is known from the late Chadronian Peanut Peak Member, about 34-35 Ma, according to revised correlations (Prothero and Swisher, 1992, fig. 2.7). In Europe, amphicyonids (including *Cynodictis*) appear in deposits predating the Grande Coupure (Remy et al., 1987). Ursoids (*Amphicynodon*) and musteloids (*Mustelictis*) first appear in the early Oligocene, immediately following the Grande Coupure (Remy et al., 1987) in MP21, approximately 33 Ma (Berggren and Prothero, 1992). In Europe, *Palaeogale* also appears after the Grande Coupure, in MP22. The Mongolian Hsanda Gol Formation which produced *Amphicynodon teilhardi*, *A. mongoliensis*, *Palaeogale sectoria*, and the musteloid *Amphicticeps* is correlated with MP 23 (Lange-Badré and Dashzeveg, 1989), which may be equivalent to the Whitneyan NALMA, approximately 31 Ma.

The abundance and diversity of amphicynodontines in Europe has suggested (e.g., Clark, 1937) that ursoids originally evolved in Europe, in spite of their Chadronian record in North America. The biostratigraphic record (e.g., Remy et al., 1987) indicates that amphicynodontines and musteloids do not appear in Europe until after the Grande Coupure and therefore must have immigrated there in the Oligocene. The earliest records for amphicynodontines and musteloids in Asia are also in the Oligocene. However the Asian record of carnivores is too incomplete to determine whether arctoids immigrated there as well or perhaps evolved in Asia. Ursoids and musteloids are first known from the Chadronian (late Eocene) of North America. This suggests that ursoids and musteloids share a pre-Chadronian common ancestor. It is possible that they are derived from Duchesnian amphicyonids in North America. *Palaeogale* also makes its first appearance in the Chadronian. However, the ancestry of *Palaeogale* remains an enigma.

ACKNOWLEDGMENTS

Support for the senior author was provided by Texas A&I University and Mr. and Mrs. Jerome Madans. L. Ginsburg of the MNNH graciously allowed the senior author to study his collections. Figure 1 was drawn by K. LaRue. C. Tipton of Texas A&I assisted with the preparation of the figures.

LITERATURE CITED

Baskin, J. A. 1987. The phylogeny of North American Musteloidea and Procyonoidea (Mammalia, Carnivora, Mustelida). Journal of Vertebrate Paleontology, 7 (supplement to number 3):10A-11A.

Baskin, J. A. in press. Musteloidea. *in* C. M. Janis, K. M. Scott, and L. L. Jacobs (eds.). Evolution of Tertiary Mammals of North America. Cambridge University Press, Cambridge.

Berggren, W. A., and D. R. Prothero. 1992. Eocene-Oligocene climatic and biotic evolution: an overview; pp. 1-28 *in* D. R. Prothero and W. A. Berggren (eds.). Eocene-Oligocene Climatic and Biotic Evolution. Princeton University Press, Princeton, N. J.

Bonis, L. de. 1969. Remarques sur la position systématique des Amphicyon. Académie des Sciences de Paris, Comptes-rendus (D) 269:1748-1750.

Bonis, L. de. 1981. Contribution à l'étude du genre *Palaeogale* Meyer (Mammalia, Carnivora). Annales de Paléontologie (Vertébrés) 67:37-56.

Bryant, H. N. 1993. Carnivora and Creodonta of the Calf Creek Local Fauna (late Eocene, Chadronian), Cypress Hills Formation, Sakatchewan. Journal of Paleontology 67:1032-1046.

Chaffee, R. G. 1954. *Campylocynodon personi*, a new carnivore from the Beaver Divide, Wyoming. Journal of Paleontology 28:43-47.

Cirot, E., and L. de Bonis. 1992. Revision du genre *Amphicynodon*, carnivore de l'Oligocene. Palaeontographica, Abteilung A 220:103-130.

Clark, J. 1937. The stratigraphy and paleontology of the Chadron Formation in the Big Badlands of South Dakota. Annals of the Carnegie Museum 25:261-350.

Clark, J., J. R. Beerbower, and K. K. Kietzke. 1967.

Oligocene sedimentation, stratigraphy, paleoecology, and paleoclimatology in the Big Badlands of South Dakota. Fieldiana: Geology Memoir 5:1-158.

Clark, J., and T. E. Guensburg. 1972. Arctoid genetic characters as related to the genus *Parictis*. Fieldiana: Geology 26:1-76.

Cope, E. D. 1873. Synopsis of new Vertebrata from the Tertiary of Colorado, obtained during the summer of 1873. Washington, Government Printing Office, 19 pp.

Emry, R. J. 1975. Revised Tertiary stratigraphy and paleontology of the western Beaver Divide, Fremont County, Wyoming. Smithsonian Contributions to Paleobiology 25:1-20.

Emry, R. J., P. R. Bjork, and L. S. Russell. 1987. The Chadronian, Orellan, and Whitneyan land mammal ages; pp. 118-152 in M. O. Woodburne (ed.), Cenozoic Mammals of North America. University of California Press, Berkeley and Los Angeles.

Flynn, J. J., and H. Galiano. 1982. Phylogeny of early Tertiary Carnivora, with a description of a new species of *Protictis* from the middle Eocene of northwestern Wyoming. American Museum Novitates 2725:1-64.

Flynn, J. J., N. A. Neff, and R. H. Tedford. 1988. Phylogeny of the Carnivora; pp. 73-116 in M. J. Benton (ed.), The Phylogeny and Classification of the Tetrapods, Volume 2, Mammals. Systematics Association Special Vol. 35B, Clarendon Press, Oxford.

Galbreath, E. C. 1953. A contribution to the Tertiary geology and paleontology of northeastern Colorado. University of Kansas Paleontological Contributions 13:1-220.

Ginsburg, L. 1966. Les Amphicyons des Phosphorites du Quercy. Annales de Paléontologie (Vertébrés) 67:37-56.

Hall, E. R. 1931. Description of a new mustelid from the later Tertiary of Oregon, with assignment of *Parictis primaevus* to the Canidae. Journal of Mammalogy 12:156-158.

Hunt, R. M., Jr. 1974a. *Daphoenictis*, a cat-like carnivore (Mammalia, Amphicyonidae) from the Oligocene of North America. Journal of Paleontology 48:1030-1047.

Hunt, R. M., Jr. 1974b. The auditory bulla in Carnivora: an anatomical basis for reappraisal of carnivore evolution. Journal of Morphology 143:21-76.

Hunt, R. M., Jr. 1977. Basicranial anatomy of *Cynelos* Jourdan (Mammalia: Carnivora), an Aquitanian amphicyonid from the Allier Basin, France. Journal of Paleontology 51:826-843.

Hunt, R. M., Jr. 1989. Evolution of the aeluroid Carnivora: Significance of the ventral promontorial process of the petrosal, and the origin of the basicranium in the living families. American Museum Novitates 2930:1-30.

Lange-Badré, B., and D. Dashzeveg. 1989. On some Oligocene carnivorous mammals from Central Asia. Acta Palaeontologica Polonica 34:125-148.

Matthew, W. D. 1899. A provisional classification of the freshwater Tertiary of the West. American Museum of Natural History Bulletin 12:19-75.

Matthew, W. D. 1903. The fauna of the *Titanotherium* beds of Pipestone Springs, Montana. American Museum of Natural History Bulletin 19:197-226.

Pilgrim, G. E. 1931. Catalogue of the Pontian Carnivora of

Europe in the Department of Geology. British Museum (Natural History). London, 174 pp.

Prothero, D. R., and C. C. Swisher. 1992. Magnetostratigraphy and geochronology of the terrestrial Eocene-Oligocene transition in North America; pp. 46-73 in D. R. Prothero and W. A. Berggren (eds.), Eocene-Oligocene Climatic and Biotic Evolution. Princeton University Press, Princeton, N. J.

Remy, J. A., J.-Y. Crochet, B. Sige, J. Sudre, L. de Bonis, M. Vianey-Liaud, M. Godinot, J.-L. Hartenberger, B. Lange-Badré, and B. Comte. 1987. Biochronologie des phosphorites du Quercy: Mise à jour des listes fauniques et nouveaux gisements de mammifères fossiles. Abhandlugen der Münchner geowissschaften (A) 10:169-188.

Schlosser, M. 1911. Grundzüge de Paläontologie (Paläozoologie) von K. A. von Zittel, II Abteilung - Vertebrata. Neuarbeitet von F. Broili und M. Schlosser. Munich and Berlin. R. Oldenbourg, 598 pp.

Schmidt-Kittler, N. 1981. Zür Stammesgeschichte der marderverwandten Raubtiergruppen (Musteloidea, Carnivora). Eclogae Geologicae Helvetiae 74:753-801.

Scott, W. B. 1893. On a new musteline from the John Day Miocene. American Naturalist, 27:658-659, 767.

Scott, W. B., and G. L. Jepsen. 1936. The mammalian fauna of the White River Oligocene. Part I: Insectivora and Carnivora. Transactions of the American Philosophical Society, new series 28:1-153.

Simpson, G. G. 1945. The principles of classification and a classification of mammals. American Museum of Natural History Bulletin 85:1-350.

Simpson, G. G. 1946. *Palaeogale* and allied early mustelids. American Museum Novitates 1320:1-14.

Tedford, R. H. 1976. Relationship of pinnipeds to other carnivores (Mammalia). Systematic Zoology 25:363-374.

Tedford, R. H., L. G. Barnes, and C. E. Ray. 1994. The early Miocene littoral ursoid Carnivoran *Kolponomos*: systematics and mode of life; in A. Berta and T. A. Deméré (eds.), Contributions in Marine Mammal Paleontology Honoring Frank C. Whitmore, Jr. Proceedings of the San Diego Society of Natural History 29:11-32.

Tedford, R. H., M. F. Skinner, R. W. Fields, J. M. Rensberger, D. P. Whistler, T. Galusha, B. E. Taylor, J. R. Macdonald, and S. D. Webb. 1987. Faunal succession and biochronology of the Arikareean through Hemphillian interval (late Oligocene through earliest Pliocene epochs) in North America; pp. 153-210 in M. O. Woodburne (ed.), Cenozoic Mammals of North America. University of California Press, Berkeley.

Teilhard de Chardin, P. 1915. Les carnassiers des phosphorites du Quercy. Annales de Paléontologie 9:103-192.

Thorpe, M. R. 1921. Two new fossil Carnivora. American Journal of Science, series 5, 1:477-483.

Van Houten, F. R. 1964. Tertiary geology of the Beaver Rim area, Fremont and Natrona Counties, Wyoming. U.S. Geological Survey Bulletin 1164:1-99.

Wang, X., and R. H. Tedford. 1992. The status of genus *Nothocyon* Matthew, 1899 (Carnivora): an arctoid not a canid. Journal of Vertebrate Paleontology 12:223-229.

25. Merycoidodontinae and Miniochoerinae

MARGARET SKEELS STEVENS AND JAMES BOWIE STEVENS

ABSTRACT

This report is a phylogenetic revision of some members of the family Merycoidodontidae. Reinterpretation primarily scrutinizes probable intraspecific morphologic variation of skulls and mandibles with special attention to post-mortem deformation, and investigation of 18 measurements made uniformly on all suitable specimens. We conclude that the subfamily Merycoidodontinae (late early or middle Chadronian to early Arikareean) contains two mesocephalic genera, *Merycoidodon* and *Mesoreodon*. *Merycoidodon* (middle Chadronian through Whitneyan, about 7 million years) contains four species, including one newly described from the middle Chadronian of the Vieja, Trans-Pecos Texas. *Merycoidodon* is ancestral to Eporeodontinae (later Whitneyan–early Arikareean), and *Mesoreodon* (latest Whitneyan or earliest Arikareean). *Mesoreodon* lasted for about 3.5 million years, and contains two species. *Mesoreodon* is ancestral to the Promerycochoerinae (middle–late Arikareean) and the Merycochoerinae (middle Arikareean–middle Hemingfordian).

The Miniochoerinae contain one genus, *Miniochoerus*. Four of the five species of *Miniochoerus* form a single series that extends from the middle Chadronian to the earlier Whitneyan, approximately 5 million years. Two species of *Miniochoerus* coexisted in the early Orellan. Miniochoerinae are not ancestral to any other group.

INTRODUCTION

Some of the commonest fossil mammals in the Tertiary beds of North America are oreodonts. Joseph Leidy (1848) described the first species, *Merycoidodon culbertsoni*, and other studies by Leidy were followed quickly by those of O. C. Marsh and E. D. Cope. At the turn of the century, Earl Douglass studied oreodonts from Montana, and J. C. Merriam and C. Stock, among others, reported species from California. Frederick B. Loomis, O. A. Peterson, and E. M. Schlaikjer published on oreodonts from Wyoming and Nebraska, and M. R. Thorpe named or characterized several from Oregon and elsewhere. Most recently, oreodonts have been investigated extensively by C. B. Schultz and C. H. Falkenbach (see Schultz and Falkenbach, 1968, for references).

Thorpe (1937a) was the first to evaluate oreodonts comprehensively as a family. Thorpe characterized the known taxa, transferred some species to different genera, presented interpretations, and outlined a generalized phylogeny, but suggested few synonymies. Scott (1940) followed Thorpe (1937a) when he reviewed the oreodonts from the White River Group.

The second attempt to evaluate oreodonts systematically came about as a direct result of the collection of large numbers of specimens by field parties from the Frick Laboratory, American Museum of Natural History. C. Bertrand Schultz, University of Nebraska State Museum, and C. H. Falkenbach, Frick Laboratory, studied these fossils. Over a period of 28 years, they collaborated on eight lengthy reports which contain many specific descriptions and phylogenetic conclusions, and initiated the practice of dividing the Merycoidodontidae into subfamilies. Oreodont taxonomy and concepts of phylogeny became chaotic. One helpful aspect of Schultz and Falkenbach's investigations (various papers, especially 1968, p. 425) has been the definition of oreodont faunal zones which, in spite of taxonomic oversplitting, are mostly valid.

The senior author of this chapter is responsible for the measurement and evaluation of the fossils, deformation experiments with plaster casts of given oreodont skulls, application of ratio diagrams to oreodonts, the systematic paleontology including the naming of the species of *Merycoidodon* described in this report, stratigraphic interpretations, synonymies, phylogenetic conclusions, study of various modern suiform artiodactyls, and for preparing Figures 1, 3, 5, 7, and 8. The junior author is responsible for statistical investigations of, and quantitative statements about, the 5,957 measurements on 756 fossil specimens and 991 measurements on 67 modern suiforms. Because this report has been in preparation for many years, the original ratio diagrams were produced by a FORTRAN program written by the junior author. Later, ratio diagrams were recalculated, and statistical analysis, in part leading to preparation of Figures 2, 4, and 6, tables, and appendices A-C, were done with a spreadsheet (Microsoft Excel) on Macintosh personal computers.

Oreodonts, as suiform artiodactyls, have comparatively durable crania that are preserved relatively more frequently than crania of most other mammals, and cranial appearance has become the basis for much of their taxonomy. Observation of skull shape has tended to be casual and subjective. Even the sturdy crania of oreodonts can be subject to post-mortem deformation. Specimen orientation within sediments is often determined by whether or not the jaws remained attached to the skull during burial, and in turn, largely determines the direction of crushing. Grain size of surrounding and infilling sediment and the completeness of infilling, influence the amount of crushing. Skull length is affected less by crushing than skull width (Phleger and Putnam, 1942, p. 548). Although anteroposterior compression can shorten the basicranial diameter, taphonomy makes this orientation of compression less probable. Latex molds of some of the least deformed (mesocephalic) oreodont skulls available were made. Then dorsoventral or lateral compression was applied to the rubber molds as plaster of Paris set within the molds. The deliberately deformed casts are remarkably similar to crushed oreodont skulls reported in the literature as representing "dolichocephalic" or "brachycephalic" species. Therefore, the terms "narrow," "high," "low," "broad," "brachycephalic," and "dolichocephalic" that dominate the recent literature must be viewed with caution. Phleger and Putnam (1942) note that crushing of oreodont crania can bias measurements and omitted deformed skulls from their analysis. Lander (1978) also noted this problem.

When oreodont skull width was measured, we were careful to use only crania which have at least one intact undeformed zygomatic arch. Similarly, the diameter of the easily deformed postorbital constriction (hollow) was not measured on obviously laterally or dorsoventrally crushed skulls. Oreodont teeth and lower jaws are less subject to post-mortem deformation, because they are more solid.

The morphology of the auditory bulla has played a role in oreodont taxonomy, especially among early Oligocene species. Intraspecific variation of the auditory bulla was studied in modern suiform and other artiodactyls to aid in the evaluation of the fossils. Much variation was found in the size and shape of the bulla. For example, a large Trans-Pecos Texas sample of the collared peccary *Dicotyles tajacu* showed auditory bullae that vary from small, medium, to large-sized, and within these sizes, from longer-than-wide to wider-than-long, rounded, pinched, and L-shaped. Although size and shape of oreodont bullae serve in a general way to reflect bullar stage of evolution, care must be taken to utilize what can be established as the probable mean condition for the given populations, and not to place too much taxonomic emphasis on slight variation between otherwise morphologically similar specimens.

Another factor which has influenced oreodont taxonomy has been the tendency to minimize the variability of a species. This is particularly true when large

numbers of more or less contemporary individuals with varying modes of preservation are known. Phleger (1940) and Phleger and Putnam (1942) gave attention to the effects of decisions about oreodont taxonomy on the variability of the taxa erected. However, since most oreodont samples have never been evaluated statistically in print, population variability has seldom been quantified.

The purpose of this report is to take another look at the taxonomy of the Merycoidodontinae and Miniochoerinae. We take into account specimen deformation, attribute to the fossil species a range of morphologic and size variation that is consistent with that established for living suiform artiodactyls (see Woodburne, 1968), and provide descriptive and analytical statistical information for oreodonts. The length of this paper requires condensation of the appendices. The original data are available from the authors upon request. Our synonymies greatly simplify the taxonomy of the two subfamilies. We also suggest lineages and a probable phylogeny of the group, particularly in the interrelationships between some of the subfamilies.

METHODS

Eighteen different measurements are defined for oreodonts (Fig. 1). Similar, or slightly modified measurements were obtained for peccary specimens. These are listed below and this sequence order follows throughout the ratio diagrams and the appendices of this report. All measurements were made with metric dial calipers and are in millimeters.

P^1-M^3, inclusive, labial, from the anterior edge of P^1 to the posterior edge of M^3, parallel with the alveolar border (P^2-M^3 for peccaries).

P^1-P^4, inclusive, labial, from the anterior edge of P^1 to the posterior edge of P^4, parallel with the alveolar border (P^2-P^4, for peccaries).

M^1-M^3, inclusive, labial, from the anterior edge of M^1 as it overlaps the P^4, to the posterior edge of M^3, parallel with the alveolar border.

APM^3, maximum anteroposterior labial diameter of M^3.

TM^3, maximum transverse diameter of M^3, across the anterior pair of selenes.

Ht. M^3, maximum labial crown height of M^3, from the unworn apex of the metacone to the base of the enamel, along the rib trace (paracone for peccaries).

Nasal L., maximum length of the nasal bone from the anterior tip to the posterior process.

Subnasal L., length of the premaxillae from the naso-premaxillary notch to the anterior alveolar margin of I^1.

Malar D., maximum dorsal-ventral depth of the malar below the ventral border of the orbit, and normal to the ventral margin.

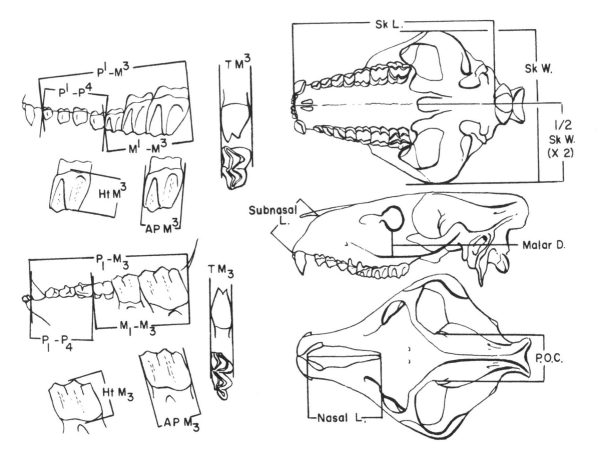

Figure 1. Illustration of definitions of measurements on an oreodont skull, and upper and lower teeth.

P.O.C., minimum diameter of the postorbital constriction.

Sk. L., basicranial length of the skull from the anterior edge of the premaxilla at I^1, to the condylar notch.

Sk. W., maximum bizygomatic diameter of the skull (or an estimation of skull width, doubling the measurement from the midline of the palate to the outside of one zygomatic arch).

P_1-M_3, inclusive, lingual, from the anterior edge of P_1 to the posterior edge of M_3, parallel with the alveolar border (P_2-M_3 for peccaries).

P_1-P_4, inclusive, lingual, from the anterior edge of P_1 to the posterior edge of P_4, parallel with alveolar margin (P_2-P_4 for peccaries).

M_1-M_3, inclusive, lingual, from the anterior edge of M_1 to the posterior margin of M_3, parallel with alveolar margin.

APM_3, anteroposterior lingual diameter of M_3.

TM_3, maximum transverse diameter across the anterior pair of selenes of M_3.

Ht. M_3, maximum lingual crown height of M_3, from the unworn apex of the entoconid to the base of the enamel, along the rib trace (metaconid for peccaries).

Three different measurements were attempted for the upper and lower third molars because these teeth are identified with confidence, because the M3s are still unworn when the individual nears maturity, and because the anteroposterior diameter changes less with wear than for the other molars.

Care was taken first to consider type specimens, then morphologically comparable, topotypic (primary) samples. Once morphology and a range of variation were established for the topotypic population, then morphologically comparable, stratigraphically equivalent specimens or subsamples were compared to the primary sample. Primary and secondary samples, combined with due regard for morphology and stratigraphy, constitute what is believed to be an oreodont species. Statistical investigations of measurements were undertaken to see if the combinations seemed reasonable.

Description of the data began with the preparation of ratio diagrams (Simpson, 1941) as an aid to judgment objectivity and visualization of measurement sets for various taxa and combinations of taxa. Ratio diagrams, ideally based on topotypic character means, provide a quick and standardized way of comparing one

Merycoidodontinae

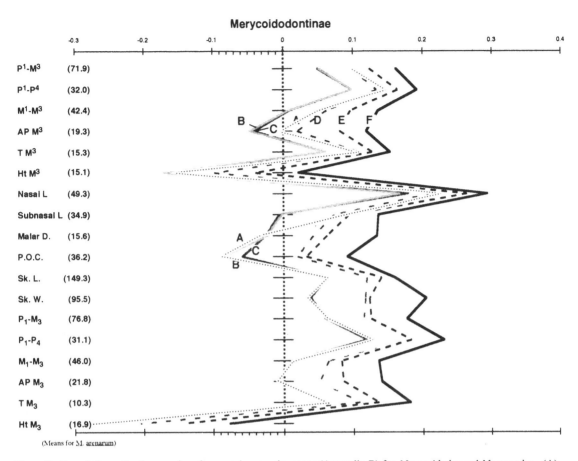

(Means for M. arenarum)

Figure 2. Cumulative ratio diagrams based on species sample means (Appendix B) for *Merycoidodon* and *Mesoreodon*. (A), *Merychyus* (*Merycoidodon*) *presidioensis* sp. nov; (B), *M.* (M.) *culbertsoni*; (C),*M.* (*Otarohyus*) *bullatus*; (D), *M.* (*O.*) *major*; (E), *Mesoreodon chelonyx*; and (F), *?M. minor*.

population with another, or, less ideally, one specimen with another. The diagrams graphically show the relative proportions and cumulative size trends between the selected characteristics of a given specimen or population (species), and provide standardization because two or more species can be compared at the same time on the same diagram.

A sample of *Merychyus arenarum* Cope (1884b), an Arikareean species (vertical dashed line, Figs. 2, 4) provides the means (shown in parentheses, Figs. 2, 4) that (logarithmically transformed) are the constants used for the construction of the oreodont ratio diagrams. *M. arenarum* was chosen because it is represented by a large primary sample, and because it is intermediate in size between the smallest and the larger oreodonts. Initially, ratio diagrams were prepared for individual type specimens, then for the topotypic samples, and then for the stratigraphically equivalent secondary sub-samples. The type specimens and sub-samples for the oreodont species in this report are listed in the hypodigms as modified from Schultz and Falkenbach's listings. Once a single species of

oreodont was identified, a summary diagram for that species was prepared (Figs. 2, 4), based on sample means for measured characteristics. The observed range and standard deviations for the given characteristics have been omitted from the ratio diagrams as a simplification. These are shown in appendices B and C. Because the taxa studied in this report are thought to be mainly members of single biostratigraphic clines, the ratio diagram for the species of each subfamily is included on a single diagram to show the cumulative relative size trends between the progressively younger species of that subfamily. The younger the taxon, the heavier the line that represents it. Ratio diagrams alone, however, cannot provide evidence for an unequivocal identification because it is not feasible to quantify all aspects of morphology. Although the morphological trends undertaken by the upper and lower teeth of a given species are closely similar, both dentitions are included on the ratio diagrams and in the appendices because type specimens often do not associate them.

The extreme similarity of the ratio diagrams of the

hypodigm suites resulted in other statistical investigations of the possible synonymies. As the appendices show, descriptive statistics were calculated if they could be. However, difference in sample size is a substantial problem, and some samples are clearly too small for some purposes. After considering the sizes of the available data sets, and desirable levels of accuracy and confidence, it was decided, somewhat arbitrarily, to call samples of 30>N≥10 small. Measurement sets of N>30 were considered adequate, and those with N<10, very small (Table 1). Adequacy of samples is not quite that simple and is discussed further below.

The usual tests of significance, and some approaches to analysis (Scheffé-Box or log-anova test for homogeneity of variance, analysis of covariance, bivariate and multiple regressions, linear and non-linear) were employed to investigate the effects of combinations of data sets, and differences between data sets. The taxonomic revisions and stratigraphy used in this report make it easy to organize "t" tests, and of course F tests, into square diagonal matrices where "acceptances" (failure to reject the H_0) close to the diagonal suggest progressive change, and those far from the diagonal suggest stereotypy. Tests of significance were evaluated for an a value of 0.05, unless otherwise noted.

Examination of histograms led us to suspect that many of the data sets are not close approximations of normal distributions. Modeling of χ^2 tests for departure from normal distribution suggests that odd numbers of classes produce more conservative results, particularly for bimodal and smaller data sets. For adequate samples, we used 11 (five for smaller samples) equal area subdivisions of the normal distribution over a range slightly greater than double the absolute value of the greatest deviation from the mean as the expected values. For adequate samples where the H_0 was rejected, this method of testing has the advantage that the probability differences from the normal distribution can be examined in detail. Because of the doubtful normality of a majority of the available distributions (discussed below), nonparametric tests, particularly, Kolmogorov-Smirnov, Kruskal-Wallis, and Hotteling's T tests, were also used. Standard references consulted include Agterberg (1974), Davis (1986), Dixon and Massey (1969), Meyer (1975), Miller and Kahn (1962), Snedecor and Cochran (1967), and Sokal and Rohlf (1981).

The Scheffé-Box test is sensitive to the input order of the data, and although it is not true to say that if one doesn't like the results, one can try again until the desired results are obtained, nevertheless trials with different orderings will yield different results, and occasionally substantially different results. Data were shuffled a minimum of 30 times by sorting on the basis of random numbers, since the original order of measurements (as seen in appendices A, B, C) was not random, and tested, reshuffled, retested, and so on for a minimum of three tests. Since it was not certain that a fully random order was achieved, the results were regarded as offering approximate probabilities that some orders existed for which the data sets would appear to have homogeneous variances. For some groups of data sets it appeared that the chances of finding such an order were very small, for others it appeared that most orders would be homogeneous, and for still others, probabilities of finding such an order were merely small, on the order of 0.01-0.10. Data sets in the first category were rejected, those in the second category were accepted, and those in the last category were accepted if orders were found such that the H_0 (no difference in variances) was not rejected.

Many modes of investigation require that substantial amounts of data be discarded, and the discarding of data changes the nature of the distributions considerably. Consequently, we used methods requiring minimal data loss. Analysis of covariance can be done with paired data, and this commonly reduced the data by only 40-50%. Data sets produced for this purpose are not random selections from the main data set, and were reinvestigated for normality, homogeneity of variances, and skewness. We took the position that analysis could proceed even when data appear non-normal, if variances are homogeneous, and the data were not significantly skewed. In a few instances it was easy to identify one or two specimens as the sources of skewness or other departure from normality. If removal of these did not change the significance of the analysis, the analysis was accepted. For methods involving regression, there is commonly no intuitive basis for assigning independent or dependent status to variables, and particularly for illustrative purposes, y-on-x, x-on-y, and reduced major axis regressions were examined. "Goodness of fit" (r^2) was rarely noticeably improved by use of power-function or multiple regressions. For the latter, substantial reduction in available data is a further consideration.

The senior author (MSS) considered many species of living artiodactyls with emphasis on suiforms, to aid in the evaluation of intraspecific variation in the oreodont taxa. Phleger (1940) and Phleger and Putnam (1942) used the felids *Smilodon californicus* (now *S. fatalis*), *Panthera atrox*, and *P. onca* to provide a guide for the evaluation of variability of various oreodont taxa as recognized by Thorpe (1937a) and Scott (1940). We feel that, despite imperfections of our samples (Table 1 and Appendix A), use of modern suiform artiodactyls, particularly *Dicotyles tajacu*, is more satisfactory.

Statistical summaries for two peccary species, *Dicotyles tajacu*, the collared peccary, and *Tayassu pecari*, the white-lipped peccary (identifications follow Woodburne, 1968, p. 48) are presented in Appendix A. We collected the sample of *D. tajacu* over a period of 30 years from a small geographic area. The complication of the distributions of measurements for this sample, mostly not approximately normal (bimodal or polymodal), may result from the fact that during this time interval, individuals were collected from local populations that experienced birth and growth within periods that ranged from optimal to minimal in available moisture and food in what is largely a desert setting of varying elevation. Most vertebrate fossil "populations" even if collected from a small area and a limited stratigraphic range must have been similarly influenced, perhaps more widely. Modern samples of known species affiliation, and among the peccaries, gender for *T. pecari*, provide information about the relative variability of characters. The most variable cranial feature among numerous species of living artiodactyls belonging to various families is the depth and robustness of the malar and the length of the nasal bones, with coefficients of variation ranging from about 9–18%. Most of the remaining cranial-dental features of the living species have coefficients of variation between 4–8%. The distribution of Malar D.s for our sample of *Dicotyles tajacu*, although bimodal, could be a sample drawn from a normally distributed population. In contemporaneous, geographically very restricted groups, modern female suiforms are commonly smaller than their male counterparts. If the geographic range of the sample is expanded to include a range of conditions for growth and maturation, overlap within cranial-dental measurements made for this paper removes size as a basis of gender distinction. We are inclined to think that sampling through time would have the same effect, and as a result, we view the importance, or at least the perceptibility, of this source of variability with doubt. We hope that ontogeny is largely removed as a source of variation by the method of selecting specimens measured, but provenance remains a source at least equal with gender for all of the samples, even one as restricted as our sample of *D. tajacu*. For fossil taxa, taphonomy must be added as a source of variation within samples, and, because of stratigraphic imprecisions, variations in climate through time is probably a part of provenance. No attempt was made here to evaluate oreodonts subspecifically, because we lack sufficient stratigraphic and paleoenvironmental control and, for most taxa, large enough samples, to make the evaluation of possible subspecies meaningful.

Although the stratigraphy utilized in this report (Fig. 8) is of necessity generalized, it forms a framework for approximate and informal lithostratigraphic organization of oreodont localities. The stratigraphic nomenclature presented in Figure 8 for the classic area of the White River Group reflects that used by the field parties of the Frick Laboratory because the Frick collectors followed the work of Wortman (1893), Osborn and Wortman (1894, 1895), and Osborn (1907), and because the bulk of the fossils studied are Frick specimens. Schultz and Stout (1955) provide the basis for the stratigraphy of the University of Nebraska State Museum specimens. For more information concerning the stratigraphy of the White River Group, see Emry et al. (1987), and Prothero (this volume, Chapter 13).

ABBREVIATIONS

Institutions are abbreviated as follows: American Museum of Natural History, New York, Department of Vertebrate Paleontology (AMNH); Academy of Natural Sciences, Philadelphia (ANSP); Carnegie Museum of Natural History, Pittsburgh (CMNH); Frick Collection, American Museum of Natural History (F:AM); Field Museum of Natural History, Chicago (FMNH); Princeton University, Princeton (PU); Museum of Comparative Zoology, Harvard University, Cambridge (MCZ); Montana State University, Bozeman (MSU); Texas Memorial Museum, The University of Texas at Austin (TMM); University of Nebraska State Museum, Lincoln (UNSM); United States National Museum, Washington, D. C. (USNM); and the Peabody Museum of Natural History, Yale University, New Haven (YPM). The following additional abbreviations also are used: number of observations (N); minimum (Min); maximum (Max); Standard Error of the Mean (SEM); standard deviation (SD); coefficient of variation (CV); Megayears before the present (Ma); and throughout this chapter quotation marks ("—") are used to denote what is regarded as an invalid taxon, an informal rock stratigraphic name, or for emphasis.

NUMERICAL DATA

Table 1 includes results of χ^2 tests of major data sets against the normal distribution. Measurement sets for the most poorly represented taxa, *Merycoidodon presidioensis, Mesoreodon chelonyx,* and *Miniochoerus forsythae* are not included. Of fossil taxa investigated, approximation of normal distribution can be rejected (a = 0.01) for 58% of the investigated distributions (62% if distributions of modern taxa are included) and is marginal (accepted at a = 0.01, but not at a = 0.05) for a further 13% of the measurement sets (12%, for all taxa). If very small samples are neglected, the percentages are 50 (57) and 15 (13), respectively. Variations on the described procedures were tried, but although they changed specific acceptances of the H_0, they did not change the overall view of the data for species as defined in this report as equivocal approximations of normally distributed sets. It appears that departure from normality was most commonly occasioned either by irregular distribution (polymodal), or by bimodal distribution. Most distributions of P.O.C. measurements are, for example, very irregular. Non-normal distributions produced by the combination of *Miniochoerus affinis* and *M. gracilis*, and many non-normal distributions of measures of *M. gracilis* as we define the taxon, are bimodal. Skewness and kurtosis, as individual measures of departure from normality, are not commonly significant.

Table 1. Coefficients of variation, well-represented taxa, ordered by descending taxon-average distances from the mean CV for the character.

Name	P^1-M^3	P^1-P^4	M^1-M^3	APM^3	TM^3	Ht. M^3	Nasal L.	Sub-nasal L.	Malar D.
Combination, *Miniochoerus affinis, M. gracilis*	9.17	9.16	9.77	10.81	11.47	10.57	12.09	13.43	12.98
	A,◊,*	A,n,*	A,n,*	A,n,*	A,◊,*	S,r,*	S,n,*	S,r,*	A,n,*
Merycoidodon culbertsoni	6.44	6.56	7.20	7.70	7.52	7.18	9.52	13.24	11.91
	A,n,L	A,n,L	A,n,I	A,n,I	A,n,L	A,r,I	A,n,L	A,n,vL	A,n,I
Merycoidodon major	5.52	5.49	6.69	8.06	6.44	6.61	10.40	7.69	11.95
	A,n,I	A,n,I	A,n,I	A,n,I	A,n,I	A,r,I	S,n,L	S,r,I	A,n,I
Miniochoerus bullatus	5.97	6.63	6.50	7.66	7.36	7.51	6.37	7.75	12.57
	A,r,I	A,r,L	A,n,I	A,n,I	A,n,L	S,r,I	vS,r,I	S,r,I	A,r,I
§*Dicotyles tajacu*	4.69	6.05	6.17	6.65	6.59	12.53	7.66	7.78	13.01
	A,r,I	A,r,I	A,n,I	A,r,I	A,◊,I	vS,r,vL	S,r,I	A,r,I	A,n,I
Miniochoerus gracilis	5.11	5.50	5.40	6.78	6.76	6.78	9.37	10.39	9.88
	A,r,I	A,r,I	A,r,I	A,r,I	A,r,I	vS,r,I	S,r,I	vS,r,I	A,n,P
Miniochoerus affinis	4.41	5.77	4.43	5.84	6.13	6.23	7.10	10.25	9.59
	A,n,I	A,r,I	A,r,P	A,n,I	A,◊,I	S,r,I	S,n,I	vS,r,I	A,n,P
Merycoidodon chadronensis	3.70	4.79	4.34	4.50	5.79	6.42	8.50	7.27	12.54
	A,r,P	A,r,P	A,r,P	A,◊,P	S,r,P	S,r,I	S,r,I	S,r,I	A,n,I

Name	P.O.C.	Sk. L.	Sk. W.	P_1-M_3	P_1-P_4	M_1-M_3	APM_3	TM_3	Ht. M_3
Combination, *M. affinis, M. gracilis*	10.33	8.28	8.83	8.45	8.46	9.05	10.80	13.39	22.21
	A,n,*	A,r,*	A,◊,*	A,◊,*	A,r,*	A,n,*	A,r,*	S,r,*	vS,r,*
M. culbertsoni	11.40	6.97	8.26	6.16	7.13	6.57	7.53	7.32	7.13
	A,n,L	A,n,I	A,◊,I	A,n,I	A,n,I	A,◊,I	A,n,I	A,r,I	S,◊,I
M. major	9.89	6.21	8.93	5.98	7.28	6.29	7.29	8.65	8.85
	A,n,I	A,◊,I	A,n,I	A,r,I	A,r,I	A,r,I	A,r,I	S,◊,I	vS,r,I
M. bullatus	9.92	6.38	5.45	5.68	6.16	5.87	6.55	7.26	6.48
	A,n,I	S,r,I	S,r,P	A,◊,I	A,r,I	A,n,I	A,r,I	A,n,I	vS,n,I
§*D. tajacu*	4.80	5.41	8.13	3.85	4.99	6.67	6.07	5.65	9.84
	A,◊,P	A,r,I	A,n,I	A,r,P	A,r,I	A,r,I	A,r,I	A,r,P	vS,r,L
M. gracilis	7.35	5.86	6.42	5.66	6.58	5.93	7.31	7.16	1.02
	A,r,I	S,r,I	S,r,I	S,r,I	S,r,I	S,r,I	S,r,I	S,r,I	vS,r,P
M. affinis	8.85	3.91	7.11	4.37	5.17	5.15	6.27	8.09	2.54
	A,◊,I	S,n,P	S,r,I	S,n,I	S,r,I	A,n,I	S,◊,I	S,n,I	vS,r,P
M. chadronensis	8.29	4.02	8.96	4.34	4.91	5.40	5.58	6.24	7.64
	A,r,I	S,r,P	S,r,I	S,n,I	S,r,I	S,r,I	S,n,P	S,r,I	vS,r,I

Abbreviations and symbols in the table:

Sample size: A $N \geq 30$ S $10 \leq N < 30$ vS $N<10$

Nature of Distribution:

n H_0, No difference from normal distribution, not rejected for $c^2_{(0.05, 8)}$ [$c^2_{(0.05, 2)}$ if N<30].

◊ H_0 rejected for $c^2_{(0.05, 8)}$ [$c^2_{(0.05, 2)}$ if N<30], not rejected for $c^2_{(0.01, 8)}$ [$c^2_{(0.01, 2)}$ if N<30].

r H_0 rejected for both a values.

Relative Size, Coefficient of Variation:

vL Coefficient of Variation (CV) is very large compared to others for that measurement (see text).

L CV is large compared to others for that measurement (see text).

I CV is intermediate compared to others for that measurement (see text).

P CV is small compared to others for that measurement (see text).

* CVs of the combined samples not included in the comparison (see text).

§ For *D. tajacu*, cheek tooth row measurements are P2-M3, and premolar measurements, P2-P4.

Scheffé-Box tests suggest that variances are generally homogeneous within, but not among the Miniochoerinae and Merycoidodontinae as we understand them. There are two kinds of exceptions to this pattern: the variance of M^1-M^3 is not homogeneous within *Miniochoerus*; and variance is homogeneous among the species of *Merycoidodon, Mesoreodon*, and *Miniochoerus* for the P^1-P^4/P^1-M^3 and APM^3/Ht. M^3 ratios. *Miniochoerus starkensis* appears to be the source of variance inhomogeneity in *Miniochoerus*. *Merycoidodon major* appears to be less similar to *M. culbertsoni* and *M. bullatus* than these are to each other.

The senior author proposes extensive taxonomic simplification on the basis of morphology and descriptive statistics. Consequently, we are interested in measures of variability. Means of different measured features differ by slightly more than an order of magnitude within a taxon, and maximum differences among taxa in means for measurements of one kind range from 28–52%. The coefficient of variation (CV) is an appealing dimensionless measure of variation. Coefficients of variation within measurement categories were normalized across taxa and distance from the means (SDs), was averaged. Kruskal-Wallis tests suggest real differences among taxa in the distributions of CVs. The arrangement of Table 1, descending average distances from the mean CV for all measurement categories, approximates descending order of variability of the best represented taxa. A CV two or more SDs below the mean is considered very small; one to two SDs below the mean, small; and within one SD of the mean, intermediate. Limits for large and very large CVs mirror those for small and very small CVs. None of the CVs for the fossil taxa that we recognize lie as much as three SDs on either side of the mean. About 69% of larger and smaller CV values are associated with small or very small samples, with a modest majority of smaller CVs. An intermediate calculation in Kruskal-Wallis tests, mean squared sums of ranks, suggests a variability ordering of taxa that differs from that used in Table 1 by interchanging the positions of *Dicotyles tajacu* and *Miniochoerus gracilis*.

The CVs produced by combining data sets of *Miniochoerus affinis* and *M. gracilis* were not part of calculations producing the ordering in Table 1, but are included for purposes of comparison. Our interpretation of *Merycoidodon culbertsoni*, which has an extensive hypodigm, has one very large and five large CVs, and, among the taxa that we recognize, is exceeded only by *Miniochoerus starkensis* (7 and 2, respectively), represented entirely by small and very small samples. By comparison with taxa that we accept, the combination of *Miniochoerus affinis* and *M. gracilis* would have about 15 large to extremely large CVs, eight of them more than three SDs from the CV mean. Although *Miniochoerus chadronensis* is a

combination of all or parts of nine previously recognized "species" (and two "geologic varieties") from four "genera," it has seven small CVs, and generally the lowest variability of fossil taxa discussed in this report.

Normalization within taxa (across the columns of measurement CVs) allows inspection and ordering of variability associated with individual measurements. A series of Kruskal-Wallis tests suggests that differences are significant, both at the high and low variability ends of the range, and produces an ordering that varies slightly from the order produced by distance from CV means for intermediately variable measures. However, the four highest variability measures, Malar D., Subnasal L., P.O.C., Nasal L., and two lowest variability measures, P_1-M_3, P^1-M^3 (descending order) are reproduced. Estimated sample size to achieve given levels of confidence and precision produces an order similar to the other orderings and identical for the four most variable characteristics. Several measures share two detrimental features, high variability and a tendency to be represented by small or very small samples. Nasal L. and Subnasal L. fall into this category, and their ratio is used as part of the diagnoses of the Merycoidodontinae and Miniochoerinae for possible future reference to taxa that are not part of this report.

Although adequate sampling of Malar D. does require large numbers of measurements, it is a durable structure far more often available than Subnasal L. or Nasal L., and usually, for our data, is approximately normally distributed (Table 1; appendices B, C). Malar D., and even P.O.C., a more fragile structure with less commonly normally distributed measurements, are generally regarded as useful measurements for the body of data under discussion. Tooth row measurements, except TM_3, are generally intermediate to low in variability, fairly well represented for most taxa, and their CVs are generally smaller than expectation values, with those of the lower teeth less predictably and regularly so than for the uppers. Seventy-five percent of CVs associated with non-normal distributions for measurements are intermediate.

Our sample of *Dicotyles tajacu* has, in comparison to similar measures on fossil taxa, very large and large CVs for Ht. M3s. This is probably due to difficulties in judging lack of wear and determining measurement landmarks. These characters are represented by very small samples. The small sample of *Tayassu pecari* is of considerable morphological interest because it is sexed, but distributions of most measurement data are strongly bimodal for reasons apparently unrelated to gender, and these data were not used for most size comparisons. Unworn heights of M3s could not be measured on *T. pecari*. "Goodness of fit" for regressions of CVs of fossil taxa on those of *D. tajacu* are poor, and mostly not significant, but if the suspect Ht. M3s are removed, the regressions are substantially improved, and if P.O.C. is also removed, the regressions are again improved, so that r^2 values are all significant and fall in the range from 0.34 (*Miniochoerus starkensis*) to 0.84 (*Miniochoerus chadronensis*). *Miniochoerus starkensis* has by far the smallest value, and the next smallest value of r^2 is 0.52 (*Mesoreodon minor*). Removal of Nasal L. and Subnasal L. data makes a further small improvement, but removal of other data categories, singly or in combinations, does not have the same effect. Parallel trials of fossil taxon data with that of *Tayassu pecari* were made, but the regressions are poorer, except for those with *Merycoidodon bullatus* and *Mesoreodon minor*.

Similarly, regressions of the CVs of the fossil taxa on each other are improved by removal of Ht. M3s. Removal of P.O.C. data improves 8 of 36 regressions, and all the improvements involve *Miniochoerus gracilis*. Removal of Malar D. data makes regressions much poorer, except for the regression of *Merycoidodon bullatus* on ?*Miniochoerus forsythae*. Removal of CV data for Nasal L. and Subnasal L. in addition to that for Ht. M3s makes a substantial improvement in the regressions, and all but three of the r^2 values are significant and lie in the range from 0.32 to 0.83. Comparisons of CVs of *Miniochoerus starkensis* with other fossil taxa include the six poorest regressions, three that are not significant (?*M. forsythae, M. bullatus, ?Mesoreodon minor*), and the three with the smallest significant r^2 values [*M. affinis* (0.34), *M. gracilis* (0.41), and *Merycoidodon major* (0.53)].

Size is a major aspect of change in the group of oreodonts with which this report is concerned. The majority of recognized taxa, even where there is little stratigraphic, and in our opinion, taxonomic separation, have significantly different means for most measurements. There are three notable exceptions. Size alone would not separate *Merycoidodon culbertsoni* from *M. bullatus* or *Miniochoerus starkensis* from *M. chadronensis*. And although ?*M. forsythae* differs from other taxa in what are probably the most useful measures of gross size, Sk. L., P^1-M^3, P_1-M_3, it differs significantly from younger *Merycoidodon culbertsoni* and *M. bullatus* mainly in measurements involving the premolar tooth row. Size variation appears to be reasonably consistent within the oreodont species (appendices B, C) as identified in this report, and to be consistent in many respects with that seen for the modern *D. tajacu*.

Skull length is an intuitive indicator of general size, and the data, available on about a third of the specimens, are of reasonably good quality. Throughout this report reference is made to the approximate size of crania as small (Sk. L. 80-150 mm or 3-6 inches), medium (Sk. L. 150-250 mm or 6-10 inches), or large (Sk. L. exceeds 250 mm or 10 inches). P^1-M^3 is a set of data of high quality, available for 69% of the specimens, and more clearly related to Sk. L. in fossil taxa that we discuss than it is in the modern peccaries (Table 2). "Goodness of fit" for the least squares regression of

Sk. L. on P^1-M^3 for all specimens in the group of oreodonts we discuss is the best that we have found for any data pairing available for these oreodonts, and this relationship illustrates the variation in gross size. Slopes of the reduced major axes (Table 2) suggest the expected numerical dominance of Sk. L. variance over that of P^1-M^3 for the combination of all taxa, but for species of *Merycoidodon* the dominance is weaker than the overall trend. Sk. L. and P^1-M^3 for *Mesoreodon* form a wider scatter at the large end of the Merycoidodontinae trend on scatter diagrams, and have a regression that is not significant. *Miniochoerus*, and its species (except *M. gracilis*, and ?*M. forsythae*) have lower than expected values for slopes of the reduced major axes despite a substantial degree of stereotypy in this proportion for the oreodonts studied.

The slope of the y/x regression line, Sk. L./P^1-M^3, for *Merycoidodon* is significantly different neither from that of the regression line for all fossil specimens discussed in this report, nor from slopes of lines for its species, but does differ significantly from the slope of the line for *Miniochoerus*. Regression lines for species of *Miniochoerus* are not significantly different from that for the genus. Species of *Miniochoerus* from ?*M. forsythae* to *M. gracilis* are progressively smaller, but *M. starkensis*, the youngest species, returns, approximately along the original trend, to a size comparable to *M. chadronensis*. *Merycoidodon culbertsoni* and *M. bullatus* differ little in size, though *M. culbertsoni* has a substantially greater range, but *Merycoidodon* increase in size overall through time. A matrix of student's "t" tests for Sk. L. shows the size similarity of *Merycoidodon culbertsoni* and *M. bullatus*, suggests progressive change in size associating *Merycoidodon major* with *Mesoreodon*, and suggests little association within *Miniochoerus*. The same kind of matrix for P^1-M^3 data shows only the similarity of *Merycoidodon culbertsoni* and *M. bullatus*. The Sk. L./P^1-M^3 ratio which should relate these measures of size produces matrices of student's "t" tests that suggest some specific differences, but supports general stereotypy in proportions within *Miniochoerus* and *Merycoidodon*.

In preliminary investigation of these and other data, many sets of regressions were calculated, and there seems to be a pattern to "goodness of fit" values associated with them. As noted above, P^1-M^3 is a fairly good guide to Sk. L. for oreodonts discussed in this report (Table 2). If genera are considered separately, *Miniochoerus* still shows a fairly good regression, and *Merycoidodon* is weaker (Table 2). Prediction of either quantity on the basis of the other is poor at the specific level in most instances, and decreased in all instances. Of our samples, only *Merycoidodon culbertsoni* has the number of specimens sufficient to examine a kind of hierarchically smaller unit, sets of specimens originally considered to be species in their own right, but assigned to *M. culbertsoni* in this report. None of the sets shows significant Sk. L./P^1-M^3 regressions (Table 2).

Table 2

"Goodness of fit" for regressions of Sk. L. on P^1-M^3, for taxa and some combinations of taxa. Very small samples are omitted.					
Taxonomic Level	N	r^2	$s_{y/x}$	r	RMA $b1$**
Taxonomically high groups:					
All oreodonts for this report	227	0.93	8.32	0.96	2.32
Merycoidodontinae	152	0.79	8.63	0.89	2.23
Genera and combinations:					
Merycoidodon	139	0.75	7.99	0.87	2.44
Miniochoerus (and Miniochoerinae)	75	0.88	4.52	0.94	1.77
Mesoreodon	13	0.22*	12.57*	0.47*	2.02*
Miniochoerus chadronensis-M. affinis-M. gracilis combination	64	0.87	4.23	0.93	1.75
Miniochoerus chadronensis-M. affinis combination	50	0.66	4.17	0.81	1.75
Miniochoerus affinis-M. gracilis combination	38	0.82	4.47	0.90	1.85
Species					
Merycoidodon culbertsoni	92	0.71	5.94	0.84	2.00
Merycoidodon bullatus	19	0.72	6.22	0.85	1.82
Merycoidodon major	28	0.54	9.39	0.73	1.81
Miniochoerus chadronensis	26	0.55	4.00	0.74	1.62
Miniochoerus affinis	24	0.33	4.37	0.57	1.06
Miniochoerus gracilis	14	0.70	4.19	0.84	2.04
§*Dicotyles tajacu*	33	0.16	10.19	0.41	3.55
§*Tayassu pecari*	24	0.38	5.60	0.62	2.16
Hierarchically lower groups					
†*M. culbertsoni* of Schultz and Falkenbach	22	0.03*	4.85*	0.18*	2.12*
†*M. culbertsoni* "osborni" of Schultz and Falkenbach	32	0.07*	6.10*	0.27*	2.11*
† "*Otionohyus wardi*" of Schultz and Falkenbach	13	0.26*	6.60*	0.51*	1.38*

*Regression not significant   §The measurement is P^2-M^3
**Slope, Reduced Major Axis   †Data from *M. culbertsoni*, Appendix B

"Goodness of fit" values for regressions of APM^3 on TM^3 (illustrated in part by Fig. 6D and Table 9), previously discussed, show the same general pattern, as do other sets of paired data investigated. Clearly species are not, and are certainly not intended to be, random subsets of the family, but nonetheless we regard this hierarchal deterioration of "goodness of fit" as interesting.

Even when a trend is strongly expressed at a familial or generic level, it may be that the smaller the stratigraphic and geographic range, or narrower the ecologic range of a sample, the weaker the regression appears. To be perceptible, stochastic change (means increasing or decreasing, CVs approximately constant) requires a substantial range in some such dimension. As a corollary, slow change under little pressure would be poorly expressed. We are inclined to doubt the importance of gender as a source of variability for our samples, and hope to have eliminated ontogeny. Most of the taxa that we synonymize have very limited stratigraphic ranges. The taxa recognized in this report, despite our efforts to constrain them stratigraphically, generally are drawn from wider geographic, stratigraphic, and thus probably ecologic ranges. Taxa that we recognize at the specific level have metrical variabilities comparable to modern groups, but distinctly greater variabilities than taxa that we propose to replace (Table 1).

Table 3

Summary of upper tooth row ratios $(P^1-P^4/P^1-M^3)*$ as percentages

	N	Mean	Standard Deviation	Coefficient of Variation	Maximum	Minimum	SEM ¥ t(0.05, N-1)
?*Miniochoerus forsythae*	13	47.12	1.78	3.77	50.2	44.5	1.07
Miniochoerus							
M. chadronensis	42	45.42	1.22	2.68	48.1	42.9	0.38
M. affinis	64	45.94	1.56	3.40	51.7	42.5	0.39
M. gracilis	43	46.80	1.22	2.60	49.7	43.9	0.37
M. starkensis	20	44.04	1.34	3.04	46.4	40.6	0.63
Merycoidodon							
M. presidioensis	2	50.12	1.61	3.21	51.3	49.0	14.47
M. culbertsoni	179	49.67	1.35	2.71	55.5	46.2	0.20
M. bullatus	55	49.40	1.64	3.32	55.8	43.4	0.44
M. major	70	48.40	1.60	3.31	51.5	44.3	0.38
Mesoreodon chelonyx	9	48.78	0.80	1.64	49.9	47.5	0.62
?*Mesoreodon minor*	18	47.65	1.47	3.09	50.0	45.6	0.73
Dicotyles tajacu	38	42.39	1.62	3.83	46.96	39.51	0.53

$*$ P^2-P^4/P^2-M^3 for *Dicotyles tajacu*

Thorpe (1937a) was the first to recognize and make use of the variation in the amount of space the premolars occupy within the cheek tooth row, between the various oreodont genera. Thorpe expressed this as a percentage, as we do in Table 3 and the amended diagnoses. Phleger and Putnam (1942, p. 552) noted a decrease in this percentage in younger taxa, for species they regarded as *Merycoidodon* (which then included *Miniochoerus*). Table 3 shows a similar decrease for the species which we would place in the Merycoidodontinae, but a less clear trend for the Miniochoerinae. Schultz and Falkenbach (various papers) did not attribute much importance to this ratio. The terms unreduced (upper premolars occupy 48% or more of the cheek tooth row), little reduced, or reduced (upper premolars occupy less than 45% of the tooth row), are used in this report to describe the amount of space the four premolars occupy within the P^1-M^3 series. The ratio calculations provide data that are approximately normally distributed for well-represented species except *Merycoidodon bullatus* and *Miniochoerus affinis*.

Examination of P^1-P^4/P^1-M^3 ratio data structure by regression methods is not meaningful, but comparison of P^1-P^4 with M^1-M^3 produces information similar in most respects. Matrices of student's "t" tests for P^1-P^4/P^1-M^3 and P^1-P^4/M^1-M^3 ratios suggest closely similar associations of taxa for these measure of proportions, although the P^1-P^4/M^1-M^3 ratio is slightly less illustrative of stereotypy in the Merycoidodontinae. Variance of P^1-P^4 slightly dominates that of the M^1-M^3, but this occurs in only two subordinate groups,

Mesoreodon chelonyx (very small sample), and *Miniochoerus affinis*. Variance in M^1-M^3 is substantially dominant in most taxonomic groups (slopes of Reduced Major Axes, Table 9, bottom, suggest the nature of the differences). Least squares regression lines (P^1-P^4 on M^1-M^3) for genera are not significantly different from parallel, but r^2 values and slopes of the regression lines and the reduced major axes occur in descending order from *Miniochoerus* to *Merycoidodon* to *Mesoreodon*. The P^1-P^4/M^1-M^3 regression lines for *Merycoidodon*, *M. culbertsoni* and *M. bullatus* do not differ significantly, whereas the line for *M. major* is not significantly different from parallel to that for *Mesoreodon*. Species of *Miniochoerus*, except *M. gracilis*, produce regression lines that are generally approximately parallel to each other, but not to the trend for the genus (Table 9 and Fig. 6D).

In some oreodont genera, another adaptive feature is change in dental crown height. Table 4 shows this change as a ratio of APM^3 to Ht. M^3. Heights of the M3s are potentially valuable characteristics that appear to be intermediately variable, but wear is very critical to the measurements. To eliminate this source of ontogenetic bias, care was taken to measure the crown height of only unworn or very little worn M3s. Height of the M^3 is only adequately represented for *Merycoidodon*, and especially for *M. culbertsoni*. Averages for the APM^3/Ht. M^3 ratios of Miniochoerinae and most Merycoidodontinae are regarded as brachydont. The single measurable M^3 of *Mesoreodon chelonyx* at 1.53 times longer than tall and the low extremes of species

Table 4

Summary of the upper third molar length-to-height ratios (APM^3/Ht. M^3)

	N	Mean	Standard Deviation	Coefficient of Variation	Maximum	Minimum	SEM ¥ t(0.05, N-1)
?Miniochoerus forsythae	4	1.86	0.11	5.94	1.95	1.71	0.18
Miniochoerus							
M. chadronensis	16	1.85	0.14	7.74	2.25	1.67	0.08
M. affinis	15	1.89	0.14	7.37	2.19	1.66	0.08
M. gracilis	8	1.91	0.09	4.61	2.07	1.81	0.07
M. starkensis	5	2.07	0.23	11.29	2.46	1.84	0.29
Merycoidodon							
M. presidioensis	1	1.80	—	—	—	—	—
M. culbertsoni	68	1.72	0.13	7.47	2.02	1.50	0.03
M. bullatus	26	1.67	0.09	5.14	1.94	1.54	0.03
M. major	32	1.68	0.13	7.77	1.91	1.41	0.05
Mesoreodon chelonyx	1	1.53	—	—	—	—	—
?Mesoreodon minor	3	1.74	0.07	4.03	1.79	1.66	0.17
Dicotyles tajacu	6	1.62	0.18	11.11	1.89	1.40	0.19

of *Merycoidodon* might best be described as sub-mesodont. There is a small to moderate increase in APM^3/Ht. M^3 with time in *Miniochoerus*, and a decrease through time in this ratio for *Merycoidodon*. *Mesoreodon chelonyx* would appear to continue the trend, but a very small sample for *?Mesoreodon minor* does not do so. A matrix of student's "t" tests, supported by nonparametric tests, suggests stereotypy within *Merycoidodon* and *Miniochoerus* in this proportion, but separation of brachydont *Miniochoerus* from brachydont-submesodont *Merycoidodon*. Species of *Mesoreodon*, very poorly represented (Table 4), show no significant differences from other species discussed in this report, but more data are required to reach any conclusions about that group.

Primitive oreodont M^{3}s are wider transversely than long anteroposteriorly, but the molars become not only higher crowned but longer than wide in some species as the premolars become less dominant in the tooth row and more molarized. Table 5 summarizes APM^3/TM^3 ratios for the Merycoidodontinae and Miniochoerinae. The number of available measurements is more satisfactory for this ratio than for APM^3/Ht. M^3, but species of *Mesoreodon* are not well represented. Distributions of the ratios, with the exception of *Merycoidodon culbertsoni* and *?Mesoreodon minor*, depart significantly from the normal distribution.

There is a general increase with time in both APM^3 and TM^3 for species of *Merycoidodon*, although *M. presidioensis* is similar in both respects to *M. major*, and *M. culbertsoni* and *M. bullatus*, both smaller, show little change. Our data show slight but stratigraphically ordered increase in the APM^3/TM^3 ratio (Table 4),

although the only significant change in this stereotyped proportion is between *Merycoidodon* and *Mesoreodon*. Variation in APM^3 slightly dominates that in TM^3 for *Merycoidodon*, and its species. *Miniochoerus* and most of its species, including *?M. forsythae*, show nearly equal decrease in APM^3 and TM^3 with time, and little change in the ratio except for *M. starkensis*, which shows a significant increase in the ratio in comparisons with *M. affinis* and *M. chadronensis*. Although one regression line would be satisfactory for the two divergent genera (*Merycoidodon culbertsoni-bullatus* on one hand, and *Miniochoerus chadronensis-affinis* on the other), species groups that are hard to distinguish purely on the basis of measurements require separate least squares regression lines for APM^3 and TM^3 data. Relationships of the best represented Miniochoerinae are illustrated in Table 9 (upper part), and Figure 6B. "Goodness of fit" values and slopes of the least squares regression lines and reduced major axes occur in descending order from *Miniochoerus* to *Merycoidodon* to *Mesoreodon*.

"Height" of oreodont M^{3}s is an absolute measurement, but it is also part of a three-dimensional concept, proportional to both length (Table 4) and width. Table 6 completes a series by relating width of merycoidodontine and miniochoerine M^{3}s to height, and includes the peccary, *Dicotyles tajacu*, as a modern control. Close inspection of sparse and unevenly distributed data (142 of 756 specimens provide all three data; 74% of these are *Merycoidodon* specimens; Appendices B, C) suggests the possibility that, with the exception of *Mesoreodon*, width, proportionally, has low variability. For the Merycoidodontinae, there is a

Table 5

Summary of the upper third molar length-to-width ratios (APM^3/TM^3)

	N	Mean	Standard Deviation	Coefficient of Variation	Maximum	Minimum	SEM ¥ t(0.05, N-1)
?*Miniochoerus forsythae*	12	0.98	0.04	4.07	1.04	0.91	0.03
Miniochoerus							
M. chadronensis	25	1.01	0.05	4.67	1.12	0.93	0.02
M. affinis	40	1.00	0.05	5.20	1.09	0.88	0.02
M. gracilis	31	1.01	0.04	3.72	1.11	0.94	0.01
M. starkensis	16	1.05	0.03	3.07	1.10	0.99	0.02
Merycoidodon							
M. presidioensis	2	0.98	0.02	2.37	1.00	0.97	0.21
M. culbertsoni	173	0.99	0.04	4.54	1.15	0.86	0.01
M. bullatus	59	1.00	0.05	4.62	1.11	0.92	0.01
M. major	70	1.02	0.05	5.27	1.16	0.92	0.01
Mesoreodon chelonyx	7	1.13	0.05	4.67	1.20	1.05	0.05
?*Mesoreodon minor*	18	1.18	0.06	4.93	1.32	1.08	0.03
Dicotyles tajacu	40	1.16	0.06	5.56	1.27	1.01	0.02

progressive decrease in the width, relative to height, but for the Miniochoerinae, stereotyped in this respect, a less pronounced trend is for the reverse.

SYSTEMATIC PALEONTOLOGY
Class MAMMALIA
Order ARTIODACTYLA
Family MERYCOIDODONTIDAE

Merycoidodontidae Hay, 1902:665
Merycoidodontidae Hay: Thorpe, 1937a
Merycoidodontidae Hay: Schultz and Falkenbach, 1968

Type species — *Merycoidodon culbertsoni* Leidy, 1848, p. 47.

Included subfamilies — Oreonetinae Schultz and Falkenbach (1956); Leptaucheniinae Schultz and Falkenbach (1940; 1968); Miniochoerinae Schultz and Falkenbach (1956); Merycoidodontinae (Hay, 1902; Schultz and Falkenbach, 1968); Eporeodontinae Schultz and Falkenbach (1940, 1968); Promerycochoerinae (Schultz and Falkenbach (1940, 1949); Merycochoerinae Schultz and Falkenbach (1940), in part; Merychyinae Simpson (1945), Schultz and Falkenbach (1947); Ticholeptinae Schultz and Falkenbach (1940, 1941), in part; Phenacocoelinae Schultz and Falkenbach (1950), in part; and the undescribed subfamilies defined by *Ustatochoerus-Mediochoerus* (the Ustatochoerinae, new subfamily) and *Brachycrus* (the Brachycrurinae, new subfamily).

Amended diagnosis — Oreodonts are suiform, selenodont artiodactyls that differ from most Agriochoeridae by lack of a divided P^4 para-metacone, usual absence of molar labial ribs, the posterior arm of the protoselene usually directed medially and deeply into the median valley between the protocone and hypocone, usual absence of protoconules, and a usually closed postorbital. The Merycoidodontidae usually lack diastema (except those to receive the caniform teeth), have a full dental series of I3/3, C1/1, P4/4, and M3/3, an incisiform lower canine, caniform P$_1$, and molars with W-shaped ectolophs.

Oreodont crania primitively have rather steep premaxillae, usually short Subnasal L. and a strongly convex nasal profile, but the profile becomes progressively straighter or more concave, and the Subnasal L. increases in some genera as the nasal notch migrates posteriorly. One genus firmly coalesces the rostrum ventral to the nasal aperture which causes the external nares to migrate posteriorly, and three genera greatly shorten the length of the nasal bones. Oreodont skulls are usually mesocephalic but certain groups become more brachycephalic, and one group becomes quite dolichocephalic. The orbits are situated in the mid-cranial region, and are primitively located relatively low on the sides of the face. Facial vacuities are absent in primitive species but are characteristic of numerous derived genera. The zygoma are usually unspecialized but become massively upturned in several taxa. The nuchal crest is primitively narrow and protracted strongly posteriorly, but the crest becomes less protracted and considerably expanded laterally in some specialized forms and a sagittal table may develop. Some genera progressively either develop or reduce the depth of the preorbital fossae, and auditory bullae vary from small to notably enlarged. Oreodont lower jaws consistently have a greatly enlarged, rounded angular region that is not much expanded below the ventral border of the horizontal ramus.

Distribution—As currently known, from the later Duchesnean to the latest Clarendonian of North America, and the Hemingfordian of Central America; early late Eocene to late Miocene.

Discussion—There has been much speculation about the possible relationships of oreodonts to other artiodactyls. Although a discussion of the suprafamilial affinities of archaic artiodactyls is beyond the scope of this chapter, the Merycoidodontidae surely are related closely to the Agriochoeridae. Both share a full dental series, lack diastemas, have W-shaped molar ectolophs, a caniform P_1, and an enlarged, rounded angular region of the lower jaw which is not directed below the ventral margin of the horizontal ramus. These characteristics define a clade that unites them at a rank higher than family.

Most investigators have associated artiodactyl selenodonty with rumination, undoubtedly because all living selenodont artiodactyls ruminate, and because all living suiform (bunodont) artiodactyls do not. The Merycoidodontidae and Agriochoeridae are selenodont, yet suiform artiodactyls. Selenodonty, however, has developed independently in other mammalian herbivores such as the Perissodactyla, Notoungulata, and others, undoubtedly to increase the efficiency of teeth, structures responsible for the primary breakdown of vegetation. Even within the Artiodactyla, selenodonty probably has occurred independently within some of the families as evidenced by the fact that crests and ridges which produce the selenodonty do not develop in the same way or from precisely the same structures (Wilson, 1974, pp. 4-5). Rumination developed as a means of increasing the efficiency of the digestive system, responsible for the further breakdown of cellulose. The fact that the Merycoidodontidae were in existence for 30 million years side-by-side with probable ruminants certainly suggests that they were not in direct competition with them, and suggests indirectly that oreodonts had different feeding strategies. Although oreodonts historically have been called "ruminating hogs," we should at least entertain the notion that they may not have ruminated (see Janis, 1989), and we regard the inclusion of oreodonts among ruminants as unsubstantiated.

Traditionally, the Merycoidodontidae have been thought to have had a monophyletic origin from within *Protoreodon*. *Protoreodon* was often included in the Merycoidodontidae (see Thorpe, 1937a; Scott, 1945; Simpson, 1945). Peterson (1919), however, suggested that *Protoreodon* is an agriochoerid, a view upheld by Gazin (1955). In spite of the fact that knowledge about oreodont origins has been eagerly sought, the fossil record is not very informative. Gazin concluded that all known species of *Protoreodon* are more or less unsuitable direct ancestors, but among them, *P. minor* and *P. petersoni* seem less unlikely than the others. Schultz

and Falkenbach (1968, p. 398) also believed that oreodonts originated from within the Agriochoeridae.

Wilson (1971, p. 9) regards the Merycoidodontidae as polyphyletic, as descended from two lineages of *Protoreodon*. Wilson would derive the Oreonetinae *fide* Schultz and Falkenbach (1956) and Wilson (1971) from the *P. minor-P. petersoni* species group, small protoreodonts with an undivided P^4 para-metacone. Wilson thought that the larger, early Merycoidodontinae and the "Desmatochoerinae" Schultz and Falkenbach (1954), were derived from a lineage of large protoreodonts of the *P. pumilus* species group, also with an undivided P^4 para-metacone. Most recently Lander (1978) argued that the Merycoidodontidae ("Oreodonta") are monophyletic and descended directly from *P. petersoni*.

Early attempts to clarify the origin of the Merycoidodontidae were in part influenced by the assumption that the morphology of *Merycoidodon culbertsoni* Leidy, on which the Merycoidodontidae is based, represents the primitive condition for the family. The very archaic oreodonts reported by Wilson (1971) from the Vieja, Trans-Pecos Texas, provide information which more accurately defines the kind of *Protoreodon*-like (not *Merycoidodon*-like) morphology that is primitive for the family.

Vieja oreodonts often have remnants of protoconules, incompletely formed protoselenes, protoreodont-like P4s, and often incompletely closed postorbital bars. The main obstacle in deriving the Merycoidodontidae from the known species of *Protoreodon* lies in the observation that the oldest Vieja oreodonts are already notably differentiated into at least three subfamilies: Oreonetinae, Leptaucheniinae (if *Limnenetes* Douglass, a probable leptauchenine, is included), Merycoidodontinae, and *Aclistomycter*, in spite of their Duchesnean and/or early Chadronian age. It was this differentiation which caused Wilson (1971) to regard the Merycoidodontidae as polyphyletic. There is not a single suite of morphological features common to the pioneering oreodonts except that all have a full dentition which lacks diastema, W-shaped ectolophs, caniform P_1, an undivided P^4 para-metacone, a lower jaw with an enlarged angle that does not extend much below the ventral border of the horizontal ramus, and when known, suiform unguals. These characteristics serve to define a clade which unites the Merycoidodontidae.

If the Merycoidodontidae is monophyletic, its origin has to be more remote than the late Duchesnean or earliest Chadronian in order to accommodate the adaptive radiation seen in the Oreonetinae (*Bathygenys* and *Oreonetes*), Leptaucheniinae (*Limnenetes*), Merycoidodontinae (*Merycoidodon*), and in *Aclistomycter*. A remote origin for the Merycoidodontidae would be, incidentally, consistent with the suggestion that oreodonts were not ruminants. Although this chapter does not specifically address the problem of the origin of the Merycoidodontidae, the family probably is derived from a very primitive, non-ruminating early agriochoerid. This view is consistent with that of Gazin (1955), partially consistent with that of Wilson (1971), but conflicts with Lander (1978). Whether the Merycoidodontidae is regarded as polyphyletic or monophyletic depends in part on temporal perspective and taxonomy. With time, the Agriochoeridae became or remained semiarboreal browsers, and the Merycoidodontidae differentiated into low level, terrestrial, stream border, semiaquatic, savanna, or upland browsers, or ?grazers, depending on time and place.

The Merycoidodontidae experienced two major episodes of adaptive radiation during their 30-million-year history. The first radiation must have occurred before the late Eocene in order to accommodate the differentiation seen by the Chadronian, and the second major episode occurred during the Whitneyan-Arikareean, middle and late Oligocene. During the Miocene several minor differentiations also occurred. The prolonged biological success of oreodonts shows that they were not in direct competition with taller, speedier, ruminating artiodactyls until at least the times of termination of most of the primitive selenodont lineages, including oreodonts, discussed by Janis (1989). Oreodont crania and teeth are superficially similar to hyraxes, and oreodonts may have had much the same ecology.

Subfamily MERYCOIDODONTINAE

Merycoidodontinae Hay, 1902:665
Merycoidodontinae Hay: Schultz and Falkenbach, 1968:24
Promerycochoerinae Schultz and Falkenbach, 1949:140, in part
Desmatochoerinae Schultz and Falkenbach, 1954:163, in part

Type species — *Merycoidodon culbertsoni* Leidy, 1848, p. 24.

Included genera — *Merycoidodon* and *Mesoreodon.*

Amended diagnosis — Medium-sized oreodonts primitively with narrow, high, and arched muzzles, steep premaxillae, Subnasal L./Nasal L. ratios averaging from 53% to 47%, deep preorbital fossae, infraorbital foramina above P^2, and small auditory bullae each with a deep hyoidal groove. Later species have lower and broader muzzles, more inclined premaxillae, smaller preorbital fossae, infraorbital foramina above P^4, broad frontals, and progressively more inflated auditory bullae which lose the hyoidal groove. Dentally, the Merycoidodontinae remained conservative throughout their history, with unreduced premolars 50-48% of the P^1-P^4/P^1-M^3 ratio (rounded averages; Table 3), an APM^3/TM^3 ratio that is primitively 1.0 or

slightly less but increases to 1.18 in the youngest species (averages; Table 5), and with persistently brachydont-submesodont dentitions where the APM^3/Ht. M^3 ratio averages 1.80 to 1.53 (Table 4). Over time, the TM^3/Ht. M^3 ratio progressively decreases from 1.80 to 1.36 (Table 6). The Merycoidodontinae differ from Miniochoerinae by progressively larger size, lack of thinned enamel on the premolar-molar middle crests and selenes, and by inflated auditory bullae in species younger than earlier Orellan; and differ by grade only from Eporeodontinae by a usually larger size (species younger than Orellan), an arched, instead of a concave nasal profile, and complete absence of facial vacuities.

Distribution — Chambers and Capote Mountain Tuff formations, Vieja Group, Presidio County, and undifferentiated Vieja Group, Jeff Davis County, Texas; Chadron and Brule formations, White River Group or its equivalents, Montana, Colorado, Nebraska, Wyoming, North and South Dakota; John Day Formation, Oregon; Fort Logan and Cabbage Patch formations and "Canyon Ferry beds," Montana; and the Gering Formation of Wyoming and Nebraska, or its equivalents elsewhere; late early Chadronian to early Arikareean, late Eocene to late middle Oligocene.

Discussion — Early authors who studied oreodonts that we consider members of the Merycoidodontinae identified them either as *Merycoidodon,* "*Oreodon,*" "*Cotylops,*" *Eporeodon,* or "*Eucrotaphus.*" Most investigators subsequent to and including Hay (1902) regarded "*Oreodon*" and "*Cotylops*" as congeneric with *Merycoidodon* Leidy (1848), because these names are based on species deemed conspecific within *M. culbertsoni,* the type species of *Merycoidodon.* Allocation of Merycoidodontinae to *Eporeodon* stems from Marsh's (1875) suggestion that the oreodonts of *Merycoidodon* aspect that have enlarged auditory bullae should be separated generically from the *M. culbertsoni* with minute bullae. Although this may have been reasonable, Marsh selected as the type species a taxon from the John Day Formation that is more derived than the merycoidodonts with large bullae from the White River Group that are best included in the Merycoidodontinae. Scott (1890), and later Schultz and Falkenbach (1968) thought it best to reserve the name *Eporeodon* for certain John Day species, and this is followed here.

Cope (1884a) considered Leidy's (1850) taxon "*Eucrotaphys jacksoni*" to be an oreodont (which it may or may not be), and by so doing, Cope could deny Marsh (1875) valid authorship of the name *Eporeodon.* Cope (1884a) used the name "*Eucrotaphus*" in the same sense that Marsh had intended *Eporeodon,* to distinguish oreodonts of *Merycoidodon* aspect that have well inflated auditory bullae from *M. culbertsoni.* The type, however, is a braincase from the John Day Formation that cannot be distinguished from *Eporeodon* or *Agriochoerus;* therefore "*Eucrotaphus jacksoni*" is

Table 6

Summary of M^3 width to height (TM^3/Ht. M^3) ratios

	N	Mean	Standard Deviation	Coefficient of Variation	Maximum	Minimum	SEM ¥ $t_{(0.05, N-1)}$
?*Miniochoerus forsythae*	4	1.89	0.15	8.06	1.68	2.04	0.24
Miniochoerus							
M. chadronensis	8	1.82	0.11	5.90	1.71	2.05	0.09
M. affinis	10	1.85	0.12	6.59	1.71	2.09	0.09
M. gracilis	6	1.88	0.04	2.24	1.82	1.95	0.04
M. starkensis	5	1.95	0.20	9.98	1.83	2.29	0.24
Merycoidodon							
M. presidioensis	1	1.8	—	—	—	—	—
M. culbertsoni	59	1.75	0.13	7.31	1.48	2.01	0.03
M. bullatus	18	1.66	0.06	3.86	1.54	1.78	0.03
M. major	29	1.63	0.12	7.29	1.37	1.84	0.05
Mesoreodon chelonyx	1	1.46	0.00	0.00	1.46	1.46	0.00
?*Mesoreodon minor*	3	1.36	0.05	3.38	1.33	1.41	0.11
Dicotyles tajacu	6	1.38	0.17	12.38	1.18	1.62	0.18

unidentifiable at the family level. Most later authors have regarded *"Eucrotaphus"* as a *nomen dubium*. Because the morphologic differences between the *M. culbertsoni*, with small bullae, and other slightly more derived members of the Merycoidodontinae with inflated bullae are not great, the taxonomic separation between them at a rank higher than species can be handled best by using subgenera.

Schultz and Falkenbach (1968) introduced chaos into the Merycoidodontinae by splitting them into numerous genera, subgenera, and species, or by referring some of them to other subfamilies. Schultz and Falkenbach gave taxonomic importance to variations in skull shape that are the product of crushing. They (1968) established different genera for each of their presumed "brachycephalic," mesocephalic, and "dolichocephalic" lineages, or tended to assign a new generic name to notably robust or otherwise "unusual" specimens. Further, Schultz and Falkenbach failed to attribute enough size variation to the fossil populations in spite of the fact that they attempted to allow for such variation.

Schultz and Falkenbach (1949) allocated the so-called "brachycephalic" Merycoidodontinae to the Promerycochoerinae as *"Promesoreodon,"* a genus they believed was ancestral to *Mesoreodon* of the Arikareean. Schultz and Falkenbach (1968) identified certain dorsoventrally crushed skulls that are slightly smaller but otherwise similar to *"Promesoreodon,"* as *"Otionohyus"* if auditory bullae were small, or as *"O. (Otarohyus)"* if bullae were inflated. Schultz and

Falkenbach (1968) consistently identified essentially uncrushed (hence mesocephalic) larger Merycoidodontinae as *Merycoidodon* if bullae are small, or as *M. (Anomerycoidodon)* if the specimens have inflated bullae. They identified uncrushed, smaller Merycoidodontinae as *"Genetochoerus"* if without inflated bullae, or as *"G. (Osbornohyus)"* if bullae are inflated. Presumed "dolichocephalic" large Merycoidodontinae Schultz and Falkenbach (1954) referred to the "Desmatochoerinae" as *"Prodesmatochoerus"* or *"Subdesmatochoerus,"* genera erroneously presumed to be sequentially ancestral to *Desmatochoerus* of the Arikareean.

Careful consideration of measurement data for crania that elsewhere have been referred to presumed "dolichocephalic" or "brachycephalic" species and which here are either retained or placed within the Merycoidodontinae, shows that the crania are remarkably similar in proportions to the type species of the subfamily, *Merycoidodon culbertsoni* (see hypodigms for the species and Table 7). *M. culbertsoni* was defined by Schultz and Falkenbach (1968) as mesocephalic. Table 7 shows that Miniochoerinae are as mesocephalic as the Merycoidodontinae, and show similar patterns of change through time in that respect.

Schultz and Falkenbach (1968) regarded what appeared to them to be unusually robust Merycoidodontinae as *"Paramerycoidodon"* if the specimens have small auditory bullae, or as *"P. (Gregoryochoerus)"* if they have inflated bullae. Other specimens that are slightly larger still but which have inflated

Table 7
Summary of Skull Width to Skull Length ratios (Sk. W./Sk. L.) as percentages

	N	Mean	Standard Deviation	Coefficient of Variation	Maximum	Minimum	SEM ¥ $t_{(0.05, N-1)}$
?*Miniochoerus forsythae*	4	59.16	2.01	3.40	56.4	61.2	3.20
Miniochoerus							
M. chadronensis	14	61.67	3.44	5.58	56.0	67.0	1.99
M. affinis	13	63.42	4.07	6.41	57.0	69.2	2.46
M. gracilis	8	65.22	2.95	4.52	60.3	69.1	2.47
M. starkensis	4	64.42	8.45	13.11	52.5	70.8	13.44
Merycoidodon							
M. presidioensis	0	—	—	—	—	—	—
M. culbertsoni	42	61.45	3.54	5.75	53.0	67.9	1.10
M. bullatus	6	62.34	2.16	3.46	60.1	66.3	2.27
M. major	17	65.46	5.16	7.89	57.7	77.2	2.66
Mesoreodon chelonyx	4	62.12	2.42	3.90	59.3	64.6	3.85
?*Mesoreodon minor*	4	72.34	8.53	11.79	67.1	85.0	13.58
Dicotyles tajacu	37	52.12	3.22	6.17	46.2	58.7	1.07

bullae were called "*P.* (*Barbourochoerus*)." Schultz and Falkenbach (1968) reserved the name *Merycoidodon* "(*Blickohyus*)" for mesocephalic Merycoidodontinae that have inflated bullae and a seemingly foreshortened basicranium. Additionally, Schultz and Falkenbach (1968) included in the Merycoidodontinae, as "*Pseudogenetochoerus*" or "*Epigenetochoerus*," several species that are best placed in the Eporeodontinae.

Schultz and Falkenbach (1949) referred *Mesoreodon* Scott (1893) to the Promerycochoerinae as the ancestral taxon. Our revision concludes that the mesocephaly and morphologic similarity of the type species, *Mesoreodon chelonyx*, to *Merycoidodon* warrants the placement of *Mesoreodon* in the Merycoidodontinae.

The Merycoidodontinae remained remarkably stereotyped in morphology and proportions during their approximate 11-million-year history. Their stereotypy coincides remarkably well with the relatively stable ecosystem alluded to by Emry et al. (1987, p. 121) for the faunas of the classic White River Group. The only osteological innovations were the progressive enlargement of the auditory bullae, inflation of the frontals, decrease in the size and depth of the preorbital fossae, the slight posterior migration of the infraorbital foramina, slight reduction in the dominance of the premolars within the cheek tooth row (Table 3), slight narrowing and increase in the crown height of the M3s (Table 4), slight cranial broadening (Table 7), and a generally progressive increase in size of species younger than the late Chadronian (Appendix B). Size increase is sufficient to preclude close similarity of measurements between stratigraphically or taxonomically distant species, but proportions, particularly P^1-

P^4/P^1-M^3 (Table 3) and Sk. W./Sk. L. (Table 7), show little change. As an exception, the APM_3/APM^3 ratio, while generally stereotyped, separates *Mesoreodon* from the Merycoidodontinae except *Merycoidodon major*, and suggests change in this proportion for later Merycoidodontinae. The Merycoidodontinae are ancestral to the Promerycochoerinae, the Merycochoerinae, and the Eporeodontinae (Fig. 7).

Merycoidodon Leidy

Type species — *Merycoidodon culbertsonii* Leidy, 1848, p. 47.

Included species — *Merycoidodon presidioensis* sp. nov., *M. culbertsoni*, *M. bullatus*, and *M. major*.

Amended diagnosis — Medium-sized oreodonts with steep premaxillae and arched nasal profiles; infraorbital foramen primitively located above P^2 that migrates to above P^4 in later species; the preorbital fossae, enlarged and deep in early species, becomes more pit-like and shallower in late examples; the zygomatic arch is unspecialized; the braincase is elongated with a narrow, protracted nuchal crest; the auditory bullae are initially small but become more inflated and lose their hyoidal grooves in later species; the premolars are unreduced with P^1-P^4/P^1-M^3 ratio from 50% to 48% (rounded averages; Table 3); the average APM^3/TM^3 ratios are near 1.00 (Table 5); M^3s are brachydont with $APM^3/Ht.$ M^3 averages from 1.80 to 1.67 (Table 4) though low extremes (1.4-1.5) are submesodont, and $TM^3/Ht.$ M^3 ratios progressively decrease (Table 6) from 1.80 (oldest) to 1.63 (youngest), as M^3 becomes higher crowned relative to the width. Of pertinent genera, *Merycoidodon* differs from *Miniochoerus* Schultz and Falkenbach (1956) by a progressively larger size and lack of thinned enamel on the premolar-molar middle crests and selenes. *Merycoidodon* differs from *Mesoreodon* Scott (1893) in its smaller size, with slightly less reduced premolars, significantly broader molars in proportion to height (Table 6), APM$_3$ 22-23% longer than APM^3 (compared to 14% for *Mesoreodon*; *Miniochoerus* resembles *Merycoidodon* in this stereotypical aspect), and a more arched nasal profile. *Merycoidodon* differs from *Eporeodon* by its slightly larger size, relatively smaller auditory bullae, and higher, more convex nasal profile and broader frontals.

Distribution—Late early or middle Chadronian to the late Whitneyan, late Eocene to middle Oligocene.

Discussion—Early students of *Merycoidodon* tended to identify them either as *Merycoidodon*, "*Oreodon*," "*Cotylops*," *Eporeodon*, or "*Eucrotaphus*" (see discussion above). Thorpe (1937a) followed Marsh's (1875) lead, as did Scott (1940), and placed the White River oreodonts of *Merycoidodon* aspect with large bullae in *Eporeodon*. Thorpe (1937a), Scott (1940), and others, however, also included in *Merycoidodon* certain species that are now placed in *Miniochoerus*, because Miniochoerinae, a sister-group, generally resemble *Merycoidodon* cranially, and have a small auditory bulla that is similar to that of *M. culbertsoni*.

Schultz and Falkenbach (1968) divided animals that are best called *Merycoidodon* between numerous genera, subgenera, and species, but the types that define this taxonomy usually do not withstand scrutiny. Much of this chapter will deal with the sorting out of this taxonomy. Lander (1978) viewed *Merycoidodon* as a "*nomen vanum*" because it is "based on an incomplete species" [*sic*—he means specimens] and Lander resurrected the junior name "*Oreodon*," hence once again used the name "Oreodonta" for the family. Genera, however, are based on species, not specimens, and the types which define most fossil species consist of only minor parts of the original organism. Although Leidy's type specimen is poor, he designated enough cotypes of *M. culbertsoni* to justify the genus. *Merycoidodon* is the valid generic name for this and other comparable taxa, and we recommend abandonment of the obsolete term "Oreodonta."

Merycoidodon (*Merycoidodon*), new combination

Type species—*Merycoidodon* (*Merycoidodon*) *culbertsoni* (Leidy), 1848, p. 47.

Included species—*Merycoidodon* (*?Merycoidodon*) *presidioensis* sp. nov., and *M.* (*M.*) *culbertsoni*.

Diagnosis—The cranial characterization of the subgenus is the same as for the genus as defined by

Merycoidodon culbertsoni, a species with a small auditory bulla that has a deep hyoidal groove (see Thorpe, 1937a, fig. 7A). It differs from *M.* (*Otarohyus*) by a small auditory bulla with a deep hyoidal groove.

Distribution—Late early or middle Chadronian through the early Orellan, late Eocene to early Oligocene.

Discussion—When phylogenetic significance of the morphology of the auditory bulla in fossil artiodactyls is considered, we should emphasize the average condition within the given species because the intraspecific variation in size and shape of the bulla is known to vary considerably in living suiform, and other artiodactyls (see Introduction). Oreodonts were probably no less variable. There are, however, more or less predictable stratigraphic trends in the bullae of the Merycoidodontinae.

As pointed out by Schultz and Falkenbach (1968), it is useful to be able to distinguish taxonomically the primitive from a more derived stage at a rank higher than species, if the oreodonts are placed in the same genus. Schultz and Falkenbach (1968) indicated the appearance of Merycoidodontinae with well inflated auditory bullae by the use of subgenera. However, they never defined the primitive condition. It is proposed here that *Merycoidodon* with small auditory bullae with deep hyoidal grooves shall be called *Merycoidodon* (*Merycoidodon*).

Merycoidodon (*?Merycoidodon*) *presidioensis* sp. nov.
Figures 2A, 3A; Tables 3-6; Appendix B

?Prodesmatochoerus cf. *meekae* Schultz and Falkenbach: Wilson, 1971:42
?Prodesmatochoerus cf. *meekae* Schultz and Falkenbach: Wilson, 1978:22
Merycoidodon lewisi Clark and Beerbower: Emry, 1992:110

Etymology—*presidioensis*, from Presidio County, Texas.

Type—TMM 40505-2, the anterior part of a skull and lower jaws from the "Valley of 10,000 skulls locality," lower middle part of the Capote Mountain Tuff Formation, Airstrip Local Fauna, Presidio County, Texas.

Hypodigm—Type, and possibly USNM 244306, from Dry Hole Quarry, Flagstaff Rim area below ash "B," and an unnumbered Frick specimen from 45 feet (13.7 m) below ash "B," Flagstaff Rim area, Natrona County, Wyoming.

Diagnosis—The cranial characterization is the same as for the genus, as the most primitive known species. The Subnasal L. is 53% of the Nasal L., premolars are unreduced with a mean P^1-P^4/P^1-M^3 of 50% (Table 3), values for the APM^3/TM^3 and APM^3/Ht. M^3 ratios are 0.98 (mean) and 1.80 (sole) respectively (Tables 4, 5),

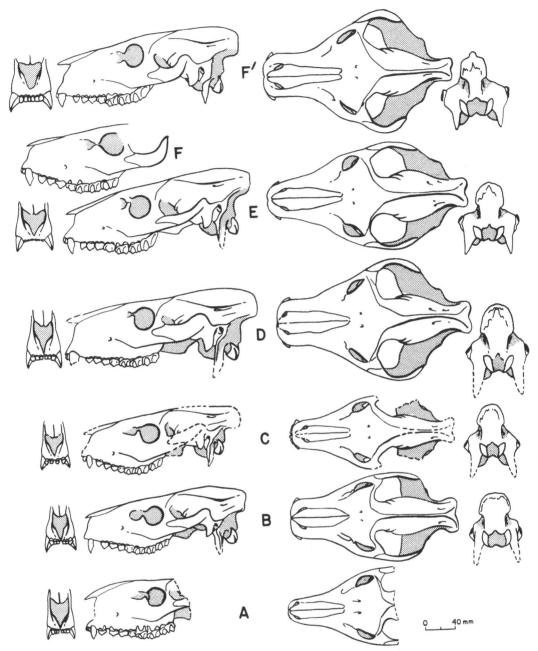

Figure 3. Selected, somewhat schematic representations of crania of species of *Merycoidodon* and *Mesoreodon*. (A), *Merycoidodon* (?*Merycoidodon*) *presidioensis* sp. nov., three views of type, TMM 40505-2, middle Chadronian, Capote Mountain Tuff, Presidio County, Texas; (B), *M.* (*Merycoidodon*) *culbertsoni*, later Chadronian to earlier Orellan, anterior view of F:AM 45155, "lower oreodon beds," Pennington County, South Dakota, and three views of AMNH 12319, "lower oreodon beds," South Dakota (nasals restored from F:AM 72150); (C), *M.* (*Otarohyus*) *bullatus*, later Orellan to earlier Whitneyan, four views of F:AM 45267, "upper nodules," "middle oreodon beds," Shannon County, South Dakota, as slightly restored from F:AM 7200, from ?equivalent rocks in Stark County, North Dakota ; (D), *M.* (*O.*) *major*, middle and late Whitneyan, anterior view of F:AM 45307, and three views of F:AM 45298, both from the "lower *Leptauchenia* beds," Washabaugh County, South Dakota; (E), *Mesoreodon chelonyx*, latest Whitneyan-earliest Arikareean, three views of PU 10418 (teeth restored from PU 11769), and occipital view of PU 10410, Fort Logan Formation, Meagher County, Montana; (F), ?*Mesoreodon minor*, early Arikareean, type, MCZ 769, adapted from Douglass (1903), reversed, from east of Drummond, Granite County, Montana ; and (F'), four views of F:AM 45430 (identified as *M.* "*cheeki*" by Schultz and Falkenbach, 1949), from Goshen County, Wyoming.

and TM^3/Ht. M^3 is 1.80 (sole; Table 6). *M. presidioensis* differs from *M. culbertsoni* by a narrower, deeper, and more convex nasal profile, steeper premaxillae, longer face, a more anterior infraorbital foramen, slightly thinned enamel on the middle surfaces of the premolar-molar selenes, and a significantly larger size [Subnasal L., tooth row and tooth measurements, except APM^3, Ht. M3s; 11 of 15 measurement categories available for comparison for *M. presidioensis* (Appendix B) lie within observed ranges of those of *M. culbertsoni*; exceptions: P1-M3, M1-M3, APM3, TM3] as known from a very small sample which does not include a measurable Sk. L.

Distribution—Lower middle part of the Capote Mountain Tuff Formation, Presidio County, Texas; tentatively identified from the basal part of the White River Formation, Flagstaff Rim area, Natrona County, Wyoming; questionably from oreodont faunal zone Chadron "A," and from Chadron "B" of the northern Plains; late early or middle Chadronian, late Eocene.

Description—The type of *Merycoidodon presidioensis* is a well-preserved skull (Fig. 3A) without the brain case, associated with lower jaws that lack the posterior part of the ascending rami. The dentition is slightly worn but is well-preserved and morphologically similar to that of other *Merycoidodon*, although the premolars are less reduced, the middle surfaces of the premolar-molar selenes have enamel that is slightly thinner than that on the external parts, and a short diastema (of dubious significance) occurs between P^1 and P^2. The skull has a very narrow face, a strongly convex nasal profile, a steep nasal aperture, a moderately deep preorbital fossa, incisors that extend in a rounded arcade ahead of the canines, and an infraorbital foramen that is situated above the posterior part of P^2 (Fig. 3A, also see Wilson, 1971, figs. 29, 30).

Discussion—The type specimen of *Merycoidodon presidioensis* is most notable for its facial narrowness. Most oreodont specimens that are said to have narrow rostra owe their facial narrowness to lateral crushing. Invariably when oreodont skulls are crushed laterally, the rostrum becomes high and narrow, the palate either becomes gable-shaped ("pinched" upward), or the palate ruptures and the broken bones shingle over one another, and the zygomatic arches become offset, tilted upward, or broken. The lack of fractures and the essentially horizontal palate of the Capote specimen show that the facial narrowness of this type is not due to deformation. Facial narrowness, however, does not necessarily imply that the cranium of *M. presidioensis* was dolichocephalic because the bizygomatic diameter, the measure of oreodont skull width, for TMM 40502-2 is unknown.

The type came from approximately 570 feet (173.7 m) above the Bracks Rhyolite which underlies the Capote Mountain Tuff Formation, and about 1,000 feet (305 m) below the overlying Mitchell Mesa Rhyolite (=

"Brite Ignimbrite," a junior name). Although the radiochronology of the Buckshot, which occurs below the Bracks, and the Bracks lava has been in flux because the first obtained radiometric age determinations violated superposition (see Wilson et al., 1968), recent compilations which utilize more accurately determined ^{40}K/^{40}Ar ratio decay constants, provide ages of about 36.2-37.7 Ma for the Buckshot Ignimbrite and about 37.5 to 37.9 Ma for the Bracks Rhyolite (Henry et al., 1986). Prothero and Swisher (1992, p. 50) note that work in progress by Swisher and Henry which utilizes ^{40}Ar/^{39}Ar ratios from single crystals of sanidine from the Bracks Rhyolite and Buckshot Ignimbrite give an average age of 36.67 Ma for the Bracks, and 37.80 Ma for the Buckshot. The proximity of these age determinations implies that the Buckshot Ignimbrite, the intervening Chambers Tuff Formation, and the Bracks Ignimbrite are bracketed at about 37-38 Ma.

A series of 18 radiometric age determinations for the Mitchell Mesa Rhyolite give an age of about 32.3 Ma for that ignimbrite (Henry et al., 1986). The type of *M. presidioensis* is bracketed at 37-32.3 Ma. Based on the distinctly primitive morphology of *M. presidioensis* and low stratigraphic position within the Capote Mountain Tuff, the actual age of *M. presidioensis* is probably nearer the older age determination.

Wilson (1971), working with the literature and the type of "*Prodesmatochoerus meekae*," identified TMM 40505-2 as "*P. cf. meekae*," because of the facial narrowness. The type of "*P. meekae*" Schultz and Falkenbach (1968) from the lowest part, Orella Member, Brule Formation, Seaman Hills, Wyoming, however, is severely laterally crushed as shown by numerous fractures and the pinched palate, as are most of the referred specimens. Furthermore, the rostrum of the type of "*P. meekae*" does not closely resemble that of *Merycoidodon presidioensis*; the premaxillae are more inclined, the infraorbital foramen is located more posteriorly, and the premolars are more reduced. In size, *M. presidioensis* resembles the type of "*P. meekae*," one of the larger specimens that we refer to *M. culbertsoni*. *M. presidioensis* is the most primitive *Merycoidodon* of which we have adequate knowledge, and its morphology probably defines the features that are primitive for the Merycoidodontinae. Although the basicranium is unknown, we would expect the species to have small auditory bullae in keeping with its antiquity relative to *M. culbertsoni*. The ancestry of *M. presidioensis* is not known.

There apparently existed during the middle Chadronian, and earlier, a lineage of medium-sized, dentally *Merycoidodon*-like oreodonts for which we have little information. Clark and Beerbower (Clark et al., 1967, p. 55) report CMNH 9391 from the early Chadronian Ahearn Member of the Chadron Formation. The specimen may belong with this group. Wilson (1971) reports as "Merycoidodontidae genus and species indeterminate, no. 1," a specimen of *Merycoididon* aspect, FMNH PM 153, from very low in the upper Duchesnean Chambers Tuff Formation, which underlies the Capote Mountain Tuff Formation. FMNH PM 153, however, is a maxilla fragment with DP^4, and provides little information about these very ancient and elusive oreodonts. USNM 244306, a well-preserved lower jaw of *Merycoididon* aspect, came from Dry Hole Quarry located 110 feet (33.5 m) below ash "B," low in the White River Formation of the Flagstaff Rim area, Natrona County, Wyoming. K/Ar age determinations for Ash "B" are 33.3 and 35.2 Ma. (Evernden et al., 1964; see Emry, 1973). Swisher and Prothero (1990) obtained a $^{40}Ar/^{39}Ar$ age determination of 35.9 Ma for biotite in Ash "B." Recent $^{40}Ar/^{39}Ar$ ratios on crystals of sanidine produced a slightly older average age determination of about 36 Ma for Ash B (Prothero and Swisher, 1992, p. 72). USNM 244306 is morphologically indistinguishable from the jaw of the type of *M. presidioensis*, but whether or not this fossil (and the unnumbered Frick specimen also from below ash "B"; see hypodigm) properly belong with this species cannot be verified without cranial material from the Flagstaff Rim area. Emry (1992, p. 110) identified this *Merycoididon* as *M. lewisi*, in part because of its relatively large size. The early Chadronian age at 36+ Ma, however, for USNM 244306, probably mitigates against its referral to a late Chadronian population (see *M. culbertsoni*, this report). Because USNM 244306 and the unnumbered Frick specimen are only tentatively referred to *M. presidioensis*, only the type is shown in Figure 2.

Merycoididon (Merycoididon) culbertsoni (Leidy)
Figures 2B, 3B; Tables 1-7; Appendix B

Merycoididon culbertsonii Leidy, 1848:48
Merycoididon culbertsonii Leidy: Schultz and Falkenbach, 1968:38
Merycoididon culbertsoni Leidy: Evanoff et al., 1992:123
Merycoididon culbertsonii browni Schultz and Falkenbach, 1968:55
Merycoididon culbertsonii osborni Schultz and Falkenbach, 1968:57
Oreodon culbertsonii periculorum Cope, 1884a:513
Genetochoerus periculorum (Cope): Schultz and Falkenbach, 1968:136
Oreodon macrorhinus Douglass, 1903:163
Merycoididon macrorhinus (Douglass): Douglass, 1907a:821
Merycoididon macrorhinus (Douglass): Schultz and Falkenbach, 1968:72
Prodesmatochoerus meekae Schultz and Falkenbach, 1968:72
Merycoididon georgei Schultz and Falkenbach, 1968:86
Otionohyus wardi Schultz and Falkenbach, 1968:107

Otionohyus vanderpooli Schultz and Falkenbach, 1968:115
Otionohyus (Otarohyus) bullatus (Leidy): Schultz and Falkenbach, 1968:118, in part
Genetochoerus (Osbornohyus) norbeckensis Schultz and Falkenbach, 1968:143, in part
Merycoididon lewisi Clark and Beerbower, *in* Clark et al., 1967:53
Merycoididon, Harris, 1967b:3
Merycoididodontidae, genus and species indeterminate, no. 2, Wilson, 1971:46
Merycioidodontid, gen. et sp. indet., Wilson, 1978:23
Merycoididon culbertsoni Leidy: Prothero, 1982b:406, in part

Cotypes—ANSP 10727, a partial jaw with M_1-M_3, and ANSP 10728, a maxilla with M^2-M^3, from the Big Badlands, South Dakota.

Hypodigm—Cotypes, and Schultz and Falkenbach's (1968) sample of this species such as AMNH 9793, F:AM 45240, 72150, 72154, 45234, 45155, 39461, and 45160, from "lower oreodon beds"; type, F:AM 72286, and Schultz and Falkenbach's (1968) sample of *M. culbertsoni* "browni," such as F:AM 72287, 72288, and 49765, from just below the "upper purplish white layer," Seaman Hills, Wyoming; Schultz and Falkenbach's (1968) type, F:AM 49668, and sample of *Merycoididon culbertsoni* "osborni," such as AMNH 595, 604, 609, 610, 39002, 39425, 39426, 39414, 39430, F:AM 72269, 72262, 45070, and 45217A, from the "lower oreodon beds," South Dakota, F:AM 72246A, 72246B, 49776, 49777, 49778, 49779, 49780, 49782, 49784, 45084, and 49646, from about 25 feet (7.6 m) above the "upper purplish white layer," to just below the "channel," Seaman Hills, Niobrara County, Wyoming, and UNSM 28077, 28104, 28283, 28150, and F:AM 72238 and F:AM 72239 from the lower part of the Orella Member, Brule Formation, Sioux County, Nebraska; type, AMNH 6397, Logan County, Colorado, and Schultz and Falkenbach's (1968) sample of "*Genetochoerus periculorum*," such as UNSM 28337, 28338, 28475, 28085, 28168, 28133, 28073, AMNH 1291, and F:AM 49730, from the lower part of the Brule Formation, western Nebraska, F:AM 45082, 49741, 49724, 49725, 49726, 49727, 49728, from about 25 feet above the "upper purplish white layer," to just below the "channel," Seaman Hills, Wyoming, and F:AM 49743, 45019, 45021, 45022, 49742, and AMNH 39462 and 48820, from the "lower oreodon beds," South Dakota; type, CMNH 767, of *Merycoididon* "*macrorhinus*," "Toston beds," Broadwater County, Montana; type, F:AM 45023, and Schultz and Falkenbach's (1954) sample of "*Prodesmatochoerus meekae*," about 25 feet above the "upper purplish white layer" to just below the "channel," Seaman Hills, such as F:AM 45024, 45025, 45026, 45332, and 49548; type, CMNH 9105, of *M.* "*lewisi*," from 12 feet (3.7 m) below the Chadron-Brule contact, Shannon County, South Dakota; type, F:AM 45143, and Schultz and Falkenbach's (1968) sample of *"Paramerycoidodon georgei"* from the "lower oreodon beds" or its equivalents, such as F:AM 45230, 45238, 45239, 45245, and 72210; type, F:AM 49662, and the samples of *"Otionohyus wardi,"* such as F:AM 72037, 72039, UNSM 28188, 28059, UNSM 28196, 28288, 28179, and USM 28198, from the lower part of the Brule Formation, Sioux County, Nebraska, F:AM 72041, 72042, 45015, 45027, 45094, 45015, 45102, 45103, 72043, 72313, 45029, and 72044, from about 25 feet above the "upper purplish white layer" to just below the "channel" in the Seaman Hills, Wyoming, F:AM 45255, 45043, 72276, 72220, 72218, 72050, 49693, AMNH 39118, 39434, and 39436, from the "lower oreodon beds," South Dakota; type, F:AM 49766 of *"Otionohyus vanderpooli"* Schultz and Falkenbach (1968), from the upper part of the Chadron Formation, Sioux County, Nebraska; specimens from the "lower oreodon beds" of South Dakota, F:AM 49795 and F:AM 45277, that Schultz and Falkenbach (1968) referred to *"Otionohyus (Otarohyus)" bullatus;* type, F:AM 45278, of *"Genetochoerus (Osbornohyus) norbeckensis,"* "lower oreodon beds," South Dakota; and TMM 40283-66, Merycoidodontidae, genus and species indeterminate, undifferentiated Vieja Group, Ash Spring, Jeff Davis County, Texas.

Amended diagnosis—Cranial characterization defines the genus-subgenus. The quite variable Subnasal L. averages 49% of the Nasal L., the premolars are unreduced with an average P^1-P^4/P_1-M_3 of 50% (Table 3), the APM^3/TM^3 ratio averages 0.99 (Table 5), the M^3 is brachydont-submesodont with an average APM^3/Ht. M^3 ratio of 1.72 (Table 4), and TM^3/Ht. M^3 averages 1.75 (Table 6). *M. culbertsoni* differs from *Merycoidodon presidioensis* by a significantly smaller size (based on P^1-M^3, P_1-M_3, and associated measurements), more inclined premaxillae, wider, lower muzzle, and more posterior infraorbital foramina. *M. culbertsoni* differs from *M. bullatus* by small, instead of inflated auditory bullae, and a slightly but significantly narrower P.O.C. and lower crowned M^3, otherwise similar.

Distribution—Upper part of the Chadron Formation and the "lower nodules," "lower oreodon beds," the lower part of the Brule Formation and its equivalents, Nebraska, Wyoming, Colorado, South Dakota, and Montana; and from the undifferentiated Vieja Group (Ash Spring), Jeff Davis County, Texas; oreodont faunal zones Chadron "C" and Brule "A," of the northern High Plains; later Chadronian-early Orellan, latest Eocene to earliest Oligocene.

Discussion—*Merycoidodon culbertsoni*, the type species of the type genus of the family Merycoidodontidae, is based on a maxilla fragment, ANSP 10728, with M^2-M^3, and a jaw fragment, ANSP 10727, with M_1-M_3 (broken), probably from the same subadult individual and presumably from the "lower oreodon beds," South Dakota. After Leidy's (1848) publication

of the name, *M. culbertsoni* was characterized thoroughly by Thorpe (1937a), Whitmore (1953), Scott (1940), and Schultz and Falkenbach (1968), among others.

Leidy (1848) used an incorrect Latinization with the double "i" ending of the species name. Since the stem is "culbertson," the proper genitive singular ending is a single "i"; a double "i" would be appropriate only if the stem ended in "i." Most authors have followed this incorrect usage ever since. According to Article 32c, section iii of the International Code of Zoological Nomenclature (1985), this incorrect Latinization should be corrected, rather than perpetuating mistakes established by past usage.

Much of the recent oreodont literature taxonomically oversplit essentially similar fossils. *Merycoidodon culbertsonii "osborni"* is based on a type and sample that is stratigraphically contemporary with *M. culbertsoni*. Schultz and Falkenbach (1968) distinguished the subspecies from the species on the basis of a slightly smaller size relative the morphologically comparable specimens these authors referred to *M. culbertsonii*. Schultz and Falkenbach (1968) also note that *M. culbertsonii "osborni"* have wider skulls with slightly shorter and smaller faces. Comparison of Schultz and Falkenbach's (1968) *M. culbertsoni* sample with *M. culbertsonii "osborni"* suggests that *M. c. "osborni"* is significantly smaller than their version of *M. culbertsoni* in Sk. L. and tooth row measurements. Upper tooth row, tooth, and skull measurements of Schultz and Falkenbach's (1968) sample of *M. culbertsoni* (except Ht. M^3, Nasal L., Sk. L.) have observed ranges that fall within those of *M. c. "osborni,"* and no overlap in other measurements is less than 35% (Sk. L.) of the combined ranges. The Sk. L./P^1-M^3 proportion of Schultz and Falkenbach's (1968) sample of *M. culbertsoni* is slightly but significantly greater than that of *M. c. "osborni."*

"Oreodon culbertsonii periculorum" Cope (1884a) is based on a skull from the lower part of the Brule Formation, Logan County, Colorado. The type and sample as identified by Schultz and Falkenbach (1968) are approximate contemporaries of other specimens here considered to be *Merycoidodon culbertsoni*. Cope did "not detect any difference between *Oreodon culbertsonii periculorum*" and the Nebraska form [*Merycoidodon culbertsoni*] other than "a slightly diminutive size." Hay (1902) and Thorpe (1921a, 1937a) regarded *"O. periculorum"* either conspecific with or subspecific within *M. culbertsoni*. Phleger and Putnam (1942, p. 556) on the basis of their statistical study concluded that *M. "periculorum"* was separable from *M. culbertsoni*. Schultz and Falkenbach (1968) also regarded the "smaller . . . skull, . . . lighter teeth, and . . . smaller" limbs "diagnostically different from those of *Merycoidodon*," and they designated *"O. periculorum"* the type species of *"Genetochoerus."*

Our comparison of Schultz and Falkenbach's (1968) *M. culbertsoni* and "*G. periculorum*" samples shows significant differences in means of all measurements except Ht. M₃, but not in proportions. However, we consider Schultz and Falkenbach's (1968) *M. culbertsoni* to be near the large end of a gradational series with *M. "periculorum"* at the small end. When comparisons are made with intervening "taxa" the differences become less important. Seven of 18 measurement categories for *M. "periculorum"* have observed ranges that lie within the observed range for other specimens that we assign to *M. culbertsoni*, and the observed ranges of the remainder overlap by an average of 40% of the combined ranges, with a minimum of 16% (Sk. W.). The observed range for Sk. L. for all of the earlier named taxa that we consider together as *M. culbertsoni*, is 54.5 mm (95 skulls; mean, 181.2 mm; Appendix B), and the observed range for a sample of 38 skulls of *Dicotyles tajacu* (mean, 195.7 mm; Appendix A) is 54 mm. Although means of measurements for the type and assigned sample of "*G. periculorum*" are smaller than those of most other specimens here identified as *M. culbertsoni*, the size difference occurs within an intraspecifically acceptable observed range, and there are no known morphologic, stratigraphic, or geographic differences that might otherwise separate them.

Merycoidodon "macrorhinus" (Douglass, 1903) is based on a robust partial skull, CMNH 767, from rocks in Montana that are probably correlative with the lower part of the Brule Formation on the northern High Plains. Originally, Douglass (1903) thought that the robustness was not due to sexual dimorphism, but later (Douglass, 1907a) concluded that *M. "macrorhinus"* "may turn out to be a very robust variety or individual of "*M. culbertsoni*. Thorpe (1937a) followed Douglass's concept of *M. "macrorhinus"* as being "a robust species (or subspecies), living in . . . western Montana, but not far removed from the Great Plains *M. culbertsonii*." Although Schultz and Falkenbach (1968) carried the name *M. "macrorhinus"* along in print, these authors noted that the wide muzzle of the type "could be individual variation." The type is larger than most, but not all, of the fossils that Schultz and Falkenbach (1968) allocated to *M. culbertsoni*. Morphology and size provide no convincing evidence to support the separation of CMNH 767 from *M. culbertsoni*.

The type of "*Prodesmatochoerus meekae*" is F:AM 45023. The type, and all of the referred specimens are crushed laterally, usually severely so, which makes their skulls artificially dolichocephalic. The type and referred specimens came from rocks that are stratigraphic equivalents of those which yielded most of the other suites of specimens that are here referred to *Merycoidodon culbertsoni*. Schultz and Falkenbach (1954) neither compared F:AM 45023 with *Merycoidodon*, a mesocephalic genus they (1968) regarded as representing a different oreodont "phylum," nor

adequately characterized the taxon except to say that it is slightly larger than "*P. natronensis*" (see ?*Miniochoerus forsythae*, this chapter), and that it has a small auditory bulla. This collection of crushed specimens is similar in morphology, size, and proportions to the other specimens here identified as *M. culbertsoni*, particularly, insofar as measurements can reveal similarity, to Schultz and Falkenbach's (1968) sample of *M. culbertsonii "osborni*."

The type of *Merycoidodon "lewisi"* is a skull, jaw, and a partial skeleton, CMNH 9105, from the upper part of the Peanut Peak Member of the Chadron Formation, 12 feet below the overlying "lower oreodon beds," Shannon County, South Dakota. Clark and Beerbower (Clark et al., 1967) characterized the species as "typically *Merycoidodon* in basicranial structures [lacks inflated auditory bullae, brackets ours], but large, robust, with heavy canines, long palate, almost straight transverse row of incisors, strong postorbital constriction, and flaring zygoma." The robustness led Clark and Beerbower to suggest that *M. "lewisi"* had a relationship with *Eporeodon*, but these authors also suggested that the robustness may be a "function of size or very possibly of sex rather than indicating" an affinity with that genus. Prior to Schultz and Falkenbach's (1968) report, oreodonts from the White River Group with inflated auditory bullae, some of which were rather large, were called "*Eporeodon*." Clark and Beerbower must have been comparing "*E. lewisi*" with "*E." major (= M. major),* the youngest and largest *Merycoidodon*. More important, the genus *Eporeodon* as originally defined (Marsh, 1875) differs from *Merycoidodon* by a very large auditory bulla, which *M. culbertsoni* and *M. "lewisi"* lack.

Schultz and Falkenbach (1968), unaware of the pending description of *Merycoidodon "lewisi*," designated a contemporary taxon as *M. culbertsonii "browni*," and noted the great similarity of the subspecies with the species. Specimens of *M. c. "browni"* came from just below the "upper purplish white layer," Seaman Hills area, Wyoming. The "upper purplish white layer" [Persistent White Layer (PWL) of the collectors for the F:AM; Purple-white Layer of Schultz and Stout, 1955, 1961] has long been a stratigraphic marker for fossils in eastern Wyoming and northwestern Nebraska. Recently the age of this layer [5 tuff of Evanoff, 1990; Glory Hole Ash (Prothero, 1982a, 1985)] in the region southeast of Douglas, Wyoming, has been estimated to be 33.7 Ma by the ^{40}Ar/^{39}Ar technique. Presumably, the slightly greater antiquity of the subspecies "*browni*" relative to the earlier Orellan sample of *M. culbertsoni* prompted the taxonomic separation. There can be no doubt that the names *M. "lewisi"* and *M. culbertsoni "browni"* are based on types that came from the same late Chadronian population that inhabited the relatively small geographic region between eastern Wyoming and southwestern South Dakota. No convincing morphologic evidence exists to warrant the specific separation of the latest Chadronian from the earlier Orellan specimens being discussed. In measurements, the combined samples of *M. "lewisi"* and *M. culbertsonii "browni"* average slightly larger than (significant: P^1-M^3, M^1-M^3, APM^3), but proportionally similar to those of Schultz and Falkenbach's (1968) sample of *M. culbertsoni*, and are similar to *"Paramerycoidodon georgei,"* discussed below. Our conclusion that late Chadronian *Merycoidodon* is *M. culbertsoni* is consistent with the opinion of Emry et al. (1987) that the late Chadronian faunas are generally similar to those of the lower (early Orellan) part of the Brule Formation.

Harris (1967b) and Harris and Wood (1969) described a rodent and *Toxotherium* (a possible lophiodont—see Schoch, 1989) from the Ash Spring Local Fauna, northern Vieja, Jeff Davis County, Texas. Another Ash Spring taxon is a medium-sized oreodont best represented by TMM 40283-66, a slightly dorsoventrally crushed partial skull, jaws, and some other bones. The rocks at Ash Spring form a lithologically isolated outcrop of reworked, tuffaceous sediment that has been regarded as undifferentiated Vieja Group in the literature, but which has no equivalent in the southern Vieja. The sediments of the southern Vieja are distinctly pink or reddish brown, whereas those at Ash Spring are nearly white, suggesting they were deposited under a different climate (Harris, 1967a, p. 134). Furthermore, the Ash Spring sediment is not associated with flow rocks which are the principal marker-beds in the southern Vieja.

Harris (1967a, b), Wilson et al. (1968), and Wilson (1971, 1978, 1986) all agree that the Ash Spring Local Fauna is the most advanced and probably the youngest of the Vieja faunas. The assemblage, however, has tended to be regarded as early Chadronian. Wilson (1978, table 14) implies that the Ash Spring assemblage is younger than "31.4" Ma [the presumed age for the latest Orellan-earliest Whitneyan 32.3-Ma (Henry et al., 1986) Mitchell Mesa Rhyolite] but there is no radiochronology for the Ash Spring deposits. A latest Orellan-earliest Whitneyan age for the Ash Spring Local Fauna is unacceptable. Wilson was simply trying to show on a generalized diagram that the Ash Spring fossils are younger than those from the Airstrip Local Fauna, and based the position of the "31.4" Ma age determination on the fact that the Mitchell Mesa Rhyolite marks the top of the Capote Mountain Tuff Formation in which the Airstrip fossils occur.

Emry et al. (1987, fig. 5.3) believe that the Ash Spring Local Fauna is more or less correlative with other 33-34 Ma Chadronian faunas, which contain *Toxotherium*. This view was altered slightly by Emry (1992), when newly obtained $^{40}Ar/^{39}Ar$ age determinations placed an older age of about 36 Ma on ash "B" of the Flagstaff Rim area. The Ash Spring *Merycoidodon* is virtually indistinguishable from *Merycoidodon*

culbertsoni of the late Chadronian and earlier Orellan of the northern High Plains, near 34 Ma. The Ash Spring species neither resembles very closely its larger geographic neighbor, *M. presidioensis* of the ?35-36 Ma Airstrip Local Fauna, nor smaller *"Merycoidodon" forsythae* (= ?*Miniochoerus forsythae*, this chapter) of the Flagstaff Rim area, about 36 Ma.

Many genera were geographically widespread during the Chadronian. The oreodont *Bathygenys* was common to both Trans-Pecos Texas and the northern High Plains during the early and middle Chadronian. Wilson (1986, p. 358) indicates that *Bathygenys* occurs in the Ash Spring Local Fauna. A search through the Ash Spring collection found no *Bathygenys* specimens, and Harris (1967a, p. 133, 1967b) and Harris and Wood (1969) do not report *Bathygenys* as a member of the Ash Spring assemblage. The stage of evolution of TMM 40283-66, along with the lack of *Bathygenys*, suggests that the Ash Spring Local Fauna lived after *Bathygenys* had become extinct.

Emry (1992) shows that *Toxotherium* in the Flagstaff Rim area occurs only below Ash "B," whereas *Bathygenys* occurs from below Ash "B" to Ash "G." By itself this may seem significant, but based on the most recent $^{40}Ar/^{39}Ar$ age determinations for ashes "B" (at about 36 Ma), "F" (at 35.8 Ma), and "G" (at 35.6 Ma) (Prothero and Swisher, 1992) the approximately 380 meters of White River Formation under consideration appears to span little time. The absence of *Toxotherium* above ash B may have been due to ecologic removal of a rare species, and not to actual extinction as implied by Emry (1992). *Bathygenys* as a common and geographically widespread genus seems not to occur in rocks that are younger than about 35 Ma. The last occurrence of *Bathygenys* in the Vieja is in the Airstrip Local Fauna, some 1000 feet (305 m) below the Mitchell Mesa Rhyolite, where it is rare. In the Agua Fria-Green Valley area, *Bathygenys* is known but again is rare in the Coffee Cup Local Fauna (Red Hill level, see Wilson, 1986), about 789 feet below the Mitchell Mesa Rhyolite. Wilson regarded the Red Hill assemblage as somewhat older than the Airstrip fossils. Because the ubiquitous *Merycoidodon culbertsoni* has been identified for the Ash Spring assemblage but not in the Flagstaff Rim area, it seems more reasonable to assign an age of about 34 Ma to the Ash Spring assemblage, than to consider the fauna as approaching the 36 Ma suggested by Emry (1992). The dual evidence of the occurrence of *M. culbertsoni* and lack of *Bathygenys* in the Ash Spring deposits, is used here to justify the 34 Ma age attributed to the Ash Spring Local Fauna. However, this suggests that *Toxotherium* survived in West Texas some 2 million years after its last known occurrence on the northern High Plains.

Schultz and Falkenbach (1968) allocated large individuals that are here called *Merycoidodon culbertsoni*, to *"Paramerycoidodon georgei,"* the type species of

"*Paramerycoidodon*." They regarded "*Paramerycoido-don*" as an Orellan to Whitneyan lineage of robust oreodonts of *Merycoidodon* aspect. Because Schultz and Falkenbach recognized *M. "macrorhinus"* as a more robust species than its contemporary, *M. culbertsoni* as identified by Schultz and Falkenbach, one wonders why the sample designated as "*P. georgei*" was not identified as *M. "macrorhinus."* The rather minor geographic separation of the type localities must have been a consideration. Additionally, Schultz and Falkenbach thought that "*P. georgei*" was more brachycephalic than the mesocephalic *M. culbertsoni*, but this is unsupported by measurements. The only way that "*P. georgei*" differs from the fossils that Schultz and Falkenbach (1968) identified as *M. culbertsoni* is that the sample of "*P. georgei*" averages slightly to significantly larger (P^1-M^3, P^1-P^4, M^1-M^3, APM^3, TM^3, and Subnasal L.). However, observed ranges of nine of 17 mutually available measurement categories for "*P. georgei*" lie within the ranges observed for Schultz and Falkenbach's (1968) sample of *M. culbertsoni*, seven other available measurements overlap substantially (average, 38% of combined ranges; minimum, P^1-M^3, 19%), and only the single TM_3 available for "*P. georgei*" lies outside of an observed range. Available proportions do not appear to differ substantially between these groups. In terms of measurements and proportions, "*P. georgei*" is also similar to *M. "lewisi"* and *M. culbertsoni "browni."* Because this revision attempts to provide the given fossil populations (i.e., species) with a realistic geographic distribution, and a range of intraspecific variation that is comparable with that seen for modern peccary suiform artiodactyls, the slightly larger size of "*P. georgei*" relative to *M. culbertsoni* as identified by Schultz and Falkenbach does not warrant a specific, let alone a generic, separation.

"*Otionohyus wardi*" Schultz and Falkenbach (1968), the type species of "*Otionohyus,*" is based on a skull, F:AM 49662, from the lower part of the Brule Formation, Sioux County, Nebraska. Schultz and Falkenbach viewed "*O. wardi*" as similar to but more brachycephalic than "*Genetochoerus periculorum,*" but all of the specimens so identified are crushed dorsoventrally which has lowered and broadened the crania. Specimens identified as "*O. wardi*" and "*G. periculorum*" are here regarded as smaller than average individuals of *M. culbertsoni*. Measurements on these specimens are more similar to each other than they are to other subsamples of *M. culbertsoni*, and the overlap of ranges between "*O. wardi*" and the next larger groups, "*P. meekae*" and *M. c. "osborni,"* is less than is seen for other adjacent comparisons of size-sequenced subsamples within *M. culbertsoni*.

The type of "*Otionohyus vanderpooli,*" F:AM 49766, is a poorly preserved skull and some postcranial bones from 8 feet (2.4 m) below the "upper purplish white layer" which marks the approximate top of the Chadron

Formation (see Evanoff et al., 1992, fig. 6.4), Sioux County, Nebraska. Thus the type is a contemporary of *Merycoidodon "lewisi"* and *M. culbertsonii "browni,"* discussed above. The type lacks the basicranium and zygoma, and is crushed severely dorsoventrally as shown by the compressed orbits, nasals, and frontals. Because Schultz and Falkenbach (1968) viewed "*Otionohyus*" as representing a lineage of phylogenetically brachycephalic oreodonts, F:AM 49766 was questionably referred to "*Otionohyus*" as a new, late Chadronian taxon. In size, F:AM 49766 compares more closely with "*Prodesmatochoerus meekae*" than any other subsample of *Merycoidodon culbertsoni*. From a reexamination of the type, the name "*O. vanderpooli*" appears to be based on a somewhat small crushed specimen of *M. culbertsoni*.

Certain earlier Orellan Merycoidodontinae are, in the morphology of their auditory bullae, somewhat more derived toward stratigraphically younger *Merycoidodon* than their contemporaries. F:AM 49795 and F:AM 45277 from the "lower oreodon beds," have small, but nevertheless slightly inflated bullae. Schultz and Falkenbach (1968) refused to accept the stratigraphic data on these specimens, and referred these skulls to the younger species, "*O. (Otarohyus)" bullatus*, but they are probably progressive individuals of *M. culbertsoni*. The stratigraphic provenance of these fossils as written on the specimens, probably is correct. Additionally, Schultz and Falkenbach (1968) referred F:AM 45278, a partial skull from the middle part of the "lower nodules," early Orellan "lower oreodon beds," to the late Orellan population called "*Genetochoerus (Osbornohyus) norbeckensis*" (= *Merycoidodon bullatus*, this report), presumably also because the specimen has a slightly inflated auditory bulla. F:AM 45278 otherwise is indistinguishable from contemporaries here called *M. culbertsoni*. Certain earlier Orellan *Merycoidodon* thus are known to have had a slightly more inflated bulla than their conspecific contemporaries. This, in turn, suggests that the inflation of the bulla within *Merycoidodon* was not so "sudden" as the literature implies. The "sudden" appearance has been amplified by taxonomic over-splitting.

The later Chadronian-early Orellan subsamples discussed in this chapter which span about 2.5 million years surely do not represent five genera, nine species, and three subspecies of oreodonts, divisible between two subfamilies. The combined sample is believed to represent a single species, *Merycoidodon culbertsoni*, that displays size, but little morphologic, variation (Table 1; Appendix B). It is interesting to note that the large sample produced by the combination of taxa that we include in *M. culbertsoni* does not produce remarkably large CVs, but does produce the only measurement data set for which the distributions are, for the most part, acceptably close to normal (13 of 18 data distributions). The exceptions include: Ht. M3s, Sk. W., M_1-M_3, T. M_3; Table 1). Our *M. culbertsoni* perhaps more closely approximates a sampling of an actual "species longevity," than any other taxon here studied.

It is possible to derive *Merycoidodon culbertsoni* (Fig. 3B) of the later Chadronian-early Orellan from *M. presidioensis* (Fig. 3A) of the middle Chadronian. The transition would necessitate only a slight posterior migration of the infraorbital foramina, a lowering of the nasal profile, broadening of the frontal and muzzle region, and a slight decrease in mean size (based on limited information for *M. presidioensis*, Appendix B). Means of most of the few available measurements on *M. presidioensis* lie in the range from about one to slightly more than two (Subnasal L., P_1-M_3) SDs (of *M. culbertsoni* samples; average, 1.3 SDs) above *M. culbertsoni* means, and are significantly larger than means found for *M. culbertsoni*. Exceptions which fall close to the *M. culbertsoni* means include single measurements on Ht. M^3, and P.O.C., and two measurements of Malar D. Measurement ratios calculated for *M. presidioensis* do not differ significantly from those of *M. culbertsoni*. More information about *M. presidioensis* is needed to clarify the relationship between the two taxa.

Merycoidodon culbertsoni, in turn, appears to be directly ancestral to *M. bullatus* (Fig. 3C) of the late Orellan-earlier Whitneyan. The only changes between these similar-sized species are persistence of an inflated auditory bulla, and, for *M. bullatus*, significantly broader postorbital constriction (based on 138 specimens of *M. culbertsoni*, 40 of *M. bullatus*), and M^3s that are proportionally slightly narrower and significantly higher crowned (smaller APM^3/TM^3 ratio, Table 5; larger TM^3/Ht. M^3 ratio, Table 6). Other differences in means (<0.6 SDs, *M. culbertsoni;* Tables 3, 4, 7, and Appendix B) are not significant.

Merycoidodon (Otarohyus)
(Schultz and Falkenbach)

Type species—*Merycoidodon bullatus* (Leidy): Leidy, 1869:109.

Included species—*Merycoidodon (Otarohyus) bullatus* and *M. (O.) major*.

Amended diagnosis—The subgenus differs from the genus by progressive inflation of auditory bullae and reduction of hyoidal grooves (see Thorpe, 1937a, fig. 7C, D); differs from *M. (Merycoidodon)* by inflated auditory bullae with shallow hyoidal grooves.

Distribution—Late Orellan through the Whitneyan, late early and early middle Oligocene.

Discussion—Historically, certain species which have inflated auditory bullae (which this chapter places in *Merycoidodon*) have been allocated either to "*Eucrotaphus*" Leidy, or *Eporeodon* Thorpe. The type specimen of the type species of "*Eucrotaphus*" is a braincase with an enlarged bulla that cannot be distinguished from *Merycoidodon* with inflated bullae. In addition, the inflated bulla of the Merycoidodontinae is independently developed from that seen in the Agriochoeridae. For these reasons, most workers have preferred to regard "*Eucrotaphus*" as a *nomen vanum*. *Eporeodon* Marsh (1875), a genus of general *Merycoidodon* aspect, also has a large bulla. Based on this single feature, the name *Eporeodon* could be used here for the Merycoidodontinae with large bullae, as it has been in the past (see Thorpe, 1937a). Schultz and Falkenbach (1968), however, agreed with Scott (1890) that the name *Eporeodon* should be reserved for certain oreodonts that are stratigraphically younger and more derived than most *Merycoidodon* being discussed in this chapter. In spite of the fact that *Eporeodon* differs from *Merycoidodon* in grade only, we have followed Scott, and Schultz and Falkenbach in reserving *Eporeodon* for some of the John Day oreodonts.

This revision concurs with Schultz and Falkenbach (1968) that it is useful to separate the Merycoidodontinae with small bullae from those which have enlarged bullae by the use of subgenera. However, because Schultz and Falkenbach (1968) divided the Merycoidodontinae into several lineages, they logically had to define a new subgenus for each "lineage" when that "lineage" developed the derived bullae. Of the available names, "*Anomerycoidodon*," "*Blickohyus*," "*Barbourochoerus*," "*Gregoryochoerus*," and "*Otarohyus*," the last one is the appropriate name, following the principle of first reviser (Articles 24A and 69B of the International Code of Zoological Nomenclature, 1985). *Otarohyus* was designated the subgenus for the species *bullatus* by the first revisers, Schultz and Falkenbach (1968, p. 118).

Merycoidodon (Otarohyus) bullatus (Leidy)
Figures 2C, 3C; Tables 1-7; Appendix B

Oreodon bullatus Leidy, 1869:106
Otionohyus (Otarohyus) bullatus (Leidy): Schultz and Falkenbach, 1968:118
Eporeodon socialis Marsh, 1885:299
Subdesmatochoerus socialis (Marsh): Schultz and Falkenbach, 1954:220
Subdesmatochoerus socialis dakotensis Schultz and Falkenbach, 1954:223
Eucrotaphus helenae Douglass, 1901:265
Otionohyus (Otarohyus) hybridus helenae (Douglass): Schultz and Falkenbach, 1968:131
Eporeodon major, var. *cedrensis* Matthew, 1901:396
Otionohyus (Otarohyus) cedrensis (Matthew): Schultz and Falkenbach, 1968:127
Eucrotaphus dickinsonensis Douglass, 1907b:99
Genetochoerus (Osbornohyus) dickinsonensis (Douglass): Schultz and Falkenbach, 1968:154,

tentatively referred

Merycoidodon (Anomerycoidodon) dani Schultz and
Falkenbach, 1968:80

Merycoidodon (Blickohyus) galushai Schultz and
Falkenbach, 1968:80

Paramerycoidodon (Barbourochoerus) bacai Schultz
and Falkenbach, 1968:91

Otionohyus wardi Schultz and Falkenbach, 1968:112,
113, in part

Otionohyus (Otarohyus) hybridus (Leidy): Schultz and
Falkenbach, 1968:129, in part

Genetochoerus (Osbornohyus) norbeckensis Schultz
and Falkenbach, 1968:143, in part

Genetochoerus (Osbornohyus) geygani Schultz and
Falkenbach, 1968:149

Merycoidodon culbertsoni Leidy: Prothero, 1982b:406,
in part

Type—ANSP 10681, Big Badlands, South Dakota.

Hypodigm—Type, and Schultz and Falkenbach's
(1968) samples of this species such as F:AM 45268,
72056, 45275, 45270, 45271, 45176E, 49788, 45276,
49790, 72147, 49798, 49796, and AMNH 611, "middle
oreodon beds," South Dakota, F:AM 72005, 72006,
72007, from the "upper nodules," Orella Member,
Brule Formation, Nebraska; cotypes of "*Eporeodon
socialis*," YPM 13118 and YPM 13119, Brule Formation, Scotts Bluff promontory; Schultz and Falkenbach's (1954) sample of "*Subdesmatochoerus socialis*,"
such as F:AM 45333, 45179, 45181, 45177, and 49640,
from near or within the "upper nodules," Brule
Formation; type, F:AM 45351, of "*S. socialis dakotensis*"; type, CMNH 765, of "*Eucrotaphus helenae*,"
Toston beds," Montana; type, AMNH 8949, and
Schultz and Falkenbach's (1968) sample of "*Otionohyus (Otarohyus) cedrensis*," such as AMNH 8946A,
8946B, and F:AM 45272, Brule Formation, northeastern Colorado; type, CMNH 1584, of "*Eucrotaphus
dickinsonensis*," Little Badlands, North Dakota; type,
F:AM 72132, and Schultz and Falkenbach's (1968)
sample of *Merycoidodon (Anomerycoidodon) "dani*,"
such as F:AM 72136, 72133, 72137, 72236, 45262,
72138, 45265, and UNSM 28061, from near or within
the "upper nodules," Brule Formation, Nebraska; type,
F:AM 45279, and F:AM 72059 of *M. "(Blickohyus)
galushai*," upper part of the "upper oreodon beds,"
South Dakota; type, UNSM 28191, and Schultz and
Falkenbach's (1968) sample of "*Paramerycoidodon
(Barbourochoerus) bacai*," such as UNSM 28469,
F:AM 45264; specimens F:AM 45099, "middle
oreodon beds," and F:AM 45100, 20 feet (6.1 m) above
the "lower nodules," "middle oreodon beds," which
Schultz and Falkenbach (1968) identified as "*Otionohyus wardi*"; Schultz and Falkenbach's (1968) referred
specimens of "*O. (Otarohyus) hybridus*," such as F:AM
72009 and F:AM 72011; type, F:AM 49733, and
Schultz and Falkenbach's (1968) sample of "*Genetochoerus (Osbornohyus) norbeckensis*," such as F:AM
49759, 45359, 45353, and 45358, "middle oreodon

beds," South Dakota; and the type, F:AM 49734, and
Schultz and Falkenbach's (1968) sample of
"*Genetochoerus geygani*," "upper oreodon beds," 35
feet (10.7 m) below the "*Leptauchenia*-like" nodules,
South Dakota, such as F:AM 49746, 49749, 45280, and
49752.

Amended diagnosis—The cranial characterization of
the species is the same as for the genus-subgenus.
Auditory bullae vary from slightly inflated with a
moderate hyoidal groove to well inflated with a shallow
groove. The Subnasal L. averages 46.5% of the Nasal
L. (7 specimens; Appendix B), P^1-P^4/P^1-M^3 averages
49.4% (Table 3), APM^3/TM^3 ratio averages 1.00
(Table 5), APM^3/Ht. M^3 ratio averages 1.67 (Table 4),
and TM^3/Ht. M^3 averages 1.66 (Table 6). *Merycoidodon bullatus* differs from *Merycoidodon (Merycoidodon) culbertsoni* by variously inflated auditory bullae
and slightly but significantly broader P.O.C. and higher
crowned M^3, otherwise similar. *Merycoidodon bullatus*
differs from *M. (Otarohyus) major* by significantly
(except Nasal L.) smaller size, larger and deeper
preorbital fossae, less expanded frontals, and smaller
auditory bullae with deeper hyoidal grooves.

Distribution—From the "middle oreodon beds,"
"upper nodules," Scenic Member, and the upper part of
the "upper oreodon beds," lowermost Poleslide
Member, Brule Formation of South Dakota; upper part
of the Orella Member, and the lowest part of the
Whitney Member of the Brule Formation, western
Nebraska; "Toston beds," Montana; and approximate
equivalents in North Dakota and northeastern Colorado;
oreodont faunal zones Brule "B" and "C"; later Orellan-
earlier Whitneyan, late early to early middle Oligocene.

Discussion—*Merycoidodon bullatus* is based on a
partial skull, ANSP 10681, that was mixed (surely after
recovery) with a collection of specimens of *M. culbertsoni* obtained by the Hayden Expedition of 1866 into
the Big Badlands of what is now South Dakota. The
small but nevertheless inflated auditory bullae of
ANSP 10681 relative to the small bullae characteristic
of *M. culbertsoni* surprised Leidy, and the specimen
became the type of "*Oreodon*" *bullatus*. Although
ANSP 10681 lacks precise stratigraphic documentation,
a large sample of comparable specimens has been
collected subsequently above the occurrence of *M.
culbertsoni* in the type area from near or within the
"upper nodules," upper part of the "middle oreodon
beds," Big Badlands of South Dakota.

Marsh (1885, p. 299) introduced the name "*Eporeodon socialis*" in a caption for comparative illustrations
of an oreodont front and hind foot but he gave no
description. The illustrations of the top of an associated
skull, and the feet again, of "*E. socialis*" were published
later (Marsh, 1886, pp. 64, 187), but still there was no
description. The cotypes of "*E. socialis*," two nearly
complete and associated skeletons from the Brule
Formation at Scott's Bluff, Scottsbluff County,
Nebraska, were not characterized until Thorpe (1921a,

1931) studied them. Most authors after Thorpe, including Schultz and Falkenbach (1954), have regarded YPM 13118 and YPM 13119 as cotypes of "*E. socialis*."

Schultz and Falkenbach (1954) gave YPM 13118 and YPM 13119 a middle Orellan age because their auditory bullae are similar to numerous Frick specimens from the "upper nodules" of the "middle oreodon beds" of South Dakota, and this is accepted here. Schultz and Falkenbach (1954) placed "*Eporeodon socialis*" in "*Subdesmatochoerus*" of the "Desmatochoerinae" because they thought that "*Eporeodon socialis*" was dolichocephalic, and thus unrelated to the mesocephalic *Merycoidodon*. The cotypes are somewhat crushed laterally, especially YPM 13119, and the zygomatic arches are not preserved. Crushing and inaccurate restoration with plaster of Paris combine to make the crania appear narrow. All of the Frick specimens that Schultz and Falkenbach (1954) allocated to "*S. socialis*" are crushed laterally, usually severely so. In most measurements, the type of *M. bullatus* is close to the means for the species as we understand it, and measurements available for YPM 13118 and YPM 13119 average, respectively, about 1.2 and 1.4 SDs larger than specimens we assign to *M. bullatus* (excluding the sample of "*S. socialis*"). Except for two measurements on YPM 13119 (APM^3, above the observed range, and Subnasal L., below), measurements lie within ranges observed for our sample of *M. bullatus* excluding "*S. socialis*," though several measurements (four of 14 available, YPM 13118; 8 of 12 available, YPM 13119) are significantly large. If the sample of "*S. socialis*" is compared to other specimens that we assign to *M. bullatus*, the situation is similar. Morphologically, including various measurement ratios, the collection of specimens hitherto identified as "*S. socialis*" does not differ from, and should be placed within, *Merycoidodon bullatus* as a set of somewhat larger specimens.

"*Subdesmatochoerus socialis dakotensis*" is based on, and known only from, an incomplete, laterally crushed skull, F:AM 45351, from 20 feet (6.1 m) below the top of the "upper nodules," "middle oreodon beds" of South Dakota. Schultz and Falkenbach (1954) separated F:AM 45351 from "*S. socialis*" on the basis of some minor dental differences, smaller auditory bullae, and smaller size (average of 2.3 SDs for 14 available measurements; significant: P1-M3s, P^1-P^4, TM3s, Malar D., and M$_1$-M$_3$), but in comparison to other specimens that we assign to *Merycoidodon bullatus*, F:AM 45351 is only slightly smaller than the mean condition (<0.7 SDs on the average; significant: P.O.C.; Appendix B).

Morphologically comparable to *Merycoidodon bullatus* is the species that Douglass (1901) named "*Eucrotaphus helenae*." The type, CMNH 765, came from the "Toston beds" northeast of Toston, Broadwa-

ter County, Montana, from a layer of "cracklin," nodular, sandy clay. Douglass (1901, 1903) regarded the "Toston beds" as mainly equivalent with the "oreodon division" of the White River Group on the northern High Plains, but the precise provenance of CMNH 765 is not known. Of the pertinent rocks, Chadronian to later Orellan deposits occur in the Toston area. Pardee (1925) detailed a stratigraphic section in an area 3 miles (4.8 km) northeast of Toston where Douglass had obtained fossils. Pardee notes that the upper two-thirds of the section is either covered, hard dendritic clay, pink clay, or conglomerate with a red clay matrix. It is unlikely that the type of "*E. helenae*" came from this upper interval on the basis of Douglass's comments. Presumably the type came from the lower third of the section. Pertinent to this discussion is the fact that Douglass (1901), while naming "*E. helenae*," also named "*Oreodon robustum*" ["*O. robustum*," preoccupied, = "*O. macrorhinus*" Douglass, 1903 (see *Merycoidodon culbertsoni*, this chapter)], also from the "Toston beds." The small auditory bullae of "*O. macrorhinus*" show that some of the rocks (?the lower "Toston beds") are more or less equivalent with the late Chadronian or earlier Orellan deposits which yield *M. culbertsoni* on the northern High Plains. Douglass's (1901) remarks seem to imply that "*E. helenae*" and "*O. robustum*" came from the same general area. The relatively small, hence primitively inflated auditory bullae of "*E. helenae*" is consistent with specimens from the later Orellan, and this, along with Douglass's comments, casts doubt on Schultz and Falkenbach's (1968) belief that "*E. helenae*" was Whitneyan in age.

Schultz and Falkenbach (1968) regarded "*Eucrotaphus helenae*" as a subspecies of the large, middle and later Whitneyan species, "*Oreodon hybridus*" Leidy (1869). Although Schultz and Falkenbach used the identification "*Otionohyus (Otarohyus) hybridus helenae*" to justify attributing a later Whitneyan age to the type of "*E. helenae*" (CMNH 765), there is little morphologic or size similarity between "*E. helenae*" as based on CMNH 765, and "*Oreodon hybridus*" as based on ANSP 10860. Schultz and Falkenbach erred in their concept of "*Oreodon hybridus*." These authors incorrectly viewed Leidy's taxon "*hybridus*" as a relatively small species, and only referred stratigraphically older (hence smaller) crania to that species. The type of "*Oreodon hybridus*" (ANSP 10860), which Schultz and Falkenbach (1968) transferred to "*Otionohyus (Otarohyus),*" is the facial part of a skull, with damaged, unmeasurable teeth, that is much larger than the facial portion of CMNH 765, or the faces of skulls from the "upper oreodon beds" of South Dakota that Schultz and Falkenbach referred to "*O. (O.) hybridus.*" The referred specimens of "*O. (O.) hybridus*" are morphologically unlike the type (here referred to *M. major*), but resemble *M. bullatus* with their relatively deep, elongated, not pit-like, preorbital fossae, only modestly inflated frontal sinuses, and smaller size. "*O. (O.) hybridus helenae*" is probably conspecific with *Merycoidodon bullatus* of the later Orellan and earlier Whitneyan.

Matthew (1901) described a damaged juvenile partial skull, jaws, and a partial skeleton, AMNH 8949, from the Brule Formation, Cedar Creek, Logan County, Colorado, as "*Eporeodon major* var. *cedrensis*," and noted that the fossil came from Hayden's and Leidy's "Horizon C," the "*Leptauchenia* beds." "Horizon C," however, as used by Matthew certainly contains a mixed fauna. Schultz and Falkenbach (1968, p. 127) questionably referred "*E. cedrensis*" to their faunal zone "Brule C," regarded as earliest Whitneyan, but elsewhere (1968, p. 128) state that the type belongs to "faunal zone Brule B," which they regarded as later Orellan. Although the taxonomic history of "*E. cedrensis*" is complex, the type specimen (AMNH 8949) was eventually called "*Otionohyus (Otarohyus) cedrensis*." Based on a juvenile individual, the type specimen lacks most of the frontals, the zygoma, and the nuchal crest, which further minimize its appearance. Furthermore, the dorsoventrally crushed type appears "low." The 7 available measurement categories of the type and referred specimens of "*O. (O.) cedrensis*" lie within or extensively overlap (average of 60% of the combined range) the range of the part of Schultz and Falkenbach's (1968) sample of *Merycoidodon bullatus* that we also refer to *M. bullatus* (Appendix B). The means of measurements for the two groups show no systematic differences. AMNH 8949 is remarkably similar to the type of "*Eucrotaphus helenae*," which Schultz and Falkenbach (1968) also called "*Otarohyus*," due in part to shared immaturity. Additionally, both these specimens are not very different from, and have approximately the same stratigraphy as *Merycoidodon bullatus*, with which they are probably conspecific.

"*Eucrotaphus dickinsonensis*" is based on a skull, mandible, and partial skeleton (CMNH 1584), from the Little Badlands of North Dakota. Douglass (1907a) reports that the type specimen came from "near the top of the thick nodular beds of the Middle White River ("Oreodon") horizon of the Little Badlands," but he was unsure of the precise stratigraphic provenance because the bones were not found in place and Douglass acknowledgde that "it is barely possible that [the type specimen] came from the upper beds." Schultz and Falkenbach (1968) thought that the type specimen came from the highest (Whitneyan) rocks in the area based on the specimen's much enlarged auditory bullae. The relative completeness of the type suggests to us that the specimen probably has not migrated downward very far. Douglass (1909, p. 286) presents a generalized measured section of the rocks in the Little Badlands where he collected. Several sets of exposures with different thicknesses, preserved sections, and lithologies are present in the Little Badlands area. Skinner (1951, p. 57) also shows a measured section, and Murphy et al. (1993, p. 35) provide information for three areas within the Little Badlands. Douglass's (1909) section is remarkably consistent with the main Little Badlands section of Murphy et al. (1993), and presumably this was where Douglass collected.

Based on Douglass's (1909) published thicknesses, the type of "*Eucrotaphus dickinsonensis*" came from about 90 feet above the base of the section, near the top of the thick nodular zone in his "Middle White River," and just below what he termed the "Greenish band." This greenish layer surely corresponds with Skinner's "White marker" (the Antelope Creek tuff of Murphy et al., 1993, p. 35, unit 1), a useful marker bed in the area. The type specimen seems to have come from near the top of Skinner's (1951) nodular layer, unit 4A, and near the top of unit 7 of Murphy et al. (1993).

Douglass (1909, p. 286) mentions that he found only oreodonts with small auditory bullae in the more fossiliferous, lower third of his "Middle White River" section. This is consistent with the recovery of Merycoidodontinae with only small bullae from the lower part of Skinner's nodular zone, as reported by Schultz and Falkenbach (1968, p. 54). Specimens with inflated bullae have been found higher in this unit (Schultz and Falkenbach, 1968, p. 76). This, however, conflicts with Hoganson and Lammers's (1992, p. 253) statement that remains of *Merycoidodon culbertsoni* (?this species), a taxon with small bullae, occur in the Fitterer bed (see Stone, 1973; Skinner's, 1951, Fitterer channel), which overlies the "White marker" bed. Hoganson and Lammers (1992, fig. 4) have placed the discontinuous and stratigraphically somewhat variable channels which constitute the Fitterer bed, below, instead of above the "White marker," which may explain their identification. Hoganson and Lammers (1992) do not clarify the bulla morphology of their specimens of *M. ?culbertsoni*. Furthermore, in spite of much renewed interest in, and the publication of numerous faunal lists about, the fossils from the Brule Formation of North Dakota, the taxa have not been documented biostratigraphically. This in turn probably obscures the recognition of the faunal change, however subtle, that is present.

Based on the observation that no Merycoidodontinae with notably inflated auditory bullae from South Dakota or Nebraska are associated with those that have small bullae, and based on Douglass's (1907a, 1909) and Schultz and Falkenbach's (1968) biostratigraphy of oreodonts with larger bullae in the North Dakota section(s), we agree with Skinner (1951) that the rocks from which the type of "*Eucrotaphus dickinsonensis*" came are late Orellan in age.

Schultz and Falkenbach (1968) placed "*Eucrotaphus dickinsonensis*" in "*Genetochoerus (Osbornohyus)*." In available measurements and proportions, "*E. dickin-

sonensis" is comparable to, though slightly smaller than, means of the subset of *M. bullatus* called *"Genetochoerus (Osbornohyus) geygani"* (Schultz and Falkenbach, 1968; discussed below). *"E. dickinsonensis"* is tentatively placed within *Merycoidodon bullatus* on the basis of size and comparable cranial morphology. CMNH 1584, however, does have a larger and more globular auditory bulla than is found in other specimens of *M. bullatus*.

Critical to this discussion is the fact that the divergence of *Eporeodon* and the Eporeodontinae (characterized by large auditory bullae among other features) from *Merycoidodon* must have occurred before *M. bullatus* evolved into *M. major*. The possibility exists that CMNH 1584 was a very early representative of the line which led to *Eporeodon* (Fig. 7), and belonged to a population that was separating from *M. bullatus*. However, because the size and shape of auditory bullae is so variable in modern suiform artiodactyls (see Introduction), the large and well-rounded bulla of a single specimen, CMNH 1584, relative to *M. bullatus* as a whole does not justify by itself the taxonomic separation of this specimen from *M. bullatus*. More specimens from the type area of CMNH 1584 are needed to document whether or not the North Dakota population consistently has larger bullae than *M. bullatus*.

Schultz and Falkenbach (1968) designated a sample of well-preserved and uncrushed (thus mesocephalic) crania that are more or less stratigraphically equivalent with the others discussed in this section, as *Merycoidodon (Anomerycoidodon) "dani."* The type specimen, F:AM 72132, came from 23 feet (7 m) below the "white zone" (see Skinner, 1951, p. 57, zone 4), Stark County, North Dakota, from rocks that are probably equivalent with the "middle oreodon beds" of South Dakota. The referred specimens of *M. "dani"* mostly came from 0-15 feet (0-4.6 m) below the "upper nodules" of the "middle oreodon beds" of South Dakota. Schultz and Falkenbach noted that the crania of *M. "dani"* are within the size range of *M. culbertsoni*, an earlier species, but have inflated auditory bullae. As identified by those authors, means of *M. "dani"* measurements do not differ systematically from those of Schultz and Falkenbach's (1968) sample of *M. culbertsoni*, and most ranges for *M. "dani"* lie within the range observed for that sample of *M. culbertsoni* (12) or overlap substantially (the rest; average overlap, 58%). Of 13 metric categories where means of *M. "dani"* are larger, 6 (P^1-M^3, M1-M3s, APM^3, TM^3, and Ht. M^3) are significantly so. Schultz and Falkenbach regarded other contemporary subsamples (here equated with *M. bullatus*) as members of separate "phyla" and did not compare *M. "dani"* with them. Groups that we consider to occupy the central to large part of the general size range of *M. bullatus* include, in order of ascending size, *"Otionohyus cedrensis,"*

"Subdesmatochoerus socialis," and *M. "dani."* Measurements of the limited samples of *M. "dani"* and *"S. socialis"* show no significant differences, and although 14 means for measurements of *M. "dani"* are significantly greater than those for the rest of the specimens in *M. bullatus* as we use the name, *M. "dani"* overlaps substantially other subsamples that lie in the intermediate (*"O. cedrensis"*) to large part of the range of specimens that we have assigned to *M. bullatus*. The gradational differences and morphology do not justify taxonomic separation.

Merycoidodon "(Blickohyus) galushai" is based on a partial skull, F:AM 45279, which lacks the premaxillae, parts of the zygoma, and the occiput, associated with a partial jaw, from the earliest Whitneyan "upper oreodon beds" of South Dakota. *M. "galushai"* originally was compared only to *M. "(Blickohyus) lynchi"* (= *M. major*, this report). It was regarded as a middle and late Whitneyan member of the same lineage by Schultz and Falkenbach (1968). They thought that *M. "(B.) galushai"* differs from *M. "(B). lynchi"* by a relatively larger orbit and smaller auditory bulla with a deeper hyoidal groove. The alleged larger orbit of *M. "galushai"* does not survive scrutiny, and the smaller bulla is in keeping with the older stratigraphy of *M. "galushai"* relative to *M. "lynchi."*

Although both *Merycoidodon "(Anomerycoidodon)"* and *M. "(Blickohyus)"* were defined to contain *Merycoidodon* with inflated auditory bullae, Schultz and Falkenbach (1968) believed that *M. "(Blickohyus)"* had a shorter braincase than *M. "(Anomerycoidodon)."* Just how this was determined, however, is unclear because all of the skulls identified in the literature as *M. "(Blickohyus)"* are poorly preserved and extensively restored in the basicranial region with plaster. Although the type and the single referred specimen (dentally unmeasurable) of *M. "galushai"* are slightly younger (Brule faunal zone "C," earlier Whitneyan) than most of the other specimens here equated with *M. bullatus* (Brule faunal zone "B"), the specimens are morphologically more similar to *M. bullatus* than they are to *M. major* of the later Whitneyan, Brule faunal zone "D." Measurements on the type of *M. "galushai"* compare closely with subsamples toward the larger end of the range for *M. bullatus* (particularly the sample of *"Subdesmatochoerus socialis"*; P.O.C. of *M. "galushai"* noticeably but not significantly small), and, with few exceptions, are smaller than means of those of *M. major* and its subsamples, except *"Genetochoerus (Osbornohyus) chamberlaini."* Because there is no convincing information to suggest that *M. "galushai"* should be regarded as a valid taxon, *M. "galushai"* here is placed within *M. bullatus* as a late representative of that species.

"Paramerycoidodon (Barbourochoerus) bacai" Schultz and Falkenbach (1968), the type species of the subgenus, is based on UNSM 28191, a partial skull

from upper Orellan rocks, Toadstool Park area, northwest of Crawford, Sioux County, Nebraska. The UNSM referred specimens have the same stratigraphy as the type, and F:AM 45264 came from 10 feet (3 m) below the top of the "upper nodules," "middle oreodon beds" of Shannon County, South Dakota. Most specimens from the Toadstool Park area are severely crushed and thus poorly preserved. The incomplete and distorted type of "*P. (B.) bacai*" is typical. Many wide, matrix-filled fractures in the teeth have increased the apparent size of the dentition. Nevertheless, Schultz and Falkenbach (1968) viewed the teeth of the type as "enormous." Based on morphology and general size of the crania and the dentally homologous parts (thus excluding the fractures), "*P. bacai*," although larger than average, falls within the range of size (eight of nine categories available for comparison; exception: M^1-M^3) and morphological variation expected for the population best interpreted as *M. bullatus*.

Schultz and Falkenbach (1968) identified two crania [F:AM 45099, from the lower part of the "middle oreodon beds," and F:AM 45100, from 20 feet (6.1 m) above the "lower nodules," also in the lower part of the "middle oreodon beds"] as "*Otionohyus wardi*" (the type specimen of "*O. wardi*" is referred to *M. culbertsoni* in this chapter), because the fossils have small auditory bullae. Rather than refer these two specimens to an older species, it would seem more reasonable to view them as slightly atavistic (bullar morphology) individuals of *M. bullatus*, especially in view of our knowledge about intraspecific variation of bullae, and the observation that *M. culbertsoni* and *M. bullatus* seem to be linked in a single phyletic series.

"*Genetochoerus (Osbornohyus) norbeckensis*" is based on a skull, F:AM 49733, which lacks the premaxillae, occipital crests, and zygoma, associated with a partial lower jaw and some other bones from the top of the "middle oreodon beds" of South Dakota. Schultz and Falkenbach (1968) referred several stratigraphically mixed and poorly preserved specimens to "*G. norbeckensis*." F:AM 45278 is a skull from the "lower nodules." These rocks yield *Merycoidodon culbertsoni*, to which F:AM 45278 is here referred. But F:AM 45278 was referred to "*G. norbeckensis*" by Schultz and Falkenbach (1968) because it has slightly inflated, although still quite small, auditory bullae. UNSM 28624, a maxilla and dentary fragment also referred to "*G. norbeckensis*," came from about 10 feet (3 m) above the base of the "*Leptauchenia* beds" ($\approx$ "lower *Leptauchenia* beds"; see Schultz and Falkenbach, 1968, p. 425), and F:AM 49745, a skull referred to "*G. norbeckensis*," came from stratigraphically high rocks in North Dakota. The Whitneyan specimens probably should be referred to *M. major* as smaller than average individuals. Although the breadth of the P.O.C. is more like that of *Merycoidodon bullatus*, the limited set of measurements for "*G. norbeckensis*" provide little support for its inclusion with, or separa-

tion from, either *M. culbertsoni* or *M. bullatus*. The type of "*G. norbeckensis*" is not morphologically or stratigraphically separable from *M. bullatus*, to which it is here referred.

"*Genetochoerus (Osbornohyus) geygani*" is based on F:AM 49734, a skull, jaw, and some limb bones from the upper part of the "upper oreodon beds," Brule faunal zone "C," some 35 feet (10.7 m) below the "*Leptauchenia*-like nodules," Shannon County, South Dakota. The original comparisons (Schultz and Falkenbach, 1968) were between "*G. geygani*" and "*G. (O.) norbeckensis*," on the one hand, and "*G. (O.) chamberlaini*" (= *M. major*, this report) on the other; "*G. geygani*" was considered intermediate. Comparison of measurement means for "*G. geygani*" with those for other specimens in our sample of *Merycoidodon bullatus* shows "*G. geygani*" to be slightly smaller in most respects (average: about 0.5 SDs for 15 of 18 measurement means; exceptions: P.O.C., TM_3, and Ht. M_3 are slightly larger). Of the intermediate- to smaller-sized groups that we have placed mainly in *M. bullatus*, specimens of "*Otionohyus*" *bullatus*, "*O. wardi*," and "*Genetochoerus dickinsonensis*" show only slight differences from "*G. geygani*," and "*G. norbeckensis*" differs only by a significantly smaller Malar D. and TM_3. Considering their morphologic similarity and stratigraphic positions, the slightly smaller size of "*G. geygani*" does not warrant separating it from the population that is best referred to *M. bullatus* (see Appendix B).

The oreodont samples discussed in this section surely do not represent seven genera, five subgenera, and twelve species divisible between two subfamilies. Disregarding the influence that crushing has had on taxonomy, the size differences between the otherwise morphologically similar specimens surely must represent a sorting of smaller than average-, average-, or larger than average-sized individuals of the same species, *Merycoidodon bullatus*. Furthermore, size distribution of the auditory bullae, from small (rare), small-medium, large-medium, to large (rare), is similar to that seen for the population of modern peccary, *Dicotyles tajacu* from Trans-Pecos Texas (see Introduction). Other morphological variability is also similar. *M. bullatus* (Fig. 3C) was in existence for a modest 2 million years, and averages the same size as its probable ancestor, *M. culbertsoni*. The most important way that *M. (Otarohyus) bullatus* differs from *M. (M.) culbertsoni*, is that most *M. bullatus* have inflated although not particularly large auditory bullae, whereas most *M. culbertsoni* do not. This reinforces the observed stereotypy of *Merycoidodon*. *M. bullatus* probably was ancestral to both *M. major*, middle and late Whitneyan, and *Eporeodon occidentalis* of the late Whitneyan or earliest Arikareean.

Merycoidodon (Otarohyus) major (Leidy)
Figures 2D, 3D; Tables 1-7; Appendix B

Oreodon major Leidy, 1853:392
Paramerycoidodon (Barbourochoerus) major (Leidy):
   Schultz and Falkenbach, 1968:92
Oreodon hybridus Leidy, 1869:105
Otionohyus (Otarohyus) hybridus (Leidy): Schultz and
   Falkenbach, 1968:129, in part
Promesoreodon scanloni Schultz and Falkenbach,
   1949:152
Subdesmatochoerus shannonensis Schultz and
   Falkenbach, 1954:223
Merycoidodon (Anomerycoidodon) lambi Schultz and
   Falkenbach, 1968:77
Merycoidodon (Blickohyus) lynchi Schultz and
   Falkenbach, 1968:81
Paramerycoidodon (Gregoryochoerus) wanlessi
   Schultz and Falkenbach, 1968:99
Otionohyus (Otarohyus) alexi Schultz and Falkenbach,
   1968:132
Genetochoerus (Osbornohyus) chamberlaini Schultz
   and Falkenbach, 1968:52
?Desmatochoerus (Paradesmatochoerus) anthonyi
   Schultz and Falkenbach, 1968:381, in part
Type — USNM 19099, Big Badlands, South Dakota.
Hypodigm — Type, and Schultz and Falkenbach's
(1968) sample of this species such as AMNH 612,
1038, F:AM 45298, 45306, 45307, 45310, 45312,
45315, 45303, 45309, 45313, 45314, 72117, 45305,
72118, 72119, 72027, 72123, 45316, 45306, 72304, and
72305, from the "*Leptauchenia* beds"; type, ANSP
10860, of "*Oreodon hybridus*"; type, F:AM 45329, and
Schultz and Falkenbach's (1949) sample of "*Promesoreodon scanloni*" from the "lower *Leptauchenia*
beds"; type, AMNH 1310, and Schultz and Falkenbach's (1954) sample of "*Subdesmatochoerus shannonensis*," such as F:AM 45183, 45184, 45327, 45328,
49550, and 45460 from the "lower *Leptauchenia* beds";
type, F:AM 72139, and Schultz and Falkenbach's
(1968) sample of *Merycoidodon (Anomerycoidodon)*
"*lambi*," such as F:AM 72097, 72098, 72099, 72084,
45295, 72128, F:AM 72105, 72124, and 72108 from
the "lower *Leptauchenia* beds"; type, F:AM 45297, and
Schultz and Falkenbach's (1968) sample of *Merycoidodon* "*(Blickohyus) lynchi*," such as AMNH 1288,
614, 619, F:AM 45322, 72086, 72083, 72087, 72091,
72114, 72106, 72141, and UNSM 28067 from the
"lower *Leptauchenia* beds"; type, F:AM 72014, and
Schultz and Falkenbach's (1968) sample of "*Paramerycoidodon (Gregoryochoerus) wanlessi*," such as
F:AM 45288, 72022, 72023, 72078, 72280, 72016,
72024, 72026, 72279, 72109, 72032, 72033, 72149, and
45299 from the "lower *Leptauchenia* beds"; type,
F:AM 72060, and Schultz and Falkenbach's (1968)
sample of "*Otionohyus (Otarohyus) alexi*," such as
F:AM 72113, 72062, 72063, 72095, 72067, 72068,

72069, 72072, 72061, 72070, 72071, 72089, 72064,
72146, and 72013 from the "lower *Leptauchenia* beds";
type, UNSM 28340, and Schultz and Falkenbach's
(1968) sample of "*Genetochoerus (Osbornohyus)
chamberlaini*," such as F:AM 49753, 49735, 45283,
72080, and 49740 from the lower part of the "*Leptauchenia* beds"; and Schultz and Falkenbach's (1968)
referred specimen of "*?Desmatochoerus (Paradesmatochoerus) anthonyi*," F:AM 72393, from the Whitney
Member of the Brule Formation.

Amended diagnosis — The cranial characterization is
the same as for the genus-subgenus, but the skull trends
toward massiveness, with broad frontals and muzzle,
widely spread zygomatic arches, well inflated auditory
bullae which lack distinct hyoidal grooves, and small
and generally pit-like preorbital fossae. The Subnasal
L. averages 51.5% of the Nasal L. (13 specimens), the
P^1-P^4/P^1-M^3 averages 48.4% (Table 3), the APM^3/
TM^3 ratio averages 1.02 (Table 5), the mean APM^3/Ht.
M^3 ratio is 1.68 (Table 4), and TM^3/Ht. M^3 averages
1.63 (Table 6). *Merycoidodon major* differs from
Merycoidodon bullatus by a substantially (on the
average, means of *M. bullatus* lie 1.7 SDs below those
of *Merycoidodon major*; basis: SDs of *M. major*), and
in terms of measurement means, systematically, larger
size (significant for all measurement means except that
of Nasal L.), a usually broader P.O.C., slightly but
significantly more reduced premolars, and the cranial
features noted above, otherwise similar. *Merycoidodon
major* differs from *Mesoreodon chelonyx* by a systematically smaller size (means average about 0.9 SDs of
M. major smaller; significant: P^1-M^3, P^1-P^4, M^1-M^3,
APM3s, Ht. M^3, and Nasal L.), significantly less
anteroposteriorly attenuated M^3s, and slightly lower
crowned teeth, otherwise very similar.

Distribution — Known from the "lower *Leptauchenia*
beds," Poleslide Member of the Brule Formation of
South Dakota and the Whitney Member of the Brule
Formation, western Nebraska; oreodont faunal zone
Brule "D," middle to late Whitneyan, middle Oligo-
cene.

Discussion — Leidy (1853) in a communication to the
Academy of Natural Sciences of Philadelphia, applied
the name "*Oreodon*" *major* to a maxillary fragment
with M^1-M^3, USNM 19099, from the Big Badlands of
South Dakota (then Nebraska Territory), but Leidy gave
no description. Later, the specimen was described and
figured by Leidy (1854) where he noted that "*O.*"
major is similar to "*O.*" *culbertsoni* except that its teeth
are "much larger than any of the specimens which have
been attributed" to "*O.*" *culbertsoni* (= *Merycoidodon
culbertsoni*). Leidy (1869) then reported on a large and
well-preserved cranium, ANSP 10863, from South
Dakota and noted that the teeth equal those of the type
of "*O.*" *major* in size, and that ANSP 10863 has an
inflated auditory bulla which *M. culbertsoni* lacks.
Leidy at this time was just discovering that certain

Merycoidodon-like oreodonts have inflated bullae, and Leidy used the large size and the presence of the enlarged bulla of ANSP 10863 to convince himself further that "*O.*" *major* is separable from *M. culbertsoni.* Thorpe (1937a, p. 74) designated ANSP 10863 an allotype of *M. major.*

Leidy (1869, pp. 105, 457) studied another specimen, ANSP 10860, from the Big Badlands of South Dakota, which he variously called either "*Oreodon hybridus,*" or a "Variety of *O. major?*" ANSP 10860 has broken and well worn teeth that are unsuitable for measurement. Meaningful dental comparisons between this specimen and the type of "*O.*" *major* cannot be made, but the muzzle is well-preserved. ANSP 10860 has the great size, the breadth, and the distinctively shallow and pit-like preorbital fossa that is characteristic of ANSP 10863 (the well-preserved specimen noted above that has teeth that are similar in morphology and size to those of the type of "*O.*" *major*). Leidy (1869) suspected that ANSP 10860 is conspecific with "*O.*" *major,* but Thorpe (1937a), although of similar opinion, carried the name "*hybridus*" along in print.

Schultz and Falkenbach (1968) identified ANSP 10860 as "*Otionohyus (Otarohyus) hybridus,*" and referred several stratigraphically older specimens to "*O. hybridus.*" The referred specimens, however, including F:AM 72009 illustrated as a representative example, are considerably smaller than ANSP 10860. Furthermore, the referred specimens differ from ANSP 10860 by a deeper preorbital fossa, in keeping with their greater antiquity, and lack the broad frontal region that is characteristic of later Whitneyan *Merycoidodon.* The referred specimens are probably members of *Merycoidodon bullatus* (this chapter).

Schultz and Falkenbach (1949) selected a dorsoventrally crushed, incomplete skull, F:AM 45329, from the Whitney Member of the Brule Formation, "*Leptauchenia* beds" (≈"lower *Leptauchenia* beds"; see Schultz and Falkenbach, 1968, p. 425), Washabaugh County, South Dakota, which lacks the premaxillae, sagittal crest, and part of the dentition, to serve as the type of "*Promesoreodon scanloni.*" The distortion renders the type specimen "brachycephalic" as currently preserved. Schultz and Falkenbach viewed the relatively "low" and "broad" type specimen as truly brachycephalic, and regarded the species as in phylogenetic line with *Mesoreodon* of the Arikareean, a presumed brachycephalic genus. However, the type species of *Mesoreodon, M. chelonyx,* has Sk. W./Sk. L. ratios that are similar to those of *Merycoidodon culbertsoni* (Fig. 2; Table 7; Appendix B), and is thus as mesocephalic as the type species of *Merycoidodon.* This proportion is so stereotyped among the taxa that we accept for use in this report, that *M. chelonyx* (and *Miniochoerus affinis* and *M. starkensis*) shows no significant difference in this respect from any other species. Among the groups that we assign to *Merycoidodon major,* available measurements for the type and referred specimens of

"*P. scanloni*" are particularly close to Schultz and Falkenbach's (1954) sample of "*Subdesmatochoerus shannonensis*" and their 1968) sample of *M.* "*(Blickohyus) lynchi.*" As far as can be determined from study of a small and not very well-preserved sample, "*P. scanloni*" is similar to *M. major,* a contemporary.

Before Schultz and Falkenbach (1968) studied *Merycoidodon* in detail, they (1954) had named "*Subdesmatochoerus shannonensis.*" This species was based on a fairly complete but laterally crushed subadult skull, AMNH 1310, from the Poleslide Member, Brule Formation, Shannon County, South Dakota. Because Schultz and Falkenbach (1954) thought AMNH 1310 was truly dolichocephalic, "*S. shannonensis*" was placed within the "Desmatochoerinae" and the original comparisons were made with only "*S. socialis*" (see *Merycoidodon bullatus,* this chapter) and "*S. montanus*" (see *Mesoreodon chelonyx,* this chapter). Just as Schultz and Falkenbach tended to place any seemingly "low" and "broad" skull (among specimens pertinent to this discussion) within the Promerycochoerinae, these authors tended to place any pertinent "high" and "narrow" skull in the "Desmatochoerinae." Re-examination of AMNH 1310 convinces us that the "high" and "narrow" appearance of the type is the result of lateral distortion. AMNH 1310 shows a morphology, specifically a distinctively pit-like preorbital fossa, that is comparable with that seen for its contemporary, *Merycoidodon major.* AMNH 1310, in keeping with its immaturity, is probably a small individual of *Merycoidodon major.* Measurements for the type and referred specimens of "*Promesoreodon scanloni*" and "*Subdesmatochoerus shannonensis*" are closely comparable, and "*S. shannonensis*" differs little from the means for other specimens that we include in *M. major,* except for a smaller TM^3, and a significantly larger APM^3/TM^3 ratio.

The type of *Merycoidodon* "*(Anomerycoidodon) lambi*" Schultz and Falkenbach (1968) is a slightly laterally skewed and fractured skull with jaws and a partial skeleton, F:AM 72139, from the upper part of the Brule Formation, lower part of the "*Leptauchenia*" beds, 2 miles (3.2 km) west of Cedar Pass, Jackson County, South Dakota. Schultz and Falkenbach's referred specimens have a similar provenance. Schultz and Falkenbach (1968) note the shallow and small (pit-like) preorbital fossa, the large size, and the lack of a hyoidal groove in the auditory bulla of *M.* "*lambi.*" Although these are features which also characterize *M. major,* a senior taxon, *M.* "*lambi*" was not compared to *M. major* by its authors because *M. major* was regarded as a member of a separate lineage. However, the specimens of *M.* "*lambi,*" including the type, are similar to Leidy's type, USNM 19099 (teeth only) and the referred specimen, ANSP 10863 (teeth poorly preserved, but the skull well-preserved) of *Merycoidodon major.* Most measurements for *M.* "*lambi*" lie nearly centrally within distributions formed by other specimens that we refer to *M. major*, and although the P_1-M_3, M_1-M_3, and Sk. W./Sk. L. (available for three *M. "lambi"* skulls) are significantly smaller, *M. "lambi"* probably is conspecific with *M. major*.

Other specimens that are stratigraphically contemporary with *Merycoidodon major* concern *M. "(Blickohyus) lynchi"* Schultz and Falkenbach. *M. "lynchi"* is based on a skull and jaws, F:AM 45297, from high in the Brule Formation, Pennington County, South Dakota. Schultz and Falkenbach distinguished *M. "lynchi"* from *M. "lambi"* on the basis of a slight size difference. That *M. "lynchi"* averages slightly smaller than *M. "lambi"* is not supported by measurements. For the very limited samples available, Sk. L. of *M. "lynchi"* is significantly smaller, but for most other measurements average differences are trivial, with *M. "lynchi"* slightly the larger of the two, particularly for the tooth row. For our sample of *M. major*, Sk. L. is not strongly associated with P^1-M^3, though more so than is true for the sample of *Dicotyles tajacu* (Table 2). No persuasive evidence exists that would separate *M. "lynchi"* from either *M. "lambi,"* or *M. major*, the senior taxon.

Another population that is probably best referred to *M. major* was called *"Paramerycoidodon (Gregoryochoerus) wanlessi"* by Schultz and Falkenbach (1968). They regarded *"Paramerycoidodon (Gregoryochoerus)"* as a lineage of oreodonts distinct from *"P. (Barbourochoerus),"* the group in which they had placed *M. major*. Schultz and Falkenbach separated *"P. (Gregoryochoerus)"* from *"P. (Barbourochoerus)"* on the basis of a slightly smaller size, less prominent sagittal crest, and "lighter" limbs. In spite of the fact that Schultz and Falkenbach did not compare *"P. (G.) wanlessi"* with either *M. "lambi"* or *M. major*, all of the specimens representative of this taxonomy are very similar except for slight variations in size and robustness. With the exception of Sk. L. (slightly smaller), measurements on *"P. (G.) wanlessi"* average slightly larger (Malar D., significantly larger) than those of *M. "lambi,"* and *"P. (G.) wanlessi"* has a proportionally broader skull. Differences between means of measurements for *"P. (G.) wanlessi"* and those of other *M. major* specimens (as we use the term) are not systematic or large (<0.9 SDs, relative to other *M. major*), but a majority of means of measurement categories for *"P. (G.) wanlessi"* are slightly (to significantly: TM^3, Malar D.) larger. The observed variation when *"P. (G.) wanlessi"* is included in *M. major* is well within the limits expected for a single species of suiform artiodactyl (see Appendix A), best called *M. major*.

Schultz and Falkenbach (1968) went on to segregate another sample of Whitneyan oreodonts from *Merycoidodon major*, under the name of *"Otionohyus (Otarohyus) alexi."* The type specimen, F:AM 72060, a skull which lacks the premaxillae and the occipital crests,

came from talus within the "lower *Leptauchenia* beds," high in the Brule Formation, Washabaugh County, South Dakota. Schultz and Falkenbach compared F:AM 72060 with neither contemporary *M. major* nor with any of the other samples discussed in this section. Schultz and Falkenbach (1968) believed that *"Otionohyus"* represented a lineage of small, brachycephalic oreodonts that had been long separated from, and by their logic could not be related to, other Merycoidodontinae. In size, *"O. alexi"* is smaller than the majority of specimens we refer to *M. major*, (significant: M^1-M^3, TM^3, Malar D., and Sk. L.), and differs significantly from *"Paramerycoidodon wanlessi"* in M^1-M^3, TM^3, Malar D., and from *M. "lambi"* in Malar D., and Sk. L., a larger ratio of Sk. W. /Sk. L. (single *"O. alexi"* skull), and a smaller ratio of Sk. L./P^1-M^3 (two *"O. alexi"* skulls). Skulls of *"O. alexi"* are "low" as the result of physical dorsoventral deformation. Considering the limited sample of *"O. alexi,"* the size difference with other specimens of *M. major* is unimpressive. Disregarding the effects of fractures, the suite of specimens identified in the literature as *"O. alexi"* have the same derived features seen in *M. major*, such as a relatively broad frontal region, a shallow and pit-like preorbital fossa, and a greatly inflated auditory bulla which lacks a distinct hyoidal groove.

"Genetochoerus (Osbornohyus) chamberlaini" is based on a partial skull and jaws, UNSM 28340, from above the "lower ash" (see Schultz and Stout, 1955), in the Whitney Member of the Brule Formation, east of Broadwater, Morrill County, Nebraska. $^{40}Ar/^{39}Ar$ age determinations on the "lower ash" (Lower Whitney Ash, LWA) suggest an age of about 31.8 Ma for that part of the Whitney Member (Swisher and Prothero, 1990, p. 760). Schultz and Falkenbach's (1968) F:AM referred specimens of *"G. (O.) chamberlaini"* came from near the "lower ash," Sioux County, Nebraska, and from the middle part of the Poleslide Member of the Brule Formation, near or within the *"Leptauchenia* nodules" of the *"Leptauchenia* beds" (≈"lower *Leptauchenia* beds"; see Schultz and Falkenbach, 1968, p. 425) of South Dakota. Averages of available measurements on the very small sample of *"G. chamberlaini"* are significantly smaller than those for other specimens that we assign to *M. major* (exceptions: Nasal L., Subnasal L.) and significantly smaller than *"Otionohyus alexi"* in means of 12 of 16 mutually available measurements (exceptions: Nasal L., Subnasal L., Malar D., Sk. L.), but the proportions that can be determined, and the morphology do not differ from *M. major*.

Schultz and Falkenbach's (1968) referred specimen of *"?Desmatochoerus anthonyi,"* F:AM 72393, is pertinent to this discussion (the type of *D. anthonyi* is not a part of the *Merycoidodon-Mesoreodon* lineage). This fossil came from the north side of 66 Mountain, Goshen County, Wyoming, and although originally tentatively equated with an Arikareean type, F:AM

72393 actually came from 60 feet (18.3 m) below the "upper ash," in the middle part of the Whitney Member of the Brule Formation (Morris F. Skinner, personal communication, October, 1972), and within an area where the Whitney Member is characterized by several distinctive ash layers. An ^{40}Ar/^{39}Ar age determination on the "upper ash" (Upper Whitney Ash, UWA) suggests an age of about 30.6 Ma for this part of the Whitney Member (Swisher and Prothero, 1990, p. 760). Schultz and Falkenbach (1968) mistook these ashes for ash deposits in the Arikareean Gering Formation. Specimens of "*G. chamberlaini*" and F:AM 72393 are not unlike the other Whitneyan fossils discussed in this section as probably conspecific within *Merycoidodon major*, to which they are referred.

Schultz and Falkenbach (1968) skewed *Merycoidodon* (*Otarohyus*) *major* toward a large size by denying to the species its medium and smaller fractions. Because these authors believed that large Merycoidodontinae belonged to a separate lineage from other oreodonts of *Merycoidodon* aspect, the species *major* was placed within "*Paramerycoidodon*," a genus specifically created to represent this lineage. The youngest *Merycoidodon*, *M. major* (as identified in this report) is in fact substantially larger than the others (Fig. 2; Appendix B). "*Paramerycoidodon*," however, is based on "*P. georgei*," a taxon that is deemed conspecific within *M. culbertsoni* (as a subsample of larger than average individuals), and *M. culbertsoni* is a small merycoidodontine. The name and the concept of "*Paramerycoidodon*" are not available because of synonymy. Even if the name were available, there is not enough morphologic difference between the Orellan and Whitneyan Merycoidodontinae to warrant a generic separation.

Merycoidodon major, the youngest *Merycoidodon*, is not a suitable ancestor for *Eporeodon* of the latest Whitneyan because *M. major* is too large, has a more reduced preorbital fossa, and has a broader and higher frontal region due to inflation of the frontal sinuses. The separation of *Eporeodon* from *Merycoidodon* probably occurred before the occurrence of *M. major* (see *M. bullatus*). *M. major*, however, probably is the direct ancestor of *Mesoreodon chelonyx*.

Mesoreodon Scott

Type species — *Mesoreodon chelonyx* Scott, 1893, p. 659, 661.

Included species — *Mesoreodon chelonyx*, and tentatively *?M. minor*.

Amended diagnosis — Primitively, *Mesoreodon* has a generalized, mesocephalic cranium of *Merycoidodon* aspect, but *?M. minor* is sub-brachycephalic (Table 7). *Mesoreodon* has inclined premaxillae, Subnasal L. averages about 47-52% of the Nasal L., a shallow pit-like preorbital fossae, an infraorbital foramen above P^3-P^4, usually a generalized zygomatic arch but the

posterior part of the arch in robust individuals of *?M. minor* is distinctly upturned, a relatively narrow braincase and posteriorly protracted nuchal crest, little reduced premolars with average P^1-P^4/P^1-M^3 of 49-48% (Table 3), the APM^3/TM^3 averages 1.13 to 1.18 (Table 5), and the APM^3/Ht. M^3 averages 1.53 to 1.74 (Table 4), and the TM^3/Ht. M^3 ratio decreases (Table 6) from 1.46 (single measurement; oldest) to 1.36 (mean; youngest) as M^3 becomes higher crowned relative to the width. *Mesoreodon* differs from *Merycoidodon* (*Otarohyus*) by larger size and relatively narrower and higher crowned teeth (particularly noticeable for our data in the significant difference between species of *Mesoreodon* and other species examined in this report in the APM^3/TM^3 ratio, a proportion distinctly stereotyped among the other taxa), otherwise very similar. *Mesoreodon* differs from *Eporeodon* by larger size, smaller and more pit-like preorbital fossae, higher muzzle, more reduced premolars, more anteroposteriorly attenuated and slightly higher crowned molars, relatively smaller auditory bullae, and sometimes a more specialized, more upturned zygomatic arch. *Mesoreodon* differs from *Desmatochoerus* Thorpe (1921a) by smaller size. *Mesoreodon* differs from *Desmatochoerus* and *Hypsiops* Schultz and Falkenbach (1950) by a less specialized, less upturned zygomatic arch. *Mesoreodon* differs from *Hypsiops* by relatively greater Nasal L., a shorter Subnasal L., less reduced premolars, and a broader M^3.

Distribution — Turtle Cove Member of the John Day Formation, Oregon; Cabbage Patch and Fort Logan formations and "Canyon Ferry beds," Montana; deposits above the Whitney Member of the Brule Formation but below the Gering Formation (brown siltstone beds, Brule Formation), Nebraska; and the Gering Formation of Wyoming and Nebraska, and equivalents in South Dakota; latest Whitneyan to late early Arikareean, late middle Oligocene.

Discussion — The taxonomy of *Mesoreodon* is complex. There has been a trend, apparently starting with Schlaikjer (1934, 1935), followed by Thorpe (1937a) and Koerner (1940), and championed by Schultz and Falkenbach (1949), to identify only what appeared to be oreodonts with "wide" and "low" skulls (of pertinent specimens) as *Mesoreodon*. Additionally, the species that hitherto has been called "*Mesoreodon*" *megalodon* has been viewed as the characteristic *Mesoreodon*.

The type species of *Mesoreodon*, *M. chelonyx*, is based on and is augmented by undeformed topotypic specimens that are as mesocephalic as *Merycoidodon culbertsoni* (Table 7). The middle Arikareean "brachycephalic" crania from Wyoming and Nebraska which have come to typify *Mesoreodon*, such as skulls of "*M.*" *megalodon*, are typically dorsoventrally crushed. Conversely, restudy of the specimens identified in the literature as *Desmatochoerus* (or versions of this name), convinces us that this name mostly has been applied to laterally crushed skulls. These skulls are the counterparts of the "brachycephalic" crania mentioned above. *Mesoreodon* in this revision has been redefined as a mesocephalic genus. Based on the striking similarities between *M. chelonyx*, and *M. major* and other *Merycoidodon* (Figs. 2, 3), there can be no doubt that *Mesoreodon* is derived directly from *M. major*.

Thorpe (1937a) believed that *Mesoreodon* was allied with *Promerycochoerus*. Schultz and Falkenbach (1949) viewed *Mesoreodon* as immediately ancestral to *Promerycochoerus*. Thus, they considered it a member of the Promerycochoerinae, because these authors believed that the broad-headed *Promerycochoerus* had to be derived from a brachycephalic ancestor, and because the species *megalodon* was allocated to *Mesoreodon*. Metrically, *Promerycochoerus* is brachycephalic on the basis of its great bizygomatic diameter, yet the remainder of the skull, although larger, has the same proportions, and is otherwise similar to *Merycoidodon* and *Mesoreodon sensu stricto*. Schultz and Falkenbach (1949), like earlier workers, dismissed the very *Merycoidodon*-like *M. chelonyx* as incidental to the genus and placed too much taxonomic emphasis on younger crushed fossils erroneously called *Mesoreodon*. These younger "*Mesoreodon*" specimens are best placed in *Desmatochoerus* when that genus is defined more in keeping with its type species.

Thorpe (1921a) originally described *Desmatochoerus* as a subgenus of *Promerycochoerus*, for *P. (Desmatochoerus) curvidens*. Schultz and Falkenbach (1954), however, believed that *Desmatochoerus*, as a member of the "Desmatochoerinae," was not related to *Promerycochoerus*. The taxonomy of certain of the dorsoventrally or laterally crushed middle Arikareean crania and the taxa they represent, alluded to above, is very complex and will not be dealt with here. However, we have concluded that *P. "curvidens"* is conspecific with the species *megalodon* Peterson (1907), which requires placement of the species *megalodon* in *Desmatochoerus*. Through the necessary shifts in taxonomy consequent on reevaluation of morphology and consideration of the effects of deformation on types and samples, we conclude that *Mesoreodon* is ancestral directly to *Desmatochoerus*, and that *Desmatochoerus* is the direct ancestor for *Promerycochoerus* (and *Megoreodon*). Thus Thorpe's (1921a) opinion that *Desmatochoerus* is a promerycochoerine is correct. Although Schultz and Falkenbach (1954) did not accept this, they (1949) are correct in regarding the species *megalodon* as the ancestor for *Promerycochoerus*. The placement of *Desmatochoerus* in the Promerycochoerinae requires the abandonment of the Subfamily Desmatochoerinae Schultz and Falkenbach (1954).

Mesoreodon, in addition to being ancestral to *Desmatochoerus-Promerycochoerus*, also is ancestral directly to *Hypsiops* Schultz and Falkenbach (1950), the progenitor of *Submerycochoerus-Merycochoerus*.

Mesoreodon of the Merycoidodontinae, thus is ancestral to the Promerycochoerinae (probably upland forms) and Merycochoerinae (probably semiaquatic, lowland forms based on their progressively more tapiroid rostral morphology). These two derived subfamilies often paralleled each other in the trend for gigantism and zygomatic specialization. The origin of *Eporeodon* of the Eporeodontinae, has to be more remote within *Merycoidodon* (*Otarohyus*) than the occurrence of *M. (O.) major* and *Mesoreodon chelonyx*. *Eporeodon*, approximately a contemporary of *Mesoreodon chelonyx*, initiated the late Whitneyan-early Arikareean adaptive radiation of oreodonts that tended toward phyletic dwarfism, hence remotely was responsible for the formation of the Merychyinae, Ticholeptinae, Ustatochoerinae (new rank, see family Merycoidodontidae, this chapter), and certain other groups.

Mesoreodon chelonyx Scott
Figures 2E, 3E; Tables 3-7; Appendix B

Mesoreodon chelonyx Scott, 1893:659, 661
Mesoreodon chelonyx Scott: Schultz and Falkenbach, 1949:140
Eucrotaphus montanus Douglass, 1907b:100
Subdesmatochoerus montanus (Douglass): Schultz and Falkenbach, 1954:219
Eporeodon meagherensis Koerner, 1940:845
Paramerycoidodon (Gregoryochoerus) meagherensis (Koerner): Schultz and Falkenbach, 1968:103
Mesoreodon danai Koerner, 1940:847

Type—PU 10425, Fort Logan Formation, Smith River Valley, Meagher County, Montana.

Hypodigm—Type, and Schultz and Falkenbach's (1949) sample of *Mesoreodon chelonyx* such as PU 10418, 10410, 11769, and 10443A; type, CMNH 907, of "*Eucrotaphus montanus*"; type, YPM 13948, of "*Eporeodon meagherensis*"; type, YPM 13949, of *Mesoreodon "danai"*; type, F:AM 45443, and Schultz and Falkenbach's (1954) sample of "*D. sandfordi*," such as F:AM 47580 and F:AM 49636; and F:AM 24429.

Amended diagnosis—The cranial characterization defines the genus. The Subnasal L. is 47% of the Nasal L. (three specimens), the mean P^1-P^4/P^1-M^3 is about 49% (Table 3), the mean APM^3/TM^3 ratio is 1.13 (Table 5), the APM^3/Ht. M^3 ratio is 1.53 (Table 4, one specimen), and the sole value for TM^3/Ht. M^3 is 1.46 (Table 6). *Mesoreodon chelonyx* differs from *Merycoidodon major* by systematically slightly to significantly (P^1-M^3, P^1-P^4, M^1-M^3, APM3s, Ht. M^3, Nasal L) larger size and by a significantly more anteroposteriorly attenuated M^3, otherwise similar. *Mesoreodon chelonyx* differs from *Eporeodon occidentalis* by systematically larger size (significant except for P.O.C., Sk. W., Ht. M_3; 11 of 18 observed ranges do not overlap, and range overlaps do not exceed 35%), slightly more reduced premolars, significantly narrower M^3, and narrower skull in proportion to Sk. L. *Mesoreodon chelonyx* differs from *?Mesoreodon minor* by a systematically slightly (not significant: Ht. M3s, Nasal L., Sk. L.) to significantly smaller size, slightly less anteroposteriorly attenuated M3s, slightly lower crowned teeth (observed ranges, Ht. M3s, do not overlap), slightly but significantly larger P^1-P^4/P^1-M^3 ratio, significantly smaller Subnasal L./Nasal L. ratio, a less robust, less upturned zygomatic arch, and by a mesocephalic instead of a sub-brachycephalic cranium (Table 7). Metrical and morphologic distinctions between *M. chelonyx* and *?M. minor* are more impressive than those between *Merycoidodon major* and *Mesoreodon chelonyx*.

Distribution—Fort Logan Formation and probably from some of the Canyon Ferry "beds," Montana; and from the brown siltstone beds of the Brule Formation, Nebraska; latest Whitneyan or earliest Arikareean, late middle Oligocene.

Discussion—*Mesoreodon chelonyx*, the type species, is based on a well-preserved, mesocephalic, slightly fractured but undistorted skull, jaws, and a partial skeleton. The species is augmented by a topotypic sample of several undeformed but fractured, variously preserved mesocephalic skulls, and jaws. Scott (1893, 1895) remarks on the similarity of *M. chelonyx* to *Eporeodon* (*?sensu lato*). The type came from rocks discovered by Grinnell and Dana (1876) along what was then called "Deep River," now the Smith River, Meagher County, Montana. Grinnell and Dana recognized a "Miocene" and a "Pliocene" horizon in the deposits along the Smith River Valley. Cope (1878) called these rocks the "*Ticholeptus* beds," and believed them to be intermediate in age between the John Day Formation and the "Loup Fork."

Scott (1893) and Douglass (1903) agreed with Grinnell and Dana (1876) that Cope's (1878) "*Ticholeptus* beds" do not represent a single biostratigraphic unit. Scott, however, continued to refer to the rocks singularly, but introduced a new name "Deep River beds." Douglass (1903) then restricted Scott's name, "Deep River beds," to the upper, younger (early Barstovian) sediments because Scott had discussed more fully the fauna from that unit. Douglass (1903) then introduced a new term, Fort Logan beds, to designate the older so-called John Day equivalent. This was accepted by Koerner (1940).

Schultz and Falkenbach (1949, 1954) thought that the rocks along the Smith River Valley represented "a group of formations." One, a Gering equivalent based on the stage of evolution of *Mesoreodon chelonyx*, one equivalent with the Harrison Formation based on Schultz and Falkenbach's identification of an oreodont from these deposits as "*Promerycochoerus*" *latidens*, and a third formation comparable with the "Sheep Creek" (including the "Lower Snake Creek") because the highest rocks contain *Ticholeptus*. This is poor lithostratigraphy.

Douglass (1907a) named as "*Eucrotaphus montanus*" a skull, jaws, and some associated bones, CMNH 907, from unindurated sediment that "evidently [overlies] the Lower White River," 10 miles (16.1 km) northeast of Helena, Lewis and Clark County, Montana. The skull is slightly laterally crushed, and appears "high" and "narrow" as currently preserved. The Tertiary rocks along the Missouri River from Helena down along and under the present-day Canyon Ferry Lake have been grouped collectively as yielding the Canyon Ferry faunas (see Kay et al., 1958, p. 33). These faunas range in age from Chadronian to Hemingfordian. Douglass, however, did not say that "*E. montanus*" came from his "Canyon Ferry Beds," which he regarded as "Miocene"; presumably it came from older rocks. Schultz and Falkenbach (1954) believed that "*Eucrotaphus montanus*" was Whitneyan in age on its morphologic stage of evolution, which is not unreasonable. Additionally, the great similarity of "*E. montanus*" to *Mesoreodon chelonyx* supports a latest Whitneyan or earliest Arikareean age assessment for the species.

Because the type skull of "*Eucrotaphus montanus*" appears narrow as now preserved, Schultz and Falkenbach (1954) identified "*E. montanus*" as "*Subdesmatochoerus*" and a member of the "Desmatochoerinae," but the type must have been mesocephalic in life. Disregarding the cranial deformation, the type has a P.O.C., Sk. W., and significantly smaller Nasal L. that fall below the observed range for other *Mesoreodon chelonyx*, but otherwise closely resembles those specimens (averages about 0.3 SDs smaller; SDs of *M. chelonyx*). "*E. montanus*" is referred to *M. chelonyx* here.

When Koerner (1940) reviewed the faunas from the Fort Logan and Deep River formations, he named "*Eporeodon meagherensis*." "*E. meagherensis*" is based on a well-preserved skull, jaws, and associated bones, YPM 13948, from the Fort Logan Formation. Koerner noted the type's similarity with *Eporeodon pacificus*, and based (correctly) a specific separation of "*E. meagherensis*" from *E. pacificus* on the shallower, more pit-like preorbital fossa and the more reduced premolars of "*E. meagherensis*." Koerner did not compare "*E. meagherensis*" with *Mesoreodon chelonyx*, which is curious because Koerner knew that *M. chelonyx* is also like *Eporeodon* (and *Merycoidodon*), except for size. Presumably the lack of comparison results from the fact that *Mesoreodon* is not *Eporeodon* (or *Merycoidodon*) by definition, and also because Koerner (1940) erroneously considered *Mesoreodon* to be brachycephalic. The characteristics which Koerner used, however, to separate "*E. meagherensis*" from *E. pacificus*, are precisely those which distinguish *Mesoreodon* from *Eporeodon*, and phylogenetically tie *Mesoreodon chelonyx* to *Merycoidodon major*.

Schultz and Falkenbach (1968) identified "*Eporeo-*

don meagherensis" as "*Paramerycoidodon (Gregoryo-choerus) meagherensi.*" These names show that they recognized that the species "*meagherensis*" is *Merycoi-dodon*-like, and placed the taxon in the Merycoido-dontinae. "*E. meagherensis,*" however, is *Merycoido-don*-like in the same way that *Mesoreodon chelonyx* is *Merycoidodon*-like. Schultz and Falkenbach, like Koerner, did not compare *M.* "*(G.) meagherensis*" with *M. chelonyx*, probably for the same reasons. Yet except for a size slightly smaller than average (average: about 0.6 SDs; most notably: P$_1$-M$_3$, and the significantly smaller Sk. L.), *M.* " *(G.) meagherensis,*" is indistinguishable from *M. chelonyx*, to which it is here referred.

Mesoreodon "danai" Koerner (1940) is based on a skull and partial mandible, YPM 13949, also from the Fort Logan Formation. Koerner distinguished *M. "danai"* from *M. chelonyx* by several trivial features such as the absence of a pit on the anteroexternal corner of P^4, and more widely separated supraorbital foram-ina. Schultz and Falkenbach (1949) referred *M. "danai"* to *M. chelonyx*, and, although YPM 13949 has significantly larger Malar D. and Sk. L. (upper limits of the observed ranges of these measurement categories in *M. chelonyx* as we understand the species), we concur with this synonymy.

The type of "*Desmatochoerus sandfordi*" is F:AM 45433, from the Gering Formation, Sioux County, Nebraska, a poorly preserved, deformed, and almost unmeasurable skull and jaws of an aged individual. As noted by Schultz and Falkenbach (1954) the type is small relative to the taxa to which the specimen was compared. Of the 11 measurement categories available for comparison, means of eight are slightly smaller than, but within the observed ranges of, those of other *M. chelonyx* (exceptions: more noticeably larger Subnasal L., P$_1$-M$_3$, and P$_1$-P$_4$). The diminutive appearance of the type is exaggerated by incomplete-ness and the fact that the teeth are worn to their roots. In morphology, determined from the broken and crushed type, and size, determined mainly from referred specimens, the sample of "*Desmatochoerus sandfordi*" agrees better with *Mesoreodon chelonyx* than ?*M. minor*, and is tentatively referred to *M. chelonyx*.

The late Morris F. Skinner, formerly of the Frick Laboratory, American Museum of Natural History, collected a skull and jaws, F:AM 24429, from rocks that occur below the unconformity which underlies the Gering Formation of Darton (1899) and overlies deposits that usually have been called Brule Formation, in an area one quarter mile (0.4 km) west of the Redington Gap Road, Morrill County, Nebraska (Skinner, personal communication, April, 1970; see Tedford et al., 1987, p. 165). These rocks have been designated informally as the brown siltstone beds of the Brule Formation (see Tedford et al., 1985, p. 341; Swinehart et al., 1985; Tedford et al., this volume,

Chapter 15). Radiochronology on the Nonpareil ash (NPAZ) gives an age of about 30 Ma (Swisher and Prothero, 1990, p. 760) for the lower part of the brown siltstone. F:AM 24429, although laterally crushed and "narrow" as now preserved, is similar morphologically to Schultz and Falkenbach's (1949) very small sample of *Mesoreodon chelonyx*, and compares closely in size except for noticeably greater Subnasal L., Malar D. (significant), and P$_1$-P$_4$. Comparison to specimens in our sample of *M. chelonyx* is similar except that Malar D. is unremarkable, but Sk. L., though within the range observed for other *M. chelonyx*, is significantly large. The known stratigraphy of F:AM 24429 relative to the older, Whitney Member of the Brule Formation, and younger (Arikaree) rocks is in agreement with the latest Whitneyan or earliest Arikareean age and morphologic intermediacy (between *Merycoidodon major* of the Whitneyan and ?*M. minor* of the early Arikareean) here attributed to *M. chelonyx*.

Mesoreodon chelonyx as redefined more consistent with its type specimen and topotypic sample, is a mesocephalic species (Fig. 3E) that is more similar to *Merycoidodon major* than it is to any of the younger species that have here, or traditionally, been placed in *Mesoreodon*. *M. chelonyx* is a phylogenetic continua-tion of *Merycoidodon* (Fig. 2D-E). Because *M. chelonyx* is so much like *Merycoidodon*, *Mesoreodon* must be placed in the Merycoidodontinae. *M. chelonyx*, in turn, is thought to be the immediate ancestor for ?*M. minor* (Figs. 2F, 3F) of the early Arikareean.

?*Mesoreodon minor* (Douglass)
Figures 2F, 3F, F'; Tables 3-7; Appendix B

Promerycochoerus minor Douglass, 1903:158
Promerycochoerus (Pseudopromerycochoerus) minor (Douglass): Schultz and Falkenbach, 1949:124
Eporeodon cheeki Schlaikjer, 1934:220
Mesoreodon cheeki (Schlaikjer): Schultz and Falkenbach, 1949:135
Mesoreodon scotti Schlaikjer, 1934:223
Mesoreodon cheeki scotti (Schlaikjer): Schultz and Falkenbach, 1949:137
Desmatochoerus (Paradesmatochoerus) grangeri Schultz and Falkenbach, 1954:194
Desmatochoerus (Paradesmatochoerus) monroecreekensis Schultz and Falkenbach, 1954:195
Desmatochoerus (Paradesmatochoerus) wyomingensis Schultz and Falkenbach, 1954:199
Pseudodesmatochoerus wascoensis Schultz and Falkenbach, 1954,:210
Desmatochoerus macrosynaphus Riel, 1964:3

Type—CMNH 769, Cabbage Patch Formation, Granite County, Montana.

Hypodigm—Type, and the type, MCZ 17765, and Schultz and Falkenbach's (1949) sample of *Mesoreodon "cheeki,"* such as F:AM 45430, 37567,

33359, and 42309; type, MCZ 17480, and Schultz and Falkenbach's (1949) sample of *M.* "*cheeki scotti,*" such as F:AM 33532, 44924, 44916, and 44939; type, F:AM 33303, and UNSM 28500 of *Desmatochoerus* "*Paradesmatochoerus grangeri*"; type, F:AM 37551, and Schultz and Falkenbach's (1954) sample of *D. (P.) monroecreekensis,* such as F:AM 44915, 45439, 45438, and 33682; type, F:AM 33312, and Schultz and Falkenbach's (1954) sample of *D.* "*wyomingensis,*" such as F:AM 37571, 49637A and 49638; type, AMNH 7827, and AMNH 7634 of "*Pseudodesmatochoerus wascoensis*"; and the type, MSU 0940, of "*Desmatochoerus macrosynaphus.*"

Amended diagnosis — The cranial characterization is that for a derived *Mesoreodon,* in which the Subnasal L. is 52% of the Nasal L., P^1-P^4/P^1-M^3 averages 48% (Table 3), the mean APM^3/TM^3 is 1.18 (Table 5), the APM^3/Ht. M^3 is 1.74 (one specimen; Table 4), and TM^3/Ht. M^3 averages 1.36 (Table 6). ?*Mesoreodon minor* differs from *Mesoreodon chelonyx* by a slightly to significantly (length and width aspects of tooth rows, Subnasal L., Malar D., P.O.C., Sk. W.) larger average size, with M3s higher crowned (no range overlap) as well as significantly narrower, slightly but significantly smaller P^1-P^4/P^1-M^3 ratio, significantly larger Subnasal L./Nasal L. ratio, a more robust and posteriorly more upturned zygomatic arch, and a broader skull (Table 7), otherwise similar. ?*Mesoreodon minor* differs from *Desmatochoerus megalodon* by smaller size and a less specialized, less robust, and less upturned zygomatic arch, otherwise similar.

Distribution — Cabbage Patch Formation, Montana; Gering Formation, Nebraska; and John Day Formation, Oregon; early Arikareean, early late Oligocene.

Discussion — Douglass (1903) designated a poorly preserved and greatly restored partial skull with jaws, CMNH 769, as the type of "*Promerycochoerus*" *minor.* The fossil came from what now is called the University of Kansas-Montana State University locality KU-MT-33, Cabbage Patch no. 1 (Rasmussen, personal communication, 1977), 3.5 miles (5.6 km) northeast of Drummond, Montana. This site is located in the basal part of the type section of the Cabbage Patch Formation (also see Macdonald, 1956; Konizeski and Donohoe, 1958; and Rich and Rasmussen, 1973).

Konizeski and Donohoe (1958) regarded the Cabbage Patch Formation as more or less correlative with the lower and middle parts (presumably the "middle," or Turtle Cove Member) of the John Day Formation of Oregon. The type of ?*Mesoreodon minor,* however, has commonly been thought to be a small species of *Promerycochoerus* because of its robust and upturned zygomatic arch. This suggested to Schultz and Falkenbach (1949) that "*P.*" *minor* was late Arikareean in age, because these authors correctly attributed a late Arikareean age to *Promerycochoerus.* Rasmussen's study (personal communication, 1977) of the stratigraphy of the Cabbage Patch Formation indicated to him

that the rocks are equivalents of the Gering and Monroe Creek formations, thus also equivalent with a part of the middle John Day, and this is accepted here.

Schlaikjer (1934) designated an almost complete skeleton, MCZ 17765, as the type of "*Eporeodon cheeki,*" and noted that the fossil came from the "lower Harrison beds," 200 feet (61 m) above the Whitney Member of the Brule Formation, Goshen County, Wyoming. Schlaikjer (1934, 1935) did not subdivide Peterson's (1907) "lower Harrison" beds, rocks that are now called the Gering, Monroe Creek, and Harrison formations. Based on the stated position above the Brule Formation and the specimen's stage of evolution, the type may have come from the upper part of the Gering Formation. Schlaikjer characterized the type as having broad, widely spread zygoma, a rather low facial region, not particularly large auditory bullae (for the size of the skull), and noted that the specimen resembles *Eporeodon.* In Schlaikjer's time, specimens of *Merycoidodon*-aspect from the Brule Formation were placed in *Eporeodon* if they had inflated auditory bullae. Schlaikjer viewed "*E. cheeki*" as "the last known survivor of '*Eporeodon*' existing as a 'living fossil' in the early Miocene along with such genera as *Mesoreodon* and *Promerycochoerus* that are descendants of the Oligocene [i.e., White River] members of this genus." "*E. cheeki,*" however, while slightly larger in P^1-M^3, M^1-M^3, and P_1-P_4 (exception: slightly smaller M_1-M_3), is morphologically the same kind of oreodont as that described earlier as "*P.*" *minor* by Douglass (1903), and except for a significantly larger P.O.C., and significantly smaller Sk. L., compares closely with other specimens that we call ?*Mesoreodon minor.*

Schlaikjer (1934) went on to name another nearly complete skeleton, MCZ 17480, as *Mesoreodon* "*scotti.*" The type had been collected by F. B. Loomis in the Muddy Creek area, Niobrara County, Wyoming, from rocks then regarded as "lower Harrison." Schultz and Falkenbach (1949) believed that *M.* "*scotti*" came from the Monroe Creek Formation (the middle sediments of the so-called "lower Harrison beds"), and that the taxon was only subspecifically different from *M.* "*cheeki.*" Unfortunately, the Gering (the lower sediments of the "lower Harrison beds") and Monroe Creek formations in the Muddy Creek area cannot be distinguished lithologically (M. F. Skinner, personal communication, 1970). The type of *M.* "*scotti*" is morphologically very similar to *M.* "*cheeki,*" but both types in turn are not very different from the type of ?*M. minor.* Measurements on specimens of *M.* "*scotti*" average slightly larger than those for *M.* "*cheeki,*" and are significantly larger than those of the type of ?*M. minor* for measurements on P^1-M^3 and M^1-M^3. We believe that the taxa under discussion, drawn from a limited geographic area, are based on types which belong to a single population, and that the observed minor differences between them are due to slight intraspecific variation in robustness and size, especially of the zygomatic arch. The amount of intraspecific variation in the robustness of the zygomatic arch becomes progressively greater in the Promerycochoerinae (and Merycochoerinae) as this group evolves. We begin to see this variation in *?M. minor.*

"*Desmatochoerus grangeri*" is based on a skull, jaws, and other bones, F:AM 33303, again from the Muddy Creek area, Niobrara County, Wyoming. Schultz and Falkenbach (1954) believed that F:AM 33303 came from the Gering Formation, but the same stratigraphic uncertainties seen for *M.* "*scotti*" apply to F:AM 33303. "*D. grangeri*" as characterized by its authors has a long face, robust zygoma, and relatively retracted nasals. The degree of retraction of the nasals, however, cannot be verified from the type specimen because the cranium is skewed anteroposteriorly by crushing which has pushed the dorsal surface of the skull backward relative to the palate. When the affects of physical deformation of the type are considered, "*D. grangeri*" cannot be distinguished from *?Mesoreodon minor* on the basis of morphology or size, although the Sk. L. is a significantly larger maximum among *?M. minor* specimens, P1-M3s and M^1-M^3 measurements are significantly large, M_1-M_3 is a significantly small minimum, and the Sk. L./P^1-M^3 (2.30) is a maximum for our sample of *?M. minor*, while Sk. W./Sk. L. is a minimum (67.1%; Table 7).

"*Desmatochoerus monroecreekensis*" is based on a skull with jaws and some other bones, F:AM 37551, again from the Muddy Creek area, Niobrara County, Wyoming. Schultz and Falkenbach (1954) thought that the type came from the Monroe Creek Formation. The type of "*D. monroecreekensis*" and most of its referred specimens are severely laterally crushed which makes the crania appear relatively "long" and "narrow." Because Schultz and Falkenbach believed that *Desmatochoerus* was phylogenetically dolichocephalic (an observation derived mainly from the study of laterally crushed referred specimens), F:AM 37551 was not compared with *Mesoreodon*. Mean values of measurements on specimens of "*D. monroecreekensis*" compare closely with those of other specimens that we include in *?M. minor*, although "*D. monroecreekensis*" means are usually slightly larger (for a very small sample, average of 0.8 SDs, relative to other *?M. minor*; exceptions: Malar D., P.O.C., P_1-P_4). "*D. monroecreekensis*" is larger than the type of *?M. minor*, but significantly so only for P^1-M^3. Based on morphology and size, the specimens of "*D. monroecreekensis*" are probably deformed crania of *?Mesoreodon minor*.

Another suite of specimens that is morphologically comparable to those being discussed in this section was described by Schultz and Falkenbach (1954) as "*Desmatochoerus wyomingensis*." This taxon is based on F:AM 33312, a skull with jaws from the ?Gering Formation, Willow Creek area, Niobrara County,

Wyoming. The same stratigraphic uncertainties which affect the Muddy Creek area, apply to Willow Creek. The type and sample of "*D. wyomingensis*" is similar to the sample of "*D. monroecreekensis*" of the Muddy Creek area in size as well as morphology, and these in turn are little different from (in size, somewhat larger than) the type of *?Mesoreodon minor*. Size differences between "*D. monroecreekensis*" and "*D. wyomingensis*" for 15 available comparisons are nonsystematic and minor, and do not warrant specific separation. Schultz and Falkenbach's (1954) distinction rests mainly on a belief that "*D. wyomingensis*" was early, and "*D. monroecreekensis*" was middle Arikareean in age, but there is no lithostratigraphic evidence on which to base this conclusion.

"*Pseudodesmatochoerus wascoensis*" is based on a severely fractured cranium from the John Day Formation, Oregon. The basicranial length of the type, AMNH 7827, has been increased by poor restoration as evidenced by a wide, plaster-filled fracture between the braincase and the rest of the skull, and the cranium is otherwise much fractured and restored. Schultz and Falkenbach (1954) did not compare "*P. wascoensis*" with any early or middle Arikareean oreodont because these authors maintained throughout their oreodont revisions that all John Day taxa (middle, Turtle Cove, and upper, Kimberly members of the John Day Formation) were late Arikareean in age. This belief has had an enormous impact on oreodont taxonomy because it excluded the John Day taxa from possible phylogenetic relationship with oreodonts elsewhere known to be older than late Arikareean. Most of the John Day species were regarded by Schultz and Falkenbach as relict, late survivors of primitive groups (an opinion that can be traced back to Schlaikjer, 1935), or as taxa endemic to the far northwest. Four of six measurement categories available for comparison of "*P. wascoensis*" with the type of *?Mesoreodon minor* are smaller (exceptions: P^1-M^3, P^1-P^4), but for five of nine mutually available characters, specimens referred to "*P. wascoensis*" fall well within the observed ranges for other specimens that we assign to *?M. minor*. Compared to other *?Mesoreodon minor*, "*P. wascoensis*" has significantly smaller M^1-M^3, and APM^3, and noticeably but not significantly smaller TM^3, while the Sk. W. is slightly above the range observed for other *?M. minor*. Based on gross morphology, the type of "*P. wascoensis*" appears similar to *?M. minor*, to which it is referred.

Riel (1964) designated a jaw with worn teeth, MSU 0940, associated with several worn upper teeth from locality no. 1, KU-Mt-31 (University of Kansas-Montana State University locality, the type locality of the Cabbage Patch Formation), the type of "*Desmatochoerus macrosynaphus*." The type has the same stratigraphy as the type of *?Mesoreodon minor* (Rasmussen, personal communication, 1977). Riel's taxonomy is based on the view that "*D. macrosynaphus*" had a long mandibular symphysis, but the symphysis of the type is severely crushed dorsoventrally and the fragments are separated. Riel also thought that the generalized P3 and P4 relative to other *Desmatochoerus* as identified in the literature, represented a primitive condition. The generalized premolars, however, are in keeping with the stage of evolution expected for an early Arikareean oreodont relative to middle Arikareean descendants. The TM3 measurement of MSU 0940 is significantly smaller than the average, and below the observed range, for that character in other specimens that we refer to ?*M. minor*, M^1-M^3 is slightly but significantly smaller, and P1-P4, above the range observed for other ?*M. minor,* is significantly larger. Otherwise, undeformed parts of the type of "*D. macrosynaphus*" are comparable in morphology and size to homologues in ?*M. minor*, to which MSU 0940 is referred.

?*Mesoreodon minor* is a derived species that is transitional between *M. chelonyx* on the one hand, and *Desmatochoerus megalodon* of the Promerycochoerinae on the other. ?*Mesoreodon minor* also has an affinity with *Hypsiops* of the Merycochoerinae.

Subfamily MINIOCHOERINAE

Miniochoerinae Schultz and Falkenbach, 1956:391
Miniochoerinae Schultz and Falkenbach: Schultz and
    Falkenbach, 1968:382
    Type species—*Miniochoerus gracilis* (Leidy), 1851,
p. 239.
    Included genera—*Miniochoerus* only.
    Amended diagnosis—Medium-small to small, mesocephalic (Table 7) oreodonts that have skulls with steep premaxillae, convex nasal profiles, Subnasal L. averages 50-63% of the Nasal L., relatively small orbits located rather low on the sides of the face, persistently small auditory bullae with deep hyoidal grooves, unspecialized zygoma, generally with only a moderately prominent, little protracted, and somewhat broadened nuchal crest, and teeth that have thinned and progressively thinner enamel on the middle surfaces of the crests and selenes, a distinctive feature first noted by Leidy (1852, p. 551). Mean P^1-P^4 values range from 47% to 44% of P^1-M^3 (Table 3), mean APM^3/TM^3 ratios range from 0.98 to 1.05 (Table 5), the M^3 is brachydont and mean APM^3/Ht. M^3 ratios range from 1.85 to 2.07 (Table 4), and the TM^3/Ht. M^3 ratios increase slightly but regularly (except ?*M. forsythae*, 1.89) with time from 1.82 to 1.95 (youngest) as the M^3 becomes lower crowned relative to the width (Table 6). Primitive Miniochoerinae have less reduced enamel, longer and narrower occiputs, and deeper preorbital fossae than younger species. Of pertinent subfamilies, Miniochoerinae differ from Oreonetinae Schultz and Falkenbach (1956) and Merycoidodontinae

Hay (1902) by notably thinned enamel on the premolar-molar crests and selenes; differ from Oreonetinae and early Leptaucheniinae Schultz and Falkenbach (1940, 1968) (*Limnenetes*) by much larger size; differ from *Limnenetes* by lower crowned teeth; and differ from post-early Orellan Merycoidodontinae and Leptaucheniinae by retaining small auditory bullae.

Distribution—White River Formation, Flagstaff Rim area, Natrona County, and Chadron Formation, White River Group, Niobrara and Converse counties, Wyoming; "lower," "middle," and "upper oreodon beds," Scenic and Poleslide members of the Brule Formation, South Dakota; Orella and Whitney members of the Brule Formation, Nebraska; and presumed equivalents in North Dakota; middle Chadronian to earlier Whitneyan, late Eocene to middle Oligocene.

Discussion—Historically, *Merycoidodon culbertsoni*, the type species of the family Merycoidodontidae, was thought to be related closely to certain small oreodonts that Leidy had named as "*Oreodon*" *gracile* and "*O.*" *affinis*. Hay (1902, p. 665) transferred "*O.*" *gracile* and "*O.*" *affinis* to *Merycoidodon*, the senior name, and this was accepted by Thorpe (1921a, 1937a, b), Bump and Loomis (1930), and others. Schultz and Falkenbach (1956) believed that the differences between *M. culbertsoni* and "*M.*" *gracilis* and its allies are sufficient to warrant the removal of them from *Merycoidodon*, hence from the Merycoidodontinae (see Schultz and Falkenbach, 1968), and placed them in a new genus, *Miniochoerus*, basis for a new subfamily, Miniochoerinae.

Schultz and Falkenbach (1956, pp. 384, 387-388, 391) characterized the Miniochoerinae to contain generally small "Oligocene" oreodonts that retain persistently small auditory bullae with deep hyoidal grooves, and have relatively short and broad occiputs. A small bulla in oreodonts is probably primitive for the family, contrary to Scott's (1940) interpretation. The lack of prominent occipital crests and the wide occiput of Miniochoerinae (relative the width of the braincase) results from the fact that Miniochoerinae are fetalized due to phylogenetic dwarfism relative to their sister group, the Merycoidodontinae. As a result the brain dominates more of the skull than in larger oreodonts. One aspect of regressions of Sk. L. on P^1-M^3 (Table 2) discussed above, is that in *Miniochoerus*, unlike *Merycoidodon*, P^1-M^3 changes with (generally decreasing) size more rapidly than does Sk. L. (Table 8). The difference between extremes (*Miniochoerus chadronensis*, *M. gracilis*) is significant.

Just as Schultz and Falkenbach (1968) believed that they could recognize "brachycephalic," mesocephalic, and "dolichocephalic" lineages among the Merycoidodontinae, they (1956) divided the Miniochoerinae into several lineages on the basis of apparent skull shape. Schultz and Falkenbach (1956) designated the "brachycephalic" Miniochoerinae as "*Platyochoerus*," the mesocephalic species as *Miniochoerus*, and the

Table 8

Summary of skull length/upper tooth row length (Sk. L./P^1-M^3) ratios for Miniochoerinae

	N	Mean	Standard Deviation	Coefficient of Variation	Maximum	Minimum	SEM ¥ $t_{(0.05, N-1)}$
?*Miniochoerus forsythae*	4	1.98	0.079	3.98	2.088	1.903	0.030
Miniochoerus							
M. chadronensis	26	1.98	0.057	2.86	2.07	1.84	0.008
M. affinis	24	2.02	0.076	3.78	2.14	1.87	0.011
M. gracilis	14	2.07	0.075	3.62	2.23	1.94	0.014
M. starkensis	7	2.02	0.098	4.86	2.15	1.86	0.027

"dolichocephalic" representatives as "*Stenopsochoerus*." Re-examination of the specimens which are the basis for this taxonomy shows that the "brachycephalic" Miniochoerinae are represented by dorsoventrally, usually severely, crushed specimens; the mesocephalic species are based on essentially uncrushed fossils; and the "dolichocephalic" forms are based on laterally, usually severely, crushed skulls. Measurements on undeformed (as nearly as could be determined) skulls suggest that Miniochoerinae are somewhat more stereotyped in the Sk. W./Sk. L. ratio (Table 7) than the Merycoidodontinae, with the possible exception of *M. gracilis* (significantly different in this respect from *M. chadronensis* and ?*M. forsythae*), and species of the two subfamilies show few significant differences in Sk. W./Sk. L. ratios.

Schultz and Falkenbach (1956) further divided *Miniochoerus* between *Miniochoerus* and *M.* "(*Paraminiochoerus*)," and "*Stenopsochoerus*" between "*Stenopsochoerus*" and "*S. (Pseudostenopsochoerus).*" Although poorly diagnosed, *M.* "(*Paraminiochoerus*)" was considered by its authors to be slightly more gracile and to have narrower frontals than *Miniochoerus*. The minor differences, however, can be accounted for by intraspecific variation and minor deformation bias. The distinction between "*Stenopsochoerus*" and "*S. (Pseudostenopsochoerus)*" rests on a presumed broader postorbital constriction for "*S. (Pseudostenopsochoerus),*" which in effect nullifies "*Stenopsochoerus*" relative to *Miniochoerus*. Schultz and Falkenbach (1956) designated as "*Parastenopsochoerus*" any Miniochoerinae with a seemingly reduced sagittal crest (probably a function of ontogeny).

Schultz and Falkenbach (1956, p. 387) believed that "The remains of Miniochoerinae do not suggest close relationship with any other subfamily of oreodonts." The Miniochoerinae, however, probably are more closely related to Merycoidodontinae than they are to any of the other pioneering oreodont subfamilies. The oldest Miniochoerinae are larger, have deeper and larger preorbital fossae with well marked anterior margins, relatively less reduced premolars, premolar-molar crests and selenes that have less thinned enamel, and more attenuated and narrower nuchal crests than younger specimens. Also, Miniochoerinae have small auditory bullae that are indistinguishable from those of Chadronian-early Orellan Merycoidodontinae. These features suggest a close relationship between these subfamilies. The divergence of Miniochoerinae from near or within the Merycoidodontinae must have occurred very early in the Chadronian, if not earlier. After divergence, the Miniochoerinae remained remarkably stereotyped in proportions during their five-million-year history. Clark and Beerbower (in Clark et al., 1967, p. 107) suggested that a "*Merycoidodon* with small bullae might well have mated with a *Miniochoerus* with large bullae." *Miniochoerus*, however, never developed enlarged bullae, and that *Merycoidodon* could have interbred with *Miniochoerus* seems improbable.

Miniochoerus Schultz and Falkenbach

Type species—*Miniochoerus gracilis* (Leidy), 1851, p. 239.

Included species—?*Miniochoerus forsythae*, *M. chadronensis*, *M. affinis*, *M. gracilis*, and *M. starkensis*.

Amended diagnosis—Same as for the subfamily.

Distribution—Same as for the subfamily.

Discussion—Oreodonts that are now regarded as members of the Miniochoerinae cannot be allocated to *Merycoidodon* or to any of its synonyms, to "*Oreodon*" [based on "*O. priscum*" (Leidy, 1851, p. 238), a juvenile of *M. culbertsoni* (Thorpe, 1937a, p. 47)], to *Oreonetes* Loomis (1924), to *Limneneetes* Douglass (1901), or to *Bathygenys* Douglass (1901) due to morphologic and/or great size dissimilarity. Schultz and Falkenbach (1956) have provided a valid name, *Miniochoerus*, for these animals as first revisers (Articles 24A and 69B of the International Code of Zoological Nomenclature, 1985).

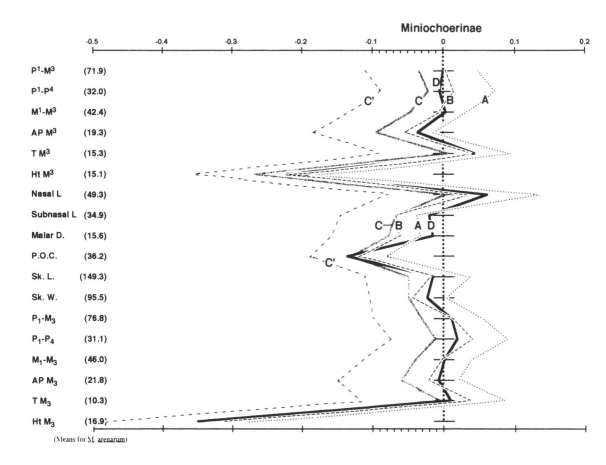

Figure 4. Cumulative ratio diagrams based on species sample means for *Miniochoerus* (Appendix C). (A), *?Miniochoerus forsythae*; (B), *M. chadronensis*; (C), *M. affinis*; (C'), *M. gracilis*; and (D), *M. starkensis*.

Miniochoerus remained stereotyped during its known history, as shown by its persistent brachydonty, cranial generality, and lack of bullar inflation. The only visible changes through time as inferred from stratigraphy, other than shifts in size, is a decrease in the depth of the preorbital fossa, premolar reduction, a slight narrowing of M_3, and cumulative thinning of the enamel on the premolar-molar internal crests and selenes. *Miniochoerus* is not a likely ancestor for any other group.

?Miniochoerus forsythae (Schultz and Falkenbach)
Figures 4A, 5A; Tables 3-8; Appendix C

Prodesmatochoerus natronensis Schultz and
    Falkenbach, 1954:228, a *nomen vanum*
Prodesmatochoerus natronensis Schultz and
    Falkenbach, 1954: Emry, 1992:110
Merycoidodon sp., Clark and Beerbower, *in* Clark et al.,
    1967:54, tentatively referred

Merycoidodon forsythae Schultz and Falkenbach,
    1968:36
Merycoidodon forsythae Schultz and Falkenbach, 1968:
    Emry, 1992:110
Otionohyus wardi degrooti Schultz and Falkenbach,
    1068:114

Type—F:AM 72303, from 65 feet (19.8 m) above ash "D," White River Formation, Flagstaff Rim area, Natrona County, Wyoming.

Hypodigm—Type, and Schultz and Falkenbach's (1968) sample of this species such as F:AM 72308, 72322, 72324, and 45194 (type of "*Prodesmatochoerus natronensis*"), between ash B-G, Flagstaff Rim area, Natrona County, Wyoming; type, F:AM 49760, of "*Otionohyus wardi degrooti*," 30 feet below the "upper purplish white layer," Converse County, Wyoming; possibly FMNH PM 16276, Ahearn Member, Chadron Formation; F:AM 24504 from 65-70 feet below the

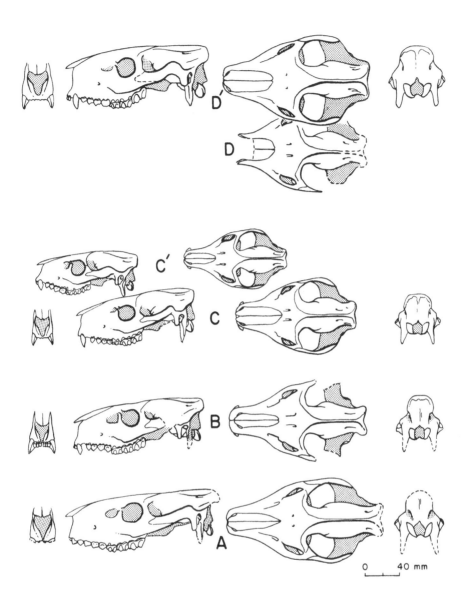

Figure 5. Selected, somewhat schematic crania of species of *Miniochoerus*. (A), *?Miniochoerus forsythae*; four views of type, F:AM 72303, from 65 feet below ash "D," White River Formation, Flagstaff Rim area, Natrona County, Wyoming; (B), *M. chadronensis*, four views of type, F:AM 45489, upper part of the Chadron Formation, Niobrara County, Wyoming; (C), *M. affinis*, four views of F:AM 44977, lower nodules, Scenic Member of the Brule Formation, Shannon County, South Dakota; (C'), lateral and dorsal view of *M. gracilis*, F:AM 45363, from high, (about 100 feet) above the "upper purplish white" layer, Seaman Hills, Niobrara County, Wyoming; (D), dorsal view of the type of *M. starkensis*, F:AM 45497, south of South Heart, North Dakota, and (D'), four views of F:AM 49585, type of *M. "nicholsae,"* Slim Buttes, Harding County, South Dakota

"upper purplish white layer," Converse County, Wyoming; and the following uncatalogued F:AM specimens with accession numbers 032-752, 032-751, 033-777, 047-1154, 047-1094, 035-806, 048-1100, 054-1234, 054-1292, 054-1251, and 055-1310 from 15-30 feet below the "upper purplish white layer," Seaman Hills, Niobrara County, Wyoming.

Amended diagnosis—Cranial characterization is that of a large and primitive member of the genus. Cranium with a prominent sagittal crest, narrow braincase, protracted and narrow nuchal crest, relatively deep preorbital fossae, teeth with slightly thinned enamel on medial crests and selenes, and the Subnasal L. is 50% of the Nasal L. (the single type specimen, F:AM 72303). The P^1-P^4 measurements average 47% of the P^1-M^3 (Table 3), the mean APM^3/TM^3 ratio is 0.98

(Table 5), the mean APM^3/Ht. M^3 ratio is 1.86 (Table 4), and TM^3/Ht. M^3 averages 1.89 (Table 6). *?Miniochoerus forsythae* differs from *Merycoidodon presidioensis* by systematically smaller size, significant for means of P1-M3s, P1-P4s, M1-M3s, Subnasal L., APM_3, TM_3 (no overall skull measures available for comparison). *?Miniochoerus forsythae* differs from *M. culbertsoni* by significantly smaller averages for most measures of gross size (P1-M3s, P1-P4s, Ht. M^3, Malar. D., Sk. L., Sk. W.; within observed ranges for *M. culbertsoni*, except P^1-P^4, Ht. M^3, ranges for which overlap by about a third of the combined ranges), significantly larger TM^3/Ht. M^3 ratio, significantly smaller P^1-P^4/P^1-M^3, Subnasal L./Nasal L., and Sk. L./P^1-M^3 ratios, and distinctly thinned enamel on the premolar-molar medial crests and selenes. *?Miniochoerus forsythae* differs from *Miniochoerus chadronensis* by systematically larger size, significant for all measurement categories for our samples except Ht. M3s, Subnasal L., Malar D. (observed ranges overlap from 6% to completely, except Nasal L., Sk. L., P_1-M3, P_1-P4), deeper preorbital fossae, more attenuated braincase and nuchal crest, significantly less reduced premolars, and thicker enamel on the medial dental crests and selenes. *?Miniochoerus forsythae* differs from *Oreonetes anceps* (Douglass) and *Bathygenys* Douglass by thinned enamel on dental crests and selenes. *?Miniochoerus forsythae* differs from *Limnenetes platyceps* Douglass by lack of early loss of fossettes with wear, relatively lower crowned teeth, deeper preorbital fossae, and lack of inflated auditory bullae. *?Miniochoerus forsythae* differs from *O. anceps, L. platyceps, B. reevesi* Wilson, and *B. alpha* Douglass by much larger size.

Distribution—White River Formation, near "D," (see Emry, 1973, fig. 16; Emry et al., 1987; Emry, 1992, p. 110, between 240 and 460 m of section), Flagstaff Rim area, Natrona County, Wyoming; possibly from the Ahearn Member, Chadron Formation, Pennington County, South Dakota; from 15-70 feet (4.6-21.3 m) below the "upper purplish white layer," Chadron Formation, Converse County, Wyoming; middle Chadronian, faunal zone Chadron A-B, late Eocene.

Discussion—A probable individual of *?Miniochoerus forsythae* originally was named as "*Prodesmatochoerus natronensis*" by Schultz and Falkenbach (1954). The name is based on a skull steinkern with some bone attached, and fragmentary jaws, with eroded and fractured teeth that are so worn that their morphology and dimensions cannot be determined. F:AM 45194 was recovered at a time when the collection of fossil vertebrates in the Flagstaff Rim area by the Frick Laboratory had just begun. The specific description was premature because much better preserved specimens soon became available. "*P. natronensis*" seems to be the same kind of oreodont as that later named as "*Merycoidodon*" *forsythae* Schultz

and Falkenbach (1968), based on a well-preserved and informative type specimen. "*P. natronensis*" here is regarded a *nomen vanum.*

The type of *?Miniochoerus forsythae* is a skull from the North Fork of Lone Tree Gulch, 65 feet above ash "D," Flagstaff Rim area. Ash "B," below and ash "F" above have produced radiometric age determinations of about 35.9 and 35.8 Ma respectively (Swisher and Prothero, 1990; Emry, 1992). Schultz and Falkenbach (1968) referred specimens from near ash "D," as well as some fragmentary remains from the Pipestone Springs Fauna, Jefferson County, Montana, and the Cypress Hills in Canada, to the species *forsythae* at the time of description. The Pipestone Spring specimens, about the same age (see Prothero, 1984), and the Cypress Hills fossils have not been re-examined.

Miniochoerinae from stratigraphically low in the Chadron Formation or its equivalents are not common. Clark and Beerbower (in Clark et al., 1967, p. 54; fig. 19, no. 2), however, report as *Merycoidodon* sp. a maxillary fragment with P^4-M^3, FMNH PM 16276, from the Ahearn Member of the Chadron formation. The fossil is probably referable to *?Miniochoerus forsythae* on the basis of the thinned enamel on the premolar-molar medial crests and selenes, and stratigraphy. Unfortunately, Clark and Beerbower give no measurements.

"*Otionohyus wardi*" Schultz and Falkenbach (1968) is based on a medium-sized skull from the lower part of the Brule Formation, Sioux County, Nebraska, with thick enamel on the medial premolar-molar crests and selenes. The type of "*O. wardi*" has been transferred to *Merycoidodon culbertsoni* (this chapter). "*O. wardi degrooti*" Schultz and Falkenbach (1968, p. 14) is based on a small-medium skull, jaws, and partial skeleton, F:AM 49760, from 30 feet (9.1 m) below the "white layer," ("upper purplish white" layer, about 34 Ma, Swisher and Prothero, 1990; also see Evanoff et al., 1992, fig. 6.2, 5 tuff, fig. 6.4), 8 miles (12.9 km) southeast of Douglas, Wyoming. The type of the presumed subspecies, thus, is a contemporary or near contemporary of *?Miniochoerus forsythae*. F:AM 49760 has thinned premolar-molar enamel of the kind that is characteristic of the Miniochoerinae, but not seen in even the earliest *Merycoidodon*. The type of "*O. wardi degrooti*" is morphologically similar to *M. forsythae* as we understand the species, and these two have similar metric parameters and ratios. Comparison of the two specimens of "*O. wardi degrooti*" with other specimens which we refer to *M. forsythae* shows no marked difference except in the Ht. M^3, a variable to highly variable character. There is little doubt that *?M. forsythae* can be identified in the stratigraphically lower deposits southeast of Douglass, Wyoming. This extends the geographic distribution of the species slightly eastward from the type area.

F:AM 24505, a slightly crushed skull, came from 65-70 feet (19.8-21.3 m) below the "upper purplish white layer" [Persistent White Layer, PWL, of F:AM collectors; Glory Hole Ash (Prothero, 1982a, 1985); 5 tuff of Evanoff, 1990; Evanoff et al., 1992], Chadron Formation, Irvine Bridge area, Converse County, Wyoming. The age of the Glory Hole Ash is about 34 Ma (Swisher and Prothero, 1990; Prothero and Swisher, 1992, p. 50). Teeth of F:AM 24505 have thinned enamel of the kind that is characteristic of *Miniochoerus*. Thus the fossil probably is not *Merycoidodon* in spite of its moderate size, narrow and protracted nuchal crest, and the relatively deep preorbital fossae. In size, F:AM 24505 compares well with the type of ?*M. forsythae*. Both specimens lie at or near the high end of most observed ranges for specimens which we refer to ?*M. forsythae*, but F:AM 24505 shows only two significant (though numerically small) differences (Sk. W., P_1-M_3).

The Frick Laboratory also has collected several specimens from 15-30 feet below the "upper purplish white" layer, Seaman Hills area, Niobrara County, Wyoming, after publication of Schultz and Falkenbach's (1956) report. Fossils in this very small sample are similar both morphologically and in their measurements to the type of ?*Miniochoerus forsythae* (no systematic differences in size; type is mainly larger, but significantly so only for P^1-M^3, and more noticeably, P^1-P^4, and Sk. W.; and is significantly smaller for P.O.C.), and more similar to other specimens that we assign to ?*M. forsythae* (significant difference, based on very small samples, only in lesser means for P^1-P^4, Ht. M^3, and P^1-P^4/P^1-M^3 ratio). These specimens extend the geographic distribution of ?*M. forsythae* slightly to the southeast of the type area. The specimens referred to ?*Miniochoerus forsythae* from the 15-30 foot interval below the "upper purplish white layer" may be slightly younger than the ≈36-million-year-old type specimen, but this is dependent on the rate of deposition of that part of the Chadron Formation. ?*M. forsythae* outside of the type area, as tied to the "upper purplish white layer," does much to clarify the biostratigraphy of ?*M. forsythae* relative to its presumed descendant, *M. chadronensis*.

The "upper persistent white layer" has not been identified in the Flagstaff Rim area. The youngest ash here, ash "J," has an ^{40}Ar/^{39}Ar age determination of about 34.5 Ma (Swisher and Prothero, 1990). Ash "J" would appear to be a slightly older ash than the "upper purplish white layer" of adjacent regions. Above this level, the White River Formation in the Flagstaff Rim area was eroded prior to deposition of the overlying middle Miocene Split Rock Formation.

Schultz and Falkenbach (1968) regarded ?*Miniochoerus forsythae* as a small, early species of *Merycoidodon*. This was probably because they were looking for an ancestral *Merycoidodon*, and because the occurrence of ?*M. forsythae* is known to be older than later Chadronian, the time when *Merycoidodon*

becomes common on the northern High Plains. Schultz and Falkenbach were unaware of the late Duchesnean and earlier Chadronian oreodonts of *Merycoidodon* aspect from the Vieja Group, Trans-Pecos Texas. Several oreodonts from the Vieja Group, such as Wilson's (1971) *Merycoidodon dunagani*, (upper part of the Chambers Tuff Formation), and *M. presidioensis* (Capote Mountain Tuff Formation), do in fact have enamel on their premolar-molar middle crests and selenes that is somewhat thinner than that on the external parts of the teeth, and enamel that is slightly thinner than that in late Chadronian and younger *Merycoidodon*. Although the systematics of these very early oreodonts, especially ?*M. dunagani*, remains in doubt, it is becoming apparent that the earliest Merycoidodontinae resemble the earliest Miniochoerinae (except for size) because they share slight enamel thinning on the medial dental crests and selenes, and have similarly narrowly protracted occiputs and small auditory bullae, among other features. We suggest the trend for enamel thinning was exaggerated progressively in *Miniochoerus* but disappeared in *Merycoidodon* as the Miniochoerinae and the Merycoidodontinae diverged.

Schultz and Falkenbach (1956; 1968) failed to relate the Miniochoerinae to the Merycoidodontinae, and thus missed the phylogenetic significance of ?*Miniochoerus forsythae* as intermediate between *Merycoidodon* and *Miniochoerus*. Because ?*M. forsythae* is intermediate between *Merycoidodon* and *Miniochoerus* and because we believe it is the ancestor of *Miniochoerus*, ?*M. forsythae* is removed from *Merycoidodon* and questionably referred to *Miniochoerus*. It is possible that ?*M. forsythae* was derived from Wilson's (1971) *Merycoidodon dunagani*. However, ?*M. dunagani* has a much more primitive dentition, is considerably smaller, and is substantially older than ?*M. forsythae*. Transitional species are needed to clarify the affinities of these two taxa. However, ?*M. forsythae* is a suitable ancestor for *Miniochoerus chadronensis* of the late Chadronian and earliest Orellan.

Miniochoerus chadronensis (Schultz and Falkenbach)
Figures 4B, 5B; Tables 1-9; Appendix C

Stenopsochoerus (*Pseudostenopsochoerus*)
   chadronensis Schultz and Falkenbach, 1956:445
Miniochoerus chadronensis (Schultz and Falkenbach):
   Evanoff et al., 1992:123
Stenopsochoerus (*Pseudostenopsochoerus*)
   chadronensis geologic variety, Schultz and
   Falkenbach, 1956:446
Miniochoerus battlecreekensis Schultz and Falkenbach,
   1956:395, in part
Miniochoerus (*Paraminiochoerus*) *affinis* (Leidy):
   Schultz and Falkenbach, 1956:405, in part

Table 9
Descriptive statistics for regressions shown in Fig. 6

$x = TM^3$ $y = APM^3$	Miniochoerus chadronensis		Miniochoerus affinis		Miniochoerus gracilis		All Miniochoerus
Regression Method: Least Squares	y/x	x/y	y/x	x/y	y/x	x/y	y/x
b_0	9.29	2.91	6.30	5.75	2.85	0.90	1.32
b_1	0.46	0.82	0.59	0.63	0.79	0.92	0.92
Standard error, b_{0m}	2.07	3.74	1.93	2.05	1.15	1.35	0.47
Standard error, b_{0y}	2.15	3.82	2.07	2.18	1.22	1.43	0.87
Standard error, b_1	0.12	0.22	0.12	0.13	0.09	0.11	0.03
Goodness of fit	0.37		0.37		0.72		0.89
Mean SS Error	0.34	0.61	0.53	0.56	0.18	0.21	0.53
Regression SS	4.70	8.39	11.92	12.62	13.74	16.05	503.07
Degrees of Freedom	23		38		29		122
F	13.68		22.55		74.76		946.79
Critical value, F, a=0.05	4.28		4.10		4.18		3.92
Regression Method: Reduced Major Axis							
Rb_0	4.41		0.44		1.09		0.43
Rb_1	0.75		0.97		0.92		0.98
Dispersion, RMA	1.35		1.46		0.88		1.46
Standard error, Rb_0	1.54		0.91		0.61		0.09
Standard error, Rb_1	0.12		0.12		0.09		0.03

$x = M^1\text{-}M^3$ $y = P^1\text{-}P^4$							
Regression Method: Least Squares	y/x	x/y	y/x	x/y	y/x	x/y	x/y
b_0	20.22	28.98	14.88	27.62	8.48	11.76	6.49
b_1	0.30	0.39	0.41	0.35	0.56	0.76	0.63
Standard error, b_{0m}	5.61	5.70	4.93	3.38	3.31	3.71	1.14
Standard error, b_{0y}	5.82	5.95	5.20	3.70	3.50	3.94	2.16
Standard error, b_1	0.13	0.17	0.13	0.11	0.10	0.14	0.03
Goodness of fit	0.12		0.14		0.43		0.73
Mean SS Error	2.32	3.00	2.68	2.31	1.27	1.73	3.36
Regression SS	11.81	15.25	26.52	22.82	36.82	49.88	1543.15
Degrees of Freedom	38		60		39		173
F	5.09		9.89		28.90		459.21
Critical value, F, a=0.05	4.10		4.00		4.09		3.90
Regression Method: Reduced Major Axis							
Rb_0	-3.93		-10.82		-0.98		2.30
Rb_1	0.88		1.08		0.86		0.74
Dispersion, RMA	3.22		3.13		2.42		4.37
Standard error, Rb_0	3.46		2.08		1.67		0.19
Standard error, Rb_1	0.13		0.13		0.10		0.03

Platyochoerus platycephalus (Thorpe): Schultz and
Falkenbach, 1956:427, in part
Stenopsochoerus sternbergi Schultz and Falkenbach,
1956:438, in part
Stenopsochoerus (*Pseudostenopsochoerus*)
douglasensis Schultz and Falkenbach, 1956:447
Stenopsochoerus (*Pseudostenopsochoerus*)
douglasensis geologic variety, Schultz and
Falkenbach, 1956:449
Stenopsochoerus (*Pseudostenopsochoerus*) *reideri*
Schultz and Falkenbach, 1956:449
Parastenopsochoerus conversensis Schultz and
Falkenbach, 1956:451

Type—F:AM 45489, upper part of the Chadron
Formation, Seaman Hills area, Niobrara County,
Wyoming.

Hypodigm—Type, and Schultz and Falkenbach's
(1956) sample of "*Stenopsochoerus chadronensis*"
geologic variety, such as F:AM 45168, 49508, 44991,
44993, 45012, and 49586; specimens from Shack Draw,
40 feet above "upper purple white" layer, Seaman Hills
area, Wyoming, that Schultz and Falkenbach (1956, p.
398) called *Miniochoerus* "*battlecreekensis*," such as
F:AM 45343, 44982, and 49579; several specimens
from the top of the nodular layer, southeast of Douglas,
Wyoming, that Schultz and Falkenbach (1956, p. 412)
called *M.* "*affinis*," such as F:AM 45481 and F:AM
45010; several specimens from the nodular layer,
southeast of Douglas, Wyoming, that these authors
(1956, p. 430) identified as "*Platyochoerus platycephalus*," such as F:AM 45471, 45472, 45476, 45169,
45007, 45488, 45008, 45171, and 45093; several
specimens from the nodular layer, southeast of Douglas, Wyoming, that Schultz and Falkenbach (1956, p.
439) identified as "*Stenopsochoerus sternbergi*," such
as F:AM 45478, 45479, 44992, 49607, and 49612; type,
F:AM 45492, and Schultz and Falkenbach's (1956, p.
448) sample of "*S. douglasensis*," such as F:AM 45496,
44989, 44990, 45493, 45494, 45495, 49517, 49538,
49613, and 45330; Schultz and Falkenbach's (1956, p.
449) geologic variety of "*S. douglasensis*," such as
F:AM 49621; type, F:AM 49620, of "*S. reideri*" from
below the "upper purple white" layer, Seaman Hills
area, Wyoming; type, F:AM 45011, of "*Parastenopsochoerus conversensis*" from Converse County, Wyoming; and 11 previously reported specimens with field
numbers such as Lusk-O-32-352, 48-1100, and 54-
1234, from just below the "upper purplish white" layer,
Seaman Hills, Wyoming.

Amended diagnosis—Cranium is typical for the
subfamily, but is relatively large, the Subnasal L.
averages 57% of the Nasal L., the preorbital fossae are
relatively deep and the sagittal-nuchal crest is moderately prominent, the P^1-P^4 averages 45% of the P^1-M^3
(P^1-P^4/P^1-M^3; Table 3), the mean APM^3/TM^3 is very
close to 1 (Table 5), the APM^3/Ht. M^3 ratio averages
1.85 (Table 4), and TM^3/Ht. M^3 averages 1.82 (Table

6). *Miniochoerus chadronensis* differs from ?*Miniochoerus forsythae* by smaller size (significant for means
of all measurement categories for our samples except
Ht. M3s, Subnasal L., Malar D.; observed ranges
overlap from 6% to completely, except Nasal L., Sk. L.,
P_1-M_3, P_1-P_4), shallower preorbital fossae, less
protracted and slightly broader nuchal crest, thinner
enamel on the premolar-molar middle crests and
selenes, and significantly more reduced premolars.
Miniochoerus chadronensis differs from *M. affinis* by
significantly larger size [for our samples, means of all
measurements except Subnasal L., Malar D., P.O.C.,
Sk. W., Ht. M3; 14 measurement categories overlap by
an average of 41% of combined observed ranges with a
maximum of 54% (APM_3) and a minimum of 23% (P_1-
M_3); ranges of *M. chadronensis* are included within
those of *M. affinis* for Malar D., Sk. W., and Ht. M3,
whereas the reverse is true for P.O.C.], and significantly
smaller Sk. L./P^1-M^3 ratio, otherwise similar. *Miniochoerus chadronensis* differs from *M. gracilis* by
systematically and significantly larger mean sizes (only
ranges of Subnasal L., Malar D., P.O.C., Sk. L., Sk. W.
overlap for our samples) and significantly smaller mean
P^1-P^4/P^1-M^3 (see Tables 3, 9), and Sk. W./Sk. L.
(stereotyped among other Miniochoerinae; Table 7)
ratios. *Miniochoerus chadronensis* differs from *M.
affinis* and *M. gracilis* by a less fetalized cranium,
deeper preorbital fossae, more attenuated nuchal crest,
and significantly smaller mean Sk. L./P^1-M^3
(stereotyped among other Miniochoerinae), Sk. W./Sk.
L., and Sk. L./P^1-M^3 ratios, otherwise similar. *Miniochoerus chadronensis* differs from *M. starkensis* by a
slight tendency to be smaller (for our samples: means of
ten of 18 measurements; significantly smaller for
APM^3, Subnasal L., Malar D., but significantly larger
for P^1-P^4, and TM3; *M. chadronensis* observed ranges
are included by those of *M. starkensis* for P1-M3s, M1-
M3s, TM3, Ht. M3, APM3, but the range for P.O.C.
includes that of *M. starkensis*; other ranges overlap by
an average of 62% of the combined ranges), and in
proportions by a significantly larger P^1-P^4/P^1-M^3 ratio,
and significantly smaller APM^3/Ht. M3 and APM^3/
TM^3 ratios.

Distribution—Upper part of the Chadron Formation,
nodular layer below the "upper purplish white" layer
(persistent white ash, 5 tuff of Evanoff et al., 1992),
Converse County, and slightly below or above the
"upper purplish white" layer but below the "channel,"
Niobrara County, Wyoming, and below the "Amelia
channel," lowest Brule Formation, Sioux County,
Nebraska; late Chadronian to earliest Orellan, faunal
zone Chadron C-earliest Brule A, latest Eocene to
earliest Oligocene.

Discussion—*Miniochoerus chadronensis* (as we con-
ceive it) has a large number of size distributions that do
not appear to be normally distributed (14 and one
doubtful; Table 1). Our sample of the modern peccary,

Dicotyles tajacu, from Trans-Pecos Texas has 15 non-normal or dubious distributions (Table 1). Inspection of probability deficiencies, particularly for adequately sampled measurements (on *M. chadronensis* specimens), suggests that an important part of the problem is a tendency for distributions to be bimodal or more complex. However, complexity does not seem to arise from any one or combination of the possible subordinate groups (taxa that we synonymize with *M. chadronensis*), but instead appears to be characteristic of the groups of the sample of *M. chadronensis* as we understand the species. Distribution complexity may be tied to the sampling of specimens from "good" versus "bad" years (see Introduction). The results are in a sense parallel to those for the sample of *Tayassu pecari*, where the expected source of distribution complexity (gender), does not appear to be the actual cause. We note that asymmetry does not appear to be a problem, and proceed on the assumption that robust tests of significance apply reasonably to the data.

Miniochoerus chadronensis is based on a slightly laterally crushed skull lacking the zygoma, and a jaw from just below the "upper purplish white layer" [Persistent White Layer of the F:AM collectors; 5 tuff (Evanoff et al., 1992)], upper part of the Chadron Formation, Seaman Hills area, south-southwest of Lusk, Wyoming. A single referred specimen was identified at the time of description. Schultz and Falkenbach (1956) regarded *Miniochoerus chadronensis* as a species of "*Stenopsochoerus*," a genus which designated a presumed lineage of dolichocephalic Miniochoerinae. "*S.*" *chadronensis* then was declared the type species of "*S. (Pseudostenopsochoerus)*" because the type of "*S. (P.)*" *chadronensis* is less "dolichocephalic" (actually, less laterally crushed) than the type species of "*Stenopsochoerus*." The characteristics of the subgenus, such as a more mesocephalic skull with a broader postorbital constriction, however, nullify the distinction of "*Stenopsochoerus (Pseudostenopsochoerus)*" from *Miniochoerus* and allow for the placement of "*S. (P.)*" *chadronensis* in *Miniochoerus*.

Miniochoerus chadronensis "geologic variety" Schultz and Falkenbach (1956) is based on several specimens from low in the "nodular layer" (? = the "fossiliferous carbonate nodular layers," Evanoff et al., 1992: fig. 6.4, just below the "upper purplish white layer" or 5 tuff) southeast of Douglas, Wyoming. The "upper purplish white layer" serves as a useful marker bed in the Seaman Hills area, and elsewhere in eastern Wyoming and NW Nebraska. Although there has been some controversy about the lateral continuity and correlation of the "upper purplish white" ash layer, it corresponds regionally with the approximate last occurrence of titanotheres (M. F. Skinner, personal communication, 1970; see also Emry et al., 1987, p. 138; Evanoff et al., 1992, p. 120; figs. 6.2, 6.4), and has recently been estimated to be about 34 Ma (Swisher and Prothero, 1990). Schultz and Falkenbach (various

papers), and others, have considered titanothere-bearing rocks for about 7.6 m (25 feet) above the "upper purplish white layer," to be basal Brule Formation, and early Orellan, a part of oreodont faunal zone Brule "A." These rocks, however, are somewhat older than undoubted Brule Formation of the Amelia area "channel" above the pronounced disconformity. The Amelia "channel" in western Nebraska has cut 100 feet down into the rocks in question. Schultz and Falkenbach's (1956) specific separation of *M. chadronensis* "geologic variety" from *M. chadronensis* rests on the fact that the type of *M. chadronensis* came from the Chadron Formation whereas those of the "geologic variety" came from rocks regarded as yielding the Brule A fauna.

Miniochoerus chadronensis "geologic variety" Schultz and Falkenbach (1956), though generally slightly smaller, compares closely in size to the type of *M. chadronensis*. Measurements of *M. chadronensis* "geologic variety," compared to those of other specimens which we refer to *M. chadronensis*, average slightly smaller except that P^1-M^3 and M^1-M^3 are significantly smaller. The observed ranges of measurements on *M. chadronensis* "geologic variety" fall within those for other specimens of *Miniochoerus chadronensis* for 12 of the 16 measurement categories available for comparison, and M^1-M^3, APM3s, P_1-P_4 measurements overlap by an average of 68% of the combined ranges. The "geologic variety" of *M. chadronensis* does not compare as well with *M. affinis* in size. Although the differences between the late Chadronian-earliest Orellan and the younger Orellan Miniochoerinae are not great, it is more reasonable to equate stratigraphically comparable specimens with their respective populations, than to regard them as "geologic varieties" (*fide* Schultz and Falkenbach, 1956) of younger (or older) species, or to maintain them as distinct species on the basis of trivial differences in geography. The synonymy of the geologic variety within *M. chadronensis* requires that the "geologic variety" designation be abandoned.

Miniochoerus "*battlecreekensis*" is based on a skull and jaw, F:AM 45001, from the early Orellan "lower oreodon beds," Scenic Member of the Brule Formation, Shannon County, South Dakota. Most of the specimens referred to this species came from early Orellan rocks in Nebraska that occur above the unconformity produced by the "Amelia" channel, but the sample from the Seaman Hills area, Wyoming, came from just above the "upper purplish white" layer and below the channel, and is thus older. The Seaman Hills specimens of *M.* "*battlecreekensis*" are more similar in size and morphology to the late Chadronian-earliest Orellan Miniochoerinae, *M. chadronensis*, than they are to Orellan specimens. Compared to other fossils that we assign to *M. chadronensis*, those of *M.* "*battlecreekensis*" (mostly the very small Seaman Hills sample) show little difference [means <1.3 SDs (other

M. chadronensis) from, and are generally smaller than (14 of 16 mutually avalable categories), means of other *M. chadronensis*] except for a single unusually and significantly small value of Ht. M^3, which produces a significantly large $APM^3/Ht. M^3$ ratio.

Schultz and Falkenbach (1965, p. 405) referred two specimens, F:AM 45481 and F:AM 45010, from the nodular layer 8 miles southeast of Douglas, Wyoming, to *Miniochoerus affinis*. These specimens came from the same rocks as other specimens that have been identified in the literature as *M.* "*battlecreekensis*" and "*Stenopsochoerus sternbergi*," but are here transferred to *M. chadronensis*. F:AM 45481 and F:AM 45010 yield measurements in only eight categories, and tend to be slightly smaller in these respects than "*S. sternbergi*." However, two of the available measurements most useful for judging size, P^1-M^3 and P^1-P^4, fall within the observed range of those measures for the very small sample of "*S. sternbergi*," and the third, M^1-M^3, is barely (0.6 mm) below the range of "*S. sternbergi*." The two specimens were probably separated from "*S. sternbergi*" because they are undeformed, and are here placed in *M. chadronensis* on the basis of stratigraphy, morphology, and size.

Schultz and Falkenbach (1956) thought that, in addition to "dolichocephalic" and mesocephalic lineages of Miniochoerinae, there existed a "brachycephalic" line within this subfamily. The presumed brachycephalic species were placed in "*Platyochoerus*." Schultz and Falkenbach referred several specimens from the "nodular layer" [? = the "fossiliferous carbonate nodule layers, Evanoff et al., 1992, fig. 6.4, just above the "white layer" (= "upper purplish white" layer; 5 tuff, Evanoff, 1992, fig. 6.4)], southeast of Douglas, Wyoming, to "*P. platycephalus*," a species based on a type specimen from the early Orellan. As with the deposits just above the "upper purplish white" layer in the Seaman Hills area, the deposits under discussion 6-9 miles southeast of Douglas, Wyoming, are latest Chadronian-earliest Orellan in age, hence are slightly older than the rocks in western Nebraska regarded as early Orellan. Our re-examination leaves little doubt that the specimens from southeast of Douglas referred to "*P. platycephalus*," are dorsoventrally deformed specimens of *Miniochoerus chadronensis*.

"*Stenopsochoerus (Pseudostenopsochoerus) douglasensis*" is based on a skull, jaws, and a partial skeleton from 8 miles (12.9 km) southeast of Douglas, Wyoming, from the middle of the "nodular layer" [? = the fossiliferous carbonate nodule layers, Evanoff et al., 1992,.fig. 6.4)] that directly overlies the "white layer" which corresponds with the "upper purplish white" layer of adjacent regions (see Prothero, 1982a, 1985; Swisher and Prothero, 1990; Prothero and Swisher, 1992; 5 tuff, Evanoff et al., 1992) on the basis of stratigraphy and its association just below the youngest

titanothere occurrences. "*S. douglasensis*," thus, has the same stratigraphy as the sub-samples of "*Platyochoerus platycephalus*," "*Stenopsochoerus sternbergi*," and Schultz and Falkenbach's (1956) geologic variety of "*S. (Pseudostenopsochoerus)*" *chadronensis*, discussed in this section. The specimens of "*S. (Pseudostenopsochoerus) douglasensis*" are somewhat crushed laterally. Three of the referred specimens of Schultz and Falkenbach's "geologic variety" of "*S. (Pseudostenopsochoerus) douglasensis*" came from the Chadron Formation below the "upper purplish white" layer. Both F:AM 45491 and F:AM 49621 ["*S. (P.) douglasensis* geologic variety"] are poorly preserved and laterally crushed, so their skulls are narrow as preserved. Schultz and Falkenbach (1956), however, note that these Chadronian specimens are larger than Miniochoerinae from the typical early Orellan. In fact, more measurements of F:AM 49621 are slightly smaller than the averages for "*S. (P.) douglasensis*" than the reverse. The P^1-M^3 and P^1-P^4 of F:AM 49621 are slightly but significantly larger than averages for specimens referred to "*S. (P.) douglasensis*," but other differences are not systematic or important.

The type of "*Stenopsochoerus (Pseudostenopsochoerus) reideri*" is a well-preserved but slightly laterally crushed skull and jaws from an unknown distance below the "upper purplish white layer," Seaman Hills, Niobrara County, Wyoming. Schultz and Falkenbach (1956) amended the implied dolichocephaly of the species "*reideri*," as placed within "*Stenopsochoerus*," again by use of the subgenus "*Pseudostenopsochoerus*." As noted above, the concept of the subgenus as defined on the basis of greater mesocephaly relative to "*Stenopsochoerus*," nullifies the concept of the genus compared to *Miniochoerus*, and allows the placement of this taxon in *Miniochoerus*. In size, F:AM 49620 approaches *?M. forsythae* (see below), but it is more similar in proportions to *M. chadronensis* as we understand the species, to which F:AM 49620 is referred.

"*Parastenopsochoerus conversensis*" is another taxon based on and known from specimens from the Douglas, Wyoming, area. "*Parastenopsochoerus*" according to Schultz and Falkenbach (1956) has certain features in common with *Miniochoerus*, "*Stenopsochoerus*," and "*Platyochoerus*," "but differs diagnostically in possessing a short sagittal crest." The type, F:AM 45011, and a referred specimen, F:AM 49642, do in fact have short and low sagittal crests, but they also have shallow malars and unworn third molars. Measurements for "*P. conversensis*" (no mandibular material available) are generally slightly smaller, but lie entirely within the ranges observed for other specimens of our sample of *M. chadronensis*, and means, including those of ratios, except P.O.C. (larger for "*P. conversensis*") are not significantly different. These specimens are probably subadults of *M. chadronensis*.

Comparison of measurements for the type of *Minio-choerus chadronensis* with those of samples of taxa that we synonymize with *M. chadronensis* show the following features. The type of *M. chadronensis* is systematically smaller than, but proportionally similar to "*S. (P.) reideri*." Measurements of the type of *M. chadronensis* lie within the observed ranges of a large majority of measurements for a group of specimens not previously reported, but tend to be slightly to noticeably (Nasal L., Sk. L., M_1-M_3, TM3) smaller. The type of *M. chadronensis* has measurements that are, in most instances, within the observed ranges of those for "*P. platycephalus*" in part, "*S. (P.) douglasensis*," and "*S. (P.)*" *chadronensis* "geologic variety," slightly above observed ranges for a Subnasal L. (significantly larger than those of "*P. platycephalus*" in part, "*S. (P.) douglasensis*") and Malar D. (significantly larger than that of "*S. (P.) douglasensis*"), and slightly below the observed range of that set of specimens for scattered categories (particularly TM3). Measurements for the type fall within the observed range, or slightly exceed single measurements for *M.* "*battlecreekensis*" in part (the Seaman Hills sample), for a large majority of available measurement categories, and only the P.O.C. differs by being below the range for *M.* "*battle-creekensis*" in part. Measurements for the type of *M. chadronensis* slightly exceed the ranges for all measurements on "*S. sternbergi*" in part, except that TM3 lies below the range.

Comparisons of averages of measurements of each of these groups to averages of measurements of other specimens that we refer to *Miniochoerus chadronensis*, and averages for each other, make the relationships somewhat clearer. In comparisons to other specimens of *M. chadronensis*, "*Stenopsochoerus (Pseudostenop-sochoerus) reideri*" is significantly larger in many respects, and means for measurements on "*Stenopso-choerus sternbergi*" in part, are smaller (except a single TM3), significantly so for Sk. L. and P_1-M_3. When the single specimen of "*S. (P.) reideri*" is compared to the very small sample of "*S. sternbergi*" the size difference is systematic, but only P^1-M^3 and M^1-M^3 means differ significantly. These are the extremes of a size series that does not appear to require subdivision. The following generalizations can be made about the specimens previously assigned to taxa that we synonymize within *M. chadronensis*. In order of descending size, the larger *M. chadronensis* include "*S. (P.) reideri*," the 11 uncatalogued specimens, the type of *M. chadronensis*, and "*S. (P.) douglasensis*." The central parts of most character distributions are occupied by "*P. platycephalus*" in part, and "*S. (P.)*" *chadronensis* "geologic variety," and the smaller *M. chadronensis* include the Seaman Hills sample of *M.* "*battlecreekensis*," "*Parastenopsochoerus conversen-sis*" (no discernible size difference), "*Stenopsochoerus sternbergi*," and two specimens of *M.* "*affinis*" that we reassign to *M. chadronensis* (no reliable difference in size, in an unusually limited set of comparisons). In size, extremes of this series are not much different from small ?*M. forsythae*, or from large *M. affinis*, but as a group, *M. chadronensis* seems sufficiently distinct.

The late to latest Chadronian-earliest Orellan samples here united within *Miniochoerus chadronensis* show only modest size and morphologic variation that is consistent with that seen for various living suiform Artiodactyla (Appendix A, C). There is no evidence that *M. chadronensis*, as we understand the species, should be divided between four genera, two subgenera, and nine species. Additionally, in spite of our taxo-nomic simplification, the possibility still exists that *M. chadronensis* may be conspecific with *M. affinis*. Means of the 18 basic measurements of *M. chadronen-sis*, except Ht. M3, are larger, 13 significantly so, than those of *M. affinis*, but the two populations overlap substantially. Three of the measurement sets for *M. affinis*, Malar D., Sk. W., and Ht. M3, lie within the ranges of measurements on *M. chadronensis*, whereas the range of P.O.C. measurements on *M. affinis* includes the range for that measurement on *M. chadronensis*. Other measurements overlap an average of 41% of the combined ranges. ?*M. forsythae*, *M. chadronensis*, and *M. affinis* represent segments of a chronocline that experienced a progressive decrease in average size, progressive decrease in the size and depth of the preorbital fossae, and increased cranial fetaliza-tion, from the middle Chadronian to the Orellan. In spite of the fact that *M. chadronensis* is closely related to *M. affinis*, *M. chadronensis* is distinguished specifi-cally from *M. affinis* on the basis of its older stratigra-phy, persistently slightly larger size, and deeper preorbital fossae.

Miniochoerus affinis (Leidy)
Figures 4C, 5C, 6; Tables 1-9; Appendix C

Oreodon affinis Leidy, 1869:105
Miniochoerus (*Paraminiochoerus*) *affinis* (Leidy):
    Schultz and Falkenbach, 1956:405, in part
Merycoidodon platycephalus Thorpe, 1921b:339
Platyochoerus platycephalus (Thorpe): Schultz and
    Falkenbach, 1956:427, in part
Miniochoerus battlecreekensis Schultz and Falkenbach,
    1956:395, in part
Stenopsochoerus sternbergi Schultz and Falkenbach,
    1956:438, in part

Type—ANSP 10679, the facial part of a skull with some teeth, White River drainage, South Dakota.

Hypodigm—Type, and most of Schultz and Falken-bach's (1956) sample of this species such as AMNH 1316, 1290, 6407, 8872, F:AM 44977, 44974, 44974, 49533, and 49575; type, YPM 12752, and most of Schultz and Falkenbach's (1956) sample of "*Platyochoerus platycephalus*," such as F:AM 45338, 49536, 45000, 49576, 44970, 45089, 45475, 45004, 45470, and 49551; type, F:AM 45001, and most of

Schultz and Falkenbach's (1956) sample of *Miniocho-erus* "*battlecreekensis*," such as F:AM 45344, 45345, 45347, 44976, 49582, and 49583; and the type, F:AM 44980, and most of the referred specimens of "*Stenopsochoerus sternbergi*," such as F:AM 44981, 45078, 45079, 44983, 45083, 45050, 49567, and 49606.

Amended diagnosis—Cranial characterization is the same as for a derived member of the genus, the Subnasal L. averages 63% of the Nasal L. (three specimens), the P^1-P^4 averages of 46% of the P^1-M^3 series (Table 3), the APM^3/TM^3 averages 1.00 (Table 5), the APM^3/Ht. M^3 averages 1.89 (Table 4), and TM^3/Ht. M^3 averages 1.85 (Table 6). *Miniochoerus affinis* differs from *Miniochoerus chadronensis* by shallower preorbital fossae, slightly more fetalized cranium, significantly smaller size [for our samples (Appendix C), means of all measurements except Subnasal L., Malar D., P.O.C., Sk. W., Ht. M_3; 14 characteristics overlap by an average of 41% of combined observed ranges with a maximum of 54% (APM_3) and a minimum of 23% (P_1-M_3); ranges of *M. chadronensis* are included within those of *M. affinis* for Malar D., Sk. W., and Ht. M_3, whereas the reverse is true for P.O.C.], and significantly larger Sk. L./P^1-M^3 ratio, otherwise similar. *Miniochoerus affinis* differs from *M. gracilis*, an approximate contemporary by systematically and significantly larger size [for our data, overlap of combined observed ranges is from 4% (P^1-M^3) to 67% (Subnasal L.) with an average of 24%; no overlap for Ht. M3s], and significantly smaller P^1-P^4/P^1-M^3 ratio (Tables 3, 9), otherwise very similar. *Miniochoerus affinis* differs from *M. starkensis* by slightly but significantly smaller size [significant differences of means for our samples except Subnasal L., P.O.C. (larger), Sk. W., TM_3, and Ht. M_3 (larger; very small samples); observed ranges overlap substantially, from 27% to 76% of the combined ranges, averaging 44%], significantly less reduced premolars, M^3s slightly but significantly broader, and significantly higher in proportion to length, otherwise similar.

Distribution—"Lower nodules," "lower oreodon beds," Scenic Member of the Brule Formation, South Dakota; lower part of the Orella Member, Brule Formation, western Nebraska; from stratigraphically high (approximately 100 feet above the "upper purplish white" layer; also see Evanoff et al., 1992, fig. 6.5), Seaman Hills area, Niobrara County, Wyoming, and equivalents elsewhere; early Orellan, faunal zone Brule A, early Oligocene.

Discussion—While characterizing a referred skull of "*Oreodon*" *gracile*, Leidy (1869) applied the name "*O.*" *affinis* to an incomplete skull, ANSP 10679, from the Big Badlands of South Dakota. Leidy was inclined to view "*O.*" *affinis* as a "doubtful species" because he thought the type might have been a large individual of "*O.*" *gracile*. Thorpe (1937a) and Schultz and Falkenbach (1956) thought that "*O.*" *gracile* and "*O.*" *affinis*

represented two populations. During the preparation of this report we were faced with the same question, and have concluded that two populations in fact are present. When the samples are combined and evaluated graphically [histograms, or scatter diagrams (Fig. 6)], the populations are separable, and descriptive statistics of the combination produce unacceptably high CVs (Table 1; Appendix C). Analysis of covariance for various data pairings (including Sk. L./P^1-M^3, APM^3/TM^3, APM_3/TM_3, P^1-P^4/M^1-M^3) suggests differences in size, some differences in proportional variability, but generally no difference in the style or nature of the variability. Nonparametric tests further confirm these findings.

Hay (1902), Bump and Loomis (1930), Thorpe (1937a), and Phleger and Putnam (1942), among others, believed that the small oreodonts that are now regarded as Miniochoerinae were *Merycoidodon* on the basis of their small auditory bullae and general cranial similarity with *M. culbertsoni*. Schultz and Falkenbach (1956) transferred "*M.*" *affinis* to their new genus, *Miniochoerus*, then went on to designate *M. affinis* the type species of the subgenus "*Paraminiochoerus*." *M.* "*Paraminiochoerus*" was distinguished from *Miniochoerus* on the basis of smaller size relative to other taxa that Schultz and Falkenbach had placed in *Miniochoerus*. The diminutive size of *M. affinis* is due to fetalization brought about by phyletic dwarfism, which renders the skull and mandible as "smaller," "lighter," or "shallower" than homologues in older *Miniochoerus*. One wonders why Schultz and Falkenbach did not designate *M. gracilis*, a truly small and senior species, as the type species of *M.* ("*Paraminiochoerus*"). And as stated by Schultz and Falkenbach, "examples of the two forms [*Miniochoerus* and *M.* "(*Paraminiochoerus*)"] cannot always be readily separated." We do not recognize the subgenus "*Paraminiochoerus*."

Thorpe (1921b) named as "*Merycoidodon platy-cephalus*" a well-preserved skull with jaw fragments, YPM 12752, from near the Scotts Bluff promontory, Scotts Bluff County, Nebraska. Although the precise stratigraphic provenance of the type is not known, it probably came from the lower part of the Brule Formation. These rocks are exposed well in the badlands to the northeast of the landmark (see Darton, 1903, figs. 18-21). The type belonged to an aged and robust individual as shown by the very worn dentition, the rugose sagittal crest, and the unusually robust malar and zygoma. Thorpe recognized that the robustness is the result of either advanced ontogeny or pathology. Thorpe (1921b, 1937a), however, made the comment that the cranium is low and broad. It must be kept in mind that Thorpe was comparing YPM 12752 to *Merycoidodon culbertsoni*, a much larger animal, and not to any species now a part of the Miniochoerinae.

Schultz and Falkenbach (1956) altered the concept of

"*Merycoidodon platycephalus*," as brachycephalic relative to *Merycoidodon culbertsoni*, when they used the species to define a presumed brachycephalic genus, "*Platyochoerus*," within the Miniochoerinae. The type specimen, YPM 12752, however, is metrically no more brachycephalic than *Miniochoerus* as the genus was defined by Schultz and Falkenbach, based on cranial ratios. It is the usually severely dorsoventrally crushed crania that Schultz and Falkenbach (1956) referred to "*Platyochoerus*" that was the basis for their concept. Measurements on the type of "*P. platycephalus*" with the exception of the significantly large Malar D., but including the slightly but significantly large P.O.C., and Sk. W., fall within, though generally on the large side of, the observed range of other specimens that we refer to *M. affinis* (excluding all "*P. platycephalus*" specimens). Interestingly, comparison of the type with other specimens referred to "*P. platycephalus*" is somewhat similar. In comparison of means of measurements of specimens of "*P. platycephalus*" to those of the part of Schultz and Falkenbach's (1956) sample of *M. affinis* that we retain, "*P. platycephalus*" is slightly, to slightly but significantly (P^1-M^3, P^1-P^4, M^1-M^3, TM^3, Sk. W.) larger with the exception of APM_3. The comparison of mean measurements of specimens of "*P. platycephalus*" to those that we refer to *M. affinis*, is strikingly similar, though the size difference is smaller and less regular. The only significant proportional difference is the smaller Sk. L./P^1-M^3 ratio observed in "*P. platycephalus*." Based on size, morphology, and apparent stratigraphy, the type of "*P. platycephalus*" is believed to be an aged individual of *M. affinis*. The samples identified in the literature as "*P. platycephalus*," except as noted elsewhere in this report, are placed easily within *M. affinis* as dorsoventrally crushed individuals.

Miniochoerus as defined by Schultz and Falkenbach (1956), represents mesocephalic Miniochoerinae. Only specimens that are uncrushed, or nearly so, have been identified as *Miniochoerus* in the literature. *M.* "*battlecreekensis*" is based on F:AM 45001, a well-preserved skull and jaws from the "lower nodules," "lower oreodon beds," Shannon County, South Dakota. The published sample of this species, excluding specimens from southeast of Douglas, Wyoming (see *M. chadronensis*, this chapter), contains specimens that are comparable with, but, for most available measurement categories, average slightly to significantly (P^1-M^3, P1-P4, P.O.C., Sk. W., TM_3) larger than the stratigraphically restricted subset which we retain from specimens Schultz and Falkenbach (1956) identified as *M. affinis*. Observed ranges of this restricted sample of *M.* "*battlecreekensis*" overlap substantially with (average of 48% of the combined ranges for 9 of the 17 mutually available measurements), include (M^1-M^3, APM^3, Subnasal L.), or are included within (Ht. M^3, Malar D., Sk. L.), ranges for measurements on the restricted subset of the Schultz and Falkenbach (1956)

version of *M. affinis*. Ranges do not overlap for Nasal L. and Sk. W. Greater similarity exists in comparisons of means and observed ranges of the restricted sample of *M.* "*battlecreekensis*" (Seaman Hills subsample excluded) to those of the sample of "*Platyochoerus platycephalus*." Comparison of the restricted sample of *M.* "*battlecreekensis*" with other specimens that we refer to *M. affinis* is similar to that with Schultz and Falkenbach's (1956) sample of *M. affinis*, but only one measurement set (P^1-P^4) is significantly larger, and inclusion of the ranges of *M.* "*battlecreekensis*" within the ranges other specimens that we refer to *M. affinis* is usual (11 of 17). Two ratios, P^1-P^4/P^1-M^3 and APM^3/Ht. M^3, average significantly larger for *M.* "*battlecreekensis*" than those for others of our version of *M. affinis*. The size difference is not great and *M.* "*battlecreekensis*" is a slightly large *M. affinis*.

Causes and effects of crushing of skulls are discussed in the introduction. A search through the literature and collections shows that the specimens identified as "*Stenopsochoerus*" invariably are laterally crushed, usually severely. F:AM 44980, type specimen of "*S. sternbergi*," (type species of "*Stenopsochoerus*") came from high above the "upper purplish white" layer, Seaman Hills area, from rocks that are stratigraphically comparable with the lower part of the Orella Member of the Brule Formation in northwestern Nebraska. The type and most of the referred specimens (excluding those from 6-9 miles southeast of Douglas, referred to *Miniochoerus chadronensis* above) are thus contemporaries of the subsamples being discussed in this section.

Schultz and Falkenbach (1956) characterized "*Stenopsochoerus sternbergi*" as having a long, narrow skull, with a narrow postorbital constriction. They (1956, p. 438) also state that it "is of interest but not of particular significance to note that examples of *S. sternbergi* have not been recovered from South Dakota." Most of the oreodont skulls from South Dakota are much less crushed, and thus more mesocephalic than contemporary fossils from northwestern Nebraska which form much of the "*sternbergi*" sample. Averages of measurements for the part of "*S. sternbergi*" that we refer to *Miniochoerus affinis* tend to be slightly larger than those of the rest of our sample of *M. affinis*, but the average for P.O.C. (suspect for crushing bias, despite conservative acceptance of specimens for measurement) is significantly smaller. Measurements for the subset of "*S. sternbergi*" specimens are comparable to those of *M.* "*battlecreekensis*," but do not show the degree of proportional variation from other specimens that we refer to *M. affinis* that specimens of *M.* "*battlecreekensis*" do, and are closely comparable to the sample of "*Platyochoerus platycephalus*." When allowances are made for crushing, the early Orellan collection of "*S. sternbergi*" cannot be separated from the other subsamples discussed in this section as conspecific with *M. affinis* on the basis of morphology, or size differences.

Miniochoerus gracilis (Leidy)
Figures 4, 5, 6; Tables 1-9; Appendix C

Oreodon gracile Leidy, 1851:239.
Miniochoerus (Paraminiochoerus) gracilis (Leidy):
Schultz and Falkenbach, 1956:413.
Miniochoerus (Paraminiochoerus) affinis (Leidy):
Schultz and Falkenbach, 1956:405, in part,
tentatively referred.

Type—ANSP 10692 (lectotype, Schultz and Falkenbach, 1956, p. 415), P^4-M^3, Mauvaises Terres, "lower oreodon beds" of South Dakota.

Hypodigm—Type, and Schultz and Falkenbach's (1956) sample of this species such as ANSP 10685, F:AM 49521, 49624, 45363, 49562, 49627, 49556; 44962, 45189A, AMNH 38947, 39431, and 39432; and some of Schultz and Falkenbach's (1956) sample of Miniochoerus affinis, such as AMNH 8875, F:AM 44969, 44966, 49622, 49510, 49534, 44972, and 49574.

Amended diagnosis—Cranial characterization is the same as for a derived member of the genus as the most fetalized species, the Subnasal L. is 62% of the Nasal L., the P^1-P^4 averages 47% of the P^1-M^3 (Table 3), the APM^3/TM^3 ratio averages 1.01 (Table 5), the mean APM^3/Ht. M^3 ratio is 1.91 (Table 4), and TM^3/Ht. M^3 averages 1.88 (Table 6). M. gracilis differs from Miniochoerus chadronensis by more fetalized cranium, systematically and significantly smaller size [for our data, means for all measurements significantly smaller; observed ranges of Subnasal L., Malar D., P.O.C., and Sk. L. and Sk. W. overlap by from 3% (Sk. L.) to 53% (Malar D.) of the combined ranges for our data sets], and significantly larger P^1-P^4/P^1-M^3, Sk. W./Sk. L., Sk. L./P^1-M^3 (ranges overlap substantially; Tables 3, 7-9). M. gracilis differs from M. affinis, an approximate contemporary, by systematically and significantly smaller size [overlap of combined observed ranges is from 4% (P^1-M^3) to 67% (Subnasal L.) with an average of 24%; no overlap for Ht. M3s], and significantly larger P^1-P^4/P^1-M^3 ratio (Tables 3 and 9). M. gracilis differs from M. starkensis by more fetalized cranium, and systematically significantly smaller size [observed ranges of only eight characteristic overlap: P1-P4s, TM^3 (5% of the combined ranges), Subnasal L., Malar D., P.O.C. (35%), Sk. L., Sk. W., with an average overlap of 17%], significantly less reduced premolars, significantly broader M3s and noticeably higher M^3s (proportions that are stereotyped in comparisons with other Miniochoerinae).

Distribution—High above the "upper purplish white" layer, Seaman Hills area, Niobrara County, Wyoming; the lower part of the Orella Member of the Brule Formation, western Nebraska; the "lower oreodon beds," lower part, Scenic Member, Brule Formation, South Dakota, and equivalents in northeastern Colorado; early Orellan, oreodont faunal zone Brule A, early Oligocene.

Discussion—Miniochoerus gracilis has the largest number of distributions that do not appear to be normally distributed (17; Table 1) of any of the better represented fossil taxa. Comments on normality of distribution made at the beginning of the discussion of M. chadronensis apply equally to M. gracilis. We proceed with discussion of M. gracilis with the same assumptions.

Leidy (1851) named several specimens, among them a maxillary fragment with well-preserved P^4-M^3, ANSP 10692, as "Oreodon" gracile. Although he noted the small size of these specimens, Leidy did not describe them in detail. Later, Leidy (1852, p. 551) more fully characterized "O." gracile and commented on the fact that the enamel on the medial shelves of the proto-, and hyposelenes of the molars is very thin. This enamel thinning is the most notable innovation of the Miniochoerinae. Thorpe (1937a) regarded the suite of specimens that had been listed as "Oreodon" gracile by Leidy (1869, pp. 94-96), along with certain other specimens, as cotypes of "O." gracile when the species was transferred to "Merycoidodon." Schultz and Falkenbach (1956) designated the P^4-M^3, mentioned by Leidy (1851), ANSP 10692, as the lectotype of Miniochoerus gracilis when the species was placed in Miniochoerus.

Leidy's (1851) specimens of "Oreodon" gracile came from what now is known, among other names, as the "lower oreodon beds," near the head of Bear Creek, north of Scenic, South Dakota. Many other specimens from these or comparable rocks are attributable to Leidy's species. Hay (1902), Bump and Loomis (1930), Thorpe (1937a), Scott (1940), Phleger and Putnam (1942), and other authors regarded "O." gracile as a very small Merycoidodon. Thorpe and Scott distinguished "M." gracilis from M. culbertsoni on the basis of its relatively broad and little posteriorly protracted occiput. The gracile occiput is consistent with that expected for a dwarfed species relative to larger oreodonts.

Schultz and Falkenbach (1956) attributed nine specimens from different geographic areas to Miniochoerus affinis that are here tentatively placed in M. gracilis. They (1956, pp. 409, 413) especially note that two of the nine, F:AM 44969 and AMNH 8875, may be large individuals of M. "(P.)" gracilis. The other seven specimens of this questionable "affinis" sample closely match F:AM 44969 and AMNH 8875 in size. In overall size, this group is systematically smaller than the part of Schultz and Falkenbach's (1956) sample of M. affinis that we retain in M. affinis [significant differences for seven (P1-M3s, P^1-P^4, M1-M3s, APM3s) of 16 measurements means available for comparison; overlap of combined observed ranges averages 24% for seven characteristics; ranges for M^1-M^3, Ht. M^3 do not overlap]. This subsample is problematic, in that the specimens are almost perfectly

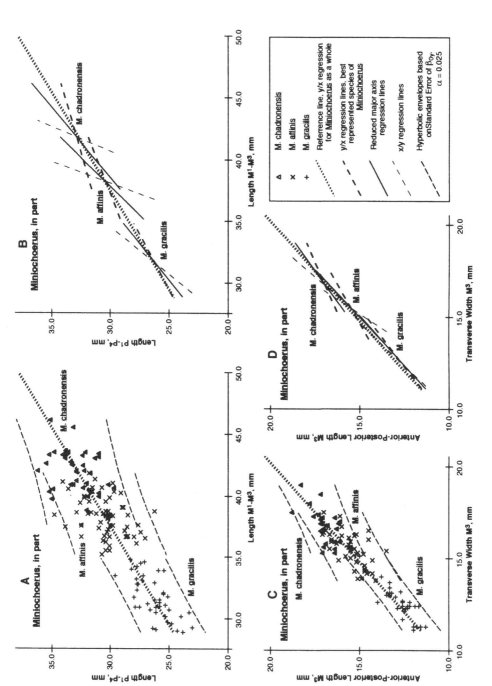

Figure 6. Scatter diagrams, least squares regression lines, and reduced *major* axis lines for selected characteristics of *Miniochoerus chadronensis*, *M. affinis* and *M. gracilis*; (A) scatter plots of P^1-P^4, M^1-M^3 data, and y/x hyperbolic confidence envelopes; (B) regression lines for P^1-P^4, M^1-M^3 data; (C) scatter plots of APM^3, TM^3 data, and y/x hyperbolic confidence envelopes; (D) regression lines for APM^3, TM^3 data. Descriptive statistics are presented in Table 9.

intermediate between *M. affinis*, as we understand that group, and Schultz and Falkenbach's (1956) sample of *M. "(P.)" gracilis*. This subsample, nine specimens out of 66, however, does not account for the distribution complexity of our version of *M. gracilis* (Table 1), although they do add to it. In one important respect, a larger P^1-P^4/P^1-M^3 (available for all 9 specimens; approximated by regressions and scatter-plots of P^1-P^4, M^1-M^3, Fig. 6A, Table 9) which allows the clearest distinction of our versions of *M. gracilis* and *M. affinis*, the part of Schultz and Falkenbach's (1956) sample of *M. affinis* that we here tentatively refer to *M. gracilis* is distinctly more like *M. gracilis*.

Miniochoerus gracilis seems a contemporary of *M. affinis*, as noted by various authors. Schultz and Falkenbach (1956, p. 414) are probably correct in the view that *M. gracilis* represents a distinct, but short-lived lineage within the Miniochoerinae, restricted to oreodont faunal zone Brule A. We agree with them that the specimens of *M. gracilis* cannot be regarded as females of *M. affinis* partly for reasons discussed above (Numerical Data), and partly because other oreodont species samples that surely must contain both males and females on the basis of the high number of observations, are not as variable as the combination of these two taxa (Table 1, and Appendices B, C). As previously noted, data for *M. gracilis* (Appendix C) is commonly bimodal, but we hesitate to attribute this to gender differences. The CVs obtained for the combination of *M. affinis* and *M. gracilis* exceed the observed range of CVs of fossil species discussed in this report for all measurements except for some of the most variable ones, Malar D., P.O.C., Sk. W.; and an anomalously high CV for P_1-P_4 of *M. starkensis* (Table 1 and Appendix C). Thirteen of 18 CVs of the combination of *M. affinis*-*M. gracilis* data are also larger than CVs obtained by grouping together all of the species of *Merycoidodon*. However, it is interesting to note that, the modern peccary, *Dicotyles tajacu*, has two CVs (Ht. M^3, Malar D.) that are larger than the corresponding CVs for combined *M. affinis*-*M. gracilis* data. Scatter diagrams (Fig. 6A, C) graphically show better separation of data sets (*M. affinis*, *M. gracilis*) than is usual for data for the Miniochoerinae or Merycoidodontinae. Plots usually show a lack of separation more closely resembling that seen for the *M. chadronensis*-*M. affinis* data sets (Fig. 6A, C). As previously noted, analysis of covariance suggests significant size differences between all three data sets illustrated. Table 9 provides descriptive statistics for the regression lines shown (Fig. 6C, D). Nevertheless, *M. gracilis* is very closely related to *M. affinis*. *M. gracilis* may have evolved through phylogenetic dwarfing from *M. chadronensis*, the probable ancestor for *M. affinis*, or more rapidly from early *M. affinis*. Schultz and Falkenbach (1956, p. 414) note that *M. gracilis* is not ancestral to any younger oreodont, and this may be true.

Miniochoerus starkensis Schultz and Falkenbach
Figures 4D, 5D, D'; Tables 3-8; Appendix C

Miniochoerus starkensis Schultz and Falkenbach, 1956:398

Miniochoerus nicholsae Schultz and Falkenbach, 1956:400

Miniochoerus cheyennensis Schultz and Falkenbach, 1956:401

Miniochoerus (Paraminiochoerus) helprini Schultz and Falkenbach, 1956:422

Miniochoerus (Paraminiochoerus) ottensi Schultz and Falkenbach, 1956:424

Platyochoerus heartensis Schultz and Falkenbach, 1956:432

Platyochoerus hatcreekensis Schultz and Falkenbach, 1956:433

Stenopsochoerus joderensis Schultz and Falkenbach, 1956:441

Stenopsochoerus berardae Schultz and Falkenbach, 1956:441

Type—F:AM 45497, from 5 feet (1.5 m) below the top of the "nodular zone," just below the "white zone" (see Skinner, 1951, p. 57), Fitterer Badlands, 7 miles (11.3 km) south of South Heart, Stark County, North Dakota.

Hypodigm—Type, and Schultz and Falkenbach's (1956) sample of this species such as F:AM 45498; type, F:AM 49585, and Schultz and Falkenbach's (1956) sample of *Miniochoerus "nicholsae,"* such as AMNH 1322; type, AMNH 9797, and F:AM 34336 of *M. "cheyennensis"*; type F:AM 49501, and Schultz and Falkenbach's (1956) sample of *M. "helprini,"* such as F:AM 49502, 49504, 49506; type, UNSM 28336, and presumably the referred specimens (not seen) of *M. "ottensi"*; type, F:AM 45476, and Schultz and Falkenbach's (1956) sample of *"Platyochoerus heartensis,"* such as F:AM 45469, F:AM 49545; type, F:AM 45463, and Schultz and Falkenbach's (1956) sample of *"P. hatcreekensis,"* such as F:AM 45465A, 45465B, 45459, and 49560; type, F:AM 45483, of *"Stenopsochoerus joderensis"*; and the type, F:AM 49617, and Schultz and Falkenbach's (1956) sample of *"S. berardae,"* such as F:AM 45484, 45486, 49570, 49618A, and 49618B.

Amended diagnosis—Cranial characterization is the same as for the genus, the Subnasal L. is 58% of the Nasal L., the P^1-P^4 averages 44% of the P^1-M^3 (Table 3), the mean APM^3/TM^3 is 1.05 (Table 5), the mean APM^3/Ht. M^3 is 2.07 (APM^3 absolutely increased; Table 4), and TM^3/Ht. M^3 averages 1.95 (Table 6). *M. starkensis* differs from *Miniochoerus affinis* by slightly but significantly larger size [for our samples (Appendix C): significant differences except Subnasal L., P.O.C. (smaller), Sk. W., TM_3, and Ht. M_3 (smaller; very small samples); observed ranges overlap substantially, from 27% to 76% of the combined ranges, averaging 44%], significantly more reduced premolars, M^3s slightly but significantly narrower, and significantly lower in proportion to length (proportions that are stereotyped among Miniochoerinae). *M. starkensis* differs from *M. gracilis* by systematically significantly larger mean size [for our samples: observed ranges of only eight characteristic overlap: P1-P4, TM^3 (5% of the combined ranges), Subnasal L., Malar D., P.O.C. (35%), Sk. L., Sk. W., with an average overlap of 17%], significantly more reduced premolars, significantly narrower M3s and noticeably lower M^{3}s in proportion to length (proportions that are stereotyped among other Miniochoerinae). *M. starkensis* differs from *M. chadronensis* by a slight tendency to be larger (for our samples: means of ten of 18 measurements; significantly larger for: APM^3, Subnasal L., and Malar D., but significantly smaller for P^1-P^4, and TM3; *M. starkensis* observed ranges include those of *M. chadronensis* for P1-M3s, M1-M3s, TM^3, Ht. M^3, APM3, but the range for P.O.C. is included in that of *M. chadronensis*; other ranges overlap by an average of 62% of the combined ranges), and in proportions by significantly larger APM^3/Ht. M^3 and APM^3/TM^3 ratios and a significantly smaller P^1-P^4/P^1-M^3 ratio.

Distribution—The Fitterer Badlands, North Dakota; "upper nodules," "middle oreodon beds" and the "upper oreodon beds," Scenic Member, and the lower part of the "lower *Leptauchenia* beds," Poleslide Member of the Brule Formation, South Dakota; the upper part of the Orella Member, and the lower part, "K" or crumbly zone, of the Whitney Member of the Brule Formation, western Nebraska; and presumed equivalents in adjacent regions; late Orellan to early Whitneyan, early middle Oligocene.

Discussion—*Miniochoerus starkensis* is based on a partial skull, jaws, and some vertebrae from the Fitterer Badlands of North Dakota. The fossil came from 5 feet below the top of the nodules of Skinner (1951, p. 58), unit 4A, and just below the "White marker bed" (Antelope Creek Tuff of Hoganson and Lammers, 1992; Murphy et al., 1993). These rocks are probably correlative with the upper part of the Orella and Scenic members of the Brule Formation (see Skinner, 1951, p. 58; also Douglass, 1909, pp. 285-288) of South Dakota and Nebraska. Although the type, F:AM 45497, is relatively undistorted and thus mesocephalic, the muzzle anterior to P^1, most of the zygoma, and the occipital crests are missing. Schultz and Falkenbach (1956) did not refer any contemporary fossils from the type area to *M. starkensis*; the two referred specimens came from the Toadstool Park area, western Nebraska. Additionally, the only comparisons made at the time of description were with taxa that Schultz and Falkenbach had identified as *Miniochoerus*.

The type of *Miniochoerus* "*nicholsae*," F:AM 49585, is a skull and some postcranial elements, from a pinkish clay in the middle of the exposures on the east side of Slim Buttes, Harding County, South Dakota. This clay layer is equivalent with either the upper part of the "upper oreodon beds" or the lower part of the "*Leptau-*

chenia beds" elsewhere in South Dakota. F:AM 49585 and the several referred specimens slightly larger than the type of *M. starkensis*, and size are the only features originally used to distinguish *M. starkensis* from *M.* "*nicholsae*." Fourteen of 16 available measurement comparisons of the type of *M.* "*nicholsae*," with other specimens that we refer to *M. starkensis* lie within observed ranges (exceptions: Ht. M^3, Sk. W.), and four measurements of the type are slightly but significantly (P^1-M^3, P.O.C., and Sk. L.) to significantly (Sk. W.) larger. A significantly large Subnasal L./Nasal L. ratio of the type of *M.* "*nicholsae*," is the only proportion more than slightly different from other *M. starkensis*. These taxa are otherwise very similar, and are probably members of the same species.

The type of *Miniochoerus* "*cheyennensis*," AMNH 9797, a partial cranium, came from either the top of the "upper oreodon beds" or the base of the "*Leptauchenia* beds," Shannon County, South Dakota. Schultz and Falkenbach (1956) referred two comparable specimens from presumably earlier Whitneyan rocks west of Pawnee Buttes, Weld County, Colorado, and a jaw, F:AM 45916, from South Dakota to *M.* "*cheyennensis*." The jaw, however, probably is a *Merycoidodon* on the basis of its substantially larger size, and lack of premolar-molar enamel thinning. The type of *M.* "*cheyennensis*" is significantly larger than other specimens that we refer to *M. starkensis* in four of five measurement categories available for comparison (exception, APM^3). A referred specimen, F:AM 45336 is not as large, and none of the proportions available on either specimen are very different from those of *M. starkensis*. Schultz and Falkenbach (1956) are correct to note the great similarity between the species "*cheyennensis*" and "*nicholsae*," and these in turn are very similar to the type of *M. starkensis*.

Miniochoerus "*helprini*" is based on a poorly preserved, fractured skull and partial jaws, F:AM 49501, from talus, 20 feet (6.1 m) below the "K" zone, a unit used as a field marker by Skinner and Mefferd when they collected fossils in the Toadstool Park area, western Nebraska. F:AM 49501 came either from the upper part of the Orella Member, or from the lowest part of the Whitney Member of the Brule Formation. Specimens from Stark County, North Dakota, came from the top of the nodules, unit 4A, of Skinner (1951). Measurements on the type of *M. starkensis*, F:AM 45497, lie above the range observed for the very small sample of *M.* "*helprini*" for nine of ten available measurement sets (Malar D. of F:AM 45497 is smaller than the mean for *M.* "*helprini*"), but only the difference in P.O.C. is significant. Means of measurements for the set of specimens referred to *M.* "*helprini*" are systematically smaller than those for other specimens that we refer to *M. starkensis*, significantly so for P^1-M^3, M1-M3s, APM3s, and TM3s, but the proportions are not significantly different. The type of *M.* "*helprini*" is slightly smaller than the type of *M.*

starkensis (Appendix C), but is otherwise similar, and it seems most useful to consider *M.* "*helprini*" as small individuals of *M. starkensis*.

Miniochoerus "(*Paraminiochoerus*) *ottensi*" is based on UNSM 28336, a skull and jaws from the "K" zone of Skinner and Mefferd, lower part of the Whitney Member of the Brule Formation, Sioux County, Nebraska. The UNSM referred specimens have not been re-examined. F:AM 49546 came from 23 feet (7 m) above the base of the "K" zone. Schultz and Falkenbach (1956) identified this limited suite of specimens as *Miniochoerus* "(*Paraminiochoerus*)," a subgenus used to designate certain *Miniochoerus* of small size, and excluded *M. starkensis*, from the subgenus. For the nine available measurements, the type of *M.* "*ottensi*" is similar in size (slightly above average size in most respects) and P^1-P^4/P^1-M^3 proportion to the type of *M. starkensis*, and to other specimens that we refer to *M. starkensis*. There seems to be no evidence to support the separation of *M.* "(*P.*) *ottensi*" from the approximately contemporary populations discussed in this section as probably conspecific within *M. starkensis*.

Schultz and Falkenbach (1956) regarded what they thought were phylogenetically brachycephalic Miniochoerinae as species of "*Platyochoerus*." Presumed late Orellan-earlier Whitneyan members of this allegedly brachycephalic line were identified either as "*P. heartensis*" or "*P. hatcreekensis*." "*P. heartensis*" is based on F:AM 45467, an almost complete cranium with a jaw fragment and some other bones from the top of the nodules, unit 4A of Skinner (1951), Stark County, North Dakota. These are the same rocks which yielded the type of *Miniochoerus starkensis*. The identification of F:AM 45467 as "*Platyochoerus*" results from the fact that the type is crushed dorsoventrally as shown by the compressed frontals and fractured and laterally spread palate and occiput. As a result, the skull is broader as now preserved than it was in life. Means of measurements for the very small sample of "*P. heartensis*" are mostly smaller (13 of 16 characters available for comparison; significant: very small single measurements for Subnasal L. and Sk. L., and P_1.M_3, P_1.P_4; exceptions: APM^3, Sk. W., TM_3) but ranges observed for "*P. heartensis*" are mainly (10 of 16) within those of other specimens that we refer to *M. starkensis*, and proportions do not differ significantly.

"*Platyochoerus hatcreekensis*" is based on F:AM 45463, from about 15 feet (4.6 m) above the base of the "K" zone, 10 miles (16.1 km) north of Harrison, Sioux County, Nebraska. The type lacks the premaxillae, zygomatic arches, and nuchal crest, and the dorsoventral deformation that has altered the shape of the type is shown by the broken and laterally spread-apart palate. After accounting for deformation, F:AM 45463 (slightly larger than the type of *Miniochoerus starken-*

sis) is similar morphologically to the other specimens discussed in this section, and the few (6; Appendix C) measurements available for comparison do not differ significantly from the means for other specimens, except for a large P.O.C., largest among specimens referred to "*P. hatcreekensis*." The dorsoventral crushing that deformed the type of *P. hatcreekensis* would tend to enlarge the P.O.C., and the measurement may be suspect. The sample of *P. hatcreekensis*, in comparison to other specimens that we refer to *M. starkensis*, is slightly larger than average for most characters (15 of 16; exception: Malar D.), most noticeably but not significantly so for M_1-M_3 and TM_3. Proportions show only trivial differences from other specimens that we refer to *M. starkensis*.

The laterally crushed, hence "narrow-headed" more or less contemporary counterparts of the "brachycephalic" Miniochoerinae discussed above, appear in the literature as "*Stenopsochoerus joderensis*" and "*S. berardae*." "*S. joderensis*" is based on and known only from F:AM 45483, a very poorly preserved and laterally crushed skull from ?late Orellan rocks in the Toadstool Park area, Sioux County, Nebraska. "*S. berardae*" is based on F:AM 49617, a moderately large skull from the east side of Slim Buttes, Harding County, South Dakota, and is a contemporary of at least the type of *Miniochoerus* "*nicholsae*." The sample of "*S. berardae*" is neither systematically, nor, in any available characteristic, significantly different from types of "*S. joderensis*" or *M.* "*nicholsae*." Measurement means for the species "*joderensis*" and "*berardae*" are mostly larger (exceptions: "*berardae*" APM^3, TM^3 measurements) than averages for other specimens that we refer to *M. starkensis*, but not significantly so (except "*joderensis*" Malar D.), and the morphology of these groups does not differ appreciably from other *M. starkensis*.

Miniochoerus starkensis differs from *M. affinis* in size and proportions, and from *M. chadronensis* mainly in proportions. Taxa previously identified as "*Stenopsochoerus berardae*," "*S. joderensis*," "*Platyochoerus hatcreekensis*," "*P. heartensis*," *M. starkensis*, *M.* "*nicholsae*," *M.* "*cheyennensis*," *M.* "*helprini*," and *M.* "*ottensi*," were separated from each other on the basis of differences in preservation and size, and the size differences are significant only when extremes are isolated and compared. Analyses used to establish the order in Table 1 suggest that our version of *M. starkensis* is more variable as a taxon, than *Merycoidodon culbertsoni*, but only slightly so, and *M. starkensis* is not nearly as well represented. Based on morphology and descriptive statistics for *M. starkensis* (Appendix C), it is unlikely that the subsamples which now constitute the taxon as discussed in this section contain three genera and nine species of Miniochoerinae. *Miniochoerus* declined during the latest Orellan-earliest Whitneyan, and *M. starkensis* is the last species.

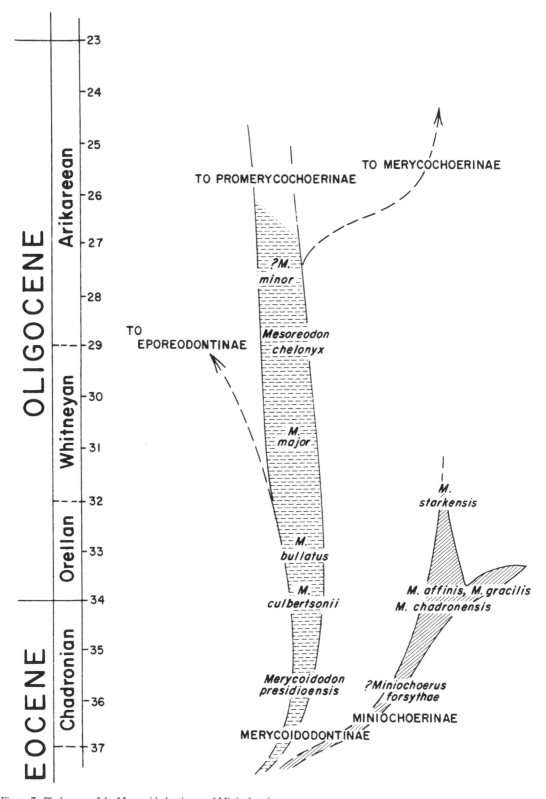

Figure 7. Phylogeny of the Merycoidodontinae and Miniochoerinae.

SUMMARY AND CONCLUSIONS

Descriptive statistics are presented for the Merycoidodontinae and Miniochoerinae (appendices B and C) recognized in this report. For comparison, similar statistical information for the peccaries *Dicotyles tajacu* and *Tayassu pecari*, modern suiform artiodactyls, is also presented (Appendix A, Table 1). Ratio diagrams based on sample means graphically illustrate the gradual but cumulative trends that have been discussed in the text (Figs. 2, 4). Illustrations of four views of a cranium for each oreodont species (Figs. 3, 5), a phylogenetic summary chart (Fig. 7), and biostratigraphic correlations (Fig. 8) are presented.

The Merycoidodontinae contains *Merycoidodon* and *Mesoreodon* and spanned some 10 million years. *Merycoidodon* (late early or middle Chadronian to the later Whitneyan as known) is divisible among four mesocephalic species, *M. presidioensis* sp. nov, *M. culbertsoni* Leidy, *M. bullatus* (Leidy), and *M. major* (Leidy), and these are linked in a single series. *M. major*, the terminal species of *Merycoidodon*, was ancestral to *Mesoreodon chelonyx* Scott (latest Whitneyan and/or earliest Arikareean), the type species of *Mesoreodon*. Average species longevity for the Merycoidodontinae is about 1.7 million years.

Auditory bullae of the Merycoidodontinae show progressive change from small, with a deep hyoidal groove in *Merycoidodon (Merycoidodon)* new combination (Chadronian-earlier Orellan) to increasingly inflated with loss of the hyoidal groove in *Merycoidodon (Otarohyus)* (Schultz and Falkenbach) from the later Orellan and Whitneyan. The size and morphology of bullae thereafter became stereotyped in *Mesoreodon chelonyx* and *?M. minor*, and the inflated bullae tend to look "small" on the relatively large skulls of these species. During the time of the transition from a dominantly small (*M. culbertsoni*) to inflated bullae (*M. bullatus*), certain individuals of either species were either more, or less derived in bullar morphology than their conspecific contemporaries. This is to be expected as intraspecific variation within a single evolving lineage. The Merycoidodontinae show a progressive trend for slight premolar reduction and slight anteroposterior attenuation of the third molars, more pronounced in *Mesoreodon* than in *Merycoidodon*, a decrease in the size and depth of the preorbital fossae, slight posterior migration of the infraorbital foramen, slight broadening of the muzzle and cranium, and a lowering of the nasal profile.

If *Merycoidodon presidioensis* is representative of middle Chadronian Merycoidodontinae, there may be a middle and late Chadronian trend for size reduction from *M. presidioensis* to *M. culbertsoni*. Better supported is a general increase in size for the Merycoidodontinae from the Orellan through early Arikareean, particularly from the Whitneyan onward. The latest species, *?Mesoreodon minor*, initiates the trend for an increase in the bizygomatic diameter of the skull, and for a notable upturning of the posterior margin of the zygomatic arch.

Mesoreodon (latest Whitneyan through the early Arikareean) is redefined to be more consistent with the mesocephalic type species, *Mesoreodon chelonyx*. *Mesoreodon* contains two species, *M. chelonyx* and *?M. minor*. *?M. minor* is essentially morphologically intermediate between *M. chelonyx* and *Desmatochoerus megalodon* (Peterson). It is questionably referred to *Mesoreodon* because it is more brachycephalic and has more specialized zygomatic arches than *Mesoreodon sensu stricto*. *Mesoreodon* is ancestral to the Promerycochoerinae and the Merycochoerinae (Fig. 7). *?M. minor* probably is ancestral directly to *Desmatochoerus-Promerycochoerus* (and *Megoreodon*) of the Promerycochoerinae, and to *Hypsiops* of the Merycochoerinae, the middle-early late Arikareean ancestor of *Submerycochoerus-Merycochoerus*.

The origin of *Eporeodon*, and thus the Eporeodontinae, also lies within the Merycoidodontinae. *Eporeodon*, however, must have originated earlier than the time of occurrence of both *Merycoidodon major* (later Whitneyan) and *Mesoreodon chelonyx* (latest Whitneyan or earliest Arikareean) because these are too specialized in the direction of *?M. minor*, to be ancestral. We believe that *Eporeodon* was derived from within late Orellan members of *M. bullatus*.

Miniochoerus and the monogeneric Miniochoerinae existed for approximately 5 million years, and remained morphologically conservative throughout their history. At least four of the five species of Miniochoerinae, *?Miniochoerus forsythae*, *M. chadronensis*, *M. affinis*, and *M. starkensis* are linked in a single phyletic series, and have a species longevity of about 1.25 million years. *?M. forsythae*, the largest and most primitive species, is structurally and presumably phylogenetically intermediate between early *Merycoidodon* and *Miniochoerus*. *M. chadronensis* and *M. affinis* are very closely related, but *M. chadronensis* is recognized as valid on the basis of a somewhat larger size and deeper preorbital fossae, than *M. affinis*. Species within Miniochoerinae are distinguishable on the basis of morphology, but the fact remains that in many aspects of size and proportion they are gradational or stereotyped.

The most notable innovation within the Miniochoerinae was the progressive thinning of enamel on the premolar-molar medial crests or selenes, but there were also trends toward premolar reduction, and reduction in preorbital fossae. The Miniochoerinae, as with the Merycoidodontinae, show a clear decrease in size from middle through late Chadronian, a trend reversed without much change in proportional stereotypy in the Orellan. The Miniochoerinae, except for *M. starkensis*, illustrate a group diminishing in size, whose species appear to be samples that could be drawn from a set

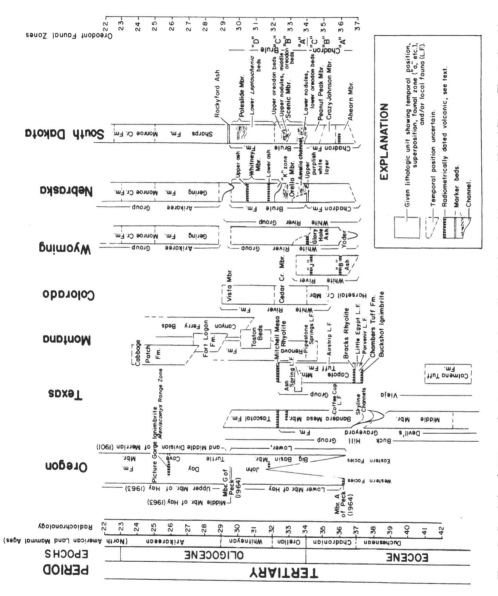

Figure 8. Correlations between pertinent oreodont-bearing and/or related rocks of the western United States. The stratigraphic nomenclature discussed in this report reflects that used by authors who were affiliated with the various institutions whose specimens were used in this study, for example Wortman (1893), Osborn and Wortman (1894, 1895), Merriam (1901), Osborn (1907), Osborn and Matthew (1909), Schultz and Stout (1955), Hay (1963), Peck (1964), Schultz and Falkenbach (various papers, butespecially, 1968), and Wilson (1986), among others.

generated by linear stochastic variation, a chronocline. Although *M. affinis* and *M. gracilis* are statistically separable, and have been recognized as separate but ?contemporary species in this report, an appealing possibility is that they may represent part of the chronocline that begins with ?*M. forsythae*. Stratigraphy of *M. gracilis* and *M. affinis* localities deserves reexamination and recollection. Together with the Merycoidodontinae, the Miniochoerinae form a larger group that also appears to illustrates gradual evolution, in different, but compatible directions. The impetus for change in size may have been middle through late Chadronian imposition and Orellan through early Arikareean relaxation of environmental stress.

Species of Chadronian to Whitneyan Merycoidodontinae and Miniochoerinae conform reasonably well with the various "Oligocene" oreodont faunal zones designated by Schultz and Falkenbach (various papers, but see 1968), although faunal zone Chadron "C" probably merges with Brule "A," and faunal zone Brule "B" probably blends with Brule "C." As known, *Merycoidodon presidioensis* sp. nov., and ?*Miniochoerus forsythae* are restricted to Chadron faunal zone "A-?B." *Merycoidodon culbertsoni*, Chadron "C" through Brule "A," and *Miniochoerus chadronensis*, Chadron "C" through earliest Brule "A," are approximate contemporaries, but *M. affinis* and *M. gracilis* are confined to Brule "A." *Merycoidodon bullatus* occurs in Brule faunal zones "B" and "C," *Miniochoerus starkensis* has been identified from Brule "C" and lower "D," and *Merycoidodon major* is restricted to faunal zone Brule "D." *Mesoreodon chelonyx* lived during the latest Whitneyan or earliest Arikareean, and ?*M. minor* lived during the early Arikareean.

ACKNOWLEDGMENTS

We thank Drs. Malcolm C. McKenna and Richard H. Tedford of the American Museum of Natural History, for allowing the senior author (MSS) to study the F:AM and AMNH oreodonts between 1967-1970, for their support, and/or for freely giving of their extensive knowledge about the stratigraphy of the deposits in which the oreodont specimens were found. Drs. McKenna and Tedford, and Drs. Michael O. Woodburne, University of California, Riverside, J. A. Wilson, retired, and Melissa Winans, of The University of Texas at Austin, and Donald R. Prothero, Occidental College, painstakingly and critically read parts or all of earlier versions of this manuscript, and gave advice. We thank J. A. Wilson for making the oreodonts from the Vieja available for study. We thank the late Morris F. Skinner, formerly of the Frick Laboratory, American Museum of Natural History, for sharing with the senior author the geologic provenance of specimens he had the privilege to collect, and for many stratigraphic discussions. We thank Dr. Ted Daeschler, Academy of Natural Sciences, Philadelphia, for measuring some oreodonts under his care, and Dr. Donald L. Rasmussen for providing the senior author (MSS) with locality and stratigraphic information for oreodonts from the Cabbage Patch Formation. We thank Dr. C. Bertrand Schultz, University of Nebraska, retired, for permitting the senior author (MSS) to examine some of the oreodonts then under his care. We thank Dr. Robert Emry, U.S. National Museum, for providing loans and valuable points of information about the stratigraphy of the Flagstaff Rim area, Natrona County, Wyoming. We thank Dr. H. G. Richards, Chairman, Department of Geology, Academy of Natural Sciences, Dr. Mary R. Dawson and Ms. Helen McGinnis, Carnegie Museum, and Messers. Raymond J. Gooris, and George Krochak, and Ms. Charlotte Holton of the American Museum of Natural History, for providing loans or casts, or for measuring specimens under their care. We thank Dr. Richard G. Van Gelder, Department of Mammalogy, American Museum of Natural History, for making Recent specimens of *Tayassu pecari* available for study. We thank Dr. Carl C. Swisher, Berkeley Geochronology Center, Berkeley, California, for providing the senior author (MSS) with then unpublished radiometric age determinations.

We thank Dr. H. E. Eveland, retired, Dr. Donald E. Owen, and the late Mr. William H. Matthews III, former heads, and the current Chair of the Department of Geology, Lamar University-Beaumont, for providing office and research space necessary for the preparation of this report. We thank President W. Sam Monroe, Lamar University-Port Arthur, and former presidents Joe Ben Welch and Steve Maradian, Lamar University-Orange, Texas, for providing necessary materials. Processing of the statistical information and hard copy printings of the manuscript and computer-generated Figures were performed in the Geology Statistics and Mapping Laboratory, Department of Geology, Lamar University-Beaumont. That laboratory was made possible by NSF grant EAR-8851521.

LITERATURE CITED

Agterberg, F. P. 1974. Geomathematics, Mathematical Background and Geoscience Applications. Elsevier Scientific Publishing Company, New York.

Bump, J. D., and F. B. Loomis. 1930. Variation in the species of *Merycoidodon*. American Journal of Science, series 5, 20:17-21.

Clark, J., J. R. Beerbower, and K. K. Kietzke. 1967. Oligocene sedimentation, stratigraphy, paleoecology and paleoclimatology in the Big Badlands of South Dakota. Fieldiana, Geological Memoir 5:1-158.

Cope, E. D. 1878. A new genus of Oreodontidae. American Naturalist 12:129.

Cope, E. D. 1884a. Synopsis of the species of Oreodontidae. Proceedings of the American Philosophical Society 21:503-572.

Cope, E. D. 1884b. The history of the Oreodontidae. American Naturalist 18:280-282.

Darton, N. H. 1899. Preliminary report on the geology and water resources of Nebraska west of the one hundred and third meridian. 19th Annual Report, United States Geological Survey 1897-1898, 4:719-814.

Darton, N. H. 1903. Description of the Scotts Bluff Quadrangle. United States Geological Survey, Geologic Atlas, Folio 88:1-4.

Davis, J. C. 1986. Statistics and Data Analysis in Geology, second edition. John Wiley and Sons, New York.

Dixon, W. J. and Massey, F. J., Jr. 1969. Introduction to Statistical Analysis, third edition. McGraw-Hill Book Company, New York.

Douglass, E. 1901. Fossil Mammalia of the White River beds of Montana. Transactions, American Philosophical Society, new series 20:237-279.

Douglass, E. 1903. New vertebrates from the Montana Tertiary. Annals of the Carnegie Museum 11:145-200.

Douglass, E. 1907a. New merycoidodonts from the Miocene of Montana. Bulletin of the American Museum of Natural History 23:809-822.

Douglass, E. 1907b. Some new merycoidodonts. Annals of the Carnegie Museum 4:99-109.

Douglass, E. 1909. A geological reconnaissance in North Dakota, Montana, and Idaho; with notes on Mesozoic and Cenozoic geology. Annals of the Carnegie Museum 4:211-288.

Emry, R. J. 1973. Stratigraphy and preliminary biostratigraphy of the Flagstaff Rim area, Natrona County, Wyoming. Smithsonian Contributions to Paleobiology 18:1-43.

Emry, R. J. 1992. Mammalian range zones in the Chadronian White River Formation at Flagstaff Rim, Wyoming; pp. 106-115 in D. R. Prothero and W. A. Berggren (eds.), Eocene-Oligocene Climatic and Biotic Evolution. Princeton University Press, Princeton, N. J.

Emry, R. J., L. S. Russell, and P. R. Bjork. 1987. The Chadronian, Orellan, and Whitneyan North American Land Mammal Ages; pp. 118-152 in M. O. Woodburne (ed.), Cenozoic Mammals of North America, Geochronology and Biostratigraphy. University of California Press, Berkeley.

Evanoff, E. 1990. Late Eocene and early Oligocene paleoclimates as indicated by the sedimentology and nonmarine gastropods of the White River Formation near Douglas, Wyoming. Ph.D. dissertation, University of Colorado, Boulder, 440 pp.

Evanoff, E., D. R. Prothero, and R. H. Lander. 1992. Eocene-Oligocene climatic change in North America: the White River Formation near Douglas, east-central Wyoming; pp. 116-130 in D. R. Prothero and W. A. Berggren (eds.), Eocene-Oligocene Climatic and Biotic Evolution. Princeton University Press, Princeton, N. J.

Evernden, J. F., D. E. Savage, G. H. Curtis, and G. T. James. 1964. Potassium-argon dates and the Cenozoic mammalian chronology of North America. American Journal of Science 262:145-198.

Gazin, C. L. 1955. A review of the upper Eocene Artiodactyla of North America. Smithsonian Miscellaneous Collections 128:1-96.

Grinnell, G. B., and E. S. Dana. 1876. On a new Tertiary lake basin. American Journal of Science, series 3, 11:125-128.

Harris, J. M. 1967a. Oligocene vertebrates from western Jeff Davis County, Trans-Pecos Texas. M.A. thesis, University of Texas at Austin, 164 pp.

Harris, J. M. 1967b. Toxotherium (Mammalia: Rhinocerotoidea) from western Jeff Davis County, Texas. The Pearce-Sellards Series, Texas Memorial Museum, The University of Texas at Austin 9:1-7.

Harris, J. M., and A. E. Wood. 1969. A new genus of eomyid rodent from the Oligocene Ash Spring local fauna of Trans-Pecos Texas. The Pearce-Sellards Series, Texas Memorial Museum, The University of Texas at Austin 14:1-7.

Hay, O. P. 1902. Bibliography and catalogue of the fossil Vertebrata of North America. Bulletin, United States Geological Survey 179:1-868.

Hay, R. L. 1963. Stratigraphy and zeolite diagenesis of the John Day Formation of Oregon. University of California Publications in the Geological Sciences 42:199-262.

Henry, C. D., F. W. McDowell, J. G. Price, and R. C. Smyth. 1986. Compilation of potassium-argon ages of Tertiary igneous rocks, Trans-Pecos Texas. The University of Texas at Austin, Bureau of Economic Geology, Geological Circular 86-2:1-34.

Hoganson, J. W., and G. E. Lammers. 1992. Vertebrate fossil record, age and depositional environments of the Brule Formation (Oligocene) in North Dakota; pp. 243-255 in J. M. Erickson, and J. W. Hoganson (eds.), Frank D. Holland, Jr. Symposium. North Dakota Geological Survey Miscellaneous Series 76:1-318.

Janis, C. M. 1989. A climatic explanation for patterns of evolutionary diversity in ungulate mammals. Palaeontology 32(3) 463-481.

Kay, J. L., R. W. Fields, and J. B. Orr. 1958. Fauna lists of Tertiary vertebrates from western and southwestern Montana; pp. 33-39 in R. W. Fields (ed.), Guidebook, Eighth Field Conference, Society of Vertebrate Paleontology, western Montana. Montana State University Press.

Koerner, H. E. 1940. The geology and vertebrate paleontology of the Fort Logan and Deep River formations of Montana. American Journal of Science 238:837-862.

Konizeski, R., and J. C. Donohoe. 1958. Faunal and stratigraphic relationships of the Cabbage Patch beds, Granite County, Montana; pp. 45-49 in R. W. Fields (ed.), Guidebook, Eight Field Conference, Society of Vertebrate Paleontology, western Montana. Montana State University Press.

Lander, E. B. 1978. A review of the Oreodonta (Mammalia, Artiodactyla), Parts I, II, and III. Dissertation Abstracts International:38.

Leidy, J. 1848. On a new fossil genus and species of ruminatoid Pachydermata, Merycoidodon culbertsonii. Proceedings, Academy of Natural Sciences, Philadelphia 4:47-50.

Leidy, J. 1850. Observations on two new genera of fossil Mammalia, Eucrotaphus jacksoni, and Archaeotherium mortoni. Proceedings, Academy of Natural Sciences, Philadelphia 5:90-93.

Leidy, J. 1851. Descriptions of fossil ruminant ungulates from Nebraska. Proceedings, Academy of Natural Sciences, Philadelphia 4:237-239.

Leidy, J. 1852. Description of the remains of extinct Mammalia and Chelonia from Nebraska Territory, collected during the geological survey under the direction of Dr. D. D. Owen; pp. 534-572 in D. D. Owen, Report of a geological survey of Wisconsin, Iowa, and Minnesota and incidentally a portion of Nebraska Territory. Lippincott, Grambo and Company, Philadelphia, 638 pp.

Leidy, J. 1853. [Remarks on a collection of fossil Mammalia from Nebraska]. Proceedings, Academy of Natural Sciences, Philadelphia 6:392-394.

Leidy, J. 1854. The ancient fauna of Nebraska, or a description of remains of extinct Mammalia and Chelonia from the Mauvaises Terres of Nebraska. Smithsonian Contributions to Knowledge 6:1-126.

Leidy, J. 1869. The extinct mammalian fauna of Dakota and Nebraska, including an account of some allied forms from other localities, together with a synopsis of the mammalian remains of North America. Journal, Academy of Natural Sciences, Philadelphia, series 2, 7:1-472.

Loomis, F. B. 1924. The oreodonts of the lower Oligocene. Annals of the Carnegie Museum 15:369-378.

Macdonald, J. R. 1956 The North American anthracotheres.

Journal of Paleontology 30:615-645.

Marsh, O. C. 1875. Notice of new Tertiary mammals, Part 4. American Journal of Science, series 3, 5:239-250.

Marsh, O. C. 1885. The gigantic mammals of the Dinocerata. Fifth. Annual Report, United States Geological Survey:243-302.

Marsh, O. C. 1886. Dinocerata, a monograph of an extinct order of gigantic mammals. Monograph, United States Geological Survey 10:1-237.

Matthew, W. D. 1901 Fossil mammals of the Tertiary of northeastern Colorado. Memoir, American Museum of Natural History 1(7):353-447.

Merriam, J. C. 1901. A contribution to the geology of the John Day Basin. Bulletin, University of California Department of Geology 2(9):269-314.

Meyer, S. L. 1975. Data Analysis for Scientists and Engineers. John Wiley and Sons, New York, 513 pp.

Miller, R. L., and J. S. Kahn. 1962. Statistical Analysis in the Geological Sciences. John Wiley and Sons, New York.

Murphy, E. C., J. W. Hoganson, and N. F. Forsman. 1993. The Chadron, Brule, and Arikaree formations in North Dakota: the buttes of southwestern North Dakota. Report of Investigation. North Dakota Geological Survey 96:1-144.

Osborn, H. F. 1907. Tertiary mammal horizons of North America. Bulletin of the American Museum of Natural History 23:237-254.

Osborn, H. F., and W. D. Matthew. 1909. Cenozoic mammal horizons of western North America, with faunal lists of the Tertiary Mammalia of the West. Bulletin, United States Geological Survey 361:1-138.

Osborn, H. F., and J. L. Wortman. 1894. Fossil mammals of the lower Miocene White River beds, collection 1892. Bulletin of the American Museum of Natural History 6:199-228.

Osborn, H. F., and J. L. Wortman. 1895. Perissodactyls of the lower Miocene White River Beds. Bulletin of the American Museum of Natural History 7:343-375.

Pardee, J. T. 1925. Geology and ground-water resources of Townsend Valley, Montana. United States Geological Survey, Water Supply Paper 539:1-61.

Peck, D. L. 1964. Geologic reconnaissance of the Antelope-Ashwood area, north-central Oregon, with emphasis on the John Day Formation of late Oligocene and early Miocene age. Bulletin, United States Geological Survey 1161-D:1-26.

Peterson, O. A. 1907. The Miocene beds of western Nebraska and eastern Wyoming and their vertebrate faunae. Annals, Carnegie Museum 4:21-72.

Peterson, O. A. 1919. Report upon the material discovered in the upper Eocene of the Uinta Basin by Earl Douglass in the years 1908-1909, and by O. A. Peterson in 1912. Annals, Carnegie Museum 12:40-168.

Phleger, F. B. 1940. Relative growth and vertebrate phylogeny. American Journal of Science 238:643-667.

Phleger, F. B., and W. S. Putnam. 1942. Analysis of *Merycoidodon* skulls. American Journal of Science 240:547-566.

Prothero, D. R. 1982a. Medial Oligocene magnetostratigraphy and mammalian biostratigraphy: testing the isochroneity of mammalian biostratigraphic events. Ph. D. dissertation, Columbia University, New York, 284 pp.

Prothero, D. R. 1982b. How isochronous are mammalian biostratigraphic events? Third North American Paleontological Convention, Proceedings 2:405-409.

Prothero, D. R. 1984. Magnetostratigraphy of the early Oligocene Pipestone Springs locality, Jefferson County, Montana. Contributions to Geology, University of Wyoming 23(1):33-36.

Prothero, D. R. 1985. Chadronian (early Oligocene) magnetostratigraphy of eastern Wyoming: implications for the age of the Eocene-Oligocene boundary. Journal of Geology 93:555-565.

Prothero, D. R., and C. C. Swisher. 1992. Magnetostratigraphy and geochronology of the terrestrial Eocene-Oligocene transition in North America; pp. 46-73 in D. R. Prothero and W. A. Berggren (eds.), Eocene-Oligocene Climatic and Biotic Evolution. Princeton University Press, Princeton, N. J.

Rich, T. H. V., and D. L. Rasmussen. 1973. New World American erinaceine hedgehogs (Mammalia: Insectivora). Occasional Papers, Museum of Natural History, University of Kansas 21:1-54.

Riel, S. J. 1964. A new oreodont from the Cabbage Patch local fauna, western Montana. Postilla, Peabody Museum of Natural History 85:1-10.

Schlaikjer, E. M. 1934. Three new oreodonts. Proceedings, Boston Society of Natural History 40:219-231.

Schlaikjer, E. M. 1935. Contributions to the stratigraphy and paleontology of the Goshen Hole area, Wyoming, Part 4, New vertebrates and the stratigraphy of the Oligocene and early Miocene. Bulletin, Museum of Comparative Zoology, Harvard University 76:97-189.

Schoch, R. M. 1989. A review of the tapiroids; pp. 298-320 in D. R. Prothero and R. M. Schoch (eds.), The Evolution of Perissodactyls. Oxford Clarendon Press, New York.

Schultz, C. B., and C. H. Falkenbach. 1940. Merycochoerinae, a new subfamily of oreodonts. Bulletin of the American Museum of Natural History 77:213-306.

Schultz, C. B., and C. H. Falkenbach. 1941. Ticholeptinae, a new subfamily of oreodonts. Bulletin of the American Museum of Natural History 79:1-105.

Schultz, C. B., and C. H. Falkenbach. 1947. Merychyinae, a subfamily of oreodonts. Bulletin of the American Museum of Natural History 88:157-286.

Schultz, C. B., and C. H. Falkenbach. 1949. Promerycochoerinae, a new subfamily of oreodonts. Bulletin of the American Museum of Natural History 93:69-198.

Schultz, C. B., and C. H. Falkenbach. 1950. Phenacocoelinae, a new subfamily of oreodonts. Bulletin of the American Museum of Natural History 95:91-149.

Schultz, C. B., and C. H. Falkenbach. 1954. Desmatochoerinae, a new subfamily of oreodonts. Bulletin of the American Museum of Natural History 105:143-256.

Schultz, C. B., and C. H. Falkenbach. 1956. Miniochoerinae and Oreonetinae, two new subfamilies of oreodonts. Bulletin of the American Museum of Natural History 109:373-482.

Schultz, C. B., and C. H. Falkenbach. 1968. The phylogeny of the oreodonts, Part 1, Merycoidodontinae, Eporeodontinae, and Leptaucheniinae, three subfamilies of oreodonts, with an appendix to the revision of the Merycoidodontidae, Part 2, summary and conclusions concerning the Merycoidodontidae. Bulletin of the American Museum of Natural History 139:1-498.

Schultz, C. B., and T. M. Stout. 1955. Classification of Oligocene sediments in Nebraska. Bulletin of the Nebraska State Museum 4:16-52.

Schultz, C. B., and T. M. Stout. 1961. Field conference on the Tertiary and Pleistocene of western Nebraska. Special Publication of the University of Nebraska State Museum 2:1-54.

Scott, W. B. 1890. Beiträge zur Kenntniss der Oreodontidae. Morphologische Jährbuch 6:319-395.

Scott, W. B. 1893. The mammals of the Deep River beds. American Naturalist 27:659-662.

Scott, W. B. 1895. The Mammalia of the Deep River beds.

Transactions, American Philosophical Society 18:55-185.

Scott, W. B. 1940. Artiodactyla; pp. 363-746 *in* W. B. Scott and G. L. Jepsen (eds.), The mammalian fauna of the White River Oligocene. Transactions, American Philosophical Society, new series 28.

Scott, W. B. 1945. The mammalian fauna of the Duchesne River Oligocene. Transactions, American Philosophical Society, new series 34:209-252.

Simpson, G. G. 1941. Large Pleistocene felines of North America. American Museum of Natural History, Novitates 1136:1-27.

Simpson, G. G. 1945. The principles of classification and a classification of mammals. Bulletin of the American Museum of Natural History 85:1-350.

Skinner, M. F. 1951. The Oligocene of western North Dakota; pp. 51-58 *in* J. D. Bump (ed.), Guidebook, Fifth Field Conference, Society of Vertebrate Paleontology, western South Dakota. South Dakota School of Mines and Technology, Museum of Geology.

Snedecor, G. W., and Cochran, W. G. 1967. Statistical Methods (sixth ed.). Iowa State University Press, Ames, IA.

Sokal, R. R., and F. J. Rohlf. 1981. Biometry, second edition. W. H. Freeman and Company, San Francisco.

Stone, W. J. 1973. Stratigraphy and sedimentary history of middle Cenozoic (Oligocene and Miocene) deposits in North Dakota. Ph.D. dissertation, University of North Dakota, Grand Forks, 217 pp.

Swinehart, J. B., V. L. Souders, H. M. DeGraw, and R. F. Diffendal, Jr. 1985. Cenozoic paleogeography of western Nebraska. Society of Economic Paleontologists and Mineralogists, Rocky Mountain Section, Third Rocky Mountain Paleography Symposium, Denver Colorado:460.

Swisher, C. C., III, and D. R. Prothero. 1990. Single-crystal ^{40}Ar/^{39}Ar dating of Eocene-Oligocene transition in North America. Science 249:760-762.

Tedford, R. H., J. B. Swinehart, R. M. Hunt, Jr., and M. R. Voorhies. 1985. Uppermost White River and lowermost Arikaree rocks and faunas, White River Valley, northwestern Nebraska, and their correlation with South Dakota; *in* J. E. Martin (ed.), Fossiliferous Cenozoic deposits of western South Dakota and northwestern Nebraska. Dakoterra 2(2):335-352.

Tedford, R. H., M. F. Skinner, R. W. Fields, J. M. Rensberger, D. P. Whistler, T. Galusha, B. E. Taylor, J. R. Macdonald, and S. D. Webb. 1987. Faunal succession and biochronology of the Arikareean through Hemphillian interval (late Oligocene through earliest Pliocene epochs) in North America; pp. 153-210 *in* M. O. Woodburne (ed.), Cenozoic Mammals of North America, Geochronology and Biostratigraphy. University of California Press, Berkeley.

Thorpe, M. R. 1921a. A newly mounted *Eporeodon*. American Journal of Science, series 5, 2:309-313.

Thorpe, M. R. 1921b. A new *Merycoidodon*. American Journal of Science, series 5, 2:334-342.

Thorpe, M. R. 1931. The osteology of *Eporeodon socialis* Marsh. Bulletin, Peabody Museum of Natural History 2:1-43.

Thorpe, M. R. 1937a. The Merycoidodontidae, an extinct group of ruminant mammals. Memoir, Peabody Museum of Natural History 3:1-428.

Thorpe, M. R. 1937b. Phylogeny of the merycoidodonts. American Journal of Science 33:252-259.

Whitmore, F. C. 1953. Cranial morphology of some Oligocene Artiodactyla. United States Geological Survey, Professional Paper 243-H:117-159.

Wilson, J. A. 1971. Early Tertiary vertebrate faunas, Vieja Group, Trans-Pecos Texas, Agriochoeridae and Merycoidodontidae. Bulletin, Texas Memorial Museum, The University of Texas at Austin 18:1-83.

Wilson, J. A. 1974. Early Tertiary vertebrate faunas, Vieja Group and Buck Hill Group, Trans-Pecos Texas, Protoceratidae, Camelidae, Hypertragulidae. Bulletin, Texas Memorial Museum, The University of Texas at Austin 23:1-34.

Wilson, J. A. 1978. Stratigraphic occurrence and correlation of early Tertiary vertebrate faunas, Trans-Pecos Texas. Bulletin, Texas Memorial Museum, The University of Texas at Austin 25:1-42.

Wilson, J. A. 1986. Stratigraphic occurrence and correlation of early Tertiary vertebrate faunas, Trans-Pecos Texas, Agua Fria-Green Valley areas. Journal of Vertebrate Paleontology 6:350-373.

Wilson, J. A., P. C. Twiss, R. K. DeFord, and S. E. Clabaugh. 1968. Stratigraphic succession, potassium-argon dates and vertebrate faunas, Vieja Group, Rim Rock Country, Trans-Pecos Texas. American Journal of Science 266:590-604.

Wolfe, J. A. 1974. A paleobotanical interpretation of Tertiary climates in the Northern Hemisphere. American Scientist 66:694-703.

Wood, H. E., II 1961. *Toxotherium hunteri*, a peculiar new Oligocene mammal from Saskatchewan. Natural History Papers, National Museum of Canada 13:1-3.

Woodburne, M. O. 1968. The cranial myology and osteology of *Dicotyles tajacu*, the collared peccary, and its bearing on classification. Memoirs of the Southern California Academy of Science 7:1-48.

Wortman, J. L. 1893. On the divisions of the White River or lower Miocene of Dakota. Bulletin of the American Museum of Natural History 5:95-105.

APPENDICES

Descriptive statistics for two species of modern suiform artiodactyls,
and for the Merycoidodontinae and Miniochoerinae.

Appendix A

Tayassu pecari, males (13 specimens)

	P^2-M^3	P^2-P^4	M^1-M^3	AP M^3	T M^3	Ht. M^3	Nasal L.	Sub-nasal L.	Malar D.
N	13	13	13	13	13	N/A	13	13	13
Min	73.9	29.2	45.0	16.5	13.7		140.3	43.7	22.7
Max	83.1	34.2	50.6	19.0	16.5		159.9	51.2	30.1
Mean	78.61	31.75	47.47	17.47	14.98		149.00	46.18	26.53
SEM ¥ t(0.05)	1.67	0.91	1.07	0.53	0.48		3.64	1.16	1.47
s	2.76	1.50	1.77	0.88	0.80		6.03	1.91	2.43
CV	3.52	4.73	3.74	5.01	5.34		4.04	4.14	9.16

Tayassu pecari, females (12 specimens)

	P^2-M^3	P^2-P^4	M^1-M^3	AP M^3	T M^3	Ht. M^3	Nasal L.	Sub-nasal L.	Malar D.
N	12	12	12	10	10	N/A	11	12	12
Min	73.00	29.00	44.50	15.40	13.50		138.20	42.10	20.40
Max	86.30	35.50	51.10	19.10	16.50		164.40	52.50	29.70
Mean	78.28	31.93	47.18	17.28	15.11		149.22	46.73	23.63
SEM ¥ t(0.05)	2.44	1.27	1.27	0.91	0.81		5.79	1.95	1.61
s	3.84	2.00	1.99	1.28	1.13		8.62	3.07	2.54
CV	4.91	6.26	4.22	7.39	7.48		5.78	6.57	10.75

Tayassu pecari, combined male and female samples (25 specimens)

	P^2-M^3	P^2-P^4	M^1-M^3	AP M^3	T M^3	Ht. M^3	Nasal L.	Sub-nasal L.	Malar D.
N	25	25	25	23	23	N/A	24	25	25
Min	73.0	29.0	44.5	15.4	13.5		138.2	42.1	20.4
Max	86.3	35.5	51.1	19.1	16.5		164.4	52.5	30.1
Mean	78.45	31.84	47.33	17.39	15.03		149.10	46.44	25.14
SEM ¥ t(0.05)	1.35	0.71	0.76	0.45	0.40		3.02	1.03	1.17
s	3.26	1.72	1.85	1.05	0.94		7.16	2.49	2.84
CV	4.15	5.41	3.90	6.02	6.22		4.80	5.37	11.31

Dicotyles tajacu, males and females (40 specimens)

	P^2-M^3	P^2-P^4	M^1-M^3	AP M^3	T M^3	*Ht. M^3	Nasal L.	Sub-nasal L.	Malar D.
N	40	38	38	40	40	7	23	40	39
Min	57.10	24.00	27.90	11.70	10.00	7.10	100.50	27.70	15.00
Max	69.90	30.80	40.80	15.70	12.90	10.20	133.70	41.80	25.40
Mean	62.89	26.75	36.87	13.30	11.50	8.39	118.77	36.05	20.37
SEM ¥ t(0.05)	0.94	0.53	0.75	0.28	0.24	0.97	3.93	0.91	0.86
s	2.95	1.62	2.27	0.88	0.76	1.05	9.09	2.83	2.65
CV	4.69	6.05	6.17	6.65	6.59	12.53	7.66	7.85	13.01

* Measurement made on the paracone   § Measurement made on the metaconid

P. O. C.	Sk. L.	Sk. W.	P_2-M_3	P_2-P_4	M_1-M_3	AP M_3	T M_3	Ht. M_3
13	13	13	12	12	13	13	13	N/A
60.9	231.7	103.7	79.9	30.0	50.5	20.8	13.8	
70.0	249.8	125.5	90.0	34.0	55.8	23.9	16.9	
66.01	239.85	119.19	84.30	31.93	52.65	22.00	14.65	
1.65	3.09	3.45	2.06	0.90	1.05	0.62	0.54	
2.73	5.12	5.71	3.24	1.42	1.73	1.02	0.90	
4.13	2.13	4.79	3.85	4.46	3.29	4.65	6.13	

P. O. C.	Sk. L.	Sk. W.	P_2-M_3	P_2-P_4	M_1-M_3	AP M_3	T M_3	Ht. M_3
12	11	12	10	12	10	10	10	N/A
62.30	230.20	102.10	77.40	29.30	48.40	19.70	13.40	
70.80	251.80	129.90	93.20	35.60	57.70	24.80	16.30	
66.43	240.42	116.96	84.28	32.48	52.32	21.72	14.63	
1.94	5.98	4.12	3.31	1.13	1.98	1.14	0.68	
3.06	8.90	6.48	4.62	1.77	2.77	1.59	0.96	
4.60	3.70	5.54	5.48	5.46	5.30	7.31	6.54	

P. O. C.	Sk. L.	Sk. W.	P_2-M_3	P_2-P_4	M_1-M_3	AP M_3	T M_3	Ht. M_3
25	24	25	22	24	23	23	23	N/A
60.9	230.2	102.1	77.4	29.3	48.4	19.7	13.4	
70.8	251.8	129.9	93.2	35.6	57.7	24.8	16.9	
66.21	240.11	118.12	84.29	32.20	52.50	21.88	14.64	
1.17	2.93	2.51	1.70	0.67	0.95	0.55	0.39	
2.84	6.94	6.07	3.83	1.60	2.19	1.27	0.90	
4.29	2.89	5.14	4.54	4.96	4.18	5.82	6.16	

P. O. C.	Sk. L.	Sk. W.	P_2-M_3	P_2-P_4	M_1-M_3	AP M_3	T M_3	§ Ht. M_3
38	38	38	32	35	35	37	40	8
48.10	169.00	80.00	63.20	23.80	38.50	15.00	9.70	8.40
61.60	223.00	120.00	74.50	29.80	55.40	19.60	12.60	11.00
54.27	195.74	102.14	68.68	27.05	42.25	17.60	11.06	9.38
0.86	3.48	2.73	0.95	0.46	0.97	0.36	0.20	0.77
2.60	10.58	8.30	2.64	1.35	2.82	1.07	0.62	0.92
4.80	5.41	8.13	3.85	4.99	6.67	6.07	5.65	9.84

Merycoidodon presidioensis (3 specimens)

Type Code	Catalog No.	P²-M³	P²-P⁴	M¹-M³	AP M³	T M³	Ht. M³	Nasal L.	Sub-nasal L.	Malar D.
	TMM 40505-2	95.6	49.0	50.0	20.5	21.2	-	87.2	46.2	15.7
	N	2	2	2	2	2	1	1	1	2
	Min	92.9	45.5	50.0	19.8	19.8	11.0	87.2	46.2	14.5
	Max	95.6	49.0	50.6	20.5	21.2	11.0	87.2	46.2	15.7
	Mean	94.25	47.25	50.30	20.15	20.50	-	-	-	15.1
	SEM ¥ $t_{(0.05)}$	17.15	22.24	3.81	4.45	8.89	-	-	-	7.62
	s	1.91	2.47	0.42	0.49	0.99	-	-	-	0.85
	CV	2.03	5.24	0.84	2.46	4.83	-	-	-	5.62

Merycoidodon culbertsoni (253 specimens)

Type Code	Catalog No.	P²-M³	P²-P⁴	M¹-M³	AP M³	T M³	Ht. M³	Nasal L.	Sub-nasal L.	Malar D.
mac	CMNH 767	90.5	45.0	47.5	19.0	20.0	-	96.0	-	21.0
geo	F:AM 45143	94.6	47.0	51.4	21.1	21.3	-	-	-	-
osb	F:AM 49668	88.9	43.9	47.9	18.6	18.4	10.3	-	39.9	14.5
cul	ANSP 10728	-	-	-	19.5	17.0	-	-	-	-
mek	F:AM 45023	93.3	45.5	51.0	19.7	18.8	11.2	77.1	37.0	16.8
per	AMNH 6397	75.9	38.0	39.8	15.0	15.3	10.0	-	34.6	12.8
wrd	F:AM 49662	87.1	41.9	46.5	18.1	18.9	-	67.3	30.1	16.3
lew	CMNH 9105	96.3	46.4	50.5	20.2	20.2	-	-	-	-
bwn	F:AM 72286	90.4	43.2	47.9	20.6	19.9	10.2	74.9	40.9	17.1
van	F:AM 49766	84.4	41.1	45.7	17.5	18.9	10.1	-	35.6	12.2
	N	182	181	205	208	175	69	53	76	170
	Min	72.2	36.2	38.1	14.2	15.3	9.5	63.2	25.0	11.3
	Max	97.4	50.3	52.7	22.2	21.8	12.6	97.2	48.2	21.0
	Mean	85.11	42.22	45.38	18.19	18.43	10.69	76.12	36.50	15.53
	SEM ¥ $t_{(0.05)}$	0.80	0.41	0.45	0.19	0.21	0.18	2.00	1.10	0.28
	s	5.48	2.77	3.27	1.40	1.39	0.77	7.25	4.83	1.85
	CV	6.44	6.56	7.20	7.70	7.52	7.18	9.52	13.24	11.91

Type specimen codes (No. of specimens examined):

bwn, *Merycoidodon culbertsoni "browni"* (6)
cul, *Merycoidodon culbertsoni* (59)
geo, *Paramerycoidodon georgei* (8)
lew, *Merycoidodon "lewisi"* (1)
mac, *Merycoidodon "macrorhinus"* (1)

Merycoidodon bullatus (91 specimens)

Type Code	Catalog No.	P²-M³	P²-P⁴	M¹-M³	AP M³	T M³	Ht. M³	Nasal L.	Sub-nasal L.	Malar D.
dak	FAM 45351	81.8	41.4	42.8	17.4	16.9	10.2	-	37.9	14.5
bul	ANSP 10681	84.5	42.5	45	-	19	-	-	-	15
nor	FAM 49733	78.3	40.1	41.1	16.1	17.1	-	-	-	13.6
hel	CMNH 765	-	-	47.4	17.5	18.4	10.9	-	-	17.8
ced	AMNH 8949	-	46.2	-	16.9	-	-	-	-	14.4
soc	YPM 13119	91.9	45.3	-	21.4	20	-	-	32.3	-
soc	YPM 13118	90.5	44	49.5	20.2	20.6	12.5	-	-	18.3
dan	FAM 72132	93.1	46.4	49.2	20.3	19.7	-	81.5	38.5	16.6
bac	UNSM 28191	-	-	-	-	-	-	-	-	-
gal	FAM 45279	90.1	45.7	47.7	20.1	20.2	-	-	-	14.9
dic	CMNH 1584	80.1	39.4	41.1	17	16.5	10.4	-	-	14.4
gey	FAM 49734	81.5	39.4	44.6	17.5	18.8	-	76.5	35.4	-
	N	56	56	68	72	63	27	9	12	47
	Min	73.3	36.0	40.4	15.2	15.6	9.5	72.8	32.3	12.5
	Max	96.3	48.0	52.5	21.4	21.5	12.5	89.5	40.9	20.2
	Mean	84.86	42.09	45.46	18.43	18.48	10.84	78.32	36.24	15.48
	SEM ¥ $t_{(0.05)}$	1.36	0.75	0.72	0.33	0.34	0.32	3.84	1.78	0.57
	s	5.07	2.79	2.95	1.41	1.36	0.81	4.99	2.81	1.95
	CV	5.97	6.63	6.50	7.66	7.36	7.51	6.37	7.75	12.57

Type specimen codes (No. of specimens examined):

bac, *"Paramerycoidodon (Barbourochoerus) bacai"* (3)
bul, *"Otionohyus (Otarohyus)" bullatus* (37)
ced, *"Otionohyus (Otarohyus) cedrensis"* (4)
dak, *"Subdesmatochoerus socialis dakotensis"* (1)
dan, *Merycoidodon (Anomerycoidodon) "dani"* (13)
dic, *"Genetochoerus (Osbornohyus) dickinsonensis"* (1)

P. O. C.	Sk. L.	Sk. W.	P₂-M₃	P₂-P₄	M₁-M₃	AP M₃	T M₃	Ht. M₃
-	-	-	103.0	50.0	51.5	25.4	14.1	-
1	0	0	2	2	2	2	2	0
31.0	-	-	103.0	47.5	51.5	25.4	14.1	-
31.0	-	-	107.5	50.0	58.4	26.0	14.7	-
-	-	-	105.25	48.75	54.95	25.70	14.40	-
-	-	-	28.59	15.88	43.84	3.81	3.81	-
-	-	-	3.18	1.77	4.88	0.42	0.42	-
-	-	-	3.02	3.63	8.88	1.65	2.95	-

P. O. C.	Sk. L.	Sk. W.	P₂-M₃	P₂-P₄	M₁-M₃	AP M₃	T M₃	Ht. M₃
-	200.0	-	-	-	-	-	-	-
34.2	-	-	-	-	-	-	-	-
34.2	183.7	114.4	97.9	46.2	51.7	22.5	12.6	9.5
-	-	-	-	-	-	-	-	-
29.5	182.1	-	-	-	52.5	23.8	12.0	10.1
30.2	160.5	-	83.6	38.4	45.4	19.8	10.9	9.2
30.0	163.8	105.0	-	40.1	-	-	-	-
-	206.5	-	100.0	46.8	52.3	23.6	14.4	-
33.2	193.3	102.4	99.4	47.6	52.6	25.1	13.1	9.4
29.9	-	-	-	-	-	-	-	-
138	95	75	84	85	106	96	79	27
22.2	152.0	82.8	79.9	36.3	41.1	18.7	10.6	8.2
41.3	206.5	129.0	106.4	51.5	56.6	25.9	14.5	10.8
31.14	181.19	108.25	93.36	43.81	49.77	22.53	12.53	9.41
0.60	2.57	2.06	1.25	0.67	0.63	0.34	0.21	0.27
3.55	12.64	8.94	5.75	3.12	3.27	1.70	0.92	0.67
11.40	6.97	8.26	6.16	7.13	6.57	7.53	7.32	7.13

mek, *Prodesmatochoerus meekae* (10)
osb, *Merycoidodon culbertsoni "osborni"* (73)
per, *"Genetochoerus periculorum"* (31)
van, *"Otionohyus vanderpooli"* (3)
wrd, *"Otionohyus wardi"* (47)

P. O. C.	Sk. L.	Sk. W.	P₂-M₃	P₂-P₄	M₁-M₃	AP M₃	T M₃	Ht. M₃
29.2	-	-	89.5	43.2	47.4	21.3	11.2	-
-	-	-	-	-	-	-	-	-
32	-	-	83.3	37.8	46.2	19.4	10.5	-
35.5	-	-	-	-	51.5	22	12.6	9.8
-	-	104	-	-	-	-	-	-
38.5	190.8	-	100.3	46.2	55.9	24.8	13.2	-
32.6	185.5	-	98.6	44.3	52.6	24.2	13.5	-
33.6	195.5	-	102.2	49.8	54.4	25.6	14.3	9.4
37	194	116.5	-	-	-	-	-	-
25.3	-	-	91.1	41.6	52.1	23.5	13.6	-
33.3	170.4	-	86.9	41.6	45.4	21	11.4	9.4
-	171.7	-	88.6	40.9	49.4	22	13.5	-
40	20	16	31	32	41	42	38	9
25.3	162.0	102.5	82.3	37.8	43.3	19.4	10.5	8.6
41.5	198.7	122.8	106.0	49.8	55.9	25.6	14.9	10.7
33.29	181.10	109.52	92.54	43.05	49.61	22.53	12.55	9.46
1.06	5.41	3.18	1.93	0.96	0.92	0.46	0.30	0.47
3.30	11.56	5.97	5.26	2.65	2.91	1.48	0.91	0.61
9.92	6.38	5.45	5.68	6.16	5.87	6.55	7.26	6.48

gal, *Merycoidodon "(Blickohyus) galushai"* (1)
gey, *"Genetochoerus (Osbornohyus) geygani"* (12)
hel, *"Otionohyus (Otarohyus) hybridus helenae"* (1)
nor, *"Genetochoerus (Osbornohyus) norbeckensis"* (7)
soc, *"Subdesmatochoerus socialis"* (7)

Merycoidodon major (130 specimens)

Type Code	Catalog No.	P²-M³	P²-P⁴	M¹-M³	AP M³	T M³	Ht. M³	Nasal L.	Sub-nasal L.	Malar D.
lyn	FAM 45297	98.9	49.4	53.0	20.7	21.1	13.5	-	-	18.2
lmb	FAM 72139	93.8	46.9	50.1	20.6	20.7	12.2	-	-	17.8
wan	FAM 72014	90.9	42.9	50.3	21.4	21.7	-	72.6	42.5	21.0
alx	FAM 72060	91.1	44.3	49.3	19.9	19.7	11.9	71.8	-	14.5
chm	UNSM 28340	82.8	39.4	45.2	18.9	-	-	-	-	15.8
scn	FAM 45329	93.1	47.5	49.5	20.9	20.1	-	-	-	17.2
maj	VSUM19099	-	-	58.0	24.0	22.0	-	-		
shn	AMNH 1310	96.2	44.5	54.3	22.2	19.1	-	-	-	17.3
	N	74	72	91	98	73	34	22	22	74
	Min	79.8	38.5	42.7	17.3	17.3	10.4	71.8	37.3	12.5
	Max	105.0	49.9	60.7	26.0	24.0	14.4	105.8	48.7	23.2
	Mean	94.73	45.73	51.52	21.29	20.70	12.63	84.61	43.41	18.11
	SEM ¥ t(0.05)	1.21	0.59	0.73	0.35	0.31	0.29	3.90	1.48	0.50
	s	5.23	2.51	3.49	1.73	1.33	0.83	8.80	3.34	2.16
	CV	5.52	5.49	6.77	8.11	6.43	6.61	10.40	7.69	11.95

Type specimen codes (No. of specimens examined):

alx, "*Otionohyus (Otarohyus) alexi*" (17)
chm, "*Genetochoerus (Osbornohyus) chamberlaini*" (9)
lmb, *Merycoidodon (Anomerycoidodon) "lambi"* (16)
lyn, *Merycoidodon "(Blickohyus) lynchi"* (15)

Mesoreodon chelonyx (12 specimens)

Type Code	Catalog No.	P²-M³	P²-P⁴	M¹-M³	AP M³	T M³	Ht. M³	Nasal L.	Sub-nasal L.	Malar D.
chx	PU 10425	101.3	50.1	55.5	24.0	22.2	-	94.0	-	17.5
dan	YPM 13949	99.6	49.0	55.0	-	-	-	-	-	22.0
mgr	YPM 13948	99.6	48.0	55.1	23.1	20.8	-	98.1	45.8	18.0
mnt	CMNH 907	99.8	49.0	56.5	25.9	22.1	-	88.9	43.7	21.5
san	FAM 45443	-	49.0	-	-	-	-	-	49.1	19.4
	N	9	11	9	8	7	1	5	6	11
	Min	95.4	45.3	52.0	22.2	20.8	14.5	88.9	42.0	15.2
	Max	108.2	52.9	58.5	26.3	22.7	14.5	100.5	49.3	22.0
	Mean	100.57	49.06	55.23	24.18	21.51	14.5	95.24	46.15	18.50
	SEM ¥ t(0.05)	2.55	1.22	1.46	1.24	0.73	-	5.48	3.07	1.36
	s	3.32	1.82	1.91	1.48	0.79	-	4.41	2.92	2.03
	CV	3.30	3.71	3.45	6.14	3.69	-	4.63	6.33	10.95

Type specimen codes (No. of specimens examined):

chx, *Mesoreodon chelonyx* (5)
dan, "*Mesoreodon danai*" (1)
mgr, "*Paramerycoidodon (Gregoryochoerus) meagherensis* (1)

?*Mesoreodon minor* (26 specimens)

Type Code	Catalog No.	P²-M³	P²-P⁴	M¹-M³	AP M³	T M³	Ht. M³	Nasal L.	Sub-nasal L.	Malar D.
cki	MCZ 17765	110.0	50.9	61.0	28.0	23.5	-	94.2	45.3	26.5
sti	MCZ 17480	110.0	51.0	61.5	-	-	-	-	-	-
mro	FAM 37551	110.6	50.8	63.4	30.1	25.4	-	100.8	51.8	26.4
wyo	FAM 33312	113.4	55.3	60.9	25.7	22.8	-	106.5	52.5	18.5
min	CMNH 769	104.0	50.0	58.3	25.6	21.5	-	-	-	23.0
gng	FAM 33303	105.5	52.4	56.7	25.1	22.0	-	102.5	52.8	24.1
was	AMNH 7827	108.3	53.0	56.7	23.8	20.8	-	-	47.0	22.8
	N	18	19	21	20	18	3	9	10	16
	Min	104.0	49.0	55.3	23.8	20.8	16.2	92.7	44.5	18.5
	Max	119.5	59.1	68.8	30.7	25.5	17.4	114.1	56.1	28.7
	Mean	110.94	52.79	61.60	27.08	23.04	16.70	102.02	50.35	23.30
	SEM ¥ t(0.05)	2.42	1.26	1.81	1.02	0.71	1.55	4.90	2.70	1.58
	s	4.87	2.61	3.98	2.18	1.44	0.62	6.38	3.77	2.96
	CV	4.39	4.94	6.47	8.05	6.24	3.74	6.25	7.49	12.72

Type specimen codes (No. of specimens examined):

cki, *Mesoreodon "cheeki"* (5)
gng, "*Desmatochoerus (Paradesmatochoerus) grangeri*" (1)
min, "*Promerycochoerus (Pseudopromerycochoerus)" minor* (1)
mro, "*Desmatochoerus (Paradesmatochoerus) monroecreekensis*" (6)

P. O. C.	Sk. L.	Sk. W.	P_2-M_3	P_2-P_4	M_1-M_3	AP M_1	T M_1	Ht. M_1
41.5	-	115.2	-	48.0	-	-	-	-
36.7	202.7	121.9	104.1	50.8	54.7	26.0	13.3	11.0
39.4	191.1	131.4	101.5	47.5	55.8	26.4	14.6	-
37.9	-	122.7	-	-	-	-	-	-
-	-	-	88.8	37.2	51.2	22.8	12.5	-
37.5	186.9	132.9	102.5	47.4	54.6	26.1	14.2	-
-	195.8	-	-	-	-	-	-	-
60	35	34	32	34	47	39	26	8
29.0	180.8	113.9	88.8	37.2	46.7	20.9	11.2	9.0
47.4	235.7	162.3	115.2	55.6	62.1	28.9	16.5	12.2
39.78	206.19	131.35	105.23	48.87	56.08	25.76	13.91	11.04
1.02	4.40	4.09	2.27	1.24	1.04	0.61	0.49	0.82
3.93	12.80	11.73	6.30	3.56	3.53	1.88	1.20	0.98
9.89	6.21	8.93	5.98	7.28	6.29	7.29	8.65	8.85

maj, *"Paramerycoidodon (Barbourochoerus)" major* (35)
scn, *"Promesoreodon scanloni"* (4)
shn, *"Subdesmatochoerus shannonensis"* (7)
wan, *"Paramerycoidodon (Gregoryochoerus) wanlessi"* (28)

P. O. C.	Sk. L.	Sk. W.	P_2-M_3	P_2-P_4	M_1-M_3	AP M_1	T M_1	Ht. M_1
39.6	216.7	140.0	106.9	48.0	59.1	28.3	14.5	-
-	230.0	-	-	-	60.0	-	-	-
39.0	208.5	127.2	106.2	50.2	56.9	26.1	13.9	-
38.0	210.5	124.8	106.6	49.8	57.7	27.7	14.6	-
-	-	-	-	-	-	-	-	-
5	7	4	6	6	8	6	7	1
38.0	202.3	124.8	106.2	48.0	55.4	25.8	13.7	12.7
44.6	230.0	141.8	110.3	51.1	61.0	30.1	15.7	12.7
41.04	216.51	133.45	108.33	50.03	58.46	27.88	14.67	12.70
3.77	9.18	13.83	2.05	1.18	1.54	1.80	0.66	-
3.04	9.93	8.69	1.95	1.12	1.85	1.71	0.72	-
7.40	4.59	6.51	1.80	2.24	3.16	6.14	4.89	-

mnt, *"Subdesmatochoerus montanus"* (1)
san, *"Desmatochoerus (Paradesmatochoerus) sandfordi"* (3)

P. O. C.	Sk. L.	Sk. W.	P_2-M_3	P_2-P_4	M_1-M_3	AP M_1	T M_1	Ht. M_1
52.8	218.8	148.0	121.9	58.2	65.7	32.0	16.6	-
-	219.0	-	-	54.0	65.0	-	-	-
44.3	231.6	161.2	122.5	55.1	68.1	35.1	17.0	-
-	234.5	-	124.0	57.9	65.0	30.4	17.0	-
-	-	-	117.5	54.0	65.0	-	-	-
43.7	242.8	162.9	117.8	56.5	62.6	30.2	14.7	-
50.0	-	166.6	-	-	-	-	-	-
7	8	5	15	17	17	14	14	2
42.1	194.5	148.0	116.9	53.6	62.6	28.9	14.0	14.0
52.8	242.8	166.6	129.0	59.0	73.1	35.1	18.9	15.7
47.09	226.39	160.82	121.75	55.84	66.64	31.72	16.66	14.85
4.02	12.94	9.28	1.97	0.89	1.62	1.27	0.94	10.80
4.35	15.47	7.47	3.56	1.73	3.14	2.19	1.64	1.20
9.24	6.83	4.64	2.93	3.10	4.72	6.91	9.82	8.09

sti, *Mesoreodon "cheeki scotti"* (6)
was, *"Pseudodesmatochoerus wascoensis"* (2)
wyo, *"Desmatochoerus (Paradesmatochoerus) wyomingensis"* (4)

Appendix C

?Miniochoerus forsythae (18 specimens)

Type Code	Catalog No.	P^2-M^3	P^3-P^4	M^1-M^3	AP M^3	T M^3	Ht. M^3	Nasal L.	Sub-nasal L.	Malar D.
nat	F:AM 45194	80.0	38.5	42.0	-	-	-	-	-	-
for	F:AM 72303	84.9	41.4	47.2	19.2	20.2	-	62.6	31.4	13.8
deg	F:AM 49760	79.6	37.8	45.1	18.1	17.8	10.6	70.8	-	14.1
	N	13	14	13	12	4	3	3	11	9
	Min	35.5	42.0	16.8	16.1	9.3	62.6	31.0	12.2	23.9
	Max	41.4	51.9	21.0	20.5	10.6	70.8	33.4	15.9	33.7
	Mean	37.88	45.95	18.70	18.84	9.75	67.67	31.93	14.41	30.08
	SEM ¥ $t_{(0.05)}$	1.01	1.48	0.72	0.84	0.92	11.00	3.19	0.82	2.20
	s	1.67	2.57	1.19	1.32	0.58	4.43	1.29	1.22	2.86
	CV	4.40	5.59	6.37	7.01	5.95	6.54	4.03	8.50	9.52

Type specimen codes (No. of specimens examined):

deg,   *"Otionohyus wardi degrooti"* (2)
for,   *"Merycoidodon" forsythae* (1)

Miniochoerus chadronensis (59 specimens)

Type Code	Catalog No.	P^2-M^3	P^3-P^4	M^1-M^3	AP M^3	T M^3	Ht. M^3	Nasal L.	Sub-nasal L.	Malar D.
chd	F:AM 45489	72.2	33.4	42.7	17.2	16.9	-	56.4	29.9	15.8
dug	F:AM 45492	76.4	33.2	45.5	18.8	17.5	9.8	51.5	-	14.1
con	F:AM 45011	70.5	31.3	43.0	16.9	15.8	8.9	51.2	32.3	11.3
reid	F:AM 49620	78.5	35.1	46.1	18.3	19.0	-	62.3	33.2	14.7
	N	44	43	45	47	25	16	21	15	50
	Min	66.6	29.7	37.6	15.2	15.3	7.9	47.8	27.0	9.0
	Max	78.5	36.2	46.1	18.8	19.0	10.1	62.3	34.6	17.2
	Mean	72.44	32.93	41.87	17.00	16.73	9.23	53.57	29.85	13.48
	SEM ¥ $t_{(0.05)}$	0.81	0.49	0.55	0.22	0.40	0.32	2.07	1.20	0.48
	s	2.68	1.58	1.82	0.77	0.97	0.59	4.56	2.17	1.69
	CV	3.70	4.79	4.34	4.50	5.79	6.42	8.50	7.27	12.54

Type specimen codes (No. of specimens examined):

chd,   *"Stenopsochoerus (Pseudostenopsochoerus)" chadronensis* (1)
con,   *"Parastenopsochoerus conversensis"* (2)

Miniochoerus affinis (69 specimens)

Type Code	Catalog No.	P^2-M^3	P^3-P^4	M^1-M^3	AP M^3	T M^3	Ht. M^3	Nasal L.	Sub-nasal L.	Malar D.
bat	F:AM 45001	69.3	32.9	38.6	15.1	15.3	-	-	-	13
pty	YPM 12752	66	30.4	35.5	14.5	16.5	-	52	-	16.5
stn	F:AM 44980	69	31.8	39.5	16	16.7	-	48.6	29.2	12.2
aff	ANSP 10679	62.5	27	36.5	-	13.5	-	-	-	13
	N	64	64	64	58	45	16	13	8	60
	Min	60.6	26.2	34.5	13.5	13.5	7.2	43.7	23.0	10.6
	Max	73.4	34.8	41.8	17.6	17.5	9.1	55.5	33.0	16.5
	Mean	66.16	30.40	38.13	15.48	15.44	8.13	49.68	29.81	13.03
	SEM ¥ $t_{(0.05)}$	0.73	0.44	0.42	0.24	0.28	0.27	2.13	2.56	0.32
	s	2.92	1.75	1.69	0.90	0.95	0.51	3.53	3.06	1.25
	CV	4.41	5.77	4.43	5.84	6.13	6.23	7.10	10.25	9.59

Type specimen codes (No. of specimens examined):

aff,   *Miniochoerus "(Paraminiochoerus)" affinis* (29)
bat,   *Miniochoerus "battlecreekensis"* (9)

Miniochoerus gracilis (66 specimens*)

Type Code	Catalog No.	P^2-M^3	P^3-P^4	M^1-M^3	AP M^3	T M^3	Ht. M^3	Nasal L.	Sub-nasal L.	Malar D.
grc	ANSP 10692	-	-	30.4	11.9	12.4	-	-	-	-
grc	ANSP 10694*	-	-	30.6	11.7	12.4	-	-	-	-
	N	43	45	46	43	33	8	10	9	41
	Min	51.2	23.0	28.9	11.3	11.1	5.8	34.9	21.9	8.3
	Max	61.5	29.6	35.1	15.0	15.1	7.2	46.6	30.4	13.7
	Mean	55.96	26.20	31.73	12.75	12.55	6.78	41.29	25.07	10.92
	SEM ¥ $t_{(0.05)}$	0.88	0.43	0.51	0.27	0.30	0.38	2.77	2.00	0.34
	s	2.86	1.44	1.71	0.87	0.85	0.46	3.87	2.60	1.08
	CV	5.11	5.50	5.40	6.78	6.76	6.78	9.37	10.39	9.88

Type specimen codes (No. of specimens examined):

grc,   *Miniochoerus ("Paraminiochoerus") gracilis* (55*)

P. O. C.	Sk. L.	Sk. W.	P_2-M_3	P_2-P_4	M_1-M_3	AP M_3	T M_3	Ht. M_3
-	-	-	-	-	-	-	-	-
23.9	167.3	99.6	-	-	-	-	-	-
-	166.2	93.7	86.6	38.0	50.5	23.5	11.4	9.4
4	9	6	6	9	11	9	4	13
155.3	88.4	85.8	37.0	48.4	21.9	11.4	7.6	44.49
167.3	103.4	89.8	40.0	54.3	24.4	14.0	9.5	50.19
162.45	96.10	87.28	38.38	50.37	22.90	12.61	8.93	47.12
8.75	3.73	1.45	1.17	1.32	0.50	0.60	1.42	1.07
5.50	4.86	1.38	1.11	1.72	0.74	0.79	0.89	1.78
3.39	5.05	1.58	2.89	3.42	3.24	6.24	10.00	3.77

nat,   "*Prodesmatochoerus natronensis*"

P. O. C.	Sk. L.	Sk. W.	P_2-M_3	P_2-P_4	M_1-M_3	AP M_3	T M_3	Ht. M_3
27.2	141.5	-	80.4	34.4	46.3	20.5	10.4	-
-	144.6	84.0	83.3	34.5	49.2	22.6	11.4	7.3
31.9	138.2	81.1	-	-	-	-	-	-
31.8	155.3	96.0	85.0	35.1	50.6	22.7	12.6	-
34	27	20	21	20	23	19	10	5
22.6	128.7	73.2	72.1	31.0	41.8	18.5	10.3	7.3
33.0	155.3	98.6	85.0	36.6	51.0	22.7	12.6	9.0
27.68	142.27	86.07	79.15	34.06	45.84	20.70	11.23	8.26
0.80	2.26	3.61	1.56	0.78	1.07	0.56	0.50	0.78
2.29	5.71	7.71	3.44	1.67	2.48	1.15	0.70	0.63
8.29	4.02	8.96	4.34	4.91	5.40	5.58	6.24	7.64

dug,   "*Stenopsochoerus (Pseudostenopsochoerus) douglasensis*" (12)
reid,   "*Stenopsochoerus (Pseudostenopsochoerus) reideri*" (1)

P. O. C.	Sk. L.	Sk. W.	P_2-M_3	P_2-P_4	M_1-M_3	AP M_3	T M_3	Ht. M_3
29.2	139	96.2	76.9	34	43.4	19.6	10.4	-
30.3	-	89.5	-	-	-	-	-	-
22.2	131	80.2	74.2	31.8	43	19.5	10.7	-
-	-	-	-	-	-	-	-	-
46	24	25	26	28	33	29	19	2
22.2	118.5	74.5	64.2	27.3	35.3	16.6	8.6	8.2
33.6	145.0	96.2	76.9	34.0	46.5	21.8	11.9	8.5
26.77	133.37	85.31	71.37	30.25	41.92	18.97	10.16	8.35
0.70	2.20	2.51	1.26	0.61	0.77	0.45	0.40	1.91
2.37	5.22	6.07	3.12	1.56	2.16	1.19	0.82	0.21
8.85	3.91	7.11	4.37	5.17	5.15	6.27	8.09	2.54

pty,   "*Platyochoerus platycephalus*" (19)
stn,   "*Stenopsochoerus sternbergi*" (10)

P. O. C.	Sk. L.	Sk. W.	P_2-M_3	P_2-P_4	M_1-M_3	AP M_3	T M_3	Ht. M_3
-	-	-	-	-	-	-	-	-
-	-	-	-	-	-	-	-	-
33	19	15	17	19	18	15	11	3
20.6	105.3	66.4	55.2	23.0	31.7	14.1	7.2	5.6
26.7	130.4	86.2	66.1	29.4	38.7	17.6	9.3	5.7
23.68	116.28	75.85	61.40	26.45	35.54	15.52	7.95	5.63
0.62	3.29	2.70	1.79	0.84	1.05	0.63	0.38	0.14
1.74	6.82	4.87	3.47	1.74	2.11	1.14	0.57	0.06
7.35	5.86	6.42	5.66	6.58	5.93	7.31	7.16	1.02

*,   ANSP 10694 may be a part of the same individual as the type of
      M. gracilis, ANSP 10692.

Miniochoerus affinis and *Miniochoerus gracilis*, combined (134 specimens)

	P²-M³	P³-P⁴	M¹-M³	AP M³	T M³	Ht. M³	Nasal L.	Sub-nasal L.	Malar D.
N	107	109	110	101	78	24	23	17	101
Min	51.2	23.0	28.9	11.3	11.1	5.8	34.9	21.9	8.3
Max	73.4	34.8	41.8	17.6	17.5	9.1	55.5	33.0	16.5
Mean	62.06	28.67	35.45	14.32	14.22	7.68	46.03	27.30	12.17
SEM ¥ t(0.05)	1.11	0.50	0.68	0.32	0.38	0.34	2.41	1.88	0.31
s	5.79	2.64	3.59	1.62	1.70	0.81	5.57	3.67	1.57
CV	9.33	9.20	10.14	11.31	11.95	10.57	12.09	13.43	12.91

Miniochoerus starkensis (29 specimens)

Type Code	Catalog No.	P²-M³	P³-P⁴	M¹-M³	AP M³	T M³	Ht. M³	Nasal L.	Sub-nasal L.	Malar D.
str	F:AM 45497	72.4	31.8	42.5	18.1	16.9	-	-	-	13.3
hrt	F:AM 45467	70.8	32	41.1	17.1	16.2	-	57.6	-	13.4
hlp	F:AM 49501	69	31.2	39.7	15.7	14.7	-	-	-	15.2
jod	F:AM 45483	68	30.5	43	-	18	-	57	-	20
ott	UNSM 28336	73.5	32.5	44	-	16.5	-	-	-	16
hat	F:AM 45463	-	-	46	18.9	18	-	59.4	-	13.5
nic	F:AM 49585	77.2	33.8	46.1	19	17.6	9.5	53.6	34.8	15.4
chy	AMNH 9797	78	34.4	46.5	19.2	19.2	-	-	-	-
ber	F:AM 49617	76.3	35	44.2	18.4	16.8	-	60.2	35.5	16.4
N		20	20	23	19	19	5	9	4	21
Min		64.7	28.1	37.6	15.3	14.7	7.8	50.3	28.6	12.6
Max		83.0	35.0	49.8	19.2	19.2	10.2	63.0	35.5	20.0
Mean		71.86	31.62	42.61	17.72	16.89	9.06	56.82	33.33	15.06
SEM ¥ t(0.05)		2.17	0.84	1.35	0.61	0.63	1.10	3.04	5.06	0.76
s		4.63	1.80	3.12	1.26	1.31	0.89	3.96	3.18	1.68
CV		6.45	5.70	7.33	7.09	7.78	9.80	6.97	9.55	11.13

Type specimen codes (No. of specimens examined):

ber,   "*Stenopsochoerus berardae*" (6)
chy,   *Miniochoerus "cheyennensis"* (2)
hat,   "*Platyochoerus hatcreekensis*" (7)
hlp,   *Miniochoerus "(Paraminiochoerus) helprini"* (5)
hrt,   "*Platyochoerus heartensis*" (3)

Appendix C

P. O. C.	Sk. L.	Sk. W.	P_2-M_3	P_2-P_4	M_1-M_3	AP M_3	T M_3	Ht. M_3
79	43	40	43	47	51	44	30	5
20.6	105.3	66.4	55.2	23.0	31.7	14.1	7.2	5.6
33.6	145.0	96.2	76.9	34.0	46.5	21.8	11.9	8.5
25.48	125.82	81.77	67.43	28.71	39.67	17.80	9.35	6.72
0.59	3.21	2.32	1.81	0.73	1.05	0.61	0.49	1.85
2.61	10.42	7.26	5.89	2.49	3.74	2.02	1.31	1.49
10.25	8.28	8.88	8.74	8.66	9.42	11.35	13.95	22.21

P. O. C.	Sk. L.	Sk. W.	P_2-M_3	P_2-P_4	M_1-M_3	AP M_3	T M_3	Ht. M_3
29.4	-	-	76.5	29.7	46.9	21.8	10.8	-
26.4	-	100.6	-	-	44.8	20.2	9.9	-
22.8	142	74.5	-	-	-	-	-	-
-	140	-	-	-	-	-	-	-
-	-	86.5	80.5	35	48	-	-	-
31.8	-	-	-	-	-	-	-	-
30.9	153	107	-	-	44.5	20.3	10.7	8
-	-	-	-	-	-	-	-	-
24.4	-	-	85.9	33.9	50.7	23.6	10.9	-
14	7	11	9	10	15	15	13	2
22.8	125.7	74.5	66.1	26.3	40.4	18.0	9.3	7.1
31.8	154.0	107.0	85.9	35.8	51.9	24.5	11.6	8.0
26.49	144.11	90.52	78.79	32.50	46.07	21.48	10.52	7.55
1.69	9.27	6.13	4.81	2.37	1.98	0.99	0.39	5.72
2.92	10.03	9.13	6.26	3.31	3.57	1.79	0.65	0.64
11.04	6.96	10.08	7.95	10.20	7.75	8.32	6.21	8.43

jod, *"Stenopsochoerus joderensis"* (1)
nic, *Miniochoerus "nicholsae"* (2)
ott, *Miniochoerus "(Paraminiochoerus) ottensi"* (1)
str, *Miniochoerus starkensis* (2)

26. Leptaucheniinae

EMILY A. COBABE

ABSTRACT

The Leptaucheniinae previously have been split into 7 genera and 31 species. Using a combination of bivariate and multivariate statistics and shared derived characters, the species of Leptaucheniinae from the northern Great Plains are re-evaluated, and a new classification is suggested. This classification recognizes the genera *Leptauchenia* Leidy and *Sespia* Stock. The two genera can be distinguished on the basis of skull size, dental features, and orbit size. *Leptauchenia* contains two species, *L. decora* and *L. major*. These species are separated by differences in dental formula, auditory bullae volume, orbit area, and orbital bridge width. The only *Sespia* from the Great Plains is *Sespia nitida*. This interpretation of the Leptaucheniinae restricts the use of these taxa in the biostratigraphy of the Northern Great Plains.

INTRODUCTION

The subfamily Leptaucheniinae of the family Merycoidodontidae is a small group of Oligocene oreodonts characterized by the presence of a long nasal-facial vacuity, high crowned teeth and large auditory bullae (Fig. 1). They have been known from the northern Great Plains since the 1850s (Leidy, 1856). Since that time, the group has undergone several systematic revisions (Thorpe, 1921, 1937; Scott, 1940). The most recent treatment, Schultz and Falkenbach (1968), left the group oversplit into 7 genera and 31 species. In this chapter, I suggest that many of the characters used by Schultz and Falkenbach and others to define groups within the Leptaucheniinae are strongly correlated to skull size and are therefore not useful for taxonomic purposes. However, several characters, such as orbit area and auditory bullae volume, do allow some taxa to be distinguished.

MATERIALS AND METHODS

A series of 8 measurements (Figure 1) and dental features were taken from approximately 1000 censused specimens at the American Museum of Natural History of New York, the University of Nebraska State Museum, the South Dakota School of Mines, and the Academy of Natural Sciences in Philadelphia. [Specimens referred to in this chapter are labeled with the following abbreviations: F:AM = Frick Collection of the American Museum of Natural History, New York; AMNH = American Museum of Natural History, New York; ANSP = Academy of Natural Sciences, Philadelphia.] Of these specimens, 127 skulls were complete enough to be used in this study.

DISCUSSION

An initial separation of taxa was made on the basis of dental characters. This created three natural groups, *Sespia* Stock, *Leptauchenia* Leidy and "*Cyclopidius*" (now *Leptauchenia major*). Factor analysis substantiated the validity of *Sespia*, but could not distinguish *Leptauchenia* from *Cyclopidius* (Fig. 2). K-mean cluster analysis failed to distinguish between any groups. Bivariate plots do, however, provide additional evidence to support the validity of *Leptauchenia decora* and *L. major* (Figs. 3 and 4).

SYSTEMATIC PALEONTOLOGY
Order ARTIODACTYLA
Family MERYCOIDODONTIDAE (Hay, 1902)
Subfamily LEPTAUCHENIINAE
(Schultz & Falkenbach, 1940 and 1968)

Included genera—*Sespia* and *Leptauchenia*.
Distribution—Brule, Gering, and Monroe Creek formations and their equivalents in Montana, Nebraska, South Dakota, Wyoming, Colorado, and California.
Diagnosis—Small to moderate-sized oreodonts with an extended pair of nasal-facial vacuities enclosed by the nasal, frontal, maxillary, and premaxillary bones; fan-shaped occipital region; inflated auditory bullae; malar moderate to very deep; teeth subhypsodont to extremely hypsodont; dental formula: I 2-3/3; C 1/1; P 4/4; M 2-3/3.

Leptauchenia Leidy 1856

Leptauchenia Leidy 1856:88; Cope 1884:546; Sinclair 1910:196; Thorpe 1921:406; Loomis 1925:245; Schultz and Falkenbach 1968:272; Macdonald 1970; Lander 1977:81

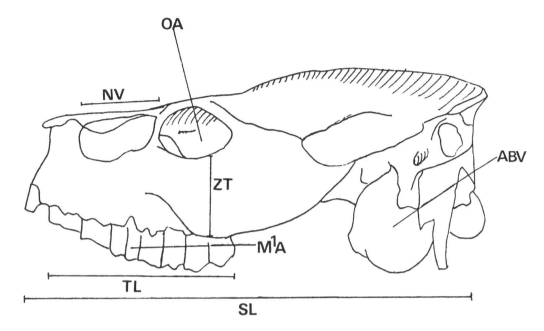

Figure 1. Measured characters on leptauchenin skull. ABV = Auditory Bullae Volume; M1A = Area of the Upper First Molar; NV = Nasal-Facial Vacuity Length; OA = Orbit Area; SL = Skull Length; TL = Tooth row Length; ZT = Zygomatic Arch Thickness. Orbital Bridge measured on the top of the skull from the midpoint of the left orbit to the midpoint of the right.

Cyclopidius Cope 1877:221 and 1884:553; Thorpe 1921:412 and 1937:252; Loomis 1925:247; Schlaikjer 1935:161; Schultz and Falkenbach 1968:296

Pithecistes Cope 1877:219, and 1884:557; Scott 1890:357; Schultz and Falkenbach 1968:258; Lander 1977:81

Hadroleptauchenia Schultz and Falkenbach 1968:306

Pseudocyclopidius Schultz and Falkenbach 1968:327

Brachymeryx Cope 1877:220

Type species—*Leptauchenia decora* (Leidy).

Distribution—Brule, Gering, and Monroe Creek Formations and their equivalents in Nebraska, South Dakota, Colorado, Montana, and Wyoming.

Revised Diagnosis—skull size larger than *Sespia*; malar moderately wide and thick; teeth subhypsodont; dental formula: I 3/3; C 1/1; P 4/4; M 2-3/3.

Discussion—*Leptauchenia* can be distinguished from *Sespia* on the basis of having subhypsodont teeth rather than very hypsodont teeth. Additionally, *Leptauchenia* has three upper incisors, whereas *Sespia* has two. *Leptauchenia* has a skull that averages larger than *Sespia*. Factor analysis demonstrated a slight segregation of *Sespia* from *Leptauchenia* (Fig. 2). Cluster analysis yielded no significant groupings.

Leptauchenia decora Leidy 1856

Leptauchenia decora Leidy 1856:88; 1869:127; Sinclair 1910:196; Thorpe 1921: 405, and 1937:235; Scott 1940, fig. 134; Schultz and Falkenbach 1968:277;

Lander 1977: 86

Pithecistes brevifacies Cope 1878: 219; Scott 1890, Plate 15; Schultz and Falkenbach 1968:268

Cyclopidius heterodon Cope 1877:22

Pithecistes heterodon Cope 1884:559

Pithecistes decedens Cope 1884:558

Cyclopidius decedens (Cope) Matthew 1899:73; Loomis 1925:248

Cyclopidius (Chelonocephalus) schucherti Thorpe 1921:415

Pithecistes breviceps (Cope) Loomis 1925: 248

Cyclopidius loganensis Koerner 1940: 856

Pithecistes tanneri Schultz and Falkenbach 1968: 261

Pithecistes altageringensis Schultz and Falkenbach 1968:265

Pithecistes copei Schultz and Falkenbach 1968:269

Pithecistes mariae Schultz and Falkenbach 1968:264

Leptauchenia harveyi Schultz and Falkenbach 1968: 276

Leptauchenia martini Schultz and Falkenbach 1968: 288

Leptauchenia parasimus Schultz and Falkenbach 1968:291

Leptauchenia margeryae Schultz and Falkenbach 1968:294

Hadroleptauchenia eiselyi Schultz and Falkenbach 1968: 306

Pseudocyclopidius orellaensis Schultz and Falkenbach 1968:327

Holotype—ANSP 10878.

Type Locality and Horizon—Zone `D' of the

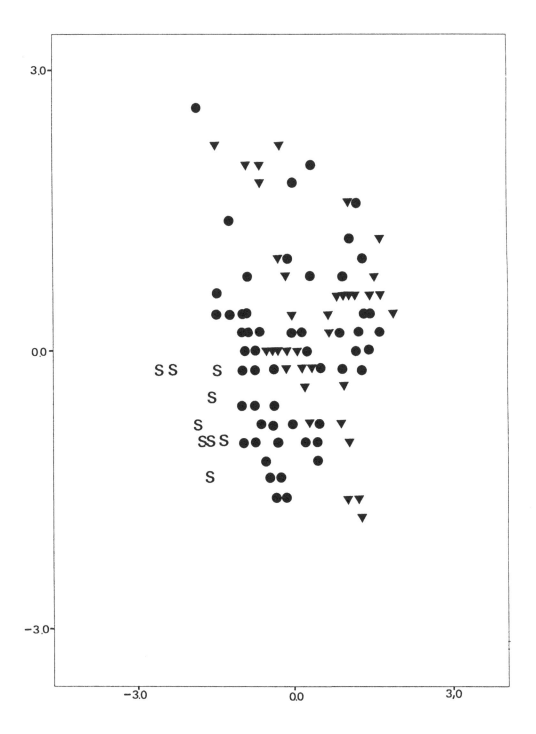

Figure 2. Factor analysis of leptauchenins. S = *Sespia*; dots = *Leptauchenia decora*; triangles = *Leptauchenia major*. Factor 1 (horizontal axis) loaded zygomatic arch thickness, nasal facial vacuity length, auditory bullae volume, and tooth row length. Factor 2 (vertical axis) loaded orbit area.

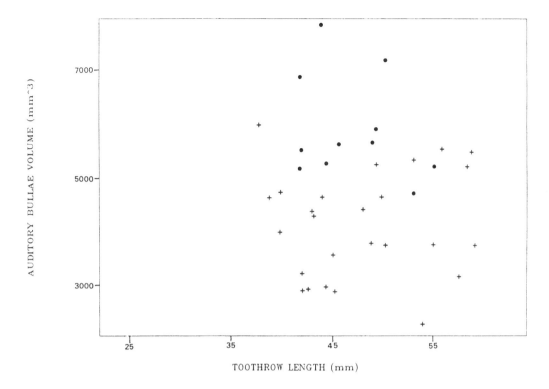

Figure 3. Bivariate plot of auditory bullae volume and tooth row length of *Leptauchenia decora* (+) and *Leptauchenia major* (dots).

Brule Formation; "Valley of the White River," South Dakota.

Distribution—Same as for genus.

Revised Diagnosis—Skull smaller than *Leptauchenia major*; malar shallow to moderately thick; dental formula: I 2/3; C 1/1; P 4/4; M 2/3.

Referred Specimens—Academy of Natural Sciences, Philadelphia: ANSP 10878. American Museum of Natural History: F:AM 114263, 34483, 45682, 45633, 56824B, 57150, 57139, 56831, 56824B, 45685, 45680, 45679, 45632, 45593, 45588, 45570, 45571A, 45571B, 45574, 45564, 45565, 45567, 45561, 45546, 45508, 45505, 45502, 45503, 45504, 114263. University of Nebraska State Museum: UNSM 28255, 28238, 28256, 28430, 2285-73, 29531, 28538, 28539, 28540, 28484, 28492, 2843, 28451, 28488, 28425, 28426, 28594, 28428, 28491, 28609, 28496, 28497, 28450, 28482. South Dakota School of Mines: SDSM 347, 562, 2446, 2449, 2451, 2856, 2577, 2578, 3131, 3214, 3679, 4028, 4076, 5772, 5954, 6123, 6211, 35532, 53533, 54159, 54171, 54172, 54181, 54182, 54191, 54183, 54210, 54227, 55166, 55168, 55183, 56126, 62462, 62465, 2993, 53531, 54211, 2447, 69118.

Discussion—*Leptauchenia decora* differs from *L. major* by having one less upper molar. Bivariate plots indicate that *L. decora* has a slightly smaller orbit area, a smaller auditory bullae volume, and a smaller orbital bridge than *L. major* skulls of similar size (Figs. 3 and

4). Factor and cluster analyses could not distinguish any further groupings.

Leptauchenia major (Leidy)

Leptauchenia major Leidy 1856:163; Thorpe 1937:238; Lander 1977:88
Cyclopidius simus Cope 1877:211; Schultz and Falkenbach 1968:298
Brachymeryx feliceps Cope 1977:220
Cyclopidius emydinus Cope 1884:553; Thorpe 1937:252; Schultz and Falkenbach, 1968:302
Cyclopidius incisivus Scott 1893:661, and 1895:163; Thorpe 1937:255
Cyclopidius lullianus Thorpe 1921:413
Leptauchenia densa Loomis 1925:245
Cyclopidius densa (Loomis) Schlaikjer 1935:161; Thorpe 1937:243
Cyclopidius quadratus Koerner 1940:857
Hadroleptauchenia densa Schultz and Falkenbach 1968:316
Hadroleptauchenia primitiva Schultz and Falkenbach 1968:307
Hadroleptauchenia shanafeltae Schultz and Falkenbach 1968:311
Hadroleptauchenia extrema Schultz and Falkenbach 1968:322
Pseudocyclopidius frankforteri Schultz and Falkenbach 1968:328

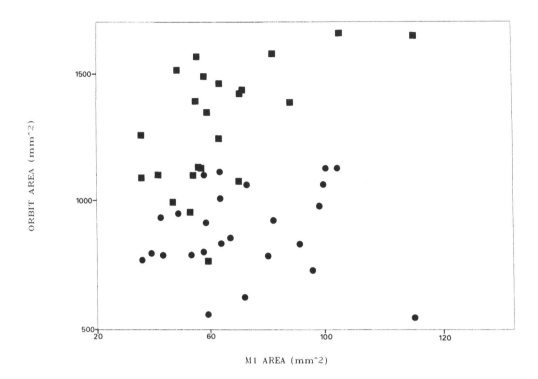

Figure 4. Bivariate plot of orbit area and M1 area of *Leptauchenia decora* (dots) and *Leptauchenia major* (squares).

Pseudocyclopidius major Schultz and Falkenbach
    1968:331
Pseudocyclopidius lullianus Schultz and Falkenbach
    1968:336
Pseudocyclopidius lullianus expiratus Schultz and
    Falkenbach 1968:342
Pseudocyclopidius quadratus Schultz and Falkenbach
    1968:345
    Holotype—AMNH 8115.
    Type Locality and Horizon—Deep River beds,
Smith Creek, Montana.
    Distribution—Upper Brule (Zone `D'), Gering and
Monroe Creek formations and their equivalents in
Montana, Nebraska, South Dakota, Wyoming, and
Colorado.
    Revised Diagnosis—Skull larger than *Sespia* or
Leptauchenia decora; dental formula: I 2/3; C 1/1; P
4/4; M 3/3.
    Referred Specimens—Academy of Natural
Science, Philadelphia: ANSP 10871, 10872, 10874,
10940, 10941. American Museum of Natural History:
AMNH 8115, 8116, F:AM 34484, 34486, 37525,
45500, 45501, 45517, 45528, 45534, 45555,
45566, 45572, 4573, 45576, 45577, 45597,
45598, 45599, 45600, 4602, 56860, 56876,
56886, 56890, 56891, 56899, 56959, 56963,
56978, 56990B, 56994, 56995, 56997, 57005,
57013, 57014, 57015, 57018, 57020, 57022,
57082, 57083, 57088, 57093, 57094, 57111,
57118, 57137, 57138, 57154. University of

Nebraska State Museum: UNSM 1080, 1081, 1082,
28401, 28431, 28567. South Dakota School of Mines:
SDSM 53437, 54217, 54218, 54232, 54345, 62245,
52246.
    Discussion—*Leptauchenia major* has three upper
molars, a slightly larger orbit area, greater auditory
bullae volume and a broader orbital bridge than skulls of
L. decora of similar size (Figs. 3 and 4).

Sespia Stock (1930)

Leptauchenia? (*Sespia*) Stock 1930:38
Leptauchenia (in part) Leidy 1869:129; Cope
    1884:546; Thorpe 1921:410
Sespia (Stock) Schultz and Falkenbach 1968:239;
    MacDonald 1970:77; Lander 1977:94
Megasespia Schultz and Falkenbach 1968:255
    Type species—*Sespia californica* Stock 1930.
    Included species—*Sespia nitida* Leidy 1869.
    Distribution—Gering and Monroe Creek forma-
tions and their equivalents of Nebraska, Montana,
Wyoming, South Dakota, and Colorado.
    Revised Diagnosis—Skull size is significantly
smaller; the malar is narrower and thinner; and the teeth
are much more hypsodont than other leptauchenins;
dental formula: I 2-3/3; C 1/1;P 4/4; M 2-3/3.

Sespia nitida (Leidy, 1869)

Leptauchenia nitida Leidy 1869:129; Thorpe 1921:410

Sespia nitida (Leidy) Schultz and Falkenbach 1968:247;
Lander 1977:96

Cyclopidius heterodon Cope 1878:222

Leptauchenia minora Schlaikjer 1935:164

Sespia heterodon (Cope) Schultz and Falkenbach
1968:244

Sespia marianae Schultz and Falkenbach 1968: 244

Sespia ultima Schultz and Falkenbach 1968:254

Megasespia middleswarti Schultz and Falkenbach
1968:256

Holotype—ANSP 10870.

Type Locality and Horizon—Sharps Formation, White Earth Creek, White River Drainage,
Washabaugh County, South Dakota.

Revised Diagnosis—Same as for genus.

Referred Specimens—Academy of Natural
Sciences–Philadelphia: ANSP 10870. American
Museum of Natural History: F:AM 45606, 45686,
45622, 57144. University of Nebraska State Museum:
UNSM 28409, 28455, 28408, 28498, 28420A,
28420B, 28442.

Discussion—*Sespia* has extremely hypsodont teeth
with a skull that is smaller than in *Leptauchenia*. Factor
analysis distinguishes *Sespia* from other leptauchenins,
as discussed above.

CONCLUSION

Many of the traditional taxonomic characters used to
classify the Leptaucheniinae, such as the zygomatic arch
thickness, the nasal-facial vacuity length and the orbital
bridge width, appear not to be useful in distinguishing
groups. Variations in these characters are due either to
allometry related to skull size or to post-depositional
deformation. Several characters, however, do allow for
the Leptaucheniinae to be divided into three taxa, *Sespia
nitida*, *Leptauchenia decora* and *L. major*. The reduction
of the number of taxa from 31 to 3 limits the use of the
Leptaucheniinae in biostratigraphy.

ACKNOWLEDGMENTS

The author thanks Drs. D. Prothero for his suggestions and advice; R. Tedford for supervising this
summer research project; M. Voorhies, C. Byers, and P.
Morris for helpful comments on early drafts of this
manuscript; Jon Garbisch and Jon Christianson for their
patience and good humor during a long week of skull
measuring. Thanks also to Drs. M. Stevens and R.
Joeckel for their reviews of the manuscript and very
helpful comments. This work was funded by the
American Museum of Natural History Undergraduate
Research Fellowship and by the Department of Geology
and Geophysics at the University of Wisconsin at
Madison.

LITERATURE CITED

Cope, E. D. 1877. Description of new Vertebrata from the
upper Tertiary formations of the West. Proceedings of
the American Philosophical Society 17:219-231.

Cope, E. D. 1878. A new genus of Oreodontidae. American
Naturalist 12:129.

Cope, E. D. 1884. Synopsis of the species of the Oreodontidae. Proceedings of the American Philosophical
Society 21:503-572.

Hay, O. P. 1902. Bibliography and Catalogue of Fossil
Vertebrates of North America. U.S.G.S. Bulletin 179.
868 pp.

Koerner, H. E. 1940. The geology and paleontology of the
Fort Logan and Deep River Formations of Montana.
American Journal of Science 238:837-862.

Lander, E. B. 1977. A review of the Oreodonta (parts I, II,
and III). Ph.D. dissertation. University of California,
Berkeley,. 474pp.

Leidy, J. 1856. Notices of remains in extinct Mammalia,
discovered by Dr. F. V. Hayden, in Nebraska Territory.
Proceedings of the Academy of Natural Sciences,
Philadelphia 8:88-90.

Leidy, J. 1856. 1858. Notice of remains of extinct
Vertebrata, from the valley of the Niobrara River,
collected during the expedition of 1857, in Nebraska,
under the command of Lieut. G. K. Warren. Proceedings
of the Academy of Natural Sciences, Philadelphia 10:20-
29.

Leidy, J. 1869. The mammalian fauna of Dakota and
Nebraska, including an account of some allied forms
from other localities, together with a synopsis of the
mammalian remains of North America. Proceedings of
the Academy of Natural Sciences, Philadelphia, series 2,
7:1-472.

Loomis, F. G. 1925. *Leptauchenia* and *Cyclopidius*.
American Journal of Science, series 5, 9:241-249.

Macdonald, J. R. 1970. Review of the Miocene Wounded
Knee area of southwestern South Dakota. Bulletin of the
Los Angeles County Museum of Natural History 8:1-82.

Matthew, W. D. 1899. A provisional classification of the
fresh-water Tertiary of the West. Bulletin of the
American Museum of Natural History 12:19-75.

Matthew, W. D. 1909. Cenozoic mammal horizons of
western North America. U.S.G.S. Bulletin 361:91-138.

O'Harra, C. C. 1920. The White River Badlands. South
Dakota School of Mines Bulletin 13:1-173.

Osborn, H. F. 1910. The Age of Mammals in Europe, Asia,
and North America. Macmillan, New York. 250 pp.

Schlaikjer, E. M. 1935. New vertebrates and the stratigraphy of the Oligocene and early Miocene. Bulletin of the
Museum of Comparative Zoology 76:97-189.

Schultz, C. B. 1955. Classification of Oligocene sediments of Nebraska. Bulletin of the University of
Nebraska State Museum 4:2.

Schultz, C. B., and C. H. Falkenbach. 1940. Merycochoerinae, a new subfamily of oreodonts. Bulletin of the
American Museum of Natural History 77:213-306.

Schultz, C. B., and C. H. Falkenbach. 1968. The phylogeny of the oreodonts, parts 1 and 2. Bulletin of the
American Museum of Natural History 139:1-498.

Scott, W. B. 1890. Beiträge zur Kenntniss der Oreodontidae. Morphologische Jarhbuch 16:319-395.

Scott, W. B. 1893. The mammals of the Deep River Beds.
American Naturalist 27:659-662.

Scott, W. B. 1895. The Mammalia of the Deep River Beds.
Transactions of the American Philosophical Society
18:55-185.

Scott, W. B. 1940. Artiodactyla; pp. 363-746 *in* W. B.
Scott and G. L. Jepsen (eds.), The Mammalian Fauna of
the White River Oligocene. Transactions of the
American Philosophical Society 28.

Sinclair, W. J. 1910. The restored skeleton of *Leptauchenia decora*. Proceedings of the Academy of Natural

Sciences, Philadelphia, 49:196-199.

Stock, C. 1930. Oreodonts from the Sespe deposits of South Mountain, Ventura County, California. Publications of the Carnegie Institute of Washington 404:27-42.

Tedford, R. H., J. B. Swinehart, R. M. Hunt, Jr., and M. R. Voorhies. 1985. Uppermost White River and lowermost Arikaree rocks and faunas, White River Valley, north western Nebraska and their correlation with South Dakota. Dakoterra 2:335-352

Thorpe, M. R. 1921. *Leptauchenia* Leidy and *Cyclopidius* (*Pithecistes*) Cope, with descriptions of new and little known forms from the Marsh collection. American Journal of Science, series 5, 1:405-419.

Thorpe, M. R. 1937. The Merycoidodontidae, an extinct group of ruminant mammals. Memoirs of the Peabody Museum of Natural History 3:1-428.

27. Leptomerycidae

Timothy H. Heaton and Robert J. Emry

ABSTRACT

Leptomerycids appear in North America in the late Duchesnean and extend into the Arikareean. The earliest grade, called *Hendryomeryx* by some workers, differs from later *Leptomeryx* in having lower crowned cheek teeth. *Leptomeryx yoderi* appeared in the early Chadronian and became a widespread species. This small but high crowned species probably gave rise to all later species of *Leptomeryx*. At Flagstaff Rim, Wyoming, *L. yoderi* underwent a gradual anagenetic size increase to become *L. mammifer* of the middle to late Chadronian. *Leptomeryx speciosus* probably also descended from *L. yoderi*. *Leptomeryx yoderi* and *L. speciosus* are of similar size and differ only in the shape of the entoconulid on the last molar. Both *L. speciosus* and *L. mammifer* became widespread species and are found at Pipestone Springs, Montana, and in the Cypress Hills, Saskatchewan. *Leptomeryx mammifer* went extinct before the end of the Chadronian, leaving *L. speciosus* as the only known surviving species. At Douglas, Wyoming, *L. speciosus* can be seen undergoing gradual anagenesis across the Chadronian-Orellan boundary to become *L. evansi*, the type species of the genus. *Leptomeryx evansi* is probably the only valid Orellan species, and it is found in great numbers in the Big Badlands of South Dakota and at other Orellan exposures of the Great Plains.

In the past, the taxonomy of the leptomerycids has been confused by the enormous variation in the ridge configuration of the lower premolars, especially P_3. Our statistical study of large samples demonstrates that these are merely individual variations within all species of *Leptomeryx* and therefore have little taxonomic value. Individual specimens cannot be identified based on premolar morphology, but, using statistical samples, the proportion of different character states can help in establishing ancestor-descendant relationships.

INTRODUCTION

Leptomeryx, one of the smaller and more common artiodactyls of the Chadronian and Orellan land mammal "ages," comprises a cohesive group of about seven species, the more primitive of which have been placed in the genus *Hendryomeryx* by some workers (e.g., Black, 1978; Storer, 1981). The type species, *L. evansi*, is common in the Big Badlands of South Dakota and at other Orellan localities in the Great Plains. Chadronian forms are more diverse and are found in deposits from Saskatchewan to Texas, but the most

abundant material with the best stratigraphic control comes from eastern Wyoming.

Because *Leptomeryx* is a hornless artiodactyl, only the teeth have been used to distinguish species. Dentaries are the most common elements recovered, and many of these show signs of having been eaten by scavengers. Complete dentaries include four incisiform teeth (I_{1-3} and C_1), a peg-like first premolar (P_1) within the diastema, three anteroposteriorly elongate premolars (P_{2-4}), and three selenodont molars (M_{1-3}). Most commonly the anterior part with incisors is missing, and the preserved part contains P_2-M_3 (Fig. 1B). Less commonly the P_1 is also preserved (Fig. 1D). The lower cheek teeth form the basis of most studies. Upper dentitions are less common and include three elongate premolars and three molars.

The sporadic distribution of fossil deposits makes environmental and evolutionary reconstruction difficult because geographic and chronologic variations cannot always be distinguished and because important transitions are not always recorded (see Heaton, this volume, Chapter 18). This problem is exacerbated for the Chadronian-Orellan transition because the most fossiliferous Chadronian deposits are located in isolated basins of the Rocky Mountains while the most productive Orellan localities occur in the more climatically uniform Great Plains. Fortunately, *Leptomeryx* material from central and eastern Wyoming spans the late Duchesnean through early Orellan, and the species found there occur in deposits of similar age throughout the region.

During its existence *Leptomeryx* underwent several important morphological transformations that were interspersed with periods of stasis. The timing and nature of these transformations is the subject of this chapter. The evolutionary history of *Leptomeryx*, as we believe it can best be reconstructed, is discussed in three parts. In chronological order they are: (1) the development of *Leptomeryx* from *Hendryomeryx* in the late Duchesnean or earliest Chadronian; (2) the apparent cladogenesis of *L. yoderi* into *L. speciosus* and *L. mammifer* in the early to middle Chadronian; and (3) the

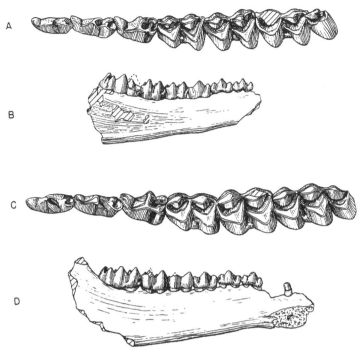

Figure 1. Comparison of two dentaries of *Leptomeryx* from Flagstaff Rim, Wyoming, in occlusal and lingual view showing cheek tooth series. A and B are of *L. speciosus* (AMNH, Frick Collection Field No. B.H. 43-859-AN). C and D are of *L. mammifer* (AMNH, Frick Collection Field No. B.H. 57-1427-1).

development of *L. evansi* from *L. speciosus* during the late Chadronian to early Orellan. The later history of leptomerycids is described by Taylor and Webb (1976).

ABBREVIATIONS

AMNH, American Museum of Natural History, New York, New York; CM, Carnegie Museum of Natural History, Pittsburgh, Pennsylvania; MCZ, Museum of Comparative Zoology, Harvard University, Cambridge, Massachusetts; MPUM, University of Montana Museum of Paleontology, Missoula, Montana; ND, North Dakota Geological Survey, Bismarck, North Dakota; NMC, National Museums of Canada, Ottawa, Ontario; PTM, Pioneer Trails Museum, Bowman, North Dakota; ROM, Royal Ontario Museum, Toronto, Ontario; SDSM, South Dakota School of Mines and Technology, Rapid City, South Dakota; SMNH, Royal Saskatchewan Museum, Regina, Saskatchewan; USNM, U.S. National Museum, Smithsonian Institution, Washington, D.C.

THE DEVELOPMENT OF *LEPTOMERYX* FROM *HENDRYOMERYX*

The ancestry of the leptomerycids is uncertain. Emry et al. (1987) noted their sudden appearance in North America and suggested the Eocene genus *Archaeomeryx* of Asia as a possible ancestor.

We have not studied the earliest specimens of *Leptomeryx* (e.g., *Hendryomeryx*) in detail, and they are less relevant to this volume than the later forms. All

are small in size. Most of the material is incomplete and difficult to evaluate, especially when one considers the enormous amount of individual variation found in later species. We will therefore limit our coverage of the earliest specimens to a review of the literature, a report of new material, and some comments on taxonomy.

Wilson (1974) described a new species, *Leptomeryx defordi*, based on what was then the earliest known leptomerycid material, from the Porvenir local fauna of Trans-Pecos Texas. He described it as a primitive species, "upper molars with incomplete selenes," and "in lower molars the metaconid is a distinct cusp but joined to the posterior internal crest by a ridge at right angles to the tooth."

Black (1978) reported even earlier leptomerycid dentitions from Badwater Creek in central Wyoming and chose to give them, as well as the Porvenir material and two specimens from McCarty's Mountain, Montana, the name *Hendryomeryx*. According to Black "*Hendryomeryx* is less advanced than the Oligocene *Leptomeryx* of the plains and intermontane regions in having lower crowned, less selenodont molars, and a weak metaconid-entoconid lophid." Black gave his Badwater Creek material the specific name *H. wilsoni*.

Storer (1981), in his review of leptomerycids from the Cypress Hills, Saskatchewan, assigned Cope's (1889) small species *Leptomeryx esulcatus* to *Hendryomeryx* and gave the genus a fuller description. Storer described the molars of *Hendryomeryx*, in

Table 1. Tooth length measurements (mm) for the three species assigned to "*Hendryomeryx*": *H. deford* from Trans-Pecos Texas (late Duchesnean), *H. wilsoni* from Badwater Creek, Wyoming (early Duchesnean), and *H. esulcatus* from the Cypress Hills, Saskatchewan (middle Chadronian). All Trans-Pecos Texas and Badwater leptomerycid material is included here, whereas the Cypress Hills measurements represent those specimens assigned by Storer (1981) to *Hendryomeryx* rather than *Leptomeryx*.

Tooth	*H. defordi* Trans-Pecos Texas (from Wilson, 1974)			*H. wilsoni* Badwater Creek (from Black, 1978)			*H. esulcatus* Cypress Hills (from Storer, 1981)		
	Range	Mean	N	Range	Mean	N	Range	Mean	N
P_2	4.8	4.8	1	3.5	3.5	1			
P_3	5.3	5.3	1						
P_4	5.4-5.8	5.6	4	4.9	4.9	1	5.3	5.3	1
M_1	5.7-6.3	6.1	3	4.9	4.9	2	6.0-6.8	6.4	2
M_2	6.4-6.8	6.5	3	5.1-5.4	5.2	4	5.6-7.8	6.5	*26
M_3	9.4	9.4	1						
M_{1-3}	21.4	21.4	1						

*These 26 specimens were listed by Storer (1981) as M_1 or M_2.

Table 2. Tooth length measurements (mm) for small leptomerycids from two early Chadronian localities in Wyoming and one locality of less certain age in Montana. All West Canyon Creek and Canyon Ferry leptomerycid material is included here, whereas the Yoder measurements represent only those specimens considered by Kihm (1987) to be too small for *Leptomeryx yoderi*. The West Canyon Creek and Canyon Ferry material was measured by Heaton.

Tooth	Yoder local fauna (from Kihm, 1987)			West Canyon Creek			Canyon Ferry		
	Range	Mean	N	Range	Mean	N	Range	Mean	N
P_2							4.1-4.9	4.5	2
P_3	5.6	5.6	1				4.9-6.5	5.6	7
P_4	5.3-5.5	5.4	2	6.9	6.9	1	5.0-6.5	5.8	10
M_1	5.6-6.3	6.0	2	6.0	6.0	1	5.3-6.6	6.0	14
M_2	6.2-7.3	6.8	2	5.8-6.5	6.2	2	6.0-7.5	6.7	11
M_3	9.5-9.7	9.6	2	9.2-10.0	9.6	2	9.5-10.7	10.2	7
M_{1-3}	21.3-22.0	21.7	2				20.9-22.7	22.0	5

addition to being lower crowned, as "much shallower-patterned than in *Leptomeryx*, with the bottoms of the fossettes and fossettids easily visible; in *Leptomeryx*, valley walls are much steeper, and valley bottoms are not easily visible except in very worn specimens." Tooth length measurements for the three species assigned by Storer (1981) to *Hendryomeryx* are reported in Table 1. *Hendryomeryx esulcatus* occurs with a middle Chadronian fauna, and Storer indicated that its teeth were higher crowned than in *H. wilsoni*. The inclusion of *L. esulcatus* in *Hendryomeryx* would extend this genus into the middle Chadronian when more derived species were also present.

We consider Black's (1978) decision to separate *Hendryomeryx* from *Leptomeryx* to be unfortunate because the differences between the species assigned to this genus and the Chadronian species retained in *Leptomeryx* are hardly worthy of generic distinction. *Leptomeryx evansi* of the Orellan is more distinct from

the Chadronian leptomerycids than any of the Chadronian and Duchesnean species are from one another, and even these differences are only of specific value. On the whole, the species of *Leptomeryx* exhibit remarkable uniformity, so there is no utility in dividing the genus.

Undescribed material from several Duchesnean and early Chadronian localities will need to be considered when unraveling early *Leptomeryx* evolution. Black (1978) reported that "in the McCarty's Mountain fauna in Montana an undescribed species of *Hendryomeryx* appears to be present (CM 1057 and 31397) together with a species of *Leptomeryx*," though Storer (1984) considered both of these to represent *Leptomeryx*. The West Canyon Creek fauna of late Duchesnean or early Chadronian age in central Wyoming (under study by Emry) includes several *Leptomeryx* specimens, possibly of *L. (H.) wilsoni*. Tooth lengths for the West Canyon Creek sample are listed in Table 2. A new sample has also been found in the Medicine Pole Hills of North Dakota (discussed below).

Because late Duchesnean and early Chadronian faunas are few in number and do not contain abundant leptomerycid material, tracing the origin and early development of *Leptomeryx* is a difficult task. All we know at present is that the transition to a later grade involved an increase in crown height and a deepening of valleys in the molars (Storer, 1981).

THE CLADOGENESIS OF *LEPTOMERYX YODERI* INTO *L. SPECIOSUS* AND *L. MAMMIFER*

The main thrust of our study is a detailed statistical analysis of Chadronian *Leptomeryx* dentaries from Flagstaff Rim and other localities in central Wyoming. Emry (1970, 1973, 1992) supervised the collection of nearly all this material, keeping careful stratigraphic records, and Heaton (1989) measured and analyzed these specimens as a Smithsonian Institution postdoctoral research project. The sample size (1400 dentaries from Flagstaff Rim alone), the number of characters measured (89 on complete specimens), and the stratigraphic control allow us to examine individual vs. specific variation and evolutionary change at a level of detail that is rarely possible in paleontological studies.

As noted by Emry (1970, 1973), there are several species of *Leptomeryx* at Flagstaff Rim, and at least one lineage clearly represents an evolutionary transition from one species to another. Figures 2 through 4 are bivariate plots of *Leptomeryx* showing the size of specimens from different stratigraphic levels. The size is based on length of the cheek tooth series (P_2-M_3) in Figure 2, length of the molars (M_{1-3}) in Figure 3, and length of M_2 (the most abundantly represented tooth) in Figure 4. Figure 5 is a similar plot using M_2 length for *Leptomeryx* from the Ledge Creek locality, 11 miles SSE of Flagstaff Rim. In each plot a unimodal size distribution is found at the lowest level while a bimodal distribution, which includes a larger species, is found throughout the upper part of the section. Levels in between exhibit an intermediate amount of size variation.

At Flagstaff Rim the single tooth measurement (Fig. 4) provides the largest sample size, while the full cheek tooth row measurement (Fig. 2) provides the cleanest separation between two coexisting species in the upper part of the section. The length of the molars (Fig. 3) makes a good compromise and is used in a series of size histograms in Figure 6. Because the two species from the upper portion of the Flagstaff Rim section are in stasis (or become only slightly larger up section), they are grouped into a single histogram (Fig. 6D). Specimens from the lower part of the section are concentrated in three quarries that exhibit different size distribution patterns, so they are shown (together with specimens from nearby stratigraphic levels) in three separate histograms (Figs. 6A-C). We will refer to these samples as Quarry A ("Low Pocket"; oldest), Quarry B ("Dry Hole Pocket"; middle), and Quarry C ("B-44 Pocket"; youngest).

Of the three quarries, Quarry A has the largest sample size, shows the greatest morphologic uniformity, and comes closest to exhibiting a normal size distribution (Fig. 6A). Emry (1973) referred this population to *Leptomeryx yoderi* because of its similarity in morphology and age to the type specimen from the Yoder local fauna, located 125 miles ESE of Flagstaff Rim. *Leptomeryx yoderi* was described by Schlaikjer (1935), and more material from the type locality was referred to this species by Kihm (1987). Statistical data on the Quarry A sample and on *L. yoderi* from the Yoder local fauna are presented in Table 3.

The upper portion of the Flagstaff Rim section exhibits a distinctly bimodal size distribution that is best interpreted as representing two species. Similar bimodal size distributions are found in samples of equivalent age from Ledge Creek, Wyoming (Fig. 5); Pipestone Springs, Montana (Matthew, 1903; Tabrum and Fields, 1980); and Calf Creek in the Cypress Hills, Saskatchewan (Storer, 1981). Emry (1973) considered the smaller of the two species from Flagstaff Rim to be the same as the smaller species from Pipestone Springs, which Matthew (1903) had provisionally referred to *Leptomeryx esulcatus*. However, Storer (1981) instead assigned this material to *L. speciosus*, a species originally described from the Cypress Hills by Lambe (1908), and this name is used here (see Emry et al., 1987). Statistical data on this sample and on similar samples from Ledge Creek and Pipestone Springs are shown in Table 4.

Emry (1973) referred the larger of the two species from the upper part of the Flagstaff Rim section to *Leptomeryx mammifer* because of its similarity to the sample of similar size from Pipestone Springs, which Matthew (1903) referred to that species. The type specimen of *L. mammifer* was described by Cope (1885) from the Cypress Hills, but it is very fragmentary and may not be diagnostic of this species (Storer, 1981). We have not examined the type but will use the name *L. mammifer* until this taxonomic problem is resolved. Table 5 presents statistics for this larger species from Flagstaff Rim, Ledge Creek, and Pipestone Springs.

The samples from Quarries B and C (Figs. 6B-C) are more difficult to interpret because of their skewed size distributions, but they seem to exhibit intermediacy between the Quarry A sample and the sample from the upper part of the Flagstaff Rim section. Emry (1973) suggested that *Leptomeryx* from Flagstaff Rim represents two coexisting lineages, both of which underwent anagenetic increases in size. In particular, he considered *L. mammifer* to be a descendant of *L. yoderi*. Even more significantly, this sequence of four chronologically distinct samples may represent a case of sympatric speciation, and this is the reason for our unusually detailed analysis.

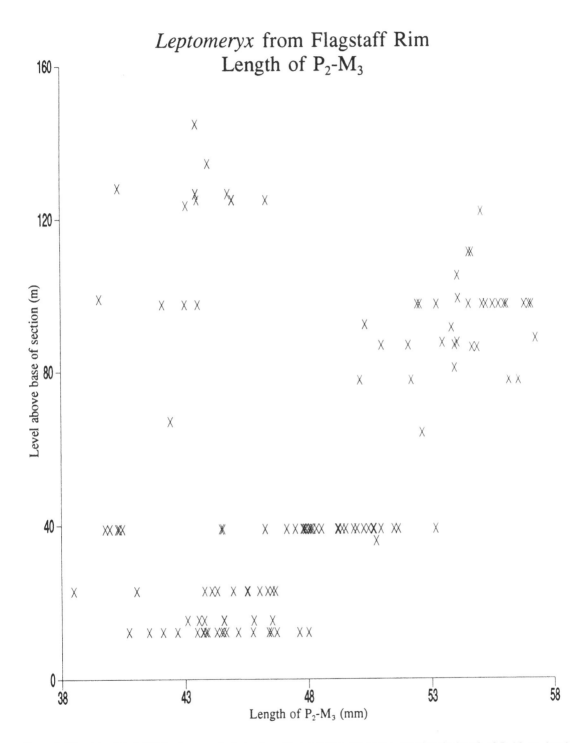

Figure 2. Bivariate plot of 129 *Leptomeryx* dentaries from Flagstaff Rim, Wyoming, showing the length of P_2-M_3 vs. level in the stratigraphic section. Specimens from the lower part of the section are concentrated in three quarry deposits, the lowest of which contains only *L. yoderi*. This measurement of six consecutive teeth allows for excellent resolution between *L. speciosus* (small) and *L. mammifer* (large) in the upper part of the section, but the sample size is small. See Tables 3-6 for tooth size statistics on each species.

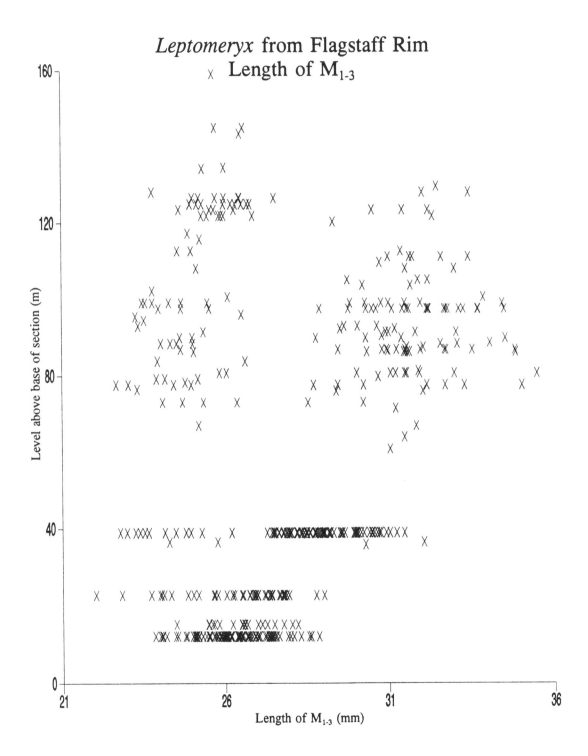

Figure 3. Bivariate plot of 466 *Leptomeryx* dentaries from Flagstaff Rim, Wyoming, showing the length of M_{1-3} vs. level in the stratigraphic section. The measurement of three teeth increases the sample size over that seen in Figure 2 and still allows good resolution between *L. speciosus* and *L. mammifer*. See histograms of four stratigraphic samples based on this measurement in Figure 6.

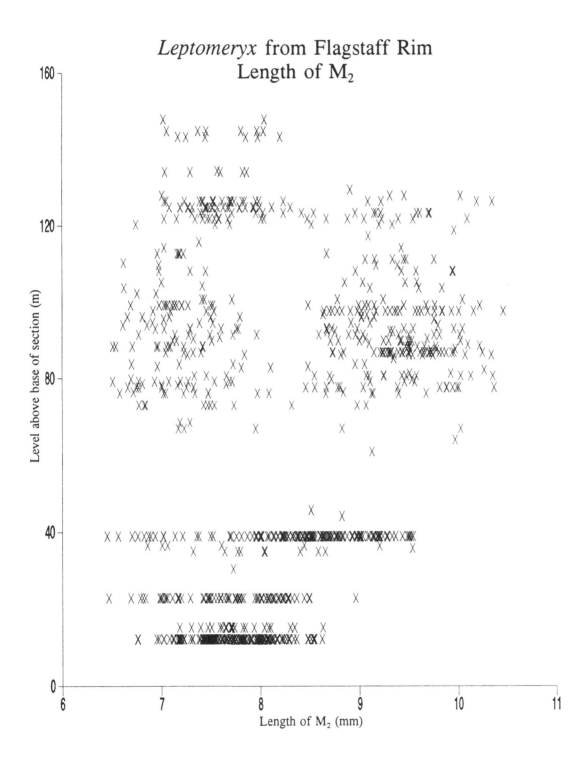

Figure 4. Bivariate plot of 928 *Leptomeryx* dentaries from Flagstaff Rim, Wyoming, showing the length of M_2, the most abundantly preserved tooth, vs. level in the stratigraphic section. The measurement of a single tooth provides an enormous sample size but reduces the resolution between *L. speciosus* and *L. mammifer* seen in Figures 2-3.

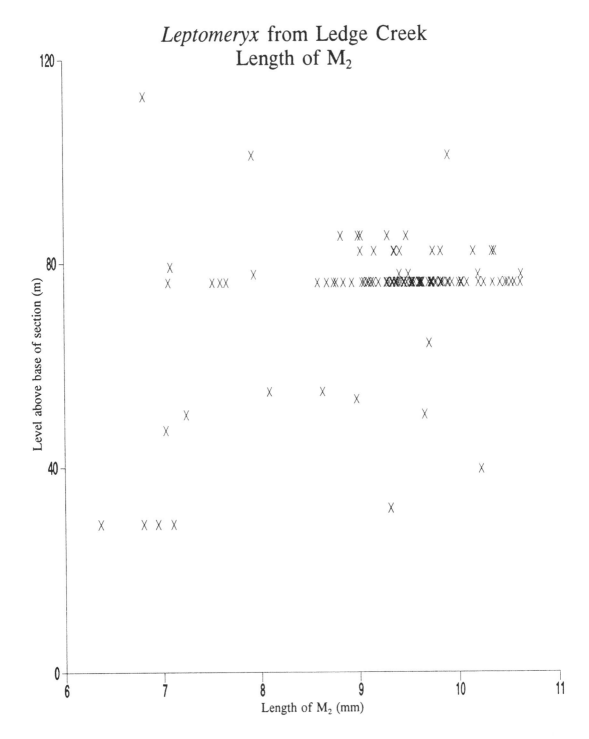

Figure 5. Bivariate plot of 111 *Leptomeryx* dentaries from Ledge Creek, Wyoming, showing the length of M_2 vs. level in the stratigraphic section. *Leptomeryx mammifer* (large) greatly outnumbers *L. speciosus* (small) in the upper part of the section. Comparisons for the lower part of the section are obscured by the small sample size.

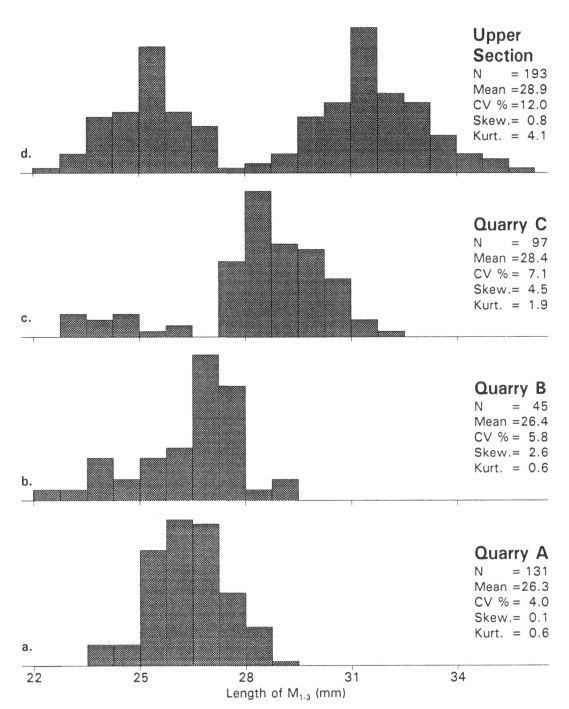

Figure 6. Size histograms, based on M_{1-3} length, of four stratigraphic samples of *Leptomeryx* from Flagstaff Rim, Wyoming. Quarry A ("Low Pocket") is a localized deposit (probably a single carnivore den) containing *Leptomeryx* almost exclusively. Quarry B ("Dry Hole Pocket") and Quarry C ("B-44 Pocket") are also localized deposits; a few specimens from above and below Quarry C are included in that sample. Because *L. speciosus* and *L. mammifer* remain in relative stasis throughout the upper part of the section (Figs. 2-4), all specimens from above 50 m have been grouped into histogram D. Only *L. yoderi* is found in Quarry A, whereas two species coexist in the other three samples. *Leptomeryx yoderi* can be traced through size increase to *L. mammifer* in this diagram. *Leptomeryx speciosus* remains about the same size in samples B, C, and D. See Tables 3-6 for tooth size statistics on each species.

Table 3. Tooth length measurements (mm) for two populations and the type specimen of *Leptomeryx yoderi*, all from eastern Wyoming. The type is a distorted dentary from the Yoder local fauna collected by Schlaikjer (1935). All leptomerycid material from Quarry A at Flagstaff Rim (Figure 6a) and from the Yoder local fauna is included except for two Yoder dentaries considered by Kihm (1987) to be too small to be *L. yoderi* (see Table 2). The Flagstaff Rim and type material was measured by Heaton.

Tooth	Flagstaff Rim Quarry A			Yoder local fauna (from Kihm, 1987)			*Leptomeryx yoderi* Type specimen
	Range	Mean	N	Range	Mean	N	Value
P_2	4.9-6.4	5.6	43	6.2	6.2	1	
P_3	5.8-7.0	6.4	111	6.0-6.8	6.4	5	6.5
P_4	5.5-7.3	6.6	187	6.1-7.2	6.7	10	6.8
M_1	5.9-8.1	7.0	227	6.2-8.1	7.2	16	6.5
M_2	6.8-8.6	7.7	229	7.5-8.6	8.1	14	7.4
M_3	10.1-13.0	11.7	178	10.3-12.5	11.5	8	11.1
M_{1-3}	23.8-28.9	26.3	131	23.8-28.5	26.4	8	
P_2-M_3	40.7-48.0	44.5	29	41.5	41.5	1	

Table 4. Tooth length measurements (mm) for three populations of *Leptomeryx speciosus* from Flagstaff Rim (Figure 6d) and Ledge Creek (Figure 5), Wyoming, and Pipestone Springs, Montana. Included are all specimens judged on size and morphology to be *L. speciosus* rather than *L. mammifer*. All measurements were made by Heaton.

Tooth	Flagstaff Rim upper section			Ledge Creek upper section			Pipestone Springs		
	Range	Mean	N	Range	Mean	N	Range	Mean	N
P_2	4.8-6.4	5.7	21				5.0-5.8	5.5	9
P_3	5.4-7.0	6.3	53	6.0-6.3	6.2	3	5.9-6.6	6.2	10
P_4	5.3-6.9	6.3	99	6.4-6.8	6.6	3	6.0-7.0	6.3	13
M_1	5.7-7.9	6.7	212	5.8-7.5	6.8	12	5.7-7.4	6.6	21
M_2	6.5-8.4	7.4	222	6.8-7.9	7.5	8	6.3-7.8	7.3	24
M_3	9.8-12.6	11.3	161	11.0-12.2	11.6	4	9.5-12.0	11.2	24
M_{1-3}	22.6-27.5	25.1	81	24.3-25.8	25.1	4	24.9-26.8	25.4	5
P_2-M_3	39.6-46.3	43.3	16				42.5-46.0	44.3	2

Table 5. Tooth length measurements (mm) for three populations of *Leptomeryx mammifer* from Flagstaff Rim (Figure 6d) and Ledge Creek (Figure 5), Wyoming, and Pipestone Springs, Montana. Included are all specimens judged on size and morphology to be *L. mammifer* rather than *L. speciosus*. All measurements were made by Heaton. The type specimen of *L. mammifer* from the Cypress Hills of Saskatchewan (NMC 6278) has an M_2 length of 10.2 mm; an M_3 from the Calf Creek local fauna (SMNHP1585.1230) has a length of 15.2□mm (Storer, 1981).

Tooth	Flagstaff Rim upper section			Ledge Creek upper section			Pipestone Springs		
	Range	Mean	N	Range	Mean	N	Range	Mean	N
P_2	6.5-7.9	7.2	44	7.0-8.3	7.6	26	6.4-7.7	6.9	3
P_3	7.0-8.9	7.9	82	7.5-8.8	8.2	43	7.4-7.8	7.6	3
P_4	6.9-9.4	7.9	158	7.1-8.7	8.0	61	7.4-8.4	7.9	4
M_1	7.3-9.8	8.4	262	7.6-9.8	8.7	89	8.1-8.5	8.4	4
M_2	8.0-10.5	9.4	239	8.6-10.6	9.6	90	8.7-9.9	9.5	3
M_3	12.1-15.9	13.9	177	12.8-16.0	14.4	74	13.5-14.0	13.9	4
M_{1-3}	28.5-35.5	31.7	112	29.7-36.3	32.5	58			
P_2-M_3	50.1-57.3	54.3	34	52.4-60.1	56.4	25			

Table 6. List of objective (44) and subjective (45) measurements taken on *Leptomeryx* dentaries used in this study, including summary statistics of four species. See Figure 7 for illustrations. The *L. yoderi* sample is from Quarry A, and the *L. mammifer* and *L. speciosus* samples are from the upper part of the section at Flagstaff Rim, Wyoming. The *L. evansi* sample is from above the PWL at Douglas, Wyoming.

OBJECTIVE MEASUREMENTS

		L. evansi			*L. yoderi*			*L. mammifer*			*L. speciosus*		
Serial Tooth Lengths		**Min.**	**Mean**	**Max.**	**Min.**	**Mean**	**Max.**	**Min.**	**Mean**	**Max.**	**Min.**	**Mean**	**Max.**
1	Length of P_2	4.9	5.6	6.4	6.5	7.2	7.9	4.8	5.7	6.4	4.6	5.0	5.5
2	Length of P_{2-3}	10.9	11.8	12.8	13.9	15.1	16.2	10.2	11.9	13.0	10.4	11.0	12.0
3	Length of P_{2-4}	16.5	18.2	19.5	20.9	22.7	24.7	16.2	17.9	19.5	15.6	16.9	18.3
4	Length of P_2-M_1	22.8	25.1	26.4	28.4	31.0	33.0	22.0	24.6	26.5	21.5	22.9	24.8
5	Length of P_2-M_2	29.7	32.8	35.0	37.2	40.5	42.7	29.0	32.0	34.2	27.7	29.2	31.5
6	Length of P_2-M_3	40.7	44.5	48.0	50.1	54.3	57.3	39.6	43.3	46.3	36.6	38.6	41.1
7	Length of P_3	5.8	6.4	7.0	7.0	7.9	8.9	5.4	6.3	7.0	4.9	6.1	6.8
8	Length of P_{3-4}	11.4	12.7	14.0	14.2	15.6	17.1	11.0	12.3	13.5	10.6	12.0	13.4
9	Length of P_3-M_1	17.3	19.5	21.2	21.6	23.9	26.3	17.2	18.9	20.6	16.5	17.8	19.4
10	Length of P_3-M_2	24.1	27.2	29.5	30.5	33.2	35.8	23.8	26.2	28.3	22.3	24.4	28.0
11	Length of P_3-M_3	35.8	38.9	42.1	43.1	47.0	50.0	34.7	37.5	41.0	31.7	34.0	37.4
12	Length of P_4	5.5	6.6	7.3	6.9	7.9	9.4	5.3	6.3	6.9	4.7	6.0	7.0
13	Length of P_4-M_1	11.7	13.4	14.8	14.3	16.3	18.0	11.6	12.9	14.2	9.7	11.9	14.0
14	Length of P_4-M_2	18.1	21.0	23.2	23.5	25.6	28.2	18.6	20.1	21.9	16.4	18.2	20.9
15	Length of P_4-M_3	29.2	32.8	35.7	36.2	39.5	44.0	29.0	31.5	34.3	26.0	28.2	32.6
16	Length of M_1	5.9	7.0	8.1	7.3	8.4	9.8	5.7	6.7	7.9	4.7	6.0	6.9
17	Length of M_{1-2}	12.8	14.7	16.6	15.9	17.9	20.2	12.4	14.0	15.6	10.7	12.4	14.3
18	Length of M_{1-3}	23.8	26.3	28.9	28.5	31.7	35.5	22.6	25.1	27.5	20.1	22.3	25.5
19	Length of M_2	6.8	7.7	8.6	8.0	9.4	10.5	6.5	7.4	8.4	5.6	6.5	7.3
20	Length of M_{2-3}	17.7	19.4	21.6	20.7	23.3	26.1	16.5	18.5	20.5	14.5	16.5	18.6
21	Length of M_3	10.1	11.7	13.0	12.1	13.9	15.9	9.8	11.3	12.6	8.7	10.1	11.6
	P_2 Measurements												
22	Length anterior to protoconid	3.0	3.4	4.0	3.5	4.4	5.0	3.1	3.6	4.0	2.7	3.1	3.5
23	Width of tooth anteriorly	1.3	1.6	1.8	1.5	1.9	2.3	1.2	1.5	1.8	1.1	1.3	1.6
24	Width of tooth posteriorly	1.8	2.2	2.5	2.3	2.8	3.2	1.9	2.1	2.6	1.6	1.9	2.1
	P_3 Measurements												
25	Length anterior to protoconid	3.2	4.1	4.7	4.5	5.1	5.8	2.7	4.1	4.5	3.5	4.0	4.5
26	Width of tooth anteriorly	1.7	2.1	2.6	1.8	2.6	3.0	1.7	2.1	2.3	1.6	2.0	2.6
27	Width of tooth posteriorly	2.2	2.8	3.7	2.5	3.5	4.1	2.4	2.8	3.3	2.1	2.6	3.3
	P_4 Measurements												
28	Length anterior to protoconid	3.7	4.4	5.2	4.5	5.4	6.1	3.7	4.4	5.2	3.5	4.1	4.9
29	Width of tooth anteriorly	1.9	2.5	3.0	2.3	3.0	3.7	2.0	2.5	3.1	2.0	2.3	2.8
30	Width of tooth posteriorly	2.9	3.6	4.5	3.9	4.6	5.5	2.9	3.6	4.8	2.7	3.3	3.8
	M_1 Measurements												
31	Length of anterior selene	2.9	3.5	4.0	3.5	4.2	4.8	2.8	3.4	4.0	2.3	3.0	3.5
32	Width of tooth at ant. selene	4.0	4.8	5.5	4.7	6.0	7.0	3.5	4.5	5.5	3.7	4.3	5.3
33	Width of tooth at post. selene	4.6	5.3	6.1	5.1	6.6	7.7	3.8	5.1	5.7	4.0	4.7	5.5

Table 6. Continued.

OBJECTIVE MEASUREMENTS

	M₂ Measurements	L. evansi			L. yoderi			L. mammifer			L. speciosus		
		Min.	Mean	Max.	Min.	Mean	Max.	Min.	Mean	Max.	Min.	Mean	Max.
34	Length of anterior selene	3.1	3.8	4.3	3.8	4.6	5.2	3.0	3.7	4.3	2.8	3.3	3.8
35	Width of anterior selene	4.7	5.6	6.6	5.8	7.0	8.6	4.2	5.3	6.1	4.4	5.0	5.8
36	Width of posterior selene	5.1	5.8	6.7	5.9	7.1	8.0	4.4	5.5	6.5	4.4	5.1	5.9
	M₃ Measurements												
37	Length of anterior selene	3.2	3.9	4.5	3.8	4.6	5.5	3.2	3.8	4.4	2.8	3.3	3.7
38	Length of two selenes	6.6	7.6	8.5	7.9	9.1	10.1	6.5	7.3	8.3	5.6	6.5	7.6
39	Width of anterior selene	4.9	5.6	6.5	5.7	6.9	8.4	4.6	5.4	6.5	4.4	5.0	5.6
40	Width of posterior selene	4.7	5.3	6.3	5.1	6.4	7.5	4.4	5.0	6.0	4.2	4.8	5.5
41	Width of posterolophid	2.1	2.9	3.8	2.7	3.5	5.9	2.3	2.9	3.7	2.3	2.7	3.3
	Jaw (lingual side)												
42	Depth at P₂-P₃ junction	6.8	9.5	11.5	9.1	11.9	14.7	7.9	9.7	11.7	7.9	8.9	10.7
43	Depth at P₄-M₁ junction	8.5	10.9	13.3	9.1	12.9	16.2	8.0	10.8	13.6	7.6	10.1	12.5
44	Depth at M₂-M₃ junction	10.1	14.3	17.0	13.0	16.9	21.2	10.6	13.1	15.9	11.1	12.7	14.6

SUBJECTIVE MEASUREMENTS

	P₂ Measurements	L. yoderi			L. mammifer			L. speciosus			L. evansi		
		Min.	Mean	Max.	Min.	Mean	Max.	Min.	Mean	Max.	Min.	Mean	Max.
1	Lateral position of main protoconid ridge	1	2.7	4	1	1.8	4	1	1.3	4	1	3.8	6
2	Strength of ridge connection to entoconid	0	1.7	3	1	2.0	3	1	2.6	3	1	2.3	3
3	Size of lingual ridge behind protocone	2	3.8	5	2	4.1	5	4	4.9	5	0	3.5	5
4	Size of medial ridge behind protocone	0	2.1	4	0	1.6	5	0	0.8	4	0	3.2	5
5	Size of labial ridge behind protocone	3	3.9	4	3	4.0	4	3	4.0	5	0	1.8	5
6	Size of posterolingual cingulum	0	0.8	3	0	0.2	1	0	0.2	2	0	0.5	2
	P₃ Measurements												
7	Lateral position of main protoconid ridge	1	2.2	6	1	1.3	4	1	1.3	5	1	4.5	7
8	Strength of ridge connection to entoconid	1	2.6	3	3	3.0	3	2	3.0	3	2	3.0	3
9	Size of lingual ridge behind protocone	2	4.4	5	0	4.8	5	4	5.0	5	0	4.0	5
10	Size of medial ridge behind protocone	0	2.2	5	0	2.0	5	0	2.4	5	0	3.0	5
11	Size of labial ridge behind protocone	3	3.7	5	0	4.1	5	2	3.9	5	0	2.3	5
12	Size of posterolingual cingulum	0	1.3	4	0	0.5	3	0	0.9	3	0	0.9	2
13	Size of anterior fork	0	0.0	1	0	0.0	0	0	0.1	2	0	0.2	2
14	Size of accessory ridge 1	0	0.8	3	0	0.7	3	0	0.6	2	0	0.1	2
15	Size of accessory ridge 2	2	3.1	4	2	3.0	4	0	2.8	4	2	3.6	5
16	Size of accessory ridge 3	0	0.0	1	0	0.0	1	0	0.0	1	0	0.0	1
17	Size of accessory ridge 4	0	0.0	0	0	0.0	2	0	0.0	0	0	0.1	2
18	Size of accessory ridge 5	0	0.0	0	0	0.0	2	0	0.0	0	0	0.0	0
19	Size of accessory cusp 1	0	0.1	3	0	0.1	2	0	0.0	0	0	0.0	0
20	Size of accessory cusp 2	0	0.4	3	0	0.1	2	0	1.2	4	0	0.2	3

Table 6. Continued.

		SUBJECTIVE MEASUREMENTS										
	P₄ Measurements	*L. yoderi*			*L. mammifer*			*L. speciosus*			*L. evansi*		
	Max.	Min.	Mean	Max.	Min.	Mean	Max.	Min.	Mean	Max.	Min.	Mean
21	Size of anterior fork	0	0.2	3	0	0.2	4	0	0.2	3	0	0.5	2
22	Size of accessory ridge 1	0	0.2	3	0	0.2	3	0	0.0	2	0	0.0	2
23	Size of accessory ridge 2	0	0.0	2	0	0.1	1	0	0.1	2	0	0.2	2
24	Size of accessory ridge 3	0	0.3	2	0	0.8	4	0	0.4	4	0	0.8	3
25	Size of accessory ridge 4	0	0.9	4	0	1.8	4	0	2.2	4	0	0.7	5
26	Size of accessory ridge 5	0	0.2	3	0	1.0	4	0	0.1	2	0	0.1	2
27	Size of accessory cusp 1	0	0.5	3	0	0.4	3	0	0.0	1	0	0.2	3
28	Size of accessory cusp 2	0	0.0	2	0	0.1	3	0	0.0	1	0	0.0	0
29	Size of accessory cusp 3	0	0.0	3	0	0.0	1	0	0.0	0	0	0.0	1
30	Size of accessory cusp 4	0	0.6	4	0	0.9	4	0	0.7	4	0	1.1	4
31	Size of accessory cusp 5	0	0.0	2	0	0.0	0	0	0.0	1	0	0.0	1
32	Size of anterolingual cingulum	0	1.5	3	0	2.4	3	0	1.5	3	0	1.1	3
	M₁ Measurements												
33	Size of Paleomeryx fold	0	0.0	0	0	0.1	3	0	0.0	1	0	2.1	4
34	Size of labial accessory cusp	2	3.0	4	0	2.9	4	0	2.6	4	0	0.8	3
35	Size of post. accessory cusp or ridge	0	1.5	3	0	1.5	4	0	1.1	4	0	0.7	4
	M₂ Measurements												
36	Size of Paleomeryx fold	0	0.0	0	0	0.0	0	0	0.0	1	0	1.8	3
37	Size of labial accessory cusp	0	2.7	3	0	2.2	4	0	1.2	3	0	0.9	3
38	Size of post. accessory cusp or ridge	0	1.6	3	0	1.1	3	0	0.6	3	0	0.7	4
	M₃ Measurements												
39	Size of Paleomeryx fold	0	0.0	0	0	0.0	2	0	0.0	0	0	2.0	4
40	Size of anterior labial accessory cusp	0	2.3	3	0	1.4	3	0	0.8	3	0	1.1	3
41	Size of posterior labial accessory cusp	0	0.3	2	0	0.2	4	0	0.7	4	0	0.7	3
42	Shape of lingual valley on posterolophid	0	1.6	7	1	1.3	3	1	1.9	5	1	2.1	4
43	Depth of lingual valley on posterolophid	1	3.5	8	1	3.4	7	0	4.0	8	1	4.3	7
44	Classification of entoconulid		Y/M			Y/M			S/E			S/E	
	Tooth Row												
45	Wear stage	1	5.5	9	0	5.2	9	0	5.1	9	1	4.9	8

a. CUSP NAMES

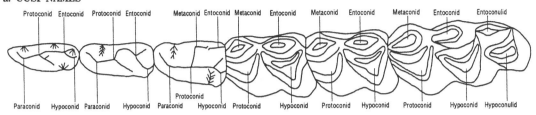

b. OBJECTIVE MEASUREMENTS   Lingual

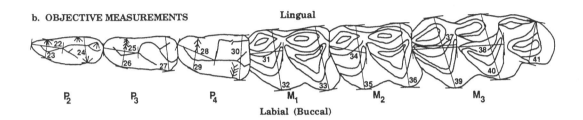

Labial (Buccal)

c. SUBJECTIVE MEASUREMENTS

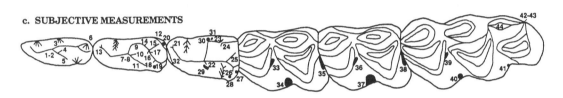

Figure 7. Three diagrams of the lower dentition of *Leptomeryx* in occlusal view: a. tooth cusps referred to in the text, b. objective measurements made with calipers, and c. subjective measurements (scores) given to various accessory cusps, ridges, and other variable features. See Table 6 for a list of the measurements illustrated and summary statistics.

Methods

In analyzing *Leptomeryx* from Flagstaff Rim, we first sought to measure the range of variation in each supposed species to consider the possibility that two or more species of similar size could be mixed. Specimens from all levels at Flagstaff Rim and samples from several other Chadronian and Orellan localities were examined to assess the total range of variation, and measurable characters were chosen to evaluate this variation quantitatively. Objective measurements chosen were the anteroposterior length and the anterior and posterior transverse width of each cheek tooth, the pre-protoconid length of the premolars, the length of the anterior selene of the molars, and the depth of the jaw below the tooth row in three places (Table 6, Figs. 7A-B). Variations within the occlusal surfaces of the teeth were harder to quantify, so a series of subjective scores were developed to facilitate this (Fig. 7C).

Six subjective measurements were made on P_2, with corresponding measurements on P_3. The first five refer to the configuration of one or more ridges that extend posteriorly from the protoconid (discussed below). The sixth measurement is the size of the cingulum on the posterolingual corner of the tooth. Eight additional measurements were made on P_3. The first refers to a

forking at the tip of the ridge extending anteriorly from the paraconid. The others refer to five accessory ridges and two accessory cusps found on some specimens. Twelve subjective measurements were made on P_4: the anterior fork (as in P_3), five accessory ridges, five accessory cusps, and an anterolingual cingulum (Table 6, Fig. 7C).

Three subjective measurements were made on M_1 and M_2: (1) development of the "*Palaeomeryx* fold," which extends posteriorly off the protoconid to meet the anterior extension of the hypoconid; (2) presence of an accessory cusp (referred to as a "buccal column" by Storer, 1981) in the valley between the protoconid and hypoconid on the labial margin of the tooth; and (3) presence of an accessory cusp or transverse ridge at the posterior end of the tooth. M_3 has a posterolophid behind the two selenes. Six subjective measurements were made on this tooth, two of which are development of the "*Palaeomeryx* fold" and presence of the labial accessory cusp as in M_1 and M_2. The third is the presence of an additional labial accessory cusp that is sometimes found in the valley between the hypoconid and the hypoconulid (Fig. 7C). The final three subjective measurements relate to the shape of the entoconulid and the valley posterior to it on the lingual

Table 7. Lateral position of largest ridge extending posteriorly from the protoconid of P_3 for ten samples of *Leptomeryx*. Some variations are illustrated in Figure 8. The sample from Quarry B at Flagstaff Rim is a mixture of *L. yoderi* and *L. speciosus*, most of which cannot be distinguished. The sample from below the PWL at Douglas is a species intermediate between *L. speciosus* and *L. evansi* (Table 12).

Locality			Flagstaff Rim				Pipestone		Douglas	
	Quar. A	Quar. B	Quarry C		Upper section		Springs		-PWL	+PWL
Species	*yoderi*	mix	*speciosus*	*mamm.*	*speciosus*	*mamm.*	*speciosus*	*mamm.*	int.	*evansi*
Mean	2.2	1.7	2.1	1.1	1.3	1.3	1.6	1.0	1.0	4.5
Lingual 1	42	18	3	48	39	60	13	5	4	4
2	29	6	2	8	3	9	1	0	0	3
3	26	6	1	0	1	5	0	0	0	0
Medial 4	19	1	0	0	1	2	1	0	0	7
5	1	0	1	0	1	0	0	0	0	12
6	1	0	0	0	0	0	1	0	0	7
Labial 7	0	0	0	0	0	0	0	0	0	3
Total	118	31	7	56	45	77	16	5	4	36

Table 8. Size of secondary ridges extending posteriorly from the protoconid of P_3 of *Leptomeryx*. Some variations are illustrated in Figure 8. Each P_3 was given a score for a labial, medial, and lingual ridge from 0 (absent) to 5 (large and connecting to entoconid and/or hypoconid). In this table, scores for two ridges are given for the sample in which the third ridge (lingual or medial) was given a score of 5. Only four samples were considered large enough for inclusion: *L. yoderi* (Y), *L. mammifer* (M), and *L. speciosus* (S) from Flagstaff Rim, and *L. evansi* (E) from Douglas, Wyoming.

		Lingual Ridge = 5								Medial Ridge = 5							
		Medial Ridge				Labial Ridge				Lingual Ridge				Labial Ridge			
Species		Y	M	S	E	Y	M	S	E	Y	M	S	E	Y	M	S	E
None 0		35	17	16	8	0	0	0	5	0	1	0	2	0	1	0	17
1		17	12	1	0	0	0	0	0	0	0	0	0	0	0	0	0
Small 2		10	17	1	0	0	0	1	0	0	0	0	0	0	0	0	2
3		4	9	2	0	19	2	13	2	12	2	0	3	7	2	1	1
4		0	6	19	1	47	49	22	3	10	3	1	12	15	5	2	1
Large 5		0	2	3	4	0	14	8	3	0	2	3	4	0	0	1	0
Total		66	63	42	13	66	65	44	13	22	8	4	21	22	8	4	21

side of the posterolophid (discussed below).

Dentaries from Flagstaff Rim (Chadronian) and Douglas (Orellan) were studied extensively prior to the selection of measurements in order to include all possible variations. Very few variants were found that could not be measured with this set of objective and subjective measurements. All measurements of accessory cusps, ridges, folds, cingula, and forks were scored with an integer from zero (feature absent) to five (feature exceptionally large or well-developed), though some did not exhibit this entire range of variation (Table 6). The only subjective measurements with a different scoring system are the position of the primary ridge behind the protoconid on P_2 and P_3, the last three

measurements on M_3, and the wear stage.

All our examinations of *Leptomeryx* teeth were conducted using a binocular microscope. Objective measurements were made using a single pair of fine-tip metal calipers. The bulk of material was measured at the Smithsonian's National Museum of Natural History, though some specimens were measured by Heaton at the University of South Dakota or on visits to the American Museum of National History in New York City.

Premolar Variations

The greatest morphological variation is exhibited in P_3. One, two, or three ridges extend posteriorly from

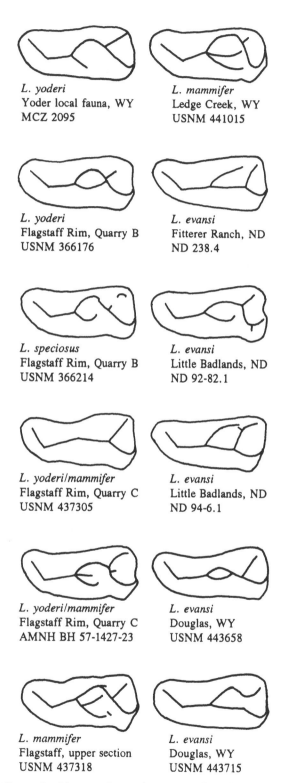

L. yoderi
Yoder local fauna, WY
MCZ 2095

L. mammifer
Ledge Creek, WY
USNM 441015

L. yoderi
Flagstaff Rim, Quarry B
USNM 366176

L. evansi
Fitterer Ranch, ND
ND 238.4

L. speciosus
Flagstaff Rim, Quarry B
USNM 366214

L. evansi
Little Badlands, ND
ND 92-82.1

L. yoderi/mammifer
Flagstaff Rim, Quarry C
USNM 437305

L. evansi
Little Badlands, ND
ND 94-6.1

L. yoderi/mammifer
Flagstaff Rim, Quarry C
AMNH BH 57-1427-23

L. evansi
Douglas, WY
USNM 443658

L. mammifer
Flagstaff, upper section
USNM 437318

L. evansi
Douglas, WY
USNM 443715

Figure 8. Diagram of ten *Leptomeryx* P_3s showing variability of the ridge(s) extending posteriorly from the protoconid. See Tables 7-8 for statistics concerning these ridges.

the protoconid and approach or connect with the entoconid and/or hypoconid (Fig. 8). In most cases a single ridge dominates, but the position of this ridge varies continuously from a lingual to a labial position on the tooth. We chose to score the position of the primary ridge from 1 (fully lingual) to 7 (fully labial), and we also assigned a score from 0 to 5 to the strength of the connection between the primary ridge and the entoconid (Table 6). Because as many as three ridges can be present, we also gave a size score from 0 to 5 to the lingual, medial, and labial ridges: 0, absent, to 5, large and connected to the entoconid and/or hypoconid. In specimens with fewer than three ridges the assignment was sometimes arbitrary. These two independent scoring systems helped us to assess the enormous variety of configurations found in the posterior ridges of P_3. P_2 was also scored using this system. It has the same ridge configuration as P_3 but exhibits much less variation.

This effort served only to document that the ridges on P_3 are highly variable in all the species studied and therefore have little taxonomic value. We base this conclusion on several observations: (1) roughly the same range of variation is found in all species studied; (2) this range of variation exhibits nearly every conceivable intermediate condition; and (3) the configuration of the ridges, within any sample, does not correlate well with any other characters on P_3 or any other tooth. Several variations in the P_3 ridges are shown in Figure 8.

Table 7 shows counts of specimens with the main ridge behind the P_3 protoconid in various lateral positions for samples from Flagstaff Rim, Pipestone Springs, and Douglas. The vast majority of Chadronian *Leptomeryx* have the primary ridge in a lingual position, whereas most Orellan specimens (*L. evansi*) have it in a medial to labial position. But the range of variation covers nearly the entire spectrum in most large samples with no indication of bimodality, so this is not a reliable character for distinguishing any two species. It is only useful in comparing species with large samples for evidence of relatedness. Of the ten samples compared in Table 7, the earliest (*L. yoderi*) and the latest (*L. evansi*, discussed later) exhibit the widest and most curious ranges of variation.

The primary P_3 ridge in *L. yoderi* from Quarry A very commonly occurs in a fully lingual to a fully medial position (Table 7), and this same variation is seen in other early Chadronian populations of *Leptomeryx* from the Yoder local fauna of Wyoming (Kihm, 1987), the Medicine Pole Hills local fauna of North Dakota (Pearson and Hoganson, 1995), and the Southfork local fauna of Saskatchewan (Storer, 1984). The medial position of the ridge becomes less common up section from Quarry A and is unusual in *L. speciosus* and *L. mammifer* from Flagstaff Rim, Ledge Creek, and Pipestone Springs (Table 7). In this character *L. speciosus* and *L. mammifer* are more similar to one another than either is to *L. yoderi*.

Table 7 considers only the most prominent ridge behind the protoconid of P_3, which in most cases is much larger than any others. In some cases two ridges connect the protoconid with the entoconid and/or hypoconid and are of nearly equal size, but only the larger one was considered for this measurement. Because as many as three ridges commonly extend posteriorly from the protoconid, separate scores were given for their sizes (Table 6). A presentation of these statistics comparable to that in Table 7 would not be useful because it would fail to account for the relationship among the ridges in individual specimens. Table 8, instead, shows statistics for two ridges based on samples for which the third ridge is fully connected to the entoconid (scored as 5). Only four samples were deemed large enough to be considered: *L. yoderi* from Quarry A, *L. speciosus* and *L. mammifer* from the upper section at Flagstaff Rim, and *L. evansi* (discussed below) from above the persistent white layer (PWL) at Douglas. These are the best tabular data we can provide to show the enormous individual variation in the ridges of P_3.

Though not included in Table 8, we have found that P_3s of *L. speciosus* from Pipestone Springs exhibit the same range of variation seen at Flagstaff Rim, including multiple ridges connecting the protoconid with the entoconid and/or hypoconid. This variation was noted by Tabrum and Fields (1980). The same is true for a population with a smaller mean size from Canyon Ferry, Montana (White, 1954; Table 2), where three of six P_3s have two connecting ridges. Clark (1937) noted variation in the size of the accessory ridges behind the P_3 protoconid in small *Leptomeryx* from the Chadronian of South Dakota. The large sample of *L. mammifer* from Ledge Creek exhibits the same variations seen at Flagstaff Rim, including multiple ridges in various configurations. Extreme variation in P_3 ridges is the rule rather than the exception in populations of *Leptomeryx*.

The Chadronian *Leptomeryx* species more commonly have a fully connected lingual ridge on P_3 while Orellan *L. evansi* more commonly has a fully connected medial ridge, and this is reflected in the totals at the bottom of Table 8. Although *L. yoderi* exhibits more variation than the other Chadronian populations in the position of the primary ridge (Table 7), it shows less variation in the additional ridges (Table 8). Not a single specimen from Quarry A at Flagstaff Rim has two ridges that connect the protoconid with the posterior end of the tooth, though three such specimens have been found in Quarry B, one of which (USNM 366176, Figure 8) is positively identified as *L. yoderi*. All Quarry A specimens with a fully connected lingual ridge have a large but non-connected labial ridge and, in some cases, a smaller medial ridge. All specimens with the connecting ridge in a more medial position have moderate-sized labial and lingual ridges. No bimodality

is exhibited in any of the *L. yoderi* columns of the table. Specimens with two connecting ridges are much more common in the other three species, as are bimodal distributions. This again suggests that *L. speciosus* and *L. mammifer* are more similar to each other than either one is to *L. yoderi*.

Leptomeryx evansi exhibits the greatest bimodality, with additional ridges being either absent or fairly large, so this species displays the greatest variability in both Table 7 and Table 8. As discussed below, *L. evansi* is easily recognized based on several characters and is most certainly a single species, so premolar variations should not be used to suggest that any of these samples represent coexisting species of the same size. What Table 8 shows, instead, is that for every specimen with a second ridge connecting the protoconid with the entoconid and/or hypoconid, there are many specimens in the population with a second ridge that is large and nearly makes such a connection. The only bimodality exhibited is in the presence/absence of secondary ridges, and since secondary ridges take on an almost infinite variety of shapes and orientations (Fig. 8), their absence in some specimens is hardly surprising.

This study demonstrates that previous attempts to use the ridges on P_3 for taxonomic distinction are completely invalid. Matthew (1903) stated:

> In the third lower premolar the protoconid has two posterior ridges, of which [in *Leptomeryx speciosus* and *L. mammifer*] the internal one connects with the heel, and the external one does not; while in *L. evansi* and the other species from the *Oreodon* and *Leptauchenia* Beds, the external ridge connects with the heel, and the internal one does not. In the lower jaw I have observed no entirely constant distinctions, except [this one] in P_3.

Matthew, using limited samples, failed to note that P_3 has from one to three ridges extending posteriorly from the protoconid and that the position of the connecting ridge (or ridges) can be found in a labial, medial, or lingual position in all of the species he studied. Although his observations are true for the majority of specimens, this is clearly not a distinguishing feature even between the rather distinct *Leptomeryx* species of the Chadronian and the Orellan.

Attempts have also been made to distinguish the various Chadronian species of *Leptomeryx* using premolars. Storer (1981, 1984) listed three possible characters to distinguish *L. yoderi* from *L. speciosus*: (1) small paraconid on P_2; (2) less prominent labial "heel" on the P_4 hypoconid; and (3) lack of an enclosed basin behind the P_3 protoconid (as is occasionally found in *L. speciosus* when two ridges connect the protoconid and entoconid). Kihm (1987), in his study of *L. yoderi* from the Yoder local fauna, dismissed the first two of Storer's distinctions as individual variation, which is

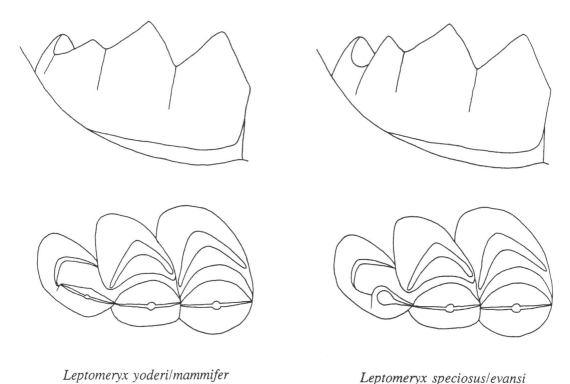

Leptomeryx yoderi/mammifer   *Leptomeryx speciosus/evansi*

Figure 9. Diagrams of *Leptomeryx* M$_3$s in lingual and occlusal view showing the different configuration of the entoconulid in *L. yoderi/mammifer* and *L. speciosus/evansi*. See Table 10 for counts of specimens with these conditions from several samples.

also confirmed by our sample, but he found the distinction in P$_3$ to be valid. We found the same to be true for the large sample from Quarry A at Flagstaff Rim as indicated by the lack of any *L. yoderi* specimens with two fully connected ridges as discussed above. However, one positively identified specimen (USNM 366176, Fig. 8) and two probable specimens (USNM 366201 and 366203) of *L. yoderi* from Quarry B have the enclosed basin. This distinction appears to be of little value, anyway, because the vast majority of *L. speciosus* P$_3$s also lack the enclosed basin between the two ridges (Table 8).

Kihm (1987) proposed two additional distinguishing characters for *L. yoderi*, bifurcation of the talonid crest on P$_2$ and presence of a labial cingulum on P$_3$, but as he suspected, larger samples demonstrate that these characters are variable in all species. We have found many premolar variations that have not been previously reported, and it is our firm opinion that no two species of *Leptomeryx* can be positively distinguished using characters of these teeth. Some character traits are found more commonly in one species than another, however, and these may be useful in tracing the evolution of certain lineages.

Molar Variations

Storer (1981, 1984) stated that an accessory labial

cusp (buccal column) in the valley between the protoconid and hypoconid of the molars is more common and larger in *Leptomeryx speciosus* than in *L. yoderi* or *L. mammifer*. We found the opposite to be true (Table 9), though the range of variation is roughly the same for each species measured. This cusp is always present on the M$_1$ of *L. yoderi* from Quarry A. A corresponding accessory cusp between the hypoconid and hypoconulid of M$_3$ is more common and generally larger in *L. speciosus* (and *L. evansi*), however (Table 9). As with premolar variations, large samples demonstrate that these characters of the molars are not species diagnostic.

Storer (1981) and Kihm (1987) provisionally accepted *Leptomeryx yoderi* as a species distinct from *L. speciosus*, though all of the differences they listed are actually individual variations. As can be seen in Figures 2-6 and Tables 3, 4, and 6, these two species have very similar size ranges. Our study was unable to identify any distinguishing character between the two except one that Emry discovered prior to our joint work (Emry, 1970; cited by Storer, 1984). This involves a difference in the configuration of a small cusp (entoconulid) and valley on the lingual side of the posterolophid of M$_3$ (Fig. 9; Table 10). In this character *L. yoderi* and the much larger *L. mammifer* are identical while *L. speciosus* is distinct from both. This

Table 9. Size of accessory cusps (buccal columns) on the labial margin of the tooth in the valley between the protoconid and hypoconid of the molars (hypoconid and hypoconulid of M_3 posterior; Figure 7) for four species of *Leptomeryx*: *L. yoderi* (Y), *L. mammifer* (M), and *L. speciosus* (S) from Flagstaff Rim, and *L. evansi* (E) from Douglas, Wyoming. Each cusp was given a score from 0 (absent) to 4 (large).

Species		M_1				M_2				M_3 Anterior				M_3 Posterior			
		Y	M	S	E	Y	M	S	E	Y	M	S	E	Y	M	S	E
None	0	0	6	10	64	3	19	93	54	5	54	103	28	126	120	87	30
	1	0	6	8	3	2	36	35	15	12	47	34	13	27	12	17	13
Small	2	19	12	37	11	62	63	43	12	105	54	38	16	11	7	20	4
	3	193	205	149	16	159	110	50	15	75	42	17	11	0	1	10	5
Large	4	12	18	3	0	0	1	0	0	0	0	0	0	0	1	1	0
Total		224	247	207	94	226	229	221	96	197	197	192	68	164	141	135	52

Table 10. Classification of *Leptomeryx* M_3s from Flagstaff Rim and several other localities that contain *L. yoderi* based on the configuration of the entoconulid as illustrated in Figure 9. The categories are: 1) *L. speciosus* configuration (S), 2) *L. yoderi* and *L. mammifer* configuration (Y/M), 3) intermediate or mixed characters of *L. speciosus* and *L. yoderi/mammifer* (I), and 4) odd configuration not matching either type (O). The two samples from the upper section at Flagstaff Rim include only USNM specimens. Size ranges (length of M_3 in mm) are given for the "S" and "Y-M" samples. The two "O" specimens from the Yoder local fauna are the two small specimens listed in Table 2.

M_3 Sample	S	Y-M	I	O	S (size)	Y-M (size)
Flagstaff Rim, Wyoming						
Upper section (large)	0	15	1	1		12.3-14.7
Upper section (small)	7	0	3	0	10.5-12.0	
Quarry C	9	88	1	3	9.9-11.2	10.9-14.5
Quarry B	3	37	1	1	10.3-11.2	10.7-13.2
Quarry A	0	132	3	2		10.5-13.0
Yoder, Wyoming	2	6	0	2	12.0-12.2	10.6-11.5
Raben Ranch, Nebraska	2	2	0	0	9.3-9.7	9.2-9.5
Medicine Pole, North Dakota	0	2	0	0		10.3-10.3
Southfork, Saskatchewan	0	2	0	0		9.2-11.6

similarity, together with populations of intermediate size from Quarries B and C at Flagstaff Rim (Figs. 2-4 and 6), demonstrate beyond reasonable doubt that *L. mammifer* is a direct descendant of *L. yoderi*.

In the *Leptomeryx yoderi/mammifer* lineage the entoconulid is long and narrow, and the anterior and posterior ridges extending from its peak slope downward at about the same angle when viewed lingually. In *L. speciosus* the entoconulid is much more rounded and pronounced, and it slopes steeply downward posteriorly (Fig. 9). We assigned subjective scores for the shape (V-shaped vs. U-shaped) and depth of the valley behind the entoconulid (Subjective Measurements 42 and 43 in Table 6), but these did not help in distinguishing the two lineages. Ultimately we simply assigned each M_3 to the lineage to which it belongs based on the shape of the entoconulid (Subjective Measurement 44, Table 6).

This distinguishing character on M_3 is not perfect because a few specimens exhibit an intermediate condition or an odd condition that cannot be classified.

There are also many specimens that cannot be classified due to excessive wear or damage to the posterolophid. Nevertheless, the distinctive shape of the M_3 entoconulid exhibits remarkable consistency in middle Chadronian populations where size can be used to identify species independently. The large sample of *Leptomeryx speciosus* from Pipestone Springs and the smaller sample from Canyon Ferry are identical to *L. speciosus* from Flagstaff Rim in this respect. For early Chadronian samples, this is the only character we have found useful for distinguishing the similar-sized *L. yoderi* and *L. speciosus*. It may also hold the clue to their relationship.

As can be seen in Table 10, the large sample from Quarry A at Flagstaff Rim consists entirely of *Leptomeryx yoderi*-type M_3s with the exception of three intermediate and two odd cases. Not a single specimen matches *L. speciosus*. This uniformity, combined with the normal size distribution at Quarry A (Fig. 6A), strongly suggests that only a single species is represented (*L. yoderi*). In the smaller sample from Quarry B, three M_3s match *L. speciosus*, and they are among the smallest specimens in this left-skewed sample (Fig. 6B). The same is true for Quarry C (Fig. 6C). Only in the upper part of the section at Flagstaff Rim are M_3s of *L. speciosus* and *L. mammifer* non-overlapping in size (Tables 4, 5, 10).

Kihm (1987) assigned all *Leptomeryx* from the Yoder fauna to *L. yoderi* except for two unusually small, low crowned specimens (Tables 2, 3). Based on this, he correlated the Yoder local fauna with the lower quarries at Flagstaff Rim. Kihm (1987) did not mention the M_3 posterolophid configuration of his sample, so this will be described here. The majority of M_3s from the Yoder local fauna, including the type specimen of *L. yoderi* (MCZ 2095), match the *L. yoderi/mammifer* entoconulid pattern of Figure 9 (SDSM 5343, 5346, 5348, 53295, 53304). The two smallest specimens (SDSM 8659 and 53300) do not match either configuration and may be a different species as Kihm (1987) noted. The two largest of Kihm's "*L. yoderi*" (SDSM 5345 and 8435) actually match *L. speciosus*. This suggests that Kihm's correlation between the Yoder local fauna and the Flagstaff Rim section may be suspect, especially when one considers that the localities are only 125 miles apart and that *Leptomeryx* must have been a highly mobile animal.

The absence of the small Yoder leptomerycid from all levels at Flagstaff Rim might suggest that the Yoder local fauna is older, predating the extinction of this species. But the joint occurrence of M_3s matching both *Leptomeryx yoderi* and *L. speciosus* correlates the Yoder local fauna only to Quarry B at Flagstaff Rim. (The large sample from Quarry A lacks *L. speciosus* whereas the *L. yoderi/mammifer* specimens from Quarry C are too big to match *L. yoderi* from the Yoder local fauna.) We considered the possibility that the Yoder local fauna contains a mixture of several ages, but this seems unlikely because (1) the fossils come from a relatively small area; (2) the fossils have excellent preservation; and (3) the three "morphs" are sometimes found together in small quarry samples.

The M_3s of *Leptomeryx speciosus* from the Yoder local fauna differ from those of Quarry B in being larger than *L. yoderi* specimens from the same localities (Table 10). Though it is statistically improbable, this could be due to sampling error as all of the Yoder M_3s fall within the size range of *L. speciosus* from the upper part of the Flagstaff Rim section (Table 4).

The situation becomes more complex when one considers the Raben Ranch local fauna of northwest Nebraska, 160 miles east of Flagstaff Rim and 75 miles NNE of Yoder. Ostrander (1980) recovered four isolated M_3s of "*Leptomeryx* sp." from Raben Ranch, two of which match *L. yoderi* (both SDSM 10191) and two of which match *L. speciosus* (SDSM 10193 and 10210) in the configuration of the M_3 entoconulid (Table 10). The odd thing is that all these teeth are relatively small

(M_3 length of 9.2 to 9.7 mm). The shortest M_3 from Flagstaff Rim is 9.6 mm long (from Quarry B), and the shortest M_3s from Pipestone Springs and Canyon Ferry are 9.5 mm long. In size the Raben Ranch specimens match the small Yoder species (Table 3), but in cusp height and other morphological characters they clearly match *L. yoderi* and *L. speciosus*. It is peculiar that three geographically close Chadronian localities are known to contain a mixture of the two M_3 types (Fig. 9), but that none of the three exhibits the same size distribution.

Localities in Saskatchewan and North Dakota also contain *Leptomeryx* with probable affinities to *L. yoderi*. Storer (1984) reported a dentary with P_3-M_3 (ROM 23207) and an isolated M_2 (SMNH P1276.3) of *L.* cf. *yoderi* from the Southfork local fauna of Saskatchewan, together with several smaller specimens that he referred to *L.* cf. *blacki* after a species from California named by Stock (1949). The single complete M_3 of each species (ROM 23207 and SMNH P1185.8, respectively) matches *L. yoderi* in entoconulid pattern (Table 10; Storer, 1984, fig. 2) as well as in crown height and other features. The smallest specimen that Storer referred to *L.* cf. *blacki* is no smaller than the Raben Ranch *Leptomeryx* and may represent the same species (i.e., a small variant of *L. yoderi* or a closely related species). We find no justification based on morphology or size range to regard the Southfork *Leptomeryx* as two different species. If a single species is represented, however, it would have a smaller mean size than *L. yoderi* of Wyoming.

Storer (1984) based his identification of *Leptomeryx* cf. *blacki* in part on the lack of a P_1-P_2 diastema in SMNH P1185.7. The position of P_1 is highly variable in all Flagstaff Rim species. For example, in the 31 specimens of *L. yoderi* from Quarry A in which the P_1-P_2 diastema can be measured, its length ranges from 2.7 to 6.5 mm with a mean value of 4.7 mm and a coefficient of variation of 18.3% (Emry, 1970). Because this character is so variable, we do not consider the lack of a P_1-P_2 diastema to be significant.

Leptomeryx has recently been discovered from the Medicine Pole Hills of southwestern North Dakota (Pearson and Hoganson, 1995). In size, the teeth match *L. yoderi* from the Yoder local fauna and from Quarry A at Flagstaff Rim (Table 10). Of four M_3s discovered so far, two (PTM 663 and 1525) match *L. yoderi* in entoconulid pattern, and the others are too heavily worn to be certain. The posterolophid of M_3 in both unworn specimens is smaller relative to the rest of the tooth than in *L. yoderi* and has a less developed entoconulid with a shallower valley behind it. The molars are very high crowned like *L. yoderi*, however. The Medicine Pole Hills *Leptomeryx* may represent an early grade of *L. yoderi* or possibly a new species.

Evolutionary Relationships

The fact that *Leptomeryx yoderi* and *L. mammifer* share the same M_3 posterolophid pattern, are distinguishable only by size, and show a gradual size increase up section at Flagstaff Rim suggests that they are chronospecies resulting from anagenetic evolution. Although Emry (1973) referred to the intermediate sample from Quarry C as an unnamed species, we have decided not to establish a species name for this sample. Such a species would differ from *L. yoderi* and *L. mammifer* only in its intermediate mean size, and there is so much overlap in the size ranges that this species could be identified only in samples large enough to define statistically. Therefore, we simply recognize the Quarry C form as a temporally restricted sample of the anagenetic continuum linking *L. yoderi* with *L. mammifer*.

The relationship of *Leptomeryx yoderi* to *L. speciosus* is more problematic. Only a single character (shape of the M_3 entoconulid, Fig. 9) reliably distinguishes between them, and that not in every case (Table 10). *Leptomeryx speciosus*, like *L. mammifer*, is abundant in the middle Chadronian, and the most likely early Chadronian ancestor for both species is *L. yoderi*. This raises the possibility that the Flagstaff Rim section illustrates a case of sympatric speciation in *Leptomeryx*. The principal questions are: (1) whether the Quarry B specimens with *L. speciosus* type M_3 entoconulids descended from the *L. yoderi* population represented by the Quarry A sample; and (2) whether the Quarry B sample represents two coexisting species (*L. yoderi* and *L. speciosus*) or a single species (*L. yoderi*) with newly acquired variation (incipient *L. speciosus*). An alternative explanation is that *L. speciosus* evolved elsewhere (still probably from *L. yoderi*) and migrated into the Flagstaff Rim region after the deposition of the Quarry A deposit.

We will not review the literature on speciation except to state that many biologists discount the possibility of sympatric speciation (Cracraft, 1989; Lynch, 1989). The division of an interbreeding population into two species would seem particularly unlikely in a prolific and mobile artiodactyl like *Leptomeryx*, though it might be possible for distinct herds to develop genetic differences while still occupying the same region. This would be a borderline case of allopatry/sympatry, sometimes called microallopatry. Genetically isolated sympatric populations, or sibling species, have been found in less mobile vertebrates such as salamanders (Larson, 1989). Such cases are usually attributed to habitat fragmentation (another form of microallopatry) which would be less likely to effect an artiodactyl.

The fact that the *Leptomeryx* sample from Quarry B, like the other samples shown in Table 10, contains few intermediate or odd M_3s compared to the number of easily-classified specimens argues for two distinct, coexisting species rather than a single, morphologically diverse species. The asymmetry of the size distribution

(Fig. 6) and the fact that all of the *L. speciosus* type M_3s are among the smallest specimens in the long left tail of that distribution (Table 10) also suggest the presence of two species. If the Quarry B and Yoder samples are combined (on the assumption of similar age), then M_3s of the two types have similar size ranges, but as stated above, correlation of the Yoder fauna with any level at Flagstaff Rim is problematic.

We do not know the duration of the hiatus between Quarry A and B at Flagstaff Rim because no datable ashes exist below Quarry C, so it is not possible to know the length of time available for evolutionary change to take place (Emry, 1973, 1992). *Leptomeryx yoderi* underwent only a slight increase in mean size between Quarry A and B, much less than between Quarry B and C or between Quarry C and the upper portion of the Flagstaff Rim section (Fig. 6). Sympatry of two similar-sized species may have driven or accelerated the increase in size of the *L. yoderi/mammifer* lineage, however. The only hope of unraveling this relationship is to compare the three Flagstaff Rim leptomerycids to see if *L. speciosus* shares any characters with *L. yoderi* or *L. mammifer* that the other lacks.

The most obvious similarity between *Leptomeryx yoderi* and *L. speciosus* is their size, although the latter is slightly smaller on average (Tables 3-4, Figure 6). If Quarry B contains two species that can be distinguished based on the shape of the the the M_3 entoconulid, as suspected, then a size comparison can be made between these two species and *L. yoderi* from Quarry A. As can be seen in Table 10, *L. yoderi* from Quarry B ranges to a larger size than in Quarry A, whereas *L. speciosus* ranges smaller than Quarry A *L. yoderi* (0.2 mm difference in length of M_3 in each case). These size range extensions are significant because the Quarry B sample contains so few specimens compared to the Quarry A sample. Above Quarry B, *L. speciosus* remains roughly the same size whereas the *L. yoderi/mammifer* lineage undergoes a marked increase in size. But between Quarries A and B the divergence is nearly symmetric. The only other unique similarity between *L. yoderi* (Quarry A) and *L. speciosus* (and *L. evansi*) is that some specimens have the main ridge behind the P_3 protoconid in a more labial than lingual position; no specimen of *L. mammifer* with this condition has yet been found (Table 7).

Leptomeryx speciosus and *L. mammifer* share several unique similarities. Unlike *L. yoderi* from Quarry A, both have relatively fewer P_3s with the main ridge in a medial position (Table 7), and both include P_3s with two ridges connecting the protoconid with the entoconid and/or hypoconid (Table 8). Both these characters first appear in Quarry B, so they could be shared, derived characters (though they are relatively uncommon in both species). Rarely *L. speciosus* and *L. mammifer* show some development of the *Palaeomeryx* fold in the molars (Table 11, discussed below), but no cases of this

Table 11. Degree of development of the "*Paleomeryx* fold" (Figure 10) on the molars of eight populations of *Leptomeryx* ranging from 0 (absent) to 4 (large). All specimens from Quarries A and B from Flagstaff Rim are combined, as are the two species from Quarry C. Specimens from the upper part of the section at Flagstaff Rim and from Pipestone Springs are separated by size into *L. speciosus* (S) and *L. mammifer* (M). Specimens from Douglas are separated into those below the persistent white layer (PWL) and those above it, which are *L. evansi* (E). Counts are shown separately for each molar and combined.

| | | Flagstaff Rim | | | | Pipestone Springs | | Douglas | |
| | | Quar. A/B | Quar. C | Upper section | | | | -PWL | +PWL |
Species		mix	mix	S	M	S	M	int	E
M₁									
None	0	206	157	197	227	21	4	12	15
	1	0	1	2	4	0	1	1	11
Small	2	0	0	0	2	0	1	1	10
	3	0	0	0	2	0	0	3	46
Large	4	0	0	0	0	0	0	0	1
Total		206	158	199	235	21	6	17	83
M₂									
None	0	212	156	227	231	25	2	18	18
	1	0	0	1	0	0	0	3	16
Small	2	0	0	0	0	0	0	1	25
	3	0	0	0	0	0	1	3	34
Large	4	0	0	0	0	0	0	0	0
Total		212	156	228	231	25	3	25	93
M₃									
None	0	189	141	199	206	26	3	16	12
	1	0	0	0	0	0	0	1	12
Small	2	0	0	0	1	0	0	0	10
	3	0	0	0	0	0	1	4	29
Large	4	0	0	0	0	0	0	1	5
Total		189	141	199	207	26	4	22	68
M₁₋₃ combined									
None	0	607	454	623	664	72	9	46	45
	1	0	1	3	4	0	1	5	39
Small	2	0	0	0	3	0	1	2	45
	3	0	0	0	2	0	2	10	109
Large	4	0	0	0	0	0	0	1	6
Total		607	455	626	673	72	13	64	244

have been found below Quarry C.

In contrast to these somewhat dubious similarities, *Leptomeryx yoderi* and *L. mammifer* share the same unique configuration of the M_3 entoconulid (Fig. 9, Table 10) in nearly every specimen. They also have a similar distribution pattern of accessory labial cusps in the molars (Table 9) and are less likely than *L. speciosus* to have a large medial ridge behind the protoconid of P_3 when a fully connected lingual ridge is

present (Table 8). The only significant difference between *L. yoderi* and *L. mammifer* is size, and at Flagstaff Rim the size transition is well documented. No comparable graded link has been found between *L. yoderi* and *L. speciosus*. Although *L. yoderi* remains the most likely ancestor for *L. speciosus*, it is uncertain whether *L. yoderi* speciated sympatrically or, alternatively, whether *L. speciosus* developed from *L. yoderi* or a related species in another location.

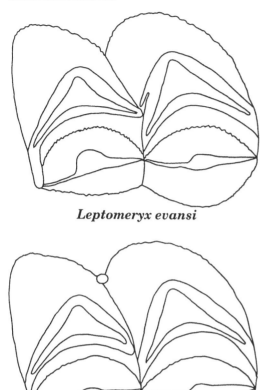

Leptomeryx evansi

Leptomeryx speciosus

Figure 10. Diagrams of *Leptomeryx* M₂s in occlusal view showing the development of the *Palaeomeryx* fold and the greater incidence of crenulations in *L. evansi* than in *L. speciosus*. Statistics concerning the *Palaeomeryx* fold can be found in Table 11. The accessory cusp in the labial valley is found more commonly in *L. speciosus* than in *L. evansi* (Table 9).

In conclusion, the dominant Chadronian species of *Leptomeryx* are closely related and represent a grade of evolution beyond their Duchesnean counterparts. Six stages in the evolution of Chadronian *Leptomeryx* seem to be indicated: (1) earliest Chadronian: *Leptomeryx* from the Medicine Pole Hills of North Dakota may represent an early grade of *L. yoderi* or its immediate ancestor; (2) middle early Chadronian: *L. yoderi* became a widespread species and is the only species found in Quarry A at Flagstaff Rim; (3) late early Chadronian: *Leptomeryx speciosus* appeared, probably a descendant of *L. yoderi*, and these two similar-sized species are found together in Quarry B; the Yoder local fauna, considered early Chadronian by Kihm (1987), and the Raben Ranch local fauna, considered middle Chadronian by Ostrander (1980), seem to best fit this stage, though the match is imperfect in both cases; (4) early middle Chadronian: *L. yoderi* increased in size, approaching *L. mammifer*, and is found with *L. speciosus* in Quarry C;

(5) middle middle to early late Chadronian: *L. mammifer* developed as a species distinctly larger than *L. speciosus*, and the two species became widespread and stable; both are found throughout the upper part of the Flagstaff Rim section, at Pipestone Springs, and in the Calf Creek local fauna of the Cypress Hills; and (6) latest Chadronian: *L. mammifer* became extinct, and *L. speciosus* underwent some modifications approaching the Orellan *L. evansi* grade (discussed below).

THE DEVELOPMENT OF *LEPTOMERYX EVANSI* FROM *LEPTOMERYX SPECIOSUS*

The highly fossiliferous Orellan deposits of the Great Plains contain large numbers of a small leptomerycid that includes the type specimen of the type species of the genus, *Leptomeryx evansi* Leidy (1853). Clark and Kietzke (1967) found *L. evansi* to be the most abundant animal in the open plains fauna of the South Dakota badlands. Emry et al. (1987) believed this species to be restricted to the Orellan, attributing reports in older and younger rocks to misidentifications.

Leptomeryx evansi is slightly smaller on average than *L. speciosus* and has a number of unique characters. First, the primary ridge behind the P_3 protoconid is usually more labial than lingual (Table 7), though as in the Chadronian species, there is considerable variation in this feature. It commonly has two ridges connecting the P_3 protoconid with the entoconid and/or hypoconid, one medial and one lingual (Table 8). Second, most specimens have a well-developed *Palaeomeryx* fold on each of the lower molars (Table 11). *Leptomeryx evansi* also tends to have more strongly developed vertical crenulations around the tooth crowns, especially on the lingual side of the lower teeth, though this is a variable feature. Crenulations on the lingual side of the metaconid and entoconid of the lower molars tend to be better developed in *L. evansi* than in other species (Fig. 10). In the critical feature of the M_3 posterolophid, *L. evansi* matches *L. speciosus* rather than *L. yoderi* and *L. mammifer* (Fig. 9). In fact the M_3 entoconulid tends to be even broader and more rounded posteriorly in *L. evansi* than in *L. speciosus*, or at least more consistently so.

We have not found any variation that would suggest that more than a single species of *Leptomeryx* existed in the Orellan of the Great Plains. The distinguishing characters are variable in their degree of development, just as in the Chadronian species, but the range of variation is generally continuous rather than bimodal, and the degree of development of the various characters on individuals is not well correlated. For example, there are specimens with the major ridge on P_3 in a labial position that have little or no development of the *Palaeomeryx* fold on the molars (USNM 443715), and there are specimens with the P_3 ridge in the lingual position with strongly developed *Palaeomeryx* folds (AMNH 127017). The *Palaeomeryx* fold can even be

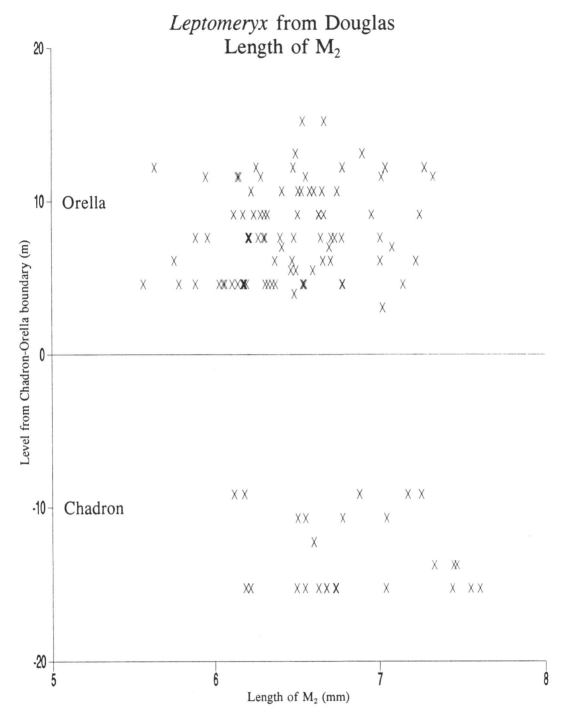

Figure 11. Bivariate plot of 116 *Leptomeryx* dentaries from Douglas, Wyoming, showing the length of M₂ vs. level in the stratigraphic section. The vertical axis is based on the Chadronian/Orellan boundary, marked by the Persistent White Layer (PWL). The Orellan specimens are *L. evansi* while the Chadronian specimens are a population showing characters intermediate between *L. speciosus* and *L. evansi*. See Table 12 for size statistics on these two stratigraphic populations.

Table 12. Tooth length measurements (mm) for two stratigraphically restricted populations of *Leptomeryx* from the Douglas area, Wyoming, and the type specimen of *L. evansi* from the Big Badlands of South Dakota. The population from above the persistent white layer (PWL) is *L. evansi*, whereas the population from below the PWL appears to be ancestral to that species. All measurements were made by Heaton.

Tooth	Douglas Below PWL			Douglas Above PWL			Leptomeryx evansi Type specimen
	Range	Mean	N	Range	Mean	N	value
P$_2$				4.6-5.5	5.0	12	
P$_3$	5.6-8.1	6.3	4	4.9-6.8	6.1	35	
P$_4$	5.4-7.0	6.2	15	4.7-7.0	6.0	66	6.2
M$_1$	5.6-7.4	6.3	21	4.7-6.9	6.0	90	6.5
M$_2$	6.1-7.6	6.8	26	5.6-7.3	6.5	90	6.9
M$_3$	9.9-11.7	10.7	11	8.7-11.6	10.1	52	10.0
M$_{1-3}$	22.3-23.9	23.3	5	20.1-25.5	22.3	36	23.2
P$_2$-M$_3$	38.7	38.7	1	36.6-41.1	38.6	6	

strongly developed on some molars of a jaw and entirely absent on adjacent ones (AMNH 127019, USNM 443664), though this is unusual. The type specimen of *L. evansi* (USNM 157) lacks the P$_3$ but has molars with prominent *Palaeomeryx* folds. Several other Orellan and Whitneyan species were named by Cook (1934) and Frick (1937). We have not examined the type specimens of these species but suspect that they are synonyms of *L. evansi*.

South and east of Douglas, Wyoming, about 90 miles east of Flagstaff Rim, White River deposition is apparently continuous from middle Chadronian through Orellan time (Evanoff et al., 1992). The Chadronian/Orellan transition is approximately at a volcanic ash bed referred to as the persistent white layer (PWL), "Glory Hole Ash," purplish-white ash, 100-foot white ash, or number 5 tuff (Evanoff et al., 1992, p. 117). Except for the uppermost part of the Chadron Formation, the entire section is fossiliferous (Kron, 1978; Evanoff et al., 1992).

The lower part of the Chadron Formation at Douglas is temporally equivalent to at least part of the upper portion of the section at Flagstaff Rim (Prothero and Swisher, 1992; Evanoff et al., 1992), though the exact correlation is uncertain due in part to the paucity of fossils in the uppermost levels at Flagstaff Rim. *Leptomeryx speciosus* appears to persist later at Flagstaff Rim than does *L. mammifer* (Figs. 2-4; see also Emry, 1992), and a population with a slightly smaller mean size than *L. speciosus* of Flagstaff Rim is found below the PWL at Douglas (Fig. 11, Table 12). The *L. evansi* population above the PWL at Douglas has a mean size that is slightly smaller yet, and the size ranges are similar. A comparison of these three successive populations suggests the strong possibility that *L. evansi* descended from *L. speciosus* in the latest Chadronian. This relationship has also been proposed by Prothero (1985a, b, and unpublished data).

Our study of Chadronian *Leptomeryx* has shown that the most diagnostic character for distinguishing species of similar size is the configuration of the M$_3$ entoconulid. Because the variant found in *L. speciosus* is a derived character that first appears in that species and is shared by *L. evansi*, a strong case can be made for an ancestor/descendant relationship. The few M$_3$s found below the PWL at Douglas all share this condition as well. Another similarity between the two species is found in the labial accessory cusps on the molars (Table 9). Both have less frequent or smaller cusps between the protoconid and hypoconid than are seen in *L. yoderi* and *L. mammifer* (being even more uncommon in *L. evansi*, especially on M$_1$), and both more commonly have an accessory cusp between the hypoconid and hypoconulid of M$_3$ (Table 9).

If the *Leptomeryx* population from below the PWL at Douglas is intermediate between *L. speciosus* from Flagstaff Rim and *L. evansi*, then an even stronger case can be made that these are chronospecies. Unfortunately this population is very small, but it does exhibit remarkable intermediacy. In mean size it is intermediate between Flagstaff Rim *L. speciosus* and *L. evansi* (Tables 4 and 12), though somewhat closer to *L. evansi*. Only four specimens include P$_3$, all of which have a fully lingual main ridge behind the protoconid as in *L. speciosus* (Table 7). In terms of the *Palaeomeryx* fold this population is perfectly intermediate, having the fold much more commonly than in *L. speciosus* but much less commonly than in *L. evansi* (Table 11). The folds even become more common higher in the Douglas section (less common 12 to 16 m below the PWL; more common 9 to 12 m below the PWL). Because three of the four P$_3$s come from the lowest level at Douglas (16 m below the PWL), the distinguishing characters of this tooth may also undergo gradual change. Evidence for such a conclusion comes from specimens of *L. evansi* found above the PWL at Douglas. Those P$_3$s that have the dominant ridge in the lingual position (as in *L. speciosus*) occur more frequently near the Chadronian/Orellan boundary than they do higher in the Douglas section.

Assigning the population from below the PWL at Douglas to a species would be a judgment call between *Leptomeryx speciosus* and *L. evansi*. Taxonomically the same sort of dilemma is faced as with the *L. yoderi/mammifer* population from Quarry C at Flagstaff Rim. We therefore conclude that *L. speciosus* and *L. evansi* are chronospecies exhibiting gradual anagenesis across the Chadronian/Orellan boundary.

Chadronian *Leptomeryx* resembling *L. evansi*

Before leaving the topic of *L. evansi*, we need to address some unusual leptomerycid specimens from the Pipestone Springs area that resemble this species in several characters. If *L. evansi* or a likely precursor distinct from *L. speciosus* could be found in the early or middle Chadronian, then the genealogy presented above would be undermined.

MPUM 3296 is a dentary from Little Pipestone Creek (near Pipestone Springs) that is as large as the largest *L. mammifer* specimens from Flagstaff Rim and exhibits several *L. evansi* features: (1) the main ridge behind the P_3 protoconid is labial rather than lingual; (2) the *Palaeomeryx* fold is moderately developed on the molars; (3) the enamel is strongly crenulated (though no more than in some *L. mammifer* specimens); and (4) the M_3 posterolophid matches *L. speciosus* and *L. evansi* more closely than *L. mammifer*. This specimen differs from *L. evansi* and the other species of *Leptomeryx*, however, in lacking a deep and relatively wide depression in the posterolophid of M_3. MPUM 3296 also has an unusual accessory ridge extending posteriorly from the metaconid on M_1.

USNM 19001, a large M_3 from Pipestone Springs, has stronger development of the *Palaeomeryx* fold than any M_3 from Flagstaff Rim and has strongly crenulated enamel. The entoconulid of USNM 19001 has a rounded posterior edge as in *L. speciosus* and *L. evansi*, but unlike all other *Leptomeryx* specimens we have examined, the entoconulid has a small ridge projecting into the medial basin toward the hypoconulid. So while the posterolophids of these molars match *L. evansi* in some respects, in other ways they are entirely unique.

The P_3 ridge configuration on MPUM 3296 and the crenulated enamel on both specimens can be dismissed as individual variations in *L. mammifer*, as can the *Palaeomeryx* fold because these are occasionally found on *L. mammifer* molars from Flagstaff Rim (Tables 7 and 11). The M_3 entoconulid is more problematic, as is its presence in specimens that possess several other *L. evansi* characters. There are three possible explanations: (1) these large specimens are merely unusual variants of *L. mammifer*, (2) there is a second large species of *Leptomeryx* from the Pipestone Springs area which has developed features paralleling *L. evansi*; or (3) this species shares a unique common ancestry with *L. evansi* independent of *L. speciosus* and *L. mammifer*. We

view the third possibility as very unlikely. *Leptomeryx evansi* has never been found in the Chadronian, and we have provided substantial evidence that it evolved from *L. speciosus* across the Chadronian/Orellan boundary. The *L. evansi* features seen in MUPM 3296 and USNM 19001 can all be found as variants in *L. speciosus* and/or *L. mammifer* (Tables 8-11). The vast size difference and lack of intermediate populations argue against a close relationship with *L. evansi*.

The sample of large *Leptomeryx* specimens from Pipestone Springs and other Montana localities is very small compared with the sample from Flagstaff Rim (see Table 5, though not all Pipestone Springs material is included). The presence of greater morphological variation in this much smaller sample is perhaps the strongest evidence that more than one species is represented in the Montana material. However, larger samples are needed to determine for certain whether these specimens with unusual combinations of characters comprise a new species or merely represent individual or geographic variations in *L. mammifer*.

SUMMARY AND CONCLUSIONS

The leptomerycids, a cohesive group of about seven species, first appear in North America in the Duchesnean, become abundant in the Chadronian and Orellan, are less common in the Whitneyan (Cook, 1934; Frick, 1937), and are present but rare in the Arikareean (Taylor and Webb, 1976). Individual variation is often more pronounced than differences between species, and this has long caused taxonomic confusion. Our study of large samples from Flagstaff Rim, Wyoming, has succeeded in determining which characters are useful for identification of species, and this information can now be applied to smaller samples. Most important, we have documented that premolar ridge configurations, especially on P_3, are highly variable in each species studied and therefore offer little help in identification, their prominence notwithstanding.

The early evolution of the group is obscured by small sample sizes and a predominance of isolated teeth rather than full dentitions. Duchesnean forms, called *Hendryomeryx* by some authors, exhibit a slightly lower-crowned condition than their Chadronian and Orellan counterparts. These lower-crowned forms appear to extend into the Chadronian in the Cypress Hills, Saskatchewan (Storer, 1981) and at Yoder, Wyoming (Kihm, 1987).

Leptomeryx yoderi became the dominant species of the early Chadronian and is the only species found in the enormous sample from Quarry A at Flagstaff Rim. It is slightly larger than its predecessors and has higher-crowned, steeper-walled teeth. This species underwent a gradual size increase and evolved into *L. mammifer* of the middle Chadronian. This anagenetic shift is well documented at Flagstaff Rim. Prior to this increase in size, *L. yoderi* apparently gave rise to *L. speciosus*, a species of similar size that can only be distinguished from it by a change in the shape of the entoconulid on the last molar. *Leptomeryx speciosus* and *L. mammifer* are found together in middle Chadronian deposits from Wyoming to Saskatchewan. Large samples exhibit clear bimodality in size, so that individual specimens can be distinguished by size as well as by M_3 entoconulid configuration.

Leptomeryx mammifer became extinct before the end of the Chadronian, leaving *L. speciosus* as the only known surviving species of the genus. *Leptomeryx speciosus* underwent a size reduction at the end of the Chadronian and gradually evolved into *L. evansi*, apparently the only *Leptomeryx* species of the Orellan. This anagenetic transition is documented by an intermediate population from the latest Chadronian of Douglas, Wyoming, where *L. evansi* characters become more common stratigraphically higher in the section. The most distinctive characters of *L. evansi*, the labial ridge behind the protoconid on P_3 and *Palaeomeryx* fold on the molars, are only rarely found in *L. speciosus* and almost never in the same specimens.

Changes in *Leptomeryx* across the Duchesnean/ Chadronian boundary can be accounted for by normal evolutionary adaptation. *Leptomeryx yoderi*, being larger and possessing higher-crowned teeth than its contemporaries, may well have out-competed them. The Chadronian/ Orellan transition, however, is marked by the extinction of a large species and the size reduction and morphological modification of a smaller species. No new predators or competitors appear at this boundary, so these changes must be attributable to climatic factors. Evanoff et al. (1992) provide evidence for a fairly rapid change from moist subtropical to semiarid warm temperate climate at the Chadronian/ Orellan boundary. The standardization in *L. evansi* of characters that are rare in its predecessors, as well as the decrease in size, suggests strong selection pressure, a population bottleneck, or both. Nevertheless, *L. evansi* survived this transition to become one of the most common and widespread species of the Orellan.

ACKNOWLEDGMENTS

We thank Robert W. Purdy of the Smithsonian Institution, Richard H. Tedford of the American Museum of Natural History, James E. Martin and Janet L. Whitmore of the South Dakota School of Mines and Technology, Charles R. Schaff of the Harvard University Museum of Comparative Zoology, John W. Hoganson of the North Dakota Geological Survey, and Dean A. Pearson of the Pioneer Trails Museum (Bowman, North Dakota) for providing us with *Leptomeryx* specimens and associated data in their care. John E. Storer and Alan R. Tabrum provided helpful critiques of the manuscript even though they didn't always agree with our conclusions. Kay L. Kassube assisted with the illustrations.

Partial funding for this project was provided by a Smithsonian Institution postdoctoral fellowship to the senior author.

LITERATURE CITED

Black, C. C. 1978. Paleontology and geology of the Badwater Creek Area, central Wyoming. Part 14. The Artiodactyls. Annals of Carnegie Museum 47:223-259.

Clark, J. 1937. The stratigraphy and paleontology of the Chadron Formation in the Big Badlands of South Dakota. Annals of Carnegie Museum 25:261-350.

Clark, J., and K. K. Kietzke. 1967. Paleoecology of the Lower Nodular Zone, Brule Formation, in the Big Badlands of South Dakota; pp. 111-137 *in* J. Clark, J. R. Beerbower, and K. K. Kietzke (eds.), Oligocene Sedimentation, Stratigraphy, Paleoecology and Paleoclimatology of the Big Badlands of South Dakota. Fieldiana: Geology Memoir 5:1-158.

Cook, H. J. 1934. New artiodactyls from the Oligocene and lower Miocene of Nebraska. American Midland Naturalist 15:148-165.

Cope, E. D. 1885. The Vertebrata of the Swift Current Creek region of the Cypress Hills. Annual Report of the Geological and Natural History Survey of Canada 1:79-85.

Cope, E. D. 1889. The Vertebrata of the Swift Current River. American Naturalist, 23:151-155.

Cracraft, J. 1989. Speciation and its ontogeny: the empirical consequences of alternative species concepts for understanding patterns and processes of differentiation; pp. 28-59 *in* D. Otte and J. A. Endler (eds.), Speciation and its Consequences. Sinauer Associates, Sunderland, MA.

Emry, R. J. 1970. Stratigraphy and Paleontology of the Flagstaff Rim Area, Natrona County, Wyoming. Ph.D. dissertation, Columbia University, New York, 195 pp.

Emry, R. J. 1973. Stratigraphy and preliminary biostratigraphy of the Flagstaff Rim area, Natrona County, Wyoming. Smithsonian Contributions to Paleobiology 18:1-43.

Emry, R. J. 1992. Mammalian range zones in the Chadronian White River Formation at Flagstaff Rim, Wyoming; pp. 106-115 *in* D. R. Prothero and W. A. Berggren (eds.), Eocene-Oligocene Climatic and Biotic Evolution. Princeton University Press, Princeton, N. J.

Emry, R. J., P. R. Bjork, and L. S. Russell. 1987. The Chadronian, Orellan, and Whitneyan Land Mammal Ages; pp. 119-152 *in* M. O. Woodburne (ed.), Cenozoic Mammals of North America: Geochronology and Biostratigraphy. University of California Press, Berkeley.

Evanoff, E., D. R. Prothero and R. H. Lander. 1992. Eocene-Oligocene climatic change in North America: the White River Formation near Douglas, east-central Wyoming; pp. 116-130 *in* D. R. Prothero and W. A. Berggren (eds.), Eocene-Oligocene Climatic and Biotic Evolution. Princeton University Press, Princeton, NJ.

Frick, C. 1937. Horned ruminants of North America. American Museum of Natural History Bulletin 69:1-669.

Heaton, T. H. 1989. Cladogenesis in a lineage of *Leptomeryx* (Artiodactyla, Mammalia) from the Chadronian of Flagstaff Rim, Natrona County, Wyoming (Abstract). Journal of Vertebrate Paleontology 9:24A-25A.

Kihm, A. J. 1987. Mammalian paleontology and geology of the Yoder Member, Chadron Formation, east-central Wyoming. Dakoterra, 3:28-45.

Kron, D. G. 1978. Oligocene vertebrate paleontology of the Dilts Ranch area, Converse County, Wyoming. M.S.

thesis, University of Wyoming, Laramie, 185 pp.

Lambe, L. M. 1908. The Vertebrata of the Oligocene of the Cypress Hills, Saskatchewan. Contributions to Canadian Paleontology 3:1-65.

Larson, A. 1989. The relationship between speciation and morphological evolution; pp. 579-598 in D. Otte and J. A. Endler (eds.) Speciation and its Consequences. Sinauer Associates, Sunderland, MA.

Leidy, J. 1853. Remarks on a collection of fossil Mammalia from Nebraska. Proceedings of the Academy of Natural Sciences of Philadelphia 6:392-394.

Lynch, J. D. 1989. The gauge of speciation: on the frequencies of modes of speciation; pp. 527-553 in D. Otte and J. A. Endler (eds.), Speciation and its Consequences. Sinauer Associates, Sunderland, MA.

Matthew, W. D. 1903. The fauna of the *Titanotherium* beds of Pipestone Springs, Montana. Bulletin of the American Museum of Natural History 19:197-226.

Ostrander, G. 1980. Mammalia of the early Oligocene (Chadronian) Raben Ranch local fauna. M.S. thesis, South Dakota School of Mines and Technology, Rapid City, 288 pp.

Pearson, D. A., and J. W. Hoganson. 1995. The Medicine Pole Hills local fauna: Chadron Formation (Eocene: Chadronian), Bowman County, North Dakota. Proceedings of the North Dakota Academy of Science 49:65.

Prothero, D. R. 1985a. Mid-Oligocene extinction event in North American land mammals. Science 229:550-551.

Prothero, D. R. 1985b. North American mammalian diversity and Eocene-Oligocene extinctions. Paleobiology 11(4):389-405.

Prothero, D. R., and C. C. Swisher III. 1992. Magneto-stratigraphy and geochronology of the terrestrial Eocene-Oligocene transition in North America; pp. 46-73 in D. R. Prothero and W. A. Berggren (eds.), Eocene-Oligocene Climatic and Biotic Evolution. Princeton University Press, Princeton, N. J.

Schlaikjer, E. M. 1935. Contributions to the stratigraphy and paleontology of the Goshen Hole Area, Wyoming. III. A new basal Oligocene formation. Bulletin of the Museum of Comparative Zoology 76:71-93.

Stock, C. 1949. Mammalian fauna from the Titus Canyon Formation, California. Contributions to Paleontology 8:231-244.

Storer, J. E. 1981. Leptomerycid Artiodactyla of the Calf Creek Local Fauna (Cypress Hills Formation, Oligocene, Chadronian), Saskatchewan. Saskatchewan Museum of Natural History Contributions 3:1-32.

Storer, J. E. 1984. Fossil mammals of the Southfork local fauna (early Chadronian) of Saskatchewan. Canadian Journal of Earth Sciences 21:1400-1405.

Tabrum, A. R., and R. W. Fields. 1980. Revised mammalian faunal list for the Pipestone Springs local fauna (Chadronian, early Oligocene), Jefferson County, Montana. Northwest Geology 9:45-51.

Taylor, B. E., and S. D. Webb. 1976. Miocene Leptomerycidae (Artiodactyla, Ruminantia) and their relationships. American Museum Novitates 2596:1-22.

White, T. E. 1954. Preliminary analysis of the fossil vertebrates of the Canyon Ferry Reservoir area. Proceedings of the U.S. National Museum 103:395-438.

Wilson, J. A. 1974. Early Tertiary vertebrate faunas, Vieja Group and Buck Hill Group, Trans-Pecos Texas: Protoceratidae, Camelidae, Hypertragulidae. Texas Memorial Museum Bulletin 23:1-34.

28. Camelidae

DONALD R. PROTHERO

ABSTRACT

Our understanding of the early evolution of the Camelidae has long been confused by poor specimens, bad taxonomy, and dubious methodology, but the excellent specimens in the Frick Collection have clarified much of the confusion. The earliest known camelid is *Poebrodon*, from the early and late Uintan of Utah, Wyoming, and California. *Hidrosotherium* is not a camel at all, but a leptomerycid. Four valid species of *Poebrotherium* are recognized: *P. chadronense*, n. sp., and *P. franki* from the Chadronian of Texas; *P. eximium* from the Chadronian and early Orellan of the High Plains; and *P. wilsoni*, from the Chadronian to Whitneyan of the High Plains. "*Poebrotherium*" *labiatum*, from the early Orellan, is transferred to *Paratylopus*, which previously included only the type species, the Whitneyan camel *Paratylopus primaevus*. The long misunderstood camels known as "*Protomeryx cedrensis*," "*Protomeryx campester*," and "*Paralabis matthewi*" are now combined as *Paralabis cedrensis*. The highly specialized, extremely hypsodont, gazelle-like stenomyline camels of the late Arikareean-Barstovian have long been phylogenetically isolated, but Frick specimens show that *Pseudolabis dakotensis* (from the Whitneyan-Arikareean) and *Miotylopus* (from the Arikareean) are sister-taxa to the Stenomylini. *Miotylopus* includes three species: the small, *Stenomylus*-like *M. leonardi*; the medium-sized *M. gibbi* (including *Dyseotylopus*); and a large new species, *M. taylori*. *Gentilicamelus sternbergi* is the only valid species of this wastebasket genus, and is the sister-taxon of *Nothokemas*. The bizarre, long-snouted floridatragulines are closely related to higher camels.

INTRODUCTION

The first fossil vertebrate described from the western United States was the Oligocene camel *Poebrotherium wilsoni* (Leidy, 1847). Since that time, many more Eocene and Oligocene camels have been collected and described, particularly from the White River Group. Despite the abundance of specimens, early camel evolution remains poorly understood. Their taxonomy is one of the most confused among fossil mammals, with numerous invalid, misassigned, and "wastebasket" taxa (Table 1). This taxonomic confusion has made it necessary to reassign nearly every taxon in the group, and to resurrect several taxa that had been unjustly forgotten. Several new taxa are also named and described

below. Some of this taxonomic confusion can be attributed to the poor quality of the material, but much is due to inaccurate descriptions, incompetence, or bad methodology. The Oligocene Camelidae demonstrate how excessive reliance on stratigraphic sequence, "ancestor worship," and primitive characters can completely obscure an hierarchical pattern of relationships. The new material in the Frick Collection made it possible to incorporate characters of the facial region, the basicranium, the auditory bulla, and the skull sutures into a phylogenetic analysis, and reduce the over-reliance on teeth and metapodials.

Because the systematic papers in this book focus on the White River Chronofauna, this paper covers only the Camelidae (predominantly middle Eocene to late Oligocene) that are primitive sister-taxa to the clade that includes miolabines, protolabines, *Oxydactylus sensu stricto* and higher camels (Fig. 1). These higher camels first appear in the Harrison Formation, which is probably early Miocene in age (Honey et al., in press). This provides a convenient cut-off point at the base of a monophyletic group. The relationships of Miocene through Recent camels are also indicated in Figure 1, but their systematics will require much further work (currently under study by J. Honey; see Honey et al., in press).

ABBREVIATIONS

AC, Amherst College Museum, Amherst, Massachusetts; AMNH, Department of Vertebrate Paleontology, American Museum of Natural History, New York; ANSP, Academy of Natural Sciences, Philadelphia, Pennsylvania; CM, Carnegie Museum of Natural History, Pittsburgh, Pennsylvania; F:AM, Frick Collection, American Museum of Natural History, New York; KU, University of Kansas Museum of Natural History, Lawrence, Kansas; LACM(CIT), California Institute of Technology collection, now at the Los Angeles County Museum of Natural History; MCZ, Museum of Comparative Zoology, Harvard University, Cambridge, Massachusetts; SDSM, South Dakota

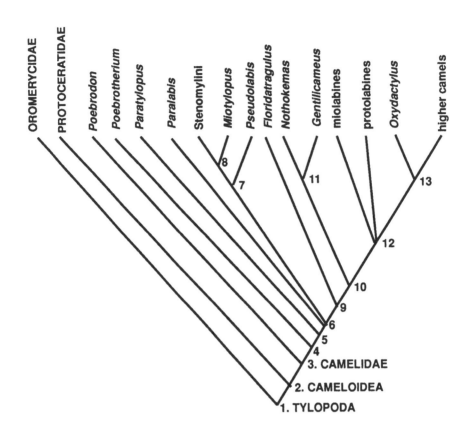

Figure 1. Phylogenetic relationships of the primitive Camelidae and their sister taxa (modified from Honey et al., in press). Character states as follows: 1) Tylopoda: camel condyle; metatarsals II, V reduced to nodules; 2) Cameloidea: long slender metapodials; fibula incomplete, becomes malleolar bone; wing-like basioccipital processes covering hypoglossal foramen; vertebrarterial canal passes through transverse processes of anterior cervical vertebrae; 3) Family Camelidae: highly hypsodont, dorsoventrally narrow upper molars, with straight ectolophs, and fossettes that close with wear; 4) higher-crowned teeth; increased size; M2 wide posteriorly; lingual hypoconulid lobe of m3 reduced; lingual entoconid and metaconid more flattened; hooked angular process; inflated auditory bulla with deep tympanohyal groove, filled with spongy bone; metacarpals II and V reduced to nodules; divergent distal metapodials; 5) reduced inner lobe of bulla; reduced premolars; 6) stronger sagittal crest; smaller lacrimal vacuity; posterior extension of inner portion of bulla; 7) Stenomylinae: premaxillary extended posteriorly; longer rostrum; deep, elongate maxillary fossa; elongated upper molars, which are laterally compressed and more hypsodont; weak mesostyles; partially fused metapodials; closed orbit; 8) highly reduced p2-3; 9) lower molars with metastylids, formed by overlap of metaconid on entoconid; heavier premaxilla; large postglenoid foramen; diastemal crest on mandible; 10) lower incisors more spatulate; P1 shortened; P3 internal cingulum stronger; orbits closed; 11) mandibular angle enlarged; p1 roots closely appressed; 12) medial plate of auditory bulla transversely compressed; rostrum lengthened; P2-3 reduced; molars more hypsodont; upper molars with weaker ribs; 13) cervical vertebrae elongate; metapodials slender, elongate, longer than basal length of skull; metatarsals and metacarpals approximately equal in length.

School of Mines Museum of Geology, Rapid City, South Dakota; TMM, Texas Memorial Museum, University of Texas, Austin, Texas; YPM, Yale Peabody Museum, New Haven, Connecticut; YPM-PU, Princeton University collection, now stored at the YPM.

SYSTEMATIC PALEONTOLOGY
Class MAMMALIA Linnaeus, 1758
Order ARTIODACTYLA Owen, 1848
Suborder TYLOPODA Illiger, 1811

Superfamily CAMELOIDEA Gill, 1872
Family CAMELIDAE Gray, 1821

Revised Diagnosis—The Family Camelidae is united by the following shared characters that distinguish it from the remaining Tylopoda, and all other artiodactyls: fully selenodont, transversely compressed upper molars that are relatively high-crowned (compared to other primitive Tylopoda), with straight ectolophs, and with fossettes that are closed

Table 1. Chronological list of camelid names discussed in this study.

NAME	AUTHOR	PRESENT REFERENCE	COMMENT/LAST REVISER
Poebrotherium wilsoni	Leidy, 1847	*Poebrotherium wilsoni*	Type species of genus
Protomeryx halli	Leidy, 1856	indeteminate	*nomen dubium*
Poebrotherium sternbergi	Cope, 1879	*Gentilicamelus sternbergi*	Only species of genus
Poebrotherium labiatum	Cope, 1881	*Paratylopus labiatus*	this paper
Gomphotherium	Cope, 1886	preoccupied	McKenna, 1966, note 1
Gomphotherium cameloides	Wortman, 1898	oxydactyline	McKenna, 1966
Gomphotherium serus	Douglass, 1900	oxydactyline	this paper
Protomeryx campester	Matthew, 1901	*Paralabis cedrensis*	*lapsus calami*
Protomeryx cedrensis	Matthew, 1901	*Paralabis cedrensis*	this paper
Poebrotherium eximium	Hay, 1902	*Poebrotherium eximium*	this paper
Oxydactylus longipes	Peterson, 1904	*Oxydactylus longipes*	Type species of genus
Miolabis (Paratylopus) primaevus	Matthew, 1904	*Paratylopus primaevus*	Type species of genus
Protomeryx leonardi	Loomis, 1911	*Miotylopus leonardi*	this paper
Oxydactylus gibbi	Loomis, 1911	*Miotylopus gibbi*	McKenna and Love, 1972
Poebrotherium andersoni	Troxell, 1917	*Poebrotherium wilsoni*	this paper
Pseudolabis (Paralabis) matthewi	Lull, 1921	*Paralabis cedrensis*	this paper
Paratylopus wortmani	Lull, 1921	oxydactyline	McKenna, 1966
Miotylopus bathygnathus	Schlaikjer, 1935	*Miotylopus gibbi*	McKenna and Love, 1972
Dyseotylopus migrans	Stock, 1935	*Miotylopus gibbi*	this paper
Gentilicamelus wyomingensis	Loomis, 1936	juvenile oxydactyline	McKenna, 1966
Gentilicamelus campestris	Loomis, 1936	*Paralabis cedrensis*	this paper
Gentilicamelus cederensis [sic]	Loomis, 1936	*Paralabis cedrensis*	*lapsus calami*
Miotylopus brachygnathus [sic]	McKenna, 1966	*Miotylopus gibbi*	*lapsus calami*
Paratylopus matthewi	McKenna and Love, 1972	*Paralabis cedrensis*	this paper
Miotylopus gibbi	McKenna and Love, 1972	*Miotylopus leonardi*	(in part)
Poebrotherium franki	Wilson, 1974	*Poebrotherium franki*	this paper
Hidrosotherium transpecosensis	Wilson, 1974	*Hendryomeryx defordi*	this paper
Miotylopus wilsoni	Dalquest and Mooser, 1974	floridatraguline	Stevens, 1977

anteriorly and posteriorly after moderate wear. Relatively long, narrow rostrum. Postorbital processes well developed, and nearly or completely closed. Inflated tympanic bulla filled with cancellous bone and indented by a deep tympanohyal groove. Angular process on mandible with a distinctive dorsal "hook." Long, unfused to fused, distally divergent middle metapodials with metacarpals II and IV and metatarsals II and V reduced to nodules. Metatarsals III and IV have flattened dorsal surfaces and their distal keels do not extend to the dorsal side. Fibular facet on the calcaneum has a proximal convexity and a dorsal concavity.

Poebrodon Gazin, 1955

Type Species—*Poebrodon kayi* Gazin, 1955
Included Species—The type and *Poebrodon californicus* Golz, 1976
Range—Washakie Formation, Adobe Town Member unit B, Washakie Basin, Wyoming (McCarroll et al., this volume, Chapter 2) (early Uintan); Myton Pocket, Uinta Formation C, Uinta Basin, Utah; Laguna Riviera Quarry, Santiago Formation, San Diego Co., California (late Uintan).

Discussion—The known material of *Poebrodon* was described by Gazin (1955) and Golz (1976). No additional material of *Poebrodon* has since been reported, except for the specimens mentioned by McCarroll et al. (this volume, Chapter 2), which will be described by them.

Wilson (1974, p. 24) placed *Poebrodon* in its own subfamily, the Poebrodoninae, along with his new taxon, "*Hidrosotherium*," because he felt that *Poebrodon* could not be ancestral to *Poebrotherium*. He argued that some primitive Chadronian *Poebrotherium* had bifurcate protocones on their upper molars, and since it was not clear that *Poebrodon* did also (the relevant specimens are too worn to determine this), *Poebrodon* was considered too advanced to be ancestral to camels. However, I have examined all the relevant specimens discussed by Wilson (1974), and I find only one or two with bifurcate protocones that are clearly referable to *Poebrotherium*; many are not camels at all, but oromerycids. In fact, some of these reports are erroneous. Wilson

(1974, p. 25) says that TMM 40504-22 (here referred to *Poebrotherium chadronense*, new species, described below) has bifurcate protocones on M3, but the specimen is too worn to determine whether this is so. Besides, this single highly variable character does not invalidate the large number of unique synapomorphies that ally *Poebrodon* with the Camelidae. If *Poebrodon* lacks bifurcate protocones and some *Poebrotherium* have them (neither of which is established yet), this would make *Poebrodon* too autapomorphous to be an ancestor, but would not prevent it from being closest sister-taxon to *Poebrotherium*.

A Note on "*Hidrosotherium*"—Wilson (1974) described a skull, jaws, and additional specimens from the Porvenir l.f. (late Duchesnean) of Trans-Pecos Texas, and named it "*Hidrosotherium transpecosensis*." Wilson (1974, p. 29) commented that "the general appearance of the skull resembles a large *Leptomeryx* but the premolar and molar pattern in no way resembles that genus." Wilson compared the teeth with those of camels, and decided that the specimen was a primitive camelid related to *Poebrodon*, justifying the new genus.

As Wilson correctly noted, the skull is indeed leptomerycid, right down to the distinctive diamond shape of the prelacrimal vacuity, the reduced anterior dentition, and the lack of a camelid auditory bulla. However, I examined the cheek teeth closely, and they are leptomerycid in every feature. The upper molars are not as transversely narrow, high-crowned, or selenodont as true camelids, but they are a good match for many primitive leptomerycids in the Frick Collection. In fact, "*Hidrosotherium*" is only slightly larger than "*Leptomeryx*"(= ?*Hendryomeryx*) *defordi* from the same deposits, and I suspect that they are the same animal. Therefore, I remove "*Hidrosotherium*" from the Camelidae, and synonymize it with "*Leptomeryx*"(= ?*Hendryomeryx*) *defordi*.

Poebrotherium Leidy, 1847

Type Species—*Poebrotherium wilsoni* Leidy, 1847

Included Species— *P. chadronense*, new species; *P. eximium* Hay, 1902; *P. franki* Wilson, 1974.

Range— Early Chadronian to middle Whitneyan, High Plains and Texas.

Diagnosis— Small to medium-sized camels (M1-3 length = 28-37 mm), with anterior dentition becoming differentiated and developing diastemata. Relatively low-crowned molars with strong styles. Distinguished from *Poebrodon* by: larger size; greater hypsodonty; M2 wider posteriorly; lingual hypoconulid lobe of M3 reduced; lingual surface of metaconid and entoconid more flattened.

Discussion— The genus *Poebrotherium* is highly variable in both size and in the differentiation of its anterior dentition. The most primitive species, *P.*

franki, has a moderately long rostrum with undifferentiated anterior teeth and short diastemata. The typical Chadronian camels, *P. chadronense* and *P. eximium*, are considerably larger, but still retain the primitive condition in their anterior teeth. *P. wilsoni* is within the size range of *P. eximium*, but develops a large P1/P1-P2/P2 diastema and a smaller C-P1/P1 diastema. *P. wilsoni* frequently develops a caniniform I3. The size of the upper canine in *P. wilsoni* appears to be sexually dimorphic, with larger canines in males. The longer rostrum is also correlated with a larger, more attenuated mandibular symphysis, which is frequently ventrally deflected. The metapodials of *P. wilsoni* are shortened relative to the rest of the skeleton. In associated material of *P. eximium*, for example, the length of the metatarsals is equal to the distance from the foramen magnum to the canine on the skull. In comparably-sized *P. wilsoni*, however, the metatarsal length is equivalent to the distance from the foramen magnum to the anterior part of P2. This change in ratios is partly due to relative limb shortening, but due also to the lengthening of the rostrum. Although *P. wilsoni* is clearly more derived than *P. eximium* or more primitive species of the genus, the differences are too minor and too variable to justify erection of a separate genus for primitive *Poebrotherium*.

Poebrotherium franki Wilson, 1974
Figure 2, Tables 2-3

Type—TMM 40504-149, a skull with I1-M3, and lower jaw with p2-m1, and fragments of vertebrae (Fig. 2). Airstrip l.f. (early Chadronian), Vieja Group of Texas.

Referred Specimens—see Wilson (1974)

Range—Early-middle Chadronian (Airstrip and Ash Springs l.f.), Vieja Group of Texas.

Diagnosis—Smallest species of *Poebrotherium* (length M1-3 = 28-29 mm). Elongate rostrum with simple, equal-sized bladelike I1-3 and canine, and no diastema. P3 with no lingual cingulum or cusp.

Description—*P. franki* was fully described by Wilson (1974). No new material of *P. franki* has been reported.

Discussion—*P. franki* is the oldest known species that shows the skull features characteristic of camels (since *Poebrodon* is known only from teeth). *P. franki* has the elongate rostrum (derived for camels) with the primitive, undifferentiated anterior dentition. The premolars and molars are fully camelid, and the bulla is fully inflated. Most of these features are not yet known for *Poebrodon*, but they would be predicted to occur in more complete material. *Poebrotherium franki* (M1-3 length 28-29 mm) is considerably larger than *Poebrodon kayi* (M1-3 length 20.2 mm). *Poebrodon californicus* (consisting only of a dP4-M1 at present) is slightly larger than *Poebrodon kayi*, and much smaller than *P.*

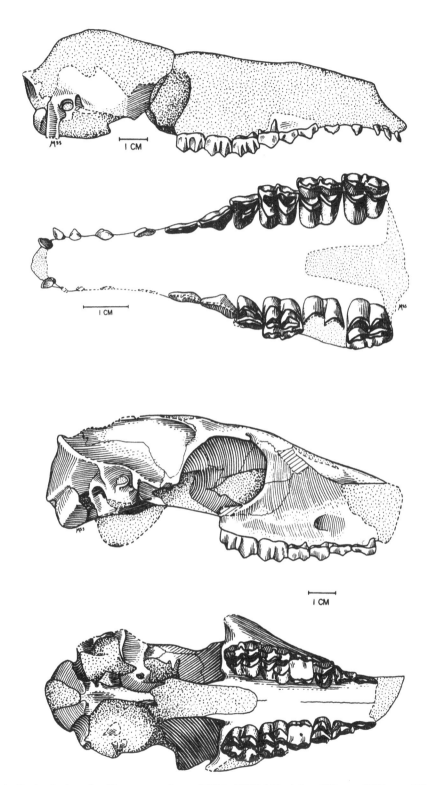

Figure 2. (Top). *Poebrotherium franki*, type specimen, TMM 40504-149 (after Wilson, 1974, p. 23). (Bottom) *Poebrotherium chadronense*, new species, TMM 40504-22 (after Wilson, 1974, p. 26).

Table 2. Camelid skull and upper teeth measurements (in mm)

	franki	*chadro-nense*	*Poebrotherium* *eximium*			*wilsoni*		*Paratylopus* *labiatus*			*primaevus*	*Paralabis* *cedrensis*
	TMM 40504-149 (type)	TMM 40504-22 (type)	AMNH 632 (type)	F:AM 47118	F:AM 47091	ANSP 1101 (type)	F:AM 47130	F:AM 42284	AMNH 6520 (type)	F:AM 42278	AMNH 9806 (type)	YPM 10167 (type)
CHARACTER												
Skull L (occ. cond>pmax)	131.5	—	163.6	161.5	161.4	—	187.5	189.0	—	182.3	188.0	—
I1-M3 L	78.4	—	100.2	97.0	97.8	—	109.1	108.1	—	106.0	103.0	—
L M3-occ. cond.	52.9	64.5	68.5	67.9	66.0	—	82.6	81.9	95.0	82.0	84.1	—
Skull W at M3	49.4	55.3	62.3	66.0	58.5	—	55.3	57.5	64.0	65.6	71.5	63.0
Rostral W at P1	13.4	17.0	23.2	20.7	15.8	—	17.7	18.0	—	22.0	21.4	18.0
P1-M3 L	57.8	—	74.7	70.3	72.5	—	77.8	76.7	—	88.3	81.5	84.8
P1-4 L	30.1	—	39.0	35.3	37.5	—	42.2	40.3	—	51.0	42.0	43.3
P1-2 diastema	2.6	—	2.0	1.5	3.3	—	10.8	8.3	—	6.0	9.0	10.5
P1L	5.1	—	7.9	6.8	7.8	—	8.5	7.5	—	7.5	8.1	—
P1W	1.9	—	2.5	3.1	2.0	—	3.0	2.2	—	3.0	—	—
P2L	7.3	10.4	10.4	9.5	8.3	—	8.8	8.8	9.1	9.0	9.7	8.7
P2W	27.	3.0	3.1	3.2	3.0	—	3.7	3.1	3.0	3.3	—	3.5
P3L	8.2	9.4	11.7	9.7	11.0	—	9.3	9.5	9.0	10.5	9.0	9.2
P3W	4.5	7.1	4.7	5.0	4.7	—	4.9	4.5	4.0	4.1	—	5.5
P4L	7.1	8.4	10.0	9.0	8.0	—	7.9	9.8	8.9	8.5	8.7	9.5
P4W	7.0	9.3	7.0	8.4	7.7	—	7.5	9.3	6.9	7.5	—	7.6
M1-3 L	29.1	34.4	36.8	36.3	37.5	39.5	35.7	36.3	39.0	37.8	39.5	43.5
M1L	9.5	10.2	10.6	11.4	11.0	11.5	10.8	10.3	11.5	11.8	11.6	12.6
M1W	9.1	10.9	11.3	10.2	10.3	10.3	10.5	10.2	10.1	10.7	10.8	12.5
M2L	10.5	12.7	12.7	12.7	14.3	13.5	12.8	12.9	14.0	13.0	13.4	15.9
M2W	10.6	12.6	12.0	10.5	11.2	11.5	11.5	12.1	12.1	11.7	12.8	14.6
M3L	10.6	13.3	14.7	14.1	15.0	14.5	14.9	16.0	16.1	15.0	17.0	18.0
M3W	11.0	13.4	10.9	11.2	10.5	12.5	12.2	12.1	11.8	12.8	14.6	15.0

franki, but it is intermediate in all its preserved features. The biggest morphologic gap is between *Poebrotherium franki* and *Poebrodon californicus*, so the generic distinction can still be justified.

Poebrotherium chadronense, new species
Figure 2, Table 2

Poebrotherium sp. Wilson, 1974

Type and only known specimen—TMM 40504-22, a skull and partial skeleton (Fig. 2). Airstrip l.f. (early Chadronian), Vieja Group of Texas.

Etymology—In reference to the Chadronian occurrence of this species.

Diagnosis—Medium-small sized *Poebrotherium* (M1-3 length = 33-35 mm) with a distinct posterolingual cusp on P3 and a wider P4 than *P. eximium*. Intermediate in size between *P. franki* and *P. eximium* or *P. wilsoni*.

Discussion—Wilson (1974) has adequately described and figured this species. He refrained from naming it "until more is known of the variation in species of *Poebrotherium*" (Wilson, 1974:26). It is clear from the presently known variation in *P. eximium* and

P. wilsoni that *P. chadronense* can be distinguished on the basis of size (Table 2). Although the size distinction is not obvious in the length of M1-3 (which I have used throughout as a convenient size measure), it is very noticeable in the rest of the skull measurements (Table 2). Since this species does not fit within the normal range of variation of *P. eximium*, and is much too large for *P. franki*, a new species is justified.

P. chadronense is known only from the early Chadronian Airstrip l.f. of Texas. It is very similar in overall morphology to *P. eximium* from the early and middle Chadronian of the High Plains, except that it is significantly smaller with a shorter rostrum.

Poebrotherium eximium Hay, 1902
Figure 3, Tables 2, 3, 5

Poebrotherium wilsoni Wortman, 1898
Poebrotherium eximium Hay, 1902

Type—AMNH 632, skull, mandible, parts of skeleton; supposedly from "Lower Oreodon Beds" (early Orellan), Big Badlands of South Dakota (Fig. 3), although all subsequently found specimens from South Dakota are known only from the Chadron Formation.

Table 2 (continued).

CHARACTER	Pseudolabis dakotensis			Miotylopus			Gentilicamelus sternbergi
				leonardi	gibbi	taylori	
	AMNH 9807 (type)	F:AM 36469 (male)	F:AM 41687 (female)	F:AM 36655	F:AM 36446	F:AM 36459 (type)	AMNH 7970 (type)
Skull L (occ. cond>pmax)	—	265.0	—	—	250.0	310.0	—
I1-M3 L	158.0	157.1	129.6	128.0	150.0	179.5	—
L M3-occ. cond.	—	110.3	90.0	—	106.5	136.6	99.5
Skull W at M3	—	86.6	66.5	—	81.5	74.0	87.4
Rostral W at P1	—	29.5	21.5	—	28.1	24.5	29.2
P1-M3 L	107.4	104.8	91.1	83.0	98.8	109.7	87.3
P1-4 L	61.4	57.8	45.0	41.6	52.3	60.4	47.7
P1-2 diastema	16.0	12.6	10.4	9.0	18.5	20.1	12.2
P1L	9.2	10.8	7.0	7.0	7.5	9.7	6.7
P1W	3.8	3.6	3.2	2.1	3.1	4.2	—
P2L	12.5	11.8	10.5	7.9	10.5	9.1	10.3
P2W	4.5	4.5	5.0	3.0	4.5	5.1	4.7
P3L	13.3	9.7	10.0	8.6	8.5	11.9	9.8
P3W	5.3	6.0	7.0	4.3	4.0	6.0	5.4
P4L	12.0	10.5	9.9	9.4	9.2	10.0	9.1
P4W	10.0	8.8	10.0	7.5	8.1	8.2	9.0
M1-3 L	50.9	48.0	47.2	42.2	46.8	51.4	40.4
M1L	16.0	12.9	14.6	12.5	13.1	14.2	11.9
M1W	14.2	13.1	13.5	10.5	12.8	12.5	11.2
M2L	17.5	15.3	16.7	14.5	15.7	16.3	13.5
M2W	15.3	15.3	15.4	12.5	13.7	14.0	12.1
M3L	20.1	20.3	18.6	16.3	20.3	21.1	16.4
M3W	16.9	17.1	14.0	13.1	15.5	16.0	13.5

Referred Specimens—From 290 feet on zonation section, Ledge Creek, Natrona Co., Wyoming (early Chadronian): F:AM 47396, rami, limb elements. From McCarty's Mountain l.f., Madison Co., Montana (early Chadronian): F:AM 47446, mandible. From Flagstaff Rim, Natrona Co., Wyoming (local range—from 15 feet below Ash F—middle Chadronian): F:AM 47066; mandible. Chadron Formation, Shannon Co., South Dakota (late Chadronian): F:AM 42241, skull and jaws; F:AM 42240, skull and jaws; F:AM 42252, skull and jaws. From 2.5 miles north of Chadron, Dawes Co., Nebraska (Brecht Ranch, Morris Ranch—late Chadronian): F:AM 47116, skull and partial skeleton; F:AM 49119, skull and jaws; F:AM 17115, skull and jaws; F:AM 17143, left ramus; F:AM 17118, right ramus; F:AM 47105, skull and partial skeleton. From Geike Ranch, Sioux Co., Nebraska: F:AM 47427, mandible; F:AM 47426, skull and mandible; F:AM 47428, right ramus; F:AM 47429, skull, jaws, and partial skeleton. From the Douglas area, Converse Co., Wyoming (stratigraphic range 60 below to 100 feet above the 5 tuff of Evanoff et al., 1992): F:AM 47103, skull and complete, articulated skeleton; F:AM 47194, left and right rami; F:AM 47222, left and right rami; F:AM 63834, right ramus; F:AM 47194, palate; F:AM 47029, skull and jaws; F:AM 63832, mandible; F:AM 63836, left ramus; F:AM 63835, mandible; F:AM 63837, right ramus; F:AM 63833, right ramus; F:AM 47006, palate and mandible; F:AM 47001, skull and mandible; F:AM 47007, skull; AMNH 22467, skull and complete, articulated skeleton; F:AM 47002, skull and mandible; F:AM 47008, skull and jaws; F:AM 47023, skull and jaws; F:AM 47371, mandible; F:AM 47196, right ramus; F:AM 47009, juvenile palates, rami; F:AM 47197, mandible; F:AM 47104, juvenile skull; F:AM 42244 juvenile palate; F:AM 47191 partial skull; F:AM 47195, right ramus; F:AM 47319, skull and rami; F:AM 47318, juvenile skull and rami; F:AM 47192, skull; F:AM 47032, skull and jaws; F:AM 42242, skull and jaws; F:AM 47316, skull and partial skeleton; F:AM 47193, skull and partial skeleton; F:AM 47091, skull and partial skeleton; F:AM 42293, skull and partial skeleton; F:AM 42300, skull and jaws; F:AM 42298, juvenile skull; F:AM 47229, skull and jaws; F:AM 63830, skull and jaws; F:AM 42243, skull and jaws; F:AM 63831, skull and jaws; F:AM 47224, skull; F:AM 42297, skull and partial skeleton; F:AM 47021, skull. From the Lusk Area, Niobrara Co., Wyoming (stratigraphic range 0-40 feet above PWL): F:AM 63822, skull; F:AM 47092, skull and jaws; F:AM 63820, palate and partial skull; F:AM 63823, skull and rami; F:AM 47094, mandible;

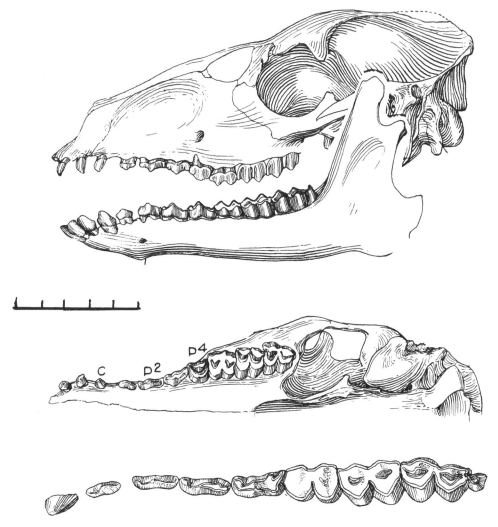

Figure 3. *Poebrotherium eximium*, type specimen, AMNH 632. Left lateral view of skull and mandible; palatal view; crown view of lower dentition. Scale in cm.

F:AM 47035, juvenile skull; F:AM 63821, mandibular and maxilliary fragments; F:AM 63824, left ramus; F:AM 47077, skull and complete skeleton; F:AM 47030, skull and mandible; F:AM 47036, skull; F:AM 47038, mandible; F:AM 47087, skull and partial skeleton.

Diagnosis—Medium-sized (M1-3 length 36-38 mm) camelid with undifferentiated anterior dentition and no diastemata. The m3 hypoconulid is a narrow crest, rather than a horn. Metapodials are longer than *P. wilsoni* or *Paratylopus labiatus*. Slightly larger than *P. chadronense*, with no posterolingual cusps on P3 and a narrower P4.

Description—The type skull was originally figured by Wortman (1898, figure 7), but never adequately described. The rest of the animal has never been described, since it was unknown until the Frick Collection became available.

Except for the anterior dentition, *P. eximium* differs very little from *P. wilsoni* or *Paratylopus labiatus*. As a consequence, some postcranial material of *P. eximium* very closely resembles that of *P. wilsoni*. Most specimens of *P. eximium* have primitive camelid skeletal morphology, also seen in *Paratylopus labiatus*

The skull of *P. eximium* (Fig. 3) differs from *P. wilsoni* in being slightly broader and shorter. The main differences between the two species are in the anterior dentition. Most specimens of *P. eximium* have equal sized, leaf-shaped upper incisors spaced evenly across the top of the rostrum. The canine is also incisiform and equal to the incisors in size, although some specimens (such as the type of *P. eximium*) have slightly more caniniform canines. P1 is usually a long, two-rooted blade which is separated from the canine by a short diastema. P2 is also a narrow, two-rooted blade with a distinct central cusp. In some advanced specimens of *P. eximium*, the P1-2 diastema that characterizes *P. wilsoni* is beginning to develop. P3 is also bladelike, but there is usually a small discontinuous lingual cingulum such as on TMM 40504-22. P4 is completely selenodont. The molars are typically camelid, with distinct styles and moderate to weak ribs. Mesostyles are generally well developed on the upper molars, although they can be weak in some specimens. Otherwise, the skull and dentition differ little from *P. wilsoni*.

The mandible (Fig. 3) has equal-sized, leaf-shaped lower incisors. Unlike *P. wilsoni*, the incisors are not arranged in a closed "fan" at the tip of the symphysis. The lower canine is completely incisiform and continuous with the incisor row. The p1 is two-rooted, bladelike, and lies immediately behind the canine. There may be a small diastema between p1 and p2, although these teeth are normally not separated. The p2 and p3 are both two-rooted blades, with a distinct anterior cusp. The p4 has a small anterolingual spur. Of the two posterior crests on p4, the labial crest is longer than the lingual crest, and lingually inflated at its posterior end. The m1-3 show the typical selenodont camel condition. The hypoconulid of m3 in *P. eximium* is usually a single narrow crest, while in most *P. wilsoni* it is basin-like. There are some specimens that show exceptions in this feature, however. The posterior symphysis extends to the level of the p1-2 diastema. The symphysis in *P. eximium* is generally not as attenuated and ventrally deflected as it is in *P. wilsoni*, but has only a slight ventral curvature. The coronoid process is slender and straight. The condyle shows the typical camel condition, and is anteromedially inflected. The angular process has the typical camelid "hook" on the dorsal edge. This is below the level of the tooth row. The ventral portion of the angle extends slightly below the level of the ventral margin of the ramus.

The postcranial skeleton (Figs. 19-26) differs little from that of *Paratylopus labiatus* except in the limbs. Scott's (1940) descriptions apply to most of the known elements except the metapodials. As mentioned above, the metapodials in *P. eximium* are relatively longer than they are in *P. wilsoni* or *Paratylopus labiatus*. In length, they equal the distance from the foramen magnum to the canine in their associated skulls. Later individuals of *P. eximium* (for example, F:AM 47077) have a postcranial skeleton (including metapodials) that closely approaches the condition found in *P. wilsoni* although their skulls still have a dentition like *P. eximium*. Thus, the only really diagnostic elements of *P. eximium* are the skull and mandible.

Discussion — Wortman (1898, p. 111, fig. 7) first figured a lateral view of a skull of "*P. wilsoni*." He considered *P. wilsoni* to be the precursor of *Paratylopus labiatus* and stated that *P. wilsoni* lacked diastemata between its anterior teeth. However, Leidy's (1847) type specimen of *P. wilsoni* has diastemata between P1/1 and P2/2, so Wortman's characterization of *P. wilsoni* is incorrect. Hay (1902) apparently recognized this, and so gave the name *P. eximium* to the specimen figured by Wortman. Hay gave no further discussion, description, or characterization of the new species.

Matthew (1904) and Loomis (1928) subsequently adopted the name *P. eximium* for primitive poebrotheres with no caniniform teeth or diastemata. Scott (1940, p. 645) commented that: "This species, if really distinct, is of the sort that one would expect to find in the Chadron, for it much more nearly approximates *Protylopus*, of the Uinta Eocene [this was before *Protylopus* was recognized as an oromerycid by Gazin, 1955], than do the other species of *Poebrotherium*. In size, it about equals *P. wilsoni*, but the five anterior upper teeth (I1-P1) are almost equal in size and equally spaced apart, without distinct diastemata. If these characteristics should prove to be constant, the propriety of recognizing it as a separate species will be clear, especially if it should turn out to be a survivor from the Chadron substage."

Further material, particularly from the Chadronian, has confirmed Scott's suggestions. The skull morphology of *Poebrotherium eximium* does indeed seem to be remarkably consistent in the Chadronian. At the end of its range (early Orellan), *P. eximium* was apparently sympatric with *P. wilsoni* and *Paratylopus labiatus*. Although the latter two species intergrade with *P. eximium* in some features, the appearance of the characteristic anterior dentition of *P. wilsoni* and *Paratylopus labiatus* is quite sudden. Cranial material of all three species can be unambiguously distinguished where they occur together. *P. wilsoni* and *Paratylopus labiatus* both appear to have speciated from *P. eximium*, which persisted unchanged well into the Orellan. The youngest known specimen of *P. eximium* occurs 40 feet above the "Persistent White Layer" (PWL) in the Seaman Hills, near Lusk, Niobrara County, Wyoming (early Orellan) (see this volume, Chapter 14).

Poebrotherium wilsoni Leidy, 1847
Figure 4, Tables 2, 3, 5

Poebrotherium andersoni Troxell, 1917

Type — ANSP 11012, a skull and attached jaws of a juvenile individual (Fig. 4).

Referred Specimens — *P. wilsoni* is by far the most abundant camel from the Oligocene. Most smaller camel specimens in the many White River collections around the world can be referred to this species. Since there are over 500 catalogued specimens of *P. wilsoni* in the Frick Collection alone, and nearly as many uncatalogued specimens, it is clearly impractical to give a complete hypodigm here. The partial list given below includes specimens that had diagnostic portions of the skull or jaw preserved, and were identified to species.

Table 3. Camelid mandible and lower tooth measurements (in mm)

| | Poebrotherium | | | | | | | Paratylopus | | |
| | *franki* | *eximium* | | | *wilsoni* | | | *labiatus* | | *primaevus* |
CHARACTER	TMM 40504-149 (type)	AMNH 632 (type)	F:AM 47118	F:AM 47002	ANSP 11012 (type)	F:AM 47130	F:AM 39086	AMNH 6520 (type)	F:AM 42278	AMNH 9806 (type)
Symphysis L	—	29.0	21.2	25.5	—	40.0	40.0	—	31.9	40.0
p1-m3 L	—	81.5	73.6	73.8	—	83.3	83.3	89.0	81.7	—
p1-4 L	—	41.5	35.0	38.0	—	40.1	44.5	42.7	42.4	—
p1-2 diastema	—	3.0	4.0	1.0	—	13.0	13.2	14.1	6.8	—
p1L	—	7.0	6.6	7.2	—	6.0	5.7	6.0	7.1	—
p1W	—	2.4	2.6	2.6	—	3.8	1.9	2.3	2.9	—
p2L	7.2	11.1	9.4	8.1	—	8.6	8.0	9.1	11.8	—
p2W	2.0	2.7	2.8	2.5	—	3.3	3.5	—	2.5	—
p3L	7.9	11.1	9.0	9.6	—	9.5	9.7	8.5	9.9	—
p3W	2.4	2.8	3.7	3.6	—	3.1	2.8	2.9	3.5	—
p4L	8.1	12.0	10.0	9.0	—	9.0	8.8	9.5	9.8	—
p4W	3.3	4.0	4.5	4.7	—	4.5	4.0	4.5	4.5	—
m1-3 L	—	42.0	3.70	37.2	39.5	39.1	38.8	43.3	42.2	44.2
m1L	9.7	11.3	10.6	10.7	10.0	10.3	11.0	11.8	11.8	11.3
m1W	6.5	6.4	7.0	6.4	—	6.5	7.0	6.6	6.6	—
m2L	—	13.4	11.8	11.4	13.5	11.9	11.3	12.5	12.7	13.2
m2W	—	8.1	7.0	7.5	—	8.3	7.7	8.4	7.4	8.9
m3L	—	18.1	16.8	16.5	18.0	18.0	17.0	20.1	17.9	19.0
m3W	—	8.6	7.1	7.4	—	8.2	8.1	8.6	8.2	8.8
jaw depth at p2	—	14.9	16.0	14.8	14.5	17.6	16.6	19.1	18.7	—
jaw depth at m2	11.0	18.0	20.3	17.8	14.5	20.7	19.5	19.8	25.7	—

From Pipestone Springs, Jefferson Co., Montana (middle Chadronian): F:AM 47445, skull and partial skeleton; F:AM 47446, left ramus with p2-m3. From Flagstaff Rim, Natrona Co., Wyoming (local range—from 25 feet below Ash D to 15 feet below Ash G—middle Chadronian): F:AM 47065, maxilla and ramus; F:AM 47413, ramus with right m3; F:AM 47414, ramus with right p4-m3; F:AM 47070, maxilla, right ramus, partial skeleton; F:AM 47409, right ramus with m2-3; F:AM 47406, right M2-3; F:AM 47067, right maxilla; F:AM 47412, left ramus; F:AM 47411, right ramus with m2-3; F:AM 47407, left maxilla with M1-3; F:AM 47359, right ramus with p2-m3. There is much additional fragmentary uncatalogued material in the Frick Collection. From Devil's Gap, Beaver Divide, Fremont Co., Wyoming (Chadronian?): AMNH 14590, skull and jaws; AMNH 14592, mandible; AMNH 14593, right ramus. From the Lusk area, Niobrara Co., Wyoming (late Chadronian-early Orellan): F:AM 63719, left ramus and metapodial fragment; F:AM 47103, skull, mandible, and partial skeleton; F:AM 47077, associated male and female partial skeletons; F:AM 63704, partial skull, rami, partial skeleton; F:AM 63725, right and left rami; F:AM 63742, left partial ramus; F:AM 63724, right ramus; F:AM 47053, partial skull and mandible, vertebrae; F:AM 47059, right partial ramus; F:AM 63730, partial mandible; F:AM 47090, right and left rami; F:AM 47098, skull

and mandible; F:AM 63723, left ramus; F:AM 47051, partial skull, left ramus, partial skeleton; F:AM 63741, right partial ramus; F:AM 63715, left ramus; F:AM 47335, partial skull and skeleton; F:AM 47097, partial skull and mandible, partial skeleton. From the Douglas area, Converse Co., Wyo. (late Chadronian-early Orellan): F:AM 47220, left maxilla and partial mandible; F:AM 47332, palate and partial mandible; F:AM 47370, partial skull and ramus, atlas; F:AM 47015, partial mandible; F:AM 47016, skull and mandible; F:AM 47022, partial skull and mandible; F:AM 47011, skull and mandible; F:AM 63761, right partial ramus; F:AM 47328, partial mandible; F:AM 47331, right ramus; F:AM 47373, anterior partial skull and mandible; F:AM 42248, right ramus. From the "lower nodular zone" (early Orellan), Cottonwood Pass area, Big Badlands, Shannon Co., S.D.: F:AM 42279, skull, mandible, partial skeleton; AMNH 39085, skull, cervicals; AMNH 28841, mandible, partial skeleton; AMNH 39086, skull and jaws; AMNH 38992, mandible; AMNH 38943, left ramus. From Geike Ranch, Sioux Co., Neb. (early Orellan): F:AM 47263, right ramus; F:AM 47271, right ramus; F:AM 47279, partial maxilla and femur; F:AM 47282, partial skull, right ramus; vertebrae; F:AM 47285, partial skull and mandible; F:AM 47276, right and left rami, articulated forelimb. From Munson Ranch, Sioux Co., Neb. (early Orellan): F:AM 47273, right ramus; F:AM 47262,

Table 3 (continued).

CHARACTER	Paralabis cedrensis AMNH 8969 (type)	Paralabis cedrensis SDSM 4097	Pseudolabis dakotensis F:AM 36469 (male)	Pseudolabis dakotensis F:AM 41687 (female)	Miotylopus leonardi F:AM 36796 (female)	Miotylopus gibbi F:AM 36446 (male)	Miotylopus taylori F:AM 36459 (male)	Gentilicamelus sternbergi AMNH 7970 (type)
Symphysis L	—	39.3	51.0	33.2	31.5	38.4	58.1	40.0
p1-m3 L	92.0	90.1	107.2	99.4	88.1	108.2	127.3	99.1
p1-4 L	48.7	49.5	51.5	47.3	43.2	59.4	71.7	50.2
p1-2 diastema	20.3	16.2	13.8	10.7	16.7	26.7	31.8	16.0
p1L	6.3	—	8.2	6.7	6.0	6.3	7.9	6.7
p1W	2.7	—	2.6	3.3	2.6	2.5	2.7	—
p2L	7.5	6.3	11.2	9.4	5.1	8.4	10.3	9.0
p2W	2.9	2.4	3.9	3.2	2.0	2.6	2.8	3.3
p3L	8.6	8.5	11.3	10.2	8.2	9.3	12.0	11.8
p3W	2.9	3.2	4.5	3.7	3.0	2.9	4.2	4.2
p4L	9.1	9.1	12.1	11.5	10.5	10.5	12.9	11.6
p4W	4.1	4.7	5.2	3.5	3.5	3.5	4.5	5.0
m1-3 L	—	47.1	52.4	52.2	46.5	49.7	55.6	47.9
m1L	10.0	12.8	12.4	13.7	12.9	12.5	13.9	13.3
m1W	6.6	8.4	10.0	8.1	7.4	8.2	9.8	8.4
m2L	13.3	14.7	15.1	17.1	14.0	15.1	17.0	16.4
m2W	7.7	8.9	11.8	8.9	7.0	10.3	10.6	9.0
m3L	18.4	201.5	15.0	22.8	18.3	23.2	24.8	19.2
m3W	7.5	9.3	11.8	8.3	6.1	9.5	11.0	8.5
jaw depth at p2	15.5	18.1	24.0	16.6	15.6	23.8	34.5	21.0
jaw depth at m2	19.4	22.6	28.1	20.5	16.4	24.0	30.5	26.1

partial skull and mandible; F:AM 47269, left partial ramus; F:AM 47284, partial skull and mandible; F:AM 47260, partial maxilla and rami; F:AM 47266, left ramus; F:AM 47264, left ramus; F:AM 47257, partial skull, left ramus, partial skeleton; F:AM 47259, right partial ramus; F:AM 47263, right ramus. From the area north and west of Chadron, Dawes Co., Neb. (late Chadronian-early Orellan): F:AM 47144, right ramus; F:AM 47145, right ramus; F:AM 47140, partial skull and mandible; F:AM 47125, partial palate and mandible; F:AM 47141, skull and mandible; F:AM 47235, right partial ramus; F:AM 47253, right ramus. From Kostelecky Ranch, Stark Co., N.D. (early Orellan): F:AM 47228, right P2-M2. From the Little Badlands, Stark Co., N.D. (early Orellan): F:AM 47229, M2; F:AM 47230, proximal metatarsal III-IV. From the "Middle Oreodon Beds" (late Orellan), Big Badlands, Shannon Co., S.D.: F:AM 42277, skull, jaws, partial skeleton; F:AM 47170, left ramus; F:AM 42265, right radius-ulna; F:AM 47185, partial skeleton; F:AM 47185, right ramus; F:AM 47173, left maxilla with M2-3; F:AM 47181, right ramus with p3-m3; F:AM 47176, left ramus with p3-m3; F:AM 47178, right ramus with m1-3; F:AM 47190, skull, jaws, partial skeleton; F:AM 47183, skull and jaws; F:AM 47177, right ramus with p4-m3; F:AM 47173, left ramus with m2-3; F:AM 47174, left maxilla with M1-

3; F:AM 47172, right ramus with m1-3; F:AM 47169, left ramus with m1-3; F:AM 47171, right ramus with p2-m2. From the *Leptauchenia* nodules (Whitneyan), Big Badlands, S.D.: AMNH 39082, right M3 (tentatively referred). From the west end of Eagle Nest Butte, Scottsbluff Co., Neb. (middle Whitneyan): F:AM 47425, right ramus.

Diagnosis—Medium to small camels (length of M1-3 = 30-37 mm), with a long diastema between P1/1 and P2/2, and between the canine and P1/1. I3 enlarged. Lower incisors closely appressed in a fanlike arrangement. Lower canine larger, particularly in males. The m3 hypoconulid usually basined. Skeleton with relatively long and slender limbs, and short, arched back. Metapodials relatively shorter than in *P. eximium*.

Description—*P. wilsoni* was described by Scott (1940). Figures 19-26 shows the material referred to *P. wilsoni* in comparison with the other species. Measurements are given in Tables 2, 3, and 5.

Discussion—*Poebrotherium wilsoni* (Leidy, 1847) was based on an attached skull and mandible of a juvenile individual, with the rostrum anterior to P1 broken off (Fig. 4). The diagnostic diastemata between P1/1 and P2/2 can be clearly seen. The measurements of the specimen indicate that *P. wilsoni* is the smallest of the White River camels with long diastemata. Galbreath

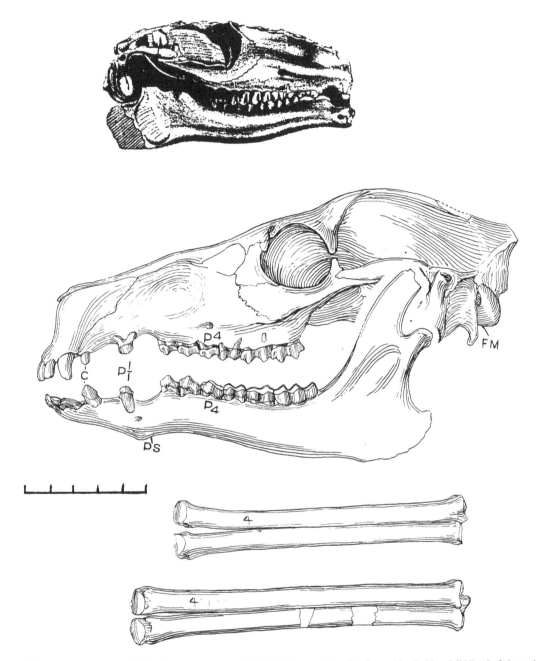

Figure 4. *Poebrotherium wilsoni*. (Top) Type specimen, ASNP 11012, as originally figured by Leidy. (Middle) Left lateral view of referred skull and mandible, F:AM 47130, with associated metacarpals and metatarsals (bottom). Scale in cm.

(1953, pp. 88-89) found that the Orellan camels of Colorado separated into two distinct clusters: the small, abundant, long-ranging *P. wilsoni*, and the larger, less abundant, and shorter-ranging *Paratylopus labiatus*. The camels in the Frick Collection show a similar pattern. *Poebrotherium wilsoni* is very abundant, and shows a moderately large range of size variation. It is similar in size to the contemporaneous specimens of *P. eximium*. Loomis (1928) thought that *P. wilsoni* could not be

distinguished from *Paratylopus labiatus*, attributing the size differences to sexual dimorphism. As Scott (1940) and Galbreath (1953) have pointed out, there are several arguments against this idea:

1) Scott (1940) showed that *P. wilsoni* has a very derived, gazelle-like skeleton, very different from the primitive skeletal proportions of *Paratylopus labiatus* or *P. eximium*.

2) The relative scarcity of *Paratylopus labiatus* was

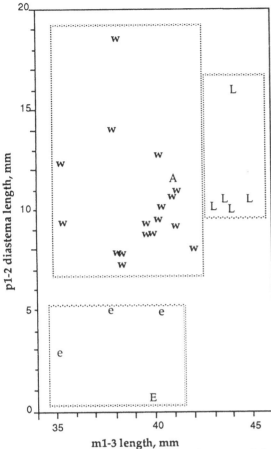

Figure 5. Bivariate plots of m1-3 length versus p1-2 diastema length in Orellan camels. A = type specimen of "*Poebrotherium andersoni*"; E = type specimen of *P. eximium*; e = referred specimens of *P. eximium*; L = specimens of *Paratylopus labiatus*; w = specimens of *Poebrotherium wilsoni*. *P. eximium* is clearly separated by its short diastema (less than 6 mm); *P. labiatus* by its larger size (m1-3 length greater than 42 mm). The type of *P. andersoni* falls within the range of variation of *P. wilsoni*.

used by Galbreath (1953) as evidence that *Paratylopus labiatus* was not the male of *P. wilsoni*. This argument is invalid, however, since some wild camelids have herds consisting of a single male and up to a dozen or more females (Koford, 1957; Franklin, 1983).

3) As Galbreath (1953) has shown, *Paratylopus labiatus* has a shorter stratigraphic range than *P. wilsoni*. There are no *Paratylopus labiatus* in rocks younger than the late early Orellan, whereas there are many examples of *P. wilsoni* in rocks correlated with the early and middle Whitneyan. In addition, *P. wilsoni* appears in the middle Chadronian, but *P. labiatus* is first known from the early Orellan.

4) There is sexual dimorphism in the upper canines of *P. wilsoni*, and several males showing this feature are distinctly smaller than *Paratylopus labiatus*.

5) "*Poebrotherium*" *labiatum*, but not *P. wilsoni*, share derived features with *Paratylopus primaevus*, and

thus I remove it from *Poebrotherium* entirely and place it with *Paratylopus* (see below).

Troxell (1917) erected the species *Poebrotherium andersoni* for a skull and mandible from the "Oreodon zone" near Harrison, Nebraska. Troxell compared this skull to *P. eximium* and *Paratylopus labiatus*, but not to *P. wilsoni*. The characters he cited as diagnostic are variable features of the anterior dentition that also occur in *P. wilsoni*. *Poebrotherium andersoni* fits entirely within the size range of *P. wilsoni* (Fig. 5). Loomis (1928) suggested that *P. andersoni* was a synonym of "*Poebrotherium*" *labiatum*, and Scott (1940) followed this synonymy. However, Loomis did not make adequate comparisons to *P. wilsoni*, either. The enlargement of the I3 and reduction of the canine cited by Loomis (1928) are typical of many specimens of *P. wilsoni*. The variability of the anterior dentition apparently confused many authors. *Paratylopus labiatus* and *P. wilsoni* are most easily distinguished by size, since they have similar development of the anterior dentition and diastemata.

P. wilsoni, with its unusual derived gazelle-like skeleton, is probably the terminal member of the *Poebrotherium* lineage. There seem to be no derived features which unite it with any later camel taxon. *P. wilsoni* persists into the mid-Whitneyan, but is relatively rare after the Orellan.

Poebrotherium wilsoni also shows a strong geographic gradient in abundance. It is extremely scarce in the Orellan deposits of North Dakota and northwest South Dakota, and also rare in the Big Badlands of South Dakota (less than 3% of the fauna, according to Clark et al., 1967). *Poebrotherium* is quite common in the samples collected along the Pine Ridge (particularly in eastern Wyoming and western Nebraska), and Galbreath (1953, p. 88) reports that they are more common than oreodonts in northeastern Colorado. There seems to be a clear north-south gradient in abundance of Orellan camels, which is paralleled by a similar gradient in abundance of *Hypertragulus* and *Miniochoerus* (Prothero, 1982). According to Clark et al. (1967), *Poebrotherium* samples were too small to determine if they showed a significant preference for near-stream versus more open habitat in the Big Badlands; they were found in both environments.

There is clearly sexual dimorphism in *Poebrotherium wilsoni*, as there is in *Pseudolabis*. The best evidence of this is a block containing two partially articulated skeletons (F:AM 47077 A and B) from 40 feet above the PWL ("Persistent White Layer"), Seaman Hills, Niobrara Co., Wyoming (earliest Orellan). The larger individual (F:AM 47077B) has a slightly longer and more pointed upper canine than the smaller individual (F:AM 47077A). The latter individual has a shorter, blunter upper canine with weak anterior and posterior cingula. Presumably, the larger specimen was a male. Sexual dimorphism may explain the large range in size variation in *P. wilsoni*. Complete skulls with good upper canines are rare in the sample, but most show a rough correlation between larger body size and larger, more pointed upper canines. The number of skulls that can be sexed is too small to determine if the male/ female ratios common in camelid harems (Koford, 1957; Wilson, 1975, p. 480; Franklin, 1983) occur in *Poebrotherium*.

Paratylopus Matthew, 1904

Poebrotherium Cope, 1881 (in part)
Protomeryx Hay, 1902 (in part)
Miolabis (Paratylopus) Matthew, 1904
Paratylopus Matthew, 1909
Gentilicamelus Loomis, 1936 (in part)
    Type Species—*Paratylopus primaevus* (Matthew, 1904)
    Included Species—*Paratylopus labiatus* (Cope, 1881), new combination.
    Revised Diagnosis—Medium-sized camels (M1-3 length = 37-41 mm) with brachydont teeth, reduced premolars, and large skulls relative to their tooth size. Otherwise, skull and skeletal proportions remain primitive. Distinguished from *Poebrotherium* by larger size and reduced premolars. Distinguished from *Paralabis* by its smaller size, and slightly reduced premolars. Distinguished from other Whitneyan-Arikareean camels by its brachydont teeth, which are relatively small for their skull size.
    Discussion—The nomenclature of Whitneyan and Arikareean camels is one of the most confused in the literature. McKenna (1966) cleared up much of this confusion, but did not attempt a formal revision. He also did not have access to the Frick Collection at the time, so many specimens that are now available for study were unknown to him.
    The first name applied to a Whitneyan-Arikareean camel was *Protomeryx* (Leidy, 1856). The type specimen of *Protomeryx halli* (ANSP 11011) is a left mandibular fragment with the canine, p1-3, and the root of i3 (Fig. 6). In the size of the canine, the length of the p1-2 diastema, and in overall size, it could belong to *Paratylopus primaevus*, *Paratylopus labiatus* (for example, compare F:AM 42278 with ANSP 11011), and some specimens of *Poebrotherium wilsoni*. The type specimen of *Protomeryx halli* is clearly indeterminate, as McKenna (1966) showed, and the name *Protomeryx* is a *nomen dubium*.
    As the first generic name applied to a Whitneyan camel, *Protomeryx* became a wastebasket taxon for many Whitneyan and Arikareean camels. Yet even the supposed Whitneyan age of the type specimen of *Protomeryx halli* is uncertain. Leidy (1856) states only that the type came from "Bear Creek, Nebraska Territory." Leidy (1869) wrote that the specimen came from Hayden's "Level D," which Matthew (1901)

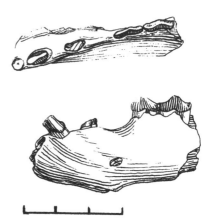

Figure 6. ANSP 1101, anterior ramal fragment, type of *"Protomeryx halli."* Scale in cm.

interpreted as Whitneyan in age. Cope (1873) paid little attention to its stratigraphic position, but correctly noted that its morphology was nearly indistinguishable from *Poebrotherium wilsoni*. Scott (1891) and Wortman (1898) considered *Protomeryx halli* to be from the "Loup Fork" beds (late Miocene-Pliocene), contemporary with *Procamelus*, *Homocamelus*, *Pliauchenia*, and *Megalomeryx*. This confused many later paleontologists. Matthew's (1899, 1901) Whitneyan age assignment for *Protomeryx halli* influenced most subsequent authors. These age uncertainties are irrelevant, however, since the specimen is not diagnostic in the first place.
    The next Whitneyan-Arikareean camel to be described was *Poebrotherium sternbergi* (Cope, 1879), from the ?upper John Day (?late Arikareean) of Oregon (Fig. 18). In 1886, Cope renamed this animal *Gomphotherium*. He was apparently unaware that the name was preoccupied by a proboscidean named by Burmeister in 1837. *Gomphotherium* then became another wastebasket taxon used by many authors for Arikareean and younger camels. Matthew (1901) placed *Gomphotherium* in synonymy with *Protomeryx*. Hay (1902) finally realized that *Gomphotherium* was preoccupied, and placed most Whitneyan-Arikareean camels in *Protomeryx*.
    The third name to be applied to Whitneyan-Arikareean camels was *Paratylopus* (Matthew, 1904). It was originally proposed as a subgenus of *Miolabis* (a middle Miocene genus) and then raised to generic level by Matthew in 1909. *Paratylopus* was based on *P. primaevus* from the early Whitneyan of South Dakota. Matthew (1904) and Cope and Matthew (1915) referred *"Gomphotherium" sternbergi* to *Paratylopus*.
    Since *Protomeryx* and *Gomphotherium* were both invalid, Loomis (1936) decided that no name was available for most Arikareean camels that were more derived than *Paratylopus primaevus*. He created the genus *Gentilicamelus*, based on *"Gomphotherium"*

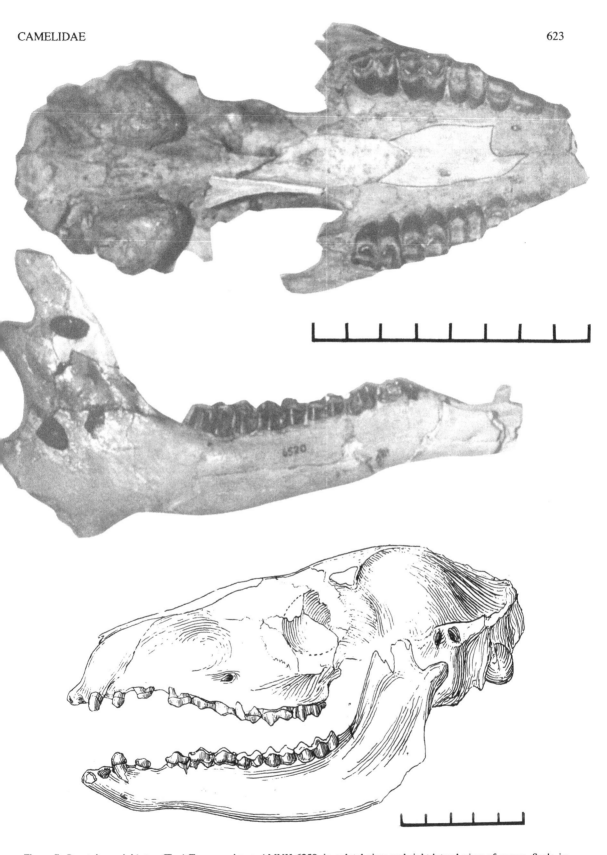

Figure 7. *Paratylopus labiatus*. (Top) Type specimen, AMNH 6250, in palatal view and right lateral view of ramus. Scale in cm. (Bottom) Referred specimen, F:AM 42278, in left lateral view. Scale bar in cm.

sternbergi. Unfortunately, he included within the genus a number of forms that have nothing to do with the type species, creating yet another taxonomic wastebasket. *Gentilicamelus* included such forms as the oxydactylines "*Gomphotherium*" *cameloides*, and "*Gentilicamelus*" *wyomingensis* (this taxon is based on an immature individual), and the enigmatic form "*Protomeryx*" *campestris* (here considered to be *Paralabis*), and "*Protomeryx*" *leonardi* (here considered to be a species of *Miotylopus*). Nothing unites this heterogeneous assemblage of camels except their supposed Arikareean age, and even that is open to question in some cases (McKenna, 1966). If *Gentilicamelus* is valid, it applies only to "*Gomphotherium*" *sternbergi*. As discussed below, *G. sternbergi* has a more derived dentition than *Paratylopus* or most of the other camels discussed here, so it appears to be a sister-taxon of the nothokematines (Honey et al., in press).

Paratylopus labiatus (Cope, 1881), new combination
Figure 7, Tables 2, 3, 5

Poebrotherium labiatum Cope, 1881
Poebrotherium wilsoni Loomis, 1928 (in part)

Type—AMNH 6520, a partial skull, associated mandible, and partial skeleton (Fig. 7).

Referred Specimens—*Paratylopus labiatus* is not as common as the contemporary form *Poebrotherium wilsoni*, but it is still quite abundant. Most large Orellan camels can be referred to *P. labiatus*. A partial listing of specimens in the Frick Collection includes the following: From the Lusk area, Niobrara Co., Wyoming (stratigraphic range = 40-80 feet above the PWL): F:AM 63743, right ramus; F:AM 63711, skull and jaws; F:AM 47334, partial skull and skeleton; F:AM 53733, right ramus; F:AM 63731, right ramus. From the Douglas area, Converse Co., Wyoming (stratigraphic range = 30-100 feet above the 5 tuff—Evanoff et al., 1992): F:AM 63837, juvenile right ramus; F:AM 63831, skull, jaws, and partial skeleton. From Geike Ranch, Sioux Co., Neb. (stratigraphic range = 0-30 feet above the PWL): F:AM 47436, skull and partial skeleton; F:AM 47283, skull and jaws. From Munson Ranch, Sioux Co., Neb. (stratigraphic range = 30-90 feet above PWL): F:AM 47274, left ramus; F:AM 47261, skull and jaws. From the Chadron area (Morris, Brecht, Bartlett, Schmechel Ranches), Dawes Co., Neb.: F:AM 47235A, quarry block of associated skulls, jaws and partial skeletons; F:AM 47147, right ramus; F:AM 47146, mandible; F:AM 47142, palate and mandible. From the "lower Oreodon beds," Sheep Mountain Table, Big Badlands, Shannon Co., S.D.: F:AM 42278, skull and jaws.

Diagnosis—Medium-sized camel (length M1-3 = 37.5-40 mm). Distinguished from *Poebrotherium* by its slightly reduced premolars (P2-4/M1-3 ratio is

Table 4. P2-4/M1-3 ratios, emphasizing the reduction of premolars in *Paralabis cedrensis*

SPECIMEN	P2-4/M1-3
Poebrotherium eximium	
AMNH 632 (type)	0.749
F:AM 47150	0.714
F:AM 47118	0.776
Poebrotherium wilsoni	
F:AM 42284	0.773
F:AM 47130	0.693
Paratylopus labiatus	
AMNH 6520 (type)	0.646
Paratylopus primaevus	
AMNH 9806	0.582
Paralabis cedrensis	
YPM 10167 (type)	**0.506**
Pseudolabis dakotensis	
AMNH 9807 (type)	0.709
F:AM 41687	0.587
F:AM 36469	0.684
Miotylopus gibbi	
F:AM 36446	0.629
Gentilicamelus sternbergi	
AMNH 7970 (type)	0.709

approximately 0.65—see Table 4) and its larger size. Distinguished from *Paratylopus primaevus* by its slightly smaller size, and its teeth, which are primitively large in relation to skull size.

Description—*Paratylopus labiatus* was described by Scott (1940), and figured by Cope and Matthew (1915, plate 115). In most skeletal features, *Paratylopus labiatus* resembles *Poebrotherium eximium*, except that it is slightly larger (Figs. 19-26). It has essentially primitive poebrotherine skeletal proportions (Scott, 1940, p. 640, fig. 131), which differs greatly from the gazelle-like proportions of *Poebrotherium wilsoni*. The most diagnostic features are size and the anterior part of the cranium. *Paratylopus labiatus* has many of the derived features found in *Poebrotherium wilsoni*: a large p1-2 diastema, large canines in males, enlarged I3, and i1-3 arranged in a closed "fan."

Discussion—I transfer "*Poebrotherium*" *labiatum* to the genus *Paratylopus* because it shares the following derived features with *Paratylopus primaevus*: larger size, more reduced premolars, and larger, more robust skull in relation to its tooth size (Tables 2, 3). *Paratylopus labiatus* seems to be clearly part of the monophyletic lineage that includes *Paratylopus primaevus*, so it is best referred to *Paratylopus*. It is not as hypsodont as the *Paralabis-Oxydactylus* lineages, or the stenomylines. If *P. labiatus* were retained in *Poebrotherium*, the latter genus would become paraphyletic.

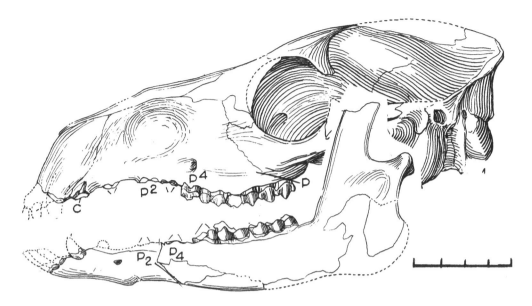

Figure 8. *Paratylopus primaevus*. Type specimen, AMNH 9806, in left lateral view of skull and mandible. Scale in cm.

Paratylopus primaevus (Matthew, 1904)
Figure 8, Tables 2, 3, 5

Miolabis (Paratylopus) primaevus Matthew, 1904
Paratylopus primaevus Matthew, 1909

Type—AMNH 9806, a skull with right ramus, partial skeleton (Fig. 8); Whitneyan (upper Oreodon beds), South Dakota.

Referred Specimens—From the Whitneyan (upper Oreodon beds), South Dakota: AMNH 17358, skull and jaws; USNM 14763, skull and jaws; AMNH 9803, skull, jaws, and partial skeleton. From Wetzel Ranch, 6 miles north of Mitchell, Sioux Co., Neb. (early Whitneyan): F:AM 47248, right ramus; F:AM 47247, right ramus. From the east side of Chimney Rock, 25 feet above lower Whitney Ash, Morrill Co., Neb. (early Whitneyan): F:AM 47237, fragmentary palate, right ramus, badly distorted partial skeleton. From H. A. Blackburn Ranch, 7 miles north of Mitchell, Sioux Co., Neb. (early Whitneyan): F:AM 47249, right ramus. From the Sherrill Hills, Niobrara Co., Wyoming (210 feet above PWL, ?early Whitneyan): F:AM 47076, left ramus.

Diagnosis—Medium-sized camel (M1-3 length = 39-41 mm, P3-M3 length = 64-65 mm), slightly larger than *Paratylopus labiatus*, but with more robust skull and more reduced premolars. Differs from *Gentilicamelus sternbergi* in that it is slightly smaller with less reduced premolars and less robust skull. Differs from species of *Paralabis* and *Pseudolabis* in having much lower-crowned teeth which are small relative to the size of the skull.

Description—Matthew (1904) and Scott (1940) thoroughly described this species. Measurements are given in Table 2, 3 and 5. Postcranial elements are shown in Figures 19-26.

Discussion—Very little new material of *Paratylopus primaevus* has appeared since its original description. If a larger sample of this species were known, it would probably overlap considerably with *Paratylopus labiatus;* the differences between the two are very slight.

Several authors (for example, Matthew, 1904; McKenna, 1966) have considered *Paratylopus* to represent the "central camelid lineage" which gave rise to *Oxydactylus*, *Miotylopus*, and *Gentilicamelus*. Although *Paratylopus primaevus* is a very primitive camel, it has autapomorphies (such as the skull which is large relative to the dentition) which exclude it from close relationship with *Paralabis*, *Oxydactylus*, or with the stenomylines.

Paralabis (Lull, 1921), new rank

Protomeryx Matthew, 1901 (in part)
Pseudolabis (Paralabis) Lull, 1921
Pseudolabis Hay, 1902 (in part)
Protomeryx Hay, 1930 (in part)
Gentilicamelus Loomis, 1936 (in part)
Paratylopus (="*Paralabis*") McKenna, 1966

Type and only species—*Paralabis cedrensis* (Matthew, 1901)

Diagnosis—Medium-sized camel (length of M1-3 = 43-45 mm) with moderately hypsodont teeth and highly reduced premolars (P2-4/M1-3 ratio = 0.506—see Table 4). Distinguished from *Poebrotherium*, *Paratylopus*, and *Gentilicamelus* by its larger size, more hypsodont teeth, and more reduced premolars. Distinguished from

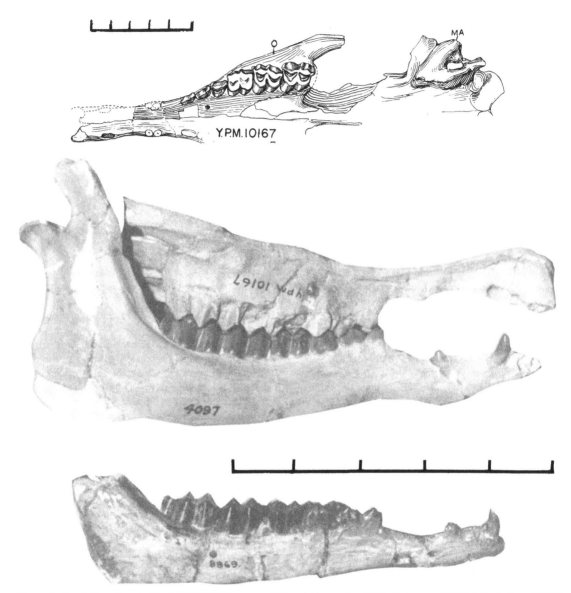

Figure 9. *Paralabis cedrensis* (Top). YPM 10167, type of "*Paralabis matthewi*." Scale in cm. (Middle) Occlusion of YPM 101067 with SDSM 4097, a lower jaw referred to "*Protomeryx cedrensis*," showing their close match in occlusion and premolar reduction. (Bottom) AMNH 8969, type of "*Protomeryx cedrensis*" (= "*P. campester*"). Scale: 1 bar = 2 cm.

Pseudolabis or *Miotylopus* by less hypsodont teeth and presence of mesostyle. Smaller than *Oxydactylus*, but otherwise quite similar.

Discussion—Whitneyan camels with extremely reduced premolars were first referred to the wastebasket taxon *Protomeryx cedrensis* by Matthew (1901). At that time, the species was known from a single lower jaw (AMNH 8969) from Colorado (Fig. 9). In 1921, Lull described a skull (YPM 10167) from the Whitneyan "*Protoceras* beds" of South Dakota, which showed a similar extreme reduction of the premolars. Lull compared it to *Pseudolabis* and *Poebrotherium*, but not to "*Protomeryx cedrensis*." On this basis, he concluded that it was referable to *Pseudolabis*, and created a new subgenus, *Paralabis*. Apparently, no one has noticed that the skull and jaw seem to match in size and morphology. As a result, the lower jaw taxon ("*Protomeryx cedrensis*") was placed in one wastebasket genus after another, and the skull has languished in obscurity, incorrectly referred to *Pseudolabis*.

McKenna (1966) did not associate the skull and jaw, but he did realize that they were both placed in the wrong genera. He recognized that "*Protomeryx campester*" was a taxon completely distinct from other Whitneyan camels, but he considered *Paralabis matthewi* to be an advanced form of *Paratylopus* that might be ancestral to *Gentilicamelus sternbergi*. Apparently he did not realize how extreme the premolar reduction was in *Paralabis*, or he would not have associated it with *Paratylopus* and *Gentilicamelus*.

In comparing the types of "*Protomeryx*" *cedrensis* and *Paralabis matthewi*, it seems clear that they belong to the same genus and species. Both specimens have more extreme premolar reduction than any other Whitneyan or Arikareean camel. The P2-4/M1-3 ratio of *Paralabis matthewi* is 0.506, the lowest of any camel in this study (Table 4). The type lower jaw of "*Protomeryx*" *cedrensis* is too broken to measure accurately, but a referred lower jaw (SDSM 4097) from the *Protoceras* channels of South Dakota has a p2-4/m1-3 ratio of 0.540, lower than any other camel. *Paralabis matthewi* and "*Protomeryx*" *cedrensis* are the same size, and, allowing for post-mortem distortion, they even occlude quite well (Fig. 9). The type of "*Protomeryx*" *cedrensis* comes from Colorado, but the referred specimen (SDSM 4097) comes from the same beds as the type of *Paralabis matthewi*. Although we still do not have very complete material, it is clear from the present evidence that they are the same taxon, and I synonymize them. Since *Paralabis* is the first valid genus-rank name applied to this taxon, I raise it from subgeneric rank to generic rank (since it is clearly not a subgenus of *Pseudolabis*).

The relationships of *Paralabis* are somewhat ambiguous. The extreme reduction of premolars seems to be an autapomorphy, and thus does not clarify relationships. The increased hypsodonty strongly argues against *Paralabis* having affinities to *Paratylopus* or *Gentilicamelus* (*contra* McKenna, 1966). Instead, the hypsodonty suggests that *Paralabis* is an advanced camel, related to either the stenomyline group or to the higher camels (Fig. 1). Unfortunately, *Paralabis* is too poorly known at present to resolve its relationships further.

Paralabis cedrensis (Matthew, 1901), new combination
Figure 9, Tables 2, 3, 4

Protomeryx cedrensis Matthew, 1901: 358, fig. 29
Protomeryx campester Matthew, 1901: 422
Protomeryx campester Matthew, 1904
Protomeryx cedrensis Matthew, 1909
Pseudolabis (Paralabis) matthewi Lull, 1921
Pseudolabis matthewi Hay, 1930
Protomeryx campester Hay, 1930
Protomeryx cedrensis Hay, 1930
Protomeryx campester Stock, 1935

Gentilicamelus (Protomeryx) cederensis [*sic*] Loomis, 1936
Gentilicamelus (Protomeryx) campestris Loomis, 1936
Paratylopus (= "*Paralabis*") *matthewi* McKenna, 1966
Protomeryx campester (= *cedrensis*) McKenna, 1966

Type—AMNH 8969 (Fig. 9), a lower mandible from the Whitneyan (Vista Member) of northern Colorado (Galbreath, 1953, p. 89).

Referred Specimens—KU 133, mandible from the Vista Member, Colorado; YPM 10167, skull, Whitneyan, *Protoceras* beds, South Dakota (Fig. 9, type of *Paralabis matthewi*); SDSM 4097, right ramus, *Protoceras* beds, South Dakota (Fig 9); F:AM 47310, skull, middle Whitney Member (5 feet below Upper Whitney Ash), 1 mile northwest of Pussy Springs, Ruby Ranch, Morrill Co., Nebraska. The postcranial skeleton of *Paralabis cedrensis* is presently unknown.

Diagnosis—Same as for genus.

Description—Since the only new specimen (F:AM 47310) is less well preserved than the type of *P. matthewi* (YPM 10167), nothing new can be added to Lull's (1921) original description of the skull. The referred lower jaws are all very similar to the type, so Matthew's (1901) original description is sufficient. In most features, *Paralabis* is very similar to *Paratylopus*, except that it is much more hypsodont and has much more reduced premolars.

Discussion—The correct species name of this taxon has been greatly confused in the literature. Matthew (1901) gave two different names to the same specimen. In the text, the name *Protomeryx cedrensis* appears first on p. 358 and in fig. 29 (p. 422), and in a subsequent table (p. 423). But at the bottom of p. 422, below fig. 29, is the formal descripton under the heading "*Protomeryx campester* n. sp." This mistake misled several authors (including Hay, 1930, and Loomis, 1936), who thought that the two names represented different specimens. In his next mention of the species, Matthew (1904) used *Protomeryx campester*. However, in the subsequent literature, Matthew (1909) went back to using the name *cedrensis*. In a manuscript in the American Museum archives that was unpublished at the time of his death, Matthew continued to use *cedrensis*.

Stock (1935, p. 122, footnote 1) decided that Matthew originally intended to use the name *campester*, even though the name *cedrensis* appears many times, and *campester* is used only once. McKenna (1966) followed Stock's determination. However, the name *campester* cannot be used. First, it is clear that Matthew intended to use *cedrensis*, since the name *campester* appears only once, and must have been a *lapsus calami*. Secondly, as first reviser, Matthew (1909, MS) persisted in using *cedrensis* in all other publications (except Matthew, 1904). This was apparently his final intent, since he had referred the species "*Protomeryx cedrensis*" to *Oxydactylus*, which already had a species named

Table 5. Camelid postcranial measurements (in mm)

	Poebrotherium			Paratylopus		Pseudolabis			Miotylopus		Gentili-
	eximium	wilsoni		labiatus	primaevus	dakotensis		leonardi	gibbi	taylori	camelus
	F:AM	F:AM	AMNH	AMNH	AMNH	F:AM	F:AM	F:AM	F:AM	F:AM	AMNH
	47118	47130	1364	6520	9806	41687	41942	36655	36446	41829	7910
		(female)	(male)	(type)	(type)	(female)	(male)	(female)	(male)	(male)	(type)
CHARACTER											
Axis centrum L	—	49	—	61	56	58	—	—	67	73	—
Axis centrum dors/vent ht	—	37	—	41	48	—	—	—	45	50	—
Atlas W, at transv. proc.	49	40	—	55	54	37	—	—	53	68	—
Atlas L, dorsal centrum	25	30	—	25	31	25	—	—	30	39	—
Scapula, length	—	140	—	—	127	—	163	—	170	167	—
W of neck	—	18	—	18	18	19	22	—	28	23	—
W of scapula	—	98	—	—	77	88	88	—	103	90	—
Humerus length	150	147	150	—	—	—	—	—	195	—	177
Midshaft W	14	16	15	19	—	18	—	—	14	18	22
Radius-ulna L	177	188	182	214	—	258	310	—	309	—	259
Midshaft radial W	15	14	13.	16	—	22	23	—	20	21	21
Metacarpus L	—	117	127	120	—	183	214	180	238	298	178
MCIII-IV W	—	15	13	15	—	19	22	13	13	17	29
Femur L	185	182	161	—	—	—	264	—	250	—	223
Midshaft W	18	12	14	29	—	—	21	—	19	—	20
Tibia L	193	—	180	222	—	254	292	—	289	335	256
Midshaft W	16	15	13	16	—	20	21	—	29	22	20
Metatarsus L	142	133	129	152	—	185	216	—	244	—	186
MTIII-IV W	16	16	13	18	—	18	22	—	13	—	28
Astragalus L	27	25	28	28	—	29	38	—	30	—	34
Astragalus W	16	15	18	17	—	20	21	—	17	—	20
Calcaneum L	52	—	55	55	57	63	54	—	76	82	74
Calcaneal tuber L	34	—	31	35	48	37	54	—	51	61	52

campestris. Although I do not agree that this taxon is synonymous with *Oxydactylus*, a serious problem arises if the name *campester* is used for *Paralabis*. The feminine form of the adjective, *campestris*, is required when it modifies the suffix *-labis* (Latin, feminine, "forceps"). (The ending of *cedrensis* is correct for either masculine or feminine nouns.) If the name became *Paralabis campestris*, it could be easily confused with a similar camel, *Oxydactylus campestris* Cook, 1909, from the late Arikareean of Nebraska. Since the author's original intent and later revisions favored *cedrensis*, and reviving the name *campester* could create a rare case of secondary homonymy (and certainly a lot of confusion), I designate *cedrensis* as the valid species name for the only species of *Paralabis*.

Subfamily STENOMYLINAE Matthew, 1910
(= Pseudolabidinae Simpson, 1945)

Known Distribution—Whitneyan-early Barstovian (early Oligocene to middle Miocene) of the North American High Plains; early Arikareean (late Oligocene) of California.

Diagnosis— Medium to large camels (P2-M3 length = 65-82 mm) with long rostra and a deep, anterioposteriorly elongate maxillary fossa. Teeth are more hypsodont and transversely narrow than *Poebrotherium*, *Paratylopus*, *Paralabis*, or oxydactylines. Very weak or no mesostyle on the upper teeth, or metastylid on the lower teeth. The premaxilla is extended posteriorly at least to the level of P1.

Included Taxa—*Pseudolabis* Matthew, 1909; *Miotylopus* Schlaikjer, 1936; and the Stenomylini Matthew, 1910 [here reduced to tribe rank], which include *Stenomylus* Peterson, 1906; *Blickomylus* Frick and Taylor, 1968; *Rakomylus* Frick, 1937.

Discussion—As listed above, there are a number of unique shared derived characters that unite a clade consisting of *Pseudolabis*, *Miotylopus*, and the stenomylines. Despite the striking similarities between these taxa, most authors have postulated that these similarities arose independently. Since the Stenomylini appear suddenly in the late Arikareean with all their bizarre specializations, several authors (for example, Peterson, 1906; Frick and Taylor, 1968) postulated that they diverged very early in camel evolution, possibly as early as the Eocene. McKenna (1966, p. 4) was the first

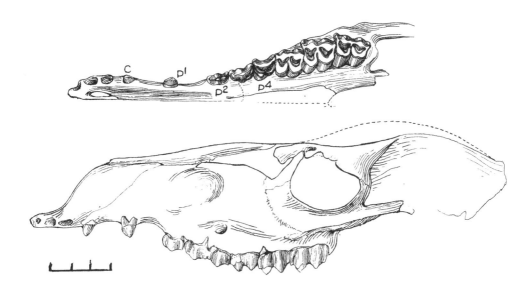

Figure 10. *Pseudolabis dakotensis*. AMNH 9807, type specimen in left lateral and palatal view. Scale in cm.

to note the similarities between stenomylines, *Pseudolabis*, and *Miotylopus (= Dyseotylopus)*, but he did not pursue these questions further. Honey and Taylor (1978, p. 419) pointed out that *Stenomylus, Pseudolabis*, and *Miotylopus* share the derived character of a weak or absent mesostyle. In addition to this feature, the peculiar maxillary fossa, the precocious elongation and hypsodonty of the teeth, and the posterior extension of the premaxillary are all good synapomorphies which corroborate this hypothesis of relationships. Therefore, I expand the subfamily Stenomylinae to include not only the traditional stenomylines of Frick and Taylor (1968), but also their primitive sister-taxa, *Pseudolabis* and *Miotylopus*. As a consequence, the traditional taxa of the Subfamily Stenomylinae are grouped under the Tribe Stenomylini, to reflect this hierarchical relationship (Fig. 1).

The expanded taxon Stenomylinae is an important, previously unrecognized monophyletic group that was the dominant group of camels during the Whitneyan and Arikareean. By contrast, the *Paratylopus* lineage was relatively rare in the Whitneyan, and extinct by the Arikareean. The oxydactylines, paralabines, and miolabines did not diverge until the late Arikareean, when the more primitive stenomylines, such as *Pseudolabis* and *Miotylopus*, declined and only the hyperspecialized Stenomylini remained.

Pseudolabis Matthew, 1904

Pseudolabis Matthew, 1904
non Pseudolabis Lull, 1921
   Type and only species—*Pseudolabis dakotensis* Matthew, 1904

Range—Whitneyan (*Protoceras* beds) to late Arikareean (Harrison Formation), High Plains of Nebraska, Wyoming, and South Dakota.

Diagnosis—Medium to large camels (length of M1-3 = 47-59 mm) with a slight flexure of the P4 lingual selene. Like all stenomylines, *Pseudolabis* has a deeply depressed maxillary fossa, a posteriorly elongated premaxilla, and relatively high-crowned teeth. It is further distinguished from *Miotylopus* in having a slightly shorter rostrum and less reduced premolars. *Pseudolabis* can be distinguished from all non-stenomyline camels by its weak mesostyles, deep maxillary fossa, and posteriorly extended premaxilla.

Discussion—From his initial descriptions, Matthew (1904) recognized that *Pseudolabis* was a precociously specialized form in having hypsodont teeth and a closed orbit. However, *Pseudolabis* has always been relegated to a "side branch" of the Camelidae, since it is clearly too specialized to be ancestral to any other camel. McKenna (1966, p. 4) was the first to point out dental similarities between *Pseudolabis, Dyseotylopus*, and the stenomylines. In addition to these dental similarities, several other unique features of the skull strongly corroborate the hypothesis that *Pseudolabis* is part of a monophyletic group including *Miotylopus* and the stenomylines (see above).

Pseudolabis dakotensis Matthew, 1904
Figures 10-11, Tables 2, 3, 5

   Type—AMNH 9807, badly crushed female skull lacking basicranium (Fig. 10), and an associated atlas, from the Whitneyan "*Protoceras* beds" of South Dakota.
   Referred Specimens—From the Whitneyan of

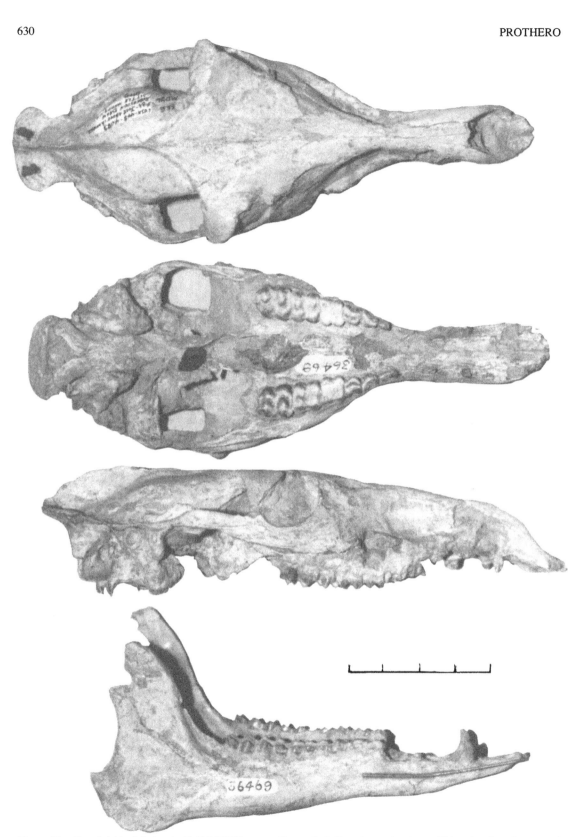

Figure 11. *Pseudolabis dakotensis*. F:AM 36469, an undistorted skull and associated mandible, showing the normal morphology of the species. Scale: 1 bar = 2 cm.

Nebraska (5 feet below the base of the Gering Formation, 155 feet above the Upper Whitney Ash, northeast corner of Castle Rock, southeast of Bayard, Scottsbluff Co., Nebraska): F:AM 41687, female skull and jaws, three partial skeletons. From the Sharps Formation, Shannon Co., S.D. (latest Whitneyan or Arikareean, *fide* Tedford et al., 1985): F:AM 47188, male skull. Gering Formation and equivalents (early Arikareean): From south side of 66 Mountain, 15 feet below white layer, Goshen Co., Wyoming: F:AM 31867, left male maxilla; F:AM 36793, tibia and calcaneum; From Horse Creek, Willow Creek area, Goshen Co., Wyoming: F:AM 36798, femur, tibia, tarsals and metatarsals; F:AM 36795, metacarpals and metatarsals; F:AM 36802, metatarsals and phalanges; F:AM 41800, female skull and jaws; Little Muddy Creek area, Goshen Co., Wyoming: F:AM 41807, partial male skull and jaws; F:AM 41815, partial skull and jaws; F:AM 36432, right ramus; F:AM 36469, male skull and jaws (Fig. 9B); F:AM 41942, male skull, jaws, partial skeleton. Monroe Creek Formation and equivalents (middle Arikareean): Muddy Creek area, Goshen Co., Wyoming: F:AM 41700, female mandible; F:AM 36642, male mandible and partial skeleton; F:AM 36791, juvenile left ramus; F:AM 41814, juvenile skull and partial skeleton; F:AM 41948, complete forelimb; F:AM 36470, male palate; F:AM 41943, female left ramus; F:AM 41811, palate and partial skeleton; F:AM 36475, male right ramus. Harrison Formation, Niobrara Co., Wyoming (late Arikareean): F:AM 36487, female mandible; F:AM 41845, humerus and ulna, metacarpals; F:AM 36547, humerus, metacarpals; F:AM 36812, metacarpals and phalanges. Harrison Formation, Sioux Co., Nebraska: F:AM 41844, femur, tibia, metatarsals, phalanges; F:AM 36647, femur, tibia, metatarsals, tarsals, phalanx.

Diagnosis—Same as for genus.

Description—A beautifully uncrushed skull and mandible (F:AM 36469) gives a much better conception of *Pseudolabis* than does the badly crushed and distorted type specimen (Fig. 11). The rostrum is more elongate than *Paratylopus*, but not as narrow as seen in *Miotylopus*. The posterodorsal extension of the premaxilla terminates above the diastema between P1 and P2. There is a deep, anteroposteriorly elongated maxillary fossa. The lacrimal vacuity is small and slit-like, contrary to the condition in the type specimen (which is badly crushed and distorted). Between the supraorbital canals is a slight midline concavity on the frontals. The postorbital processes are very broad and fused to the zygomatic arch. They form a definite "hood" around the posterodorsal rim of the orbit. The braincase is relatively small, but the sagittal crest and lambdoid crests are very large, sharp, and prominent. The zygomatic arches are slender; their maximum lateral width is at the level of the orbit.

The basicranium is well preserved, and shows the typical camelid condition. The wing-like basioccipital processes are present, although weak. The postero-internal (paroccipital) extension of the tympanic bulla is very prominent because the lateral portion of the bulla is broken. The glenoid surfaces are broad and slightly dorsolaterally inclined (as is typical for camels). The external auditory meatus is strong, but the postglenoid foramen is nearly absent. The well preserved pterygoids do not show the peculiar bifurcate tips unique to *Pseudolabis*. The secondary palate terminates at the level of M2. The infraorbital foramen lies above P4, as is typical of camels.

The anterior dentition of F:AM 36469 is poorly preserved (as is the anterior dentition of the type), but F:AM 41942 has a good anterior dentition. I1-2 are small, chisel-shaped teeth spaced out along the tip of the premaxilla. I3 is greatly enlarged, conical, and recurved, as in most primitive camels, including *P. wilsoni* (but not *P. eximium* or *P. franki*), *Paratylopus*, *Miotylopus*, and *Oxydactylus*. There is a large I3-canine diastema. The canines in males are also large, conical and recurved, although not so large as I3. The canines of females are much smaller, and in many cases barely protrude from the maxilla. There is a larger canine-P1 diastema, and a slightly smaller P1-P2 diastema. P1 is a narrow, two-rooted blade. P2 is also bladelike, but P3 is more triangular, with a strong parastyle and weak lingual cusps and cingula. The P3 ectoloph is concave. This condition is primitive for camels, and differs markedly from the condition seen in higher camels. P4 shows the diagnostic *Pseudolabis* flexure of the lingual selene that was a part of Matthew's (1904) definition of the genus. However, only the type shows the extreme development of two lingual crests. Most other specimens have just a slight kink in the posterior half of the lingual selene.

M1-3 show the typical camelid selenodont condition, but are higher crowned than the upper molars of *Poebrotherium*, *Paratylopus*, or *Oxydactylus*. All of the styles are relatively weak, but the mesostyle is completely absent or weakly developed near the base of the tooth. The ribs are stronger than the styles, but not as marked as in most other camels. There are no lingual cingula or cusps on the molars.

The mandible is unknown in the type specimen, but F:AM 36469 is nearly complete and undistorted (Figure 9B). I1-3 are leaf-shaped and arranged into a continuous fan-like pattern. These teeth wear flat at the tips in older individuals. The lower canines follow the incisors without a diastema, and show sexual dimorphism (as do the uppers). Males have canines which are large, conical and recurved, but females have smaller, laterally flattened canines about the size of p1. The c-p1 and p1-p2 diastemata are of about equal length. p1 is a small, simple blade. p2 is also bladelike, but with a parastylid. p3 has a weak parastylid, and two posterior crests which are closely appressed. As usual, the labial crest is lingually recurved, and longer than the lingual crest. p4 has a strong parastylid and a short posterolingual crest.

The lower molars are high crowned, anteroposteriorly elongate, and transversely compressed. Their lingual borders are smooth, with no metastylids. m3 is particularly narrow, with a basined hypoconulid.

The symphysis is unusually long and narrow, and fused to the level of posterior p1. It shows relatively little ventral deflection. The coronoid process is long and slender, rising from the tooth row at the typical camelid angle. There is a deep elongate fossa running along the dorsolateral edge of the coronoid process immediately posterior to m3. It is deeper in *Pseudolabis* than in any other camelid; this fossa condition seems to be unique to this genus. The condyle shows the typical camelid condition, with a straight posterior border. The angular hook is on the level of the crowns of the lower molars. The angular process protrudes very slightly ventrally.

Most of the best preserved elements described below are from a nearly complete skeleton, F:AM 41942. Of the axial skeleton, only the cervical vertebrae are known. They are similar in proportion to those of *Paratylopus*, except for their larger size (Fig. 19). The neck of *Pseudolabis* was apparently no longer than that of any other primitive camel.

The scapula is best known from F:AM 41942 (Fig. 20). It is narrower dorsally than *Paratylopus*, with a smaller infraspinous fossa. The spine is very prominent and posteroventrally recurved at its margin. There is a small acromion process, but it does not extend as far as the glenoid. The coracoid process is a distinct knob.

The humerus (Fig. 21) is relatively short and slender compared to *Paratylopus*. It is otherwise very similar except for a distally extended entepicondyle. The fused radius-ulna is also relatively long and slender, with a less pronounced olecranon (Fig. 22). No carpals are known for *Pseudolabis*.

Metacarpals III-IV are unfused, and slightly shorter than the metatarsals (Fig. 23). Absolute metapodial length seems quite variable, probably because of the strong sexual size dimorphism. The metacarpal length is usually equal to the distance between the foramen magnum and the P2 on a male skull.

The poorly preserved pelves are similar to those of other camels in parts that remain. The femur has a deep digital fossa and a prominent greater trochanter that is slightly above the level of the head (Fig. 24). The lesser trochanter is a distinct process on the plantar side, as in *Paratylopus* and most tylopods. The tibia (Fig. 25) is relatively long and slender compared to *Paratylopus*, with a very strong cnemial crest. The tarsals are typically camelid (Fig. 26). Metatarsals III and IV are fused (unlike the metacarpals), and equal in length to the distance between the occipital condyle and P1. The phalanges are unremarkable.

In general, *Pseudolabis* is a slightly longer-limbed, more gracile animal than *Paratylopus*, but its proportions are very similar to those of *Oxydactylus*. The metacarpals are slightly shorter than the metatarsals; the former are still unfused, but the latter are completely fused.

Discussion — Although the sample of *Pseudolabis* is still small, there seems to be an unusually large variation in size. This size variation persists throughout the history of *Pseudolabis*, since both small and large individuals can be found at each level in the Whitneyan and Arikareean. I suspect that this size difference is due to sexual dimorphism, since the larger forms invariably have much larger canines than the smaller forms. Both large and small morphs have the same stratigraphic range, which also suggests that they are sexual dimorphs, rather than two species.

Miotylopus Schlaikjer, 1935

Oxydactylus Loomis, 1911 (in part)
Protomeryx Loomis, 1911 (in part)
Miotylopus Schlaikjer, 1935
Dyseotylopus Stock, 1935
Gentilicamelus Loomis, 1936 (in part)

Type Species — *Miotylopus gibbi* (Loomis, 1911)

Included Species — *Miotylopus leonardi* (Loomis, 1911), new combination; *Miotylopus taylori*, new species.

Range — Arikareean (Gering to Harrison formations), Wyoming, Nebraska, and South Dakota; early Arikareean, southern California.

Revised Diagnosis — Stenomyline camels with highly reduced premolars and elongate rostra. Differs from *Pseudolabis* in these two features and in the lack of a pseudolabine flexure on P4 lingual selene. Differs from the Stenomylini in having lower-crowned teeth; P1/1-C1/1 diastemata still present; and no diastemata between P2/2 and P3/3. Differs from all other camels in having high-crowned teeth with very weak or no mesostyles, and dorsal premaxilla extended posterior to the level of P1. Size quite variable. P2-M3 length ranges from 65 to 84 mm.

Discussion — Two small, primitive oxydactyline-like camel jaws were described by Loomis (1911). The larger jaw was named *Oxydactylus gibbi* (p. 67) and the smaller jaw was called *Protomeryx leonardi* (p. 68). Loomis thought that these specimens were from the "Upper Harrison Formation" in the Muddy Creek area, Goshen County, Wyoming, but McKenna and Love (1972, p. 26) believed that they were from much lower in the Arikaree Group. In the Frick Collection, camels from the Muddy Creek area are from sediments equivalent to the Gering and Monroe Creek formations, and occasionally equivalent to the Harrison Formation — but none are equivalent to the "Upper Harrison" rocks of Nebraska. Apparently, Loomis's Muddy Creek

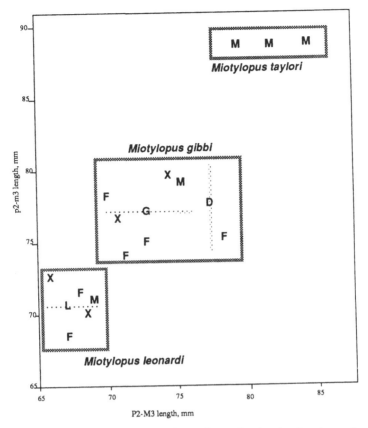

Figure 12. Plot of p2-m3 vs. P2-M3 of selected specimens of *Miotylopus*, showing the size clusters here recognized as *M. leonardi* (small species), *M. gibbi* (medium species), and *M. taylori* (large species). "M" and "F" indicate specimens that can be sexed as male or female; "X" indicates a specimen whose sex is indeterminate. "L" = type of "*Protomeryx*" *leonardi* (a lower jaw, so P2-M3 length is unknown); "G" = type of "*Oxydactylus*" *gibbi* (a lower jaw, so P2-M3 length is unknown); "D" = type of "*Dyseotylopus migrans*" (a partial skull, so p2-m3 length is unknown).

specimens are early or middle Arikareean in age.

In May, 1935, Schlaikjer described the skull and jaws of a camel from the "Lower Harrison" on the south side of 66 Mountain, Goshen County, Wyoming. According to McKenna and Love (1972, p. 28), this specimen was actually from Monroe Creek Formation or older rocks. Schlaikjer named this form *Miotylopus bathygnathus*, and compared it only to "*Paratylopus*" *sternbergi*, *Poebrotherium*, and "*Protomeryx*." He recognized that the mesodont molars without a mesostyle were a new combination for camels, and this was part of his diagnosis for the genus. In July of the same year, Stock (1935) described a camel, *Dyseotylopus migrans*, from the early Arikareean Kew Quarry in the Sespe Formation, Las Posas Hills, Ventura County, California (see this volume, Chapter 8). It also had mesodont teeth and weak mesostyles on the upper molars. Schlaikjer (1935, p. 176) briefly discussed this specimen, and considered it distinct from his *Miotylopus*, or possibly intermediate between *Miotylopus* and *Poebrotherium*.

Loomis (1936) placed the enigmatic *Protomeryx leonardi* in his new genus *Gentilicamelus*. As discussed above, nothing united the different species that Loomis assigned to *Gentilicamelus* except their Whitneyan-Arikareean age. Most of the contents of *Gentilicamelus* are here referred to *Paralabis*, *Oxydactylus*, and *Miotylopus*, except for the type species, *G. sternbergi*.

McKenna (1966) straightened out much of the confusion regarding small Arikareean camels with weak mesostyles. He transferred "*Oxydactylus*" *gibbi* to *Miotylopus*, although he did not formally synonymize it with *M. bathygnathus* (consistently misspelled "*brachygnathus*" by McKenna, 1966, and McKenna and Love, 1972). McKenna (1966) also noted the similarities of *Miotylopus* with *Dyseotylopus*, although he felt they represented separate lineages. In his 1966 paper, McKenna suggested that "*Protomeryx*" *leonardi* might be referable to *Dyseotylopus*. But in 1972, McKenna and Love formally synonymized *leonardi*, "*brachygnathus*" [*sic*], and some other specimens with *M. gibbi* (new combination). They also described a new specimen from the Arikareean of Darton's Bluff, Johnson County, Wyoming, which was considerably smaller than other specimens of *M. gibbi*.

McKenna and Love (1972) referred specimens from the Castolon l.f. of Texas described by Stevens et al. (1969) to *M. gibbi*. Stevens (1977, p. 48) has since referred these camels to *Michenia*, based on more complete material, including upper molars with strong mesostyles. Frick and Taylor (1971) suggested that *Dyseotylopus* might be related to *Michenia* and the protolabidines, but Taylor (cited in Stevens, 1977, p. 51) later rejected this idea.

In my studies of the large samples of Arikareean camelids with weak mesostyles in the Frick Collection, I found considerable variation in size. Specimens with relatively broad rostra, unreduced premolars, and the typical *Pseudolabis* P4 were all referred to *Pseudolabis* (see above). The remaining specimens were all united by the derived condition of reduced premolars and relatively slender rostra. All had weak or absent mesostyles, and their dorsal premaxillae are always extended posterior to the level of P1, so they are clearly stenomylines. Most specimens have the deep, elongate premaxillary fossa characteristic of stenomylines, although there is some variation in this feature (see *M. taylori* below). After sorting specimens by size and canine development (which seem to be the best indicator of sexual dimorphism in camels), I found no strong size dimorphism in most *Miotylopus* (unlike in *Poebrotherium* or *Pseudolabis*). For example, one of the smallest jaws, F:AM 36427, has male-shaped canines, yet is the same size as a jaw with presumed female canines, F:AM 36806, from the same deposits (Fig. 13D). In sorting the sample, there seemed to be three distinct size clusters (Fig. 12) with both males and females represented. The small form includes the type of "*Protomeryx*" *leonardi*. The medium-sized form includes the types of "*Oxydactylus*" *gibbi*, *Miotylopus bathygnathus*, and *Dyseotylopus migrans*. A very large form was also found that has never been named or described, and must be a new species. These size clusters are much more apparent in overall proportions of skull and mandible than they are in tooth dimensions (Fig. 12).

The largest forms are a new species, described as *Miotylopus taylori* below. The medium- and small-sized forms are more difficult to separate. McKenna and Love (1972) lumped them together as *Miotylopus gibbi*. However, I am not comfortable with such a wide difference in size and morphology (not attributable to sexual dimorphism) in a single species. The differences are clearly not due to ontogeny, either. F:AM 36441 (a small male) and F:AM 36446 (a medium-sized female) are strikingly different in size, yet their M3's are fully erupted and show comparable wear. I find that the specimens can easily be sorted by size into small and medium-sized *Miotylopus*. Therefore, I recognize two species: the medium-sized *M. gibbi* and the smaller *M. leonardi* (new combination).

Miotylopus is the first valid generic name for this group of camels. It is the senior synonym of *Dyseotylopus* by two months. The name *Miotylopus* was originally chosen because these camels are typical of the Arikareean, which was then considered early Miocene. Ironically, nearly all of the early and middle Arikareean (and thus nearly all *Miotylopus*) are now considered late Oligocene (Tedford et al., 1987; Prothero and Rensberger, 1985), so the name *Miotylopus* has become a misnomer.

Miotylopus leonardi new combination
Figures 13-14, Tables 2, 3, 5

Protomeryx leonardi Loomis, 1911
Gentilicamelus leonardi Loomis, 1936
"*Protomeryx leonardi*" Skinner et al., 1968
Miotylopus gibbi McKenna and Love, 1972 (in part)

Type—AC 2004, a complete right ramus (Fig. 13), from the early Arikareean (Monroe Creek or Harrison formations), Muddy Creek area, "3 miles below Spanish Diggings Spring" (Loomis, 1911, p. 68), Goshen County, Wyoming.

Referred Specimens—The listing below includes catalogued specimens which have the necessary diagnostic elements (usually skulls or jaws). Much catalogued and uncatalogued material in the Frick Collection is referable to this species based on size, but is too incomplete or fragmentary to be certain of this.

From the Gering Formation and equivalents, early Arikareean, Horse Creek-Tremaine area, Sioux County, Nebraska: F:AM 36486, left ramus, right maxilla; F:AM 36463, left and right rami; F:AM 36660, left female ramus; F:AM 36658, male skull and mandible; F:AM 41819, skull; F:AM 36797, juvenile skull, mandible, partial skeleton. Little Muddy Creek area, Niobrara County, Wyoming: F:AM 36806, female mandible; F:AM 36665, female rostrum and mandible; F:AM 36464 partial skull, left ramus; F:AM 36820, female right ramus; F:AM 36442, right ramus; F:AM 36427, female right ramus; F:AM 41944, male left ramus, partial skeleton; F:AM 36441, male mandible; F:AM 36450, partial skull; F:AM 36806, female rostrum; F:AM 36443, female palate; F:AM 36807, female left ramus; F:AM 36473, left ramus; F:AM 41810, palate; F:AM 36825, left ramus; F:AM 41946, female left ramus; F:AM 36452, left ramus; F:AM 36440, left ramus; F:AM 36451, female skull, tarsus; F:AM 36448, juvenile skull; F:AM 36435, left ramus; F:AM 41945, right ramus, partial skeleton; F:AM 36425, skull; F:AM 36655, male skull, mandible, partial skeleton; F:AM 36796, female skull, mandible, partial skeleton (Fig. 13B-C); F:AM 36441, mandible; F:AM 36810, female partial skull and skeleton; F:AM 36846, female skull and partial mandible; F:AM 36447, female skull, mandible, and partial skeleton. From the Monroe Creek Formation and equivalents, Muddy Creek area, Niobrara County, Wyoming (middle Arikareean):

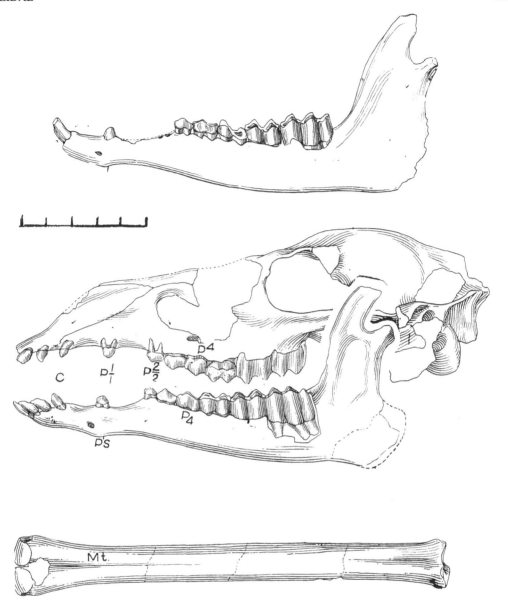

Figure 13. *Miotylopus leonardi*. (Top) AC 2004, type of *"Protomeryx" leonardi* (reversed for comparison) (Middle) F:AM 36796, skull and jaws referred to *M. leonardi*. (Bottom) F:AM 36796, associated metatarsals. Scale in cm.

F:AM 36821, partial skeleton; F:AM 36823, female skull, mandible, and partial skeleton; F:AM 41995, partial skeleton; F:AM 36436, right maxilla, partial skeleton; F:AM 41949, female right ramus; F:AM 41950, palate; F:AM 36445, male mandible; F:AM 36483, left ramus; F:AM 36484, left ramus; F:AM 36434, female left ramus, partial skeleton; F:AM 41953, female mandible; F:AM 41831, male mandible. From the Harrison Formation and equivalents, near Node, Niobrara County, Wyoming (late Arikareean): F:AM 36785, female mandible; F:AM 36826, female left ramus; F:AM 36437, left ramus; F:AM 36426, male right ramus.

Diagnosis—Smallest species of *Miotylopus* (P2-M3 length = 65-70 mm).

Description—The type specimen (Fig. 13) of *M. leonardi* is a right ramus. Much more complete material of the species is represented in the Frick Collection. The skull (as shown by F:AM 36425, F:AM 36796, and F:AM 36655) has a rather delicate premaxilla similar to that seen in *Stenomylus*. The nasal process of the premaxilla extends well posterior to P1. There is a deep, elongate maxillary fossa. The lacrimal vacuity was apparently small and slit-like, although most specimens are damaged in this area. The postorbital bar is complete. The frontal portion of the postorbital bar has an almost pedicellate contact with the braincase, since the orbits project laterally from the side of the head. The cranium is small and bulbous, as in *Stenomylus*, with very small sagittal and lambdoid crests. The basicranium is very poorly preserved in most specimens, but what can be seen is typically camelid. The bulla is relatively small for the size of the skull.

The upper incisors are small and chisel-like. They are arranged continuously around the tip of the rostrum. I3 is not enlarged. A small incisiform canine (slightly larger in males) lies immediately posterior to I3. There is a short C-P1 diastema and the P1-P2 diastema is about the same length. P1 is incisiform, but P2 is a narrow blade. P3 is blade-like, with a weak postero-lingual cingulum. P4 is fully selenodont, with no flexure of the lingual selene (contrary to the condition in *Pseudolabis*). P2-4 are considerably reduced in comparison to the molars. The molars are relatively elongate, narrow, and subhypsodont, with very weak mesostyles. However, they do not show the extreme elongation and hypsodonty seen in *Stenomylus*.

The mandible (as seen in F:AM 36796 or F:AM 36441) is slender and gracile, with a very narrow symphysis. The i1-3 are leaf-shaped, with a strong lingual rib. They are arranged in a continuous "fan" around the symphysis. The lower canine shows marked sexual dimorphism. In presumed females, it is almost incisiform, and only slightly separated from the incisors. In presumed males, however, it is a much larger, more robust caniniform tooth, lying much further posterior from the rostrum. The c-p1 and p1-p2 diastemata are of about equal length. The symphysis terminates at the level of p1, which is a small, simple, bladelike tooth, as is p2. The p3 has a lingually inflected parastyle, but is otherwise bladelike. The p4 has a strong parastyle, but is so severely compressed that the posterior ridges are fused. The p2-4 are noticeably reduced in size relative to the size of the molar row, a feature of *Miotylopus* that distinguishes it from many other camels. The lower molars are transversely narrow, subhypsodont, and m3 is particularly elongated. There is a bladelike hypoconulid on m3. The posterior ramus shows the typical slender recurved coronoid, high condyle, and distinct hooked angle. The dorsal edge of the angular "hook" is well above the tooth row.

Although there is much postcranial material from the localities that produce *M. leonardi*, little of it is associated with cranial material that can be identified with *M. leonardi*. The metapodials associated with F:AM 36655 and F:AM 36796 are long and slender, and both metacarpals and metatarsals are fused. The metacarpals are approximately equal in length to the metatarsals, and both are equal to the total length of the skull. The smaller unassociated postcranial elements which may belong to *M. leonardi* are similarly long and slender compared to other camels.

Discussion — As discussed above, *Miotylopus* can be divided into three distinct size groups. The smallest species includes the ramus that was originally named *Protomeryx leonardi* (Loomis, 1911). This ramus best matches the small species of *Miotylopus*, although it shows few uniquely derived features of that genus. It is clear, however, that it can be referred to no other taxon, because: 1) it is too large and too hypsodont for *Poebrotherium*; 2) it is too small, too hypsodont, and its premolars are too reduced for *Paratylopus*; 3) it is similar in size and premolar reduction to *Paralabis*, but the ramus is much shallower dorsoventrally; 4) it is too small for *Pseudolabis*, *Oxydactylus*, or any other species of *Miotylopus*; and 5) it does not have the extremely hypsodont teeth of *Stenomylus*. "*Protomeryx*" *leonardi* also comes from the same deposits as the Frick Collection samples which are here referred to *M. leonardi*.

Miotylopus leonardi shows some striking resemblances to *Stenomylus* (Fig. 14A-C). It is nearly identical in size, and has similarly slender premaxillae, flaring pedicellate postorbital processes, bulbous braincase with weak sagittal crests, and relatively short basicranium. The anterior dentitions are very similar, particularly in females, with their small canines (male *Stenomylus* show no noticeable dimorphism in the canines). The major difference between the two taxa is that *Stenomylus* has shifted an incisiform p1 into the anterior cropping battery, and developed a p2-p3 diastema. The cheek teeth of *Stenomylus*, of course, are much more derived, with their extreme anteroposterior lengthening and hypsodonty of M1-3/1-3 (particularly M3/3), and their highly reduced premolars. *Stenomylus* also has a smooth, rounded angular process, and short coronoid process not seen in most other camels.

Nevertheless, the similarities between the two are very impressive, although there are few features which are unique to these two taxa. Both taxa are similar-sized members of the Stenomylinae. The skull similarities are partly related to their similarity in size, and most of the dental similarities are primitive for the Steno-mylinae. The reduction of I3 may unite the two (other species of *Miotylopus* have a larger I3). However, this same reduction occurs independently in several other camelids, and may not be very reliable. No unique synapomorphies unite *Miotylopus leonardi* and *Stenomylus*, so the species *leonardi* cannot be referred to *Stenomylus*. In non-cladistic terms, *Stenomylus* more closely approaches *M. leonardi* than any other known camel, and *M. leonardi* has no known features which would rule out ancestry of the Stenomylini. It is also of the right age to represent the lineage which gave rise to the Stenomylini. Such hypotheses, however, are speculative.

However, there is certainly no reason to believe (as did Peterson, 1908, and Frick and Taylor, 1968) that the

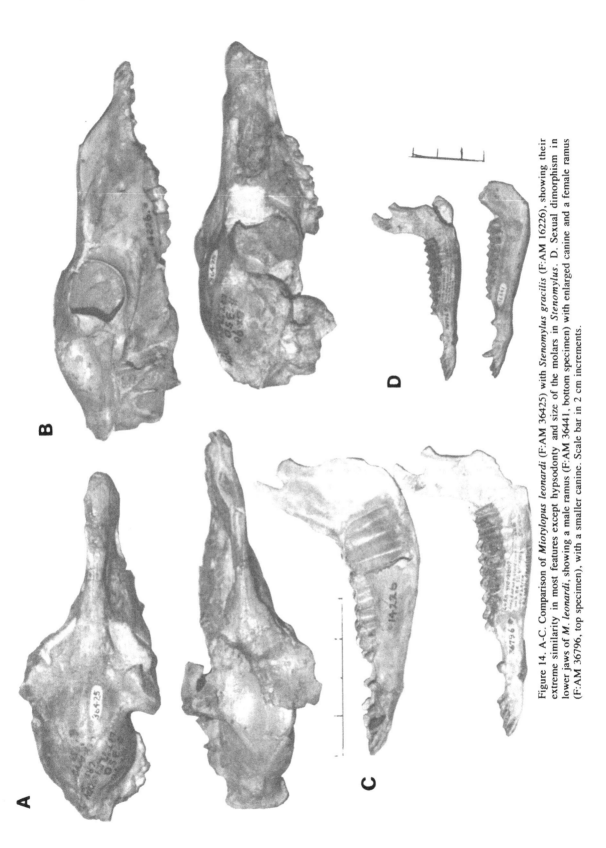

Figure 14. A-C. Comparison of *Miorylopus leonardi* (F:AM 36425) with *Stenomylus gracilis* (F:AM 16226), showing their extreme similarity in most features except hypsodonty and size of the molars in *Stenomylus*. D. Sexual dimorphism in lower jaws of *M. leonardi*, showing a male ramus (F:AM 36441, bottom specimen) with enlarged canine and a female ramus (F:AM 36796, top specimen), with a smaller canine. Scale bar in 2 cm increments.

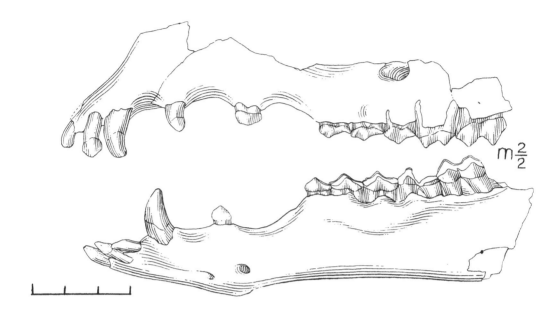

Figure 15. *Miotylopus gibbi*. YPM 10328, type specimen of *"Oxydactylus" gibbi*, including partial upper and lower jaws. Scale in cm.

ancestry of stenomylines must go back to the Eocene. Although stenomylines are very specialized, relatively little change (increased hypsodonty and the anterior shift of p1 and p2) is required to make a *Stenomylus* out of *M. leonardi*. Frick and Taylor (1968, pp. 6-7) argued against the relationship of stenomylines with other Oligocene camelids because "it would require a more rapid rate of evolution for *Stenomylus* to be derived from *Poebrotherium* between Whitneyan and Arikareean time than is known to occur elsewhere in the Camelidae." It is fallacious to argue that because some camelids show relatively slow, stereotyped evolution, all of them must. Stenomylines were highly specialized for an ecological niche very different from their contemporaries. In occupying this new niche, they could have evolved very rapidly.

Miotylopus gibbi (Loomis, 1911)
Figures 15-16, Tables 2, 3, 5

Oxydactylus gibbi Loomis, 1911
Miotylopus bathygnathus Schlaikjer, 1935
Dyseotylopus migrans Stock, 1935
Miotylopus gibbi McKenna, 1966
Miotylopus brachygnathus [*sic*] McKenna, 1966
Miotylopus gibbi McKenna and Love, 1972 (in part)

Type—YPM 10328, palate and lower jaws missing M3/3 (Fig. 13A); from the Gering or Monroe Creek formations ("Upper Harrison" according to Loomis, 1911), "on Muddy Creek, about opposite to the spring associated with the 'Spanish Diggings'" (Loomis, 1911, p. 67).

Referred Specimens—As was the case with *M. gibbi*, the list below includes only a part of the Frick Collection that bears the diagnostic features of this species (mainly skulls and jaws). Many more catalogued and uncatalogued specimens are probably referable to this species, but will not be listed below.

From Gering Formation and equivalents, Little Muddy Creek, Niobrara County, Wyoming (early Arikareean): F:AM 36477, left ramus; F:AM 36454, skull, mandible, partial skeleton; F:AM 41803, male partial skull, mandible; F:AM 36446, male skull, mandible, partial skeleton (Fig. 16); F:AM 41826, male rostrum. From the Monroe Creek Formation and equivalents, Muddy Creek, Niobrara County, Wyoming (middle Arikareean): F:AM 41689, skull; F:AM 41695, female left ramus; F:AM 41805, partial skeleton; F:AM 41830, partial mandible and skeleton; F:AM 36457, partial skeleton; F:AM 36453, male rostrum; F:AM 41808, male mandible; F:AM 48133, female mandible; F:AM 41809, skull and ramus; F:AM 41831, partial skull; F:AM 41812, male mandible; F:AM 41854, right ramus. Head of Warbonnet Creek, Sioux County, Nebraska: CM 1329, mandible, right maxillary fragment. From the Harrison Formation and equivalents (late Arikareean), near Node, Niobrara County, Wyoming: F:AM 36436, female mandible and humerus; F:AM 36431, female left ramus and symphysis. From the early-middle Arikareean, 150 feet

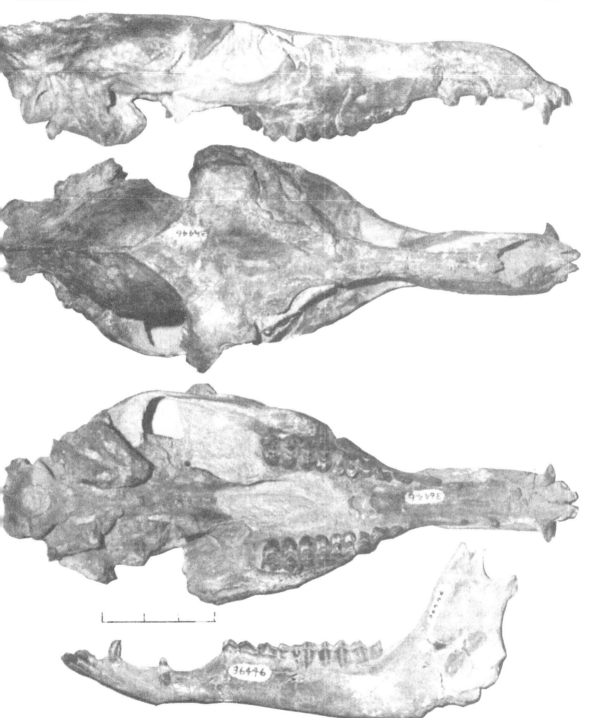

Figure 16. *Miotylopus gibbi*. F:AM 36446, referred male specimen of *M. gibbi*, showing complete skull and mandible. Scale in 2 cm increments.

above the top of the Brule Formation, Goshen Hole, Goshen County, Wyoming: MCZ 2924, partial skull and jaws, pelvis (type of *Miotylopus bathygnathus*). From the ?early Arikareean Kew Quarry, Sespe Formation, Las Posas Hills, Ventura County, Califor-

nia: LACM(CIT) 1721 (type of *Dyseotylopus migrans*).

Diagnosis—Medium-sized species of *Miotylopus*. Length of P2-M3 = 71-79 mm. M3 more elongate relative to the size of the animal than in any other species of *Miotylopus*.

Description—The type of *M. gibbi* consists of a palate and jaws, and the types of *M. bathygnathus* and *Dyseotylopus migrans* are poorly preserved. A skull and jaw, F:AM 36446 (referred to *M. gibbi*) shows all the known features of these three type specimens and is much better preserved, with only slight dorsoventral flattening (Fig. 16). The rostrum is longer and narrower than in *Pseudolabis* or *Oxydactylus*. The nasal process of the premaxilla extends almost to the level of the P2. There is a deep, elongate maxillary fossa, slightly exaggerated by the crushing. The frontals are very broad, with almost no medial depression between the supraorbital canals. The lacrimal vacuity is narrow and slit-like. The postorbital processes are broad and flaring, with a prominent lateral flange. They are well sutured to the zygomatic arch, closing the orbit. The braincase is relatively small, with moderately weak sagittal and lambdoid crests. The basicranium is of the normal camelid type, with moderately large bullae. The pterygoids flare widely. The secondary palate terminates at the level of M2, and the internal nares are broadly open.

I1-2 are small, chisel-like, and evenly spaced around the tip of the rostrum. I3 is very large and caniniform. There is a long I3-C diastema, and shorter C-P1 diastema, and a moderately long P1-P2 diastema. The upper canine is small and incisiform. P1 is a small, blunt blade. P2 is bladelike, but with a weak lingual cingulum. P3 has a strongly convex ectoloph and a weak lingual cingulum. P4 is fully selenodont, with a faint cusp on the posterolabial rim of the lingual selene. M1-3 are fully selenodont, with weak mesostyles and moderately strong ribs. M3 is more elongate relative to the size of the animal than is found on other species of *Miotylopus*.

The lower i1-3 are chisel-like, with faint lingual ribs. The lower canine follows after a short diastema. It is large, upright, and caniniform in males, but nearly incisiform in females. The lower c-p1 and p1-p2 diastemata are quite large, with sharp diastematal crests on the dorsal edge of the ramus. The p1 is a small, blunt crest. p2-4 are blade-like, very narrow, and small relative to the size of the molars. m1-3 are mesodont with a basined m3 hypoconulid. The posterior ramus is typically camelid in all its preserved parts.

The postcranial skeleton of F:AM 36446 is relatively complete. The scapula (Fig. 20) is slender, with a relatively small supraspinous fossa. The acromion is long and delicate, and the coracoid process is unusually thick, partly because the neck above the glenoid is so narrow. The humerus (Fig. 21) has a very shallow bicipital groove and a relatively weak lateral tuberosity. The condyles are quite narrow, and the epicondyles are very weak. The radius-ulna (Fig. 22) is long and slender, with a prominent olecranon. The metacarpals are fused and approximately the same length as the skull. They are only slightly shorter than the fused metatarsals. The femur (Fig. 24) is remarkably long and slender, with a pronounced curvature that is convex dorsally. The greater trochanter is less robust than in *Paratylopus*, but the lesser trochanter is in the typical plantar position. The tibia (Fig. 25) is as long as the ulna. It is unusually slender at midshaft, but quite robust at the proximal end, with a prominent cnemial crest. The calcaneum has an unusually long tuber, and a very prominent convex proximal fibular facet. The phalanges are long and slender.

Discussion—Medium-sized *Miotylopus* have been referred to two different genera and three species, but there seems no justification for more than one species. The first name applied to this group was *Oxydactylus gibbi* (Loomis, 1911). Although this specimen consists only of a palate and mandible, the teeth are typical of *Miotylopus*, including the weak mesostyles and large I3.

Schlaikjer's (1935) *Miotylopus bathygnathus* is a camel of the same size and morphology as "*Oxydactylus*" *gibbi*, and this provides the first valid generic name for the group. In all of its preserved features, it falls within the normal range of variation of medium-sized *Miotylopus*. Schlaikjer considered the orbits open, but the specimen is too fragmentary to determine this. The only anomalous feature is the slightly smaller bullae with a shallow tympanohyal groove, but this is within the range of variation seen in the Frick Collection sample.

Dyseotylopus migrans (Stock, 1935) is a camel similar in size, age, and preserved morphology to those referred here to *M. gibbi*. The only significant difference between *Dyseotylopus* and the other specimens of *M. gibbi* is that the mesostyles are stronger at the base of the molars of *Dyseotylopus* than they are in most stenomylines. However, I have seen a similar condition in some Frick Collection *M. gibbi*. This feature is not of generic or specific significance, so I place *Dyseotylopus migrans* in synonymy with *Miotylopus gibbi*. Demére (1988) reports specimens of both cf. *Dyseotylopus* and *Miotylopus* from the early Arikareean Eastlake l.f. in southern San Diego County. However, based on more extensive collections, Demére (personal communication) now recognizes only a single camelid taxon at Eastlake. It is neither *Dyseotylopus* nor *Miotylopus*, but a new stenomyline, which will be described by Deméré in the near future.

McKenna (1966) suggested that *Miotylopus* was an intermediate between *Paratylopus* and *Oxydactylus*. However, the anatomy and relationships of *Miotylopus* are much better known now. It is clear that *Miotylopus* is a highly derived camel closely related to *Pseudolabis* and the Stenomylini. If one seeks a sister-taxon for the oxydactylines, *Paralabis* makes a much better candidate than does *Miotylopus*.

Miotylopus taylori, new species
Figure 17, Tables 2, 3, 5

Type—F:AM 36459, male skull, mandible (Fig. 17), atlas, and axis. From the early Arikareean (Gering Formation correlative) rocks of the Willow Creek area, Goshen County, Wyoming.

Referred Specimens—From the Muddy Creek area (middle Arikareean, Monroe Creek equivalent), Niobrara County, Wyoming: F:AM 36824, male skull and mandible; F:AM 41855, partial skeleton; F:AM 41829, partial skeleton; F:AM 36461, male palate, mandible, and partial skeleton; F:AM 34460, skull, mandible, partial skeleton.

Etymology—In honor of Beryl Taylor, who devoted most of his career to the study of the Frick Collection camels, and whose hard work and insights made this research possible.

Diagnosis—Largest species of *Miotylopus*. P2-M3 length = 81-85 mm. The p4 posterior crests are separate, unlike other species in the genus.

Description—The type specimen (Fig. 17) is a nearly complete skull and jaws missing only the jugal portion of the zygomatic arches. The skull is unusually large and narrow for *Miotylopus*, although this may be partly due to lateral crushing. The nasal process of the premaxilla extends to the level of the center of the P1-P2 diastema. The maxillary fossa is relatively shallow in the type specimen, although it shows the more typical condition in another referred specimen, F:AM 36824. The rostrum is long and slender, but the palate is not as wide as is typical of other *Miotylopus*. This, too, may be an artifact of crushing, since F:AM 36824 shows more normal *Miotylopus* proportions in the palate. The lacrimal vacuity is ovoid in shape, with a tapered anterior end. There is a marked midline depression in the frontals between the supraorbital canals. The postorbital processes of the zygomatic arches are broken, so it is impossible to determine if the orbits were closed in the type. They are closed in F:AM 36824, however. The braincase in the type is relatively narrow, again due to deformation, since the braincase in F:AM 36824 looks more normal in proportions. There is a very high, thin sagittal crest, and a long lambdoid crest which flares laterally from a narrow base.

The basicranium is well preserved, although it has not been sufficiently prepared to see many foramina or other fine details. It does show the typical higher camelid bulla, posteriorly projecting paroccipitals, wing-like basioccipital processes anterior to the occipital condyles, and broad shelf-like glenoids. The glenoid foramen was apparently quite small. The pterygoids are long and slender, and slightly pinched posterior to the palate. The secondary palate terminates at the posterior end of M2.

I1-2 are leaf-shaped, without a lingual rib, and spaced evenly across the front of the rostrum. I3 is very large and caniniform. The upper canine is also fairly large and caniniform, although smaller than I3; apparently the type specimen was a male (as is evident from the lower canine as well). There are long diastemata between the I3-C, the C-P1, and the P1-P2. P1 is a two-rooted, pointed blade. P2 is blade-like and simple. P3 has a convex ectoloph, but no lingual cingula or cusps. P4 is selenodont with a small labial space in the posterior lingual selene (as in *M. gibbi*). The molars are mesodont, with strong ribs and weak styles. The mesostyles are completely absent. M2 and M3 are also relatively elongate anteroposteriorly.

The mandible is slender, particularly anterior to the cheek teeth. The lower incisors are leaf-shaped and arranged in the typical "fan." The lower canine in the type is large and caniniform, so this specimen appears to be a male. There is a short i3-c diastema, and long, convex c-p1 and p1-p2 diastemata. p1 is a small, single-rooted triangular blade. p2 and p3 are also bladelike and transversely compressed. p4 is unlike that in *M. gibbi* in that the posterior crests are still separate, although short and nearly connected. m1-3 are mesodont and anteroposteriorly elongate. The m3 hypoconulid is only partially basined.

The coronoid process slopes backward much more than is typical for camels, and the tip is shorter and less slender. This is in part due to the unusually high condyle, which resembles the condition seen in some higher camelids. The angular process is badly broken, but its ventral extension projects sharply below the ventral border of the ramus.

Two cervical vertebrae were associated with F:AM 364591. The transverse processes of the atlas (Fig. 19) are unlike most other camels because they have long posterior extensions that are sharply separated from the posterior face of the centrum. The intervertebral and alar foramina were apparently quite large, although this has been exaggerated by breakage of the lateral part of the transverse processes. The axis is long with a short odontoid process and a long dorsal spine. The spine has a nearly horizontal dorsal border, and a much broader anterior end than *Oxydactylus campestris*. It is also laterally narrower than the condition seen in *O. campestris*, but otherwise similar in size and proportion.

No other postcranial remains are associated with the type specimen, but parts of the skeleton are found in association with other fragmentary referred dentitions. A scapula (F:AM 36461) is very similar to that of *M. gibbi*, except that it is proportionally larger (Fig. 20). It has a relatively small supraspinous fossa, a long curved acromion, and a thick blunt coracoid process. The humerus (based on F:AM 41829) is relatively short and slender, with a posteriorly extended entepicondyle (Fig. 21). It is articulated with the proximal end of the radius-ulna, which has a relatively short, robust olecranon. The fused metacarpals are equal to the skull in length,

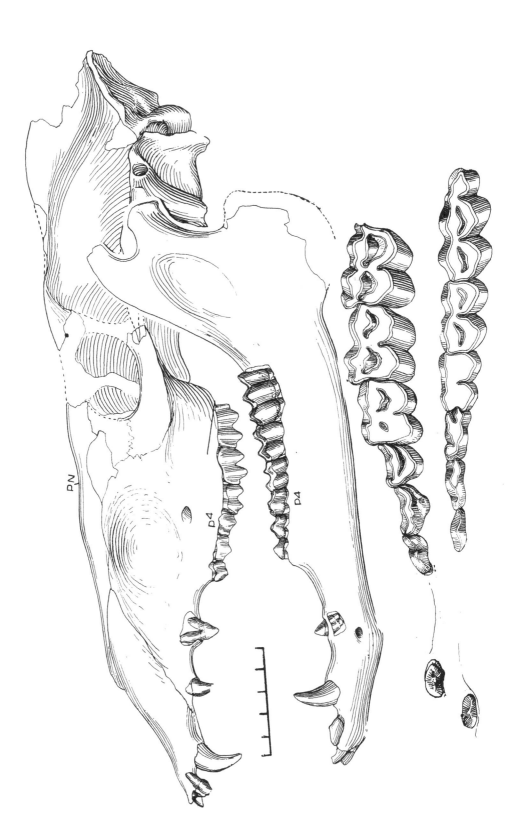

Figure 17. *Miotylopus taylori*, new species. F:AM 36459, type specimen, showing left lateral view of skull and mandible, and crown views of upper and lower dentitions. Scale in cm.

and approximately as long as the metatarsals.

The pelvis (F:AM 46829) is known from a fragment, which shows that it was relatively small and gracile, with a deep acetabulum. The femur is unknown. The tibia (F:AM 41829) is exceptionally long and slender, with a weaker cnemial crest and proximal end than *M. gibbi*. It does show a similar dorsal curvature. This specimen of a tibia is articulated with the tarsals and metatarsals (Fig. 26). Although poorly preserved, the striking feature of these specimens is the length of the calcaneal tuber, which has a knob-like, rugose end, and a plantar keel.

Discussion—The striking feature of the skull and skeleton of *M. taylori* is its more giraffe-like proportions, with elongate cervical vertebrae and distal limb elements. In this respect, it converges on many of the oxydactylines and aepycamelines. However, the diagnostic *Miotylopus* skull features clearly show that these features are convergent (as happens many times in camels).

The difference in overall proportions may also explain why three closely related species of *Miotylopus* could live sympatrically through most of the Arikareean in Wyoming and Nebraska (Fig. 12). Apparently, they were ecologically differentiated. *M. gibbi* maintains the more primitive proportions seen in many camelids. *M. leonardi* has the gazelle-like size and build that eventually was the niche of the stenomylines. *M. taylori* had the longer neck and legs of the giraffe-camel niche, which was later occupied by *Oxydactylus*, the aepycamelines, and other later camels.

Although sparse material referable to *M. gibbi* and *M. leonardi* is known from the upper Arikareean Harrison Formation and equivalents, it appears that the heyday of *Miotylopus* was over by the late Arikareean. Instead, the Harrison and younger formations are characterized by the oxydactylines (which were also highly differentiated in size and ecological adaptations) and by the highly specialized stenomylines.

Subfamily "NOTHOKEMATINAE" Honey et al.
(in press)
Gentilicamelus Loomis, 1936

Poebrotherium Cope, 1879 (in part)
Gomphotherium Cope, 1886 (*non* Burmeister, 1837)
Protomeryx Hay, 1902 (in part)
Miolabis (Paratylopus) Matthew, 1904 (in part)
Paratylopus Cope and Matthew, 1915 (in part)
Type and Only Species—*Gentilicamelus sternbergi* (Cope, 1879).

Revised Diagnosis—Large camel (P2-M3 length = 70.0 mm), distinguished from contemporary *P. primaevus* by its larger size, and by the presence of metastylids on the lower molars. Distinguished from all other Eocene/Oligocene camels by its advanced features, including lower molar metastylids, strong P3 lingual

cingulum, more spatulate lower incisors, and shortened P1. Distinguished from Miocene camels by its primitive skull proportions and by its brachydont teeth which are small relative to the skull size (Table 2).

Discussion—As described above, the generic affinities of this specimen have been disputed ever since it was first described. Although it shares many primitive similarities with *Paratylopus*, the presence of many derived oxydactyline features, such as the lower molar metastylids, high angular process, strong P3 lingual cingulum, shortened P1, and more spatulate lower incisors show that it is probably a sister-taxon of the later Arikareean oxydactylines. Since its stratigraphic position within the John Day Formation is unknown, it is possible that it comes from the late Arikareean, early Miocene portion of the section.

Honey et al. (in press) presented evidence that *G. sternbergi* is a much more advanced camel than the primitive poebrotheres and stenomylines discussed so far. In particular, it shares the enlarged mandibular angle and closely appressed P1 roots of *Nothokemas*, so it is tentatively placed in the "Nothokematinae." This clade of camels is also closely related to the bizarre floridatragulines, although in the cladogram (Fig. 1) of Honey et al. (in press), the "Nothokematinae" are slightly closer to the advanced camelids.

Gentilicamelus sternbergi (Cope, 1879)
Figure 18, Tables 2, 3, 5

Poebrotherium sternbergi Cope, 1879
Gomphotherium sternbergi Cope, 1886
Protomeryx sternbergi Hay, 1902
Miolabis (Paratylopus) sternbergi Matthew, 1904
Paratylopus sternbergi Cope and Matthew, 1915
Paratylopus sternbergi Lull, 1921
Gentilicamelus sternbergi Loomis, 1936

Type—AMNH 7910, skull, mandible, and partial skeleton (Fig. 18); from the ?Upper John Day beds (?late Arikareean), Oregon.

Referred Specimens—AMNH 7913, mandible and limb bones; AMNH 7911, mandible; both from the type locality.

Diagnosis—Same as for genus.

Description—Cope's (1879, 1886) and Wortman's (1898) descriptions of *P. sternbergi* are very sketchy, and also inaccurate. The skull of *P. sternbergi* has a very robust appearance seen in no other camel (Fig. 18). This is due to the relatively short, broad rostrum and the disproportionately small teeth. The rostrum is broken anterior to the canine alveolus, so the condition of the upper incisors and canines is unknown. The maxilla has a broad, shallow, ovoid fossa, characteristic of primitive camels. There is a slight concavity along the anterior midline of the frontals, just anterior to the supraorbital canals. The skull is broken in the region of the lacrimals on both sides, so it is impossible to determine

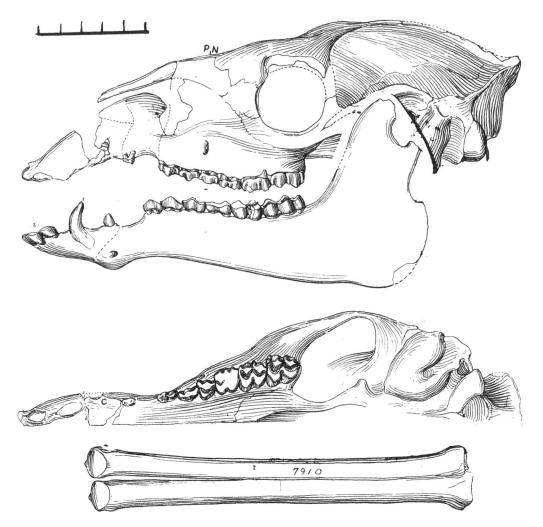

Figure 18. *Gentilicamelus sternbergi*. AMNH 7910, type specimen, showing left lateral view of skull, mandible, palatal view of skull, and associated metatarsals. Scale in cm.

the original size and shape of the lacrimal vacuity. As McKenna (1966) noted, the postorbital region is badly broken, so the original continuity of the postorbital bar is impossible to determine. Neither Wortman's (1898, p. 114) nor Cope's (Cope and Matthew, 1915, p. 116) figures are accurate in this respect. The braincase is disproportionately large by comparison with *P. primaevus*. The sagittal and lambdoid crests are weaker than in *P. primaevus,* but they are partially broken.

The zygomatic arches are poorly preserved, but the left side is more complete and shows that the arch flares much less than in most other camels. The basicranium is only crudely prepared, so few details of the basicranial structure can be determined. However, the bullae are even more inflated and bulbous than would be expected for the skull size, and the portion of the bulla medial to the tympanohyal groove is considerably smaller than

the lateral portion. The pterygoids are poorly prepared, but show the primitive, broadly separated condition (in contrast to the pinched condition seen in stenomylines). The secondary palate terminates at the level of M2 (the primitive condition).

The upper dentition is well worn, but reasonably well preserved. I1-3 and the canines are missing, as mentioned above. Both first upper premolars are broken, although from their remnants, it appears that they have the shape of double-rooted blades. They appear to be disproportionately small in comparison to the first upper premolars of *P. primaevus* and other camels. There is a large diastema between P1 and P2, which is a bladelike tooth with a strong parastyle and lingual cingulum. P3 is more triangular, with an even stronger parastyle and a labially-inflected metastyle. A distinct lingual cingulum with a small cusp is present on P3.

P4 shows the typical camelid selenodont condition, although with a stronger parastyle. M1-3 are fully selenodont, with strong styles and even stronger ribs. The overall impression of the teeth is that they are unusually brachydont for a camel, and disproportionately small in comparison to the size of the skull.

The mandible is complete but poorly preserved and not very well prepared. The ramus is relatively robust, yet has a more ventrally deflected symphysis than does *P. primaevus*. Lower i1-3 are wedgelike and closely associated, fanning out from the symphysis. The lower canine is an unusually large, posterolabially-curved tusk. Apparently this individual was a male, since the canine seems to be sexually dimorphic in many camels. There are short diastemata before and after the canine, and a longer diastema between p1 and p2. The p1 is a small, narrow wedge with two roots. p2 is bladelike, but broader and less compressed than is typical for camels. The same is true of p3, which also has a strong, lingually-inflected parastylid. The parastylid of p4 is even larger and more lingually inflected. The posterolabial crest of p4 is much larger than the posterolingual crest, and both are connected to form a small fossette. Lower m1-2 are too badly worn and broken to describe, other than that they are low-crowned and selenodont. The m3, however, shows a faint metastylid, a derived character seen only in higher camels. The hypoconulid of m3 is broad and basined.

The posterior ramus shows the typical slender camel coronoid, with small "pillow"-shaped condyles which are relatively high on the jaw. The dorsal "hook" of the angular process is very high—well above the level of the tooth row, and much higher than in primitive camels. This, too, appears to be a derived feature that unites it with oxydactylines. There is a slight ventral extension of the angular process below the level of the ventral margin of the ramus. The whole posterior portion of the jaw appears to be more robust than is typical, but this may be an artifact of preservation.

Except for the larger size, the postcranial skeleton (Figs. 19-26) is very similar to *P. primaevus*, and requires no additional description.

Camelidae *incertae sedis*

(sister taxon to the Aepycamelinae and higher camels)

Oxydactylus Peterson, 1904

Discussion—The genus *Oxydactylus* has long been a taxonomic wastebasket for a variety of Arikareean camels (McKenna, 1966; Honey et al., in press). Most do not belong to the genus, and some have been reassigned in Table 1. In addition, the genus *Oxydactylus* as presently consituted is probably a paraphyletic assemblage that includes the sister-taxa of aepycamelines, protolabines, miolabines, and other higher camels (Fig. 1). Honey and Taylor (1978) have already shown that some specimens referred to *Oxydactylus* are

actually primitive protolabines. The same may also be true for primitive aepycamelines labeled "*Oxydactylus*." Honey (personal communication) is currently revising the taxonomy of oxydactyline camels, some of which is summarized in Honey et al. (in press). Since *Oxydactylus s.s.* first appears in the early Miocene (latest Arikareean, Harrison Formation and younger beds), it is beyond the scope of the present study. When comparisons to *Oxydactylus* are made in this document, I am referring to the type species, *O. longipes*, and to the most primitive species, *O. campestris*.

A NOTE ON THE FLORIDATRAGULINES

The affinities of the bizarre, long-snouted floridatragulines (*Floridatragulus, Aguascalientia*) from the early Hemingfordian of Florida and Texas have mystified many scientists (for example, White, 1940, 1941, 1942; Simpson, 1945; Romer, 1948; Ray, 1957; Olsen, 1959, 1962; Patton, 1964, 1966, 1969; Maglio, 1966; McKenna, 1966). They were originally placed in the hypertragulines by White (1940), but Ray (1957) and most subsequent workers have realized that they were camels. Their peculiar, crocodile-like rostrum, strong intervallic cingular cusps on the upper molars, and divided hypoconulids on m3 are autapomorphic and do not clarify their relationships. However, the rest of their anatomy is clearly camelid, as several recent workers (Patton, 1966, 1969; Maglio, 1966) have shown in detail. Some (for example, Patton, 1969; Maglio, 1966) thought that floridatragulines were an early side branch of the Camelidae, with no apparent close relatives. Stevens (1977, p. 56) states that the floridatragulines are descended from oromerycids, but she cites no derived oromerycid characters in support of this contention. All of the characters she does mention are primitive for tylopods, or autapomorphic, and she ignores the great number of camelid synapomorphies seen in floridatragulines. Apparently, she has changed her mind, because in her contribution to Honey et al. (in press) she agrees floridatragulines belong in the Camelidae.

Honey et al. (in press) cited a number of derived characters to show that floridatragulines are camelids, more derived than the stenomyline-pseudolabine clade, but more primitive than miolabines, protolabines, or oxydactylines (Fig. 1). An undescribed specimen of *Floridatragulus* (F:AM 31864) shows an apparently continuous postorbital bar, so it may be even more derived than postulated by Honey et al. (in press). Stevens (in Honey et al., in press) argues that *Poebrotherium franki* is the sister-taxon of the floridatragulines, since it has a slightly elongate rostrum with diastemata between some of the anterior teeth. Elongated rostra and diastemata betweeen the anterior teeth occur several times in the Camelidae, and this feature is so prone to parallelism that I see no reason for attempting to push the ancestry of floridatragulines back

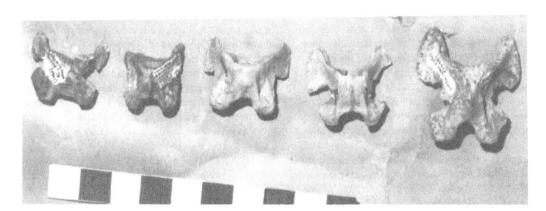

Figure 19. Atlas vertebrae. From left to right: *Poebrotherium eximium*, F:AM 47118; *Poebrotherium wilsoni*, F:AM 47130; *Paratylopus labiatus*, AMNH 6520; *Paratylopus primaevus*, AMNH 9806; *Miotylopus taylori*, F:AM 36459. Scale bar in 2 cm increments.

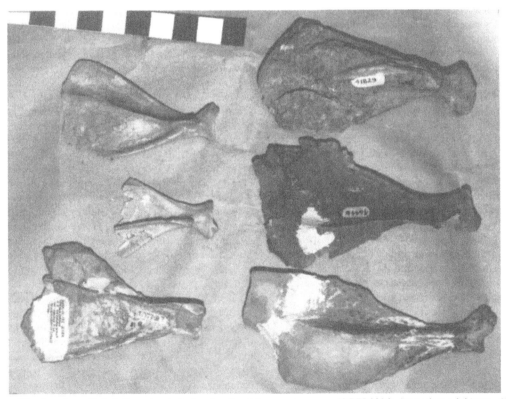

Figure 20. Scapulae. On left (from top to bottom): *Paratylopus primaevus*, AMNH 9806; *Paratylopus labiatus*, AMNH 6520; *Poebrotherium wilsoni*, F:AM 47077. On right (from top to bottom): *Miotylopus taylori*, F:AM 41829; *Miotylopus gibbi*, F:AM 36446; *Pseudolabis dakotensis*, F:AM 41942. Scale bar in 2 cm increments.

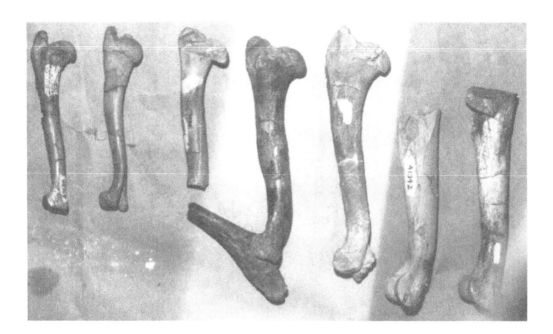

Figure 21. Humeri. From left to right: *Poebrotherium eximium*, F:AM 47118; *Poebrotherium wilsoni*, AMNH 1364; *Paratylopus labiatus*, AMNH 6520; *Gentilicamelus sternbergi*, AMNH 7910; *Miotylopus gibbi*, F:AM 36446; *Pseudolabis dakotensis*, F:AM 41942; *Miotylopus taylori*, F:AM 41855. Scale bar in 2 cm increments.

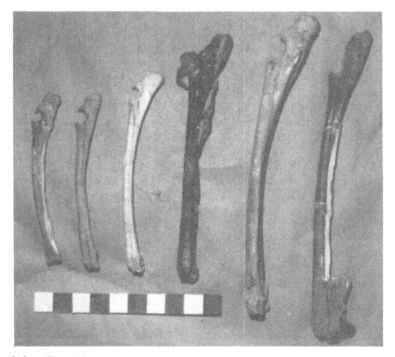

Figure 22. Radii and ulnae. From left to right: *Poebrotherium eximium*, F:AM 47118; *Poebrotherium wilsoni*, AMNH 1364; *Paratylopus labiatus*, AMNH 6520; *Gentilicamelus sternbergi*, AMNH 7910; *Pseudolabis dakotensis*, F:AM 41942; *Miotylopus gibbi*, F:AM 36446. Scale bar in 2 cm increments.

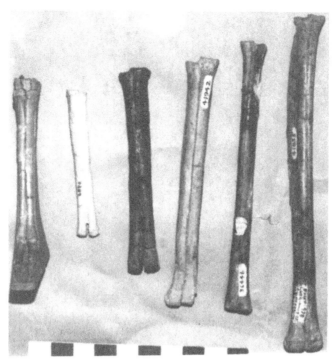

Figure 23. Metacarpals. From left to right: *Poebrotherium wilsoni*, AMNH 1364; *Paratylopus labiatus*, AMNH 6520; *Gentilicamelus sternbergi*, AMNH 7910; *Pseudolabis dakotensis*, F:AM 41942; *Miotylopus gibbi*, F:AM 36446; *Miotylopus taylori*, F:AM 41855. Scale bar in 2 cm increments.

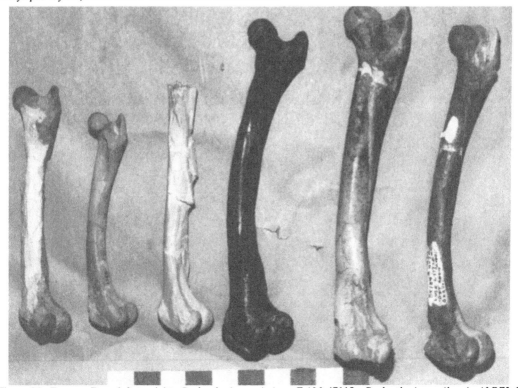

Figure 24. Femora. From left to right: *Poebrotherium eximium*, F:AM 47118; *Poebrotherium wilsoni*, AMNH 1364; *Paratylopus labiatus*, AMNH 6520; *Gentilicamelus sternbergi*, AMNH 7910; *Pseudolabis dakotensis*, F:AM 41942; *Miotylopus gibbi*, F:AM 36446. Scale bar in 2 cm increments.

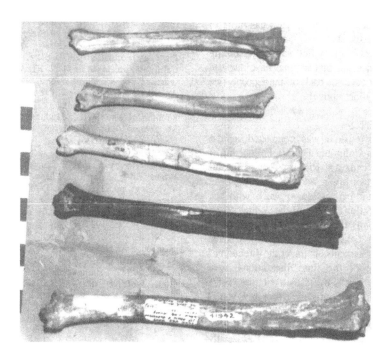

Figure 25. Tibiae. From top to bottom: *Poebrotherium eximium*, F:AM 47118; *Poebrotherium wilsoni*, AMNH 1364; *Paratylopus labiatus*, AMNH 6520; *Gentilicamelus sternbergi*, AMNH 7910; *Pseudolabis dakotensis*, F:AM 41942. Scale bar in 2 cm increments.

Figure 26. Tarsals and metatarsals. From left to right: *Poebrotherium eximium*, F:AM 47118; *Poebrotherium wilsoni*, AMNH 1364; *Paratylopus labiatus*, AMNH 6520; *Gentilicamelus sternbergi*, AMNH 7910; *Pseudolabis dakotensis*, F:AM 41942; *Miotylopus gibbi*, F:AM 36446 (laterally crushed); *Miotylopus taylori*, F:AM 41855. Scale bar in 2 cm increments.

to the Chadronian. Instead, I suspect that it has more to do with the gradualistic bias of many mammalian paleontologists, who, when they find something bizarre that appears suddenly, insist that it must have diverged from some unknown ancestor back in the Eocene (see discussion of stenomylines above).

ACKNOWLEDGMENTS

The bulk of this research was completed in the summer of 1982 on a Columbia Faculty Fellowship. I thank Malcolm McKenna and Richard Tedford for all their years of support of my research at the American Museum, and for their permission to work with the fabulous Frick Collection. Most of the White River specimens described in this study were collected by Morris Skinner and field parties, and the Arikaree specimens by Charles Falkenbach, under the direction of Childs Frick. Many of the line illustrations were prepared for Childs Frick by artists of the Frick Laboratory, including Ray Gooris and Hazel de Berard. I thank Ray Gooris and Richard Tedford for access to these unpublished illustrations. I thank Clifford Prothero for his help with darkroom work. I especially thank Beryl Taylor for all his help with the camels, and for his years of work cataloguing and organizing the collection for Childs Frick. Many of the insights in this paper came from his understanding of the fossils, and were conveyed in the way he curated them. I also thank Jim Honey for all his generous help in recent years, especially in our collaboration on a summary of camel evolution (Honey et al., in press). I thank Tom Deméré, Jim Honey, Tom Kelly, Malcolm McKenna, Margaret Stevens, Beryl Taylor, and Jack Wilson for their helpful comments on this manuscript.

LITERATURE CITED

Burmeister, K. H. K. 1837. Handbuch der Naturgeschichte. Berlin, Abteilung I and II.

Clark, J., J. R. Beerbower, and K. K. Kietzke. 1967. Oligocene sedimentation, stratigraphy, paleoecology and paleoclimatology of the Big Badlands of South Dakota. Fieldiana: Geology Memoir 5:1-158.

Cope, E. D. 1873. [On *Menotherium lemurinum, Hypisodus minimus, Hypertragulus calcaratus, H. tricostatus, Protohippus,* and *Procamelus heterodontus.*] Proceedings of the Academy of Natural Sciences of Philadelphia 25:419-420.

Cope, E. D. 1879. Observations on the faunae of the Miocene Tertiaries of Oregon. Bulletin of the U. S. Geological and Geographical Survey of the Territories 5:55-69.

Cope, E. D. 1881. On the origin of foot structures of the ungulates. American Naturalist 15:269-273.

Cope, E. D. 1886. The phylogeny of the Camelidae. American Naturalist 20:611-624.

Cope, E. D., and W. D. Matthew. 1915. Hitherto unpublished plates of Tertiary Mammalia and Permian Vertebrata. Monographs of the American Museum of Natural History 2:1-154.

Dalquest, W. W., and O. Mooser. 1974. Miocene vertebrates from Aguascalientes, Mexico. The Pearce-Sellards Series, Texas Memorial Museum 31:1-10.

Deméré, T. A. 1988. Early Arikareean (late Oligocene) vertebrate fossils and biostratigraphic correlations of the Otay Formation at Eastlake, San Diego County, California; pp. 35-44 in M. V. Filewicz and R. L. Squires (eds.), Paleogene Stratigraphy, West Coast of North America. Pacific Section SEPM Publication 58.

Eldredge, N., and S. J. Gould. 1972. Punctuated equilibria: an alternative to phyletic gradualism; pp. 82-115 in T. J. M. Schopf (ed.), Models in Paleobiology. Freeman Cooper, San Francisco.

Evanoff, E., D. R. Prothero, and R. H. Lander. 1992. Eocene-Oligocene climatic change in North America: the White River Formation near Douglas, east-central Wyoming; pp. 116-130 in D. R. Prothero and W. A. Berggren (eds.), Eocene-Oligocene Climatic and Biotic Evolution, Princeton University Press, Princeton, N.J.

Franklin, W. L. 1983. Contrasting socioecologies of South America's wild camelids: the vicuña and the guanaco. American Society of Mammalogists Special Publication 7:573-629.

Frick, C. 1937. Horned ruminants of North America. Bulletin of the American Museum of Natural History 69:1-669.

Frick, C., and B. E. Taylor. 1968. A generic review of stenomyline camels. American Museum Novitates 2353:1-51.

Galbreath, E. C. 1953. A contribution to the Tertiary geology and paleontology of northeastern Colorado. University of Kansas Paleontological Contributions, Vertebrata 4.

Gazin, C. L. 1955. A review of the upper Eocene Artiodactyla of North America. Smithsonian Miscellaneous Collections 128(8):1-96.

Golz, D. J. 1976. Eocene Artiodactyla of southern California. Los Angeles County Natural History Museum Bulletin 26, 85 pp.

Hay, O. P. 1902. Bibliography and catalogue of the fossil Vertebrata of North America. Bulletin of the U.S. Geological Survey 179:1-868.

Hay, O. P. 1930. Second bibliography and catalogue of the fossil Vertebrata of North America. Carnegie Institute of Washington Publication 390(2):1-1074.

Honey, J. G., and B. E. Taylor. 1978. A generic revision of the Protolabidini (Mammalia, Camelidae) with a description of two new protolabidines. Bulletin of the American Museum of Natural History 161:371-425.

Honey, J., J. A. Harrison, D. R. Prothero, and M. S. Stevens. In press. Camelidae; in C. Janis, K. M. Scott, and L. Jacobs (eds.), Tertiary Mammals of North America, Cambridge University Press, Cambridge (in press).

Koford, C. B. 1957. The vicuña and the puna. Ecological Monographs 27:153-219.

Leidy, J. 1847. On a new genus and species of fossil Ruminantia. Proceedings of the Academy of Natural Sciences of Philadelphia 3:322-326.

Leidy, J. 1856. Notice of some remains of extinct vertebrated animals. Proceedings of the Academy of Natural Sciences of Philadelphia 8:163-165.

Leidy, J. 1869. The extinct mammalian fauna of Dakota and Nebraska. Journal of the Academy of Natural Sciences of Philadelphia 7:1-472.

Loomis, F. B. 1911. The camels of the Harrison beds, with three new species. American Journal of Science 31:65-70.

Loomis, F. B. 1928. *Poebrotherium.* American Journal of Science 16:137-142

Loomis, F. B. 1936. The skeleton of a new fossil camel from Wyoming. University of Wyoming Publications 2(5):59-64.

Lull, R. S. 1921. New camels in the Marsh collection. American Journal of Science 1:392-404.

Maglio, V. J. 1966. A revision of the fossil selenodont artiodactyls from the middle Miocene Thomas Farm, Gilchrist County, Florida. Breviora 255:1-27.

Matthew, W. D. 1899. A provisional classification of the freshwater Tertiary of the west. Bulletin of the American Museum of Natural History 12(3):19-75.

Matthew, W. D. 1901. Fossil mammals of the Tertiary of northeastern Colorado. Memoirs of the American Museum of Natural History 1:355-447.

Matthew, W. D. 1904. Notice of two new Oligocene camels. Bulletin of the American Museum of Natural History 20:211-215.

Matthew, W. D. 1909. Faunal lists of the Tertiary Mammalia of the West. U. S. Geological Survey Bulletin 361:91-138.

Matthew, W. D. 1910. On the skull of *Apternodus* and the skeleton of a new artiodactyl. Bulletin of the American Museum of Natural History 28:33-42.

McKenna, M. C. 1966. Synopsis of Whitneyan and Arikareean camel phylogeny. American Museum Novitates 2253:1-11.

McKenna, M. C., and J. D. Love. 1972. High-level strata containing early Miocene mammals on the Bighorn Mountains, Wyoming. American Museum Novitates 2490:1-31.

Olsen, S. J. 1959. Fossil mammals of Florida. Special Publications of the Florida Geological Survey 6:1-74.

Olsen, S. J. 1962. The Thomas Farm fossil quarry. Quarterly Journal of the Florida Academy of Sciences 25(2):142-146.

Patton, T. H. 1964. The Thomas Farm fossil vertebrate locality; pp. 12-20 in W. Auffenberg, W. T. Neill, T. H. Patton, and S. D. Webb, Guidebook to the 1965 field trip, Society of Vertebrate Paleontology, central Florida. Gainesville, Florida.

Patton, T. H. 1966. Revision of the selenodont artiodactyls from Thomas Farm. Quarterly Journal, Florida Academy of Sciences 29:179-190.

Patton, T. H. 1969. Miocene and Pliocene artiodactyls, Texas Gulf Coastal Plain. Bulletin of the Florida State Museum 14:115-227.

Peterson, O. A. 1904. Osteology of *Oxydactylus*, a new genus of camels from the Loup Fork of Nebraska, with descriptions of two new species. Annals of the Carnegie Museum of Natural History 2(3):434-476.

Peterson, O. A. 1906. The Miocene beds of western Nebraska and eastern Wyoming and their vertebrate faunas. Annals of the Carnegie Museum 4:21-72.

Peterson, O. A. 1908. Description of the type specimen of *Stenomylus gracilis* Peterson. Annals of the Carnegie Museum 4:286-300.

Prothero, D. R. 1982. Medial Oligocene magnetostratigraphy and mammalian biostratigraphy: testing the isochroneity of mammalian biostratigraphic events. Ph.D. dissertation, Columbia University, New York.

Prothero, D. R. 1994. Mammalian evolution; pp. 238-270 in D. R. Prothero and R. M. Schoch (eds.), Major Features of Vertebrate Evolution, Paleontological Society Short Courses in Paleontology 7.

Prothero, D. R., and J. M. Rensberger. 1985. Magnetostratigraphy of the John Day Formation, Oregon, and the North American Oligocene-Miocene boundary. Newsletters in Stratigraphy 15(2):59-70.

Ray, C. E. 1957. A list, bibliography, and index of fossil vertebrates of Florida. Special Publication of the Florida Geological Survey 3:1-175.

Romer, A. S. 1948. The fossil mammals of Thomas Farm, Gilchrist County, Florida. Quarterly Journal of the Florida Academy of Sciences 10(1):1-11.

Schlaikjer, E. M. 1935. Contributions to the stratigraphy and paleontology of the Goshen Hole area, Wyoming. IV. New vertebrates and the stratigraphy of the Oligocene and early Miocene. Bulletin of the Museum of Comparative Zoology, Harvard, 76(4):97-189.

Scott, W. B. 1940. The mammalian fauna of the White River Oligocene. Part 4. Artiodactyla. Transactions of the American Philosophical Society 28(4):363-746.

Simpson, G. G. 1945. The principles of classification and a classification of mammals. Bulletin of the American Museum of Natural History 85:1-350.

Skinner, M. F., S. M. Skinner, and R. J. Gooris. 1968. Cenozoic rocks and faunas of Turtle Butte, south-central South Dakota. Bulletin of the American Museum of Natural History 138(7):381-436.

Stevens, M. S. 1977. Further study of the Castolon local fauna (early Miocene), Big Bend National Park, Texas. The Pearce-Sellards Series, Texas Memorial Museum 28:1-69.

Stevens, M. S., J. B. Stevens, and M. R. Dawson. 1969. New early Miocene formation and vertebrate local fauna, Big Bend National Park, Brewster County, Texas. The Pearce-Sellards Series, Texas Memorial Museum 15:3-52.

Stock, C. 1935. Artiodactyla from the Sespe of Las Posas Hills, California. Publications of the Carnegie Institute of Washington 453:119-125.

Tedford, R. H., J. B. Swinehart, R. M. Hunt, Jr., and M. R. Voorhies. 1985. Uppermost White River and lowermost Arikaree rocks and faunas, White River Valley, northwestern Nebraska and their correlation with South Dakota. Dakoterra 2:335-352.

Tedford, R. H., T. Galusha, M. F. Skinner, B. E. Taylor, R. W. Fields, J. R. Macdonald, J. M. Rensberger, S. D. Webb, and D. P. Whistler. 1987. Faunal succession and biochronology of the Arikareean through Hemphillian interval (late Oligocene through earliest Pliocene Epochs) in North America; pp. 152-210 in M. O. Woodburne (ed.), Cenozoic Mammals of North America, Geochronology and Biostratigraphy, Univ. Calif. Press, Berkeley.

Troxell, E. L. 1917. An Oligocene camel, *Poebrotherium andersoni*, n. sp. American Journal of Science 43:381-389.

White, T. E. 1940. New Miocene vertebrates from Florida. Proceedings of the New England Zoological Club 18:31-38.

White, T. E. 1941. Additions to the Miocene fauna of Florida. Proceedings of the New England Zoological Club 18:91-98.

White, T. E. 1942. The lower Miocene mammal faunas of Florida. Bulletin of the Museum of Comparative Zoology, Harvard, 92(1):1-49.

Wilson, J. A. 1974. Early Tertiary vertebrate faunas, Vieja Group and Buck Hill Group, Trans-Pecos Texas: Protoceratidae, Camelidae, Hypertragulidae. Bulletin of the Texas Memorial Museum 23:1-34.

Wilson, E. O. 1975. Sociobiology: a synthesis. Harvard University Press, Cambridge, Massachusetts.

Wortman, J. D. 1898. The extinct Camelidae of North America and some associated forms. Bulletin of the American Museum of Natural History 10:93-142.

29. Hyracodontidae

DONALD R. PROTHERO

ABSTRACT

The taxonomy of *Hyracodon* has long been confused by the variability of the cusps and crests on the upper premolars. Some authors have oversplit the genus into as many as 11 species, while others recognized only a single species. Based on much larger collections now available, I recognize five valid species of *Hyracodon*: *H. primus* from the Duchesnean; *H. petersoni* from the early Chadronian; *H. priscidens* from the early and middle Chadronian; the highly variable type species, *H. nebraskensis*, from the late Chadronian to early Arikareean; and the larger *H. leidyanus* from the Whitneyan and early Arikareean. Both *H. nebraskensis* and *H. leidyanus* last appear in the lower Sharps Formation, and their extinction coincides with the early Arikareean faunal reorganization.

INTRODUCTION

In 1850, Joseph Leidy published the first notice of North American rhinoceroses. In a few brief paragraphs, he described (but did not illustrate) some teeth he called *Rhinoceros occidentalis* (now referred to the rhinocerotid taxon *Subhyracodon occidentalis*) and *Rhinoceros nebraskensis*. In 1851, Leidy placed both *R. occidentalis* and *R. nebrascensis* (misspelled by Leidy and most later authors from this point onward with a "c" instead of a "k") in the European Oligócene-Miocene genus *Aceratherium*, because their teeth had cingula, unlike living *Rhinoceros*. Finally, in 1852 Leidy gave a full description and illustration of *R. nebrascensis*. He figured and discussed these specimens at length in his 1853 monograph on the "Ancient Fauna of Nebraska." In 1856, Leidy transferred *R. nebrascensis* to a new genus, *Hyracodon*, without a generic description or diagnosis, other than that it possessed "a greater number of teeth than any other known member of the *Rhinoceros* family." This refers to the fact that *Hyracodon* still has the incisors and canines that have been lost in most of the Rhinocerotidae.

Edward Drinker Cope described hyracodontids in 1873, when he proposed the species *Hyracodon arcidens* for fragments of two or three individuals from the Orellan Cedar Creek beds of Colorado.

While they were still undergraduates at Princeton, Scott and Osborn (1887) described some White River

collections at the Museum of Comparative Zoology of Harvard University. They named two new species of *Hyracodon*. One was *H. major*, which was based on a partial skeleton of a large individual that cannot be compared to any taxon based on teeth. Sinclair (1922) recommended that this taxon be abandoned. The other was *H. planiceps*, which was actually a calf of *Subhyracodon*. Wood (1961) applied to the International Commission of Zoological Nomenclature for suppression of the name *planiceps*, and the name has been suppressed.

In the early part of this century, species names of rhinocerotoids continued to proliferate because most paleontologists considered every slight difference in premolars to be grounds for a new taxon. Lambe (1905) described *H. priscidens* from the Chadronian of Saskatchewan. Troxell (1921) proposed three new species of *Hyracodon*: *H. leidyanus*, *H. selenidens*, and *H. arcidens mimus* (misspelled "*minimus*" by some authors). Sinclair (1922) reviewed all the known species of *Hyracodon*, and recognized that all the Orellan species intergraded in size and morphology. Unfortunately, he did not use this argument to reduce the number of invalid names. Instead, he recognized four species groups: *H. nebrascensis*, *H. arcidens* (including *H. priscidens* and *H. selenidens*), *H. leidyanus*, and a new species, *H. apertus*. Abel (1926) reviewed the molarization of premolars in *Hyracodon*, showing their continuous intergradation. His diagram (Fig. 1) is misleading, because it gives the impression of a continuous chronocline in premolar crests. In fact, the first four specimens (I-V) all come from the early Orellan, and are thus part of a nearly contemporaneous population sample; the lower four specimens (VII-X) are also from the same level (the late Whitneyan) and represent variability within a single time frame. Based on this intergradation, he considered them to be "races" or subspecies of *H. nebrascensis*.

Horace Wood's first (1926) publication on rhinocerotoids erected yet another new species, *H. petersoni*. In 1927, Wood published his landmark

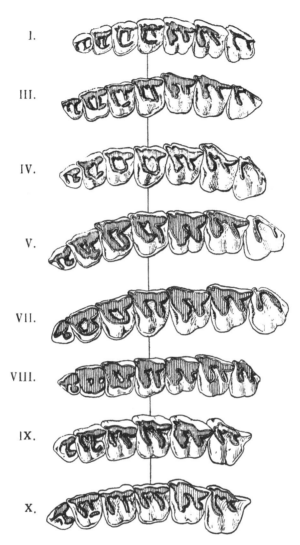

Figure 1. The molarization of upper premolars in *Hyracodon* cheek teeth, as originally shown by Abel (1926). The vertical line runs through the P4 in each specimen for reference. I is the type of *H. selenidens* (YPM 11173), as illustrated by Troxell (1921). III was referred to *H. nebraskensis* (YPM-PU 12687) by Sinclair (1922). IV was referred to *H. arcidens* (YPM-PU 12518) by Sinclair (1922). V was referred to *H. nebraskensis* (YPM-PU 12563) by Sinclair (1922). Although I-V are drafted to give the impression of a chronocline, all are from the early Orellan. VII is the type specimen of *H. apertus* (YPM-PU 10803) figured by Sinclair (1922). VIII was also referred to *H. apertus* (YPM-PU 12705) by Sinclair (1922). IX is the type specimen of *H. leidyanus* (YPM 11169) after Troxell (1921). X was also referred to *H. leidyanus* (YPM-PU 10802) by Sinclair (1922). VII-X also appear to be a chronocline, but all four are from the late Whitneyan.

review of Tertiary rhinoceroses and hyracodonts, but did not add any new species of *Hyracodon*.

The discovery of a large quarry sample of the early rhinocerotid *Trigonias* from the Chadronian Davis Ranch Quarry, Weld County, Colorado, provided the first evidence that primitive rhinocerotoid upper premolars were highly variable within a single population, and might not be reliable criteria for biological species. These specimens were described by Gregory and Cook (1928), who demonstrated that the entire sample was homogeneous in size and in all morphological features except for the molarization of upper premolars. Nevertheless, they followed the typological traditions of their time and recognized six species of *Trigonias* (four of them new) and called one specimen ?*Caenopus premitis*. In their paper, Gregory and Cook (1928, p. 4) commented on how plastic the variability of the population was, and suggested that this might be due to "hybridization of originally distinct races." It did not occur to them that the upper premolars might be meaningless in recognizing species. Apparently, the belief in the utility of premolar characters was too deeply entrenched to be uprooted by this evidence.

William Diller Matthew's (1931) reaction to the Davis Ranch *Trigonias* sample was surprisingly modern. His critique of typological species concepts sounds as if it could have been written in the last decade. Matthew (1931, pp. 5-6) argued that, on ecological grounds, such a large number of closely related species in a single quarry was unlikely, and thus they were all the same species.

Despite Matthew's arguments, his contemporaries continued to name new species of rhinocerotoids based on slight variations of premolars. Peterson (1934) named *H. primus* for a specimen from the Duchesnean of Utah. Russell (1934) reviewed the Cypress Hills fauna first described by Lambe (1905, 1908). He reduced *H. priscidens* to synonymy with *H. arcidens*, and created a new species, *H. browni*. Schlaikjer (1935) described a number of new species from the early Chadronian Yoder l.f. of Goshen County, Wyoming, including *Caenopus yoderensis* and *Hyracodon ischyrolophus*.

In the White River monographs, Scott (1941) acknowledged the problem of the variability of upper premolars in rhinocerotoids, and reduced most of the species of *Hyracodon* (including *arcidens, selenidens, leidyanus,* and *apertus*) to synonymy. He did the same for *Trigonias*. Unfortunately, he still followed Wood in retaining the multiple invalid species of *Subhyracodon*.

After the White River monographs, almost 20 years passed before further research on hyracodonts was published. Green (1958) reported *Hyracodon* from the early Arikareean Sharps Formation of South Dakota. Radinsky (1967) published the first comprehensive review of the Hyracodontidae, revising the definition of the family and making comparisons to the Asian taxa for the first time. He agreed with Scott (1941) that the only valid species of *Hyracodon* was *H. nebraskensis*, and described a new genus and species, *Triplopides rieli*. However, Tanner and Martin (1976) disagreed, and recognized not only *H. nebraskensis* but also *H. priscidens, H. selenidens* (because of its supposed Chadronian provenience), *H. petersoni, H. primus, H. arcidens minimus* [*sic*], and a new species, *Hyracodon doddi*, from the late Chadronian of Sioux County, Nebraska.

In their review of *Forstercooperia*, Lucas, Schoch, and Manning (1981) also discussed the phylogeny of the rest of the Hyracodontidae. Russell (1982) reviewed the Cypress Hills rhinocerotoids, but failed to discuss several long-neglected taxa, including his own (1934) *Hyracodon browni*. Wilson and Schiebout (1984) described new material from the Duchesnean and Chadronian of the Trans-Pecos Texas region. On the basis of these additional specimens, they argued for the validity of *H. primus* and *H. petersoni*. Emry and Purdy (1984) showed that the type specimen of *Hyracodon nebraskensis* had long been misidentified, and determined which specimen was Leidy's original type.

The most recent research on North American hyracodontids was the rhino phylogeny of Prothero, Manning, and Hanson (1986), and the papers published in the perissodactyl symposium volume (edited by Prothero and Schoch, 1989). Prothero, Guérin, and Manning (1989) summarized the viewpoint on *Hyracodon* presented here. Lucas and Sobus (1989) reviewed the indricotheres. Heissig (1989) rejected the long-established view that indricotheres are distinct from rhinocerotids, and placed indricotheres within the Rhinocerotidae. In the American Museum Mongolian collections, he found a foot of *Juxia* that showed that primitive indricotheres had a functional fifth metacarpal. Apparently, he was unaware that the fifth metacarpal is a highly variable digit, disappearing and then re-appearing in several rhino genera (Prothero et al., 1986, p. 359). As a character, it is not sufficient evidence to outweigh all the other synapomorphies that support the distinction between the two families.

ABBREVIATIONS

AMNH, Department of Vertebrate Paleontology, American Museum of Natural History, New York; CM, Carnegie Museum of Natural History, Pittsburgh, PA; F:AM, Frick Collection in the American Museum of Natural History, NY; MCZ, Museum of Comparative Zoology, Harvard University, Cambridge, MA; NMC, National Museum of Canada, Ottawa, Ontario, Canada; ROM, Royal Ontario Museum, Toronto, Ontario, Canada; SDSM, South Dakota School of Mines Museum, Rapid City, SD; SMNH P, Saskatchewan Museum of Natural History, Regina, Saskatchewan, paleontological collections; TMM, University of Texas Memorial Museum, Austin, TX; UNSM, University of Nebraska State Museum, Lincoln, NE; USNM, National Museum of Natural History, Smithsonian Institution, Washington, DC; YPM, Yale Peabody Museum, New Haven, CT; YPM-PU, Princeton University collection, now housed at the YPM.

METHODS

The literature review above shows how taxonomic practices have changed over the years. Crucial to the controversy is the significance of the crests on the upper premolars. Before the advent of population thinking in the 1930s and 1940s, nearly all paleontologists were typologists who used almost any minor variation as the basis for a new taxon. As we have seen, however, Matthew (1931) and Scott (1941) were ahead of their time in realizing that rhinocerotoid upper premolars were highly variable within populations (first established by the Davis Ranch *Trigonias* sample). However, a thorough analysis of the variability of the upper premolars of *Hyracodon* has never been published. Radinsky (1967) assumed without discussion that all the species of *Hyracodon* were synonyms of *H. nebraskensis*; he did no detailed analysis of the large

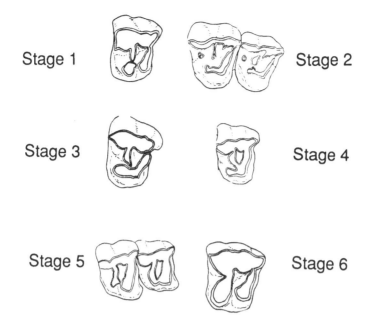

Stage 1   Stage 2

Stage 3   Stage 4

Stage 5   Stage 6

Figure 2. Coding system for the six standard stages of premolar molarization in *Hyracodon*. Figures modified from Gregory and Cook (1928).

samples of *Hyracodon* then available. In addition, most of the specimens discussed above come from the early Orellan. Relatively few analyses have been made of the variability of Duchesnean, Chadronian, or Whitneyan *Hyracodon* upper premolars.

Unlike the large quarry samples of *Trigonias* and *Subhyracodon*, however, there is apparently no similar quarry sample of *Hyracodon* than I have been able to locate. Although *Hyracodon* is very abundant in many museum collections from the late early Orellan "lower nodular zone" of the Big Badlands, they apparently did not die off in large numbers in a single place. Perhaps they lived in the open plains (as suggested by Clark et al., 1967), and so did not accumulate large masses of bones in river channels, as the larger-bodied, possibly more aquatic *Trigonias* and *Subhyracodon* did. Nixon and LaGarry-Guyon (1993) have described trackways from Toadstool Park, Nebraska, which seem to show that *Hyracodon* was solitary and *Subhyracodon* traveled in herds. Perhaps this explains the lack of quarry samples of *Hyracodon*. In the absence of a quarry, the best available sample is the late early Orellan "lower nodular zone" collection from a single area. For this study, I used the large sample from the western Big Badlands (Cottonwood Pass-Big Corral Draw-Quinn Draw area), collected for the Frick Laboratory by Morris Skinner and crews in the 1950s, and now curated in the American Museum of Natural History. Magnetic stratigraphy (Prothero and Swisher, 1992) shows that the lower nodular zone in this area is late early Orellan and spans the latter third of Chron C13n (33.1-33.5 Ma), and thus represents about 100,000-200,000 years

of accumulation (not the 1100 to 11,000 years suggested by Clark et al., 1967).

To quantify the variability in upper premolars, I coded the different cusp and crest configurations into six standard stages of increased molarization (Fig. 2). In stage 1, the protocone and hypocone are completely separated from the ectoloph, and merge with it only in the latest wear stages. In stage 2, the protoloph curves around the anterolingual margin of the tooth, but does not connect to the ectoloph except in the latest wear stages, and remains distinct from the hypocone. The metaloph, however, is larger than in stage 1, and may contact the hypocone. In stage 3, the protoloph connects to the ectoloph, and curves completely around the anterolingual margin of the tooth crown, merging with the hypocone. In stage 4, the protoloph contacts the metaloph from the early wear stages, but still has a posterolingual crest remaining in the hypocone position. In stage 5, the protoloph loses its lingual extension, forming two parallel crests (the protoloph and metaloph), which are still in contact at the lingual end. Finally, stage 6 premolars are fully molarized in the typical rhinocerotoid "π"-shaped pattern, with short, parallel protoloph and metaloph, and no closure of the valley between them.

Gregory and Cook (1928) developed their own scheme of coding the premolar variability in the Davis Ranch *Trigonias* sample. Although they coded the characters differently, their figures 4 and 5 show that all six of these premolar stages occur in this single population sample, and their Graph 4 shows that all six stages form an almost continuous frequency distribution.

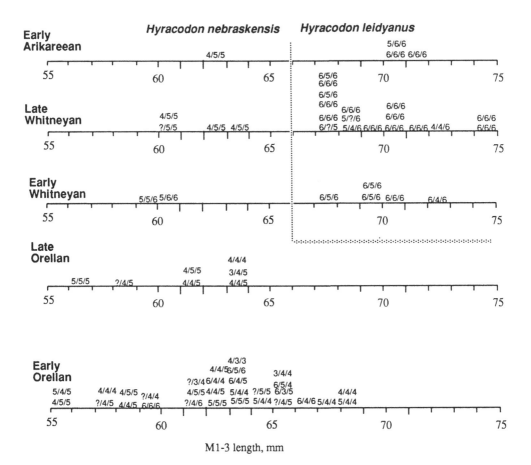

Figure 3. Size-frequency distributions (as measured by length of M1-3) of different specimens of *Hyracodon*, with their premolar stages (coded P2/P3/P4 in sequence) following the system shown in Figure 2. A question mark indicates that a particular premolar crown was missing, too damaged, or too worn to be assessed. Note that through the Orellan to early Arikareean, there is a series of samples with highly variable premolars, whose mean size is about 63 mm; these are referred to *H. nebraskensis*. In the Whitneyan and early Arikareean, there is a much larger lineage (mean size = 70 mm) with more advanced premolars; these are referred to *H. leidyanus*.

Thus, this range of upper premolar variability can be documented from a single homogeneous population, and clearly has less taxonomic significance than previously thought.

Each specimen was coded by stage of molarization for each of the three upper premolars (P2/P3/P4 in sequence) and plotted by size (as indicated by M1-3 length), as shown in Figure 3. The large sample from the lower nodular zone of the western Big Badlands forms a single unimodal size-frequency distribution. Inspection of the premolar stages within the sample shows extreme and continuous variability in each tooth. P2's in stages 4, 5 and 6 are equally common. P3's are typically stage 4 or 5, although there are few in stage 3 and 6. P4's are usually stage 4 or 5, with only a few in stage 6. This large range of variability includes not only the type specimen of *H. nebraskensis* (coded 6/4/5) from the ?early Orellan of South Dakota, but also the types

of *H. arcidens* (3/3/5), and *H. selenidens* (4/3/3), which are from the early Orellan of Colorado (Galbreath, 1953).

Figure 3 shows, however, that the early Orellan to early Arikareean *Hyracodon* sample is not completely homogeneous or monospecific, contrary to Scott (1941) and Radinsky (1967). For example, in the Whitneyan and early Arikareean, there is a disjunctly larger (M1-3 length = 68-76 mm) species which is distinct from contemporary specimens referable to *H. nebraskensis*. These larger specimens nearly all have stage 6 P2s, stage 5-6 P3s, and stage 6 P4s. Many are coded 6/6/6 (completely molarized), a condition found in only one of the *H. nebraskensis* specimens. This sample seems to be referable to Troxell's (1921) species *H. leidyanus* from the Whitneyan of South Dakota (premolar stages of the type = 4/6/6). *H. apertus* (Sinclair, 1922) from the late Whitneyan *Protoceras* channels in Big Corral

Draw in the Big Badlands, is also within this range of variation (premolar stages of the type = 4/5/6).

Although there are fewer Chadronian specimens that can be plotted in this fashion, many of these fall outside the range of variation of *H. nebraskensis,* and so some of the Chadronian taxa are considered valid as well (as discussed further below).

SYSTEMATIC PALEONTOLOGY
Class MAMMALIA Linnaeus, 1758
Order PERISSODACTYLA Owen, 1848
Superfamily RHINOCEROTOIDEA Owen, 1845
Family HYRACODONTIDAE Cope, 1879

Type Genus—*Hyracodon* Leidy, 1856
Included Genera—*Forstercooperia* Wood, 1939; *Juxia* Chow and Chiu, 1963; *Urtinotherium* Chow and Chiu, 1963; *Paraceratherium* Forster Cooper, 1911; *Triplopus* Cope, 1880; *Epitriplopus* Wood, 1927; *Ardynia* Matthew and Granger, 1923; *Triplopides* Radinsky, 1967; *Prohyracodon* Koch, 1897; *Ilianodon* Chow and Xu, 1961; *Eggysodon* Roman, 1911.

Range—Middle Eocene to early Miocene, Eurasia and North America.

Diagnosis—Small to gigantic cursorial hornless rhinocerotoids with long, slender metapodials. I3-1/3-1 C1-0/1-0 P4-3/4-3 M3/3. Incisors conical and pointed, approximately equal in size with each other. Premolars non-molariform to molariform. M3 triangular, retaining a lingually deflected metacone at the end of the ectoloph.

Discussion—Since the focus of most of the systematic papers in this volume is the White River Chronofauna, and since the Asian taxa were thoroughly reviewed by Radinsky (1967), I will not discuss them further here. *Triplopus* is currently being reviewed by R. Stucky (personal communication) *Forstercooperia* was discussed by Lucas et al. (1981) and is now being restudied by L. Holbrook (personal communication). No new material of *Epitriplopus* or *Triplopides* has been reported, so there is little to add to Radinsky's (1967) discussion. Only the genus *Hyracodon* is overdue for a revision.

Hyracodon Leidy, 1856

Rhinoceros Leidy, 1850 (*non* Linnaeus)(in part)
Hyracodon Leidy, 1856
Caenopus Schlaikjer, 1935

Type Species—*Hyracodon nebraskensis* (Leidy, 1850).
Included Species—Type and *H. leidyanus* Troxell, 1921; *H. priscidens* Lambe, 1905; *H. petersoni* Wood, 1926; *H. primus* Peterson, 1934.
Range—Late middle Eocene (late Duchesnean) to late early Oligocene (earliest Arikareean), North America.

Diagnosis—Small hyracodontids (length of P1-M3 = 108-149 mm) with posterolingually creased P3-4 protocone; reduced canines; deep rostrum; dp1 lost in adults; deciduous premolars retained into young adult life; molarization in upper premolars variable; coronoid process anteriorly inclined.

Discussion—As reviewed above, *Hyracodon* has been oversplit into 11 different species based on highly variable upper premolars. Most of the latest Chadronian and Orellan specimens fall within a single range of continuous variation that can be called *H. nebraskensis.* A larger, more advanced Whitneyan-Arikareean taxon seems distinct and is referred to *H. leidyanus.* Three Duchesnean and Chadronian taxa (*H. primus, H. petersoni,* and *H. priscidens*) are smaller than *H. nebraskensis,* and fall outside its range of premolar variability. For this reason, they are considered valid taxa.

Hyracodon primus Peterson, 1934

Hyracodon primus Peterson, 1934
Type—CM 11914, right maxillary fragment with P4-M2, Lapoint Member, Duchesne River Formation, Uinta Basin, Utah (Duchesnean).
Referred Material—Extensive sample from the late Duchesnean Porvenir l.f., Trans-Pecos Texas, listed by Wilson and Schiebout (1984, pp. 20, 22).
Diagnosis—Smallest species of *Hyracodon* (M1-3 length = 47-49 mm; see Table 1). No molarization of premolars; P2 in stage 1, P3 in stage 2-3, and P4 in stage 3. Prominent cristae on P3-4 and M1-2. Constricted protocone on upper molars. Differs from *Epitriplopus* in having shorter and wider M1, less molariform P4, and molar cristae. Differs from *H. petersoni* in its smaller size and much more primitive and unmolarized premolars.
Discussion—Wilson and Schiebout (1984) provided a thorough description of the material of *H. primus* from Texas. Although Radinsky (1967) considered the taxon to be indeterminate based on the type specimen, Wilson and Schiebout (1984) have clearly shown that *H. primus* is valid. Despite its primitive features, it also has the derived features of *Hyracodon,* including conical incisors, constricted symphysis, and strong premolar and molar cristae.

Hyracodon petersoni Wood, 1926

Hyracodon petersoni Wood, 1926
Hyracodon ischyrolophus Schlaikjer, 1935
Hyracodon petersoni Russell, 1982
Type—CM 3572, maxilla and premaxilla with most of the upper dentition, from the "*Titanotherium* beds, Badland Creek, Sioux County, Nebraska" (Chadronian).
Referred Material—From the early Chadronian

Table 1. Measurements of *Hyracodon* dentitions (in mm) *H. primus* after Wilson and Schiebout (1984, Tables 5-7); *H. petersoni* upper teeth based on type (CM 3572) and TMM 42156-4 (Wilson and Schiebout, Table 4), lower teeth after Russell (1982:13); *H. priscidens* based on type (NMC 6564) and four referred specimens; *H. nebraskensis* based on type and five referred specimens; *H. leidyanus* based on type (YPM 11173) and five referred specimens; *H. selenidens* type (YPM 11169) after Troxell (1921); *H. doddi* type (UNSM 1106) after Tanner and Martin (1976); *H. arcidens mimus* type after Troxell (1921); *H. apertus* type (YPM-PU 10803) after Sinclair (1922).

	primus	petersoni	priscidens	nebraskensis	leidyanus	selenidens type	doddi type	arcidens type	apertus type
P2-M3	—	97.0	106.0-116.0	115.0±5.2	123.3±4.2	—	—	—	144
M1-3	47.7	52.0-55.0	59.3-63.0	62.5±1.4	69.5±1.9	—	—	—	72
P2L	10.0	13.6-14.5	14.6-16.0	17.0±1.2	19.4±1.1	14	—	16	20
P2W	11.4	15.6-16.5	17.5-21.0	20.3±1.8	21.9±0.9	16	—	19	21
P3L	10.6	14.4-16.2	15.6-18.1	18.0±0.2	19.7±1.1	15	—	18	21
P3W	15.0	18.5-18.8	20.7-23.3	23.1±1.5	24.8±2.1	19	—	23	24
P4L	13.0	14.7-17.4	17.3-18.5	18.2±0.9	21.9±1.5	16	15.3	20	23
P4W	17.5	19.3-20.6	19.1-25.3	25.5±2.6	27.8±2.6	21	22.1	26	28
M1L	14.8-17.0	17.3-19.6	19.9-22.0	20.3±1.2	21.8±1.8	19	16.4	—	24
M1W	16.5-17.3	20.2-20.6	22.2-24.5	23.6±1.6	26.2±2.1	21	20.5	—	27
M2L	16.9-18.3	18.5-22.7	21.1-23.5	22.0±1.8	18.5±0.9	20	—	—	24
M2W	18.5-20.4	21.7-23.6	23.0-28.0	25.2±2.0	27.8±1.6	21	22.6	—	27
M3L	14.4-16.3	16.2-17.0	19.1-21.1	20.5±1.3	22.3±1.4	17	—	—	23
M3W	17.8-19.2	22.7-22.7	20.4-25.1	22.9±1.2	25.2±0	21	—	—	28
p2-m3	80.0	—	106.0-114.3	117.9±1.9	128.8±5.4	—	—	—	—
m1-3	46.0-49.7	—	64.0-66.0	62.9±1.8	68.2±1.2	—	—	—	—
p2L	10.8	12.6	14.2-15.7	17.0±2.0	18.9±0.5	—	—	—	—
p2W	7.4	8.8	9.3-10.0	11.8±0.5	12.4±1.6	—	—	—	—
p3L	12.5	14.9	15.3-19.0	19.1±0.9	22.8±0.7	—	—	—	—
p3W	8.9	12.0	11.6-13.1	14.9±0.9	15.5±2.1	—	—	—	—
p4L	12.6-14.3	15.3	16.5-20.0	19.6±0.4	23.2±1.1	—	—	—	—
p4W	9.6-10.9	11.9	12.0-14.0	15.5±1.3	16.9±1.0	—	—	—	—
m1L	13.7-16.0	13.8	17.7-20.0	19.4±0.4	22.8±1.3	—	—	—	—
m1W	9.9-10.2	12.1	12.4-15.0	14.1±1.4	14.9±0.9	—	—	—	—
m2L	15.0-18.9	19.5	20.0-22.8	22.5±0.6	24.3±0.5	—	—	—	—
m2W	10.0-12.5	12.9	13.0-14.0	14.0±1.4	15.5±1.3	—	—	—	—
m3L	17.3-18.3	19.9	24.2-24.6	23.5±0.9	23.0±1.9	—	—	—	—
m3W	10.3-11.2	13.1	13.1-13.7	13.1±1.2	14.0±0.9	—	—	—	—

Southfork l.f., Cypress Hills, Saskatchewan (Russell, 1982; this volume, Chapter 12): SMNH P1179.1, right maxillary fragment with P3-M3; SMNH P1179.2, right maxillary fragment with P3-M3; SMNH P1204.1, mandible. From the early Chadronian "Yoder Formation," Goshen County, Wyoming: SDSM 8748, type of *H. ischyrolophus* Schlaikjer, 1935. From the early Chadronian Coffee Cup and Airstrip l.f., Trans-Pecos Texas (Wilson and Schiebout, 1984, pp. 30-32): TMM 42153-4, skull fragment with P1-M3, and associated lower jaw fragment with partial m2 and m3; TMM 40501-1, M1.

Range—Early Chadronian of Wyoming, Nebraska, Saskatchewan, and Texas.

Diagnosis—Small species of *Hyracodon* (M1-3 length = 52-55 mm), slightly larger than *H. primus*, but smaller than *H. priscidens*. Differs from *H. priscidens* in that the P3-4 metaloph curves posteriorly, keeping the medial valley of P3-4 open posteriorly even in advanced stages of wear. Differs from *H. primus* in its larger size and more advanced upper premolars

(4/3/3, except for the posteriorly-curved metaloph).

Discussion—*H. petersoni* is clearly distinct from the *H. priscidens* in its smaller size and peculiarly deflected P3-4 metalophs (not seen to this degree in any specimen of *H. priscidens*). The validity of *H. petersoni* was reaffirmed by Russell (1982), who showed that the Cypress Hills Chadronian *Hyracodon* could be separated into two groups, *H. petersoni* and *H. priscidens*.

Hyracodon ischyrolophus was named by Schlaikjer (1935) for an isolated left M3 from the early Chadronian Yoder l.f., Goshen County, Wyoming. As Schlaikjer (1935, fig. 8) shows, it is identical in size to *H. petersoni*, and distinguished only in having a slightly larger metaloph. However, this condition is seen in some of the referred material of *H. petersoni* (e.g., TMM 42153-4, shown in Wilson and Schiebout, 1984, fig. 20), so it is clearly within the normal range of variation of *H. petersoni*. Schlaikjer (1935, p. 86) named a new species on such an inadequate specimen because it allegedly filled the gap between *Triplopus* and *Hyracodon*.

Hyracodon priscidens Lambe, 1905

Hyracodon priscidens Lambe, 1905
Hyracodon browni Russell, 1934

Type—NMC 6564, left and right maxillae missing left P4 and right P3; plesiotype: NMC 66561, mandibular symphysis with left ramus. Both are from "Bone Coulee" (Conglomerate Creek Valley), now called the Calf Creek l.f., middle Chadronian, Cypress Hills, Saskatchewan (see Storer, this volume, Chapter 12).

Referred Material—From the Calf Creek l.f., Saskatchewan: SMNH P1634.1, mandible; ROM 23195, mandible (Russell, 1982). From the Pipestone Springs l.f., Jefferson County, Montana: AMNH 9708, mandible with right p1-3; 9710, right m3; 9855; isolated teeth; 9711, isolated teeth; 9709, right ramus with m1-3. From the McCarty's Mountain l.f., Madison County, Montana: F:AM 112166, left lower molar. From the Flagstaff Rim area, Natrona County, Wyoming (Emry, 1973, 1992): F:AM 111850, from Dry Hole Quarry; F:AM, 111849, 44 feet below Ash B; 111853, from 75 feet above Ash B, and over a dozen uncatalogued specimens from levels between these specimens. From the Ledge Creek area, Natrona County, Wyoming: F:AM 112167, 111852, 111854, from the "Red Fauna." Additional uncatalogued specimens ranging from 30 feet below the 160 Ash to 50 feet below the 310 Ash (Skinner and Gooris, 1968; Prothero, 1985). Much additional material from Chadron Formation collections in many museums could be referred to this taxon.

Range—Early to middle Chadronian, Wyoming, Nebraska, Montana, and Saskatchewan.

Diagnosis—Medium-sized *Hyracodon* (length M1-3 = 59-63 mm) with unmolarized upper premolars (P2 typically in stage 3-4; P3 in stage 2-3; P4 in stage 3-4). Strong protoloph on P2-4 curves posterolingually to hypocone. Metaloph on P3-4 does not connect to protoloph, leaving median valley open posteriorly. Posterior extension of metaloph on M3 is short and reflected posterolabially.

Discussion—*H. priscidens* is the common early-middle Chadronian hyracodont found in most localities of this age. It is slightly larger than *H. petersoni*, and lacks the strong posterior deflection of the metalophs on P3-4. Although it overlaps the size range of *H. nebraskensis*, *H. priscidens* has much less molarized premolars (typically stage 3 in P3-4), and never connects the protoloph or metaloph to enclose the medial valley (as is typical of *H. nebraskensis*).

H. priscidens was erroneously synonymized with *H. arcidens* by Sinclair (1922), who misinterpreted the morphology shown in Lambe's clear illustration. Wood (1926, 1927) followed Sinclair in this synonymy. Apparently neither saw the type specimen, but relied on figures. Scott (1941) and Radinsky (1967) considered *H.*

priscidens to be a synonym of *H. nebraskensis*. But Tanner and Martin (1976) considered *H. priscidens* a valid species, and Russell (1982) showed that it was distinct from all other species of *Hyracodon*.

Hyracodon browni was named by Russell (1934) based on two P3's from the Cypress Hills Chadronian. These teeth differ from *H. priscidens* from the same beds only in their lack of a metaloph, a feature seen in no other rhinocerotoid. In size and in the rest of the morphology, *H. browni* cannot be distinguished from *H. priscidens*. Curiously, in his review of the Cypress Hills rhinos, Russell (1982) did not comment on *H. browni*, even though he had originally described it from the same fauna 48 years earlier. Given the enormous variability of upper premolars in *Hyracodon* documented so far, I see no reason to base a separate species on such limited material of what is almost certainly an aberrant individual.

Hyracodon nebraskensis (Leidy, 1850)

Rhinoceros nebraskensis Leidy, 1850 (*non* Linneaus)
Aceratherium nebrascense Leidy, 1851 (*non* Kaup)
Hyracodon nebrascensis Leidy, 1856
Hyracodon arcidens Cope, 1873
Hyracodon selenidens Troxell, 1921
Hyracodon arcidens mimus Troxell, 1921
Hyracodon apertus Sinclair, 1922 (in part, not type)
Hyracodon apertus Green, 1958
Hyracodon doddi Tanner and Martin, 1976

Type—USNM 336207, partial skull and mandible missing the rostrum, the incisors and canines, and the braincase (see Emry and Purdy, 1984), from the "Mauvaises Terres, Nebraska Territory" (presumably from the Orellan, Big Badlands, South Dakota). Figured by Emry and Purdy (1984, figs. 1 and 2); original figures in Leidy (1852, Plate 12A, fig. 6; 1853, Plate 14, figs. 1, 2).

Referred Material—Virtually all the abundant *Hyracodon* specimens from the late Chadronian and Orellan are referable to this species. In addition to the type and other specimens described by Leidy (e.g., USNM 138, 336208), they include a large number of specimens from the late Chadronian *Chadronia* Pocket (see Ostrander, 1985), Sioux County, Nebraska: AMNH 82925, 82595, 82923, 82910, 82924, 82985, 92984, 86201, 86200; over a hundred catalogued specimens from the late early Orellan "lower nodular zone," Scenic Member of the Brule Formation, Big Badlands, South Dakota, in the American Museum collections; dozens of catalogued specimens, including the types of *H. "arcidens"* (AMNH 6318) and *H. "selenidens"* (YPM 11173) from the Orellan Cedar Creek Member of the Brule Formation, Logan and Weld Counties, Colorado, in the American Museum Collection, and in the University of Kansas collections (see Galbreath, 1953). From the late Orellan "middle Oreodon beds," Big

Badlands of South Dakota, including AMNH 12305, 12306, 9783, 1165, and 1176. From the late Orellan Frank Kostelecky Ranch, Stark County, North Dakota: F:AM 112168. From the Whitneyan Poleslide Member, Big Badlands, South Dakota: AMNH 560. From the late Whitneyan Pussy Springs locality, Morrill County, Nebraska: F:AM 111866. From the late Whitneyan Redington Gap area, Scottsbluff County, Nebraska: F:AM 111805. From the early Arikareean Sharps Formation, Washabaugh County, South Dakota: F:AM 111869 (50 feet above the Rockyford Ash).

Range—Late Chadronian to early Arikareean of the High Plains (the Dakotas, Nebraska, Colorado, Wyoming).

Diagnosis—Medium-sized *Hyracodon* (M1-3 length = 55-70 mm). Molarization of upper premolars is highly variable. P2 ranges from stages 4-6; P3 typically stage 4-5; P4 typically stage 4-5. Most specimens are larger than the sample of *H. priscidens* and have more advanced premolars. Distinguished from *H. leidyanus* by its smaller size and less advanced premolars.

Discussion—As discussed above, the *nebraskensis-arcidens-apertus-selenidens* morphotypes intergrade completely in size and morphology. As Scott (1941, p. 841) put it, "It is quite out of the question that four separate species of *Hyracodon* should have lived together in the same area which is now South Dakota and Nebraska, nor is the problem rendered any less difficult by calling these variations subspecies. Several Recent subspecies do not occur together, but each one has its own range, though with some overlapping on the borders of the ranges. Nor is a stratigraphic separation possible . . . For zonal purposes the species are of little value, as they range through considerable thickness of beds, and the same is true of size variants."

Tanner and Martin (1976) described a new species, *H. doddi* (UNSM 1106) from the late Chadronian (18 feet below the Upper Purplish White layer), Sioux County, Nebraska. In having a fully molarized (= stage 6) P4, it is virtually identical in morphology with the type of *H. apertus* from the Whitneyan, and differs only in its slightly smaller size and its Chadronian age. However, there are numerous specimens in the late Chadronian and early Orellan collections in the American Museum collections (e.g., AMNH 9788) which also show the morphology of *H. doddi*, so there is no reason to separate it from the highly variable sample of *H. nebraskensis*. Tanner and Martin (1976) justified the distinction of *H. doddi* from *H. nebraskensis* and *H. apertus* on the grounds that there were no specimens in the Nebraska collections older than the Whitneyan with a fully molarized P4. The larger collections in the American Museum, however, have completely encompassed this range of variation, so there is no longer any basis for retaining *H. doddi*.

Tanner and Martin (1976, p. 211) also argued for the validity of *H. selenidens* distinct from *H. arcidens* and

H. nebraskensis because the types of the latter species were both "about 20% larger than *H. selenidens* and both are geologically younger (Orellan?)." They argued that the type of *H. selenidens* is Chadronian because it was apparently associated with brontothere material in the Yale collection. Tanner and Martin did not examine the full range of sizes shown by *H. nebraskensis*, or they would have seen that *H. selenidens* is at the small end of the frequency distribution of *H. nebraskensis* (which also ranges into the late Chadronian). As Sinclair (1922) showed, *H. selenidens* cannot be distinguished from *H. arcidens*, and the fact that one is late Chadronian and the other early Orellan should have no bearing on systematic decisions.

Hyracodon leidyanus Troxell, 1921

Hyracodon leidyanus Troxell, 1921
Hyracodon leidyanus Sinclair, 1922
Hyracodon apertus Sinclair, 1922 (in part)
Hyracodon leidyanus Wood, 1927
Hyracodon nebraskensis Scott, 1941 (in part)
Hyracodon nebraskensis Radinsky, 1967 (in part)

Type—YPM 11169, left maxilla with P1-M3, from "Crow Buttes, South Dakota." Sinclair (1922) and Wood (1926, 1927) argued that it is probably from the "*Protoceras* beds" (Whitneyan) in age.

Referred Material—From the Poleslide Member of the Brule Formation, Big Badlands, South Dakota: YPM 11168 (paratype), lower jaw and some postcranial fragments; YPM-PU 10802, 10144, two skulls; F:AM 111856, skull and jaws; F:AM 111857, skull; F:AM 111858, palate; F:AM 111859, partial skull and jaws; YPM-PU 10803, type of *H.* "*apertus.*" From the Vista Member, Logan County, Colorado: AMNH 8821, palate and mandible; 8813, skull and jaws. From the Harris Ranch-Slim Buttes area, Fall River County, South Dakota (see Simpson, 1985): F:AM 111860, 111861, 111862, 111863, all skulls with jaws and associated postcranials. From Roberts Ranch, Scottsbluff County, Nebraska: F:AM 111864, skull and jaws. From Redington Gap, Morrill County, Nebraska: F:AM 111867, skull. From the basal Sharps Formation, Shannon and Washabaugh Counties, South Dakota: F:AM 111870, maxilla with right P1-4 (15 feet above the Rockyford Ash); F:AM 111868, maxilla with M1-3, and SDSM 54141 (referred by Green, 1958, to *H. apertus*), both found 5 feet above the Rockyford Ash.

Range—Whitneyan to earliest Arikareean of South Dakota, Nebraska, and Colorado.

Diagnosis—Largest species of *Hyracodon* (M1-3 length = 66-75 mm). P3-4 completely molariform or nearly so (mostly stage 6, a few stage 5), with parallel protoloph and metaloph and no connection blocking the valley between them. P2 not completely molariform, but the hypocone is connected to the metaloph, and separate from the protoloph. Crochets and cristae are occasionally present in the upper molars. Teeth slightly more hypsodont than in *H. nebraskensis*. Skull and skeletal features are very robust, reflecting its large size.

Discussion—The large sample of Whitneyan *Hyracodon* in the Frick Collection (Fig. 3) separates into two clusters of morphology: a smaller *H. nebraskensis-H. apertus* group that remains stable in size and range of morphologies throughout the late Chadronian to early Arikareean, and a disjunctly larger group that first appears in the early Whitneyan. In most skulls the upper premolars are fully molarized, as in the type of *H. leidyanus*. It is unlikely that these larger *Hyracodon* represent sexual dimorphs, since earlier *Hyracodon* show no such dimorphism, and the premolars of these larger specimens are consistently more advanced than contemporary small *H. nebraskensis*. These morphological and size gaps are adequate justification for recognizing *H. leidyanus*. This species apparently split off from *H. nebraskensis* in the early Whitneyan and co-existed with it until they both became extinct in the early Arikareean.

SUMMARY

The variability of the upper premolars in *Hyracodon* has led to tremendous taxonomic confusion. In the last century, researchers have gone back and forth from oversplitting (e.g., Troxell, Wood, Tanner and Martin) to excessive lumping (e.g., Scott, Radinsky). I believe that the large sample now available in the Frick Collection suggests an intermediate solution. There are five valid species of *Hyracodon*. The Duchesnean *H. primus* and early Chadronian *H. petersoni* are clearly valid taxa, although their available sample is still small. In the early and middle Chadronian, the common hyracodont was *H. priscidens*. Most of the specimens of *Hyracodon* from the late Chadronian-early Arikareean are encompassed within the highly variable type species, *H. nebraskensis*. These include the types and most specimens referred to *H. apertus*, *H. selenidens*, *H. arcidens*, and *H. doddi*. However, in the Whitneyan and early Arikareean, another larger species, *H. leidyanus*, persists alongside *H. nebraskensis*. *Hyracodon* disappears from the fossil record in the early Arikareean along with a number of other taxa characteristic of the White River Chronofauna. This is part of the "faunal reorganization" of the late early Arikareean discussed by Tedford et al. (1985, 1987; this volume, Chapter 15).

ACKNOWLEDGMENTS

This research was supported by a Columbia University Faculty Fellowship, and by a Guggenheim Fellowship, at the American Museum of Natural History. I thank Malcolm McKenna and Richard Tedford for access to the AMNH and F:AM collections, and Farish Jenkins and Chuck Schaff (MCZ) and Bob Emry and Bob Purdy (USNM) for loans of specimens in their care. I thank John Storer (SMNH) for the opportunity to examine the Cypress Hills material, John A. Wilson (TMM) for access to the Texas specimens, and Phil Bjork and Jim Martin (SDSM) for the chance to see the Yoder material. I thank Earl Manning for curating and figuring out the AMNH and F:AM *Hyracodon* material, and teaching me so much about rhinos, and for critiquing this manuscript in detail. I thank Steve King for entering the statistical data on a spreadsheet. I thank Bruce Hanson, Luke Holbrook, Allen Kihm, Spencer Lucas, Larry Martin, Judith Schiebout, and Jack Wilson for reviewing the manuscript.

LITERATURE CITED

Abel, O. A. 1926. Die Molarisierung der Oberen Prämolaren von *Hyracodon nebrascensis* Leidy. Palaeontologische Zeitschrift 8(3): 224-245.

Clark, J., J. R. Beerbower, and K. K. Kietzke. 1967. Oligocene sedimentation, stratigraphy, paleoecology and paleoclimatology of the Big Badlands of South Dakota. Fieldiana: Geology Memoir 5:1-158.

Cope, E. D. 1873. Second notice of extinct Vertebrata from the Tertiary of the Plains. Paleontological Bulletin 15: 1-2.

Emry, R. J. 1973. Stratigraphy and preliminary biostratigraphy of the Flagstaff Rim area, Natrona County, Wyoming. Smithsonian Contributions to Paleobiology 18.

Emry, R. J. 1992. Mammalian range zones in the Chadronian White River Formation at Flagstaff Rim, Wyoming; pp. 106-115 *in* D. R. Prothero and W. A. Berggren (eds.), Eocene-Oligocene Climatic and Biotic Evolution, Princeton University Press, Princeton, N.J.

Emry, R. J., and R. W. Purdy. 1984. The holotype and would-be holotypes of *Hyracodon nebraskensis* (Leidy 1850). Notulae Naturae of the Academy of Natural Sciences, Philadelphia 460: 1-18.

Galbreath, E. C. 1953. A contribution to the Tertiary geology and paleontology of northeastern Colorado. University of Kansas Paleontological Contributions, Vertebrata 4.

Green, M. 1958. Arikareean rhinoceroses from South Dakota. Journal of Paleontology 32:587-594.

Gregory, W. K., and H. J. Cook. 1928. New material for the study of evolution: a series of primitive rhinoceros skulls (*Trigonias*) from the lower Oligocene of Colorado. Proceedings of the Colorado Museum of Natural History 8(1): 3-39.

Hanson, C. B. 1989. *Teletaceras radinskyi*, a new primitive rhinocerotid from the late Eocene Clarno Formation, Oregon; pp. 379-398 *in* D. R. Prothero and R. M. Schoch (eds.), The Evolution of Perissodactyls. Oxford University Press, New York.

Heissig, K. 1989. The Rhinocerotidae; pp. 399-417 *in* D. R. Prothero and R. M. Schoch (eds.), The Evolution of Perissodactyls. Oxford University Press, New York.

Kihm, A. J. 1987. Mammalian paleontology and geology of the Yoder Member, Chadron Formation, east-central Wyoming. Dakoterra 3:28-45.

Lambe, L. M. 1905. A new species of *Hyracodon* (*H. priscidens*) from the Oligocene of the Cypress Hills,

Assiniboia. Transactions, Royal Society of Canada, Series 2 11(Section 4):37-42.

Lambe, L. M. 1908. The Vertebrata of the Oligocene of the Cypress Hills, Saskatchewan. Contributions to Canadian Palaeontology 3(4):5-64.

Leidy, J. 1850. [First descriptions of *Rhinoceros nebraskensis, Agriochoerus antiquus, Palaeotherium proutii*, and *P. bairdii*.] Proceedings of the Academy of Natural Sciences of Philadelphia 5: 119.

Leidy, J. 1851. [Remarks on *Aceratherium occidentale* and *A. nebrascense*.] Proceedings of the Academy of Natural Sciences of Philadelphia 5:331.

Leidy, J. 1852. Description of the remains of extinct Mammalia and Chelonia, from Nebraska Territory; pp. 535-572 in D. D. Owen, Report on the geological survey of Wisconsin, Iowa, and Minnesota; and incidentally of a portion of Nebraska Territory. Lippincott, Grambo & Co., Philadelphia.

Leidy, J. 1853. The ancient fauna of Nebraska; or a description of remains of extinct Mammalia and Chelonia, from the Mauvaises Terres of Nebraska. Smithsonian Contributions to Knowledge 6(7): 1-126.

Leidy, J. 1856. Notices of several genera of extinct Mammalia, previously less perfectly characterized. Proceedings of the Academy of Natural Sciences of Philadelphia 8: 91-92.

Lucas, S. G., R. M. Schoch, and E. Manning. 1981. The systematics of *Forstercooperia*, a middle to late Eocene hyracodontid (Perissodactyla, Rhinocerotoidea) from Asia and western North America. Journal of Paleontology 55: 826-841.

Lucas, S. G., and J. Sobus. 1989. The systematics of indricotheres; pp. 358-378 in D. R. Prothero and R. M. Schoch (eds.), The Evolution of Perissodactyls. Oxford University Press, New York.

Matthew, W. D. 1931. Critical observations on the phylogeny of rhinoceroses. University of California Publications in Geological Sciences 20: 1-8.

Nixon, D. A., and H. E. LaGarry-Guyon. 1993. New trackway sites in the White River Group type section at Toadstool Park, Nebraska: paleoecology of an Oligocene braided stream, riparian woodland, and adjacent grassland. Journal of Vertebrate Paleontology 13 (supplement to number 3):51A.

Ostrander, G. E. 1985. Correlation of the early Oligocene (Chadronian) in northwestern Nebraska; pp. 205-231 in J. E. Martin (ed.), Fossiliferous Cenozoic deposits of western South Dakota and northwestern Nebraska. Dakoterra, Museum of Geology, South Dakota School of Mines 2.

Peterson, O. A. 1934. List of species and descriptions of new material from the Duchesne River Oligocene, Uinta Basin, Utah. Annals of the Carnegie Museum 23: 373-389.

Prothero, D. R. 1985. Chadronian (early Oligocene) magnetostratigraphy of eastern Wyoming: implications for the Eocene-Oligocene boundary. Journal of Geology 93: 555-565.

Prothero, D. R., C. Guérin, and E. Manning. 1989. The history of the Rhinocerotoidea; pp. 322-340 in D. R. Prothero and R. M. Schoch (eds.), The Evolution of Perissodactyls. Oxford University Press, New York.

Prothero, D. R., E. Manning, and C. B. Hanson. 1986. The phylogeny of the Rhinocerotoidea (Mammalia, Perissodactyla). Zoological Journal of the Linnean Society of London 87:341-366.

Prothero, D. R., and R. M. Schoch (eds.) 1989. The Evolution of Perissodactyls. Oxford University Press, New York.

Prothero, D. R., and C. C. Swisher III. 1992. Magnetostratigraphy and geochronology of the terrestrial Eocene-Oligocene transition in North America; pp. 46-74 in D. R. Prothero and W. A. Berggren (eds.), Eocene-Oligocene Climatic and Biotic Evolution, Princeton University Press, Princeton, N.J.

Radinsky, L. B. 1967. A review of the rhinocerotoid family Hyracodontidae (Perissodactyla). Bulletin of the American Museum of Natural History 136(1): 1-45.

Russell, L. S. 1934. Revision of the Lower Oligocene vertebrate fauna of the Cypress Hills, Saskatchewan. Transactions, Royal Canadian Institute 29(1), no. 43:49-67.

Russell, L. S. 1982. Tertiary mammals of Saskatchewan. Part VI: The Oligocene rhinoceroses. Life Sciences Contributions, Royal Ontario Museum 133:1-58.

Schlaikjer, E. M. 1935. Contributions to the stratigraphy and paleontology of the Goshen Hole area, Wyoming. III. A new basal Oligocene formation. Bulletin of the Museum of Comparative Zoology, Harvard University 74(3): 71-93.

Scott, W. B. 1941. The mammalian fauna of the White River Oligocene. Part V. Perissodactyla. Transactions of the American Philosophical Society 28(5): 747-980.

Scott, W. B., and H. F. Osborn. 1887. Preliminary account of the fossil mammals from the White River Formation contained in the Museum of Comparative Zoology. Bulletin of the Museum of Comparative Zoology 13(5): 151-171.

Simpson, W. F. 1985. Geology and paleontology of the Oligocene Harris Ranch badlands, southwestern South Dakota; pp. 303-333 in J. E. Martin (ed.), Fossiliferous Cenozoic deposits of western South Dakota and northwestern Nebraska. Dakoterra, Museum of Geology, South Dakota School of Mines 2.

Sinclair, W. J. 1922. Hyracodons from the Big Badlands of South Dakota. Proceedings of the American Philosophical Society 61: 65-79.

Skinner, S. M., and R. J. Gooris. 1966. A note on *Toxotherium* (Mammalia, Rhinocerotidae) from Natrona County, Wyoming. American Museum Novitates 2261: 1-12.

Tanner, L. G., and L. D. Martin. 1976. New rhinocerotoids from the Oligocene of Nebraska; pp. 210-219 in C. S. Churcher (ed.), Athlon: Essays in paleontology in honour of Loris Shano Russell. Royal Ontario Museum, Life Sciences, Miscellaneous Publications.

Tedford, R. H., T. Galusha, M. F. Skinner, B. E. Taylor, R. W. Fields, J. R. Macdonald, J. M. Rensberger, S. D. Webb, and D. P. Whistler. 1987. Faunal succession and biochronology of the Arikareean through Hemphillian interval (late Oligocene through earliest Pliocene epochs) in North America; pp. 153-210 in M. O. Woodburne (ed)., Cenozoic Mammals of North America, Geochronology and Biostratigraphy, University of

California Press, Berkeley, California.

Tedford, R. H., J. B. Swinehart, R. M. Hunt, Jr., and M. R. Voorhies. 1985. Uppermost White River and lowermost Arikaree rocks and faunas, White River Valley, northwestern Nebraska and their correlation with South Dakota; pp. 335-352 *in* J. E. Martin (ed.), Fossiliferous Cenozoic deposits of western South Dakota and northwestern Nebraska. Dakoterra, Museum of Geology, South Dakota School of Mines 2.

Troxell, E. L. 1921. New species of *Hyracodon*. American Journal of Science II: 34-40.

Wilson, J. A., and J. A. Schiebout. 1984. Early Tertiary vertebrate faunas, Trans-Pecos Texas: Ceratomorpha less

Amynodontidae. Pearce-Sellards Series, Texas Memorial Museum 39: 1-47.

Wood, H. E., II. 1926. *Hyracodon petersoni*, a new cursorial rhinoceros from the Lower Oligocene. Annals of the Carnegie Museum 16(7): 315-318.

Wood, H. E., II. 1927. Some early Tertiary rhinoceroses and hyracodonts. Bulletins of American Paleontology 13(50): 5-105.

Wood, H. E., II. 1961. *Planiceps* Scott & Osborn, 1887 (*Hyracodon*): Proposed suppression under the plenary powers as a *nomen dubium* (Mammalia). Bulletin of Zoological Nomenclature 18(3):28.

30. Summary

DONALD R. PROTHERO AND ROBERT J. EMRY

CHRONOSTRATIGRAPHY

The chronostratigraphic age assignments of the major late middle Eocene through early Oligocene terrestrial deposits in western North America are shown in Figures 1-5. Although not all areas are dated and correlated with equal precision, we feel that these correlations are considerably more accurate and highly resolved than those presented by Prothero (1985), Emry et al. (1987), and Krishtalka et al. (1987). These major improvements in geochronology are due primarily to the development of $^{40}Ar/^{39}Ar$ dating, providing both new dates, and the correction and rejection of certain K-Ar dates. In addition, magnetic stratigraphy provides even finer-scale correlation of localities whose general age is already known from biostratigraphy or radiometric dating.

Uintan

When Prothero and Swisher (1992) summarized the chronostratigraphic controls on the Uintan known at that time, there were relatively few constraining data points, and few grounds for controversy. The primary data base consisted of Flynn's (1986) magnetic stratigraphy of selected Bridgerian-Uintan sections, plus recently obtained $^{40}Ar/^{39}Ar$ dates and the magnetic polarity pattern of the Uinta Formation in the Uinta Basin (this volume, Chapter 1). Based on the evidence available at the time, Prothero and Swisher (1992) suggested that the Bridgerian/Uintan boundary occurred early in Chron C20r (based largely on Flynn's data), and that the early/late Uintan boundary (Uinta B2/C boundary in the Uinta Basin) occurred in Chron C20n. They also discussed another possibility—that the early/late Uintan boundary occurred in Chron C19n— but rejected that alternative as less likely for the reasons discussed below.

Since that time, the interpretation of Prothero and Swisher (1992) has been challenged, and new data have emerged. McCarroll et al. (1993) moved the early/late Uintan boundary to C19n, based on previously unpublished data from the Washakie Basin of Wyoming. This interpretation was followed in several early drafts of Prothero's chapters in this volume until new data from

the San Diego section (this volume, Chapters 4, 5, and 6) forced yet another re-examination of the evidence (compare Chapters 2 and 4 of this volume). Rather than repeat all the intricate arguments detailed in those chapters, we will summarize the chronostratigraphic constraints on the problem, and assess the reliability of the various sources of data.

Geochronological constraints

Despite many attempts, there are few reliable radioisotopic dates. Prothero and Swisher (1992) gave reasons for rejecting much of the old K-Ar data base in favor of the new $^{40}Ar/^{39}Ar$ dates, and this argument is still valid. We will mention K-Ar dates as appropriate below, but many are discordant with all other sources of data. We feel that the most reliable dates are the following:

—an $^{40}Ar/^{39}Ar$ date by Carl Swisher of 47.3 ± 0.05 Ma on the Henry's Fork tuff, high in the upper Bridgerian Bridger C (E. Evanoff, personal communication). This suggests that the late Bridgerian is at least as young as 47.3 Ma, which would place it early in Chron C21n (Berggren et al., 1995).

— an $^{40}Ar/^{39}Ar$ date by Chris Henry of 46.29 ± 0.05 Ma on a tuff just below the Alamo Creek basalt in the lower member of the Devil's Graveyard Formation of Trans-Pecos Texas (Henry, personal communication; see this volume, Chapter 9). This date is just below the early Uintan Whistler's Squat l.f. (in rocks of reversed polarity—Walton, 1992) and just above the late Bridgerian Junction and 0.6 local faunas, which suggests that the Bridgerian/Uintan boundary occurs in late Chron C21n.

—an $^{40}Ar/^{39}Ar$ date by John Obradovich of 42.83 ± 0.24 Ma from the late Uintan Mission Valley Formation (Berry, 1991; Prothero, 1991; Obradovich and Walsh, in prep; this volume, Chapters 4 and 6). On the Berggren et al. (1995) time scale, this date falls within Chron C20n, and the date comes from rocks of normal magnetic polarity.

—an $^{40}Ar/^{39}Ar$ date by Carl Swisher of 39.74 ± 0.07 Ma on the Lapoint Tuff at the base of the late Duchesnean Lapoint Member of the Duchesne River Formation, Uinta Basin, Utah (Prothero and Swisher, 1992). This date seems to place the early Duchesnean in Chron C18n, and relegates the latest Uintan Brennan Basin and Dry Gulch Creek members of the Duchesne River Formation (Andersen and Picard, 1972) to early Chron C18n or earlier time.

In addition to the ^{40}Ar/^{39}Ar dates, there are several K-Ar dates of varying reliability:

—a K-Ar date of 46.9 ± 1.1 Ma on the Quarry Tuff, in rocks of reversed magnetic polarity just below the early Uintan Whistler Squat l.f. of West Texas (McDowell, 1979; Walton, 1992). Given the large error estimates, this seems to place early Uintan faunas early in Chron C20r.

—a K-Ar date of 43.9 ± 0.9 Ma on a tuff just above the early Uintan Whistler Squat l.f. of West Texas in rocks of both normal and reversed polarity (McDowell, 1979; Walton, 1992). This seems to place early Uintan mammals late in Chron 20r, based on the Berggren et al. (1995) time scale.

—a K-Ar date of 42.7 ± 1.6 Ma on the Skyline tuff in rocks of normal magnetic polarity above the late Uintan Serendipity l.f., and just below the early Duchesnean Skyline Channels l.f. of West Texas (Stevens et al., 1984; Walton, 1992; see this volume, Chapter 9). This date has been widely regarded as anomalous, because it has large error bars and seems to conflict with other data. It was based on a few tiny crystals of biotite (M. Stevens, personal communication). It seems to place the late Uintan Serendipity and Purple Bench l.f. in Chron C20r if taken at face value. Given the large error estimates, however, these faunas could also be placed in Chron C19r.

—a series of K-Ar dates averaging 42.2 Ma associated with NP16 calcareous nannofossils and late Uintan mammals from the Casa Blanca l.f. of the Texas Gulf Coastal Plain (Westgate, 1988). Although NP16 spans late Chron C20n to early Chron C18n (Berggren et al., 1995), the date would tend to support the correlation of the late Uintan with Chrons C20n-C19r.

As discussed by Prothero and Swisher (1992), all the K-Ar dates given by Mauger (1977) for the Uinta Basin have been rejected; none seem to be consistent with the framework of dates outlined above. Mauger's date of 43.1 ± 1.3 Ma on rocks of normal polarity correlative with Uinta A seems much too young, given that those rocks seem to be associated with Chron C20r elsewhere. The date of 42.8 ± 1.0 Ma in rocks of reversed polarity (placing them in Chron C20n or at best C19r) laterally correlative with the Uinta-Duchesne River contact is much too old, given its apparent correlation with Chron C18r. In addition, there are three long magnetic polarity zones in the Indian Canyon section between the 43.1 and 42.8 dates spaced only 0.3 million years apart (Prothero and Swisher, 1992; this volume, Chapter 1). Likewise, many of the old K-Ar dates from Wyoming used by Flynn (1986) were rejected by Swisher, as has

been the K-Ar date of 42.3 Ma from Duchesnean Badwater locality 20 (Prothero and Swisher, 1992).

Marine biostratigraphic constraints

The San Diego section is the only place where well-dated strata with Bridgerian and Uintan mammals interfinger with marine strata bearing planktonic microfossils. The most important datum is the occurrence of coccolith subzones CP12b, CP13a, and CP13b, and planktonic foraminiferan Zone P10 (Flynn, 1986; Walsh et al., this volume, Chapter 6) in the Ardath Shale. Based on these microfossil zones, all authors (Flynn, 1986; Bottjer et al., 1991; Walsh et al., this volume, Chapter 6) agree that the normal magneto-zone that encompasses the Ardath, Scripps, and lower Friars formations correlates with Chron C21n (46.3-48.0 Ma in the Berggren et al., 1995, time scale). Bridgerian faunas occur beneath this level in rocks of reversed polarity (= Chron C21r) in the upper part of the Delmar Formation (Swami's Point l.f.; see Walsh, this volume, Chapter 5). Bridgerian or early Uintan mammals occur above the Ardath Shale in rocks of normal magnetic polarity (= C21n) in the lower part of the Scripps Formation ("Horizon A" of Flynn, 1986, fig. 9; Black's Beach l.f. of Walsh, this volume, Chapter 5). Undoubted early Uintan faunas occur in rocks of normal magnetic polarity (= C21n) at the base of the Friars Formation. As discussed by Walsh (this volume, Chapters 4 and 6), these data would place the Bridgerian/Uintan boundary in Chron C21n.

The Mission Valley Formation in San Diego produces a late Uintan mammalian fauna in rocks of variable magnetic polarity. Steineck et al. (1972) also reported planktonic foraminifera correlative with Zone P13, which would imply a correlation of these strata with Chron C18n (Berggren et al., 1995). However, this age assignment is based on only two poorly preserved taxa, *Globorotaloides suteri* and *Truncorotaloides collacteus*, and there are problems with the quality of these identifications, and with the supposed restriction of these taxa to Zones P13-P14 (McWilliams, 1972; Philips, 1972; Flynn, 1986, p. 350; W. A. Berggren, personal communication). However, the ^{40}Ar/^{39}Ar date of 42.83 Ma mentioned above rules out correlation with Chrons C18n or C19n for the late Uintan strata of normal magnetic polarity in the Mission Valley Formation. Correlation with Chron C20n is the only reasonable alternative.

Magnetic Stratigraphy and Mammalian Biostratigraphy

Given the constraints outlined above, we can now examine the numerous magnetic polarity stratigraphies for Bridgerian-Uintan-Duchesnean strata that have been reported and see if we can make sense of the pattern (Fig. 1). Three primary constraints seem to limit our range of possible interpretations:

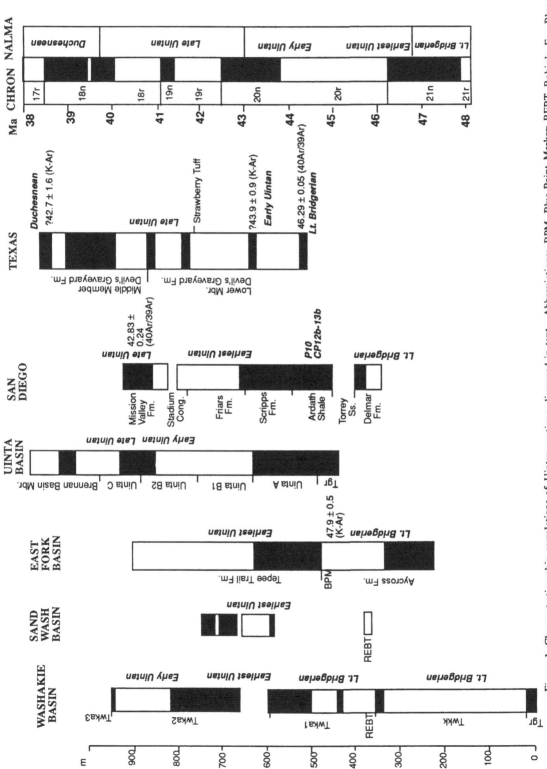

Figure 1. Chronostratigraphic correlations of Uintan sections discussed in text. Abbreviations: BPM, Blue Point Marker; REBT, Robin's Egg Blue Tuff; Tgr, Green River Formation; Twk, Kinney Rim Member, Washakie Formation; Twka1, Twka2, Twka3, units 1, 2, and 3, Adobe Town Member, Washakie Formation.

1. The San Diego magnetic and marine microfossil data seem to place the Bridgerian/Uintan boundary in Chron C21n. Only in San Diego is there a superposed sequence of Bridgerian mammals (Delmar Formation, Swami's Point l.f.) underlying planktonic microfossils of Zones P10 and CP12b-CP13b (Ardath Shale), which are in turn overlain by early Uintan mammals (lower Friars Formation); the Ardath, Scripps, and lower Friars formations are apparently all within the same zone of normal magnetic polarity. This interpretation is also consistent with the ^{40}Ar/^{39}Ar dates of 47.3 Ma for the late Bridgerian in the Bridger Basin, and 46.29 Ma for the Bridgerian/Uintan transition in Trans-Pecos Texas.

2. The ^{40}Ar/^{39}Ar date of of 42.83 on the late Uintan Mission Valley Formation places the early/late Uintan boundary in Chron C20n, and does not allow it to be as young as C19n.

3. The ^{40}Ar/^{39}Ar date of 39.74 Ma on the Lapoint Tuff places the Dry Gulch Creek and Lapoint Members of the Duchesne River Formation in middle Chron C18n in the time scale of Berggren et al. (1995). Any interpretation which forces the Uintan/Duchesnean boundary younger than this would be contradicted by this date (and also by other dates on the Duchesnean discussed below).

Uinta Basin, Utah: Prothero and Swisher (1992) and Prothero (this volume, Chapter 1) reported the results of magnetostratigraphic analysis in the Uinta Basin of Utah, the "type areas" for the Uintan and Duchesnean land mammal "ages." Unfossiliferous Uinta A is of normal polarity. Uinta B1 and lower Uinta B2 are of reversed polarity. Upper B2 and lower Uinta C are of normal magnetic polarity. Upper Uinta C and the lower Brennan Basin Member of the Duchesne River Formation are in a zone of reversed polarity. Another zone of normal and then of reversed polarity is found in the lower Brennan Basin Member. Sampling was discontinued above this level because of problems with diagenetic hematite, and because the rocks were either too hard or too crumbly for sampling.

Prothero and Swisher (1992) and Prothero (this volume, Chapter 1) correlated the Uinta A normal magnetozone with Chron C21n, the upper B2-lower C normal magnetozone with Chron C20n, and the lower Duchesne River normal magnetozone with Chron C19n. McCarroll et al. (1993) suggested that this entire sequence should be shifted up one magnetic chron, placing Uinta A in Chron C20n, Uinta B2-C in Chron C19n, and the lower Brennan Basin Member in Chrons C18n and C17r. However, this interpretation conflicts with several of the data constraints outlined above. In particular, it contradicts the 42.83 Ma date on the late Uintan in San Diego, and it would force the Lapoint Tuff into Chron C17n or younger, which is inconsistent with its ^{40}Ar/^{39}Ar date of 39.74 Ma.

Thus, the Uinta Basin section is best correlated with Chrons C21n-C18r, which places the early/late Uintan boundary in Chron C20n, and the Bridgerian/Uintan

boundary at some time earlier than mid-Chron C20r.

Washakie Basin, Wyoming: This section was originally reported by Flynn (1986), and has since been modified by McCarroll et al. (1993; this volume, Chapter 2). The late Bridgerian Kinney Rim Member (Twkk) was mostly of reversed polarity except for a few sites below the unconformity at the top. The lower half of unit 1 of the Adobe Town Member (Twka1) was of reversed polarity, with a short (2-site) zone of normal polarity; the upper third was of normal polarity. There is a 90-m (300-foot) sampling gap in the basal part of member 2 (Twka2); the middle third of this unit was of normal polarity, and produces earliest Uintan ("Shoshonian") and early Uintan faunas. The upper third of Twka2 was of reversed polarity except for a single normal site at the top. Unit 3 (Twka3) has not yet been sampled magnetically.

Flynn (1986, fig. 9) originally correlated the reversed magnetozone in the late Bridgerian Kinney Rim Member to Chron C21r, and the upper part of the Kinney Rim Member to Chron C21n. The reversed magnetozone spanning the lower half of Twka1 was correlated to Chron C20r, and the long normal magnetozone (C+ of Flynn, 1986) spanning upper Twka1 and lower Twka2 was correlated with Chron C20n. The upper part of Twka2 was correlated with Chron C19r. McCarroll et al. (1993) also preferred this correlation, although McCarroll et al. (this volume, Chapter 2) were equivocal. However, it presents several problems. It places late Bridgerian faunas in Chron C20r; that conflicts with the marine planktonic evidence which places early Uintan mammals within Chron C21n. It also forces the early/late Uintan boundary up into Chron C19n, contradicting the 42.83 Ma date on the late Uintan in San Diego. This also implies that the Brennan Basin Member of the Duchesne River Formation correlates with Chron C18n and C17r (McCarroll et al., this volume, Fig. 5). This interpretation conflicts with the ^{40}Ar/^{39}Ar date of 39.74 Ma on the Lapoint Tuff, as discussed above.

An alternative correlation, as suggested by Prothero and Swisher (1992), places the late Bridgerian faunas of Twwk and lower Twka1 in Chron C21n-C21r and earlier (how much earlier depends on how one interprets the many short zones of normal polarity in this interval). "Shoshonian" and early Uintan faunas of Twka2 are then correlated with Chron C21n and C20r, which is consistent with all the known data.

Sand Wash Basin, Colorado: Stucky et al. (this volume, Chapter 3) outline the magnetic stratigraphy and biostratigraphy of the Washakie Formation in the Sand Wash Basin of Colorado, just to the south of the Washakie Basin of Wyoming. Certain key beds, such as the "Robin's Egg Blue Tuff," allow correlation of these two sections across the state line. Given these constraints, Stucky et al. (this volume, Chapter 3) found that the most reasonable correlation matched the lower (earliest Uintan, or "Shoshonian") reversed magnetozone with the 300-foot (90-m) with either the reversed rock of middle Twka1 (which may have Bridgerian small mammals, but lacks the key large mammal taxa), or with the sampling gap within magnetozone C+ in the lower part of Twka2. This latter interpretation is consistent with the level from which McCarroll et al. (this volume, Chapter 2, Fig. 3) report earliest Uintan faunas in Twka2. The upper half of the Sand Wash section is of normal polarity, and probably correlates with the rest of magnetozone C+ in Twka2. If magnetozone C+ is correlated with Chron C21n (as suggested above), these correlations would place "Shoshonian" and early Uintan faunas in Chron C21n or possibly C21r.

East Fork Basin, Wyoming: Flynn (1986, fig. 6) reported magnetostratigraphic results from the East Fork Basin of the Absaroka Mountains of northwest Wyoming. The Aycross Formation is characterized by a basal normal magnetozone (containing an early Bridgerian fauna, "Horizon A" of Flynn, 1986, fig. 9), a long reversed interval (containing the late Bridgerian "Horizon B" of Flynn, 1986, fig. 9), followed by an upper normal magnetozone that continued into the basal Tepee Trail Formation. The rest of the Tepee Trail Formation is of reversed polarity, and includes ?early Uintan "Horizon C" of Flynn (1986, fig. 9) and Bone Bed A of McKenna (1980; "Horizon D" of Flynn, 1986), a definite early Uintan assemblage. Flynn (1986, Fig. 9) interpreted this sequence as Chrons C22n-C20r, placing the Bridgerian/Uintan boundary at the base of C20r (although the evidence could also place the boundary in Chron C21n). These results are consistent with the interpretations advocated above and with those of Prothero and Swisher (1992), but not with the interpretation of McCarroll et al. (1993), which would place the Bridgerian/Uintan boundary in Chron C20n.

As Sundell et al. (1984) point out, however, Flynn (1986) apparently missed some polarity zones in the Absaroka sequence. They agreed with Flynn that the normal magnetozone at the base of the Tepee Trail Formation is Chron C21n, but showed that Flynn missed a 300-m interval in the upper Aycross and lower Tepee Trail Formations. Thus, the normal magnetozone at the top of the Aycross Formation is Chron C22n, not C21n.

Trans-Pecos Texas: Walton (1992) described the magnetostratigraphy of the lower and middle members of the Devil's Graveyard Formation in Trans-Pecos Texas (Stevens et al., 1984; Wilson, 1984, 1986). Walton (1992, fig. 3.8) correlated the Bridgerian 0.6 l.f. and Junction l.f. with Chron C21n and correlated the early Uintan Whistler Squat l.f. with C20r (consistent with our preferred correlations discussed above, and with all the available dates). However, she also placed the late Uintan Serendipity l.f. in Chron C20r, which conflicts with the data discussed above. As discussed in Chapter 9 of this volume, the normal magnetozone immediately above the Strawberry Tuff is probably

Chron C20n, placing these late Uintan faunas in C20n, C19r, and C19n, and making them consistent with the rest of the available data. The long normal magnetozone in the middle member of the Devil's Graveyard Formation probably correlates with Chron C18n (this volume, Chapter 9), as it is overlain by the early Duchesnean Skyline and Cotter Channels faunas (see discussion below).

Summary

Despite the apparently conflicting mass of data and interpretations discussed by McCarroll et al. (this volume, Chapter 2) and Walsh (this volume, Chapter 4), there does seem to be a solution to this problem that conforms to the constraints of all the reliable data. Several sections (primarily San Diego, but also the Washakie Basin, East Fork Basin, and Trans-Pecos Texas) are consistent with the interpretation that the Bridgerian/Uintan boundary occurs in Chron C21n (and not in C20n, as suggested by McCarroll et al., 1993). The Uinta Basin section, and the date of 42.83 Ma on the late Uintan Mission Valley Formation in San Diego, establishes that the early/late Uintan transition occurs within Chron C20n (and not C19n, as suggested by McCarroll et al., 1993). The magnetostratigraphy of the Uinta Basin section, the date on the Lapoint Tuff, and the pattern of other late Uintan-Duchesnean sections (such as the Sespe Formation; see this volume, Chapter 8) place the Uintan/Duchesnean boundary in Chron C18n (and not in Chron C17r or younger, as suggested by McCarroll et al., 1993).

Of course, this interpretation is subject to further testing and falsification. It predicts, for example, that the 90-m sampling gap in the lower part of unit 2 of the Adobe Town Member between the Adobe Town and Skull Creek sections of Flynn (McCarroll et al., this volume, Chapter 3, Fig. 3) will be found to have reversed polarity, as is the Shoshonian lower half of the Sand Wash Basin section (this volume, Chapter 4). It also predicts that all of Bridger C and at least the lower part of Bridger D will be of normal polarity, corresponding to Chron C21n. Because Bridger E may be early Uintan (Evanoff et al., 1994), the Chron C21n-C20r transition will probably occur in either upper Bridger D or in Bridger E. Flynn and McCarroll (personal communication) have resampled the Bridger Formation, so it will soon will be possible to test this prediction.

Duchesnean

Emry (1981), Wilson (1978, 1984, 1986), Krishtalka et al. (1987), Kelly (1990), and Lucas (1992) have discussed the basis for the Duchesnean. The main problem is that the "type area" for this land mammal "age," part of the Duchesne River Formation, is very sparsely fossiliferous, and most of the rare "type" Duchesnean fossils from the formation come from unknown levels within the Lapoint Member. The lowest two members (Brennan Basin and Dry Gulch

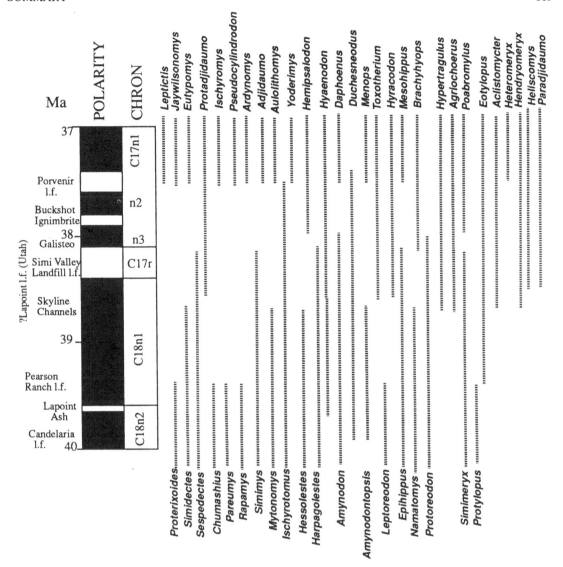

Figure 2. Ranges of key taxa and chronostratigraphy of Duchesnean localities discussed in text.

Creek Members of Andersen and Picard, 1972) are apparently late Uintan. The "type Duchesnean fauna" from the Lapoint Member has been considered so depauperate, and so similar to the the the faunas of the Chadronian, that Emry (1981) and Wilson (1984, 1986) considered the Duchesnean a "subage" of the Chadronian. Kelly (1990; Kelly, 1992; Kelly et al., 1991; Kelly and Whistler, 1994), however, further documented the Duchesnean faunas of California, and Storer (1987, 1990, 1993, this volume, Chapter 12) described Duchesnean faunas from Saskatchewan. Lucas (1992) summarized the Duchesnean localities from throughout western North America, and argued that there was more to the interval than just the faunas of the Lapoint Member. Since these studies, the validity of the Duchesnean as a discrete land mammal "age" has regained popularity.

The advent of ^{40}Ar/^{39}Ar dating and magnetic stratigra-phy have improved the chronostratigraphy of key Duchesnean sections. Although most of the type Duchesne River Formation proved intractable to magnetostratigraphy (this volume, Chapter 1), Prothero and Swisher (1992) demonstrated that the lower part of the Brennan Basin Member was correlative with parts of Chrons C19r-C18r. In addition, Swisher obtained a ^{40}Ar/^{39}Ar date of 39.74 ± 0.07 Ma from the Lapoint Ash, at the base of the Lapoint Member, about 50 m (150 feet) below the Carnegie Museum "Teleodus" (= Duchesneodus) Quarry, the principal Duchesnean locality in the Lapoint Member. This information suggests that at least some part of the Duchesnean correlates with the long Chron C18n. Combining this fact with the magnetic stratigraphy of the late Uintan and early Duchesnean (Pearson Ranch l.f.) faunas in the Sespe Formation of California (this volume, Chapter 8), it appears that the Uintan/Duchesnean boundary

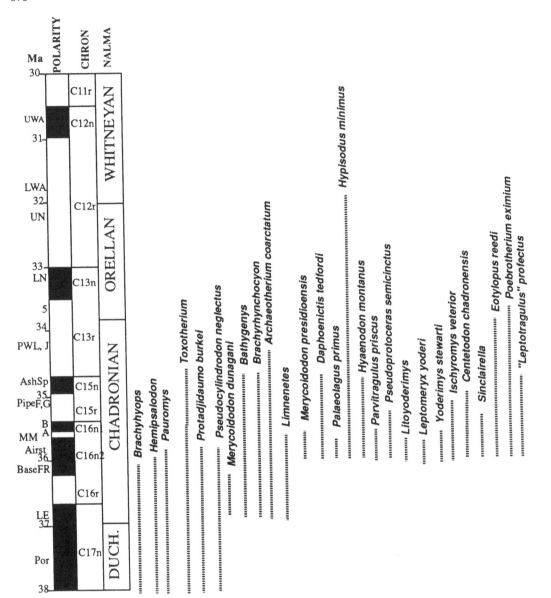

Figure 3. Chronostratigraphy of key early Chadronian localities and taxa discussed in text. Abbreviations: NALMA = North American land mammal "age"; l.f. = local fauna; LE = Little Egypt l.f.; BaseFR = Base of Flagstaff Rim section; Airst. = Airstrip l.f.; Pipe = Pipestone Springs l.f.; AshSp = Ash Springs l.f.; PWL = Persistent White Layer; A, B, G, J = Marker ashes at Flagstaff Rim; LN = lower nodular zone in the Big Badlands; UN = upper nodular zone; LWA = Lower Whitney Ash; UWA = Upper Whitney Ash. Time scale after Berggren et al. (1995).

occurs early in Chron C18n, and that early Duchesnean faunas are correlative with Chron C18n (38.3-40.1 Ma). The slightly younger Duchesnean Simi Valley Landfill l.f. of Kelly et al. (1991) is correlative with C17r (38.1-38.3 Ma).

The other constraint on the Duchesnean comes from the Trans-Pecos Texas region (this volume, Chapter 9). The late Duchesnean Porvenir l.f. occurs just above the Buckshot Ignimbrite, which has been $^{40}Ar/^{39}Ar$ dated at 37.8 ± 0.15 Ma (Prothero and Swisher, 1992). Within the Chambers Tuff in the Capote Creek drainage, the

Porvenir l.f. occurs below the Chadronian Little Egypt l.f. This fauna, in turn, is overlain by the Bracks Rhyolite, which has been $^{40}Ar/^{39}Ar$ dated at 36.7 ± 0.07 Ma. Combining the magnetic stratigraphy with the stratigraphic position (this volume, Chapter 9) suggests that the Porvenir l.f. is about 37.5 Ma in age (Chron C17n2). Similarly, the Duchesnean Skyline and Cotter channels of the Agua Fria area (Stevens et al., 1984; Wilson, 1986) appear to correlate with Chron C18n-C17r (this volume, Chapter 9). Based on magnetostratigraphy and faunal similarities, the Duchesnean faunas of the Galisteo Formation of New Mexico are correlated with C17n3-C17r (this volume, Chapter 10).

Although this does not provide precise chronostratigraphy of all Duchesnean faunas, a framework for subdividing the Duchesnean can be developed (Fig. 2). The Pearson Ranch l.f. is early Duchesnean (C18n2, 39-40 Ma), and the Simi Valley Landfill l.f. and Skyline-Cotter Channels l.f. are "middle" Duchesnean (C17r-C18n1, 38.2-38.6 Ma). The Povenir l.f. is late Duchesnean, because it overlies a date of 37.8 Ma and is overlain by a Chadronian fauna constrained by a date of 36.7 Ma. The Galisteo fauna also appears to be late Duchesnean (38.0-38.3 Ma). The chronostratigraphic position of the Lapoint fauna is not well constrained, but based on faunal evidence, it is most likely late Duchesnean (Wilson, 1984, 1986; Kelly, 1990; Lucas, 1992). Lucas (1992) discussed the possible correlation of other Duchesnean localities, but presently none of these has ^{40}Ar/^{39}Ar dates or magnetic stratigraphy to date them more precisely.

Chadronian

The Duchesnean/Chadronian boundary is best constrained in Trans-Pecos Texas, where the late Duchesnean Porvenir l.f. and early Chadronian Little Egypt l.f. are bracketed by ^{40}Ar/^{39}Ar dates of 37.8 ± 0.15 and 36.7 ± 0.07 Ma (Prothero and Swisher, 1992; this volume, Chapter 9). Based on the magnetic stratigraphy and the stratigraphic position of the faunas, the Reeves Bonebed (part of the Little Egypt l.f.) probably correlates with part of Chron C17n1. This would place the Duchesnean/Chadronian boundary at about 37.0 Ma.

The best chronostratigraphic data for the remainder of the Chadronian come from the Flagstaff Rim section in Wyoming. ^{40}Ar/^{39}Ar dating and magnetic stratigraphy (Prothero and Swisher, 1992) shows that the sampled part of the section spans Chrons C16n to C13r (Fig. 3). However, the lowest part of the section (from the base to about 60 feet on the composite section of Emry, 1973, 1992) was not sampled paleomagnetically. As indicated by Emry (1992), this section may be much older than Chron C16n because at the very base Duchesnean taxa such as *Brachyhyops* and *Hemipsalodon* are associated with earliest Chadronian taxa similar to those found in the Yoder l.f. of Wyoming. Following Emry (1992, fig. 5.3), we correlate the base of the Flagstaff Rim section with the earliest Chadronian (36.0-36.5 Ma). This suggests that the lower 60 feet of the section spans Chrons C16r and C16n2, with either very low sedimentation rates, or else significant hiatuses through this interval.

Emry et al. (1987, p. 136) gave evidence that the Airstrip l.f. of Trans-Pecos Texas was early Chadronian, correlative with that part of the section below Ash B at Flagstaff Rim. Ash B has been ^{40}Ar/^{39}Ar dated at 35.9 ± 0.2 Ma (Prothero and Swisher, 1992) and 35.41 ± 0.14 Ma (Obradovich et al., 1995). Either date places Ash B

within Chron C16n of the Berggren et al. (1995) time scale. The normal magnetic polarity of the main Airstrip section also suggests a correlation with Chron C16n (this volume, Chapter 9). Similarly, the McCarty's Mountain l.f. of Montana is early Chadronian (Emry et al., 1987; Tabrum et al., this volume, Chapter 14), and its magnetostratigraphy (this volume, Chapter 14) suggests a correlation with Chrons C16n1 to C16r.

Typically middle Chadronian faunas occur between Ash B and Ashes F and G at Flagstaff Rim (Emry et al., 1987; Ostrander, 1985). Based on the magnetic stratigraphy (Prothero and Swisher, 1992), the middle Chadronian spans Chron C16n1 and early Chron C15r (35.0-35.5 Ma). The faunas and magnetic stratigraphy of the Pipestone Springs area (this volume, Chapter 14) also suggest a correlation with Chron C13r-C15r. The magnetic stratigraphy of the middle Chadronian Raben Ranch l.f. of Nebraska (this volume, Chapter 13) is consistent with this correlation.

Emry et al. (1987, p. 136) suggested that the Ash Springs l.f. of Trans-Pecos Texas was also middle Chadronian, based on the presence of *Toxotherium* and the stage of evolution of *Meliakrounomys*. However, Stevens and Stevens (this volume, Chapter 25) argue that Ash Springs is slightly younger than other middle Chadronian localities based on the oreodonts. The Ash Springs l.f. contains *Merycoidodon culbertsoni* rather than *Merycoidodon presidioensis* or *Miniochoerus forsythae*, and lacks *Bathygenys* and other characteristic middle Chadronian taxa. Unfortunately, there are no magnetostratigraphic or ^{40}Ar/^{39}Ar data to resolve this question, but we place the Ash Springs l.f. slightly later than Ash G to reflect this interpretation (Fig. 3).

Above Ash G (440 feet on the composite section of Emry, 1973), fossils become very scarce at Flagstaff Rim. A few taxa are known from the upper part of the section that might characterize the late Chadronian (Emry, 1992), but a better record of this time interval comes from the base of the Pine Ridge escarpment in Wyoming and Nebraska. Recent tephrochronology (Larson and Evanoff, personal communication) has shown that the widespread marker ash known as the "Persistent White Layer," "Purplish-White Layer," or "PWL" in Niobrara County, Wyoming, and Sioux County, Nebraska, is not the same as the "100 foot white layer" or "5 tuff" of the Douglas area, Converse County, Wyoming (Evanoff et al., 1992), as has long been assumed. Instead, the PWL appears to correlate geochemically with Ash J at Flagstaff Rim, which has been ^{40}Ar/^{39}Ar dated at 34.7 ± 0.04 Ma (Prothero and Swisher, 1992) or 34.36 ± 0.11 Ma (Obradovich et al., 1995). According to Larson and Evanoff (personal communication), the PWL and Ash J also appear to correlate with the "4 tuff" at Douglas (Evanoff et al., 1992), rather than the 5 tuff, which has been ^{40}Ar/^{39}Ar dated at 33.91 ± 0.06 Ma (Prothero and Swisher, 1992). Ash J at Flagstaff Rim and the 4 and 5 tuffs at Douglas

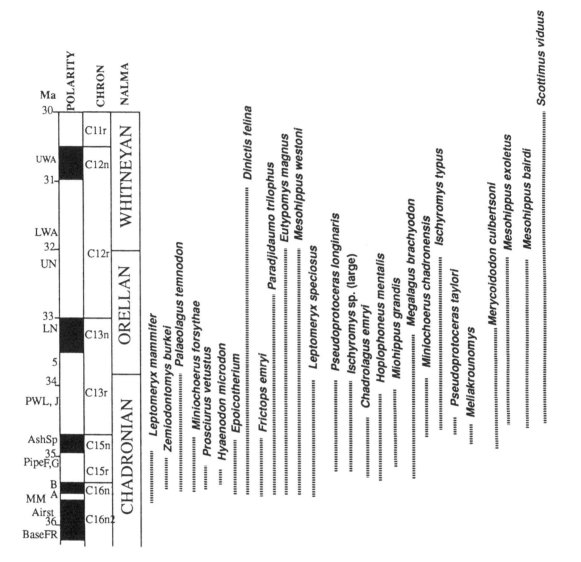

Figure 4. Chronostratigraphy and ranges of key middle-late Chadronian taxa. Abbreviations as in Figure 3.

all occur in Chron C13r (Prothero and Swisher, 1992; Evanoff et al., 1992).

These new data require some slight adjustments in the correlations within Wyoming (this volume, Chapter 13). More important, they provide the first direct tie between the upper Flagstaff Rim section and the slightly more fossiliferous sections below the PWL in Niobrara County, Wyoming. For this reason, we will use the sections in the Lusk area to characterize the biostratigraphy of the late Chadronian.

Orellan-Whitneyan

Prothero and Swisher (1992) and Prothero (this volume, Chapter 13) outlined the new ^{40}Ar/^{39}Ar dates and magnetics within the White River Group, which places chronostratigraphic controls on the Orellan and

Whitneyan (Figs. 3-5). The Chadronian/Orellan boundary (see Prothero and Whittlesey, in press) lies within Chron C13r and just above the "5 tuff" at Douglas (Evanoff et al., 1992), which has been ^{40}Ar/^{39}Ar dated at 33.91 ± 0.06 Ma (Prothero and Swisher, 1992) or 33.59 ± 0.02 Ma (Obradovich et al., 1995). Earliest Orellan faunas occur in the latest part of Chron C13r, and late early Orellan faunas are found in strata correlative with Chron C13n. Late Orellan mammals occur throughout the early part of Chron C12r. The Orellan/Whitneyan boundary occurs in the middle of Chron C12r, just below the Lower Whitney Ash, which has been ^{40}Ar/^{39}Ar dated at 31.8 ± 0.15 Ma (Prothero and Swisher, 1992). This places the Orellan/ Whitneyan boundary at about 32.0 Ma. Late Whitneyan faunas first appear in the later part of Chron C12r.

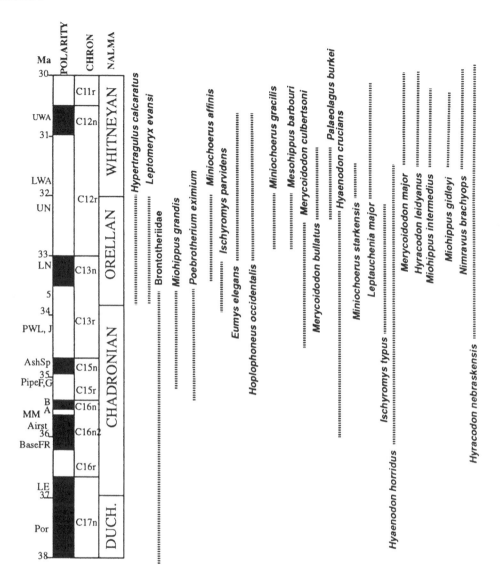

Figure 5. Chronostratigraphy and temporal ranges of key Orellan and Whitneyan taxa. Abbreviations as in Figure 3.

The Whitneyan/Arikareean transition is discussed by Tedford et al. (1985, 1987; this volume, Chapter 15). Characteristic Arikareean taxa occur in the base of the Sharps Formation in South Dakota, and in the brown siltstone beds of the Brule Formation in Nebraska in early Chron C11n (30.0 Ma). Because the Berggren et al. (1995) time scale places the early-late Oligocene boundary at 28.5 Ma, this makes the Orellan, Whitneyan, and earliest Arikareean (including the Sharps and Gering formations) early Oligocene.

BIOCHRONOLOGY AND BIOSTRATIGRAPHY

In the past, mammalian paleontologists rarely attempted to resolve the age of assemblages more closely than to a single North American land mammal "age." In many cases, this was because few of the specimens had precise enough biostratigraphic data to determine where they came from within a given rock unit. As discussed by Emry et al. (1987), some paleontologists used lithostratigraphic units as bio-stratigraphic units (e.g., Schultz and Falkenbach, 1968). In recent years, however, a number of authors have begun to emphasize the exact stratigraphic position of their fossils within lithostratigraphic units. It is now possible to subdivide the Uintan through Whitneyan North American land mammal "ages" into smaller units, and ultimately into biostratigraphic zones based on measured sections in the field.

In this regard, the Uintan through Whitneyan have lagged behind the rest of the mammalian time scale. For example, Archibald et al. (1987) suggested biostratigraphic "zones" for much of the Paleocene, and Gingerich (1983) and Gunnell (1989) have erected biostratigraphic "zones" for part of the Paleocene and Eocene. However, in both of these instances, these are not true biostratigraphic zones in the sense of the 1983 North American Code of Stratigraphic Nomenclature, because they lack type sections and other features specified by the Code (see Prothero, 1995). Tedford et al. (1987) proposed biochronological criteria for subdividing the late Oligocene-Miocene land mammal "ages," but did not attempt to create a range-zone biostratigraphy for the entire interval.

The detailed biostratigraphic data for the Uintan through early Arikareean interval has progressed considerably since the reports of Krishtalka et al. (1987) and Emry et al. (1987). Yet because key parts of the data base are still incomplete as this volume goes to press, we will not propose formal biostratigraphic range zones at this time. Instead, we will indicate the criteria that we would recommend for such a zonation, and likely areas that might serve as zonal stratotypes for each unit.

Uintan

Biostratigraphic zonation of the Uintan is complicated by several factors. The conflicts over chrono-stratigraphy are the foremost problem. Assuming that the correlation shown in Figure 1 is approximately correct, there is also a problem with the biostratigraphic data base itself. Until the summary by McCarroll et al. (this volume, Chapter 2), there was no detailed biostratigraphic information for the Washakie Basin. The biostratigraphic summary of the Uinta Basin (this volume, Chapter 1) was the first in over 30 years (since Black and Dawson, 1966), and the first to tie the specimens to specific levels within Uinta B or C. A third problem is the high degree of endemism of Uintan faunas (Lillegraven, 1979; Walsh, this volume, Chapter 5). In particular, the smaller mammals of the Rocky Mountain region are less well studied, and their distribution in the Uinta Basin is restricted to a few well sampled levels (such as White River and Myton Pockets). Consequently, the Rocky Mountain zonation is dominated by larger mammals. By contrast, the Uintan localities in California produce mostly small mammals, and only a few of the large mammal taxa represented in the Rocky Mountains are also present in the San Diego or Sespe sections.

The Uintan was originally based on the faunas of the Uinta Formation in Utah, so this is the obvious place to begin. The available biostratigraphic data for the collections of the Uinta Basin were summarized in Chapter 1 of this volume. Combining this information with the biostratigraphy of the Washakie Basin (this volume, Chapter 2) and the faunal list of the Sand Wash Basin (this volume, Chapter 3), it is possible to suggest a biostratigraphy for the Rocky Mountains.

Early Uintan

Flynn (1986, p. 380) defined the beginning of the Uintan (and the beginning of his "Shoshonian Subage")

on the first appearance of *Amynodon*, and characterized it by the first appearances of *Leptoreodon*, *Protoreodon*, *Protylopus*, *Epihippus* and *Achaenodon*. Krishtalka et al. (1987, p. 89) gave a similar list: *Amynodon*, *Protoreodon*, *Triplopus*, *Forstercooperia*, *Metarhinus*, *Dolichorhinus* (= *Sphenocoelus*, according to Mader, 1989), *Epihippus*, and *Achaenodon*. Stucky et al. (1989, p. 38) gave a slightly longer list, with many of the same taxa: *Amynodon*, *Oligoryctes*, *Eobasileus*, *Protoreodon*, *Achaenodon*, *Macrotarsius*, *Protoptychus*, *Protylopus*, *Ourayia*, *Forstercooperia*, *Triplopus*, *Epihippus*, and *Isectolophus*.

If the fauna from the middle Adobe Town Member (unit 2, or Twka2) of the Washakie Formation in the Washakie Basin, and from the "Locality 1—Turtle Hill" collections in the Sand Wash Basin, are taken as typical of the earliest Uintan, then not all the taxa listed above define the beginning of the Uintan. *Amynodon*, *Protoptychus*, *Eobasileus*, *Metarhinus*, *Sphenocoelus*, *Triplopus*, *Forstercooperia*, *Achaenodon*, and *Protylopus* first occur in Twka2 in the Washakie Basin; *Epihippus* first occurs in Twka3 (McCarroll et al., this volume, Chapter 2, Table 1). In the Sand Wash Basin, *Amynodon*, *Eobasileus*, *Protoreodon*, *Epihippus*, *Metarhinus*, and *Triplopus* are recorded from the Turtle Hill-Locality 1 levels (Stucky et al., this volume, Chapter 3). Combining the common elements of these lists, only *Protoreodon*, *Eobasileus*, *Metarhinus*, and *Triplopus* are consistently recorded at the beginning of the Uintan in more than one location. *Amynodon*, *Ourayia*, *Macrotarsius*, *Epihippus*, *Protoreodon*, *Leptoreodon*, and *Protylopus* make their first appearance in Texas in the earliest Uintan Whistler's Squat l.f. (Wilson, 1986). *Amynodon*, *Leptoerodon* and *Protoreodon* also occur in the earliest Uintan part of the Friars Formation in San Diego (Walsh, this volume, Chapter 5), so these seem to be the only elements that can be used across North America. Apparently, Flynn's (1986) recommendation of *Amynodon* as the defining taxon of the beginning of the Uintan still works.

Although the large number of first occurrences makes the Bridgerian/Uintan distinction fairly easy to make, there is some controversy over whether the early Uintan can be further subdivided. Flynn (1986) proposed a "Shoshonian Subage" for the earliest Uintan, and this has been followed by a number of authors. The Shoshonian was defined on the overlapping first occurrences of the early Uintan taxa listed above, and the last occurrence of Bridgerian holdover taxa, such as *Uintasorex*, *Notharctus robustior*, *Trogolemur myodes*, *Microsyops annectens*, and *Hemiacodon*. *Notharctus robustior*, *Microsyops annectens*, and *Hemiacodon* also occur in the Sand Wash Basin faunas. However, as Walsh (this volume, Chapter 4) points out, *Uintasorex* and *Microsyops* are now known from the late Uintan, and *Uintasorex* even ranges into the late Duchesnean (Storer, personal communication), leaving only three taxa which last appear in the earliest Uintan. To that list could be added *Hyopsodus paulus* and *Dilophodon minusculus*. So far, none of these taxa are known from Uinta B, although this may not be very significant, since the existing collections from Uinta B are almost entirely large mammals (except for those from White River Pocket, at the very top of Uinta B2).

A better way to subdivide the early Uintan would be based on first occurrences which separate the Twka2-Sand Wash faunas from Uinta B. Stucky et al. (1989) listed the first appearances of *Hessolestes, Auxontodon, Bunomeryx, Diplobunops*, and *Leptotragulus* as defining the late early Uintan (although *Hessolestes* is known from the earliest Uintan Whistler's Squat l.f. of Texas). Unfortunately, the remaining list of artiodactyls (plus many other taxa restricted to White River Pocket—see this volume, Chapter 1, Fig. 8) are not known from any early Uintan localities outside the Rocky Mountains. As Walsh (this volume, Chapter 4) points out, these taxa are unknown from San Diego, the only place where a possible "late early Uintan" assemblage (Murray Canyon l.f.) overlies a presumed "Shoshonian" fauna (Poway assemblage). The Badwater area of central Wyoming produces several superposed early Uintan localities (Krishtalka et al., 1987), but so far it is not clear whether they represent both "Shoshonian" and "Uinta B" equivalents, because a comprehensive biostratigraphy of the Wagonbed Formation has never been published.

Thus, there is no direct evidence from superposed assemblages for a faunal distinction between "Shoshonian" and "late early Uintan." Instead, the evidence must come from faunal differences and the chronostratigraphic correlations shown in Figure 1, which are controversial, as we have seen. Nevertheless, the fact that the Twka2 faunas are in a zone of normal polarity, and Uinta B is almost entirely of reversed polarity, shows that they are different in age, no matter how one juggles the correlations.

Late Uintan

The distinction between early and late Uintan is easy to make in the Uinta Basin, since there are a large number of taxa which last appear in Uinta B2, and many more which first appear in Uinta C (this volume, Chapter 1, Fig. 8). Krishtalka et al. (1987, p. 89) list *Protoptychus, Metarhinus, Telmatherium*, taeniodonts, and uintatheres as last appearing in Uinta B, and eomyids, *Domnina, Thylacaelurus, Colodon, Poebrodon, Prodaphoenus, Simidectes, Procynodictis*, and *Epitriplopus* as first occurring in the late Uintan. *Telmatherium*, as now defined (Mader, 1989), is restricted to the Bridgerian, but the rest of the last occurrences are still valid. *Simidectes* (Whistler's Squat l.f.) and *Poebrodon* (Twka2) are now known from the early Uintan, and *Epitriplopus* first occurs in the latest Uintan Randlett horizon of the lower Duchesne River Formation, but the rest of the list of late Uintan first appearances remains valid. Stucky et al. (1989, p. 38)

gave a slightly different list of taxa defining the late Uintan: *Domnina, Thylacaelurus, Chumashius, Mytonius, Tapocyon, Colodon, Simimeryx, Mytonolagus*, and *Pseudocylindrodon*. Again, this list works for the Uinta Basin and possibly Badwater, but not for San Diego or the Sespe Formation. *Tapocyon* occurs in the early Uintan Friars Formation (this volume, Chapter 5) but in the late Uintan in the Sespe Formation. *Simimeryx* first appears in the late Uintan in the Sespe Formation (Mason, 1988). *Domnina, Thylacaelurus, Mytonius, Prodaphoenus, Colodon*, and *Pseudocylindrodon* are unknown from the San Diego area, and all but *Pseudocylindrodon* is unknown from the Sespe Formation as well.

The chronostratigraphic controversies, faunal endemism, and lack of superposed faunas in more than one region have created problems for any biostratigraphic zonation of the Uintan. Since the San Diego region (this volume, Chapter 5) has more superposed Uintan faunas than any other, it may eventually provide the standard on which a Uintan biostratigraphy could be based. In the Rocky Mountain region, the faunal evidence is insufficient to subdivide the Uintan beyond "Shoshonian," "Uinta B," and "Uinta C," since in no place do all three "subages" occur in superposition.

Duchesnean

Lucas (1992) summarized the faunal lists for most Duchesnean localities, and that information is presented graphically in Figure 2. Wilson (1984, 1986) and Kelly (1990) suggested that distinctions could be made between early and late Duchesnean faunas, but Lucas (1992) questioned whether we have enough information to make that assessment. However, with the new chronostratigraphic data outlined above, both the relative and numerical ages of Duchesnean faunas are becoming better known.

Between the Pearson Ranch and Porvenir local faunas, the distinction between early and late Duchesnean is apparent (Fig. 2). Indeed, as Wilson (1978, 1984, 1986) and Emry (1981) noted, the Pearson Ranch l.f. has much more in common with the late Uintan, and the Porvenir l.f. with the Chadronian, than either share with each other as parts of the Duchesnean. Only the brontothere *Duchesneodus* unites the various Duchesnean faunas (Lucas, 1992). Early Duchesnean local faunas (primarily Pearson Ranch and Skyline-Cotter Channels) are dominated by Uintan holdovers, including the genera *Proterixoides, Simidectes, Sespedectes, Chumashius, Pareumys, Rapamys, Simimys, Mytonomys, Ischyrotomus, Hessolestes, Harpagolestes, Amynodon, Leptoreodon, Epihippus, Protoreodon, Simimeryx*, and *Protylopus*. Some of the species of these genera, however, are unique to the Duchesnean. Only a few genera, such as *Amynodontopsis* and *Mahgarita*, are restricted to the early Duchesnean.

Similarly, late Duchesnean local faunas (primarily Porvenir, Simi Valley Landfill, Galisteo, and possibly

Lapoint) are dominated by genera which continue into the Chadronian. These include *Leptictis, Jaywilsonomys, Eutypomys, Ischyromys, Pseudocylindrodon, Ardynomys, Protadjidaumo, Adjidaumo, Aulolithomys, Yoderimys, Hyaenodon, Daphoenus, Menops, Toxotherium, Hyracodon, Mesohippus, Agriochoerus, Poabromylus, Eotylopus, Aclistomycter, Heteromeryx,* and *Hendryomeryx. Hemipsalodon* and *Brachyhyops* continue into the Chadronian in a number of places. The genera *Rooneyia, Ischognathus,* and *Haplohippus,* and some of the species of the genera listed above, are restricted to the late Duchesnean.

From these data, a subdivision and biostratigraphic zonation of the Duchesnean is within reach. Since the exposures of the "type" Duchesnean in the Uinta Basin are so poorly fossiliferous, they are inadequate as a basis for biostratigraphic zonation. Instead, we would recommend that the early Duchesnean be based on the exposures of the middle member of the Sespe Formation in western Simi Valley, which produce the Pearson Ranch l.f. Unfortunately, the exposures are located in Simi Valley Landfill, and they are vanishing as the landfill is modified and filled in. *Amynodontopsis* would probably be the best candidate for a zonal indicator of the early Duchesnean. The best candidate for a late Duchesnean type section would be the exposures of the Chambers Tuff, such as Big Cliff (Wilson, 1978, Fig. 7), that produce the Porvenir l.f. The zonal name-bearers might be *Hemipsalodon,* or possibly *Brachyhyops,* since they are very distinctive and virtually restricted to this interval, although many of the genera that carry over into the Chadronian are much more common.

Storer (1987; this volume, Chapter 12) described the Duchesnean Lac Pelletier Lower Fauna of Saskatchewan. Storer was unsure whether it was early or late Duchesnean, although he thought it was older than the Porvenir l.f. This uncertainty is due to the high endemism of the fauna, with only 51% of the genera shared with other Duchesnean faunas. The fact that Lac Pelletier has many taxa in common with the Chadronian, and few in common with the Pearson Ranch l.f., suggests that it is later Duchesnean. In fact, the most distinctive feature is the large number of taxa which do not appear until the Chadronian or later in the White River Group. They include *Sinclairella, Hesperocyon, Heptacodon, Pseudoprotoceras, "Leptotragulus"* (=*Trigenicus*) *profectus, Adjidaumo,* and *Heliscomys. Hesperocyon* and *Adjidaumo* (known from the Porvenir l.f.) and *Heptacodon* (known from the Duchesnean of the Gulf Coast), however, do occur in the Duchesnean outside the White River Group.

Chadronian
Earliest Chadronian (36.5-37.0 Ma)

Based on the chronostratigraphic data discussed above, the earliest Chadronian (Chron C16r) may be incompletely represented at Flagstaff Rim. The only early

Chadronian fauna that clearly falls within the interval between 36.5 and 37.0 Ma is the Little Egypt l.f. of Texas, which lies below a date of 36.7 ± 0.07 Ma (the Rancho Gaitan l.f. of Chihuahua, Mexico, may be a correlative). Thus, the taxa that define the beginning of the Chadronian and the earliest Chadronian time interval (C16r-C17n1, 36.5-37 Ma) are known mainly from Texas (Fig. 3). They include the first appearance of *Bathygenys, Merycoidodon dunagani, Brachyrhynchocyon* (formerly *Daphoenocyon*) *dodgei,* and *Archaeotherium* (if the upper Porvenir entelodont is *Brachyhyops,* rather than *Archaeotherium,* as suggested by Storer *in* Emry, 1992, p. 114). On this basis, the Duchesnean/Chadronian boundary could be indicated by the first appearance of *Bathygenys, Merycoidodon, Brachyrhynchocyon,* and *Archaeotherium.* This list is less extensive than the early Chadronian indicators suggested by Lucas (1992, p. 98), but these are the only taxa that can be clearly shown to first occur immediately above the Duchesnean/Chadronian boundary. Most of the rest have not been documented until strata correlative with Chron C16n (36.0 Ma).

Of these taxa, we suggest *Bathygenys* as the best candidate for zonal indicator of the beginning of the Chadronian, since it is extremely distinctive and abundant in the earliest Chadronian. The sections of Chambers Tuff that include the Reeves Bonebed (the main locality in the Little Egypt l.f.) might serve to typify the Duchesnean/Chadronian boundary and the earliest Chadronian.

Late early Chadronian (35.7-36.5 Ma)

Strata which are correlative with Chron C16n2 (35.7-36.5 Ma) are much better known, and might be considered late early Chadronian. They include the interval from 60-120 feet above base of the section (to 50 feet below Ash B) at Flagstaff Rim, as well as the Airstrip l.f. of Texas and the McCarty's Mountain l.f. of Montana. They may also include the Yoder l.f. of Wyoming (Kihm, 1987), although there are no radiometric dates or magnetics to test this correlation.

Based primarily on the lower Flagstaff Rim section, a large number of taxa make their first appearance in the late early Chadronian, and are more or less restricted to that interval (Fig. 3). These include *Palaeolagus primus, Merycoidodon presidioensis, Parvitragulus priscus, Pseudoprotoceras semicinctus, Litoyoderimys lustrorum, Yoderimys stewarti,* and *Leptomeryx yoderi.* In addition, a number of taxa first appear in the late early Chadronian and range into the middle Chadronian. These include *Daphoenictis tedfordi, Hyaenodon montanus, Ischyromys veterior, Centetodon chadronensis, "Leptotragulus"* (=*Trigenicus*) *profectus,* and *Sinclairella dakotensis.* As noted above, however, the Duchesnean Lac Pelletier Lower Fauna of Saskatchewan includes a number of taxa that are otherwise known from strata no earlier than the late early Chadronian in the White River Group, Montana, or Texas. These include *Sinclairella, Pseudoprotoceras, "Leptotragulus"* (*=Trigenicus*) *profectus*, and *Heliscomys*. This shortens the list of late early Chadronian first appearances somewhat, although it remains valid south of the Canadian border.

A number of Duchesnean taxa last appear in the late early Chadronian, including *Ardynomys* and *Protadjidaumo*. The oreodont *Limnenetes* appears to be restricted to this interval.

Of these taxa, *Leptomeryx yoderi* might be used as the defining taxon of the late early Chadronian, since it is common, distinctive, and found not only at Flagstaff Rim, but also at many other late early Chadronian localities. The type section for this interval should also be established at Flagstaff Rim, which has the most complete section in terms of both fossils and chronostratigraphy.

Middle Chadronian (34.7-35.7 Ma)

As discussed above, the middle Chadronian is typified by the strata from about 50 feet below Ash B to Ash G at Flagstaff Rim (120-440 feet on the zonation section of Emry, 1973). The principal correlatives are the Pipestone Springs l.f. and Little Pipestone Creek l.f. of Montana (this volume, Chapter 14), the Raben Ranch l.f. of Nebraska (Ostrander, 1985; this volume, Chapter 13), and fauna of the Crazy Johnson Member of the Chadron Formation in South Dakota (Fig. 4). These strata all appear to correlate with Chron C15n-C15r (34.7-35.7 Ma).

A large number of taxa first appear in this interval, and are more or less restricted to it. They include *Leptomeryx mammifer, Zemiodontomys burkei, Miniochoerus forsythae, Prosciurus vetustus, Hyaenodon microdon, Chadrolagus emryi*, and *Frictops emryi*. Many other taxa first appear at this level and range into younger strata, including *Palaeolagus temnodon, Dinictis felina, Eutypomys magnus, Mesohippus westoni, Leptomeryx speciosus, Pseudoprotoceras longinaris*, large *Ischromys* (see Heaton and Emry, this volume, Chapter 27), *Hoplophoneus mentalis, Miohippus grandis*, and *Megalagus brachyodon*. Of the taxa listed above, Emry et al. (1987) recommended *Leptomeryx mammifer* as the best indicator of the middle Chadronian. The best candidate for a type section would be Flagstaff Rim from about 50 feet below Ash B to Ash G.

Late Chadronian (34.7-33.7 Ma)

As discussed above, the Flagstaff Rim section above Ash G is too poorly fossiliferous to characterize the late Chadronian, so the section below the PWL (= Ash J at Flagstaff Rim, 4 tuff at Douglas) in the Seaman Hills, north of Lusk, Niobrara County, Wyoming, better typifies the late Chadronian. The Ash Springs l.f. in Texas might also correlate with these strata, as might the Douglas section below the 4 tuff (Evanoff et al., 1992), the late Chadronian strata in western Nebraska

(Ostrander, 1985; Gustafson, 1986; this volume, Chapter 13), and the Peanut Peak Member of the Chadron Formation in South Dakota. All of these strata appear to correlate with late Chron C13r-C15n (33.7-34.7 Ma).

A few taxa, such as *Meliakrounomys* and *Pseudoprotoceras taylori*, are restricted to the higher levels at Flagstaff Rim (Emry, 1992). Other late Chadronian sections, however, produce the first occurrences of *Merycoidodon culbertsoni, Poebrotherium franki* (both found at Ash Springs), *Miniochoerus chadronensis, Ischyromys typus, Mesohippus exoletus, Mesohippus bairdi*, and *Scottimus viduus*, among many others (Fig. 4). *Pseudocylindrodon* and *Toxotherium* are among the Duchesnean taxa that last appear at this level in the Ash Springs l.f. *Palaeolagus temnodon, Hoplophoneus mentalis, Miohippus grandis, Mesohippus westoni, Eotylopus reedi, Poebrotherium eximium*, and *Archaeotherium coarctatum* are among the taxa that last occur in the late Chadronian (most terminate near the Chadronian/Orellan boundary).

Of these taxa, *Miniochoerus chadronensis* might be the best candidate for a zonal indicator, since it is abundant, distinctive, and restricted to this interval. An appropriate type section could be designated in the Seaman Hills, Niobrara County, Wyoming.

Orellan

Prothero and Whittlesey (in press) reviewed the biostratigraphy of the Orellan and Whitneyan strata in the White River Group (Fig. 5). The key points are summarized below.

The Chadronian/Orellan boundary occurs late in Chron C13r, above the 5 tuff at Douglas, which has been ^{40}Ar/^{39}Ar dated at 33.91 ± 0.06 Ma (Prothero and Swisher, 1992) and at 33.59 ± 0.02 Ma (Obradovich et al., 1995). As discussed by Prothero and Whittlesey (in press), there are at least four separate occurrences of brontotheres in strata correlative with the Brule Formation, so this creates problems for the Wood Committee's (1941) definitions of the Chadronian based on the last appearance of brontotheres and the top of the Chadron Formation. Prothero and Whittlesey (in press) recommended that the first appearance of the ruminant *Hypertragulus calcaratus* be used to recognize the beginning of the Orellan, because it is distinctive, abundant, and appears suddenly at the beginning of the Orellan. There are no Chadronian species of *Hypertragulus* in the White River Group with which it might be confused. *Hypertragulus heikeni* from the early Chadronian Rancho Gaitan l.f. of Mexico is very different, both in time and morphology, and may not be *Hypertragulus* at all.

Korth (1989) called the Orella A strata in Toadstool Park, Nebraska, the "*Palaeolagus hemirhizis* zone." As discussed by Prothero and Whittlesey (in press), this zone is difficult to use because *P. hemirhizis* may be a taxonomic composite of specimens of *P. temnodon* and

P. haydeni, and also because *P. hemirhizis*, if valid, may also occur in the Chadronian of Saskatchewan. In addition, the zone was based on a lithostratigraphic unit, and not on detailed biostratigraphic data within Orella A, so it unacceptably mixes lithostratigraphy and biostratigraphy. Finally, Korth (1989) designated no type section for this or any other zone in his scheme.

The earliest Orellan can also be recognized by the first appearance of *Leptomeryx evansi*, although as shown by Heaton and Emry (this volume, Chapter 27), the transformation from *L. speciosus* to *L. evansi* is subtle and hard to recognize. The earliest Orellan is also marked by the first appearance of *Palaeolagus intermedius* and *Paratylopus labiatus*, and by the last appearance of *Poebrotherium eximium*, *Miohippus grandis*, and brontotheres. At this level, there is also a small species of *Miniochoerus* (mean M^{1-3} length = 42 mm; observed range = 39-45 mm) between *M. chadronensis* and *M. affinis* in size. It was originally called *M. "douglasensis"* by Stevens (MS), but is now referred to *M. chadronensis* (Stevens and Stevens, this volume, Chapter 26). This oreodont is also a good indicator of the earliest Orellan, along with the overlapping ranges of *H. calcaratus*, *L. evansi*, *M. grandis* and brontotheres. On this basis, Prothero and Whittlesey (in press) recognized the earliest Orellan *Hypertragulus calcaratus* Interval Zone. Its type section is the Reno Ranch East section at Douglas, Wyoming, 20-50 feet above the 5 tuff (Evanoff et al., 1992).

Several different biostratigraphic events mark the late early Orellan (Fig. 5), which appears to span most of Chron C13n (33.0-33.5 Ma). Specimens referable to *Miniochoerus affinis* (mean M^{1-3} length = 38 mm; observed range = 36-41 mm) first appear at the beginning of Chron C13n, as well as *Eumys elegans*, *Pelycomys brulanus*, *Adjidaumo minutus*, *Cedromus wardi*, and *Hoplophoneus occidentalis*. Korth (1989) named the Orella B strata at Toadstool Park the "*Eumys elegans* zone," but we hesitate to use this taxon as name-bearer. Although it first appears at this level in the White River Group, *Eumys* may appear earlier in Montana and Saskatchewan (this volume, Chapters 12 and 14). The late early Orellan would be better defined by the overlapping ranges of *Miniochoerus affinis* and *Ischyromys parvidens*, taxa which are much more abundant and easy to recognize. Prothero and Whittlesey (in press) named the *Miniochoerus affinis* Interval Zone for the late early Orellan. This Zone is typified by the strata between 50-80 feet above the PWL in Boner Ranch section in the Seaman Hills, north of Lusk, Niobrara County, Wyoming.

Early late Orellan strata are correlative with the earliest part of Chron C12r (32.5-33.0 Ma), and are marked by slightly fewer distinctive biostratigraphic events. However, the first appearance of the dwarfed oreodont *Miniochoerus gracilis* (mean M^{1-3} length = 34 mm; observed range = 31-37 mm) and the advanced horse *Mesohippus barbouri* are both unique to this

interval. *Ischyromys parvidens* last appears at this level as well (see this volume, Chapter 18). Korth (1989) showed that *Agnotocastor readingi*, *Paradjidaumo validus*, *Eutypomys thomsoni*, and *Eumys parvidens* are retricted to the early late Orellan (Orella C). Prothero and Whittlesey (in press) named this interval the *Miniochoerus gracilis* Interval Zone, and designated its type section as the strata from 80-150 feet above the PWL in the Boner Ranch section of the Seaman Hills.

Latest Orellan strata (early Chron C12r, 32.0-32.5 Ma) are marked by many distinctive taxa. The most useful is the appearance of *Merycoidodon* with large auditory bullae, now referred to *Merycoidodon bullatus* by Stevens and Stevens (this volume, Chapter 26). The advanced rabbit *Palaeolagus burkei* first appears at this level, as does the last of the miniochoeres, *Miniochoerus starkensis*, along with many other taxa listed by Prothero and Whittlesey (in press). Korth (1989) named a late Orellan zone based on Orella D in Toadstool Park, Nebraska, after the rare rodent *Diplolophus insolens*. Although this taxon is apparently unique to the latest Orellan, it is so rare that it would not be a very useful range-zone indicator. In addition, Korth (1989) did not designate a type section for this zone. Prothero and Whittlesey (in press) named the *Merycoidodon bullatus* Interval Zone for the latest Orellan. The "Upper Nodular zone" of the Big Badlands of South Dakota in the Cottonwood Pass-Sheep Mountain Table area was chosen as the stratotype.

Many taxa disappear at or near the Orellan/Whitneyan boundary, including *Ischyromys*, *Mesohippus*, *Subhyracodon*, *Prosciurus*, *Pelycomys*, *Protosciurus*, *Oligospermophilus*, *Eutypomys*, *Adjidaumo*, *Heliscomys*, *Wilsoneumys*, *Eoeumys*, *Tenudomys*, *Pipestoneomys*, *Megalagus*, *Palaeolagus intermedius*, *Leptictis haydeni*, and *Hyaenodon crucians*.

Whitneyan

Unlike the detailed biostratigraphic records for the Orellan, few collections have very detailed records for the Whitneyan. Even in the Frick Collection, specimens are only recorded as "upper Oreodon beds," "*Leptauchenia* beds" or "*Protoceras* channels" with little indication of exactly how many feet they occurred above or below a given horizon. In the University of Nebraska State Museum collections, specimens are only recorded as derived from "Whitney A," "Whitney B," or "Whitney C." Because the primary data base is so low in resolution, the Whitneyan cannot be so finely subdivided as the Chadronian or Orellan at the present time (Fig. 5).

A number of distinctive biostratigraphic events mark the Orellan/Whitneyan boundary (mid-Chron C12r, about 32.0 Ma), although their apparent coincidence may be an artifact of the coarse resolution of the biostratigraphic data. The earliest Whitneyan is marked by abundant *Leptauchenia decora* and the first appearance of *Leptauchenia* (formerly *Cyclopidius*) *major*, and by the first appearances of *Hyracodon leidyanus, Paratylopus primaevus, Paralabis cedrensis, Diceratherium tridactylum, Protapirus obliquidens, Ectopocynus antiquus, Oxetocyon cuspidatus, Cynodesmus thooides, Agnotocastor praetereadens,* and *Oropyctis pediasius.* This zone also yields the last *Miniochoerus* and *Hyaenodon horridus.* Prothero and Whittlesey (in press) named this the *Leptauchenia major* interval zone, with a stratotype in the "upper Oreodon beds" on the south side of Sheep Mountain Table in the Big Badlands. It correlates with mid-Chron C12r (32.0-31.4 Ma).

The late Whitneyan (late Chron C12r–early-C11n, 31.4 Ma–30.0 Ma) can also be distinguished by a number of distinctive taxa. They include the first appearance of the large oreodont *Merycoidodon major,* the tylopod *Protoceras celer,* the horses *Miohippus intermedius, Miohippus equinanus, Miohippus annectens,* and *Miohippus gidleyi,* the nimravids *Hoplophoneus dakotensis, Eusmilus cerebralis,* and *Nimravus brachyops,* the rodents *Eumys brachyodus* and *Scottimus lophatus,* and the creodont *Hyaenodon brevirostrus.* Prothero and Whittlesey (in press) designated the late Whitneyan as the *Merycoidodon major* Interval Zone, with its stratotype in the "*Leptauchenia-Protoceras* beds" in the Sheep Mountain Table area of the Big Badlands.

Tedford et al. (1985, 1987; this volume, Chapter 15) noted a number of taxa that mark the Whitneyan/ Arikareean boundary. According to their definition, this boundary occurs in early Chron C11n (about 30.0 Ma), very near the Rockyford Ash at the base of the Sharps Formation in South Dakota, and near Nonpareil Ash 2 in the brown siltstone beds in Nebraska. Earliest Arikareean first occurrences include the rabbits *Palaeolagus hypsodus* and *P. philoi,* the beaver *Palaeocastor nebrascensis,* the canid *Shunkehetanka geringensis,* the ruminant *Nanotragulus loomisi,* and the oreodonts *Sespia nitida* and *?Mesoreodon minor.* In addition, the rhinos *Diceratherium armatum* and *D. annectens,* and the rodents *Leidymys blacki, Sanctimus stuartae, Geringia mcgregori, Tenudomys, Plesiosminthus,* and several other taxa first occur in the earliest Arikareean as now defined. Of these taxa, *Sespia* would make the most distinctive and abundant biostratigraphic indicator, especially since it also marks the early Arikareean in areas outside the High Plains, such as California. The best candidate for a type section might be the Sharps Formation above the Rockyford Ash in either the Cedar Pass or Wolff Table-Wanblee areas of the Big Badlands of South Dakota, since these produce the richest faunas. In addition, many taxa last occur in the late Whitneyan and are currently unknown from Arikareean strata, including *Leptomeryx, Merycoidodon, Paratylopus, Paralabis, Perchoerus, Heptacodon, Leptochoerus, Hyracodon, Colodon, Protapirus, Hesperocyon, Osbornodon, Dinictis, Paradjidaumo, Eumys,* and *Scottimus.*

Tedford et al. (1987; this volume, Chapter 15) also note that the late early Arikareean (early Chron C9n, 27.8 Ma) is marked by the "enrichment phase" of Arikareean faunas, including the addition of such taxa as *Paciculus, Gregorymys, Pseudotheridomys, Archaeolagus,* and *Enhydrocyon crassidens.* These taxa are all first reported from the "Monroe Creek Formation" and equivalents. This zone falls very near the early/late Oligocene boundary as currently defined (Berggren et al., 1995), and concludes our discussion of the early Oligocene zonation of the North American terrestrial record.

Although much work remains to be done, we have made considerable progress in our understanding of the chronostratigraphy and biostratigraphy of the Uintan through Arikareean interval since the last summaries were published in 1987. For the Orellan and Whitneyan, range zone biostratigraphy has already been proposed. In the Uintan, Duchesnean, and Chadronian, much needs to be resolved before formal biostratigraphic zones can be proposed, and they may not be applicable beyond their local area. Nevertheless, we hope this discussion has laid the foundation for future work, and before yet another decade has passed, we may have a standardized, well-dated chronostratigraphy and biostratigraphy for the entire middle Eocene through early Oligocene.

THE EOCENE-OLIGOCENE CLIMATIC TRANSITION

Berggren and Prothero (1992) and Prothero (1994a, 1994b) reviewed the evidence of global climatic changes that accompanied the Eocene-Oligocene transition. Although most of the conclusions presented in those papers are still valid, it is worthwhile to re-examine them in the light of the correlations discussed above.

According to Stucky (1990, 1992), North American land mammal faunas reached a peak of diversity in the late Uintan, and began to decline in overall diversity through the Duchesnean and Chadronian. Wolfe (1978; 1994, fig. 3) also reported a drop in mean annual temperatures of 10°C as indicated by floras from mid-latitudes of North America at about 40 Ma (the Uintan/Duchesnean boundary). According to Boersma et al. (1987) and Aubry (1992), the Uintan/Duchesnean boundary (early Chron C18n, early Bartonian, planktonic foraminiferan Zone P13/P14 boundary, nannoplankton Zone NP16/NP17 boundary) apparently corresponds to a number of global climatic events and oceanographic changes (cooling of bottom waters, which become decoupled from surface waters; increased thermal isolation of Antarctica; increased oxygenation of surface waters; cooling in mid-latitiudes and extinction of their warm-water plankton).

The most dramatic change in global faunas occurred at the end of the middle Eocene (about 37-38 Ma, according to Berggren et al., 1995), when mass extinctions decimated the marine invertebrates and plankton. Tropical taxa and warm-adapted organisms were the most hard-hit, suggesting that this extinction was caused by global cooling. Based on the dating outlined above, this event probably corresponds to the middle Duchesnean transition, when the typically Uintan-early Duchesnean faunas (such as the Pearson Ranch l.f.) were replaced by elements of the White River Chronofauna (such as the Porvenir l.f.). Although total mammalian diversity did not change between the early and late Duchesnean (Stucky, 1990, 1992), there was a much higher rate of turnover during this interval.

The Eocene-Oligocene boundary itself was not a major extinction horizon in the global marine record (Berggren and Prothero, 1992; Prothero, 1994a), nor was it a major event on the land. Based on the chrono-stratigraphy outlined above, the Eocene/ Oligocene boundary (late Chron C13r, 33.7 Ma) falls very near the Chadronian/Orellan boundary, which saw the disappearance of a few archaic groups (especially bronto-theres, oromerycids, and cylindrodonts), but relatively little in the way of extinction in the remaining taxa (Stucky, 1992, fig. 24.4; Prothero and Heaton, in press).

Based on many climatic indicators, the most dramatic event of all should have occurred in the earliest Oligocene (about 33 Ma, middle Chron C13n). This was when the first major ice sheets appeared in Antarctica, global temperatures dropped 5-6°C (as indicated by oxygen isotopes), and many marine invertebrates suffered another episode of mass extinc-tion. In Europe, a major change took place in land faunas, as the archaic Eocene endemic mammals were replaced by Eurasian immigrants in an event known as the "Grande Coupure" (Hooker, 1992).

In North America, the climatic signals are even more dramatic. Wolfe (1978) reported a 13°C drop in mean annual temperatures based on land floras, with a great increase in seasonality during what he then called the "Terminal Eocene Event" (now dated as earliest Oligocene). Paratropical floras now found in central America were replaced by broadleaved deciduous forests or northern hardwood forests, now found in New England. His more recent estimates based on more sophisticated methods (Wolfe, 1994, fig. 3) are consistent with this interpretation. According to Retallack (1983, 1992), the paleosols in the Big Badlands show a transformation from dry woodland to wooded grassland in the earliest Orellan. In places such as Douglas, Wyoming, late Chadronian floodplain deposits are replaced by early Orellan eolian deposits in mid Chron C13n (Evanoff et al., 1992). Land snails from these same beds are consistent with this interpreta-tion. Late Chadronian land snails are large-shelled, tropical forms typical of the modern southern Rocky Mountains or central Mexican Plateau, and indicate a mean annual temperature of 16.5°C and a mean annual precipitation of 450 mm (Evanoff et al., 1992). During Chron C13n, these snails were replaced by small-shelled, drought-tolerant taxa indicative of a open woodlands habitat with a pronounced dry season, such

as are found today in Baja California. In addition, aquatic reptiles and amphibians (especially salamanders, crocodilians, and pond turtles) disappear by the early Orellan, replaced by land tortoises (Hutchison, 1982, 1992).

Given all this striking climatic evidence, the response by the mammalian fauna was remarkably mild. As reviewed by Prothero and Heaton (in press), the vast majority of mammalian lineages (62 out of 70) found in the earliest Orellan continued into the late early Orellan with no observable morphological change worthy of species distinction. The few extinctions and originations that did occur were remarkably minor, and about the only noticeable change was the dwarfing in the oreodont *Miniochoerus*, which was already underway in the late Chadronian. The implications of this discordance between the climatic evidence and mammalian faunal data is analyzed in Prothero and Heaton (in press), so it will not be discussed further here.

ACKNOWLEDGMENTS

We thank Tom Kelly, Spencer Lucas, Steve McCarroll, John Storer, Richard Stucky, Alan Tabrum, Steve Walsh, and Mike Woodburne for helpful reviews of this manuscript.

LITERATURE CITED

Andersen, D. W., and M. D. Picard. 1972. Stratigraphy of the Duchesne River Formation (Eocene-Oligocene?), northern Uinta Basin, northeastern Utah. Utah Geologi-cal and Mineralogical Survey Bulletin 97:1-23.

Archibald, J. D., W. A. Clemens, P. D. Gingerich, D. W. Krause, E. H. Lindsay, and K. D. Rose. 1987. First North American land mammal ages of the Cenozoic Era; pp. 24-76 in M. O. Woodburne (ed.), Cenozoic Mammals of North America, Geochronology and Biostratigraphy, University of California Press, Berkeley.

Aubry, M.-P. 1992. Late Paleogene calcareous nan-noplankton evolution: a tale of climatic deterioration; pp. 272-309 in D. R. Prothero and W. A. Berggren (eds.), Eocene-Oligocene Climatic and Biotic Evolution. Princeton University Press, Princeton, N. J.

Berggren, W. A., D. V. Kent, C. C. Swisher III, and M.-P. Aubry. 1995. A revised Cenozoic geochronology and chronostratigraphy. SEPM Special Publication 54:129-212.

Berggren, W. A., and D. R. Prothero. 1992. Eocene-Oligocene climatic and biotic evolution: an overview; pp. 1-28 in D. R. Prothero and W. A. Berggren (eds.), Eocene-Oligocene Climatic and Biotic Evolution. Princeton University Press, Princeton, N. J.

Berry, R. W. 1991. Deposition of Eocene and Oligocene bentonites and their relationship to Tertiary tectonics, San Diego County; pp. 107-113 in P. L. Abbott, and J. A. May (eds.), Eocene Geologic History San Diego Region, Pacific Section SEPM Guidebook 68.

Black, C. C., and M. R. Dawson. 1966. A revew of the late Eocene mammalian faunas from North America. American Journal of Science 264:321-349.

Boersma, A., I. Premoli-Silva, and N. J. Shackleton. 1987. Atlantic Eocene planktonic foraminiferal paleohydro-graphic indicators and stable isotope paleoceanography. Paleoceanography 2(3):287-331.

Bottjer, D. J., S. P. Lund, J. E. Powers, and M. C. Steele.

1991. Magnetostratigraphy of Paleogene strata in San Diego and the Simi Valley, southern California; pp. 115-124 in P. L. Abbott and J. A. May (eds.), Eocene Geologic History San Diego Region. Pacific Section SEPM Volume 68.

Emry, R. J. 1973. Stratigraphy and preliminary biostratigraphy of the Flagstaff Rim area, Natrona County, Wyoming. Smithsonian Contributions to Paleobiology 18.

Emry, R. J. 1981. Additions to the mammalian fauna of the type Duchesnean, with comments on the status of the Duchesnean "Age." Journal of Paleontology 55:563-570.

Emry, R. 1992. Mammalian range zones in the Chadronian White River Formation at Flagstaff Rim, Wyoming; pp. 106-115 in D. R. Prothero and W. A. Berggren (eds.), Eocene-Oligocene Climatic and Biotic Evolution. Princeton University Press, Princeton, N. J.

Emry, R. J., P. R. Bjork, and L. S. Russell. 1987. The Chadronian, Orellan, and Whitneyan land mammal ages; pp. 118-152 in M. O. Woodburne (ed.), Cenozoic Mammals of North America, Geochronology and Biostratigraphy, University of California Press, Berkeley.

Evanoff, E., D. R. Prothero, and R. H. Lander. 1992. Eocene-Oligocene climatic change in North America: the White River Formation near Douglas, east-central Wyoming; pp. 116-130 in D. R. Prothero and W. A. Berggren (eds.), Eocene-Oligocene Climatic and Biotic Evolution, Princeton University Press, Princeton, N. J.

Evanoff, E., P. Robinson, P. Murphey, D. G. Kron, D. Engard, and P. Monaco. 1994. An early Uintan fauna from Bridger E. Journal of Vertebrate Paleontology 14, supplement to no. 3:24A.

Flynn, J. J. 1986. Correlation and geochronology of middle Eocene strata from the western United States. Palaeogeography, Palaeoclimatology, Palaeoecology 55: 335-406.

Gingerich, P. D. 1983. Paleocene-Eocene faunal zones and a preliminary analysis of the Laramide structural deformation of the Clark's Fork Basin, Wyoming. Wyoming Geological Association Guidebook 34: 185-195.

Gunnell, G. F. 1989. Evolutionary history of the Microsyopoidea (Mammalia, ?Primates) and the relationship between Plesiadapiformes and Primates. University of Michigan Papers in Paleontology 27: 1-157.

Gustafson, E. P. 1986. Preliminary biostratigraphy of the White River Group (Oligocene, Chadron and Brule Formations) in the vicinity of Chadron, Nebraska. Transactions of the Nebraska Academy of Sciences, XIV: 7-19.

Henry, C. D., M. Kunk, and W. McIntosh. 1994. ^{40}Ar/^{39}Ar chronology and volcanology of silicic volcanism in the Davis Mountains, Trans-Pecos Texas. Geological Society of America Bulletin 106: 1359-1376.

Hooker, J. J. 1992. British mammalian paleocommunities across the Eocene-Oligocene transition and their environmental implications; pp. 494-515 in D. R. Prothero and W. A. Berggren (eds.), Eocene-Oligocene Climatic and Biotic Evolution. Princeton University Press, Princeton, N. J.

Hutchinson, J. H. 1982. Turtle, crocodilian, and champsosaur diversity changes in the Cenozoic of the north-central region of the western United States. Palaeogeography, Palaeoclimatology, Palaeoecology 37: 149-164.

Hutchison, J. H. 1992. Western North American reptile and amphibian record across the Eocene/Oligocene boundary

and its climatic implications; pp. 451-463 in D. R. Prothero and W. A. Berggren (eds.), Eocene-Oligocene Climatic and Biotic Evolution. Princeton University Press, Princeton, N. J.

Kelly, T. S. 1990. Biostratigraphy of Uintan and Duchesnean land mammal assemblages from the middle member of the Sespe Formation, Simi Valley, California. Contributions to Science of the Los Angeles County Museum 419: 1-43.

Kelly, T. S. 1992. New Uintan and Duchesnean (middle and late Eocene) rodents from the Sespe Formation, Simi Valley, California. Bulletin of the Southern California Academy of Sciences 91(3): 97-120.

Kelly, T. S., E .B. Lander, D. P. Whistler, M. A. Roeder, and R. E. Reynolds. 1991. Preliminary report on a paleontologic investigation of the lower and middle members, Sespe Formation, Simi Valley Landfill, Ventura County, California. PaleoBios 13 (50): 1-13.

Kelly, T. S., and D. P. Whistler. 1994. Additional Uintan and Duchesnean (middle and late Eocene) mammals from the Sespe Formation, Simi Valley, California. Contributions to Science of the Los Angeles County Museum 439: 1-29.

Kihm, A. J. 1987. Mammalian paleontology and geology of the Yoder Member, Chadron Formation, east-central Wyoming. Dakoterra 3:28-45.

Korth, W. W. 1989. Stratigraphic occurrence of rodents and lagomorphs in the Orella Member, Brule Formation (Oligocene), northwestern Nebraska. Contributions to Geology, University of Wyoming 27(1): 15-20.

Krishtalka, L., R. K. Stucky, R. M. West, M. C. McKenna, C. C. Black, T. M. Bown, M. R. Dawson, D. J. Golz, J. J. Flynn, J. A. Lillegraven, and W. D. Turnbull. 1987. Eocene (Wasatchian through Duchesnean) biochronology of North America; pp. 77-117 in M. O. Woodburne (ed.), Cenozoic Mammals of North America, Geochronology and Biostratigraphy, University of California Press, Berkeley.

Lillegraven, J. A. 1979. A biogeographical problem involving comparisons of later Eocene terrestrial vertebrate faunas of western North America; pp. 333-347 in J. Gray and A. J. Boucot (eds.), Historical biogeography, plate tectonics, and the changing environment. Oregon State University Press, Corvallis.

Lucas, S. G. 1992. Redefinition of the Duchesnean land mammal "age," late Eocene of western North America; pp. 88-105 in D. R. Prothero and W. A. Berggren (eds.), Eocene-Oligocene Climatic and Biotic Evolution. Princeton University Press, Princeton, N. J.

Mader, B. J. 1989. The Brontotheriidae: a systematic revision and preliminary phylogeny of North American genera; pp. 458-484 in D. R. Prothero and R. M. Schoch (eds.), The Evolution of Perissodactyls. Oxford University Press, New York.

Mason, M. A. 1988. Mammalian paleontology and stratigraphy of the early to middle Tertiary Sespe and Titus Canyon Formations, southern California. Ph.D. dissertation, University of California, Berkeley, 257 pp.

Mauger, R. L. 1977. K-Ar ages of biotites from tuffs in Eocene rocks of the Green River, Washakie, and Uinta basins, Utah, Wyoming, and Colorado. Contributions to Geology of the University of Wyoming 15: 17-41.

McCarroll, S. M., J. J. Flynn, and W. D. Turnbull. 1993. Biostratigraphy and magnetic polarity correlations of the Washakie Formation, Washakie Basin, Wyoming. Journal of Vertebrate Paleontology 13(3): 49A.

McDowell, F. W. 1979. Potassium-argon dating in the Trans-Pecos Texas volcanic field. Bureau of Economic Geology of the University of Texas Guidebook 19: 10-

18.

McKenna, M. C. 1980. Late Cretaceous and early Tertiary vertebrate paleontological reconnaissance, Togwotee Pass area, northwestern Wyoming; pp. 321-343 in L. L. Jacobs (ed.), Essays in Honor of Edwin Harris Colbert. Museum of Northern Arizona Press, Flagstaff.

McWilliams, R. G. 1972. Age and correlation of the Eocene Ulatisian and Narizian Stages, California: Discussion. Geological Society of America Bulletin 83:533-534.

Obradovich, J. D., E. Evanoff, and E. E. Larson. 1995. Revised single-crystal laser-fusion $^{40}Ar/^{39}Ar$ ages of Chadronian tuffs in the White River Formation of Wyoming. Geological Society of America, Abstracts with Programs 27(3): 77-78.

Ostrander, G. E. 1985. Correlation of the early Oligocene (Chadronian) in northwestern Nebraska; pp. 205-231 in J. E. Martin (ed.), Fossiliferous Cenozoic deposits of western South Dakota and northwestern Nebraska. Dakoterra, Museum of Geology, South Dakota School of Mines 2.

Philips, F. J. 1972. Age and correlation of the Eocene Ulatisian and Narizian Stages, California: Discussion. Geological Society of America Bulletin 83:2217-2224.

Prothero, D.R. 1985. North American mammalian diversity and Eocene-Oligocene extinctions. Paleobiology 11(4): 389-405.

Prothero, D. R. 1991. Magnetic stratigraphy of Eocene-Oligocene mammal localities in southern San Diego County; pp. 125-130 in P. L. Abbott and J. A. May (eds.), Eocene Geologic History San Diego Region. Pacific Section SEPM Volume 68.

Prothero, D. R. 1994a. The late Eocene-Oligocene extinctions. Annual Reviews of Earth and Planetary Sciences 22: 145-165.

Prothero, D. R. 1994b. The Eocene-Oligocene Transition: Paradise Lost. Columbia University Press, New York.

Prothero, D. R. 1995. Geochronology and magnetostratigraphy of Paleogene North American land mammal "ages": an update. SEPM Special Publication 54:305-315.

Prothero, D. R., and T. H. Heaton. (in press). Speciation and faunal stability in mammals during the Eocene-Oligocene climatic crash. Palaeogeography, Palaeoclimatology, Palaeoecology.

Prothero, D. R., and C. C. Swisher III. 1992. Magnetostratigraphy and geochronology of the terrestrial Eocene-Oligocene transition in North America; pp. 46-73 in D. R. Prothero and W. A. Berggren (eds.), Eocene-Oligocene Climatic and Biotic Evolution. Princeton University Press, Princeton, N. J.

Prothero, D. R., and K. E. Whittlesey. (in press). Magnetostratigraphy and biostratigraphy of the Orellan and Whitneyan land mammal "ages" in the White River Group. Geological Society of America Special Paper.

Retallack, G. J. 1983. Late Eocene and Oligocene paleosols from Badlands National Park, South Dakota. Geological Society of America Special Paper 193.

Retallack, G. J. 1992. Paleosols and changes in climate and vegetation across the Eocene/Oligocene boundary; pp. 382-398 in D. R. Prothero and W. A. Berggren (eds.), Eocene-Oligocene Climatic and Biotic Evolution. Princeton University Press, Princeton, N. J.

Schultz, C. B., and C. H. Falkenbach. 1968. The phylogeny of the oreodonts, Parts 1 and 2. Bulletin of the American Museum of Natural History 139: 1-498.

Steineck, P. L., J. M. Gibson, and R. W. Morin. 1972. Foraminifera from the middle Eocene Rose Canyon and Poway formations, San Diego, California. Journal of Foraminiferal Research 2: 137-144.

Stevens, M. S. (MS) Re-evaluation of taxonomy and phylogeny of some oreodonts (Arctiodactyla, Merycoidodontidae). Unpublished manuscrript in the Osborn Library, American Museum of Natural History, 1977.

Stevens, J. B., M. S. Stevens, and J. A. Wilson. 1984. Devil's Graveyard Formation (new), Eocene and Oligocene age, Trans-Pecos Texas. Texas Memorial Museum Bulletin 32: 1-21.

Storer, J. E. 1987. Dental evolution and radiation of Eocene and Early Oligocene Eomyidae (Mammalia, Rodentia) of North America, with new material from the Duchesnean of Saskatchewan. Dakoterra 3:108-117.

Storer, J. E. 1990. Primates of the Lac Pelletier Lower Fauna (Eocene: Duchesnean), Saskatchewan. Canadian Journal of Earth Sciences 27:520-524.

Storer, J. E. 1993. Multituberculates of the Lac Pelletier Lower Fauna, Late Eocene (Duchesnean), of Saskatchewan. Canadian Journal of Earth Sciences 30:1613-1617.

Stucky, R. K. 1990. Evolution of land mammal diversity in North America during the Cenozoic. Current Mammalogy 2: 375-432.

Stucky, R. K. 1992. Mammalian faunas in North America of Bridgerian to early Arikareean "Ages" (Eocene and Oligocene); pp. 464-493 in D. R. Prothero and W. A. Berggren (eds.), Eocene-Oligocene Climatic and Biotic Evolution. Princeton University Press, Princeton, N. J.

Stucky, R. K., L. Krishtalka, and M. R. Dawson. 1989. Paleontology, geology, and remote sensing of Paleogene rocks in the northeastern Wind River Basin, Wyoming, USA; pp. 34-44 in J. J. Flynn and M. C. McKenna (eds.), Mesozoic/Cenozoic Vertebrate Paleontology: Classic Localities, Contemporary Approaches. 28th International Geological Congress Field Trip Guidebook T322.

Sundell, K. A., P. N. Shive, and J. G. Eaton. 1984. Measured sections, magnetic polarity, and biostratigraphy of the Eocene Wiggins, Tepee Trail, and Aycross Formations within the southeastern Absaroka Range, Wyoming. Earth Science Bulletin of the Wyoming Geological Association 17: 1-48.

Tedford, R. H. 1970. Principles and practices of mammalian geochronology in North America. Proceedings of the North American Paleontological Convention 2F: 666-703.

Tedford, R. H., J. B. Swinehart, R. M. Hunt, Jr., and M. R. Voorhies. 1985. Uppermost White River and lowermost Arikaree rocks and faunas, White River Valley, north western Nebraska and their correlation with South Dakota. Dakoterra 2:335-352.

Tedford, R. H., T. Galusha, M. F. Skinner, B. E. Taylor, R. W. Fields, J. R. Macdonald, J. M. Rensberger, S. D. Webb, and D. P. Whistler. 1987. Faunal succession and biochronology of the Arikareean through Hemphillian interval (late Oligocene through earliest Pliocene Epochs) in North America; pp. 152-210 in M. O. Woodburne (ed.), Cenozoic Mammals of North America, Geochronology and Biostratigraphy, University of California Press, Berkeley.

Walton, A. H., 1992. Magnetostratigraphy and the ages of Bridgerian and Uintan faunas in the lower and middle members of the Devil's Graveyard Formation, Trans-Pecos Texas; pp. 74-87 in D. R. Prothero and W. A. Berggren (eds.), Eocene-Oligocene Climatic and Biotic Evolution, Princeton University Press, Princeton, N. J.

Westgate, J. W. 1988. Biostratigraphic implications of the first Eocene land-mammal fauna from the North American coastal plain. Geology 16: 995-998.

Wilson, J. A. 1978. Stratigraphic occurrence and correlation of early Tertiary vertebrate faunas, Trans-Pecos Texas, Part 1: Vieja area. Texas Memorial Museum Bulletin 25: 1-42.

Wilson, J. A. 1984. Vertebrate fossil faunas 49 to 36 million years ago and additions to the species of *Leptoreodon* found in Texas. Journal of Vertebrate Paleontology 4: 199-207.

Wilson, J. A. 1986. Stratigraphic occurrence and correlation of early Tertiary vertebrate faunas, Trans-Pecos Texas: Agua Fria-Green Valley areas. Journal of Vertebrate Paleontology 6: 350-373.

Wolfe, J. A. 1978. A paleobotanical interpretation of Tertiary climates in the Northern Hemisphere. American Scientist 66: 694-703.

Wolfe, J. A. 1994. Tertiary climatic changes at middle latitudes of western North America. Palaeogeography, Palaeoclimatology, Palaeoecology 108: 195-205.

Wood, H. E., R. W. Chaney, J. Clark, E. H. Colbert, G. L. Jepsen, J. B. Reeside, Jr., and C. Stock. 1941. Nomenclature and correlation of the North American continental Tertiary. Geological Society of America Bulletin 52:1-48.

Index